Invasive Stink Bugs and Related Species (Pentatomoidea)

Biology, Higher Systematics, Semiochemistry, and Management

CONTEMPORARY TOPICS in ENTOMOLOGY SERIES

THOMAS A. MILLER Editor

Insect Sounds and Communication: Physiology, Behaviour, Ecology, and Evolution
Edited by Sakis Drosopoulos and Michael F. Claridge

Insect Symbiosis, Volume 2
Edited by Kostas Bourtzis and Thomas A. Miller

Insect Symbiosis, Volume 3
Edited by Kostas Bourtzis and Thomas A. Miller

Food Exploitation by Social Insects: Ecological, Behavioral, and Theoretical Approaches
Edited by Stefan Jarau and Michael Hrncir

Molecular Biology and Genetics of the Lepidoptera
Edited by Marian R. Goldsmith and František Marec

Honey Bee Colony Health: Challenges and Sustainable Solutions
Diana Sammataro and Jay A. Yoder

Forensic Entomology: International Dimensions and Frontiers
Edited by Jeffery Keith Tomberlin and Mark Eric Benbow

Greenhouse Pest Management
Edited by Raymond A. Cloyd

Cerambycidae of the World: Biology and Pest Management
Edited by Qiao Wang

Molecular Biology and Genetics of the Lepidoptera
Edited by Marian R. Goldsmith and Frantisek Marec

Invasive Stink Bugs and Related Species (Pentatomoidea): Biology, Higher Systematics, Semiochemistry, and Management
Edited by J.E. McPherson

Invasive Stink Bugs and Related Species (Pentatomoidea)

Biology, Higher Systematics, Semiochemistry, and Management

Edited by
J. E. McPherson

CRC Press
Taylor & Francis Group
Boca Raton London New York

CRC Press is an imprint of the
Taylor & Francis Group, an **informa** business

CRC Press
Taylor & Francis Group
6000 Broken Sound Parkway NW, Suite 300
Boca Raton, FL 33487-2742

First issued in paperback 2020

© 2018 by Taylor & Francis Group, LLC
CRC Press is an imprint of Taylor & Francis Group, an Informa business

No claim to original U.S. Government works

ISBN 13: 978-0-3675-7034-7 (pbk)
ISBN 13: 978-1-4987-1508-9 (hbk)

Library of Congress Cataloging-in-Publication Data

Names: McPherson, J. E. (John Edwin), 1941- , author.
Title: Invasive stink bugs and related species (Pentatomoidea) : biology, higher systematics, semiochemistry, and management / J.E. McPherson.
Description: Boca Raton : Taylor & Francis, 2017. | Includes index.
Identifiers: LCCN 2016048509 | ISBN 9781498715089 (hardcover : alk. paper)
Subjects: LCSH: Stink bugs. | Hemiptera. | Agricultural pests.
Classification: LCC QL523.P5 M37 2017 | DDC 595.7/54--dc23
LC record available at https://lccn.loc.gov/2016048509

Visit the Taylor & Francis Web site at
http://www.taylorandfrancis.com

and the CRC Press Web site at
http://www.crcpress.com

This book is dedicated to the late Carl W. Schaefer, Professor Emeritus at the University of Connecticut, Storrs. Dr. Schaefer was a friend and colleague to many of us and a recognized world leader in the study of Hemiptera-Heteroptera. He was a prolific researcher and writer, authoring over 240 journal articles and editing or co-editing seven books. He began "The Heteropterists' Newsletter" in 1973, which he edited and produced for over 20 years. He also was editor of the Annals of the Entomological Society of America (ESA) for 25 years (1973–1998) and served in several additional leadership positions in the Society, including President of the Eastern Branch. In recognition of his outstanding career, he was elected by the ESA as an Honorary Member in 1996 and as a Fellow in 2006. In 2014, he was elected for Honorary Life Membership by the International Heteropterists' Society, one of the first two individuals ever to have received this award. Truly, his contributions during his career were outstanding and certainly are worthy of this dedication.

Contents

Preface ix
Acknowledgments xi
Editor xiii
Contributors xv

Section I Introduction

1. **Overview of the Superfamily Pentatomoidea** 3
   J. E. McPherson, C. Scott Bundy, and Alfred G. Wheeler, Jr.

Section II Systematics

2. **Higher Systematics of the Pentatomoidea** 25
   David A. Rider, Cristiano F. Schwertner, Jitka Vilímová, Dávid Rédei, Petr Kment,
   and Donald B. Thomas

Section III Invasive Pentatomoidea

3. *Bagrada hilaris* **(Burmeister)** 205
   C. Scott Bundy, Thomas M. Perring, Darcy A. Reed, John C. Palumbo, Tessa R. Grasswitz,
   and Walker A. Jones

4. *Halyomorpha halys* **(Stål)** 243
   George C. Hamilton, Jeong Joon Ahn, Wenjun Bu, Tracy C. Leskey, Anne L. Nielsen,
   Yong-Lak Park, Wolfgang Rabitsch, and Kim A. Hoelmer

5. *Megacopta cribraria* **(F.)** 293
   Joe E. Eger, Wayne A. Gardner, Jeremy K. Greene, Tracie M. Jenkins, Phillip M. Roberts,
   and Dan R. Suiter

6. *Murgantia histrionica* **(Hahn)** 333
   J. E. McPherson, C. Scott Bundy, and Thomas P. Kuhar

7. *Nezara viridula* **(L.)** 351
   Jesus F. Esquivel, Dmitry L. Musolin, Walker A. Jones, Wolfgang Rabitsch,
   Jeremy K. Greene, Michael D. Toews, Cristiano F. Schwertner, Jocélia Grazia,
   and Robert M. McPherson

8. *Piezodorus guildinii* **(Westwood)** 425
   C. Scott Bundy, Jesus F. Esquivel, Antônio R. Panizzi, Joe E. Eger, Jeffrey A. Davis,
   and Walker A. Jones

Section IV Potentially Invasive Pentatomoidea

9. *Oebalus* spp. and *Arvelius albopunctatus* (De Geer)..455
   J. E. McPherson and C. Scott Bundy

Section V A Noninvasive Group (Antestia Complex)

10. The Antestia Bug Complex in Africa and Asia 465
    Régis Babin, Pierre Mbondji Mbondji, Esayas Mendesil, Harrison M. Mugo, Joon-Ho Lee,
    Mario Serracin, N. D. T. M. Rukazambuga, and Thomas A. Miller

Section VI Diapause and Seasonal Cycles of Pentatomoidea

11. Diapause in Pentatomoidea 497
    Dmitry L. Musolin and Aida Kh. Saulich

12. Seasonal Cycles of Pentatomoidea...565
    Aida Kh. Saulich and Dmitry L. Musolin

Section VII Vectors of Plant Pathogens

13. Pentatomoids as Vectors of Plant Pathogens 611
    Paula Levin Mitchell, Adam R. Zeilinger, Enrique Gino Medrano, and Jesus F. Esquivel

Section VIII Symbiotic Microorganisms

14. Symbiotic Microorganisms Associated with Pentatomoidea 643
    Yoshitomo Kikuchi, Simone S. Prado, and Tracie M. Jenkins

Section IX Semiochemistry

15. Semiochemistry of Pentatomoidea...677
    Donald C. Weber, Ashot Khrimian, Maria Carolina Blassioli-Moraes, and Jocelyn G. Millar

Section X Management

16. General Insect Management 729
    Jeremy K. Greene, James A. Baum, Eric P. Benson, C. Scott Bundy, Walker A. Jones,
    George G. Kennedy, J. E. McPherson, Fred R. Musser, Francis P. F. Reay-Jones,
    Michael D. Toews, and James F. Walgenbach

Insects and Spiders Index...775

Plants Index 801

Microorganisms and Plant Diseases Index 817

Preface

This book began with Tom Miller, University of California-Riverside, in his position as editor of the series "Contemporary Topics in Entomology," produced by CRC Press and Taylor & Francis Group, with John Sulzycki as senior editor with the publisher. The series results from people and symposium titles Tom encounters at various meetings and is centered on topics of current interest. Thus far, the series has published the following titles: *Insect Symbiosis* (3 volumes); *Insect Sounds and Communication: Physiology, Behaviour, Ecology, and Evolution; Food Exploitation by Social Insects: Ecological, Behavioral, and Theoretical Approaches; Molecular Biology and Genetics of Lepidoptera; Honey Bee Colony Health: Challenges and Sustainable Solutions; Forensic Entomology: International Dimensions and Frontiers;* and *Greenhouse Pest Management*.

The idea for this book began with Tom's involvement with Global Knowledge Initiative, a Washington D.C.-based non-profit organization dedicated to helping developing countries address science and technology problems. The first project of this organization in 2012 was to help solve a defect associated with Arabica coffee from East Africa. This defect is called "Potato Taste Defect" (PTD) because a few out of 100 cups of brewed coffee taste and smell like rotten potatoes. PTD is attributed to contamination of coffee cherries by microbes left from feeding by stink bugs of the genus *Antestiopsis*; occurrence of these bugs is centered in Rwanda and Burundi and, to a lesser extent, in Uganda, Kenya, Congo, and Tanzania. Coffee exports represent about 25% of Rwanda's income, and although the coffee industry is thriving there, solving PTD would be of great benefit.

During this same period, Tom also became familiar with the biology of the brown marmorated stink bug and the southern green stink bug and learned of their association with microbes. He found the information so interesting that he organized a symposium, "Pentatomids and Microbes," for the European Congress of Entomology that was held in York, United Kingdom, 3–8 August 2014. The contributors for that symposium constituted the starting core for designing the book on stink bug biology. Shortly thereafter, Tom asked me to serve as editor of the section on invasive stink bug species plus the *Antestiopsis* complex (antestia bugs).

I soon realized that serving as editor for the invasive stink bugs and *Antestiopsis* was going to be difficult because the other chapters involved Pentatomoidea in general (e.g., Higher Classification, Pathogens, Semiochemistry) and there was no underlying theme to pull the information together. After consulting with Tom about my concerns, he asked me to serve as editor for the entire book. Oh, well!

My first task was to select an individual for each chapter who, based on my knowledge or on recommendations from others, was highly qualified and willing to serve as chair. Then, I asked each chair to select contributors for his/her chapter who also were highly qualified and enthusiastic about contributing their expertise. The net result was the selection of 60 contributors from 13 countries (15, if you include the native countries of two authors whose current locations are elsewhere), giving the book a cosmopolitan flavor.

The contributors for each chapter were asked to submit the most recent information of which they were aware and to treat it as though they were preparing an article for the *Annual Review of Entomology*. All were highly enthusiastic about the project, each group of contributors considering their chapter to be an excellent opportunity to present the most comprehensive and up-to-date treatment of their specialties. As a result, this book encompasses a wide-ranging series of topics that are connected, sometimes loosely, by the underlying theme of the invasive species (even with the inclusion of the *Antestiopsis* complex, **Chapter 10**).

The book is divided into **10 sections**, reflecting the breadth of coverage among the **16 chapters**. **Section I (Chapter 1)** discusses introductory information including a brief classification overview of the Pentatomoidea, general biology of this superfamily, predators, parasites, chemical defenses, and brief information on the invasive species. **Section II (Chapter 2)** presents a thorough treatment of the higher classification of the Pentatomoidea, primarily at the tribal level. **Section III (Chapters 3–8)** provides a detailed discussion of each of the six recognized invasive species, including three that recently have been introduced into the United States [*Bagrada hilaris* Burmeister, *Halyomorpha halys* (Stål), and *Megacopta cribraria* (F.)]. **Section IV (Chapter 9)** deals with potentially invasive species in the United States and includes only three species, *Oebalus insularis* (Stål), *O. ypsilongriseus* (De Geer), and *Arvelius albopunctatus* (De Geer). The two species of *Oebalus*, which are noted pests of rice, are widely distributed in South and Central America but have been recorded in the United States only from Florida; potentially, they appear capable of spreading to adjacent states but, thus far, have not done so. *A. albopunctatus* apparently prefers wild and cultivated species of Solanaceae. It occurs throughout most of South America north through Baja California (Mexico) to Arizona, Texas, and Florida; it has not, as yet, become a pest in the United States. **Section V (Chapter 10)** discusses the *Antestiopsis* complex in Africa and Asia, a taxon noted as a pest of coffee that has not spread further although other pests of coffee are present today wherever coffee is grown. **Section VI (Chapters 11–12)** discusses diapause and seasonal cycles of Pentatomoidea. **Section VII (Chapter 13)** deals with Pentatomoidea as vectors of plant pathogens and **Section VIII (Chapter 14)**, symbiotic microorganisms associated with this superfamily. **Section IX (Chapter 15)** presents a detailed discussion of the semiochemistry of Pentatomoidea including pheromones, allomones, and kairomones. And, finally, **Section X (Chapter 16)** considers general management practices (both historical and current controls) from a broad perspective.

Serving as editor of this book was a challenging task, and I agreed to do it with some trepidation. Dealing with numerous contributors from several countries could have proven to be an unpleasant experience. But, such was not the case. All participants were so cooperative and enthusiastic about their contributions that my role was both exciting and rewarding. I want to extend my thanks to all of the contributors and hope that readers of this book will find it beneficial, rewarding, and a great tool for their own research and teaching and for increasing their own general knowledge of these areas of entomology.

J. E. McPherson
Carbondale, Illinois, U.S.A.

Acknowledgments

With a book of this size and breadth of topics, it is obvious that many individuals had to be involved in its development. Those who were not authors but contributed valuable information are acknowledged after the various chapters. However, others were involved with the development of the book as a whole. I am grateful to John Sulzycki, our editor at CRC Press, who provided me with his expertise and encouragement throughout the development of the book and always was willing to listen to my suggestions and ideas, adopting them when possible or carefully explaining why they were not possible. I am grateful to Jill Jurgensen, Project Coordinator, formerly at CRC Press, who worked with me on a day-to-day basis and was a joy to deal with. Her expertise on preparation of the chapters, tables, figures, and copyrights was outstanding and made my work much easier. Jill's responsibilities for the book were assumed by Jennifer Blaise, Editorial Assistant, CRC Press, who continued the same high quality of its preparation for publication. Jill and Jennifer's outstanding professionalism was continued by Marsha Hecht, Project Editor, CRC Press; and Adel Rosario, Project Manager, Manila Typesetting Company, both of whom were responsible for preparation of the galley proofs and final copyediting. I feel fortunate to have had the opportunity to work with such a dedicated group of individuals. I also owe a special thanks to Dmitry Musolin, co-chair of Chapters 11 and 12 and a contributing author to Chapter 7, who volunteered to help proofread the entire book manuscript and did an outstanding job.

I thank Tom J. Henry (Systematic Entomology Laboratory, U.S. Department of Agriculture-Agricultural Research Service [USDA-ARS], c/o National Museum of Natural History, Washington, DC) for his advice on various taxonomic problems. I also thank Michael T. Madigan (Department of Microbiology, Southern Illinois University, Carbondale) who provided invaluable expertise in the preparation of the microorganism index and assistance with the proper citing of these organisms in the text.

Doreen S. Hees (Office Administrator, Department of Zoology, SIUC) spent considerable time photocopying literally thousands of manuscript pages of this book for my copyediting (because I do not like to edit on the screen) and completed each job quickly. Her help was invaluable.

Lastly, I am pleased to acknowledge my wife, Jean, who tolerated a moody husband who used her as a sounding board for 4 years as the book was evolving. She provided the calming atmosphere that I needed, and I want her to know how much I appreciate her understanding and patience.

Editor

J. E. McPherson is Professor Emeritus of Zoology at Southern Illinois University, Carbondale (SIU). He obtained his Ph.D. in entomology from Michigan State University in 1968 and joined SIU in 1969 as Assistant Professor. He was promoted to Associate Professor in 1974 and to Professor in 1979. He retired in July 2012 but has continued to conduct research, maintaining the same office and laboratory space he had during his employment, and still is managing the entomology collection.

Dr. McPherson has written broadly on the ecology and systematics of the Heteroptera, particularly the Pentatomoidea, Reduvioidea, and various aquatic and semiaquatic taxa. He has authored or co-authored 200 refereed journal articles and three books, presented papers at both national and regional meetings, and given invited lectures at various universities. He has received several research grants, primarily from the USDA Forest Service.

Dr. McPherson is a member of the Entomological Society of America (ESA) and has served on numerous ESA national and branch committees. He was the recipient of an ESA National Service Award in 1991 for his work on the Editorial Board of the *American Entomologist*, the society's flagship publication, and served as editor of that publication from 1993 through 2001. He was the 1993 recipient of the Distinguished Achievement Award in Teaching, the 1997 recipient of the C. V. Riley Achievement Award, and the 2006 Award of Merit, all from the ESA North Central Branch. He also was the 1996 recipient of the Outstanding Teacher in the College of Science, SIU. He served 6 years on the ESA Governing Board, 3 (1994–1996) as Section A (now Systematics, Evolution, and Biodiversity) representative and 3 as an officer (Vice-President, 2001; President, 2002; Past-President, 2003). He was elected an Honorary Member in 2004 and a Fellow in 2007 of the ESA. Recently, he received the 2017 Michigan State University Entomology Distinguished Alumnus Award.

Dr. McPherson is a member of several additional societies including the Entomological Society of Washington, Florida Entomological Society, Michigan Entomological Society, and the New York Entomological Society. A Festschrift issue of the *Great Lakes Entomologist* was dedicated to him in 2012 by the Governing Board of the Michigan Entomological Society and was comprised of a series of papers, primarily on the Pentatomidae, contributed by colleagues including some former graduate students.

Contributors

Jeong Joon Ahn
National Institute of Horticultural and Herbal
  Science
Rural Development Administration
Jeju, 690-150
REPUBLIC OF KOREA
j2ahn33@korea.kr

Régis Babin
International Centre of Insect Physiology and
  Ecology (ICIPE)
P.O. Box 30772-00100
Nairobi
KENYA
regis.babin@cirad.fr

James A. Baum
Monsanto Company
700 Chesterfield Parkway West, BB2A
Chesterfield, Missouri 63017
U.S.A.
James.a.baum@monsanto.com

Eric P. Benson
Clemson University
130 McGinty Court
266 P&AS Building
Clemson, South Carolina 29634
U.S.A.
ebenson@clemson.edu

Maria Carolina Blassioli-Moraes
Laboratório de Semioquímicos
Embrapa Recursos Genéticos e Biotecnologia
Brasília
BRAZIL
carolina.blassioli@embrapa.br

Wenjun Bu
Institute of Entomology
College of Life Sciences
Nankai University
Tianjin 300071
CHINA
wenjunbu@nankai.edu.cn

C. Scott Bundy
Department of Entomology, Plant Pathology,
  and Weed Science
New Mexico State University
Las Cruces, New Mexico 88003
U.S.A.
cbundy@nmsu.edu

Jeffrey A. Davis
Department of Entomology
Louisiana State University Agricultural Center
Baton Rouge, Louisiana 70803
U.S.A.
jeffdavis@agcenter.lsu.edu

Joe E. Eger
2606 S. Dundee St.
Tampa, Florida 33629
U.S.A.
1-813-294-9467
Jeeger811@gmail.com

Jesus F. Esquivel
USDA, ARS, SPARC
2765 F&B Road
Insect Control and Cotton Disease Research Unit
College Station, Texas 77845
U.S.A.
Jesus.Esquivel@ars.usda.gov

Wayne A. Gardner
Department of Entomology
University of Georgia
Griffin Campus
Griffin, Georgia 30223
U.S.A.
wgardner@uga.edu

Tessa R. Grasswitz
Cornell University
Lake Ontario Fruit Team
12690 State Route 31
Albion, New York 14411
U.S.A.

Jocélia Grazia
Departamento de Zoologia
Instituto de Biociências
Universidade Federal do Rio Grande do Sul
    (UFRGS)
Av. Bento Gonçalves 9500, prédio 43435 Bairro
    Agronomia
Porto Alegre, RS 91501-970
BRAZIL
jocelia@ufrgs.br

Jeremy K. Greene
Clemson University
64 Research Road
Blackville, South Carolina 29817
U.S.A.
greene4@clemson.edu

George C. Hamilton
Department of Entomology
Rutgers University
96 Lipman Drive
New Brunswick, New Jersey 08901
U.S.A.
ghamilto@njaes.rutgers.edu

Kim A. Hoelmer
Beneficial Insects Introduction Research Unit
USDA ARS
501 S. Chapel St.
Newark, Delaware 19713
U.S.A.
kim.hoelmer@ars.usda.gov

Tracie M. Jenkins
Department of Entomology
University of Georgia
Griffin Campus
Griffin, Georgia 30223
U.S.A.
75meadow@gmail.com

Walker A. Jones
Biological Control of Pests Research Unit
National Biological Control Laboratory
ARS USDA
P.O. Box 67
Stoneville, Mississippi 38776
U.S.A.
walker.jones@gmail.com

George G. Kennedy
Department of Entomology and Plant Pathology
North Carolina State University
Box 7630
Raleigh, North Carolina 27695
U.S.A.
gkennedy@ncsu.edu

Ashot Khrimian
USDA Agricultural Research Service
Invasive Insect Biocontrol and Behavior
    Laboratory
BARC-West
Beltsville, Maryland 20705
U.S.A.
ashot.khrimian@ars.usda.gov

Yoshitomo Kikuchi
Bioproduction Research Institute (BPRI)
National Institute of Advanced Industrial Science
    and Technology (AIST) Hokkaido
2-17-2-1 Tsukisamu-higashi, Toyohira-ku
Sapporo 062-8517
JAPAN
y-kikuchi@aist.go.jp

Petr Kment
Department of Entomology
National Museum
Cirkusova 1740
193 000 Praha 9 – Horni Porcernice
CZECH REPUBLIC
sigara@post.cz

Thomas P. Kuhar
Department of Entomology
Virginia Tech
Blacksburg, Virginia 24061
U.S.A.
tkuhar@vt.edu

Joon-Ho Lee
Department of Agricultural Biotechnology
Seoul National University
Seoul, 08826
REPUBLIC OF KOREA
Jh7lee@snu.ac.kr

Tracy C. Leskey
USDA-ARS
Appalachian Fruit Research Laboratory
Kearneysville, West Virginia 25430
U.S.A.
Tracy.leskey@ars.usda.gov

Pierre Mbondji Mbondji
Laboratory of Entomology
Institute of Agricultural Research
Agriculture and Public Health Advisory Group
P.O. Box 8206 Yaoundé
CAMEROON
pmbondji.aphag@yahoo.fr

J. E. McPherson
Department of Zoology
Southern Illinois University
Carbondale, Illinois 62901
U.S.A.
mcpherson@zoology.siu.edu

Robert M. McPherson
Department of Entomology
University of Georgia
42 No Point Lane
Blairsville, Georgia 30512
U.S.A.
pherson@uga.edu

Enrique Gino Medrano
USDA-ARS
Insect Control and Cotton Disease Research Unit
2765 F&B Road
College Station, Texas 77845
U.S.A.
Gino.Medrano@ars.usda.gov

Esayas Mendesil
Department of Horticulture and Plant Sciences
Jimma University
Jimma, P.O. Box 307
ETHIOPIA
emendesil@yahoo.com

Jocelyn G. Millar
Department of Entomology
University of California
Riverside, California 92521
U.S.A.
jocelyn.millar@ucr.edu

Thomas A. Miller
2180 Prince Albert Drive
Riverside, California 92507
U.S.A.
chmeliar@gmail.com

Paula Levin Mitchell
Department of Biology
Winthrop University
Rock Hill, South Carolina 29733
U.S.A.
mitchellp@winthrop.edu

Harrison M. Mugo
Coffee Research Institute
Ruiru, P.O. Box 4-00232
KENYA
mugohmu@yahoo.com

Dmitry L. Musolin
Department of Forest Protection, Wood Science
    and Game Management
Saint Petersburg State Forest Technical University
Institutskiy per., 5
St. Petersburg 194021
RUSSIA
musolin@gmail.com

Fred R. Musser
Department of Biochemistry, Entomology,
    and Plant Pathology
Mississippi State University
P.O. Box 9775, 100 Old Hwy 12
Mississippi State, Mississippi 39762
U.S.A.
fm61@msstate.edu

Anne L. Nielsen
Department of Entomology
Rutgers University
96 Lipman Drive
New Brunswick, New Jersey 08901
U.S.A.
nielsen@njaes.rutgers.edu

John C. Palumbo
Department of Entomology
Yuma Agricultural Center
University of Arizona
Yuma, Arizona 85364
U.S.A.
jpalumbo@ag.arizona.edu

Antônio R. Panizzi
Laboratory of Entomology
Embrapa National Wheat Research Center
P.O. Box 3081
Passo Fundo, RS 99001-970
BRAZIL
antonio.panizzi@embrapa.br

Yong-Lak Park
Division of Plant and Soil Sciences
West Virginia University
Morgantown, West Virginia 26506
U.S.A.
Yopark@mail.wvu.edu

Thomas M. Perring
Department of Entomology
University of California
Riverside, California 92521
U.S.A.
thomas.perring@ucr.edu

Simone S. Prado
Laboratório de Quarentena "Costa Lima"
Embrapa Meio Ambiente
Rodovia SP 340 – Km 127,5 – Tanquinho Velho
Jaguariúna, SP, 13820-000
BRAZIL
simone.prado@embrapa.br

Wolfgang Rabitsch
Environment Agency Austria
Spittelauer Lände 5
1090 Vienna
AUSTRIA
wolfgang.rabitsch@umweltbundesamt.at

Francis P. F. Reay-Jones
Clemson University
2200 Pocket Road
Florence, South Carolina 29506
U.S.A.
freayjo@clemson.edu

Dávid Rédei
Institute of Entomology
College of Life Sciences
Nankai University
Tianjin 300071
CHINA
david.redei@gmail.com

Darcy A. Reed
Department of Entomology
University of California
Riverside, California 92521
U.S.A.
darcy.reed@ucr.edu

David A. Rider
Department of Entomology
North Dakota State University, Dept. 7650
P.O. Box 6050
Fargo, North Dakota 58108
U.S.A.
david.rider@ndsu.edu

Phillip M. Roberts
Department of Entomology
University of Georgia
2360 Rainwater Rd.
Tifton, Georgia 31794
U.S.A.
proberts@uga.edu

N. D. T. M. Rukazambuga
School of Agriculture, Rural Development and
    Agricultural Economics
College of Agriculture, Animal Sciences and
    Veterinary Medicine
University of Rwanda
Butare, P.O. Box 117
RWANDA
dnrukazambuga@gmail.com

Aida Kh. Saulich
Department of Entomology
Saint Petersburg State University
Universitetskaya nab., 7/9
St. Petersburg 199034
RUSSIA
325mik40@gmail.com

Cristiano F. Schwertner
Departamento de Ciências Biológicas
Universidade Federal de São Paulo–Campus
    Diadema
Rua Artur Riedel 275, 09972-270 Diadema, SP
BRAZIL
schwertner@unifesp.br

Mario Serracin
Rogers Family Company
Huye
RWANDA
mserracin@rogersfamilyco.com

Dan R. Suiter
Department of Entomology
University of Georgia
Griffin Campus
1109 Experiment Street
Griffin, Georgia 30223
U.S.A.
dsuiter@uga.edu

Donald B. Thomas
USDA-ARS Cattle Fever Tick Research
    Laboratory
Moore Air Base
Edinburg, Texas 78541
U.S.A.
donald.thomas@ars.usda.gov

Michael D. Toews
Department of Entomology
University of Georgia
Tifton, Georgia 31793
U.S.A.
mtoews@uga.edu

Jitka Vilímová
Department of Zoology
Charles University
Vinicna 7
128 44 Praha 2
CZECH REPUBLIC
vilim@natur.cuni.cz

James F. Walgenbach
Department of Entomology
North Carolina State University
MHCREC/455 Research Drive
Mills River, North Carolina 28759
U.S.A.
Jim_walgenbach@ncsu.edu

Donald C. Weber
USDA Agricultural Research Service
Invasive Insect Biocontrol and Behavior
    Laboratory
BARC-West
Beltsville, Maryland 20705
U.S.A.
Don.Weber@ars.usda.gov

Alfred G. Wheeler, Jr.
Department of Agricultural and Environmental
    Sciences
277 Poole Agricultural Center
Clemson University
Clemson, South Carolina 29634
U.S.A.
awhlr@clemson.edu

Adam R. Zeilinger
Department of Environmental Science, Policy,
    and Management
University of California
Berkeley, California 94720
U.S.A.
arz@berkeley.edu

Section I

Introduction

1

Overview of the Superfamily Pentatomoidea[1,2]

J. E. McPherson, C. Scott Bundy, and Alfred G. Wheeler, Jr.

CONTENTS

1.1 General Information.. 3
1.2 Classification Overview.. 4
1.3 Biology.. 4
1.4 General Life History.. 5
1.5 Predators and Parasitoids.. 7
1.6 Chemical Defenses of Pentatomoids.. 7
1.7 Management Practices 8
1.8 Pentatomoid Fauna: Potential Impact of Invasive Insects on Noninvasive Species 8
    1.8.1 Terminology.. 8
    1.8.2 What Factors Allow Species to Become Invasive? 9
    1.8.3 Harmful Effects of a Successful Invasion by Insects.. 9
1.9 Invasive Pentatomoids... 9
    1.9.1 History of Invasive Pentatomoids in America North of Mexico........................... 10
        1.9.1.1 *Bagrada hilaris* (Burmeister), Bagrada Bug or Painted Bug............ 10
        1.9.1.2 *Halyomorpha halys* (Stål), Brown Marmorated Stink Bug.............. 10
        1.9.1.3 *Megacopta cribraria* (F.), Kudzu Bug... 10
        1.9.1.4 *Murgantia histrionica* (Hahn), Harlequin Bug.................................. 10
        1.9.1.5 *Nezara viridula* (L.), Southern Green Stink Bug.............................. 11
        1.9.1.6 *Piezodorus guildinii* (Westwood), Redbanded Stink Bug................ 11
1.10 Potentially Invasive Pentatomoids... 11
    1.10.1 *Oebalus* spp. 11
    1.10.2 *Arvelius albopunctatus* (De Geer), Tomato Stink Bug 11
    1.10.3 Other Species... 11
1.11 Noninvasive Group (A Comparison).. 12
1.12 Key to Families of Pentatomoidea in America North of Mexico 12
1.13 Chapters 2–16.. 17
1.14 Acknowledgments.. 17
1.15 References Cited.. 17

1.1 General Information

The superfamily Pentatomoidea (stink bugs and their relatives) comprises 18 families worldwide (including two fossil families) with over 8,000 species, the largest of which is the Pentatomidae (about 5,000 species) (**Table 2.2**). Six families are represented in America north of Mexico: Acanthosomatidae (acanthosomatids or parent bugs), Cydnidae (burrower bugs), Pentatomidae (stink bugs), Scutelleridae (shieldbacked

---

[1] This chapter was modified and updated (in part) from Stink bugs of economic importance in America north of Mexico by J. E. McPherson and R. M. McPherson. Copyright 2000 CRC Press.

[2] Statements describing content of **Chapters 2 and 9–16** were contributed by one or more authors of those chapters.

or jewel bugs), Thyreocoridae (black bugs or ebony bugs), and, recently, Plataspidae (plataspids). Most species in these families are phytophagous, the major exception being the asopine pentatomids, which are predaceous. Within each of these families are species that cause economic injury to crops in the New World, Old World, and worldwide (Schuh and Slater 1995, Schaefer and Panizzi 2000, Eger et al. 2010, Ruberson et al. 2013). The Pentatomidae, largest of the six pentatomoid families, contains the highest number of economically important species (Schuh and Slater 1995, Schaefer and Panizzi 2000).

1.2 Classification Overview

The Pentatomoidea are members of the order Hemiptera and suborder Heteroptera (true bugs). The Heteroptera are recognized by a segmented beak that arises from the front of the head; and wings that, when present and well developed, have the first pair leathery basally and membranous distally (**Figure 1.1A**), the second pair membranous, with both pairs lying flat on the abdomen. The other suborders (i.e., Auchenorrhyncha, Sternorrhyncha, Coleorrhyncha) have the segmented beak appearing to arise ventrally from the rear of the head or between the front coxae; and wings, that when present and well developed, usually are held rooflike over the abdomen with both pairs of uniform texture throughout.

As with most other insect groups, the higher classification of the Pentatomoidea has changed considerably over time. Various families and subfamilies have been raised, lowered, then raised again between those two categories on a number of occasions. With few exceptions, the family and subfamily classification now seems relatively stable (see Grazia et al. 2008). However, the tribal classification remains in a state of chaos and requires a thorough phylogenetic analysis. The higher classification of the Pentatomoidea is discussed in **Chapter 2**. A list of the families, subfamilies, and tribes currently in use is presented along with notes concerning how these taxa are defined. The generic groups of Gross (1975, 1976) and Linnavuori (1982) also are discussed. Keys are provided to the families of the Pentatomoidea and subfamilies and tribes of the Pentatomidae. Some preliminary speculation is given on the validity of these taxa and their phylogenetic relationships.

1.3 Biology

As noted above, most pentatomoids are phytophagous. Phytophagous pentatomoids feed on a wide variety of fruit, vegetable, nut, and grain crops as well as wild hosts. Generally, the bugs feed on roots, stems, and leaves but most often are associated with developing seeds, fruits, or growing shoots. Adults and nymphs obtain nutrients by piercing the plant tissues with their mandibular and maxillary stylets.

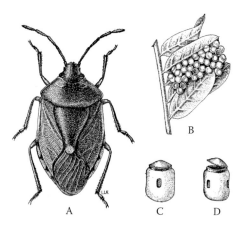

FIGURE 1.1 Adult and eggs of a typical stink bug, *Chlorochroa ligata*. A, Adult; B, Egg mass; C, Individual egg; D, Empty egg showing displaced pseudoperculum. (Modified from Morrill 1910).

Although all stages feed on plant material (often excluding the first instars), adults and/or fifth instars cause the most injury, at least to soybeans. Often, the bug leaves a salivary sheath at the feeding site (see discussion in McPherson and McPherson 2000, p. 3).

Many phytophagous insects, including aphids, sharpshooters, and weevils, possess symbiotic microorganisms; these symbionts play major roles in the lives of their hosts including provision of essential nutrients, digestion of food materials, and protection of their hosts from parasites, pathogens, and/or chemical pesticides. Among phytophagous insects, the Pentatomoidea exhibit extraordinary diversity in their symbiotic system, including morphology of the symbiotic organ, mechanism of symbiont transmission, and host-symbiont interdependency. The current biological knowledge of the diverse stink bug-microbe associations is reviewed in **Chapter 14**, highlighting the role of symbiotic microorganisms in the evolution of pentatomoid species. Injury can be associated with microbial symbionts, which can affect the taste of the crop (e.g., coffee, tea; see **Chapter 10**).

Compounding the detrimental effects of feeding is the ability to transmit plant pathogens. Pentatomids are known or suspected to transmit a variety of disease-causing plant pathogens including the causal pathogens for seed and boll rot, yeast spot, leaf spot and vein necrosis, stem canker, stigmatomycosis, panicle and shoot blight, witches' broom, hartrot, and marchitez. Affected crops range from pistachio and oil palms to cotton, soybean, and cowpea. Organized by vector-borne diseases, the relationships between various pentatomid species and fungi, bacteria, phytoplasmas, viruses, and trypanosomatids are explored and reviewed in **Chapter 13**, including an examination of feeding, transmission, and vector-pathogen interactions.

1.4 General Life History

Adults (sometime nymphs, rarely eggs) overwinter beneath leaf litter and other ground debris, usually remaining inactive. Some species [e.g., *Murgantia histrionica* (Hahn), *Nezara viridula* (L.)] can become active during milder temperatures with feeding, copulation, and oviposition possible. In fact, *N. viridula* will even feed when in reproductive diapause and, apparently, feeding during this time seems to enhance overwintering survival (see McPherson and McPherson 2000). In some instances, when winters are mild, species remain active throughout the winter. Examples include *Bagrada hilaris* (Burmeister) (Taylor al. 2015) (**Chapter 3**) and *Mecidea minor* Ruckes (Bundy and McPherson 2011), which are found in southern North America.

Diapause and related phenomena in the Pentatomidae and other pentatomoid families are reviewed in **Chapter 11**. Using pentatomoids as examples, the consecutive stages of the complex dynamic process of diapause (such as diapause preparation, induction, initiation, maintenance, termination, post-diapause quiescence, and resumption of direct development) are described and discussed.

Adults emerge in the spring as temperatures rise and begin feeding and reproducing on grasses, herbaceous plants, shrubs, and trees, depending on the species. As noted above, they are attracted most often to the developing seeds, fruits, or growing shoots. In fact, they will move from host to host as earlier hosts pass peak suitability and that of later hosts approaches (McPherson and McPherson 2000). Reproduction of the bugs begins shortly thereafter. For those few species that overwinter as eggs or nymphs, the patterns of their life cycles in spring are somewhat different.

Precopulatory and copulatory behaviors have been reported for several species (e.g., McPherson 1982), and certain patterns are apparent. Mating usually begins with the male antennating various parts of the female's body but eventually concentrating on or near the tip of her abdomen. If she is receptive, she will raise the tip of her abdomen for aedeagal insertion. If she is not receptive, the male may replace or combine antennating with head butting and may attempt to lift her abdomen with his head. If he is successful in stimulating the female to lift the tip of her abdomen, he will turn 180°, elevate his abdomen, and attempt to insert his aedeagus. If he is successful in doing so, copulation may last for several hours in this end-to-end position. Both adults may feed during this time, the female sometimes dragging the male along. When the female is not receptive, she may not elevate her abdomen, or may kick at the male with her hind legs, or simply walk away (McPherson 1982, McPherson and McPherson 2000).

Eggs of the Pentatomoidea, though superficially similar as a group, are variable in their morphology depending upon the family, features of which are helpful in their identification (see Southwood 1956, Hinton 1981, Javahery 1994). A ring of micropylar processes of various shapes and sizes is present at the cephalic end of the pentatomoid egg. Members of the Pentatomidae, Plataspidae, Scutelleridae, and Tessaratomidae have a thick chorion (thinner in the Plataspidae) with a pseudoperculum, a caplike structure through which the hatching nymph emerges (Hinton 1981, McPherson 1982) (**Figure 1.1C, D**). However, members of the Acanthosomatidae, Cydnidae, and Thyreocoridae have a thin chorion, which splits irregularly when the nymph hatches (McPherson 1982). An egg burster, which aids in emergence from the egg, is present in embryos of at least some species in all these families (Hinton 1981, McPherson 1982).

In most Pentatomoidea, eggs are laid on the host plants in round or subhexagonal clusters on the leaves or in longitudinal rows generally on the leaves and stems; the eggs adhere to the plant and to each other by a sticky secretion (**Figure 1.1B**). Members of the Plataspidae deposit their eggs in two rows along with fecal pellets containing bacterial symbionts that enhance survival of the nymphs (Hosokawa et al. 2007, Ruberson et al. 2013; see **Chapter 5**). In rare instances, the eggs are laid singly in the soil although the sticky secretion is still evident (e.g., *Bagrada hilaris*, Taylor et al. 2014b; see **Chapter 3**). In the Thyreocoridae, eggs are laid singly on the host plant; and in the Cydnidae, eggs are laid in loose clusters in the soil (McPherson 1982). In the closely-related Parastrachiidae, *Parastrachia japonensis* Scott lays its eggs in clusters of up to 100 in shallow nests in the leaf litter (Filippi et al. 2001). Recently, Cervantes et al. (2013) reported that the cydnid *Melanaethus crenatus* (Signoret), lays its eggs singly in the soil, usually close together.

The eggs usually are abandoned immediately after oviposition but members of most pentatomoid families contain species in which females exhibit parental care (e.g., Acanthosomatidae, Pentatomidae, Scutelleridae, Cydnidae, and Parastrachiidae), guarding the eggs and young nymphs (e.g., Eberhard 1975, McPherson 1982, Sites and McPherson 1982, Tallamy and Schaefer 1997, Peredo 2002, Costa 2006).

Pentatomoids have five nymphal instars (**Figure 1.2**). The first instars generally are gregarious, inactive, and remain atop or near the egg shells during the stadium. If they are disturbed, the cluster begins to break up and the individuals seem unable to reaggregate. Although first instars of pentatomids generally are thought not to feed (McPherson 1982), they may acquire symbionts by sucking the secretions covering the shells of the unhatched eggs (McPherson 1982, McPherson and McPherson 2000). In some species

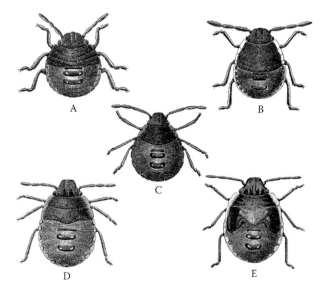

FIGURE 1.2 Nymphal instars of a typical stink bug, *Chlorochroa ligata*. A-E, First through fifth instars, respectively. (Modified from Morrill 1910).

[e.g., *Halyomorpha halys* (Stål)], the female defecates on the egg, and the first instars acquire symbionts by feeding on her feces (Taylor et al. 2014a). Also, the first instars of *Nezara viridula* apparently do feed (Esquivel and Medrano 2014), although the probing may be for water rather than nutrient uptake. Among the other families of pentatomoids, the first instars of at least some species of Thyreocoridae and Cydnidae are active and feed on the host plants (McPherson 1982, Sites and McPherson 1982, Bundy and McPherson 2009, Cervantes et al. 2013).

For those species that aggregate during the first stadium, the later instars begin to gradually and progressively disperse. This tendency to remain clustered in earlier instars may provide some protection from predation (Lockwood and Story 1986).

As the nymphs pass through the various stadia, subtle morphological changes are evident, particularly in the fourth and fifth instars. These two older instars can be distinguished from younger instars by the lengths of the wing pads, which are longest in the fifth instars. The eggs and nymphal instars have been described for many pentatomoid species, including the six invasive species discussed in this book (e.g., Moizuddin and Ahmad 1975, McPherson 1982, McPherson and McPherson 2000, Hoebeke and Carter 2003, Zhang et al. 2012, Leslie et al. 2014, Taylor et al. 2015).

Species of pentatomoids in northern and central North America generally are uni- or bivoltine with the number increasing to five in the extreme south (McPherson 1982, McPherson and McPherson 2000). The diversity of seasonal cycles known in the Pentatomoidea, mostly from the Temperate Zone, including those that are uni-, bi-, multi-, and semivoltine, is reviewed in **Chapter 12**. Further discussion focuses on the ecological importance of photoperiodic and thermal responses in natural or artificial expansions of pentatomids outside their original ranges.

1.5 Predators and Parasitoids

Pentatomoids are attacked by numerous invertebrate and vertebrate species, both parasitoids and predators. The parasitoids consist primarily of hymenopteran egg parasitoids (e.g., Scelionidae) and tachinid flies (Tachinidae). Numerous predators have been recorded, including predaceous stink bugs (e.g., *Podisus*). Among vertebrates, birds often have been reported as predators of pentatomoids (McPherson 1982). Native fungal pathogens, nematodes, and other parasitic organisms also have been recorded from invasive stink bugs (e.g., Sosa-Gómez and Moscardi 1998).

As with most other invasive insect pests, alien pentatomoids have arrived in new geographic areas without their most important coevolved natural enemies, and, thus, some have been targets for classical biological control. One of the best-known projects against an invasive pest insect was the introduction of the egg parasitoid *Trissolcus basalis* (Wollaston) (Platygastridae) for management of the southern green stink bug, *Nezara viridula*, across five continents (Walker A. Jones, personal communication).

Tachinid flies attacking pentatomids include certain parasitic species that have been subjects of classical projects. The recent invasions of the brown marmorated stink bug, *Halyomorpha halys*; the painted bug, *Bagrada hilaris* (Pentatomidae); and the bean plataspid or kudzu bug, *Megacopta cribraria* (F.) (Plataspidae) into North America have triggered projects that are in various stages of progress (Walker A. Jones, personal communication).

1.6 Chemical Defenses of Pentatomoids

Pentatomoids, like many other heteropterans, have a rich array of semiochemical compounds that function as pheromones, allomones, synomones, and kairomones. Although probably best known for their allomonal scent-gland secretions, pentatomoids have many semiochemical intra- and interspecific interactions that are just being uncovered. Among the increasingly evident complex chemistry and relationships are defenses against parasitoids and predators, eavesdropping by these natural enemies, cross-species attraction, and intraspecific variation in production of and response to semiochemicals based on life stage and physiology. The exciting discoveries in this area and the potential and actual uses of these chemicals in pest detection, monitoring, and management are addressed in **Chapter 15**.

1.7 Management Practices

Management tactics are considered in detail in **Chapter 16**. Beginning with a brief overview of the types of control, the history of these tactics is discussed. The earliest control practice apparently was the use of sulfur by the Sumerians in 2500 B.C. to control insects and mites. Following were reports of cultural, botanical, and biological control, the practices of which moved from the Old World to the New World and continued to become more sophisticated. The dramatic effect of the introduction of the synthetic organic insecticides for insect control in the 1940s, the resulting detrimental effects of their widespread use, the resurgence of research in biological control, and the development of integrated pest management are discussed in detail. Following is a detailed discussion of control practices in the modern era ending with a discussion of future management practices.

1.8 Pentatomoid Fauna: Potential Impact of Invasive Insects on Noninvasive Species

1.8.1 Terminology

Invasion biology (ecology, science), the study of organisms that become established in areas outside their native ranges, assumed prominence in the 1980s. Since the late 1990s, the number of books and journal articles on nonnative species has exploded. Those who consider themselves invasion biologists represent numerous disciplines and work with disparate taxa. The controversies that pervade the literature on nonnative species (e.g., Simberloff 2012, Richardson and Ricciardi 2013, Valéry et al. 2013) might have been anticipated in a field characterized by practitioners from dissimilar backgrounds and interests. Some of invasion biology's critics represent disciplines other than biology or ecology, such as history, philosophy, and sociology (Simberloff 2003). An unfounded criticism is that invasion biologists are xenophobes who regard all nonnative species as "bad" when, in fact, the benefits of such species often are mentioned (e.g., Simberloff 2003, Wheeler and Hoebeke 2009). Helping to fuel controversy is the emotionalism that infects the language of invasion biology, which shares certain military metaphors (e.g., invader, invasion) used by L. O. Howard in the late nineteenth and early twentieth century to promote the importance of economic entomology (Russell 1999). Especially contentious among invasion biologists has been use of the terms "invasion" and "invasive species" (e.g., Colautti and MacIsaac 2004, Colautti and Richardson 2009).

Attempts to standardize terminology involving plant and animal invasions, as noted by Davis (2009), Blackburn et al. (2011), and Heger et al. (2013), have met with minimal success. The term "invasive species" continues to be used inconsistently; it can refer to any nonnative organism (Wheeler and Hoebeke 2009, Simberloff 2011) and be used with or without consideration of impact. Simberloff (2013) prefers a biologically based definition of organisms that become invasive: "species that arrive with human assistance [intentionally, as well as inadvertently], establish populations, and spread." Similarly, Pyšek and Richardson (2006) would restrict invasive species to those that spread rapidly, regardless of any ecological harm or economic loss that might accrue. Definitions that incorporate harm or impact are subjective because they introduce human perception and values, which can vary regionally (Lodge et al. 2006). Moreover, the addition of impact in defining "invasive" tends to obscure an ecological and evolutionary appreciation of the invasion process (Colautti and Richardson 2009).

Yet, it is unrealistic for entomologists to avoid using the value-based term "pests" in referring to insects that adversely affect human well-being. In addition, reference to harm or impacts of invasive species seems unavoidable when considering the potential risks that nonnative species pose for natural ecosystems (Ward et al. 2008) or assessing the ecological and economic consequences of alien insects (e.g., Pimentel et al. 2005, Kenis et al. 2009, Kenis and Branco 2010, Aukema et al. 2011, Vinson 2013, Herms and McCullough 2014). Davis (2009), even though he preferred other terminology, acknowledged the absurdity of omitting the word "invasion" from his book on invasion biology. Regulatory agencies likewise are compelled to introduce human values in defining impacts (Jeschke et al. 2014). Similarly, we need to include harm in our definition of invasive species, which, as Davis (2009) pointed out, is consistent with usage adopted by the Global Invasive Species Program, International Convention on Biological Diversity, and the United States National Invasive Species Council. Thus, contributors throughout this book will use the following definition for their chapters: ***invasive pentatomoids are nonnative species whose populations become established and spread, sometimes exhibiting uncontrolled population growth and displacing native populations, and cause adverse socioeconomic, environmental, or human-health effects***.

Species not considered invasive can be termed adventive, immigrant, introduced, and native (Wheeler and Hoebeke 2009). Adventive is an inclusive term that refers to any species that is not native. Adventives can be immigrants, that is, species not deliberately or intentionally introduced; or they can be purposeful introductions, such as natural enemies used in biological control. All remaining species, then, represent the native fauna. Native species also can become invasive (Buczkowski 2010, Davis et al. 2011), but for the purposes of this book, all invasive species are adventive, but not all adventive species are invasive.

1.8.2 What Factors Allow Species to Become Invasive?

Researchers long have attempted to identify attributes that enable plants and animals to be successful invaders and, independently, to determine environmental conditions that favor invasion. The invasion process, however, is best understood by appreciating the interconnectedness of environmental conditions and organisms' traits as well as the significance of propagule pressure (Lockwood et al. 2005, Pyšek and Richardson 2006, Davis 2009, Su 2013, Jeschke 2014). Several characteristics, however, tend to facilitate invasiveness, although not all insect species possessing these traits become invasive (Su 2013). Traits that can favor invasiveness include: (1) ability to move readily in commerce; (2) tolerance of multiple habitat conditions, such as tolerance of temperature and humidity; (3) ability to reproduce rapidly with a concomitant increase in population size; (4) ability to compete successfully for resources; (5) lack of natural enemies; and (6) ability to fly.

1.8.3 Harmful Effects of a Successful Invasion by Insects

Wheeler and Hoebeke (2009) listed several harmful effects including (1) transmission of animal (and plant) diseases, (2) extreme crop damage, (3) severe damage to forests, homes, and gardens, (4) elimination of competing native species, and (5) effects on evolution. To these can be added effects on recreational areas and ecosystem processes and the insects becoming nuisance urban pests.

1.9 Invasive Pentatomoids

Invasive insect species worldwide consist primarily of Hymenoptera (Formicidae, Vespidae) but also many include other groups such as Coleoptera, Isoptera, Lepidoptera, and Hemiptera (e.g., Anonymous 2014). Within the Hemiptera, one such group is the Pentatomoidea. In Europe and Asia, only two pentatomoid species are listed as invasive, *Halyomorpha halys* and *Nezara viridula* (Rabitsch 2008). However, six invasive species are found in North America, including *H. halys* and *N. viridula*. The additional four species include *Bagrada hilaris*, *Murgantia histrionica*, *Piezodorus guildinii* (Westwood), and *Megacopta cribraria*.

Certain biological characteristics and environmental factors enhance the ability of pentatomoids to become invasive species, Panizzi (2015) discusses both topics under the following categories: (1) polyphagy (feeding on a wide range of host plants), (2) ability to survive unfavorable conditions, and (3) climate change. In addition, specifically concerning the Neotropics (but also the impact of pentatomoids on a worldwide basis), other factors include changes in cultivation practices, growth of agribusiness, and increased trade.

In the United States, injury caused by stink bugs in the native fauna has increased for one major reason, at least for cotton – "a reduction in the frequency of foliar, broad-spectrum insecticide applications" (Greene et al. 2006). This reduction in pesticide use may have played a role in the dramatic increase in populations of the three most recent invasive pentatomoids, *Bagrada hilaris*, *Halyomorpha halys*, and *Megacopta cribraria*.

1.9.1 History of Invasive Pentatomoids in America North of Mexico

1.9.1.1 *Bagrada hilaris* (Burmeister), Bagrada Bug or Painted Bug

This species (**Chapter 3**) was described by Burmeister as *Cimex hilaris* in 1835 (p. 368). Although the locality was not given, and the type was lost, Fabricius had described the same species in 1775. But the name he proposed, *Cimex pictus*, was invalid because it was a primary homonym of *Cimex pictus* Drury (1770). The locality of this specimen was given as India.

This stink bug, which somewhat resembles a small harlequin bug, has an Old World distribution of Asia (including India), Africa, southern Europe, and the Middle East (Taylor et al. 2015). First reported in the United States from southern California in June 2008, *Bagrada hilaris* has extended its range north and south in California and east to Nevada, Utah, Arizona, New Mexico, and western Texas (Reed et al. 2013).

Bagrada hilaris prefers cruciferous crops (Halbert and Eger 2010, Reed et al. 2013) and has reached economic importance in California and Arizona (Palumbo and Natwick 2010). It attacks broccoli, cabbage, cauliflower, kale, collards, and radish but also will injure sunflower, corn, and cotton, among others (Reed et al. 2013).

1.9.1.2 *Halyomorpha halys* (Stål), Brown Marmorated Stink Bug

This Asian species (**Chapter 4**) was described by Stål as *Pentatoma halys* in 1855 from China (p. 182) and now is considered a recent invasive in Europe (Rabitsch 2008, Milonas and Partsinevelos 2014) and North America (Hoebeke and Carter 2003, Fogain and Graff 2011). It first was reported in the United States by Hoebeke and Carter in 2003 from sightings in Allentown, PA, in fall 1996 (Adams Island), September 1998, and January 1999. It has spread rapidly in the intervening years and now occurs in more than 41 states and the District of Columbia (Leskey et al. 2012, Wallner et al. 2014) and in Ontario and Quebec, Canada (Fogain and Graff 2011). It is highly polyphagous, feeding on a wide variety of agricultural and nonagricultural plants including ornamentals, hardwood trees, field crops, tree and small fruits, vegetables, and wild plants (Nielsen and Hamilton 2009, Wallner et al. 2014). It also is considered a significant nuisance because it overwinters in houses, garages, offices, and other similar enclosures (Inkley 2012, Leskey et al. 2012).

1.9.1.3 *Megacopta cribraria* (F.), Kudzu Bug

This species (**Chapter 5**), although not a stink bug, is member of the pentatomoid family Plataspidae. Described by Fabricius as *Cimex cribraria* in 1798 from India (p. 531), it later was reported from various localities in Asia and the Indian subcontinent (Eger et al. 2010). It first was reported in the United States in 2009 from Georgia and, as of 2012, had spread to seven states in the Southeast (Ruberson et al. 2013) and, as of 2013, as far north as Maryland and the District of Columbia (Leslie 2014). The preferred host plants are kudzu and soybeans, both of which are legumes.

1.9.1.4 *Murgantia histrionica* (Hahn), Harlequin Bug

This species (**Chapter 6**) apparently was the first invasive stink bug for the United States and certainly the first for which substantial records are available. Described by Hahn as *Strachia histrionica* in 1834 from Mexico (p. 116), it first was reported in the United States by Walsh (1866) from specimens collected in 1864 from Washington Co., TX. Its spread was monitored closely after its detection because of its potential to injure crucifers. Today, it ranges in the continental United States primarily from New England south to Florida and west to Minnesota, South Dakota, Nebraska, and California but occurs primarily in the South (McPherson 1982). It is an established immigrant in Hawaii (Froeschner 1988).

1.9.1.5 *Nezara viridula* (L.), Southern Green Stink Bug

This species (**Chapter 7**) was described by Linnaeus as *Cimex viridulus* in 1758 from India (p. 444). It subsequently was reported from other parts of Asia and Europe and the New World including the West Indies, Jamaica, St. Domingo, Cuba, and Venezuela (DeWitt and Godfrey 1972). The earliest record for the United States was by Distant (1880) (p. 78), who reported it from the southern states, including Texas. Today, it occurs primarily from Virginia to Florida west to Texas and Oklahoma and also occurs in California (McPherson and McPherson 2000) and Washington (see **Chapter 7**). It feeds on a wide variety of plants including soybeans, tomatoes, vegetables, row crops, cruciferous vegetation, and leguminous weeds (McPherson and McPherson 2000).

1.9.1.6 *Piezodorus guildinii* (Westwood), Redbanded Stink Bug

This species (**Chapter 8**) was described by Westwood as *Rhaphigaster guildinii* in 1837 from St. Vincent Island (p. 31) and today ranges from the West Indies to South America and north to the southern United States (Panizzi et al. 2000). The earliest records for the United States of which we are aware are southern Florida (Uhler 1894) and New Mexico (Van Duzee 1904). This species previously was collected primarily in Florida but now has extended its range to include South Carolina, Tennessee, Georgia, Alabama, Mississippi, Arkansas, Missouri, Louisiana, Texas, and New Mexico (Van Duzee 1904, McPherson and McPherson 2000, Bundy 2012, Davis 2012, Temple et al. 2013, Vyavhare et al. 2014) (Bundy 2012 noted that the bug's presence in New Mexico needs to be verified). It seems to prefer soybeans but will attack many other plants such as alfalfa, clover, cotton, kidney bean, lentil, peanut, strawberry, and others (McPherson and McPherson 2000, Panizzi et al. 2000). It has reached economic importance in Louisiana and Texas (Temple et al. 2013, Vyavhare et al. 2014). The important question is why does it now appear to be expanding its range and increasing in numbers in the southern states.

1.10 Potentially Invasive Pentatomoids

1.10.1 *Oebalus* spp.

The genus *Oebalus* Stål (1862) (**Chapter 9**) contains three species that occur in the United States including *O. pugnax* (F.), *O. insularis* Stål (1872), and *O. ypsilongriseus* (De Geer) (1773). Although *O. pugnax* is widely distributed in the United States (Froeschner 1988), *O. insularis* and *O. ypsilongriseus* have been recorded only from Florida. All three species are noted pests of rice.

1.10.2 *Arvelius albopunctatus* (De Geer), Tomato Stink Bug

This species occurs from South America north to Arizona, Texas, and Florida in the United States. It is a pest of several economically important crops in Brazil and Mexico but has been reported to feed only on solanaceous weeds in the United States.

1.10.3 Other Species

Recently, Panizzi (2015) authored a paper on the invasive species of the pentatomids in the United States, which includes a section on species of potential invaders from the Neotropics. This list includes the following five species, all of which are of major economic importance in South America: *Dichelops furcatus* (F.), *Dichelops melacanthus* (Dallas), *Edessa meditabunda* (F.), *Euschistus heros* (F.), and *Tibraca limbativentris* Stål.

1.11 Noninvasive Group (A Comparison)

A group of pentatomids, commonly known as antestia bugs, has been the object of many studies as important pests of Arabica coffee in Africa. They currently attract interest because they are supposed to be the cause of potato taste defect (PTD), which significantly downgrades the value of the coffee crop, especially in the Great Lakes region of Africa. The specific linkage between PTD and stink bugs is not known, although symbiotic microorganisms might be involved. The current knowledge of the distribution, life history, and natural enemies of antestia bugs, as well as their injury, role in coffee potato taste defect, economic impact, and control is reviewed in **Chapter 10**. Also discussed are their current ecological range and distribution on coffee. These bugs are particularly interesting because they appear capable of becoming invasive species wherever coffee is grown but have remained limited to Africa and Asia. Yet other major coffee pests, such as the coffee berry borer, which also originated in Africa, are found wherever coffee is grown. Possible reasons are offered to explain the geographical differences between antestia bugs and other coffee pests.

1.12 Key to Families of Pentatomoidea in America North of Mexico[3]

1. Tarsi 2-segmented (**Figure 1.3**)... 2
1'. Tarsi 3-segmented (**Figure 1.4**)... 3
2. Scutellum U-shaped, greatly enlarged, covering wings and most of abdomen; abdomen widest subapically (**Figure 1.5**)...Plataspidae
2'. Scutellum subtriangular, small, not covering wings and most of abdomen; abdomen not widest subapically (**Figure 1.6**).. Acanthosomatidae
3. Pronotum expanded posteriorly, covering base of scutellum (**Figure 1.7**).......Tessaratomidae[4]
3'. Pronotum not expanded posteriorly, not covering base of scutellum....................................... 4
4. Scutellum U-shaped, greatly enlarged, covering wings and most of abdomen (**Figures 1.8 and 1.9**).. 5
4'. Scutellum usually subtriangular, not greatly enlarged, not covering most of abdomen (**Figures 1.10 and 1.11**); if scutellum large and U-shaped, then colors bright and contrasting or prominent tooth present either side of each anterolateral angle of pronotum.................................. 6
5. Tibiae with strong spines (**Figure 1.12**); color basically shiny black (**Figure 1.9**).................... ....Thyreocoridae
5'. Tibiae without strong spines (**Figure 1.13**); color variable but never shiny black (**Figure 1.8**)... ....Scutelleridae
6. Tibiae with strong spines (**Figure 1.10**); front legs fossorial (**Figure 1.14**) or cultrate (**Figure 1.15**)..Cydnidae
6'. Tibiae without strong spines (**Figure 1.16**); front legs not fossorial or cultrate ... Pentatomidae

---

[3] Modified from Eger et al. (2010).

[4] Represented by *Piezosternum subulatum* (Thunberg), which occurs from Mexico to Brazil and the West Indies but has not yet been found in the United States (Froeschner 1988).

FIGURE 1.3 Two-segmented tarsus of *Megacopta cribraria* (Plataspidae). (Courtesy of J. E. Eger)

FIGURE 1.4 Three-segmented tarsus of *Apoecilus cynicus* (Pentatomidae). (Courtesy of J. E. Eger)

FIGURE 1.5 Habitus of *Megacopta cribraria* (Plataspidae). (Courtesy of J. E. Eger)

FIGURE 1.6   Habitus of *Elasmucha* sp. (Acanthosomatidae). (Courtesy of J. E. Eger)

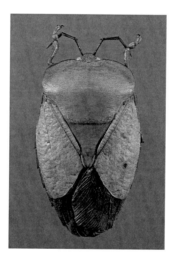

FIGURE 1.7   Habitus of *Catacanthus* sp. (Tessaratomidae). (Courtesy of C. Scott Bundy)

FIGURE 1.8   Habitus of *Augocoris* sp. (Scutelleridae). (Courtesy of J. E. Eger)

FIGURE 1.9 Habitus of *Corimelaena* sp. (Thyreocoridae). (Courtesy of J. E. Eger)

FIGURE 1.10 Habitus of *Pangaeus* sp. (Cydnidae). (Courtesy of J. E. Eger)

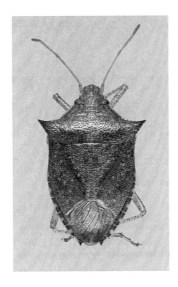

FIGURE 1.11 Habitus of *Euschistus quadrator* (Pentatomidae). (Courtesy of J. E. Eger)

FIGURE 1.12   Hind tibia of *Corimelaena lateralis* (Thyreocoridae). (Courtesy of C. Scott Bundy)

FIGURE 1.13   Hind tibia of *Tetyra robusta* (Scutelleridae). (Courtesy of C. Scott Bundy)

FIGURE 1.14   Fossorial foretibia of *Cyrtomenus* sp. (Cydnidae). (Courtesy of J. E. Eger)

FIGURE 1.15   Cultrate foretibia of *Atarsocoris* sp. (Cydnidae). (Courtesy of J. E. Eger)

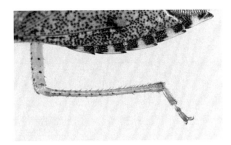

FIGURE 1.16 Hind tibia of *Euschistus servus* (Pentatomidae). (Courtesy of J. E. Eger)

1.13 Chapters 2–16

In the following chapters, we summarize and update the knowledge of the extent of problems associated with invasive pentatomoids worldwide including economic importance (feeding injury, nuisance problems, transmission of pathogens, role of symbionts); research on their biology (including diapause and voltinism), behavior, chemical ecology, monitoring, and control. We discuss why these species have become invasive and what the future holds for their continued geographic expansion and the resulting direct and indirect effects of these bugs on the human population.

1.14 Acknowledgments

We thank J. E. Eger (Dow Agrosciences, Tampa, FL) for providing many of the images used in this chapter.

1.15 References Cited

Anonymous. 2014. 100 of the world's worst invasive alien species. Global Invasive Species Database, http://www.issg.org/database/species/search.asp?st=100ss.

Aukema, J. E., B. Leung, K. Kovacs, C. Chivers, K. O. Britton, J. Englin, S. J. Frankel, R. G. Haight, T. P. Holmes, A. M. Liebhold, D. G. McCullough, and B. Von Holle. 2011. Economic impacts of non-native forest insects in the continental United States. PLoS ONE 6(9): e24587. http://dx.doi.org/10.1371/journal.pone.0024587.

Blackburn, T. M., P. Pyšek, S. Bacher, J. T. Carlton, R. P. Duncan, V. Jarošík, J. R. U. Wilson, and D. M. Richardson. 2011. A proposed unified framework for biological invasions. Trends in Ecology and Evolution 26: 333–339.

Buczkowski, G. 2010. Extreme life history plasticity and the evolution of invasive characteristics in a native ant. Biological Invasions 12: 3343–3349.

Bundy, C. S. 2012. An annotated checklist of the stink bugs (Heteroptera: Pentatomidae) of New Mexico. The Great Lakes Entomologist 45: 196–209.

Bundy, C. S., and J. E. McPherson. 2009. Life history and laboratory rearing of *Corimelaena incognita* (Hemiptera: Heteroptera: Thyreocoridae), with descriptions of immature stages. Annals of the Entomological Society of America 102: 1068–1076.

Bundy, C. S., and J. E. McPherson. 2011. Life history and laboratory rearing of *Mecidea minor* (Hemiptera: Heteroptera: Pentatomidae), with descriptions of immature stages. Annals of the Entomological Society of America 104: 605–612.

Burmeister, H. 1835. Handbuch der Entomologie. Volume 2: 1–400. T. Enslin, Berlin.

Cervantes, L., C. Mayorga, and M. Lopez Ortega. 2013. Description of immature stages of *Melanaethus crenatus* (Hemiptera: Heteroptera: Cydnidae: Cydninae: Geotomini), with notes on oviposition, seed-carrying and feeding behaviors. Florida Entomologist 96: 1434–1441.

Colautti, R. I., and H. J. MacIsaac. 2004. A neutral terminology to define 'invasive' species. Diversity and Distributions 10: 135–141.

Colautti, R. I., and D. M. Richardson. 2009. Subjectivity and flexibility in invasion terminology: too much of a good thing? Biological Invasions 11: 1225–1229.

Costa, J. T. 2006. The other insect societies. The Belknap Press of Harvard University Presss, Cambridge, MA. 767 pp.

Davis, J. 2012. Identifying host plant resistance to redbanded stink bug. 2012 Soybean Breeders/Entomologists Workshop. St. Louis, MO. February 27, 2012. http://soybase.org/meeting_presentations/soybean_breeders _workshop/SBW_2012/Davis.pdf (accessed 4 June 2015).

Davis, M. A. 2009. Invasion biology. Oxford University Press, New York. 244 pp.

Davis, M. A., M. K. Chew, R. J. Hobbs, A. E. Lugo, J. J. Ewel, G. J. Vermeij, J. H. Brown, M. L. Rosenzweig, M. R. Gardener, S. P. Carroll, K. Thompson, S. T. A. Pickett, J. C. Stromberg, P. Del Tredici, K. N. Suding, J. G. Ehrenfeld, J. P. Grime, J. Mascaro, and J. C. Briggs. 2011. Don't judge species on their origins. Nature 474: 153–154.

De Geer, C. 1773. Mémoires pour servir à l'histoire des insectes. Volume III. P. Hesselberg, Stockholm. ii + 696 pp, 44 plates.

DeWitt, N. B., and G. L. Godfrey. 1972. The literature of arthropods associated with soybeans. II. A bibliography of the southern green stink bug, *Nezara viridula* (Linneaus) (sic) (Hemiptera: Pentatomidae). Illinois Natural History Survey. Biological Notes 78: 1–23.

Distant, W. L. 1880. Insecta. Rhynchota. Hemiptera-Heteroptera, Volume I, pp. 1–88. *In* F. D. Godman and O. Salvin (Eds.), Biologia Centrali-Americana, London. 462 pp.

Drury, D. 1770. Illustrations of natural history. Volume 1. B. White, London. i–xxvii, 130 pp., plates 1–50.

Eberhard, W. G. 1975. The ecology and behavior of a subsocial pentatomid bug and two scelionid wasps: strategy and counterstrategy in a host and its parasites. Smithsonian Contributions to Zoology 205: 1–39.

Eger, J. E., Jr., L. M. Ames, D. R. Suiter, T. M. Jenkins, D. A. Rider, and S. E. Halbert. 2010. Occurrence of the Old World bug *Megacopta cribraria* (Fabricius) (Heteroptera: Plataspidae) in Georgia: a serious home invader and potential legume pest. Insecta Mundi 121: 1–11.

Esquivel, J. F., and E. G. Medrano. 2014. Ingestion of a marked bacterial pathogen of cotton conclusively demonstrates feeding by first instar southern green stink bug (Hemiptera: Pentatomidae). Environmental Entomology 43: 110–115.

Fabricius, J. C. 1775. Systema entomologiae sistens insectorum classes, ordines, genera, species, adjectis synonymis, locis, descriptionibus, observationibus. Flensburgi et Lipsiae, Kortii. xxvii + 832 pp.

Fabricius, J. C. 1798. Supplementum entomologiae systematicae. Proft et Storch, Hafniae. 572 pp. (Classis XII, Rhyngota, pp. 511–546).

Filippi, L., M. Hironaka, and S. Nomakuchi. 2001. A review of the ecological parameters and implications of subsociality in *Parastrachia japonensis* (Hemiptera: Cydnidae), a semelparous species that specializes on a poor resource. Population Ecology 43: 41–50.

Fogain, R., and S. Graff. 2011. First records of the invasive pest, *Halyomorpha halys* (Hemiptera: Pentatomidae), in Ontario and Quebec. Journal of the Entomological Society of Ontario 142: 45–48.

Froeschner, R. C. 1988. Family Pentatomidae Leach, 1815. The stink bugs, pp. 544–597. *In* T. J. Henry and R. C. Froeschner (Eds.), Catalog of the Heteroptera, or true bugs, of Canada and the continental United States. E. J. Brill, New York. 958 pp.

Grazia, J., R. T. Schuh, and W. C. Wheeler. 2008. Phylogenetic relationships of family groups in Pentatomoidea based on morphology and DNA sequences (Insecta: Heteroptera). Cladistics 24(6): 932–976.

Greene, J. K., C. S. Bundy, P. M. Roberts, and B. R. Leonard. 2006. Identification and management of common boll-feeding bugs in cotton. Clemson University Extension Bulletin 158: 1–28.

Gross, G. F. 1975. Handbook of the flora and fauna of South Australia. Plant-feeding and other bugs (Hemiptera) of South Australia. Heteroptera – Part 1. Handbooks Committee, South Australian Government, Adelaide, pp. 1–250, 4 color plates.

Gross, G. F. 1976. Handbook of the flora and fauna of South Australia. Plant-feeding and other bugs (Hemiptera) of South Australia. Heteroptera – Part 2. Handbooks Committee, South Australian Government, Adelaide, pp. 251–501.

Hahn, C. W. 1834. Die Wanzenartigen Insecten. C. H. Zeh'chen Buchhandlung, Nürnberg. Volume 2: 33–120.

Halbert, S. E., and J. E. Eger. 2010. Bagrada bug (*Bagrada hilaris*) (Hemiptera: Pentatomidae) an exotic pest of Cruciferae established in the western USA. Pest Alert, Florida Department of Agriculture and Consumer Services, Division of Plant Industry. http//www.freshfromflorida.com/pi/pest-alerts/pdf/bag rada-bug-pest-alert.pdf (accessed 27 June 2012).

Heger, T., A. T. Pahl, Z. Botta-Dukát, F. Gherardi, C. Hoppe, I. Hoste, K. Jax, L. Lindström, P. Boets, S. Haider, J. Kollmann, M. J. Wittmann, and J. M. Jeschke. 2013. Conceptual frameworks and methods for advancing invasion ecology. Ambio 2013, 42:527–540. http://dx.doi.org/10.1007/s13280-012-0379-x.

Herms, D. A., and D. G. McCullough. 2014. Emerald ash borer invasion of North America: history, biology, ecology, impacts, and management. Annual Review of Entomology 59: 13–30.

Hinton, H. E. 1981. Biology of insect eggs, Vols. 1-3. Pergamon, New York, NY. 1125 pp.

Hoebeke, E. R., and M. E. Carter. 2003. *Halyomorpha halys* (Stål) (Heteroptera: Pentatomidae): a polyphagous plant pest from Asia newly detected in North America. Proceedings of the Entomological Society of Washington 105: 225–237.

Hosokawa, T., Y. Kikuchi, M. Shimada, and T. Fukatsu. 2007. Obligate symbiont involved in pest status of host insect. Proceedings of the Royal Society B 274: 1979–1984.

Inkley, D. B. 2012. Characteristics of home invasion by the brown marmorated stink bug (Hemiptera: Pentatomidae). Journal of Entomological Science 47: 125–130.

Javahery, M. 1994. Development of eggs in some true bugs (Hemiptera-Heteroptera). Part I. Pentatomoidea. Canadian Entomologist 126: 401-433.

Jeschke, J. M. 2014. General hypotheses in invasion ecology. Diversity and Distributions 20: 1229–1234.

Jeschke, J. M., S. Bacher, T. M. Blackburn, J. T A. Dick, F. Essl, T. Evans, M. Gaertner, P. E. Hulme, I. Kühn, A. Mrugala, J. Pergl, P. Pysek, W. Rabitsch, A. Ricciardi, D. M. Richardson, A. Sendek, M. Vilà, M. Winter, and S. Kumschick. 2014. Defining the impact of non-native species. Conservation Biology 28: 1188–1194.

Kenis, M., and M. Branco. 2010. Impact of alien terrestrial arthropods in Europe. BioRisk 4: 51–71.

Kenis, M., M.-A. Auger-Rozenberg, A. Roques, L. Timms, C. Péré, M. J. W. Cock, J. Settele, S. Augustin, and C. Lopez-Vaamonde. 2009. Ecological effects of invasive alien insects. Biological Invasions 11: 21–45.

Leskey, T. C., G. C. Hamilton, A. L. Nielsen, D. F. Polk, C. Rodriguez-Saona, J. C. Bergh, D. A. Herbert, T. P. Kuhar, D. Pfeiffer, G. P. Dively, C. R. R. Hooks, M. J. Raupp, P. M. Shrewsbury, G. Krawczyk, P. W. Shearer, J. Whalen, C. Koplinka-Loehr, E. Myers, D. Inkley, K. A. Hoelmer, D.-H. Lee, and S. E. Wright. 2012. Pest status of the brown marmorated stink bug, *Halyomorpha halys* in the USA. Outlooks on Pest Management. doi.org/10.1564/23oct07.

Leslie, A. W., C. Sargent, W. E. Steiner, Jr., W. O. Lamp, J. M. Swearingen, B. B. Pagac, Jr., G. L. Williams, D. C. Weber, and M. J. Raupp. 2014. A new invasive species in Maryland: the biology and distribution of the kudzu bug, *Megacopta cribraria* (Fabricius) (Hemiptera: Plataspidae). The Maryland Entomologist 6(2): 2–23.

Linnaeus, C. 1758. Systema naturae per regna tria naturae, secundum classes, ordines, genera, species, cum characteribus, differentiis, synonymis, locis. Editio decima, reformata. Laurentii Salvii, Holminae. Volume I: 1–823 + 1 p.

Linnavuori, R. E. 1982. Pentatomidae and Acanthosomatidae (Heteroptera) of Nigeria and the Ivory Coast, with remarks on species of the adjacent countries in West and Central Africa. Acta Zoologica Fennica 163: 1–176.

Lockwood, J. A., and R. N. Story. 1986. Adaptive functions of nymphal aggregation in the southern green stink bug, *Nezara viridula* (L.) (Hemiptera: Pentatomidae). Environmental Entomology 15: 739–749.

Lockwood, J. L., P. Cassey, and T. Blackburn. 2005. The role of propagule pressure in explaining species invasions. Trends in Ecology and Evolution 20: 223–228.

Lodge, D. M., S. Williams, H. J. MacIsaac, K. R. Hayes, B. Leung, S. Reichard, R. N. Mack, P. B. Moyle, M. Smith, D. A. Andow, J. T. Carlton, and A. McMichael. 2006. Biological invasions: recommendations for U.S. policy and management. Ecological Applications 16: 2035–2054.

McPherson, J. E. 1982. The Pentatomoidea (Hemiptera) of northeastern North America with emphasis on the fauna of Illinois. Southern Illinois University Press, Carbondale and Edwardsville. 240 pp.

McPherson, J. E., and R. M. McPherson. 2000. Stink bugs of economic importance in America north of Mexico. CRC LLC, Boca Raton, FL. 253 pp.

Milonas, P. G., and G. K. Partsinevelos. 2014. First report of brown marmorated stink bug *Halyomorpha halys* Stål (Hemiptera: Pentatomidae) in Greece. EPPO Bull, 44: 183–186. doi:10.1111/epp.12129

Moizuddin, M., and I. Ahmad. 1975. Eggs and nymphal systematics of *Coptosoma cribrarium* (Fabr.) (Pentatomoidea: Plataspidae) with a note on other plataspids and their phylogeny. Records, Zoological Survey of Pakistan 7(1–2): 93–100 (1979).

Morrill, A. W. 1910. Plant-bugs injurious to cotton bolls. United States Department of Agriculture Bureau of Entomology Bulletin 86: 1–110.

Nielsen, A. L., and G. C. Hamilton. 2009. Life history of the invasive species *Halyomorpha halys* (Hemiptera: Pentatomidae) in northeastern United States. Annals of the Entomological Society of America 102: 608–616.

Palumbo, J. C., and E. T. Natwick. 2010. The bagrada bug (Hemiptera: Pentatomidae): a new invasive pest of cole crops in Arizona and California. Online. Plant Health Progress. 3 pp. http://dx.doi.org/10.1094/PHP -2010-0621-01-BR

Panizzi, A. R. 2015. Growing problems with stink bugs (Hemiptera: Heteroptera: Pentatomidae): species invasive to the U. S. and potential Neotropical invaders. American Entomologist 61: 223–233.

Panizzi, A. R., J. E. McPherson, D. G. James, M. Javahery, and R. M. McPherson. 2000. Chapter 13. Stink bugs (Pentatomidae), pp. 421–474. *In* C. W. Schaefer and A. R. Panizzi (Eds.), Heteroptera of economic importance. CRC Press LLC, Boca Raton, FL. 828 pp.

Peredo, L. C. 2002. Description, biology, and maternal care of *Pachycoris klulgii* (Heteroptera: Scutelleridae). Florida Entomologist 85: 464–473.

Pimentel, D., R. Zuniga, and D. Morrison. 2005. Update on the environmental and economic costs associated with alien-invasive species in the United States. Ecological Economics 52: 273–288.

Pyšek, P., and D. M. Richardson. 2006. The biogeography of naturalization in alien plants. Journal of Biogeography 33: 2040–2050.

Rabitsch, W. 2008. Alien true bugs of Europe (Insecta: Hemiptera: Heteroptera). Zootaxa 1827: 1–44.

Reed, D. A., J. C. Palumbo, T. M. Perring, and C. May. 2013. *Bagrada hilaris* (Hemiptera: Pentatomidae), an invasive stink bug attacking cole crops in the southwestern United States. Journal of Integrated Pest Management 4(3): 1–7.

Richardson, D. M. and A. Ricciardi. 2013. Misleading criticisms of invasion science: a field guide. Diversity and Distributions 19: 1461–1467.

Ruberson, J. R., K. Takasu, G. D. Buntin, J. E. Eger, Jr., W. A. Gardner, J. K. Greene, T. M. Jenkins, W. A. Jones, D. M. Olson, P. M. Roberts, D. R. Suiter, and M. D. Toews. 2013. From Asian curiosity to eruptive American pest: *Megacopta cribraria* (Hemiptera: Plataspidae) and prospects for its biological control. Applied Entomology and Zoology 48: 3–13.

Russell, E. P. III. 1999. L. O. Howard promoted war metaphors as a rallying cry for economic entomology. American Entomologist 45: 74–78.

Schaefer, C. W., and A. R. Panizzi (Eds.). 2000. Heteroptera of economic importance. CRC Press LLC, Boca Raton, FL. 828 pp.

Schuh, R. T., and J. A. Slater. 1995. True bugs of the world (Hemiptera: Heteroptera). Classification and natural history. Cornell University Press, Ithaca, NY. 336 pp.

Simberloff, D. 2003. Confronting introduced species: a form of xenophobia? Biological Invasions 5: 179–192.

Simberloff, D. 2011. How common are invasion-induced ecosystem impacts? Biological Invasions 13: 1255–1268.

Simberloff, D. 2012. Nature, natives, nativism, and management: worldviews underlying controversies in invasion biology. Environmental Ethics 34: 5–25.

Simberloff, D. 2013. Invasive species: what everyone needs to know. Oxford University Press, New York. 329 pp.

Sites, R. W., and J. E. McPherson. 1982. Life history and laboratory rearing of *Sehirus cinctus cinctus* (Hemiptera: Cydnidae), with descriptions of immature stages. Annals of the Entomological Society of America 75: 210–215.

Southwood, T. R. E. 1956. The structure of the eggs of the terrestrial Heteroptera and its relationship to the classification of the group. Transactions of the Royal Entomological Society of London 108: 163–221.

Sosa-Gómez, D. R., and F. Moscardi. 1998. Laboratory and field studies on the infection of stink bugs, *Nezara viridula*, *Piezodorus guildinii*, and *Euschistus heros* (Hemiptera: Pentatomidae) with *Metarhizium anisopliae* and *Beauveria bassiana* in Brazil. Journal of Invertebrate Pathology 71: 115–120.

Stål, C. 1855. Nya Hemiptera. Öfversigt af Kongliga Svenska Vetenskaps-Akademiens Förhandlingar 12(4): 181–192.

Stål, C. 1862. Hemiptera Mexicana enumeravit speciesque novas descripsit. Stettin Entomologische Zeitung (Entomologische Zeitung Herausgegeben von dem Entomologischen Vereine zu Stettin) 23(1–3): 81–118.

Stål, C. 1872. Enumeratio Hemipterorum. Bidrag till en förteckning öfver alla hittills kända Hemiptera, jemte systematiska meddelanden. Parts 1–5. Konglilga Svenska Vetenskaps-Akademiens Handlingar, 1872, part 2, 10(4): 1–159.

Su, N.-Y. 2013. How to become a successful invader. Florida Entomologist 96: 765–769.

Tallamy, D. W., and C. W. Schaefer. 1997. Maternal care in the Hemiptera: ancestry, alternatives, and current adaptive value, pp. 94–115. *In* J. C. Choe and B. J. Crespi (Eds.), The evolution of social behaviour in insects and arachnids. Cambridge University Press, New York, NY. 541 pp.

Taylor, C. M., P. L. Coffey, B. D. DeLay, and G. P. Dively. 2014a. The importance of gut symbionts in the development of the brown marmorated stink bug, *Halyomorpha halys* (Stål). PLoS ONE 9(3): e90312. http://dx.doi.org/10.1371/journal.pone.0090312.

Taylor, M. E., C. S. Bundy, and J. E. McPherson. 2014b. Unusual ovipositional behavior of the stink bug *Bagrada hilaris* (Hemiptera: Heteroptera: Pentatomidae). Annals of the Entomological Society of America 107: 872–877.

Taylor, M. E., C. S. Bundy, and J. E. McPherson. 2015. Life history and laboratory rearing of *Bagrada hilaris* (Hemiptera: Heteroptera: Pentatomidae) with descriptions of immature stages. Annals of the Entomological Society of America 108: 536–551.

Temple, J. H., J. A. Davis, S. Micinski, J. T. Hardke, P. Price, and B. R. Leonard. 2013. Species composition and seasonal abundance of stink bugs (Hemiptera: Pentatomidae) in Louisiana soybean. Environmental Entomology 42: 648–657.

Uhler, P. R. 1894. On the Hemiptera-Heteroptera of the island of Grenada, West Indies. Proceedings of the Zoological Society of London, pp. 167–224.

Valéry, L., H. Fritz, and J.-C. Lefeuvre. 2013. Another call for the end of invasion biology. Oikos 122: 1143–1146.

Van Duzee, E. P. 1904. Annotated list of the Pentatomidae recorded from America north of Mexico, with descriptions of some new species. Transactions of the American Entomological Society 30: 1–80.

Vinson, S. B. 2013. Impact of the invasion of the imported fire ant. Insect Science 20: 439–455.

Vyavhare, S. S., M. O. Way, and R. F. Medina. 2014. Stink bug species composition and relative abundance of the redbanded stink bug (Hemiptera: Pentatomidae) in soybean in the upper Gulf Coast Texas. Environmental Entomology 43: 1621–1627.

Wallner, A. M., G. C. Hamilton, A. L. Nielsen, N. Hahn, E. J. Green, and C. R. Rodriquez-Saona. 2014. Landscape factors facilitating the invasive dynamics and distribution of the brown marmorated stink bug, *Halyomorpha halys* (Hemiptera: Pentatomidae), after arrival in the United States. PloS ONE 9(5): e95691. 12 pp. http://dx.doi.org/10.1371/journal.pone.0095691.

Walsh, B. D. 1866. The Texan cabbage-bug. (*Strachia histrionica* Hahn.). *In* The Practical Entomologist I(11): 1–116. (p. 110).

Ward, D. F., M. C. Stanley, R. J. Toft, S. A. Forgie, and R. J. Harris. 2008. Assessing the risk of invasive ants: a simple and flexible scorecard approach. Insectes Sociaux 55: 360–363.

Westwood, J. O. 1837, 1842. A catalogue of Hemiptera in the collection of the Rev. F. W. Hope, with short Latin descriptions of the new species. J. C. Bridgewater, London. 1837, Part I: 1–46; 1842, Part II, 1–26 (stating that descriptions are by J. O. Westwood).

Wheeler, A. G., Jr., and E. R. Hoebeke. 2009. Adventive (non-native) insects: importance to science and society, pp. 475–521. *In* R. G. Foottit and P. H. Adler (Eds.), Insect diversity. Wiley-Blackwell, John Wiley & Sons, Ltd., Hoboken, NJ. 632 pp.

Zhang, Y., J. L. Hanula, and S. Horn. 2012. The biology and preliminary host range of *Megacopta cribraria* (Heteroptera: Plataspidae) and its impact on kudzu growth. Environmental Entomology 41: 40–50.

Section II

Systematics

2

Higher Systematics of the Pentatomoidea

David A. Rider, Cristiano F. Schwertner, Jitka Vilímová,
Dávid Rédei, Petr Kment, and Donald B. Thomas

CONTENTS

2.1   Introduction 27
2.2   Pentatomoidea 29
    2.2.1   Key to the Families of Pentatomoidea..29
    2.2.2   Acanthosomatidae Signoret, 1863 34
        2.2.2.1   Key to the Subfamilies of Acanthosomatidae 38
    2.2.3   Canopidae Amyot and Serville, 1843 38
    2.2.4   Cydnidae Billberg, 1820...39
        2.2.4.1   Key to the Subfamilies of Cydnidae 41
    2.2.5   Dinidoridae Stål, 1868...42
        2.2.5.1   Key to the Subfamilies of Dinidoridae.............................44
    2.2.6   Lestoniidae China, 1955 44
    2.2.7   Megarididae McAtee and Malloch, 1928 46
    2.2.8   Mesopentacoridae Popov, 1968 47
    2.2.9   Parastrachiidae Oshanin, 1922 47
    2.2.10 Pentatomidae Leach, 1815 48
        2.2.10.1   Key to the Subfamilies of Pentatomidae 51
        2.2.10.2   Aphylinae Bergroth, 1906 54
        2.2.10.3   Asopinae Amyot and Serville, 1843 54
        2.2.10.4   Cyrtocorinae Distant, 1880 57
        2.2.10.5   Discocephalinae Fieber, 1860..57
            2.2.10.5.1   Key to the Tribes of Discocephalinae 59
        2.2.10.6   Edessinae Amyot and Serville, 1843 59
        2.2.10.7   Pentatominae Leach, 1815 60
            2.2.10.7.1   Tentative Key to the Tribes of Pentatominae 61
            2.2.10.7.2   Aeliini Douglas and Scott, 1865....................................69
            2.2.10.7.3   Aeptini Stål, 1871 69
            2.2.10.7.4   Aeschrocorini Distant, 1902 70
            2.2.10.7.5   Agaeini Cachan, 1952..71
            2.2.10.7.6   Agonoscelidini Atkinson, 1888......................................72
            2.2.10.7.7   Amyntorini Distant, 1902 72
            2.2.10.7.8   Antestiini Distant, 1902 73
            2.2.10.7.9   Axiagastini Atkinson, 1888 74
            2.2.10.7.10 Bathycoeliini Atkinson, 1888..75
            2.2.10.7.11 Cappaeini Atkinson, 1888...76
            2.2.10.7.12 Carpocorini Mulsant and Rey, 1866..............................76
            2.2.10.7.13 Catacanthini Atkinson, 1888...79
            2.2.10.7.14 Caystrini Ahmad and Afzal, 1979..................................80
            2.2.10.7.15 Chlorocorini Rider, Greve, Schwertner, and Grazia, New Tribe....80

2.2.10.7.16 Coquereliini Cachan, 1952 81
2.2.10.7.17 Degonetini Azim and Shafee, 1984 81
2.2.10.7.18 Diemeniini Kirkaldy, 1909 82
2.2.10.7.19 Diplostirini Distant, 1902 82
2.2.10.7.20 Diploxyini Atkinson, 1888 83
2.2.10.7.21 Eurysaspini Atkinson, 1888 83
2.2.10.7.22 Eysarcorini Mulsant and Rey, 1866 84
2.2.10.7.23 Halyini Amyot and Serville, 1843 85
2.2.10.7.24 Hoplistoderini Atkinson, 1888 86
2.2.10.7.25 Lestonocorini Ahmad and Mohammad, 1980 87
2.2.10.7.26 Mecideini Distant, 1902 87
2.2.10.7.27 Memmiini Cachan, 1952 88
2.2.10.7.28 Menidini Atkinson, 1888 88
2.2.10.7.29 Myrocheini Stål, 1871 89
2.2.10.7.30 Nealeriini Cachan, 1952 90
2.2.10.7.31 Nezarini Atkinson, 1888 91
2.2.10.7.32 Opsitomini Cachan, 1952 92
2.2.10.7.33 Pentamyrmecini Rider and Brailovsky, 2014 92
2.2.10.7.34 Pentatomini Leach, 1815 92
2.2.10.7.35 Phricodini Cachan, 1952 94
2.2.10.7.36 Piezodorini Atkinson, 1888 94
2.2.10.7.37 Procleticini Pennington, 1920 95
2.2.10.7.38 Rhynchocorini Stål, 1871 96
2.2.10.7.39 Rolstoniellini Rider, 1997 96
2.2.10.7.40 Sciocorini Amyot and Serville, 1843 97
2.2.10.7.41 Sephelini Breddin, 1904 98
2.2.10.7.42 Strachiini Mulsant and Rey, 1866 98
2.2.10.7.43 Triplatygini Cachan, 1952 99
2.2.10.7.44 The Generic Groups of Gross (1975-1976) 100
            2.2.10.7.44.1 *Ochisme* Group 101
            2.2.10.7.44.2 *Kitsonia* Group 101
            2.2.10.7.44.3 *Kumbutha* Group 101
            2.2.10.7.44.4 *Macrocarenus* Group 101
            2.2.10.7.44.5 *Menestheus* Group 102
            2.2.10.7.44.6 *Ippatha* Group 102
            2.2.10.7.44.7 *Dictyotus* Group 102
            2.2.10.7.44.8 *Tholosanus* Group 103
            2.2.10.7.44.9 *Poecilotoma* Group.. 103
            2.2.10.7.44.10 *Kapunda* Group 103
            2.2.10.7.44.11 *Cephaloplatus* Group 104
            2.2.10.7.44.12 *Mycoolona* Group.. 104
2.2.10.7.45 The Generic Groups of Linnavuori (1982) 104
            2.2.10.7.45.1 *Tyoma* Group 104
            2.2.10.7.45.2 *Acoloba* Group.. 105
            2.2.10.7.45.3 *Aeliomorpha* Group.. 105
            2.2.10.7.45.4 *Carbula* Group.. 105
            2.2.10.7.45.5 *Veterna* Group 105
            2.2.10.7.45.6 *Halyomorpha* Group 105
            2.2.10.7.45.7 *Banya* Group.. 106
            2.2.10.7.45.8 *Eipeliella* Group 106
            2.2.10.7.45.9 *Kelea* Group.. 106

		2.2.10.7.46	Other Unplaced or Questionably Placed Genera	106
		2.2.10.7.47	Fossil Genera	111
	2.2.10.8	Phyllocephalinae Amyot and Serville, 1843		112
		2.2.10.8.1	Cressonini Kamaluddin and Ahmad, 1991	113
		2.2.10.8.2	Megarrhamphini Ahmad, 1981	113
		2.2.10.8.3	Phyllocephalini Amyot and Serville, 1843	115
		2.2.10.8.4	Tetrodini Ahmad, 1981	115
	2.2.10.9	Podopinae Amyot and Serville, 1843		115
		2.2.10.9.1	Key to the Podopinae Genus Groups	118
		2.2.10.9.2	*Podops* Group	119
		2.2.10.9.3	*Deroploa* Group	120
		2.2.10.9.4	*Graphosoma* Group	120
		2.2.10.9.5	*Tarisa* Group	121
		2.2.10.9.6	*Brachycerocoris* Group	121
		2.2.10.9.7	Genera *incertae sedis*	121
		2.2.10.9.8	Phylogenetic Notes	122
	2.2.10.10	Serbaninae Leston, 1953		124
	2.2.10.11	Stirotarsinae Rider, 2000		124
2.2.11	Phloeidae Amyot and Serville, 1843			125
2.2.12	Plataspidae Dallas, 1851			125
	2.2.12.1	Key to the Subfamilies and Genus Groups of Plataspidae		127
2.2.13	Primipentatomidae Yao, Cai, Rider, and Ren, 2013			128
2.2.14	Saileriolidae China and Slater, 1956			128
2.2.15	Scutelleridae Leach, 1815			129
	2.2.15.1	Key to the Subfamilies of Scutelleridae		131
2.2.16	Tessaratomidae Stål, 1865			132
	2.2.16.1	Key to the Subfamilies of Tessaratomidae		134
2.2.17	Thaumastellidae Seidenstücker, 1960			134
2.2.18	Thyreocoridae Amyot and Serville, 1843			135
	2.2.18.1	Key to the Subfamilies of Thyreocoridae		137
2.2.19	Urostylididae Dallas, 1851			137
2.3	Conclusions			138
2.3.1	Size of Project			139
2.3.2	Interpretation of Characters			139
2.3.3	Lack of Specimens			139
2.3.4	Solutions			140
2.4	Acknowledgments			140
2.5	References Cited			141

2.1 Introduction

The Pentatomoidea – What a wonderful and diverse assemblage of insects! Members of this superfamily occur in nearly all parts of the World and occupy virtually all terrestrial habitats. They range greatly in size with the smallest species not much larger than a pin head (e.g., *Megaris* Stål in Megarididae and *Sepontia* Stål in the Pentatomidae) to some of the largest and most robust species in the Heteroptera (e.g., many Tessaratomidae). Most species are phytophagous, and several are known to cause economic damage to various crops; however, some species are predatory (Pentatomidae: Asopinae), preying mainly on other insects, and a few groups are at least suspected to be fungivores (Canopidae, Megarididae, some Plataspidae). Nearly all species have scent glands, both in the immatures and the adults, which emit a foul odor, presumably to ward off predators. Even so, many species also rely on crypsis to avoid predators as many are various shades of greens, tans, and browns, allowing them to blend in with the vegetation, soil, or whatever substrate they live on. There are, however, some species that seem to advertise their repugnant qualities; that is, they are aposematically colored with reds, yellows, oranges, and blues, sometimes with a brilliant metallic sheen. Certain groups have evolved elaborate stridulatory mechanisms involving the legs, abdomen, and/or wings. Additionally, some groups (e.g., some Acanthosomatidae; some Cydnidae; some Dinidoridae; Parastrachiidae; Phloeidae; some Discocephalinae, Edessinae, and Pentatominae in the Pentatomidae; some Plataspidae; some Scutelleridae; and some Tessaratomidae) exhibit parental care with the females, and, perhaps, sometimes the males, standing guard over the egg masses and early instars.

In this chapter, we provide a taxonomic history for each family and subfamily (if applicable) and, in addition, for the tribes and generic groups within the Pentatomidae, that, perhaps, will give some perception of the taxonomic position of each group. We also discuss the systematic classification within each of these groups and provide general diagnoses and descriptions for all families within the Pentatomoidea and for the subfamilies, tribes, and generic groups within the Pentatomidae. We also have summarized some of the biological information for each group. Keys to aid in the identification of the various taxa are included, but it should be understood that to build keys that work for all taxa all the time would become long and tedious. We have chosen to present keys that will work for most of the commonly encountered taxa but, perhaps, not for some of the exceptional taxa. Most of these exceptional taxa are discussed as they arise in the body of the chapter. The present treatise is not meant as the last word in the phylogenetic study of this superfamily. Rather, it is a work in progress and reflects the current state of knowledge on the higher classification of the Pentatomoidea. We discuss potentially important characters and point out problems in the existing classification. It is our intention to provide information on the currently recognized or disputed taxa, their defining characters, and their relationships. We recognize that with further study the classification presented herein will change greatly.

There are several characteristics that we believe are of paramount importance for the classification of the Pentatomoidea. For example, in females belonging to several pentatomoid families, the spermathecal duct has a large membranous dilation around its middle. In members of the Pentatomidae, there is an elongate, sclerotized, double-walled tube (sclerotized rod) projecting from the distal orifice of the dilation into its lumen. All known members of the family Pentatomidae have this structure, save one, the genus *Trichopepla* Stål (**Figure 2.20G**) (currently a member of the Carpocorini), which we consider to be a secondary loss. Although a similar structure is also found in some members of other pentatomoid families (e.g., Cydnidae, Thyreocoridae) (Štys and Davidová-Vilímová 1979, Pluot-Sigwalt and Lis 2008), sometimes with an analogous projection of the proximal orifice (e.g., some Scutelleridae) (Tsai et al. 2011), its presence in the Pentatomidae can be regarded as a synapomorphy of the family (Gapud 1991, Grazia et al. 2008). Accordingly, we tentatively include the Aphylinae, Cyrtocorinae, and Serbaninae as subfamilies within the Pentatomidae, and exclude the Lestoniidae and the Phloeidae, based on the presence or absence of the dilation, respectively, but stress that the evolution of this character is not fully understood. Additionally, there are, undoubtedly, several important characters in the male genitalia that have phylogenetic significance. For example, Gross (1976) stated that the presence of the so-called "median penial plates" (mesal, sclerotized portions of the second pair of conjunctival processes fused along their midline and closely associated with the distal portion of the vesica) is strictly a pentatomid character. Further study of the phallus and aedeagus of the males will be necessary to determine which characters are informative and which are homoplastic.

Another character that we feel may be of primary importance is the structure of the thoracic sterna (Gross 1975b), at least for the classification within the Pentatomidae. Most pentatomid genera have the mesosternum medially carinate, or at least with a weak, raised line. There are, however, several small groups of genera that have the mesosternum medially sulcate without any sign of a medial carina. The significance of this character is yet to be determined, but it could lend support to an eventual splitting of the Pentatominae as presently conceived. The absence of the mesosternal carina is especially common in certain Australian and African groups of genera. We also believe the structure of the base of the abdomen (rounded or produced), and the structure of the ostiole and its associated structures, will be important. They are difficult to interpret, however, as certain characteristics have arisen multiple times, and some have been lost secondarily. The form of the ostiolar structure has been especially well-studied (Kment and Vilímová 2010b).

2.2 Pentatomoidea

As with many other superfamilies, heteropteran or not, the Pentatomoidea has a long and complicated taxonomic history. Through time, each of the included families, subfamilies, and tribes has
been transferred across the family level ranks as often as there have been heteropterists working
on the classification of these taxa (see **Table 2.1**). Although the recognition of the most important
families (and subfamilies) has more or less stabilized, their evolutionary relationships are still poorly
understood; therefore, the current classification is certainly tentative even at the family and subfamily levels. The tribal classification of most families, and particularly within the Pentatomidae remains
chaotic at best.

The superfamily Pentatomoidea is a member of the insect order Hemiptera, suborder Heteroptera,
and infraorder Pentatomomorpha. Typically, this infraorder has been divided into six superfamilies:
Aradoidea, Idiostoloidea, Pentatomoidea, Lygaeoidea, Pyrrhocoroidea, and Coreoidea. The Aradoidea
lack the trichobothria that are present in the other superfamilies; as such, those five superfamilies are
grouped together as the Trichophora, a group first proposed by Tullgren (1918). Most workers since
that time have supported the monophyly of the Trichophora (Schuh and Slater 1995). Although it was
not a focus of their study, the phylogenetic analysis of Grazia et al. (2008) also generally supported the
Trichophora as a monophyletic group. Their study also supported the monophyly of the Pentatomoidea
based primarily on four characters: (1) the enlarged scutellum, the apex of which usually reaches or
surpasses an imaginary line uniting the posterolateral angles of abdominal connexiva III, (2) the claval
commissure that is usually obsolete (i.e, the clavi usually do not surpass the apex of the scutellum),
(3) the trichobothria (usually paired but sometimes singular) located laterally near the spiracular line
on abdominal sternites II through VII, and (4) tergite VIII covering tergite IX in females. Additionally,
the five-segmented antennae may be a significant synapomorphy. Several other recent studies based on
molecular markers, including Grazia et al. (2008), also support the monophyly of the Pentatomoidea
and most of the included families. Relationships among families, subfamilies, and tribes within the
Pentatomoidea are mostly unsettled, and the available information will be discussed under each corresponding section. The pentatomoid family and subfamily classification used in this chapter is outlined in
Table 2.2. We also are developing a catalog of the Pentatomoidea of the World; this work has led directly
to a series of nomenclatural changes throughout the superfamily (Rider and Rolston 1995; Rider 1998a,b,
2007; Rider and Fischer 1998; Rider and Kment 2015). At present, there are 1410 genera and 8042 species described in the Pentatomoidea (**Table 2.2**). A tentative list of pentatomoid genera, and their placement to family, subfamily, and tribe can be found on the internet (Rider 2015b).

2.2.1 Key to the Families of Pentatomoidea

1	Large to medium-sized species with body strongly dorsoventrally flattened; lateral mar-gins of juga, pronotum, and base of coria greatly enlarged, foliaceous (**Figures 2.1A, 2.16F**); each compound eye divided into a dorsal and a ventral section by foliaceous lateral margin of head, cryptic, bark-dwelling species (**Figures 2.25G, 2.27K**)............ 2
1'	Size and shape variable, usually not conspicuously flattened; lateral margins of juga, pronotum, and base of coria not foliaceous; if body somewhat flattened, then compound eyes not divided into two sections 3
2(1)	Antennae 3-segmented (**Figure 2.1A**); spermatheca lacking a sclerotized rod; South America (**Figures 2.16F, 2.25G**) .... Phloeidae
2'	Antennae 4-segmented; spermatheca with a sclerotized rod; Southeast Asia (Borneo) (**Figure 2.27K**).. Pentatomidae (part: Serbaninae)
3(1)	Body tortoise-shaped, greatly convex dorsally, flat ventrally; margins of head, lateral margins of pronotum, part of costal margins of coria, and lateral margins of abdomen laminately produced ventrally (**Figure 2.25D**); abdominal venter of female with one or two pairs of disc-shaped organs; Australia .... Lestoniidae

TABLE 2.1

Historical Classifications

Amyot & Serville 1843	Stål, 1870s	Atkinson, Late 1880s	Distant, 1900s
Family Longiscutes	Subfamily Cimicina	Subfamily Plataspina	Subfamily Plataspidinae
Tribe Orbiscutes	Division Elvisuraria	Subfamily Cydnina	Subfamily Scutellerinae
Race Anguleux	Division Sphaerocoraria	Section Cydnides	Division Elvisuraria
Group Scutellérides	Division Scutelleraria	Section Séhirides	Division Sphaerocoraria
Group Pachycorides	Division Tetyraria	Subfamily Scutellerina	Division Scutelleraria
Group Tétyrides	Division Odontotarsaria	Division Elvisuraria	Division Tetyraria
Group Eurygastrides	Division Eurygastraria	Division Sphaerocoraria	Division Odontotarsaria
Group Podopides	Division Odontoscelaria	Division Scutellaria	Division Eurygastraria
Group Oxynotides	Subfamily Asopina	Division Tetyraria	Division Odontoscelaria
Race Globuleux	Subfamily Tessaratomina	Division Odontotarsaria	Subfamily Graphosomatinae
Group Thyréocorides	Division Oncomerina	Division Eurygastraria	Subfamily Cydninae
Group Odontoscelides	Division Tessaratomina	Division Odontoscelaria	Subfamily Pentatominae
Group Canopides	Division Eusthenina	Subfamily Pentatomina	Division Halyaria
Tribe Coniscutes	Division Prionogastrina	Division Podoparia	Division Sciocoraria
Libertirostres	Division Cyclogastrina	Division Halyaria	Division Dorpiaria
Longirostres	Subfamily Dinidorina	Division Sciocoraria	Division Dymantaria
Race Spissirostres	Subfamily Oxynotina	Division Myrocharia	Division Mecidaria
Group Stirétrides	Subfamily Phloeina	Division Odiaria	Division Amyntaria
Group Asopides	Subfamily Discocephalina	Division Tropicorypharia	Division Carpocoraria
Ténuirostres	Subfamily Pentatomina	Division Cappaearia	Division Aeschrocoraria
Race Spinipèdes	Division Tarisaria	Division Carpocoraria	Division Eusarcocoriaria
Group Cydnides	Division Trigonosomaria	Division Diploxyaria	Division Hoplistoderaria
Group Séhirides	Division Graphosomaria	Division Eysarcoriaria	Division Antestiaria
Group Pododides	Division Podoparia	Division Agonosceliaria	Division Eurydemaria
Race Nudipèdes	Division Halyaria	Division Strachiaria	Division Compastaria
Inermiventres	Division Sciocoraria	Division Hoplistoderaria	Division Tropicaria
Sulciventres	Division Aeliaria	Division Catacantharia	Division Rhynchocoraria
Group Halydes	Division Eysarcoraria	Division Nezaria	Division Nezaria
Group Phléides	Division Pentatomaria	Division Hyllaria	Division Menidaria
Pléniventres	Division Strachiaria	Division Plautiaria	Division Diplostiraria
Group Sciocorides	Subfamily Acanthosomina	Division Axiagastaria	Division Euryaspisaria
Group Pentatomides	Subfamily Plataspina	Division Euryasparia	Subfamily Asopinae
Armiventres	Subfamily Cydnina	Division Menidaria	Subfamily Tessaratominae
Group Rhaphigastrides	Subfamily Urolabidina	Division Piezodoraria	Division Tessaratomaria
Race Brévirostres	Subfamily Phyllocephalina	Division Bathycoeliaria	Division Eusthenaria
Group Edessides		Division Rhynchocoraria	Subfamily Dinidorinae
Group Phyllocéphalides		Division Tropicoraria	Subfamily Phyllocephalinae
Canalirostres		Subfamily Asopina	Subfamily Urostylinae
Mégyménides		Subfamily Acanthosomina	Subfamily Acanthosomatinae
		Subfamily Urostylina	
		Subfamily Tessaratomina	
		Division Tessaratomaria	
		Division Eusthenaria	
		Division Oncomeraria	
		Subfamily Dinidorina	
		Subfamily Phyllocephalina	

3'	Body not as in 3(1), if somewhat tortoise-shaped with convex dorsum and flat venter, then margins of head, pronotum, coria, and abdomen not laminately produced ventrally, abdominal venter of both sexes lacking disc-shaped organs... 4
4(3)	Scutellum subtriangular, leaving apices of clavi exposed at rest; apices of clavi either meeting in a single point or forming a distinct claval commissure (**Figures 2.3C; 2.15F; 2.16J; 2.25I, L; 2.31B**)... 5
4'	Scutellum or at least its extreme tip projecting posteriad beyond frena, concealing apices of clavi at rest; apices of clavi separated 9
5(4)	Length less than 3.5 mm; body elongate; scutellum short, not surpassing posterior margin of metanotum (**Figure 2.16J**); scent gland ostiole situated close to lateral margin of metapleuron, with long vestibular scar; ocelli, if present, placed relatively far from each other; resembling lygaeoids in appearance; southern and northern Africa, southwest Asia (**Figure 2.16J**) .... Thaumastellidae
5'	Size usually greater, body broader; scutellum usually longer, surpassing posterior margin of metanotum; scent gland ostiole, if present, situated close to acetabula; ocelli, if present, placed close to each other 6
6(5)	Ant-mimic (**Figures 2.14E, 2.31B**); black, marked with white, with large spines; clavi extending beyond apex of scutellum and meeting at a single point; Thailand .... Pentatomidae (part: Pentamyrmecini)
6'	Not an ant mimic; color variable, but usually not black with white markings, may have bristles or small spines or pegs, but not large spines; clavi extend beyond apex of scutellum and form a claval commissure... 7
7(6)	Lateral and anterior margins of head with strong, peg-like setae (**Figure 2.1B**); tibiae provided with spines along length (**Figure 2.1C**); ocelli placed closer to compound eyes than to each other; reddish-brown to blackish-brown species; Western Hemisphere except for one species introduced into Iran (**Figure 2.15F**)......................... Cydnidae (part: Amnestinae)
7'	Head and tibiae unarmed; ocelli, if present, placed closer to each other than to compound eyes; color variable, but usually not red-brown or black-brown 8
8(7)	Larger (usually more than 10 mm in length), variously colored (**Figure 2.25L**); scent gland ostiole present; spiracles III-VII situated far from lateral margins of abdomen; Eastern Hemisphere (Oriental region and neighboring areas of temperate East Asia)...... ....Urostylididae
8'	Smaller (usually less than 4.5 mm); generally of pale color (**Figure 2.25I**); scent gland ostiole strongly reduced or lacking; spiracles III-VII situated close to lateral margins of abdomen; Oriental region ....Saileriolidae
9(4)	Coxal combs (a row of strong, frequently broadened and flattened setae along apical margin of coxa) present on all legs (**Figure 2.1D**); tibiae frequently with distinct spines along their length (**Figures 2.1E-H**)... 10
9'	Coxal combs absent; tibiae may have hairs, but not distinctly spinose (except in a few exceptional cases)...12
10(9)	Scutellum usually subtriangular (**Figures 2.15K, L; 2.25E**), sometimes enlarged but if so, not nearly covering all of abdomen (**Figures 2.15G, I**).. 11
10'	Scutellum greatly enlarged, nearly covering entire abdomen; worldwide (**Figure 2.25K**)..... ....Thyreocoridae
11(10)	Tibiae without distinct spines; medium-sized, aposematically black and red species; Africa, India through China to Japan (**Figures 2.25E, F**) .... Parastrachiidae
11'	Tibiae with distinct, frequently strong spines; of various colors but not aposematically black and red; worldwide...Cydnidae (part)

TABLE 2.2

Pentatomoidea Classification and Diversity (as of April 2017)

Family	Subfamily	Tribe	Genera[1]	Species[1]
Acanthosomatidae			**57**	**287**
	Acanthosomatinae		17	211
	Blaudusinae		23	47
		Blaudusini	10	22
		Lanopini	13	25
	Ditomotarsinae		17	29
		Ditomotarsini	13	22
		Laccophorellini	4	7
Canopidae			**1**	**9**
Cydnidae			**111**	**852**
	Amaurocorinae		3	5
	Amnestinae		8	59
	Cephalocteinae		8	31
		Cephalocteini	2	4
		Scaptocorini	6	27
	Cydninae		75	659
		Cydnini	11	160
		Geotomini	64	499
	Garsauriinae		5	16
	Sehirinae		12	82
Dinidoridae			**17**	**109**
	Dinidorinae		13	82
		Amberianini	1	2
		Dinidorini	9	76
		Thalmini	3	4
	Megymeninae		4	27
		Byrsodepsini	1	2
		Megymenini (including Eumenotini)	3	25
Lestoniidae			**1**	**2**
Megarididae			**2**	**18**
Mesopentacoridae[3]			**2**	**3**
Parastrachiidae			**2**	**8**
Pentatomidae[2]			**940**	**4949**
	Aphylinae		2	3
	Asopinae		63	303
	Cyrtocorinae		4	11
	Discocephalinae		81	325
	Edessinae		15	338
	Pentatominae		660	3484
	Phyllocephalinae		45	214
	Podopinae		68	269
	Serbaninae		1	1
	Stirotarsinae		1	1

(Continued)

TABLE 2.2 (CONTINUED)

Pentatomoidea Classification and Diversity

Family	Subfamily	Tribe	Genera[1]	Species[1]
Phloeidae			**3**	**4**
Plataspidae			**66**	**606**
Primipentatomidae[3]			**4**	**5**
Saileriolidae			**3**	**4**
Scutelleridae			**100**	**531**
	Elvisurinae		7	29
	Eurygastrinae		10	37
		Eurygastrini	2	21
		Psacastini	8	16
	Hoteinae		7	24
	Odontoscelinae		5	29
	Odontotarsinae		11	78
		Odontotarsini	6	39
		Phimoderini	5	39
	Pachycorinae		26	120
	Scutellerinae		33	212
		Scutellerini	29	196
		Sphaerocorini	4	16
	Tectocorinae		1	2
Tessaratomidae			**62**	**252**
	Natalicolinae		9	18
	Oncomerinae		18	62
	Tessaratominae		35	172
		Prionogastrini	1	1
		Sepinini	5	17
		Tessaratomini	29	154
Thaumastellidae			**1**	**3**
Thyreocoridae			**30**	**223**
	Corimelaeninae		28	217
	Thyreocorinae		2	6
Urostylididae			**8**	**172**
TOTAL			**1410**	**8042**

[1] Numbers of genera and species are compiled from the World Catalog of Pentatomoidea (David A. Rider, unpublished), and include subgenera, subspecies, and fossil genera and species.

[2] For a list of the tribes in the pentatomid subfamilies with numbers of included genera and species, see **Table 2.3**.

[3] Fossil family.

12(9)	Tarsi 2-segmented; antennae clearly 5-segmented (Plataspidae) or appear to be 4-segmented (subdivision of pedicel indistinct) (Megarididae); scutellum greatly enlarged, nearly covering entire abdomen (**Figures 2.16E, G, I; 2.25H**) 13
12'	Not with the above combination of characters 14
13(12)	Evaporatoria of scent glands covering nearly entire ventral surfaces of prothorax, meso-thorax, and metathorax (**Figure 2.1I**); Eastern Hemisphere, except two introduced species (southeastern North America, and Panama) (**Figures 2.16G-I; 2.25H**) .... Plataspidae
13'	Evaporatoria of scent glands small, never on ventral surface of prothorax; Neotropics (**Figure 2.16E**) .... Megarididae
14(12)	Scutellum greatly enlarged, nearly covering entire abdomen (**Figures 2.16A; 2.26A-L**) ... 15

14'   Scutellum usually subtriangular in shape, sometimes enlarged, but usually leaving lateral and sometimes apical portions of abdomen uncovered (exceptions: Pentatomidae: Aphylinae, *Sepontia* and *Sepontiella*, and several other Oriental and Austro-Papua genera) 16

15(14)   Anterior margin of pronotum meeting lateral margins in rounded arc; forewing with secondary transverse fold at apex of corium; ventral abdominal intersegmental sutures obscured laterally; dorsum strongly convex, venter flat, body uniformly black, usually shining; Neotropics (**Figure 2.16A**)..Canopidae

15'   Anterior and lateral margins of pronotum angulate; forewing not folded transversely; ventral abdominal intersegmental sutures complete, reaching lateral margins; shape and color variable, but if dorsum strongly convex and venter flat, then body usually not totally black; worldwide (**Figures 2.26A-L**) .... Scutelleridae

16(14)   Tarsi 2-segmented; mesosternum often with a large compressed median keel (**Figures 2.1J, K**) posterior margin of abdominal sternite VII deeply excised in males, leaving abdominal segment VIII exposed (**Figure 2.1L**); worldwide (**Figures 2.15A-E; 2.25A**) .... Acanthosomatidae

16'   Tarsi 3-segmented (exceptions: Dinidoridae: Dinidorinae: Thalmini; Tessaratomidae: Natalicolinae; Pentatomidae: Pentatominae: Nealeriini, Opsitomini, *Phalaecus*, *Rolstoniellus*, *Prionocompastes*); abdominal sternite VII of male not as above 17

17(16)   Bucculae elongate, uniformly elevated along length, forming ridge on each side of first rostral segment (**Figures 2.1M, N**); worldwide .... Pentatomidae (in part)

17'   Bucculae much shorter, arcuately elevated, forming short, flap-like structure on each side along anterior portion of first rostral segment (**Figures 2.1O, 2.2A**)............................ 18

18(17)   Spiracles on abdominal segment II (first visible segment) partially or fully exposed, not completely concealed by posterior margin of metapleuron (**Figure 2.2B**) 19

18'   Spiracles on abdominal segment II (first visible segment) usually completely concealed by posterior margin of metapleuron; Old World, Neotropics (**Figures 2.18E-H; 2.27F-H**) .... Pentatomidae (part: Edessinae and Phyllocephalinae)

19(18)   Metasternum usually produced laterally between coxae, and anteriorly onto mesosternum (**Figure 2.2C**); antennae usually 4-segmented, if 5-segmented, then segment III quite short, segments more or less cylindrical (**Figure 2.2D**); pronotum often extended posteriorly over base of abdomen (**Figure 2.2E**); ostiolar ruga in form of anterior and posterior lobes (Natalicolinae and Tessaratominae) (**Figure 2.4C**) or spout-shaped (Oncomerinae) (**Figure 2.4D**); veins in hemelytral membrane usually subparallel (**Figure 2.2F**); Old World, except one genus in Neotropics (**Figures 2.16K, L; 2.25J**) .... Tessaratomidae

19'   Metasternum usually not produced laterally or anteriorly; antennae 4- or 5-segmented, but usually with segment III not distinctively short, some segments often flattened (**Figure 2.2F**); pronotum not produced posteriorly over base of abdomen (**Figure 2.25C**); ostiolar ruga always spout-shaped (**Figures 2.4A, B**); veins of hemelytra may be reticulate; Old World with one genus in Neotropics (**Figures 2.16B-D; 2.25C**) ....Dinidoridae

2.2.2 Acanthosomatidae Signoret, 1863

The family Acanthosomatidae has had a typical pentatomoid taxonomic history. It, as well as many other families of Pentatomoidea, has been treated by most early workers as a subfamily of a broadly-defined Pentatomidae. Although China (1933) accorded it family status, it would be another 20-30 years before most workers accepted it as a valid family. In his generic revision of the group, Kumar (1974a) recognized this taxon as a valid family.

Fischer (1994) and Grazia et al. (2008) provided support for the monophyly of the family; however, the phylogenetic relationships of the Acanthosomatidae within the Pentatomoidea are still unsettled (Kment 2006, Carvajal and Faúndez 2013). Gapud (1991) established the family as an intermediate lineage

FIGURE 2.1 A, *Phloeophana longirostris*, antennae, ventral view; B, *Amnestus pusio*, head, dorsal view; C, *Amnestus pusio*, anterior tibia, inner view; D, *Cyrtomenus mirabilis*, middle and posterior coxae, ventral view; E, *Cyrtomenus mirabilis*, anterior tibia, inner view; F, *Galgupha* sp., anterior tibia, inner view; G, *Sehirus cinctus*, anterior tibia, inner view; H, *Sehirus cinctus*, posterior tibia, inner view; I, *Coptosoma* sp., ostiole and evaporatoria, ventral view; J, *Elasmostethus ligatus placidus*, mesosternum, lateral view; K, *Elasmostethus ligatus placidus*, mesosternum, ventral view; L, *Planois gayi*, posterior part of the abdomen, lateral view; M, *Dolycoris baccarum*, head, ventral view; N, *Plautia affinis*, head, ventral view; O, *Basycriptus distinctus*, head, lateral view.

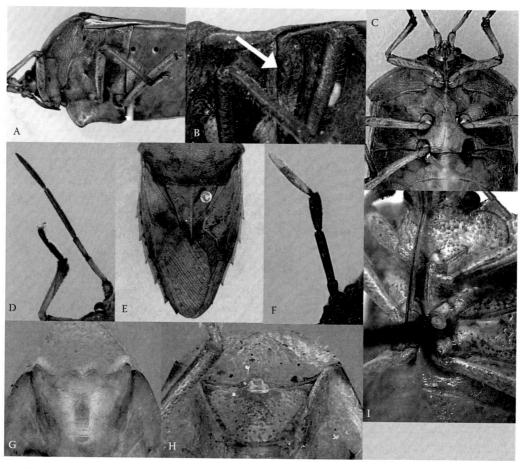

FIGURE 2.2 A, *Piezosternum thunbergi*, thoracic pleura and base of abdomen, lateral view; B, *Dinidor mactabilis*, thorax and anterior abdomen, lateral view; C, *Piezosternum thunbergi,* head and thorax, ventral view; D, *Piezosternum thunbergi*, antennae; E, *Piezosternum thunbergi*, thorax and abdomen, dorsal view; F, *Dinidor mactabilis*, antennae; G, *Nopalis sulcatus*, posterior abdomen, ventral view; H, *Elasmostethus ligatus placidus*, posterior abdomen, ventral view; I, *Lanopis rugosus*, thorax, ventral view.

within the Pentatomoidea. Results in Grazia et al. (2008) varied according to the analysis performed; the morphological evidence pointed to a basal position, whereas molecular data suggested that the family is a derived taxon related to Pentatomidae. Both analyses, as well as another molecular study by Wu et al. (2016) suggested a sister group relationship between the Acanthosomatidae and the Lestoniidae.

Members of this family, in general appearance, resemble pentatomids, being approximately the same size (5-20 mm) and shape and having a more or less triangular scutellum (with the frena extending about three-fourths the lateral distance). They differ from most species of Pentatomidae, however, by their having two-segmented tarsi, the males have the posterior margin of abdominal segment VII broadly emarginated, leaving segment VIII broadly exposed (**Figure 2.1L**), the females usually have Pendergrast's organs [i.e., glandular abdominal organs that are usually rounded or oval in shape and are located on abdominal sternites V through VII, or only on VII in some species; these are usually lacking in species that exhibit maternal care (Fischer 2006, Tsai et al. 2015)], and they lack a dilation and sclerotized rod in the spermathecal duct (Schuh and Slater 1995, Schwertner and Grazia 2015). Except for a small group of Madagascan genera that has four-segmented antennae (Kment 2006), they have five-segmented antennae. The mesosternum is often strongly carinate medially (**Figures 2.1J, K**) (subfamily Acanthosomatinae), and abdominal sternite III is often armed with an anteriorly directed spine or tubercle. The trichobothria on abdominal sternites III through VII are transverse in position. The tibiae lack distinct spines or bristles. Females usually have the posterior margin of abdominal sternite VII deeply emarginated medially. Twelve species of acanthosomatids have been karyotyped, of which nine have a diploid number of 10 + XY (Ueshima 1979, Kerzhner et al. 2004, Rebagliati et al. 2005).

Kumar (1974a) divided the Acanthosomatidae into three subfamilies: the Blaudusinae (spelled Blaudinae by some authors), the Ditomotarsinae, and the Acanthosomatinae, with the Blaudusinae further subdivided into two tribes (the Blaudusini and Lanopini), and the Ditomotarsinae also divided into two tribes (the Ditomotarsini and Laccophorellini) (see **Table 2.2**). Separation into subfamilies was based primarily on differences in the sternal structure of the thorax and the presence or absence of a spine or tubercle at the base of the abdomen. The resulting classification is controversial and leaves the character evolution unclear; as Kment (2006) pointed out, even Kumar (1974a) realized that there were many exceptions and that some genera were not easily placed in these subfamilies or tribes. Currently, there are 57 genera and 287 species in the Acanthosomatidae (**Table 2.2**).

The distribution of the Acanthosomatidae is predominantly in the Southern Hemisphere, especially Argentina, Chile, South Africa, and Australia. Only a few genera of Acanthosomatinae (i.e., *Acanthosoma* Curtis, *Cyphostethus* Fieber, *Elasmostethus* Stål, *Elasmucha* Stål, and *Sastragala* Amyot and Serville) have representatives in the Northern Hemisphere but there has been remarkable species radiation in East and Southeast Asia. Kumar (1974a) monographed the World fauna, Kment (2006) reviewed the fauna of Madagascar, and Rolston and Kumar (1974) provided keys to the genera occurring in the Western Hemisphere. The Australian, Iranian, Nearctic, and Palearctic species have been catalogued by Cassis and Gross (2002), Ghahari et al. (2014), Froeschner (1988a), and Göllner-Scheiding (2006a), respectively. Knowledge on the Neotropical species was summarized by Schwertner and Grazia (2015). There have been five fossil species named in the genus *Acanthosoma* (Heer 1853, Förster 1891, Piton and Théobald 1935), and one named in *Elasmostethus* (Popov 1968a). Also, Fujiyama (1987) recognized but did not name a fossil species in *Acanthosoma* and another in *Elasmucha*. Additionally, another fossil genus and species, *Suspectocoris grandis* Jordan (1967), were originally described in the Acanthosomatidae, but was later transferred to the Pentatomidae (Popov 2007).

Ahmad and Moizuddin (1990) reviewed the family for the Indo-Pakistan region, and Kment (2006) reviewed those species occurring in Madagascar. Recent revisions, either in part or in whole, include *Acanthosoma* (Tsai and Rédei 2015a,b,c), *Acrophyma* Bergroth (Faúndez 2009), *Archaeoditomotarsus* Faúndez, Carvajal, and Rider (Faúndez et al. 2014a), *Cyphostethus* (Ahmad and Önder 1993), *Duadicus* Dallas (Wang et al. 2014), *Elasmostethus* (Thomas 1991, Ahmad 1997, Yamamoto 2003), *Elasmucha* (Thomas 1991), *Eupolemus* Distant (Jensen-Haarup 1931b), *Hellica* Stål (Froeschner 2000), *Lindbergicoris* Leston (Zheng and Wang 1995), *Mahea* Distant (Kment 2006), *Noualhieridia* Breddin (Kment 2007), *Panaetius* Stål (Wang et al. 2015), *Rhopalimorpha* Dallas (Pendergrast 1950, 1952), and *Tolono* Rolston and Kumar (Carvajal et al. 2015b).

Acanthosomatids are herbivorous bugs that are usually polyphagous with some species being monophagous or oligophagous (Schaefer and Ahmad 1987; Faúndez 2007a, 2009). Host plants includes trees and shrubs (Kumar 1974a; Schaefer and Ahmad 1987; Faúndez 2007b, 2009), and feeding sites include young tissues and reproductive parts of the hosts (Schaefer and Ahmad 1987; Faúndez 2007a,b). Casual records of feeding on decaying organic matter and predation are known (Miller 1971) and are probably related to a shortage of suitable host plants (Schaefer and Ahmad 1987).

Maternal care is relatively common in species of the subfamily Acanthosomatinae (Schuh and Slater 1995, Hanelová and Vilímová 2013, Tsai et al. 2015). Faúndez and Osorio (2010) described maternal care behavior for *Sinopla perpunctatus* Signoret (subfamily Blaudusinae); they also reported coloration change in the female associated with the reproductive period and the guarding of eggs and nymphs. Maternal care in the Acanthosomatidae does not represent phylogenetic conservatism; the results of a recent study (Tsai et al. 2015) do not support the hypothesis of Tallamy and Schaefer (1997) that parental care is a plesiomorphic relict in Hemiptera, which has been lost repeatedly due to high carrying cost. Maternal care has arisen four times independently in the subfamily Acanthosomatinae: common ancestor of the genus *Elasmucha*, common ancestor of *Sastragala*, in *Acanthosoma firmatum* (Walker), and in *Sinopla perpunctatus* (Tsai et al. 2015).

2.2.2.1 Key to the Subfamilies of Acanthosomatidae (modified from Kumar 1974a)

1   Abdominal spine absent; posterolateral angles of abdominal segment VII rounded or angulate, not spinose (**Figure 2.2G**); anterolateral pronotal margins thick; Australia, Africa, South America (**Figures 2.15D, E**) .... Ditomotarsinae

1'   Abdominal spine usually present but if absent, then either posterolateral angles of last connexival segments spinose (**Figure 2.2H**) or anterolateral pronotal margins thin 2

2(1)   Mesosternal carina usually absent, but if present, it is only a short raised wedge at juncture of pro- and mesosterna (may extend slightly forward and backward); when both pro- and mesosternal carinae present then invariably poorly developed (**Figure 2.2I**) and never continuous (in such cases abdominal spine well-developed and sometimes reaching anterior end of prosternal carina); Australia, southern Africa, South America (**Figures 2.15C, 2.25A**)....
....Blaudusinae

2'   Mesosternal carina usually well-developed and receiving at its posterior end the generally distally concave abdominal spine, latter closely apposed to sternal carina on left-hand side (or extending over it, or completely fused with it) (**Figures 2.1J,K**); worldwide (**Figures 2.15A, B**) .... Acanthosomatinae

2.2.3 Canopidae Amyot and Serville, 1843

The family Canopidae is a small family, both in size (5-7 mm) and number of included taxa with one genus, *Canopus* F., and nine species (**Table 2.2**). Interestingly, twice this many species have been described, but most were based on nymphal specimens that are now considered to be of unknown identity (McAtee and Malloch 1928). As with many pentatomoid families, this group was treated as a subfamily of an inclusive Pentatomidae. Horváth (1919) proposed tribal status for this group (as Canoparia), and described two new species. The first major monograph of the group (McAtee and Malloch 1928) treated it as such. It was elevated to family status by McDonald (1979) where it has remained. The only included genus was described by Fabricius in 1803 to include the species *C. obtectus* F., the type specimens of which are all nymphs, thus the type species for *Canopus* is now considered to be *incertae sedis* (McAtee and Malloch 1928). A key to the species was provided by McAtee and Malloch (1928).

Diagnostic characters (Schuh and Slater 1995, Schwertner and Grazia 2015) for this family include the following: the bugs have a circular or obovate shape, are strongly convex dorsally, rather flattened ventrally, and generally are shiny black in color, sometimes with purple or greenish reflections (**Figure 2.16A**). The length of the head anterior to the compound eyes is almost as long as the width between the compound eyes, and the jugal margins are narrowly reflexed. The antennae are five-segmented with segment II short, subequal to its diameter. The scutellum is greatly enlarged, nearly covering the entire abdominal dorsum; only a small fraction of the hemelytra is exposed. The hemelytra are quite elongate, usually twice the length of the abdomen, with a line of weakness for folding the wing near the apex of the costa; the wing membrane has at least five parallel veins. The hind wings have lobate posterior margins; they possess a stridulatory mechanism composed of a strigil on the posterior anal vein (postcubitus in several earlier papers) of the hind wing and a plectrum located on abdominal tergum I. The tibiae may be setose, but they are not distinctly bristled or spined; the tarsi are three-segmented. The abdominal sutures become obsolete laterad of the spiracles; the abdominal trichobothria are longitudinally placed mesad of the spiracular line on abdominal sterna III through VII. The nymphs are strongly convex and sclerotized, with three pairs of dorsal abdominal scent gland openings between terga III-IV, IV-V, and V-VI; the anterior gland openings are twice the width of the other two; sterna II and III are divided mesally.

The general habitus, development of the scutellum, and some abdominal and wing characters may suggest a relationship with the Megarididae and Plataspidae; however, other morphological and molecular evidence are conflicting (Grazia et al. 2008). Schaefer (1981a, 1988) considered the families Canopidae, Cydnidae, Cyrtocoridae, Lestoniidae, Megarididae, Plataspidae, and Thaumastellidae as primitive groups within the Pentatomoidea because of the plesiomorphic characters they have in common (e.g., structure of the spermatheca, stridulatory structures).

The biology and ecology of the canopids are poorly known. McHugh (1994) found two species of *Canopus* on sporophores of certain fungi in different localities in Bolivia and Costa Rica, giving evidence that canopids may be mycetophagous in habit. Schaefer (1988) provided some information on plant associations of the Canopidae.

2.2.4 Cydnidae Billberg, 1820

This family, in contrast to most of the pentatomoid families, generally has been considered to be a valid family almost from its conception. Billberg (1820) proposed the group Cydnides with a similar concept as the family Pentatomidae of Leach (1815). Amyot and Serville (1843) recognized the 'race Spinipèdes' to include the groups Cydnides, Séhirides, and Pododides, the latter currently included in the Pentatomidae under the tribe Sciocorini. Dallas (1851) treated the Cydnides and Séhirides of Amyot and Serville in his concept of the Cydnidae. The main instability pertaining to this family has been associated with its contents; that is, the taxa included within the Cydnidae have changed often. Both Fieber (1860) and Stål (1876) considered the genera *Corimelaena* White and *Thyreocoris* Schrank, both herein included in the Thyreocoridae, to be members of the Cydnidae. Uhler (1872) and Lethierry and Severin (1893) treated the Cydnidae and Thyreocoridae as separate entities. More recently, however, Dolling (1981) held a rather broad view of the family, defined by the presence of setal combs on the coxae and a strigil on the ventral surface of the hind wings; he included eight subfamilies within the Cydnidae, three of which are not included within the family in this treatise (Corimelaeninae, Thaumastellinae, and Thyreocorinae), and at least one or two others may be raised to family status in the future (see below).

Important monographs of the Cydnidae include that of Froeschner (1960) for the New World; Linnavuori (1993) for west, central, and northeast Africa; and Lis (1994) for the Oriental Region. Lis provided a catalog of the Old World taxa (1999a) as well as an updated version of the Palearctic species (2006a). Recent regional catalogs or checklists of North American (Froeschner 1988b), Old World (Lis 1999a), Palearctic (Lis 2006a), Austro-Papuan (Lis 1995) and Australian (Cassis and Gross 2002) faunas are available. Information regarding the Neotropical members of this family has been provided by Schwertner and Nardi (2015).

The Cydnidae is arguably the morphologically most diverse family of the Pentatomoidea. Accordingly, although it is easy to recognize small, apparently monophyletic subgroups within it, the monophyly of the family as a whole is questionable, and its phylogenetic relationships within Pentatomoidea are unknown (Grazia et al. 2008, Pluot-Sigwalt and Lis 2008, Lis 2010). Most modern authors exclude Thyreocoridae, *Thaumastella* Horváth, *Parastrachia* Distant, and *Dismegistus* Amyot and Serville from the Cydnidae, but there is little doubt that all of these taxa (particularly the latter two) are closely related to some cydnid subgroups; a comprehensive study of the whole complex is needed to elucidate their relationships. All of these groups possess coxal combs, and Thyreocoridae and *Thaumastella* have spine-like setae on the tibiae (albeit very weak in the latter). Some or all of these taxa are sometimes considered as subfamilies in a more broadly defined Cydnidae (Dolling 1981, Schuh and Slater 1995).

Any morphological definition of the Cydnidae will greatly overlap at least with *Parastrachia* and *Dismegistus*. The diagnostic characters (Schuh and Slater 1995, Schwertner and Nardi 2015) for the Cydnidae, in a narrow sense, include small to medium in size (2-25 mm), usually somewhat ovoid in shape, and usually glossy or shiny black or brown (some sehirines are bluish with white markings) in color. Cydnids are usually somewhat convex dorsally and distinctly convex ventrally. The head may be somewhat quadrate or more often semi-circular, relatively wide, often somewhat explanate. The antennae are five-segmented, rarely four-segmented (e.g., *Schiodtella* Signoret, *Geopeltus* Lis, and *Adrisa* Amyot and Serville). The scutellum is usually subtriangular in shape, but sometimes may be somewhat enlarged, apically broadly rounded, and usually reaching less than three quarters of the length of the abdomen. The subfamily Amnestinae has the clavi extending beyond and meeting caudad of the scutellar apex, forming a distinct claval commissure (**Figures 2.3C, 2.15F**). Most cydnids possess a stridulatory mechanism composed of a strigil on the posterior anal vein (postcubitus in several earlier papers) of the hind wing and a plectrum located on abdominal tergum I. The distal margins of the coxae are provided with a row of setae or bristles, known as coxal combs (**Figure 2.1D**). The legs are adapted for digging in soil and leaf litter, having strong spine-like setae on the tibiae (**Figures 2.1C, E-H**), sometimes set on distinct wart-like projections; the anterior tibiae often are compressed. The tarsi are three-segmented except in the Cephalocteinae where the anterior and posterior tibiae are strongly modified and, thus, the accompanying tarsi may be reduced or absent (**Figures 2.3D, E**). Abdominal trichobothria usually are present on segments III through VII and arranged longitudinally or transversely (**Figures 2.3A, B**), usually laterad of the spiracular line. Nymphal scent glands are present between abdominal terga III and IV, IV and V, and V and VI. The female spermatheca is diverse, and the spermathecal duct is either simple or is provided with a dilation of various sizes, shapes, and inner structure; sclerotized tubular projections of the distal orifice are also not rare (Pluot-Sigwalt and Lis 2008). Fourteen species of Cydnidae have been karyotyped, of which eight and four have a diploid number of 10 + XY, and 12 + XY, respectively (Ueshima 1979, Kerzhner et al. 2004, Rebagliati et al. 2005).

The Cydnidae currently contains 111 genera and 852 species. They are classified into six subfamilies (**Table 2.2**), but the Amaurocorinae are sometimes considered to be a tribe of the Sehirinae (e.g., Lis 1994). The Amaurocorinae and Garsauriinae are restricted to the Eastern Hemisphere (Lis 1999a, 2002), the Amnestinae is predominantly Neotropical, and the Sehirinae is predominantly Palearctic (Froeschner 1960, Lis 1999a, Mayorga 2002); species of the subfamilies Cephalocteinae and Cydninae are distributed worldwide. As the family is potentially non-monophyletic, it may be necessary to remove some of the included taxa.

The classification of this group has been studied thoroughly, but general conclusions remain unsettled (Pluot-Sigwalt and Lis 2008, Lis 2010). Concerning the classification at the subfamilial and tribal level, only Cephalocteinae is supported in a phylogenetic context (Lis 1999b), whereas the non-monophyly of the Cydninae and Sehirinae is suggested by some authors (Pluot-Sigwalt and Lis 2008, Lis 2010).

The biology is well-known for some members of this family and not so well-known for others (Schwertner and Nardi 2015). Species that live above ground, especially those in the subfamily Sehirinae [e.g., *Sehirus cinctus* (Palisot de Beauvois)] have been the subject of a number of studies. Others, however, that spend most of their time underground as excavators (subfamilies Cephalocteinae and Cydninae) are more difficult to study in their subterranean habitats. Nymphs and adults of the subterranean species are thought to feed on the sap of roots, although some species [e.g., *Pangaeus bilineatus* (Say)] have been observed feeding on ground pods of *Arachis hypogaea* L. (Chapin et al. 2006). Sehirines and amnestines usually feed on above-ground structures of their host plants. For example, *Sehirus cinctus cinctus* feeds on mature seeds that have fallen from several different host plants (Froeschner 1960, Sites and McPherson 1982), whereas species of *Amnestus* Dallas feed on the fruits and seeds of *Ficus colubrinae* Standley in Mexico (Mayorga and Cervantes 2001).

The females of Cephalocteinae and Cydninae lay eggs singly below the ground surface (García and Belotti 1980, Riis et al. 2005a,b). In contrast, sehirine females lay egg masses in shallow cracks in the soil surface (Sites and McPherson 1982). At least one species of Amnestinae, *Amnestus ficus* Mayorga and Cervantes, lays eggs inside the fruits of *Ficus* L. (Moraceae) (Mayorga and Cervantes 2001). Maternal care has been observed in some of those species that lay egg masses [e.g., *Sehirus cinctus cinctus* and *Adomerus triguttulus* (Motschulsky)]. The females guard the eggs until the nymphs disperse after hatching (Southwood and Hine 1950, Nakahira et al. 2013). Within the entire Pentatomoidea, an active provisioning of food by females to the nymphs is known only in the Cydnidae and the Parastrachiidae. When oviposition occurs in the soil, eggs usually are deposited near the host plant. The moisture and soil texture seem to determine the depth of oviposition (Willis and Roth 1962, Riis and Esbjerg 1998a,b, Riis et al. 2005a,b). In times of prolonged drought in the Brazilian Cerrado, eggs of *Scaptocoris* Perty species have been found as deep as 1.5 meters (Nardi et al. 2008). The eggs of cydnid species are characterized by a smooth corium, a uniform creamy coloration, and no conspicuous projections (García and Belotti 1980, Mayorga and Cervantes 2001, Vivan et al. 2013). Incubation time can vary from 1 week (García and Belotti 1980, Sites and McPherson 1982, Riis et al. 2005a,b) up to 4 weeks (Sales and Medeiros 2001). Several sehirine species (e.g., *Adomerus triguttulus* and *Canthophorus niveimarginatus* Scott), are known to produce trophic eggs (i.e., inviable eggs that are usually used for food for offspring) (Nakahira 1994; Kudô and Nakahira 2004, 2005; Kudô et al. 2006; Filippi et al. 2008).

Burrower bug nymphs live within or near the surface of the soil, feeding on the roots or fallen seeds of their hosts. They are typically oligophagous or polyphagous, feeding on the plants near the site where they hatched from the eggs. As with most terrestrial Heteroptera, cydnids go through five nymphal instars. The majority of cydnid species seems to be polyphagous with several plant families reported as hosts (Timonin 1958, Becker 1967a, Mayorga and Cervantes 2001, Riis et al. 2005a,b, Chapin et al. 2006, Schwertner and Nardi 2015). On occasion, the bugs are reported as agricultural pests (Gallo et al. 2002, Riis et al. 2005a,b, Schwertner and Nardi 2015).

Some species, and some individuals within other species, are brachypterous and, thus, are unable to fly. Those that can fly seem to take wing mostly for dispersal, but flying could also be for colonization of new areas, location of food, and finding new mates (Willis and Roth 1962, Oliveira and Malaguido 2004, Nardi et al. 2008). Little is known about cydnid reproductive behavior. Soil-dwelling species mate in the soil (Willis and Roth 1962, Nardi 2005). Similar to other Heteroptera, it appears that copulation in the Cydnidae is mediated by chemicals and sound communication (Gogala et al. 1974, Gogala 1984, Čokl et al. 2006, Pluot-Sigwalt 2008).

As expected from a group whose members live primarily on or near the ground, there are several fossil genera and species known. Also, at least three family-level names have been proposed within the Cydnidae. For example, Pinto and Ornellas (1974) erected two fossil families, the Pricecoridae for *Pricecoris beckerae* Pinto and Ornellas and the Latiscutellidae for *Latiscutella santosi* Pinto and Ornellas. Additionally, Popov (1986) proposed the subfamily Clavicorinae for two fossil genera, *Clavicoris* Popov and *Cretacoris* Popov. Later, Popov and Pinto (2000) placed all three of these family-groups as junior synonyms of the Amnestinae. Within the subfamily Amnestinae, there are six genera and nine fossil species (Pinto and Ornellas 1974, Popov 1986, Thomas 1988, 1994a, Yao et al. 2007); within the subfamily Cydninae, there are three genera and 41 fossil species (Heer 1853; Oustalet 1874; Novák 1877; Scudder 1878, 1890; Förster 1891; Cockerell 1909; Henriksen 1922; Piton 1933; Théobald 1937; Statz and Wagner 1950; Jordan 1967; Kinzelbach 1970; Popov 1986, 2007; Schaefer and Crepet 1986; Thomas 1994a); and within the Sehirinae, there are five fossil species (Statz and Wagner 1950, Vršanský et al. 2015). There are currently six genera and 25 fossil species that have not been placed to subfamily. Another fossil genus and species, *Ovalocoris parvis* Jordan, was originally described in the Lygaeidae but later (Popov 2007) transferred to the Cydnidae. Additionally, there are three fossil species that were originally placed in the Cydnidae that have now been placed elsewhere. That is, *Cydnopsis affinis* Jordan and *C. ventralis* Jordan have both been transferred to the nepomorphan family Aphelocheiridae (Popov 2007), and the third species, *Cydnopsis nigromembranacea* Jordan (1967) has been transferred to the Coleoptera (Popov 2007).

2.2.4.1 Key to the Subfamilies of Cydnidae (modified from Schuh and Slater 1995)

1   Clavi reaching beyond and meeting in a straight line beyond apex of scutellum, forming claval commissure (**Figure 2.3C**); New World (except one species introduced into Old World) (**Figure 2.15F**) ....Amnestinae

1'   Claval commissure absent 2

2(1)   Fore tibiae falcate or cultrate, much produced beyond tarsal insertion, therefore tarsi appearing to arise at middle of tibial length (except in *Cephalocteus* where tarsus is inserted apically or subapically) (**Figures 2.3D, E**); posterior tibiae strongly broadened; worldwide (**Figures 2.15G, H**) .... Cephalocteinae

2'   Fore tibiae not cultrate, tarsi arising at or near apices of tibiae (**Figures 2.1E, G**).............. 3

3(2)   A submarginal row of setigerous punctures present along each lateral pronotal margin (**Figure 2.3F**); diameter of tarsal segment II subequal to diameters of tarsal segments I and III (**Figure 2.1E**); worldwide (**Figures 2.15J, K; 2.25B**) ....Cydninae

3'   Pronotum lacking submarginal row of setigerous punctures along lateral margins; diameter of tarsal segment II distinctly narrower than diameters of tarsal segments I and III (**Figures 2.1G, H**).. 4

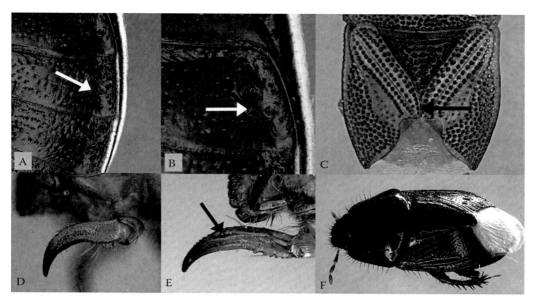

FIGURE 2.3 A, *Sehirus cinctus*, lateral abdomen, ventral view; B, *Sehirus cinctus*, detail of the second abdominal segment, ventral view; C, *Amnestus pusio*, scutellum and hemelytra, dorsal view; D, *Scaptocoris castaneus*, anterior tibia, lateral view; E, *Scaptocoris castaneus*, anterior tibia, inner view; F, *Cyrtomenus mirabilis*, habitus, dorsolateral view.

4(3)   Pronotum with a fine, distinctly impressed subapical groove paralleling anterior margin; abdominal trichobothria on segments III through VII arranged in longitudinal pairs; body conspicuously flattened and coarsely punctured; Old World .... Garsauriinae

4'   Pronotum without distinctly impressed subapical groove paralleling anterior margin; abdominal trichobothria on segments III through VII arranged in transverse pairs (**Figures 2.3A, B**); body not conspicuously flattened... 5

5(4)   Long setae present at least on lateral margins of head, pronotum, and corium (**Figure 2.15I**); female spermatheca composed of a long, coiled, simple, non-differentiated tube; Old World (**Figure 2.15I**) .... Amaurocorinae

5'   Long setae not present along lateral margins of head, pronotum and corium (**Figure 2.15L**); female spermatheca more complex, possessing a spermathecal bulb and pump region; worldwide (**Figure 2.15L**)... Sehirinae

2.2.5 Dinidoridae Stål, 1868

The Dinidoridae represents another family that was treated by most early workers as a subfamily of the Pentatomidae. Amyot and Serville (1843) recognized the group Mégyménides, including only *Megymenum* Guérin-Méneville, as a group within their tribe Coniscutes. Stål (1868) proposed the group Dinidorida to include the genera *Aspongopus* Laporte (= *Coridius* Illiger), *Atelides* Dallas (= *Sagriva* Spinola), *Dinidor* Latreille, and *Megymenum*; he later (Stål 1870) gave the group subfamilial status, a classification that most workers followed until the 1950s. The main exception was Lethierry and Severin (1893) who catalogued this group as a family. Cachan (1952), in his monograph of the Madagascar fauna, treated this group as a family. Although the name Megymeninae has priority over the Dinidoridae, the latter family has been well established and should not be supplanted (see article 35.5 of the International Code of Zoological Nomenclature [ICZN]); the Megymeninae is retained as a subfamily within the Dinidoridae.

The most recent authoritative work on the family was by Durai in 1987 who also treated this group at the family level; earlier, this family was treated in a still useful paper by Schouteden (1913). Durai (1987) established the presently used classification (with a couple updates - see below), and provided diagnoses and keys for identifying all genera and species. A checklist of Old World taxa was provided by Lis (1990), a World catalog was provided by Rolston et al. (1996), and the Palearctic species also have been catalogued (Lis 2006b). Information on the Neotropical species have been compiled by Schwertner and Grazia (2015).

There are at least two fossil pentatomoids possibly belonging to this family: *Dinidorites margiformis* described from Eocene deposits in Colorado (Cockerell 1921) and an unnamed fossil reported from deposits in British Columbia (Archibald and Mathewes 2000), Canada, which was originally placed in the subfamily Megymeninae.

Diagnostic characters of the family (Schuh and Slater 1995, Schwertner and Grazia 2015) include the lateral margins of the head, which are usually carinate; and the bucculae, which are short and elevated, almost flap-like. The antennae may be four or five-segmented, some segments may be flattened (**Figure 2.2F**), and the rostrum usually reaches to or beyond the middle coxae. The humeral angles of the pronotum are rounded, almost never developed, but the lateral margin usually bears an anteriorly directed projection in the Megymeninae (**Figure 2.16D**). The scutellum is usually somewhat triangular in shape, with the basal width subequal to the medial length, and it never covers the corium; the apex is usually rounded. The hemelytral membrane usually has reticulate venation. The tarsi may be two- or three-segmented; the tibiae lack distinct bristles or spines, and the coxae lack the coxal combs evident in the Cydnidae and related families. The female spermatheca lacks a dilation and sclerotized rod. Eight of twelve species that have been karyotyped have a diploid number of 12 + XY (Ueshima 1979, Kerzhner et al. 2004, Rebagliati et al. 2005).

Typically, this family is divided into two subfamilies (i.e., Dinidorinae and Megymeninae) (**Table 2.2**), each with quite different facies. The dinidorines are somewhat rounded to oval, colored in browns, blacks, and tans, occasionally with red or yellow, but are relatively smooth surfaced with typical pentatomoid punctures. The megymenines are usually gray to dark grey or almost black, not quite so smoothly ovoid in shape, and their dorsal surface is rather rough or granulated (**Figure 2.16D**). *Eumenotes* Westwood (**Figure 2.16C**) and *Afromenotes* Kment and Kocorek, which superficially resemble members of the Megymeninae, are of uncertain placement. It may be justified to place them in a third subfamily, the Eumenotinae, or keep them as a tribe in either of the other two subfamilies.

The family Dinidoridae includes 17 genera and 109 species in two subfamilies and five tribes (**Table 2.2**), distributed mostly in the Old World. The only representative in the New World is the nominotypical genus, *Dinidor*, with six endemic species. The classification used in this chapter was established by Durai (1987), followed by Rolston et al. (1996) in their World catalog of the group, and includes updates from Kocorek and Lis (2000) and Lis et al. (2015). In the first work, Kocorek and Lis (2000) proposed a new tribe (Byrsodepsini [**Figure 2.16B**]) within the Megymeninae and placed the Eumenotini as a junior synonym of the Megymenini. The Eumenotini has had a complex taxonomic history, at one time or another, having been classified within the Aradidae, Pentatomidae, Tessaratomidae, Dinidoridae, or as a distinct family, the Eumenotidae (see Kment and Kocorek 2014). The molecular analysis by Lis et al. (2012a) revealed that the Eumenotini is more closely related to the Dinidorinae than to the Megymeninae, which resulted in its removal from the synonymy of the Megymeninae and restoring it as a tribe without subfamily assignment. In the second work, Lis et al. (2015) conducted a molecular investigation of the Madagascan genus *Amberiana* Distant. Interestingly, the molecular data indicated that *Amberiana* was related to the cydnid subfamily Sehirinae or the family Parastrachiidae, but the authors stated that the morphology was typical dinidorid; they ultimately decided to leave *Amberiana* in the Dinidoridae, but they erected a new tribe, the Amberianini, for this single genus.

Gapud (1991) considered the Dinidoridae and Tessaratomidae as sister groups based on two synapomorphies: the spiracles on abdominal segment II at least partially exposed (**Figure 2.2B**) and laterotergites IX quite large in females. Grazia et al. (2008) found similar results, although they felt that the Dinidoridae could be paraphyletic or monophyletic according to the analyses they performed (morphological, molecular, or combined analyses). The sister group relationship of Dinidoridae and Tessaratomidae was supported morphologically (Kment and Vilímová 2010a) and by molecular data (Lis et al. 2012a,c). The two families together certainly form a single monophyletic assemblage, but the monophyly of each of them, particularly the Tessaratomidae, needs further support.

Only a few species of this family have had their biology studied; most of the available information is limited to label data or other field observations (Schaefer and Ahmad 1987, Schaefer et al. 2000). All species studied to date are exclusively phytophagous, feeding on both the reproductive and vegetative parts of the host plant. The data indicate that polyphagy is widespread, but oligophagy in some species is likely. Some species seem to show a preference for certain plant families (e.g., *Coridius* Illiger species prefer plants of the family Cucurbitaceae and can be pests on melons and squash; *Eumenotes* species are frequently found on Convolvulaceae). All species studied are univoltine and usually use more than one host plant during their life cycles (Schaefer et al. 2000). Some species exhibit gregarious behavior at certain times of the year, being found in large quantities together on their host plants. Dinidorids only lay one or two clutches of eggs per female, each containing 14-28 eggs (Schaefer et al. 2000). At least one species, *Cyclopelta parva* Distant, has been reported to exhibit parental care of the immature forms (Hoffmann 1936).

Various species of dinidorids have been used for human consumption (Strickland 1932, Hoffmann 1947). For example, in the Sudan, gelatin has been extracted from *Coridius viduatus* (F.), which has been used in making ice cream (Mariod and Fadul 2014). Dinidorids also are thought to have medicinal value, especially in oriental regions. For example, *C. chinensis* (Dallas), once popular in China as an aphrodisiac (Hoffman 1947), still it is in use as a traditional Chinese medicine; it is believed that it can regulate breath and relieve pain (Zhang and You 2002, Yao 2006); Hoffmann (1947) stated that "this species is very commonly used in China in an aphrodisiacal medicine and is on sale in Chinese medicine shops throughout China. It is called 'Chu Shan Chung' or 'Hai Tao Chung' and was written about in 1590 by Li Shih Chen and in 1890 by Fang Shui." *Coridius nepalensis* (Westwood) is used similarly in India; again Hoffmann (1947) stated "the natives of Assam are very fond of these bugs which they pound up and mix with foods that are made of rice to improve the taste." In Sudan, oils have been extracted from *C. viduatus*, which were found to have anti-bacterial properties (Mustafa et al. 2008).

2.2.5.1 Key to the Subfamilies of Dinidoridae (modified from Schuh and Slater 1995)

1	Basal angles of scutellum without fovea; posterolateral angles of abdominal connexiva neither tuberculate nor lobed; Old World except one genus (*Dinidor*) in New World tropics (**Figures 2.16B, 2.25C**)..Dinidorinae
1'	Basal angles of scutellum each with a fovea; posterolateral angles of abdominal connexiva either tuberculate or lobed.. 2
2(1)	Abdominal spiracles arranged along a line; trichobothria paired; Old World (**Figure 2.16D**)... Megymeninae
2'	Spiracles of abdominal segment II situated close to lateral margin, not in line with spiracles of remaining segments (**Figure 2.4E**); trichobothria unpaired; Old World (**Figure 2.16C**)..Eumenotini (of uncertain placement)

2.2.6 Lestoniidae China, 1955

This endemic Australian family contains a single genus (*Lestonia* China [**Figure 2.25D**]) and two species (*L. haustorifera* China and *L. grossi* McDonald) (**Table 2.2**). This group originally was described as a subfamily of the Plataspidae (China 1955) but subsequently was elevated to family status by China and Miller (1959). Because of the enlarged scutellum, this family has been associated at one time or another with the Plataspidae, Scutelleridae, or the pentatomid subfamily Aphylinae. Fischer (2000) proposed a possible relationship with the Acanthosomatidae, based on the two-segmented tarsi, and the supposed homology of the Pendergrast's organs in the Acanthosomatidae with the pair of unusual, small, disc-shaped organs located on the abdominal venter of female specimens of this family. A relationship with the Aphylinae can be dismissed by the presence of a sclerotized rod in the female spermatheca of the aphylines (lacking in the Lestoniidae). Studies by McDonald (1969a, 1970), Gross (1975b), Gapud

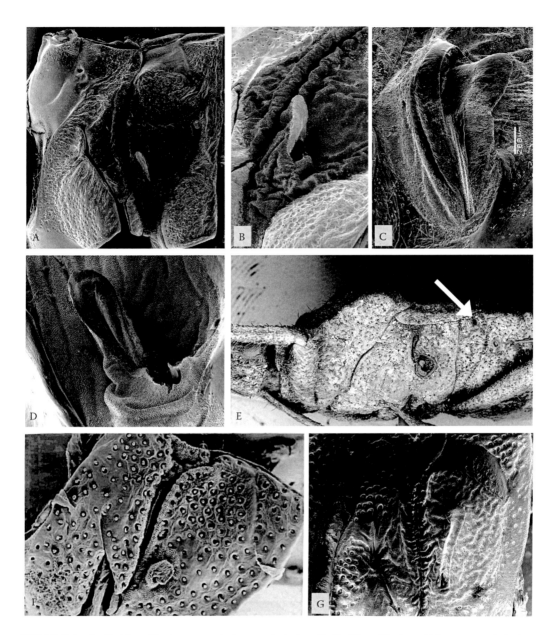

FIGURE 2.4 A, *Coridius viduatus*, evaporatorium, ventral view; B, *Coridius viduatus*, ostiolar ruga, ventral view; C, *Tessaratoma papillosa*, ostiolar ruga, ventral view; D, *Oncomeris flavicornis*, ostiolar ruga, ventral view; E, *Eumenotes obscura*, thorax and anterior abdomen, lateral view; F, *Kayesia parva*, evaporatorium, ventral view; G, *Eurygaster maura*, evaporatorium, ventral view.

(1991), and Schaefer (1993) seemed to support a relationship with the Plataspidae, but they ultimately decided that there were enough differences to maintain separate family status. The phylogenetic studies by Grazia et al. (2008) and Wu et al. (2016) provide further support for the recognition of Lestoniidae as the sister group of the Acanthosomatidae.

Members of this family are relatively small (5-6 mm), ovoid, dorsally convex, and ventrally flattened (Schuh and Slater 1995). The scutellum is large in *Lestonia haustorifera* (**Figure 2.25D**) covering much, but not all, of the dorsum but usually reaching to near the apex of the abdomen; it is much shorter and more triangular in *L. grossi*. The margins of the head, pronotum, and basal areas of the coria, in both species, are laminately produced, giving the insect the appearance of a small tortoise beetle or scale. The pronotum is relatively large with the anterior angles reaching beyond the anterior margin of the compound eyes. The ocelli are quite small and widely separated. The antennae are relatively short, four-segmented. They have two-segmented tarsi, which has been used to link them with the Acanthosomatidae. Similarly, the females have a pair of small disclike organs located on abdominal sternite VI just anterior to and laterad of the external genitalia; again, these have been considered to be homologous to the Pendergrast's organs in the Acanthosomatidae; China (1955) speculated that they may help specimens adhere to the substrate. The two species possess a pair of small, transversely arrayed trichobothria (initial studies by China 1955 indicated that there was only a single trichobothrium on each segment; this was later corrected by China 1963) with the ental trichobothrium of each pair in line with the spiracular line. The female spermathecal duct lacks a dilation and a sclerotized rod; the spermathecal bulb is ball-shaped, without diverticula; Schuh and Slater (1995) indicated that *L. haustorifera* lacks flanges near the bulb, but they indicated such a flange is present near the spermathecal bulb in *L. grossi*.

Little is known about the biology of lestoniids. They have only been found near the growing tips of trees of the genus *Callitris* Ventenat (Cupressaceae) in mostly arid conditions, where they resemble small scales or tortoise beetles (Gross 1975b, Cassis and Gross 2002).

2.2.7 Megarididae McAtee and Malloch, 1928

Members of the Megarididae are rare in collections, probably due to their small size (less than 5 mm), and their secretive habits. It is also a small family in number of taxa, including a single extant genus (i.e., *Megaris*) and currently 18 species (**Table 2.2**), all occurring in the Neotropics (McDonald 1979). A second genus with one species (*Minysporops dominicanus* Poinar and Heiss) has been described from Dominican amber (Poinar and Heiss 2013). This group was not recognized as a family-level taxon until McAtee and Malloch (1928) proposed the subfamily Megaridinae within the Pentatomidae; they also provided a key to species known at that time. Kormilev (1954) elevated it to family status, a position that nearly all recent workers have recognized.

Members of this family, in general appearance, are quite similar to the Canopidae and some members of the Plataspidae, only much smaller (**Figure 2.16E**). Diagnostic characters (Schuh and Slater 1995, Schwertner and Grazia 2015) include the rounded shape, the convex dorsum, the flattened venter, and the usual shiny black color, occasionally with reddish spots or markings. The anterolateral margins of the head and pronotum are carinate; the bucculae are undeveloped. The antennae are four-segmented, with many long setae that are about the same length as the diameter of the antennal segment in females, and much longer in males. The scutellum is enlarged and nearly covers the abdominal dorsum (**Figure 2.16E**). Similar to the Canopidae, the forewings are longer than the abdomen, but they have a thin, weak area at about the middle of the costa, thus allowing the wings to fold up underneath the scutellum. The wing membrane lacks veins or has a single longitudinal vein. The tibiae lack spines or bristles; the tarsi are two-segmented. The coxal combs are absent. The female spermathecal duct lacks a dilation and sclerotized rod.

In the study by Grazia et al. (2008), the Megarididae are placed near the Plataspidae in their morphological studies. They did indicate that this placement was inconclusive because it was based heavily on the similar shape and structure of the scutellum. They stated that "body shape alone is misleading with regard to the establishment of phylogenetic affinities." Unfortunately, they were not able to obtain fresh material to do DNA studies, so the phylogenetic placement of this family within the Pentatomoidea remains unclear.

The biology and ecology of the megaridids are poorly known. It is believed that all species are exclusively phytophagous. *Megaris puertoricensis* Barber and *M. semiamicta* McAtee and Malloch have been recorded feeding on flowers of *Eugenia* L. species (Myrtaceae) (Schuh and Slater 1995). More recently, a specimen of an unknown species of *Megaris* was collected in a light trap placed inside a plot of cultivated species of *Eucalyptus* L'Héritier de Brutelle (Myrtaceae) in the state of São Paulo, Brazil (Cristiano F. Schwertner, unpublished data).

2.2.8 Mesopentacoridae Popov, 1968

Popov (1968b) erected a new fossil family, the Mesopentacoridae, to contain a single genus and species, *Mesopentacoris costalis* Popov. Superficially, this genus and species are similar in appearance to modern day urostylidids or the fossil coreoid family Pachymeridiidae (Yao et al. 2008). Popov (1989) added a second species, *M. orientalis*, and then in 1990, he added another genus and species, *Corienta transbaicalica*. Finally, in 1996, Ren et al. added a third genus with another new species, *Pauropentacoris macruratus*.

2.2.9 Parastrachiidae Oshanin, 1922

This family presently contains two genera (and eight species), *Dismegistus* (**Figure 2.25F**) and *Parastrachia* (**Figure 2.25E**) (**Table 2.2**), which have had similar but separate taxonomic histories. Distant (1883) originally placed *Parastrachia* in the Pentatomidae "somewhere between the genera *Strachia* Hahn [currently Strachiini] and *Catacanthus* Spinola [currently Catacanthini]." Oshanin (1922) placed *Parastrachia* in its own tribe, but still within the Pentatomidae. In the first thorough study of the group, Schaefer et al. (1988) established the subfamily Parastrachiinae within the family Cydnidae solely for the genus *Parastrachia*. Schaefer later (Sweet and Schaefer 2002) admitted that he had "private reservations" about this placement and felt that the Parastrachiinae probably deserved family status, a move that was made by Sweet and Schaefer (2002). This position has been supported by more recent studies (Grazia et al. 2008, Lis 2010).

Originally, *Dismegistus* was placed in the cydnid subfamily Sehirinae (Amyot and Serville 1843), a position that was supported by Stål (1876). Signoret (1880) transferred this genus to the Pentatomidae, placing it near the genus *Strachia* (currently Strachiini). Interestingly, Bergroth (1923) moved *Dismegistus* to the pentatomid subfamily Asopinae. Leston (1956a) returned the genus to its original placement as a member of the Sehirinae in the Cydnidae. Dolling (1981), in his thorough study of the Cydnidae and related families, removed *Dismegistus* from the Cydnidae, but he did not know where to place it. It remained in limbo until Pluot-Sigwalt and Lis (2008) noticed similarities with the genus *Parastrachia* (mainly in the structure of the spermatheca); at about the same time, Grazia et al. (2008) determined that *Dismegistus* and *Parastrachia* had similar DNA sequences. Consequently, *Dismegistus* only recently has been transferred into the Parastrachiidae.

A couple of factors may have contributed to the two genera having separate taxonomic histories. For example, they have different distributions, the two species of *Parastrachia* occurring from India through China, and into Japan, and the six species of *Dismegistus* being confined to the African continent. Furthermore, although members of both genera are usually red and black in coloration, their size and shape are different. Species of *Parastrachia* are somewhat larger, more slender, and somewhat similar in shape to some Largidae or Pyrrhocoridae (**Figure 2.25E**). Species of *Dismegistus* are smaller and much more ovoid in shape (**Figure 2.25F**). This also may help explain why few papers have been recently published on *Dismegistus* as compared to *Parastrachia*. Schaefer et al. (1991) reviewed the genus *Parastrachia*. The species have recently been catalogued (Lis 2006c).

Members of this family are small to medium in size and usually colored red and black; species of *Parastrachia* (**Figure 2.25E**) are larger and more slender, whereas species of *Dismegistus* (**Figure 2.25F**) are smaller and more broadly oval. The bucculae meet posteriorly, similar to that seen in the pentatomid subfamily Asopinae, but the rostrum is not particularly crassate, and these species are not typically predatory (united bucculae are also found elsewhere in the Pentatomoidea, including the pentatomid subfamily Edessinae, see **Section 2.2.10.6**). The antennae are five-segmented. The prosternum is medially sulcate; the sternal thoracic structure is similar to that seen in the Sehirinae, but parastrachiines have tibiae that are not adorned with bristles or spines. Both *Parastrachia* and *Dismegistus* possess coxal combs composed of an irregular series of long, narrow setae. The ostiolar rugae and associated evaporative areas are well developed in *Parastrachia*, but strongly reduced, obsolete in *Dismegistus*. Abdominal segment VIII is broadly exposed, similar to that seen in the Acanthosomatidae and the Urostylididae, but the bugs lack the claval commissure of the urostylidids; also, they have three-segmented tarsi, thus differing from the acanthosomatids. The female spermatheca is relatively simple, the duct lacking the dilation and sclerotized rod; the spermathecal bulb is simple, ball-shaped, with a pair of flanges present.

Virtually nothing is known about the biology of members of *Dismegistus*. On the other hand, much is known about *Parastrachia japonensis* Scott. Females of this species exhibit parental care as they will excavate an egg chamber in the soil or leaf litter, provision it with seeds, and stand guard over the eggs and early instars (Nomakuchi et al. 1998, 2001, 2005; Filippi et al. 2000a,b, 2001, 2002, 2005; Hironaka et al. 2003a,b, 2007a-c, 2008a,b). They are known to produce trophic eggs, that is eggs that are not viable but are used for food by offspring (Hironaka et al. 2005). The known host plant of *P. japonensis* is *Schoepfia jasminodora* Siebold and Zuccarini (Schoepfiaceae).

2.2.10 Pentatomidae Leach, 1815

The Pentatomidae is the largest family in the Pentatomoidea, containing 940 genera and 4,949 species in ten subfamilies (**Table 2.2**) with many new taxa still awaiting description. The total number of species undoubtedly will reach 5,000 in the near future and perhaps even 6,000 eventually. As with any large and diverse group, it is difficult to provide defining characters that work for all members. The Pentatomidae was proposed by Leach (1815) as a family group; he included it together with the Scutelleridae in a taxon currently equivalent to the Pentatomoidea. Burmeister (1835) referred to this group as the Scutata.

Amyot and Serville (1843) presented one of the first classifications of the Pentatomoidea (**Table 2.1**). Their "famille Longiscutes" (long scutellum) is essentially equivalent to our present day Pentatomoidea. Amyot and Serville divided this 'family' into two large groups, the Orbiscutes for those taxa with a scutellum that covered most of the abdomen, and the Coniscutes for those taxa in which the scutellum did not nearly cover the entire abdomen. They further divided the Orbiscutes into two "races" based on body shape: the Anguleux contained those taxa that were somewhat more flattened (our present day scutellrids, podopines, and cyrtocorines), and the Globuleux contained those taxa that were rounded and more globular in shape (our present day thyreocorids, canopids, and some scutellerids). The remaining Coniscutes were separated into smaller groups using some of the same characters we use today (e.g., size and length of rostrum, the armature of the venter, and whether the legs had spines or not).

Dallas (1851) recognized many of the same taxa but now at the family level (Asopidae, Edessidae, Halydidae, Oxynotidae, Pentatomidae, Phyllocephalidae and Podopidae) within the group Scutelleroidea. Fieber (1861) recognized two more family groups, the Macropeltidae and Discocephalidae. Stål (1865) considered only the subfamilies Asopida, Pentatomida, and Phyllocephalida in a more inclusive Pentatomidae. He refined his classification in later works (Stål 1870-1876), treating also Discocephalina and Oxynotina as subfamilies and dividing the Pentatomina into several groups of genera (see comment below in the Pentatominae section); current genera included in Podopinae were recognized as a distinct group within Pentatomina. The taxa recognized in the Stål-based classification (**Table 2.1**) were largely followed by subsequent workers (e.g., Lethierry and Severin 1893, Kirkaldy 1909, Oshanin 1912, Leston 1952a, China and Miller 1955, Linnavuori 1982). The works of Singh-Pruthi (1925) and Pendergrast (1957) helped establish the current limits of the family. Based on the male and female genitalia, the results of these two authors supported the recognition of a single group including five of Stål's subfamilies (Asopinae, Discocephalinae, Pentatominae, Phyllocephalinae, and Podopinae). This arrangement has been refined and expanded (Leston 1958; McDonald 1966; Gross 1975b; Rolston and McDonald 1979; Gapud 1991; Rider 2006a, 2015c; Cassis and Gross 2002; Grazia et al. 2008).

The family Pentatomidae represents a well-supported group among the Pentatomoidea, based on both morphology and molecular data (Gapud 1991; Henry 1997; Li et al. 2005; Xie et al. 2005; Grazia et al. 2008; Li et al. 2012; Yao et al. 2012). Morphological synapomorphies include gonapophyses 8 and first rami lost (non-homoplastic); gonapophyses 9 reduced, fused to gonocoxites 9, second rami lost (non-homoplastic); and spermathecal duct dilated, its distal orifice provided with a sclerotized invagination (slerotized rod) (homoplastic) (Gross 1975b, Gapud 1991, Grazia et al. 2008). However, with such a large group, there is much variation in size, shape, and coloration, thus making it difficult to give a precise diagnosis for the family.

Some of the smallest species (*Sepontia*) are only a couple of millimeters in length, whereas some species of *Alcaeorrhynchus* Bergroth (**Figure 2.17B**), *Catacanthus* (**Figure 2.29C**), *Porphyroptera* China, *Mustha* Amyot and Serville, and *Xiengia* Distant are quite large, approaching or surpassing 25-35 millimeters. The body is usually broad and ovate, but some grass-feeding species are quite long and slender (e.g., *Acoloba* Spinola [**Figure 2.21D**], *Aelia* F. [**Figure 2.28A**], *Mecidea* Dallas [**Figure 2.30F**]). All variations of coloration exist, from browns and greens allowing the individuals to blend in with their surroundings, to bright reds, oranges, and blues, sometimes in metallic hues (**Figures 2.27A-2.32L**). The family is characterized further (Schuh and Slater 1995, Grazia et al. 2015) by the antennae usually having five segments, although some subfamilies (Cyrtocorinae, Serbaninae) and species only have four, and at least one halyine genus, *Omyta* Spinola, has only three. The scutellum is usually large and subtriangular, although it may be enlarged in some groups, and even cover a large portion of the abdomen in some asopines, podopines, and pentatomines. Frena are present, and they usually extend beyond the middle of the scutellar margins; the clavi usually do not extend beyond the apex of the scutellum, so the claval commissure is lacking. In many groups, the thoracic sterna are sulcate medially, whereas in more advanced groups, the sulcus has been replaced by a medial carina. The tarsi are usually three-segmented, but they are only two-segmented in the Cyrtocorinae, Stirotarsinae, and in some minor pentatomine tribes or isolated genera. The abdominal trichobothria are usually transverse, and located near the spiracular line. In some species, there are areas on the abdominal venter on each side that appear to be thinner, somewhat opaque; these "cuticular patches" were studied in a series of papers (Staddon 1992, 1998, 2000; Staddon and Ahmad 1994; Staddon et al. 1994). The female spermathecal duct has a dilation with a long, slender, sclerotized rod evaginated from the distal orifice; the spermathecal pump is well-developed with both a proximal and a distal flange; the spermathecal bulb may be simple, digitoid, or ball-shaped, often with one to three tubular diverticula. The eggs are usually barrel-shaped with a pseudoperculum. Nearly 300 species of Pentatomidae have been karyotyped; the diploid number varies from 10 + XY up to 24 + XY, but the most common diploid number by far is 12 + XY (Ueshima 1979, Kerzhner et al. 2004, Rebagliati et al. 2005).

Almost all subfamilies of Pentatomidae are well-defined by unique apomorphies, supporting the monophyly of these taxa (Rolston and McDonald 1979, Gapud 1991, Konstantinov and Gapon 2005, Campos and Grazia 2006, Gapon and Konstantinov 2006). The only exception is the subfamily Pentatominae, a 'catch-all' taxon that includes several genera and groups of genera not recognized in any of the other subfamilies (Cassis and Gross 2002).

The phylogenetic relationships among pentatomid lineages have been almost completely ignored. Leston (1953a, 1954a) was the first to document and discuss the monophyly and phylogenetic relationships within the Pentatomidae. McDonald (1966), Gross (1975b), and Linnavuori (1982) expanded these studies and suggested good apomorphic characters for the recognition of several monophyletic groups within the family. Gross (1975b) and Linnavuori (1982) also discussed possible phylogenetic relationships among those groups. Gapud (1991) and Hasan and Kitching (1993) were the first authors to discuss phylogenetic relationships with the Pentatomidae based on phylogenetic trees.

More recently, studies under a phylogenetic framework have been conducted at different taxonomic levels. The monophyly of the subfamily Edessinae was tested in part by Barcellos and Grazia (2003b); they also proposed hypotheses concerning the relationships among the included genera. The work of Campos and Grazia (2006) supported the monophyly of the tribe Ochlerini (Discocephalinae); Garbelotto et al. (2013) and Roell and Campos (2015) expanded the knowledge about phylogenetic relationships within this tribe. Concerning the subfamily Pentatominae, phylogenetic studies at the tribal (Schaefer and Ahmad 1987, Memon et al. 2011, Schwertner and Grazia 2012), groups of genera (Grazia 1997, Bernardes et al. 2009), and genera levels (Thomas 1985, Fortes and Grazia 2005, Ferrari et al. 2010, Greve et al. 2013) have been published. But most of the subfamilies, tribes, and groups of genera have never been studied in a phylogenetic framework, and the relationships within the Pentatomidae remain unknown. The results of Gapud (1991) and Grazia et al. (2008) suggest that the Aphylinae and Cyrtocorinae may be basal lineages, but a definitive conclusion and better resolution at the family and tribal levels in the phylogenetic hypothesis presented were limited by the scope of the study conducted. The pentatomid subfamilial and tribal classification used in this book is outlined in **Table 2.3**.

We will not go into the biology of the Pentatomidae here. Some notes on biology, habitats, etc., will be discussed under the subfamily and tribal sections, and also in many of the chapters throughout this book. The Australian, Iranian, North American, and Palearctic species have recently been catalogued by Cassis and Gross (2002), Ghahari et al. (2014), Froeschner (1988c), and Rider (2006a), respectively. Salini and Viraktamath (2015) provided keys for the identification of the South Indian genera. Grazia et al. (2015)

TABLE 2.3

Pentatomidae Classification and Diversity

Subfamily	Tribe	Genera[1]	Species[1]	Tribe	Genera[1]	Species[1]	Total Genera[1]	Species[1]
Aphylinae							2	3
Asopinae							63	303
Cyrtocorinae							4	11
Discocephalinae							81	325
	Discocephalini	46	195	Ochlerini	35	130		
Edessinae							15	338
Pentatominae							660	3484
	Aeliini	4	28	Hoplistoderini	7	35		
	Aeptini	8	19	Lestonocorini	4	20		
	Aeschrocorini	8	20	Mecideini	1	16		
	Agaeini	2	11	Memmiini	1	7		
	Agonoscelidini	1	26	Menidini	28	164		
	Amyntorini	6	8	Myrocheini	25	75		
	Antestiini	29	171	Nealeriini	2	4		
	Axiagastini	4	12	Nezarini	26	272		
	Bathycoeliini	1	32	Opsitomini	1	1		
	Cappaeini	24	151	Pentamyrmecini	1	1		
	Carpocorini	127	503	Pentatomini	56	316		
	Catacanthini	11	63	Phricodini	1	7		
	Caystrini	14	66	Piezodorini	5	21		
	Chlorocorini	10	77	Procleticini	11	36		
	Coquereliini	4	5	Rhynchocorini	18	106		
	Degonetini	1	2	Rolstoniellini	6	19		
	Diemeniini	14	55	Sciocorini	16	126		
	Diplostirini	2	2	Sephelini	10	34		
	Diploxyini	5	29	Strachiini	20	142		
	Eurysaspini	3	25	Triplatygini	4	9		
	Eysarcorini	19	230	Unplaced	29	108		
	Halyini	91	430					
Phyllocephalinae							45	214
	Cressonini	6	12	Phyllocephalini	32	167		
	Megarrhamphini	3	12	Tetrodini	4	23		
Podopinae							68	269
	Brachycerocoris group	3	13	*Podops* group	28	135		
	Deroploa group	9	20	*Tarisa* group	3	22		
	Graphosoma group	19	70	Unplaced	6	9		
Serbaninae							1	1
Stirotarsinae							1	1
TOTAL							940	4949

[1] Numbers of genera and species are compiled from the World Catalog of Pentatomoidea (David A. Rider, unpublished, as of April, 2017) and include subgenera, subspecies, and fossil genera and species.

compiled information for the Neotropical species. A key to the subfamilies and tribes of American Pentatomidae was provided by Rolston and McDonald (1979). Local monographs and checklists are available for several countries or larger geographic areas.

2.2.10.1 Key to the Subfamilies of Pentatomidae (modified from Schuh and Slater 1995)

1   Tarsi 2-segmented, longitudinally carinate dorsally (**Figures 2.14A, B**); rostrum apparently 3-segmented, flattened (**Figure 2.14C**); antennae 5-segmented, with antennal segments II and V enlarged, inflated, segment III small; Neotropics (**Figures 2.13F, 2.14D, 2.27L**)....Stirotarsinae

1'   Tarsi usually 3-segmented, if 2-segmented then lacking longitudinal carina dorsally; rostrum clearly 4-segmented, rounded (not flattened); antennae may be 3 to 5 segmented, but never with both segments II and V enlarged or inflated and segment III small 2

2(1)   Scutellum enlarged, covering most of abdomen (lateral margins uncovered) (**Figure 2.27A**); pronotum curves downward and posteriorly, forming a large posterolateral lobe on each side, with a large, distinct posterolateral notch that leaves a 3-4 sclerite portion of the mesopleural region exposed (exponium); head rather short, not extending much beyond eyes (**Figure 2.5A**); Australia (**Figures 2.17A, 2.27A**) ....Aphylinae

2'   Scutellum usually subtriangular, sometimes more expansive; pronotum not as described above, exponium absent; head usually more elongate.. 3

3(2)   Rostrum rather short, exceeding at most slightly beyond procoxae (**Figures 2.1O, 2.5B**); Old World (**Figures 2.18H; 2.27G, H**)..Phyllocephalinae

3'   Rostrum more elongate, extending well beyond procoxae.. 4

4(3)   Rostrum distinctly thickened, especially segment I which does not lie entirely between bucculae (**Figures 2.5C-E**); worldwide (**Figures 2.17B-F; 2.27B**) ....Asopinae

4'   Rostrum not distinctly thickened, segment I lying entirely between bucculae or only slightly beyond posterior buccal margins of bucculae 5

5(4)   Antennae 3- or 4-segmented 6

5'   Antennae 5-segmented 8

6(5)   Bucculae obsolete, shorter than first rostral segment; body extremely flattened; Borneo (**Figure 2.27K**).. Serbaninae

6'   Bucculae well developed, nearly as long, subequal, or longer than first labial segment; body not extremely flattened, more robust 7

7(6)   Scutellum produced dorsad into a robust thorn-like projection medially (**Figure 2.17G**); tarsi two-segmented; body usually gray to dark gray, covered by a very dense pubescence composed of short and thick setae, dorsal surface rather rough; New World tropics (**Figures 2.17G, 2.27C**) ....Cyrtocorinae

7'   Scutellum usually not produced into a robust thorn-like projection, if projection is present, then tarsi three-segmented, coloration and vestiture different and dorsal surface not roughened (the New World edessine genus *Peromatus* will also key here, but it has the mesosternum produced anteriorly as a bifurcate process [**Figure 2.5G**]); worldwide ...... Pentatominae (in part)

8(5)   Metasternum produced anteriorly onto mesosternum or rarely prosternum (**Figures 2.5F-H**); rostrum not surpassing mesocoxae (**Figures 2.5F-H**); New World (**Figures 2.18E-G; 2.27F**) ....Edessinae

8'   Metasternum rarely produced anteriorly onto mesosternum but if so, then rostrum extending onto abdomen or at least to metacoxae.. 9

9(8)   Trichobothrium nearest spiracle on sternum 7 laterad of spiracular line by distance at least equal to greatest diameter of spiracular opening (**Figures 2.5I, J**) 10

9'   At least one trichobothrium on sternum 7 on or mesad of spiracular line (**Figure 2.5K**) 11

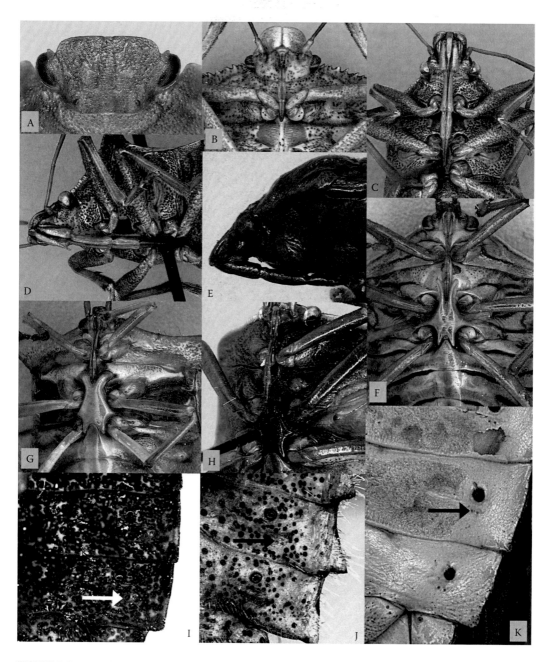

FIGURE 2.5   A, *Aphylum syntheticum*, head, anterodorsal view; B, *Basycriptus distinctus*, head and anterior thorax, ventral view; C, *Picromerus bidens*, head and thorax, ventral view; D, *Picromerus bidens*, head and thorax, ventrolateral view; E, *Zicrona caerulea*, head and thorax, lateral view; F, *Edessa rufomarginata*, thoracic sterna, ventral view; G, *Peromatus notatus*, head and thorax, ventrolateral view; H, *Brachystethus geniculatus*, thorax, ventrolateral view; I, *Macropygium reticulare*, lateral margins of abdominal segments VI and VII, ventral view; J, *Dryptocephala spinosa*, lateral margins of abdominal segments VI and VII, ventral view; K, *Pellaea stictica*, lateral margins of abdominal segments VI and VII, ventral view.

10(9) Base of abdominal venter with mesal tubercle; metasternum produced, flattened; worldwide ....
.... Pentatominae (in part)

10' Base of abdominal venter rarely tuberculate but if so, then metasternum thinly carinate mesally; New World .... Discocephalinae (in part)

11(9) Rostrum arising on or posterior to an imaginary line traversing head at anterior limit of eyes (**Figures 2.6A, B**) and/or superior surface of tarsal segment III of hind legs shallowly excavated in females (**Figure 2.6C**); New World .... Discocephalinae (part)

11' Rostrum arising anterior to an imaginary line traversing head at anterior limit of eyes (**Figures 2.1M, N; 2.6E**); superior surface of tarsal segment III of hind legs convex or flattened in both sexes 12

12(11) Tibiae usually sulcate on outer surface (**Figure 2.6D**); rostral segment I often longer than bucculae (**Figure 2.6E**); trichobothria paired; scutellum not reaching apex of abdomen; frena one-third or more length of scutellum (**Figure 2.6F**) worldwide .... Pentatominae (in part)

12' Tibiae not sulcate on outer surface; rostral segment I usually not longer than bucculae; trichobothria single (**Figure 2.6G**) or paired; scutellum usually reaching apex of abdomen (**Figure 2.6H**); frena short, less than one-third length of scutellum (**Figure 2.6H**), but if one-half length of scutellum, then scutellum not reaching apex of abdomen; worldwide (**Figures 2.18I-L; 2.27I, J**) .... Podopinae

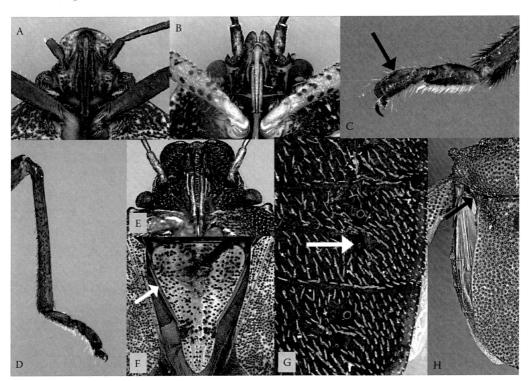

FIGURE 2.6 A, *Antiteuchus mixtus*, head, ventral view; B, *Macropygium reticulare*, head, ventral view; C, *Miopygium cyclopeltoides*, tarsus, dorsolateral view; D, *Pentatoma rufipes*, tibia, inner view; E, *Eysarcoris aeneus*, head, ventral view; F, *Rhaphigaster nebulosa*, scutellum, dorsal view; G, *Amaurochrous dubius*, lateral margins of abdominal segments V, VI and VII, ventral view; H, *Amaurochrous dubius*, detail of scutellum, dorsal view.

2.2.10.2 Aphylinae Bergroth, 1906

This is a small group (two genera, three described and several undescribed species) endemic to Australia (**Tables 2.2, 2.3**). As with other pentatomine taxa, it has had an interesting taxonomic history. Originally, it was described as a subfamily of the Pentatomidae (Bergroth 1906b); that same year, Schouteden published a detailed monograph of the subfamily (Schouteden 1906). Reuter (1912) and China and Miller (1959) treated this group as a family; Gross (1975b) lowered it back to a subfamily, indicating that it might be related to the podopine genus *Tarisa* Amyot and Serville. In their review of the World Heteroptera, Schuh and Slater (1995) once again treated this group as a family, a position which was followed by Cassis and Gross (2002) and Kment et al. (2012). Recently, as a result of their cladistic analysis, Grazia et al. (2008) concluded that for now, it should be treated as a pentatomid subfamily. The genitalia of both sexes, particularly the presence of a distinct pentatomid-type sclerotized rod within the dilation of the female spermathecal duct, support this placement.

Members of this subfamily are relative small (usually less than 5 mm), oval, with the dorsal surface strongly convex and the pleural and abdominal areas flattened or concave (Schuh and Slater 1995) (**Figure 2.17A**). The scutellum, somewhat enlarged, usually extends to the apex of the abdomen but leaves the connexiva and lateral portions of the coria exposed (**Figure 2.27A**). The pronotum curves downward and posteriorly forming a large posterolateral lobe on each side, with a large, distinct posterolateral notch, which in turn leaves three to four sclerites of the mesopleural region exposed (normally covered by the forewings in other pentatomoids); this region has been given the term exponium by Štys and Davidová-Vilímová (2001). Its function is to draw the defensive secretion from the metathoracic scent glands upwards to the outer surface of the body (Kment et al. 2012). The head is relatively short and broad, barely extending anteriorly beyond the compound eyes. The parameres, at least in the genus *Aphylum* Bergroth, have two articulating sections. The female spermathecal duct has a distinct sclerotized rod that is narrowed apically and hooked; the spermathecal bulb is globose with two finger-like diverticula (McDonald 1970).

This subfamily contains two genera, *Aphylum* (**Figures 2.17A, 2.27A**) with two described species distributed in eastern Australia, and the recently described *Neoaphylum* Štys and Davidová-Vilímová with a single species only known from western Australia. The only known biological information is that some specimens have been collected from under the bark of the River Red Gum, *Eucalyptus camaldulensis* (Myrtaceae) (Cassis and Gross 2002). Štys and Davidová-Vilímová (2001) stated that the spiracular structure in *Neoaphylum* (with a "stopper"-like structure located in each spiracle) indicated that this species might be adapted to live in dry, arid regions.

2.2.10.3 Asopinae Amyot and Serville, 1843

The asopines are considered and recognized as a natural (e.g., monophyletic) taxon, having as a defining synapomorphy a crassate rostrum (**Figures 2.5C-E**) (Gapud 1991). An adaptation for predation, the "beak" consists of barbed stylets strengthened by the thickened rostrum are used to "spear" their prey and hold it at beaks-length, thus limiting potential injury when the prey, typically larvae of holometabolous insects, thrashes in an effort to escape. Additionally, the bucculae are united behind the oral groove, not open as in the Pentatominae. United bucculae is likely a plesiomorphic character as it is found in other pentatomid groups (e.g., Edessinae) as well. In many, but not all, genera, the forelegs seem to be adapted for predation with spurs on the profemora and a foliate expansion of the protibiae (also seen in several phytophagous species) (**Figure 2.27B**), possibly to ward off counter-attacks by the prey. Barão et al. (2013) cite the protibiae of asopines as unique and a defining apomorphy for the subfamily. Many asopines have the evaporatoria reduced compared to the typical non-predatory stink-bugs and such reduction would presumably minimize the problem of the scent giving away their presence to potential prey.

Predatory behavior is not limited to the asopines. Insectivory by species in the halyine genus *Brochymena* Amyot and Serville has been observed with sufficient frequency to conclude that they are at least adventitious predators. They have a non-crassate rostrum and, thus, lack specific adaptations for predation unless one counts the foliate protibiae and reduction in the external structures of the metathoracic scent gland apparatus. Reductions in the scent gland apparatus occurs in some unrelated pentatomines and podopines (e.g., *Braunus* Distant, *Murgantia* Stål, *Tornosia* Bolívar) but is likely a convergent adaptation for avoidance of parasitoids that are known to key on the scent to find their hosts. Moreover, at least one genus of asopines, *Tylospilus* Stål, has species that are at least partially, and presumably secondarily, herbivorous (Stoner et al. 1974), and this latter genus has the least crassate rostrum in the subfamily.

Although predation, and its accompanying adaptations, could result from convergence, important synapomorphies of the genitalia strongly support asopine monophyly. Although neither feature, the thecal shield (apical part of theca *sensu* Gapon and Konstantinov 2006) or the genital plates (parandria *sensu* Tuxen 1970) is unique to the asopines, these bugs are the only pentatomids that have both in combination (McDonald 1966).

Members of the Asopinae are found in both the Old and New World and also in the Pacific islands. The Old World fauna consists of 40 genera (Thomas 1994b) and the New World 26 (Thomas 1992b), with four genera found in both. Gapon (2010a) added a new New World genus bringing the total to 63 genera, with 303 species worldwide (**Tables 2.2, 2.3**). The exact number of species is in doubt because many of the larger Old World genera have not been revised recently. Nomenclatural notes and a checklist of the Chinese species was provided by Rider and Zheng (2002). When making identifications of asopine specimens, the two monographs by Thomas should be consulted. His 1992b publication provides keys to the genera and species of the New World taxa; his 1994 work provides keys to the Old World genera and checklists of the known species. Also, the Euro-Mediterranean taxa were treated by Péricart (2010). Two older papers that still provide very useful information for identifying taxa are those by Schouteden (1905a, 1907). The Chinese species of the genera *Amyotea* Ellenrieder and *Picromerus* Amyot and Serville have been reviewed by Zhao et al. (2011, 2013) respectively. Other genera that have been reviewed, either in part or in whole, include *Anasida* Karsch (Ahmad and Rana 1991), *Arma* Hahn (Ahmad and Önder 1990a, Zheng 1981), *Blachia* Walker (Ahmad and Rana 1994), *Canthecona* Amyot and Serville (Ahmad and Rana 1988, Khuong and Lam 2001), *Oechalia* Stål (Usinger 1941, 1942), *Perillus* Stål (Knight 1952), *Picromerus* (Ahmad and Önder 1990b), and *Rhacognathus* Fieber (Josifov and Kerzhner 1978).

The tribal classification is unresolved. Various suprageneric names have been proposed but not in the context of a formal classification. The oldest name applied to the group, Spissirostres, by Amyot and Serville (1843) was not based on a genus name and, therefore, has no significance in tribal classification. Kirkaldy (1909) used the name "Cimicinae" for the group based on the genus *Cimex*, but a subsequent ICZN decision restricted this name to the bed bugs and, thus, cannot apply to Asopinae. The name Asopinae, based on the genus *Asopus* Burmeister, was proposed by Amyot and Serville (1843) as "Asopides." Tribal names that have been proposed, and their type genera, are provided in **Table 2.4**.

When Amyot and Serville (1843) divided their "Spissirostres" into two subordinate taxa (i.e., Asopides and Stirétrides), they were following Burmeister's (1835) Handbuch that recognized three genera: *Stiretrus* Laporte, *Discocera* Laporte, and *Asopus*. Their "Asopides" included only *Asopus*, and their "Stirétrides" included *Stiretrus* and *Discocera*. Thus, Kirkaldy's (1909) selection of *Asopus gibbus* Burmeister, a junior synonym of *Discocera cayennensis* Laporte, as the type-species of *Asopus*, was unfortunate as it rendered Asopini, Stiretrini, and Discocerini as synonymous. Kirkaldy was a strict priorist and his selection serves as an excellent example for not always adhering to strict priority.

TABLE 2.4

Proposed Asopine Tribal Names, Authority and Nominate Genus

Asopini Amyot & Serville (1843) based on *Asopus* Burmeister

Stiretrini Amyot & Serville (1843) based on *Stiretrus* Laporte

Armini Bergroth (1904) based on *Arma* Hahn

Discocerini Schouteden (1907) based on *Discocera* Laporte

Amyotini Schouteden (1907) based on *Amyotea* Ellenrieder

Jallini Dupuis (1949) based on *Jalla* Hahn

Stilbotini Gapud (2015) based on *Stilbotes* Stål

As Amyot and Serville (1843) and Schouteden (1907) perceived, there does appear to be a natural dichotomy within the asopines. Although many of the genera are coarsely punctate and cryptically colored in earth-tones (i.e., brown, tan, gray or black), about an equal number are glabrous and aposematic (i.e., brightly colored or even metallic). There is circumstantial evidence that the bright colors are meant to mimic the colors of the prey, typically chrysomelid beetles (van Doesburg 1970, Schaefer 1996); this could be aggressive mimicry allowing the pentatomids to move closer to their prey without being noticed, but it could also be a form of protective mimicry as both the pentatomids and the chrysomelids probably taste bad to predators. But there is only weak support from the other character states that would suggest that aposematic coloration is synapomorphic. Given the limited material at hand, it is likely that Amyot and Serville's concept of the "Asopides" was meant for the dull-colored forms and would be most exemplified by *Asopus argus* (F.), [now *Amyotea malabarica* (F.)], which Bergroth (1911) argued should be the type species for the asopines. This led to Leston's (1953a) proposal to replace the subfamily name with Amyotinae, which would have had as its nominate tribe Amyotini. But the oldest name for this tribe is Armini, based on the genus *Arma*, as proposed by Bergroth (1904), leaving the nominate tribe Asopini typified by the genus *Discocera*.

The most striking apomorphy in the group is the presence of a pair of pilose glands on the abdominal sternum of males. It has been shown that this gland secretes an aggregative pheromone that is released when the males find prey. This attracts other members of the species including potential mates (Aldrich 1988, Aldrich and Lusby 1986). Of the 62 genera for which males are known (*Australojalla* Thomas is known from a single female), 23 have the abdominal glands. The glands are present in *Afrius* Stål (except the subgenus *Subafrius* Schouteden), *Andrallus* Bergroth, *Apateticus* Dallas, *Apoecilus* Stål, *Blachia* Walker, *Bulbostethus* Ruckes, *Canthecona*, *Cazira* Amyot and Serville, *Coryzorhaphis* Spinola, *Discocera*, *Eocanthecona* Bergroth, *Heteroscelis* Latreille, *Hoploxys* Dallas, *Leptolobus* Signoret, *Macroraphis* Dallas (except the subgenus *Megarhaphis* White), *Mecosoma* Dallas, *Montrouzierellus* Kirkaldy, *Oplomus* Spinola, *Perillus*, *Platynopiellus* Thomas, *Platynopus* Amyot and Serville, *Stilbotes* Stål, and *Stiretrus*. Because of its complexity, function, and, most importantly, the constant nature of its morphology among genera, there is little likelihood that its presence is the result of convergence rather than relationship. Its absence in two subgenera of genera that possess them would be explainable as a secondary loss. Because the above 23 genera include *Asopus*, they would constitute the nominotypical tribe Asopini.

Species with reduced or suppressed glands are known at least in the genera *Afrius*, *Cazira* (**Figure 2.27B**), and *Macroraphis*. This indicates that the glands could have been lost multiple times. The non-glandular genera would constitute a sister clade, an arrangement consistent with the dichotomy envisioned by early workers, Amyot and Serville (1843), Schouteden (1907) and Bergroth (1904, 1911). But, inasmuch as it is defined by a plesiomorphy, the resulting clade, functionally the Armini, may well be polyphyletic. Within this group, a second tribe has been proposed by Dupuis (1949). The Jallini was erected to include *Jalla* Hahn and *Zicrona* Amyot and Serville, but there is no reason to suspect that *Jalla* and *Zicrona* are related. Dupuis cited the extension of the seventh abdominal tergite; however, the extension is found also in the genus *Dorycoris* Mayr, which shares no apomorphic characters with either of the two genera included by Dupuis and thus there is minimal support for the tribe as proposed. We suspect that *Jalla* more likely is related to the genera *Amyotea*, *Anasida*, and *Pseudanasida* Schouteden (but not *Jalloides* Schouteden or *Australojalla*) based on a similarly broad but flattened body form and the scent-gland ruga is gutter-like. Without a formal phylogenetic analysis to define its relationships, we consider that Dupuis' Jallina is no more than a subclade of Armini, moreover with *Zicrona* excluded.

Similarly, within the clade of male-gland possessing genera, the monotypic tribe Stilbotini has been proposed by Gapud (2015) based on *Stilbotes*, from the Philippines. Based on a suite of shared characteristics, *Stilbotes* is related to the African genus *Leptolobus*. Both genera have pedunculated eyes distant from the cervix (i.e., a membranous area between the head and thorax), a bilobed thorax, and spinous humeri. Both genera also have the male abdominal glands and therefore fall within the Asopini; thus, Stilbotini is probably a subclade of the latter tribe. Until a formal phylogenetic analysis can confirm its validity, we propose that Gapud's name should be treated as a synonym of the Asopinae, being a subclade of Asopina, including *Leptolobus*.

In a study of the aedeagus of the Asopinae, Gapon and Konstantinov (2006) found three more or less distinct patterns; however, one genus (*Euthyrhynchus* Dallas) was difficult to place into any of the categories. As only 18 of the 63 asopine genera were studied, and no studied genus included more than one species (which might have provided information on how plastic or consistent the characteristics are), the results of this work should be tested on a larger sample. The three groups outlined by Gapon and Konstantinov did not correspond to the groups possessing or lacking male abdominal glands. This lack of congruent pattern might suggest that either the male genital characters, or the abdominal glands, or perhaps both are subject of convergent/homoplastic evolution. Gapon and Konstantinov wisely eschewed proposing tribal designations for their schemes, a decision we support.

Only two fossil species have been described in the Asopinae, one of which has been transferred into the Pentatominae. The species still treated as an asopine is *Asopus puncticollis* Piton (1940). The other species, *Arma contusa* Förster (1891), was later placed as a synonym of *Eysarcoris mammillata* Förster (Théobald 1937), which is a member of the Pentatominae: Eysarcorini.

2.2.10.4 *Cyrtocorinae Distant, 1880*

The Cyrtocorinae is a small subfamily containing four genera and eleven species, all restricted to the New World tropics (**Tables 2.2, 2.3**). They have bounced back and forth in being considered a subfamily within the Pentatomidae or being a separate family. For example, historically, Dallas (1851) and Walker (1867a) (using the synonymic preoccupied name Oxynotidae) treated this group as a family, Distant (1880) as a subfamily, Lethierry and Severin (1893) as a family, Kirkaldy (1909) as a subfamily, and so on. More recently, Schuh and Slater (1995) treated it as a subfamily, but Packauskas and Schaefer (1998), in their revision of the group, treated it as a family. The cladistic study of Grazia et al. (2008) supported the treatment of this group as a subfamily within the Pentatomidae. We also treat it as a subfamily, supported heavily by its members having a dilation and a sclerotized rod in the female spermathecal duct. Amyot and Serville (1843) first recognized this taxon as a family level group under the name Oxynotides; this name, however, was based on a preoccupied generic name, *Oxynotus* Laporte. Family names based on preoccupied generic names are not allowed, so White (1842) proposed *Cyrtocoris* as a replacement generic name, but it was Distant (1880) who first used Cyrtocorinae as a family-level name.

Members of this family are small to medium in size (6-10 mm), somewhat squarish in shape (**Figure 2.27C**), sometimes with long spine-like projections in various directions. They are usually dark grey to black in color, but often appear somewhat lighter because of the presence of many small brown-tan or white scale-like setae (Rolston and McDonald 1979, Packauskas and Schaefer 1998). Their dorsal surface is quite roughened, irregular, and almost granulate (**Figures 2.17G, 2.27C**). They are characterized by the flattened expansions of the juga; the form of the scutellum, which usually reaches the apex of the abdomen; and the presence of a rather robust dorsal thorn-like protuberance on the scutellum (**Figure 2.17G**). The antennae are four-segmented. The tarsi are two-segmented.

Brailovsky et al. (1988) described life history traits of two species, *Cyrtocoris egeris* Packauskas and Schaefer and *C. trigonus* (Germar), which are summarized here. The eggs are deposited in grooves in the bark of hosts in masses of variable number (usually more than ten eggs). The nymphs are gregarious and remain on the same host during their entire development, which can last about 45 days. There is one generation per year, and the insects hibernate as adults. These species are exclusively phytophagous, and feeding sites mainly include the branches of the hosts. Host plant associations include species of Araceae, Euphorbiaceae, Fabaceae, Malvaceae, and Piperaceae (Brailovsky et al. 1988, Schaefer et al. 2005). The eggs and first-fifth nymphal instars of *C. egeris* have been recently studied by Bianchi et al. (2011).

2.2.10.5 *Discocephalinae Fieber, 1860*

The Discocephalinae presently contains 81 genera and 325 species, all endemic to the New World tropics (**Tables 2.2, 2.3**). Fieber (1860) proposed Discocephalidae to include *Discocephala* Laporte, *Dryptocephala* Laporte, and *Platycarenus* Fieber. The taxon was treated as a subfamily by Stål (1868), a classification followed by most subsequent authors (Stål 1872, Distant 1880, Lethierry and Severin 1893, Schuh and Slater 1995), although Kirkaldy (1909) and McDonald (1966) treated the group as a tribe (Discocephalini). The current classification was defined by Rolston and McDonald (1979), transferring to the Discocephalinae most of the New World genera formerly placed in the Halyini (Pentatominae). Rolston (1981) divided this subfamily into two tribes (i.e., Discocephalini and Ochlerini), erecting the Ochlerini for those genera that have the third tarsal segment of the hind legs in females excavate or flattened. More recently, Campos and Grazia (2006) supported the monophyly of this subfamily. Grazia et al. (2015) presented an updated checklist for the group.

Most members are yellows, greys, browns, and/or blacks, often mottled, helping them blend in with their environment. The body is sometimes somewhat flattened (many genera in the Discocephalini) or more robust and convex ventrally (many genera in the Ochlerini). The labium arises on or posterior to an imaginary line traversing the head at the anterior margins of the eyes (**Figures 2.6A, B**); the first rostral segment usually extends onto the anterior region of the prosternum. The antennae may be four or five-segmented. The metasternum is not produced anteriorly onto the mesosternum. The tarsi are three-segmented; in females in the Ochlerini, the dorsal surface of tarsal segment III of the hind legs usually is excavated and concave (**Figure 2.6C**). The trichobothrium nearest the spiracle on abdominal sternite VII usually arises laterad of the spiracular line (**Figures 2.5I, J**). Males of both tribes have the basal portion of abdominal segment X membranous; the conjunctiva are reduced; the phallotheca, ejaculatory duct, median penial lobes, and conjunctival appendages (when present) are strongly sclerotized; and the vesica is undifferentiated from the conjunctiva (Rolston and McDonald 1979, Schuh and Slater 1995, Konstantinov and Gapon 2005, Campos and Grazia 2006).

As mentioned above, this subfamily is endemic to the New World. There are, however, several Old World genera that are quite similar in habitus to the discocephalines. For example, the Australian genus *Cephaloplatus* White (**Figure 2.19I**) (currently in Caystrini) and *Discimita* Kment and Garbelotto (2016) from Central Africa (tentatively placed in the Myrocheini) are similar in appearance to the discocephaline genus *Dryptocephala*. Also, several genera from Southeast Asia and New Guinea [e.g., *Aednus* Dallas (currently in Myrocheini), *Goilalaka* Ghauri (currently in Halyini), and *Mimikana* Distant (currently in Halyini)] are quite similar in general appearance to some ochlerine genera. These similarities however, probably are due to convergence.

All species, as far as known, are phytophagous. There are not many host plant records available. The dull black ochlerines have been found under dead tree trunks where they may be feeding on fungi. Other ochlerine species, especially in the genus *Lincus* Stål, are known to vector a flagellate disease of cultivated palms (Dolling 1984, Asgarali and Ramkalup 1985, Louise et al. 1986, Resende et al. 1986, Resende and Bezerra 1990, Alvarez 1993, Mitchell 2004; also see **Chapter 13**). Several species, especially from the discocephaline genus *Antiteuchus* Dallas (**Figure 2.17H**), exhibit maternal care with the adults guarding the eggs and early instar nymphs (Rau 1918, Fennah 1935, Callan 1944, Eberhard 1975, Melber and Schmidt 1977, Santos and Albuquerque 2001a,b). In the Discocephalini, maternal care also has been recorded for species of *Dinocoris* Burmeister (**Figure 2.17I**), *Eurystethus* Mayr (**Figure 2.17J**), and *Mecistorhinus* Dallas (Tallamy and Schaefer 1997). More recently, Guerra et al. (2011) recorded for the first time trophobiosis between ants and a pentatomid species, *Eurystethus microlobatus* Ruckes. Rolston (1990, 1992) provided keys to the "broad-headed" genera of the Discocephalini (**Figures 2.12F, 2.17K, L; 2.27D**) and the genera of Ochlerini (**Figures 2.18A-D; 2.27E**), respectively. There is one fossil genus and species described in the Discocephalinae: *Acanthocephalonotum martinsnetoi* Petrulevičius and Popov (2014), however, its placement is based merely on external characters, and the genus could fit into the Pentatominae: Triplatygini as well.

Recently revised Discocephalini genera include: *Abascantus* Stål (Becker 1977), *Ablaptus* Stål (Rolston 1988a, Becker and Grazia 1989a), *Agaclitus* Stål (Becker and Grazia 1992), *Alcippus* Stål (Becker and Grazia 1989b), *Alveostethus* Ruckes (Ruckes 1966b), *Antiteuchus* (Ruckes 1964, Engleman and Rolston 1983, Rolston 1993, Fernandes and Grazia 2006), *Callostethus* Ruckes (Fernandes et al. 2011), *Cataulax* Spinola (Grazia et al. 2000), *Dinocoris* (Becker and Grazia 1985), *Dryptocephala* (Ruckes 1966c), *Eurystethus* (Ruckes 1966a), *Lineostethus* Ruckes (Ruckes 1966b), *Mecistorhinus* (Ruckes 1961, 1966e), *Parantiteuchus* Ruckes (Fernandes and Grazia 2002), *Pelidnocoris* Stål (Ruckes 1966d), and *Priapismus* Distant (Rolston 1984a). *Colpocarena* Stål (**Figures 2.12F, 2.27D**) and *Phoeacia* Stål (**Figures 2.13B,**

2.17L) are currently being revised (David A. Rider, unpublished data). Also, the genus *Anhanga* Distant recently has been transferred from the Discocephalini to the Pentatominae: Carpocorini, near the genus *Galedanta* Amyot and Serville (Bianchi et al. 2016a). The Discocephalini presently contains 46 genera and 195 species (**Table 2.3**).

Recent taxonomic work in the Ochlerini includes: Dellapé and Dellapé (2016), including a key to the species of *Adoxoplatys* Breddin, modified from Kormilev (1955); *Alitocoris* Sailer (Sailer 1950, Garbelotto et al. 2013); *Lincus* (Rolston 1983c, 1989); *Ochlerus* Spinola (Simões and Campos 2014, 2015); and *Paralincus* Distant (Rolston 1983d). Recent work by colleagues in Brazil and North America also has resulted in the descriptions of several new ochlerine genera: *Candeocoris* (Roell and Campos 2015), *Hondocoris* (Thomas 2003), *Ocellatocoris* (Campos and Grazia 2001), *Parastalius* (Matesco et al. 2007), *Pseudocromata* (Ortego-León and Thomas 2016), *Stapecolis* (Garbelotto and Campos 2016), and *Xynocoris* (Garbelotto and Campos 2014). The Ochlerini presently contains 35 genera and 130 species (**Table 2.3**).

2.2.10.5.1 Key to the Tribes of Discocephalinae

1 Dull black or fuscous coloration; dorsal surface of hind tarsal segment concave (**Figure 2.6C**) or flattened in females and sometimes in males (**Figures 2.18A-D; 2.27E**)................Ochlerini

1' Brown, often mottled with black, or shiny black; hind tarsal segment cylindrical in both sexes, not concave or flattened (**Figures 2.12F, G; 2.13B; 2.17H-L**)...........................Discocephalini

2.2.10.6 Edessinae Amyot and Serville, 1843

The exclusively New World subfamily Edessinae has undergone significant changes of late. Until recently, this subfamily was considered to contain only four genera: *Edessa* F. (**Figure 2.27F**), *Olbia* Stål, *Pantochlora* Stål, and *Peromatus* Amyot and Serville (**Figure 2.18G**). Now, several genera usually placed in the Pentatominae have been transferred to the Edessinae (e.g., *Brachystethus* Laporte [**Figure 2.18E**], *Lopadusa* Stål [**Figure 2.18F**]) (Barcellos and Grazia 2003b, Rider 2015a), and the extremely speciose genus *Edessa* is being split into several smaller genera (e.g., *Ascra* Say, *Grammedessa* Correia and Fernandes) (Santos et al. 2015, Correia and Fernandes 2016). Currently, this subfamily contains 15 genera and about 338 species (**Tables 2.2, 2.3**), but this undoubtedly will increase as there are still many known species of *Edessa* that are undescribed.

In its early taxonomic history, this group was often included in, or at least confused with, the family Tessaratomidae. Amyot and Serville (1843) included in Edessides the genera related to *Edessa* and also some genera currently included in the Tessaratomidae. Dallas (1851) treated the group as a family, but the recognition of edessines as a distinct group was virtually ignored after the proposition of the Tessaratomidae by Stål (1865). For example, Horváth (1900) erected the family-level name Pantochloraria within the Tessaratomidae to hold the edessine genus *Pantochlora*. Kirkaldy (1909) recognized *Edessa* and related genera in the tribe Edessini, but left the Pantochlorini in the Tessaratomidae. McDonald (1966) also treated this group as a tribe, the Edessini, within the Pentatomidae. Rolston and McDonald (1979) raised the group to subfamily level, which has been followed by subsequent workers (Gapud 1991, Schuh and Slater 1995, Grazia and Schwertner 2008b, Grazia et al. 2015). Even though Leston (1955b) indicated that *Pantochlora* might be related to the edessines, it was not formally transferred to the Edessinae until 1969 by Kumar (1969a), but he said that *Pantochlora* might deserve its own tribal status. Barcellos and Grazia (2003b) supported the placement of *Pantochlora* within the Edessinae.

Members of this subfamily are medium to large in size, often green in color with dark markings, and, occasionally some may have some brighter markings; they tend to be ovoid to ovaloid but narrowing posteriorly. The bucculae are rather short, arcuately elevated, almost flap-like. The most diagnostic character is the prominantly elevated, tumid metasternum, that in many species is produced forward onto or beyond the mesosternum (**Figure 2.5H**) and, in the genus *Edessa* (and related genera), bifurcates anteriorly (**Figures 2.5F, G**). The rostrum is short and, usually, in *Edessa*, the apex fits into the notched metasternal process (Rolston and McDonald 1979, Barcellos and Grazia 2003a,b). Similar appearing processes are known from several pentatomine groups, but they usually are abdominal or mesosternal in origin. The humeral angles are sometimes rather prominent, spinously or truncately produced.

At quick glance, the edessines have a similar appearance to the rhynchocorines (Pentatominae), but they are actually quite different. In the rhynchocorines, the base of the abdomen, the metasternum, and the mesosternum are all produced ventrad and tightly contact each other (nearly appearing fused); it is the mesosternum that protrudes forward over the prosternum and often onto the base of the head (**Figure 2.7E**). In the edessines, only the base of the abdomen and the metasternum are produced ventrad; it is the metasternum that is produced forward over the mesosternum, often becoming bifid anteriorly over the prosternum (**Figures 2.5F-H**). Additionally, the rostrum is much shorter in the edessines, and the bucculae are shorter and flap-like.

Several edessine genera have been reviewed recently, either in part or in whole: *Ascra* (Santos et al. 2015), *Brachystethus* (Barcellos and Grazia 2003a), *Doesburgedessa* Fernandes (Fernandes 2010), *Grammedessa* (Correia and Fernandes 2016), *Lopadusa* (Becker and Grazia 1970), and *Paraedessa* Silva and Fernandes (Silva et al. 2013). Several species groups within *Edessa* have been the subject of revisionary work (Fernandes and van Doesburg 2000a-c; Fernandes et al. 2001; Silva et al. 2004, 2006; Fernandes and Campos 2011; Silva and Fernandes 2012; and Santos et al. 2014).

Various species may feed on a variety of plant hosts, but there seems to be some preference for members of the plant family Solanaceae among the economically important species (Panizzi et al. 2000). Several members of this subfamily (e.g., *Lopadusa augur* Stål and *Edessa nigropunctata* Berg) have been reported to exhibit parental care of the young (Requena et al. 2010). There is one fossil species described in the Edessinae, *Edessa protera* Poinar and Thomas (2012).

2.2.10.7 *Pentatominae Leach, 1815*

This is, by far, the most diverse subfamily in the Pentatomidae, containing 660 genera and 3,484 species (**Tables 2.2, 2.3**). Its members occur worldwide. A taxon similar to our current Pentatominae was proposed by Stål (1865) and included genera of the groups Halydes, Pentatomides, Pododides, Podopides, Rhaphigastrides, and Sciocorides of Amyot and Serville. Later, the subfamily was divided into groups of genera (Stål 1872, 1876). These groups eventually were rearranged and named by different authors (Atkinson 1888, Distant 1902, Cachan 1952), and were followed partially until recently (Uhler 1886; Kirkaldy 1909; Van Duzee 1917; Cachan 1952; Gross 1975b; Rolston and McDonald 1979; Rider 2006a, 2015a). However, depending on the characters considered important for the classification proposed, the organization and hierarchy of the groups differ considerably (**Table 2.1**).

The lack of unique diagnostic characters hampers the identification of this subfamily, making it difficult to construct a useful and stable classification. As with any large group, it is difficult to find characters that all members possess. In general, pentatomines can be quite small (*Sepontia* and *Sepontiella* Miyamoto species are only a few mm in length) to quite large (*Catacanthus* species are nearly as large as most tessaratomids). Many of its members are dull yellows, tans, greens, browns, and blacks, colored so as to blend in with their surroundings; others, however, may be brilliantly colored in reds, yellows, and oranges, sometimes with a metallic sheen, perhaps aposematically colored to advertise their repugnant odors. Most species have five-segmented antennae, but some species only have four segments, and at least one Australian genus (*Omyta*) has only three segments. The humeral angles are often simple and rounded, but, in some species, they are quite prominent and spinuously produced. The scutellum is usually subtriangular but, in a few species, it is more spatulate; if spatulate, it usually does not reach the apex of the abdomen. The frena usually extend at least two-fifths the length of the scutellum (**Figure 2.6F**). The tarsi are usually three-segmented, but a couple of tribes (Nealeriini, Opsitomini) and at least two genera included in other separate tribes (*Phalaecus* Stål, *Rolstoniellus* Rider) have two-segmented tarsi.

The classification within this subfamily has been chaotic at best. The number of tribes recognized has varied dramatically from worker to worker. For example, various workers from various parts of the world still recognize over 40 different tribes (e.g., Cassis and Gross 2002, Derjanschi and Péricart 2005, Rider 2006a, Salini and Viraktamath 2015), and another 15-20 generic groups have been proposed (Gross 1975b, 1976; Linnavuori 1982). And yet, Schuh and Slater (1995) only recognized eight valid tribes.

There are no recent keys to all of the known tribes or genera, except on a regional basis. For example, there are several important works covering the fauna of various Old World regions: West Palearctic

(Derjanschi and Péricart 2005, Ribes and Pagola-Carte 2013), Central Asia (Putshkov 1965), Far East of Russia (Vinokurov et al. 1988), China (Hsiao et al. 1977), South India (Salini and Viraktamath 2015), West and Central Africa (Linnavuori 1975, 1982), Madagascar (Cachan 1952), and South Australia (Gross 1975b, 1976). Rider et al. (2002) provided nomenclatural notes and a checklist of the Chinese species. In the New World, Lawrence H. Rolston headed up a small group who authored a series of four papers that provided keys for all known pentatomine genera (Rolston and McDonald 1979, 1981, 1984; Rolston et al. 1980). The focus of this series of papers was not to reflect phylogeny or even to hint at a practical classification, but was more utilitarian - it was meant to give workers a means to identify their specimens to genus. This series set the stage for further work in the New World including a number of generic revisions, a book on the Pentatomoidea of northeastern North America (McPherson 1982), a set of regional treatments for various states in the United States in honor of J. E. McPherson's contributions (Bundy 2012, O'Donnell and Schaefer 2012, Packauskas 2012, Rider 2012, Sites et al. 2012, Swanson 2012, Zack et al. 2012), and a separate contribution on the fauna of Ontario (Paiero et al. 2013). More recently, Barão et al. (2017) published a review of the metathoracic scent glands in the Carporcorini; this information will undoubtedly be useful in future phylogenetic studies.

Because one of the primary purposes of this chapter is to provide a framework for future phylogenetic studies, we have chosen to discuss every tribe unequivocally that is still considered valid by someone somewhere, and also any generic groups proposed by Gross (1975b, 1976) and Linnavuori (1982) that do not fit easily into one of the existing tribes. As such, the following key is only tentative at best, with some tribes keying out multiple times. Hopefully, this will still be of some benefit to some workers, but the user should be aware of many exceptional genera (and some genera that have never been placed in a tribe), many of which are discussed under the tribal headings. Obviously, there will be drastic changes in our treatment of the various taxa, particularly the tribal classification, but this chapter should help facilitate the efficiency of future phylogenetic studies as well as provide basic information for identification.

Nearly all Pentatominae are phytophagous (see comments in Asopinae section), usually oligophagous, but a few species are known from many host plants. There are some reports of facultative feeding on dung or carrion (Adler and Wheeler 1984, Eger et al. 2015b). A number of species are known economic pests of various crops (e.g., *Nezara viridula* L. [**Figure 2.22K**], *Euschistus* Dallas spp. [**Figure 2.20A**], *Eurydema* Laporte spp. [**Figure 2.32E**], etc.) (McPherson and McPherson 2000, Panizzi et al. 2000). In some parts of the world (e.g., southern Mexico), pentatomines are used as human food.

2.2.10.7.1 Tentative Key to the Tribes of Pentatominae

1 Sublateral, lunate stridulatory structure present on each side of the abdominal venter (**Figures 2.7A, B**), corresponding stridulatory pegs on inner surface of hind femora....... 2

1' Abdominal venter lacking lunate stridulatory structures (*Neomazium* [Carpocorini] has stridulatory structures, but they form a rather narrow band along lateral margins of abdominal venter), hind femora lacking pegs 3

2(1) Elongate, slender, coloration pale yellowish to tan; occurs in arid areas of the New World and in the Middle East, Africa, and India (**Figure 2.30F**) .... Mecideini

2' Variously shaped, but never elongate and slender; coloration usually not yellowish); occurring in Australia and adjacent areas (**Figures 2.12D; 2.21B, C; 2.29I**)............ Diemeniini

3(1) Ant mimic, black with white markings on corium giving it a constricted appearance; pronotum elongate, distinctly constricted in middle; many spines along edges of pronotum, on base and apex of scutellum, and one on each side of abdomen, many of which are oriented dorsad; Southeast Asia (**Figures 2.14E, F; 2.31B**) .... Pentamyrmecini

3' Not an ant mimic; coloration variable, not as above; usually lacking numerous spines along pronotal margins and on scutellum and abdomen, if some spines are present then light brown in color (*Phricodus*) or quite larger (*Mustha*)... 4

4(3) Small, pale brownish species, with numerous spines along margins of pronotum, and base of coria; antennae 4-segmented, segment II long and gradually thickened towards apex, the distal two segments thickened, spindle-shaped; Old World (**Figure 2.31I**) .... Phricodini

4'   Size and coloration variable, but not a combination of pale brownish with numerous spines on pronotum and coria; antennae usually 5-segmented, if 4-segmented, then they are not as above 5

5(4)   Ostiole obsolete (**Figure 2.9A**), at most, represented by small opening between the bases of the mid- and hind-coxae, rugae short, merging with surrounding smooth pleuron, associated evaporative area also obsolete, V-shaped, confined to metapleuron (black brachypterous *Trochiscocoris* [Strachiini] has ostioles completely reduced; *Otantestia* [Antestiini] will key here) 6

5'   Ostiole and associated external structures may be reduced and small, but always distinct, sometimes ostiole, rugae, and evaporative areas quite large (**Figures 2.9B-F**) 7

6(5)   Large species, body length usually 20 mm or greater; body elongate-oval; head narrowly triangular, tylus distinctly surpassing apices of juga; Afrotropical and Oriental (**Figure 2.28D**) .... Agaeini

6'   Usually smaller and broader; head much broader than its length, tylus never surpassing apices of juga; worldwide (**Figures 2.12H; 2.32D-G**) .... Strachiini

7(5)   Mesosternum longitudinally carinate, sometimes carina is low and indistinct (**Figure 2.7C**) or in a shallow sulcus, but is usually visible 8

7'   Mesosternum longitudinally sulcate, not carinate (**Figures 2.7F, G**), occasionally a low, indistinct carina may be visible just near anterior margin 39

8(7)   Mesosternal carina large and robust, produced anteriorly over prosternum and often onto base of head as a spine or flattened wedge (**Figures 2.7E, H**) 9

8'   Mesosternal carina usually less elevated, not robust, at most protruding just onto posterior margin of the prosternum (**Figure 2.7D**); may have large spine or wedge produced forward, but these originate from metasternum or abdomen 10

9(8)   Usually green in color when alive, but often fades to yellowish after death; India through China, Southeast Asia, and Australia to Oceania (**Figures 2.13D, E; 2.23L; 2.24A, B; 2.31L**) .... Rhynchocorini

9'   Usually brown or mottled brown and pale; New World (**Figures 2.13A, 2.31E**) .... Pentatomini (part, *Evoplitus* group)

10(8)   Antennae 4-segmented; abdominal venter in females nearly covered with large, opaque areas which are provided with a dense layer of hairs (several genera of Halyini may key here); Madagascar (**Figure 2.30G**) .... Memmiini

10'   Antennae with 3-5 segments, but usually 5-segmented; abdominal venter in females lacking large opaque areas, not covered with hairs 11

11(10) Abdominal venter with distinct medial, longitudinal sulcus, the rostrum often elongate with at least the apex lying in sulcus (**Figure 2.8A**); usually green or greyish-green in color; Old World tropics (**Figures 2.19F, 2.28J**) .... Bathycoeliini

11'   Abdominal venter usually lacking medial sulcus, or if sulcus present, then usually brown or grey in color 12

12(11) Abdominal venter armed basally with a forwardly directed spine or tubercle (**Figures 2.8B, C**) 13

12'   Abdominal venter unarmed basally 18

13(12) Large species, body length usually around or above 20 mm; body colorful, marked with reds, yellows, oranges, and sometimes metallic blues or greens; dorsal surface of head often nearly impunctate, but with oblique transverse ridges or wrinkles (**Figure 2.8F**); worldwide (except Europe) (**Figures 2.20I, J; 2.29C, D**) .... Catacanthini (part)

13'   Small to medium sized, if larger, then usually not colorfully marked, if colorfully marked, then head usually distinctly punctate 14

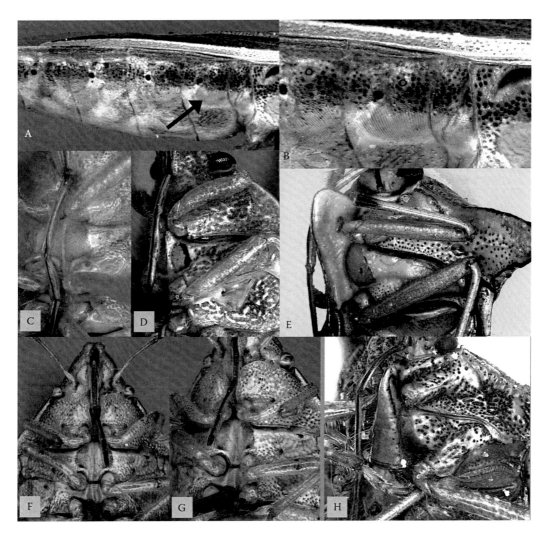

FIGURE 2.7 A, *Mecidea major*, abdomen, lateral view; B, *Mecidea major*, detail of abdominal segments II, III and IV, lateral view; C, *Thyanta acuminata*, thorax, ventrolateral view; D, *Piezodorus guildiniii*, thorax, ventrolateral view; E, *Rhynchocoris humeralis*, thorax, ventrolateral view; F, *Aelia acuminata*, head and thorax, ventral view; G, *Aelia acuminata*, thorax, ventrolateral view; H, *Evoplitus humeralis*, thorax, ventrolateral view.

14(13) Green in color when alive, and keeping the green color after death (note that some species have seasonal or semi-permanent yellowish or brownish forms, but the typical form is green); worldwide (**Figures 2.12E; 2.22I-L; 2.30K, L**).......................................Nezarini

14' Coloration variable, but if predominately green, this green fades to yellow after death.... 15

15(14) Usually small to medium in size; usually tans, browns and/or blacks, sometimes with pale marks; head relatively broad, usually semi-circular; abdominal spine may be relatively long, but usually terete, slender; ostiolar rugae elongate, reaching near to lateral margin of metapleuron, apex acute, attached to metapleuron; worldwide (except Europe) (**Figures 2.22C-E; 2.30H**) ....Menidini

15' Usually medium to large in size; color variable, but if tan, brown, or black, then head is not so broad and/or semicircular; abdominal spine usually large, robust; ostiolar rugae variable.... 16

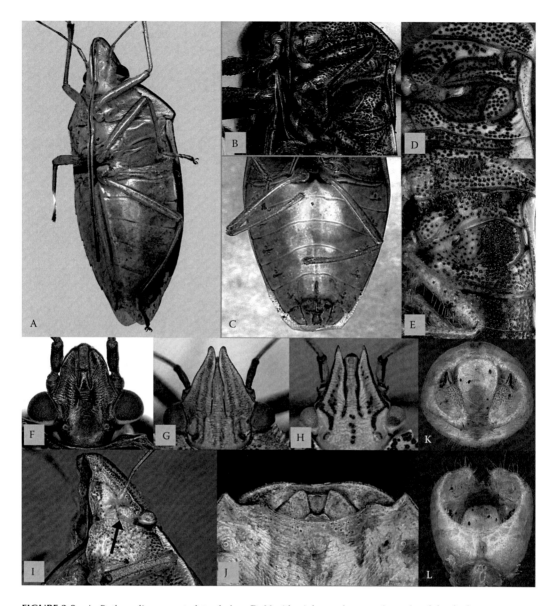

FIGURE 2.8   A, *Bathycoelia* sp., ventrolateral view; B, *Menida violacea*, thorax and anterior abdominal segments, ventrolateral view; C, *Chinavia viridans,* abdomen, ventral view; D, *Tibraca limbativentris*, evaporatorium, ventral view; E, *Eysarcoris aeneus*, evaporatorium, ventral view; F, *Catacanthus incarnatus*, head, dorsal view; G, *Loxa viridis*, head, dorsal view; H, *Arvelius albopunctatus*, head, dorsal view; I, *Aelia acuminata*, head and anterior part of thorax, ventrolateral view; J, *Odmalea concolor*, female genital plates, ventral view; K, *Odmalea concolor*, pygophore, posterior view; L, *Odmalea concolor*, pygophore, dorsal view.

16(15)   Usually tans, browns, and/or blacks in color, occasionally with metallic green, if green in
        life, does not fade to yellow after death; worldwide (**Figures 2.23A-H; 2.31C-H**)............
        ....Pentatomini (part)

16'   Usually yellowish-green fading to yellow after death (exception South African *Flaminia*
        [**Figure 2.21F**] which is black with red and white markings)... 17

17(16)   Scutellum usually large, spatulate, apex broadly rounded; Old World (**Figures 2.21F,
        2.29L**)..Eurysaspini

17'   Scutellum subtriangular in shape, apex narrowly rounded; worldwide (**Figures 2.23K, 2.31J**) .... Piezodorini

18(12) Relatively large and colorful, marked with reds, yellows, oranges, and sometimes metallic blues or greens; dorsal surface of head often nearly impunctate, but with oblique transverse ridges or wrinkles (**Figure 2.8F**); worldwide (except Europe) (**Figures 2.20I, J; 2.29C, D**) ....Catacanthini (part)

18'   Small to medium sized, if larger, then usually not colorfully marked, if colorfully marked, then dorsal surface of head usually distinctly punctate 19

19(18) Ostiolar rugae relatively short, often auriculate in form (**Figure 2.9F**), if more spout-like, still not reaching beyond middle of metapleuron (**Figure 2.9D**) 20

19'   Ostiolar rugae relatively long, reaching beyond middle of metapleuron, usually acute or acuminate apically, curving anteriad laterally (**Figures 2.9B, C**) 29

20(19) Occurs in the New World... 21

20'   Occurs in the Old World 24

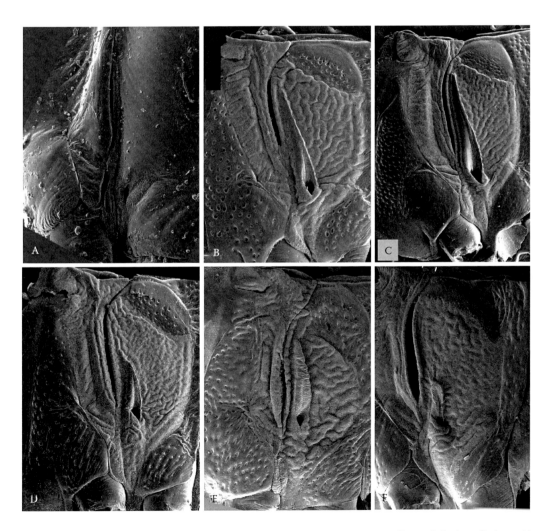

FIGURE 2.9   Evaporatoria, ventral view. A, *Murgantia histrionica*; B, *Acrosternum millierei*; C, *Bathycoelia horvathi*; D, *Palomena prasina*; E, *Carpocoris purpureipennis*; F, *Pentatoma rufipes*.

21(20) Yellowish-green in life (*Rhyncholepta* may be reddish brown with yellowish-green areas), usually fading to yellow after death; dorsal surface of head relatively flat, often subtriangular in shape (**Figures 2.8G, H**; **2.20K, L**; **2.21A**; **2.29F**)..Chlorocorini

21'   Color variable, but usually not yellowish-green in life fading to yellow after death; characters of head variable, but usually not with the combination of flat dorsal surface and subtriangular shape 22

22(21) Black to grayish-black, dorsal surface coarsely punctate, roughened; juga somewhat foliaceous or at least dentate near apex; North and Central America (**Figure 2.21K**)........... Halyini (part)

22'   Color variable; puncturing and dorsal surface variable, but usually with coloration not black to grayish-black with dorsal surface roughened; juga not foliaceuos nor dentate near apex 23

23(22) Dorsal surface covered with elongate hairs; veins in wing membrane marked with fuscous or black; rostrum usually elongate, reaching onto base of abdomen or more; one introduced species occurring in southern North America and the West Indies (**Figures 2.19B**; **2.28E**)..Agonoscelidini (part)

23'   Dorsal surface usually not covered with elongate hairs, if elongate hairs present, then veins in wing membrane are not marked with color; length of rostrum variable, but usually reaching at most onto base of abdomen (**Figures 2.12A, I**; **2.19J-L**; **2.20A-H**; **2.29A, B**) ....Carpocorini (part)

24(20) Small to medium in size; scutellum enlarged, spatulate in shape; ostiolar rugae short, usually auriculate (**Figures 2.8E**; **2.30A**)..Eysarcorini (part)

24'   Size variable, but if small then scutellum not usually enlarged; ostiolar rugae variable..... 25

25(24) Usually medium to large in size, gray to grayish-black mottled; head often with long preocular portion, narrowed, slender, lateral margins sometimes foliaceous or dentate (**Figures 2.21J, L**; **2.22A, B**; **2.30B, C**).. Halyini (part)

25'   Size variable but usually small to medium, color variable; head shape variable, but usually not as above 26

26(25) Body shape somewhat elongate, slender, narrowing posteriorly; humeral angles usually prominent, rounded or often spinose; female genital plates sharply produced posteriorly as an ovipositor; Africa to India (**Figure 2.30E**)..Lestonocorini

26'   Body shape variable, but if somewhat elongate-slender, narrowing posteriorly, then female genital plates typically pentatomid, not produced posteriorly as an ovipositor 27

27(26) Body often covered with elongate hairs; head tends to be elongate, slender; rostrum usually elongate, usually reaching onto base of abdomen or more; Africa, India, Southeast Asia to Australia (**Figures 2.19B, 2.28E**) ....Agonoscelidini (part)

27'   Body usually not covered with elongate hairs, if elongate hairs present, head not particularly elongate and slender; length of rostrum variable, but usually reaching at most onto base of abdomen.. 28

28(27) Ostiolar rugae short, usually auriculate (**Figures 2.8E**; **2.21G-I**; **2.30A**).... Eysarcorini (part)

28'   Ostiolar rugae longer, usually spout-like (**Figures 2.9E**; **2.12A**; **2.19J, L**; **2.29A**) ....
        ....Carpocorini (part)

29(19) Uniformly light brown to dark brown in color; juga longer than and meeting anterior to tylus, head distinctly triangular in shape; China, Southeast Asia, Madagascar (**Figure 2.28F**) .... Amyntorini

29'   Coloration variable but if brown, usually with pale markings; juga and tylus usually subequal in length, if juga longer than tylus, then head not distinctly triangular in shape ... 30

30(29) General coloration green to dark green, often with some pale or dark markings, sometimes with bright orange or reddish markings; ostiolar rugae elongate, often reaching to near lateral margin of metapleuron, apex acute, attached to metapleuron; worldwide (**Figures 2.19D, E**; **2.28G, H**) .... Antestiini

30'   General coloration light brown to dark fuscous, but usually not greenish, if some green color present, it is usually a metallic green; ostiolar rugae variable 31

31(30) Small to medium sized, very robust (i.e., venter strongly convex), head and anterior disk of pronotum strongly declivent; scutellum somewhat enlarged, spatulate in shape; China, Southeast Asia, Madagascar (**Figure 2.30D**)..Hoplistoderini

31'   Medium to large in size, usually not so robust, head and anterior disk of pronotum not strongly declivent; scutellum usually subtriangular, if enlarged or spatulate in shape, then body size much larger and not so robust.. 32

32(31) From Madagascar.. 33

32'   From regions outside of Madagascar 35

33(32) Anterolateral pronotal margins usually strongly convex, reflexed; relative large in size (14-25 mm) (**Figure 2.29G**)..Coquereliini

33'   Anterolateral pronotal margins usually not strongly convex, at most narrowly reflexed; size variable.. 34

34(33) Usually brown to dark brown in color; head with juga and tylus subequal in length, occasionally with juga longer than tylus (*Tripanda*), but apex still evenly rounded, not dentate ....Cappaeini (part)

34'   Usually gray to dark gray; head variable, but often elongate, slender, sometimes foliaceous or provided with teeth near apex.. Halyini (part)

35(32) Generally ovoid in shape, brown, sometimes with darker brown stripes or spots, somewhat flattened, depressed; lateral margins of head and pronotum sharply edged, somewhat laminate; Old World (**Figures 2.12C; 2.19I; 2.29E**) ....Caystrini

35'   Shape and coloration variable, but not usually both ovoid and brown with darker brown markings, more robust; lateral margins of head and pronotum may be edged, but not sharply so, not laminate... 36

36(35) Scutellum tends to be enlarged, spatulate in shape (more triangular in *Acesines*); each buccula in males provided with a large, curved tooth (*Axiagastus*), smaller, strongly angled tooth (*Oncotropis*), or normally shaped (*Acesines*); ostiolar rugae elongate, each reaching to near lateral margin of metapleuron, apex acute, attached to metapleuron; India to Australia (**Figure 2.28I**).. Axiagastini

36'   Scutellum subtriangular in shape; each buccula unarmed, or at most obtusely angulate; ostiolar rugae variable 37

37(36) General coloration usually gray to dark gray; head often elongate (especially in preocular portion), sometimes foliaceous or provided with teeth near apex; worldwide (except South America) (**Figures 2.21J, L; 2.22A, B; 2.30B, C**).......................................Halyini (part)

37'   General coloration brown to dark brown; head more typically pentatomoid in shape, not usually elongate, foliaceous, or dentate...38

38(37) Medium to large in size; anterolateral pronotal margins often provided with several strong teeth; abdominal spiracles relatively large; Southeast Asia (**Figure 2.32A**) .... Rolstoniellini

38'   Small to medium in size; anterolateral margin usually lacking denticles although humeral angles may be spinose; spiracles of typical size; worldwide (**Figures 2.19G, H; 2.28K, L**)..Cappaeini (part)

39(7)   Base of abdomen provided with an anteriorly directed spine or tubercle 40

39'   Base of abdomen unarmed.. 41

40(39) Posterior margin of pronotum distinctly concave; antennae 5-segmented; rostral groove sometimes margined on each side with sharply edged ridge (*Diplostira*) or abdominal spine long, nearly reaching base of head (*Ambiorix*); India, China (**Figure 2.29J**)............ .... Diplostirini

40'   Posterior margin of pronotum not distinctly concave; antennae 4-segmented; rostral groove not margined by distinct ridges, and abdominal spine shorter; India (**Figure 2.29H**)........ .... Degonetini

41(39) Propleura expanded forwards and upwards, covering or nearly covering base of each antenna; Holarctic (**Figures 2.8I; 2.19A; 2.28A**) .... Aeliini

41'   Propleura not expanded forwards and upwards, base of antennae exposed 42

42(41) Female terminalia with basal plates quite small, nearly obscured under posterior margin of last abdominal segment (**Figure 2.8J**); posterior margin of male pygophore usually with a distinct, medial slit, sometimes becoming more rounded (**Figures 2.8K, L**); New World (**Figures 2.23I, J; 2.31K**)... Procleticini

42'   Female terminalia with basal plates larger, more conspicuous; posterior margin of male pygophore usually entire, not with deep slit or rounded emargination 43

43(42) Veins in hemelytral membrane usually reticulate; spiracles usually located laterally, just below the connexival margin; African, Oriental (**Figures 2.19C, 2.28C**)... Aeschrocorini

43'   Veins in hemelytral membrane usually subparallel; spiracles not located distinctly laterad, usually remote from connexival margin 44

44(43) Tarsi either two-segmented or appearing to be two-segmented (distal two segments nearly fused); Madagascar 45

44'   Tarsi distinctly three-segmented 46

45(44) Smaller, usually less than 15 mm in length; second antennal segment quite small, antennae appearing to be four-segmented (**Figure 2.31A**).. Opsitomini

45'   Larger, usually greater than 15 mm in length; antennae five-segmented (**Figure 2.30J**) ... . Nealeriini

46(44) Head large and wide, nearly quadrangular in front of eyes; humeri of pronotum prominent, spinose or lobate; Madagascar (**Figure 2.32H**) .... Triplatygini

46'   Head of different shape; humeri of pronotum usually rounded, never greatly spinose or lobate (occasionally with a short, acute spine).. 47

47(46) Lateral margins of head, pronotum, and bases of coria sharply edged, at least somewhat laminate; general shape more ovoid.. 48

47'   Lateral margins of head, pronotum, and bases of coria less sharply edged, not laminate; general shape variable but usually more slender, parallel-sided.................................... 49

48(47) Size generally larger, body somewhat more robust; often with a few small teeth on inferior surface of front femur; tibiae not finely denticulate; Africa, south Palearctic, Oriental region, Australia (**Figures 2.22F-H; 2.30I**)... Myrocheini

48'   Size generally smaller, body more flattened; front femora unarmed; tibiae finely denticulate, rarely spinose; widespread (**Figures 2.24C, D; 2.32B**)................................. Sciocorini

49(47) Australia (Gross's *Menestheus* group).. Aeptini (part)

49'   Africa, India, Oriental region... 50

50(49) Tendency for apices of juga, humeral angles, apices of femora, and last abdominal connexiva to be produced, spinose, or at least angulate; Africa (**Figures 2.21D, E; 2.29K**) .... Diploxyini

50'   Not with above set of characters; juga may be produced beyond apex of tylus, but not spinose.. 51

51(50) Smaller, more obovate; ostiolar rugae quite reduced; Africa, Madagascar (**Figures 2.12B; 2.28B**).. Aeptini (part)

51'   Longer, more slender, African species often parallel sided; ostiolar rugae distinct, medium in length; Africa, India, Oriental region (**Figures 2.13C, 2.32C**)....................... Sephelini

2.2.10.7.2 Aeliini Douglas and Scott, 1865

Douglas and Scott (1865) first proposed this taxon as a family to include two genera, *Aelia* and *Aelioides* Dohrn (now considered to be a junior synonym of *Neottiglossa* Kirby). Puton (1869) lowered it to tribal rank, a taxonomic position in which nearly all subsequent workers have treated it. At present, four genera (or subgenera) and 28 species are included in the Aeliini (**Table 2.3**). *Aelia* (**Figure 2.28A**) and *Neottiglossa* (**Figure 2.19A**) (the latter divided into two subgenera, *Neottiglossa* s. str. and *Texas* Kirkaldy) are both definitely members of this tribe. The North African genus *Aeliopsis* Bergevin is included only because Bergevin (1931) initially compared his new genus with *Aelia*. A careful study of *Aeliopsis* undoubtedly will result in its removal from this tribe. Bergevin provided a lateral view of the head and pronotum. In this illustration, the propleura are not expanded anteriorly, and the posterior margins of the bucculae are greatly expanded, lobate, a character not seen in *Aelia* or *Neottiglossa*. The original description of the genus and its only species, *Aeliopsis bucculata* Bergevin, leave little doubt that the genus is closely related and probably identical with *Gomphocranum* Jakovlev (Carpocorini, see **Section 2.2.10.7.12**).

Zaidi (1996) provided a different classification of the Aeliini, including five genera with an elongate body from the Indo-Pakistan region: *Aelia*, *Adria* Stål, *Aeliomorpha* Stål, *Bonacialus* Distant, and *Gulielmus* Distant. This classification is highly inconsistent with opinions of other workers. Linnavuori (1982) gave conclusive evidence for excluding *Aeliomorpha* from the Aeliini. He, however, did not place it in any known tribe, but, rather, treated it in its own generic group. *Aeliomorpha*, *Bonacialus*, and *Gulielmus* have tentatively been placed in the Carpocorini and *Adria* in the Eysarcorini. These genera, however, have the mesosternum medially sulcate and, as such, are probably more closely related to the Aeptini, Myrocheini, or Diploxyini. More recently, Rider (2016) described a new genus, *Aeliavuori*, as superficially similar to *Aelia* (**Figure 2.12A**). He, however, noted that it had a carinate mesosternum, and it lacked the prosternal 'flaps' characteristic of the Aeliini, and tentatively placed *Aeliavuori* in the Carpocorini.

Members of this tribe are characterized by the propleura being greatly expanded anteriorly, covering or nearly covering the antenniferous tubercles (**Figure 2.8I**). They are small to medium sized, somewhat elongate-cylindrical in shape, and usually tan to dark brown in color with a few pale markings on the dorsum. Species of *Neottiglossa* (**Figure 2.19A**) tend to be smaller than species of *Aelia* (**Figure 2.28A**). The head is usually somewhat elongate, triangular in shape with the apices of the juga sometimes swollen (note: *Neottiglossa cavifrons* Stål in North America has the apex of the head much more rounded and concave). The prosternum is deeply sulcate between the propleural lobes, the mesosternum is distinctly sulcate (**Figures 2.7F, G**), and the metasternum is weakly concave. The ostiolar rugae are rather short, nearly obsolete. The abdominal venter is unarmed. The female spermathecal bulb is low and broadly ball-shaped, with two long, robust, finger-like projections.

Most members are known to feed on various grasses, with several species of *Aelia* considered to be pests on wheat (Sunn pest) (Panizzi et al. 2000). The genus *Aelia* is most speciose in the Mediterranean and the Middle East, with only one species (*A. americana* Dallas) occurring in North America. The genus *Neottiglossa* has a similar number of species (10-15) in the Palearctic and Nearctic. Most of the *Aelia* species were revised by Wagner (1960) and Brown (1962); Old World species of *Neottiglossa* were treated by Tamanini (1988); species of both genera occurring in the West Palearctic were keyed by Derjanschi and Péricart (2005); and the New World species of *Neottiglossa* were revised by Rider (1990). Staddon and Abdollahi (1999) dealt with the comparative morphology of the phallus of *Aelia* and *Neottiglossa*.

2.2.10.7.3 Aeptini Stål, 1871

The type genus, *Aeptus* Dallas, at present contains only a single species, *A. singularis* Dallas, but at least two or three undescribed species are known (**Figures 2.12B, 2.28B**), including one from Madagascar. Stål (1871) originally included the genera *Dymantis* Stål, *Aeptus*, *Menestheus* Stål, and *Eribotes* Stål in his Aeptini, and placed this tribe near the Myrocheini. Bergroth (1920) further refined the tribe, including only *Aeptus* from Africa, and several small genera from Australia. *Dymantis* generally is placed in the

Myrocheini; and *Paramecocoris* Stål, as a replacement name for the preoccupied genus *Paramecus* Fieber, usually is placed in the Caystrini. The remaining Australian genera often have been included in this tribe, but Gross (1975b) placed them in their own group (the *Menestheus* group), and this will be discussed under that heading. At present, this tribe contains eight genera and 19 species (**Table 2.3**), but this is likely to change drastically in the near future.

Aeptus species tend to be small to medium in size, elongate to pear-shaped (**Figures 2.12B, 2.28B**). The juga extend well beyond the apex of the tylus, the lateral margins are sharply edged and not reflexed, and the length of the head is usually longer than the median length of the pronotum. The eyes and ocelli are reduced. The first antennal segment does not reach the apex of the head, and the antenniferous tubercles are small, not visible from above. The rostrum reaches to or beyond the hind coxae, the second segment is longer than the last two segments together. The prosternum is bordered by a pair of erect lobes in front of the fore coxae. The hemelytra, in *Aeptus* species, are usually brachypterous with the coria somewhat squarish; the scutellum is more spatulate, reaching beyond the coria (**Figures 2.12B, 2.28B**). The thoracic sterna are sulcate without median carinae; the scent efferent system and evaporatoria are quite reduced. The abdominal venter is unarmed basally. The male pygophore is robust with the apical margin deeply insinuated, and the ventral surface is depressed medially. The parameres are long and slender with numerous hairs apically. In the female, the spermathecal bulb is simple, ball-shaped, and lacks diverticula.

Linnavuori (1982) indicated that this tribe was probably related to *Dymantis* (now considered a member of the Myrocheini), and speculated that *Aeptus* might belong in the Myrocheini (see **Section 2.2.10.7.29**). He further indicated that its members occur on grasses in mesic and moist habitats. Our initial studies indicate that one of a couple scenarios may be possible, depending on whether one is a lumper or a splitter. On one hand, it appears that the Australian genera often placed in this tribe may not be closely related to the type genus *Aeptus* and may need a new tribe of their own. On the other hand, the Australian "aeptine" genera show a marked resemblance to African genera now placed in the tribe Diploxyini, which in turn may be con-tribal with *Aeptus*. Finally, there are at least another three genera currently placed in the Carpocorini that may belong with these genera. Two of these genera occur in India (*Bonacialus* and *Gulielmus*), and one is South American in distribution (*Poriptus* Stål). All three have the mesosternum distinctly sulcate without a medial carina, which should preclude their membership in the Carpocorini; they are also elongate, slender with the juga longer than the tylus, and at least *Poriptus* is known to feed on grasses (similar to other "aeptines" whose host plants are known).

2.2.10.7.4 *Aeschrocorini Distant, 1902*

Distant (1902) originally proposed the Aeschrocoraria to hold two genera, *Aeschrocoris* Bergroth and *Scylax* Distant. This tribe was not mentioned again until Cachan (1952) added *Aeschrus* Spinola and his new genus, *Chraesus*. This tribe is still relatively small, tentatively containing eight genera (Oriental Region: *Aeschrocoris* [**Figure 2.28C**], *Scylax*; Afrotropical Region: *Aeschrus*, *Chraesus*, *Geomorpha* Bergroth, *Risbecella* Schouteden, *Tyoma* Spinola [**Figure 2.19C**], and *Tyomana* Miller) and 20 species (**Table 2.3**). Most of the included genera are dark brown to blackish with a strongly gibbose body variously outfitted with tubercles, spines, and/or other elevations. The head is somewhat elongate, and the juga may or may not extend beyond the apex of the tylus. The thoracic sterna are sulcate, usually lacking a medial carina. The ostiolar rugae are small and auriculate; and the evaporative area is also reduced. The scutellum is spatulate to subtriangular and, in some cases, the coria do not reach the apex of the scutellum. The hemelytral membranes are often reticulate (except in *Tyomana*). The first visible abdominal sternite usually has a deep medial depression, which often is delimited by an elevated lobe on each side. The position of the spiracles is unusual in that they are situated more lateral than in most pentatomines, located just below the connexival margin and with a distinct tubercle between each pair of spiracles (a similar condition is found in the South American genus *Caonabo* Rolston, but that genus has the thoracic sterna carinate medially).

The male pygophore usually has a small genital opening; the parameres are long and slender, sometimes enlarged apically, and sometimes bilobed. According to Linnavuori (1982), the penisfilum in *Tyomana* is of the usual pentatomine type, but in *Aeschrus*, it is quite distinctive with the theca broad and flattened, and the vesica is rather long and coiled. The female spermathecal bulb in *Tyoma* is elongate basally, rounded apically, with three medium-length, somewhat robust, diverticula arising from near the apex of the bulb; the spermathecal duct below the proximal flange is not swollen.

We have examined specimens of *Aeschrocoris, Aeschrus, Geomorpha, Risbecella, Scylax,* and *Tyoma.* At present, we believe that *Aeschrocoris, Geomorpha,* and *Tyoma* all belong to a single tribe. *Aeschrus,* as noted by Linnavuori (1982), is quite different, and may not belong in this tribe. It looks superficially similar in general appearance to the members of this tribe, but the spiracles are not located laterally, the pronotum has a distinct rounded flap extending posteriorly on each side over the base of each corium, and the scutellum is expanded laterally beyond the apex of the frenum on each side, also covering part of each corium. It appears that *Chraesus* and *Risbecella* are also members of this tribe; *Tyomana* is highly similar to and is probably a junior synonym of *Risbecella.* We also have serious doubts about including *Scylax* in this tribe. In *Scylax,* the characters of the head, sternum, and scutellum are similar to those described above for *Aeschrus,* but the general habitus of the animal is markedly different (dorsum weakly convex, integument relatively smooth); the hemelytral membrane is not reticulate; the base of the abdominal venter is neither grooved nor spinose; the spiracles are in a more typical position, not located laterad; and there are no tubercles between the spiracles.

Linnavuori (1982) indicated that there were similarities between this group and the *Diploxys* group (= Diploxyini); that is, they have a similar body form, humeral spines on the pronotum, apical spines on the femora, posterolateral angles of the last abdominal segment often spinose, and a similar structure of the thoracic sterna. Hassan et al. (2016) provided a review of the Indian species of *Aeschrocoris.*

Other than a few host plant records, there is little biological information available for members of this tribe. Linnavuori (1982) recorded *Aeschrus inaequalis* Spinola from *Croton zambesicus* Müller Argoviensis (Euphorbiaceae) and *Tyoma verrucosa* Montandon from eggplant (Solanaceae). Also, *Tyoma cryptorhyncha* (Germar) has been recorded from cowpea (Fabaceae) (Phelps and Oosthuizen 1958).

2.2.10.7.5 Agaeini Cachan, 1952

This is a monotypic tribe, containing only the genus *Agaeus* Dallas (**Figure 2.28D**). Members of this tribe occur in Madagascar, Africa, through the Indian subcontinent, China, into southeast Asia. Recently, *Agaeus* has been divided into two genera with the erection of a new genus, *Jostenicoris* Arnold for the Afrotropical species (Arnold 2011a). However, as Arnold (2011a) failed to mention any single differentiating character for his new genus, *Jostenicoris* is unavailable (Kment 2013). *Agaeus* currently contains eleven species (**Table 2.3**), and it has had a chaotic taxonomic history. Stål (1865) treated this genus between *Afrania* Stål and *Stenozygum* Fieber, both of which are now in the Strachiini, thus indicating that Stål may have believed *Agaeus* also belonged in the Strachiini. Cachan (1952) erected the Agaeini based on his study of Madagascan material. Linnavuori (1982) included this genus in the Eurydemini (= Strachiini). More recently, Azim (2002) placed this genus in the Halyini, a placement supported by Memon et al.'s (2011) phylogenetic study of the South Asian halyines, but their methodology has been criticized (Barão et al. 2012).

Members of this genus are medium to relatively large in size; they are generally dark (sometimes reddish) in coloration with pale yellowish to orangish stripes or mottled areas forming distinct patterns on the dorsum (**Figure 2.28D**). The head is somewhat elongate triangular with the lateral margins narrowly reflexed; the tylus is longer than the juga. The rostrum extends well onto the abdomen; the abdominal venter may be shallowly to distinctly sulcate medially. The anterolateral pronotal margins are distinctly reflexed. The thoracic sterna are not sulcate, the mesosternum has a medial carina, and the metasternum is flattened. The ostiole is quite small, lying mesad between the middle leg and hind leg; the ostiolar rugae are reduced to a short, rod-like, shiny area, each narrowly surrounded by the much reduced, V-shaped evaporatoria (very similar to that seen in the Strachiini). The scutellum is triangular in shape with the coria reaching well beyond the scutellar apex. The abdominal venter is unarmed basally. The male parameres are rather robust, hatchet-shaped, often with a large, distinct tooth near the base of the parameral stem. The female basal plates in some species are relatively small, well separated, with a large medial plate between them.

The elongate head with the tylus somewhat longer than the juga, and the medially sulcate abdominal venter might ally this genus with the Halyini. The much reduced ostiolar apparatus could indicate a relationship with the Strachiini. Until a thorough phylogenetic study can be completed, we prefer to leave these species in their own tribe.

There are a few host plant records known for several species of *Agaeus*. Zhang et al. (1995) recorded *Agaeus mimus* Distant from two different species of Verbenaceae: *Clerodendron villosum* Blume and *Gmelina arborea* Roxburgh; they also recorded *Agaeus tessellatus* Dallas from *Tabebuia rosea* de Candolle (Bignoniaceae). Additionally, *Agaeus pavimentatus* Distant has been recorded from *Tecoma stans* (L.) Jussieu ex Kunth (Bignoniaceae) (Golding 1931) and *Cordia millenii* Baker (Boraginaceae) (Golding 1927).

2.2.10.7.6 Agonoscelidini Atkinson, 1888

This tribe contains a single genus, *Agonoscelis* Spinola, with 26 species total, most of which occur in Africa (**Table 2.3**) but several others extend into Madagascar and through the Indian subcontinent to Australia. One species, *A. puberula* Stål, has been introduced accidentally into North America; it has become established there and spread throughout much of the southern and western United States. Some workers credit Stål (1876) as the author of this tribal name, but Stål did not officially use a name in the family sense; he did, however, key out *Agonoscelis* in its own generic group.

Species of *Agonoscelis* tend to be small to medium in size; they are usually pale in color with numerous black punctures that form irregular patterns (**Figures 2.19B, 2.28E**); in some species, the coria are reddish; and in *A. rutila*, the background color is black with red markings. There is a tendency for *Agonoscelis* species to possess long, erect hairs; in fact, several species can be quite hairy. The head may be typically shaped or elongate and slender; the juga and tylus are usually subequal in length; the lateral margins of the head are usually not reflexed or only feebly reflexed. The rostrum reaches to and usually beyond the hind coxae. The posterolateral margins of the pronotum are usually not reflexed. The scutellum is typically triangular in shape. The thoracic sterna are not distinctly sulcate - the prosternum is flattened to shallowly concave, the mesosternum is carinate medially, and the metasternum is flattened to weakly convex (the metasternum is relatively large due to the widely separated placement of the mid and hind coxae). The ostiole and associated evaporative area are relatively small and reduced, and the ostiolar rugae are obsolete. The hemelytral veins are usually distinctly brown to black (**Figures 2.19B, 2.28E**). The abdominal venter is unarmed basally. According to Linnavuori (1982), the male pygophore has special sclerifications on the basal lobes; the parameres are arcuate, usually long and slender, with the apex sometimes triangularly expanded or with a subapical tooth. The female spermathecal bulb is ball-shaped, somewhat low and broad, with two rather long finger-like diverticula; the spermathecal duct is slightly, conically swollen just below the proximal flange.

There is a great similarity between *Agonoscelis* species and those in the North American genus *Trichopepla* (currently placed in the Carpocorini). With the accidental introduction of a species of *Agonoscelis* into North America, identification of the two genera has become complicated. Members of the genus *Agonoscelis* usually have the veins in the hemelytra distinctly marked with brown or black (**Figures 2.19B, 2.28E**); in *Trichopepla*, the venation is usually concolorous with the wing membrane (**Figure 2.20G**). Species of *Agonoscelis* also resemble *Dolycoris* Mulsant and Rey (**Figure 2.19L**) and other related carpocorine genera (see **Section 2.2.10.7.12**) both in somatic and genital characters (Gross 1975a, 1976), and it is possible that this tribe will eventually be subsumed into the Carpocorini.

In the Sudan, gelatin is extracted from specimens of *Agonoscelis versicoloratus* (Turton) and used to make ice cream (Mariod and Fadul 2014).

2.2.10.7.7 Amyntorini Distant, 1902

Distant (1902) originally proposed the Amyntaria for five genera: *Amyntor* Distant (which is a junior synonym of *Bolaca* Walker), *Belopis* Distant, *Halyabbas* Distant, *Ochrophara* Stål, and *Sennertus* Distant. *Ochrophara* is considered to be a member of the Sephelini, a position first indicated by Stål (1871, 1876) when he described this genus. Cachan (1952) added two new genera (*Ambohicorypha* and *Madecorypha*), bringing the total of included genera to six. The two Cachan genera are endemic to Madagascar, *Sennertus* has only been recorded from Myanmar, and the remaining three genera are more widespread, occurring throughout India, China, and Southeast Asia. There are currently eight species in this tribe (**Table 2.3**). Singh et al. (2013) recently reported *Halyabbas unicolor* Distant from India, but their description and illustrations clearly indicate that they had a species of *Palomena* Mulsant and Rey (Nezarini).

In spite of the small number of included taxa, this is a morphologically rather heterogenous group difficult to characterize. Specimens of this tribe tend to be medium to large in size; they are usually pale yellowish-brown to dark brown (**Figure 2.28F**). The head is relatively large and usually distinctly triangular in shape; the juga usually extend beyond the apex of the tylus and usually meet anterior to the tylus; the lateral margins of the head are not reflexed. The rostrum reaches to and usually beyond the hind coxae. The thoracic sterna are not distinctly sulcate, the prosternum is flat or shallowly concave, the mesosternum is distinctly carinate medially, and the metasternum is flat (note that the metasternum is relatively small due to the nearly contiguous placement of the mid and hind coxae). The ostioles are relatively large; the ostiolar rugae are quite long and slender, curving slightly cephalad, with the apices acuminate; the evaporative area is relatively large. The anterolateral pronotal margins are often edged (obscurely so in *Halyabbas*) and not reflexed; they are minutely toothed in *Sennertus*. The tarsi are three-segmented. The abdominal venter is unarmed basally, although this area may appear somewhat swollen or hump-like.

Very little biological information is available for this tribe. *Bolaca unicolor* (Walker) has been recorded from *Pinus yunnanensis* Franchet (Pinaceae) (Shi and Xu 2006) and *Halyabbas unicolor* has been recorded from bamboo (Poaceae) (Hoffmann 1931, Zheng and Zou 1982, Zheng 1994, Wang et al. 2002) and sandalwood (Santalaceae) (Chatterjee 1934).

2.2.10.7.8 Antestiini Distant, 1902

Atkinson (1888) is actually the first worker to recognize this group at the family level, proposing two names that currently belong here. His Hyllaria was based on the genus *Hyllus* Stål, 1868, which is preoccupied by an arachnid genus of the same name described by Koch in 1846. A family-group name cannot be based upon a preoccupied generic name, so the Hyllini is invalid. Bergroth (1891) recognized this homonymy and proposed the generic name *Anaca* as a replacement name. The second family-level name proposed by Atkinson was the Plautiaria, based on the genus *Plautia* Stål. As we discuss later in this section, the tribal placement for *Plautia* is not certain, but if it does belong here, it could compete for priority with Antestiaria Distant (1902). However, due to the uncertain placement of *Plautia*, we prefer the continued usage of Antestiini for this tribe. Distant (1902) erected the Antestiaria to hold four genera, two of which are the two discussed above (*Anaca* and *Plautia*); the other two were *Apines* Dallas and *Antestia* Stål. The genus *Apines* is currently considered to be a member of the Menidini.

This is a relatively large tribe containing 29 genera and 171 species (**Table 2.3**). Linnavuori (1982) divided the African genera into six subgroups. There is considerable variation in the color and structure of members of this tribe. Usually they are small to medium in size, occasionally they are quite large (e.g., *Porphyroptera*); they are for the most part, robust, broadly ovate in shape. They tend to be somewhat shiny dorsally; punctures are present, usually dark and may be coarse, dense or not. The coloration is usually green or greenish (**Figures 2.19E, 2.28G**), but some species may have bright orange or reddish markings on the dorsum (**Figures 2.19D, 2.28H**). The lateral margins of the head are margined, often narrowly reflexed; the juga and tylus are usually subequal in length. The anterior pronotal margins and/ or the posterolateral pronotal margins are narrowly, but distinctly, reflexed. The scutellum is somewhat broad at the base (basal width and medial length are subequal) but narrows to a narrowly rounded apex. The mesosternum has a low but distinct medial carina. The ostiolar rugae are elongate, each reaching beyond the middle of the metapleuron, and are usually acuminate apically; the associated evaporative area is relatively extensive, distinct (except *Bergrothina* Schouteden). The base of the abdomen is unarmed or at most slightly humped.

According to Gross (1976), the male parameres are C-shaped or T-shaped. The phallosoma of the aedeagus may be sclerotized or not, and a narrow collar-like thecal shield may be present or absent; the medial penial plates are simple and may be attached to the conjunctiva. The spermathecal bulb in females is rounded with two long tubular processes (tubules are lacking in a few species of several different genera).

The placement of the genus *Plautia* Stål has been problematic and may bridge the gap between this tribe and the Nezarini. Gross (1976) stated that "on the shape of the claspers and aedeagus I am convinced that *Plautia* belongs, with a series of other genera from our region, to a later grouping here called the *Pentatoma* group and that *Antestia* should be placed in its own group which is, however, allied to the genera of the *Pentatoma* group." Gross' *Pentatoma* group was broadly conceived containing mainly those genera we recognize as belonging to the Nezarini. We believe the general facies, the dark coloration on the hemelytra of some species, and the ostiolar structure place *Plautia* in the Antestiini, but a thorough phylogenetic study is needed to solve this problem. Gross (1976) also indicated that the genus *Axiagastus* Dallas might belong in this tribe. It currently is treated as a member of its own tribe, Axiagastini.

Otantestia Breddin differs from other antestiines by having the ostiole (and associated structures) nearly obsolete; other than that, the remaining morphology and coloration clearly places this genus in the Antestiini. *Porphyroptera* also differs from most antestiines by its much larger size. In fact, China (1929) originally placed *Porphyroptera* near the catacanthine genera *Catacanthus* and *Chalcocoris* Dallas, but Linnavuori (1982) transferred it to the Antestiini. The overall color of *Porphyroptera* is green with some dark markings. More recently, Schwertner (2005) based primarily on characters of the male and female genitalia, concluded that *Porphyroptera* is related to *Nezara*, and probably should be transferred to the tribe Nezarini. Finally, Bergroth (1891) placed the Australian genus *Novatilla* Distant as a junior synonym of *Anaxilaus* Stål. This was disputed by Distant (1900a), but Bergroth (1906a) reiterated his opinion. Kirkaldy (1909) followed Bergroth, but more recent workers (Gross 1976, Cassis and Gross 2002) have treated the two as separate and valid genera. The Chinese and Indian species of *Plautia* were reviewed by Liu and Zheng (1994) and Ahmad and Rana (1996), respectively. Linnavuori (1975, 1982) treated most of the African genera of this tribe.

Species of the genus *Antestiopsis* Leston are considered to be major pests of coffee in Africa (Taylor 1945a,b; Michelmore 1949; Greathead 1966, 1969; Abebe 1987; van der Meulen and Schoeman 1990; also **see Chapter 10**). Several species of *Plautia* are known to cause economic damage to various fruit crops in eastern Asia (Moriya et al. 1987, Nakamura and Nishino 1999) and often are considered to be nuisance pests because they overwinter in people's homes.

2.2.10.7.9 Axiagastini Atkinson, 1888

Although Atkinson (1888) erected this tribe in a nomenclaturally available way, Stål (1876) first recognized this group of genera as "*Axiagastus* et affinia." The Axiagastini at present contains four genera (**Table 2.3**): *Acesines* Stål, *Axiagastus* (**Figure 2.28I**), *Indrapura* China, and *Oncotropis* Stål; the latter two are monotypic with *Indrapura klossi* China known only from Sumatra and *Oncotropis carinatus* (Stål) known only from the Philippines. *Acesines* contains five species, three of which are known only from China, one is only known from India, and the last is known from both India and Thailand. *Axiagastus* also contains five species, two of which occur in China and Southeast Asia and Indonesia, and the remaining three occur in Indonesia and New Guinea.

Species of *Axiagastus* are rather robust and medium-sized, dark brown in color with some pale mottling (**Figure 2.28I**). The head is typically shaped with the juga and tylus subequal in length; the lateral margins of the head are not or only feebly reflexed. The rostrum reaches to or beyond the hind coxae. The bucculae in male specimens are each provided with a large, curved tooth anteriorly. The thoracic sterna are not sulcate, but, rather, the prosternum is weakly carinate medially, the mesosternum is distinctly carinate medially, and the metasternum is flat (it is rather small due to the mid and hind coxae being nearly contiguous). The anterior margin of each propleuron is developed into a relatively wide, obtuse, oblique ridge on each side just behind head. The ostioles are relatively large; the ostiolar rugae are long and slender, curving anteriorly, and becoming acuminate apically; and the evaporatoria are well-developed. The anterolateral margins of the pronotum are distinctly, but narrowly, reflexed, this reflexion continues as a series of punctures or sulcus along the anterior margin, setting off a collar-like structure just behind head. The abdominal venter is unarmed basally.

Species of *Oncotropis* differ slightly from species of the other three genera in that they possess a distinct line of punctures submarginally along the anterior and lateral margins of the head and pronotum, and the bucculae in males are angulate anteriorly but not spinose. In *Acesines*, the head is quite short and broad, and the submarginal line of punctures on the head and pronotum are lacking. The rostrum reaches between the meso- and metacoxae, and the bucculae are simple, not spined anteriorly. The anterior margins of the propleura are not swollen; the mesosternum is distinctly carinate medially; and the metasternum is swollen and produced posteriorly, meeting with the basal abdominal spine. The ostiolar rugae are short and spout-like.

In general, axiagastine species have some superficial resemblance to several genera placed in the Menidini or Eurysaspini, but they lack the apically free, basal spine on the abdomen. In fact, Distant (1902) treated *Acesines* in the Menidaria and *Axiagastus* in the Hoplistodaria. And as mentioned previously, Gross (1976) indicated that *Axiagastus* might be related to the Antestiini.

Axiagastus cambelli Distant (**Figure 2.28I**) is considered to be an occasional pest on coconut palm in the Bismarck Archipelago (Smee 1965, Lever 1969, Baloch 1973).

2.2.10.7.10 *Bathycoeliini Atkinson, 1888*

Stål (1876), without formally naming it, first recognized this group at a higher level when he referred to it as "*Bathycoelia* et affinia." Atkinson (1888), following Stål, established the Bathycoeliaria for three genera: *Abeona* Stål, *Bathycoelia* Amyot and Serville (**Figures 2.19F, 2.28J**), and *Jurtina* Stål. Bergroth (1913) placed *Jurtina* as a junior synonym of *Bathycoelia*. *Abeona* is a junior homonym of *Abeona* Girard, 1855, a genus in the Pisces; Bergroth (1891) recognized this homonymy and proposed *Amblycara* as a replacement name. *Amblycara* currently is considered to be a member of the Pentatomini, but Tsai and Rédei (2014) speculated that *Amblycara* and *Bathycoelia* are closely related (see discussion later in this section) and should probably be placed together in the same tribe.

This tribe at present contains a single genus, *Bathycoelia*, and 32 species (**Table 2.3**). Most species occur in Africa and adjacent islands (Madagascar, Mascarenes, Seychelles, Socotra), but other species occur in India, China, Southeast Asia, and many islands of Malesia and Oceania. The ability of species of this genus to colonize remote islands and evolve endemic species is quite remarkable. Two species were described from South America (Jensen-Haarup 1931a, 1937), but these are probably mislabeled; Rolston et al. (1980) examined the types for both species and confirmed that they belong in the genus *Bathycoelia*.

Most species are greenish with black or red spots or lines on the connexiva and along the margins of the head and anterolateral margins of the pronotum (**Figure 2.19F**); the basal angles of the scutellum are often black, foveate, or sometimes have a pale yellowish callus (**Figure 2.28J**). The head is usually triangular to subtriangular in shape, flattish, with the juga and tylus subequal in length. The anterolateral pronotal margins are usually narrowly to broadly reflexed, and the anterior margin also is beaded, forming a collar posterior to the vertex that disappears laterally behind the eyes. The humeral angles are usually rounded to angulate, never spinose. The prosternum is shallowly concave, the mesosternum is distinctly carinate medially, and the metasternum is usually elevated, somewhat tectiform. The ostiolar rugae are quite elongate, curving cephalad, extending over three-quarters of the way to the lateral metapleural margin, becoming sharply acuminate apically; the associated evaporative area is extensive (**Figure 2.9C**). The connexiva are often distinctively patterned. The abdominal venter is usually distinctly furrowed longitudinally along the median with the elongate rostrum fitting into the abdominal groove (**Figure 2.8A**). The posterior margin of the male pygophore is usually emarginate medially, usually with variously shaped or produced lobes on each side of emargination; dorsal plates are often present, sometimes well-developed. The parameres are somewhat T-shaped. In the females, the spermathecal bulb is usually broader than long and provided with two diverticula of unequal lengths.

Linnavuori (1982) indicated a possible relationship with the antestiines with which they share the greenish coloration and the elongate ostiolar rugae. *Bathycoelia horvathi* Schouteden (western Africa) seems to represent a separate lineage, being smaller and more robust; its coloration is a deeper green, and the abdominal sulcation is less developed. It may not be congeneric with other *Bathycoelia* species. More recently, Tsai and Rédei (2014) discussed the relationship between the genera *Amblycara* and *Bathycoelia*. *Amblycara* typically has been considered to be a member of the Pentatomini, but they proposed that *Amblycara* might be closely related to *Bathycoelia* based on the genitalia and the fact that both genera have a ventral sulcus on the abdomen. They stated "*Bathycoelia* is not of such an isolated position which would justify a placement in a tribe of its own, or even with *Amblycara*," but they chose to not formally synonymize the Bathycoeliini with the Pentatomini until a more conclusive study could be conducted.

Bathycoelia thalassina (Herrich-Schäffer) is considered to be a major pests of cacao in tropical Africa (Gerald 1965; Lodos 1967; Owusu-Manu 1971, 1976, 1977, 1990), and *B. natalicola* Distant (= *B. distincta* Distant) has been recorded as a pest on both avocado (Erichsen and Schoeman 1992; van den Berg et al. 1999b, 2000) and macadamia nut in southern Africa (van den Berg et al. 1999a, Schoeman 2013).

2.2.10.7.11 *Cappaeini Atkinson, 1888*

This group was recognized first, but not formally named, by Stål (1876) who referred to it as "*Cappaea* et affinia"; he originally included three genera *Cappaea* Ellenrieder (**Figure 2.28K**), *Halyomorpha* Mayr (**Figure 2.28L**), and *Tolumnia* Stål. In 1888, Atkinson formally erected two tribes, the Tropicorypharia (based on *Tropicorypha* Mayr, currently included in this tribe), and the Cappaearia (based on *Cappaea*). Distant (1902) treated both tribes as synonyms of the Carpocorini, but most recent workers have recognized the Cappaeini as a valid tribe (the exception being Zaidi and Shaukat, 1993, who also treated both of the above tribes as synonyms under the Carpocorini). Members of the Cappaeini tend to be brown in color (several species of *Caura* Stål and *Halyomorpha* are dark bluish), lack a spine or tubercle on the base of the abdomen, and have ostiolar rugae that are elongate, tapering to a point apically. This tribe contains 24 genera and 151 species (**Table 2.3**), all of which are restricted to the Old World with the exception of *Halyomorpha halys* (Stål) (**Figure 2.28L**), which has been introduced accidentally into and is spreading throughout North America, and also has become established in Chile, South America (Faúndez and Rider 2017). Linnavuori (1982) dealt with the African genera included here in the *Veterna* group and the *Halyomorpha* group.

The head is typically shaped, usually subtriangular (elongate, spatulate in *Tropicorypha*), with the juga and tylus subequal in length (juga longer than and meeting in front of tylus in *Benia* Schouteden and *Tripanda* Berg). The body shape is often broadly rounded, somewhat robust, convex below. The anterolateral margins of the pronotum may be narrowly reflexed or not, and the anterior margin is usually flat behind the head (a row of punctures may form a transverse ridge behind the vertex in some species of *Tolumnia*). The humeral angles are usually rounded (**Figures 2.28K, L**) but may be weakly spined (some species of *Caura*) to strongly spinose (some species of *Veterna* Stål [**Figure 2.19H**]). The prosternum is shallowly to distinctly sulcate medially, the mesosternum is distinctly carinate medially, and the metasternum is generally flat. The ostiolar rugae tend to be quite elongate, usually curving cephalad, and sharply acuminate apically (more rounded apically in *Benia*, *Tripanda*, and *Tropicorypha*, and relatively short in *Veterna*). The abdominal venter is unarmed.

Linnavuori (1982) described the male pygophore as having excavations and lobe-like processes, and the parameres are rather short, incrassate, and variously modified. The female spermatheca is ball-shaped without diverticula (*Benia*, *Mabusana* Distant, *Tripanda* subgenus *Tenerva* Cachan), or with two diverticula (*Boerias* Kirkaldy, *Paralerida* Linnavuori, *Tripanda* sensu stricto, *Veterna*) or three diverticula (*Caura*, *Lerida* Karsch); or not ball-like but rather with an irregular bulb (*Lokaia* Linnavuori). There seems to be some overlap in characters with members of the Carpocorini.

Several cappaeinine genera have received attention in recent years. For example, *Cappaea* was studied by Zaidi (1993), *Caura* was revised by Leston and Dutton (1957), a second species was added to *Massocephalus* Dallas (Rider 2008), *Prytanicoris* Gross was recently described and reviewed (Gross 1978), the Indian species of *Tolumnia* have been treated (Zaidi 1995), and the genus *Tripanda* recently has been revised (Kment and Jindra 2009).

One species in particular has been the focus of much study recently - *Halyomorpha halys* (**Figure 2.28L**). This species, native to eastern Asia, has been introduced accidentally into Europe, North America, and South America and is spreading quickly through all three continents. It attacks several fruit trees and vegetable crops and is considered to be a nuisance because it congregates around and enters homes in the fall for overwintering (see **Chapter 4**).

2.2.10.7.12 *Carpocorini Mulsant and Rey, 1866*

Mulsant and Rey (1866) proposed three family-level names that we now treat under the tribe Carpocorini. The three were the Rubiconiaires, based on the genus *Rubiconia* Dohrn; Aulacetraires, based on the genus *Aulacetrus* Mulsant and Rey; and Carpocorates, based on the genus *Carpocoris* Kolenati. *Rubiconia* was transferred to the Eysarcorini (Puton 1869), but most current workers treat it as a member of the Carpocorini. *Aulacetrus* is an objective synonym of *Holcogaster* Fieber (**Figure 2.20B**); that is, both *Aulacetrus* and *Holcogaster* have the same type species, *Pentatoma fibulatus* Germar. Aulacetrini was considered valid for quite some time but now has been synonymized under the Carpocorini (but see later discussion in this section). Oshanin (1906) also proposed a separate tribe for *Holcogaster*, calling it the Holcogastraria. So, the two names, Carpocorini and Rubiconiini (and even the Aulacetrini if *Holcogaster* is determined to belong in the Carpocorini) would compete for priority. The first reviser (Atkinson 1888) used the name Carpocoraria, which now seems to be in prevailing use. Additionally, Yang (1962) proposed the tribal name Dolycorini, based on the genus *Dolycoris* (**Figure 2.19L**), but all known workers consider *Carpocoris* and *Dolycoris* to be closely related, definitely belonging to the same tribe.

This is the largest tribe in the Pentatominae, containing 127 genera and 503 species (**Table 2.3**). Its members occur worldwide. As would be expected in a large taxon, there is considerable variation in structure and color in this group. In general, genera of this tribe are yellowish to brownish (but other colors including metallic greens and blues are also known), sometimes mottled, and they lack a spine or tubercle at the base of the abdomen. The juga and tylus are usually subequal in length; occasionally the juga are a little longer but usually not meeting in front of the tylus. The prosternum is usually flat to shallowly concave (in *Holcostethus* Fieber [**Figure 2.20C**] and *Peribalus* Mulsant and Rey, the prosternum is inflated, forming an obtuse ridge on either side of the rostral canal), the mesosternum is not sulcate and usually has a well-defined medial carina, and the metasternum is usually flat to shallowly concave. The scutellum is usually triangular in shape, occasionally more spatulate (e.g., *Coenus* Dallas).

According to Gross (1976), the most striking characteristic of this group is "the strong and thick F-shaped clasper with sculptured lateral surfaces," but even this can vary. He goes on to describe the internal male genitalia: "The structure of the aedeagus is somewhat variable but there is a strong tendency for the vesica to be robust, often sclerotized, and sometimes quite long, if the latter then not infrequently reflexed, curved or wound in a spiral (see McDonald re cit of *Euschistus* or my figure of *Notius*). The medial penial plates are not infrequently absent." The spermathecal bulb in the female tends to be simple, ball-shaped, and without diverticula; the sclerotized rod may or may not be swollen near its apex.

Most of the New World genera presently considered to be in this tribe may not actually belong here. There seems to be a fundamental difference in the shape and length of the ostiolar rugae between the two geographical areas. In the Old World, the ostiolar rugae tend to be somewhat longer (still usually less than half the width of the metapleuron), and angulate apically, with the apex attached to or detached from the metapleuron (**Figure 2.9E**). In most New World genera, the ostiolar rugae are much shorter, more auriculate in form, and the apex is often detached from the metapleuron (**Figure 2.8D**). The exceptions are *Antheminia*, which has a couple species in the New World, but its members are distinctly related to *Carpocoris* and has the Old World ostiolar ruga; and the genus *Trichopepla* (**Figure 2.20G**), which may also be more closely related to the Old World carpocorines than those presently placed there from the New World. There is a group of New World genera related to the well-known genus *Euschistus* (**Figure 2.20A**) that have distinctly auriculate ostiolar rugae, and, in the male genitalia, the penisfilum is distinctly elongate and coiled. It is possible that these genera will be found to be a monophyletic group and warrant a tribe of their own.

It is impossible to find clear characters that universally separate the Carpocorini from the Pentatomini. The structure of the thoracic venter is roughly the same in both tribes. The rugae associated with the metathoracic scent gland ostioles are usually relatively short in the Carpocorini, but more or less shortened, sometimes very strongly reduced, rugae occur in several genera that otherwise clearly belong in the Pentatomini. The base of the abdominal venter is generally unarmed in the Carpocorini, and it can be unarmed or armed with a tubercle or spine of various lengths in the Pentatomini; this character is occasionally variable even within a single genus (e.g., *Pentatoma* Olivier) or among closely related genera. As a consequence, several genera (particularly those that have reduced rugae and an unarmed abdominal venter) are difficult to place into either of these two tribes.

The Carpocorini presently contains several genera that exhibit some variable characters that may preclude them from belonging in this tribe. Some of those are noted here. As mentioned above, Mulsant and Rey (1866) proposed the tribe Aulacetrini for the single genus, *Aulacetrus*, which is an objective synonym of *Holcogaster*. Cachan (1952) noted a similarity of this genus with members of the Antestiini. Several modern workers (Stichel 1961, Putshkov 1965, Wagner 1966) treated *Holcogaster* as a member of the Carpocorini. Our studies agree with Cachan; that is, species of *Holcogaster* (**Figure 2.20B**) have elongate, apically acute ostiolar rugae, more characteristic of the antestiines. Members of the European genera *Rubiconia* and *Staria* Dohrn have the ostiolar rugae elongate, reaching beyond the middle of the metapleuron; additionally, the inner mesial angles of the propleura are produced anteriorly (similar to the aeliines) in species of *Staria*. Both genera need further study, but they could possibly be related to the eysarcorines.

Cnephosa Jakovlev has the inner, mesial margins of the propleura lobed and produced anteriorly, similar to that seen in the Aeliini, but they do not cover the antenniferous tubercles. Members of the genus *Rhombocoris* Mayr have the posterior half of each buccula greatly expanded, lobe-like; otherwise, the ostiolar and sternal structures are similar to the carpocorines. In our discussion of the tribes Aeliini and Aeptini (see **Sections 2.2.10.7.2 and 2.2.10.7.3**, respectively), we mentioned that two genera, *Bonacialus* and *Gulielmus*, both have the mesosternum distinctly sulcate, not carinate; they probably need to be transferred out of the Carpocorini but their placement is questionable (perhaps they should be included in the Aeptini, Diploxyini, Myrocheini, or Sephelini). Members of the Australian genus *Pseudapines* Bergroth look extremely similar to members of the tribe Menidini, but they lack the small, anteriorly directed spine on the base of the abdomen. This could be a secondary loss of the spine, and this genus probably should be transferred to the Menidini. In fact, Gross (1976) stated that he was hesitant in treating *Pseudapines* as a member of the Carpocorini. The general habitus of members of the genus *Kamaliana* Ahmad and Zaidi is very similar to that found in many genera in the Cappaeini.

The Australian Region has a number of genera that are questionably placed. For example, the New Zealand genus *Hypsithocus* Bergroth has the prosternum distinctly sulcate and the metasternum shallowly sulcate; the mesosternum has a medial carina but only on the anterior half. The ostiolar apparatus is extremely small and the wings are brachypterous. Its general facies appears to be similar to the myrocheines. The ostiolar rugae are quite elongate in the genus *Monteithiella* Gross; otherwise, this genus seems to fit well into the Carpocorini. Gross (1976) stated that he was placing *Monteithiella* in the Carpocorini with hesitation. Similarly, members of the genus *Eurinome* Stål have the ostiolar rugae quite elongate, acuminate apically, and are brownish in color, so they may need to be transferred to the Cappaeini or Rolstoniellini. The genus *Mycoolona* Distant lacks a medial carina on the metasternum, and so its proper placement is in need of further study. Another interesting Australian genus is *Neomazium* Distant. It has the ostiolar rugae slightly elongate but still not reaching halfway to the lateral margin of the metapleuron. It has a series of short ridges forming a narrow band around the lateral margins of the abdomen; these ridges are for stridulation. We know of no other pentatomoid genus possessing a stridulatory mechanism similar to this. The genus *Notius* Dallas does not resemble the typical carpocorines; its members have the hind coxae widely separated, leaving a broad, shallowly sulcate metasternum. The ostiolar rugae are slightly elongate. Interestingly, the ventral margins of the bucculae curve or are rolled mesad, almost forming a tube that covers the first rostral segment.

The South American genus *Poriptus* has the mesosternum distinctly sulcate without a medial carina; the elongate juga meeting anterior to the tylus is similar to that found in the African members of the tribe Diploxyini; otherwise the remaining characters are typical of the Carpocorini. Members of the genus *Prionotocoris* Kormilev have the mesosternum and metasternum shallowly sulcate with a medial carina weakly evident only on the anterior third of the mesosternum. Faúndez and Rider (2014) previously noted some similarities between their new South American genus *Thestral* and the Australian genus *Poecilotoma* Dallas (**Figure 2.24G**). Because of this, they tentatively transferred *Poecilotoma* into the Carpocorini. In truth, *Poecilotoma*, *Thestral*, and the related *Acledra* Signoret all have sulcate, non-carinate mesosterna, and most likely none of them belong in the Carpocorini. Rolston (1987b) placed the genus *Calagasma* Bergroth as a junior synonym of *Epipedus* Spinola. Recently, Lupoli (2016) re-established *Calagasma* as a valid genus, and described a new species within each of the two genera.

There are also several questionably placed genera from Africa. Members of the genus *Steleocoris* Mayr have the prosternum and metasternum distinctly sulcate with the metasternum forming relatively sharp ridges, one on each side of the rostral groove; the mesosternum is much more shallowly sulcate, and it has a weak, medial carina. In *Theloris* Stål, the sternal structures are typical carpocorine, but the ostiolar rugae are quite elongate, apically acuminate, and so it may need to be transferred to the Cappaeini. The genus *Kahlamba* Distant appears to be closely related to *Acoloba*, presently considered to be a member of the Diploxyini; both genera may have ties to the myrocheine genus *Neococalus* Bergroth.

As one would expect from a large tribe, many of the included genera have been treated systematically, either in part or as a whole. To help the reader locate these papers, we have provided a list of some of the more important works: *Acledra* (Faúndez et al. 2014b); *Adustonotus* Bianchi (Bianchi et al. 2017); *Agatharchus* Stål (Belousova 1999); *Agroecus* Dallas (Rider and Rolston 1987); *Amauromelpia* Fernandes and Grazia (Fernandes and Grazia 1998a); *Braunus* (Thomas 1997a, 2001; Barão et al. 2016); *Caribo* Rolston (Rolston 1984b, Rider 1988); *Carpocoris* (Belousova 2004, Ribes et al. 2007, Lupoli et al. 2013); *Coenus* (Rider 1996); *Cosmopepla* Stål (McDonald 1986); *Curatia* Stål (Barcellos and Grazia 1998); *Dichelops* Spinola (Grazia 1978); *Euschistus* (Rolston 1974, 1985; Bianchi et al. 2017); *Euschistus (Lycipta)* (Rolston 1982, Weiler et al. 2016); *Euschistus (Mitripus)* (Rolston 1978b); *Galedanta* (Grazia 1967, 1981); *Glyphepomis* Berg (Campos and Grazia 1998, Bianchi et al. 2016b); *Gulielmus* (Zaidi and Ahmad 1989); *Holcostethus* (McDonald 1974, 1982; Ribes and Schmitz 1992; Belousova 2007); *Hymenarcys* Amyot and Serville (Rolston 1973a); *Hypanthracos* Grazia and Campos (Grazia and Campos 1996); *Hypatropis* Bergroth (Fernandes and Grazia 1996); *Ladeaschistus* Rolston (Rolston 1973b, Cioato et al. 2015); *Luridocimex* Grazia, Fernandes, and Schwertner (Grazia et al. 1998); *Mcphersonarcys* Thomas (Thomas 2012); *Mecocephala* Dallas (Schwertner et al. 2002); *Mimula* Jakovlev (Belousova 1997); *Mormidea* Amyot and Serville (Rolston 1978c, Rider and Rolston 1989); *Ochyrotylus* Jakovlev (Belousova 2003); *Oebalus* Stål (Sailer 1944); *Ogmocoris* Mayr (Frey-da-Silva et al. 2002a); *Paramecocephala* Benvegnú (Frey-da-Silva et al. 2002b); *Parentheca* Berg (Campos and Grazia 1999); *Pentatomiana* Barcellos and Grazia (Barcellos and Grazia 2004); *Peribalus* (Belousova 2007); *Poriptus* (Barcellos and Grazia 2008); *Sibaria* Stål (Rolston 1976); *Spinalanx* Rolston and Rider (Rolston and Rider 1988, Thomas 1997b); *Stysiana* Grazia, Schwertner, and Fernandes (Grazia et al. 1999); *Tibraca* Stål (Fernandes and Grazia 1998b); and *Trichopepla* (McDonald 1976).

Several species currently placed in the Carpocorini are known to be pests of various crops. For example, species of the genus *Euschistus* (**Figure 2.20A**) will attack legume (e.g., soybean) and other crops (Panizzi et al. 2000). Also, species in the genera *Oebalus* (Pantoja et al. 1995, 1999, 2000) and *Tibraca* (**Figure 2.20H**) (Pantoja et al. 2005, 2007) are considered pests of rice.

At least one species of *Euschistus* is used for human consumption in southern Mexico. Hoffmann (1947) stated "*Euschistus zopilotensis* Distant [a junior synonym of *E. strenuus* Stål]. Ancona, writing in Ann. Inst. Biol. (Mex.) in 1933 states that this pentatomine species, popularly known as 'jumiles' as are many other species of Heteroptera, is extensively used as food in Cuautla, State of Morelos, Mexico. Mr. N. L. H. Krauss, in a letter to Dr. R. I. Sailer in 1945 states that he has seen them sold alive in buckets at Cuernavaca, Morelos, by the indians who say that they scrape the bugs off trees in the nearby mountains. He further states that they are usually sold in paper cones, in handful lots, by venders, the price being about two pesos (about 42 cents U.S. cy.) per kilo. They are eaten alive or dropped into stew just before serving. Only a few are used since they have a strong taste." He also noted that they were supposed to cure kidney, liver, and stomach ailments.

2.2.10.7.13 *Catacanthini Atkinson, 1888*

Although he did not formally refer to it by name, Stål (1876) was the first to recognize this group at a higher level. He included six genera in his "*Catacanthus, Arocera, Vulsirea* et affinia": *Catacanthus, Chalcocoris, Commius* Stål, *Coquerelia* Signoret, *Anaxilaus,* and *Hyrmine* Stål. Atkinson (1888) originally proposed the Catacantharia for a single genus, *Catacanthus*. Bergroth (1908) catalogued *Chalcocoris* directly after the genus *Catacanthus*, inferring that the two might be related. Until recently, most workers have included only these two Old World genera in the Catacanthini. Rider (2015a), based on Stål's (1876) inclusion of *Arocera* Spinola and *Vulsirea* Spinola in his "*Catacanthus, Arocera, Vulsirea* et affinia" transferred five New World genera (*Arocera, Boea* Walker, *Rhyssocephala* Rider, *Runibia* Stål, and *Vulsirea*) into this tribe. This brings the total number of genera and species included in this tribe to 11 and 63, respectively (**Table 2.3**).

Members of all included genera tend to be medium to large in size and are usually quite colorful, ranging from yellows, reds, and oranges to metallic blues and greens (**Figures 2.20I, J; 2.29C, D**). They may or may not have the abdomen armed basally, the ostiolar rugae tend to be elongate (in *Boea, Runibia,* and *Vulsirea,* the ostiolar rugae are somewhat shorter), and the dorsum of the head is nearly impunctate but has oblique striae or wrinkles (**Figure 2.8F**). The anterolateral and anterior margins of the pronotum are usually distinctly reflexed, as is the lateral margins of the head.

There may be more New World genera related to this group; for example, the genus *Pharypia* Stål (**Figure 2.31G**) recently has been included in the Pentatomini, separate from the other similar New World genera because its members usually have a spine or tubercle at the base of the abdomen. This character is variable within the Catacanthini, so it is possible, based on the similar structure of the head and the colorful coloration, that this genus may also belong in the Catacanthini. A phylogenetic analysis by Fürstenau and Grazia (2014) indicates that members of the tribe Coquereliini may also be related to, if not belonging in, the Catacanthini (see **Section 2.2.10.7.16**). Several New World genera currently placed in the Catacanthini have been revised recently: *Arocera* (Rider 1992b), *Rhyssocephala* (Rider 1992a), and *Runibia* (Zwetsch and Grazia 2001).

Catacanthus incarnatus (Drury), the so-called man-faced stink bug, has been recorded attacking cashew in south India (Davis 1949, Bhat and Srikumar 2013, Waghmare et al. 2015).

2.2.10.7.14 Caystrini Ahmad and Afzal, 1979

Credit is usually given to Stål (1876) for first recognizing this group when he keyed "*Odius* et affinia" (*Odius* is a synonym of *Caystrus* Stål), but no one actually referred to this group using a family-group name until Ahmad and Afzal (1979). This tribe contains 14 genera and 66 species (**Table 2.3**) all of which are restricted to the Old World. They tend to be small to medium in size, brown to black in color, and have a striking resemblance to members of the Myrocheini.

There appear to be three characters, none of which is totally reliable when used individually, that are used to separate the Caystrini from the Myrocheini. The caystrines have the mesosternum medially carinate (sulcate in the myrocheines), the ostiolar rugae are usually longer (short to obsolete in myrocheines) with conspicuous evaporative areas, and the front femora lack spines or teeth (at least small spines usually present in myrocheines). Most caystrines will exhibit all three characters, but occasionally only two (or rarely one) are present. Both the myrocheines and caystrines have the body somewhat depressed and broadly ovate (**Figure 2.29E**), and the lateral margins of the head and pronotum are sharp and somewhat laminate; the juga are often longer than the tylus, and meet anterior to the tylus. The spermathecal bulb in females is ball-shaped with two to five finger-like diverticula.

Ahmad and Kamaluddin (1989b) revised the tribe for the Indo-Pakistan subcontinent. Three genera, *Surenus* Distant (from Halyini), *Agathocles* Stål and *Exithemus* Distant (both from Rolstoniellini) will be transferred to the Caystrini, based primarily on the study of their morphology (Salini Shivaprakash and Petr Kment, unpublished data). Linnavuori (1982) stated that most caystrines are apparently grass feeders in moist habitats.

2.2.10.7.15 Chlorocorini Rider, Greve, Schwertner, and Grazia, New Tribe

Type genus: *Chlorocoris* Stål, here designated.
This tribe contains 10 New World genera (or subgenera) and 77 species (**Table 2.3**) that usually are included in the tribe Pentatomini (Gapud 1991, Grazia and Schwertner 2008b, Grazia et al. 2015). Different authors suggested, at different times, a possible relationship among at least some of these genera (Grazia 1968, Becker and Grazia-Vieira 1971, Grazia-Vieira 1972, Grazia 1976, Eger 1978, Thomas 1992a, 1998). However, this taxon has been recognized only recently as a probable monophyletic group (Greve 2010, Rider 2015a) but, as yet, has not been formally named. As this group of genera appears to form a monophyletic group, it seems desirable to formally establish this tribe at this time with the type genus *Chlorocoris* Stål. The members of this tribe tend to be medium to large in size; they are usually green in life but fade to yellow after death (**Figure 2.29F**). They tend to be somewhat depressed. The head is usually rather flat and often triangular in shape with the apices of the juga acute to spinose (**Figures 2.8G, H**). The humeral angles are often prominently spined; the anterolateral pronotal margins are edged and usually provided with a row of small to fairly large denticles. The ostiolar rugae are usually short and linear, reaching to middle of metapleuron or less (longer in *Chloropepla* Stål), and the associated evaporatorium is large and extensive. The mesosternum is medially carinate, sometimes strongly so (*Arvelius* Spinola). The base of the abdomen is usually shallowly sulcate and receives the apex of the rostrum (exception is *Arvelius* in which the base of the abdomen has a forwardly projecting spine, the tip of which meets with the elevated metasternal carina). The tarsi are three-segmented.

The included genera are *Arvelius*, *Chlorocoris* Spinola (**Figure 2.29F**) (with 3 subgenera including *Arawacoris* Thomas, *Chlorocoris*, and *Monochrocerus* Stål), *Chloropepla*, *Eludocoris* Thomas, *Fecelia* Stål, *Loxa* Amyot and Serville (**Figure 2.20K**), *Mayrinia* Horváth (**Figure 2.20L**), and *Rhyncholepta* Bergroth (**Figure 2.21A**), although *Arvelius*, by virtue of the armed abdominal venter, may not belong here. Several of the above listed genera have been reviewed recently: *Arvelius* (Brailovsky 1981), *Chlorocoris* (Thomas 1985, 1998), *Chloropepla* (Greve et al. 2013), *Fecelia* (Grazia 1976, 1980b, Eger 1980), *Loxa* (Eger 1978), and *Mayrinia* (Grazia-Vieira 1972). Host plant preferences can vary, but at least species of *Arvelius* seem to favor members of the plant family Solanaceae. *Arvelius albopunctatus* (DeGeer) has been recorded on tomato (Basso et al. 1974) in Brazil, and, more recently, has been reported as becoming a problem on cherry tomatoes in Baja California (Panizzi 2015). It also is known to vector trypanosomatid protozoans (Kastelein and Camargo 1990). They also have been studied as possible biological control agents for weedy species of *Solanum* in South Africa (Siebert 1977, Olckers and Zimmermann 1991).

2.2.10.7.16 *Coquereliini Cachan, 1952*

Cachan (1952) proposed the Coquereliaria for two previously described genera (*Coquerelia* and *Coquerelidea* Reuter) and two new genera (*Cleoqueria* Cachan and *Neocoquerelidea* Cachan), all of which are endemic to Madagascar. *Coquerelia* (**Figure 2.29G**) contains two species; the remaining three genera are monotypic (**Table 2.3**). From the descriptions, it appears that this group could be related to the Catacanthini or possibly the Antestiini. The abdominal venter is described as being unarmed, but it is armed with a small tubercle in *Neocoquerelidea*. There is a tendency for the anterolateral pronotal margins to be strongly convex, expanded, and distinctly reflexed (**Figure 2.29G**). The ostiolar rugae are usually elongate and apically acute (except shorter in *Cleoqueria*). Some of the markings on the dorsum are similar to those seen in *Catacanthus* (e.g., *Coquerelia*, *Coquerelidea*), *Antestia* (e.g., *Neocoquerelidea*), or the Madagascan acanthosomatine *Noualhieridia ornatula* Breddin (e.g., *Cleoqueria*). The Coquereliini is potentially closely related to the Catacanthini (Fürstenau and Grazia 2014) and there is the possibility that it will be proven to be a synonym of that tribe.

2.2.10.7.17 *Degonetini Azim and Shafee, 1984*

Distant (1902) treated the genus *Degonetus* Distant in the Tropicoraria (= Pentatomini). Azim and Shafee (1984a) noted that this genus exhibited several characters that would preclude it from being a member of that tribe, but they were unable to place it in any other known tribe; thus, they proposed the tribe Degonetini.

Members of this tribe are medium in size, ovate, with prominent acutely produced humeral angles (**Figure 2.29H**). The antennae are four-segmented in both sexes. The juga are longer than the tylus and curve mesad apically. The anterolateral pronotal margins are distinctly dentate. The prosternum, mesosternum, and metasternum are sulcate medially, with no sign of a medial carina. The ostioles are relatively large, but unattended; that is, there is no ostiolar auricle or ruga extending laterally; the associated evaporative area is relatively large and shiny. The base of the abdomen is provided with a stout, conically shaped tubercule reaching the middle of the hind coxae (this is the only known pentatomoid taxon that has the combination of a sulcate mesosternum and an armed abdominal venter). According to Azim and Shafee (1984a), the spermathecal bulb has tubular outgrowths, and the proximal part of the sclerotized rod is long and narrow.

This tribe contains a single genus, *Degonetus* (**Figure 2.29H**), and two species [*D. serratus* (Distant) and *D. sikkimensis* Mathew] (**Table 2.3**), both only known from India. Based on the illustrations provided by Mathew (1969), *D. sikkimensis* is not conspecific with *D. serratus*, but it probably belongs to or it is related to *Prionaca* Dallas (currently in the Pentatomini, see **Section 2.2.10.7.34**). Little is known about the biology of these species. There are two known plant associations: Chatterjee (1934) recorded *D. serratus* (Distant) from sandalwood, *Santalum album* L. (Santalaceae), and there have been three records from teak, *Tectona grandis* Loureiro (Lamiaceae) (Chatterjee 1934, Azim 2011, and Roychoudhury and Chandra 2011).

2.2.10.7.18 Diemeniini Kirkaldy, 1909

Bergroth (1905) first proposed the Platycoraria, based on the genus *Platycoris* Guérin-Méneville, for this group. There was some confusion, however, surrounding the publication dates of Guérin-Méneville's paper. In 1831, two of the plates (the text was not published until 1838) were published out of order, which affected the type species for *Platycoris* (see Dupuis 1952 for more details). The result is that *Platycoris* is actually a member of the Halyini, not the Diemeniini; Bergroth's Platycoraria is based on a misidentified type genus and, therefore, a junior synonym of Halyini. Kirkaldy (1909) proposed Diemeniini to replace Platycorini, based on the genus *Diemenia* Spinola.

This tribe includes 14 genera, of which 12 are endemic to Australia; *Oncocoris* Mayr (**Figure 2.21C**) is mostly distributed in Australia, but a few species are found in New Guinea and the Solomon Islands; *Caridophthalmus* Assmann (**Figures 2.12D, 2.21B**) also has species in Australia, but its center of diversity is Papua New Guinea. There are about 55 described species in the tribe (**Table 2.3**). This group has been associated with the Halyini by Dallas (1851) and Stål (1876). As might be expected, due to the presence of abdominal stridulatory structures in both tribes, the Diemeniini occasionally has been lumped with the Mecideini. Gross (1975b) indicated that he found, especially in the structure of the male genitalia, that the Diemeniini may have some relationship with the Halyini, but he further stated that the Diemeniini is not identical with the Halyini, and that both were distinct from the Mecideini.

Members of this tribe are small to medium in size (5-14 mm), and exhibit several different forms and colors; in fact, the variability might make one question whether they all belong in the same tribe or not (**Figures 2.21B, 2.29I**). The single, unifying character is that all included genera have a distinct, lunately shaped, stridulatory area located on each side of the abdominal venter. Only one other pentatomid genus (*Mecidea*, presently in the Mecideini) has a similar stridulatory area, but the included species are quite elongate, slender, and pale, and they occur in dry, arid areas outside of the Australian region. There is another genus (*Neomazium*, presently placed in the Carpocorini) that has a stridulatory mechanism on the abdominal venter, but this mechanism is placed along the margins of the abdomen and is not homologous with that found in the Diemeniini. The antennae may be four- or five-segmented. Often, the juga are produced beyond the tylus, and the lateral margins are usually reflexed, often with a small process in front of each eye. According to Gross (1975b), the male genitalia are quite variable with the parameres either Y-shaped (*Diemenia* and *Caridophthalmus*), thin and elongate (*Aplerotus* Dallas), or flattened or plate-like (*Kalkadoona* Distant and *Oncocoris*). The spermathecal bulb is simple, ball-shaped, without finger-like diverticula.

Both *Kalkadoona* and *Oncocoris* (**Figure 2.21C**) have the prosternum, mesosternum, and metasternum medially carinate; and the ostiolar rugae are relatively elongate and straight. As mentioned above, *Caridophthalmus* (**Figures 2.12D, 2.21B**) is the only diemeniine genus with greater diversity outside of Australia (but most species remain undescribed). The species also differ significantly from the Australian members of this tribe. The stridulatory ridges in *Caridophthalmus* are more widely spaced and somewhat obtuse. In fact, one might question whether they are actually involved in stridulation except that there are stridulatory pegs on the inner surface of the hind femora. Additionally, members of this genus have the surface of the head distinctly horizontal, facing forward, and the dorsal surface is armed with various distinct spines (**Figures 2.12D, 2.21B**). It is possible that once a thorough phylogenetic analysis is completed, this genus will be placed in its own tribe. There are some general similarities between this genus and the newly described *Pentamyrmex* Rider and Brailovsky, an ant-mimic genus for which a new tribe, Pentamyrmecini, has been erected (see discussion under Pentamyrmecini, **Section 2.2.10.7.33**).

Relatively recent revisions have been provided for *Diemenia* (Ahmad and Kamaluddin 1989a) and *Oncocoris* (McDonald and Edwards 1978).

2.2.10.7.19 Diplostirini Distant, 1902

This is a small tribe, containing two monotypic genera (**Table 2.3**); they are confined to India and the Orient. Distant (1902) originally based this tribe on the distinctly concave posterior margin of the pronotum (conversely, the basal margin of the scutellum is distinctly convex). In fact, the two included genera have little in common, except for the general elongate shape, and there is little doubt that they are not closely related. The only species of the genus *Diplostira* Dallas, *D. valida* Dallas (**Figure 2.29J**), is quite large (body length about 25 mm) and has the base of the abdominal venter obtusely spined;

the mesosternum and metasternum have two ridges, one on each side of the rostral channel. The other included genus, *Ambiorix* Stål (type species: *A. aenescens* Stål) is much smaller (about 10 mm), lacks the ridges on each side of the rostral groove, but the base of the abdominal venter has an anteriorly directed spine that reaches to the base of the head. *Ambiorix* is apparently closely related to *Iphiarusa* Breddin (currently Pentatomini, see **Section 2.2.10.7.34**) and it likely will be necessary to be remove it from the proximity of *Diplostira*.

Interestingly, the unusual structure of the mesosternum and metasternum in *Diplostira* is similar to that found in the asopine genus *Ealda* Walker.

2.2.10.7.20 Diploxyini Atkinson, 1888

Stål (1876) was the first to formally recognize this group at a higher level, but he did not formally name it. He included three genera in his "*Diploxys* et affinia": *Adria*, *Diploxys* Amyot and Serville, and *Steleocoris*. Atkinson (1888) proposed Diploxyaria, based on the African genus *Diploxys*. As his paper concerned the Indian fauna, he did not specifically mention *Diploxys*. Interestingly, he originally included four additional genera in this tribe, all of which have been transferred to other tribes. The four genera he specifically included are *Adria*, *Scylax*, *Aeschrocoris*, and *Aeliomorpha*. *Adria* has, at times, been included in the Aeliini, but we have tentatively placed it in the Eysarcorini (but see discussion on this genus in the Eysarcorini section, **Section 2.2.10.7.22**). *Scylax* and *Aeschrocoris* are both now considered members of the Aeschrocorini although the placement of *Scylax* is questionable (see discussion under Aeschrocorini, **Section 2.2.10.7.4**). And *Aeliomorpha* has had a confused history, having been included in the Aeliini, the Carpocorini, and in its own group (see discussion within the *Aeliomorpha* group of Linnavuori, **Section 2.2.10.7.45.3**).

This is a relatively small tribe, containing four genera (and one subgenus) and 29 species (**Table 2.3**), but some of the included species are relatively common. The four genera are *Acoloba* (**Figure 2.21D**), *Aeladria* Cachan, *Coponia* Stål, and *Diploxys* (**Figures 2.21E, 2.29K**). All known species occur in Africa or Madagascar with *Acoloba lanceolata* (F.), *D. orientalis* Linnavuori, and *D. pulchricornis* Linnavuori also known from Yemen, and *D. cordofana* (Mayr) also known from Saudi Arabia (Linnavuori 1975, 1982, 1986, 1989). As a group, they are medium in size, elongate and slender, and usually brownish or yellowish in color. There is a tendency for the formation of spines on the apices of the juga, humeral angles, posterolateral angles of the last connexival segment, and the apices of the femora (**Figure 2.29K**). The posterior angles of the bucculae are triangularly produced. The prosternum, mesosternum, and metasternum are medially sulcate, lacking carinae. The ostiolar rugae are short, sometimes obsolete, and curve cephalad; the associated evaporative areas are also reduced. The pygophore is somewhat robust with the apical margin weakly sinuate. The spermathecal bulb may be simple and ball-shaped, or it may have a couple of finger-like tubules.

There is some general resemblance between these genera and the Australian genera presently included in the Aeptini. The situation is complicated because the Australian genera may not actually be related to *Aeptus*, the type species of the Aeptini, so they may deserve a new tribal name; however, they may be related to the diploxyine genera. Interestingly, there are two genera that occur in the Indian subregion that presently are placed in the Carpocorini but may be related to these genera. In both of these (*Bonacialus* and *Gulielmus*), the mesosternum is distinctly sulcate. The South American genus *Poriptus* is also quite similar in general appearance (and it has a sulcate mesosternum) to the diploxyines.

Linnavuori (1982) stated that members of this tribe are "grass feeders in mesic or moist savannah habitats, marshes and shore meadows."

2.2.10.7.21 Eurysaspini Atkinson, 1888

Atkinson (1888) erected the Euryasparia [sic] for a single genus, *Eurysaspis* Signoret (incorrectly spelled as *Euryaspis*). The grammatically correct stem derived from *Eurysaspis* would be Eurysaspid-, but as the spelling Eurysaspidini is not in prevailing usage, Eurysaspini is to be maintained (ICZN 1999, Article 29.3.1.1). This is a relatively small tribe presently containing three genera (*Eurysaspis* [**Figure 2.29L**], *Flaminia* Stål [**Figure 2.21F**], *Platacantha* Herrich-Schäffer) and 25 species (**Table 2.3**). There is some doubt about the validity of this tribe as all three genera have strikingly different morphology. The single unifying character is the enlarged, spatulate scutellum (**Figures 2.21F, 2.29L**), but this scutellar form is seen in many other groups as well. The ostiolar rugae tend to be elongate and apically acute, curving cephalad. In *Eurysaspis*, the base of the abdominal venter is armed with a robust, truncate projection that meets with an elevated metasternum that, in turn, meets with an elevated mesosternum; the mesosternum becomes narrower and flattened, and protrudes as a rounded lobe onto the prosternum, between the front coxae (much like that seen in the Rhynchocorini). In *Flaminia*, the abdominal venter is not as elevated as in *Eurysaspis*, but it is still truncately produced forward and comes into contact with the slightly elevated, but distinctly sulcate, metasternum. The mesosternum is somewhat flat posteriorly, becomes slightly carinate anteriorly, but is not produced forward onto the prosternum. In *Platacantha*, the base of the abdominal venter is armed with a robust, ventrally flattened spine that extends forward beyond the middle coxae and covers both the meta- and mesosterna. This is rather similar to the sternal structure exhibited in the Piezodorini, so *Platacantha* may need to be transferred to that tribe. In fact, Linnavuori (1982) indicated that *Platacantha* was distinctively different from *Eurysaspis* and should not be placed in the same group with that genus. He argued that there were some similarities with the genus *Acrosternum* Fieber (**Figure 2.22I**) (currently in the Nezarini), based on the sternal structures, or to members of the Antestiini, based on genitalic structures.

2.2.10.7.22 Eysarcorini Mulsant and Rey, 1866

Mulsant and Rey (1866) proposed Eysarcoriens as a new family with two branches (= tribes): Eysarcoraires and Rubiconiaires. Their Rubiconiaires originally contained two genera, *Rubiconia* and *Staria*, both of which are now considered to be members of the Carpocorini (see discussion in **Section 2.2.10.7.12**). They originally included three genera in their Eysarcoraires: *Dalleria* Mulsant and Rey, *Eysarcoris* Hahn, and *Onylia* Mulsant and Rey. *Dalleria* is currently treated as a subgenus of *Stagonomus* Gorski, and *Onylia* is a junior synonym of the same genus. Two other tribal names have been proposed that are now considered to be synonyms of the Eysarcorini. Putshkov (1961) proposed the tribe Stolliini, based on the genus *Stollia* Ellenrieder, which is considered by most workers to be a synonym of *Eysarcoris*. Fuente (1974) proposed the tribal name Stagonomini as a replacement name for Eysarcorini (at one time there was a dispute concerning the correct type species for *Eysarcoris*, which could have endangered the availability of the Eysarcorini). Finally, of minor importance, but worthy of mention is the fact that the genus *Eysarcoris* has been unjustifiably emended several times, and, so, various derivations (e.g., Eusarcorini, Eysarcocorini) frequent the literature.

This is a difficult tribe to characterize. Its members are usually small to medium in size, and tend to be somewhat robust in shape (**Figure 2.30A**). The abdominal venter is unarmed, and the scutellum tends to be enlarged and spatulate (several exceptions include but are not limited to *Aspavia* Stål [**Figure 2.21G**], *Carbula* Stål [**Figure 2.21H**], and *Durmia* Stål). They are usually brown to dark brown in color. It is a relatively large tribe containing 19 genera and 230 species (**Table 2.3**). Linnavuori (1982) stated that the lateral margins of the pronotum are usually callose, whitish, and sharply delimited, except for the genus *Adria* (but see further discussion of *Adria* below). The ostiolar rugae are usually short and auriculate (**Figure 2.8E**); the associated evaporative areas are also fairly small. The mesosternum is medially carinate. In one species of *Stagonomus*, the female abdomen and genital plates are produced caudad, becoming acutely pointed posteriorly (**Figure 2.21I**), much as is seen in the Lestonocorini. The male parameres are usually strongly F-shaped. The spermathecal bulb is simple, ball-shaped, and lacks diverticula.

There is little to separate this tribe from the Carpocorini, except for those few genera in which the scutellum is enlarged. As mentioned above, several included genera do not have the scutellum expanded, and there are some carpocorine genera (e.g., *Coenus*) that have an enlarged, spatulate scutellum. The genus *Adria* typically has been treated as a member of the tribe Aeliini, but we have tentatively included it in the Eysarcorini. It may not belong in either tribe. Its species have the mesosternum sulcate, thus allying it (along with the carpocorine genera *Gulielmus* and *Bonacialus*) more closely to the Aeptini, Diploxyini, Myrocheini, or Sephelini. In fact, Atkinson (1888) treated *Adria* as a member of the Diploxyini. Additionally, it is likely that the genus *Corisseura* Cachan should be transferred to another tribe. Its members have the ostiolar rugae quite long and acuminate apically. This might place it in the Cappaeini or Hoplistoderini.

Linnavuori (1982) stated that the species of this tribe usually are found on grasses in fields, meadows, and in forest clearings; several of the European species of *Eysarcoris* and *Stagonomus* have been associated with Lamiaceae (Derjanschi and Péricart 2005).

Recent revisions have been provided for several eysarcorine genera: *Adria* (Zaidi 1994, 1996), *Durmia* (Linnavuori 1973, 1982), Indian species of *Eysarcoris* (Azim and Shafee 1984b), Indian species of *Hermolaus* Distant (Azim and Shafee 1985), and *Pseudolerida* Schouteden (Linnavuori 1982).

2.2.10.7.23 Halyini Amyot and Serville, 1843

Amyot and Serville (1843) proposed this family-level name (as Halydes), and it has been recognized as a valid tribe by nearly all subsequent workers. The real controversy concerning the taxonomic history of this group is its composition. The reader will notice that many of the existing pentatomid tribes have been associated at one time or another with the Halyini. This is a relatively large tribe containing 91 genera and 430 species (**Table 2.3**). Members of this tribe are extremely variable in size (8-35 mm) and structure, making it rather difficult to characterize the group as a whole. In fact, Gross (1976) quoted a colleague as saying 'We all reckon we know whether a thing is a halyine or not, but what are really the distinguishing characters of this group?' They are medium to large in size, usually somewhat flattened or depressed, and usually a mottled coloration of browns, grays, and pale areas (**Figures 2.21J-L; 2.22A, B; 2.30B, C**). It is easy to envision most species occurring on the bark of trees, blending in with this cryptic coloration. It is mostly an Old World group with only two genera (*Brochymena* [**Figure 2.21K**] and *Parabrochymena* Larivière) known from the New World (Larivière 1992, 1994).

There is a tendency for the head to be long and slender (**Figures 2.30B, C**); the lateral margins of the head and pronotum are often foliate or at least dentate or crenulate. Most genera have five-segmented antennae, but a few have four segments (*Neagenor* Bergroth, some *Poecilometis* Dallas), and at least one Australian genus has only three segments (*Omyta*). The rostrum reaches to the hind coxae, and often beyond, well onto the abdominal venter. The scutellum is usually subtriangular in shape, although in at least one genus (*Mezessea* Linnavuori) it is enlarged and spatulate-like. The legs and antennae appear longer and more gracile than in members of other tribes. The femora may or may not be armed with small teeth or spines; the outer surfaces of the tibiae are usually distinctly sulcate, and, occasionally, the tibiae are dilated (e.g., *Erthesina* Spinola [**Figure 2.21L**]). The prosternum is usually weakly sulcate, the mesosternum is medially carinate, and the metasternum is usually flat to shallowly concave. In the New Guinean region, there is a group of genera (*Accarana* Distant, *Auxentius* Horváth, *Babylas* Horváth, *Brizica* Walker, *Bromocoris* Horváth, and *Coctoteris* Stål) in which the mesosternum may be more or less sulcate, but there is still at least a weak medial carina; in these genera, the metasternum is usually strongly sulcate, this sulcation continuing more or less onto the abdominal venter. *Ectenus* Dallas and *Epitoxicoris* Wall appear to be intermediate between this group and the rest of the Halyini. Also, the New World genera (*Brochymena* [**Figure 2.21K**] and *Parabrochymena*), and at least two Old World genera (*Mustha* and *Orthoschizops* Spinola) have the sterna weakly to distinctly sulcate with the medial carinae absent or obsolete; *Mustha* also has the margins of the head, pronotum, and abdomen provided with a row of spines. Regardless of the above sternal structures, the abdomen may or may not be sulcate; if sulcate, the sulcation may continue for most of the length of the abdominal venter or may occur only on the base of the abdomen. The ostiolar rugae are quite variable, ranging from short to long. The base of the abdominal venter is unarmed. Several genera (*Atelocera* Laporte [**Figure 2.30C**], *Carenoplistus* Jakovlev, and *Pseudatelus* Linnavuori [**Figure 2.22A**]), in females, have the abdominal venter nearly covered with two opaque spots (one on each side) with numerous hairs that seems to accumulate waxy secretions (also seen in the Memmiini). The spermathecal bulb is ball-shaped with two to three finger-like diverticula. For further discussion of various halyine characters, see Gross (1976), Memon et al. (2011), and Kment (2013).

Due to the variability in structure, the Halyini has served as a dumping ground for many genera that have been moved to other tribes (e.g., *Agaeus*, *Agonoscelis*, *Phricodus* Spinola, each now in its own tribe, and, at one time, all of the taxa in the New World Ochlerini were included in the Halyini). Memon et al. (2011) presented a phylogenetic analysis, based on morphology, of the South Asian halyine genera, and although the methodology has been questioned (Barão et al. 2012), their results maintained that *Agaeus* (**Figure 2.28D**) and *Phricodus* (**Figure 2.31I**) should be included in the Halyini. It should also be noted

that Wall (2007) recognized a separate clade within the Halyini, called the *Solomonius* group, character-
ized by extremely reduced parameres in the male genitalia that are fused with the proctiger complex. His
Solomonius group contains seven genera, namely *Anchises* Stål, *Auxentius*, *Babylas*, *Brizica*, *Elemana*
Distant, *Epitoxicoris*, and *Solomonius* Wall.

There are still several genera placed in the Halyini that seem to not belong. For example, *Eurus* Dallas
has a different head structure and the mesosternum is sulcate; *Goilalaka*, in general appearance, seems
quite similar to the genus *Aednus* (now in Myrocheini but probably belongs in the Caystrini or a related
tribe); *Mimikana*, which actually is reminiscent of some members of the Ochlerini; and *Tinganina*
Bergroth, which exhibits a different head structure (i.e, the bucculae are distally toothed), and many
specimens exhibit brachyptery (not seen in other halyine genera). *Polycarmes* Stål has both the mesosternum and the metasternum sharply carinate medially; this condition of the metasternum is relatively rare.
The correct placement of these genera will probably not be known until a thorough phylogenetic study is
completed. The genus *Surenus* will be transferred to the Caystrini (Salini Shivaprakash and Petr Kment,
unpublished data). On the other hand, two tribes defined mostly by reductions, Memmiini (3-segmented
tarsi and 4-segmented antennae) and Nealeriini (2-segmented tarsi and 4-segmented antennae), may be
synonyms of the Halyini (Linnavuori 1982, Kment 2013).

There may even be some question on whether the only New World representatives, *Brochymena*
(**Figure 2.21K**) and *Parabrochymena*, are true halyines. Again, this is a question that will only be
answered via a new phylogenetic study. Species of these two genera are quite similar in general appearance; in fact, the validity of *Parabrochymena* has been questioned. A relatively recent study (Ahmad and
McPherson 1998) confirmed that there are differences between these two genera; whether these differences are great enough to warrant separating this group into two genera may require further study. The
biology of a species of each genus also has been studied (Cuda and McPherson 1976).

Several halyine genera have been reviewed in recent years: *Apodiphus* Spinola (Ghauri 1977a),
Auxentius (Wall 2007), *Brizica* (Wall 2007), *Brochymena* (Ruckes 1947, Larivière 1992), *Cahara* Ghauri
(Ghauri 1978, Fan and Liu 2013), *Carenoplistus* (Memon and Ahmad 1998), *Ectenus* (Uichanco 1949),
Epitoxicoris (Wall 2007), *Faizuda* Ghauri (Ghauri 1988b), *Goilalaka* (Ghauri 1972), *Halys* F. (Ghauri
1988a), *Jugalpada* Ghauri (Ghauri 1975a), *Parabrochymena* (Larivière 1994), *Paranevisanus* Distant
(Ghauri 1975b), *Platycoris* (McDonald 1995), *Poecilometis* (Gross 1972), *Sarju* Ghauri (Ghauri 1977b,
Memon and Ahmad 2009), *Solomonius* Wall (Wall 2007), *Tachengia* China (Ahmad 2004), *Theseus* Stål
(Baehr 1989, 1991), *Tipulparra* Ghauri (Ghauri 1980), and *Zaplutus* Bergroth (Cachan 1952).

It appears that most halyine species are arboreal, residing on various trees; their dark mottled coloration allows them to blend in well with the bark of trees.

2.2.10.7.24 *Hoplistoderini Atkinson, 1888*

This group was first recognized by Stål (1876), but he did not formally name it. He included four genera
in his "*Hoplistodera, Alcimus* et affinia": *Alcimus* Dallas (= *Alcimocoris* Bergroth), *Taurodes* Dallas,
Hoplistodera Westwood, and *Stachyomia* Stål. Atkinson (1888) was the first to formally name this tribe,
and he originally included three genera in his Hoplistoderaria: *Alcimus, Hoplistodera*, and *Bolaca*.
Distant (1902) transferred *Bolaca* to the Amyntorini, where we currently treat it, and added two more
genera, *Axiagastus* (now in the Axiagastini) and *Paracritheus* Bergroth (still considered to belong in the
Hoplistoderini). Bergroth (1918) described the genus *Glottaspis*, placing it near the genus *Alcimocoris*.

Members of this tribe tend to be small to medium in size; they are somewhat robust in form with a
tendency for the humeral angles to be produced, the scutellum to be enlarged (spatulate) (**Figure 2.30D**),
and the head and anterior disk of pronotum to be quite declivent. Otherwise, however, there is some variability in structure. For example, *Hoplistodera* and *Stachyomia* both have the mesosternum distinctly
carinate medially, whereas *Alcimocoris* has a sulcate mesosternum. The ostiolar rugae are usually elongate, reaching beyond the middle of the metapleura. The abdominal venter is unarmed.

This is not a large group with most genera and species known from the Indian-China region south-
eastward into Southeast Asia with *Glottaspis* being endemic to Madagascar. There are seven genera and
35 species (**Table 2.3**). Distant (1902) indicated a possible relationship with the Eysarcorini as members
of both tribes usually have the scutellum enlarged, and they tend to have a somewhat shortened, robust

body. But the similarities seem to end there. None of the included genera has been revised recently, but Hsiao et al. (1977) provided keys for the Chinese taxa.

2.2.10.7.25 *Lestonocorini Ahmad and Mohammad, 1980*

Although the genera included in this tribe have unusual female genitalia, they were not segregated off into their own tribe until relatively recently. Ahmad and Mohammad (1980) originally included four genera, *Gynenica* Dallas, *Lestonocoris* Ahmad and Mohammad, *Neogynenica* Yang, and *Umgababa* Leston, all of which are still included in the Lestonocorini.

This is a small group, containing four genera and 20 species (**Table 2.3**), distributed primarily in Africa and the Indian subcontinent. They are small to medium in size, usually brown in color, often with some ivory markings on the dorsum; there is a tendency for the humeral angles to be produced or even spinose (**Figure 2.30E**). The primary character separating this group from other groups is the structure of the female genital plates; the plates (and the abdomen) are drawn out posteriorly, forming a sharply triangular shape (similar to the condition seen in the eysarcorine genus *Stagonomus*, the antestiine genus *Birketsmithia* Leston, and at least one species of *Banasa* Stål) (**Figure 2.30E**). Otherwise, these insects possess most of the characters seen in either the Carpocorini or Eysarcorini, especially *Carbula* or *Aspavia*. In fact, Schouteden (1910) treated *Gynenica* as being a relative of *Carbula*. The mesosternum is carinate medially, and the ostiolar rugae are quite small and auriculate. The head tends to be somewhat more slender than that seen in the Carpocorini or Eysarcorini. The outer surface of each tibia is rounded, not sulcate. The base of the abdomen is unarmed. The spermathecal bulb is ball-shaped, basally flaring outward to the edges of the proximal flange.

The genus *Gynenica* has been revised (Leston 1953d). Salini (2016) recently redescribed *Dardjilingia* Yang and discussed its taxonomic placement. She concluded that this genus probably does not belong in the Lestonocorini, but she did not place it elsewhere.

2.2.10.7.26 *Mecideini Distant, 1902*

Interestingly, Stål (1876) included *Mecidea* in the Halyini. Distant (1902) originally placed two genera in his Mecidaria, *Aenaria* Stål and *Mecidea*. Members of both genera are rather long and slender, but species of *Aenaria* lack the characteristic stridulatory structures on the abdominal venter. *Aenaria* is now considered to be a member of the Sephelini.

This is a monotypic tribe containing a single genus, *Mecidea*, and 16 species (**Table 2.3**). The mecideines are small to medium in size, elongate, slender, pale yellowish species (**Figure 2.30F**) with a lunate stridulatory region on each side of the abdominal venter. Leston (1957) placed much importance on the presence of the stridulatory structures, and treated this group as a subfamily. The prosternum is shallowly concave, punctured, and undifferentiated from the propleura; the mesosternum is carinate medially, the carina usually becoming somewhat stronger anteriorly; the metasternum has two low, but sharp ridges forming a V-shaped sulcus with the narrow part of the 'V' cephalad. The ostiolar rugae are long and slender, set at an angle, and curving cephalad laterally. The scutellum is elongate, subtriangular, with the lateral margins straight from base to apex. The abdominal venter is unarmed; stridulatory ridges are present in a small arc on each side of abdominal venter (**Figures 2.7A, B**); associated pegs are located on the inner surface of the hind femora. The spermathecal bulb is elongate oval, without any projections.

The abdominal lunate stridulatory areas are found only in this tribe and in the diemeniines; the body shape and coloration in the diemeniines, however, are quite variable but never elongate, slender, and pale. The diemeniines occur in the Australian region whereas species of *Mecidea* occur in the New World and in Africa, the West Palearctic, Central Asia, India, and Western China, usually in arid areas. Abdominal stridulatory ridges are known in one other genus, *Neomazium*, but those are confined to the extreme lateral margins of the abdomen.

The genus *Mecidea* was revised in 1952 by Sailer, but additional species were described subsequently (Villiers 1954, Wagner 1954, Ahmad and Shah 1994).

Mecideines are grass feeders, tending to occur in dry, arid regions (Sailer 1952, Linnavuori 1982). Bundy and McPherson, in a series of papers, studied the eggs, nymphs, life history, and laboratory rearing of both *Mecidea minor* Ruckes and *M. major* Sailer (Bundy and McPherson 2005, 2010, 2011; Bundy et al. 2005).

2.2.10.7.27 Memmiini Cachan, 1952

This is currently a monotypic tribe containing a single genus, *Memmia* Stål, and seven species (**Table 2.3**), all endemic to Madagascar. They are relatively large in size, oval in shape, and darkly colored or mottled (**Figure 2.30G**). The head is somewhat slender with the juga and tylus subequal in length or the juga slightly longer than the tylus. The antennae are four-segmented. The anterolateral pronotal margins are crenulate, and the scutellum is somewhat elongate, sometimes reaching beyond apex of coria. The mesosternum may or may not be sulcate, but, in either case, there is a weak but distinct medial carina. The ostiolar rugae are small, nearly obsolete, each forming a small circular elevation around the ostiole. The abdominal venter in females is nearly covered with a large opaque area on each side; these opaque areas possess many long hairs that seem to accumulate waxy secretions (this also is seen in several halyine genera including *Atelocera*, *Carenoplistus*, and *Pseudatelus*).

This genus often has been associated with the Halyini. Distant (1882) and Linnavuori (1982) both indicated a close relationship with halyine genera, *Atelocera* for Distant and *Pseudatelus* for Linnavuori. Cachan (1952) erected a new tribe (Memmiini) for this genus, but a thorough phylogenetic analysis may determine that it belongs in the Halyini. Or, alternatively, the presence of the opaque glands on the abdominal venter of females may be important, and it may be necessary to transfer those continental African and Palearctic genera (*Atelocera*, *Carenoplistus*, and *Pseudatelus*) into the Memmiini.

2.2.10.7.28 Menidini Atkinson, 1888

Although he did not formally name it, Stål (1876) was the first to recognize this group at a higher level when he referred to it as "*Menida* et affinia"; he included eleven genera in this group: *Aegaleus* Stål, *Alciphron* Stål, *Ambiorix*, *Amphimachus* Stål, *Anaximenes* Stål, *Anchesmus* Stål, *Antestia*, *Brachycoris* Stål, *Cresphontes* Stål, *Menida* Motschulsky, and *Piezodorus* Fieber. Five of the eleven are no longer included in the Menidini: *Alciphron* tentatively is treated in the Nezarini, *Ambiorix* is currently doubtedly placed in the Diplostirini, *Anaximenes* and *Piezodorus* are currently members of the Piezodorini, and *Antestia* is the type genus for the Antestiini. Atkinson (1888) formally named the group (Menidaria), placing in it three of the above genera: *Antestia*, *Cresphontes*, and *Menida*, plus a fourth genus, *Apines*.

This is a relatively large tribe with 28 genera and 164 species (**Table 2.3**). Members in general are small to medium in size (4-12 mm), usually elongate, and rounded to circular in shape (**Figures 2.22C-E; 2.30H**). The head is usually somewhat shorter and broader than in most other tribes. The base of the abdominal venter usually is armed with a forwardly projected spine (**Figure 2.8B**); the spine may be short or long, and the apex usually is not apposed by the metasternum. The humeral angles are usually rounded. In some genera (*Amphimachus*, *Aspideurus* Signoret, *Decellella* Schouteden, *Keleacoris* Rider and Rolston, *Saceseurus* Breddin), the scutellum is somewhat enlarged and spatulate-like. The mesosternum is carinate medially although the carina may be more weakly developed than in some tribes. The ostiolar rugae are relatively long, reaching beyond the middle of the metapleuron. The parameres are somewhat T-shaped. The spermathecal bulb is usually simple, ball-shaped and lacks diverticula.

As with other large and variable tribes, there seem to be several genera that have been placed in this tribe that, perhaps, do not belong here. For example, members of the genus *Dabessus* Distant are much larger than most menidines and are more cryptically colored in light browns. Their general body form is wedge-shaped with their greatest width across the humeral angles. The abdominal spine is obsolete in one species, and in two other species, the abdominal spine meets with an elevated metasternum. The genus *Lathraedoeus* Breddin may belong in the Menidini, but it differs by having the last abdominal sternite in males extending posteriorly over the genital capsule; that is, the male pygophore is hidden up inside the pocket formed by the last abdominal sternite and tergite. Gross (1976) indicated that the genus *Sciomenida* Gross differs from other menidine genera by having a much shorter ostiolar ruga.

The South American genus *Elanela* Rolston (**Figure 2.22C**) superficially resembles members of the Menidini, but the abdominal armature is short and spine-like and fits into a bifurcate notch in the metasternum. Additionally, the genus *Udonga* Distant does not appear to be related to the menidines. In fact, the pentagonal shape of the head seems to ally it with several other Old World genera (e.g., *Parvacrena* Ruckes), which may be a separate monophyletic group. Linnavuori (1982) indicated that the African genera *Decellella* and *Brachycoris* have a misleading resemblance to *Sepontia* of the

Eysarcorini. He also stated that the Palearctic genus *Actuarius* Distant had the base of the abdomen unarmed.

The genera *Keriahana* Distant and *Neostrachia* Saunders have had very confused taxonomic histories. Most recent workers have followed Bergroth (1914) in treating *Neostrachia* as a synonym of *Menida*. At the same time, other workers have considered *Neostrachia* to be a junior synonym of *Apines* (e.g., Linnavuori 1986). Leston (1955a), however, placed *K. elongata* (Distant) (the type species of *Keriahana*) as a junior synonym of *N. bisignata* (Walker) (the type species of *Neostrachia*) without commenting on either genus. More recently, Ahmad et al. (1974) resurrected the use of *Keriahana* as a valid genus; they treated *K. elongata* and *K. bisignata* as separate species but placed both in *Keriahana*. If *Keriahana* and *Neostrachia* are congeneric, and both are taxonomically different from *Menida* and *Apines*, then *Neostrachia* is the senior name. *Menida* is undoubtedly a large and diverse group of species, desperately in need of revision. The actual status of these genera will remain a question until such a revision is completed.

Another problematic genus is the genus *Dunnius* Distant. It is fairly obvious that *Dunnius* and *Acesines* are rather closely related, if not congeneric. *Acesines* is presently considered to be a member of the Axiagastini.

Specimens of *Udonga montana* (Distant), locally known as Thangnang, are used for human consumption in India (Gahukar 2012). Azad and Firake (2012) stated "Thangnang is an economically important food for the local people. The bugs appear in swarms and form hives on branches, which sometimes break due to heavy weight. Furthermore, the bugs are easy to collect after a drizzle. Groups of farmers were found collecting around 20-30 kg of the bugs in gunny bags, bamboo containers, buckets, etc. According to the residents, Thangnang is a precious food available profusely only after 50 years during such events. Local people used to fry it in oil; some locals also make 'chutney' using the bugs. Besides, oil is also extracted from the bugs using traditional processing. This oil has high market value in spite of its bad smell and the pure oil is believed to cure many health problems."

Leston (1952b) revised the genus *Aegaleus*, but there are at least one or two undescribed species remaining (David A. Rider, unpublished data). Linnavuori (1982) keyed several African genera and species; except for this work few other menidine genera have been revised recently.

2.2.10.7.29 *Myrocheini Stål, 1871*

Stål (1871) originally proposed the division Myrocheina for seven genera: *Myrochea* Amyot and Serville, *Paramecocoris*, *Ennius* Stål, *Erachtheus* Stål, *Cocalus* Stål, *Laprius* Stål, and *Aednus*. *Paramecocoris* was proposed originally by Stål (1854) as a replacement name for *Paramecus* (which was preoccupied by *Paramecus* Dejean, 1829, in the Coleoptera), and as such, these two genera must have the same type species and are synonyms (Stål, in later publications, confused the issue by treating both *Paramecus* and *Paramecocoris* as valid genera). At any rate, *Paramecocoris* now is considered to be a member of the Caystrini. *Cocalus* is preoccupied by the arachnid genus *Cocalus* Koch, 1846; Bergroth (1891) proposed the replacement name *Neococalus*; it still is considered to be a member of this tribe although some of its characters are not typical (see below).

Stål (1876) presented keys to the genera for two separate groups that he considered to be closely related: the "*Dymantis* et genera nonnula affinia" group and the "*Myrochea* et genera quaedam affinia" group. His *Dymantis* group included both *Aeptus* (the type genus for the Aeptini) and *Dymantis* (which probably = *Dymantis* + *Neodymantis* Kment and Rider in the present sense), which is now considered to be a subgenus of *Myrochea* (Kment and Rider 2015), the type genus for the Myrocheini. This may be an indication that these two tribes (Aeptini and Myrocheini) could be synonyms.

At present, this tribe contains 25 genera and 75 species (**Table 2.3**), all from the Old World. Kment (2015) and Kment and Garbelotto (2016) recently presented a diagnosis of this tribe in its broad sense. Members of this tribe are medium in size, brownish, and bear a striking resemblance to members of the Caystrini. Members of both tribes tend to be convex below, somewhat flat above, brown in color with some pale marks or dark brown streaks (**Figures 2.22G**, **2.30I**). The head is usually parabolic or sometimes triangular, with the lateral margins of the head and pronotum somewhat sharply edged, lamellate, but not reflexed; further, they both usually have the antennifer and the base of the first antennal segment hidden by the lateral margin of the head. The antennae are five-segmented. There are no prosternal lobes.

There are, however, three characters, none of which works at all times but, when used in concert, usually can separate myrocheines from caystrines. Myrocheines usually have: (1) the mesosternum distinctly sulcate, without a medial carina; (2) the ostiolar apparatus tends to be reduced (i.e., the ostiole is rather small, the rugae tend to be quite small and reduced, and the evaporative area is quite small); and (3) there often are small teeth or spines located on the inferior surface of the front femora. The female latero-tergites VIII are dorsally insinuated. The spermathecal bulb is relatively simple, ball-shaped, with 0-3 finger-like diverticula.

As inferred above, the distinction between the myrocheines and caystrines is not as distinct as one would hope. It is possible that the Myrocheini in its broadest sense is paraphyletic in relationship to the Caystrini. For example, the genus *Aednus* has the reduced ostiolar structures and the fore femora are armed with spines, but the mesosternum is carinate rather than sulcate. Gross (1975b) placed *Arniscus* Distant (**Figure 2.22F**) in his *Tholosanus* group but indicated that this group probably was related to the Myrocheini. *Arniscus*, in fact, has the mesosternum medially carinate and the fore femora are unarmed. As such, it is doubtful that *Arniscus* belongs in the Myrocheini. We also doubt the placement of the genus *Humria* Linnavuori in the Myrocheini; even though the mesosternum is medially sulcate, the ostiolar rugae are much more developed, and the fore femora are unarmed. Also, the placement of the newly described *Discimita* is rather tentative (see Kment and Garbelotto 2016). Finally, although the genus *Neococalus* has the front femora armed, the mesosternum is weakly keeled medially and the ostiolar rugae are quite elongate.

The genus *Phaeocoris* Jakovlev has had a confused taxonomic history. Horváth (1903) placed his own genus, *Timuria* Horváth, as a synonym of *Phaeocoris* and treated them both in the Sciocorini. Kirkaldy (1909) catalogued *Phaeocoris* in a broadly conceived Pentatomini, but he also catalogued *Timuria* as a valid genus in the Sciocorini. Kerzhner (1972) placed *Phaeocoris* (and both of its synonyms, *Timuria* and *Tancreisca* Jensen-Haarup) as a junior synonym of *Dymantis*, a myrocheine genus generally thought to be restricted to Africa. Ahmad and Önder (1996) described another new myrocheine genus, *Lodosia*, that has been placed as a synonym of *Phaeocoris* (Gapon and Baena 2005). Finally, Gapon and Baena (2005) after carefully studying the situation, stated "as temporary decision, we propose to retain *Phaeocoris* in Myrocheini, to which this genus is similar in external appearance."

Ahmad and Afzal (1989) reviewed the genera and species occurring in the Indo-Pakistan region, and Linnavuori (1975, 1982) covered the tropical African species. Kment and Rider (2015) corrected some nomenclatural problems involving *Dymantis* that resulted in the erection of a new genus, *Neodymantis*.

Linnavuori (1982) indicated that most species are grass feeders although members of *Humria* may be arboreal. He also indicated that this group might be related to the Aeptini.

2.2.10.7.30 Nealeriini Cachan, 1952

Stål (1876) described the genus *Aleria* and included it in his key to genera of the Halyini. Unfortunately, this generic name was preoccupied by *Aleria* Marshall, 1874, in the Hymenoptera. Bergroth (1893) proposed *Nealeria* as a replacement name but still treated this genus as a member of the Halyini. Additionally, Reuter in 1887 described another new genus, *Paraleria*, also including it in the Halyini. Both genera remained in the Halyini until Cachan (1952) transferred them into his new tribe Nealeriaria.

This tribe still contains only those two genera and only four species (**Table 2.3**), all endemic to Madagascar. They are relatively large pentatomids, elongate-oval, having the appearance of African halyines (**Figure 2.30J**). We have examined a single pair of one species. The juga are a little longer than the tylus. The mesosternum is narrowly and shallowly sulcate, not carinate medially; the proster-num and metasternum are shallowly concave. The sternal sulcus continues onto the abdomen for at least 3-4 segments. The ostiolar rugae are rather small and auriculate with a correspondingly small evapo-rative area. The anterolateral margins of the pronotum are usually concave and obtusely denticulate. The abdominal venter is covered with a relatively dense layer of short hairs over its entirety but differ from the Memmiini in that the hairs are less dense, and not as obvious; also, they are present in both sexes rather than present only in females (see **Section 2.2.10.7.27**). The antennae are four-segmented. The third rostral segment is the longest. The tarsi are two-segmented and the fore tibiae are somewhat dilated.

Some recent workers (Kment 2013) have included these genera in the Halyini, and they have a strong resemblance to various halyines. The two-segmented tarsi and the hirsute abdomen will separate them from that tribe. The hairy patches on the abdominal venter may ally this group with the Memmiini (and the few halyine genera with similar structures).

2.2.10.7.31 *Nezarini Atkinson, 1888*

Atkinson (1888), in forming his classification, relied heavily on the classification presented by Stål (1876), even though Stål did not call many of his family-groups by name. Atkinson (1888) provided names for most of Stål's groups. In this case, Stål (1876) keyed *Nezara* Amyot and Serville and *Acrosternum* together, indicating that they may form a higher group. Atkinson (1888) called this group the Nezaria. Yang (1962) proposed Palomenini based on *Palomena*, which we treat as a member of the Nezarini.

This is a relatively large group containing 26 genera and 272 species (**Table 2.3**) with its members occurring in both the New World and Old World. Members are small to medium large (usually greater than 9 mm) in length. This is another tribe that is difficult to define. Basically, these are the genera that are green in life, and they retain their green coloration after death (**Figures 2.30K, L**). This would seem to be a most inappropriate character on which to base a tribe, but, for the most part, it seems to work. The comparative lengths of the juga and tylus can vary, even within a genus. The humeral angles are usually rounded but occasionally spinose. The mesosternum is medially carinate (**Figure 2.7C**), becoming more elevated in some species of *Glaucias* Kirkaldy, and finally being quite elevated and produced forward onto the base of prosternum in *Amblybelus* Montrouzier and *Acrozangis* Breddin (indicating a possible relationship between the Nezarini and the Rhynchocorini). The ostiolar ruga can be short and auriculate (e.g., *Brachynema* Mulsant and Rey, *Nezara*) or rather long and acuminate apically (*Acrosternum*, *Chinavia* Orian) (**Figures 2.9B, D**). The base of the abdominal venter also can vary greatly from being unarmed to possessing an elongate, forwardly projecting spine (**Figure 2.8C**). The male parameres may be complex in shape, ranging from T-shaped to C-shaped. The female spermathecal bulb is ball-shaped with two to three long finger-like diverticula.

Acrozangis presents a special problem because not only is the mesosternum quite developed, the posterior surface of the mesosternum is excavated to receive the apex of the ventral abdominal spine. Its characters clearly place it in the Rhynchocorini, yet it appears to be closely related to *Glaucias*, whose characters place it in the Nezarini. As a sidenote, Fan et al. (2012) recorded *Cuspicona antica* Vollenhoven from China. We have examined the type specimens of *C. antica*, and their description and illustrations match the type fairly well. But, this species is definitely not a species of *Cuspicona* Dallas but, rather, is a species of *Acrozangis*.

It is evident that Gross (1976) did not have clear understanding of this group. He included nearly all nezarine genera in his *Pentatoma* group which would be the Pentatomini in this chapter. Beyond that, however, his descriptions and illustrations fit well with the Nezarini. He also discussed a possible relationship with the Rhynchocorini. There is also a possible relationship with the Antestiini. For example, we include the genus *Plautia* in the Antestiini. Although Gross discussed this as a possibility, he ultimately placed *Plautia* in the Nezarini. Linnavuori (1982) also indicated a resemblance between this group and the antestiines (including the male genitalia), but he indicated that the anterior margin of the pronotum is completely flat and densely punctate in the nezarines, whereas it often is set off by a suture or row of punctures in the antestiines.

A number of nezarine genera have been treated, either in part or in whole, in recent times as follows: *Acrosternum* (Ahmad and Rana 1989, Linnavuori and Al-Safadi 1993, Ribes and Pagola-Carte 2013), *Brachynema* (Zaidi and Ahmad 1988, Ribes and Schmitz 1992), *Chinavia* (Day 1965, Orian 1965, Rolston 1983a, Rider and Rolston 1986, Rider 1987, Frey-da-Silva and Grazia 2001, Schwertner and Grazia 2007), *Chlorochroa* Stål (Buxton et al. 1983, Thomas 1983, Scudder and Thomas 1987), *Glaucias* (Ruckes 1963), *Nezara* (Freeman 1940, 1946; Ferrari et al. 2010), *Palomena* (Zheng and Ling 1989, Zaidi and Ahmad 1991), and *Parachinavia* (Grazia and Schwertner 2008a).

This tribe includes several economically important species. For example, *Brachynema germarii* (Kolenati) attacks pistachio in the Middle East. Several species of *Acrosternum* (**Figure 2.22I**) and *Chinavia* (**Figure 2.30K**) are known to attack several crops. But, of course, the best known pest is *Nezara viridula* (**Figure 2.22K**), a species of cosmopolitan distribution that attacks a number of different crops (see **Chapter 7**).

2.2.10.7.32 Opsitomini Cachan, 1952

Opsitomaria was proposed by Cachan (1952) to contain his new genus *Opsitoma*. This tribe still contains just a single genus with one species (**Table 2.3**), endemic to Madagascar. It is a medium sized species, somewhat elongate rounded in shape (**Figure 2.31A**). It is brown in color, and has the head wider subapically than just in front of the eyes. The mesosternum is sulcate, with no carina medially. The ostiolar rugae are quite short and auriculate; the associated evaporative areas are also relatively small. The second antennal segment is extremely short, almost giving the appearance that the antennae are four-segmented. The juncture between the distal two tarsal segments is rather weak, again giving the impression that the tarsi are only two-segmented.

The sulcate mesosternum could ally this genus with the Aeptini, Myrocheini, or the Diploxyini. Further research is needed to see if this is a valid tribe, or whether it might fall into one of these tribes.

2.2.10.7.33 Pentamyrmecini Rider and Brailovsky, 2014

This recently described tribe contains a single genus and species, *Pentamyrmex spinosus* Rider and Brailovsky (**Figure 2.31B**) from Thailand (Rider and Brailovsky 2014, 2015) (**Table 2.3**). The general appearance of this species is so distinctive that its placement in a superfamily was initially difficult. Photos of this insect were sent to several notable heteropterists with the hope of determining the correct superfamily and then the family. The female genital plates and the five-segmented antennae place this creature in the Pentatomidae. As the scientific name implies, this species appears to be an ant mimic. We know of no other example of ant mimicry within the Pentatomoidea except possibly the genus *Caridophthalmus* (**Figures 2.12D, 2.21B**), currently placed in the Diemeniini.

The only known individual of this taxon is just over 6 mm in length, black with a few white spots. Members of this tribe have the head strongly declivent, almost vertical (**Figure 2.14E**). The ocelli are present, but are relatively small. The antennae are five-segmented, with the first segment not reaching the apex of the head. The bucculae are large, each obtusely lobed posteriorly. The pronotum has a deep transverse constricture dividing it into two equal halves; the lateral margins are not edged or reflexed but are evenly rounded. The lateral pronotal margins are provided with several large spines; each humeral angle is formed into a rather large spine, which, in turn, possesses several small spines (**Figure 2.31B**). The scutellum is elongate-triangular with an upwardly directed spine in each basal angle and a larger bifurcate spine arising from the apex (**Figures 2.14E, F**). The coria are slightly constricted medially each with a pale white spot nearby lending to the appearance of a narrow waste. There are also two strategically placed white waxy spots on the pleural regions (one on each side), also giving the appearance of constrictures (for possible ant mimicry). The clavi are quite long and slender, extending and meeting beyond apex of scutellum at a point, not really forming a claval commissure. The thoracic pleurites are nearly vertical; the ostiolar apparatus is reduced with the ostiolar rugae short and auriculate. The posterolateral connexival angles are not spinose except for those of segment V, each of which has a large upwardly directed spine (**Figure 2.14E**).

2.2.10.7.34 Pentatomini Leach, 1815

This is a large, widespread tribe with current members occurring in both the New World and Old World. It contains 56 genera and 316 species (**Table 2.3**). Again, as with many of the larger tribes, included members are quite variable. In fact, this may be one of the more poorly defined tribes within the family. Basically, members of this tribe are medium to large in size, often brown or mottled brown and white (**Figures 2.23A-H; 2.31C-F, H**) and usually have the base of the abdomen produced into an anteriorly directed spine or tubercle. It appears that this last character has arisen multiple times within the family and secondarily has been lost several times, making it quite difficult to determine its utility. This character is not even consistent within the nominate genus *Pentatoma*; most species have an armed abdominal venter, but there are several species in which the base of the abdomen is unarmed. Additionally, in some of the genera, the abdominal spine or tubercle is free apically, reaching between the hind coxae or further (Rolston et al. 1980, Pentatomini, section two); in several other genera, however, the apex of the abdominal spine is not free, but is met by the posterior margin of an elevated metasternum that may or may not be notched (Rolston et al. 1980, Pentatomini, section three). At present, taxa having either of the above two conditions are included within the Pentatomini, but this may in the future serve as way to divide this group into two better defined tribes (see later discussion under *Banasa* in the 'Other Unplaced Genera' section, **Section 2.2.10.7.46**).

Typically, once those genera showing characteristics of other tribes (e.g., some Catacanthini, Menidini, some Nezarini, Piezodorini, some Procleticini) have been removed, the remaining unplaced genera that possess an armed abdominal venter have been placed in the Pentatomini. This is not totally satisfactory. It is difficult at this time to better place these genera, and so we have chosen to continue to treat them here until a more thorough phylogenetic analysis can be completed.

This tribe can be further characterized by several general characters, most of which have exceptions (Tsai and Rédei 2014). The antennae are usually five-segmented. The mesosternum is usually carinate. Interestingly, we generally define the Pentatomini as having elongate ostiolar rugae with the apices acute to acuminate and attached to the pleura, but, in fact, this character is variable even within some genera. For example, some members of *Pentatoma* and *Priassus* have an ostiolar ruga that is somewhat shorter; even when the rugae are longer, they are not acuminate apically, and the apex appears to be detached from the pleural surface (**Figure 2.9F**). The tarsi are three-segmented (except *Phalaecus*).

Because of the lack of clear-cut diagnostic characters, it is impossible to provide a universal definition of the Pentatomini. Several genera traditionally placed into other tribes (first of all Carpocorini and Rolstoniellini) are potentially related to Pentatomini (see further discussion under the above mentioned tribes).

There is a small group of New World genera (*Adevoplitus* Grazia and Becker, *Evoplitus* Amyot and Serville [**Figure 2.31E**], *Pseudevoplitus* Ruckes, and an old genus in need of validation, *Platencha* Spinola [**Figure 2.13A**]) that have the sternal structure strikingly similar to the Old World tribe Rhynchocorini. The abdominal venter is armed at the base with an anteriorly directed spine, which, in turn, is received by the posteriorly notched metasternum; the metasternum then joins with the elevated mesosternal carina, which extends anteriorly onto the prosternum to near the base of the head (**Figure 2.7H**) (Grazia 1997). These species, however, do not seem to belong to the Rhynchocorini, mainly because they are brown and white mottled and more densely punctate. There is also a possibility that *Nocheta* Rolston, *Paratibilis* Ruckes, *Phaleacus*, *Tibilis* Stål, and related genera may belong to this group - the sternal structure is essentially the same except that the mesosternal carina is usually shorter, just reaching onto the posterior margin of the prosternum. These genera may be a monophyletic group warranting the erection of a new tribe for their placement. *Phalaecus* differs further from other pentatomines by having short flap-like bucculae (similar to that seen in other groups, e.g., Edessinae), and the tarsi are two-segmented (Bergroth 1910).

There is another small group of Old World genera (currently placed in several different tribes) that may form a monophyletic group. They have a relatively short abdominal tubercle, and the sternal characters are fairly typical of more advanced pentatomoids. The ostiolar rugae, however, are rather short, dissimilar to *Pentatoma*. Also, these taxa have a fairly distinctive head shape; it is relatively flat with the lateral margins of the juga subparallel in front of the eyes, but, more distally, they are rather sharply angled near the apex. Included genera are *Parvacrena* (currently in the Pentatomini) and *Udonga* (currently placed in the Menidini). See notes under the Menidini.

There are several other genera presently placed in the Pentatomini that have some characters that could exclude them from their placement here. For example, members of the genera *Disderia* Bergroth (Central America) and *Iphiarusa* (Indian and Oriental regions) have the abdominal spine quite long, reaching to or beyond the front coxae; it is flat ventrally, somewhat reminiscent of the Piezodorini. There are several genera in which the sternal structure is similar to the pentatomines, but the ostiolar rugae are much shorter or auriculate. These include, but are not limited to, *Elsiella* Froeschner, *Lelia* Walker (**Figure 2.31F**), *Marghita* Ruckes, *Myota* Spinola (**Figure 2.23C**), *Placosternum* Amyot and Serville, *Rhaphigaster* Laporte (**Figure 2.31H**), *Stictochilus* Bergroth (**Figure 2.23H**), and *Taurocerus* Amyot and Serville. Some of these genera (*Marghita* and *Stictochilus*) are quite similar in general appearance to species of the *Euschistus* group of the carpocorines and may more properly belong there. Members of the genus *Placosternum* have a somewhat different thoracic sternal structure, somewhat more reminiscent of the *Evoplitus* group discussed above.

The genus *Janeirona* Distant has a distinct 'gestalt', being more long and slender and parallel-sided. The bucculae are shorter and more flap-like than what is seen in most pentatomines, perhaps the reason Distant (1911) originally placed this genus in the Tessaratomidae. The sternal structure, although strictly speaking, fits into the pentatomine definition, simply looks different. In many ways, *Janeirona* is similar in appearance to some members of the discocephaline tribe Ochlerini. Kumar (1969a) stated "*Janeirona* is transferred to Pentatomidae and probably deserves a tribal status in Pentatominae close to *Menecles* (**Figure 2.29B**) and *Euschistus* (**Figure 2.20A**), which it resembles in possessing a long filiform tail." The genera *Pharypia* and *Ramosiana* Kormilev are relatively large and colorful New World genera, marked with various reds, yellows, and other colors. The species of *Pharypia* (**Figure 2.31G**) especially have a strong resemblance to species of *Arocera* (**Figure 2.29D**), but they have traditionally been separated from *Arocera* et affinia because they have an abdominal spine. Now that *Arocera* and related genera have tentatively been placed within the Catacanthini, perhaps *Pharypia* should be considered similarly. There are members of the Catacanthini that have abdominal spines.

The genus *Sabaeus* Stål is also somewhat enigmatic. Members of this genus are green and usually remain green after death. They strongly resemble members of the Rhynchocorini, but the rhynchocorines nearly always fade to yellow after death, and *Sabaeus* lacks the characteristic sternal structure of the rhynchocorines. An argument could be made to place them in the Nezarini except that their general shape is different from nezarines. There are currently other members of the Pentatomini that are green in color (e.g., the New World genus *Banasa* [**Figure 2.31D**]) but, again, their placement in this tribe is only tentative. Finally, the dorsal habitus and coloration of *Amirantea* Distant is more reminiscent of members of the tribe Antestiini than those in the Pentatomini.

Relatively recent revisionary treatments exist for the following pentatomine genera: *Adevoplitus* (Grazia and Becker 1997), *Amblycara* (Tsai and Rédei 2014), *Bifurcipentatoma* Fan and Liu (Fan and Liu 2012), *Lelia* (Fan and Liu 2010b), *Marghita* Ruckes (Grazia and Koehler 1983), *Neotibilis* Grazia and Barcellos (Grazia and Barcellos 1994, Bernardes et al. 2006), *Phalaecus* (Grazia 1983), *Pseudevoplitus* (Thomas 1980; Grazia et al. 1995, 2002, 2016b), *Ramivena* Fan and Liu (Fan and Liu 2010a), *Serdia* Stål (Becker 1967b, Thomas and Rolston 1985, Fortes and Grazia 2005), *Stictochilus* (Rolston and Rider 1986), *Taurocerus* (Grazia and Barcellos 2005), and *Tibilis* (Barcellos and Grazia 1993).

Members of the genera *Pentatoma* (**Figure 2.31C**) and *Rhaphigaster* (**Figure 2.31H**) are known to be aboreal, living on various leafy trees.

2.2.10.7.35 Phricodini Cachan, 1952

This tribe contains a single genus (*Phricodus*) and seven species (**Table 2.3**), all of which are quite different from other known pentatomine species. This genus has been moved in and out of the Halyini several times (Stichel 1961, Memon et al. 2011). Cachan (1952) was the first to place it in its own tribe. More recently, Linnavuori (1982) stated that *Phricodus* differed drastically from members of the Halyini (body structure, genitalia), and finally concluded that "the genus had an isolated position within the African Pentatominae." Memon et al. (2011) included *Phricodus* in their phylogenetic analysis of the South Asian halyine genera; they concluded that *Phricodus* should be a member of the Halyini, but their methodology has been questioned (Barão et al. 2012).

These species are relatively small in size, pale to dark brown, and profusely spinose (**Figure 2.31I**). The apices of the juga are spinose, the antennifers are spinose, there is a large anteocular spine just in front of the compound eyes, and the lateral margins of the pronotum have five to six long spines, some of which bifurcate. The antennae are four-segmented; segment II is enlarged distally, and the apical two segments are much enlarged and spindle-shaped. The dorsal surface of the body has numerous erect bristles (best seen in lateral view). The prosternum and mesosternum are shallowly sulcate with the mesosternum having a weak medial carina. The ostiolar rugae are medium in length, spout-like, and detached apically; and the associated evaporative area is relatively large and sharply delimited. The tarsi are three-segmented with segment II rather small. The scutellum is triangular and elongate, and the scutellar tongue is sharply acuminate apically. The male pygophore has a pair of long hinged processes directed caudad; the parameres are quite small. The spermathecal bulb is simple, ball-shaped, and lacks projections.

The genus was revised recently by Göllner-Scheiding (1999).

2.2.10.7.36 Piezodorini Atkinson, 1888

Stål (1876) keyed *Piezodorus* out by itself, vaguely indicating that it might deserve a higher grouping. Atkinson (1888) formally named this group the Piezodoraria, and included within it only *Piezodorus* and *Ambiorix*. *Ambiorix* is now considered to be a member of the Diplostirini.

This tribe at present contains five genera and 21 species (**Table 2.3**). Members of this tribe are small to medium in size (9-12 mm), usually pale yellowish in color (may be more greenish when alive) with concolorous punctures [an exception is *Piezodorus lituratus* (F.), which has darker punctures] (**Figures 2.23K, 2.31J**). The humeral angles are always rounded. The mesosternum is carinate medially with the anterior portion produced anteriorly onto the prosternum as a small rounded lobe (**Figure 2.7D**). The base of the abdomen is armed with a large, robust, usually ventrally flattened spine that often reaches to or beyond the middle coxae. The ostiolar rugae are elongate, apically acuminate, and curved cephalad. The posterior margin of the pygophore is simple and vertical, and the pygophoral opening is distinctly dorsad. According to Gross (1976), the aedeagus is "one of the most complicated of any seen in the Pentatomidae." The phallosoma is small, a thecal shield may or may not be present, the conjunctiva is produced into long tubular lobes, and the vesica may be long or short. There is a tendency for the basal plates of the female genitalia to have their anterior margins hidden under the last non-genital abdominal sternite. The spermathecal bulb is constricted in the middle, and the apical portion is simple or provided with a single diverticulum (Linnavuori 1982).

Gross (1976) indicated that there were some superficial similarities with members of the Menidini. On the other hand, Linnavuori (1982) stated that the Piezodorini was possibly related to the Nezarini. There also seem to be some general similarities in appearance with members of the Eurysaspini.

Recent reviews, either in whole or in part, are available for *Pausias* Jakovlev (Ahmad 1994) and *Piezodorus* (Staddon and Ahmad 1995).

Some species of *Piezodorus* are considered to be pests of various crops, especially legumes (Panizzi et al. 2000). For example, the tropical pest *Piezodorus guildinii* (Westwood) appears to have moved northwards in recent years, becoming a problem species on soybean in the southern United States (see **Chapter 8**).

2.2.10.7.37 Procleticini Pennington, 1920

Pennington (1920) was the first to recognize this group at the family-group level when he proposed the Procleticini for two monotypic genera, *Procleticus* Berg and *Lobepomis* Berg. Kormilev (1955) added the genus *Neoderoploa* Pennington, and Pirán (1963) added his new genus *Terania*. No further mention of this tribe was made until Rider (1994) provided a generic conspectus of the tribe and broadened the definition to include another seven genera: *Aleixus* McDonald, *Brepholoxa* Van Duzee, *Dendrocoris* Bergroth (**Figure 2.23I**), *Odmalea* Bergroth (**Figure 2.23J**), *Parodmalea* Rider, *Thoreyella* Spinola (**Figure 2.31K**), and *Zorcadium* Bergroth.

This tribe contains 11 genera and 36 species (**Table 2.3**). Members of this tribe are small to medium in size and usually ovate in shape (**Figures 2.23I, J; 2.31K**). This New World group can be separated into two groups based on the shape of the scutellum. In some genera (*Brepholoxa, Dendrocoris, Odmalea, Parodmalea*), the scutellum is more or less triangular in shape, whereas in other genera (*Lobepomis, Neoderoploa, Procleticus, Terania*), the scutellum is enlarged and spatulate in shape. The juga are usually longer than the tylus [except in some *Odmalea concolor* (Walker)] and often contiguous in front of the tylus. The first antennal segment does not reach the apex of the head; the bucculae are lobed posteriorly. The ostiolar rugae usually reach the middle of the metapleura and are usually acuminate apically (except *Brepholoxa*). The thoracic sterna are shallowly to deeply sulcate, and the mesosternum is not carinate medially. The abdominal venter is armed with a forwardly projecting spine or tubercle (except it is quite small in *Lobepomis* and *Procleticus* and lacking in *Parodmalea* and males of some *Dendrocoris*). The genital plates are small and, as a group, they appear to be recessed into the venter; the basal plates are quite small and are often partially or completely obscured by the last abdominal sternite (**Figure 2.8J**). The pygophore is somewhat produced posteriorly with a distinct medial emargination that may be narrow and parallel-sided (**Figures 2.8K, L**) or, often, becoming circular ventrally. Gapon (2005) studied the male genitalia of procleticines and discussed their relationship with other tribes.

Rider (1994) provided a generic conspectus of this tribe, defined the unique apomorphies that support the monophyly of this group, and established a current classification. Schwertner and Grazia (2012) corroborated the monophyly and proposed a phylogenetic classification. Several procleticine genera recently have been revised, at least in part: *Dendrocoris* (Nelson 1955, 1957; Thomas 1984; Thomas and Brailovsky 2000), *Odmalea* (Rolston 1978a), and *Thoreyella* (Rolston 1984c; Bernardes et al. 2009, 2011). A revision of *Brepholoxa* is nearly complete (David A. Rider, unpublished data).

2.2.10.7.38 Rhynchocorini Stål, 1871

Stål (1871) proposed the family-level group Rhynchocorina and provided a key to the nine genera he included within it. The nine genera were *Rhynchocoris* Westwood, *Hoffmanseggiella* Spinola, *Morna* Stål, *Pugione* Stål, *Pegala* Stål, *Vitellus* Stål, *Cuspicona*, *Ocirrhoe* Stål, and *Periboea* Stål (= *Diaphyta* Bergroth). This is a fairly well-characterized tribe; as such, it has undergone relatively few changes, other than the addition of several more genera, since Stål originally established it.

This is a fairly large tribe (18 genera and 106 species; **Table 2.3**) occurring in the Australian, New Guinean, and Oceanic regions, and marginally reaching to Southeast Asia and India. Members of this tribe range from small to large in size, most are green in color when alive (fading to yellow after death), and they often have black or reddish markings, especially along the margins of the head and pronotum (**Figures 2.23L; 2.24A, B; 2.31L**). There is a tendency for the punctures to be less dense, and as such, the body surface is somewhat shiny. The most unifying characteristic is the structure of the thoracic sterna and the abdominal venter. The base of the abdomen is armed with a relatively stout spine or tubercle that is usually rounded apically. The spine usually reaches to the hind coxae where it is received by an elevated metasternum. The caudal margin of the metasternum is either excavated or bifurcated to receive the abdominal spine; anteriorly the metasternum joins with the elevated mesosternal ridge, which is also quite elevated and produced anteriorly onto the prosternum (**Figure 2.7E**), sometimes greatly so, and, in some cases, even reaches onto the base of the head. The ostiolar rugae are usually quite elongate, reaching over half the distance to the metapleural margins. Many genera also have the humeral angles produced into conical productions or spines. The male parameres are somewhat F-shaped. The female spermathecal bulb is ball-shaped with three long, finger-like diverticula.

There are two rhynchocorine genera that possess a charater unique within the Pentatomidae. In all known pentatomids, the front wings, when at rest, lie adjacent to the lateral margins of the scutellum with the distal part of the corium lying below the apex of the scutellum (i.e., scutellar tongue). In members of the genera *Petalaspis* Bergroth and *Vitellus*, the front wings, at rest, lie in a similar position, but the distal part of the corium now lies on top of the scutellar apex. The lateral margins of the scutellar tongue have now become flattened and recessed below the coria. Gross (1975a) described this as a quadrate membranous plate around and beneath the coria; this gives the scutellar apex a decidedly acute appearance.

Gross (1975b), based on external features, indicated that the rhynchocorines are similar to members of the Catacanthini, Menidini, Nezarini, and Piezodorini but, eventually, dismissed these possible relationships base on the male genitalia. Superficially, the rhynchocorines look similar to members of the Edessinae, but their thoracic structure is quite different. They are, however, structurally quite similar to a group of South American genera tentatively placed in the Pentatomini (called the *Evoplitus* group in this chapter, see **Section 2.2.10.7.34**). The thoracic structure is extremely similar, but the coloration is quite different (the South American genera are usually mottled with brown and fuscus, never greenish or yellowish). Gross (1975a) revised some of the Australian members of this tribe (e.g., *Cuspicona* and *Ocirrhoe*). Other groups that have received recent attention include *Avicenna* Distant (Gross and McDonald 1994), *Diaphyta* (McDonald 2001), and *Vitellus* (Gross and McDonald 1994).

Biprorulus bibax Breddin is considered to be a major pest on citrus in Australia (James 1989, 1992, 1993, and numerous others).

2.2.10.7.39 Rolstoniellini Rider, 1997

Distant (1902) first proposed this group as the Compastaria, based on the preoccupied generic name *Compastes* Stål. Family-group names based on a preoccupied generic name are not allowed, so Rider (1997) proposed *Rolstoniellus* as a replacement name for *Compastes* with the resultant tribal name Rolstoniellini. This tribe contains six genera (*Agathocles*, *Amasenus* Stål, *Critheus* Stål, *Exithemus*, *Nesocoris* Bergroth, and *Rolstoniellus* [**Figure 2.32A**]) and 19 species (**Table 2.3**) distributed in the Oriental region except for *Nesocoris*, which occurs in New Caledonia. Two genera, *Agathocles* and *Exithemus* will be transferred to the Caystrini based on their morphology (Salini Shivaprakash and Petr Kment, unpublished data). Another genus originally placed in this tribe, *Homalogonia* Jakovlev, was later transferred to the Cappaeini. On the other hand, the Oriental genera *Prionocompastes* Breddin and *Zhengius* Rider (= the preoccupied *Tibetocoris* Zheng and Liu), are obviously closely related to *Rolstoniellus* as well.

Little has been written about this tribe since its inception. Distant (1902) defined the group as "possessing a greater breadth of body with the head broader at the apex and the scutellum usually less acuminate. The species are obscurely coloured." He also mentioned that the base of the abdomen is unarmed.

We have examined specimens of *Agathocles*, *Amasenus*, *Critheus*, *Exithemus*, *Prionocompastes*, *Rolstoniellus*, and *Zhengius*. Members of *Critheus* and *Exithemus* are elongate oval and punctate but without projections or protuberances. *Critheus* has a submarginal line of punctures paralleling the margins of the head and pronotum that is very similar to that seen in *Axiagastus* (Axiagastini). Members of *Amasenus* are larger with the humeri prominent and truncately produced, the scutellum has a pair of large 'bumps' basally, and there are also small teeth along the anterolateral pronotal margins. All three genera have a similar structure of the thoracic sterna and ostioles. The mesosternum is carinate, especially anteriorly. The ostiolar rugae are elongate, curving anteriorly, and then becoming rather thin, forming a narrow line-like ridge in the apical third. The species of *Amasenus* and *Critheus* have the rostrum quite elongate, reaching well onto the abdomen; in fact, in one species of *Critheus*, the rostrum reaches the base of the genital segment. In some specimens of *Exithemus* we examined, the rostrum reaches to the hind coxae. In all specimens examined, the spiracles appeared unusually large; the pair on abdominal segment II was at least partially exposed, and, in *Amasenus*, there was, at least, a pair of small spiracles on abdominal segment VIII of the females. *Prionocompastes*, *Rolstoniellus*, and *Zhengius* (the first and last currently in the Pentatomini) are characterized by serrate anterolateral margins of the pronotum and broadly expanded, apically denticulate or lobate humeri (**Figure 2.32A**); the strongly reduced ruga of the metathoracic scent gland ostiole; and two-segmented tarsi. They obviously form a complex of closely related species, and their generic level differences (particularly those of *Rolstoniellus* and *Zhengius*) are unclear. In spite of its 3-segmented tarsi, *Liicoris* Zheng and Liu (see **Section 2.2.10.7.46**) is probably related to these genera.

The tribe is morphologically very heterogenous and likely non-monophyletic. On the other hand, there is little to distinguish its type genus (*Rolstoniellus*) and related genera from certain members of the Pentatomini; therefore, it is probable that this tribe eventually will be subsumed into the Pentatomini.

The genus *Critheus* has been revised (Ghauri 1963). Two species of *Critheus*, *C. indicus* (Distant) and *C. lineatifrons* Stål, have been recorded feeding on, and sometimes damaging, bamboo (Hoffmann 1931, 1932; Cheo 1935; Zheng 1986, 1994; Zhang et al. 1995).

2.2.10.7.40 Sciocorini Amyot and Serville, 1843

Amyot and Serville (1843) proposed two family-level names that are now associated with this tribe; Pododides was based on their new genus *Pododus* Amyot and Serville, and Sciocorides was based on the genus *Sciocoris* Fallén. In 1866, Mulsant and Rey proposed the family-level name Oploscelates, based on their new genus *Oploscelis*, which now is considered to be a subgenus of *Menaccarus* Amyot and Serville. Stål (1876) essentially synonymized all three of the above tribes when he included *Pododus*, *Sciocoris*, and *Menaccarus* in his "*Sciocoris* et affinia" group.

This tribe contains 16 genera and subgenera and about 126 species (**Table 2.3**) (most of them in the genus *Sciocoris*) and occurs worldwide, although the taxa are most diverse in the Palearctic Region. Members of this tribe tend to be small to medium in size, usually somewhat flattish (except several New World species are more robust) (**Figures 2.24C, D; 2.32B**). The lateral margins of the head and especially the pronotum are broadly explanate, shelf-like. The ocelli are somewhat remote from the relatively small compound eyes, and the antenniferous tubercles are not visible from above. The scutellum is somewhat enlarged, spatulate. The prosternum and mesosternum are shallowly to deeply sulcate, not carinate medially (except a faint carina is evident in *Trincavellius* Distant [**Figure 2.24D**]), and the metasternum is flattish to slightly concave. The ostiolar rugae are usually short and auriculate (a little longer in the Australian genus *Kapunda* Distant) with the associated evaporative areas also quite small. There are at least two New World species (*S. crassus* Ruckes and *S. longifrons* Barber) presently placed in *Sciocoris* that have the inner mesial margins of the propleura produced, forming a ridge on either side of the rostral groove.

The spermathecal bulb is usually small, ball-shaped, simple, and lacks diverticula; the distal and proximal flanges tend to be quite small. Further studies are needed to determine if the Australian genera *Kapunda* (longer ostiolar ruga) and *Adelaidena* Distant (not examined by us), and the South American *Trincavellius* (less sulcate mesosternum with faint medial carina) actually belong in the Sciocorini; also, the two New World species with expanded propleura may deserve a new generic name. Additionally, recent work by Gapon and Baena (2005) seems to indicate that based on the male and female genitalia, *Dyroderes* Spinola (**Figure 2.24C**) may not belong in the Sciocorini.

Sciocorine species seem to be well-adapted to more arid regions. The Indian species of *Sciocoris* were recently treated (Azim and Shafee 1987); the Palearctic species have also been reviewed (Péricart 2002, Derjanschi and Péricart 2005).

2.2.10.7.41 *Sephelini Breddin, 1904*

Stål (1876) first recognized this group at the family-level when he provided a key for the "*Sephela, Brachymna* et affinia" group; this group contained *Sephela* Amyot and Serville, *Brachymna* Stål, *Ochrophara, Drinostia* Stål, and *Aenaria*. Breddin (1904) was the first to give this group a formal family-group name, the Sephelaria.

This tribe contains 10 genera and 34 species (**Table 2.3**). Members of this tribe are elongate, slender, pale brown species occurring in the African, Indian, and Oriental regions (**Figures 2.13C, 2.32C**). The body is somewhat depressed, the antennifers are rather small, and the first antennal segment does not reach the apex of the head. The juga are usually longer than and meet in front of the tylus (except in *Niphe* Stål, they are subequal in length); the head is usually parabolic in shape, except it is distinctly triangular in *Brachymna*. Rostral segment II is usually shorter than segments III and IV combined. The pronotum is flattish with the anterolateral margins lamellate. The scutellum is long and slender, narrowly triangular. The mesosternum is sulcate (except in *Niphe* in which a medial carina is present). The ostiolar rugae are medium in length, extending just beyond the middle of the metapleura, and usually curving cephalad; the evaporative areas are relatively large. The male pygophore usually has a deep medial emargination; the parameres are small and weakly sclerified. The female spermathecal bulb is somewhat elongate, forming a ball distally with a long fingerlike diverticulum arising from the top of the ball.

According to Linnavuori (1982), the African species feed on grasses in savanna habitats. Two sepheline genera have been studied recently: *Aenaria* (Fan and Liu 2009) and *Niphe* (Zaidi and Ahmad 1990).

2.2.10.7.42 *Strachiini Mulsant and Rey, 1866*

Mulsant and Rey (1866) proposed the Strachiaires for two genera, *Nitilia* Mulsant and Rey and *Strachia*. *Nitilia* now is considered to be a subgenus of *Bagrada* Stål, and Mulsant and Rey treated *Strachia* essentially the same as the current genus *Eurydema*; that is, all of the species they treated in their work are now placed in *Eurydema*. The Eurydemaria was proposed by Distant (1902) and, although Oshanin (1906) synonymized the two, Eurydemini continued to be used by many workers, even into the 1990s. Oshanin (1906) also proposed the Trochiscocoraria for the genus *Trochiscocoris* Reuter, but this genus now is considered to be a member of the Strachiini (see comments below).

This tribe contains 20 genera and 142 species (**Table 2.3**). Members of this tribe are small to medium in size (3-12 mm), oval to elongate oval in shape, and are often quite colorful with reds, oranges, and yellows mixed with the black patterns (**Figures 2.12H; 2.32D-G**). This tribe has members in all major regions of the world. There seems to be a general reduction in punctation, and the dorsum and especially the venter are shiny and nearly impunctate (not so much in *Eurydema*). The single most useful character for recognizing this tribe is the reduction of the ostiole and associated external structures of the scent glands. The opening has become extremely small and is situated more mesad between the middle and hind coxae; the ostiolar rugae are extremely short and merge into the surrounding pleuron; each, in turn, is flanked anteriorly and posteriorly by an extremely reduced V-shaped evaporative area (**Figure 2.9A**). A reduced ostiolar structure may appear in other tribes but usually not to the extent seen in the Strachiini; an exception is found in *Otantestia*, but this genus in all other structures and colors aligns itself with the Antestiini. The head is subtriangular, with the width usually greater than the length; the eyes are prominent; and the lateral margins of the head are narrowly reflexed (the anterolateral margins of the pronotum are also usually narrowly reflexed). In some genera (e.g., *Eurydema* [**Figure 2.32E**], *Madates*

Strand [**Figure 2.12H**]), the anterior pronotal margin is set off by a row of punctures or a suture, forming a collar behind the head. The scutellum is triangular in shape. The prosternum is shallowly sulcate, the mesosternum is distinctly carinate. The abdominal venter is unarmed. The posterior margin of the male pygophore can be rather complex; the male parameres may be fairly simple to somewhat Y-shaped; the aedeagus is equipped with a thecal shield. The female spermathecal bulb is elongate, simple, or provided with a finger-like diverticulum.

The genus *Strachia* may have a rudimentary stridulatory mechanism. There are numerous peg-like tubercles on the inner surface of the hind femora that are scattered randomly. The abdominal venter has the margins between segments slightly and obtusely elevated, and there is often a secondary transverse ridge in the middle of each segment. This is only seen in the males; the females have the abdominal ridges much reduced, and the inner surface of the hind femur is nearly smooth, not tuberculate as it is in the males. This is not seen in other members of the tribe.

There is another aberrant genus whose taxonomic placement needs further study. The species in the genus *Trochiscocoris* are quite small and usually brachypterous; they have the external scent efferent system even more reduced than in most strachiines; that is, the ostiole is obsolete. Although the general 'gestalt' of these species is quite different, the genitalia of both sexes are highly similar to other members of the Strachiini. Oshanin (1906) erected the Trochiscocoraria for this genus, but Asanova and Kerzhner (1969) placed *Trochiscocoris* back in the Strachiini.

The West Palearctic (Derjanschi and Péricart 2005) and Indian (Azim and Shafee 1986) species of this tribe were reviewed, and two strachiine genera have been recently revised: *Madates* (Rider 2006d) (**Figure 2.12H**) and *Murgantia* (Brailovsky and Barrera 1989) (**Figure 2.32F**). Also, the Oriental and Australian species of *Stenozygum* have been reviewed (Ahmad and Khan 1983).

Gross (1976) compared the strachiines with some asopines; for example, *Oechalia* (**Figure 2.17C**) has reduced scent gland structures, and the asopines possess a thecal shield in the male aedeagus (but other groups also sometimes have a thecal shield). He also discussed a possible relationship with the Antestiini, again based on male genitalia.

This tribe includes several economically important species. For example, various species of *Eurydema* are pests of cole crops in the Old World. *Murgantia histrionica* (Hahn) also attacks cruciferous plants in the New World (see **Chapter 6**). Additionally, *Bagrada hilaris* (Burmeister) was introduced accidentally into western North America recently and is considered a pest of several plant species in the Brassicaceae (see **Chapter 3**). It now also has been found in Chile, South America, and is causing extensive damage to crops there (Faúndez et al. 2016).

2.2.10.7.43 Triplatygini Cachan, 1952

Cachan (1952) proposed the family-level name Triplatyxaria for three genera endemic to Madagascar: *Anoano* Cachan, *Tricompastes* Cachan, and *Triplatyx* Horváth (**Figure 2.32H**). This tribe received little attention until Kment (2008, 2012, 2015) and Kment and Baena (2015) treated all included genera and described one new genus (see comments below).

Most of the information included here comes from papers by Kment (2008, 2012, 2015) and Kment and Baena (2015). There are currently four genera (*Anoano, Nene* Kment, *Tricompastes*, and *Triplatyx*) and nine species assigned to this tribe (**Table 2.3**). All included taxa are endemic to Madagascar. Members of this tribe have large juga that are somewhat foliaceous, the lateral margins are edged, and their apices reach beyond the apex of the tylus and either meet anteriorly or leave a small V-shaped notch. Usually there is a small anteocular spine along the jugal margin on each side. The antennae are five-segmented; antennal segment I is nearly covered by the head margins, and, therefore, only its apex is visible from above. The humeral angles are lobe-like, elongate, and usually incised apically. The scutellum is usually triangular and, often, broadly rounded apically. The mesosternum is longitudinally sulcate (except *Nene*, which has the mesosternum carinate). The ostiolar rugae are short and auriculate, and the evaporative areas also are reduced. In *Triplatyx*, the female spermathecal bulb is somewhat spherical and provided with four to five randomly directed, bifurcating diverticula (Kment 2008); in *Nene*, there are only three simple diverticula (Kment 2015); in *Tricompastes*, there are only two diverticula (Kment and Baena 2015); and in *Anoano*, there is only a single diverticulum positioned apically (Kment 2012).

Practically no biological information is available, except for one report of *Triplatyx bilobatus* Cachan being beaten from the lower parts of tree branches (Kment 2008). The known distribution of Triplatygini species correspond to forest areas of Madagascar.

2.2.10.7.44 The Generic Groups of Gross (1975-1976)

Gross (1975b, 1976), in two volumes monographed the pentatomoid fauna of South Australia. He devoted considerable space towards the higher classification of the Pentatomoidea, and his thoughts have been extremely influential in the development of this chapter. He realized, however, that building a rigorous classification based almost solely on Australian material would be impossible, and he did not want to clutter the literature with new names that might become synonyms. Even though he recognized some of the more common tribes from other geographical areas, he still placed all of the Australian genera into generic groups, even if tribal names for some groups were available. Many of these groups may end up representing monophyletic groups and deserve tribal status; others undoubtedly will not.

Gross' classification in many ways is similar to what is currently used, except that he treated several subfamilies (Asopinae, Podopinae, and Phyllocephalinae) as generic groups, equivalent in status as the other generic groups (essentially lowering them to tribal level). The following discussion includes brief notes on those groups not already covered in the discussions of currently known tribes (**Table 2.5**).

In truth, Gross (1976) presented a non-linear classification stating that "the picture which has emerged differs from that originally conceived in that our present day groups seem to form more of a radiative pattern from a centre located around *Asopus*, *Halys* and *Poecilotoma* groups." He further stated that "as in any treatment of a two dimensional concept of groups, the groups in the text must appear in a linear

TABLE 2.5

Gross and Linnavuori Classifications

The Classification of Gross (1975, 1976)	The Classification of Linnavuori (1982)
Ochisme group	Mecideini - see earlier discussion
Tarisa group - see discussion under Podopinae	*Phricodus* group - see discussion under Phricodini
Kitsonia group	Halyini - see earlier discussion
Kumbutha group	Sciocorini - see earlier discussion
Macrocarenus group	Aeptini - see earlier discussion
Menestheus group	Myrocheini - see earlier discussion
Ippatha group	*Caystrus* group - see discussion under Caystrini
Dictyotus group	*Sephela* group - see discussion under Sephelini
Tholosanus group	*Tyoma* group
Asopus group - see discussion under Asopinae	*Aeschrus* group - see discussion under Aeschrocorini
Poecilotoma group	*Diploxys* group - see discussion under Diploxyini
Halys group - see discussion under Halyini	*Acoloba* group
Carpocoris group - see discussion under Carpocorini	*Aeliomorpha* group
Kapunda group	Eysarcorini - see earlier discussion
Diemenia group - see discussion under Diemeniini	*Carbula* group
Cephaloplatus group	*Agonoscelis* group - see discussion under Agonoscelidini
Phyllocephala group - see discussion under Phyllocephalinae	*Veterna* group
Mycoolona group	*Halyomorpha* group
Eysarcoris group - see discussion under Eysarcorini	*Banya* group
Strachia group - see discussion under Strachiini	*Eipeliella* group
Antestia group - see discussion under Antestiini	*Antestia* group - see discussion under Antestiini
Menida group - see discussion under Menidini	*Nezara* group - see discussion under Nezarini
Piezodorus group - see discussion under Piezodorini	*Bathycoelia* group - see discussion under Bathycoeliini
Pentatoma group - see discussions under Nezarini and Pentatomini	*Piezodorus* group - see discussion under Piezodorini
	Eurysaspis group - see discussion under Eurysaspini
	Kelea group
Rhynchocoris group – see discussion under Rhynchocorini	*Menida* group - see discussion under Menidini
	Eurydemini - see discussion under Strachiini

sequence …". Given that, we have discussed his generic groups in the same order that he presented them, still having a general trend from more primitive groups to the more advanced groups.

2.2.10.7.44.1 Ochisme Group   Gross (1975b) segregated four genera (*Amphidexius* Bergroth [**Figure 2.24E**], *Dysnoetus* Bergroth, *Ochisme* Kirkaldy [= *Trachyops* Dallas; **Figure 2.32K**], and *Turrubulana* Distant), each with a single species, placing them in his *Ochisme* group. All four genera are endemic to Australia. Dallas (1851) and Stål (1876) both included *Trachyops* in the Halyini. Similarly, Distant (1910a) placed his genus, *Turrubulana*, in the Halyini, but this was disputed by Bergroth (1916). Bergroth (1918), based especially on the shape of the pronotum, originally placed *Amphidexius* near *Dandinus* Distant, a member of the Podopinae, or *Paranotius* Breddin (= *Lubentius* Stål), a member of the Carpocorini; he also indicated that the placement of *Dysnoetus* was questionable, although he discussed a possible relationship with the Myrocheini or Sciocorini.

Gross (1975b) considered the structure of the thoracic sterna of primary importance, which he characterized as "deeply sulcate pro-, meso- and metasterna; the edges of the prosternal sulcus are sharply raised as two prominent keels lying forward and on the inner side of the fore coxae." The rostrum is elongate, sometimes reaching to abdominal sternite IV (shorter in *Dysnoetus*). All tibiae are sulcate on superior surfaces. There may be a shallow sulcus on the first two abdominal sternites. The male parameres are C-shaped; the phallosoma sometimes ends in a narrow thecal shield which may possess a pair of horn-like processes dorsally. The female spermathecal bulb is ball-shaped with two short diverticula. The sulcate nature of the mesosternum may relate this group to other groups with a sulcate mesosternum (e.g., Aeptini and Myrocheini; see **Sections 2.2.10.7.3 and 2.2.10.7.29**, respectively).

2.2.10.7.44.2 Kitsonia Group   The generic name, *Kitsonia* Gross, upon which this group is based, is preoccupied. Rider (1998b) proposed *Kitsoniocoris* Rider as a replacement name. This group contains a single genus and species, endemic to Australia. This genus has a triangular head, the margins of which appear to be rounded from above, but the edge can be seen from lateral view. The scutellum is enlarged, reaching to the apex of the abdomen, but it does not cover the lateral margins of the coria. All thoracic sterna are sulcate; that of the prosternum is bordered on both sides by a sharply elevated carina in front of the fore coxae.

Gross (1975b), based on the shape of the scutellum and the structure of the prosternum, related this genus to the *Tarisa* group [= Podopinae] (see **Section 2.2.10.9**), but the triangular shape of the head and the outline of the pronotum and scutellum differ from most podopines. He also noted that the shape of the scutellum was similar to that seen in *Macrocarenus* Stål (but again head shape is different), and *Bachesua* Gross and *Ippatha* Distant (both of which differ again in head structure and lack the prosternal structure).

2.2.10.7.44.3 Kumbutha Group   Gross (1975b) established this group to accommodate nine genera stating that they appeared to form a natural grouping, but its relationship with other groups was far from clear. He characterized the group as having an enlarged scutellum that covered most of the dorsum or, if smaller, it was still semi-elliptical in shape (although he stated that the scutellum was about normally sized in *Cooperocoris* Gross), and by the medially sulcate prosternum and mesosternum. The prosternal sulcus is margined on both sides by a keel, or a series of coalescing granules forming a keel, this keel continues forward and upwards to form a sharp anterior margin to the propleura. The lateral angles of the pronotum are obtusely rounded.

Gross (1975b) compared this group to a number of other Australian groups, mainly those that had a similar scutellar structure or a similar structure of the male genitalia. Ultimately, he concluded that the relationships with other groups were obscure, and further study was needed. The morphology of the head and the structure of the male internal genitalia, at least of some of the genera, support a relationship with the *Macrocarenus* group, discussed below.

2.2.10.7.44.4 Macrocarenus Group   Gross (1975b) established this group for two Australian genera - *Macrocarenus*, and his new genus *Macrocarenoides* Gross (**Figure 2.32I**). Species in these two genera are over 6 mm in size. This group also has the scutellum greatly enlarged, reaching nearly to the apex of the abdomen. All three thoracic sterna are sulcate with the prosternal sulcus sharply margined on both sides. The head is acuminate and elongately triangular. The male parameres are strongly F-shaped, each with the upper ramus broad and truncate, and the middle ramus platform-shaped. The aedeagus has a small thecal shield, and the medial penial plates are thin and rod-like.

Gross (1975b) stated that this group was clearly related to the *Kumbutha* group. He also indicated that this group might be related to the *Menestheus* group or the Aeliini (See **Section 2.2.10.7.2**), based on male genital characters. He noted that some workers had treated *Macrocarenus* as a member of the Scutelleridae due to the greatly enlarged scutellum, but the male aedeagus and parameres were distinctly pentatomoid rather than scutelleroid.

2.2.10.7.44.5 *Menestheus* Group

This group was originally proposed to contain six Australian genera. Gross (1975b) characterized the group as having the head elongate, narrowed anteriorly, with the juga narrowly rounded, acuminate, and frequently longer than the tylus and often meeting anterior to the tylus. The antennifers are not visible from dorsal view, and the bucculae are strongly elevated. All thoracic sterna are sulcate with the prosternal sulcus margined on both sides by an elevated sharp ridge that projects forward as a small tooth onto the base of the head. The scutellum is subtriangular. The parameres are strongly C-shaped. The male aedeagus lacks a thecal shield but does possess a pair of thin, sclerotized rods.

This group has had an interesting taxonomic history. Stål (1876) first associated several of the included genera with the genus *Dymantis*, an African genus now placed in the Myrocheini (they share the sulcate mesosternum). Kirkaldy (1909) recognized the same general group under the name Aeptini; the type genus *Aeptus* is another African genus with a sulcate mesosternum, a somewhat elongate head, but its members usually are brachypterous. Gross (1975b) also related this group with the Aeliini, but ultimately concluded that the Australian genera had enough differences to warrant placement in their own group. Based on Kirkaldy, we treat these genera as members of the Aeptini (see **Section 2.2.10.7.3**), but we are currently studying this affiliation. Along with its possible relationship with *Aeptus*, there are also similarities with groups with the mesosternum sulcate (e.g., the African tribe Diploxyini, the Myrocheini, and, perhaps, the South American genus *Poriptus*) (see **Sections 2.2.10.7.20 and 2.2.10.7.29**, respectively). Also, there are few characters of any significance that reliably separate this group from the *Ochisme* group. Additionally, several Indian genera (*Adria*, *Bonacialus*, and *Gulielmus*) may also be related to this group. Further research is needed.

2.2.10.7.44.6 *Ippatha* Group

Gross (1975b) proposed this grouping to hold two Australian genera - *Ippatha* and *Bachesua*. Both of these genera have the scutellum enlarged, covering most of the abdominal dorsum (**Figure 2.24F**). The thoracic sterna are all sulcate, but the margins of the sulcus are not sharply keeled; the margins are either bluntly rounded or they are depressed. Gross (1975b) further indicated that these two genera were similar in appearance to members of the Scutelleridae, but their broad heads with explanate margins were more reminiscent of members of the *Kapunda* group in the Pentatomidae. The male parameres are flattened and strongly T-shaped. The female spermathecal duct has a dilation with a sclerotized rod, and the spermathecal bulb is simple, ball-shaped, and lacks diverticula.

Musgrave (1930) treated these genera as members of the subfamily Graphosomatinae (= Podopinae) (See **Section 2.2.10.9**), another group that often has enlarged scutella. Gross (1975b) treated the Podopinae as two other groups at the same level as the *Ippatha* group - the *Tarisa* group and the *Podops* group. He indicated that these two groups have more sharply raised margins to the thoracic sulci, and the male genitalia do not "reveal any particular relationships with those of other groups of Australian Pentatominae."

2.2.10.7.44.7 *Dictyotus* Group

This group was proposed to accommodate three Australian genera – *Dictyotus* Dallas (**Figure 2.22H**), *Paradictyotus* Gross, and *Utheria* Gross. Its members have the thoracic sterna sulcate, but the sulci are lacking strongly elevated margins. The head is rather broad with the apex broadly rounded, and the jugal margins are explanate and somewhat reflexed; these margins also have a blunt tooth in front of each compound eye. The anterolateral pronotal margins are straight or only feebly concave or convex (except in *Utheria*) and somewhat explanate. The scutellum is subtriangular.

The hemelytral membrane is reticulate. The rostrum reaches the middle coxae. The male genitalia possess at least feebly developed parandria (probably not true parandria, but rather simply processes on the lateral rims of the pygophore); some species have a narrow thecal shield; there is also a pair of rigid 'struts' directed upwards at the base of the conjunctiva. The female spermathecal duct has a dilation and a sclerotized rod; the spermathecal bulb is ball-shaped with a pair of tubular diverticula.

Gross (1975b) stated that "this group does not appear to have any particular close relatives in the Pentatomidae except the following *Tholosanus* group and exotic forms which may be related to that group." In his discussions of the *Tholosanus* group, he indicated that there might be a relationship with the African members of the tribe Myrocheini (see **Section 2.2.10.7.29**). So, it is possible that members of this group, as well as those of the *Tholosanus* group, could be referred to the Myrocheini.

2.2.10.7.44.8 Tholosanus Group   This group was proposed to contain two Australian genera – *Arniscus* (**Figure 2.22F**) and *Tholosanus* Distant. Gross (1975b) indicated that, based on external morphology, members of this group were difficult to distinguish from the *Dictyotus* group. He stated that *Tholosanus* group members were slightly larger, and that they lacked the reticulate venation in the hemelytral membrane. He also indicated that the male parameres of *Arniscus* are quite similar to those of *Dictyotus*, but the parameres of *Tholosanus* are more F-shaped. He further indicated that there were several differences in the structure of the aedeagus.

Gross (1975b) stated that this group was closely related to the *Dictyotus* group, but either or both may be related to a group of African genera now included in the tribe Myrocheini (see **Section 2.2.10.7.29**). He personally compared these genera with two myrocheine genera, *Laprius* and *Dorpius* Distant (= *Myrochea*) and found them to be quite similar. It is possible that the genera included in this group could be referred to the Myrocheini.

2.2.10.7.44.9 Poecilotoma Group   This group is monotypic, containing a single Australian genus, *Poecilotoma* (**Figure 2.24G**). Gross (1975b) indicated that members of this genus have a similar appearance as members of the *Halys* group (= Halyini), although they are smaller than most members of that tribe. Characters that the two groups share include the prominent humeral angles, the crenulate anterolateral pronotal margins, the antennifers forming a hook or flap laterally, the juga extending beyond the apex of the tylus, and the four-segmented antennae.

Gross (1975b) indicated that many of these same characters are found in some members of the *Diemenia* group (= Diemeniini), but both the halyines and the genus *Poecilotoma* lack the stridulatory mechanism on the abdominal venter characteristic of diemeniines. The thoracic sterna are sulcate, lacking a medial carina (which is present in the Halyini and Diemeniini). The male parameres show no particular resemblance to those of either group; the aedeagus has a thecal shield, similar to that seen in the Asopinae. More recently, Faúndez and Rider (2014) noted a resemblance between *Poecilotoma* and the South American genus *Thestral*, a genus that appears to be related to the genus *Acledra*.

2.2.10.7.44.10 Kapunda Group   Traditionally, the Australian members of this group had been placed into three genera: *Adelaidena* Distant, *Kapunda* Distant, and *Sciocoris*. Gross (1976), after studying pertinent type material, determined that *Adelaidena* was a good genus, but the Australian members of '*Sciocoris*' were congeneric with *Kapunda*. Gross (1976) further indicated that at least based on internal male genitalia, the Australian members of this group were not related to European sciocorines, although this may not be as important as Gross thought because Gapon and Baena (2005) determined that the genitalic structure of the European sciocorines was particularly heterogenous. He concluded that the male genitalia more closely resembled that of members of the *Diemenia* group or the *Poecilotoma* group. The members of the *Kapunda* group lack the abdominal stridulatory structures characteristic of the diemeniines.

Members of this group are relatively small (less than 7 mm); they are oval in shape with a relatively broad and magnate head which is flat dorsally, and the juga and tylus are subequal in length. The pronotum is relatively flat with explanate margins. The scutellum is subtriangular and relatively flat. All thoracic pleura are sulcate without sharply raised margins; there is a short medial carina on the anterior portion of the mesosternum.

2.2.10.7.44.11 Cephaloplatus Group   Gross (1976) proposed this group for a single Australian genus, *Cephaloplatus* (**Figure 2.19I**), and indicated that a second undescribed genus from northern Australia might belong here. That second genus, *Linea* McDonald (**Figure 2.12C**), has been described (McDonald 2003). Unfortunately, *Linea* was preoccupied; *Birna* McDonald has been proposed as a replacement name (McDonald 2006). Gross (1976) further indicated, however, that *Cephaloplatus* (**Figure 2.19I**) might be closely related to the genus *Caystrus*, the type genus for the tribe Caystrini (see **Section 2.2.10.7.14**).

Members of this group are oval to elongate-oval in shape, small to medium in size (7-14 mm), and a general yellowish to yellowish-tan coloration. The juga are broad, extending beyond the apex of the tylus, their apices rounded to obliquely truncate, and their lateral margins weakly reflexed. There is usually a lateral tooth-like projection anterior to each compound eye. The ocelli are widely separated. The antero-lateral pronotal margins are usually explanate, often slightly reflexed. The rostrum reaches the hind coxae. The prosternum is depressed without a longitudinal carina, whereas the mesosternum has a weak longitudinal raised line. The ostiolar rugae are reduced and auriculate in shape. The male parameres vary from F-shaped to T-shaped, usually with a corresponding ridge on the wall of the pygophore. The male aedeagus lacks a thecal shield. The female spermathecal duct has a dilation and sclerotized rod, and the spermathecal bulb is ball-shaped with two small diverticula.

Although Gross (1976) discussed the possible relationship with the genus *Caystrus* (a member of the Caystrini), he finally concluded that " in so much as the *Cephaloplatus* group show any relationship with any other group in this generic section of the family it is with the following *Phyllocephala* group" (= subfamily Phyllocephalinae). Interestingly, many of the species of *Cephaloplatus* (**Figure 2.19I**) show a general similar facies with some genera of the South American tribe Discocephalini (Discocephalinae), especially the genera *Dryptocephala* and *Pelidnocoris*, but this is probably not a sign of relationship. The genus *Cephaloplatus* has been revised recently (Gross 1970).

2.2.10.7.44.12 Mycoolona Group   This monotypic group contains a single genus and species, *Mycoolona atricornis* (Westwood), and was not formally treated by Gross (1976) as it does not occur in South Australia. He felt that it was important to discuss this group, along with the *Eysarcoris* group (= Eysarcorini), for completeness in studying relationships among the Australian pentatomines. He characterized this species as a subelongate creature (9 mm in length), greenish grey in color with coarse black punctures; antennal segments II through V are black. The male parameres are robust, triangular in shape, each with the dorsal surface concave. The male aedeagus has a thecal shield.

Distant (1910b) originally placed *Mycoolona* near the genus *Anaxarchus* Stål (**Figure 2.19J**), a genus that Gross (1976) speculated may belong in the *Carpocoris* group (= Carpocorini) (see **Section 2.2.10.7.12**). Based on general appearance, Gross (1976) stated that *Mycoolona* was similar to members of the *Halys* group (= Halyini) or the *Carpocoris* group. He also indicated that it looks similar to some species of the genus *Ocirrhoe* (Rhynchocorini), but *Mycoolona* lacks the sternal structures characteristic of the rhynchocorines. The external structures associated with the scent gland apparatus are more similar to the carpocorines than to the halyines. But he further indicated that the male genitalia was different from both groups.

2.2.10.7.45 The Generic Groups of Linnavuori (1982)
Linnavuori (1982), much like Gross (1975b, 1976), spent considerable space discussing the classification and relationships among the various pentatomine groups, focusing on the African fauna. He also did not want to add new names to the literature until a thorough phylogenetic analysis could be completed. He did, however, use some tribal names if they were already in use by others and only used generic group names for those genera that were difficult to place in the tribes known to him. His classification, in some ways, closely resembles our current classification in that he considered the Asopinae, Phyllocephalinae, and Podopinae to be subfamilies. What follows are brief discussions of his generic groups that have not already been covered in our previous discussions of known tribes. The groups are in the same order that Linnavuori (1982) treated them (**Table 2.5**).

2.2.10.7.45.1 Tyoma Group   Although Linnavuori (1982) separated this group from the *Aeschrus* group, he indicated that there was a close relationship between the two groups. Most current workers treat the genera from this group as members of the Aeschrocorini. This may or may not be valid, but see discussions under that tribe (see **Section 2.2.10.7.4**).

2.2.10.7.45.2 Acoloba Group   Linnavuori (1982) proposed this group to accommodate a single genus, *Acoloba* (**Figure 2.21D**), a genus endemic to Africa. He indicated that it was probably related to the *Diploxys* group (= Diploxyini) (see **Section 2.2.10.7.20**), but differed from that group by its greenish coloration; the body being quite elongate, narrow, and depressed; the bucculae gradually tapering basad; the anterolateral pronotal margins being narrowly reflexed; the external structures associated with the metathoracic scent glands; and by the structure of the male pygophore. The ostiolar rugae are longer, and the evaporatoria are distinct, well delimited from the shiny and punctate lateral margins of the metapleura. The male pygophore is slender, apically trilobate; the male parameres are strongly curved. The spermathecal bulb is broad and rather flat, with two tubular diverticula.

Linnavuori (1982) stated that this group displayed "several advanced characters connected with the specialized mode of life on grass leaves."

2.2.10.7.45.3 Aeliomorpha Group   This is another monotypic group, containing only the genus *Aeliomorpha* that has several species occurring in Africa, the Middle East, and into the Indian region. Linnavuori (1982) stated that this was a well-known group and noted that many previous workers had regarded it as a close relative of the genus *Aelia*, the type genus for the tribe Aeliini (see **Section 2.2.10.7.2**). In fact, in some phylogenetic papers (Ahmad et al. 1990, Zaidi 1994), *Aeliomorpha* was used to represent the tribe Aeliini, which of course may have given erroneous results as this genus is no longer accepted as a member of the Aeliini. Linnavuori (1982) recognized possible relationships with the Aeliini and possibly the genus *Acoloba* but concluded that they were not contribal. He listed a few characteristics that separated *Aeliomorpha* from the Aeliini: the body is less convex, the juga and tylus are subequal in length, the antennae are sexually dimorphic (segment II in males is relatively short), the propleura are not lobed anteriorly over the base of the antennifers, the mesosternum is carinate (not sulcate), the parameres are shaped differently, and the female spermathecal bulb is simple and ovate. Stål (1865) treated *Aeliomorpha* as a subgenus of the genus *Pentatoma*, but, at that time, his concept of *Pentatoma* was quite broad and contained a number of subgenera that are now placed in the Carpocorini (see **Section 2.2.10.7.12**) and Pentatomini (see **Section 2.2.10.7.34**). Cachan (1952) formally treated *Aeliomorpha* as a member of the Carpocorini.

Linnavuori (1982) stated that members of this genus are "grass feeders in various savanna habitats, and in meadows, fields, etc."

2.2.10.7.45.4 Carbula Group   Linnavuori (1982) included seven genera in his *Carbula* group – *Aspavia* (**Figure 2.21G**), *Carbula* (**Figure 2.21H**), *Durmia*, *Empiesta* Linnavuori, *Gynenica*, *Mulungua* Linnavuori, and *Pseudolerida*. He devoted much discussion to the interrelationships of the included genera but a much less effort to actually defining the group. He stated that the *Carbula* group was a uniform group. He characterized the group as having a carinate mesosternum; ostiolar rugae that are relatively short but with fairly large, well-delimited evaporative areas; and the male parameres that are biramous. The female spermathecal bulb is ball-shaped, lacking diverticula.

Most workers have treated these genera as members of the tribe Eysarcorini (see **Section 2.2.10.7.22**), although the scutellum is not as large as in most eysarcorine genera, with one exception – the genus *Gynenica* now is regarded as a member of the Lestonocorini (see **Section 2.2.10.7.25**).

2.2.10.7.45.5 Veterna Group   Linnavuori (1982) stated that he "studied" nine genera belonging in this group, implying that there probably were other genera that also belonged here. The nine he studied were *Benia*, *Caura*, *Lerida*, *Leridella* Jeannel, *Lokaia*, *Mabusana*, *Paralerida*, *Tripanda*, and *Veterna* (**Figure 2.19H**). Most of the included genera in this group are brownish in color, have the abdominal venter unarmed, and have the ostiolar rugae relatively long, all characters shared with the Cappaeini (see **Section 2.2.10.7.11**). As such, these genera have been discussed under that tribe.

2.2.10.7.45.6 Halyomorpha Group   Linnavuori (1982) included four genera in this group – *Adelocus* Bergroth, *Boerias* (**Figure 2.19G**), *Halyomorpha* (**Figure 2.28L**), and *Hymenomaga* Karsch. All of these genera essentially possess those characters that define the tribe Cappaeini (see **Section 2.2.10.7.11**), and it is fairly well accepted that *Halyomorpha* is closely related to *Cappaea*, the type genus of the tribe Cappaeini. So, the genera listed above have been discussed under that tribe.

2.2.10.7.45.7 Banya Group   This is a monotypic group containing the single genus and species *Banya leplaei* Schouteden (Linnavuori 1982). Schouteden (1916) originally noted that this genus had some resemblance to the genera *Benia* and *Mabusana* of the *Veterna* group (Cappaeini) (see **Section 2.2.10.7.11**) but differed strikingly by the sharp spine at the base of the abdomen, which extends onto the metasternum. Members of the Cappaeini lack an abdominal spine, but several other groups, including the Menidini (see **Section 2.2.10.7.28**), have a similar spine. In fact, Schouteden (1916) also suggested a close relationship with the Menidini. Linnavuori (1982), however, gave a number of characters in which *Banya* differs from members of the Menidini, but many of these may not be true if non-African menidines are compared. As mentioned under that tribe, the Menidini is in dire need of revision and will probably undergo many changes in the future.

2.2.10.7.45.8 Eipeliella Group   Linnavuori (1982) proposed this group for a single genus, *Eipeliella* Schumacher, which contains five African species. This genus originally was described as a member of the Graphosomatinae (= Podopinae) (see **Section 2.2.10.9**), based primarily on the large scutellum (Schumacher 1912). We now know that an enlarged scutellum has evolved a number of times in the Pentatomoidea, and that, by itself, is not enough to indicate relationship. Linnavuori (1982) indicated that the general body form and the simple spermathecal bulb resembled that found in some members of the Eysarcorini (see **Section 2.2.10.7.22**); he then provided a number of characters that separate *Eipeliella* from eysarcorines. He finally concluded that this genus might be related to the Antestiini (see **Section 2.2.10.7.8**).

2.2.10.7.45.9 Kelea Group   This is another monotypic group containing a single genus and species, *Keleacoris congolensis* (Distant). *Kelea* Schouteden is preoccupied, so Rider and Rolston (1995) proposed *Keleacoris* as a replacement name. Linnavuori (1982) indicated that this species was similar in appearance to species of *Eurysaspis*. The head is short and broad with the lateral margins of the juga distinctly notched anterior to the compound eyes; the antenniferous tubercles are visible from dorsal view; and the apex of the head is broadly truncate. The bucculae are rather broad. The rostrum extends to the mesocoxae. The pronotum is convex, and the anterolateral margins are broadly convex and narrowly reflexed. The scutellum is somewhat enlarged, ligulate, with distinct basal callosities; the frena are rather short. The base of the abdomen is armed with a tumescence that meets with the posterior margin of the elevated metasternum. The mesosternum has a medial carina only anteriorly. The female spermathecal bulb is simple without diverticula.

   Linnavuori (1982) discussed possible connections with the Eurysaspini (see **Section 2.2.10.7.21**) but ultimately concluded that this genus is probably a derivative of the *Menida* group (= Menidini) (see **Section 2.2.10.7.28**), especially the genus *Gwea* Schouteden.

2.2.10.7.46 Other Unplaced or Questionably Placed Genera

Gross (1975b, 1976) and Linnavuori (1982) dealt with nearly all of the difficult Australian and African genera, placing them into various generic groups. There are, however, a number of pentatomid genera from the Orient and Neotropics that have never been placed in any known tribes, and their current placement remains unknown. Further study is needed before proper placement can be made confidently. We provide a few comments that may help with future placement.

Araducta Walker, 1867

Walker (1867b) originally described this genus as having a pectoral keel that extended to the fore coxae, and he compared this genus with *Cuspicona* (**Figure 2.23L**), a member of the Rhynchocorini. Distant (1900b) redescribed the genus, acknowledging the pectoral keel, but made some changes in describing the abdominal spine; he placed this genus near the genus *Brachystethus*, a member of the Edessinae. Distant also described or transferred four more species in *Araducta*, all of which have now been transferred to either *Dabessus* or *Dunnius*, genera now tentatively placed in the Menidini. Kirkaldy (1909) catalogued this genus after the genus *Exithemus* and before the genus *Lelia*, members of the tribes

Rolstoniellini and Pentatomini, respectively. We have examined specimens and confirm that they possess the sternal structure that is characteristic of the rhynchocorines; although the coloration of these specimens differs from the typical, we still treat this genus as a member of the Rhynchocorini (see **Section 2.2.10.7.38**). The genus occurs in the Malay Archipelago and New Guinea.

Banasa Stål, 1860; *Disderia* Bergroth, 1910; *Elanela* Rolston, 1980; *Glaucioides* Thomas, 1980; *Grazia* Rolston, 1981; *Janeirona* Distant, 1911; *Kermana* Rolston, 1981; *Modicia* Stål, 1872; *Nocheta* Rolston, 1980; *Pallantia* Stål, 1862; *Rio* Kirkaldy, 1909; and *Vidada* Rolston, 1980

For New World genera, Rolston et al. (1980) divided his broadly conceived Pentatomini into three sections. Section one contained those genera in which the abdominal venter was unarmed basally; section two contained those genera in which the base of the abdomen was armed with a spine or tubercle that was unapposed by an elevated metasternum (i.e., the apex was free distally); and section three contained those genera in which the abdominal venter was armed with a tumescence, tubercle, or small spine that was apposed by an elevated metasternum and might be truncate or notched.

Nearly all of the Old World genera that have the abdominal venter armed would fall into Rolston's section two (Rolston and McDonald 1981); that is, the abdominal armature is unapposed by the metasternum. There are however, a number of New World genera that belong in Rolston's section three (Rolston et al. 1980). Some of these have been moved elsewhere (e.g., *Arvelius* into the Chlorocorini; *Brachystethus* and *Lopadusa* into the Edessinae), or at least have been discussed elsewhere in this chapter (e.g., *Evoplitus* and related genera). Many of the remaining genera do not fit neatly into the well-known tribes that have the abdomen armed (e.g., Catacanthini, Eurysaspini, Menidini, Nezarini, Piezodorini). Based on a similar general appearance, we treat *Elanela* (**Figure 2.22C**) and *Rio* (**Figure 2.22E**) as members of the Menidini (maybe erroneously). We believe the treatment of the remaining genera as members of the Pentatomini is suspicious, but, without further evidence, we conservatively leave them in the Pentatomini. Several of the genera treated here have been revised recently: *Banasa* (**Figure 2.31D**) (Thomas and Yonke 1981, 1988, 1990), *Disderia* (Rolston 1983b), *Elanela* (Grazia and Greve 2011, Grazia et al. 2016a), *Pallantia* (**Figure 2.23E**) (Grazia 1980a, 1986; Brailovsky 1986), and *Rio* (Grazia and Fortes 1995, Fortes and Grazia 2000). The monotypic genus *Glaucioides* Thomas (**Figure 2.23A**) was described recently (Rolston et al. 1980).

Brachyhalys Ahmad, 1981

Ahmad (1981) listed as a new genus and species *Brachyhalys pilosum*, and he provided a new family-level name, Brachyhalyini, for its inclusion. He did not, however, provide any descriptions of the tribe, genus, or species, and there were no type designations. As such, these names should all be considered as unavailable.

Brasilania Jensen-Haarup, 1931

We have not examined any specimens of this genus. Jensen-Haarup (1931a) originally placed this New World genus near the Old World genus *Bathycoelia* (currently in the Bathycoeliini or possibly Pentatomini; see **Section 2.2.10.7.10**). Jensen-Haarup (1931a) described this genus as having the abdominal venter unarmed, and Rolston (1987a) included it in an artificial group of seven South American genera of Pentatomini that have an unarmed abdominal venter and elongate ostiolar rugae. The sternum was originally described as not carinate. This genus is unusual in that the coloration is described as pitchy black all over (including the membranes), and the rostrum is extremely long, reaching well beyond the apex of the abdomen.

Cachaniellus Rider, 2007

This genus originally was described by Cachan (1952) using the preoccupied generic name *Sambirania*. He placed it in a broadly conceived Carpocorini but indicated that it was related to *Corisseura*, the placement of which also is uncertain. We tentatively treat both genera as members of the Eysarcorini. Both genera are endemic to Madagascar.

Catalampusa Breddin, 1903

Breddin (1903) described its only known species, *C. oenops* Breddin, from Bolivia, placing it between a new species of *Nezara* and a new species of *Tibilis*. In the original description, Breddin (1903) indicated that the metasternum is elevated and received the short abdominal spine, thus placing it in Rolston's section three. Kirkaldy (1909), the last worker to mention this genus, also catalogued it near the genus *Tibilis*, thus perhaps indicating a possible relationship with that genus. Because *Tibilis* is a member of the Pentatomini, we are tentatively treating this genus as a member of this tribe.

Chrysodarecus Breddin, 1903

Breddin (1903) did not place this genus in any known tribe or subfamily, nor did he mention any possible relationships. Kirkaldy (1909) catalogued it among the Discocephalinae, possibly because the next genus dealt with in Breddin's work was *Adoxoplatys*, a member of the Discocephalinae. In truth, the original description in some ways is quite similar to the genus *Placocoris* Mayr. Breddin (1903) stated that the body was quite flat and had a double row of spines on the hind femora (also found in *Placocoris*). The single species of this genus was described from Peru.

Cyptocephala Berg, 1883; *Tepa* Rolston and McDonald, 1984; and *Thyanta* Stål, 1862

These three New World genera are obviously related to each other and although they have been studied extensively by one of us (DAR), their tribal placement remains questionable. Rolston and McDonald (1984), by virtue of their having an unarmed abdominal venter, placed these genera in section one of their broadly conceived Pentatomini. These three genera also have the ostiolar rugae relatively elongate, reaching to (*Tepa*) or extending beyond (*Cyptocephala* and *Thyanta* [**Figure 2.32L**]) the middle of the metapleura, with the apices usually acuminate. Because the adults are usually green (yellowish or brown forms are present in colder months), and usually remain green after death, we have tentatively placed them in the Nezarini. Further study is needed to confirm this placement. All three genera have been revised: *Cyptocephala* (Rolston (1986), *Tepa* (Rolston 1972, Rider 1986), and *Thyanta* (**Figure 2.32L**) (Rider and Chapin 1991, 1992). As a side note, it appears that at least some of the subgenera of *Thyanta* probably should be elevated to generic rank.

Dardjilingia Yang, 1936 and *Zouicoris* Zheng, 1986

These two genera were described from India and Bhutan, and from southwestern China, respectively; both have remained monotypic so far. They share an unarmed abdominal venter; a mesosternum provided with a very low median carina; a relatively short ostiolar ruga not reaching the middle of the metapleuron; and a rostrum that is relatively short, approaching or reaching the middle coxae. The male genitalia and the female genital plates of members of the two genera are similar as well, and there is little doubt that they belong to the same tribe. It is difficult, however, to determine which tribe this should be because the general habitus of these species is quite different from the majority of pentatomids. Yang (1936) compared *Dardjilingia* with *Neogynenica*, currently in the Lestonocorini (see **Section 2.2.10.7.25**); the female genital plates, however, are much different from the condition found in members of that tribe (also see Salini 2016). Zheng (1986) indicated that the short rostrum in *Zouicoris* might ally this genus with the phyllocephalines, but he ultimately stated that it was not a member of that group and that it probably belonged in the Pentatominae; he did not, however, indicate to which tribe it might belong.

Lakhonia Yang, 1936

This genus, containing a single species occurring in Thailand and Southwest China, never has been formally placed in any pentatomid subfamily or tribe. It is obvious from the illustrations and description that the more recently described genus, *Oedocoris* Zheng and Liu, is synonymous (Rider et al. 2002), and yet its tribal placement remains unknown. The dorsal surface is covered by a thick layer of short, soft, hairs, and the head and anterior disk of the pronotum are strongly declivent. The abdominal venter is unarmed, and the ostiolar rugae are short and auriculate. This suite of characters is similar to members of the Carpocorini.

Leovitius Distant, 1900

Distant (1900b) originally related this genus to the genus *Prionochilus* (a preoccupied name now placed in the synonymy of the pentatomine genus *Lelia*). He later (1902) placed *Leovitius* in the Rhynchocorini, even though Dallas (1851) and Distant, himself (1902), described the ventral armature as consisting of a long basal abdominal spine, reaching beyond the front coxae. There is no mention of the elevated metasternum (meeting posteriorly with the abdominal spine or tubercle), or the mesosternum being elevated and produced anteriorly, characteristics that define the Rhynchocorini (see **Section 2.2.10.7.38**). Until this genus can be studied in more detail, we tentatively treat this genus as a member of the Pentatomini. This genus contains a single species from northern India.

Liicoris **Zheng and Liu, 1987**

This genus contains a single species, *L. tibetanus* Zheng and Liu, from Tibet. It was originally compared with members of the Australian *Dictyotus* group (Zheng and Liu 1987), but based on the re-examination of the holotype, it is obviously unrelated to members of that group. Its general habitus is quite similar to that of members of *Homalogonia* (currently Cappaeini, see **Section 2.2.10.7.11**). The thoracic venter is similar to the condition found generally in Carpocorini and Pentatomini; the abdominal venter is unarmed; and the rugae associated with the metathoracic scent gland ostioles are strongly reduced. Based mainly on the metapleural rugae, the general habitus, and the male genitalia, the genus probably belongs to the Pentatomini (see **Section 2.2.10.7.34**).

Manoriana **Ahmad and Kamaluddin, 1978**

This genus is monotypic; its single species *M. pakistanensis* Ahmad and Kamaluddin was described from Pakistan (Ahmad and Kamaluddin 1978a). This taxon was originally placed in the Carpocorini, and Afzal and Hasan (1988) considered *Manoriana* to be a junior synonym of *Mormidella*, also a member of the Carpocorini. Based on characters in the original description, and the illustrations, we believe this genus actually belongs in the Antestiini (see **Section 2.2.10.7.8**) and even may be synonymous with *Antestia*.

Mathiolus **Distant, 1889**

This New World genus originally was placed by Distant (1889) in the Discocephalinae. Rolston (1988b) provided evidence supporting the transferal of this genus to his broadly conceived Pentatomini; he described it as having the abdominal venter unarmed (section one of Rolston and McDonald 1984) and the ostiolar rugae short; thus, our tentative treatment of this genus in the Carpocorini (see **Section 2.2.10.7.12**).

Muscanda **Walker, 1868 and** *Xiengia* **Distant, 1921**

Both of these Old World genera, and their only included species, *M. testacea* Walker and *X. elongata* Distant, respectively, originally were placed in the Tessaratomidae. Walker (1868) compared his new genus with *Piezosternum* Amyot and Serville (Tessaratomidae), and Distant (1921b) placed his new genus near *Origanaus* Distant (Tessaratomidae). All subsequent workers continued to treat both genera as members of the Tessaratomidae until Kumar and Ghauri (1970) transferred them to the Pentatomidae. They did not place them in any known subfamily or tribe but, rather, commented that the two genera might be related and might deserve a new tribe of their own. We have examined habitus photographs of both *Muscanda testacea* and *Xiengia elongata*, and both genera appear to be members of the Phyllocephalinae (Cressonini), and they may be related to the genus *Uddmania* Bergroth (see Arnold 2011b for recent work on *Uddmania*).

Myota **Spinola, 1850**

Spinola (1850) first included this Neotropical genus (**Figure 2.23C**) in a key to genera where it keyed near several genera in the Edessinae and Tessaratomidae. Dallas (1851) described this genus (under the name *Aegius*) between the genera *Tropicoris* Hahn (= *Pentatoma*) and *Cataulax*, a member of the Discocephalinae. Lethierry and Severin (1893) catalogued it between *Abeona* (= *Amblycara*), a member of the Pentatomini, and *Placocoris*, another difficult to place genus (see below). Kirkaldy (1909) also catalogued it between these same two genera and among a number of other genera tentatively placed in the Pentatomini. We tentatively treat this genus as a member of the Pentatomini.

Neojurtina **Distant, 1921**

Distant (1921a) originally placed this Oriental genus (**Figure 2.23D**) in his broadly conceived Nezaria (= Nezarini), but he also indicated a possible relationship with the genus *Jurtina*, currently considered to be a junior synonym of *Bathycoelia*, the type genus of the Bathycoeliini. We tentatively treat this genus as a member of the Pentatomini.

Patanius Rolston, 1987

This genus and its only included species (*P. vittatus* Rolston [**Figure 2.32J**]) were described by Rolston (1987a) among an artificial group of seven South American genera that have the abdominal base unarmed and the ostiolar rugae elongate and acuminate apically. Two members of this group have been placed in tribes (e.g., *Arocera* in the Catacanthini, *Chloropepla* in the Chlorocorini), but the placement of the remaining five genera is questionable (*Brasilania, Cyptocephala, Patanius, Senectius* Rolston, and *Thyanta*). One interesting character of this genus is the apparent absence of parameres, a trait that is only shared in South American pentatomids by some carpocorines (e.g., *Luridocimex, Stysiana*), an unde-scribed genus, and the genus *Rhyncholepta* (**Figure 2.21A**), which is currently considered to be a mem-ber of the Chlorocorini. The mesosternum also was described originally as weakly carinate mesially.

Placocoris Mayr, 1864

This is another Neotropical genus that Rolston et al. (1980) placed in the Pentatomini (section three) defined by an armed abdominal venter that is apposed by an elevated metasternum (see discussion under *Banasa* above in this section). This genus with its two species, *P. albovenosus* Kormilev and *P. viridis* Mayr, however, is significantly different than the other genera in Rolston's section three. It is extremely flattened, and each of the hind femora has a double row of spines (a character also possessed by members of the genus *Chrysodarecus*). For now, this genus is left in the Pentatomini, but it may deserve a new tribe or even subfamily.

Platycoris Guérin-Méneville, 1831

There has been some confusion concerning the appropriate name for this Australian genus; this confu-sion has stemmed from the uncertainty of the dates of publication of Guérin-Méneville's plates (1831) and text (1838) for his Voyage Coquille paper, which in turn has played a role in determining the type species for this genus. To make a long story short, the proper name for this genus is *Platycoris*, and the oft-used generic name *Hypogomphus* Spinola is a synonym. Gross (1976) speculated that this genus, as *Hypogomphus*, might belong in the Carpocorini, but McDonald (1995) who revised this genus, again under the name *Hypogomphus*, placed the genus in the Halyini.

Riaziana Hasan, Afzal, and Ahmad, 1989

This genus, and its only included species, *R. serrata* Hasan, Afzal, and Ahmad, described from Pakistan, originally were not placed in any known tribe. Hasan et al. (1989) discussed its placement, indicating that it had a superficial resemblance to the halyine genera *Sarju* and *Izharocoris* Afzal and Ahmad but ulti-mately concluded that it was not a member of the Halyini. They also indicated that the serrate anterolateral pronotal margins might relate it to the Tropicorini (= Pentatomini), but they decided the unarmed abdomi-nal venter would keep it from being a member of that tribe (although we now know that some members of the Pentatomini have the abdominal venter unarmed). They left this genus unplaced, and no one since has worked on this genus. The inadequate original description and poor accompanying figures render any speculation about the identity and placement of *Riaziana* difficult, but the shapes of the paramere and phallus suggest that it might be a member of the Eysarcorini, potentially related to *Carbula* or *Durmia*.

Senectius Rolston, 1987

This South American genus, with its only included species, *S. metallicus* Rolston, is another that Rolston (1987a) included in his artificial group of seven genera that have an unarmed abdominal venter and an elongate ostiolar ruga. Rolston (1987a) did not indicate any possible relationship to any other genera. This genus has metallic green or blue markings and has the mesosternum weakly carinate.

Vitruvius Distant, 1901

Distant (1901b) originally placed this Oriental genus, and its only included species, *V. insignis* Distant (**Figure 2.24H**), in the Tessaratomidae but further indicated that it did not seem to be related to any known tessaratomine genus. This is where all other workers treated it until Kumar and Ghauri (1970) transferred it to the Pentatomidae. They did not place it in any known tribe but suggested that it might deserve a new tribe. Superficially, members of this genus resemble members of the Phyllocephalinae. The rostrum is shorter than in most pentatomines, but it is still longer than that seen in most phyllocephalines, reaching between the middle coxae. The mesosternum is longitudinally carinate, the abdominal venter is unarmed, and the ostiolar rugae are relatively short, not reaching to the middle of the metapleura.

Arrow-headed Bugs

Štys (2000), at the 21st International Congress of Entomology, reported on a "strange arrow-headed bug" from Australia. He examined both adults and immatures that had been collected under the scales of bark of a eucalyptus tree from South Australia and another adult collected in flight from Western Australia. He indicated that many of its character states were shared by other Pentatominae, but it also had many "unique" characters within the Heteroptera. Štys listed the following defining characters: (1) general facies "lygaeid" or aradid-like with many morphological characters similar to the Lygaeoidea; (2) head relatively long with lateral pre-ocular extensions; (3) bases of labrum and labium well-separated, by at least half the length of head; (4) juga much longer than tylus, apically free in immatures, but enclosed by anteriorly projecting bucculae in adults; (5) labium thin and long, reaching nearly to genitalia; (6) reduced and somewhat unique trichobothrial pattern; (7) adults with an L-shaped stridulatory area on each side of abdominal sternite III, and later instar immatures with linear stridulatory areas along lateral margins of laterotergites II and III laterad of spiracles. This last character, stridulatory areas in immatures, is only known in the Reduviidae. Štys (2000) hypothesized that the arrow-headed bugs might belong in the Pentatominae and be related to the Diemeniini or the Mecideini (probably because they share stridulatory areas), or that they may represent a new relict pentatomomorphan family. This taxon still awaits formal description.

2.2.10.7.47 Fossil Genera

Placement of fossil taxa to genus, or even to tribe, can be difficult because of the condition of most fossils and the inability to observe certain key characters. Below, we list the fossil pentatomid genera known to us, with comments on their current classification. At least three fossil families have been described in the Pentatomoidea. The Mesopentacoridae, and its included genera, is discussed earlier in this chapter (see **Section 2.2.8**); and the relatively new fossil family, Primipentatomidae, and its included genera is treated later (see **Section 2.2.13**). The third family, Spinidae, was described by Hong in 1982. The type genus of the Spinidae, *Spinus* Hong, is preoccupied so Spinidae is an invalid name. Hamilton (1992), however, placed *Spinus* as a junior synonym of the auchenorrhynchan genus *Homopterulum* Handlirsch, effectively making the Spinidae a junior synonym of the fossil membracoid family Jascopidae.

Apodiphus robustus **Jordan, 1967**
Popov (2007) placed this fossil species as a junior synonym of *Pentatomoides atratus* (also see below).

Cacoschistus **Scudder, 1890;** *Pentatomites* **Scudder, 1890;** *Polioschistus* **Scudder, 1890;** *Poteschistus* **Scudder, 1890;** *Teleoschistus* **Scudder, 1890;** *Thlimmoschistus* **Scudder, 1890;** *Thnetoschistus* **Scudder, 1890; and** *Tiroschistus* **Scudder, 1890**
Kirkaldy (1909) asserted that all of the above listed genera may possibly be related to the carpocorine genus *Euschistus*. Scudder (1890) described another fossil genus, *Mataeoschistus*, but this has been placed as a junior synonym of *Thnetoschistus* (Carpenter 1992).

Carpocoroides **Jordan, 1967;** *Halynoides* **Jordan, 1967;** *Pentatomoides* **Jordan, 1967;** *Pseudopalomena* **Jordan, 1967;** *Rhomboidea* **Jordan, 1967;** *Suspectocoris* **Jordan, 1967**
Technically speaking, these fossil generic names may not be taxonomically available. Jordan (1967) described ten new species of fossil pentatomids, placing them into six new genera. But, he did not provide a generic description for any of the six genera, nor did he designate type species. *Carpocoroides* and *Pentatomoides* originally included multiple species, so they are not taxonomically available (no type species designated). The remaining four genera were originally monotypic, and so the type species is known (by monotypy), but for these genera to be valid, ICZN Article 13.4 states that the description must be in the form of a combined description of a new genus-group taxon and new species (if marked by gen. nov., sp. nov., or equivalent expression). Jordan (1967) did not use this type of notation, but, in his discussion, he states "Trotzdem habe ich 22 Arten neu benannt, vielfach sogar neue Gattungsnamen

prägen müssen" [Despite of it I have named 22 species, many times I had also to create new generic names]. This could be interpreted as his intention to describe the new genera in combination, and may allow for *Halynoides*, *Pseudopalomena*, *Rhomboidea*, and *Suspectocoris* to be valid. At any rate, he did not indicate to what tribe these species belonged. *Suspectocoris grandis* Jordan originally was placed in the Acanthosomatidae, but Popov (2007) transferred it to the Pentatomidae. Jordan (1967) originally may have intended to show relationship by the names he chose (i.e., *Carpocoroides* might be related to *Carpocoris*, *Halynoides* might be related to *Halys*, and so on), but we do not know this for sure. Popov (2007) indicated that *Pentatomoides* might actually be a synonym of the extant genus *Carpocoris*, and he transferred one of Jordan's *Pentatomoides* species to the genus *Carpocoroides*.

Deryeuma Piton, 1940
Piton (1940) originally indicated that this genus was similar to the strachiine genus *Eurydema* (notice that *Deryeuma* is an anagram of *Eurydema*).

Mesohalys Beier, 1952
Beier (1952) originally placed this genus, and its only included species, *M. munzenbergiana* Beier, in the Halyaria (= Halyini).

Necanicarbula Zhang, 1989
This fossil genus originally was placed near the eysarcorine genera *Carbula* and *Dymantiscus* Hsiao (Zhang 1989).

Poliocoris Kirkaldy, 1910 and *Teleocoris* Kirkaldy, 1910
Kirkaldy (1910) originally indicated that *Teleocoris* had some resemblance to various tessaratomids but ultimately concluded that it probably belonged in the Pentatomidae: Halyini. He further indicated that *Poliocoris* probably was related to *Teleocoris* and, thus, it should also belong in the Halyini.

Taubatecoris Martins-Neto, 1997
This is another fossil genus, recently described from Brazil (Martins-Neto 1997). It originally was not placed in any known tribe; its correct taxonomic placement requires further study.

2.2.10.8 *Phyllocephalinae Amyot and Serville, 1843*

This interesting group seems to have received little attention in recent years. It is an Old World group, occurring in Africa, the Middle East, the Indian subregion, the Oriental region, and extending into Australia; it currently contains 45 genera and 214 species (**Tables 2.2, 2.3**).

This taxon was proposed first at the family-level (Phyllocéphalides) by Amyot and Serville (1843) and was recognized as a family (Phyllocephalidae) by Dallas (1851). This group through time has been treated at all levels from a genus group (Gross 1976), a tribe (Leston 1952a) within the Pentatominae, a subfamily (Stål, 1865, Lethierry and Severin 1893, Distant 1902, Kirkaldy 1909, China and Miller 1959, Linnavuori 1982, Schuh and Slater 1995, Rider 2006a), or a family (Vidal 1949, Miyamoto 1961). Gross (1976) treated them as a genus group at the same level as other genus groups (= pentatomine tribes) and related this group to the *Cephaloplatus* group (based on similar jugal expansions of the head, ostiolar rugae, and even male genital characters). It appears that the general consensus among recent authors is to treat it as a subfamily.

Members of this subfamily are medium to large in size, some are ovate (**Figure 2.27G**) whereas others are elongate, almost parallel-sided (**Figure 2.27H**), and the dorsal surface is often quite rugulose; their coloration is usually yellows, tans, or browns, allowing them to blend in with their surroundings. The juga are usually longer than the tylus, sometimes meeting in front of the tylus, and may be expanded or foliate; there sometimes is a tooth-like projection anterior to each compound eye. The bucculae are usually short, flap-like (**Figure 2.1O**). The single most diagnostic character is the distinctively short rostrum, which does not or only barely surpasses the procoxae (it may be a little longer in Australian genera, but it still reaches, at most, to the anterior margin of the mesocoxae) (**Figure 2.5B**). Rostral segment I and half of segment II lie between the bucculae. The humeral angles may be rounded or greatly produced. The ostiolar rugae are reduced and auriculate. The digestive system in the Phyllocephalinae is formed differently than in other pentatomine subfamilies and was referred to as "a kind of filter chamber" by Goodchild (1966). The basal plates of the phallus are usually strongly sclerotized; the phallotheca has a pair of short ventral processes that are flattened dorsoventrally, and widened posteriorly; the ventral lobe of the conjunctiva is sclerotized and projected dorsad on each side as a spoon-shaped process; and the vesica is short, never surpassing the ventral lobe of the conjunctiva (Linnavuori 1982, Hasan and Kitching 1993, Konstantinov and Gapon 2005). The female spermathecal duct has a dilation and sclerotized rod that is somewhat swollen at the base; the spermathecal bulb is ball-shaped with two short tubular diverticula (Gross 1976).

There has been little work on the higher classification within this subfamily until recently. Linnavuori (1982) presented an extensive discussion of the phylogeny of African genera, but refrained from using actual tribal names. Ahmad (1981) and Ahmad and Kamaluddin (1976; 1978b; 1988; 1990a,b; 1992; 1994), in a series of papers, studied several phyllocephaline genera and erected several tribal names to contain the Indian genera. Their classification does not totally agree with that of Linnavuori, but, between the two, and with some intuition, most phyllocephaline genera can be placed to tribe.

Linnavuori (1982) placed the African genera in three generic groups (**Table 2.6**) but indicated that there were still three genera that seemed to represent their own independent phylogenetic lineages. At the time of Linnavuori's work, only one tribal name was available, Delocephalini. Actually Delocephalini was erected by Horváth in 1900 as a tribe in the family Tessaratomidae, but Schouteden (1909) (and confirmed by Kumar 1969a) transferred the type genus, *Delocephalus* Distant, to the Phyllocephalinae of the Pentatomidae without commenting on the tribal situation.

Ahmad and Kamaluddin (1976; 1978b; 1988; 1990a,b; 1992; 1994) worked extensively on the higher classification of the subfamily, erecting three new tribal names, and assigned a number of genera to the nominate tribe (**Table 2.6**). Basically, their classification goes something like this: Those genera that have the humeral angles greatly expanded and produced anteriorly with the anterolateral margins toothed or crenulate belong in the Cressonini; the elongate slender species belong in the Megarrhamphini; the species that have the lateral margins of the head expanded and laminate belong in the Tetrodini; and all others would belong in the Phyllocephalini. By combining the above classification of Ahmad and Kamaluddin with that of Linnavuori, we have arrived at an integrated classification for this chapter (**Table 2.6**).

There have been few systematic works on phyllocephaline groups since the 1980s (Linnavuori 1982, Ahmad and Kamaluddin 1988, Kamaluddin and Ahmad 1988) or early 1990s (Ahmad and Kamaluddin 1990a,b; 1992). Kamaluddin and Ahmad (1991) discussed the taxonomic position of three cressoniine genera, *Kafubu* Schouteden, *Melampodius* Distant, and *Nimboplax* Linnavuori. Rider and Zheng (2004) provided nomenclatural notes and a checklist of the Chinese species. And Arnold (2011b) described two new species of *Uddmania*, bringing the total known species in this genus to three.

According to Gross (1976), "there is a strong tendency for members of this group to occupy desert or very dry areas though a number do live in much moister, though always tropical, areas." There are host plant records for a handful of species; the records are predominantly from various grasses (Poaceae), including a couple of records from bamboo. Linnavuori (1982) recorded an undetermined species of *Phyllocephala* Laporte as occurring on or near the ground at the base of the grass *Panicum turgidum* tussocks. He noted that most other phyllocephalines usually occur higher up on the plants.

2.2.10.8.1 *Cressonini Kamaluddin and Ahmad, 1991*

This tribe was proposed to contain those phyllocephaline genera that have the humeral angles greatly expanded and produced and have the anterolateral pronotal margins toothed or crenulate. Whether these are good phylogenetic characters remains to be determined. This tribe currently contains six genera and 12 species (**Table 2.3**).

2.2.10.8.2 *Megarrhamphini Ahmad, 1981*

This tribe has traditionally been characterized as including those genera whose species are more elongate and slender (**Figure 2.27H**). Linnavuori (1982) described his *Lobopeltista* group in a similar fashion although none of the genera that he discussed in this group is currently included in the Megarrhamphini. But he dealt primarily with the African fauna whereas *Megarrhamphus*, and relatives, are more Asian

TABLE 2.6

Recent Classifications of the Subfamily Phyllocephalinae

Linnavuori Classification	Ahmad Classification	Combined Classification
***Phyllocephala* Group**	**Cressonini** Kamaluddin & Ahmad	**Cressonini** Kamaluddin & Ahmad
Lamtoplax Linnavuori	*Cressona* Dallas	*Cressona* Dallas
Melampodius Distant	*Kafubu* Schouteden	*Kafubu* Schouteden
Nimboplax Linnavuori	*Melampodius* Distant	*Lamtoplax* Linnavuori
		Melampodius Distant
***Lobopeltista* Group**	**Megarrhamphini** Ahmad &	*Nimboplax* Linnavuori
Dichelorhinus Stål	Kamaluddin	
Frisimelica Distant	*Bakerorandolotus* Ahmad &	**Megarrhamphini** Ahmad &
Gellia Stål	Kamaluddin	Kamaluddin
Gonopsis Amyot & Serville	*Megarrhamphus* Bergroth	*Bakerorandolotus* Ahmad &
Katongoplax Linnavuori	*Randolotus* Distant	Kamaluddin
Lobopeltista Schouteden		*Megarrhamphus* Bergroth
Macrina Amyot & Serville	**Phyllocephalini** Amyot & Serville	*Randolotus* Distant
Sandehana Distant	*Dalsira* Amyot & Serville	
	Diplorhinus Amyot & Serville	**Phyllocephalini** Amyot & Serville
***Dalsira* Group** [= Delocephalini ?]	*Gonopsis* Amyot & Serville	*Basicryptus* Herrich-Schäffer
Chalcopis Kirkaldy	*Mercatus* Distant	*Borrichias* Distant
Dalsira Amyot & Serville	*Metonymia* Kirkaldy	*Chalcopis* Kirkaldy
[= *Basicryptus*]	*Nazeeriana* Ahmad & Kamaluddin	*Dalsira* Amyot & Serville
Delocephalus Distant	*Salvianus* Distant	*Delocephalus* Distant
Eonymia Linnavuori	*Schyzops* Spinola	*Dichelorhinus* Stål
Melanocryptus Linnavuori [= *Jayma*]		*Diplorhinus* Amyot & Serville
Metocryptus Linnavuori	**Tetrodini** Ahmad & Kamaluddin	*Eonymia* Linnavuori
Metonymia Kirkaldy [= *Dalsira*]	*Gellia* Stål	*Frisimelica* Distant
Penedalsira Linnavuori	*Tetroda* Amyot and Serville	*Gonopsimorpha* Yang
Schismatops Dallas	*Tetrodias* Kirkaldy	*Gonopsis* Amyot & Serville
Tantia Distant		*Jayma* Rider
Tshibalaka Schouteden		*Kaffraria* Kirkaldy
		Katongoplax Linnavuori
Independent Lineages		*Lobopeltista* Schouteden
Kafubu Schouteden		*Macrina* Amyot & Serville
Schyzops Spinola		*Magwamba* Distant
Storthogaster Karsch		*Melanocryptus* Linnavuori
		Mercatus Distant
		Metocryptus Linnavuori
		Metonymia Kirkaldy
		Minchamia Gross
		Nazeeriana Ahmad & Kamaluddin
		Neoschyzops Ahmad & Kamaluddin
		Penedalsira Linnavuori
		Phyllocephala Laporte
		Roebournea Schouteden
		Salvianus Distant
		Sandehana Distant
		Schismatops Dallas
		Schyzops Spinola
		Storthogaster Karsch
		Tantia Distant
		Tshibalaka Schouteden
		Uddmania Berg
		Tetrodini Ahmad & Kamaluddin
		Gellia Stål
		Tetroda Amyot and Serville
		Tetrodias Kirkaldy

in distribution. These two groups may be related. Linnavuori (1982) believed the narrowing of the body to be an adaptive character associated with living on narrow grass leaves. He related this group to the *Phyllocephala* group based on a common prolongation of the juga, the dorsal coloration and pattern of carinae, and the development of pale callous bands on the thoracic pleura. This tribe currently contains three genera and 12 species (**Table 2.3**).

2.2.10.8.3 *Phyllocephalini Amyot and Serville, 1843*

Under the name *Phyllocephala* group, Linnavuori (1982) treated four African genera (*Lamtoplax* Linnavuori, *Melampodius*, *Nimboplax*, and *Phyllocephala*). Kamaluddin and Ahmad (1991) transferred the first three genera listed above to the Cressonini. Linnavuori (1982) defined this group by a number of characters: body large, broadly ovate (**Figures 2.18H, 2.27G**); head shovel-shaped, with stiff hairs; of a uniform color; pronotum rugulose, the anterior margin with strong papillae bearing a setigerous apical pit; scutellum broadly ligulate; female genital plates coarsely punctate; and the mesosternum with broad, slightly elevated medial carinae. Again, it should be cautioned that these characters may apply to the Cressonini as the majority of genera treated here by Linnavuori have now been transferred to the Cressonini.

The genus *Delocephalus* has had an interesting taxonomic history, which has involved family-level nomenclature. Distant (1881) originally described the genus and species, *D. miniatus*, from Madagascar and placed the species in the family Tessaratomidae (the spiracles on abdominal segment II were exposed, a character thought to be important in defining the Tessaratomidae). Horváth (1900), realizing that *Delocephalus* was unlike any tessaratomids he knew, erected a new division, Delocephalaria, within the Tessaratomidae. Based primarily on its general facies, Schouteden (1909) transferred *Delocephalus* to the Pentatomidae (Phyllocephalinae). Cachan (1952), apparently following Horváth, listed this genus in the Delocephalaria within the Tessaratomidae. Kumar (1969a) studied the male genitalia and confirmed Schouteden's opinion that *Delocephalus* was a member of the Pentatomidae. Linnavuori (1982) included *Delocephalus* in his *Dalsira* group, but Ahmad and Kamaluddin (1990b) included *Dalsira* Amyot and Serville in the Phyllocephalini. Until further work can be conducted, we tentatively treat *Delocephalus* as a member of the Phyllocephalinae: Phyllocephalini, and Delocephalini is treated as a synonym of the Phyllocephalini.

Kamaluddin and Ahmad (1988) reviewed the genera and species from India and Pakistan. This tribe currently contains 32 genera and 167 species (**Table 2.3**).

2.2.10.8.4 *Tetrodini Ahmad, 1981*

Ahmad and Kamaluddin (1990a) defined this tribe as having the body broad and elongately ovate; the head longer than broad and distinctly longer than the pronotum; the length anterior to the compound eyes distinctly longer than the posterior part of head; the juga usually longer than and meeting in front of the tylus; the antenniferous tubercles not visible from dorsal view; the humeral angles rounded, never greatly produced; the anterolateral pronotal margins usually entire, not denticulate; and the scutellum longer than broad. The female spermathecal bulb is ball-shaped with three relatively short diverticula, and the duct between the sclerotized rod and the spermathecal pump is quite short (Linnavuori 1982). This tribe currently contains four genera and 23 species (**Table 2.3**).

2.2.10.9 Podopinae Amyot and Serville, 1843

This subfamily, described by Amyot et Serville, 1843 (as 'Podopides'), has not only been classified historically within the family Pentatomidae, but specific genera (e.g., the genus *Tarisa*) or groups of genera have at times also been placed within the family Scutelleridae (most likely due to the enlarged scutellum seen in both groups). The podopines are distributed worldwide, except in the New Zealand subregion. They represent an especially heterogenous group of species, primarily because of the difficulty in classifying some of the genera/taxa (see summary of higher classification in Davidová-Vilímová and McPherson 1995). Schouteden (1903) treated the African species of Podopinae and later (1905b) provided a summary of the Podopinae in an excellent monograph. The most recent studies concerning the higher classification of the Podopinae, including characters of the male external genitalia, were reported by Schaefer (1981b, 1983). The southern European taxa were treated by Péricart (2010).

Three main trends exist in the recent higher classification of the Podopinae (again see details in Davidová-Vilímová and McPherson 1995): (1) to demote the subfamily to the level of tribe (McDonald 1966, Ahmad 1977); (2) to view the subfamily as a monophyletic group resulting in its being divided into two or three taxa classified as tribes or generic groups (Gross 1975b, Gapud 1991); or (3) to view the subfamily as a monophyletic group resulting in its being divided into two to five tribes (Schaefer 1981b, 1983; Hasan and Kitching 1993; Davidová-Vilímová and Štys 1994 - but note that the two new tribes, Deroploini and Brachycerocorini, erected by Davidová-Vilímová and Štys (1994) are *nomina nuda*). Recently, Gapon (2008a) accepted only two sister groups in the monophyletic Podopinae: the Graphosomatini *sensu lato* and the Podopini. The latter group was divided by Gapon into three sub-tribes, two of which are new: the Podopina, the Kayesiina, and the Scotinopharina. Gapon (2008a), however, did not study all taxa in these groups, and so further study is needed before these categories are accepted conclusively.

The supposed ground plan of the ancestral podopine is as follows: The antennae are five-segmented, with two flagellar segments longer and more slender than the other segments. The labium extends to the middle or hind coxae, with segment I not extending beyond the posterior margins of the bucculae. The bucculae are relatively long, simply shaped; and the ocelli are relatively small and not protruding. The antenniferous tubercles are of the common pentatomoid pattern, located on the ventral surface of the head close to the anteroventral margin of the compound eyes. The compound eyes are large but not distinctly protruding beyond the outline of the head. The juga are relatively simple, straight and flat, without conspicuous structures, and are identically shaped in both sexes. The outline of the head anterior to the compound eyes is broadly triangular without distinct hairs or bristles and/or coatings/crust.

The pronotum is relatively simple, without a longitudinal ridge or keel and lacking distinctive chaetotaxy and/or coatings, but with a granulated area on each side anteriorly. The anterior angles of the pronotum are rounded, without processes; the lateral angles also are rounded, not protruding; and the anterolateral margins are simple and rounded. The scutellum lacks conspicuous structures or distinct hairs or coatings. The frena are developed as oblique structures on the ventral surface of the scutellar margins, reaching at most one-half of the scutellar length (**Figure 2.6H**). The apex of the scutellum reaches at least to the middle of abdominal tergite VI; it is relatively broad basally, only slightly narrowed apically, usually leaving almost all of the corium, most of the clavus, and the apex of the membrane uncovered.

The venter of the thorax has simple episterna, without conspicuous structures. The metapleural scent gland ostiole is accompanied by a small, short, and narrow ruga; and a relatively large, finely granulated, flat evaporative area. The abdominal venter is unarmed at the base and usually flat to slightly convex; and the trichobothria are paired on each side of abdominal sternites III through VII, and their alveoli are not differentiated from the surrounding area. The posterolateral angles of the connexiva are simple and rounded.

The pygophore (terminology after Schaefer 1977 and Davidová-Vilímová and McPherson 1991) has its external opening directed posteriorly. The ventral wall is smooth, simple, and lacks conspicuous structures. The ventral rim is compact; the shape ranges from nearly straight to slightly concave and may or may not have an incision and/or process. The infolding of the ventral pygophoral rim is parallel to the ventral wall and has a cup-like sclerite on the internal ventral wall that is located deep within the pygophore. The dorsal rim and its infolding lack setae, and the rim lacks processes; the infolding is simple, forming an obtuse angle to the dorsal wall. The lateral rim is simple, narrow, and sharp; its outline is rounded and lacks setae and processes. The infolding of the lateral pygophoral rim, which forms the pygophoral area, lacks setae and is simple with the dorsal processes of various shapes and sizes. The paramere has a single arm, basic in shape, roughly simple cylindrical, with fine chaetotaxy; a larger tubercle is present on the posterior margin, which lacks membranous structures. The phallus (terminology adopted from Gapon and Konstantinov 2006) has a phallotheca without a shield but with basal tubercles ventrally. The conjunctiva is larger, membranous, and has a larger number of different membranous processes. Only the paired medial penial lobes are not fused; they are developed as sclerotized structures in the conjunctiva. The vesica is straight and short.

The spermatheca (based on several species examined; see below) has the spermathecal bulb roughly globular in shape with 0 to 3 distal processes; the two flanges are developed, and the pumping region between the flanges is of various patterns, and may be straight or convoluted. The sclerotized rod and dilation are elongate, and the proximal part of the spermathecal duct is free of the rod for various lengths. Linnavuori (1982) stated that only one pentatomine type of spermatheca (bulb small and simple) occurs in the Podopinae, but he studied only a few species, none of which had developed processes on the bulb.

During the past several decades, several taxa have been removed from or transferred to the Podopinae. More recently, the podopines have been accepted by most workers as a monophyletic group defined by the following synapomorphies: hypertrophied scutellum often overlapping the apex of the abdomen (inclusive apomorphy developed several times within the Pentatomoidea); scutellum broad/wide in the anterior part, only slightly narrowed or even slightly wider toward the apex (large parts of the hemelytra still uncovered by the scutellum); frena developed on the ventral surface of the scutellum, one on each side as an oblique ledge-shaped structure reaching maximally to half of its length (autapomorphy); and ostiolar rugae relatively small, short, and narrow. Dorsal projections of various shapes and sizes are developed on the pygophoral area dorsad of parameres. Parandria (movable projections on the lateral pygophoral rim) also have been mentioned as an autapomorphy of the Podopinae, but they are developed only in some of the genera from the *Podops* group.

The following taxa were historically, at least for a short period, included in the Podopinae or classified as their close relatives. The genus *Eipeliella* originally was placed in the Graphosominen (= Podopinae), probably due to the enlarged scutellum. Schouteden (1916) tentatively placed the genus in the Pentatomidae near the genus *Eysarcoris*, which also has an enlarged scutellum. More recently, Linnavuori (1982) placed *Eipeliella* in its own group and indicated that it might be related to the Antestiini. *Eipeliella* has the frena located on the lateral margins of the scutellum, and the ostiolar rugae are long and acute apically. Distant (1910a) originally placed his Australian genus *Ippatha* (**Figure 2.24F**) in the Graphosomatinae (= Podopinae), a position followed by Musgrave (1930) in his revision of the Australian Podopinae. Gross (1975b), believing *Ippatha* to be distinct from the Podopinae, placed it in its own genus group. *Ippatha* has an enlarged evaporative area, and the frena and dorsal processes of the pygophore are not developed, thus distinguishing it from the Podopinae. The genus *Amphidexius* never has been explicitly classified in the Podopinae, but Bergroth (1918) originally indicated a possible relationship with the genus *Dandinus,* which is a member of the Podopinae. Gross (1975b) erected a new genus group, the *Ochisme* group, and placed within it *Amphidexius* and several other related genera. *Amphidexius* has the frena located laterally on the scutellum, and the scutellum is more triangular in shape. The South American genus *Prionotocoris* originally was placed near the genus *Ancyrosoma* Amyot and Serville, an Old World podopine. Grazia (1988) indicated that *Prionotocoris* was more closely related to a group of genera tentatively placed in the Carpocorini, the *Euschistus* group. All these genera can be separated from the podopines by the subtriangular scutellum and the lateral frena that are relatively long, reaching two-thirds the length of the scutellum. Another South American genus, *Glyphepomis*, that originally was touted as the only Neotropical member of the Podopini has been correctly moved to the pentatomine tribe Carpocorini; it differs from the Podopinae in having lateral frena on a short, somewhat oval, scutellum that narrows somewhat apically. Finally, the tribe Procleticini (monophyly shown by Schwertner and Grazia 2012) has been purported to be closely related to the Podopini (Kormilev 1955), but the frena are short and lateral, and some of the genera have a subtriangular scutellum.

All podopines are phytophagous, feeding on the vegetative as well as reproductive parts of plants from different families, with groups of genera showing preferences for certain habitats and hosts (Schaefer 1981b, 1983; see below for more information). The only species considered to be pests are included in the genus *Scotinophara* Stål [e.g., *S. coarctata* (F.), *S. lurida* (Burmeister) and *S. vermiculata* (Vollenhoven)]. They are known to be important pests of rice in Southeastern Asia (Panizzi et al. 2000, Joshi et al. 2007). Besides these species of *Scotinophara* (the most speciose genus with 40 described species), species of the genera *Podops* Laporte (**Figure 2.27I**) and *Graphosoma* Laporte (**Figure 2.27J**) are studied the most often (e.g., Leston 1953b; Durak and Kalender 2007, 2009; Gamberale-Stille et al. 2009).

No geographical area contains members of all the genera-groups in the Podopinae. For example, members of the *Deroploa* group are restricted to the Australian region. The Nearctic region contains members only from the *Podops* group; the Neotropical region has members from two groups (*Graphosoma* group and the *Podops* group); and the Australian region contains members from only two groups (*Deroploa* group and *Podops* group). Four groups, namely the *Brachycerocoris* group, the *Graphosoma* group, the *Podops* group, and the *Tarisa* group are distributed mainly within two regions, Afrotropical and Palearctic. In all, there are about 269 species in 68 genera (**Tables 2.2, 2.3**) currently classified within the Podopinae. Rider and Zheng (2004) provided nomenclatural notes and a checklist of the Chinese species. The podopine species have been treated or catalogued for various geographical regions: Nearctic (Froeschner 1988c), South America (Grazia et al. 2015), southern Europe (Péricart 2010), India (Salini and Viraktamath 2015), and Australia (Cassis and Gross 2002).

The position of the Podopinae within the Pentatomoidea is still not resolved. The specific structure of their frena excludes a relationship with the Scutelleridae (frena not developed), a taxon with which it was sometimes classified. Most of the pentatomoid groups have an enlarged scutellum, but it is usually triangular in outline, becoming narrowed toward its apex, with the frena lying along the lateral margins of the scutellum, not ventrally as in the Podopinae. The taxon Podopinae is defined by a specific set/group of apomorphies as a monophyletic group. In a recent phylogenetic study of Pentatomoidea, Grazia et al. (2008) included only three somewhat heterogenous members of the Podopinae, and, thus, the results are not conclusive.

The higher classification historically has varied; at present, five monophyletic genus groups are recognized.

2.2.10.9.1 Key to the Podopinae Genus Groups

1 Compound eyes pedunculate, conspicuously protruding (**Figures 2.10A, B**); humeral angles of pronotum either divided into anterior serrate section and posterior rounded section or undivided; if rounded, then ostioles of metathoracic scent glands located on distinct tubercle (**Figure 2.4F**) and frena on scutellum exceed up to 1/10 of its length; widespread (**Figures 2.18L, 2.27I**) ....*Podops* group

1' Compound eyes not pedunculate, either not or only slightly protruding; humeral angles of pronotum generally not divided into anterior serrate and posterior rounded sections 2

2(1) Each humeral angle of pronotum either with conspicuous long protuberance or rounded or divided (**Figure 2.10C**); medial margin of proepisternum flattened; longitudinal keel on pronotum extending from 2/3 its length to its posterior margin; Australia (**Figure 2.18J**)... .... *Deroploa* group

2' Each humeral angle of pronotum simply rounded but if of a different shape, then medial margin of proepisternum not flattened; longitudinal keel on pronotum developed only anteriorly.. 3

3(2) Scutellum with two large, medial tubercles, one near base and the other near middle, or basal disc of scutellum with distinct medial keel; each buccula with triangular tubercle anteriorly or medially (**Figure 2.10D**); scutellum reaching at least to middle of abdominal tergite VIII; Old World (**Figure 2.18I**) ....*Brachycerocoris* group

3' Scutellum without conspicuous medial tubercles or keel, or they are in a different pattern than above; each buccula without tubercle or if tubercle present, then scutellum shorter, reaching at most to middle of abdominal tergite VII 4

4(3) Entire surface of infoldings of lateral and dorsal pygophoral rims covered by dense, short setae; lateral rim rounded, without protuberance (**Figure 2.10E**); infolding of ventral pygophoral rim without conspicuous structures; Old World .... *Tarisa* group

4' Infoldings of lateral and dorsal pygophoral rims without dense setae, sometimes bearing only sparse, longer setae; lateral pygophoral rim with either larger protuberance or large conspicuous protuberance detached by incision (**Figures 2.10F-H**); conspicuous structures of different types developed on infolding of ventral pygophoral rim; widespread (**Figures 2.18K, 2.27J**)...*Graphosoma* group

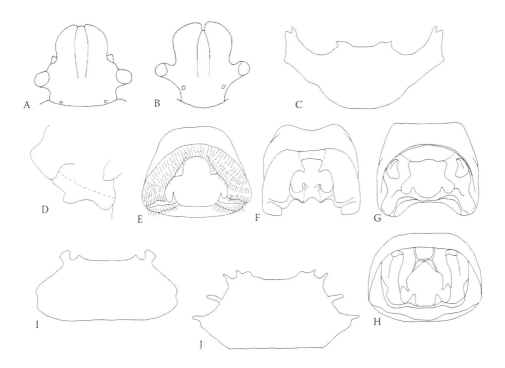

FIGURE 2.10 A, *Podops inuncta*, head, dorsal view; B, *Weda stylata*, head, dorsal view; C, *Deroploopsis trispinosus*, pronotum, dorsal view; D, *Brachycerocoris camelus*, buccula, lateral view; E, *Dybowskyia reticulata*, pygophore, posterior view; F, *Graphosoma lineatum*, pygophore, posterior view; G, *Leprosoma inconspicuum*, pygophore, posterior view; H, *Ventocoris trigonum*, pygophore, posterior view; I, *Podops inuncta*, pronotum, dorsal view; J, *Tornosia insularis*, pronotum, dorsal view.

2.2.10.9.2 Podops Group

The *Podops* group contains 28 genera and 135 species (**Table 2.3**) that are widely distributed in all zoogeographical regions, and is the only group representing Podopinae in the Nearctic and Oceanic regions (**Figures 2.18L, 2.27I**). A 29th genus, *Burrus* Distant, probably also belongs in this group. This genus has not been studied since the original description; the habitus provided in the original description appears to be similar to other members of the *Podops* group, including the stylate eyes.

The following apomorphies define the *Podops* group, the most morphologically homogeneous group within the Podopinae. They have conspicuously pedunculate compound eyes (**Figures 2.10A, B**), and the antenniferous tubercles are large and at least partially visible in dorsal view. The lateral pronotal angles are composed of an anterior, sharply toothed projection and a posterior, rounded projection (**Figures 2.10I, J**). The ventral wall of the pygophore has a lateral, transverse keel on each side, each lateral rim has a large projection or a distinct movable, distinctly delimited protuberance (parandrium). The pygophore has an infolding of the ventral rim immediately parallel to the ventral wall, and there is a cup-like sclerite near the ventral rim.

The spermathecae of five genera have been described. In *Podops* [species studied were *P. curvidens* Costa and *P. inuncta* (F.)] (Leston 1953b, Pendergrast 1957, Jitka Vilímová, unpublished data), the spermathecal bulb is roughly globular, the distal flange of the pumping region is larger and flattened, and the proximal flange is smaller and dish-shaped. The region between the flanges is straight. The spermathecal duct is slender and proceeds through the sclerotized rod; the proximal part of the duct is relatively short. In *Amaurochrous* Stål [species studied were *A. cinctipes* (Say) and *A. dubius* (Palisot de Beauvois)], the spermathecal bulb is globular with three processes; the spermathecal duct is slender, proceeding through a sclerotized dilation (McDonald 1966). *Weda parvula* (Van Duzee) is similar to *Amaurochrous* species except that it has only two processes on the spermathecal bulb (McDonald 1966). In the genus *Scotinophara* (species studied were *S. horvathi* Distant, *S. lurida*, and *S. scotti* Horváth), the spermathecal bulb is globular but without processes (*S. lurida*), or with three thornlike processes of either equal length (*S. horvathi*), or with one process distinctly longer than the other two (*S. scotti*); both flanges are of roughly equal size and shape; the region between the flanges is either medially (*S. lurida*) or distally (other species) dilated; the spermathecal duct is slender, proceeding through a sclerotized dilation, and is relatively long (*S. lurida*), short (*S. scotti*), or very short (*S. horvathi*); the proximal part of the duct is free (Kim and Lee 1994). In the genus *Kayesia* Schouteden (species studied were *K. parva* Schouteden and *K. setigera* Linnavuori), the spermathecal bulb is ball-shaped, without processes; the region between the flanges is swollen (Linnavuori 1982).

The North American genera and species were reviewed by Barber and Sailer (1953). The review of the genus *Scotinophara* by Wongsiri (1975) did not include all known species; on the other hand, in the more recent review by Barrion et al. (2007), they described 19 new species, all from the Philippines; the validity of these new species needs confirmation. Davidová-Vilímová (1999) revised the genus *Tornosia*.

The members of this group often live in damp places, mostly on the stems of their host plants from the order Poales (e.g., Cyperaceae, Juncaceae, Poaceae) (Schaefer 1981b, 1983, Linnavuori 1982, Jitka Vilímová, unpublished data).

2.2.10.9.3 *Deroploa* Group

The *Deroploa* group contains nine genera and 20 species (**Table 2.3**), which are endemic to the Australian region (**Figure 2.18J**). It is a fairly homogenous group, morphologically distinctly different from the other podopine groups.

The *Deroploa* group of genera share several apomorphies. The eyes are slightly pedunculate, and each lateral angle of the pronotum has a conspicuous, long protuberance (**Figures 2.10C, 2.18J**). The pronotum has a longitudinal keel minimally extending two-thirds of its length posteriorly; it usually extends the entire length. The scutellum is long and slightly to distinctly exceeding the apex of the abdomen. The scent gland evaporatoria are small and developed only on the metapleura; the rugae are either distinct and hemicircular or reduced. None of these apomorphies is exclusive to the *Deroploa* group, but, as a set, they unambiguously define this monophyletic group.

Nearly all members of this group, where information is known, are arboreal, living in subtropical and/or temperate habitats. Only *Jeffocoris*, described by Davidová-Vilímová (1993) is known to be an herbivore in arid areas (Cassis and Gross 2002).

2.2.10.9.4 *Graphosoma* Group

The *Graphosoma* group contains 19 genera and 70 species (**Table 2.3**) distributed in four regions: Palearctic, Afrotropical, Oriental, and Neotropical (**Figures 2.18K, 2.27J**).

In the *Graphosoma* group, the thoracic episterna are modified by the development of protuberances or flattened margins. In the male pygophore, the infolding of the ventral rim is immediately parallel to the ventral wall and medially merges into a cup-like sclerite; the lateral rim has a large or small projection on each side (**Figures 2.10F-H**); and the conjunctiva of the phallus has a high number of membranous appendices. None of these characters is exclusive to this genus group, but, as a set, they adequately define this monophyletic group.

The spermathecae of two species of *Graphosoma* have been studied, *G. lineatum* (Jitka Vilímová, unpublished data) and *G. rubrolineatum* (Kim and Lee 1994). The spermathecal bulb is broadly oval, the flanges of the pump region are similar in size and shape, the region between the two flanges is slightly convoluted, the spermathecal duct is slender and proceeds through a sclerotized dilation, and a relatively long proximal part of the duct is free.

The genus *Leprosoma* Baerensprung has been revised recently (Gapon 2008b, 2010b), and Linnavuori (1984) provided a key to species of *Tholagmus* Stål. The only genus of Podopinae recorded in South America, the monotypic *Neoleprosoma* Kormilev, is included in this group.

Members of this group occur in different types of habitat as well as on a wide spectrum of host plants from different families (Apiaceae, Brassicaceae, Poaceae, Ranunculaceae, Rosaceae, Rubiaceae). They feed mostly on the seeds (Schaefer 1981b, 1983; Jitka Vilímová, unpublished data).

2.2.10.9.5 *Tarisa* Group

The *Tarisa* group contains three genera and 22 species (**Table 2.3**) distributed in the Palearctic and Afrotropical regions.

This group is defined by a specific set or group of characters. They usually have relatively large antenniferous tubercles that are partially visible in dorsal view. The anterior part of the pronotum bears tubercles laterally and medially; the scutellum is conspicuously wide basally, completely covering the clavus and the membrane of the hemelytra; and only part of the corium is exposed. In the male pygophore, the large part of the ventral wall is either flattened or slightly concave; the infoldings of both the dorsal and lateral rims bear dense, short setae (**Figure 2.10E**).

The spermatheca of *Dybowskyia reticulata* (Dallas) has been described (Kim and Lee 1994). In this species, the spermathecal bulb is broadly oval with one long process; the distal flange of the pump region is slightly sclerotized, the proximal flange is not sclerotized, and the region between the flanges is slightly sclerotized and swollen. The spermathecal duct is slender, proceeding through a sclerotized dilation; a long proximal part of the duct is free.

All species live on different host plants, *Dybowskyia* Jakovlev on Apiaceae, and *Tarisa* on Asteraceae. Some species of this group are the only known podopines to feed on Chenopodiaceae (recently placed as a subfamily within the Amaranthaceae) (Schaefer 1981b, 1983; Jitka Vilímová, unpublished data).

2.2.10.9.6 *Brachycerocoris* Group

The *Brachycerocoris* group contains three genera and 13 species (**Table 2.3**) distributed in Palearctic, Afrotropical, and Oriental regions (**Figure 2.18I**).

This group of genera is defined by a specific set of characters. In this group, each buccula has a triangular process, either anteriorly or medially (**Figure 2.10D**). The pronotum has a longitudinal keel either anteriorly or along its entire length; the scutellum either has distinct tubercles, one basally and one on the middle in the longitudinal axis, or its basal plate continues as a longitudinal keel. The entire lateral rim of the pygophore is widened (Jitka Vilímová, unpublished data).

The spermatheca of *Bolbocoris variolosus* (Germar) has been illustrated (Linnavuori 1982). The spermathecal bulb is globular; the two flanges of the pumping region are of similar size and shape, and the region between the flanges is relatively long, longer than the spermathecal duct before it enters the sclerotized rod.

The members of the genus *Bolbocoris* Amyot and Serville are known only from Poaceae, whereas species of the genus *Brachycerocoris* Costa occur on a wide spectrum of plants including Fabaceae, Rhamnaceae, and Verbenaceae (Schaefer 1981b, 1983; Linnavuori 1982). Göllner-Scheiding (1992) provided a revision of the genus *Bolbocoris*, and Schaefer et al. (1996) revised the genus *Brachycerocoris*.

2.2.10.9.7 *Genera incertae sedis*

The six genera listed below are classified as *incertae sedis*. They do not show any characteristics (character or group of characters) that would enable us to include them in any of the monophyletic groups mentioned above. The distribution of most of the genera are summarized by Schouteden (1905b).

Cryptogamocoris Carapezza, 1997

This genus has a single species (*C. cornutus* Carapezza) known only from Tunisia where it lives on *Salsola vermiculata* L. (Amaranthaceae). In general appearance, this species is similar to *Tarisa* (Carapezza 1997), but there also appears to be some similarities with members of the *Graphosoma* group. Because the specific and important characters of the pygophore that distinguish these two groups have not been described, we cannot confidently resolve its proper placement.

Cyptocoris Burmeister, 1835

This genus, distributed in the Afrotropical region, has been included in the Graphosomatini by several authors (Schouteden 1905b, Gillon 1972, Linnavuori 1982). However, its relationship has not been resolved. In fact, several characters are not congruent with its placement in the Podopinae. Its compound eyes are not prominent, but, instead, they lie within the outline of the head. The head is strongly convex dorsally, almost semiglobular. The frena are not developed, and the pygophoral area is extremely narrow.

Eobanus Distant, 1901

This genus is distributed in the Oriental region (Distant 1901a). It has been placed at one time or another in the Podopinae near the genus *Bolbocoris* (Distant 1902), in the Graphosomatini (Schouteden 1905b), or in the Podopini (Ahmad 1977). *Eobanus* appears to be most similar to the *Deroploa* group, but this cannot be ascertained confidentally without further study.

Kundelungua Schouteden, 1951

This genus was not originally placed in any tribe (Schouteden 1951). It has been confirmed that it, distributed in the Afrotropical region in savannah type habitats, belongs in the Podopinae, but it does not seem to fit readily into any of the known generic groups. In Gapon's (2008a) classification, *Kundelungua* was placed with *Kayesia* in the subtribe Kayesiina (*Podops* group), but this is not strongly supported; many of the characters uniting the two genera are reduced in *Kundelungua* (Davidová-Vilímová 1993).

Neocazira Distant, 1883

This genus traditionally has been placed in the Podopinae with which it shares a number of important synapomorphies (e.g., pattern of frena), but it is not possible to place this genus unambiguously into any of the podopine genera groups. *Neocazira* is distributed in the Palearctic and Oriental regions. It is phenotypically most similar to the *Brachycerocoris* group.

Tahitocoris Yang, 1935

Yang (1935) erected a new family, the Tahitocoridae, for his new genus and species, *Tahitocoris cheesmani* Yang. The characters he used to differentiate this taxon from other pentatomoid families included (1) ocelli absent; (2) ostioles extremely reduced, simple and without rugae or evaporative areas; (3) scutellum quite short and broad, not extending beyond posterior margin of metathorax; and (4) wings rudimentary. China and Miller (1955) lowered this group to subfamily status within the Pentatomidae indicating that several of the characters listed by Yang also occur in various members of the Pentatomidae. Miller (1971) treated *Tahitocoris* as a member of the pentatomine subfamily Asopinae. Thomas (1994b), after conferring with W. R. Dolling, transferred *Tahitocoris* from Asopinae to the Podopinae, but kept it in its own tribe, the Tahitocorini. Dolling (1997) speculated that the broad scutellum probably evolved from a scutellerid or podopine ancestor; he further indicated that the female genitalia (presence of a triangulum with loss of distinct rami and valvulae) clearly placed this taxon within the Pentatomidae (he also described the female spermathecal duct as having a dilation with a sclerotized rod). Dolling (1997) acknowledged that the posteriorly convergent bucculae might signify a relationship with the Asopinae, but, ultimately, he indicated that the pedunculated eyes, the somber coloration, and the form of the trichobothria (single) was more indicative of the Podopinae: Podopini. Interestingly, the types for this species were reported to have been collected above 2000 ft. on the island of Tahiti from a giant fern, *Angiopteris evecta* (G. Forst.) Hoffm. (Marattiaceae) (Yang 1935).

2.2.10.9.8 Phylogenetic Notes

The *Podops* group represents the sister group to a clade including all other groups; within this clade, the *Deroploa* and *Graphosoma* groups appear as intermediate lineages, with the clade *Tarisa* + *Brachycerocoris* being the most advanced in the evolution of the Podopinae (**Table 2.7**).

Within the *Podops* group, the genus *Kayesia* is the basal taxon, with the genus *Tornosia* more advanced (its classification in the Podopinae and *Podops* group was confirmed by Davidová-Vilímová 1999), and the next group of genera related to *Podops* and finally the group related to *Scotinophara*. The specific movable protuberances of the lateral rim of the pygophore (parandria) are often erroneously mentioned as a characteristic feature of the Podopinae. They are, however, developed as a synapomorphy in the genera *Allopodops* Harris and Johnson, *Amaurochrous*, *Crollius* Distant, *Oncozygia* Stål, *Podops*, *Severinina* Schouteden, *Thoria* Stål, and *Weda* Schouteden forming a monophyletic group (subgroup Podopina). The remaining ten genera related to the genus *Scotinophara* form another monophyletic group (subgroup Scotinopharina).

TABLE 2.7

Podopinae

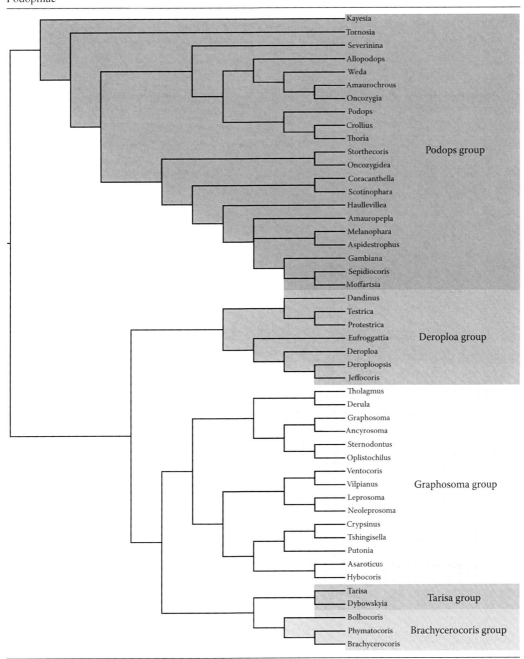

The genera *Dandinus*, *Testrica* Walker, and *Protestrica* Schouteden are basal in the *Deroploa* group, with the genera *Eufrogattia* Goding, *Deroploa* Westwood, and *Deroploopsis* Schouteden more advanced, and *Jeffocoris* the most advanced.

The genera in the *Graphosoma* group can be split into two relatively homogeneous groups, those related to *Graphosoma*: *Ancyrosoma*, *Derula* Mulsant and Rey, *Oplistochilus* Jakovlev, *Sternodontus*

Mulsant and Rey, *Tholagmus* and those related to *Ventocoris* Hahn: *Asaroticus* Jakovlev, *Crypsinus* Dohrn, *Hybocoris* Kiritshenko, *Leprosoma*, *Neoleprosoma* Kormilev and Pirán, *Putonia* Stål, *Tshingisella* Kiritshenko, *Vilpianus* Stål.

In the *Tarisa* group, the genus *Tarisa* is more basal than the genus *Dybowskyia*; in the *Brachycerocoris* group, *Brachycerocoris* is the most advanced, *Bolbocoris* is basal, and *Phymatocoris* Stål is in the middle.

2.2.10.10 Serbaninae Leston, 1953

This is a monotypic subfamily containing a single genus and species (*Serbana borneensis* Distant) (**Figure 2.27K**) (**Tables 2.2, 2.3**), which is endemic to Borneo. Members of this family are highly similar in appearance to members of the family Phloeidae, and, probably due to living in similar habitats, they have evolved (probably convergently) several novel characteristics in common with each other. For example, they both are relatively large and ovate with the margins of the head, pronotum, and abdomen produced and foliate. Also, each compound eye is divided into a dorsal and ventral section. They differ, however, in several significant characters. For example, the antennae are four-segmented in the Serbaninae (three-segmented in phloeids). The metathoracic scent glands are different in that the ostiole is placed more mesial than in the phloeids, ostiolar rugae are present (rudimentary in phloeids), and the vestibular scar is lacking (present in phloeids). Also, the serbanines possess a dilation and sclerotized rod in the female spermathecal duct (lacking in the phloeids). We believe that the spermathecal structure should place this group as a subfamily within the Pentatomidae.

Due to the similar appearance as described above, Distant (1906) originally placed the serbanines in the Phloeidae. Leston (1953c), based primarily on characters of the male genitalia, moved the serbanines to their own subfamily (within the Pentatomidae). However, Grazia et al. (2008) concluded that the serbanines should be placed in the Phloeidae based on the somatic characters of the external morphology listed above, characters that we believe may be due to convergent evolution from living in similar habitats. Further study is needed to answer this question unequivocally.

Virtually nothing is known of the biology of this group, except the supposition that their semi-unique shape and coloration is adapted for living on the bark of trees.

2.2.10.11 Stirotarsinae Rider, 2000

This relatively recently proposed subfamily was erected for a single genus and species (*Stirotarsus abnormis* Bergroth) (**Tables 2.2, 2.3**), a species only known from a couple of specimens from Peru, South America (**Figures 2.13F, 2.27L**). When originally described, Rider (2000) was somewhat hesitant in proposing a new subfamily (he had originally planned on proposing the Stirotarsini within the Pentatominae), but due to the suite of novel characters, and the encouragement of colleagues, he proposed this as a subfamily.

The known species is medium in size, is dark greyish in color, and has the body surface relatively rough and wrinkled (**Figures 2.13F, 2.14D, 2.27L**). The head is elongate, longer than wide; the antennae are five-segmented, with segment I not reaching apex of head, segments II and V are distinctly inflated, and segment III is quite short (**Figures 2.13F, 2.27L**). The rostrum is three-segmented, with only vague indication where segments III and IV have fused (**Figure 2.14C**); the distal half of the rostrum is extremely flat. The thoracic sterna are medially sulcate, without any indication of a medial carina, and the lateral margins of the sulcus are obtusely elevated. The ostiolar apparatus including the evaporatorium is reduced. The spiracles are located laterad, just below the connexival margins (**Figure 2.14D**); and the abdominal trichobothria are transverse, each pair on each side of the abdomen located mesad of the spiracular line. The tarsi are two-segmented, each with a distinct dorsal, longitudinal carina (**Figure 2.14A**); and the tibiae are foliate and dorsally concave (**Figure 2.14B**). The female spermathecal duct has a dilation with a sclerotized rod, and the spermathecal bulb is simple, ball-shaped, without fingerlike diverticula.

The genus *Stirotarsus* was proposed by Bergroth (1911) to include *S. abnormis* from Peru. Because of its aberrant features, the taxonomic placement of this genus has been ambiguous. Bergroth (1911) originally placed the genus in the Arminae (= Asopinae) near several genera that also have the tibiae foliate. Gapud (1991) removed *Stirotarsus* from the Asopinae and placed it tentatively in the Dinidoridae, probably based on the roughened, greyish dorsum, similar to members of the Megymeninae. Finally, as alluded to earlier, Rider (2000), based heavily on the presence of a dilation and sclerotized rod in the spermatheca, returned this genus to the Pentatomidae, but he felt the remaining characters were so novel as to warrant the erection of a new subfamily.

No biological information is available.

2.2.11 Phloeidae Amyot and Serville, 1843

This family contains three extant species in two genera (*Phloeophana* Kirkaldy [**Figure 2.16F**] and *Phloea* Lepeletier and Serville [**Figure 2.25G**]) and one fossil species and genus (**Table 2.2**). The present day species are endemic to Brazil, South America, with the core distribution in the Atlantic Rain Forest. The single fossil species, *Palaeophloea monstrosa* (Heer), was described based on a fragmentary impression from the Miocene of Croatia; its placement is doubtful. At times, this family has been treated as a subfamily within the Pentatomidae (e.g., Leston 1953c), and certain workers have included the Bornean genus *Serbana* within this group (e.g., Grazia et al. 2008). The phloeids lack the dilation and sclerotized rod in the female spermathecal duct; thus, we consider them to be a family separate from the Pentatomidae. On the other hand, members of the genus *Serbana* have the dilation and sclerotized rod and, so, are treated herein as a subfamily within the Pentatomidae. The two groups have a highly similar appearance though, being rather large (Phloeidae: 20-30 mm, *Serbana*: about 15 mm), ovate, and distinctively flattened with the lateral margins of the head, pronotum, and abdomen laminate or foliate; their mottled coloration allows them to easy blend in with the bark of trees upon which they occur. The taxonomic position of the Phloeidae within the Pentatomoidea still is undetermined (Grazia et al. 2008).

The phloeids are characterized (Leston 1953c, Lent and Jurberg 1965, Schuh and Slater 1995) by having the jugal foliations quite large, concealing or nearly concealing the three-segmented antennae (**Figure 2.1A**); antennae are three-segmented due to anarthrogenesis (incomplete division) between segments III and IV, segment I is rather long, and segments II and III are somewhat curved. Each compound eye is divided into dorsal and ventral sections (an apomorphy also present in *Serbana*, but probably acquired independently). The scutellum is subtriangular in shape, provided with a tongue-like apical lobe that is short in *Phloea* (**Figure 2.25G**), but greatly elongate, approaching the apex of the forewing in *Phloeophana* (**Figure 2.16F**). The hemelytral membrane has a reticulate venation, and the hind wings possess a hamus. The tarsi are three-segmented. The scent gland openings are located near the lateral margins of the metapleura, the ostiolar rugae are rudimentary, and the vestibular scar is present. Abdominal sternites III through VII, on each side, possess a pair of longitudinally arranged trichobothria that are located mesad of the spiracular line. In the female genital plates, paratergites IX are quite elongate. The egg lacks a pseudoperculum.

As mentioned previously, phloeids are well suited for living on the bark of trees; their mottled dorsal coloration looks much like lichens growing on the tree surface (Lent and Jurberg 1966). When disturbed, but also without any apparent disturbance, they are able to discharge a defensive liquid to a considerable distance from their anus. They also exhibit maternal care (Hussey 1934); the females protect the eggs and at least the first instars and possibly through the first three instars. The young nymphs can actually attach themselves to the venter of the female, and she then carries them with her (Lent and Jurberg 1965, Guilbert 2003). The life cycle occurs entirely on the trunks of their host (Salomão and Vasconcellos-Neto 2010, Salomão et al. 2012) with both nymphs and adults feeding on the vascular system (Bernardes et al. 2005). Host plants include members of the plant families Euphorbiaceae, Fabaceae, Mimosaceae, Moraceae, Myrtaceae, Rosaceae, and Urticaceae (Salomão et al. 2012).

2.2.12 Plataspidae Dallas, 1851

The family Plataspidae contains 66 genera and 606 species (**Table 2.2**), all restricted to the Old World (except for two species recently introduced into the New World). Members of this group are small to medium in size (2 to 20 mm), usually orbiculate, but occasionally more ovate or even parallel-sided; they usually are colored in blacks and browns, sometimes with pale stripes or spots, and often somewhat shiny (**Figures 2.16G-I; 2.25H**). Many species have a similar appearance of small black beetles. In a few genera, there is sexual dimorphism in the head structure; in some species, the males have elaborately produced spines or projections on the head. The margins of the juga usually are edged. The antenniferous tubercles are usually not visible in dorsal view; the antennae are either clearly five-segmented or appear to be four-segmented due to a rather weak subdivision of the pedicel, the basipedicellite (the more basal subsegment of a divided pedicel) is short. In several species, the stylets are long, and they are partially coiled inside the enlarged labrum or a lobe of labial segment II (China 1931, Rédei and Bu 2013, Rédei and Jindra 2015). The scutellum is greatly enlarged, nearly covering the entire abdomen. The hind wings are much longer than the body and folded up under the scutellum when at rest. The ostiolar evaporatoria are relatively large in comparison to the remaining ostiolar structures, covering nearly the entire ventral surface of the prothorax, mesothorax, and metathorax (**Figure 2.1I**), the extension to the prothorax is unique within the Pentatomoidea (Kment and Vilímova 2010b, Rédei and Bu 2013). The tarsi are two-segmented. The male pygophore is large, almost semiglobular, and its posterior surface has a relatively small external opening; the pygophoral rims are relatively simple without processes and/or other structures. The parameres are rather simple, stylus-like, and the hypophysis usually is bent slightly to moderately (nearly a right angle). The phallus is variably shaped with a distinct phallotheca, large membranous conjunctiva, and short, stout, sclerotized vesica. The spermathecal duct lacks a dilation and sclerotized rod; the spermathecal bulb is ovoid to globular but lacks distinct tubular diverticula; both the proximal and distal flanges are present. Sixteen different species have been karyotyped of which, 15 and 1 species had a diploid number of 10 + XY and 8 + XY, respectively (Ueshima 1979, Kerzhner et al. 2004, Rebagliati et al. 2005).

   This family was proposed by Dallas in 1851; it has gone by several other names (for example Coptosomatidae Reuter, 1912, and Brachyplatidae Leston, 1952a). In fact, the grammatically correct spelling of the family name is Plataspididae, but the spelling Plataspidae is in prevailing useage and, so, it should be conserved according to provisions of the ICZN (I. M. Kerzhner, personal communication; adopted for use by Davidová-Vilímova 2006). For the most part, most workers have accepted family status for this group. The taxonomic position of the Plataspidae within the Pentatomoidea has varied according to the study. Jessop (1983) suggested a relationship among the Plataspidae, Canopidae, Megarididae, and Lestoniidae, all families with the scutellum enlarged. Gapud (1991) supported the relationship with Lestoniidae, Cydnidae, and Thyreocoridae. There is little doubt, however, that the Plataspidae has no relationship with any of the above families. The family possesses a number of plesiomorphies that suggest that it is probably an early offshoot of the Pentatomoidea.

   Davidová-Vilímová (2006) indicated that past workers had associated the Plataspidae with these same three families but also included Aphylinae and sometimes the Scutelleridae, two more family-level groups with the scutellum enlarged. The phylogenetic results of Grazia et al. (2008) were contradictory. That is, based on molecular data, the Plataspidae was placed in a more basal clade (see also Dai and Zheng 2004; Li et al. 2005, 2006), whereas morphological data placed the Plataspidae in a more advanced clade. Many of the defining characters of this family are shared with other families (Jessop 1983). For example, the above mentioned enlargement of the scutellum (families listed above), the ability to fold the hind wings so they can fit under the scutellum (also seen in the Canopidae and Megarididae), and the two-segmented tarsi (Acanthosomatidae, Megarididae and a few pentatomid groups).

   Most plataspid species inhabit the tropics and subtropics, and are especially diverse in the Afrotropical and Oriental regions. As mentioned previously, two Oriental species have been accidently introduced into the New World. *Megacopta cribraria* (F.) was discovered in North America in 2009 (Eger et al. 2010, Suiter et al. 2010). This species attacks kudzu (Zhang et al. 2012) and could be considered beneficial, but it also attacks other legumes and has become a pest in the southeastern United States, especially on soybean (**See Chapter 5**). *Brachyplatys subaeneus* (Westwood), originally misidentified as *B. vahlii* (F.), has been recorded from several localities in Panama (Aiello et al. 2016, Rédei 2016). Additionally, *Coptosoma xanthogramma* (White) has been reported from Hawaii for quite some time (Beardsley and Fluker 1967). Actually, Froeschner (1984) discussed two other rumored records from Alaska, *Coptosoma duodecimpunctatum* (Germar) and *Coptosoma biguttulum* Motschulsky, but he concluded that these records should probably be removed from the North American fauna. Only two *Coptosoma* species are distributed in Central Europe (Davidová-Vilímová and Štys 1980), both of which hibernate in the nymphal stage, which is unusual for European Pentatomoidea. Moreover, they may overwinter in any of the first through fourth instars, depending on the weather and temperature during the preceding autumn.

All plataspids are phytophagous (mostly oligophagous or polyphagous). There seems to be a preference of many plataspids for feeding on members of the plant family Fabaceae. The long and partially coiled stylets (somewhat similar to some Aradidae) led China (1931) to suggest that plataspids might feed on fungi, but more recent evidence involving a symbiotic relationship between plataspids and ants in which the ants tended and collected honey dew from the plataspids (see below) seems to indicate that at least these species of plataspids are phloem feeders, needing the long stylets to penetrate the bark of their host plant.

Mutualistic relationship with ants has been reported in several plataspids, including *Caternaultiella rugosa* Schouteden (tropical Africa), *Tropidotylus servus* Maschwitz, Fiala, and Dolling, *T. minister* Maschwitz, Fiala, and Dolling, *Tetrisia vacca* Webb, and *Hemitrochostoma lambirense* Bergroth (Malaysia) (Maschwitz et al. 1987, Gibernau and Dejean 2001, Waldkircher et al. 2004, Tomokuni 2012). The ants build pavilions to protect and shelter the egg batches, nymphs, and adults of the bug. In at least *C. rugosa*, the adult female bugs are able to switch the maternal behavior as needed. That is, if there are more egg batches outside of the pavilions, then the females will guard the eggs; if most of the eggs are within the ant pavilions, then the adult females do not help with guarding of the eggs (Gibernau and Dejean 2001).

Similarly to most other pentatomoids, plataspid females transmit symbionts to their offspring via the eggs. However, they do it somewhat differently. In these bugs, the symbionts are contained in small, dark capsules deposited between the individual eggs of the egg batch. This phenomenon, a potentially unique apomorphy of the family, has been reported for the Palearctic species, *Coptosoma scutellatum* (Geoffroy) (Davidová-Vilímova 1987) and a number of East Asian species belonging to the genera *Coptosoma* Laporte, *Brachyplatys* Boisduval, and *Megacopta* Hsiao and Jen (e.g., Hosokawa et al. 2006, 2008). The newly hatched nymphs feed on the capsules. The egg batches of at least two species of *Libyaspis* Kirkaldy are nearly unique within the Heteroptera. The eggs are covered by an ootheca of material originating from the midgut. A similar ootheca is known in some species of Reduviidae and Urostylididae, but, in these examples, the oothecal material is a secretion produced by parts of the female genital duct (Carayon 1949, 1952).

Some plataspids (e.g., *Coptosoma lyncea* Stål, from northern Australia) also are known to form large aestivation aggregations (Monteith 1982). One such aggregation was described as covering every leaf of a small tree, including lines of individuals on the petioles. When disturbed, the individuals would take flight and disperse their defensive secretions but eventually would return to their previous roost.

At least one fossil species, *Coptosoma eocenica* Piton (1940), has been described.

The phylogeny and higher classification of the family is unresolved. Jessop (1983), based primarily on somatic characters, first classified the plataspid genera into three suprageneric categories but refrained from formally giving them tribal names (*Brachyplatys* group, *Coptosoma* group, *Libyaspis* group); he also keyed the genera of the *Libyaspis* group. Ahmad and Moizuddin (1992) split the South Asian Plataspidae into two subfamilies: Plataspinae (incorrectly as Brachyplatidinae) and Coptosomatinae (incorrectly as Plataspidinae); the names were corrected by Davidová-Vilímova (2006) in the Palearctic catalog. This classification cannot be adapted to the World fauna and it is difficult or impossible to place many genera to any of the described subfamilies (Davidová-Vilímová 2006). As such, the key presented below is based merely on the literature and will not allow placement of several described genera.

2.2.12.1 Key to the Subfamilies and Genus Groups of Plataspidae (modified from Schuh and Slater 1995)

1   Ocelli placed laterally, near compound eye or anterior angles of pronotum (**Figure 2.11A**); abdominal sterna usually clearly convex; head usually narrow, about 0.3-0.5 times width of pronotum; base of scutellum ("pseudoscutellum") usually elevated, set off from rest of scutellum by an impressed line (Coptosomatinae) .... *Coptosoma* group

1'   Ocelli placed much more mesially, near each other; abdominal sterna not or only slightly convex; head broader, usually 0.5-0.7 times width of pronotum; no distinct "pseudoscutellum" is recognizable (Plataspinae) 2

2(1)   Body relatively flat; color black, sometimes spotted with yellow, punctures variable, and often with a yellow submarginal line on head, pronotum, and scutellum.......... *Brachyplatys* group

2'   Body more convex; color pattern red, yellow, or brown, maculated with dark brown to black punctures; dark areas sometimes extensive and spotted or flecked with yellow but without a yellow submarginal line on head, pronotum, and scutellum........................ *Libyaspis* group

2.2.13 Primipentatomidae Yao, Cai, Rider, and Ren, 2013

This recently described fossil family contains four genera and five species (**Table 2.2**), all known only from Liaoning Province in China. The following descriptive notes come primarily from Yao et al. (2013). Its members are medium in size (8-15 mm), oval in shape. The head is triangular with the angles rounded; the antennae are four-segmented, with segment I extremely short. Rostrum relatively short, reaching beyond fore coxae, sometimes almost to middle coxae; segment I not reaching beyond posterior margin of bucculae. Scutellum subtriangular with angles rounded, basal disc lunately elevated; frena relatively long, reaching to apex of scutellum. Legs relatively simple, lacking distinct bristles and spines. The original description indicates that the sternum is strongly carinate medially, but, from the illustrations, it looks like it could be strongly carinate or strongly sulcate. The forewings are macropterous with the coria relatively small, not reaching beyond apex of scutellum.

This family was not described until 2013; it presently contains four genera: *Breviscutum* Yao, Cai, Rider, and Ren, *Oropentatoma* Yao, Cai, Rider, and Ren, *Primipentatoma* Yao, Cai, Rider, and Ren (with two species), and *Quadrocoris* Yao, Cai, Rider, and Ren. All are from the Yixian or Jiufotang Formations in western Liaoning Province, China. These deposits are estimated to date to the early Cretaceous.

Primipentatomidae is monophyletic and considered to be the sister group to all Pentatomoidea excluding Saileriolidae + Urostylididae (Yao et al. 2013).

2.2.14 Saileriolidae China and Slater, 1956

This group originally was described, "with some hesitation," as a subfamily of the Urostylididae (China and Slater 1956), and it only recently (Grazia et al. 2008) has been elevated to family level. This family contains three genera (*Bannacoris* Hsiao (**Figure 2.25I**), *Ruckesona* Schaefer and Ashlock, and *Saileriola* China and Slater) and four species (**Table 2.2**) known from China and Southeast Asia.

Members of this family are small (less than 5 mm) and somewhat ovate in shape (**Figure 2.25I**). The head is subtriangular, strongly declivent, and the lateral margins are rounded, not edged or reflexed. The ocelli are distinctly closer to each other than each is to their adjacent compound eye. The antenniferous tubercles are conspicuous and arise above a hypothetical line drawn through the middle of the eyes; they are easily visible from above. The antennae are five-segmented with segment I greatly elongate (longer than length of head and pronotum combined), and segment III is short. The bucculae are relatively short, not reaching beyond middle of head; the rostrum usually reaches onto abdominal segment IV. The anterolateral pronotal margins are edged. The scutellum is sharply triangular with the basal region swollen, and each frenum extends along its entire lateral margin. The scutellum leaves the clavi fully unexposed, but there is no claval commissure, the apices of the clavi meet in a single point or slightly overlap; the coria are weakly sclerotized, their apices reach well beyond the apex of the abdomen; the membrane is provided with several subparallel veins; and the hamas of the hind wings is lacking. The ostiole of the metathoracic scent gland is quite small and probably non-functional; associated external scent efferent structures are absent. The acetabula of both the mesothoracic and metathoracic legs are widely separated. The tarsi are three-segmented (erroneously indicated as two-segmented in *Bannacoris* by Hsiao 1964), with segment II shortest. Intersegmental sutures of the pregenital abdomen at least partly obscured, spiracles III-VII situated close to the lateral margin of abdomen, trichobothria are absent on segment III, single (on each side) on segment IV, and double (on each side) on segments V and VI.

Schaefer and Ashlock (1970) gave a summary of the similarities between the Urostylididae and the Saileriolidae. They stated that the two groups "have in common the placement of the antennae (dorsal to the midline of the eye), the annulate antennifers, the proximity of the ocelli, the sutures of the vertex, and the general structure of the genital capsule." China and Slater (1956) had noted some of these same similarities, and they placed "great importance" on the structure of the antennifer and the position of its insertion onto the head. Hsiao (1964) described the genus and species *Bannacoris arboreus* Hsiao, placing it in the Saileriolinae, but Schaefer and Ashlock (1970) were apparently unaware of its description as they made no mention of it. The phylogenetic study by Grazia et al. (2008) indicated that the Urostylididae in the traditional sense was not a monophyletic group, which was solved by moving the Saileriolinae to its own family. Furthermore, they concluded that the Saileriolidae is the sister group to all of the non-urostylid Pentatomoidea.

The only published biological information available comes from the type series of *Ruckesona vitrella* Schaefer and Ashlock, which Schaefer and Ashlock (1970) reported had been collected on "palm at water margin." The authors also noted that the green gut contents of specimens contained what looked like fragmented chloroplasts that led them to propose that specimens might not feed exclusively on plant sap. Specimens of *Bannacoris arboreus* (**Figure 2.25I**) were collected from leaves of banana species (D. Rédei, unpublished data).

2.2.15 Scutelleridae Leach, 1815

The Scutelleridae was proposed by Leach (1815) as a family group, and included it together with the Pentatomidae in a taxon currently equivalent to the Pentatomoidea. Amyot and Serville (1843) proposed the tribe Orbiscutes in the family Longiscutes to include all genera with the scutellum enlarged. In their classification, scutellerids were divided into five different groups (Scutellérides, Pachycorides, Tétyrides, Eurygastrides, and Odontoscélides), but they also included within these groups a number of genera now placed in other pentatomoid families. Dallas (1851) recognized many of the same taxa, but now at the family level (Eurygastridae, Pachycoridae, and Odontoscelidae) within the group Scutelleroidea, but the first and last of these also included members now placed in the Podopinae and the Thyreocoridae, respectively. Fieber (1861) recognized only one family-group, the Tetyridae, but this still contained several non-scutelleroid members. Stål (1865) was the first to treat this group in a fashion similar to our present-day classification, although he considered the scutelleroids to form a subfamily of a more inclusive Pentatomidae. He refined his classification in a later work (Stål 1873), which became the classification followed by most subsequent workers (e.g., Lethierry and Severin 1893, Kirkaldy 1909, Oshanin 1912, Leston 1952a, China and Miller 1955, Lattin 1964, Linnavuori 1982). The only major change is that Reuter (1912), without comment, elevated the scutelleroids to the family level (Tsai et al. 2011). The works of Singh-Pruthi (1925) and Pendergrast (1957) supported the recognition of this group as a valid family, which has been followed by most modern workers (e.g., Leston 1958, McDonald 1966, Gross 1975b, Rolston and McDonald 1979, Froeschner 1988d, Schuh and Slater 1995, Cassis and Gross 2002, Cassis and Vanags 2006, Göllner-Scheiding 2006b, Tsai et al. 2014, Barcellos et al. 2015).

Although limited in scope and not the direct focus of the studies, explicit phylogenetic studies have been conducted that include the Scutelleridae as a terminal taxon (e.g., Gapud 1991, Fischer 2001, Grazia et al. 2008). In all of those analyses, whether based on morphology, molecules, or a combination of the two, the monophyly of the Scutelleridae and its familial status were corroborated, but the phylogenetic relationship within the Pentatomoidea remains inconclusive. Tsai et al. (2011) argue that the monophyly of the Scutelleridae is currently weakly supported, considering no unique morphological apomorphies are recognized, and the molecular data are too fragmentary. These same arguments can be applied to most families within the Pentatomoidea.

Members of this family (**Figures 2.26A-L**) are medium to relatively large in size (5 to 20 mm in length), ovoid to elongate ovoid in shape, and the body is moderately to strongly convex. The color varies greatly among scutellerids, with some of the most brightly colored species in the Heteroptera. Many species are polymorphic. Whether the colorations represent aposematism or mimicry has been discussed (Tsai et al. 2011, Eger et al. 2015a). The juga and tylus are usually subequal in length. The antennae are usually five-segmented, but sometimes (exceptionally) only four-segmented (*Augocoris* Burmeister) or even three-segmented. The bucculae are parallel and not strongly elevated; the rostrum reaches at least to the mesocoxae. The most obvious uniting character is the enlarged scutellum that, in most species, nearly covers the entire abdominal dorsum; the exocorium and clavus are minimally exposed, and the frena are short or absent; the coria are not strongly sclerotized. The hemelytral membrane has ten or more longitudinal veins, and the hind wing has a hamus. In members of some scutellerid subgroups (Scutellerinae), a strigil is found on the posterior anal vein (postcubitus in several earlier papers) of the hind wing and a plectrum is found on abdominal tergum I. The propleura are carinate, and the prosternum is weakly to moderately sulcate. The second gonocoxae are fused, and the first and second gonapophyses are membranous. The gynatrium has a sclerotized basal groove, the female spermatheca is diverse, the spermathecal duct is with or without a dilation, if the dilation is present, its proximal and distal orifices may be provided with sclerotized tubular projections; the spermathecal bulb is simple, rounded, without tubular diverticula. The phallus usually has two or three pairs of well-developed conjunctival projections. Twenty-seven species of scutellerids have been karyotyped, all of which have a diploid number of 10 + XY (Ueshima 1979, Kerzhner et al. 2004, Rebagliati et al. 2005).

There are other families that have an enlarged scutellum (Canopidae, Megarididae, Lestoniidae, Plataspidae, Podopinae in the Pentatomidae, Thyreocoridae), but there are other characters that separate this family from those listed. For example, the scutellerids lack spines and bristles on the tibiae which is characteristic of the Thyreocoridae; the small or obsolete frena and the lack of a pentatomid-type dilation and sclerotized rod will separate scutellerids from the Podopinae and other Pentatomidae; the three-segmented tarsi will separate them from the Megarididae and Plataspidae, which have two-segmented tarsi; the jugal and pronotal margins are never laminately produced as in the Lestoniidae; and the characters of the forewing and spermatheca will separate scutellerids from the Canopidae.

The classification within the Scutelleridae has not been stable. Schuh and Slater (1995) recognized four subfamilies: Scutellerinae, Pachycorinae, Eurygastrinae, and Odontotarsinae; they treated the Elvisurinae as a tribe within the Scutellerinae, and they briefly mentioned that McDonald and Cassis (1984) had proposed a new subfamily, Tectocorinae, for the Australian genus, *Tectocoris* Hahn (**Figure 2.26E**). More recently, Carapezza (2009) erected a new subfamily, the Hoteinae (three genera). We follow Eger et al. (2015a), and currently recognize eight subfamilies, three of which are subdivided into two tribes each (**Table 2.2**). Most of the subfamilies occur primarily in the Old World (Afrotropical, Australian, Oriental, and Palearctic regions) with a few genera or species in the Western Hemisphere. The subfamily Pachycorinae is found exclusively in the New World except for one species that has been introduced into Australia (Cassis and Vanags 2006, Tsai et al. 2011, Eger et al. 2015a). There are currently 100 genera and 531 species (**Table 2.2**) placed in the Scutelleridae. There have been at least one fossil genus and ten fossil species described in the Scutelleridae: one species in the Eurygastrinae (Förster 1891); four species in the Odontotarsinae (Heer 1853, Meunier 1915) but one, *Odontotarsus archaicus* Meunier, was transferred to the Cydnidae (Théobold 1937); one genus and two species in the Pachycorinae (Heer 1853, 1865); two species in the Scutellerinae (Fujiyama 1967, Statz and Wagner 1950); and one species in the Tectocorinae (Henriksen 1922). The generic or even the subfamily placement of several of the fossil taxa is doubtful.

The Scutelleridae was reviewed in an older but excellent paper by Schouteden (1904). Since that time, there have been several regional studies on this family. For example, the North American fauna was treated in an unpublished thesis (Lattin 1964); the fauna of Australia has been thoroughly studied (McDonald and Cassis 1984, Cassis and Vanags 2006); the Taiwanese Scutellerinae taxa have been reviewed recently (Tsai et al. 2011); and the African taxa were treated in an older paper by Schouteden (1903). Parveen and Gaur (2015) presented an illustrated key to the scutellerid genera occurring in India, and Leston (1952c) treated the Scutellerinae of Angola. The Neotropical fauna was treated in Eger et al. (2015a), including a key to all genera recorded in the region. Barcellos et al. (2015) dealt with the Argentinian fauna. Some of the more notable revisionary works for various scutellerid genera include the following: *Agonosoma* Laporte (Paleari 1992), *Calidea* Laporte (Freeman 1939, 1946), *Calliphara* Germar (Lyal 1979), *Cantao* Amyot and Serville (McDonald 1988), *Deroplax* Mayr (Ahmad et al. 1988, 1998), North American species of *Eurygaster* Laporte (Vojdani 1961), *Lamprocoris* Stål (Rédei and Tsai 2016), *Odontoscelis* (Göllner-Scheiding 1986, 1987), *Odontotarsus* Laporte (Göllner-Scheiding

1990), *Poecilocoris* Dallas (Ahmad and Kamaluddin 1982), *Polytes* Stål (Eger 1990, 1992, 1994, 2015), *Promecocoris* Puton (Gapon 2007), *Sphyrocoris* Mayr (Eger 2012), and *Tiridates* Stål (Eger 1987).

All scutellerid species are considered to be phytophagous, and most are considered to be generalists (Eger et al. 2015a). Host plant preference or even specialization may occur (Javahery et al. 2000). Plant species within the families Euphorbiaceae (e.g., *Pachycoris* Burmeister spp.), Cyperaceae, Malvaceae, Myrtaceae, and Poaceae (e.g., *Eurygaster* spp.) are among the preferred hosts (Barcellos et al. 2015). Facultative feeding on non-plant material (e.g., dung and carrion) has been reported (Tsai et al. 2011, Eger et al. 2015a). Most species are not considered to be pests. Species of the genus *Eurygaster* are exceptions and, collectively, are called Sunn pests. They cause considerable damage to wheat primarily in the Middle East and Central Asia. Additionally, species in the genera *Calidea* and *Tectocoris* (**Figure 2.26E**) are minor pests of cotton in Africa and Australia, respectively. Javahery et al. (2000) reviewed the biology of some of the economically important species.

Scutellerid development and their life cycles are similar to species in the Pentatomidae (reviewed in Javahery et al. 2000, Cassis and Vanags 2006, Tsai et al. 2011, Eger et al. 2015a). Barrel-shaped eggs are deposited in or near the host plants, and, after hatching, the first instar nymphs do not feed. Nymphal aggregations occur in some species and may last until late instars; large aggregations also may include adults (Cassis and Vanags 2006). Feeding usually occurs in vegetative or reproductive parts of the plant (Javahery et al. 2000). Some adults communicate by stridulation. Mating takes place after the male approaches the female, contacting her with his antennae. The male first mounts the female, then turns around such that mating takes place end-to-end (Tsai et al. 2011). Maternal care has been reported in species of four genera from three different subfamilies: *Augocoris* and *Cantao* (Scutellerinae), *Pachycoris* (Pachycorinae), and *Tectocoris* (Tectocorinae) (Cassis and Vanags 2006, Tsai et al. 2011, Eger et al. 2015a).

2.2.15.1 Key to the Subfamilies of Scutelleridae (modified from McDonald and Cassis 1984, Schuh and Slater 1995, Tsai et al. 2011)

1 Pro-, meso- and metasterna strongly sulcate (**Figure 2.11B**); sulcus with strongly elevated sides; abdominal venter never with paired striated areas; large brown, reddish brown, or grey species; Old World (**Figure 2.26A**)..Elvisurinae

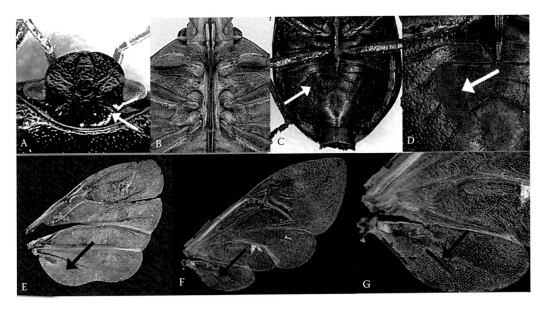

FIGURE 2.11 A, *Coptosoma* sp., head, dorsal view; B, *Solenosthedium rubropunctatum*, thorax, ventral view; C, *Pachycoris torridus*, abdomen, ventral view; D, *Pachycoris torridus*, detail of abdominal segments III–VIII, ventral view; E, *Thyreocoris scarabaeoides*, hind wing; F, *Galgupha* sp., hind wing; G, *Galgupha* sp., detail of hind wing jugal lobe.

1'   Pro-, meso- and metasterna usually, at most, with shallow sulcus not bordered with strongly elevated sides; if a deep sulcus with strongly elevated sides present, abdominal venter with paired striated areas (Pachycorinae: *Nesogenes*) 2

2(1)   Abdominal venter with striated areas present on at least sterna V and VI (**Figures 2.11C, D**), sometimes also on sterna IV.. 3

2'   Abdominal venter lacking striated areas.. 4

3(2)   Scent gland ostiole continued in a curved furrow situated on an indistinct, slightly elevated ruga; penisfilum of male phallus and female spermathecal duct long and coiled; Old World (except Europe) (**Figure 2.26F**) ....Hoteinae

3'   Scent gland ostiole different from above (except *Ephynes*); male without penisfilum, female spermathecal duct not elongate and coiled; New World (**Figures 2.26G, H**)...Pachycorinae

4(2)   Body convex dorsally and ventrally (venter usually less convex than dorsum but never flat); bases of pronotum and scutellum forming a rather continuous outline in lateral view; body not brightly colored, usually dull brown, black, or yellowish.. 5

4'   If body convex dorsally and ventrally, then dorsal outline of body distinctly emarginated between pronotum and scutellum in lateral view (Tectocorinae, Scutellerinae: Scutellerini); or dorsum strongly convex, venter flat (Scutellerinae: Sphaerocorini); body frequently vividly colored, even metallic.. 7

5(4)   Head short, subsemicircular; anterolateral margins of pronotum strongly convex; scutellum covering whole abdomen; scent gland ostioles reduced, indistinct or lacking; body with strong, dense pilosity (**Figures 2.26J, K**) .... Odontoscelinae

5'   Head longer, usually subtriangular or subquadrangular, but if (exceptionally) semicircular then scutellum not covering whole abdomen; anterolateral margins of pronotum frequently concave or straight, at most weakly convex; scent glands well developed or absent; body usually without conspicuous pilosity.. 6

6(5)   Scutellum relatively narrow, parallel-sided, leaving hemelytra and connexivum broadly exposed for nearly entire length (**Figure 2.26I**); metathoracic scent gland ostioles well developed, provided with distinct ruga (**Figure 2.4G**); Widespread (**Figure 2.26I**) .... Eurygastrinae

6'   Scutellum usually broader, leaving at most narrow lateral margin of hemelytra and connexivum exposed (**Figure 2.26L**); metathoracic scent gland ostioles lost .... Odontotarsinae

7(4)   Abdominal venter of male with a pair of androconial glands on segments IV-VI (**Figure 2.26E**).. Tectocorinae

7'   Abdominal venter without androconial glands (**Figures 2.26B-D**)................. Scutellerinae

2.2.16 Tessaratomidae Stål, 1865

Stål (1865) first recognized this as a family-group taxon. Leston (1954b) treated this group as a subfamily of the Pentatomidae and noted many similarities in genitalia with the Scutelleridae; he later (1956b), without comment, elevated the tessaratomids to family status. This was accepted by Kumar (1969a,b) and most subsequent workers (e.g., Rolston et al. 1994). Leston (1954b, 1955b) also indicated that he had removed the Natalicolini from the Tessaratominae (now Tessaratomidae) and elevated it to subfamily rank within the Pentatomidae (corresponding to a family level taxon in the current classification), but it appears that the paper in which he planned to make this change never was published.

   Diagnostic characters (Schuh and Slater 1995, Schwertner and Grazia 2015) of members of this family include their medium to large size (10-40 mm); the family includes some of the largest pentatomoids often over 20 mm in length. They are ovate to elongate-ovate, mostly in colors of yellows, greens, or browns (**Figures 2.16K, L; 2.25J**). In general, they are similar in appearance to many pentatomids. The head is subtriangular in shape, relatively small in comparison to body size; the juga often extend beyond the tylus and meet in front of it, and their lateral margins usually are edged. The antennae are usually four-segmented; if they are five-segmented, then the third segment is relatively small (**Figure 2.2D**); the antenniferous tubercles are usually not visible from dorsal view. The bucculae are relatively short, somewhat flap-like (**Figure 2.2A**); the rostrum is also short, usually not surpassing the middle coxae. The humeral angles may or may not be prominent; the posterior margin of the pronotum extends over the base of the scutellum, sometimes greatly so (**Figure 2.2E**). The scutellum is subtriangular, not covering the coria, and the apex is usually pointed. The veins of the hemelytral membrane are subparallel (**Figure 2.2E**), not reticulate, but closed cells are present along the basal margin in several species; the hind wing has a hamus. Several tessaratomids possess a stridulatory mechanism composed of a strigil on the posterior anal vein (postcubitus in several earlier papers) of the hind wing and a plectrum located on abdominal tergum I; this structure is apparently lacking in the Oncomerinae. One of the characters used most often for defining the tessaratomids is that most species have the spiracles on abdominal sternite II completely exposed (said to be covered by the posterior margin of the metapleura in other families) (**Figure 2.2B**), but this character seems to be a matter of degree. It is exposed or partially exposed in a number of species in other families and, thus, seems somewhat unreliable for defining this family. The metasternum often is produced laterad and anteriorly, sometimes reaching the fore coxae; its posterior margin is truncately abutted to the base of the abdomen (**Figure 2.2C**). The metathoracic scent gland ostiole is oval and accompanied by a spout-shaped ostiolar ruga (Oncomerinae) (**Figure 2.4D**), or it is wide and groove-shaped and accompanied by a bilobate ostiolar ruga (Tessaratominae + Natalicolinae) (**Figure 2.4C**) (Kment and Vilímová 2010a). The tarsi may be two- (Natalicolinae) or three-segmented (Oncomerinae, Tessaratominae). The abdominal trichobothria are arranged transversely on each side of abdominal sternites III through VII and are located posterior to the spiracles and mesad of the spiracular line. The ninth paratergites are greatly enlarged. The female spermatheca lacks a sclerotized rod; the spermathecal bulb is simple, ball-shaped, and lacks diverticula. Nine species of tessaratomids have been karyotyped, of which seven and two had a diploid number of 10 + XY and 12 + XY, respectively (Ueshima 1979, Kerzhner et al. 2004, Rebagliati et al. 2005).

At present, the Tessaratomidae contains 62 genera and 252 species (**Table 2.2**). The distribution is mostly Old World, with only three species of *Piezosternum* (**Figure 2.16L**) occurring in the New World. There have been four fossil species described in the Tessaratomidae. Two are members of the extant genus *Pycanum* (Piton 1940). The other two (*Latahcoris spectatus* Cockerell, *Tessaratomoides maximus* Jordan) were placed in new genera (Cockerell 1931, Jordan 1967); one of the new genera (*Tessaratomoides*) may not be available, however, as it was proposed in name only; that is, it was not formally described (see discussion of Jordan 1967 names in **Section 2.2.10.7.47**).

Based on the literature, mainly the works of Leston (1955b, 1958) and Kumar (1969a,b, 1974b), Rolston et al. (1994), in their World catalog, recognized three subfamilies (one of which was subdivided into three tribes): the Natalicolinae, Oncomerinae, and the Tessaratominae (with the tribes Prionogastrini, Sepinini, and Tessaratomini). Schuh and Slater (1995) recognized the same three subfamilies, but they divided the Oncomerinae into two tribes (Oncomerini and Piezosternini), and the Tessaratominae into five tribes (Eusthenini, Platytatini, Prionogastrini, Sepinini, and Tessaratomini). Sinclair (1989), in his Ph.D. dissertation, provided a generic revision and a phylogenetic analysis of this family. The generic revision of Oncomerinae, and the description of two new genera (one in the Oncomerinae, the other in the Tessaratominae) have been published (Sinclair 2000a,b) but the phylogenetic analysis has not. Although his phylogenetic analysis showed that the family is polyphyletic, and that the Oncomerinae should be raised to family status, he later published the generic review in which he still treated the Oncomerinae as a subfamily. In his study, the remaining members of the Tessaratomidae were placed into two subfamilies: the Tessaratominae and the Natalicolinae, the latter further divided into two tribes (Natalicolini and Prionogastrini). Grazia et al. (2008), in their analysis based only on morphology, and their analysis combining both morphological and molecular data, found that the Tessaratomidae was monophyletic, and, in all analyses, the Tessaratomidae was the sister group with the Dinidoridae, a result suggested by Gapud (1991). The intrafamiliar classification recently was discussed by Kment and Vilímová (2010a) suggesting the monophyly of Tessaratominae + Natalicolinae (= Tessaratomidae *sensu stricto*) and Dinidoridae + Tessaratomidae *sensu lato*, but the relationships among Tessaratomidae *sensu stricto*, Oncomerinae, and Dinidoridae remain unresolved. Neither the monophyly of the Tessaratomidae *sensu lato* nor the Dinidoridae + Tessaratomidae *sensu lato* clade were confirmed by Wu et al. (2016).

Other than an older work on the African Tessaratomidae (Schouteden 1905a) and the more recent comprehensive studies on the Australian oncomerines by Leston and Scudder (1957) and Sinclair (2000a,b), and the Chinese tessaratomids by Zia (1957), little recent revisionary work has been done on the family. There are a few scattered studies on the following genera: New World species of *Piezosternum* (Pirán 1971), *Pygoplatys* Dallas (Magnien 2008, 2011; Magnien et al. 2008), *Sciadiocoris* Magnien and Pluot-Sigwalt (Magnien and Pluot-Sigwalt 2016), and *Tamolia* Horváth (Carvajal et al. 2015a). There are two genera currently classified in the Tessaratominae that should be removed. *Megaedoeum* Karsch should probably be transferred to the Dinidoridae, and *Aurungabada* Distant should be placed as a junior synonym of *Halyomorpha* (Pentatomidae: Pentatominae: Cappaeini).

Only a few species of tessaratomids have had their biology studied (McDonald 1969b, Malipatil and Kumar 1975, Schaefer et al. 2000, Dzerefos et al. 2009). All species studied to date are exclusively phytophagous, feeding on both reproductive and vegetative parts of their host plants. Schaefer and Ahmad (1987) compiled a list of their host plants, and Schaefer et al. (2000) reviewed and discussed species of agricultural importance. *Musgraveia sulciventris* Stål is considered to be a minor pest on citrus in Australia (McDonald 1969b). *Tessaratoma papillosa* (Drury), commonly called the lychee stink bug, is a pest on lychee in China (Schulte et al. 2006). The Palearctic species recently have been catalogued (Rider 2006b), and there is an internet website devoted to this family (Magnien 2015).

Species are univoltine and usually use more than one host plant during their life cycle (Schaefer et al. 2000, Dzerefos et al. 2009). Females of several species lay their eggs in masses of four rows (3-4-4-3 formula), with an average of 14-28 eggs per clutch. Eggs of *Encosternum delegorguei* Spinola took an average of 18 ± 9 days to hatch, whereas the nymphs took four months to reach the adult stage (Dzerefos et al. 2009).

Maternal care has been described for the genus *Pygoplatys* (**Figure 2.25J**) of the Tessaratominae (Gogala et al. 1998) and three genera of the subfamily Oncomerinae in the Australian region (Monteith 2006, 2011). The oncomerine genera show a similar behavior as that described for species of Phloeidae, where the nymphs are carried on the modified body of the female for a period of time (Monteith 2006).

In southern Africa, specimens of *Encosternum delegorguei* (**Figure 2.16K**), commonly called Thongolifha, are collected in large numbers, dried, and used for human consumption. They are also said to cure hangovers, but in this case, they are eaten raw (Dzerefos et al. 2009, Dzerefos and Witkowski 2014).

2.2.16.1 Key to the Subfamilies of Tessaratomidae

1   Ostiole small, oval, accompanied by undivided, spout-shaped ostiolar ruga (**Figure 2.4D**); hemelytral membrane usually without closed cells basally, longitudinal veins arising from base of wing membrane (**Figure 2.2E**); Old World except one genus (*Piezosternum*) also found in Neotropics (**Figure 2.16L**)... Oncomerinae

1'   Ostiole forming a wide ostiolar groove, accompanied by a bilobate ostiolar ruga with both an anterior and a posterior lobe (**Figure 2.4C**) (strongly dorsoventrally flattened in *Platytatus*); hemelytral membrane with closed cells basally, longitudinal veins arising from basal cells (Tessaratomidae *sensu stricto*) 2

2(1)   Scutellum subequilateral; tarsi two-segmented; Old World (**Figure 2.16K**) ....Natalicolinae

2'   Scutellum distinctly longer than wide; tarsi three-segmented; Old World (**Figure 2.25J**).... .... Tessaratominae

2.2.17 Thaumastellidae Seidenstücker, 1960

This is another small family both in number (one genus and three species; **Table 2.2**) and size (less than 3.5 mm). Two of the species are known only from southern Africa, and the third species is distributed in northern Africa and the Near East. Individuals of this family closely resemble small ground-dwelling lygaeoids (**Figure 2.16J**) and, in fact, Horváth (1896) originally described *Thaumastella* as a lygaeid, a taxonomic position in which this group remained until Seidenstücker (1960) noted that it lacked the lanceolate ovipositor characteristic of most Lygaeidae. The Palearctic species recently have been catalogued (Lis 2006d).

In members of this group, the lateral margins of the head are rounded, not edged or reflexed. They have five-segmented antennae and a four-segmented rostrum with segment I usually lying entirely between the bucculae. The scutellum is subtriangular in shape, never enlarged (**Figure 2.16J**). The ostiolar rugae are disc-like, sometimes reaching to the lateral margins of the metapleura; the associated evaporative areas are relatively large. Both macropterous and brachypterous forms occur; the macropterous forms have each corium divided into an endo- and exocorium, and the claval commissure is present; the fore wings of brachypterous forms are shortened, 'staphylinoid', and distally truncate. Coxal combs, composed by broad and flattened setae, are present; and the tarsi are three-segmented.

Macropterous morphs of this family possess a stridulatory mechanism similar to that found in the Cydnidae with the stridulitrum on the ventral surface of the posterior anal vein (postcubitus in earlier papers) of the hind wing and the plectrum on the anterolateral margin of abdominal tergite I. Jacobs (1989) indicated that the South African species had a pair of trichobothria on each side of abdominal segments III through VII (*Thaumastella elizabethae* Jacobs has only a single trichobothrium on each side of segment VII); he further indicated that the pair on segment III was nearly longitudinal, and each subsequent pair became more oblique until the pair on segment VI was nearly transverse. The female spermathecal bulb is ball-shaped, lacking diverticula, and there is no dilation or sclerotized rod.

The type genus and species, *Thaumastella aradoides* Horváth (**Figure 2.16J**) was originally described as a member of the Lygaeidae *sensu lato* (Horváth 1896). Seidenstücker (1960) noted that thaumastellids lacked the characteristic lanceolate ovipositor of the lygaeoids and proposed a new subfamily for them within the Lygaeidae. This prompted Štys (1964) to elevate the group to family status. Both Štys (1964) and Dolling (1981) considered the Thaumastellidae to be related to the Cydnidae with Dolling proposing it to be the sister group of the rest of his broadly conceived Cydnidae. Jacobs (1989) indicated that many of the characters once thought to be unique to the Thaumastellidae were found in some species of Cydnidae. He did indicate that some features of the male genitalia and the presence of an m-chromosome seemed to be unique to this family. Jacobs et al. (1989) also found that the defensive secretions of thaumastellids shared more components with the lygaeoids than it did with other pentatomoids. The presence of an m-chromosome, which is present in most lygaeoids and absent in all other pentatomoids prompted Henry (1997) to suggest that the thaumastellids might not belong in the Pentatomoidea after all. Based on morphological characters, the phylogenetic study by Grazia et al. (2008) placed the Thaumastellidae within the Pentatomoidea and, more specifically, within the Cydnidae, similar to the classification proposed by Dolling (1981). Based on molecular data, however, placement of the thaumastellids was variable; as such, Grazia et al. (2008) decided to leave this group as a family until further studies could be completed.

Jacobs (1989) indicated that most of the specimens of *Thaumastella namaquensis* Schaefer and Wilcox he had collected were "under fairly large stones in cavities which are exposed when the stones are removed." He also collected individuals on the ground near the stones, especially at dusk, and noted that they seemed reluctant to leave their shelters. He speculated that they might be feeding on seeds that the wind had blown up against the stones. He indicated that nymphs were found only in April. Jacobs (1989) also provided some notes on the biology of his new species, *T. elizabethae*, including a host plant record – seeds of *Pharnaceum aurantium* (DC) Druce [Aizoaceae]. Interestingly, he noted that it appeared that individuals had difficulty freeing their stylets from the seeds; even when disturbed, the individuals could be seen scurrying off with the seeds still attached to their mouthparts.

2.2.18 Thyreocoridae Amyot and Serville, 1843

This is another family that has had a complicated taxonomic history. The Thyreocoridae was proposed (as Thyréocorides) by Amyot and Serville (1843) for a group of genera not only including *Thyreocoris* and *Strombosoma* Amyot and Serville but also *Canopus* (Canopidae) and several genera now included in the Plataspidae. Uhler (1872) proposed Corimelaenidae for the New World genera *Corimelaena* and *Galgupha* Amyot and Serville, genera that Amyot and Serville had grouped with the scutellerid genus *Odontoscelis*. Fieber (1860) and Stål (1876) treated *Corimelaena* and *Thyreocoris* as members of the family Cydnidae. Lethierry and Severin (1893) catalogued the Corimelaeninae as a subfamily of a broad Pentatomidae. Later, Horváth (1919) treated this group as a subfamily within the Cydnidae and recognized two tribes, the Canoparia and the Thyreocoraria. McAtee and Malloch (1928) excluded the Canoparia from the Thyreocorinae and, subsequently (1933), treated this group as a subfamily of the Pentatomidae. Froeschner (1960) recognized the Cydnidae and the Corimelaenidae (including both New

World and Old World species) as two separate and valid families. However, Dolling (1981) treated both the Corimelaeninae and the Thyreocorinae as subfamilies of equal status within the Cydnidae with the Corimelaeninae restricted to New World species and the Thyreocorinae restricted to Old World species. Froeschner (1988e) provided an explanation for why Thyreocoridae (or -inae) had priority over Corimelaenidae if only one family group is recognized. Štys and Davidová-Vilímová (1979) recognized this group as a family, the Thyreocoridae. Lis (1994, 1999a) treated all New World and Old World species as a single group, but used the name Corimelaeninae as a subfamily of the Cydnidae; he later (2006e) recognized this group as a family (Thyreocoridae), with two subfamilies (Corimelaeninae and Thyreocorinae). Schuh and Slater (1995) followed Dolling in recognizing this group as two subfamilies within the Cydnidae - the Thyreocorinae for Old World taxa and Corimelaeninae for New World taxa. The morphological analysis in the phylogenetic work by Grazia et al. (2008) supported the groupings of Dolling (1981), but the rest of the analysis did not. They speculated that their analysis might not have been robust enough to determine relationships or that the Cydnidae as recognized by Dolling might not be monophyletic after all. Based on mitochondrial sequences (12S and 16S), Lis et al. (2012b) did not recover the Thyreocoridae as a monophyletic group. Although the aim of their study was to test the monophyly of the Dinidoridae, they included three thyreocorid species: *Galgupha difficilis* (Breddin), *Strombosoma impictum* (Stål), and *Thyreocoris scarabaeoides* (L.). In the combined analysis, *S. impictum* and *T. scarabaeoides* were defined as sister groups, thus supporting the Thyreocorinae, whereas *G. difficilis* was the sister group of a clade including Acanthosomatidae, Cydnidae, Dinidoridae, Parastrachiidae, Pentatomidae, Scutelleridae, and Tessaratomidae. Based on morphological characters from a broad sample of included genera (12 genera of Thyreocoridae plus representatives from Canopidae, Cydnidae, Parastrachiidae, and Plataspidae), Matesco's (2014) study supported the monophyly of the Thyreocoridae, including six synapomorphies: (1) dense head punctation, (2) presence of punctation between the ocellus and adjacent compound eye, (3) presence of a pseudoruga (prolongation of the ostiolar ruga onto the mesopleuron), (4) anterior margin of abdominal sternite VII angulate in males, (5) absence of a carina on the internal surface of the dorsal rim of the pygophore, and (6) parameres partially exposed.

The relationship of the Thyreocoridae within the Pentatomoidea has been studied by several workers (Gapud 1991, Grazia et al. 2008, Lis et al. 2012b, Matesco and Grazia 2015), most of whom have indicated a close relationship with the Cydnidae. The phylogenetic study by Gapud (1991), based on morphology, suggested that the Cydnidae and Thyreocoridae were strongly related by the presence in both of coxal combs and tibial spines. Thyreocorids can be separated from the cydnids by the lack of setigerous punctures on the head and thorax (present in cydnids). The Thyreocoridae shares other important characters with some other pentatomoid families, such as the enlarged scutellum, short frenum, and ninth laterotergites fused (Matesco and Grazia 2015). For example, the enlarged scutellum is shared with the Canopidae, Lestoniidae, Megarididae, Plataspidae, Scutelleridae, and Pentatomidae (Aphylinae, Cyrtocorinae, and some members of Asopinae, Pentatominae, and Podopinae); the Thyreocoridae can be separated from all of these families by the presence of spines along the length of the tibiae.

Members of this family are relatively small (3-8 mm), oval-elongate in shape, strongly convex dorsally, flat ventrally, and usually a shiny black (**Figure 2.25K**) or dark brown color, often with some white markings. The head is declivent in lateral view and subtriangular in dorsal view. The antennae are five-segmented. The scutellum is greatly enlarged and strongly convex, nearly covering the entire abdomen; the exposed part of the corium is much reduced and often of a pale whitish color. They possess a stridulatory mechanism composed of a strigil on the posterior anal vein (postcubitus in several earlier papers) of the hind wing and a plectrum located on abdominal tergum I. In the Corimelaeninae, the hind wings have the jugal lobe perforated (**Figure 2.11F**). The tibiae are provided with numerous spines along their length, the tarsi are three-segmented, and the coxal combs are present. The abdominal trichobothria usually are arranged transversely. The female spermatheca has a short invagination basally (called a sclerotized funnel in Štys and Davidová 1979) that may be the precursor to a sclerotized rod; the spermathecal bulb is simple, ball-shaped, and lacks diverticula.

The New World taxa (nine genera and 207 species at the time) were revised and keyed by McAtee and Malloch (1933). The Old World fauna comprises three genera and six species; the European species of *Thyreocoris* were revised by Štys and Davidová (1979). The Palearctic and Nearctic species recently have been catalogued by Lis (2006e) and Froeschner (1988e), respectively. A checklist and other taxonomic and biological information available for the Neotropical species was compiled by Matesco and Grazia (2015). There are currently 30 genera and 223 species in this family (**Table 2.2**). The systematics of this family has been studied as a thesis project (Matesco 2014); a recent review of the genus *Alkindus* Distant also has been provided (Matesco and Grazia 2013).

All species of this family are phytophagous, with many different host plants reported for various species. According to Štys and Davidová (1979), the most preferred host plant for the European *Thyreocoris* species is *Viola tricolor* L. (Violaceae). Feeding, especially in *Corimelaena* spp., occurs mainly in the reproductive parts of the hosts, such as flowers and developing fruits (Schaefer 1988). McPherson (1971, 1972) observed preference of *Corimelaena lateralis* (F.) for the mature parts of *Daucus carota* L. (Apiaceae) rather than the inflorescence; however, such a preference was not observed in *C. pulicaria* (Germar) (McPherson 1972). Polyphagy seems to be the rule with some species recorded from host plant associations from more than ten plant families; yet host specialization may occur in some species.

Thyreocorids may be uni-, bi-, or multivoltine (McPherson 1972; Lung and Goeden 1982; Bundy and McPherson 1997, 2009). Adults overwinter under litter, soil, or stones. Precopulatory behavior was described by Bundy and McPherson (1997), and copulation may take several hours. Thyreocorids lay eggs singly, glued laterally to the substrate, often the reproductive parts of the host plant (less frequently on other parts of the host plant). Egg development varies from eight to eleven days, depending on the species and the temperature and humidity. First instars do not show gregarious behavior. Nymphal developmental time varies from 30 to 45 days, influenced by diet, temperature, and humidity. Several species of tachinids (Diptera) and parasitic wasps (formerly Scelionidae, now Platygastridae in the Hymenoptera) have been recorded as attacking thyreocorids. Occasionally, thyreocorids can occur in rather large numbers in their natural habitats (McPherson 1974, Mendonça et al. 2009). In field studies in southern Brazil, thyreocorids were especially abundant, and they were one of the most speciose groups collected (Schmidt and Barcellos 2007, Mendonça et al. 2009).

2.2.18.1 Key to the Subfamilies of Thyreocoridae (modified from Schuh and Slater 1995)

1 Hind wing with jugal lobe entire (**Figure 2.11E**); Eastern Hemisphere (**Figure 2.25K**)...........
.... Thyreocorinae

1' Hind wing with an oval perforation in jugal lobe (**Figures 2.11F, G**); Western Hemisphere.....
....Corimelaeninae

2.2.19 Urostylididae Dallas, 1851

Urostylidids are an Old World group occurring from India through the Oriental region and into Japan and Southeast Asia. The family currently contains eight genera and 172 species (**Table 2.2**). They are variously colored, frequently green to yellowish green (**Figure 2.25L**), but many species are contrasting brown and yellow, others are aposematically red and black, resembling pyrrhocorids, and still others have a complex color pattern. They are medium in size (8 to 15 mm), and, in general, they look more like a coreoid or pyrrhocoroid than a pentatomoid. They share several unique characters with the closely related family Saileriolidae such as more dorsal antennal insertions and ocelli that are situated close to each other. They differ, however, in several important characters. For example, in urostylidids, the abdominal spiracles are all ventral (they are situated close to the lateral margins of the abdomen in saileriolids), the scent gland rugae are distinct and spoutlike (external scent efferent structures are strongly reduced or absent in saileriolids), and a hamus is present in each hind wing (not present in saileriolids); in addition, saileriolids are much smaller (less than 5 mm in length). Most of the early work on this group was under the family name Urostylidae, which was homonymous with Urostylidae in the Ciliophora. Berger et al. (2001) emended the name of the heteropteran family to Urostylididae to remove this homonymy. The Palearctic species recently have been catalogued (Rider 2006c). At least six fossil species have been described, all in the genus *Urochela* (Zhang 1989).

Members of this family can be characterized further by a rather short, declivent head with rounded margins (not edged or reflexed). The antennae are five-segmented with segment I relatively long, reaching well beyond the apex of the head. The bucculae are small, somewhat flap-like. The clavi usually meet beyond the apex of the scutellum in a single point, sometimes forming a short claval commissure. The scutellum is triangular, the apex is acute; the frena extend the entire length, from base to apex, of the scutellum. The middle and hind coxae are widely separated; the tarsi are three-segmented. The abdominal trichobothria are paired on each side of abdominal segments III through VII and transversely oriented. In females, the second valvifers are fused, forming an M-shaped or W-shaped sclerite; the female spermathecal duct lacks a dilation or sclerotized rod. Kumar (1971) provided a detailed study of urostylidid genitalia and the alimentary canal. Five species of Urostylididae have been karyotyped, of which two and three species had a diploid number of 12 + XY and 14 + XY, respectively (Ueshima 1979, Kerzhner et al. 2004, Rebagliati et al. 2005).

The taxonomic history of this family has been even more confused than other pentatomoid families. Through time, they have been associated with a number of different families, some pentatomoid, some not. For example, Singh-Pruthi (1925) related them to the Acanthosomatidae, Yang (1938, 1939) and Pendergrast (1957) to the Pyrrhocoridae, and Miyamoto (1961) to the Pentatomidae. Most workers have, however, held the common belief that the urostylidids represent a basal position in the evolution of the pentatomoids; in fact, China and Slater (1956) considered them to represent a proto-Trichophoran group basal to the Pentatomidae, Coreidae, and Lygaeidae. The study by Grazia et al. (2008) supports a basal position, placing the urostylidids as the sister group to the rest of the Pentatomoidea.

There are no recent generic revisions available for this family, but several more comprehensive studies have been published for certain geographical areas. For example, Yang (1939), Hsiao et al. (1977), and Ren (1999) reviewed the Chinese urostylidids; Ren and Lin (2003) studied those species occurring in Taiwan; and Ahmad et al. (1992) treated the taxa from the Indian subcontinent.

The family Urostylididae is one of three heteropteran families in which females are known to form an ootheca, a structure protecting the mass/batch of eggs. Early works on the subject were in Japanese (Yamada 1914, 1915). Kobayashi (1965) and later Kobayashi and Tachikawa (2004) described the occurrence of ootheca as a gelatinous substance covering the eggs completely except for the apices of the micropylar processes. They studied four species from two genera: *Urostylis westwoodii* Scott, *U. striicornis* Scott, *Urochela luteovaria* Distant, and *U. quadrinotata* (Reuter). The egg batch usually is deposited either on the bark or in a crevice of the bark and is covered by a substance of a different color, often transparent or glossy. An interesting account of this behavior was provided by Kaiwa et al. (2014); they studied the life cycle of two species in Japan. The females have enlarged ovaries that produce a polysacharide (jelly-like) excretion. Endosymbionts from specific organs in the genital chamber are passed into the jelly during oviposition (vertical symbiont transmission). Eggs are laid in November, and nymphs hatch in February (end of winter, which is exceptional within the Heteroptera) before plant hosts (*Quercus* spp.) have formed leaves. Nymphal instars I through III remain on the egg mass, feeding on the jelly containing the symbiotic bacteria.

Additional biological information is meager. Urostylidids have been recorded from a variety of plants, but there seems to be preference for various tree species (Cornaceae, Fagaceae, Pinaceae, Rosaceae, Theaceae, and Tiliaceae). They occasionally have been considered pests.

2.3 Conclusions

We now have presented taxonomic histories, synopses, and various comments on all of the pentatomoid families and on the pentatomid subfamilies, tribes, and species groups known to us. We hope this establishes a foundation upon which new phylogenetic work can be built. In our opinion, the main barriers that must be overcome in developing such a classification include, but are not limited to (1) the size of the project (over 50 tribes and generic groups within the Pentatomidae alone); (2) the interpretation of morphological characters; and (3) the lack of specimens for molecular research.

2.3.1 Size of Project

Within the Pentatomidae, we have recognized a total of ten subfamilies, two of which have two tribes, another has four, and another has five. Within the subfamily Pentatominae, we have recognized 42 tribes, another twelve Gross genus groups and eight Linnavuori genus groups that do not fit cleanly within already established tribes. This roughly adds up to 80-85 groups that will need to be studied. A comprehensive analysis would include at least two or more taxa from each of the above groups, so we are now looking at a minimum of 160 taxa in the analysis of only the Pentatomidae. Complicating matters is the fact that some of the groups contain only a single (or few) species, and these are represented only by a few specimens in collections.

2.3.2 Interpretation of Characters

It is obvious that several important characters that have been used in past analyses have evolved multiple times. Additionally, various characters, including those that have arisen multiple times, have been secondarily lost. A good example of this is the armature of the base of the abdomen, which is a character that often has been used in establishing classifications. In the early 1980s, this character was used to segregate the majority of New World genera into three groups: (1) those lacking a spine or tubercle, (2) those having a spine or tubercle with the apex free, and (3) those with a spine or tubercle that was met in apposition with an elevated metasternum (Rolston et al. 1980; Rolston and McDonald 1981, 1984). For practical purposes, this classification worked well and allowed for further keys to be developed, thus aiding mainly in the identification of the New World genera. It did not, however, reflect phylogeny, especially once we began to integrate this with Old World classifications. It became obvious that the abdominal armature had evolved a number of times and when certain groups were separated off using other characters, those that remained began to make more sense. Similarly, it appears that the development of the structures associated with the metathoracic scent gland involves similar problems. The length and shape of the ostiolar rugae often has been used in helping develop classifications. Initial data seem to indicate that a short, auriculate ostiolar ruga may be considered primitive, and a longer, groove-shaped, apically acuminate or disc-shaped ruga may be more advanced (Schaefer 1972, Linnavuori 1982). But it also appears that this does not always support phylogenetic relationships (see Kment and Vilímová 2010a). That is, the rugae appear to have lengthened multiple times and then, in some cases, may have shortened. This makes it extremely difficult to interpret this type of character. We also believe we need to be careful in interpreting the structure of the thoracic sternal characters, and characters associated with the internal male genitalia. Again, certain aspects of these characters also appear to have evolved multiple times or have been lost secondarily.

2.3.3 Lack of Specimens

The lack of specimens has been mentioned in connection with conducting molecular research, but it is also true for morphological work. The greatest asset for conducting taxonomic research, especially phylogenetic work, is to have access to specimens for study. Although much taxonomic work on the Pentatomoidea has been completed, there is still much to do. There is still much diversity to be explored. To illustrate this, we will provide a couple examples. The Neotropical discocephaline genus *Phoeacia* (**Figures 2.13B, 2.17L**) currently has three described species. Several new species were discovered, so loans were requested to complete a review of the genus. There are now over 30 distinct species known. Similarly, the New Guinean rhynchocorine genus *Pegala* Stål currently contains six species, but at least one or two should be removed to other genera. Again, two undescribed species were discovered, so a revision was planned. Now, after accumulating museum specimens, there are over 25 new species and several new related genera in need of description. Although these two examples may not be typical, they certainly are not exceptional. Many new genera and species await discovery, hence the need for further collecting.

The lack of fresh specimens is an even more severe problem for DNA work. Molecular research requires access to relatively fresh material, preferably material collected in the last five to ten years, whereas older specimens will suffice for morphological work. Additionally, specimens earmarked for molecular studies need to be collected and preserved in particular ways so as to not destroy the DNA. We also caution those who are using sequences posted on the internet in their phylogenetic studies. Quite often, the organisms associated with the sequences are misidentified (see discussions on Panamanian plataspids earlier in this chapter, and by Rédei 2016, and Lis et al. 2016).

2.3.4 Solutions

The problems listed above are real and will be difficult, but not impossible, to overcome. Tantamount for success will be the willingness of workers on the systematics of the Pentatomoidea to form strong collaborations with each other. Collaborations will reduce the work load each worker will have to shoulder and, thus, perhaps, help overcome problem number one. Collaborations also can help to delineate a more efficient sample for analyses, thus improving our capability to deal with this problem. Interpretation of characters may be more difficult but, again, not impossible. A greater reliance on other, more useful characters may be needed initially, and then the more difficult ones will make more sense deeper into the analyses. Again, collaborations lead to discussion, and discussions on the difficult characters may result in ways to interpret these difficult characters. Furthermore, the integration of molecular data with the morphological data should also help us interpret characters that have evolved multiple times (or help us recognize secondary losses). Acquisition of specimens for molecular work, again will be bolstered by additional collaboration. A better system needs to be developed for placing sequences on the internet for other workers to use. Although this is an admirable task, we need to make sure that the organisms are correctly identified. Perhaps it should be made mandatory that the identity of all taxa whose sequences are to made accessable are authoritatively identified by an expert in the group before the sequences can be posted on the internet; it also would be helpful to require photographs of the test animals along with the sequences.

We hopefully have provided the groundwork upon which further phylogenetic work can continue. We believe a number of good collaborations among pentatomid workers already have been made; these need to be strengthened. New relationships need to be established. Perhaps a step in the right direction would be to develop an organizational workshop or summit and to encourage all serious pentatomoid taxonomists to attend. We would be able discuss ways to overcome the above listed problems (e.g., standardize character interpretation, parcel out work load, develop a molecular plan). Such a workshop could be held in conjunction with an International Heteropterists' Society Meeting, or it could be held at another time. Either way, it would need to be held near a large collection so that specimens could be examined during discussions. Determining the phylogeny of the Pentatomoidea is certainly a task for an intercontinental team!

Additionally, increased collaborations will require increased coordination. That was the impetus behind the development of the Pentatomoidea Website (Rider 2015a); it was meant to serve as a centralized spot where workers could access a bibliography on the group, to see what other workers were doing, and to gain information on various taxa and their hosts, biology, etc. This website certainly needs improvement, and to have more information added, but it also depends on colleagues to periodically update their profile pages. Those workers who do not have a profile page, but would like one, are encouraged to send their information to the webmaster (DAR). The eventual publication of the World Catalog of the Pentatomidae (and catalogs of other pentatomoid families) will also greatly aide taxonomic and phylogenetic work.

In summary, we hope the information provided herein will stimulate further work. It is now time for the fun to begin (e.g., to see how many of these taxa will hold true and which ones will fall, and which taxa will be transferred to new places and which ones will stay put). Good luck, and have fun!

2.4 Acknowledgments

The first person deserving of our sincere gratitude is Jay McPherson who has spent an uncountable number of hours reading and editing these pages; this chapter would not be nearly the quality that it is without his considerable contributions. There are many people throughout the years who have contributed greatly to the development of this chapter, simply via the numerous discussions they have had with us. Some are

FIGURE 2.25 A, Blaudusinae, Lanopini: *Sinopla humeralis* Signoret, female (Chile); B, Cydninae, Cydnini: *Cydnus aterrimus* (Förster), male (Italy); C, Dinidorinae, Dinidorini: *Coridius sanguinolentus* (Westwood), female (China: Guangxi); D, Lestoniidae: *Lestonia haustorifera* China, female (Australia); E, Parastrachiidae: *Parastrachia japonensis*, female (China: Guangxi); F, Parastrachiidae: *Dismegistus fimbriatus* (Thunberg), male, (South Africa); G, Phloeidae: *Phloea corticata* (Drury), male (Brazil); H, Plataspidae: *Libyaspis coccinelloides* (Laporte), male (Madagascar); I, Saileriolidae: *Bannacoris arboreus* Hsiao, male (Thailand); J, Tessaratominae, Tessaratomini: *Pygoplatys tenangau* Magnien, Smets, and Pluot, female (Indonesia: Sumatra); K, Thyreocorinae: *Thyreocoris scarabaeoides* (L.), male (Czech Republic); L, Urostylididae: *Urostylis annulicornis* Scott, male (North Korea).

FIGURE 2.26 A, Elvisurinae: *Coleotichus borealis* Distant, male (Philippines: Mindanao); B, Scutellerinae, Scutellerini: *Poecilocoris purpurascens* (Westwood), male (India: Sikkim); C, Scutellerinae, Scutellerini: *Calliphara caesar* (Vollenhoven), female (Indonesia: Ambon Is.); D, Scutellerinae, Sphaerocorini: *Steganocerus multipunctatus* (Thunberg), male (South Africa); E, Tectocorinae: *Tectocoris diophthalmus* (Thunberg), male (Indonesia: Selayar Is.); F, Hoteinae: *Deroplax circumducta* (Germar), female (South Africa); G, Pachycorinae: *Crathis ansata* (Distant), female (Panama); H, Pachycorinae: *Agonosoma trilineatum* (F.), female (Curaçao); I, Eurygastrinae: *Polyphyma koenigi* Jakovlev, female (Turkmenistan); J, Odontoscelinae: *Irochrotus sibiricus* Kerzhner, male (China: Inner Mongolia); K, Odontoscelinae: *Morbora australis* Distant, female (Australia: Northern Territory); L, Odontotarsinae: *Urothyreus horvathianus* (Schouteden), female (Namibia).

FIGURE 2.27 A, Aphylinae: *Aphylum syntheticum* Bergroth, female (Australia); B, Asopinae: *Cazira horvathi* Breddin, male (China: Yunnan); C, Cyrtocorinae: *Cyrtocoris gibbus* (F.), female (Brazil: São Paulo); D, Discocephalinae, Discocephalini: *Colpocarena complanatus* (Burmeister), female (Brazil: Amazonas); E, Discocephalinae: Ochlerini: *Macropygium reticulare* (F.), male (Brazil: São Paulo); F, Edessinae: *Edessa rufomarginata* (DeGeer), female (Brazil: São Paulo); G, Phyllocephalinae, Phyllocephalini: *Basycriptus distinctus* (Signoret), female (Congo); H, Phyllocephalinae, Megarrhamphini: *Megarrhamphus hastatus* (F.), female (China: Sichuan); I, Podopinae, *Podops* group: *Podops inuncta* (F.), male (Czech Republic); J, Podopinae, *Graphosoma* group: *Graphosoma lineatum* (L.), male (Czech Republic); K, Serbaninae: *Serbana borneensis* Distant, male (Borneo); L. Stirotarsinae: *Stirotarsus abnormis* Bergroth, female (Peru).

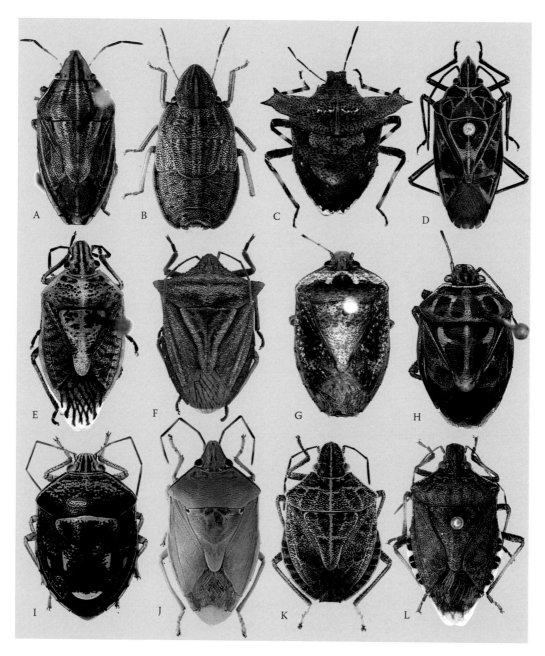

FIGURE 2.28 A, Pentatominae, Aeliini: *Aelia acuminata* (L.), female (Russia: North Caucasus); B, Pentatominae, Aeptini: *Aeptus singularis* Dallas, male (South Africa); C, Pentatominae, Aeschrocorini: *Aeschrocoris obscurus* (Dallas), female (China: Yunnan); D, Pentatominae, Agaeini: *Agaeus mimus* Distant, male (China: Yunnan); E, Pentatominae, Agonoscelidini: *Agonoscelis nubilis* (F.), male (Philippines); F, Pentatominae, Amyntorini: *Belopis unicolor* Distant, female (China: Yunnan); G, Pentatominae, Antestiini: *Antestia trispinosa* Linnavuori, female (Ghana); H, Pentatominae, Antestiini: *Antestiopsis cruciata* (F.), female (Indonesia: Sumatra); I, Pentatominae, Axiagastini: *Axiagastus cambelli* Distant, female (Solomon Islands); J, Pentatominae, Bathycoeliini: *Bathycoelia alkyone* Linnavuori, male (Yemen: Socotra Island); K, Pentatominae, Cappaeini: *Cappaea taprobanensis* (Dallas), male (Vietnam); L, Pentatominae, Cappaeini: *Halyomorpha halys* (Stål), female (South Korea).

FIGURE 2.29 A, Pentatominae, Carpocorini: *Carpocoris purpureipennis* (DeGeer), male (Czech Republic); B, Pentatominae, Carpocorini: *Menecles insertus* (Say), female (USA: Texas); C, Pentatominae, Catacanthini: *Catacanthus incarnatus* (Drury), male (India: Tamil Nadu); D, Pentatominae, Catacanthini: *Arocera apta* (Walker), female (Brazil: Amazonas); E, Pentatominae, Caystrini: *Caystrus quadrimaculatus* Linnavuori, male (Cameroon); F, Pentatominae, Chlorocorini: *Chlorocoris complanatus* (Guérin-Méneville), male (Brazil: São Paulo); G, Pentatominae, Coquereliini: *Coquerelia pectoralis* (Signoret), female (Madagascar); H, Pentatominae, Degonetini: *Degonetus serratus*, female (India: Kerala); I, Pentatominae, Diemeniini: *Diemenia rubromarginata* (Guérin-Méneville), male (Australia: Australian Capital Territory); J, Pentatominae: Diplostirini: *Diplostira valida* Dallas, female (Bangladesh); K. Pentatominae: Diploxyini: *Diploxys fallax* (Stål), male (Madagascar); L, Pentatominae, Eurysaspini: *Eurysaspis* sp., male (South Africa).

FIGURE 2.30 A, Pentatominae, Eysarcorini: *Eysarcoris aeneus* (Scopoli), female (Russia: Yaroslavl); B, Pentatominae, Halyini: *Halys sulcatus* (Thunberg), male (India: Karnataka); C, Pentatominae, Halyini: *Atelocera serrata* (F.), male (Cameroon); D, Pentatominae, Hoplistoderini: *Hoplistodera pulchra* Yang, male (China: Fujian); E, Pentatominae, Lestonocorini: *Gynenica funerea* Horváth, female (Kenya); F, Pentatominae, Mecideini: *Mecidea lindbergi* Wagner, male (Iran); G, Pentatominae, Memmiini: *Memmia femoralis* (Signoret), male (Madagascar); H, Pentatominae, Menidini: *Menida violacea* Motschulsky, female (China, Shanxi); I, Pentatominae, Myrocheini: *Myrochea cribrosa* (Klug), female (Senegal); J, Pentatominae, Nealeriini: *Nealeria asopoides* (Stål), male (Madagascar); K, Pentatominae, Nezarini: *Chinavia collis* (Rolston), male (Costa Rica); L, Pentatominae, Nezarini: *Palomena prasina* (L.), male (France).

FIGURE 2.31 A, Pentatominae, Opsitomini: *Opsitoma brunneus* Cachan, male (Madagascar); B, Pentatominae, Pentamyrmecini: *Pentamyrmex spinosus* Rider and Brailovsky, female (Thailand); C, Pentatominae, Pentatomini: *Pentatoma rufipes* (L.), male (Czech Republic); D, Pentatominae, Pentatomini: *Banasa patagiata* (Berg), female (Argentina: Formosa); E, Pentatominae, Pentatomini: *Evoplitus humeralis* (Westwood), male (Brazil: São Paulo); F, Pentatominae, Pentatomini: *Lelia octopunctata* (Dallas), male (India: Haryana); G, Pentatominae: Pentatomini: *Pharypia pulchella* (Drury), male (Venezuela); H, Pentatominae, Pentatomini: *Rhaphigaster nebulosa* (Poda), female (Bulgaria); I, Pentatominae, Phricodini: *Phricodus bessaci* Villiers, male (Iran); J, Pentatominae, Piezodorini: *Piezodorus lituratus* (F.), male (Switzerland); K, Pentatominae, Procleticini: *Thoreyella brasiliensis* Spinola, female (Brazil: São Paulo); L, Pentatominae, Rhynchocorini: *Rhynchocoris humeralis* (Thunberg), female (Myanmar).

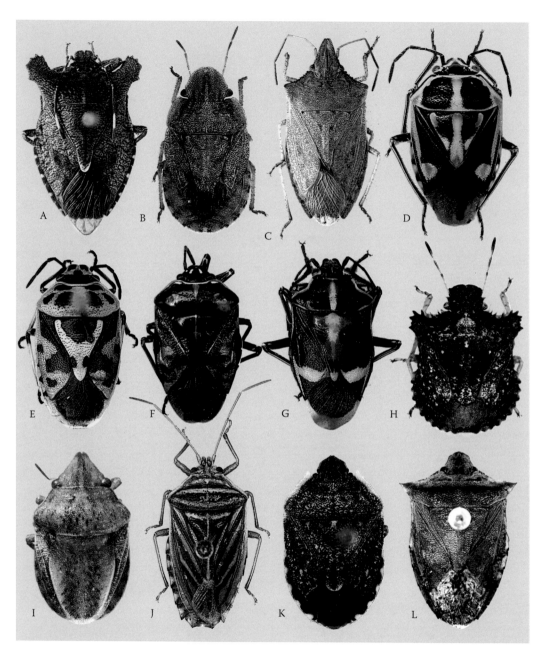

FIGURE 2.32 A, Pentatominae, Rolstoniellini: *Rolstoniellus boutanicus* (Dallas), female (India: Meghalaya); B, Pentatominae, Sciocorini: *Sciocoris cursitans* (F.), male (Czech Republic); C, Pentatominae, Sephelini: *Brachymna tenuis* (Stål), female (China: Fujian); D, Pentatominae: Strachiini: *Bagrada hilaris* (Burmeister), female (India: Meghalaya); E, Pentatominae, Strachiini: *Eurydema ornata* (L.), female (Italy); F, Pentatominae, Strachiini: *Murgantia histrionica* (Hahn), male (USA: California); G, Pentatominae, Strachiini: *Strachia crucigera* Hahn, female (Thailand); H, Pentatominae, Triplatygini: *Triplatyx bilobatus* Cachan, male (Madagascar); I, Pentatominae, unplaced: *Macrocarenoides scutellatus* (Distant), female (Australia); J, Pentatominae, unplaced: *Patanius vittatus* Rolston, male (Brazil: Amazonas); K, Pentatominae, unplaced: *Trachyops australis* Dallas, female (Australia: Queensland); L, Pentatominae, unplaced: *Thyanta perditor* (F.), female (Brazil: São Paulo).

now considered veterans in the field (Gerry Cassis, Joe Eger, Jocelia Grazia, Barbara Lis, Jerzy Lis, Philippe Magnien, J. E. McPherson, Pavel Štys), some are no longer with us (Richard Froeschner, Izya Kerzhner, Yuri Popov, Larry Rolston, Carl Schaefer), and some are relatively new (Kim Barão, Aline Barcellos, Felipe Bianchi, Luiz Campos, Mariom Carvajal, Eduardo Faúndez, José Fernandes, Dmitry Gapon, Thereza Garbelotto, Caroline Greve, Jing-Fu Tsai, Michael Wall); to all, we give thanks.

Special thanks to Gerry Cassis (University of New South Wales), Tom Henry (United States National Museum of Natural History), and Toby Schuh (American Museum of Natural History) for providing specimens for study, including specimens and photos of pertinent type material. The list of other curators of various museums (private, university, etc.) who have provided loans of specimens and types (or took photos of types) is too long to provide here, but we thank you nonetheless. We greatly appreciate the help with literature acquisition that many heteropterists have provided; DAR is especially grateful for the help received from the North Dakota State University Inter-Library Loan Service.

We also appreciate the help of Gerald Fauske and Mariom Carvajal (North Dakota State University, Department of Entomology), Petr Janšta (Charles University, Department of Zoology), and João Victor de Lima Firmino (Universidade Federal de Alagoas) for their help with some of the photographs and the podopine phylogenetic tree used in this chapter. Also, Bruno C. Genevcius (Museu de Zoologia da Universidade de São Paulo) and Jussara Suzano Bento (Universidade Federal de São Paulo) provided special help with the illustrations for the key characters. A special thanks to DAR's wife, Jayma Moore (North Dakota State University, Electron Microscopy Laboratory), who had to listen to DAR grumble throughout the process of putting this chapter together; she also helped immensely in providing word processing advice and in the assembly of the tables and plates.

Figures 2.16C and 2.32H were used by permission from the Acta Entomologica Musei Nationalis Pragae journal; **Figure 2.25D** was used by permission from the American Museum of Natural History; **Figure 2.12H** was used by permission from Denisia; **Figure 2.12A** was used by permission from Entomologica Americana; **Figures 2.13F; 2.14A-D; and 2.27L** were used by permission from the Entomological Society of America; and **Figures 2.4A-D, F; 2.14E, F; and 2.21D** were used by permission from Zootaxa. Also, all photos in **Figure 2.9** were taken at the American Museum of Natural History's Microscopy and Imaging Facility.

The work of Cristiano F. Schwertner was supported by a research grant from FAPESP (Auxílio Regular, proc 2014/00729-3). The research of Jitka Vilímová was supported by grant no. SVV 260 313/2016 of the Ministry of Education of the Czech Republic. Dávid Rédei received financial support from the National Natural Science Fundation of China (grant no. 31472024) and the One Hundred Young Academic Leaders Program of Nankai University. The work of Petr Kment was supported by the Ministry of Culture of the Czech Republic (DKRVO 2016/14 and 2017/14, National Museum, 00023272).

2.5 References Cited

Abebe, M. 1987. Insect pests on coffee with special emphasis on *Antestia*, *Antestia intricata*, in Ethiopia. Insect Science and its Application 8(4–6): 977–980.

Adler, P. H., and A. G. Wheeler, Jr. 1984. Extra-phytophagous food sources of Hemiptera-Heteroptera: Bird droppings, dung, and carrion. Journal of the Kansas Entomological Society 57(1): 21–27.

Afzal, M., and S. A. Hasan. 1988. A new species of the genus *Mormidella* Horvath (Heteroptera: Pentatomidae: Pentatominae) from Pakistan with comparative notes on the genus and the species. Annotationes Zoologicae et Botanicae, no. 183: 8 pp.

Ahmad, I. 1977. Systematics and biology of pentatomomorphous superfamilies Coreoidea and Pentatomoidea of Pakistan. Research Project No. FG-Pa-181 (A-17-ENT-37) Final Report, vii + 625 pp.

Ahmad, I. 1981. A revision of the superfamilies Coreoidea and Penttaomoidea [sic] (Heteroptera: Pentatomomorpha) from Pakistan, Azad Kashmir, and Bangladesh. Part I.: Additions and corections [sic] of coreid and pentatomid fauna with phylogenetic considerations. Entomological Society of Karachi, Supplement 4, part 1 [1979], 113 pp.

Ahmad, I. 1994. A review of taxa of pentatomine genus *Pausias* Jakovlev (Hemiptera: Pentatomidae). Pakistan Journal of Entomology 9(1): 29–42.

Ahmad, I. 1997. A revision of the genus *Elasmostethus* Fieber (Hemiptera: Acanthosomatidae) from western Palaearctic, with a new record from Turkey and their cladistics relationships. Proceedings of Pakistan Congress of Zoology 17: 63–71.

Ahmad, I. 2004. A revision of *Tachengia* China (Hemiptera: Pentatomidae: Pentatominae: Halyini) and its cladistic relationships. Proceedings of Pakistan Congress of Zoology 24: 125–130.

Ahmad, I., and M. Afzal. 1979. Resurrection of the tribe Caystrini Stål (Heteroptera, Pentatomidae, Pentatominae) with description of two new genera from Oriental region. Annotationes Zoologicae et Botanicae, no. 133: 14pp.

Ahmad, I., and M. Afzal. 1989. A revision of Myrocheini (Pentatomidae: Pentatominae) from Indo-Pakistan area. Oriental Insects 23: 243–267.

Ahmad, I., and S. Kamaluddin. 1976. A new genus and three new species of Phyllocephalinae (Pentatomomorpha: Pentatomidae) from Pakistan with notes on their zoogeography and phylogeny. Entomologische Mitteilungen aus dem Zoologischen Museum Hamburg 5(93): 81–95.

Ahmad, I., and S. Kamaluddin. 1978a. A new genus and two new species of Carpocorini (Heteroptera: Pentatomidae) from Pakistan. Pakistan Journal of Scientific and Industrial Research 21(5–6): 185–188.

Ahmad, I., and S. Kamaluddin. 1978b. A new genus and a new species of Phyllocephalinae (Hemiptera: Pentatomidae) with phylogenetic considerations. Transactions of the Shikoku Entomological Society 14(1–2): 1–6.

Ahmad, I., and S. Kamaluddin. 1982. A revision of the genus *Poecilocoris* (Pentatomoidea: Scutelleridae) from Indo-Pakistan subcontinent with descriptions of three new species. Oriental Insects 16(3): 259–295.

Ahmad, I., and S. Kamaluddin. 1988. A new tribe and a new species of the subfamily Phyllocephalinae (Hemiptera: Pentatomidae) from the Indo-Pakistan subcontinent. Oriental Insects 22: 241–258.

Ahmad, I., and S. Kamaluddin. 1989a. A revision of the Australian genus *Diemenia* Spinola (Hemiptera: Pentatomidae: Pentatominae). Records of the South Australian Museum 23(1): 21–31.

Ahmad, I., and S. Kamaluddin. 1989b. A revision of the tribe Caystrini Stal (Hemiptera: Pentatomidae: Pentatominae) from Indo-Pakistan subcontinent with description of two new species from Pakistan and their cladistic analysis. Proceedings of the Pakistan Congress of Zoology 9: 169–183.

Ahmad, I., and S. Kamaluddin. 1990a. A new tribe for phyllocephaline genera *Gellia* Stål and *Tetroda* Amyot et Serville (Hemiptera: Pentatomidae) and their revision. Annotationes Zoologicae et Botanicae 195: 1–20.

Ahmad, I., and S. Kamaluddin. 1990b. Redescription of type species of *Schyzops* Spinola, *S. aegyptiaca* (Lefebvre) and a new genus for *S. orientalis* Kamaluddin and Ahmad (Hemiptera: Pentatomidae: Phyllocephalinae) with a key to the world species and their cladistic analysis. Proceedings of the Pakistan Congress of Zoology 10: 195–208.

Ahmad, I., and S. Kamaluddin. 1992. New generic status of a rice-feeding tetrodine subgenus *Tetrodias* Kirkaldy and redescription of *Tetroda* Amyot and Serville (Hemiptera: Pentatomidae: Phyllocephalinae) and their cladistic analysis. Pakistan Journal of Zoology 24(2): 123–127.

Ahmad, I., and S. Kamaluddin. 1994. Studies on the male and female genitalia of Cressonini (Hemiptera: Pentatomidae: Phyllocephalinae) and their bearing on classification. Pakistan Journal of Entomology 9(2): 95–101.

Ahmad, I., and N. A. Khan. 1983. A revision of the genus *Stenozygum* (Pentatomidae: Strachiini) from the Oriental and Australian regions, with reference to zoogeography and phylogeny. Australian Journal of Zoology 31(4): 581–605.

Ahmad, I., and J. E. McPherson. 1998. Additional information on male and female genitalia of *Parabrochymena* Larivière and *Brochymena* Amyot and Serville (Hemiptera: Pentatomidae). Annals of the Entomological Society of America 91: 800–807.

Ahmad, I., and F. A. Mohammad. 1980. A new tribe, a new genus and two new species of the subfamily Pentatominae Amyot et Serville (Hemiptera: Pentatomidae) from Pakistan and their relattonships [sic]. Transactions of the Shikoku Entomological Society 15(1–2): 11–25.

Ahmad, I., and F. A. Mohammad. 1990. A revision of Acanthosomatidae (Hemiptera, Pentatomomorpha, Pentatomoidea) from Indo-Pakistan area with a cladistic analysis of the genera. Oriental Insects 24: 267–304.

Ahmad, I., and M. Moizuddin. 1992. Plataspidae Dallas (Hemiptera: Pentatomoidea) from Pakistan and Bangladesh with keys including Indian taxa. Annotationes Zoologicae et Botanicae, no. 208: 31 pp.

Ahmad, I., and F. Önder. 1990a. Revision of the genus *Arma* Hahn (Hemiptera: Pentatomidae: Pentatominae: Asopini) with description of two new species from Turkey. Türkiye Entomoloji Dergisi 14(1): 3–12.

Ahmad, I., and F. Önder. 1990b. A revision of the genus *Picromerus* Amyot and Serville (Hemiptera: Pentatomidae: Pentatominae: Asopini) from western Palearctic with description of two new species from Turkey. Türkiye Entomoloji Dergisi 14(2): 75–84.

Ahmad, I., and F. Önder. 1993. A review of status of *Cyphostethus* Fieber (Hemiptera: Acanthosomatidae) from western Palaearctic region. Pakistan Journal of Entomology 8: 55–62.

Ahmad, I., and F. Önder. 1996. Genus *Lodosia* Ahmad & Önder, gen. nov., p. 256. *In* Ahmad, I., C. W. Schaefer, and F. Önder. 1996. The first species of Myrocheini (Hemiptera: Pentatomidae: Pentatominae) from the Palaearctic region. European Journal of Entomology 93(2): 255–262.

Ahmad, I., and N. A. Rana. 1988. A revision of the genus *Canthecona* Amyot et Serville (Hemiptera: Pentatomidae: Pentatominae: Asopini) from Indo-Pakistan subcontinent with description of two new species from Pakistan. Türkiye Bitki Koruma Dergisi 12(2): 75–84.

Ahmad, I., and N. A. Rana. 1989. A revision of the genus *Acrosternum* Fieber (Hemiptera: Pentatomidae: Pentatominae: Pentatomini), associated with rice from Indo-Pakistan subcontinent with description of a new species from Baluchistan. Proceedings of the Pakistan Congress of Zoology 9: 185–199.

Ahmad, I., and N. A. Rana. 1991. A review and cladistic analysis of the genus *Anasida* Karsch (Pentatomidae: Pentatominae: Asopini). Proceedings of the Pakistan Congress of Zoology 10: 271–285.

Ahmad, I., and N. A. Rana. 1994. A revision of asopine stink bug genera *Blachia* Walker and *Breddiniella* Schouteden (Hemiptera: Pentatomidae: Pentatominae). Proceedings of the Pakistan Congress of Zoology 14: 137–146.

Ahmad, I., and N. A. Rana. 1996. A review of antestiine genus *Plautia* Stål (Hemiptera: Pentatomidae: Pentatominae) from Indo-Pakistan subcontinent and their cladistic relationships. Pakistan Journal of the Entomological Society of Karachi 11: 45–57.

Ahmad, I., and I.-U.-R. Shah. 1994. A revision of the stink bug genus *Mecidea* Dallas (Hemiptera: Pentatominae: Mecideini) from Indo-Pakistan subcontinent and their cladistic relationship. Proceedings of the Pakistan Congress of Zoology 14: 127–136.

Ahmad, I., Q. A. Abbasi, and A. A. Khan. 1974. Generic and supergeneric keys with reference to a check list of pentatomid fauna of Pakistan (Heteroptera: Pentatomoidea) with notes on their distribution and food plants. Proceedings of the Entomological Society of Karachi, supplement 1: 1–103.

Ahmad, I., M. Moizuddin, and S. Mushtaq. 1988. A revision of the genus *Deroplax* (Hemiptera: Scutelleridae) from Oriental region with description of two new species from Pakistan and Bangladesh. Oriental Insects 22: 259–266.

Ahmad, I., S. S. Shaukat, and R. H. Zaidi. 1990. A taxometric study of the tribe Aeliini Stal [sic] (Hemiptera: Pentatomidae: Pentatominae) from Indo-Pakistan subcontinent. Indian Journal of Zoological Spectrum 1(1): 23–30.

Ahmad, I., M. Moizuddin, and S. Kamaluddin. 1992. A review and cladistics of Urostylidae Dallas (Hemiptera: Pentatomoidea) with keys to taxa of Indian subregion and description of four genera and five species including two new ones from Pakistan, Azad Kashmir and Bangladesh. Philippine Journal of Science 121(3): 263–297.

Ahmad, I., S. A. Rizvi, and S. Kamaluddin. 1998. Review of the shesham tree shieldbug genus *Deroplax* Mayr (Hemiptera: Scutelleridae: Pachycorinae) with description of two new species from Baluchistan, Palaearctic region. Proceedings of Pakistan Congress of Zoology 18: 49–56.

Aiello, A., K. Saltonstall, and V. Young. 2016. *Brachyplatys vahlii*, an introduced bug from Asia: first in the Western Hemisphere (Hemiptera: Plataspidae: Brachyplatidinae). BioInvasions Records 5(1): 7–12.

Aldrich, J. R. 1988. Chemical ecology of the Heteroptera. Annual Revue of Entomology 33: 211–238.

Aldrich, J. R., and W. R. Lusby. 1986. Exocrine chemistry of beneficial insects: male specific secretions from predatory stinkbugs (Heteroptera: Pentatomidae). Comparative Biochemistry and Physiology 85B: 639–642.

Alvarez, F. A. 1993. Ciclo de vida de *Lincus tumidifrons* Rolston (Hemiptera: Pentatomidae), vector de la Marchitez Sorpresiva de la palma de aceite. Revista Colombiana de Entomologia 19(4): 167–174.

Amyot, C. J. B., and A. Serville. 1843. Histoire naturelle des insectes. Hémiptères. Librairie Encyclopedique de Roret ed., Paris. lxxvi + 675 pp.

Archibald, S. B., and R. W. Mathewes. 2000. Early Eocene insects from Quilchena, British Columbia, and their paleoclimatic implications. Canadian Journal of Zoology 78(8): 1441–1462.

Arnold, K. 2011a. *Jostenicoris* Arnold, n. gen., eine neue Heteropteren-Gattung aus der Äthiopischen Region (Insecta: Hemiptera: Heteroptera: Pentatomidae: Strachiini). Edessana 1: 15–18.

Arnold, K. 2011b. *Uddmania blatterti* Arnold, n. sp. von Sumatra und *Uddmania schoenitzeri* Arnold, n. sp. von Java, zwei neue Pentatomiden-Arten aus der Inselwelt Indonesiens, nebst Bemerkungen zur Gattung *Uddmania* Bergroth, 1915 (Insecta: Hemiptera: Heteroptera: Pentatomidae: Phyllocephalinae: Cressonini). Edessana 1: 19–24.

Asanova, R. B., and I. M. Kerzhner. 1969. Eine Übersicht der Gattung *Trochiscocoris* Reuter mit Beschreibung einer neuen Unterart aus dem zentralen Kasachstan. Beiträge zur Entomologie 19(1–2): 115–121.

Asgarali, J., and P. Ramkalup. 1985. Study of *Lincus* sp. (Pentatomidae) as the possible vector of hartrot in coconut. Surinam Agriculture 33: 56–61.

Atkinson, E. T. 1888. Notes on Indian Rhynchota: Heteroptera, No. 3. Journal of the Asiatic Society of Bengal 57(2): 1–72.

Azad Thakur, N. S., and D. M. Firake. 2012. *Ochrophora* [sic] *montana* (Distant): a precious dietary supplement during famine in northeastern Himalaya. Current Science 102(6): 845–846.

Azim, M. N. 2002. Studies on Indian genera of the tribe Halyini (Pentatomidae: Pentatominae). Oriental Science 2002: 41–52.

Azim, M. N. 2011. Taxonomic survey of stink bugs (Heteroptera: Pentatomidae) of India. Halteres 3: 1–10.

Azim, M. N., and S. A. Shafee. 1984a. Degonetini trib. n. (Heteroptera: Pentatomidae). Current Science 53(20): 1094–1095.

Azim, M. N., and S. A. Shafee. 1984b. Indian species of the genus *Stollia* Ellenrieder (Heteroptera, Pentatomidae). Mitteilungen der Schweizerischen Entomologische Gesellschaft 57: 291–293.

Azim, M. N., and S. A. Shafee. 1985. Studies on Indian species of *Hermolaus* (Heteroptera, Pentatomidae). International Journal of Entomology 27(4): 394–397.

Azim, M. N., and S. A. Shafee. 1986. Studies on the Indian Strachiini (Pentatomidae: Pentatominae). Journal of the Bombay Natural History Society 82(3): 586–593.

Azim, M. N., and S. A. Shafee. 1987. Studies on Indian species of *Sciocoris* Fallen (Heteroptera: Pentatomidae). Articulata 2(10): 367–370.

Baehr, M. 1989. Review of the Australian shield bug genus *Theseus* Stål. Spixiana 11(3): 243–258.

Baehr, M. 1991. A nomenclatorial note on *Theseus modestus grossi* Baehr, 1989 (Insecta, Heteroptera, Pentatomidae). Spixiana 14(3): 265.

Baloch, G. M. 1973. Natural enemies of *Axiagastus cambelli* Distant (Hemiptera: Pentatomidae) on the Gazelle Peninsula, New Britain. Papua New Guinea Agricultural Journal 24(1): 41–45.

Barão, K. R., A. Ferrari, and J. Grazia. 2012. Phylogeny of the South Asian Halyini? Comments on Memon et al. (2011): Towards a better practice in Pentatomidae phylogenetic analysis. Annals of the Entomological Society of America 105: 751–752.

Barão, K. R., A. Ferrari, and J. Grazia. 2013. Comparative morphology of selected characters of the Pentatomidae foreleg (Hemiptera: Heteroptera). Arthropod Structure and Development 42: 425–435.

Barão, K. R., T. de A. Garbelotto, L. A. Campos, and J. Grazia. 2016. Unusual looking pentatomids: reassessing the taxonomy of *Braunus* Distant and *Lojus* McDonald (Hemiptera: Heteroptera: Pentatomidae). Zootaxa 4078(1): 168–186.

Barão, K. R., A. Ferrari, C. V. K. Adami, and J. Grazia. 2017. Diversity of the external thoracic scent efferent system of Carpocorini (Heteroptera: Pentatomdae) with character selection for phylogenetic inference. Zoologischer Anzeiger 268: 102–111.

Barber, H. G., and R. I. Sailer. 1953. A revision of the turtle bugs of North America (Hemiptera: Pentatomidae). Journal of the Washington Academy of Sciences 43(5): 150–162.

Barcellos, A., and J. Grazia. 1993. Revisão de *Tibilis* Stal, 1860 (Heteroptera, Pentatomini). Anais da Sociedade Entomológica do Brasil 22(1): 183–208.

Barcellos, A., and J. Grazia. 1998. Sobre os gêneros *Curatia* e *Copeocoris* (Heteroptera, Pentatomidae, Pentatomini). Iheringia (Série Zoologia) 85: 27–46.

Barcellos, A., and J. Grazia. 2003a. Revision of *Brachystethus* (Heteroptera, Pentatomidae, Edessinae). Iheringia (Série Zoologia) 93(4): 413–446.

Barcellos, A., and J. Grazia. 2003b. Cladistic analysis and biogeography of *Brachystethus* Laporte (Heteroptera, Pentatomidae, Edessinae). Zootaxa 256: 1–14.

Barcellos, A., and J. Grazia. 2004. *Pentatomiana beckerae* gen. nov. and sp. nov., a new Neotropical Pentatomini (Hemiptera, Heteroptera, Pentatomidae). Revista Brasileira de Zoologia 21(2): 283–285.

Barcellos, A., and J. Grazia. 2008. Revision of the genus *Poriptus* Stål (Hemiptera: Heteroptera: Pentatomidae: Pentatominae). Zootaxa 1821: 25–36.

Barcellos, A., J. Eger, Jr., and J. Grazia. 2015. Scutelleridae, pp. 409–415. *In* L. E. Claps, and S. Roig-Juñent (Eds.), Biodiversidad de Artrópodos Argentinos. Volume 3. San Miguel de Tucumán, Universidad Nacional de Tucumán, Facultad de Ciencias Naturales. 546 pp.

Barrion, A. T., R. C. Joshi, A. L. A. Barrion-Dupo, and L. S. Sebastian. 2007. Systematics of the Philippine rice black bug, *Scotinophara* Stål (Hemiptera: Pentatomidae), pp. 3–179. *In* R. C. Joshi, A. T. Barrion, and L. S. Sebastian (Eds.), Rice black bugs. Taxonomy, ecology, and management of invasive species. Philippine Rice Research Institute, Science City of Muñoz, Philippines. 793 pp.

Basso, I. V., D. Link, and O. J. Lopes. 1974. Entomofauna de algumas solanáceas em Santa Maria, RS. Revista de Centro Ciencias Rurais 4: 263–269.

Beardsley, Jr., J. W., and S. Fluker. 1967. *Coptosoma xanthogramma* (White) (Hemiptera: Plataspidae), a new pest of legumes in Hawaii. Proceedings of the Hawaiian Entomological Society 19: 367–372.

Becker, M. 1967a. Estudos sôbre a subfamília Scaptocorinae na região neotropical (Hemiptera: Cydnidae). Arquivos de Zoologia 15(4): 291–325.

Becker, M. 1967b. Sôbre o gênero *Serdia* Stal, coma a descrição de uma nova espécie (Hemiptera, Pentatomidae, Pentatominae). Revista Brasileira de Biologia 27(1): 85–104.

Becker, M. 1977. The genus *Abascantus* Stal, with the description of two new species (Heteroptera, Pentatomidae, Discocephalinae). Revista Brasileira de Biologia 37(2): 385–393.

Becker, M., and J. Grazia. 1985. Revisão do gênero *Dinocoris* Burmeister, 1835 (Heteroptera, Pentatomidae, Discocephalinae). Revista Brasileira de Zoologia 3(2): 65–108.

Becker, M., and J. Grazia. 1970. Sôbre os gêneros *Lopadusa* Stal e *Bothrocoris* Mayr (Hemiptera, Pentatomidae, Pentatomini). Revista Brasileira de Biologia 30(2): 217–232.

Becker, M., and J. Grazia. 1989a. Novas contribuições ao gênero *Ablaptus* Stal, 1864 (Heteroptera, Pentatomidae, Discocephalinae). Memorias do Instituto Oswaldo Cruz 84 (supplement IV): 57–68.

Becker, M., and J. Grazia. 1989b. The discocephaline genus *Alcippus* Stal, 1867 (Heteroptera, Pentatomidae). Revista Brasileira de Entomologia 33(1): 49–55.

Becker, M., and J. Grazia. 1992. Revisão do gênero *Agaclitus* Stal (Heteroptera, Pentatomidae, Discocephalinae). Revista Brasileira de Entomologia 36(4): 831–842.

Becker, M., and J. Grazia-Vieira. 1971. Sôbre o gênero *Rhyncholepta* Bergroth, 1911, com a descrição de uma nova espécie (Hemiptera, Pentatomidae, Pentatominae). Revista Brasileira de Biologia 31(3): 389–399.

Beier, M. 1952. Miozäne und oligozäne Insekten aus Österreich und den unmittelbar angrenzenden Gebieten. Sitzungsberichte Österreichische Akademie der Wissenschaften (1)161(203): 129–134.

Belousova, E. N. 1997. Revision of shield bugs of the genus *Mimula* Jak. (Heteroptera, Pentatomidae). Entomologicheskoe Obozrenie 75(4)[1996]: 836–856. [in Russian; English translation: 1997, Entomological Review 79(9)[1996]: 1137–1157]

Belousova, E. N. 1999. Revision of shield bugs of the genus *Agatharchus* Stal (Heteroptera, Pentatomidae). Entomologicheskoe Obozrenie 78(4): 828–848. [in Russian; English translation: 1999, Entomological Review 79(7): 778–796]

Belousova, E. N. 2003. Contribution to the taxonomy of the shield-bug genus *Ochyrotylus* Jakovlev, 1885 (Heteroptera, Pentatomidae). Entomologicheskoe Obozrenie 83(3): 687–689. [in Russian; English translation: 2003, Entomological Review 83(5): 574–576]

Belousova, E. N. 2004. Use of new characters of the structure of the genitalia in identification of species of the genus *Carpocoris* Kol. (Heteroptera, Pentatomidae). Entomologicheskoe Obozrenie 83(1): 140–146. [in Russian; English translation: 2004, Entomological Review 84(1): 59–64]

Belousova, E. N. 2007. Revision of the shield-bug genera *Holcostethus* Fieber and *Peribalus* Mulsant & Rey (Heteroptera, Pentatomidae) of the Palaearctic region. Entomologicheskoe Obozrenie 86(3): 610–654. [in Russian; English translation: 2007, Entomological Review 87(6): 701–739]

Berger, H., E. Heiss, and I. M. Kerzhner. 2001. Removal of homonymy between Urostylidae Dallas, 1851 (Insecta, Heteroptera) and Urostylidae Buetschli, 1889 (Ciliophora, Hypotrichia). Annalen des Naturhistorischen Museums in Wien Serie B Botanik und Zoologie 103B: 301–302.

Bergevin, E. de. 1931. Description d'un noveau genre et d'une novelle espèces de Pentatomidae (Tribu des Pentatomaria), Hémiptère provenant des chasses de M. de Peyerimhoff au Hoggar. Bulletin de la Société d'Histoire Naturelle de l'Afrique du Nord 22(2): 77–79.

Bergroth, E. 1891. Contributions a l'etude des pentatomides. Revue d'Entomologie 10: 200–235.

Bergroth, E. 1893. On some Ethiopian Pentatomidae of the group Halyinae. Annals and Magazine of Natural History (6)12: 112–120.

Bergroth, E. 1904. Eine neue Art der Gattung *Glypsus* Dall. (Hemiptera-Heteroptera, Pentatomidae). Revue Russe d'Entomologie 4: 32–34.

Bergroth, E. 1905. On stridulating Hemiptera of the subfamily Halyinae, with descriptions of new genera and new species. Proceedings of the Zoological Society of London 2: 146–154.

Bergroth, E. 1906a. Systematische und synonymische Bemerkungen über Hemipteren. Wiener Entomologische Zeitung 25(1): 1–12.

Bergroth, E. 1906b. Aphylinae und Hyocephalinae, zwei neue Hemipteren-Subfamilien. Zoologischer Anzeiger 29: 644–649.

Bergroth, E. 1908. Enumeratio Pentatomidarum post Catalogum bruxellensem descriptarum. Mémoires de la Société Entomologique de Belgique 15(10): 131–200.

Bergroth, E. 1910. Note on the genus *Phalaecus* Stal [sic]. Entomological News 21(1): 18–21.

Bergroth, E. 1911. Zur Kentniss der neotropischen Arminen (Hem.-Het.). Wiener Entomologische Zeitung 30: 117–130.

Bergroth, E. 1913. Note on the genus *Bathycoelia* Am. S. (Hem., Pentatomidae). Annales de la Société Entomologique de Belgique 57: 150–154.

Bergroth, E. 1914. Zwei neue paläarktische Hemipteren, nebst synonymischen Mitteilungen. Wiener Entomologische Zeitung 33(5–6): 177–184.

Bergroth, E. 1916. Heteropterous Hemiptera collected by Professor W. Baldwin Spencer during the Horn expedition into Central Australia. Proceedings of the Royal Society of Victoria (ns) 29(1): 19–39.

Bergroth, E. 1918. Hendecas generum Hemipterorum novorum vel subnovorum. Annales Musei Nationalis Hungarici 16: 298–314.

Bergroth, E. 1920. New species of the genus *Eribotes* Stål (Hemiptera, Pentatomidæ). Arkiv för Zoologi 12(18): 1–5.

Bergroth, E. 1923. On the systematic position of the genera *Dismegistus* Am. S. and *Parastrachia* Dist. Annales de la Société Entomologique de Belgique 63: 70–72.

Bernardes, J. L. C., J. Grazia, A. Barcellos, and A. T. Salomão. 2005. Descrição dos estágios imaturos e notas sobre a biologia de *Phloea subquadrata* (Heteroptera: Phloeidae). Iheringia (Série Zoologia) 95(4): 415–420.

Bernardes, J. L. C., J. Grazia, and A. Barcellos. 2006. New species of *Neotibilis* Grazia & Barcellos (Hemiptera: Pentatomidae: Pentatomini). Neotropical Entomology 35(3): 344–348.

Bernardes, J. L. C., C. F. Schwertner, and J. Grazia. 2009. Cladistic analysis of *Thoreyella* and related genera (Hemiptera: Pentatomidae: Pentatominae: Procleticini). Zootaxa 2310: 1–23.

Bernardes, J. L., C. F. Schwertner, and J. Grazia. 2011. Review of *Thoreyella* Spinola with the description of two new species from Brazil (Heteroptera, Pentatomidae). Revista Brasileira de Entomologia 55(3): 299–312.

Bhat, P. S., and K. K. Srikumar. 2013. Occurrence of man-faced stink bug, *Catacanthus incarnatus* Drury on cashew in Puttar region of Karnataka. Current Biotica Insect Environment 19(1): 32–34.

Bianchi, F. M., V. C. Matesco, L. A. Campos, and J. Grazia. 2011. External morphology of the egg and the first and fifth instars of *Cyrtocoris egeris* Packauskas and Schaefer (Hemiptera: Heteroptera: Pentatomidae: Cyrtocorinae). Zootaxa 2991: 29–34.

Bianchi, F. M., A. Barcellos, and J. Grazia. 2016a. Rediscovering *Anhanga*: description and considerations on its taxonomic placement (Hemiptera: Heteroptera: Pentatomidae). Acta Entomologica Musei Nationalis Pragae 56(2): 557–566.

Bianchi, F. M., V. da Rosa Gonçalves, J. R. de Souza, and L. A. Campos. 2016b. Descriptions of three new species of *Glyphepomis* Berg (Heteroptera: Pentatomidae: Pentatominae). Zootaxa 4103(5): 443–452.

Bianchi, F. M., M. Deprá, A. Ferrari, J. Grazia, V. L. S. Valente, and L. A. Campos. 2017. Total evidence phylogenetic analysis and reclassification of *Euschistus* Dallas within Carpocorini (Hemiptera: Pentatomidae: Pentatominae). Systematic Entomology 42: 399–409.

Billberg, G. J. 1820. Enumeratio insectorum in Museo Gust. Joh. Billberg. Typis Gadelianis, Stockholm. 138 pp.

Brailovsky, H. 1981. Revisión del género *Arvelius* Spinola (Hemiptera-Heteroptera-Pentatomidae-Pentatomini). Anales del Instituto de Biología Universidad Nacional Autónoma de México (Zoológia) 51(1)[1980]: 239–298.

Brailovsky, H. 1986. Hemiptera-Heteroptera de Mexico XXXVII: Tres nuevas especies y nuevos registros de la familia Pentatomidae. Anales del Instituto de Biología Universidad Nacional Autónoma de México (Zoológia) 57(2): 281–298.

Brailovsky, H., and E. Barrera. 1989. El genero *Murgantia* Stål, con descripcion de cuatro especies nuevas y algunos registros nuevos (Hemiptera-Heteroptera-Pentatomidae-Pentatomini) de America Latina. Anales del Instituto de Biología Universidad Nacional Autónoma de México (Zoológia) 59(2): 219–244.

Brailovsky, H., L. Cervantes, and C. Mayorga. 1988. Hemiptera-Heteroptera de Mexico XL: La familia Cyrtocoridae Distant en la Estacion de Biologia Tropical "Los Tuxtlas" (Pentatomoidea). Anales del Instituto de Biología Universidad Nacional Autónoma de México (Zoológia) 58(2)[1987]: 537–560.

Breddin, G. 1903. Beiträge zur Hemipteren-fauna der Anden. Sitzungsberichte der Gesellschaft Naturforschender Freunde zu Berlin 1903: 366–383.

Breddin, G. 1904. Beschreibungen neuer indo-australischer Pentatomiden. Wiener Entomologische Zeitung 23(1): 1–19.

Brown, E. S. 1962. Notes on the systematics and distribution of some species of *Aelia* Fabr. (Hemiptera, Pentatomidae) in the Middle East, with special reference to the *rostrata* group. Annals and Magazine of Natural History (13)5: 129–145, pl. II.

Bundy, C. S. 2012. An annotated checklist of the stink bugs (Heteroptera: Pentatomidae) of New Mexico. The Great Lakes Entomologist 45(3–4): 196–209.

Bundy, C. S., and J. E. McPherson. 1997. Life history and laboratory rearing of *Corimelaena obscura* (Heteroptera: Thyreocoridae) with descriptions of immature stages. Annals of the Entomological Society of America 90: 20–27.

Bundy, C. S., and J. E. McPherson. 2005. Morphological examination of the egg of *Mecidea major* (Heteroptera: Pentatomidae). Southwestern Entomologist 30(1): 41–45.

Bundy, C. S., and J. E. McPherson. 2009. Life history and laboratory rearing of *Corimelaena incognita* (Hemiptera: Heteroptera: Thyreocoridae), with descriptions of immature stages. Annals of the Entomological Society of America 102: 1068–1076.

Bundy, C. S., and J. E. McPherson. 2010. Presence of a red morph in adult populations of *Mecidea minor* (Hemiptera: Pentatomidae: Pentatominae: Mecideini) in New Mexico. Journal of Entomological Science 45(4): 317–321.

Bundy, C. S., and J. E. McPherson. 2011. Life history and laboratory rearing of *Mecidea minor* (Hemiptera: Heteroptera: Pentatomidae), with descriptions of immature stages. Annals of the Entomological Society of America 104: 605–612.

Bundy, C. S., J. E. McPherson, and P. F. Smith. 2005. Comparative laboratory rearing of *Mecidea major* and *M. minor* (Heteroptera: Pentatomidae). Journal of Enntomological Science 40(3): 291–294.

Burmeister, H. C. 1835. Handbuch der Entomologie, vol II. Schnabelkerfe, Rhyngota. Berlin. 400 pp.

Buxton, G. M., D. B. Thomas, Jr., and R. C. Froeschner. 1983. Revision of the species of the *sayi*-group of *Chlorochroa* Stal (Hemiptera: Pentatomidae). Occasional Papers in Entomology (Sacramento, Calif.) 29: 23 pp.

Cachan, P. 1952. Les Pentatomidae de Madagascar (Hemipteres Heteropteres). Mémoires de l'Institut Scientifique de Madagascar (E) 1(2): 231–462, pls. 6–14.

Callan, E. M. 1944. Cacao stink-bugs (Hem., Pentatomidae) in Trinidad, B. W. I. Revista de Entomologia 15(3): 321–324.

Campos, L. A., and J. Grazia. 1998. Revisão de *Glyphepomis* Berg, 1891 (Heteroptera, Pentatomidae). Revista Brasileira de Entomologia 41(2–4): 203–212.

Campos, L. A., and J. Grazia. 1999. Revisão de *Parentheca* Berg (Heteroptera, Pentatomidae, Pentatomini). Revista Brasileira de Zoológia 16(3): 691–699

Campos, L. A., and J. Grazia. 2001. Um novo gênero de Ochlerini do sul do Brasil (Heteroptera, Pentatomidae, Discocephalinae). Iheringia (Série Zoologia) 90: 55–58.

Campos, L. A., and J. Grazia. 2006. Análise cladística e biogeografia de Ochlerini (Heteroptera, Pentatomidae, Discocephalinae). Iheringia (Série Zoologia) 96(2): 147–163.

Carapezza, A. 1997. Heteroptera of Tunisia. Il Naturalista Siciliano (4)21(A) (supplement): 331 pp.

Carapezza, A. 2009. On some Old World Scutelleridae (Heteroptera). Nouvelle Revue Entomologie (ns) 25(3) [2008]: 197–212.

Carayon, J. 1949. L'oothèque d'Hémiptères Plataspidès de l'Afrique tropicale. Bulletin de la Société Entomologique de France 54: 66–69.

Carayon, J. 1952. Les mécanisme de transmission héréditaire des endosymbiontes chez les Insectes. Tijdschrift voor Entomologie 95: 111–142.

Carpenter, F. M. 1992. Superclass Hexapoda. *In* Treatise on invertebrate paleontology. Part R. Arthropoda 4. Volumes 3 and 4. The Geological Society of America, Inc. and The University of Kansas, Boulder, Colorado, and Lawrence, Kansas. 655 pp.

Carvajal, M., and F. I. Faúndez. 2013. Rediscovery of *Sinopla humeralis* Signoret, 1864 (Hemiptera: Heteroptera: Acanthosomatidae). Zootaxa 3637(2): 190–196.

Carvajal, M., F. I. Faúndez, and D. A. Rider. 2015a. New data on the genus *Tamolia* Horváth, 1900 (Hemiptera: Heteroptera: Tessaratomidae), with description of a new species. Zootaxa 4052(4): 481–484.

Carvajal, M., D. A. Rider, and F. I. Faúndez. 2015b. Revision of the genus *Tolono* Rolston and Kumar, 1975 (Hemiptera: Heteroptera: Acanthosomatidae), with descriptions of two new species. Proceedings of the Entomological Society of Washington 117(1): 7–13.

Cassis, G., and G. F. Gross. 2002. Hemiptera: Heteroptera (Pentatomomorpha). *In* W. W. K. Houston and A. Wells (Eds), Zoological catalogue of Australia. Volume 27.3B. CSIRO Publishing, Melbourne, Australia. xiv + 737 pp.

Cassis, G., and L. Vanags. 2006. Jewel bugs of Australia (Insecta, Heteroptera, Scutelleridae). *In* W. Rabitsch (Ed.), Hug the bug - for love of true bugs. Festschrift zum 70. Geburtstag von Ernst Heiss. Denisia 19: 275–398.

Chapin, J. W., T. H. Sanders, L. O. Dean, K. W. Hendrix, and J. S. Thomas. 2006. Effect of feeding by a burrower bug, *Pangaeus bilineatus* (Say) (Heteroptera: Cydnidae), on peanut flavor and oil quality. Journal of Entomological Science 41(1): 33–39.

Chatterjee, N. C. 1934. Entomological investigations on the spike disease of sandal (24). Pentatomidae (Hemipt.). Indian Forest Records 20(9): 1–31.

Cheo, M.-T. 1935. A preliminary list of the insects and arachnids injurious to economic plants in China. Peking Natural History Bulletin 10(1): 5–37.

China, W. E. 1929. Notes on the genera *Glaucias*, Kirk. (*Zangis*, Stål), and *Plautia*, Stål (Hemiptera). Entomologist 62: 13–16.

China, W. E. 1931. Morphological parallelism in the structure of the labium in the hemipterous genera *Coptosomoides*, gen. nov., and *Bozius*, Dist. (Fam. Plataspidae) in connection with mycetophagous habits. Annals and Magazine of Natural History (10)7: 281–286.

China, W. E. 1933. A new family of Hemiptera-Heteroptera with notes on the phylogeny of the suborder. Annals and Magazine of Natural History (10)12: 180–196.

China, W. E. 1955. A new genus and species representing a new subfamily of Plataspidae with notes on the Aphylidae (Hemiptera, Heteroptera). Annals and Magazine of Natural History (12)8: 204–210.

China, W. E. 1963. *Lestonia haustorifera* China (Hemiptera: Lestoniidae) - A correction. Australian Journal of Entomology 2(1): 67–68.

China, W. E., and N. C. E. Miller. 1955. Check-list of family and subfamily names in Hemiptera-Heteroptera. Annals and Magazine of Natural History (12)8: 257–267.

China, W. E., and N. C. E. Miller. 1959. Check-list and keys to the families and subfamilies of the Hemiptera-Heteroptera. Bulletin of the British Museum of Natural History, Entomology 8(1): 1–45.

China, W. E., and J. A. Slater. 1956. A new subfamily of Urostylidae from Borneo (Hemiptera: Heteroptera). Pacific Science 10: 410–414.

Cioato, A., F. M. Bianchi, J. Eger, and J. Grazia. 2015. New species of *Euschistus (Euschistus)* from Jamaica, *Euschistus (Mitripus)* and *Ladeaschistus* from southern South America (Hemiptera: Heteroptera: Pentatomidae: Pentatominae). Zootaxa 4048(4): 565–574.

Cockerell, T. D. A. 1909. Fossil insects from Colorado. Entomologist 42(7): 170–174.

Cockerell, T. D. A. 1921. Some Eocene insects from Colorado and Wyoming. Proceedings of the United States National Museum 59(2358): 29–39.

Cockerell, T. D. A. 1931. Hymenoptera and Hemiptera, pp. 309-312. *In* Insects from the Miocene (Latah) of Washington. Annals of the Entomological Society of America 24: 307–323.

Čokl, A., C. Nardi, J. M. S. Bento, E. Hirose, and A. R. Panizzi. 2006. Transmission of stridulatory signals of the burrower bugs, *Scaptocoris castanea* and *Scaptocoris carvalhoi* (Heteroptera: Cydnidae) through the soil and soybean. Physiological Entomology 31(4): 371–381.

Correia, A. O., and J. A. M. Fernandes. 2016. *Grammedessa*, a new genus of Edessinae (Hemiptera: Heteroptera: Pentatomidae). Zootaxa 4107(4): 541–565.

Cuda, J. P., and J. E. McPherson. 1976. Life history and laboratory rearing of *Brochymena quadripustulata* with descriptions of immature stages and additional notes on *Brochymena arborea* (Hemiptera: Pentatomidae). Annals of the Entomological Society of America 69: 977–983.

Dai, J-x., and Z-m. Zheng. 2004. Discussion on the phylogenetic relationships of partial Pentatomidae insects based on sequences of cytochrome b gene. Zoological Research 25: 397–402.

Dallas, W. S. 1851. List of the specimens of hemipterous insects in the collection of the British Museum. Part 1. Trustees of the British Museum, London. Pp. 1–368, pls. 1–11.

Davidová-Vilímová, J. 1987. The eggs of two *Coptosoma* species, with a review of the eggs of the Plataspidae (Heteroptera). Acta Entomologica Bohemoslovaca 84: 254–260.

Davidová-Vilímová, J. 1993a. Revision of the genus *Kundelungua* (Heteroptera: Pentatomidae: Podopinae). European Journal of Entomology 90: 159–175.

Davidová-Vilímová, J. 1993b. *Jeffocoris* gen. n. - A new podopine genus from Australia (Heteroptera: Pentatomidae). Records of the South Australian Museum 26(2): 105–109.

Davidová-Vilímová, J. 1999. Review of the genus *Tornosia* Bolívar (Hemiptera: Pentatomidae: Podopinae), with the validation of *T. brevispina* Schouteden. African Entomology 7: 113–121.

Davidová-Vilímová, J. 2006. Family Plataspidae Dallas, 1851, pp. 150–165. *In* B. Aukema and C. Rieger (Eds.), Catalogue of the Heteroptera of the Palaearctic Region. Volume 5. Pentatomomorpha II. The Netherlands Entomological Society, Amsterdam. xiii + 550 pp.

Davidová-Vilímová, J., and J. E. McPherson. 1991. Pygophores of selected species of Pentatomoidea (Heteroptera) from Illinois. Acta Universitatis Carolinae Biologica 35: 143–183.

Davidová-Vilímová, J., and J. E. McPherson. 1995. History of the higher classification of the subfamily Podopinae (Heteroptera: Pentatomidae), a historical review. Acta Universitatis Carolinae Biologica 38: 99–124.

Davidová-Vilímová, J., and P. Štys. 1980. Taxonomy and phylogeny of West Palaearctic Plataspidae (Heteroptera). Studie ČSAV 4: 1–155.

Davidová-Vilímová, J., and P. Štys. 1994. Diversity and variation of trichobothrial patterns in adult Podopinae (Heteroptera: Pentatomidae). Acta Universitatis Carolinae Biologica 37(1–2)[1993]: 33–72.

Davis, T. A. 1949. An unrecorded insect pest of the cashew tree (*Anacardium occidentale* L.) in south India. Current Sciences 18(4): 133.

Day, G. M. 1965. Revision of *Acrosternum* auctt nec. Fieber from Madagascar. Annals and Magazine of Natural History (13)7[1964]: 559–565.

Dejean, P. F. M. A. 1829. Species general des coléoptères de la collection de M. le Compte Dejean. Volume 4. Chez Méquignon-Marvis, Libraire-Éditeur, Paris, France and Bruxelles, Belgium. 520 pp.

Dellapé, G., and P. Dellapé. 2016. A new species of *Adoxoplatys* Breddin (Heteroptera, Pentatomidae), Discocephalinae) from Argentina. Revista Brasileira de Entomologia 60: 15–18.

Derjanschi, V. V., and J. Péricart. 2005. Hémiptères Pentatomoidea euro-méditerranéens 1. Généralités. Systématique: Prémiere Partie. Faune de France. Volume 90. Fédération Françaice des Sociétés de Sciences Naturelles, Paris. 494 pp.

Distant, W. L. 1880. Insecta. Rhynchota, Hemiptera-Heteroptera, pp. 1–88. *In* F. D. Godman and O. Salvin (Eds.), Biologia Centrali-Americana. Volume 1, Porter, London. xx + 462 pp., 39 pls.

Distant, W. L. 1881. Descriptions of new genera and species of Rhynchota from Madagascar. Transactions of the Entomological Society of London 1881(1): 103–108.

Distant, W. L. 1882. Description of a new species of Pentatomidæ from Madagascar. Entomologist's Monthly Magazine 19: 108.

Distant, W. L. 1883. First report on the Rhynchota collected in Japan by Mr. George Lewis. Transactions of the Entomological Society of London 1883(4): 413–443, pls. 19–20.

Distant, W. L. 1889. Insecta. Rhynchota, Hemiptera-Heteroptera, pp. 305–328. *In* F. D. Godman and O. Salvin (Eds.), Biologia Centrali-Americana. Volume 1, London. Xx + 462 pp., 39 pls.

Distant, W. L. 1900a. Rhynchotal notes. – IV. Heteroptera: Pentatominæ (part). Annals and Magazine of Natural History (7)5: 386–397.

Distant, W. L. 1900b. Rhynchotal notes. – IV. Heteroptera: Pentatominae (part). Annals and Magazine of Natural History (7)5: 420–435.

Distant, W. L. 1901a. Notes and descriptions relating to some Plataspinæ and Graphosominæ (Rhynchota). Annals and Magazine of Natural History (7)8: 233–242.

Distant, W. L. 1901b. Enumeration of the Heteroptera (Rhynchota) collected by Signor Leonardo Fea in Burma and its vicinity. Part I. Family Pentatomidae. Transactions of the Entomological Society of London 1901(1): 99–114.

Distant, W. L. 1902. The Fauna of British India, including Ceylon and Burma. Published under the authority of the secretary of state for India in council. *In* W. T. Blanford (Ed.), Volume 1. Heteroptera. Taylor & Francis, London, xxxviii + 438 pp.

Distant, W. L. 1906. Oriental Heteroptera. Annales de la Société Entomologique de Belgique 50(12): 405–417.

Distant, W. L. 1910a. Rhynchotal notes. - LII. Australasian Pentatomidæ. Annals and Magazine of Natural History (8)6: 369–386.

Distant, W. L. 1910b. Rhynchotal notes. - LII. Australasian Pentatomidæ (continued). Annals and Magazine of Natural History (8)6: 465–481.

Distant, W. L. 1911. Rhynchotal notes. – LIV. Pentatomidae from various regions. Annals and Magazine of Natural History (8)7: 338–354.

Distant, W. L. 1921a. The Heteroptera of Indo-China (continued). Entomologist 54: 68–69.

Distant, W. L. 1921b. The Heteroptera of Indo-China (continued). Entomologist 54: 164–169.

Dolling, W. R. 1981. A rationalized classification of the burrower bugs (Cydnidae). Systematic Entomology 6: 61–76.

Dolling, W. R. 1984. Pentatomid bugs (Hemiptera) that transmit a flagellate disease of cultivated palms in South America. Bulletin of Entomological Research 74: 473–476.

Dolling, W. R. 1997. The systematic position of the genus *Tahitocoris* (Hemiptera: Pentatomidae: Podopinae). Journal of the New York Entomological Society 103(4)[1995]: 409–411.

Douglas, J. W., and J. Scott. 1865. The British Hemiptera. Volume I. Hemiptera-Heteroptera. Robert Hardwicke, London. 628 pp.

Dupuis, C. 1949. Les Asopinae de la faune Française [Hemiptera Pentatomidae]. Essai sommaire de synthese morphologique, systematique et biologique. Revue Française d'Entomologie 16: 233–250.

Dupuis, C. 1952. Priorité de quelque noms d'Hétéroptères de Guérin Méneville (1831). Bulletin de la Société Zoologique de France 77: 447–454.

Durai, P. S. S. 1987. A revision of the Dinidoridae of the world (Heteroptera: Pentatomoidea). Oriental Insects 21: 163–360.

Durak, D., and Y. Kalender. 2007. Fine structure and chemical analysis of the metathoracic scent glands of *Graphosoma semipunctatum* (Fabricius, 1775) (Heteroptera, Pentatomidae). Journal of Applied Biological Science 1: 43–50.

Durak, D., and Y. Kalender. 2009. Fine structure and chemical analysis of the metathoracic scent gland secretion in *Graphosoma lineatum* (Linnaeus, 1758) (Heteroptera, Pentatomidae). Comptes Rendus Biologies 332(1): 34–42.

Dzerefos, C. M., and E. T. F. Witkowski. 2014. The potential of entomophagy and the use of the stinkbug, *Encosternum delegorguei* Spinola (Hemiptera: Tessaratomidae), in sub-Saharan Africa. African Entomology 22(3): 461–472.

Dzerefos, C. M., E. T. F. Witkowski, and R. Toms. 2009. Life-history traits of the edible stinkbug, *Encosternum delegorguei* (Hem., Tessaratomidae), a traditional food in southern Africa. Journal of Applied Entomology 133(9–10): 749–759.

Eberhard, W. G. 1975. The ecology and behavior of a subsocial pentatomid bug and two scelionid wasps: Strategy and counterstrategy in a host and its parasites. Smithsonian Contribution to Zoology 205: 1–39.

Eger, J. E., Jr. 1978. Revision of the genus *Loxa* (Hemiptera: Pentatomidae). Journal of the New York Entomological Society 86(3): 224–259.

Eger, J. E., Jr. 1980. *Fecelia biorbis* n. sp. (Heteroptera: Pentatomidae), a new species from Haiti. Journal of the New York Entomological Society 88(1): 29–32.

Eger, J. E., Jr. 1987. A review of the genus *Tiridates* Stål (Heteroptera: Pentatomoidea: Scutelleridae). Florida Entomologist 70(3): 339–350.

Eger, J. E., Jr. 1990. Revision of the genus *Polytes* Stål (Heteroptera: Scutelleridae). Annals of the Entomological Society of America 83(2): 115–141.

Eger, J. E., Jr. 1992. New distribution records for *Polytes lineolatus* (Heteroptera: Scutelleridae) with description of female genitalia. Florida Entomologist 75(1): 154–155.

Eger, J. E., Jr. 1994. New synonymy in the genus *Polytes* Stal (Heteroptera, Scutelleridae). Florida Entomologist 77(3): 376–378.

Eger, J. E., Jr. 2012. The genus *Sphyrocoris* Mayr (Heteroptera: Scutelleridae: Pachycorinae). The Great Lakes Entomologist 45(3–4): 235–250.

Eger, J. E., Jr. 2015. *Polytes debra*, a new species from Peru (Heteroptera: Scutelleridae: Pachycorinae). Insecta Mundi 420: 1–6.

Eger, J. E., Jr., L. M. Ames, D. R. Suiter, T. M. Jenkins, D. A. Rider, and S. E. Halbert. 2010. Occurrence of the Old World bug *Megacopta cribraria* (Fabricius) (Heteroptera: Plataspidae) in Georgia: a serious home invader and potential legume pest. Insecta Mundi 121: 1–11.

Eger, J. E., Jr., A. Barcellos, and L. Weiler. 2015a. Shield bugs (Scutelleridae), pp. 757–788. *In* A. R. Panizzi and J. Grazia (Eds.), True bugs (Heteroptera) of the Neotropics. Springer Dordrecht Heidelberg, New York, London. 901 pp.

Eger, J. E., Jr., H. Brailovsky, and T. J. Henry. 2015b. Heteroptera attracted to butterfly traps baited with fish or shrimp carrion. Florida Entomologist 98(4): 1030–1035.

Engleman, H. D., and L. H. Rolston. 1983. Eight new species of *Antiteuchus* Dallas (Hemiptera: Pentatomidae). Journal of the Kansas Entomological Society 56(2): 175–189.

Erichsen, C., and A. Schoeman. 1992. Economic losses due to insect pests on avocado fruit in the Nelspruit/Hazyview region of South Africa during 1991. South African Avocado Growers' Association Yearbook 15: 49–54.

Fabricius, J. C. 1803. Systema Rhyngotorum secundum ordines, genera, species adjectis synonymis, locis, observationibus, descriptionibus. C. Reichard, Brunsvigae. x + 335 pp.

Fan, Z-h., and G-q. Liu. 2009. The genus *Aenaria* Stål, 1876 in China (Hemiptera, Pentatomidae). Acta Zootaxonomica Sinica 34(4): 760–765.

Fan, Z-h., and G-q. Liu. 2010a. A new genus *Ramivena* (Hemiptera: Pentatomidae), with descriptions of two new species. Oriental Insects 44: 211–223.

Fan, Z-h., and G-q. Liu. 2010b. The genus *Lelia* Walker, 1876, with the description of one new species (Hemiptera: Heteroptera: Pentatomidae: Pentatominae). Zootaxa 2512: 56–62.

Fan, Z-h., and G-q. Liu. 2012. *Bifurcipentatoma*, a new genus of Pentatomini with descriptions of two new species from China (Hemiptera: Heteroptera: Pentatomidae). Zootaxa 3274: 14–28.

Fan, Z-h., and G-q. Liu. 2013. The genus *Cahara* Ghauri, 1978 of China (Hemiptera, Heteroptera, Pentatomidae, Halyini) with descriptions of two new species. ZooKeys 319: 37–50.

Fan, Z-h., X. Xing, X. Sun, and G-q. Liu. 2012. New records of Pentatomidae (Hemiptera: Heteroptera) from China. Entomotaxonomia 34(2): 181–191.

Faúndez, E. I. 2007a. Notes on the biology of *Ditomotarsus punctiventris* Spinola, 1852 (Hemiptera: Acanthosomatidae) in the Magellan region, and comments about the crypsis in Acanthosomatidae. Anales Instituto Patagonia 35(2): 67–70.

Faúndez, E. I. 2007b. Asociación críptica entre *Sinopla perpunctatus* Signoret, 1863 (Acanthosomatidae: Hemiptera) y el ñirre *Nothofagus antarctica* (G. Forster) Oersted (Fagaceae) en la Región de Magallanes (Chile). Boletín de la Sociedad Entomológica Aragonesa 40: 563–564.

Faúndez, E. I. 2009. Contribution to the knowledge of the genus *Acrophyma* Bergroth, 1917 (Hemiptera: Heteroptera: Acanthosomatidae). Zootaxa 2137: 57–65.

Faúndez, E. I., and G. A. Osorio. 2010. New data on the biology of *Sinopla perpunctatus* Signoret, 1864 (Hemiptera: Heteroptera: Acanthosomatidae). Boletín de Biodiversidad de Chile 3: 24–31.

Faúndez, E. I., and D. A. Rider. 2014. *Thestral incognitus*, a new genus and species of Pentatomidae from Chile (Heteroptera: Pentatomidae: Pentatominae: Carpocorini). Zootaxa 3884(4): 394–400.

Faúndez, E. I., and D. A. Rider. 2017. The brown marmorated stink bug *Halyomorpha halys* (Stål, 1855) (Heteroptera: Pentatomidae) in Chile. Arquivos Entomolóxicos 17: 305–307.

Faúndez, E. I., M. A. Carvajal, and D. A. Rider. 2014a. *Archaeoditomotarsus crassitylus*, gen. and sp. nov. (Hemiptera: Heteroptera: Acanthosomatidae) from Chile. Zootaxa 3860(1): 87–91.

Faúndez, E. I., D. A. Rider, and M. A. Carvajal. 2014b. A new species of *Acledra* s. str. (Hemiptera: Heteroptera: Pentatomidae) from the highlands of Argentina and Bolivia, with a checklist and key to the species of the nominate subgenus. Zootaxa 3900(1): 127–134.

Faúndez, E. I., A. Lüer, A. G. Cuevas, D. A. Rider, and P. Valdebenito. 2016. First record of the painted bug *Bagrada hilaris* (Burmeister, 1835) (Heteroptera: Pentatomidae) in South America. Arquivos Entomolóxicos 16: 175–179.

Fennah, R. G. 1935. A preliminary list of the Pentatomidae of Trinidad, B.W.I. Tropical Agriculture 12(7): 192–194.

Fernandes, J. A. M. 2010. A new genus and species of Edessinae from Amazon region (Hemiptera: Heteroptera: Pentatomidae). Zootaxa 2662: 53–65.

Fernandes, J. A. M., and L. D. Campos. 2011. A new group of species of *Edessa* Fabricius, 1803 (Hemiptera: Heteroptera: Pentatomidae). Zootaxa 3019: 63–68.

Fernandes, J. A. M., and J. Grazia. 1996. Revisão de gênero *Hypatropis* Bergroth, 1891 (Heteroptera, Pentatomidae). Revista Brasileira de Entomologia 40(3–4): 341–352.

Fernandes, J. A. M., and J. Grazia. 1998a. *Amauromelpia*, a new northern neotropical genus (Heteroptera, Pentatomidae). Iheringia (Série Zoologia) 84: 153–160.

Fernandes, J. A. M., and J. Grazia. 1998b. Revision of the genus *Tibraca* Stål (Heteroptera, Pentatomidae, Pentatominae). Revista Brasileira de Zoologia 15(4): 1049–1060.

Fernandes, J. A. M. and J. Grazia. 2002. Contribution to the knowledge of *Parantiteuchus* (Heteroptera, Pentatomidae, Discocephalinae): description of the male of *P. hemitholus* Ruckes. Zootaxa 99: 1–4.

Fernandes, J. A. M., and J. Grazia. 2006. Revisão do gênero *Antiteuchus* Dallas (Heteroptera, Pentatomidae, Discocephalinae). Revista Brasileira de Entomologia 50(2): 165–231.

Fernandes, J. A. M., and P. H. van Doesburg. 2000a. The *E. dolichocera*-group of *Edessa* Fabricius, 1803 (Heteroptera: Pentatomidae: Edessinae). Zoologische Mededelingen 73: 305–315.

Fernandes, J. A. M., and P. H. van Doesburg. 2000b. The *E. beckeri*-group of *Edessa* Fabricius, 1803 (Heteroptera: Pentatomidae: Edessinae). Zoologische Mededelingen 74: 143–150.

Fernandes, J. A. M., and P. H. van Doesburg. 2000c. The *E. cervus*-group of *Edessa* Fabricius, 1803 (Heteroptera: Pentatomidae: Edessinae). Zoologische Mededelingen 74: 151–165.

Fernandes, J. A. M., P. H. van Doesburg, and C. Greve. 2001. The *E. collaris*-group of *Edessa* Fabricius, 1803 (Heteroptera: Pentatomidae: Edessinae). Zoologische Mededelingen 75: 239–250.

Fernandes, J. A. M., J. Grazia, and L. D. Campos. 2011. Redescription of *Callostethus* Ruckes, 1961 (Hemiptera: Heteroptera: Pentatomidae: Discocephalinae) with the description of *C. flavolineatus* sp. nov. Zootaxa 2866: 55–60.

Ferrari, A., C. F. Schwertner, and J. Grazia. 2010. Review, cladistics analysis and biogeography of *Nezara* Amyot and Serville (Hemiptera: Pentatomidae). Zootaxa 2424: 1–41.

Fieber, F. X. 1860. Die europäischen Hemiptera. Halbflügler (Rhynchota Heteroptera). Nach der analytischen Methode bearbeitet. Gerold, Wien, pp. i–vi, 1–112.

Fieber, F. X. 1861. Die europäischen Hemiptera. Halbflüger (Rhynchota Heteroptera). Nach der analytischen Methode bearbeitet. Gerold, Wien, pp. 113–444.

Filippi, L., M. Hironaka, S. Nomakuchi, and S. Tojo. 2000a. Provisioned *Parastrachia japonensis* (Hemiptera: Cydnidae) nymphs gain access to food and protection from predators. Animal Behaviour 60(6): 757–763.

Filippi, L., S. Nomakuchi, M. Hironaka, and S. Tojo. 2000b. Insemination success discrepancy between long-term and short-term copulations in the provisioning shield bug, *Parastrachia japonensis* (Hemiptera: Cydnidae). Journal of Ethology 18(1): 29–36.

Filippi, L., M. Hironaka, and S. Nomakuchi. 2001. A review of the ecological parameters and implications of subsociality in *Parastrachia japonensis* (Hemiptera: Cydnidae), a semelparous species that specializes on a poor resource. Population Ecology 43(1): 41–50.

Filippi, L., M. Hironaka, and S. Nomakuchi. 2002. Risk-sensitive decisions during nesting may increase maternal provisioning capacity in the subsocial shield bug *Parastrachia japonensis*. Ecological Entomology 27(2): 152–162.

Filippi, L., M. Hironaka, and S. Nomakuchi. 2005. Kleptoparasitism and the effect of nest location in a subsocial shield bug *Parastrachia japonensis* (Hemiptera: Parastrachiidae). Annals of the Entomological Society of America 98(1): 134–142.

Filippi, L., N. Baba, K. Inadomi, T. Yanagi, M. Hironaka, and S. Nomakuchi. 2008. Pre- and post-hatch trophic egg production in the subsocial burrower bug, *Canthophorus niveimarginatus* (Heteroptera: Cydnidae). Naturwissenschaften 96(2): 201–211.

Fischer, C. 1994. Das Pendergrast-Organ der Acanthosomatidae (Heteroptera, Pentatomoidea): Schutz des Eigeleges vor Räubern und Parasiten? Sitzungsberichte der Gesellschaft Naturforschender Freunde zu Berlin 33: 129–142.

Fischer, C. 2000. The disc-like organ of the Lestoniidae (Heteroptera: Pentatomoidea), with remarks on lestoniid relationships. Insect Systematics and Evolution 31: 201–208.

Fischer, C. 2001. Ein Beitrag zum Grundmuster, phylogenetischen System und zur Verwandtschaft der Scutelleridae (Heteroptera, Pentatomoidea). Ph.D. Dissertation, Freien Universität, Berlin, Germany. 235 pp.

Fischer, C. 2006. The biological context and evolution of Pendergrast's organs of Acanthosomatidae (Heteroptera, Pentatomoidea). *In* W. Rabitsch (Ed.), Hug the bug - for love of true bugs. Festschrift zum 70. Geburtstag von Ernst Heiss. Denisia 19: 1041–1054.

Förster, B. 1891. Die Insekten des "Plattigen Steinmergels" von Brunstatt. Abhandlungen zur Geologischen Spezialkarte von Elsass-Lothringen 3: 335–594.

Fortes, N. D. F., and J. Grazia. 2000. Novas espécies do gênero Rio (Heteroptera, Pentatomidae). Iheringia (Série Zoologia) 88: 67–102.

Fortes, N. D. F. de, and J. Grazia. 2005. Revisão e análise cladística de *Serdia* Stål (Heteroptera, Pentatomidae, Pentatomini). Revista Brasileira de Entomologia 49(3): 294–339.

Freeman, P. 1939. A contribution to the study of the genus *Calidea* Laporte (Hemipt.-Heteropt., Pentatomidae). Transactions of the Royal Entomological Society of London 88(5): 139–160.

Freeman, P. 1940. A contribution to the study of the genus *Nezara* Amyot & Serville (Hemiptera, Pentatomidae). Transactions of the Royal Entomological Society of London 90(12): 351–374.

Freeman, P. 1946. Further notes on the hemipterous genera *Calidea* LaPorte (Scutellerinae) and *Nezara* Amyot & Serville (Pentatominae). Proceedings of the Royal Entomological Society of London (B)15(3–4): 32.

Frey-da-Silva, A., and J. Grazia. 2001. Novas espécies de *Acrosternum* subgênero *Chinavia* (Heteroptera, Pentatomidae, Pentatomini). Iheringia (Série Zoologia) 90: 107–126.

Frey-da-Silva, A., J. Grazia, and J. A. M. Fernandes. 2002a. Revision of the genus *Ogmocoris* Mayr, 1864 (Heteroptera, Pentatomidae, Pentatomini). Beaufortia 52(10): 179–185.

Frey-da-Silva, A., J. Grazia, and J. A. M. Fernandes. 2002b. Revisão do gênero *Paramecocephala* Benvegnú, 1968 (Heteroptera, Pentatomidae). Revista Brasileira de Entomologia 46(2): 209–225.

Froeschner, R. C. 1960. Cydnidae of the Western Hemisphere. Proceedings of the United States National Museum 111(3430): 337–680.

Froeschner, R. C. 1984. Does the Old World family Plataspidae (Hemiptera) occur in North America? Entomological News 95: 36.

Froeschner, R. C. 1988a. Family Acanthosomatidae Signoret, 1863. The acanthosomatids, pp. 1–3. *In* T. J. Henry and R. C. Froeschner (Eds.), Catalog of the Heteroptera, or true bugs, of Canada and the continental United States. E. J. Brill, Leiden, New York. xix + 958 pp.

Froeschner, R. C. 1988b. Family Cydnidae Billberg, 1820. Burrowing Bugs, pp. 119–120. *In* T. J. Henry and R. C. Froeschner (Eds.), Catalog of the Heteroptera, or true bugs, of Canada and the continental United States. E. J. Brill, Leiden, New York. xix + 958 pp.

Froeschner, R. C. 1988c. Family Pentatomidae Leach, 1815. The stink bugs, pp. 544–607. *In* T. J. Henry and R. C. Froeschner (Eds.), Catalog of the Heteroptera, or true bugs, of Canada and the continental United States. E. J. Brill, Leiden, New York. xix + 958 pp.

Froeschner, R. C. 1988d. Family Scutelleridae Leach, 1815. The shield bugs, pp. 684–693. *In* T. J. Henry and R. C. Froeschner (Eds.), Catalog of the Heteroptera, or true bugs, of Canada and the continental United States. E. J. Brill, Leiden, New York. xix + 958 pp.

Froeschner, R. C. 1988e. Family Thyreocoridae Amyot and Serville, 1843. The negro bugs, pp. 698–707. *In* T. J. Henry and R. C. Froeschner (Eds.), Catalog of the Heteroptera, or true bugs, of Canada and the continental United States. E. J. Brill, Leiden, New York. xix + 958 pp.

Froeschner, R. C. 2000. Revision of the South American genus *Hellica* Stal (Heteroptera: Acanthosomatidae). Journal of the New York Entomological Society 107(2–3)[1999]: 164–170.

Fuente, J. A. de la. 1974. Revisión de los Pentatómidos ibéricos (Hemiptera). Parte II. Tribus Aeliini Stål, 1872, Stagonomini nov. nom. (=Eysarcorini Auct.) y Carpocorini Distant, 1902. Eos Revista Española de Entomología 48(1–4)[1972]: 115–201.

Fujiyama, I. 1967. A fossil scutellerid bug from marine deposit of Tottori, Japan (Tertiary insect fauna of Japan, 1). Bulletin of the National Museum Tokyo 10(3): 393–401.

Fujiyama, I. 1987. Middle Miocene insect fauna of Abura, Hokkaido, Japan, with notes on the occurrence of Cenozoic fossil insects in the Oshima Peninsula, Hokkaido. Memoirs of the National Science Museum of Tokyo 20: 37–44.

Fürstenau, B. B. R. J., and J. Grazia. 2014. Cladistic analysis of *Runibia* Stål, 1861 (Hemiptera, Pentatomidae), pp. 50–51. *In* T. J. Henry and K. Henderson (Eds.), Proceedings of the Fifth Quadrienniel Meeting of the International Heteropterists' Society. National Museum of Natural History, Smithsonian Institution, Washington, DC, 21–25 July 2014. 68 pp.

Gahukar, R. T. 2012. Entomophagy can support rural livelihood in India. Current Science 103(1): 10.

Gallo, D., O. Nakano, S. Silveira-Neto, R. P. L. Carvalho, G. C. Baptista, E. Berti Fo, J. R. P. Parra, R. A. Zucchi, S. B. Alves, J. D. Vendramim, L. C. Marchini, J. R. S. Lopes, and C. Omoto. 2002. Entomologia agrícola. Piracicaba, FEALQ. 920 pp.

Gamberale-Stille, G., A. I. Johansen, and B. S. Tullberg. 2009. Change in protective coloration in the striated shieldbug *Graphosoma lineatum* (Heteroptera: Pentatomidae): predator avoidance and generalization among different life stages. Evolutionary Ecology 24(2): 423–432.

Gapon, D. A. 2005. On the question of the taxonomical status of tribe Procleticini Pennington (Heteroptera: Pentatomidae). Caucasian Entomological Bulletin 1(1): 4–18. [in Russian]

Gapon, D. A. 2007. A taxonomic review of *Promecocoris* Puton, 1886 (Heteroptera, Scutelleridae), with a description of male and female genitalia. Euroasian Entomological Journal 6(1): 57–66. [in Russian]

Gapon, D. A. 2008a. New subtribes and a new genus of Podopini (Heteroptera: Pentatomidae: Podopinae). Acta Entomologica Musei Nationalis Pragae 48(2): 523–532.

Gapon, D. A. 2008b. A revision of *Leprosoma* Baerensprung, 1859 (Heteroptera: Pentatomidae), pp. 105–120. *In* S. Grozeva and N. Simov (Eds.), Advances in Heteroptera research. Festschrift in Honour of 80th Anniversary of Michail Josifov. Pensoft Publishers, Sofia-Moscow. 419 pp.

Gapon, D. A. 2010a. *Conquistator*, a new genus for *Podisus mucronatus* Uhler, 1893 (Heteroptera: Pentatomidae: Asopinae) with a redescription of the type species. Zoosystematica Rossica 18[2009]: 264–270.

Gapon, D. A. 2010b. Taxonomic notes on the genus *Leprosoma* (Heteroptera: Pentatomidae). Zoosystematica Rossica 19(2):272–276.

Gapon, D. A., and M. Baena. 2005. On the status, synonymy and tribal position of *Phaeocoris* Jakovlev, 1887 (Heteroptera: Pentatomidae). Zoosystematica Rossica 14: 61–68.

Gapon, D. A., and F. V. Konstantinov. 2006. On the structure of the aedeagus of shield bugs (Heteroptera, Pentatomidae): III. Subfamily Asopinae. Entomological Review 86: 491–507.

Gapud, V. P. 1991. A generic revision of the subfamily Asopinae, with consideration of its phylogenetic position in the family Pentatomidae and superfamily Pentatomoidea (Hemiptera-Heteroptera). Parts I and II. Philippine Entomologist 8: 865–961.

Gapud, V. P. 2015. The Philippine genus *Stilbotes* Stål and a new tribe of Asopinae (Hemiptera: Pentatomidae). Asia Life Sciences 24(2): 1–5.

Garbelotto, T. de A. and L. A. Campos. 2014. *Xynocoris* Garbelotto & Campos gen. nov., pp. 282–283. *In* T. de A. Garbelotto, L. A. Campos, and J. Grazia, *Xynocoris*, new genus of Ochlerini from Central and South America (Hemiptera: Pentatomidae: Discocephalinae). Zootaxa 3869(3): 281–305.

Garbelotto, T. de A. and L. A. Campos. 2016. *Stapecolis* Garbelotto & Campos gen. nov., pp. 546–549. *In* T. de A. Garbelotto, L. A. Campos, and J. Grazia, *Stapecolis*, new genus of Ochlerini (Hemiptera: Pentatomidae: Discocephalinae). Zootaxa 4137(4): 545–552.

Garbelotto, T. de A., L. A. Campos, and J. Grazia. 2013. Cladistics and revision of *Alitocoris* with considerations on the phylogeny of the *Herrichella* clade (Hemiptera, Pentatomidae, Discocephalinae, Ochlerini). Zoological Journal of the Linnean Society 168: 452–472.

García G., C. A., and A. C. Belotti. 1980. Estudio preliminar de la biologia y morfologia de *Cyrtomenus bergi* Froeschner, nueva plaga de la yuca. Revista Colombiana de Entomologia 6(3–4): 55–61.

Gerald, B. M. 1965. *Bathycoelia thalassina* (Herrich-Schaeffer), (Hemiptera: Pentatomidae); a pest of *Theobroma cacao* L. Nature 207: 881.

Ghahari, H., P. Moulet, and D. A. Rider. 2014. An annotated catalog of the Iranian Pentatomoidea (Hemiptera: Heteroptera: Pentatomomorpha). Zootaxa 3837(1): 1–95.

Ghauri, M. S. K. 1963. A preliminary revision of the little-known genus *Critheus* Stål (Pentatomidae, Heteroptera). Annals and Magazine of Natural History (13)5[1962]: 407–415.

Ghauri, M. S. K. 1972. New species of Heteroptera from New Guinea and New Britain. Bulletin of Entomological Research 62: 129–137.

Ghauri, M. S. K. 1975a. *Jugalpada*, a new genus of Halyini (Pentatomidae, Heteroptera). Journal of Natural History 9(6): 629–632.

Ghauri, M. S. K. 1975b. Revision of the Himalayan genus *Paranevisanus* Distant (Halyini, Pentatominae, Pentatomidae, Heteroptera). Zoologischer Anzeiger 195(5–6): 407–416.

Ghauri, M. S. K. 1977a. A revision of *Apodiphus* Spinola (Heteroptera: Pentatomidae). Bulletin of Entomological Research 67(1): 97–106.

Ghauri, M. S. K. 1977b. *Sarju* - a new genus of Halyini (Heteroptera, Pentatomidae, Pentatominae) with new species. Türkiye Bitki Koruma Dergisi 1(1): 9–27.

Ghauri, M. S. K. 1978. *Cahara* - a new genus of Halyini (Heteroptera, Pentatomidae, Pentatominae) with new species on fruit and forest trees in the Sub-Himalayan region. Journal of Natural History 12(2): 163–176.

Ghauri, M. S. K. 1980. *Tipulparra* - a new genus of Halyini with new species (Heteroptera, Pentatomidae, Pentatominae). Reichenbachia Staatliches Museum für Tierkunde in Dresden 18(21): 129–146.

Ghauri, M. S. K. 1988a. A revision of Asian species of the genus *Halys* Fabricius based on the type material (Insecta, Heteroptera, Pentatomidae, Pentatominae). Entomologische Abhandlungen Staatliches Museum für Tierkunde Dresden 51(6): 77–92.

Ghauri, M. S. K. 1988b. *Faizuda* - a new genus of Halyine with new species (Heteroptera, Pentatomidae, Pentatominae). Türkiye Entomoloji Dergisi 12(1): 3–10.

Gibernau, M., and A. Dejean. 2001. Ant protection of a heteropteran trophobiont against a parasitoid wasp. Oecologia 126: 53–57.

Gillon, D. 1972. Les Hémiptères Pentatomides d'une savane préforestière de Côte-d'Ivoire. Annales de l'Université d'Abidjan (E)5(1): 265–371, 16 pls.

Girard, C. F. 1855. Notice upon the viviparous fishes inhabiting the Pacific coast of North America, with an enumeration of the species observed. Proceedings of the Academy of Natural Sciences of Philadelphia 7: 318–323.

Gogala, M. 1984. Vibration producing structures and songs of terrestrial Heteroptera as systematic character. Biološki Vestnik 32(1): 19–36.

Gogala, M., A. Čokl, K. Drašlar, and A. Blažević. 1974. Substrate-borne sound communication in Cydnidae (Heteroptera). Journal of Comparative Physiology 94(1): 25–31.

Gogala, M., H.-S. Yong, and C. Bruehl. 1998. Maternal care in *Pygoplatys* bugs (Heteroptera: Tessaratomidae). European Journal of Entomology 95(2): 311–315.

Golding, F. D. 1927. Notes on the food plants and habits of some southern Nigerian insects. Bulletin of Entomological Research 18: 95–99.

Golding, F. D. 1931. Further notes on the food plants of Nigerian insects. Bulletin of Entomological Research 22: 221–223.

Göllner-Scheiding, U. 1986. Revision of the genus *Odontoscelis* Laporte Decastelnau, 1832 (Heteroptera, Scutelleridae). Deutsche Entomologische Zeitschrift 33(1–2): 95–127.

Göllner-Scheiding, U. 1987. Ergänzung und Korrektur zu der Revision der Gattung *Odontoscelis* Laporte de Castelnau, 1832. Deutsche Entomologische Zeitschrift 34(1–3): 217–218.

Göllner-Scheiding, U. 1990. Revision of the genus *Odontotarsus* Laporte de Castelnau, 1832 (Heteroptera: Scutelleridae). Mitteilungen aus dem Zoologischen Museum in Berlin 66(2): 333–370.

Göllner-Scheiding, U. 1992. Revision der Gattung *Bolbocoris* Amyot and Serville, 1843 (Heteroptera, Pentatomidae, Podopinae). Deutsche Entomologische Zeitschrift 39(1–3): 189–202.

Göllner-Scheiding, U. 1999. Die Gattung *Phricodus* Spinola, 1840 (Insecta: Heteroptera: Pentatomidae). Entomologische Abhandlungen Staatliches Museum für Tierkunde Dresden 58(9): 149–163.

Göllner-Scheiding, U. 2006a. Family Acanthosomatidae Signoret, 1864, pp. 166–181. *In* B. Aukema and C. Rieger (Eds.), Catalogue of the Heteroptera of the Palaearctic region. Volume 5. The Netherlands Entomological Society, Amsterdam. xiii + 550 pp.

Göllner-Scheiding, U. 2006b. Family Scutelleridae Leach, 1815 – shield bugs, pp. 190–227. *In* B. Aukema and C. Rieger (Eds.), Catalogue of the Heteroptera of the Palaearctic region. Volume 5. Pentatomomorpha II. The Netherlands Entomological Society, Wageningen. xiii + 550 pp.

Goodchild, A. J. P. 1966. Evolution of the alimentary canal in the Hemiptera. Biological Reviews 41: 97–140.

Grazia, J. 1967. Estudos sôbre o gênero *Galedanta* Amyot and Serville, 1843 (Hemiptera-Heteroptera, Pentatomidae). Iheringia (Série Zoologia) 35: 45–59.

Grazia, J. 1968. Sôbre o gênero *Chloropepla* Stal, 1867, com a descriçao de uma nova espécie (Hemiptera, Pentatomidae, Pentatominae). Revista Brasileira de Biologia 28(2): 193–206.

Grazia, J. 1976. Revisao do gênero *Fecelia* Stal, 1872 (Heteroptera, Pentatomidae, Pentatomini). Revista Brasileira de Biologia 36(1): 229–237.

Grazia, J. 1978. Revisao do gênero *Dichelops* Spinola, 1837 (Heteroptera, Pentatomidae, Pentatomini). Iheringia (Série Zoologia) (53): 3–119.

Grazia, J. 1980a. Revisão do gênero *Pallantia* Stal, 1862 (Heteroptera, Pentatomidae). Revista Brasileira de Entomologia 24(1): 15–27.

Grazia, J. 1980b. Uma nova espécie do gênero *Fecelia* Stal (Heteroptera, Pentatomidae, Pentatomini). Revista Brasileira de Biologia 40(2): 261–266.

Grazia, J. 1981. Novas consideraçoes sobre *Galedanta* Amyot & Serville, 1843 com a descriçao de duas novas espécies (Heteroptera, Pentatomini). Anais da Sociedade Entomológica do Brasil 10(1): 9–19.

Grazia, J. 1983. Sobre o gênero *Phalaecus* Stal, 1862 com a descriçao de quatro novas espécies (Heteroptera, Pentatomidae). Revista Brasileira de Entomologia 27(2): 177–187.

Grazia, J. 1986. Uma nova espécie de *Pallantia* Stal, 1862 do México (Heteroptera, Pentatomini). Anais da Sociedade Entomológica do Brasil 15(2): 343–347.

Grazia, J. 1988. Sobre o gênero *Prionotocoris* Kormilev, 1955 (Heteroptera, Pentatomidae, Pentatominae). Revista Brasileira de Entomologia 32(3–4): 493–498.

Grazia, J. 1997. Cladistic analysis of the *Evoplitus* genus group of Pentatomini (Heteroptera: Pentatomidae). Journal of Comparative Biology 2(1): 43–48.

Grazia, J., and A. Barcellos. 1994. *Neotibilis*, um novo gênero de Pentatomini (Heteroptera). Iheringia (Série Zoologia) 76: 55–94.

Grazia, J., and A. Barcellos. 2005. Revision of *Taurocerus* (Heteroptera, Pentatomidae, Pentatomini). Iheringia (Série Zoologia) 95(2): 173–181.

Grazia, J., and M. Becker. 1997. *Adevoplitus*, a new genus of neotropical Pentatomini (Heteroptera: Pentatomidae). Journal of the New York Entomological Society 103(4)[1995]: 386–400.

Grazia, J., and L. A. Campos. 1996. *Hypanthracos*, a new genus of Pentatomini Heteroptera: Pentatomidae. Iheringia (Série Zoologia) 80: 13–19.

Grazia, J., and N. D. F. de Fortes. 1995. Revisão do gênero *Rio* Kirkaldy, 1909 (Heteroptera, Pentatomidae). Revista Brasileira de Entomologia 39(2): 409–430.

Grazia, J., and C. Greve. 2011. Contributions to the knowledge of *Elanela: Elanela jordi* sp. nov., from Amazonas, Brazil (Hemiptera: Heteroptera: Pentatoidae). Heteropterus Revista de Entomología 11: 261–266.

Grazia, J. and R.T. Koehler. 1983. Revisão do gênero *Marghita* Ruckes, 1964 com a descricao de uma nova espécie (Heteroptera, Pentatomidae, Pentatomini). Iheringia (Série Zoologia) 63: 133–144.

Grazia, J., and C. F. Schwertner. 2008a. Review of *Parachinavia* Roche (Hemiptera, Pentatomidae, Pentatominae), pp. 159–169. *In* S. Grozeva and N. Simov (Eds.), Advances in Heteroptera research. Festschrift in honour of 80th anniversary of Michail Josifov. Pensoft Publishers, Sofia-Moscow. 419 pp.

Grazia, J., and C. F. Schwertner. 2008b. Pentatomidae e Cyrtocoridae, pp. 223–234. *In* L. E. Claps, G. Debandi, and S. Roig-Juñent (Eds.), Biodiversidad de Artrópodos Argentinos. Volume 2. San Miguel de Tucumán, Universidad Nacional de Tucumán, Facultad de Ciencias Naturales, 614 pp.

Grazia, J., and C. F. Schwertner. 2015. Acanthosomatidae, pp. 399–402. *In* L. E. Claps, and S. Roig-Juñent (Eds.), Biodiversidad de Artrópodos Argentinos. Volume 3. San Miguel de Tucumán, Universidad Nacional de Tucumán, Facultad de Ciencias Naturales, 546 pp.

Grazia, J., M. Becker, and D. B. Thomas. 1995. A review of the genus *Pseudevoplitus* Ruckes, (Heteroptera: Pentatomidae) with the description of three new species. Journal of the New York Entomological Society 102(4)[1994]: 442–455.

Grazia, J., J. A. M. Fernandes, and C. F. Schwertner. 1998. *Luridocimex*, um novo gênero de Pentatomini (Heteroptera, Pentatomidae) do Brasil. Iheringia (Série Zoologia) 84: 161–166.

Grazia, J., J. A. M. Fernandes, and C. F. Schwertner. 1999. *Stysiana*, a new genus and four new species of Pentatomini (Heteroptera: Pentatomidae) of the Neotropical region. Acta Societatis Zoologicae Bohemicae 63(1–2): 71–83.

Grazia, J., L. A. Campos, and M. Becker. 2000. Revision of *Cataulax* Spinola, with *Architas* Distant as a new synonym (Heteroptera: Pentatomidae: Discocephalini). Anais da Sociedade Entomológica do Brasil 29(3): 475–488.

Grazia, J., L. A. Campos, C. Greve, and F. S. Rocha. 2002. Notas sobre *Pseudevoplitus* (Heteroptera, Pentatomidae) e descrição de duas espéces novas. Iheringia (Série Zoologia) 92(1): 53–61.

Grazia, J., R. T. Schuh, and W. C. Wheeler. 2008. Phylogenetic relationships of family groups in Pentatomoidea based on morphology and DNA sequences (Insecta: Heteroptera). Cladistics 24: 932–976.

Grazia, J., A. R. Panizzi, C. Greve, C. F. Schwertner, L. A. Campos, T. de A. Garbelotto, and J. A. M. Fernandes. 2015. Stink bugs (Pentatomidae), pp. 681–756. *In* A. R. Panizzi and J. Grazia (Eds.), True bugs (Heteroptera) of the Neotropics. Springer, Dordrecht, Heidelberg, New York, London. 901 pp.

Grazia, J., L. D. de Barros, and K. R. Barão. 2016a. *Elanela* Rolston revisited (Heteroptera: Pentatomidae): new distributional records and description of new species. Zootaxa 4092(4): 561–571.

Grazia, J., G. J. Bolze, and K. R. Barão. 2016b. There and back again: contributions on *Pseudevoplitus* Ruckes (Heteroptera: Pentatomidae). Zootaxa 4078(1): 161–167.

Grazia-Vieira, J. 1972. O genero *Mayrinia* Horvath, 1925 (Heteroptera, Pentatomidae, Pentatomini). Revista Peruana de Entomologia 15(1): 117–124.

Greathead, D. J. 1966. A taxonomic study of the species of *Antestiopsis* (Hemiptera: Pentatomidae) associated with *Coffea arabica* in Africa. Bulletin of Entomological Research 56: 515–554, pls. 19, 20.

Greathead, D. J. 1969. On the taxonomy of *Antestiopsis* spp. (Hem., Pentatomidae) of Madagascar, with notes on their biology. Bulletin of Entomological Research 59[1968]: 307–315.

Greve, C. 2010. Filogenia do grupo *Chlorocoris* baseada em morfologia e evidência total, descrição de cinco novas espécies e sinopse de *Chloropepla* Stål, incluindo análise cladística e biogeográfica (Hemiptera: Heteroptera: Pentatomidae). Ph.D. dissertation. Universidade Federal do Rio Grande do Sul, Porto Alegre. 147 pp.

Greve, C., C. F. Schwertner, and J. Grazia. 2013. Cladistic analysis of *Chloropepla* Stål (Hemiptera: Heteroptera: Pentatomidae) with the description of three new species. Insect Systematics and Evolution 44: 1–43.

Gross, G. F. 1970. A revision of the Australian pentatomid bugs of the genus *Cephaloplatus* White (Hemiptera - Pentatomidae - Pentatominae). Records of the South Australian Museum 16(4): 1–58.

Gross, G. F. 1972. A revision of the species of Australian and New Guinea shield bugs formerly placed in the genera *Poecilometis* Dallas and *Eumecopus* Dallas (Heteroptera: Pentatomidae) with descriptions of new species and selection of lectotypes. Australian Journal of Zoology, Supplementary Series 15: 1–192.

Gross, G. F. 1975a. A revision of the Pentatomidae (Hemiptera-Heteroptera) of the *Rhynchocoris* group from Australia and adjacent areas. Records of the South Australian Museum 17(6): 51–167.

Gross, G. F. 1975b. Handbook of the flora and fauna of South Australia. Plant-feeding and other bugs (Hemiptera) of South Australia. Heteroptera - Part 1. Handbooks Committee, South Australian Government, Adelaide, pp. 1–250, 4 col. pls.

Gross, G. F. 1976. Handbook of the flora and fauna of South Australia. Plant-feeding and other bugs (Hemiptera) of South Australia. Heteroptera - Part 2. Handbooks Committee, South Australian Government, Adelaide, pp. 251–501.

Gross, G. F. 1978. The genus *Bathycoelia* A&S in New Guinea and *Prytanicoris* gen. nov. from the New Guinea area and the New Hebrides (Heteroptera-Pentatomidae-Pentatominae). Records of the South Australian Museum 17(29): 417–428.

Gross, G. F., and F. J. D. McDonald. 1994. Revision of *Vitellus* Stål and *Avicenna* Distant in Australia (Hemiptera: Pentatomidae). Journal of the Australian Entomological Society 33(3): 265–274.

Guérin-Méneville, F. E. 1831. Insectes, pls. 1–21. *In* M. L. I. Duperrey (Ed.), Voyage autour du monde, exécuté par ordre du Roi, sur la corvette de la majesté, la Coquille, pendant les années 1822, 1823, 1824 et 1825, Zoologie. Volume 2, partie 2. Arthus Bertrand, Libraire-Éditeur, Paris. 5 + 21 + 16 plates.

Guérin-Méneville, F. E. 1838. Crustacés, Arachnides et Insectes, pp. i–xii + 9–319. *In* M. L. I. Duperrey (Ed.), Voyage autour du monde, exécuté par ordre du Roi, sur la corvette de la majesté, la Coquille, pendant les années 1822, 1823, 1824 et 1825, Zoologie. Volume 2, partie 2. Arthus Bertrand, Libraire-Éditeur, Paris. xii + 319 + 155 pp.

Guerra, T. J., F. Camarota, F. S. Castro, C. F. Schwertner, and J. Grazia. 2011. Trophobiosis between ants and *Eurystethus microlobatus* Ruckes 1966 (Hemiptera: Heteroptera: Pentatomidae) a cryptic, gregarious and subsocial stinkbug. Journal of Natural History 45: 1101–1117.

Guilbert, E. 2003. Habitat use and maternal care of *Phloea subquadrata* (Hemiptera: Phloeidae) in the Brasilian Atlantic forest (Espirito Santo). European Journal of Entomology 100(1): 61–63.

Hamilton, K. G. A. 1992. Lower Cretaceous Homoptera from the Koonwarra fossil bed in Australia, with a new superfamily and synopsis of Mesozoic Homoptera. Annals of the Entomological Society of America 85: 423–430.

Hanelová, J., and J. Vilímová. 2013. Behaviour of the central European Acanthosomatidae (Hemiptera: Heteroptera: Pentatomoidea) during oviposition and parental care. Acta Musei Moraviae, Scientiae Biologicae 98(2): 433–457.

Hasan, S. A., and I. J. Kitching. 1993. A cladistic analysis of the tribes of the Pentatomidae (Heteroptera). Japanese Journal of Entomology 61(4): 651–669.

Hasan, S. A., M. Afzal, and I. Ahmad. 1989. Description of a new genus and species of Pentatomidae (Hemiptera: Heteroptera) from Pakistan. Pakistan Journal of Scientific and Industrial Research 32(11): 764–765.

Hassan, M. E., P. Mukherjee, and B. Biswas. 2016. A new species of *Aeschrocoris* Bergroth (Hemiptera: Heteroptera: Pentatomidae: Pentatominae) from India. Munis Entomology & Zoology 11(1): 246–249.

Heer, O. 1853. Die Insektenfauna der Tertiärgebilde von Oeningen und von Rabodoj in Croatien. Dritter Theil: Rhynchoten. Wilhelm Englemann, Leipzig. 138 pp.

Heer, O. 1865. Die Urwelt der Schweiz. Schulthess, Zürich. 622 pp.

Henriksen, K. L. 1922. Eocene insects from Denmark. Danmarks Geologiske Undersøgelse 2: 1–36.

Henry, T. J. 1997. Phylogenetic analysis of family groups within the infraorder Pentatomomorpha (Hemiptera: Heteroptera), with emphasis on the Lygaeoidea. Annals of the Entomological Society of America 90: 275–301.

Hironaka, M., H. Horiguchi, L. Filippi, S. Nomakuchi, S. Tojo, and T. Hariyama. 2003a. Progressive change of homing navigation in the subsocial bug, *Parastrachia japonensis* (Heteroptera: Cydnidae). Japanese Journal of Entomology (ns) 6(1): 1–8. (in Japanese)

Hironaka, M., S. Nomakuchi, L. Filippi, S. Tojo, H. Horijuchi, and T. Hariyama. 2003b. The directional homing behaviour of the subsocial shield bug, *Parastrachia japonensis* (Heteroptera: Cydnidae), under different photic conditions. Zoological Science 20(4): 423–428.

Hironaka, M., S. Nomakuchi, S. Iwakuma, and L. Filippi. 2005. Trophic egg production in a subsocial shield bug, *Parastrachia japonensis* Scott (Heteroptera: Parastrachiidae), and its functional value. Ethology 111(12): 1089–1102.

Hironaka, M., L. Filippi, S. Nomakuchi, H. Hiroko, and T. Hariyama. 2007a. Hierarchical use of chemical marking and path integration in the homing trip of a subsocial shield bug. Animal Behaviour 73(5): 739–745.

Hironaka, M., S. Tojo, and T. Hariyama. 2007b. Light compass in the provisioning navigation of the subsocial shield bug, *Parastrachia japonensis* (Heteroptera: Parastrachiidae). Applied Entomology and Zoology 42(3): 473–478.

Hironaka, M., S. Tojo, S. Nomakuchi, L. Filippi, and T. Hariyama. 2007c. Round-the-clock homing behavior of a subsocial shield bug, *Parastrachia japonensis* (Heteroptera: Parastrachiidae), using path integration. Zoological Science 24(6): 535–541.

Hironaka, M., L. Filippi, S. Nomakuchi, and T. Hariyama. 2008a. Guarding behaviour against intraspecific kleptoparasites in the subsocial shield bug, *Parastrachia japonensis* (Heteroptera: Parastrachiidae). Behaviour 145(6): 815–827.

Hironaka, M., K. Inadomi, S. Nomakuchi, L. Filippi, and T. Hariyama. 2008b. Canopy compass in nocturnal homing of the subsocial shield bug, *Parastrachia japonensis* (Heteroptera: Parastrachiidae). Naturwissenschaften 95(4): 343–346.

Hoffmann, W. E. 1931. Notes on Hemiptera and Homoptera at Canton, Kwangtung Province, Southern China 1924–1929. United States Department of Agriculture, Insect Pest Survey Bulletin 11(3): 138–151.

Hoffmann, W. E. 1932. Notes on the bionomics of some oriental Pentatomidae (Hemiptera). Archivio Zoologico Italiano 16(3–4): 1010–1027.

Hoffmann, W. E. 1936. Die Brutpflege bei den Wanzen. Arbeiten über Physiologische und Angewandte Entomologie aus Berlin-Dahlem 3(4): 286–288.

Hoffmann, W. E. 1947. Insects as human food. Proceedings of the Entomological Society of Washington 49(9): 233–237.

Hong, Y-c. 1982. Heteroptera, pp. 91–98. *In* Mesozoic fossil insects of Jiuquan Basin in Gansu Province. Geological Publishing House, Beijing. [in Chinese]

Horváth, G. 1896. Hemiptera nova Palæarctica. Természetrajzi Füzetek 19(3–4): 322–329.

Horváth, G. 1900. Analecta ad cognitionen Tessaratominorum. Természetrajzi Füzetek 23: 339–374.

Horváth, G. 1903. Adnotationes synonymicae de Hemipteris Palaearcticis. Annales Musei Nationalis Hungarici 1: 555–558.

Horváth, G. 1919. Analecta ad cognitionem Cydnidarum. Annales Musei Nationalis Hungarici 17: 205–273.

Hosokawa, T., Y. Kikuchi, N. Nikoh, M. Shimada, and T. Fukatsu. 2006. Strict host-symbiont cospeciation and reductive genome evolution in insect gut bacteria. PloS Biology 4(10): 1841–1851.

Hosokawa, T., Y. Kikuchi, M. Shimada, and T. Fukatsu. 2008. Symbiont acquisition alters behaviour of stink-bug nymphs. Biology Letters 4: 45–48.

Hsiao, T-y. 1964. New species and new record of Hemiptera Heteroptera from China. Acta Zootaxonomica Sinica 1(2): 283–292. [in Chinese]

Hsiao, T-y., L-y. Zheng, S-z. Ren, et al. 1977. A handbook for the determination of the Chinese Hemiptera-Heteroptera. Volume 1. Biology Department, Nankai University, Tientsin. 330 pp, 52 pls. [in Chinese; unpaginated]

Hussey, R. F. 1934. Observations on *Pachycoris torridus* (Scop.), with remarks on parental care in other Hemiptera. Bulletin of the Brooklyn Entomological Society 29(4): 133–145.

Jacobs, D. H. 1989. A new species of *Thaumastella* with notes on the morphology, biology and distribution of the two southern African species (Heteroptera: Thaumastellidae). Journal of the Entomological Society of Southern Africa 52(2): 301–316.

Jacobs, D. H., P. J. Apps, and H. W. Viljoen. 1989. The composition of the defensive secretions of *Thaumastella namaquensis* and *T. elizabethae* with notes on the higher classification of the Thaumastellidae (Insecta: Heteroptera). Comparative Biochemistry and Physiology B. Comparative Biochemistry 93(2): 459–464.

James, D. G. 1989. Population biology of *Biprorulus bibax* Breddin (Hemiptera: Pentatomidae) in a southern New South Wales citrus orchard. Journal of the Australian Entomological Society 28: 279–286.

James, D. G. 1992. Effect of citrus host variety on oviposition, fecundity and longevity in *Biprorulus bibax* (Breddin) (Heteorptera). Acta Entomologica Bohemoslovaca 89: 65–67.

James, D. G. 1993. Integrated management of spined citrus bug *Biprorulus bibax* Breddin (Hemiptera: Pentatomidae) in inland citrus of southeastern Australia, pp. 96–98. *In* S. A. Corey, D. J. Dall, and W. M. Milne (Eds.), Pest control and sustainable agriculture. CSIRO, Melbourne. xiii + 514 pp.

Javahery, M., C. W. Schaefer, and J. D. Lattin. 2000. Shield bugs (Scutelleridae), pp. 475–503. *In* C. W. Schaefer and A. R. Panizzi (Eds.), Heteroptera of economic importance. CRC, Boca Raton, Florida. 828 pp.

Jensen-Haarup, A. C. 1931a. Hemipterological notes and descriptions VI. Entomologiske Meddelelser 17: 319–336.

Jensen-Haarup, A. C. 1931b. New or little known Hemiptera-Heteroptera I. Deutsche Entomologische Zeitschrift 1930: 215–222.

Jensen-Haarup, A. C. 1937. Einige neue Pentatomidenarten aus der Sammlungen des Zoologischen Museums in Hamburg (Hem. het.). Entomologiske Rundschau 54: 169–171, 321–324.

Jessop, L. 1983. A review of the genera of Plataspidae (Hemiptera) related to *Libyaspis*, with a revision of *Cantharodes*. Journal of Natural History 17: 31–62.

Jordan, K. H. C. 1967. Wanzen aus dem Pliozän von Willershausen. Bericht der Naturhistorischen Gesellschaft zu Hannover 111: 77–90.

Joshi, R. C., A. T. Barrion, and L. S. Sebastian. 2007. Rice black bugs. Taxonomy, ecology, and management of invasive species. Philippine Rice Research Institute, Science City of Muñoz, Philippines. 793 pp.

Josifov, M. V., and I. M. Kerzhner. 1978. Heteroptera aus Korea. II. Teil (Aradidae, Berytidae, Lygaeidae, Pyrrhocoridae, Rhopalidae, Alydidae, Coreidae, Urostylidae, Acanthosomatidae, Scutelleridae, Pentatomidae, Cydnidae, Plataspidae). Fragmenta Faunistica 23(9): 137–196.

Kaiwa, N., T. Hosokawa, N. Nikoh, M. Tanahashi, M. Moriyama, X-y. Meng, T. Maeda, K. Yamaguchi, S. Shigenobu, M. Ito, and T. Fukatsu. 2014. Symbiont-supplemented maternal investment underpinning host's ecological adaptation. Current Biology 24: 2465–2470.

Kamaluddin, S., and I. Ahmad. 1988. A revision of the tribe Phyllocephalini (Hemiptera: Pentatomidae: Phyllocephalinae) from Indo-Pakistan subcontinent with description of five new species. Oriental Insects 22: 185–240.

Kamaluddin, S., and I. Ahmad. 1991. Redescription with new systematic position of *Kafubu* Schouteden, *Melampodius* Distant and *Nimboplax* Linnavuori (Hemiptera: Pentatomidae: Phyllocephalinae: Cressonini) and their cladistics analysis. Proceedings of the Pakistan Congress of Zoology 11: 259–263.

Kastelein, P., and E. P. Camargo. 1990. Trypanosomatid protozoa in fruit of Solanaceae in southeastern Brazil. Memorias do Instituto Oswaldo Cruz 85(4): 413–417.

Kerzhner, I. M. 1972. New and little-known Heteroptera from Mongolia and from adjacent regions of the USSR. I. Nasekomye Mongolii 1: 349–379. [in Russian]

Kerzhner, I. M., V. G. Kuznetsova, and D. A. Rider. 2004. Karyotypes of Pentatomoidea additional to those published by Ueshima, 1979. Zoosystematica Rossica 13(1): 17–21.

Khuong, D. D., and T. X. Lam. 2001. Species of the subfamily Asopinae (Pentatomidae-Heteroptera) in Vietnam. Tap Chi Sinh Hoc 23(2): 15–19. [in Vietnamese]

Kim, H. R., and C. E. Lee. 1994. Morphological studies on the spermathecae of Korean Podopinae and Asopinae. Korean Journal of Entomology 24(3): 217–223.

Kinzelbach, R. K. 1970. Wanzen aus dem eozänen Ölschiefer von Messel (Insecta: Heteroptera). Notizblatt des Hessischen Landesamtes für Bodenforschung zu Wiesbaden 98(4): 9–18.

Kirkaldy, G. W. 1909. Catalogue of the Hemiptera (Heteroptera) with biological and anatomical references, lists of food plants and parasites etc. Volume 1. Cimicidae. F. Dames, Berlin. xl + 392 pp.

Kirkaldy, G. W. 1910. Three new Hemiptera-Heteroptera from the Miocene of Colorado. Entomological News 21(3): 129–131.

Kment, P. 2006. Revision of *Mahea* Distant, 1909, with a review of the Acanthosomatidae (Insecta: Heteroptera) of Madagascar and Seychelles. Acta Entomologica Musei Nationalis Pragae 45[2005]: 21–50.

Kment, P. 2007. *Noualhieridia guentheri* sp. nov., a new species of Acanthosomatidae (Heteroptera) from Madagascar. Mainzer Naturwissenschaftliches Archiv, Beiheft 31: 181–185.

Kment, P. 2008. A revision of the endemic Madagascan genus *Triplatyx* (Hemiptera: Heteroptera: Pentatomidae). Acta Entomologica Musei Nationalis Pragae 48(2): 543–582.

Kment, P. 2012. Redescription of the Madagascan endemic genus *Anoano* with a new synonymy (Hemiptera: Heteroptera: Pentatomidae). Acta Entomologica Musei Nationalis Pragae 52(2): 371–382.

Kment, P. 2013. *Carduelicoris stehliki*, a new genus and species of Pentatomidae (Hemiptera: Heteroptera) from Madagascar. Acta Musei Moraviae, Scientiae Biologicae 98(2): 415–432.

Kment, P. 2015. Two new genera of Madagascan Pentatominae (Hemiptera: Heteroptera: Pentatomidae). Acta Entomologica Musei Nationalis Pragae 55(2): 591–624.

Kment, P., and M. Baena. 2015. A redescription of the endemic Madagascan genus *Tricompastes* (Hemiptera: Heteroptera: Pentatomidae). Zootaxa 4044(1): 65–78.

Kment, P., and T. de A. Garbelotto. 2016. *Discimita linnavuorii*, a new genus and species of Afrotropical Pentatominae resembling the Neotropical Discocephalinae (Hemiptera: Heteroptera: Pentatomidae). Entomologica Americana 122(1–2): 199–211.

Kment, P., and Z. Jindra. 2009. A revision of *Tripanda* and *Tenerva* (Hemiptera: Heteroptera: Pentatomidae: Pentatominae). Zootaxa 1978: 1–47.

Kment, P., and A. Kocorek. 2014. *Afromenotes hirsuta*, a new genus and species of Eumenotini from the Democratic Republic of the Congo (Hemiptera: Heteroptera: Dinidoridae). Acta Entomologica Musei Nationalis Pragae 54(1): 109–131.

Kment, P., and D. A. Rider. 2015. On the synonymy of *Dymantis* Stål, 1861 and *Eomyrochea* Linnavuori, 1982 and resulting nomenclatural changes (Hemiptera: Heteroptera: Pentatomidae). Zootaxa 4058(2): 278–286.

Kment, P., and J. Vilímová. 2010a. Thoracic scent efferent system of the Tessaratomidae *sensu lato* (Hemiptera: Heteroptera: Pentatomoidea) with implication to the phylogeny of the family. Zootaxa 2363: 1–59.

Kment, P., and J. Vilímová. 2010b. Thoracic scent efferent system of Pentatomoidea (Hemiptera: Heteroptera): a review of terminology. Zootaxa 2706: 1–77.

Kment, P., P. Štys, and J. Vilímová. 2012. Thoracic scent efferent system and exponium of Aphylidae (Hemiptera: Heteroptera: Pentatomoidea), its architecture and function. European Journal of Entomology 109: 267–279.

Knight, H. H. 1952. Review of the genus *Perillus* with description of a new species (Hemiptera, Pentatomidae). Annals of the Entomological Society of America 45: 229–232.

Kobayashi, T. 1965. Developmental stages of *Urochela* and an allied genus of Japan (Hemiptera: Urostylidae). (The developmental stages of some species of the Japanese Pentatomoidea, XIII). Transactions of the Shikoku Entomological Society 8(3): 94–104.

Kobayashi, T., and S. Tachikawa. 2004. An illustrated book of eggs and larvae of Heteroptera - morphology and ecology. Agricultural Research Series of the National Agricultural Research Center, No. 51. Yokendo Press, Tokyo. 323 pp.

Koch, C. L. 1846. Die Arachniden. Volume 13. Nürnberg. 243 pp.

Kocorek, A., and J. A. Lis. 2000. A cladistic revision of the Megymeninae of the World (Hemiptera: Heteroptera: Dinidoridae). Polskie Pismo Entomologiczne 69(1): 7–30.

Konstantinov, F. V., and D. A. Gapon. 2005. On the structure of the aedeagus in shield bugs (Heteroptera, Pentatomidae): 1. Subfamilies Discocephalinae and Phyllocephalinae. Entomologicheskoe Obozrenie 84(2): 334–352. [English translation in Entomological Review 85(3): 221–235]

Kormilev, N. A. 1954. Una familia nueva para la fauna Argentina (Hemiptera, Megaridae). Anales de la Sociedad Científica Argentina 157: 47–54.

Kormilev, N. A. 1955. Notas sobre Pentatomoidea Neotropicales II (Hemiptera). Acta Scientifica de los Institutos de Investigación de San Miguel 1: 1–16.

Kudô, S., and T. Nakahira. 2004. Effects of trophic-eggs on offspring performance and rivalry in a sub-social bug. Oikos 107(1): 28–35.

Kudô, S., and T. Nakahira. 2005. Tropic-egg production in a subsocial bug: Adaptive plasticity in response to resource conditions. Oikos 111(3): 459–464.

Kudô, S., T. Nakahira, and Y. Saito. 2006. Morphology of trophic eggs and ovarian dynamics in the subsocial bug *Adomerus triguttulus* (Heteroptera: Cydnidae). Canadian Journal of Zoology 84(5): 723–728.

Kumar, R. 1969a. Morphology and relationships of the Pentatomoidea (Heteroptera). III. Natalicolinae and some Tessaratomidae of uncertain position. Annals of the Entomological Society of America 62: 681–695.

Kumar, R. 1969b. Morphology and relationships of the Pentatomoidea (Heteroptera) IV. Oncomerinae (Tessaratomidae). Australian Journal of Zoology 17: 553–606.

Kumar, R. 1971. Morphology and relationships of the Pentatomoidea (Heteroptera) 5 - Urostylidae. The American Midland Naturalist 85(1): 63–73.

Kumar, R. 1974a. A revision of world Acanthosomatidae (Heteroptera: Pentatomoidea): Keys to and descriptions of subfamilies, tribes and genera, with designation of types. Australian Journal of Zoology, Supplemental Series no. 34: 1–60.

Kumar, R. 1974b. A key to the genera of Natalicolinae Horvath, with the description of a new species of Tessaratominae Stål and with new synonymy (Pentatomoidea: Heteroptera). Journal of Natural History 8(6): 675–679.

Kumar, R., and M. S. K. Ghauri. 1970. Morphology and relationships of the Pentatomoidea (Heteroptera) 2 – World genera of Tessaratomini (Tessaratomidae). Deutsche Entomologische Zeitschrift 17(1–3): 1–32.

Larivière, M.-C. 1992. Description of *Parabrochymena*, new genus, and redefinition and review of *Brochymena* Amyot and Audinet-Serville (Hemiptera: Pentatomidae), with considerations on natural history, chorological affinities, and evolutionary relationships. Memoirs of the Entomological Society of Canada 163: 1–75.

Larivière, M.-C. 1994. *Parabrochymena* Larivière (Hemiptera: Pentatomidae): Systematics, natural history, chorological affinities, and evolutionary relationships, with a biogeographical analysis of *Parabrochymena* and *Brochymena* Amyot and Audinet-Serville. Canadian Entomologist 126(5): 1193–1250.

Lattin, J. D. 1964. The Scutellerinae of America north of Mexico (Hemiptera: Heteroptera: Pentatomidae). Ph.D. dissertation, University of California, Berkeley. 350 pp.

Leach, W. E. 1815. Order VIII. Hemiptera, pp. 120–126. *In* D. Brewster (Ed.), The Edinburgh encyclopædia. Entomology. Volume 9. Blackwood, Edinburgh. Pp. 57–172.

Lent, H., and J. Jurberg. 1965. Contribução ao conhecimento dos Phloeidae Dallas, 1851, com um estudo sôbre genitália (Hemiptera, Pentatomoidea). Revista Brasileira de Biologia 25(2): 123–144.

Lent, H., and J. Jurberg. 1966. Os estádios larvares de *Phloeophana longirostris* (Spinola, 1837) (Hemiptera, Pentatomidae). Revista Brasileira de Biologia 26(1): 1–4.

Leston, D. 1952a. Notes on the Ethiopian Pentatomoidea (Hemiptera).-V. On the specimens collected by Mr. A. L. Capener, mainly in Natal. Annals and Magazine of Natural History (12)5: 512–520.

Leston, D. 1952b. Notes on the Ethiopian Pentatomidae (Hem.) III: The genus *Aegaleus* Stål. Entomologist's Monthly Magazine 88: 16–17.

Leston, D. 1952c. Notes on the Ethiopian Pentatomoidea (Hemiptera): VIII, Scutellerinae Leach of Angola, with remarks upon the male genitalia and classification of the subfamily. Publicaçóes Culturais Companhia de Diamantes de Angola 16: 9–26.

Leston, D. 1953a. The suprageneric nomenclature of the British Pentatomoidea (Hemiptera). Entomologist's Gazette 4: 13–25.

Leston, D. 1953b. On the wing-venation, male genitalia and spermatheca of *Podops inuncta* (F.), with a note on the diagnosis of the subfamily Podopinae Dallas (Hem., Pentatomidae). Journal of the Society for British Entomology 4(7): 129–135.

Leston, D. 1953c. "Phloeidae" Dallas: Systematics and morphology, with remarks on the phylogeny of "Pentatomoidea" Leach and upon the position of *"Serbana"* Distant (Hemiptera). Revista Brasileira de Biologia 13(2): 121–140.

Leston, D. 1953d. Notes on the Ethiopian Pentatomidae (Hem.): VII A review of *Gynenica* Dallas 1851. Revue de Zoologie et de Botanique Africaines 48(3–4): 179–195.

Leston, D. 1954a. Classification of the terrestrial Heteroptera (Geocorisae). Nature 174: 91–92.

Leston, D. 1954b. Wing venation and male genitalia of *Tessaratoma* Berthold with remarks on Tessaratominae Stål (Hemiptera: Pentatomidae). Proceedings of the Royal Entomological Society of London (A)29: 9–16.

Leston, D. 1955a. Pentatomoidea (Hemiptera): New species and synonomy. Annals and Magazine of Natural History (12)8: 698–704.

Leston, D. 1955b. A key to the genera of Oncomerini Stål (Heteroptera: Pentatomidae, Tessaratominae), with the description of a new genus and species from Australia and new synonomy. Proceedings of the Royal Entomological Society of London (B)24(3–4): 62–68.

Leston, D. 1956a. The Ethiopian Pentatomoidea (Hemiptera): XXII, on *Dismegistus* Amyot and Serville (Cydnidae). Proceedings of the Royal Entomological Society of London (A)31(7–9): 87–94.

Leston, D. 1956b. Results from the Danish expedition to the French Cameroons 1949–50. IX. Hemiptera, Pentatomoidea. Bulletin de l'Institut Français d'Afrique Noire (A)18(2): 618–626.

Leston, D. 1957. Nomenclatural changes in Mecideinae (Hem. Pentatomidae). Entomologist's Monthly Magazine 93: 179.

Leston, D. 1958. Higher systematics of shieldbugs (Hemiptera: Pentatomidae). 10th International Congress of Entomology 1[1956]: 325.

Leston, D., and P. Dutton. 1957. Ethiopian Pentatomoidea (Hem.): XXIII, A synopsis of *Caura* Stål (Pentatomidae). Entomologist's Monthly Magazine 93: 54–56.

Leston, D., and G. G. E. Scudder. 1957. The taxonomy of the bronze orange-bug and related Australian Oncomerinae (Hemiptera: Tessaratomidae). Annals and Magazine of Natural History (12)10: 439–448.

Lethierry, L., and G. Severin. 1893. Catalogue général des Hémiptères. Volume 1. Bruxelles, Pentatomidae. i–x + 286 pp.

Lever, R. J. A. W. 1969. Pests of the coconut palm. Food and Agriculture Organization of the United Nations, Agricultural Studies, no. 77: 190 pp.

Li, H-m., R-q. Deng, J-w. Wang, Z-y. Chen, F-l. Jia, and X-z. Wang. 2005. A preliminary phylogeny of the Pentatomomorpha (Hemiptera: Heteroptera) based on nuclear 18S rDNA and mitochondrial DNA sequences. Molecular Phylogenetics and Evolution 37: 313–326.

Li, H-m., X-z. Wang, and J-t. Lin. 2006. Phylogenetic relationships of the Pentatomoidea based on the mito-chondrial 16S rDNA sequences (Heteroptera: Pentatomomorpha). Journal of Huazhong Agricultural University 25: 507–511.

Li, M., Y. Tian, Y. Zhao, and W-j. Bu. 2012. Higher level phylogeny and the first divergence time estimation of Heteroptera (Insecta: Hemiptera) based on multiple genes. PLoS ONE 7(2): 1–17.

Linnavuori, R. E. 1973. Studies on African Heteroptera. Arquivos do Museu Bocage (2)4(2): 26–69.

Linnavuori, R. 1975. Hemiptera Heteroptera of the Sudan with remarks on some species of the adjacent countries. Part 5. Pentatomidae. Boletim Sociedade Portuguesa de Ciências Naturais (2)15: 5–128.

Linnavuori, R. 1982. Pentatomidae and Acanthosomatidae (Heteroptera) of Nigeria and the Ivory Coast, with remarks on species of the adjacent countries in West and Central Africa. Acta Zoologica Fennica, no. 163. 176 pp.

Linnavuori, R. 1984. New species of Hemiptera Heteroptera from Iraq and the adjacent countries. Acta Entomologica Fennica, no. 44: 59 pp.

Linnavuori, R. 1986. Heteroptera of Saudi Arabia. Fauna of Saudi Arabia 8: 31–197.

Linnavuori, R. 1989. Heteroptera of Yemen and South Yemen. Acta Entomologica Fennica 54: 1–40.

Linnavuori, R. E. 1993. Cydnidae of West, Central and North-East Africa (Heteroptera). Acta Zoologica Fennica, no. 192. 148 pp.

Linnavuori, R. E., and M. M. Al-Safadi. 1993. *Acrosternum* Fieber (Heteroptera, Pentatomidae) in the Arabian Peninsula. Entomologica Fennica 4(4): 235–239.

Lis, J. A. 1990. New genera, new species, new records and checklist of the Old World Dinidoridae (Heteroptera, Pentatomoidea). Annals of the Upper Silesian Museum, Entomology 1: 103–147.

Lis, J. A. 1994. A revision of Oriental burrower bugs (Heteroptera: Cydnidae). Upper Silesian Museum, Bytom, Poland. 349 pp.

Lis, J. A. 1995. A synonymic list of burrower bugs of the Australian Region (Heteroptera: Cydnidae). Genus 6(2): 137–149.

Lis, J. A. 1999a. Burrower bugs of the Old World - a catalogue (Hemiptera: Heteroptera: Cydnidae). Genus 10(2): 165–249.

Lis, J. A. 1999b. Taxonomy and phylogeny of Cephalocteinae with a reference to their historical biogeography (Hemiptera: Heteroptera: Cydnidae). Polskie Pismo Entomologiczne 68(2): 111–131.

Lis, J. A. 2002. Burrower bugs described after the Old World catalogue of the family (Hemiptera: Heteroptera: Cydnidae). Polskie Pismo Entomologiczne 71(1): 7–17.

Lis, J. A. 2006a. Family Cydnidae Billberg, 1820 – burrowing bugs (burrower bugs), pp. 119–147. *In* B. Aukema and C. Rieger (Eds.), Catalogue of the Heteroptera of the Palaearctic Region. Volume 5. The Netherlands Entomological Society, Amsterdam. xiii + 550 pp.

Lis, J. A. 2006b. Family Dinidoridae Stål, 1867, pp. 228–232. *In* B. Aukema and C. Rieger (Eds.), Catalogue of the Heteroptera of the Palaearctic Region. Volume 5. The Netherlands Entomological Society, Amsterdam. xiii + 550 pp.

Lis, J. A. 2006c. Family Parastrachiidae Oshanin, 1922, p. 118. *In* B. Aukema and C. Rieger (Eds.), Catalogue of the Heteroptera of the Palaearctic Region. Volume 5. The Netherlands Entomological Society, Amsterdam. xiii + 550 pp.

Lis, J. A. 2006d. Family Thaumastellidae Seidenstücker, 1960, p. 117. *In* B. Aukema and C. Rieger (Eds.), Catalogue of the Heteroptera of the Palaearctic Region. Volume 5. The Netherlands Entomological Society, Amsterdam. xiii + 550 pp.

Lis, J. A. 2006e. Family Thyreocoridae Amyot and Serville, 1843 – negro-bugs, pp. 148–149. *In* B. Aukema and C. Rieger (Eds.), Catalogue of the Heteroptera of the Palaearctic Region. Volume 5. The Netherlands Entomological Society, Amsterdam. xiii + 550 pp.

Lis, J. A. 2010. Coxal combs in the Cydnidae *sensu lato* and three other related "cydnoid" families – Parastrachiidae, Thaumastellidae, Thyreocoridae (Hemiptera: Heteroptera): Functional, taxonomic, and phylogenetic significance. Zootaxa 2476: 53–64.

Lis, J. A., M. Bulińska-Balas, P. Lis, D. J. Ziaja, and A. Kocorek. 2012a. Systematic position of Dinidoridae and Tessaratomidae within the superfamily Pentatomoidea (Hemiptera: Heteroptera) based on the analysis of the mitochondrial cytochrome oxidase II sequences. Opole Scientific Society Nature Journal 45: 43–54.

Lis, J. A., P. Lis, D. J. Ziaja, and A. Kocorek. 2012b. Systematic position of Dinidoridae within the superfamily Pentatomoidea (Hemiptera: Heteroptera) revealed by the Bayesian phylogenetic analysis of the mitochondrial 12S and 16S rDNA sequences. Zootaxa 3423: 61–68.

Lis, J. A., J. Olchowik, and M. Bulińska-Balas. 2012c. Preliminary studies on the usefulness of DNA mini-barcodes for determining phylogenetic relationships within shieldbugs (Hemiptera: Heteroptera: Pentatomidae). Heteroptera Poloniae – Acta Faunistica 4: 13–25.

Lis, J. A., A. Kocorek, D. J. Ziaja, and P. Lis. 2015. New insight into the systematic position of the endemic Madagascan genus *Amberiana* (Hemiptera: Heteroptera: Dinidoridae) using 12S rDNA sequences. Turkish Journal of Zoology 39: 610–619.

Lis, J. A., B. Lis, and D. J. Ziaja. 2016. In BOLD we trust? A commentary on the reliability of specimen identification for DNA barcoding: a case study on burrower bugs (Hemiptera: Heteroptera: Cydnidae). Zootaxa 4114(1): 83–86.

Liu, Q-a., and L-y. Zheng. 1994. On the Chinese species of *Plautia* Stål (Hemiptera: Pentatomidae). Entomotaxonomia 16(4): 235–248.

Lodos, N. 1967. Studies on *Bathycoelia thalassina* (H.-S.) (Hemiptera, Pentatomidae), the cause of premature ripening of cocoa pods in Ghana. Bulletin of Entomological Research 57(2): 289–299.

Louise, C., M. Dollet, and D. Mariau. 1986. Research into hartrot of the coconut, a disease caused by *Phytomonas* (Trypanosomatidae), and into its vector *Lincus* sp. (Pentatomidae) in Guiana. Oléagineaux 41(10): 437–449.

Lung, K. Y.-H., and R. D. Goeden. 1982. Biology of *Corimelaena extensa* on tree tobacco, *Nicotiana glauca*. Annals of the Entomological Society of America 75: 177–180.

Lupoli, R. 2016. Diagnosis of *Calagasma* Bergroth and *Epipedus* Spinola with description of *Calagasma eclipsa* sp. nov. and *Epipedus rolstoni* sp. nov. (Hemiptera: Heteroptera: Pentatomidae: Pentatominae: Carpocorini). Zootaxa 4170(2): 330–338.

Lupoli, R., F. Dusoulier, A. Cruaud, S. Cros-Arteil, and J.-C. Streito. 2013. Morphological, biogeographical and molecular evidence of *Carpocoris mediterraneus* as a valid species (Hemiptera: Pentatomidae). Zootaxa 3609(4): 392–410.

Lyal, C. H. C. 1979. A review of the genus *Calliphara* Germar, 1839 (Hemiptera: Scutelleridae). Zoologische Mededelingen 54(12): 149–181.

Magnien, P. 2008. New species and synonymy in genus *Pygoplatys* Dallas, 1851 (Heteroptera, Tessaratomidae). Nouvelle Revue d'Entomologie (ns) 24(2)[2007]: 113–124.

Magnien, P. 2011. Description of two new species of *Pygoplatys* Dallas, 1851, with a key to the species of the genus (Hemiptera: Heteroptera: Tessaratomidae). Heteropterus Revista de Entomología 11(2): 287–297.

Magnien, P. 2015. Illustrated Catalog of Tessaratomidae. http://hemiptera-databases.com/tessaratomidae/

Magnien, P. and D. Pluot-Sigwalt. 2016. *Sciadiocoris*, a new genus of Oncomerinae with three new species from Papua New Guinea (Hemiptera: Heteroptera: Tessaratomidae). Entomologica Americana 122(1–2): 294–304.

Magnien, P., K. Smets, D. Pluot-Sigwalt, and J. Constant. 2008. A new species of *Pygoplatys* Dallas (Heteroptera, Tessaratomidae) from the Damar agroforests in Sumatra: Description, immatures and biology. Nouvelle Revue d'Entomologie 24(2)[2007]: 99–112.

Malipatil, M. B., and R. Kumar. 1975. Biology and immature stages of some Queensland Pentatomomorpha (Hemiptera: Heteroptera). Journal of the Australian Entomological Society 14(2): 113–128.

Mariod, A., and H. Fadul. 2014. Extraction and characterization of gelatin from two edible Sudanese insects and its applications in ice cream making. Food Science and Technology International 2014: 1–12.

Marshall, T. A. 1874. Descriptions of a new genus and two new species of European *Oxyura*. Entomologists Monthly Magazine 10: 207–209.

Martins-Neto, R. G. 1997. A paleoentomofauna da Formacao Tremembe (Bacia de Taubate) Oligoceno do estado de Sao Paulo; descricao de novos hemipteros (insecta). Revista Universidade Guarulhos Geociencias 2(6): 66–69.

Maschwitz, U., B. Fiala, and W. R. Dolling. 1987. New trophobiotic symbioses of ants with South East Asian bugs. Journal of Natural History 21: 1097–1107.

Matesco, V. C. 2014. Sistemática de Thyreocoridae Amyot and Serville (Hemiptera: Heteroptera: Pentatomoidea): Revisão de *Alkindus* Distant, morfologia do ovo de duas espécies de *Galgupha* Amyot and Serville e análise cladística de *Corimelaena* White, com considerações sobre a filogenia de Thyreocoridae, e morfologia do ovo de 16 espécies de Pentatomidae como exemplo do uso de caracteres de imaturos em filogenias. Ph.D. dissertation, Universidade Federal do Rio Grande do Sul. 222 pp.

Matesco, V. C., and J. Grazia. 2013. Revision of the genus *Alkindus* Distant (Hemiptera: Heteroptera: Thyreocoridae: Corimelaeninae). Zootaxa 3750(1): 57–70.

Matesco, V. C., and J. Grazia. 2015. Negro bugs (Thyreocoridae), pp. 789–820. *In* A. R. Panizzi and J. Grazia (Eds.), True bugs (Heteroptera) of the Neotropics. Springer, Dordrecht, Heidelberg, New York, London. 901 pp.

Matesco, V. C., J. Grazia, and L. A. Campos. 2007. Description of new genus and species of Ochlerini from Central America (Hemiptera: Pentatomidae: Discocephalinae). Zootaxa 1562: 63–68.

Mathew, K. 1969. A new species of *Degonetus* Distant (Hem., Het., Pentatomidae) from Sikkim [India]. Oriental Insects 3(2): 197–198.

Mayorga, C. 2002. Revisión genérica de la familia Cydnidae (Hemiptera-Heteroptera) en México, con un listado de las especies conocidas. Anales del Instituto de Biología, Universidad Nacional Autónoma de México (Zoológia)73(2): 157–192.

Mayorga, C., and L. Cervantes. 2001. Life cycle and description of a new species of *Amnestus* Dallas (Hemiptera Heteroptera: Cydnidae) associated with the fruit of several species of *Ficus* (Moraceae) in Mexico. Journal of the New York Entomological Society 109(3–4): 392–402.

McAtee, W. L., and J. R. Malloch. 1928. Synopsis of pentatomid bugs of the subfamilies Megaridinae and Canopinae. Proceedings of the United States National Museum 72(25): 1–21, 2 pls.

McAtee, W. L., and J. R. Malloch. 1933. Revision of the subfamily Thyreocorinae of the Pentatomidae (Hemiptera-Heteroptera). Annals of the Carnegie Museum 21: 191–411, pls. 4–17.

McDonald, F. J. D. 1966. The genitalia of North American Pentatomoidea (Hemiptera: Heteroptera). Quaestiones Entomologicae 2: 7–150.

McDonald, F. J. D. 1969a. A new species of Lestoniidae (Hemiptera). Pacific Insects 11(2): 187–190.

McDonald, F. J. D. 1969b. Life cycle of the bronze orange bug *Musgraveia sulciventris* (Stål) (Hemiptera: Tessaratomidae). Australian Journal of Zoology 17: 817–820.

McDonald, F. J. D. 1970. The morphology of *Lestonia haustorifera* China (Heteroptera: Lestoniidae). Journal of Natural History 4: 413–417.

McDonald, F. J. D. 1974. Revision of the genus *Holcostethus* in North America (Hemiptera: Pentatomidae). Journal of the New York Entomological Society 82(4): 245–258.

McDonald, F. J. D. 1976. Revision of the genus *Trichopepla* (Hemiptera: Pentatomidae) in N. America. Journal of the New York Entomological Society 84(1): 9–22.

McDonald, F. J. D. 1979. A new species of *Megaris* and the status of the Megarididae McAtee and Malloch and Canopidae Amyot and Serville (Hemiptera: Pentatomoidea). Journal of the New York Entomological Society 87(1): 42–54.

McDonald, F. J. D. 1982. Description of the male genitalia of *Holcostethus hirtus* (Van Duzee) with a revised key to North American species (Hemiptera: Pentatomidae). Journal of the New York Entomological Society 90(1): 5–7.

McDonald, F. J. D. 1986. Revision of *Cosmopepla* Stål (Hemiptera: Pentatomidae). Journal of the New York Entomological Society 94(1): 1–15.

McDonald, F. J. D. 1988. A revision of *Cantao* Amyot and Serville (Hemiptera: Scutelleridae). Oriental Insects 22: 287–299.

McDonald, F. J. D. 1995. Revision of *Hypogomphus* Spinola in Australia (Hemiptera: Pentatomidae). Journal of the Australian Entomological Society 34(2): 169–176.

McDonald, F. J. D. 2001. Two new species of *Diaphyta* Bergroth with notes on the genus (Hemiptera: Pentatomidae). Australian Journal of Entomology 40: 17–22.

McDonald, F. J. D. 2003. A new genus and species of Pentatomidae (Hemiptera: Heteroptera) from northern Australia. Australian Entomologist 30(1): 17–20.

McDonald, F. J. D. 2006. *Birna*, a new name for *Linea* McDonald (Hemiptera: Pentatomidae). Australian Entomologist 33(1): 26.

McDonald, F. J. D., and G. Cassis. 1984. Revision of the Australian Scutelleridae Leach (Hemiptera). Australian Journal of Zoology 32: 537–572.

McDonald, F. J. D., and P. B. Edwards. 1978. Revision of the genus *Oncocoris* Mayr (Hemiptera: Pentatomidae). Australian Journal of Zoology, supplement 62: 1–53.

McHugh, J. U. 1994. On the natural history of Canopidae (Heteroptera, Pentatomoidea). Journal of the New York Entomological Society 102(1): 112–114.

McPherson, J. E. 1971. Notes on the laboratory rearing of *Corimelaena lateralis lateralis* (Hemiptera: Corimelaenidae) on wild carrot. Annals of the Entomological Society of America 64: 313–314.

McPherson, J. E. 1972. Life history of *Corimelaena lateralis lateralis* (Hemiptera: Thyreocoridae) with descriptions of immature stages and list of other species of Scutelleroidea found with it on wild carrot. Annals of the Entomological Society of America 65: 906–911.

McPherson, J. E. 1974. Three negro bug state records for Illinois (Hemiptera: Corimelaenidae). Transactions of the Illinois State Academy of Science 67(3): 361–363.

McPherson, J. E. 1982. The Pentatomoidea (Hemiptera) of northeastern North America with emphasis on the fauna of Illinois. Southern Illinois University Press, Carbondale and Edwardsville. 240 pp.

McPherson, J. E., and R. M. McPherson. 2000. Stink bugs of economic importance in America north of Mexico. CRC Press, Boca Raton, London, New York, Washington, D.C. 253 pp.

Melber, A., and G. H. Schmidt. 1977. Sozialphänomene bei Heteropteren. Zoologica, Stuttgart 43(127): 19–53.

Memon, N., and I. Ahmad. 1998. Redescription of *Carenoplistus acutus* (Signoret) (Hemiptera: Pentatomidae: Pentatominae: Halyini) with reference to its male genitalia and cladistic relationships. Pakistan Journal of Entomology 13(1–2): 5–7.

Memon, N., and I. Ahmad. 2009. A revision of halyine stink bug genus *Sarju* Ghauri (Hemiptera: Pentatomidae: Halyini) and its cladistic analysis. Pakistan Journal of Zoology 41(5): 399–411.

Memon, N., F. Gilbert, and I. Ahmad. 2011. Phylogeny of the South Asian halyine stink bugs (Hemiptera: Pentatomidae: Halyini) based on morphological characters. Annals of the Entomological Society of America 104: 1149–1169.

Mendonça, M. de S., C. F. Schwertner, and J. Grazia. 2009. Diversity of Pentatomoidea (Hemiptera) in riparian forests of southern Brazil: taller forests, more bugs. Revista Brasileira de Entomologia 53(1): 121–127.

Meunier, F. 1915. Nouvelles recherches sur quelques insects des plâtrières d'Aix en Provence. Verhandelingen der Koninklijke Akademie van Wetenschappen te Amsterdam (Sect. 2) 18(5): 1–17.

Michelmore, A. P. G. 1949. Report on coffee entomology and pathology, 1946–1948. Uganda Protectorate. Government Printer, Entebbe, Uganda. 15 pp.

Miller, N. C. E. 1971. The biology of the Heteroptera. Leonard Hill (Books) Ltd., London. 206 pp.

Mitchell, P. L. 2004. Heteroptera as vectors of plant pathogens. Neotropical Entomology 33(5): 519–545.

Miyamoto, S. 1961. Comparative morphology of the alimentary organs of Heteroptera with phylogenetic considerations. Sieboldia 2(4): 197–259.

Monteith, G. B. 1982. Dry season aggregations of insects in Australian monsoon forests. Memoirs of the Queensland Museum 20(3): 533–543.

Monteith, G. B. 2006. Maternal care in Australian oncomerine shield bugs (Insecta, Heteroptera, Tessaratomidae). *In* W. Rabitsch (Ed.), Hug the bug - for love of true bugs. Festschrift zum 70. Geburtstag von Ernst Heiss. Denisia 19: 1135–1152.

Monteith, G. B. 2011. Maternal care, food plants and distribution of Australian Oncomerinae (Hemiptera: Heteroptera: Tessaratomidae). Australian Entomologist 38(1): 37–48.

Moriya, S., M. Shiga, and M. Mabuchi. 1987. Analysis of light trap records in four major species of fruit-piercing stink bugs, with special reference to body size variation in trapped adults of *Plautia stali* Scott. Bulletin of the Fruit Tree Research Station (A)14: 79–94. (in Japanese)

Mulsant, E., and C. Rey. 1866. Histoire Naturelle des Punaises de France. Volume II. Pentatomides. F. Savy and Deyrolle, Paris. 372 pp.

Musgrave, A. 1930. Contributions to the knowledge of Australian Hemiptera. No. II. A revision of the subfamily Graphosomatinae (family Pentatomidae). Records of the Australian Museum 17: 317–341.

Mustafa, N. E. M., A. A. Mariod, and B. Matthäus. 2008. Antibacterial activity of *Aspongopus viduatus* (melon bug) oil. Journal of Food Safety 28: 577–586.

Nakahira, T. 1994. Production of trophic eggs in the subsocial burrower bug *Adomerus triguttulus*. Naturwissenschaften 81(9): 413–414.

Nakahira, T., K. D. Tanaka, and S-i. Kudo. 2013. Maternal provisioning and possible joint breeding in the burrower bug *Adomerus triguttulus* (Heteroptera: Cydnidae). Entomological Science 16(2): 151–161.

Nakamura, Y., and T. Nishino. 1999. Forecasting fruit damage caused by the brown-winged green bug, *Plautia stali* Scott, and its occurrence using synthetic aggregation pheromone traps. 1. Comparison of the seasonal prevalence of catches between an aggregation pheromone trap and a light-trap. Proceedings of the Association for Plant Protection of Kyushu 45: 119–122.

Nardi, C. 2005. Percevejos castanhos (Hemiptera, Cydnidae, *Scaptocoris*): aspectos morfológicos, ecológicos e comportamentais. Dissertação, Universidade de São Paulo-Escola Superior de Agricultura Luiz de Queiroz, Piracicaba, SP, Brasil. 67 pp.

Nardi, C., P. M. Fernandes, and J. M. S. Bento. 2008. Wing polymorphism and dispersal of *Scaptocoris carvalhoi* (Hemiptera: Cydnidae). Annals of the Entomological Society of America 101: 551–557.

Nelson, G. H. 1955. A revision of the genus *Dendrocoris* and its generic relationships (Hemiptera, Pentatomidae). Proceedings of the Entomological Society of Washington 57(2): 49–67.

Nelson, G. H. 1957. A new species of *Dendrocoris* and a new combination of *Atizies* (Hemiptera, Pentatomidae). Proceedings of the Entomological Society of Washington 59(4): 197–199.

Nomakuchi, S., L. Filippi, and S. Tojo. 1998. Selective foraging behavior in nest-provisioning females of *Parastrachia japonensis* (Hemiptera: Cydnidae): cues for preferred food. Journal of Insect Behavior 11(5): 605–619.

Nomakuchi, S., L. Filippi, and M. Hironaka. 2001. Nymphal occurrence pattern and predation risk in the subsocial shield bug, *Parastrachia japonensis* (Heteroptera: Cydnidae). Applied Entomology and Zoology 36(2): 209–212.

Nomakuchi, S., L. Filippi, S. Iwakuma, and M. Hironaka. 2005. Variation in the start of nest abandonment in the subsocial shield bug *Parastrachia japonensis* (Hemiptera: Parastrachiidae). Annals of the Entomological Society of America 98: 143–149.

Novák, O. 1877. Fauna der Cyprisschiefer des Egerer Tertiärbeckens. Sitzungsberichte der Kaiserlichen Akademie der Wissenschaften 76: 71–96.

O'Donnell, J. E., and C. W. Schaefer. 2012. Annotated checklist of the Pentatomidae (Heteroptera) of Connecticut. The Great Lakes Entomologist 45(3–4): 220–234.

Olckers, T., and H. G. Zimmermann. 1991. Biological control of silverleaf nightshade, *Solanum elaeagnifolium*, and bugweed, *Solanum mauritianum*, (Solanaceae) in South Africa. Agriculture, Ecosystems and Environment 37: 137–155.

Oliveira, L. J., and A. B. Malaguido. 2004. Flutuação e distribuição vertical da população do percevejo castanho da raiz, *Scaptocoris castanea* Perty (Hemiptera: Cydnidae), no perfil do solo em areas produtoras de soja nas regiões centro-oeste e sudeste do Brasil. Neotropical Entomology 33(3): 283–291.

Orian, A. J. E. 1965. A new genus of Pentatomidae from Africa, Madagascar and Mauritius (Hemiptera). Proceedings of the Royal Entomological Society of London (B)34(3–4): 25–30, 3 pls.

Ortega-León, G. and D. B. Thomas. 2016. *Pseudocromata*, a new genus of Ochlerini based on a new species from Ecuador (Pentatomidae: Discocephalinae). Zootaxa 4137(2): 286–290.

Oshanin, B. 1906. Verzeichnis der Paläarktischen Hemipteren mit besonderer Berücksichtigung ihrer Verteilung im Russischen Reiche. Volume 1. Heteroptera. Lieferung 1. Pentatomidae-Lygaeidae. Annuaire du Musée Zoologique de l'Académie Impériale des Sciences 11: 1–393.

Oshanin, B. 1912. Katalog der paläarktischen Hemipteren (Heteroptera, Homoptera-Auchenorhyncha und Psylloideae). R. Friedlander and Sohn, Berlin. xvi + 187 pp.

Oshanin, B. 1922. Sur les genres de la tribu des Strachiaria Put. (Heteroptera, Pentatomidae). Ezhegodnik Zoologischeskago Muzeya Rossiiskoi Akademii Nauk 23: 143–148.

Oustalet, M. 1874. Sur un Hémiptère de la famille des Pentatomides. Bulletin de la Société Philomathique de Paris (6)11: 14–16.

Owusu-Manu, E. 1971. *Bathycoelia thalassina*. Another serious pest of cocoa in Ghana. CMB Newsletter 47: 12–14.

Owusu-Manu, E. 1976. Estimation of cocoa pod losses caused by *Bathycoelia thalassina* (H.-S.) (Hemiptera, Pentatomidae). Ghana Journal of Agricultural Science 9(2): 81–83.

Owusu-Manu, E. 1977. Distribution and abundance of the cocoa shield bug, *Bathycoelia thalassina* (Hemiptera: Pentatomidae) in Ghana. Journal of Applied Ecology 14(2): 331–342.

Owusu-Manu, E. 1990. Feeding behavior and the damage caused by *Bathycoelia thalassina* (Herrich-Schaeffer) (Hemiptera, Pentatomidae). Café Cacao Thé 34(2): 97–104.

Packauskas, R. J. 2012. The Pentatomidae, or stink bugs, of Kansas with a key to species (Hemiptera: Heteroptera). The Great Lakes Entomologist 45(3–4): 210–219.

Packauskas, R. J., and C. W. Schaefer. 1998. Revision of the Cyrtocoridae (Hemiptera: Pentatomoidea). Annals of the Entomological Society of America 91: 363–386.

Paiero, S. M., S. A. Marshall, J. E. McPherson, and M.-S. Ma. 2013. Stink bugs (Pentatomidae) and parent bugs (Acanthosomatidae) of Ontario and adjacent areas: A key to species and a review of the fauna. Canadian Journal of Arthropod Identification, no. 24. 183 pp.

Paleari, L. M. 1992. Revisão do gênero *Agonosoma* Laporte, 1832 (Hemiptera, Scutelleridae). Revista Brasileira de Entomologia 36(3): 505–520.

Panizzi, A. R. 2015. Growing problems with stink bugs (Hemiptera: Heteroptera: Pentatomidae). Species invasive to the U.S. and potential Neotropical invaders. American Entomologist 61: 223–233.

Panizzi, A. R., J. E. McPherson, D. G. James, J. Javahery, and R. M. McPherson. 2000. Stink bugs (Pentatomidae), pp. 421–474. *In* C. W. Schaefer and A. R. Panizzi (Eds.), Heteroptera of economic importance. CRC Press, Boca Raton, London, New York, Washington, D.C. 828 pp.

Pantoja, A., E. Daza, C. Garcia, O. I. Mejía, and D. A. Rider. 1995. Relative abundance of stink bugs (Hemiptera: Pentatomidae) in southwestern Colombia rice fields. Journal of Entomological Science 30(4): 463–467.

Pantoja, A., E. Daza, O. I. Mejía, C. A. Garcia, M. C. Duque, and L. E. Escalona. 1999. Development of *Oebalus ornatus* (Sailer) and *Oebalus insularis* (Stal) (Hemiptera: Pentatomidae) on rice. Journal of Entomological Science 34(3): 335–338.

Pantoja, A., C. A. Garcia, and M. C. Duque. 2000. Population dynamics and effects of *Oebalus ornatus* (Hemiptera: Pentatomidae) on rice yield and quality in southwestern Colombia. Journal of Economic Entomology 93: 276–279.

Pantoja, A., M. Triana, H. Bastidas, C. García, and M. C. Duque. 2005. Development of *Tibraca obscurata* and *Tibraca limbativentris* (Hemiptera: Pentatomidae) in rice in southwestern Colombia. Journal of Agriculture of the University of Puerto Rico 89(3–4): 221–228.

Pantoja, A., M. Triana, H. Bastidas, C. García, O. I. Mejia, and M. C. Duque. 2007. Damage by *Tibraca lim-bativentris* (Hemiptera: Pentatomidae) to rice in southwestern Colombia. Journal of Agriculture of the University of Puerto Rico 91(1–2): 11–18.

Parveen, S., and A. Gaur. 2015. Illustrated key to the Indian genera of Scutelleridae (Hemiptera: Heteroptera). Indian Journal of Entomology 77(2): 169–184.

Pendergrast, J. G. 1950. The genus *Rhopalimorpha* Dallas (Hemiptera-Heteroptera) with a description of a new species. Records of the Auckland Institute and Museum 4(1): 31–34.

Pendergrast, J. G. 1952. The genus *Rhopalimorpha* Dallas (Heteroptera, Pentatomidae). Records of the Auckland Institute and Museum 4(3): 159–162.

Pendergrast, J. G. 1957. Studies on the reproductive organs of the Heteroptera with a consideration of their bearing on classification. Transactions of the Royal Entomological Society of London 109(1): 1–63.

Pennington, M. S. 1920. Lista de los Hemipteros Heteropteros de la República Argentina. Primera parte. Buenos Aires, pp. 1–16.

Péricart, J. 2002. Note sur le genre *Sciocoris* Fallén, 1829, et ses représentants euro-méditerranéens (Heteroptera, Pentatomidae). Bulletin de la Société Entomologique de France 107(4): 435–448.

Péricart, J. 2010. Hémiptères Pentatomoidea Euro-Méditerranéens 3. Podopinae et Asopinae. Faune de France Volume 93. Fédération Française des Sociétés de Sciences Naturelles, Paris. 291 pp. + 24 pls.

Petrulevičius, J. F., and Y. A. Popov. 2014. First fossil record of Discocephalinae (Insecta, Pentatomidae): a new genus from the middle Eocene of Río Pichileufú, Patagonia, Argentina. ZooKeys 422: 23–33.

Phelps, R. J., and M. J. Oosthuizen. 1958. Insects injurious to cowpeas in the Natal region. Journal of the Entomological Society of Southern Africa 21(2): 286–295.

Pinto, I. D., and L. P. Ornellas. 1974. New Cretaceous Hemiptera (Insects) from Codó Formation – Northern Brazil. Anais do XXVIII Congresso Brasileiro de Geologia 28: 289–304.

Pirán, A. A. 1963. Hemiptera neotropica. VII. Algunas especies nuevos o poco conocidas del noroeste Argentino. I. Acta Zoologica Lilloana 19: 335–341.

Pirán, A. A. 1971. La subfamilia Tessaratominae (Hemiptera-Heteroptera) en la region Neotropical. Acta Zoologica Lilloana 26: 197–208.

Piton, L. E. 1933. Insectes fossils des Cinérites et Randannites d'Auvergne. Clermont-Ferrand.

Piton, L. E. 1940. Paléontologie du gisement éocène de Menat (Puy-de-Dôme) (flore et faune). Mémoires de la Société d'Histoire Naturelle d'Auvergne 1: 1–303.

Piton, L. E. and N. Théobald. 1935. La faune entomologique des gisements mio-pliocènes du massif Central. Revue des Sciences Naturelles d'Auvergne (ns)1: 65–104.

Pluot-Sigwalt, D. 2008. A pair of basi-abdominal sex pheromone glands in the male of some burrower bugs (Hemiptera: Heteroptera: Cydnidae). Acta Entomologica Musei Nationalis Pragae 48(2): 511–522.

Pluot-Sigwalt, D., and J. A. Lis. 2008. Morphology of the spermatheca in the Cydnidae (Hemiptera: Heteroptera): Bearing of its diversity on classification and phylogeny. European Journal of Entomology 105(2): 279–312.

Poinar Jr., G., and E. Heiss. 2013. *Minysporops dominicanus* gen. n., sp. n. (Hemiptera: Pentatomoidea: Megarididae), a megaridid in Dominican amber. Historical Biology 25(1): 95–100.

Poinar Jr., G., and D. B. Thomas. 2012. A stink bug, *Edessa protera* sp. n. (Pentatomidae: Edessinae) in Mexican amber. Historical Biology 24(2): 207–211.

Popov, Y. A. 1968a. True bugs (Heteroptera) in the Holocene badger coprolite, pp. 129–132. *In* The history of vegetation cover development in the central regions of the European part of the USSR during Anthropogene. Publishing House Nauka, Moscow.

Popov, Y. A. 1968b. True bugs of the Jurassic fauna of Karatau (Heteroptera), pp. 99–113. *In* B. B. Rodendorf (Ed.), Jurassic Insects of Karatau. Nauka, Moskva.

Popov, Y. A. 1986. Peloridiina (= Coleorrhyncha) et Cimicina (= Heteroptera), pp. 50–83. *In* Insects in the early Cretaceous ecosystems of the west Mongolia. Transactions of the Joint Soviet-Mongolian Palaeontological Expedition 28: 1–213.

Popov, Y. A. 1989. New fossil Hemiptera (Heteroptera + Coleorrhyncha) from the Mesozoic of Mongolia. Neues Jahrbuch für Geologie und Paläontologie Monatshefte 3: 166–181.

Popov, Y. A. 1990. Description of fossil insects. True bugs. Cimicina. *In* Late Mesozoic insects of Eastern Transbaikalia. Transactions of the Paleontological Institute of the Academy of Sciences of the USSR 239: 20–39.

Popov, Y. A. 2007. A new notion on the heteropterofauna (Insecta: Hemiptera, Heteroptera) from the Pliocene of Willershausen (N Germany). Paläontologische Zeitschrift 81(4): 429–439.

Popov, Y. A., and I. D. Pinto. 2000. On some Mesozoic burrower bugs (Heteroptera: Cydnidae). Paleontological Journal 34, supplement 3: 298–302.

Puton, A. 1869. Catalogue des Hémiptères Hétéroptères d'Europe. Deyrolle, Paris. vii + 40 pp.

Putshkov, V. G. 1961. [Pentatomoidea.]. Fauna Ukraini. Volume 21(1). Vidavnitstvo Akademii Nauk Ukrainskoi RSR, Kiiv. 338 pp. [in Ukrainian]

Putshkov, V. G. 1965. Shield-bugs of Central Asia (Hemiptera, Pentatomoidea). Publishing House Ilim (The Academy of Sciences of Kyrgyz SSR, Institute of Biology), Frunze (Kyrgyz SSR, the USSR). 332 pp. [in Russian]

Rau, P. 1918. Maternal care in *Dinocoris tripterus* Fab. (Hemiptera). Entomological News 29(2): 75–76.

Rebagliati, P. J., L. M. Mola, A. G. Papeschi, and J. Grazia. 2005. Cytogenetic studies in Pentatomidae (Heteroptera): A review. Journal of Zoological Systematics and Evolutionary Research 43(3): 199–214.

Rédei, D. 2016. The identity of the *Brachyplatys* species recently introduced into Panama (Hemiptera: Heteroptera: Plataspidae). Zootaxa 4136: 141–154.

Rédei, D., and W. Bu. 2013. *Labroplatys*, a new genus of Oriental Plataspidae (Hemiptera: Heteroptera) with two new species from India, Laos, and China. Entomologica Americana 118[2012]: 263–273.

Rédei, D., and Z. Jindra. 2015. A revision of the genus *Hemitrochostoma* (Hemiptera, Heteroptera, Plataspidae). ZooKeys 495: 63–77.

Rédei, D. and J-F. Tsai. 2016. A revision of *Lamprocoris* (Hemiptera: Heteroptera: Scutelleridae). Entomologica Americana 122(1–2): 262–293.

Ren, D., H-z. Zhu, and Y-q. Lu. 1996. New discovery of early Cretaceous fossil insects from Chifeng City, Inner Mongolia. Acta Geoscientia Sinica 4: 432–438. [in Chinese]

Ren, S-z. 1999. Revision of the subfamily Urostylinae from China (1). Contributions from Tianjin Natural History Museum 16: 7–14. [in Chinese with English summary]

Ren, S-z., and C-s. Lin. 2003. Revision of the Urostylidae of Taiwan, with descriptions of three new species and one new record (Hemiptera-Heteroptera: Urostylidae). Formosan Entomologist 23(2): 129–143.

Requena, G. S., T. M. Nazareth, C. F. Schwertner, and G. Machado. 2010. First cases of exclusive paternal care in stink bugs (Hemiptera: Pentatomidae). Zoologia 27(6): 1018–1021.

Resende, M. L. V. de, and J. L. Bezerra. 1990. Transmissão da murcha de *Phytomonas* a coqueiros por *Lincus lobuliger* (Hemiptera: Pentatomidae). Summa Phytopatologia 16: 27.

Resende, M. L. V. de, R. E. L. Borges, J. L. Bezerra, and D. B. de Oliveira. 1986. Tranmissão da murcha de *Phytomonas* a coqueiros e dendezeiros por *Lincus lobuliger* Breddin, 1908 (Hemiptera, Pentatomidae): Resultados preliminares. Revista Theobroma 16(3): 149–154.

Reuter, O. M. 1887. Ad cognitionem Heteropterorum Madagascariensium I. Gymnocerata. Entomologisk Tidskrift 8: 77–109.

Reuter, O. M. 1912. Bemerkungen über mein neues Heteropterensystem. Öfversigt af Finska Vetenskaps-Societetens Förhandlingar 54(A)(6): 1–62.

Ribes, J., and S. Pagola-Carte. 2013. Hémiptères Pentatomoidea Euro-Méditerranéens 2. Pentatominae (suite). Faune de France. Volume 96. Fédération Française des Sociétés Sciences Naturelles. 421 pp. + 20 pls.

Ribes, J., and G. Schmitz. 1992. Révision du genre *Brachynema* Mulsant & Rey, 1852 (Heteroptera, Pentatomidae, Pentatominae). Bulletin et Annales de la Société Royale Belge d'Entomologie 128: 105–166.

Ribes, J., D. A. Gapon, and S. Pagola-Carte. 2007. On some species of *Carpocoris* Kolenati, 1846: New synonymies (Heteroptera: Pentatomidae: Pentatominae). Mainzer Naturwissenschaftliches Archiv, Beiheft 31: 187–198.

Rider, D. A. 1986. A new species and new synonymy in the genus *Tepa* Rolston and McDonald (Hemiptera: Pentatomidae). Journal of the New York Entomological Society 94(4): 552–558.

Rider, D. A. 1987. A new species of *Acrosternum* Fieber, subgenus *Chinavia* Orian, from Cuba (Hemiptera: Pentatomidae). Journal of the New York Entomological Society 95(2): 298–301.

Rider, D. A. 1988. A new species of *Caribo* Rolston from Puerto Rico (Hemiptera: Pentatomidae). Florida Entomologist 71(1): 8–11.

Rider, D. A. 1990. Review of the New World species of the genus *Neottiglossa* Kirby (Heteroptera: Pentatomidae). Journal of the New York Entomological Society 97(4)[1989]: 394–408.

Rider, D. A. 1992a. *Rhyssocephala*, new genus, with the description of three new species (Heteroptera: Pentatomidae). Journal of the New York Entomological Society 99(4)[1991]: 583–610.

Rider, D. A. 1992b. Revision of *Arocera* Spinola, with the description of two new species (Heteroptera: Pentatomidae). Journal of the New York Entomological Society 100(1): 99–136.

Rider, D. A. 1994. A generic conspectus of the tribe Procleticini Pennington (Heteroptera, Pentatomidae), with the description of *Parodmalea rubella*, new genus and species. Journal of the New York Entomological Society 102(2): 193–221.

Rider, D. A. 1996. Review of the genus *Coenus* Dallas, with the description of *C. explanatus*, new species (Heteroptera: Pentatomidae). Journal of the New York Entomological Society 103(1)[1995]: 39–47.

Rider, D. A. 1997. Rolstoniellini, replacement name proposed for Compastini Distant, 1902, a tribal name based on a generic junior homonym (Heteroptera: Pentatomidae: Pentatominae). Journal of the New York Entomological Society 103(4)[1995]: 401–403.

Rider, D. A. 1998a. Nomenclatural changes in the Pentatomoidea (Hemiptera-Heteroptera: Cydnidae, Pentatomidae). II. Species level changes. Proceedings of the Entomological Society of Washington 100(3): 449–457.

Rider, D. A. 1998b. Nomenclatural changes in the Pentatomoidea (Hemiptera-Heteroptera: Pentatomidae, Tessaratomidae). III. Generic level changes. Proceedings of the Entomological Society of Washington 100(3): 504–510.

Rider, D. A. 2000. Stirotarsinae, new subfamily for *Stirotarsus abnormis* Bergroth (Heteroptera: Pentatomidae). Annals of the Entomological Society of America 93: 802–806.

Rider, D. A. 2006a. Family Pentatomidae, pp. 233–402. *In* B. Aukema and C. Rieger (Eds.), Catalogue of the Heteroptera of the Palaearctic Region. Volume 5. The Netherlands Entomological Society, Amsterdam. xiii + 550 pp.

Rider, D. A. 2006b. Family Tessaratomidae, pp. 182–189. *In* B. Aukema and C. Rieger (Eds.), Catalogue of the Heteroptera of the Palaearctic Region. Volume 5. The Netherlands Entomological Society, Amsterdam. xiii + 550 pp.

Rider, D. A. 2006c. Family Urostylidae, pp. 102–116. *In* B. Aukema and C. Rieger (Eds.), Catalogue of the Heteroptera of the Palaearctic Region. Volume 5. The Netherlands Entomological Society, Amsterdam. xiii + 550 pp.

Rider, D. A. 2006d. Review of the genus *Madates* Strand with the description of three new species (Heteroptera, Pentatomidae). *In* W. Rabitsch (Ed.), Hug the bug - for love of true bugs. Festschrift zum 70. Geburtstag von Ernst Heiss. Denisia 19: 599–610.

Rider, D. A. 2007. *Cachaniellus*, a replacement name for the preoccupied genus *Cachanocoris* (Hemiptera: Heteroptera: Pentatomidae). Proceedings of the Entomological Society of Washington 109(1): 261.

Rider, D. A. 2008. *Massocephalus stysi*, a new species of Pentatomidae (Hemiptera: Heteroptera) from the Philippines. Acta Entomologica Musei Nationalis Pragae 48(2): 583–590.

Rider, D. A. 2012. The Heteroptera (Hemiptera) of North Dakota I: Pentatomomorpha: Pentatomoidea. The Great Lakes Entomologist 45(3–4): 312–380.

Rider, D. A. 2015a. Pentatomoidea Home Page. https://www.ndsu.edu/faculty/rider/Pentatomoidea/

Rider, D. A. 2015b. Pentatomoidea – Genus Index. https://www.ndsu.edu/faculty/rider/Pentatomoidea/Genus _Index/genus_index.htm

Rider, D. A. 2015c. Pentatomoidea – Classification. https://www.ndsu.edu/faculty/rider/Pentatomoidea /Classification/classification.htm

Rider, D. A. 2016. *Aeliavuori linnacostatus*, a new genus and species of Pentatomidae from the Democratic Republic of the Congo (Hemiptera: Heteroptera). Entomologica Americana 122(1–2): 212–219.

Rider, D. A., and H. Brailovsky. 2014. Pentamyrmexini, a new tribe for *Pentamyrmex spinosus*, a remarkable new genus and species of Pentatomidae (Hemiptera: Heteroptera) from Thailand. Zootaxa 3895(4): 595–600.

Rider, D. A., and H. Brailovsky. 2015. Pentamyrmexini, a new tribe for *Pentamyrmex spinosus*, a remarkable new genus and species of Pentatomidae (Hemiptera: Heteroptera) from Thailand. Zootaxa, 3895 (4): 595–600. Erratum. Zootaxa 3919(3): 600.

Rider, D. A., and J. B. Chapin. 1991. Revision of the genus *Thyanta* Stål, 1862 (Heteroptera: Pentatomidae) I. South America. Journal of the New York Entomological Society 99(1): 1–77.

Rider, D. A., and J. B. Chapin. 1992. Revision of the genus *Thyanta* Stål, 1862 (Heteroptera: Pentatomidae) II. North America, Central America, and the West Indies. Journal of the New York Entomological Society 100(1): 42–98.

Rider, D. A., and C. Fischer. 1998. *Zorcadium* Bergroth, an objective junior synonym of *Pseudobebaeus* Fallou [sic] (Heteroptera: Pentatomidae). Entomological News 109(4): 274–276.

Rider, D. A., and P. Kment. 2015. On the correct name of *Coptosoma pygmaeum* Jensen-Haarup, 1926 (Hemiptera: Heteroptera: Plataspidae). Proceedings of the Entomological Society of Washington 117(1): 68–69.

Rider, D. A., and L. H. Rolston. 1986. Three new species of *Acrosternum* Fieber, subgenus *Chinavia* Orian, from Mexico (Hemiptera: Pentatomidae). Journal of the New York Entomological Society 94(3): 416–423.

Rider, D. A., and L. H. Rolston. 1987. Review of the genus *Agroecus* Dallas, with the description of a new species (Hemiptera: Pentatomidae). Journal of the New York Entomological Society 95(3): 428–439.

Rider, D. A., and L. H. Rolston. 1989. Two new species of *Mormidea* from Mexico and Guatemala (Heteroptera: Pentatomidae). Journal of the New York Entomological Society 97(1): 105–110.

Rider, D. A., and L. H. Rolston. 1995. Nomenclatural changes in the Pentatomidae (Hemiptera-Heteroptera). Proceedings of the Entomological Society of Washington 97(4): 845–855.

Rider, D. A., and L-y. Zheng. 2002. Checklist and nomenclatural notes on the Chinese Pentatomidae (Heteroptera) I. Asopinae. Entomotaxonomia 24(2): 107–115.

Rider, D. A., and L-y. Zheng. 2004. Checklist and nomenclatural notes on the Chinese Pentatomidae (Heteroptera) III. Phyllocephalinae, Podopinae. Proceedings of the Entomological Society of Washington 107(1): 90–98.

Rider, D. A., L.-Y. Zheng, and I. M. Kerzhner. 2002. Checklist and nomenclatural notes on the Chinese Pentatomidae (Heteroptera). II. Pentatominae. Zoosystematica Rossica 11(1): 135–153.

Riis, L., and P. Esbjerg. 1998a. Movement, distribution, and survival of *Cyrtomenus bergi* (Hemiptera: Cydnidae) within the soil profile in experimentally simulated horizontal and vertical soil water gradients. Environmental Entomology 27(5): 1175–1181.

Riis, L., and P. Esbjerg. 1998b. Season and soil moisture effect on movement, survival, and distribution of *Cyrtomenus bergi* (Hemiptera: Cydnidae) within the soil profile. Environmental Entomology 27: 1182–1189.

Riis, L., A. C. Bellotti, and B. Arias. 2005a. Bionomics and population growth statistics of *Cyrtomenus bergi* (Hemiptera: Cydnidae) on different host plants. Florida Entomologist 88(1): 1–10.

Riis, L., P. Esbjerg, and A. C. Bellotti. 2005b. Influence of temperature and soil moisture on some population growth parameters of *Cyrtomenus bergi* (Hemiptera: Cydnidae). Florida Entomologist 88(1): 11–22.

Roell, T., and L. A. Campos. 2015. *Candeocoris bistillatus*, new genus and new species of Ochlerini from Ecuador (Hemiptera: Heteroptera: Pentatomidae). Zootaxa 4018(4): 573–583.

Rolston, L. H. 1972. The small *Thyanta* species of North America (Hemiptera: Pentatomidae). Journal of the Georgia Entomological Society 7(4): 278–285.

Rolston, L. H. 1973a. A review of *Hymenarcys* (Hemiptera: Pentatomidae). Journal of the New York Entomological Society 81(2): 111–117.

Rolston, L. H. 1973b. A new South American genus of Pentatomini (Hemiptera: Pentatomidae). Journal of the New York Entomological Society 81(2): 101–110.

Rolston, L. H. 1974. Revision of the genus *Euschistus* in Middle America (Hemiptera, Pentatomidae, Pentatomini). Entomologica Americana 48(1): 1–102.

Rolston, L. H. 1976. A new species and review of *Sibaria* (Hemiptera: Pentatomidae). Journal of the New York Entomological Society 83(4)[1975]: 218–225.

Rolston, L. H. 1978a. A revision of the genus *Odmalea* Bergroth (Hemiptera: Pentatomidae). Journal of the New York Entomological Society 86(1): 20–36.

Rolston, L. H. 1978b. A new subgenus of *Euschistus* (Hemiptera: Pentatomidae). Journal of the New York Entomological Society 86(2): 102–120.

Rolston, L. H. 1978c. A revision of the genus *Mormidea* (Hemiptera: Pentatomidae). Journal of the New York Entomological Society 86(3): 161–219.

Rolston, L. H. 1981. Ochlerini, a new tribe in Discocephalinae (Hemiptera: Pentatomidae). Journal of the New York Entomological Society 89(1): 40–42.

Rolston, L. H. 1982. A revision of *Euschistus* Dallas subgenus *Lycipta* Stål (Hemiptera: Pentatomidae). Proceedings of the Entomological Society of Washington 84(2): 281–296.

Rolston, L. H. 1983a. A revision of the genus *Acrosternum* Fieber, subgenus *Chinavia* Orian, in the Western Hemisphere (Hemiptera: Pentatomidae). Journal of the New York Entomological Society 91(2): 97–176.

Rolston, L. H. 1983b. A redefinition of *Disderia* and addition of a new species (Hemiptera: Pentatomidae). Journal of the New York Entomological Society 91(3): 246–251.

Rolston, L. H. 1983c. A revision of the genus *Lincus* Stål (Hemiptera: Pentatomidae: Discocephalinae: Ochlerini). Journal of the New York Entomological Society 91(1): 1–47.

Rolston, L. H. 1983d. The genus *Paralincus* (Hemiptera: Pentatomidae). Journal of the New York Entomological Society 91(2): 183–187.

Rolston, L. H. 1984a. A revision of the genus *Priapismus* Distant (Hemiptera: Pentatomidae). Journal of the Kansas Entomological Society 57(1): 119–126.

Rolston, L. H. 1984b. *Caribo* Rolston, new genus, pp. 80–81. *In* L. H. Rolston and F. J. D. McDonald, A conspectus of Pentatomini of the Western Hemisphere. Part 3 (Hemiptera: Pentatomidae). Journal of the New York Entomological Society 92(1): 69–86.

Rolston, L. H. 1984c. A review of the genus *Thoreyella* Spinola (Hemiptera: Pentatomidae). Proceedings of the Entomological Society of Washington 86(4): 826–834.

Rolston, L. H. 1985. Key to the males of the nominate subgenus of *Euschistus* in South America, with descriptions of three new species (Hemiptera: Pentatomidae). Journal of the New York Entomological Society 92(4)[1984]: 352–364.

Rolston, L. H. 1986. The genus *Cyptocephala* Berg, 1883 (Hemiptera: Pentatomidae). Journal of the New York Entomological Society 94(3): 424–433.

Rolston, L. H. 1987a. Two new genera and species of Pentatomini from Peru and Brazil (Hemiptera: Pentatomidae). Journal of the New York Entomological Society 95(1): 62–68.

Rolston, L. H. 1987b. Diagnosis of *Epipedus* Spinola and redescription of the type species, *E. histrio* Spinola (Hemiptera: Pentatomidae). Journal of the New York Entomological Society 95(1): 69–72.

Rolston, L. H. 1988a. The genus *Ablaptus* Stål (Pentatomidae: Discocephalinae: Discocephalini). Journal of the New York Entomological Society 96(3): 284–290.

Rolston, L. H. 1988b. The genus *Mathiolus* Distant (Hemiptera: Pentatomidae). Journal of the New York Entomological Society 96(3): 291–298.

Rolston, L. H. 1989. Three new species of *Lincus* (Hemiptera: Pentatomidae) from palms. Journal of the New York Entomological Society 97(3): 271–276.

Rolston, L. H. 1990. Key and diagnoses for the genera of 'broadheaded' discocephalines (Hemiptera: Pentatomidae). Journal of the New York Entomological Society 98(1): 14–31.

Rolston, L. H. 1992. Key and diagnoses for the genera of Ochlerini (Hemiptera: Pentatomidae: Discocephalinae). Journal of the New York Entomological Society 100(1): 1–41.

Rolston, L. H. 1993. A key and diagnoses for males of the *incurvia* species-group of *Antiteuchus* Dallas with descriptions of three new species (Hemiptera: Pentatomidae: Discocephalinae). Journal of the New York Entomological Society 101(1): 108–129.

Rolston, L. H., and R. Kumar. 1974. Two new genera and two new species of Acanthosomatidae (Hemiptera) from South America, with a key to the genera of the Western Hemisphere. Journal of the New York Entomological Society 82(4): 271–278.

Rolston, L. H., and F. J. D. McDonald. 1979. Keys and diagnoses for the families of Western Hemisphere Pentatomoidea, subfamilies of Pentatomidae and tribes of Pentatominae (Hemiptera). Journal of the New York Entomological Society 87(3): 189–207.

Rolston, L. H., and F. J. D. McDonald. 1981. Conspectus of Pentatomini genera of the Western Hemisphere - Part 2 (Hemiptera: Pentatomidae). Journal of the New York Entomological Society 88(4)[1980]: 257–272.

Rolston, L. H., and F. J. D. McDonald. 1984. A conspectus of Pentatomini of the Western Hemisphere. Part 3 (Hemiptera: Pentatomidae). Journal of the New York Entomological Society 92(1): 69–86.

Rolston, L. H., and D. A. Rider. 1986. Two new species of *Stictochilus* Bergroth from Argentina (Hemiptera: Pentatomidae). Journal of the New York Entomological Society 94(1): 78–82.

Rolston, L. H., and D. A. Rider. 1988. *Spinalanx*, a new genus and two new species of Pentatomini from South America (Hemiptera: Pentatomidae). Journal of the New York Entomological Society 96(3): 299–303.

Rolston, L. H., F. J. D. McDonald, and D. B. Thomas, Jr. 1980. A conspectus of Pentatomini genera of the Western Hemisphere. Part I (Hemiptera: Pentatomidae). Journal of the New York Entomological Society 88(2): 120–132.

Rolston, L. H., R. L. Aalbu, M. J. Murray, and D. A. Rider. 1994. A catalog of the Tessaratomidae of the World. Papua New Guinea Journal of Agriculture, Forestry and Fisheries 36(2)[1993]: 36–108.

Rolston, L. H., D. A. Rider, M. J. Murray, and R. L. Aalbu. 1996. Catalog of the Dinidoridae of the World. Papua New Guinea Journal of Agriculture, Forestry and Fisheries 39(1): 22–101.

Roychoudhury, N. and S. Chandra. 2011. *Degonetus serratus* (Distant): a new record of pentatomid bug feeding on teak. Journal of Tropical Forestry 27(4): 31–34.

Ruckes, H. 1947. Notes and keys on the genus *Brochymena* (Pentatomidae, Heteroptera). Entomologica Americana 26(4)[1946]: 143–238.

Ruckes, H. 1961. Notes on the *Mecistorhinus-Antiteuchus* generic complex of discocephaline pentatomids (Heteroptera, Pentatomidae). Journal of the New York Entomological Society 69: 147–156.

Ruckes, H. 1963. Heteroptera. Pentatomoidea. *In* Insects of Micronesia (Bernice P. Bishop Museum, Honolulu, Hawaii) 7(7): 307–390.

Ruckes, H. 1964. The genus *Antiteuchus* Dallas, with descriptions of new species (Heteroptera, Pentatomidae, Discocephalinae). Bulletin of the American Museum of Natural History 127(2): 47–102.

Ruckes, H. 1966a. The genus *Eurystethus* Mayr, with the descriptions of new species (Heteroptera, Pentatomidae, Discocephalinae). American Museum Novitates 2254: 1–37.

Ruckes, H. 1966b. An analysis and a breakdown of the genus *Platycarenus* Fieber (Heteroptera, Pentatomidae, Discocephalinae). American Museum Novitates 2255: 1–42.

Ruckes, H. 1966c. The genus *Dryptocephala* Laporte (Heteroptera, Pentatomidae, Discocephalinae). American Museum Novitates 2256: 1–31.

Ruckes, H. 1966d. A review of the bug genus *Pelidnocoris* Stål (Heteroptera, Pentatomidae, Discocephalinae). American Museum Novitates 2257: 1–8.

Ruckes, H. 1966e. A new *Mecistorhinus* from Ecuador (Heteroptera: Pentatomidae). Journal of the New York Entomological Society 73[1965]: 223–224.

Sailer, R. I. 1944. The genus *Solubea* (Heteroptera: Pentatomidae). Proceedings of the Entomological Society of Washington 46(5): 105–127.

Sailer, R. I. 1950. *Alitocoris*, a new genus of Pentatomidae (Hemiptera). Proceedings of the Entomological Society of Washington 52(2): 69–76.

Sailer, R. I. 1952. A review of the stink bugs of the genus *Mecidea*. Proceedings of the United States National Museum 102(3309): 471–505.

Sales Jr., O., and M. O. Medeiros. 2001. Percevejo castanho da 2. raiz em pastagens. *In* Reunião Sul-Brasileira sobre Pragas de Solo, 8., 2001, Londrina. Anais...Londrina: Embrapa Soja, pp. 71–79. (Embrapa Soja, Documentos 172).

Salini, S. 2016. Redescription of *Dardjilingia* (Hemiptera: Heteroptera: Pentatomidae) from India. Zootaxa 4144: 131–137.

Salini, S., and C. A. Viraktamath. 2015. Genera of Pentatomidae (Hemiptera: Pentatomoidea) from south India – an illustrated key to genera and checklist of species. Zootaxa 3924(1): 1–76.

Salomão, A. T., and J. Vasconcellos-Neto. 2010. Population dynamics and structure of the neotropical bark bug *Phloea subquadrata* (Hemiptera: Phloeidae) on *Plinia cauliflora* (Myrtaceae). Environmental Entomology 39: 1724–1730.

Salomão, A. T., T. C. Postali, and J. Vasconcellos-Neto. 2012. Bichos-cascas na Serra do Japi: história natural dos percevejos Phloeidae (Hemiptera), pp. 321–337. *In* J. Vasconcellos-Neto, P. R. Polli, and A. M. Penteado-Dias (Eds.), Novos olhares, novos saberes sobre a Serra do Japi: ecos de sua biodiversidade. Editora CRV, Curitiba.

Santos, A. V., and G. S. Albuquerque. 2001a. Custos ecofisiológicos do cuidad maternalem *Antiteuchus sepulcralis* (Fabricius) (Hemiptera: Pentatomidae). Neotropical Entomology 30(1): 105–112.

Santos, A. V., and G. S. Albuquerque. 2001b. Eficiência do cuidado maternal de *Antiteuchus sepulcralis* (Fabricius) (Hemiptera: Pentatomidae) contra inimigos naturais do estágio de ovo. Neotropical Entomology 30(4): 641–646.

Santos, B. T. S. dos, A. T. S. Nascimento, and J. A. M. Fernandes. 2014. Proposition of a new species group in *Edessa* Fabricius, 1803 (Hemiptera: Heteroptera: Pentatomidae: Edessinae). Zootaxa 3774(5): 441–459.

Santos, B. T. S. dos, V. J. da Silva, and J. A. M. Fernandes. 2015. Revision of *Ascra* with proposition of the *bifida* species group and description of two new species (Hemiptera: Pentatomidae: Edessinae). Zootaxa 4034(3): 445–470.

Schaefer, C. W. 1972. Degree of metathoracic scent-gland development in the trichophorous Heteroptera (Hemiptera). Annals of the Entomological Society of America 65: 810–821.

Schaefer, C. W. 1977. Genital capsule of the trichophoran male (Hemiptera: Heteroptera: Geocorisae). International Journal of Insect Morphology & Embryology 6(5–6): 277–301.

Schaefer, C. W. 1981a. The sound-producing structures of some primitive Pentatomoidea (Hemiptera: Heteroptera). Journal of the New York Entomological Society 88(4)[1980]: 230–235.

Schaefer, C. W. 1981b. Genital capsules, trichobothroa, and host plants of the Podopinae (Pentatomidae). Annals of the Entomological Society of America 74: 590–601.

Schaefer, C. W. 1983. Host plants and morphology of the Piesmatidae and Podopinae (Hemiptera: Heteroptera): Further notes. Annals of the Entomological Society of America 76: 134–137.

Schaefer, C. W. 1988. The food plants of some "primitive" Pentatomoidea (Hemiptera: Heteroptera). Phytophaga 2: 19–45.

Schaefer, C. W. 1993. Notes on the morphology and family relationships of Lestoniidae (Hemiptera: Heteroptera). Proceedings of the Entomological Society of Washington 95(3): 453–456.

Schaefer, C. W. 1996. Bright bugs and bright beetles: Asopine pentatomids (Hemiptera: Heteroptera) and their prey, pp. 18–56. *In* O. Alomar and R. N. Wiedemann (Eds.), Zoophytophagous Heteroptera: implications for life history and integrated pest management. Proceedings of the Thomas Say Publications in Entomology, Entomological Society of America, Lanham, Maryland. vii + 202 pp.

Schaefer, C. W., and I. Ahmad. 1987. The food plants of four pentatomoid families (Hemiptera: Acanthosomatidae, Tessaratomidae, Urostylidae, and Dinidoridae). Phytophaga 1: 21–34.

Schaefer, C. W., and P. D. Ashlock. 1970. A new genus and new species of Saileriolinae (Hemiptera: Urostylidae). Pacific Insects 12(3): 629–639.

Schaefer, C. W., and W. L. Crepet. 1986. A new burrower bug (Heteroptera: Cydnidae) from the Paleocene/Eocene of Tennessee. Journal of the New York Entomological Society 94(2): 296–300.

Schaefer, C. W., W. R. Dolling, and S. Tachikawa. 1988. The shieldbug genus *Parastrachia* and its position within the Pentatomoidea (Insecta: Hemiptera). Zoological Journal of the Linnean Society 93(4): 283–311.

Schaefer, C. W., L.-Y. Zheng, and S. Tachikawa. 1991. A review of *Parastrachia* (Hemiptera, Cydnidae, Parastrachiinae). Oriental Insects 25: 131–144.

Schaefer, C. W., J. E. O'Donnell, and J. H. Patton. 1996. A review of the Asian species of *Brachycerocoris* Costa (Heteroptera: Pentatomidae: Podopinae). Oriental Insects 30: 203–212.

Schaefer, C. W., A. R. Panizzi, and D. G. James. 2000. Several small pentatomoid families (Cyrtocoridae, Dinidoridae, Eurostylidae [sic], Plataspidae, and Tessaratomidae), pp. 505–512. *In* C. W. Schaefer and A. R. Panizzi (Eds.), Heteroptera of economic importance. CRC Press, Boca Raton, Florida. 828 pp.

Schaefer, C. W., A. R. Panizzi, and M. C. Coscarón. 2005. New records of plants fed upon by the uncommon heteropterans *Cyrtocoris egeris* Packauskas and Schaefer and *C. trigonus* (Germar) (Hemiptera: Cyrtocoridae) in South America. Neotropical Entomology 34(1): 127–129.

Schmidt, L. S., and A. Barcellos. 2007. Abundância e riqueza de espècies de Heteroptera (Hemiptera) do Parque Estadual do Turvo, sul do Brasil: Pentatomoidea. Iheringia (Série Zoologia) 97(1): 73–79.

Schoeman, P. S. 2013. Phytophagous stink bugs (Hemiptera: Pentatomidae; Coreidae) associated with macadamia in South Africa. Open Journal of Animal Sciences 3(3): 179–183.

Schouteden, H. 1903. Faune entomologique de l'Afrique tropicale. Rhynchota Æthiopica. I. Scutellerinæ et Graphosomatinæ. Annales du Musée du Congo, Zoologie (3)1(1): 1–132 + 2 pls.

Schouteden, H. 1904. Heteroptera. Fam. Pentatomidæ. Subfam. Scutellerinae. *In* P. Wytsman (Ed.), Genera Insectorum. Fascicule 24. Bruxelles. 100 pp. + 5 pls.

Schouteden, H. 1905a. Faune entomologique de l'Afrique tropicale. Rhynchota Aethiopica. II Arminae et Tessaratominae. Annales du Musée du Congo, Zoologie (3)1(2): 133–277 + 3 pls.

Schouteden, H. 1905b. Heteroptera, fam. Pentatomidae, subf. Graphosomatinae. *In* P. Wytsman (Ed.), Genera Insectorum. Fascicule 40. Bruxelles. 47 pp. + 3 pls.

Schouteden, H. 1906. Heteroptera. Fam. Pentatomidæ. Subfam. Aphylinae. *In* P. Wytsman (Ed.), Genera Insectorum. Fascicule 47. Bruxelles. 4 pp. + 1 pl.

Schouteden, H. 1907. Family Pentatomidae, Subfamily Asopinae (Amyoteinae). *In* P. Wytsman (Ed.), Genera Insectorum. Fascicule 52. Bruxelles. 82 pp. + 5 pls.

Schouteden, H. 1909. Catalogues raisonnés de la faune entomologique du Congo Belge. I. Hémiptères-Hétéroptères, Fam. Pentatomidæ. Annales du Musée Congo Belge Zoologie (3), Section II, 1(1): 1–88 + 2 pls.

Schouteden, H. 1910. 12. Hemiptera. 6. Pentatomidae, pp. 73–96, In Wissenschaftliche Ergebnisse der schwedischen zoologischen Expedition nach dem Kilimandjaro, dem Meru und den Umgebenden Massaisteppen Deutsch-Ostafrikas 1905–1906 unter Leitung von Prof. Dr. Yngve Sjöstedt. Schweischen Akademie der Wissenschaften. Band 2. Stockholm. 136 pp.

Schouteden, H. 1913. Heteroptera. Fam. Pentatomidæ. Subfam. Dinidorinae. *In* P. Wytsman (Ed.), Genera Insectorum. Fascicle 153. Bruxelles. 19 pp. + 2 pls.

Schouteden, H. 1916. Pentatomiens nouveaux du Congo Belge. Revue Zoologique Africaine 4(3): 298–313.

Schouteden, H. 1951. Un nouveaux genre de Graphosomatinae (Hemipt. Pentatom.). Revue de Zoologie et de Botanique Africaines 44(4): 295–296.

Schuh, R. T., and J. A. Slater. 1995. True bugs of the world (Hemiptera: Heteroptera). Classification and natural history. Cornell University Press, Ithaca, New York. 336 pp.

Schulte, M. J., K. Martin, and J. Sauerborn. 2006. Effects of azadirachtin injection in litchi trees (*Litchi chinensis* Sonn.) on the litchi stink bug (*Tessaratoma papillosa* Drury) in northern Thailand. Journal of Pest Science 79: 241–250.

Schumacher, F. 1912. *Eipeliella*, eine neue Gattung aus dem äthiopischen Gebeit, Vertreter der Tribus der Graphosominen. (Hemiptera, Heteroptera, Pentatomidae). Mitteilungen aus dem Zoologischen Museum in Berlin 6(1): 97–101.

Schwertner, C. F. 2005. Filogenia e classificação dos percevejos-verdes do grupo *Nezara* Amyot & Serville (Hemiptera, Pentatomidae, Pentatominae). Ph.D. dissertation, Universidade Federal do Rio Grande do Sul, Porto Alegre. xv + 246 pp.

Schwertner, C. F., and J. Grazia. 2007. O gênero *Chinavia* Orian (Hemiptera, Pentatomidae, Pentatominae) no Brasil, com chave pictórica para os adultos. Revista Brasileira de Entomologia 51(4): 416–435.

Schwertner, C. F., and J. Grazia. 2012. Review of the neotropical genus *Aleixus* McDonald (Hemiptera: Heteroptera: Pentatomidae: Procleticini), with description of a new species and cladistic analysis of the tribe Procleticini. Entomologica Americana 118(1–4): 252–262.

Schwertner, C. F., and J. Grazia. 2015. Less diverse pentatomoid families (Acanthosomatidae, Canopidae, Dinidoridae, Megarididae, Phloeidae, and Tessaratomidae), pp. 821–862. *In* A. R. Panizzi and J. Grazia (Eds.), True bugs (Heteroptera) of the Neotropics. Springer, Dordrecht, Heidelberg, New York, London. 901 pp.

Schwertner, C. F., and C. Nardi. 2015. Burrower bugs (Cydnidae), pp. 639–680. *In* A. R. Panizzi and J. Grazia (Eds.), True bugs (Heteroptera) of the Neotropics. Springer, Dordrecht, Heidelberg, New York, London. 901 pp.

Schwertner, C. F., J. Grazia, and J. A. M. Fernandes. 2002. Revisão do gênero *Mecocephala* Dallas, 1851 (Heteroptera, Pentatomidae). Revista Brasileira de Entomologia 46(2): 169–184.

Scudder, G. G. E., and D. B. Thomas, Jr. 1987. The green stink bug genus *Chlorochroa* Stål (Hemiptera: Pentatomidae) in Canada. Canadian Entomologist 119(1): 83–93.

Scudder, S. H. 1878. The fossil insects of the Green River Shales. Bulletin of the United States Geological and Geographical Survey of the Territories 4: 747–776.

Scudder, S. H. 1890. The Tertiary insects of North America. Report of the United States Geological Survey of the Territories 13: 1–734.

Seidenstücker, G. 1960. Heteropteren aus Iran 1956, III; *Thaumastella aradoides* Horv., eine Lygaeide ohne Ovipositor. Stuttgarter Beiträge zur Naturkunde 38: 1–4.

Shi, S-l., and Z-h. Xu. 2006. Taxonomy of pest insects of *Pinus yunnanensis*. Journal of Southwest Forestry College 26(1): 35–43.

Siebert, M. W. 1977. Candidates for the biological control of *Solanum elaeagnifolium* Cav, in South Africa 2. Laboratory studies on the biology of *Arvelius albopunctatus* (DeGeer) (Hemiptera: Pentatomidae). Journal of the Entomological Society of South Africa 40: 165–170.

Signoret, V. 1863. Révision des Hémiptères du Chili. Annales de la Société Entomologique de France (4)3: 541–588.

Signoret, V. 1880. Descriptions et des observations sur divers Hémiptères. Bulletin de la Société Entomologique de France 1880(22): 193–194.

Silva, E. J. E., J. A. M. Fernandes, and J. Grazia. 2004. Variações morfológicas em *Edessa rufomarginata* e revalidação de *E. albomarginata* e *E. marginalis* (Heteroptera, Pentatomidae, Edessinae). Iheringia (Série Zoologia) 94(3): 261–268.

Silva, E. J. E., J. A. M. Fernandes, and J. Grazia. 2006. Caracterização do grupo *Edessa rufomarginata* e descrição de sete novas espécies (Heteroptera, Pentatomidae, Edessinae). Iheringia (Série Zoologia) 96(3): 345–362.

Silva, V. J. da, and J. A. M. Fernandes. 2012. A new species group in *Edessa* Fabricius, 1803 (Heteroptera: Pentatomidae: Edessinae). Zootaxa 3313: 12–22.

Silva, V. J. da, B. M. Nunes, and J. A. M. Fernandes. 2013. *Paraedessa*, a new genus of Edessinae (Hemiptera: Heteroptera: Pentatomidae). Zootaxa 3716(3): 395–416.

Simões, F. L., and L. A. Campos. 2014. Taxonomic notes on *Ochlerus*: revisiting Herrich-Schäffer's species (Hemiptera: Pentatomidae: Discocephalinae: Ochlerini). Zootaxa 3774(5): 496–500.

Simões, F. L., and L. A. Campos. 2015. Breddin's types of *Ochlerus* (Hemiptera, Pentatomidae, Discocephalinae). Beiträge zur Entomologie 65(2): 213–222.

Sinclair, D. P. 1989. A cladistic, generic revision of the Oncomeridae Stål n. stat. and Tessaratomidae Schilling n. stat. (Hemiptera: Heteroptera: Pentatomoidea). PhD. dissertation, University of Sydney. 325 pp.

Sinclair, D. P. 2000a. Two new genera of Tessaratomidae (Hemiptera: Heteroptera: Pentatomoidea). Memoirs of the Queensland Museum 46(1): 299–305.

Sinclair, D. P. 2000b. A generic revision of the Oncomerinae (Heteroptera: Pentatomoidea: Tessaratomidae). Memoirs of the Queensland Museum 46(1): 307–329.

Singh, D., R. Kaur, and H. Kaur. 2013. Taxonomic studies on the genus *Halyabbas* Distant and the type-species *Halyabbas unicolor* Distant (Heteroptera-Pentatomidae-Pentatominae-Amyntorini). International Journal of Fauna and Biological Studies 1(2): 16–19.

Singh-Pruthi, H. 1925. The morphology of the male genitalia in Rhynchota. Transactions of the Entomological Society of London 1925: 127–267, pls. VI–XXXII.

Sites, R. W., and J. E. McPherson. 1982. Life history and laboratory rearing of *Sehirus cinctus cinctus* (Hemiptera: Cydnidae), with descriptions of immature stages. Annals of the Entomological Society of America 75: 210–215.

Sites, R. W., K. B. Simpson, and D. L. Wood. 2012. The stink bugs (Hemiptera: Heteroptera: Pentatomidae) of Missouri. The Great Lakes Entomologist 45(3–4): 134–163.

Smee, L. 1965. Insect pests of *Cocos nucifera* in the Territory of Papua New Guinea: their habits and control. Papua and New Guinea Agricultural Journal 17(2): 51–64.

Southwood, T. R. E., and D. J. Hine. 1950. Further notes on the biology of *Sehirus bicolor* (L.) (Hem., Cydnidae). Entomologist's Monthly Magazine 86: 299–301.

Spinola, M. 1850. Tavola sinottica dei generi spettanti alla classe degli insetti artroidignati Hemiptera Linn., Latr. – Rhyngota Fabr. – Rhynchota Burm. Modena, pp. 1–60.

Staddon, B. W. 1992. Specialized cuticular patches in Heteroptera-Pentatomidae (*Piezodorus, Aelia*) recall Slifer's patches in Acrididae. Journal of Natural History 26: 811–821.

Staddon, B. W. 1998. Colour and sternal patch pattern variation in a spring aggregation of *Nezara viridula* L. (Hem., Het., Pentatomidae) from Turkey. Entomologist's Monthly Magazine 134: 267–269.

Staddon, B. W. 2000. Surface-area variability of cryptic sternal patches of *Rhaphigaster nebulosa* (Poda) (Hem.-Het., Pentatomidae) correlated with climate. Entomologist's Monthly Magazine 136: 37–43.

Staddon, B. W., and G. A. Abdollahi. 1999. Comparative morphology and taxonomic indications of the aedeagus in the genus *Aelia* (Heteroptera: Pentatomidae). Acta Societatis Zoologicae Bohemicae 63(1–2): 209–224.

Staddon, B. W., and I. Ahmad. 1994. A further study of the sternal patches of Heteroptera-Pentatomidae with considerations of their function and value for classification. Journal of Natural History 28(2): 353–364.

Staddon, B. W., and I. Ahmad. 1995. Species problems and species groups in the genus *Piezodorus* Fieber (Hemiptera: Pentatomidae). Journal of Natural History 29(3): 787–802.

Staddon, B. W., M. J. Thorne, and I. Ahmad. 1994. Additional data on variation in the specialized cuticular patches of true bugs in the family Pentatomidae (Heteroptera). European Journal of Entomology 91(4): 391–405.

Stål, C. 1854. Nya Hemiptera från Cafferlandet. Öfversigt af Kongliga Vetenskaps-Akademiens Förhandlingar 10(9)[1853]: 209–227.

Stål, C. 1865. Hemiptera Africana. Volume 1. Norstedtiana, Stockholm. iv + 256 pp.

Stål, C. 1868. Bidrag till Hemipterernas systematik. Öfversigt af Kongliga Vetenskaps-Akademiens Förhandlingar 24(7)[1867]: 491–560.

Stål, C. 1870. Enumeratio Hemipterorum. Bidrag till en företeckning öfver alla hittils kända Hemiptera, jemte systematiska meddelanden. 1. Kongliga Svenska Vetenskaps-Akademiens Handlingar 9(1): 1–232.

Stål, C. 1871. Hemiptera insularum Philippinarum. Bidrag till Philippinska öarnes Hemipter-fauna. Öfversigt af Kongliga Vetenskaps-Akademiens Förhandlingar 27(7)[1870]: 607–776, pls. 7–9.

Stål, C. 1872. Enumeratio Hemipterorum. Bidrag till en förteckning öfver alla hittels kända Hemiptera, Jemte Systematiska meddelanden. 2. Kongliga Svenska Vetenskaps-Akademiens Handlingar 10(4): 1–159.

Stål, C. 1873. Enumeratio Hemipterorum. Bidrag till en Förteckning öfver alla hittills kända Hemiptera, Jemte Systematiska meddelanden. 3. Kongliga Svenska Vetenskaps-Akademiens Handlingar 11(2): 1–163.

Stål, C. 1876. Enumeratio Hemipterorum. Bidrag till en Förteckning öfver alla hittills kända Hemiptera, Jemte Systematiska Meddelanden. Kongliga Svenska Vetenskaps-Akademiens Handlingar 14(4): 1–162.

Statz, G., and E. Wagner. 1950. Geocorisae (Landwanzen) aus den Oberoligocäner Ablagerungen von Rott. Palaeontographica (abt. A)98: 97–136.

Stichel, W. 1961. Illustrierte Bestimmungstabellen der Wanzen. II. Europa (Hemiptera-Heteroptera Europae). Volume 4, Heft 1. Pentatomorpha, Aradoidea. W. Stichel, Berlin-Hermsdorf. 838 pp.

Stoner, A., A. M. Metcalfe, and R. A. Weeks. 1974. Plant feeding by a predaceous insect, *Podisus acutissimus*. Environmental Entomology 3: 187–189.

Strickland, C. 1932. Edible and paralysific bugs, one of which [is] a new species *Cyclopelta subhimalayensis* n. sp. (Hemipteron, Heteropteron, Pentatomida, Dinadorina [sic]). Indian Journal of Medical Research 19(3): 873–876.

Štys, P. 1964. Thaumastellidae - A new family of pentatomoid Heteroptera. Acta Societatis Entomologicae Čechosloveniae 61(3): 238–253.

Štys, P. 2000. The enigmatic "arrow-headed bug" from Australia (Heteroptera: Pentatomoidea) and its relationship. Abstract. Internationa Congress of Entomology, Brazil I: 918.

Štys, P., and J. Davidová-Vilímová. 1979. *Coptosoma sandahli* - a third European species of Plataspidae (Heteroptera). Acta Entomologica Bohemoslovaca 76(2): 140–142.

Štys, P., and J. Davidová-Vilímová. 2001. A new genus and species of the Aphylidae (Heteroptera: Pentatomoidea) from Western Australia, and its unique architecture of the abdomen. Acta Societatis Zoologicae Bohemicae 65(2): 105–126.

Suiter, D. R., J. E. Eger, W. A. Gardner, R. C. Kemerait, J. N. All, P. M. Roberts, J. K. Greene, L. M. Ames, G. D. Buntin, T. M. Jenkins, and G. K. Douce. 2010. Discovery and distribution of *Megacospta cribraria* (Hemiptera: Heteroptera: Plataspidae) in northeast Georgia. Journal of Integrated Pest Management 1: 1–4.

Swanson, D. R. 2012. An updated synopsis of the Pentatomoidea (Heteroptera) of Michigan. The Great Lakes Entomologist 45(3–4): 263–311.

Sweet, M. H., and C. W. Schaefer. 2002. Parastrachiinae (Hemiptera: Cydnidae) raised to family level. Annals of the Entomological Society of America 95: 441–448.

Tallamy, D. W., and C. Schaefer. 1997. Maternal care in the Hemiptera: ancestry, alternatives, and current adaptive value, pp. 94–115. *In* J. C. Choe and B. J. Crespi (Eds.), The Evolution of Social Behavior in Insects and Arachnids. Cambridge University Press, Cambridge, England, New York. xiii + 541 pp.

Tamanini, L. 1988. Tabelle per la determinazione dei piú comuni eterotteri Italiani (Heteroptera). Memorie della Società Entomologica Italiana 67(2): 359–471.

Taylor, T. H. C. 1945a. Recent investigations of *Antestia* species in Uganda. Part I. East African Agricultural Journal 10(4): 223–233.

Taylor, T. H. C. 1945b. Recent investigations of *Antestia* species in Uganda. Part II. East African Agricultural Journal 11(1): 47–55.

Théobald, N. 1937. Les insects fossils des terrains oligocenes de France. Mémoires de la Société de Nancy 2: 1–473.

Thomas, D. B., Jr. 1980. A new *Pseudevoplitus* Ruckes from Guatemala with a key to the species (Hemiptera: Pentatomidae). The Pan-Pacific Entomologist 56(4): 293–296.

Thomas, D. B., Jr. 1983. Taxonomic status of the genera *Chlorochroa* Stål, *Rhytidilomia* Stål, *Liodermion* Kirkaldy, and *Pitedia* Reuter, and their included species (Hemiptera: Pentatomidae). Annals of the Entomological Society of America 76: 215–224.

Thomas, D. B., Jr. 1984. A new species of *Dendrocoris* Bergroth from Mexico (Hemiptera: Pentatomidae). The Pan-Pacific Entomologist 60(1): 8–11.

Thomas, D. B., Jr. 1985. Revision of the genus *Chlorocoris* Spinola (Hemiptera: Pentatomidae). Annals of the Entomological Society of America 78: 674–690.

Thomas, D. B., Jr. 1988. Fossil Cydnidae (Heteroptera) from the Oligo-Miocene amber of Chiapas, Mexico. Journal of the New York Entomological Society 96(1): 26–29.

Thomas, D. B., Jr. 1991. The Acanthosomatidae (Heteroptera) of North America. The Pan-Pacific Entomologist 67(3): 159–170.

Thomas, D. B., Jr. 1992a. *Eludocoris*, a new genus of Pentatomidae (Insecta: Heteroptera) from Costa Rica. Annals of Carnegie Museum 61(1): 63–67.

Thomas, D. B., Jr. 1992b. Taxonomic synopsis of the asopine Pentatomidae (Heteroptera) of the Western Hemisphere. Thomas Say Foundation Monographs. Volume 16. Entomological Society of America, Lanham MD. 156 pp.

Thomas, D. B., Jr. 1994a. Fossil Cydnidae (Heteroptera) in the Dominican amber. Journal of the New York Entomological Society 102(3): 303–309.

Thomas, D. B., Jr. 1994b. Taxonomic synopsis of the Old World asopine genera (Heteroptera: Pentatomidae). Insecta Mundi 8: 145–212.

Thomas, D. B., Jr. 1997a. The anocellate, flightless genus *Lojus* McDonald (Heteroptera: Pentatomidae). Annals of the Entomological Society of America 90: 569–574.

Thomas, D. B., Jr. 1997b. A new species of *Spinilanx* [sic] Rolston and Rider from South America (Heteroptera: Pentatomidae). Journal of the New York Entomological Society 103(4)[1995]: 404–408.

Thomas, D. B., Jr. 1998. A new species of *Chlorocoris* (Heteroptera: Pentatomidae) from Jamaica. Florida Entomologist 81(4): 483–488.

Thomas, D. B., Jr. 2001. A new, ocellate species in the genus *Lojus* McDonald (Heteroptera: Pentatomidae). Proceedings of the Entomological Society of Washington 103(4): 854–857.

Thomas, D. B., Jr. 2003. *Hondocoris* Thomas, new genus, pp. 221–222. *In* N. Arismendi and D. B. Thomas, Pentatomidae (Heteroptera) of Honduras: A checklist with description of a new ochlerine genus. Insecta Mundi 17(3–4): 219–236.

Thomas, D. B., Jr. 2012. *Mcphersonarcys*, a new genus for *Pentatoma aequalis* Say (Heteroptera: Pentatomidae). The Great Lakes Entomologist 45(3-4): 127–133.

Thomas, D. B., Jr., and H. Brailovsky. 2000. Review of the genus *Dendrocoris* Bergroth with descriptions of new species (Pentatomidae: Heteroptera). Insecta Mundi 13(1–2)[1999]: 1–9.

Thomas, D. B., Jr., and L. H. Rolston. 1985. A revision of the pentatomine genus *Serdia* Stål, 1860 (Pentatomidae: Hemiptera). Journal of the New York Entomological Society 93(4): 1165–1172.

Thomas, D. B., Jr., and T. R. Yonke. 1981. A review of the Nearctic species of the genus *Banasa* Stål (Hemiptera: Pentatomidae). Journal of the Kansas Entomological Society 54(2): 233–248.

Thomas, D. B., Jr., and T. R. Yonke. 1988. Review of the genus *Banasa* 1860 (Hemiptera: Pentatomidae) for Mexico, Central America, and the Antilles. Annals of the Entomological Society of America 81: 28–49.

Thomas, D. B., Jr., and T. R. Yonke. 1990. Review of the genus *Banasa* (Hemiptera: Pentatomidae) in South America. Annals of the Entomological Society of America 83: 657–688.

Timonin, M. 1958. *Scaptocoris talpa* on roots of banana and other plants in Honduras. FAO Plant Protection Bulletin 6: 74–75.

Tomokuni, M. 2012. A new genus and species of mutualistic Plataspidae (Insecta: Heteroptera) from Sarawak, Malaysia. Memoirs of the National Museum of Natural Science 48: 39–45.

Tsai, J-F., and D. Rédei. 2014. A revision of the genus *Amblycara* (Hemiptera: Heteroptera: Pentatomidae). Acta Entomologica Musei Nationalis Pragae 54(1): 133–155.

Tsai, J-F., and D. Rédei. 2015a. The genus *Acanthosoma* in Taiwan (Hemiptera: Heteroptera: Acanthosomatidae). Acta Entomologica Musei Nationalis Pragae 55(2): 625–664.

Tsai, J-F., and D. Rédei. 2015b. The identity of *Acanthosoma vicinum*, with proposal of a new genus and species level synonymy (Hemiptera: Heteroptera: Acanthosomatidae). Zootaxa 3936(3): 375–386.

Tsai, J-F., and D. Rédei. 2015c. Redefinition of *Acanthosoma* and taxonomic corrections to its included species (Hemiptera: Heteroptera: Acanthosomatidae). Zootaxa 3950(1): 1–60.

Tsai, J-F., D. Rédei, G-f. Yeh, and M-m. Yang. 2011. Jewel bugs of Taiwan (Heteroptera: Scutelleridae). National Chung Hsing University, Taichung, Taiwan. 293 pp.

Tsai, J-F., S-I. Kudo, and K. Yoshizawa. 2015. Maternal care in Acanthosomatinae (Insecta: Heteroptera: Acanthosomatidae) – correlated evolution with morphological change. BMC Evolutionary Biology 15: 1–13.

Tullgren, A. 1918. Zur Morphologie und Systematik der Hemipteren I. I. Über das Vorkommen von s.g. Trichobothrien bei Hemiptera-Heteroptera und ihre mutmassliche Bedeutung für das Heteropterensystem. Entomologisk Tidskrift 39(2): 113–133.

Tuxen, S. L. (Ed.). 1970. Taxonomists Glossary of Genitalia in Insects (2nd edit.). Munksgaard, Copenhagen. 359 pp.

Ueshima, N. 1979. Hemiptera 2: Heteroptera. Animal Cytogenetics 3(6)(ii): 1–117.

Uhler, P. R. 1872. Notices of the Hemiptera of the Western Territories of the United States, chiefly from the surveys of Dr. F. V. Hayden, pp. 392–423. *In* F. V. Hayden (Ed.), Preliminary report of the United States Geological survey of Montana and portions of adjacent territories; being a fifth annual report of progress. Government Printing Office, Washington, D.C. 537 pp.

Uhler, P. R. 1886. Check-list of the Hemiptera Heteroptera of North America. Brooklyn Entomological Society, Henry H. Kahrs, Steam Printer, Brooklyn, N.Y. iv + 32 pp.

Uichanco, L. B. 1949. A revision of the genus *Ectenus* Dallas, with description of a new species (Hemiptera, Pentatomidæ). The Philippine Journal of Science 78(3): 285–289.

Usinger, R. L. 1941. The genus *Oechalia* (Pentatomidae, Hemiptera). Proceedings of the Hawaiian Entomological Society 11(1): 59–93.

Usinger, R. L. 1942. A new species of *Oechalia* from Oahu (Pentatomidae, Hemiptera). Proceedings of the Hawaiian Entomological Society 11(2): 217–218.

van den Berg, M. A., W. P. Steyn, and J. Greenland. 1999a. Hemiptera occurring on macadamia in the Mpumulanga Lowveld of South Africa. African Plant Protection 5: 89–92.

van den Berg, M. A., W. P. Steyn, and J. Greenland. 1999b. Monitoring stink bugs on avocado. Producers can reduce damage to fruit through monitoring and timely control. Neltropika Bulletin 1999(305): 22.

van den Berg, M. A., W. P. Steyn, and J. Greenland. 2000. Hemiptera occurring on avocado trees in the Mpumalanga Lowveld of South Africa. African Plant Protection 6(1): 29–33.

van der Meulen, H. J., and A. S. Schoeman. 1990. Aspects of the phenology and ecology of the antestia stink bug, *Antestiopsis orbitalis orbitalis* (Hemiptera: Pentatomidae), a pest of coffee. Phytophylactica 22(4): 423–426.

van Doesburg, P. H. 1970. Polymorphism in some neotropical Asopinae (Heteroptera, Pentatomidae). Actes IV Congress Latin Zoologie 1: 235–238.

Van Duzee, E. P. 1917. Catalogue of the Hemiptera of America north of Mexico, excepting the Aphididae, Coccidae and Aleurodidae. University of California Publications, Entomology 2. xiv + 902 pp.

Vidal, J. P. 1949. Hémiptères de l'Afrique du Nord et des pays circum méditerranéens. Mémoires de la Société des Sciences Naturelles du Maroc 48: 1–238.

Villiers, A. 1954. Contribution à l'étude du peuplement de la zone d'inondation du Niger (Mission G. Remaudière). X. - Hémiptères Hétéroptères. Bulletin de l'Institut Français d'Afrique Noire 16(1): 219–231.

Vinokurov, N. N., V. B. Golub, E. V. Kanyukova, I. M. Kerzhner, and G. A. Chernova. 1988. Order Heteroptera, pp. 727–930. *In* P. A. Lehr (Ed.), Keys to the insects of the Far East of the USSR. Volume 2. Nauka Publishing House, Leningrad.

Vivan, L. M., C. Nardi, J. Grazia, and J. M. S. Bento. 2013. Description of the immatures of *Scaptocoris carvalhoi* Becker (Hemiptera: Cydnidae). Neotropical Entomology 42(3): 288–292.

Vojdani, S. 1961. The Nearctic species of the genus *Eurygaster* (Hemiptera: Pentatomidae: Scutellerinae). The Pan-Pacific Entomologist 37(2): 97–107.

Vršanský, P., J. A. Lis, J. Schlögl, M. Guldan, T. Mlynský, P. Barna, and P. Štys. 2015. Partially disarticulated new Miocene burrower bug (Hemiptera: Heteroptera: Cydnidae) from Cerová (Slovakia) documents occasional preservation of terrestrial arthropods in deep-marine sediments. European Journal of Entomology 112(4): 844–854.

Waghmare, S. H., G. P. Bhawane, Y. J. Koli, and S. M. Gaikwad. 2015. A case of extensive congregation of man-faced stink bug *Catacanthus incarnatus* (Drury) (Hemiptera: Pentatomidae) together with new host records from western Maharashtra, India. Journal of Threatened Taxa 7(8): 7490–7492.

Wagner, E. 1954. Neue Heteropteren von den Kanarischen Inseln. Commentationes Biologicae 14(2): 1–27.

Wagner, E. 1960. Die paläarktischen Arten der Gattung *Aelia* Fabricius 1803 (Hem. Het. Pentatomidae). Zeitschrift für Angewandte Entomologie 47: 149–195.

Wagner, E. 1966. Wanzen oder Heteropteren I. Pentatomorpha [sic]. *In* F. Dahl (Ed.), Die Tierwelt Deutschlands und der angrenzend en Meeresteile. Volume 54. VEB Gustav Fischer Verlag, Jena. 235 pp.

Waldkircher, G., M. D. Webb, and U. Maschwitz. 2004. Description of a new shieldbug (Heteroptera: Plataspidae) and its close association with a species of ant (Hymenoptera: Formicidae) in Southeast Asia. Tijdschrift voor Entomologie 147: 21–28.

Walker, F. 1867a. Catalogue of the specimens of Hemiptera Heteroptera in the collection of the British Museum. Part I. E. Newman, London. pp. 1–240.

Walker, F. 1867b. Catalogue of the specimens of Hemiptera Heteroptera in the collection of the British Museum. Part II. E. Newman, London. pp. 241–417.

Walker, F. 1868. Catalogue of the specimens of Hemiptera Heteroptera in the collection of the British Museum. Part III: E. Newman, London. pp. 418–599.

Wall, M. A. 2007. A revision of the *Solomonius*-group of the stinkbug tribe Halyini (Hemiptera: Pentatomidae: Pentatominae). Zootaxa 1539: 1–84.

Wang, H.-j., P. Li, Y.-d. Gao, and M. Yuan. 2002. A list of insect pests of bamboos in Baishuijiang Natural Reserve. Journal of Gansu Forestry Science and Technology 27(4): 12–16.

Wang, X.-j., G-q. Liu, and G. Cassis. 2014. Systematic study of *Duadicus* Dallas, 1851 (Insecta: Hemiptera: Heteroptera: Acanthosomatidae: Blaudusinae: Blaudusini), including the description of a new species from Western Australia. Austral Entomology 53: 42–52.

Wang, X.-j., G-q. Liu, and G. Cassis. 2015. Revision of *Panaetius* Stål (Hemiptera: Heteroptera: Acanthosomatidae) from Australia, including the description of two new species and phylogenetic analysis. Austral Entomology 54: 445–464.

Weiler, L., A. Ferrari, and J. Grazia. 2016. Phylogeny and biogeography of the South American subgenus *Euschistus* (*Lycipta*) Stål (Heteroptera: Pentatomidae: Carpocorini). Insect Systematics and Evolution 47: 313–346.

White, A. 1842. Description of some hemipterous insects of the section Heteroptera. Transactions of the Entomological Society of London 3(2): 84–94.

Willis, E. R., and L. M. Roth. 1962. Soil and moisture relations of *Scaptocoris divergens* Froeschner (Hemiptera: Cydnidae). Annals of the Entomological Society of America 55: 21–33.

Wongsiri, N. 1975. A revision of the genus *Scotinophara* Stål (Hemiptera: Pentatomidae) of Southeast Asia. Department of Agriculture, Entomology and Zoology Division, Taxonomy Branch, Technology Bulletin 3: 1–19.

Wu, Y.-z., S-s. Yu, Y-h. Wang, H-y. Wu, X-r. Li, X-y Men, Y-w. Zhang, D. Rédei, Q. Xie, and W-j. Bu. 2016. The evolutionary position of Lestoniidae revealed by molecular autapomorphies in the secondary structure of rRNA besides phylogenetic reconstruction (Insecta: Hemiptera: Heteroptera). Zoological Journal of the Linnean Society 177: 750–763.

Xie, Q., W-j. Bu, and L-y. Zheng. 2005. The Bayesian phylogenetic analysis of the 18S rRNA sequences from the main lineages of Trichophora (Insecta: Heteroptera: Pentatomomorpha). Molecular Phylogenetics and Evolution 34(2): 448–451.

Yamada, Y. 1914. On *Urostylis westwoodii* Scott. Insect World 18: 138–142. [in Japanese]

Yamada, Y. 1915. On *Urostylis striicornis* Scott. Insect World 19: 313–316. [in Japanese]

Yamamoto, A. 2003. A revision of Japanese *Elasmostethus* Fieber (Heteroptera: Acanthosomatidae). Tijdschrift voor Entomologie 146(1): 49–66.

Yang, W-i. 1935. Descriptions of a new family and three new genera of heteropterous insects. Annals and Magazine of Natural History (10)16: 476–482.

Yang, W-i. 1936. Descriptions of two new genera and four new species of Pentatomidae in the collection of Paris Museum. Chinese Journal of Zoology 2: 147–156.

Yang, W-i. 1938. On a new classification of urostylid insects. Bulletin of the Fan Memorial Institute of Biology, Zoology 8(1): 35–48.

Yang, W-i. 1939. A revision of Chinese urostylid insects (Heteroptera). Bulletin of the Fan Memorial Institute of Biology, Zoology 9(1): 5–66.

Yang, W-i. 1962. Economic insect fauna of China. Fascicle 2. Hemiptera Pentatomidae. Academia Sinica, Science Press, Beijing, China. x + 138 pp.

Yao, Y. H. 2006. The biological characteristics and application value of *Aspongopus chinensis*. Journal of Southeast Guizhou National Teacher's College 24: 48–49.

Yao, Y.-z., W-z. Cai, and D. Ren. 2007. The first fossil Cydnidae (Hemiptera: Pentatomoidea) from the late Mesozoic of China. Zootaxa 1388: 59–68.

Yao, Y.-z., W-z. Cai, and D. Ren. 2008. New Jurassic fossil true bugs of the Pachymeridiidae (Hemiptera: Pentatomomorpha) from Northeast China. Acta Geologica Sinica 82(1): 35–47.

Yao, Y.-z., D. Ren, D. A. Rider, and W-z. Cai. 2012. Phylogeny of the infraorder Pentatomomorpha based on fossil and extant morphology, with description of a new fossil family from China. PLoS ONE 7(5): 1–17.

Yao, Y.-z., W-z. Cai, D. A. Rider, and D. Ren. 2013. Primipentatomidae fam. nov. (Hemiptera: Heteroptera: Pentatomomorpha), an extinct insect family from the Cretaceous of north-eastern China. Journal of Systematic Palaeontology 11(1): 63–82.

Zack, R. S., P. J. Landolt, and J. E. Munyaneza. 2012. The stink bugs (Hemiptera: Heteroptera: Pentatomidae) of Washington State. The Great Lakes Entomologist 45(3): 251–262.

Zaidi, R. H. 1993. Redescription of *Cappaea* Ellenrieder (Pentatomidae: Pentatominae: Carpocorini). Pakistan Journal of Scientific and Industrial Research 36(4): 159–161.

Zaidi, R. H. 1994. Cladistic analysis of the tribe Aeliini (Heteroptera: Pentatomidae: Pentatominae) from the Oriental region. Vingnanam Journal of Science 9: 1–14.

Zaidi, R. H. 1995. A revision of the genus *Tolumnia* Stål (Pentatomidae: Pentatominae: Carpocorini) from the Indo-Pakistan sub-continent. Pakistan Journal of Scientific and Industrial Research 37(12)[1994]: 512–515.

Zaidi, R. H. 1996. A revision of the tribe Aeliini (Hemiptera: Pentatomidae: Pentatominae) from the Indo-Pakistan region. Journal of Scientific and Industrial Research Iran 7(4): 228–238.

Zaidi, R. H., and I. Ahmad. 1988. A revision of the genus *Brachynema* Mulsant and Rey (Pentatomidae: Pentatominae: Carpocorini) from Pakistan. Pakistan Journal of Scientific and Industrial Research 31(10): 701–705.

Zaidi, R. H., and I. Ahmad. 1989. Redescription of a little known aeliine genus *Gulielmus* Distant (Hemiptera: Pentatomidae: Pentatominae) with special reference to male and female genitalia. Pakistan Journal of Scientific and Industrial Research 32(2): 95–97.

Zaidi, R. H., and I. Ahmad. 1990. A revision of the genus *Niphe* Stål (Pentatomidae: Pentatominae: Carpocorini) from Indo-Pakistan sub-continent. Pakistan Journal of Scientific and Industrial Research 33(4): 169–173.

Zaidi, R. H., and I. Ahmad. 1991. A revision of the genus *Palomena* Mulsant & Rey (Pentatominae: Carpocorini) from Oriental region. Mitteilungen der Schweizerischen Entomologischen Gesellschaft 64(3–4): 367–376.

Zaidi, R. H., and S. S. Shaukat. 1993. Taxometric studies of five closely related genera of the tribe Carpocorini Stal [sic] (Hemiptera: Pentatominae) from the Indo-Pakistan subcontinent. Bangladesh Journal of Zoology 21(1): 59–66.

Zhang, J-f. 1989. Heteroptera, pp. 71–93. *In* Fossil insects from Shanwang, Shandong. Shandong Scientific and Technological Publishing House, Jinan. 459 pp.

Zhang, S-m. *et al.* 1995. Economic insect fauna of China, Fascicle 50: Hemiptera (2). Science Press, Beijing, China. 169 pp.

Zhang, Y., J. L. Hanula, and S. Horn. 2012. The biology and preliminary host range of *Megacopta cribraria* (Heteroptera: Plataspidae) and its impact on kudzu growth. Environmental Entomology 41: 40–50.

Zhang, Y-e., and Z-l. You. 2002. Effects of Yiqihuayu capsule on the plasma level of 6-keto-PGF1alpha and TXB2 in patients with pelvis congestion syndrome (PCS). Chinese Journal of Information on Traditional Chinese Medicine 9(3): 19–20.

Zhao, Q., W-j. Bu, and G-q. Liu. 2011. The genus *Amyotea* Ellenreider from China (Heteroptera, Pentatomidae). Acta Zootaxonomica Sinica 36(4): 950–955. [in Chinese]

Zhao, Q., G-q. Liu, and W-j. Bu. 2013. A review of the Chinese species of the genus *Picromerus* Amyot and Serville, with description of a new species (Hemiptera: Heteroptera: Pentatomidae: Asopinae). Zootaxa 3613(2): 146–164.

Zheng, L-y. 1981. Chinese species of genus *Arma* Hahn (Hemiptera: Pentatomidae). Natural Enemies of Insects 3(4): 28–32. [in Chinese]

Zheng, L-y. 1986. New taxa of hemipterous insects found on bamboos from China. Journal of Bamboo Research 5(1): 1–7.

Zheng, L-y. 1994. Heteropteran insects (Hemiptera) feeding on bamboos in China. Annals of the Entomological Society of America 87: 91–96.

Zheng, L-y., and Z-p. Ling. 1989. A revision of east Asiatic species of genus *Palomena* (Hemiptera: Pentatomidae). Acta Zootaxonomica Sinica 14(3): 309–326. [in Chinese]

Zheng, L-y., and G-q. Liu. 1987. New genera, new species of Chinese Pentatomidae and a new Chinese record of Scutelleridae (Heteroptera). Acta Zootaxonomica Sinica 12(3): 286–296. [in Chinese]

Zheng, L-y., and H-j. Wang. 1995. Contribution to the taxonomy of *Lindbergicoris* Leston (Hemiptera: Acanthosomatidae). Entomologica Scandinavica 26(1): 17–25.

Zheng, L-y., and H-g. Zou. 1982. Records of heteropterous insects on bamboo from Yunnan. Zoological Research 3 (supplement): 113–120.

Zia, Y. 1957. Tessaratominae of China. Acta Entomologica Sinica 7(4): 423–448.

Zwetsch, A., and J. Grazia. 2001. Revisão do gênero *Runibia* (Heteroptera, Pentatomidae, Pentatomini). Iheringia (Série Zoologia) 91: 5–28.

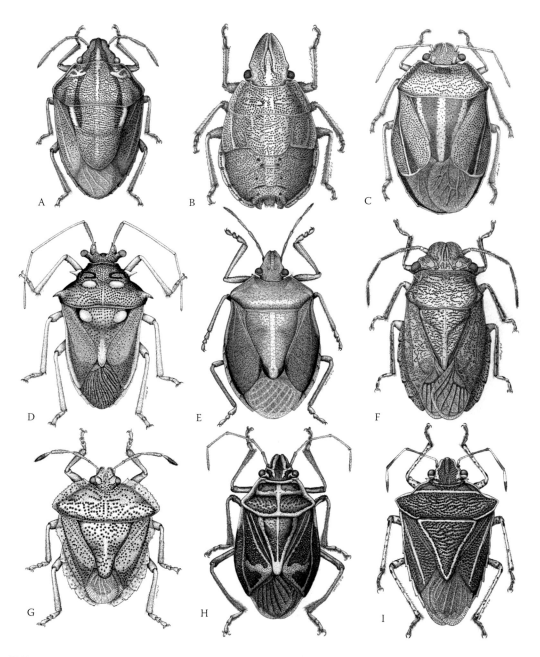

FIGURE 2.12   A, Pentatominae, Carpocorini: *Aeliavuori linnacostatus* Rider; B, Pentatominae, Aeptini: *Aeptus* sp.; C, Pentatominae, Caystrini: *Birna griggae* (McDonald); D, Pentatominae, Diemeniini: *Caridophthalmus* sp.; E, Pentatominae, Nezarini: *Chlorochroa belfragei* (Stål); F, Discocephalinae, Discocephalini: *Colpocarena* n. sp.; G, Discocephalinae, Discocephalini: n. gen., n. sp.; H, Pentatominae, Strachiini: *Madates limbata* (F.); I, Pentatominae, Carpocorini: *Mormidea* n. sp.

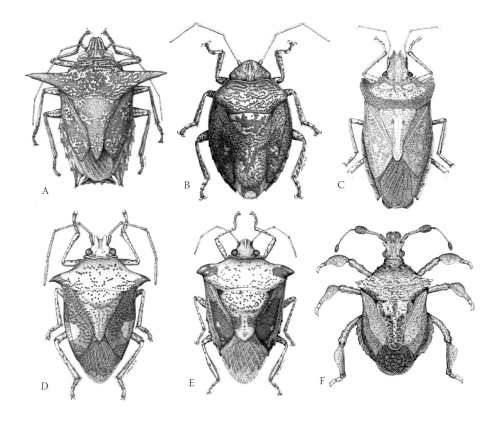

FIGURE 2.13 A, Pentatominae, Pentatomini: *Platencha* n. sp.; B, Discocephalinae, Discocephalini: *Phoeacia* n. sp.; C, Pentatominae, Sephelini: *Ochrorrhaca truncaticornis* Breddin; D, Pentatominae, Rhynchocorini n. gen., n. sp.; E, Pentatominae, Rhynchocorini n. gen., n. sp.; F, Stirotarsinae: *Stirotarsus abnormis* Bergroth.

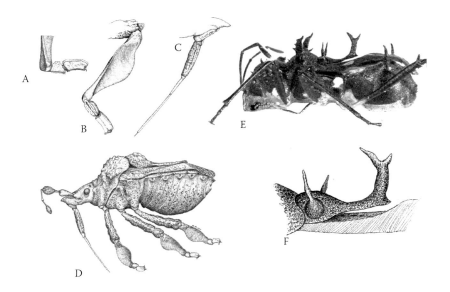

FIGURE 2.14 A, *Stirotarsus abnormis*, tarsi, lateral view; B, *Stirotarsus abnormis*, tibia and tarsi, lateral view; C, *Stirotarsus abnormis*, rostrum, lateral view; D, *Stirotarsus abnormis*, habitus, lateral view; E, *Pentamyrmex spinosus*, habitus, lateral view; F, *Pentamyrmex spinosus*, scutellum, lateral view.

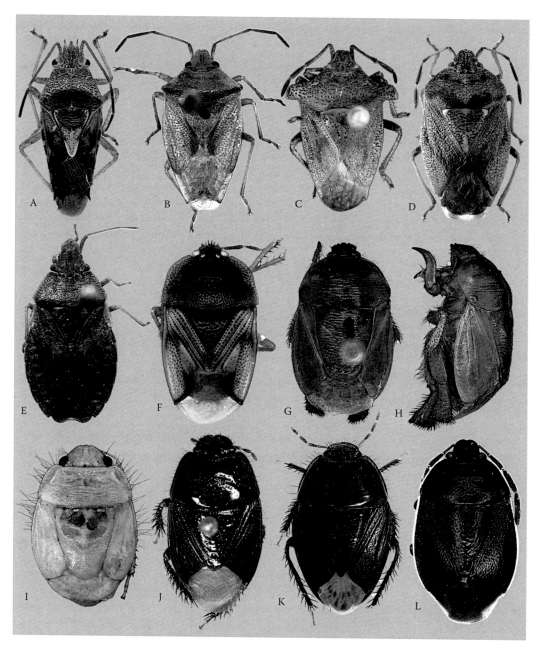

FIGURE 2.15   A, Acanthosomatinae: *Catadipson aper* Breddin, male (Uganda); B, Acanthosomatinae: *Elasmostethus ligatus placidus* (Walker), female (Australia); C, Blaudusinae, Blaudusini: *Bebaeus punctipes* Dallas, female (Venezuela); D, Ditomotarsinae, Ditomotarsini: *Ditomotarsus punctiventris*, female (Chile); E, Ditomotarsinae, Laccophorellini: *Sangarius paradoxus* Stål, female (Australia: New South Wales); F, Amnestinae: *Amnestus championi* Distant, male (Panama); G, Cephalocteinae, Scaptocorini: *Scaptocoris castaneus* Perty, male, dorsal view (Brazil: São Paulo); H, Cephalocteinae, Scaptocorini: *Scaptocoris castaneus* Perty, male, lateral view (Brazil: São Paulo); I, Amaurocorinae: *Linospa candida* (Horváth), male (Uzbekistan); J, Cydninae, Geotomini: *Cyrtomenus mirabilis* (Perty), female (Brazil: São Paulo); K. Cydninae, Geotomini: *Macroscytus brunneus* (F.), male (Croatia); L, Sehirinae: *Sehirus cinctus* (Palisot de Beauvois), male (USA: Indiana).

FIGURE 2.16 A, Canopidae: *Canopus impressus* (F.), female (Brazil); B, Megymeninae, Byrsodepsini: *Byrsodepsus* sp., female (Thailand); C, Megymeninae, Megymenini: *Megymenum brevicorne* (F.), female (China: Guangdong); D, Eumenotini: *Eumenotes obscura* Westwood, female (Malaysia); E, Megarididae: *Megaris* n. sp., male (French Guiana); F, Phloeidae: *Phloeophana longirostris* (Spinola), male (Brazil: São Paulo); G, Plataspidae, *Brachyplatys* group: *Brachyplatys hemisphaericus* (Westwood), female, dorsal view (Madagascar); H, Plataspidae, *Brachyplatys* group: *Brachyplatys hemisphaericus* (Westwood), female, lateral view (Madagascar); I, Plataspidae, *Coptosoma* group: *Coptosoma scutellatum* (Geoffroy), female (Czech Republic); J, Thaumastellidae: *Thaumastella aradoides* Horváth, female (Sudan); K, Natalicolinae: *Encosternum* sp., female (Botswana); L. Oncomerinae: *Piezosternum thunbergi* Stål, female (Brazil: São Paulo).

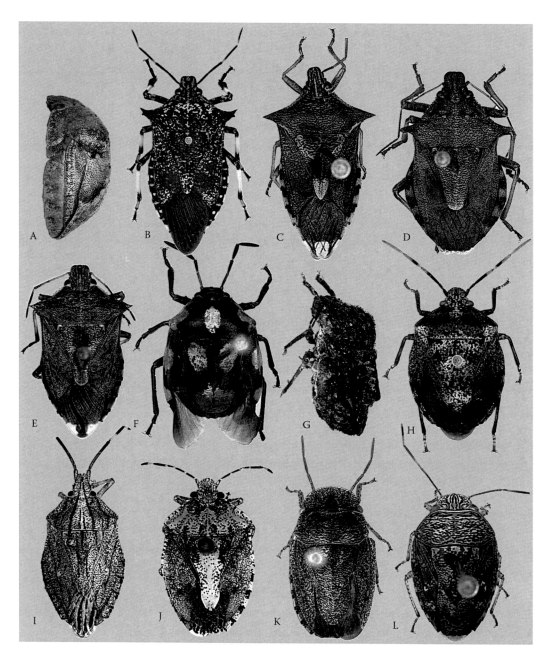

FIGURE 2.17 A, Aphylinae: *Aphylum syntheticum* Bergroth, female, lateral view (Australia); B, Asopinae: *Alcaeorrhynchus grandis* (Dallas), female (Brazil: São Paulo), C, Asopinae: *Oechalia schellenbergi* (Guérin-Méneville), female (Australia: West Australia); D, Asopinae: *Picromerus bidens* (L.), female (South Korea); E, Asopinae: *Podisus maculiventris* (Say), female (USA: Louisiana); F, Asopinae: *Stiretrus erythrocephalus* Lepeletier and Serville (Brazil: São Paulo); G, Cyrtocorinae: *Cyrtocoris gibbus* (F.), female, lateral view (Brazil: São Paulo); H, Discocephalinae, Discocephalini: *Antiteuchus mixtus* (F.), female (Brazil: Mato Grosso); I, Discocephalinae, Discocephalini: *Dinocoris (Praedinocoris) lineatus* (Dallas), female (Brazil: São Paulo); J, Discocephalinae, Discocephalini: *Eurystethus* n. sp., female (Brazil: Amazonas); K, Discocephalinae, Discocephalini: *Ischnopelta luteicornis* (Walker), female (Brazil: Maranhão); L, Discocephalinae, Discocephalini: *Phoeacia* n. sp., male (Brazil: Amazonas).

FIGURE 2.18 A. Discocephalinae, Ochlerini: *Alitocoris schraderi* (Sailer), male (Guatemala); B, Discocephalinae, Ochlerini: *Catulona lucida* Campos and Grazia, male (Brazil: São Paulo); C, Discocephalinae, Ochlerini: *Miopygium cyclopeltoides* Breddin, female (Brazil: São Paulo); D, Discocephalinae, Ochlerini: *Pseudadoxoplatys mendacis* Rolston, female (Brazil: Amazonas); E. Edessinae: *Brachystethus geniculatus* (F.), female (Brazil: São Paulo); F, Edessinae: *Lopadusa quinquedentata* (Spinola), male (Brazil: São Paulo); G, Edessinae: *Peromatus notatus* (Burmeister), female (Brazil: São Paulo); H, Phyllocephalinae, Phyllocephalini: *Diplorhinus furcatus* (Westwood), male (China: Zhejiang); I, Podopinae: *Bolbocoris inaequalis* (Germar), female (South Africa); J, Podopinae: *Deroploa parva* Westwood, male (Australia: New South Wales); K, Podopinae: *Ancyrosoma leucogrammes* (Gmelin), female (Turkey); L, Podopinae: *Amaurochrous dubius* (Palisot de Beauvois), female (USA: Louisiana).

FIGURE 2.19 A, Pentatominae, Aeliini: *Neottiglossa undata* (Say), female (USA: Washington); B, Pentatominae, Agonoscelidini: *Agonoscelis versicoloratus* (Turton), male, (Tanzania); C, Pentatominae, Aeschrocorini: *Tyoma cryptorhyncha* (Germar), male (South Africa); D, Pentatominae, Antestiini: *Antestiopsis anchora* (Thunberg), male, (Thailand); E, Pentatominae, Antestiini: *Eudryadocoris goniodes* (Dallas), male (Ethiopia); F, Pentatominae, Bathycoeliini: *Bathycoelia* sp., male (Ivory Coast); G, Pentatominae, Cappaeini: *Boerias victorini* (Stål), female (South Africa); H, Pentatominae, Cappaeini: *Veterna pugionata* (Stål), female (South Africa); I, Pentatominae, Caystrini: *Cephaloplatus granulatus* Bergroth, male (Australia: South Australia); J, Pentatominae, Carpocorini: *Anaxarchus reyi* (Montrouzier), male (New Caledonia); K, Pentatominae, Carpocorini: *Dichelops (Diceraeus) furcatus* (F.), male (Brazil: São Paulo); L, Pentatominae, Carpocorini: *Dolycoris baccarum* (L.), female (Russia: Yaroslavl).

FIGURE 2.20 A, Pentatominae, Carpocorini: *Euschistus (Euschistus) heros* (F.), male (Brazil: São Paulo); B, Pentatominae, Carpocorini: *Holcogaster exilis* Horváth, female (Iraq); C, Pentatominae, Carpocorini: *Holcostethus limbolarius* (Stål), male (USA: Alabama); D, Pentatominae, Carpocorini: *Mormidea v-luteum* (Lichtenstein), female (Brazil: São Paulo); E, Pentatominae, Carpocorini: *Prionosoma podopioides* Uhler, male (USA: Washington); F, Pentatominae, Carpocorini: *Proxys albopunctulatus* (Palisot de Beauvois), male (Brazil: São Paulo); G, Pentatominae, Carpocorini: *Trichopepla semivitatta* (Say), male (USA: Louisiana); H, Pentatominae, Carpocorini: *Tibraca limbativentris* Stål, female (Brazil: São Paulo); I, Pentatominae, Catacanthini: *Runibia perspicua* (F.), male (Brazil: Rio Grande do Sul); J, Pentatominae, Catacanthini: *Vulsirea violacea* (F.), female (Brazil: São Paulo); K, Pentatominae, Chlorocorini: *Loxa deducta* Walker, female (Brazil: São Paulo); L, Pentatominae, Chlorocorini: *Mayrinia curvidens* (Mayr), female (Brazil: São Paulo).

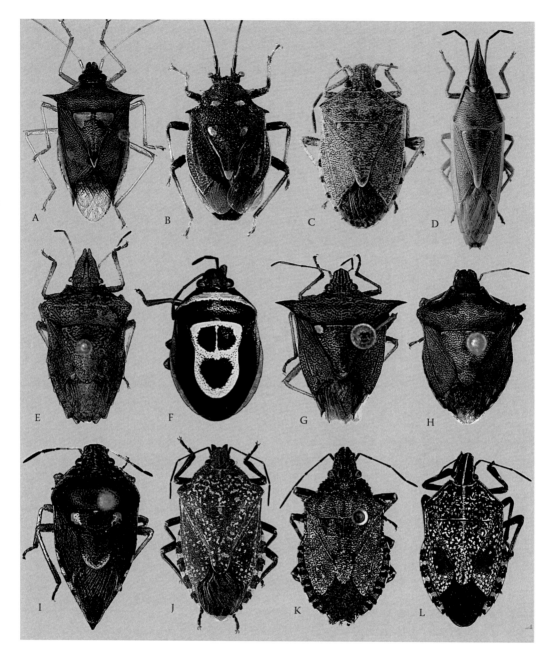

FIGURE 2.21   A, Pentatominae, Chlorocorini: *Rhyncholepta grandicallosa* Bergroth, male (Brazil: Amazonas); B, Pentatominae, Diemeniini: *Caridophthalmus* sp., female (Papua New Guinea); C, Pentatominae, Diemeniini: *Oncocoris favillaceus* (Walker), female (Australia: Northern Territory); D, Pentatominae, Diploxyini: *Acoloba lanceolata* (F.), female (Zambia); E, Pentatominae, Diploxyini: *Diploxys punctata* (Distant), female (Madagascar); F, Pentatominae, Eurysaspini: *Flaminia natalensis* (Dallas), male (South Africa); G, Pentatominae, Eysarcorini: *Aspavia armigera* (F.), male (Sierra Leone); H, Pentatominae, Eysarcorini: *Carbula putoni* (Jakovlev), male (China: Hubei); I, Pentatominae, Eysarcorini: *Stagonomus (Stagonomus) amoenus* (Brullé), female (Uzbekistan); J, Pentatominae, Halyini: *Apodiphus amygdali* (Germar), female (Croatia); K, Pentatominae, Halyini: *Brochymena quadripustulata* (F.), male (USA: Minnesota); L, Pentatominae, Halyini: *Erthesina fullo* (Thunberg), female (China: Hebei).

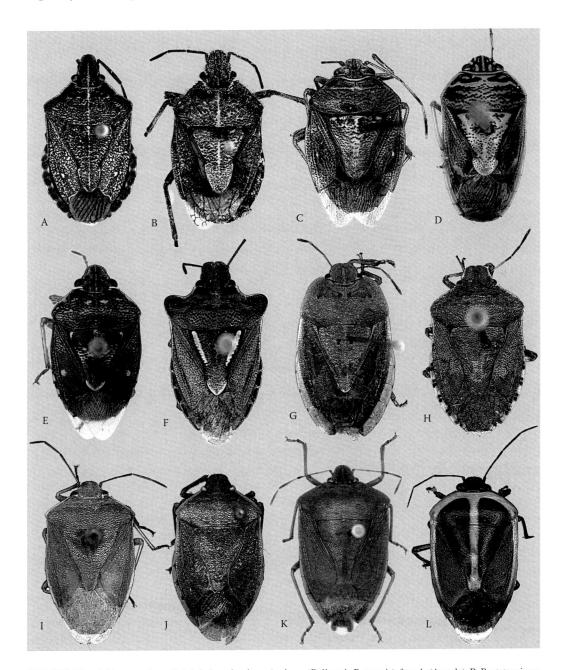

FIGURE 2.22   A, Pentatominae, Halyini: *Pseudatelus spinulosus* (Palisot de Beauvois), female (Angola); B, Pentatominae, Halyini: *Scribonia* sp., female (Mozambique); C, Pentatominae, Menidini: *Elanela hevera* Rolston, female (Surinam); D, Pentatominae, Menidini: *Menida transversa transversa* (Signoret), female (Burkina Faso); E, Pentatominae, Menidini: *Rio punctatus* Fortes and Grazia, female (Brazil: Amazonas); F, Pentatominae, Myrocheini: *Arniscus humeralis* (Dallas), male (Australia: Western Australia); G, Pentatominae, Myrocheini: *Delegorguella ventralis* (Germar), male (Mozambique); H, Pentatominae, Myrocheini: *Dictyotus caenosus* (Westwood), female (Australia: Queensland); I, Pentatominae, Nezarini: *Acrosternum millierei* (Mulsant and Rey), female (Senegal); J, Pentatominae, Nezarini: *Chlorochroa* sp., male (USA: Texas); K, Pentatominae, Nezarini: *Nezara viridula* (L.), male (Brazil: São Paulo); L, Pentatominae, Nezarini: *Roferta marginalis* (Herrich-Schäffer), female (Brazil: São Paulo).

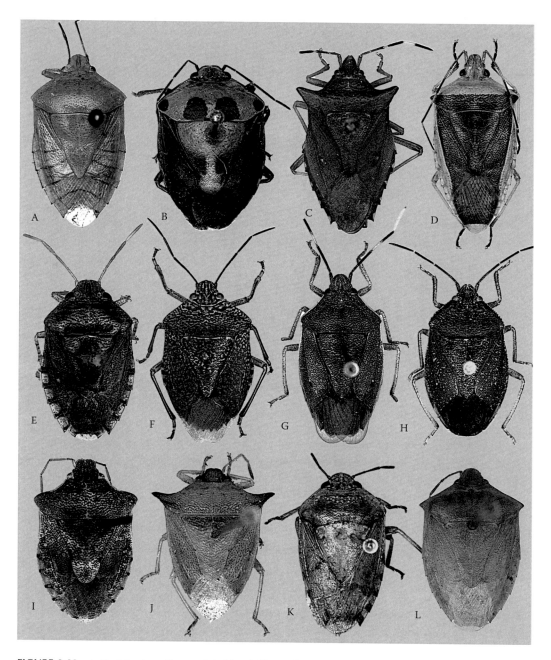

FIGURE 2.23 A, Pentatominae, Pentatomini: *Glaucioides englemani* Thomas, male (Brazil, Pará); B, Pentatominae, Pentatomini: *Hyrmine sexpunctata* (L.), male (Papua New Guinea); C, Pentatominae, Pentatomini: *Myota aerea* (Herrich-Schäffer), male (Brazil: São Paulo); D, Pentatominae, Pentatomini: *Neojurtina typica* Distant, female (China: Yunnan); E, Pentatominae, Pentatomini: *Pallantia macula* (Dallas), female (Brazil: São Paulo); F, Pentatominae, Pentatomini: *Pellaea stictica* (Dallas), female (Brazil: São Paulo); G, Pentatominae, Pentatomini: *Serdia concolor* Ruckes, female (Brazil, Rio Grande do Sul); H, Pentatominae, Pentatomini: *Stictochilus tripunctatus* Bergroth, female (Brazil: São Paulo); I, Pentatominae, Procleticini: *Dendrocoris humeralis* (Uhler), female (USA: Iowa); J, Pentatominae, Procleticini: *Odmalea concolor* (Walker), male (Brazil: Amazonas); K, Pentatominae, Piezodorini: *Pausias pulverosus* Linnavuori, male (Sudan); L, Pentatominae, Rhynchocorini: *Cuspicona simplex* Walker, female (Australia: New South Wales).

FIGURE 2.24 A, Pentatominae, Rhynchocorini: *Everardia picta* Gross, male (Australia: Western Australia); B, Pentatominae, Rhynchocorini: *Ocirrhoe unimaculata* (Westwood), male (Australia: Australian Capital Territory); C, Pentatominae, Sciocorini: *Dyroderes umbraculatus* (F.), female (Turkey); D, Pentatominae, Sciocorini: *Trincavellius galapagoensis* (Butler), female (Peru); E, Pentatominae, unplaced: *Amphidexius suspensus* Bergroth, male (Australia: Western Australia); F, Pentatominae, unplaced: *Ippatha angustilineata* Musgrave, male (Australia: Western Australia); G, Pentatominae, unplaced: *Poecilotoma grandicornis* (Erichson), female (Australia: New South Wales); H, Pentatominae: unplaced: *Vitruvius insignis* Distant, female (China: Yunnan).

FIGURE 2.25 **(See color insert.)** A, Blaudusinae, Lanopini: *Sinopla humeralis* Signoret, female (Chile); B, Cydninae, Cydnini: *Cydnus aterrimus* (Förster), male (Italy); C, Dinidorinae, Dinidorini: *Coridius sanguinolentus* (Westwood), female (China: Guangxi); D, Lestoniidae: *Lestonia haustorifera* China, female (Australia); E, Parastrachiidae: *Parastrachia japonensis*, female (China: Guangxi); F, Parastrachiidae: *Dismegistus fimbriatus* (Thunberg), male, (South Africa); G, Phloeidae: *Phloea corticata* (Drury), male (Brazil); H, Plataspidae: *Libyaspis coccinelloides* (Laporte), male (Madagascar); I, Saileriolidae: *Bannacoris arboreus* Hsiao, male (Thailand); J, Tessaratominae, Tessaratomini: *Pygoplatys tenangau* Magnien, Smets, and Pluot, female (Indonesia: Sumatra); K, Thyreocorinae: *Thyreocoris scarabaeoides* (L.), male (Czech Republic); L, Urostylididae: *Urostylis annulicornis* Scott, male (North Korea).

FIGURE 2.26 (See color insert.) A, Elvisurinae: *Coleotichus borealis* Distant, male (Philippines: Mindanao); B, Scutellerinae, Scutellerini: *Poecilocoris purpurascens* (Westwood), male (India: Sikkim); C, Scutellerinae, Scutellerini: *Calliphara caesar* (Vollenhoven), female (Indonesia: Ambon Is.); D, Scutellerinae, Sphaerocorini: *Steganocerus multipunctatus* (Thunberg), male (South Africa); E, Tectocorinae: *Tectocoris diophthalmus* (Thunberg), male (Indonesia: Selayar Is.); F, Hoteinae: *Deroplax circumducta* (Germar), female (South Africa); G, Pachycorinae: *Crathis ansata* (Distant), female (Panama); H, Pachycorinae: *Agonosoma trilineatum* (F.), female (Curaçao); I, Eurygastrinae: *Polyphyma koenigi* Jakovlev, female (Turkmenistan); J, Odontoscelinae: *Irochrotus sibiricus* Kerzhner, male (China: Inner Mongolia); K, Odontoscelinae: *Morbora australis* Distant, female (Australia: Northern Territory); L, Odontotarsinae: *Urothyreus horvathianus* (Schouteden), female (Namibia).

FIGURE 2.27   (See color insert.) A, Aphylinae: *Aphylum syntheticum* Bergroth, female (Australia); B, Asopinae: *Cazira horvathi* Breddin, male (China: Yunnan); C, Cyrtocorinae: *Cyrtocoris gibbus* (F.), female (Brazil: São Paulo); D, Discocephalinae, Discocephalini: *Colpocarena complanatus* (Burmeister), female (Brazil: Amazonas); E, Discocephalinae: Ochlerini: *Macropygium reticulare* (F.), male (Brazil: São Paulo); F, Edessinae: *Edessa rufomarginata* (DeGeer), female (Brazil: São Paulo); G, Phyllocephalinae, Phyllocephalini: *Basycriptus distinctus* (Signoret), female (Congo); H, Phyllocephalinae, Megarrhamphini: *Megarrhamphus hastatus* (F.), female (China: Sichuan); I, Podopinae, *Podops* group: *Podops inuncta* (F.), male (Czech Republic); J, Podopinae, *Graphosoma* group: *Graphosoma lineatum* (L.), male (Czech Republic); K, Serbaninae: *Serbana borneensis* Distant, male (Borneo); L. Stirotarsinae: *Stirotarsus abnormis* Bergroth, female (Peru).

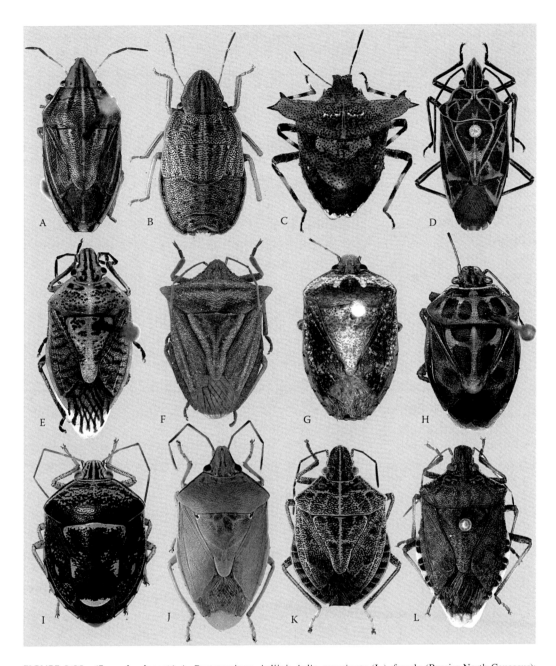

FIGURE 2.28 **(See color insert.)** A, Pentatominae, Aeliini: *Aelia acuminata* (L.), female (Russia: North Caucasus); B, Pentatominae, Aeptini: *Aeptus singularis* Dallas, male (South Africa); C, Pentatominae, Aeschrocorini: *Aeschrocoris obscurus* (Dallas), female (China: Yunnan); D, Pentatominae, Agaeini: *Agaeus mimus* Distant, male (China: Yunnan); E, Pentatominae, Agonoscelidini: *Agonoscelis nubilis* (F.), male (Philippines); F, Pentatominae, Amyntorini: *Belopis unicolor* Distant, female (China: Yunnan); G, Pentatominae, Antestiini: *Antestia trispinosa* Linnavuori, female (Ghana); H, Pentatominae, Antestiini: *Antestiopsis cruciata* (F.), female (Indonesia: Sumatra); I, Pentatominae, Axiagastini: *Axiagastus cambelli* Distant, female (Solomon Islands); J, Pentatominae, Bathycoeliini: *Bathycoelia alkyone* Linnavuori, male (Yemen: Socotra Island); K, Pentatominae, Cappaeini: *Cappaea taprobanensis* (Dallas), male (Vietnam); L, Pentatominae, Cappaeini: *Halyomorpha halys* (Stål), female (South Korea).

FIGURE 2.29   (See color insert.) A, Pentatominae, Carpocorini: *Carpocoris purpureipennis* (DeGeer), male (Czech Republic); B, Pentatominae, Carpocorini: *Menecles insertus* (Say), female (USA: Texas); C, Pentatominae, Catacanthini: *Catacanthus incarnatus* (Drury), male (India: Tamil Nadu); D, Pentatominae, Catacanthini: *Arocera apta* (Walker), female (Brazil: Amazonas); E, Pentatominae, Caystrini: *Caystrus quadrimaculatus* Linnavuori, male (Cameroon); F, Pentatominae, Chlorocorini: *Chlorocoris complanatus* (Guérin-Méneville), male (Brazil: São Paulo); G, Pentatominae, Coquereliini: *Coquerelia pectoralis* (Signoret), female (Madagascar); H, Pentatominae, Degonetini: *Degonetus serratus*, female (India: Kerala); I, Pentatominae, Diemeniini: *Diemenia rubromarginata* (Guérin-Méneville), male (Australia: Australian Capital Territory); J, Pentatominae: Diplostirini: *Diplostira valida* Dallas, female (Bangladesh); K. Pentatominae: Diploxyini: *Diploxys fallax* (Stål), male (Madagascar); L, Pentatominae, Eurysaspini: *Eurysaspis* sp., male (South Africa).

FIGURE 2.30 (See color insert.) A, Pentatominae, Eysarcorini: *Eysarcoris aeneus* (Scopoli), female (Russia: Yaroslavl); B, Pentatominae, Halyini: *Halys sulcatus* (Thunberg), male (India: Karnataka); C, Pentatominae, Halyini: *Atelocera serrata* (F.), male (Cameroon); D, Pentatominae, Hoplistoderini: *Hoplistodera pulchra* Yang, male (China: Fujian); E, Pentatominae, Lestonocorini: *Gynenica funerea* Horváth, female (Kenya); F, Pentatominae, Mecideini: *Mecidea lindbergi* Wagner, male (Iran); G, Pentatominae, Memmiini: *Memmia femoralis* (Signoret), male (Madagascar); H, Pentatominae, Menidini: *Menida violacea* Motschulsky, female (China, Shanxi); I, Pentatominae, Myrocheini: *Myrochea cribrosa* (Klug), female (Senegal); J, Pentatominae, Nealeriini: *Nealeria asopoides* (Stål), male (Madagascar); K, Pentatominae, Nezarini: *Chinavia collis* (Rolston), male (Costa Rica); L, Pentatominae, Nezarini: *Palomena prasina* (L.), male (France).

FIGURE 2.31   (See color insert.) A, Pentatominae, Opsitomini: *Opsitoma brunneus* Cachan, male (Madagascar);
B, Pentatominae, Pentamyrmecini: *Pentamyrmex spinosus* Rider and Brailovsky, female (Thailand); C, Pentatominae,
Pentatomini: *Pentatoma rufipes* (L.), male (Czech Republic); D, Pentatominae, Pentatomini: *Banasa patagiata* (Berg),
female (Argentina: Formosa); E, Pentatominae, Pentatomini: *Evoplitus humeralis* (Westwood), male (Brazil: São Paulo);
F, Pentatominae, Pentatomini: *Lelia octopunctata* (Dallas), male (India: Haryana); G, Pentatominae: Pentatomini: *Pharypia
pulchella* (Drury), male (Venezuela); H, Pentatominae, Pentatomini: *Rhaphigaster nebulosa* (Poda), female (Bulgaria);
I, Pentatominae, Phricodini: *Phricodus bessaci* Villiers, male (Iran); J, Pentatominae, Piezodorini: *Piezodorus lituratus*
(F.), male (Switzerland); K, Pentatominae, Procleticini: *Thoreyella brasiliensis* Spinola, female (Brazil: São Paulo);
L, Pentatominae, Rhynchocorini: *Rhynchocoris humeralis* (Thunberg), female (Myanmar).

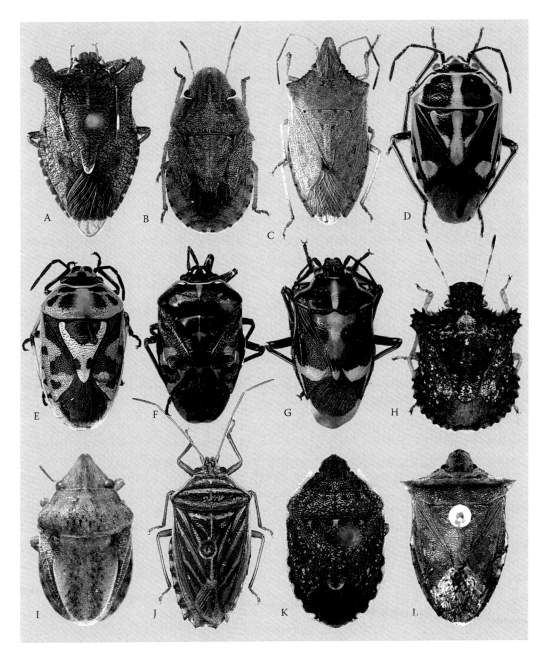

FIGURE 2.32 (See color insert.) A, Pentatominae, Rolstoniellini: *Rolstoniellus boutanicus* (Dallas), female (India: Meghalaya); B, Pentatominae, Sciocorini: *Sciocoris cursitans* (F.), male (Czech Republic); C, Pentatominae, Sephelini: *Brachymna tenuis* (Stål), female (China: Fujian); D, Pentatominae: Strachiini: *Bagrada hilaris* (Burmeister), female (India: Meghalaya); E, Pentatominae, Strachiini: *Eurydema ornata* (L.), female (Italy); F, Pentatominae, Strachiini: *Murgantia histrionica* (Hahn), male (USA: California); G, Pentatominae, Strachiini: *Strachia crucigera* Hahn, female (Thailand); H, Pentatominae, Triplatygini: *Triplatyx bilobatus* Cachan, male (Madagascar); I, Pentatominae, unplaced: *Macrocarenoides scutellatus* (Distant), female (Australia); J, Pentatominae, unplaced: *Patanius vittatus* Rolston, male (Brazil: Amazonas); K, Pentatominae, unplaced: *Trachyops australis* Dallas, female (Australia: Queensland); L, Pentatominae, unplaced: *Thyanta perditor* (F.), female (Brazil: São Paulo).

Section III

Invasive Pentatomoidea

3

Bagrada hilaris (Burmeister)[1]

**C. Scott Bundy, Thomas M. Perring, Darcy A. Reed, John C. Palumbo,
Tessa R. Grasswitz, and Walker A. Jones**

Bagrada hilaris (Burmeister, 1835)[2]

1775	*Cimex pictus* Fabricius, Syst. Ent., 715–716. (Primary junior homonym of *Cimex pictus* Drury, 1770, Coreidae).
1835	*Cimex hilaris* Burmeister, Handb. Ent. 2: 368. (not described, but refers to pl. 34 fig. 237 in Klug, 1845). (Synonymized by Horváth, 1909, Ann. Mus. Natl. Hung. 7: 631).
1837	*Pentatoma picta:* Westwood, Cat. Hope 1: 8.
1838	*Cimex hebraicus* Germar, Silb. Rev. Ent. 5: 177. (Synonymized with *hilaris* by Dallas, 1851, List Hem. Brit. Mus. 1: 259).
1844	*Eurydema picta:* Herrich-Schäffer, Wanz. Ins. 7(5): 83.
1844	*Eurydema hebraicus:* Herrich-Schäffer, Wanz. Ins. 7(5): 83.
1845	*Cimex jucundus* Klug, Symb. Phys. 5: pl. 44 fig. 6. (Synonymized with *hilaris* by Dallas, 1851, List Hem. 1: 259).
1851	*Strachia picta:* Dallas, List Hem. Brit. Mus. 1: 259.
1851	*Strachia hilaris:* Dallas, List Hem. Brit. Mus. 1: 259.
1853	*Pentatoma jucundus:* Herrich-Schäffer, Wanz. Ins. Index 53, 156.
1853	*Eurydema hilaris:* Herrich-Schäffer, Wanz. Ins. Index 93, 94.
1862	*Bagrada picta:* Stål, Stett. Ent. Zeit. 23: 105.
1865	*Bagrada hilaris:* Stål, Hem. Afr. 1: 187–188.
1909	*Bagrada cruciferarum* Kirkaldy, Cat. Hem. 1: 108, 380. (Unnecessary new name for *Cimex pictus* Fabricius, 1775).
1936	*Bagrada picta* var. *connenctens* Horváth, Ann. Mus. Natl. Hung. 30: 29. (Treated as a form after 1960 by Stichel, 1961, Ill. Bestimm. Wanz. 4(48): 757).
1936	*Bagrada picta* var. *modesta* Horváth, Ann. Mus. Natl. Hung. 30: 29. (Treated as a form after 1960 by Stichel, 1961, Ill. Bestimm. Wanz. 4(48): 757).
1961	*Bagrada hilaris* f. *connectens:* Stichel, Ill. Bestimm. Wanz. 4(48): 757. (Unjustified emendation).
1961	*Bagrada hilaris* f. *cruciferarum:* Stichel, Ill. Bestimm. Wanz. 4(48): 757.
1961	*Bagrada hilaris* f. *modesta:* Stichel, Ill. Bestimm. Wanz. 4(48): 757.

CONTENTS

3.1 Introduction 206
3.2 Taxonomy and Identification 207
    3.2.1 Taxonomy 207
    3.2.2 Identification and Comparison with Other Stink Bugs 208
3.3 Distribution 209
    3.3.1 Old World Distribution 209
    3.3.2 New World Distribution 209

---

[1] **Sections 3.4.2, 3.5, 3.6.1.1, 3.6.1.2, and 3.6.1.4** were modified and updated from Biology, ecology, and management of an invasive stink bug, *Bagrada hilaris*, in North America by J. C. Palumbo, T. M. Perring, J. G. Millar, and D. A. Reed, in Annual Review of Entomology 61: 453-473. Copyright 2016. Reprinted with permission.
[2] Synonymy adapted from David A. Rider (personal communication).

3.4   General Biology 210
     3.4.1   Life History 210
     3.4.2   Host Plants..213
3.5   Agricultural Impacts 218
     3.5.1   Feeding and Damage...218
     3.5.2   Old World 219
     3.5.3   North America..220
3.6   Management 222
     3.6.1   Conventional Cropping Systems 222
          3.6.1.1   Sampling/Monitoring 222
          3.6.1.2   Cultural Control 222
          3.6.1.3   Action Thresholds...223
          3.6.1.4   Chemical Control..223
     3.6.2   Organic Cropping Systems...224
          3.6.2.1   Planting Date and Crop Selection..224
          3.6.2.2   Trap Cropping 225
          3.6.2.3   Sanitation (Clean Culture) 225
          3.6.2.4   Crop Isolation...225
          3.6.2.5   Mechanical Control and Exclusion Techniques..226
          3.6.2.6   Organic Insecticides...226
     3.6.3   Natural Enemies 226
          3.6.3.1   Parasitoids 226
          3.6.3.2   Predators 228
          3.6.3.3   Other Natural Enemies 228
          3.6.3.4   Native Natural Enemies 228
          3.6.3.5   Classical Biological Control 229
3.7   Future Outlook...231
3.8   Acknowledgments..231
3.9   References Cited...232

3.1 Introduction

Bagrada hilaris (Burmeister), the painted bug (or bagrada bug), is known throughout the world as a serious pest of vegetable crops, particularly plants in the family Brassicaceae. Vegetable farmers in India and Pakistan have been plagued by *B. hilaris* for over a century (Vekarta and Patel 1999, Sahito et al. 2010, Malik et al. 2012), and more recent problems have occurred throughout the rest of the Middle East, Africa, Australia, southern Europe, and Southeast Asia (Lal and Singh 1993, Guarino et al. 2008, Anonymous 2012).

Bagrada hilaris is among the most recent invasive pentatomoids to become established in North America (**Figure 3.1A**). It first was identified in the New World in 2008 in California (Arakelian 2010, 2011; Garrison 2011) and, since that time, it has spread rapidly throughout the southwestern deserts of California and Arizona, Central and Coastal California, and parts of New Mexico, Arizona, and Texas (Dara 2014, Reed et al. 2013b). It recently has been detected in Hawaii (Matsunaga 2014), northern Mexico (Sánchez-Peña 2014), and Chile (Faúndez et al. 2016). In these regions, this bug has caused (or has the potential to cause) economic losses to large scale cole crops producers and to small farm and organic growers.

The rapid range expansion of *Bagrada hilaris* in the New World, combined with the continuing pest status and expansion in the Old World, has generated considerable research efforts, particularly aimed at mitigating losses in agriculture. Current and recent studies are providing useful information that is being adopted into integrated pest management programs (IPM), yet there still is a heavy reliance on repeated applications of broad-spectrum insecticides in conventional agriculture. Organic producers have limited control strategies, and both organic and conventional growers are changing their growing practices due

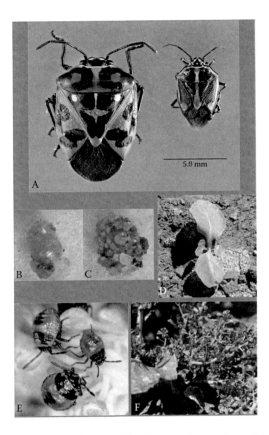

FIGURE 3.1 **(See color insert.)** A, Comparison of adult *Murgantia histrionica* and *Bagrada hilaris* (dorsal view); B, *B. hilaris* egg; C, *B. hilaris* egg covered in soil; D, "Blind" cabbage injury from *B. hilaris* feeding, Yuma, Arizona; E, 2nd and 3rd instar nymphs of *B. hilaris*; F, large cluster of *B. hilaris* nymphs feeding on broccoli seed heads. (Images A-C and D-F: courtesy of C. Scott Bundy and John C. Palumbo, respectively).

to damage by *B. hilaris*. For example, organic growers are altering planting schemes and timing to avoid painted bug infestations, and large farm growers have begun to use transplanted cole crops rather than direct seeding. These altered strategies are more costly to the growers.

The present chapter reviews the literature on *Bagrada hilaris* pertaining to its systematic and taxonomic history and identification, worldwide distribution, basic biology, host range, and impact and management in agricultural systems. Through the compilation of this information, we hope to promote the continued development of methods to reduce the damage caused by this pest.

3.2 Taxonomy and Identification

3.2.1 Taxonomy

Bagrada primarily is an Old World genus with three subgenera and 16 species (Rider 2006). The literature on the painted bug is somewhat confusing due to several alternate names and synonymies for this species. In particular, *B. pictus* and *B. cruciferarum*, synonyms of *B. hilaris*, commonly are found in the Old World literature. *B. pictus* (as *Cimex pictus*), although described prior to *B. hilaris*, is not a valid name because it previously was used to describe a leaf-footed bug (primary junior homonym); *B. cruciferarum* was an unnecessary new name for *B. pictus*. See the beginning of this chapter for details and **Chapter 2** for a review of the relationships among the pentatomoids, including *B. hilaris*. A search of the literature citing the three most common species names for this pest resulted in 42 citations of

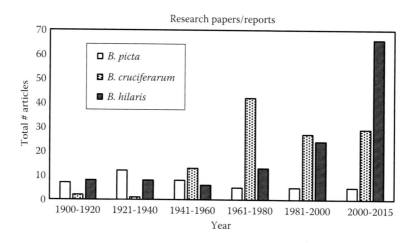

FIGURE 3.2 Citations for the painted bug broken down by species (*Bagrada cruciferarum*, *B. hilaris*, and *B. picta*), 1900 through 2015. (Courtesy of Thomas M. Perring).

B. picta, 114 citations for *B. cruciferarum*, and 125 citations for *B. hilaris*. Although *B. picta* was used more commonly in the older literature, *B. hilaris* and *B. cruciferarum* still are used today, despite the fact that *B. hilaris* is the proper scientific name (**Figure 3.2**).

The common names used for *Bagrada hilaris* also are varied (and sometimes confusing). It has been known as the "mustard bug" (Janisch 1931), the "mustard painted bug" (Gupta and Gupta 1970, Verma 1980), the "colorful bug" (Singh and Gandhi 2012), and the "caper bug" (Sozzi and Vicente 2006). It also has been called the "harlequin bug" (Hill 1975, 2008; Sánchez-Peña 2014), which is the common name reserved for the harlequin bug, *Murgantia histrionica* (Hahn) (McPherson and McPherson 2000). Most often, *B. hilaris* is labelled the painted bug, a name that describes the beautiful "painted-like" coloration of the insect that likely resulted in the species name "*pictus*" (= painted) by Fabricius (1775). More recently, the insect has been called the "bagrada bug" (Halbert and Eger 2010, Palumbo and Natwick 2010, Perring et al. 2013, Reed et al. 2013b). This common name comes from the genus *Bagrada*, which will add confusion if the insect is ever moved to another genus. For this reason, we support the use of "painted bug" for *B. hilaris*.

3.2.2 Identification and Comparison with Other Stink Bugs

Bagrada hilaris with its red, black, orange, and white markings easily is distinguished from most other pentatomids. In North America, it may be confused with *Murgantia histrionica* (see **Chapter 6**). Because both species are pests of cruciferous vegetables, this distinction is important. The most obvious difference between adults of the two species is size: the painted bug is much smaller (4–6 mm) than the harlequin bug (8–11.5 mm) (Derjanschi and Péricart 2005, McPherson 1982, respectively) (**Figures 3.1A and 3.3**). In addition, there are several morphological and color differences between species. The heads of both species are strongly declivent (sloped downward). However, each head is distinctly different (**Figure 3.3**). *B. hilaris* has a much narrower head anterior to its eyes, which are strongly pedunculate; its juga are strongly reflexed at the tip and meet beyond the tylus. *M. histrionica* has a broader head anterior to its eyes, which are more rounded; its juga are not strongly reflexed at the tip and are shorter, not meeting beyond the tylus. Although overall colors are similar (and variability does exist), there are a few distinct patterns that often distinguish the species. The scutellum of *M. histrionica* has an orange or red mediolongitudinal stripe bisected with a short transverse stripe of similar color, giving it a crosslike appearance. The scutellum of *B. hilaris* has a similar longitudinal stripe, but it lacks a transverse stripe (no cross). Further, the pronotum of *M. histrionica* often has a pair of well-developed orange or red spots, lacking or greatly reduced in *B. hilaris*.

FIGURE 3.3 Comparison of the heads of *Murgantia histrionica* and *Bagrada hilaris* (frontal view). (Courtesy of C. Scott Bundy).

3.3 Distribution

3.3.1 Old World Distribution

Bagrada hilaris is distributed widely in the Old World. It is found from Africa and southern Europe east through Pakistan, India, China, and parts of southeast Asia (Rider et al. 2002, Arakelian 2011, Reed et al. 2013a, Taylor et al. 2014, Taylor et al. 2015).

3.3.2 New World Distribution

Bagrada hilaris first was collected in North America from Los Angeles County, California, United States in 2008 (Garrison 2011), spreading quickly throughout southern California (Palumbo and Natwick 2010, Reed et al. 2013b). It rapidly spread east through Arizona (2009), Nevada (2011), Utah (2011), and New Mexico (2010) to Texas (2012) (Bealmear et al. 2012, Bundy et al. 2012, Vitanza 2012, Reed et al. 2013b, Lambert and Dudley 2014, Santa Ana 2015, Taylor et al. 2015) and south into northern Mexico (Sánchez-Peña 2014, Torres-Acosta and Sanchez-Peña 2016). There are reports, thus far, of *B. hilaris* activity from at least 22 counties in California, 5 in Arizona, 10 in New Mexico, 2 in Nevada and 6 in Texas. In 2014, a population also was found in Maui, Hawaii (Matsunaga 2014). In the continental United States, the initial spread of this bug resulted in a devastating attack on cole crop production in the desert valleys of California and Arizona during 2010–2011 (Palumbo 2015a), and now it has been found in agricultural, urban, and wild landscapes as far north as Yolo County (≈100 miles northeast of San Francisco, CA) (Dara 2014) and as far east as central Texas (Santa Ana 2015; Raul T. Villanueva, personal communication) (**Figure 3.4**).

It is likely that *Bagrada hilaris* was introduced accidentally into southern California via shipping containers and, thereby, were disseminated throughout transportation routes in the southwestern United States. These possible modes of introduction and dispersal are supported by an interception of adults of this species on a shipment of castor oil drums from India in 2010 (Gevork Arakelian, personal communication) and by 12 interceptions of these bugs by Florida agricultural inspection stations during 2011–2013 on transported plant material (LeVeen and Hodges 2015). Recent infestations of *B. hilaris* in Hawaii during 2014–2015 (HDOA 2014, Darcy E. Oishi, personal communication) also suggest commerce involvement. Movement and establishment of these bugs is further facilitated by the presence of available food sources, particularly wild and invasive weedy mustards (Suazo et al. 2012) prevalent along transportation routes and other disturbed areas.

It now has been found in South America. Faúndez et al. (2016) reported it from Estero las Cruces, Chile, where they found large numbers of nymphs and adults feeding on *Brassica rapa* L. The locality is near both an international airport and the Pan American Highway, either a possible source of introduction.

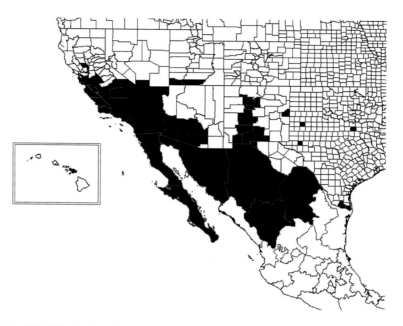

FIGURE 3.4   New World Distribution of *Bagrada hilaris*. (Courtesy of Timothy R. Lewis).

3.4 General Biology

3.4.1 Life History

Many aspects of the biology of *B. hilaris* have been described by researchers in the Old World (e.g., Howard 1906, Gunn 1918, Hutson 1935, Colazza et al. 2004, Deep et al. 2014a, b), particularly from India (e.g., Rakshpal 1949, Mukerji 1958, Atwal 1959, Batra and Sarup 1962, Azim and Shafee 1986, Singh and Malik, 1993, Rohilla et al. 2004, Ghosal et al. 2006). Its general life history is similar to most members of the Pentatomidae but is rather unique with regard to oviposition and mating behavior (Guarino et al. 2008, Taylor et al. 2014), particularly when compared with other stink bugs of North America. Both nymphs and adults exhibit warning coloration and often aggregate on host plants throughout the growing season (Reed et al. 2013b) (**Figure 3.1F**). In general, *B. hilaris* thrives under warm, dry conditions and may be active throughout the year. Life history parameters are linked closely to temperature and food source (Manzar et al. 1999, Rohilla et al. 2004) and when optimal abiotic and biotic conditions are present, this insect can reproduce extremely rapidly with populations often reaching outbreak proportions. Field abundance generally increases with increasing seasonal temperatures (Atwal 1959, Verma et al. 1993, Tiwari and Saravanan 2009, Abrol and Gupta 2010, Huang et al. 2013), but may be affected negatively by rain, irrigation, and high relative humidity (Verma et al. 1993, Nagar et al. 2011, Huang et al. 2013). Both reproduction and feeding injury inflicted upon plants diminishes during temperature extremes. Deep et al. (2014b) reported the bugs could not survive at temperatures below 16°C, and populations decreased at temperatures greater than 40°C.

Bagrada hilaris is multivoltine (Batra 1958, Hill 1975, Taylor et al. 2015), although few studies have evaluated all life stages in the field (see Taylor 2015). Batra and Sarup (1962) speculated from laboratory rearing in India that the bugs could complete as many as 10–12 generations per year. However, this is not supported by data collected from the field. In South Africa, Gunn (1918) reported four generations under caged conditions (host not given) in the field. In India, Pruthi (1946) reported four generations (life stages not given) on toria (field mustard) before the bugs moved to wild hosts. In New Mexico, Taylor et al. (2015) reported two generations with evidence for a partial third on mesa pepperwort (*Lepidium alyssoides* A. Gray), London rocket (*Sisymbrium irio* L.), and broccoli (*Brassica oleracea* L.) whose combined phenologies overlapped and spanned the year. The differences in generations reported between the Old and New Worlds may be due to the bugs' responses to differences in environmental conditions, incomplete evaluations of life stages in some studies, or a combination of these factors.

Many studies on the phenology of *Bagrada hilaris* have focused heavily on adult populations, so immature stages often have not been clearly evaluated. In both the Old and New Worlds, adult bugs generally remain active throughout the year (Batra 1958, Narayanan 1958, Atwal 1959, Sandu 1975, Singh and Malik 1993, Taylor et al. 2015). In India, Singh and Malik (1993) reported two peaks in adult populations, one in October and November during the seedling stages of rapeseed (*Brassica napus* L.) and Indian mustard [*B. juncea* (L.) Czern.] and another in March and April at harvest. Also in India, Narayanan (1958) found adults (and nymphs) most abundant from October through March. He also reported that a few adults survive the summer heat in crevices in embankments near irrigation channels. In California, Garrison (2011) reported adults and nymphs feeding on a variety of cruciferous plants in June and September to November. He also reported that the bugs had spread to the Yuma region of Arizona and were feeding on *Brassica* weeds, seed crops, canola, and cotton. In southern California, Reed et al. (2013b) reported two main population peaks, one from April to May and another from September to October, with adults overwintering in the soil near host plants. In Arizona, Palumbo (2015a) also reported similar peaks (March to April and September to October) in adult populations on hosts such as broccoli and cabbage (*B. oleracea*). In New Mexico, Bundy et al. (2012) reported nymphs and adults from several hosts in the Brassicaceae from April through September in 2010 and 2011, shortly after the bugs became established in the state. More recently (2012–2014), also in New Mexico, Taylor et al. (2015) found adults and nymphs in the field each month of the year and eggs from February through October. They also observed that, in the winter months, nymphs and adults could be found near the bases of senesced host plants in adjacent debris.

Unlike most other stink bugs, *Bagrada hilaris* appears to spend much of its time off its host plant. In India, Batra (1958) reported that the bugs moved from mustard hosts to shelter beneath weeds at night and during the heat of the day. In South Africa, Gunn (1918) reported that nymphs moved from plants in the late afternoon and during the night, hiding between leaves, in debris of fields and gardens, and under soil clumps. In southern California, Reed et al. (2013a) reported that the bugs spend more of their time off the plants in the soil during warm weather. In New Mexico, Taylor et al. (2015) found the bugs commonly on the soil near the bases of host plants throughout the year. In a field study in Arizona, Huang et al. (2013) reported that temperature had a significant effect on populations of adult bugs, with peak abundance occurring during the warmest time of the day. They also observed that environmental factors such as humidity, wind speed, and dew point were not good predictors of adult activity.

Bagrada hilaris has been reared under controlled conditions in the laboratory and the egg and nymphal stages have been described (see Taylor et al. 2015 for a summary). The subcylindrical or barrel-shaped eggs are cream-colored to light brown when deposited, turning dark pink just before hatching (**Figure 3.1B,C**). Eye spots and an egg burster are visible as the eggs reach maturity (Rakshpal 1949, Taylor et al. 2015). Eggs are laid singly or in small clusters (Howard 1906; Pruthi 1946; Reed et al. 2013a, b; Taylor et al. 2014) (See below for details on egg clutch size and ovipositional behavior). The egg incubation period lasts ≈2–6 days at 28–40°C (Atwal 1959, Batra and Sarup 1962, Azim and Shafee 1986, Singh and Malik 1993, Rohilla et al. 2004, Ghosal et al. 2006, Deep et al. 2014a). Taylor et al. (2015) reported an average incubation period of ≈7.5 days at 25°C. Wintertime incubation periods of up to 20 days have been reported (Mukerji 1958, Bhai and Singh 1961). Atwal (1959) and Deep et al. (2014b) found 100% egg mortality at 45°C. First instars hatch, with the aid of an egg burster, through the pseudopercular opening at the cephalic end of the egg (Taylor et al. 2015), a process that may take up to 20 minutes (Rakshpal 1949).

The five nymphal instars are distinguished by distinct differences in morphology and color (Taylor et al. 2015) (**Figure 3.1E**). First and second instars are reddish brown (first instar) to light brown (second instar) with a red abdomen. Taylor et al. (2015) documented two color morphs of the third through fifth instars of this species, with differing levels of melanization. These later instars often take on the bright red, black, orange, and white markings of the adults. Fourth and fifth instars are distinguished by the presence of wing pads, distinctly evident in the fifth instar.

Singh and Malik (1993) reported total nymphal stadia of 14–22 days, 14–20 days, and 29–37 days under temperatures of 25.5–34°C, 28–30°C, and 28°C, respectively. Taylor et al. (2015) reported average total nymphal stadia of ≈34 days at 25°C, with total developmental time from egg to adult averaging ≈42 days. Husain (1925) reported that the total developmental time from egg to adult took less than three weeks at temperatures greater than 35°C. Not surprisingly, the total time required for nymphs under various rearing conditions was variable, taking longer at cooler temperatures or when fed non-brassicaceous plants (Rohilla et al. 2004).

The overall aggregation and feeding behavior of *Bagrada hilaris* nymphs generally is similar to that of other pentatomids. Upon hatching, first instars do not appear to feed but remain closely associated with the empty egg choria for several hours (Atwal 1959). Nymphs and adults may remain aggregated throughout their lifetime but also disperse to locate nearby food sources (Dhiman and Gandhi 1988). Under field conditions, it is common to find all stages together on a single host plant while other nearby plants appear unoccupied and undamaged (Reed et al. 2013b). When disturbed, all stages disperse, either dropping to the ground and/or quickly crawling down the plant to soil level.

As with nymphs (Taylor et al. 2015), there is evidence of color variation among adults. Darcy A. Reed (unpublished data) has observed distinct differences in coloration of the venter of the abdomen. Specifically, the venter is much lighter in color (predominantly cream-colored) in populations found in hotter, drier areas than the venter in populations in colder, wetter areas, where it is predominantly black. Aldrich (1986) has reported that the overall coloration in other pentatomids can be darker when the bugs are reared under cooler temperatures.

The adult stage averages 8–26 days (Gunn 1918, Singh and Malik 1993, Colazza et al. 2004, Ghosal et al. 2006) when feeding on preferred hosts. Females generally have been reported to survive longer than males (Batra and Sarup 1962, Singh and Malik 1993, Ghosal et al. 2006); however, there are reports of males living slightly longer than females (Rakshpal 1949, Verma et al. 1993). Adults have been reported to live as long as 96 days (Hutson 1935), which may be a reflection of poor food quality.

Mating typically occurs within 2–6 days after eclosion (Rakshpal 1949, Batra and Sarup 1962, Azim and Shafee 1986, Verma et al. 1993, Ghosal et al. 2006); however, mating has been reported as early as 1 day after eclosion under certain conditions (Hutson 1935, Batra and Sarup 1962, Singh and Malik 1993).

Guarino et al. (2008) described three phases of courtship/mating behavior. First, upon encountering a female, the male antennates her body ("contact" phase). Second, the male mounts the female and continues to antennate her antennae and abdomen ("mount-antennation" phase). Finally, the male dismounts and initiates copulation by placing his hind legs on the dorsum of her abdomen and backing his abdomen into hers ("engagement" phase). His wings take a position above hers to allow contact between the genitalia of the two bugs. Courtship behavior may last seconds to minutes (Guarino et al. 2008), whereas copulation may be greatly prolonged (several hours). Males tend to remain with females for prolonged periods of time. Azim and Shafee (1986) reported copulation lasting 10–12 hours at a time, with pairs separating when the female prepares to oviposit. Competing males often were found in close proximity to a copulating couple and could be found atop the couple awaiting an opportunity to mate with the female (C. Scott Bundy and Darcy A. Reed, personal observations).

Following mating, there is a preovipositional period of 2–14 days before eggs are deposited (Rakshpal 1949, Batra and Sarup 1962, Azim and Shafee 1986, Singh and Malik 1993, Ghosal et al. 2006). A single female has been reported to deposit up to 19 eggs per day (Rakshpal 1949) and from 36 to 217 eggs during her lifetime (Hutson 1935, Mukerji 1958, Batra and Sarup 1962, Singh and Malik 1993, Verma et al. 1993).

Adults are active and present on plants primarily during the daytime (1300 to 1800 hours) with peak mating activity occurring between 1000 and 1600 hours (Huang et al. 2013). They spend the majority of their time *in copula*, and females exert more feeding damage than males during this time (Huang et al. 2013). Undisturbed copulation may last approximately 12 hours and repetitive and multiple matings with different males are common (Azim and Shafee 1986). Unmated females do not lay eggs and repetitive matings are not necessary for additional egg production (Rakshpal 1949). However, Thomas M. Perring (unpublished data) observed infertile egg production from unmated females within 4 days of adult emergence. He also observed that male replacement or repetitive mating increased number of progeny. The egg laying behavior of *Bagrada hilaris* appears to be extremely unusual for a stink bug and has been described in detail by Taylor et al. (2014). The majority of stink bug species deposit their eggs on host plants in clusters or distinct rows (Esselbaugh 1956, Hinton 1981, McPherson 1982). In the Old World, a few species have been reported to deposit eggs individually or in small clusters on the soil surface or on detritus (Puchkova 1961). Reports on egg laying for *B. hilaris* have been variable. Eggs have been reported to be laid singly or in small clusters on leaves or stems of its host, on the soil surface or detritus, in cracks in the soil (Hutson 1935, Pruthi 1946, Rakshpal 1949, Batra 1958, Ghosal et al. 2006, Arakelian 2010, Halbert and Eger 2010, Perring et al. 2013, Reed et al. 2013b), on artificial materials such as shade cloth or row cover (Darcy A. Reed, personal communication), or beneath the soil surface (Samuel 1942). However, Taylor et al. (2014), in documenting the ovipositional behavior of *B. hilaris*, found that the female actively buries each egg, the first known report of this behavior for a stink bug. Upon finding a suitable ovipositional site, the female anchors herself (positioning phase) to the substrate and pushes her abdomen beneath the soil surface. She then stops moving (stationary phase) as an egg apparently passes through her oviduct. Following, she then begins pivoting her abdomen back and forth and then deposits the egg in the soil (egg deposition phase). Finally, she withdraws her abdomen from the soil (abdomen withdrawal phase) and actively covers the egg with soil using her hind legs (covering phase).

3.4.2 Host Plants

Bagrada hilaris is a polyphagous insect that is known to utilize 96 hosts in 25 families (**Table 3.1**). Although most of its hosts are in the family Brassicaceae ($n = 30$), it also feeds on multiple species in the Asteraceae ($n = 8$), Fabaceae ($n = 11$), and Poaceae ($n = 16$). Because many of its brassicaceous hosts are agricultural plants, it is in this group where the majority of economic damage occurs. Recent studies by Huang et al. (2014b) demonstrated preference of *B. hilaris* for certain brassicaceous plants over others. They found that it was most attracted to radish, *Raphanus sativus* L., followed by red cabbage, *Brassica oleracea* var. capitata, sweet alyssum *Lobularia maritima* L. (Desv.), arugula, *Eruca sativa* Mill., and broccoli, *B. oleracea* var. italica Plenck. Additional work showed a preference for feeding on the younger growth of *Brassica* sp. seedlings. Feeding significantly reduced leaf area, dry weight, and chlorophyll content of all hosts fed upon by this bug (Huang et al. 2014a).

Reed et al. (2013b), in a series of greenhouse studies, showed a number of plants were not utilized as hosts by *Bagrada hilaris*. These plants were distributed across several plant families and included the following: Amaranthaceae – *Chenopodium* sp., *Spinacia oleracea* L.; Apiaceae – *Coriandrum sativum* L, *Senecio vulgaris* L., *Lactuca sativa* L., *Sonchus* sp.; Cucurbitaceae – *Cucumis melo* L., *Cucumis sativus* L, *Cucurbita foetidissima* Kunth, *Cucurbita pepo* L.; Fabaceae – *Lotus corniculatus* L., *Vicia faba* L., *Glycine max* (L.) Merrill; and Solanaceae – *Capsicum annuum* L., *Solanum nigrum* L., *S. lycopersicum* L., *Nicotiana glauca* Graham. Despite their findings, some of these plants (e.g., *C. melo*, *C. annuum*, and *S. lycopersicum*) have been reported as hosts by other authors (see **Table 3.1**). Although these results show that *B. hilaris* will infest a wide range of hosts, and even cause economic damage, we believe these situations are due to the bugs being in close proximity to more preferred brassicaceous hosts. When the non-brassica hosts *Parthenium hysterophorus*, *Pluchea lanceolata*, *Ziziphus zuzuba*, *Chenopodium album*, and *Calotropis procera* were grown near mustard, *B. hilaris* was found in high numbers. However, after the mustard was harvested, bug populations declined (Patidar et al. 2013b).

These non-brassicaceous plants also may be "bridging hosts," used by *Bagrada hilaris* to bridge the gap between the end of one season of a brassicaceous crop and the beginning of the next. This was the case for *B. hilaris* infestations in wheat, invading only after nearby crucifers had been harvested (Cheema et al. 1973). *B. hilaris* also has been observed feeding on corn as a "bridge" between spring-harvested and fall-planted cole crops in the southwestern United States. Dissections of insects collected from the corn revealed depleted fat bodies, an indication that the bugs were not obtaining adequate nutrients from that plant (Thomas M. Perring and Darcy A. Reed, personal observations).

TABLE 3.1

Plants Associated with *Bagrada hilaris*

Plant Family	Scientific Name	Common Name	Reference	Location
Amaranthaceae (Chenopodiaceae)				
	Beta vulgaris L.	Sugarbeet	Gunn 1918	South Africa
	Chenopodium album L.	Lamb's quarters	Singh and Malik 1993	India
	Spinacia oleracea L.	Spinach	Palumbo and Natwick 2010	USA (Arizona, California)
Annonaceae				
	Asimina triloba (L.) Dunal	Pawpaw	Gunn 1918	South Africa
Anacardiaceae				
	Mangifera indica L.	Mango	Tandon and Lal 1977	India
Apiaceae				
	Daucus carota (Hoffm.) Schübl and G. Martens	Carrot	Gunn 1918	South Africa
Asteraceae				
	Barbarea sp.	Winter cress	Gunn 1918	South Africa
	Bidens pilosa L.	Black-jack	Gunn 1918	South Africa
	Chrysanthemum sp. L.	Chrysanthemum	Gunn 1918	South Africa
	Cynara scolymus L.	Artichoke	Gunn 1918	South Africa
	Dahlia sp.	Dahlia	Gunn 1918	South Africa
	Geigeria alata (Benth. and Hook.)	Gud-gad	Linnavuori 1975	Sudan
	Lactuca sativa L.	Lettuce	Gunn 1918	South Africa
	Schkuhria pinnata (Lam.) Kuntze ex Thell.	Pinnate false threadleaf	Gunn 1918	South Africa
	Sonchus arvensis L.	Field sow thistle, Corn sow thistle	Singh and Malik 1993	India
Brassicaceae				
	Brassica juncea (L.) Vassiliĭ Matveievitch Czernajew	Indian mustard, Mustard greens, Raya	Bakhetia 1986, Huang et al. 2014b, Hutson 1935, Reed et al. 2013b	India, Sri Lanka, USA (Arizona, California)
	Brassica oleracea Plenk (Italica group)	Broccoli	Boopathi and Pathak 2012; Huang et al. 2014a, b; Palumbo 2015a; Reed et al. 2013b; Taylor et al. 2015	India, USA (Arizona, California)
	Brassica oleracea D.C. (Acephala group)	Kale	Huang et al. 2014a, b; Obopile et al. 2008; Reed et al. 2013b	Botswana, USA (Arizona, California)
	Brassica oleracea L. (Botrytis group)	Cauliflower, Collards	Huang et al. 2014a, b; Hutson 1935; Reed et al. 2013b; Shivalingaswamy and Satpathy 2007; Verma et al. 1993	India, Sri Lanka, USA (Arizona, California)

(Continued)

TABLE 3.1 (CONTINUED)

Plants Associated with *Bagrada hilaris*

Plant Family	Scientific Name	Common Name	Reference	Location
	Brassica oleracea L. (Capitata group)	Cabbage	Azim and Shafee 1986; Gunn 1918; Huang et al. 2014a, b; Obopile et al. 2008; Palumbo 2015a; Reed et al. 2013b; Shivalingaswamy and Satpathy 2007	Botswana, India, South Africa, USA (Arizona, California)
	Brassica oleracea L. (Gemmifera group)	Brussels sprouts	Bealmear et al. 2012	USA (Arizona)
	Brassica oleracea L. (Gongylodes group)	Kohlrabi, Knol khol	Rakshpal 1949	India
	Brassica napobrassica (L.) Mill	Rutabaga	Palumbo and Natwick 2010	USA (Arizona)
	Brassica napus L., *Brassica rapa* L. subsp. *oleifera*, *Brassica campestris* L.	Canola, Field Mustard, Rapeseed, Sarson	Bakhetia and Labana 1978, Bakhetia and Sekhan 1989, Bhati et al. 2015, Chattopadhyay et al. 2011, Cheema et al. 1973, Deep et al. 2014b, Gunn 1918, Mari and Lohar 2010, Obopile et al. 2008, Sachan and Purwar 2007, Sahito et al. 2010, Singh and Malik 1993, Singh et al. 2012	Botswana, India, Pakistan, South Africa
	Brassica rapa subsp. *chinensis* (L.) Hanelt	Bok choy, Pak choy	Grasswitz (unpublished data)	USA (New Mexico)
	Brassica rapa subsp. *narinosa* (Bailey) Hanelt	Tat-soi	Grasswitz (unpublished data)	USA (New Mexico)
	Brassica rapa subsp. *nipposinica* (Bailey) Hanelt	Mizuna	Grasswitz (unpublished data)	USA (New Mexico)
	Brassica rapa L. var *pekinensis*	Chinese Cabbage, Napa Cabbage	Huang et al. 2014b, Palumbo and Natwick 2010	USA (Arizona)
	Brassica rapa L. var *rapa*	Turnip	Gunn 1918	South Africa
	Capsella bursa-pastoris (L.) Medikus	Shepherd's purse	Reed et al. 2013b	USA (California)
	Descurainia sophia(L.) (= *Sisymbrium sophia* L.)	Flixweed, Herb Sophia, Tansy Mustard	Cheema et al. 1973	Pakistan
	Eruca sativa Miller	Arugula, Taramira	Ahuja and Joshi 1995, Huang et al. 2014b, Mahar 1974, Mari and Lohar 2010, Reed et al. 2013b	Pakistan, India, USA (Arizona, California)
	Hirschfeldia incana (L.) Lagrèze-Fossat	Shortpod Mustard	Lambert and Dudley 2014, Reed et al. 2013b	USA (California)
	Iberis sp. L.	Candytuft	Gunn 1918, Reed et al. 2013b	South Africa, USA (California)
	Lepidium alyssoides A. Gray	Mesa peperwort	Bundy et al. 2012, Taylor et al. 2015	USA (New Mexico)
	Lepidium capense Thunberg	Cape pepper-grass	Gunn 1918	South Africa
	Lepidium latifolium L.	Pepperweed	Bundy et al 2012, Lambert and Dudly 2014	USA (California, New Mexico)

(Continued)

TABLE 3.1 (CONTINUED)

Plants Associated with *Bagrada hilaris*

Plant Family	Scientific Name	Common Name	Reference	Location
	Lobularia maritima (L.) Desvaux	Sweet alyssum	Gunn 1918, Huang et al. 2014b, Reed et al. 2013b	South Africa, USA (Arizona, California, New Mexico)
	Matthiola sp. Aiton	Stock	Gunn 1918, Huang et al. 2014b, Reed et al. 2013b	South Africa, USA (Arizona, California)
	Raphanus raphanistrum L.	Wild radish, Jointed charlock	Gunn 1918, Makwali et al. 2002	Kenya, South Africa
	Raphanus sativus L.	Radish	Bath and Singh 1989; Cheema et al. 1973; Fletcher 1921; Gunn 1918; Huang et al. 2014a, b; Mahar 1974	India, Pakistan, South Africa, USA (Arizona)
	Sisymbrium irio L.	London rocket	Reed et al. 2013b, Taylor et al. 2015	USA (California)
	Sisymbrium capense Thunberg	Cape mustard	Gunn 1918	South Africa
	Tropaeolum majus L. (= *Nasturtium indicum*)	Nasturtium	Gunn 1918	South Africa
Cannabaceae				
	Cannabis sativa L.	Hemp	Cheema et al. 1973	Pakistan
Caricaceae				
	Carica papaya L.	Papaya	Daiber 1991	South Africa
Capparaceae				
	Capparis spinosa L.	Caper	Ahuja et al. 2008, Colazza et al. 2004, Guarino et al. 2007, Infantino et al. 2007, Mahar 1974, Sozzi and Vicente 2006	Pakistan, India, Italy
Convolvulaceae				
	Convolvulus arvensis L.	Field bindweed	Singh and Malik 1993	India
Cucurbitaceae				
	Citrullus lanatus (Thunberg)	Watermelon	Palumbo and Natwick 2010	USA (Arizona, California)
	Cucumis melo L. subsp. *melo* var. *cantalupo*	Cantaloupe	Palumbo and Natwick 2010	USA (Arizona, California)
	Momordica dioica Roxburgh ex. Willdenow	Spiny gourd, Kantola	Singh et al. 2009	India
Euphorbiaceae				
	Euphorbia hirta L.	Pill-pod spurge	Singh and Malik 1993	India
	Ricinus communis L.	Castor oil plant	Cheema et al. 1973	Pakistan
Fabaceae				
	Indigofera sp. L.	Indigo	Narayanan 1954, Patidar et al. 2013a, Stebbing 1903	India
	Medicago sativa L.	Alfalfa	Cheema et al. 1973	Pakistan
	Phaseolus lunatus L.	Lima bean	Reed et al. 2013b	USA (California)

(Continued)

TABLE 3.1 (CONTINUED)

Plants Associated with *Bagrada hilaris*

Plant Family	Scientific Name	Common Name	Reference	Location
	Phaseolus vulgaris L.	Snap bean	Reed et al. 2013b	USA (California)
	Pisum sativum L.	Pea	Gunn 1918, Tomar et al. 2004	South Africa
	Robinia pseudoacacia L.	Black locust	Sharma et al. 2008	India
	Trifolium alexandrinum L.	Egyptian clover	Cheema et al. 1973	Pakistan
	Vicia sp.	Vetch	Reed et al. 2013b	USA (California)
	Vigna mungo (L.) Hepper	Black gram, Black lentil	Dhuri and Singh 1983, Gupta and Gupta 1970	India
	Vigna radiata (L.) R. Wilczek	Green gram	Gupta and Gupta 1970	India
	Vigna unguiculata (L.) Walpers	Cowpea	Reed et al. 2013b	USA (California)
Malvaceae				
	Abelmoschus esculentus (L.) Moench	Okra	Sabyasachi et al. 2013, Singh and Joshi 2004	India
	Alcea sp.	Hollyhock	Gunn 1918	South Africa
	Gossypium hirsutum L.	Cotton	Harris 1937, Reed et al. 2013b	Tanzania, USA (California)
Moraceae				
	Morus alba L.	Mulberry	Sharma 2013	India
Poaceae				
	Avena sativa L.	Oats	Gunn 1918	South Africa
	Cynodon dactylon (L.) Persoon	Bermuda grass	Cheema et al. 1973	Pakistan
	Cyperus rotundus L.	Nutgrass	Singh and Malik 1993	India
	Dactyloctenum aegyptium (L.) Willdenow	Egyptian crowfoot	Sandhu 1975	India
	Hordeum vulgare L.	Barley	Gunn 1918	South Africa
	Oryza sativa L.	Rice	Patidar et al. 2013a, Rakshpal 1949	India
	Panicum miliaceum L., *Pennisetum glaucum* (L.) R. Brown, *Pennisetum typhoides* (Burm.)	Millet	Gupta and Gupta 1970, Sandhu 1975, Sandhu et al. 1974, Srivastava and Srivastava 2012, Verma 1980	India
	Pennisetum glaucum (L.) R. Br.	Pearl millet	Gupta and Gupta 1970	India
	Saccharum spontaneum L.	Kans grass (kahi)	Sandhu 1975	India
	Saccharum officinarum L.	Sugarcane	Isaac 1946, Narayanan 1954, Patidar et al. 2013a	India
	Sorghum bicolor var. *sudanense* Piper	Sudan grass	Cheema et al. 1973, Reed et al. 2013b	Pakistan, USA (California)
	Sorghum halepense (L.) Persoon	Johnson grass	Sandhu 1975	India
	Sorghum vulgare (L.) Persoon	Sorghum	Sandhu 1975	India

(Continued)

TABLE 3.1 (CONTINUED)

Plants Associated with *Bagrada hilaris*

Plant Family	Scientific Name	Common Name	Reference	Location
	Triticum aestivum L.	Wheat	Aalbersberg et al. 1989, Cheema et al. 1973, Gunn 1918, Rawat and Singh 1980, Sandhu and Deol 1976	India, Pakistan, South Africa
	Zea mays L.	Corn, Maize	Cheema et al. 1973, Halbert and Eger 2010, Narayanan et al. 1959, Reed et al. 2013b, Rizvi et al. 1986, Sandhu 1975	India, Pakistan, USA (California)
Polygonaceae				
	Polygonum plebeium R. Brown	Knotweed	Singh and Malik 1993	India
Rhamnaceae				
	Ziziphus rotundifolia (Burm. f.)	Wild jujube	Singh and Malik 1993	India
Rubiaceae				
	Coffea arabica L.	Coffee	Anderson 1917, Patidar et al. 2013a, Rakshpal 1949	India, Kenya
Rutaceae				
	Citrus sp. L.	Citrus	Gunn 1918	South Africa
Solanaceae				
	Capsicum annuum L.	Pepper	Dara 2012	USA (California)
	Physalis peruviana L.	Cape Gooseberry	Gunn 1918	South Africa
	Solanum lycopersicum L.	Tomato	Singh et al. 2011a	India
	Solanum tuberosum L.	Potato	Dharpure 2002a, b; Hill 1975	India
Theaceae				
	Camellia sinensis (L.) Kuntze	Tea	Nadda et al. 2013	India
Tropaeolaceae				
	Nasturtium integrifolium (Nuttall) Kuntze	Nasturtium	Gunn 1918	South Africa
Zygophyllaceae				
	Tribulus terrestris L.	Puncturevine	Sandhu 1975	India

3.5 Agricultural Impacts

3.5.1 Feeding and Damage

Typical feeding damage by *Bagrada hilaris* on brassicaceous plant foliage can be characterized by the presence of circular chlorotic patches or spots on leaf surfaces (**Figures 3.5 and 3.6**). Initially, these damaged areas appear as pale green, irregular-shaped spots (1–2 mm in diameter) around the point where the stylets were inserted into the leaf tissue immediately following feeding (John C. Palumbo, personal observation). After 12–24 hours, the spots became chlorotic and resemble circular or starbust

FIGURE 3.5   Fresh feeding lesions of *Bagrada hilaris* on a broccoli leaf, Yuma, Arizona. (Courtesy of John C. Palumbo).

FIGURE 3.6   Cosmetic feeding injury of *Bagrada hilaris* to kale, Yuma, Arizona. (Courtesy of John C. Palumbo).

shaped scorched areas on the leaf surface (Infantino et al. 2007, Palumbo and Natwick 2010, Reed et al. 2013b). These chlorotic spots eventually may become necrotic and lead to the desiccation and death of the plant tissue. Depending on the maturity of the plant at the time of feeding, these leaves may become distorted, resulting in malformations in leaf growth (Infantino et al. 2007, Sachan and Purwar 2007, Palumbo and Natwick 2010, Reed et al. 2013b, Sánchez-Peña 2014). Feeding also results in the removal of cell sap from plant tissues as reported on numerous brassicaceous host crops (Nyabuga 2008, Banuelos et al. 2013, Reed et al. 2013b), and non-brassicaceous host crops (Rawat and Singh 1980; Dharpure 2002a,b; Infantino et al. 2007).

Field observations of injury to cole crops in North America by *Bagrada hilaris* have suggested that the bugs feed preferentially on young, tender leaf tissue (Palumbo and Natwick 2010, Reed et al. 2013b, Sánchez-Peña 2014). Injury to cotyledons can result in rapid plant desiccation and death (Lal and Singh 1993, Makwali et al. 2002, Sachan and Purwar 2007, Hill 2008, Nyabuga 2008). Laboratory studies on broccoli confirmed that *B. hilaris* adults feed significantly more on newly emerged terminal leaves than on older, expanded true leaves (Huang et al. 2014a). There also is evidence that adult feeding differs between sexes of *B. hilaris*; females feed for longer durations and cause more feeding injury to broccoli plants than males (Huang et al. 2013).

3.5.2 Old World

Bagrada hilaris is considered one of the major important pests of oilseed brassica crops (i.e., Indian mustard, rape, and canola) in the Old World (Sachan and Purwar 2007, Abrol 2009). Oilseed brassica crops are most susceptible to feeding damage at the seedling stage and at seed maturity/harvest stage (Sachan and Purwar 2007, Singh et al. 2007, Abrol 2009). Attacks at the seedling stages can cause desiccation of terminal growing points, and feeding on foliage during the season can impact plant growth and maturity (Makwali et al. 2002, Ahuja et al. 2008, Banuelos et al. 2013). *B. hilaris* will feed readily on the developing seed and seed pods of oil-seed brassicaceous plants, which eventually can restrict the plant's overall reproductive capacity (Verma et al. 1993, Rajpoot, et al. 1996, Malik et al. 2012, Banuelos et al. 2013). Feeding on developing pods results in desiccation of seed pods and reduced oil content of the seeds (Rajpoot et al. 1996). Adults also will excrete a black, sticky resinous material (likely feces [John C. Palumbo, personal observations]) that can spoil the seed pods prior to harvest (Rajpoot, et al. 1996, Banuelos et al. 2013).

Feeding damage by *Bagrada hilaris* on other Old World host crops varies depending on crop type. *B. hilaris* recently became a major pest of caper plants, *Capparis spinosa* L., in Italy, where its feeding on the developing buds, stems, and foliage results in hollowed out plant parts, chlorotic spots, feeding punctures, and leaf deformations (Colazza et al. 2004, Infantino et al. 2007). When *B. hilaris* first was reported as a pest on potato, characteristic wilting of foliage on apical branches was attributed to the sap feeding by adults and nymphs (Dharpure 2002b). Rizvi et al. (1986) reported that maize plants infested with *B. hilaris* were wilted and stunted relative to non-infested plants. Feeding by *B. hilaris* on the leaves, spikelets, and stems of wheat resulted in characteristic white blotches on the tissue, and plants that were severely damaged, wilted, and collapsed (Rawat and Singh 1980). Feeding on pearl millet caused damage to emerging heads (Gahukart and Jotwani 1980). Although considered only an occasional pest of cotton, *B. hilaris* has been observed feeding on immature, green cotton bolls (Harris 1937).

Estimates of the impact of damage by *Bagrada hilaris* in crop losses in many cropping systems in the Old World have been subjective. For example, Colazza et al. (2004) indicated that feeding by this bug on caper plants is capable of causing significant economic crop losses. Nyabuga (2008) suggested that the supply and quality of kale available in many parts of Kenya was low during the dry season due in part to *B. hilaris* injury. Growers of rape, cabbage, and kale in Botswana considered *B. hilaris* the most serious constraint to production because of its potential to cause crop losses if not adequately controlled (Obopile et al. 2008). A few studies have provided quantitative estimates of yield reductions and crop losses associated with *B. hilaris* infestations. Joshi et al. (1989) reported that in unprotected Indian mustard crops, growers suffered reductions in seed yield as high as 70% due to *B. hilaris* damage. Similarly, Ahuja et al. (2008) reported that mustard crops experienced 32% plant mortality and 37% reduction in seed yield when *B. hilaris* was not controlled. Broccoli infested with *B. hilaris* during the vegetative and head formation growth stages in the Eastern Hill Region of India was reported to suffer damage from 20–50% (Boopathi and Pathak 2012). Studies on kale showed that increases in *B. hilaris* population density lead to corresponding increases in cosmetic damage to leaves and reductions in yield and quality (Nyabuga 2008).

3.5.3 North America

In the United States, greater than 90% of the nation's fresh-market brassicaceous crops are grown commercially in the agricultural valleys of Arizona and California. When the first outbreaks of *Bagrada hilaris* were reported in these growing locations in 2009, adults and nymphs were observed feeding on direct-seeded and transplanted broccoli, cabbage, and cauliflower cultivars that produce marketable crowns or heads as well as a number of leafy and root cultivars such as kale, radish, rutabaga, collards, arugula, turnip, and mustard crops used for salads (Palumbo and Natwick 2010). Yield losses on these crops occurred primarily from reductions in plant stand where seedling mortality exceeded 60% in some broccoli fields during crop establishment (Reed et al. 2013b). Subsequent feeding by the pest on established stands caused losses in yield and quality due to damaged terminal growing points, severely stunted plants, as well as malformed or multi-headed plants, which are not commercially marketable (Palumbo and Natwick 2010, Reed et al. 2013b, Sánchez-Peña 2014). Estimates by local growers indicated that in 2009, injury from *B. hilaris* caused 20–25% losses in cauliflower and broccoli, respectively, due to harvest delays, poor quality, and unacceptable product (blind plants or multiple crowns) (Reed et al. 2013b). Growers further estimated that from 2010 to 2014, *B. hilaris* infested more than 80% of the acreage in Arizona and southern California, resulting in excess of 10% stand losses and plant injury in direct-seeded broccoli crops (Palumbo et al. 2015).

Seedling brassicaceous crops especially are susceptible to direct feeding damage by *Bagrada hilaris* on cotyledons, newly emerged leaves, and apical meristems (Reed et al. 2013b). In Arizona and California, adults commonly move into newly established fields and begin feeding on the cotyledons and stems of direct-seeded plants immediately following emergence of the hypocotyl (Palumbo and Natwick 2010, Huang et al. 2013). Feeding on newly emerged cotyledons of brassicaceous host crops in the field can result in rapid plant desiccation and death (Palumbo and Natwick 2010). In general, larger plants with more leaf area are less susceptible to *B. hilaris* feeding injury than small seedlings. Under caged laboratory conditions, seedling (cotyledon stage) mortality in response to *B. hilaris* feeding occurred within a few hours on several brassicaceous crop species (Huang et al. 2014a). Huang et al. (2014b) showed that young brassica seedling plants (i.e., cotyledon stage) were significantly more susceptible to plant death than larger 4-leaf stage plants. Laboratory studies further demonstrated that feeding damage caused by *B. hilaris* significantly reduced leaf growth, chlorophyll content, and dry weights of broccoli, red and green cabbage, and cauliflower (Huang et al. 2014b).

In leafy, non head-forming brassicaceous plants in which the leaves are produced for fresh markets (i.e., kale, arugula, collards), feeding injury similarly can reduce plant growth on young plants (Nyabuga 2008; Huang et al. 2014a,b) and affect cosmetic quality throughout the growing season due to the chlorotic and necrotic aesthetic damage to leaves at harvest (Palumbo and Natwick 2010, Reed et al. 2013b). In head-forming brassica crops, feeding injury or death to the terminal growing point will lead to plant deformations such as adventitious bud break (plants produce multiple unmarketable heads) or 'blind' plants (no head produced) (**Figures 3.1D, 3.7, and 3.8**) (Palumbo and Natwick, 2010, Reed et al. 2013b, Sánchez-Peña 2014). Field studies in Arizona have shown that broccoli and cauliflower plants are

FIGURE 3.7   Blind cauliflower injury caused by feeding of *Bagrada hilaris*, Yuma, Arizona. (Courtesy of John C. Palumbo).

FIGURE 3.8   Unmarketable multiterminal broccoli caused by feeding of *Bagrada hilaris*, Yuma, Arizona. (Courtesy of John C. Palumbo).

susceptible to feeding injury by *Bagrada hilaris* up to the sixth leaf node stage. However, beyond this point, plants are capable of withstanding feeding injury without any significant loss in head production (John C. Palumbo, unpublished data).

3.6 Management

3.6.1 Conventional Cropping Systems

3.6.1.1 Sampling/Monitoring

Reliable sampling plans or monitoring tools for *Bagrada hilaris* in commercial cole crops have not been developed as yet. However, field observations suggest that sampling for the presence of *B. hilaris* in seedling cole crops should be conducted from mid-morning to late afternoon when ambient temperatures are warm and adults are most abundant on plants (Huang et al. 2013). University of Arizona Cooperative Extension recommendations suggest monitoring begin immediately upon seedling emergence or transplanting (Palumbo 2014). Recent research has shown a strong association between the proportion of plants with fresh feeding injury and population density of *B. hilaris*, indicating that fresh feeding damage will provide an accurate measure of *B. hilaris* abundance in commercial fields (Palumbo and Carrière 2015). Consequently, growers and consultants are encouraged to sample plants for fresh feeding signs on cotyledons and young leaves when monitoring for the presence of *B. hilaris* (Palumbo 2014).

Chemical attractants that might be used as lures in monitoring traps, such as those described by Joseph (2014), also are under investigation. For example, chemical analyses indicate that adults of both sexes of *Bagrada hilaris* release qualitatively similar volatile blends (consisting of nonanal, decanal, and (*E*)-2-octenyl acetate), but that males produce ≈2.8 times as much of (*E*)-2-octenyl acetate as do females (Guarino et al. 2008). Furthermore, in olfactometer tests, females responded positively to the odors both of live males and hexane extracts of males, but not to the odor of other females. This suggests that the elevated levels of (*E*)-2-octenyl acetate produced by males are involved in long-distance attraction of adult females to males (Guarino et al. 2008); this finding may have value in developing a pheromone-based monitoring tool. Similarly, volatile compounds derived from brassicaceous plants also are being examined for use as attractants in trapping *B. hilaris*. However, no consistently effective plant-derived attractants have been identified (Jocelyn G. Millar, personal communication).

3.6.1.2 Cultural Control

Cultural control tactics are discussed more fully in **Section 3.6.2** (Organic Cropping Systems). However, many of these tactics (e.g., trap cropping and alterations in planting date) are equally relevant to developing integrated management strategies in conventional cropping systems.

Other modifications in farming practices help prevent populations of *Bagrada hilaris* from reaching damaging levels, particularly on oilseed mustards. For example, it has been shown that nitrogen fertilization should be used conservatively, as high rates on mustards could lead to increased *B. hilaris* population development (Parsana et al. 2001). This recommendation was supported by Sachan and Purwar (2007) who reported that infestations of the bug were shown to increase as levels of nitrogen applied to mustard plants was increased. It also has been shown that timely irrigation can reduce pest damage on seedling crops (Sanchan and Purwar 2007, Singh et al. 2007, Banuelos et al. 2013). On mature crops, rapid threshing of harvested material has been suggested as a means for minimizing *B. hilaris* damage (Abrol 2009). Similarly, planting oil seed brassica crops earlier or later in the season than normal may allow emerging crops to escape damage from the pest, but results have varied among growing regions (Lal and Singh 1993, Parsana et al. 2001, Ahuja et al. 2008). In the southwestern United States, growers increasingly are trying to establish brassica crops (such as broccoli) from transplants rather than direct-seeding to avoid excessive mortality to emerging seedlings due to *B. hilaris* feeding (Palumbo and Natwick 2010, Reed et al. 2013b, Huang et al. 2014a).

3.6.1.3 Action Thresholds

Little information is available on action thresholds used for making control decisions for *Bagrada hilaris* on brassica crops. Nyabuga (2008) recommended an action threshold of three bugs per plant when initiating control of *B. hilaris* on kale. A damage threshold used for mustard crops of one bug/m² on early growth stage plants and three bugs/m² on larger plants (Daiber 1991) also has been recommended for *B. hilaris* control on caper crops (Infantino et al. 2007). Although a scientifically-validated action threshold has not yet been established for *B. hilaris* on head-forming brassica crops in Arizona, a nominal threshold has been established that recommends treating plants when numbers of adults or damaged seedlings exceed one per six row feet of seedlings or transplants (Palumbo 2014). Maintaining *B. hilaris* densities below this level can reduce stand losses and unacceptable plant damage (Reed et al. 2013b). Based on recent work by Palumbo and Carrière (2015), the nominal threshold has been modified and now triggers insecticide treatments when the number of head-forming brassica plants with fresh lesions exceeds 5%. The threshold should be used for making management decisions until plants reach the six-leaf node stage, after which plants are much less susceptible to crop losses (J. C. Palumbo, unpublished data).

3.6.1.4 Chemical Control

The predominant approach to *Bagrada hilaris* management in cole crops worldwide has been chemical control. *B. hilaris* adults can destroy seedling plants rapidly, and the lack of effective cultural and biological control alternatives have forced growers to rely on quick-acting, contact insecticides to protect their crops. Preventing adults from feeding on plant terminals and small cotyledons on commercial brassica crops in Arizona and California is critical to establishing a marketable crop (Palumbo and Natwick 2010, Reed et al. 2013b). Consequently, management of *B. hilaris* in these cropping systems has consisted of intensive insecticide usage to protect seedling crops during crop establishment (Palumbo 2015b). Insecticide usage has almost doubled on cole crops since the arrival of *B. hilaris* to the desert Southwest, and growers have indicated that over 90% of the broccoli crops produced from 2010 to 2014 received an average of four insecticide sprays per season to control *B. hilaris* (Palumbo 2015b).

Historically, growers in the Old World have relied on organochlorine, cyclodiene, organophosphate, and carbamate insecticides to control *Bagrada hilaris* (Sarop et al. 1972, Bok et al. 2006, Hill 2008). The majority of these older, highly toxic compounds that have been effective against *B. hilaris* either are outdated or banned from use in the United States (Sarop et al. 1972, Gami et al. 2004, Ahuja et al. 2008, Nyabuga 2008, Singh et al. 2011b). However, among organophosphate and carbamate compounds currently registered for use in the United States, only chlorpyrifos and methomyl have demonstrated significant contact toxicity against *B. hilaris* adults in laboratory bioassays and field trials (Palumbo et al. 2013b, Palumbo and Huang 2014a, Palumbo et al. 2015).

Presently, in many parts of the world, including North America, pyrethroids commonly are used for control of *Bagrada hilaris* because of their fast knockdown efficacy (Infantino et al. 2007, Nyabuga 2008, Obopile et al. 2008, Palumbo et al. 2015). Because of the quick-acting contact activity of pyrethroids, growers in Arizona reported that pyrethroids are the primary insecticide used to control of *B. hilaris* adults on seedling crops (Palumbo 2015b). Local laboratory bioassays and greenhouse trials have demonstrated that pyrethroids such as bifenthrin and lambda cyhalothrin cause rapid contact mortality against *B. hilaris* adults (Palumbo et al. 2015) and, in field efficacy trials, protect plants from damage for 5–7 days following application (Palumbo 2012; Palumbo et al. 2013b, d, e; Palumbo and Huang 2014a).

Among the newer classes of insecticide chemistry, neonicotinoids such as imidacloprid and acetamiprid applied as foliar sprays have been shown to control *Bagrada hilaris* (Singh et al. 2011b). Similarly, dinotefuran applied to broccoli foliage is significantly more toxic to *B. hilaris* adults and provides significantly better plant protection against *B. hilaris* adults than acetamiprid, thiamethoxam, imidacloprid and clothianidin (Palumbo 2012; Palumbo et al. 2013b, c, d; Palumbo et al. 2015). Dinotefuran also has exhibited antifeeding properties against *B. hilaris* adults (Palumbo et al. 2013d, Palumbo et al. 2015). The soil systemic toxicity of neonicotinoids to *B. hilaris* adults in laboratory bioassays has shown that dinotefuron, imidacloprid, and thiamethoxam are toxic to adults via ingestion (Palumbo et al. 2015), but their application as at-planting soil treatments in field trials did not prevent *B. hilaris* from injuring seedling broccoli plants (Palumbo et al. 2015). Neonicotinoid seed treatments showed that planting mustard seeds treated with imidacloprid resulted in significantly lower plant injury due to *B. hilaris* (Ameta et al. 2005, Ahuja et al. 2008). In Arizona, clothianidin-treated seeds protected broccoli seedlings from *B. hilaris* injury through the three-leaf plant stage under field conditions (J. C. Palumbo, unpublished data).

Several other new classes of insecticide chemistry with known activity against piercing-sucking pests have been developed (e.g., anthranillic diamides, sulfoxamines, ketoenols, spinosyns), but research with *Bagrada hilaris* has indicated that these new insecticides are only marginally effective (Palumbo 2012; Palumbo et al. 2013c, d; Palumbo and Huang 2014b; Palumbo et al. 2015).

Similarly, botanical and biorational insecticides have been evaluated for control of *Bagrada hilaris*. The evaluation of neem oil/extracts and azadirachtin products in numerous laboratory bioassays and field trials indicated that efficacy of these materials varied widely depending on the life stage of *B. hilaris* used or other experimental variables such as the extraction processes of the compound, formulation of sprayed product, concentration applied, and spray timing/frequency (Pandey et al. 1981, Charleston et al. 2006, Guarino et al. 2007, Ahuja et al. 2008, Nyabuga 2008, Chandel et al. 2011, Grasswitz 2013b, Palumbo et al. 2013a). Other botanical insecticides such as pyrethrin, rotenone, sabadilla, chili extract, garlic extract, insecticidal soaps, and numerous exotic plant extracts also have been evaluated for biological activity against *B. hilaris* with variable results (Verma and Pandey 1981, Johri et al. 2004, Charleston et al. 2006, Guarino et al. 2007, Nyabuga 2008, Chandel et al. 2011, Grasswitz 2013b, Palumbo et al. 2013a).

To date, there has been no strong evidence that *Bagrada hilaris* populations have evolved resistance to insecticides under field conditions. However, Infantino et al. (2007) suggested that increasing damage caused by *Bagrada hilaris* on insecticide-treated caper crops in Italy likely is due to the development of insecticide resistance. Furthermore, bioassay monitoring of *B. hilaris* populations in India showed that the LC_{50} values of some organophosphate and organochlorine insecticides increased significantly over a 25-year span, but corroborative evidence indicating failure of insecticide efficacy in the field has not been reported (Swaran Dhingra 1998). For North American populations, baseline susceptibility data recently generated for *B. hilaris* populations from Arizona and southern California (Palumbo et al. 2015) will become increasingly important in the future given the reliance on pyrethroids, organophosphates, and neonicotinoids for control of *B. hilaris* and other key insect pests in brassica crops (Palumbo 2015a).

3.6.2 Organic Cropping Systems

At present, there is little published research on the management of *Bagrada hilaris* in organic production systems. In the United States, management of *B. hilaris* in such systems poses particular problems because of the limited number and efficacy of organically acceptable insecticides and the lack of effective native natural enemies (Lawrence 2012). Furthermore, the seasonal phenology of this bug varies between regions in the United States, making it difficult to use recommendations broadly across these regions. Management guidelines developed in one location may not be fully transferrable to another. In New Mexico, for example, many of the state's small-scale organic farmers produce a variety of crops within the Brassicaceae (including arugula, various leafy mustard greens, broccoli, radishes, kale, and turnips). However, the timing of production of these crops varies widely depending on latitude and altitude within the state and with the use of season-extending techniques such as floating row covers and hoop-houses. Also, the phenology of *B. hilaris* may vary within the state (Tessa R. Grasswitz, personal observations), reflecting differences in seasonal temperatures and host availability. For these reasons, strategies such as adjusting planting dates to avoid peak *Bagrada* populations must be synchronized with the pest's seasonal history in the target area(s).

With this caveat in mind, the best approach to managing *Bagrada hilaris* is to use a combination of control tactics. In this respect, pest management in organic systems is similar to that in conventional farming (Dufour 2001). Possible organic management tactics for *B. hilaris* are discussed below.

3.6.2.1 *Planting Date and Crop Selection*

A combination of judicious crop choices and planting dates may allow organic growers to produce viable crops in periods when *Bagrada hilaris* is dormant or absent. In central New Mexico, for example,

B. hilaris typically does not colonize crops until mid-June to mid-July (Tessa R. Grasswitz, personal observation), allowing early-spring plantings of quickly maturing crops, such as Asian or mustard greens, arugula, and radishes, to be sown and harvested (either outdoors or under protection) prior to the first appearance of adults of this bug. The same appears to be true in parts of California (Lambert and Dudley 2014) and Arizona (Huang et al. 2013). However, in some areas (e.g., southern New Mexico), *B. hilaris* may be active all year (Taylor et al. 2015), precluding such an approach.

3.6.2.2 Trap Cropping

Throughout its range, *Bagrada hilaris* shows a distinct preference for brassicaceous crops, although there are widespread reports of its occurrence on other plants (Reed et al. 2013a; Reed et al. 2014; Deep et al. 2014a, b; Huang et al. 2014b) (see **Section 3.4.2**, Host Plants). Within the Brassicaceae, the bug shows distinct preferences for different species and varieties (Huang et al. 2014b), and these preferences may form the basis of trap-cropping systems.

Trap cropping (or diversionary cropping) for *Bagrada hilaris* first was reported over a century ago in South Africa by Howard (1906), who suggested planting mustard crops in or around cabbage and cauli-flower fields to divert the insect away from the latter crops. Similarly, trap crops of Indian mustard planted prior to cabbage and followed by insecticide sprays have been used successfully to prevent damage from a number of key cabbage pests including *B. hilaris* (Srinivasan and Moorthy 1992, Reddy 2013). In a recent study, a radish cultivar was shown to be more attractive to *B. hilaris* adults than other brassica crops (Huang et al. 2014b). Similarly, Makwali et al. (2002) found that wild radish, *Raphanus raphanistrum* L., was highly attractive to *B. hilaris* and suggested it could be used as a diversionary host to protect bras-sica crops. Aalbersberg et al. (1989) found that Japanese radish planted as a border crop around wheat to enhance aphid control unintentionally attracted high numbers of *B. hilaris* that eventually damaged the wheat. In New Mexico, a field-scale experiment involving more than 20 different host plants showed that spring raab, *Brassica ruvo* L. H. Bailey, was highly preferred (Tessa R. Grasswitz, unpublished data). Spring raab, therefore, has potential as a trap crop by sowing early to concentrate and control populations of this bug prior to establishing other brassicaceous cash crops. However, this approach is feasible only if an effective control method is available for use on the trap crop. Flame weeders or vacuum devices may be suitable options when the trap crop is relatively small. Both techniques become less effective as the trap crop grows and increases in size. Sweet alyssum, *Lobularia maritima*, an ornamental annual brassica widely grown as an 'insectary' plant to attract beneficial insects (Brennan 2013), also has been suggested as a perimeter trap crop to protect edible brassicas, using vacuuming as the control tactic (Sooby 2014). The low-growing, compact nature of this plant would make it amenable to vacuuming even when fully mature. However, there is evidence that it is not a highly preferred host (Huang et al. 2014b), which would reduce its utility as a trap crop if more favored hosts were present in the vicinity.

3.6.2.3 Sanitation (Clean Culture)

In Africa, the importance of clean culture and sanitation long has been recognized as essential for avoiding *Bagrada hilaris* outbreaks on brassicaceous crops (Lounsbury 1898, Howard 1906). Therefore, control of weedy hosts in and around the planting area and post-harvest destruction of crop residues com-monly are recommended as methods of preventing or slowing *Bagrada* population build-up (Sachan and Purwar 2007, Hill 2008, Nyabuga 2008, Abrol 2009, Anonymous 2012, Banuelos et al. 2013, Reed et al. 2014). In addition, because *B. hilaris* usually deposits its eggs in the soil (Taylor et al. 2014), it has been suggested that frequent cultivation of leafy vegetable beds during the season may help reduce *B. hilaris* populations (Bok et al. 2006, Nyabuga 2008).

3.6.2.4 Crop Isolation

In situations where multiple brassicaceous crops are grown concurrently (e.g., small-scale organic farms), spatial separation of crops is recommended for management of *Bagrada hilaris* because close proximity to a highly preferred host increases the chances of less favored crops becoming infested. Primarily, this is due to the tendency of adults and nymphs to disperse short distances by walking (Tessa R. Grasswitz, personal observation). Similarly, in the southwestern United States, larger-scale commercial growers are encouraged to avoid planting brassicaceous plants near crops considered to be alternative hosts for the bug (e.g., cotton, Sudan grass, and corn) (Palumbo 2014).

3.6.2.5 Mechanical Control and Exclusion Techniques

Historically, in Africa, gardeners and small-scale farmers resorted to mechanical control methods such as shaking infested plants and dislodging the insects into pans containing paraffin (Lounsbury 1898) or removing *Bagrada hilaris* either by hand or with sweep nets (Howard 1906, Nyabuga 2008). Early attempts to protect cabbage crops from injury by covering them with butter muslin or with frames covered with mosquito netting were unsuccessful (Howard 1906). More recently, in the United States, a combination of low metal hoops and floating row covers has been found to be effective for excluding *B. hilaris* provided that a good seal is maintained between the edge of the fabric and the soil and no holes are present in the cover (Reed et al. 2014, Tessa R. Grasswitz, personal observations). This approach has proven useful in establishing autumn-sown brassicaceous plantings in central New Mexico with the additional benefit of providing some degree of frost protection to the plants during the winter. Such an approach is useful for small-scale producers, but may not be practical on a larger scale because of the expense and labor requirements (Frederic J. Klicka, personal communication).

3.6.2.6 Organic Insecticides

Field tests of insecticides currently approved for use on brassicaceous plants in organic production systems have generated inconsistent results, indicating they generally are not very effective (Palumbo and Natwick 2010, Palumbo et al. 2013a, Grasswitz 2014) (see **Section 3.6.1.4**, Chemical Control). The variability of these results may be related to environmental conditions and the relatively rapid breakdown of most organic insecticides under intense UV radiation and/or high temperatures. For example, a 1.4% commercial formulation of pyrethrin gave much better results against *Bagrada hilaris* in New Mexico when applied in October, when temperatures were cooler, than in August (Grasswitz 2014). In India, laboratory bioassays with a neem-based product resulted in 96% mortality of a mixed population of adults and nymphs 2 days after treatment (Ghosal et al. 2006). However, in greenhouse tests in the United States, a 70% formulation of neem oil proved ineffective against both nymphs and adults (Grasswitz 2013a).

Laboratory testing of products based on various entomopathogenic fungi (Dara 2013) indicates they may have potential in situations where temperatures are sufficiently low and humidity high enough to support the development of epizootics. For example, in some climates, such materials might be appropriate for use as soil drenches or surface applications because most eggs of this bug are laid just below the soil surface (Taylor et al. 2014), and first and second instars frequently are observed aggregated on the soil surface rather than on the plant (Tessa R. Grasswitz, personal observation). If irrigation water is readily available, such applications might be potentiated by use of overhead sprinklers, although this has not yet been tested.

In general, published data on the efficacy of organically approved insecticides in the United States are too scarce and variable to support specific recommendations at this time. In certain geographic areas, some organic insecticides require an organically acceptable buffering agent to reduce the pH. An organically approved surfactant also is recommended for some brassicaceous crops. These requirements impose an additional constraint on growers. Although a few such products are on the market (e.g., Constant BupHer™ and Natural Wet® surfactant), they are not readily available in all states and, in some cases, are not sold in sizes appropriate for smaller-scale producers.

3.6.3 Natural Enemies

3.6.3.1 Parasitoids

The key natural enemies of *Bagrada hilaris* have been poorly studied in comparison to other widespread pest pentatomids. In South Africa, Howard (1906) stated "No natural enemies of the bug have been found in the Transvaal, but the fact that it is abundant some seasons in the Cape Colony, and not in others, seems to point to the work of some parasite in the Colony, and we at present are negotiating with other entomologists for the importation of such parasites whenever possible." It is not known if such a project was initiated. Later, Gunn (1918) reported that in the Pretoria area, "… a Hymenopterous parasite belonging to the family Chalcididae began to emerge [from the soil] in large numbers."

All other references to parasitoids have come from observations in India and Pakistan. Four species of parasitic Diptera have been reported from *Bagrada hilaris* including *Sarcophaga kempi* Senior-White (Sarcophagidae) (Thompson 1950) and the tachinid flies *Alophora indica* (Mesnil) (as *Parallophorella indica*) (Herting and Simmonds 1971, Cheema et al. 1973, Crosskey 1976), *Alophora* sp. (Rakshpal 1949, Cheema et al. 1973, Crosskey 1976, Azim and Shafee 1986), and *Phasia* sp. (Tachinidae) (Herting and Simmonds 1971). Little is known about the effectiveness of these parasitoids. Cheema et al. (1973) reported that only four *A.* (as *Hyalomya*) *pusilla* emerged from 554 adults from the western hills of Pakistan. In India, Rakshpal (1954) reported that only male *Alophora* sp. emerged from male *B. hilaris* and only female parasitoids emerged from female hosts. This relationship has not been reported for any other tachinid attacking a stink bug host. *A. pusilla* is known from a wide range of hosts across multiple heteropteran families. Sarcophagids are not specific to any pentatomid species or even confined to a particular insect family. All pentatomids are probably attacked by one or more tachinids (Phasiinae). Some are fairly specific but few have a significant impact on their hosts. Exceptions are species in the New World genus *Trichopoda* that have been successfully established in Hawaii and Australia for biological control of the southern green stink bug, *Nezara viridula* (L.) (Davis and Krauss 1963, Coombs and Sands 2000).

The nymphal stages of Pentatomidae are not commonly attacked. The two most common genera reported from nymphal stages are the wasps *Hexacladia* (Encyrtidae) and *Aridelus* (Braconidae). However, none has been reported from *B. hilaris*. Many tachinids attack the later nymphal instars of stink bugs but normally only emerge after the host reaches the adult stage.

Pentatomid eggs are the most susceptible stage attacked by parasitoids, and several species are known to attack *Bagrada hilaris* in Pakistan and India. Outside of the South African report of abundant "Chalcididae" emerging from *B. hilaris* eggs (Gunn 1918), no specific parasitoids have been reported from the rest of the insect's broad distribution, but certainly must occur everywhere that this pentatomid is endemic. Several genera and species of wasps in the family Platygastridae have been reported from the eggs: *Gryon* sp. (Herting and Simmonds 1971), *G. karnalensis* Chacko and Katiyar (Cheema et al. 1973), *Telenomus samueli* (Mani) (Mani 1941, Samuel 1942, Rajmohana 2006), *Typhodytes* sp. (Samuel 1942, Herting and Simmonds 1971), *Trissolcus* sp. (Ghosal 2006), and *Trissolcus hyalinipennis* Rajmohana and Narendran (Subba Rao and Chacko 1961, Herting and Simmonds 1971, Fergusson 1983, Rajmohana and Narendran 2007). *Psix* sp. nr. *striaticeps* (Dodd), reared from another pentatomid, successfully attacked *B. hilaris* eggs in the laboratory (Cheema et al. 1973) but has not been recovered from field collections. A detailed survey of egg parasitism of *B. hilaris* (as *B. cruciferarum*) on maize (Narayanan et al. 1959) concluded that *G. karnalensis* (as *Hadrophanurus* sp.) "… increases to such an extent that it exterminates the host and ultimately proves fatal to the parasite itself." Although egg parasitoids normally are the most important specialist natural enemy of stink bugs, none has been reported as this effective on any crop. Unfortunately, the illustrations clearly show that the eggs were deposited together in regular rows, as with most other stink bugs, and the eggs appear more elongate than the globular *B. hilaris* eggs, bringing into question the identity of the host species studied. The authors do refer to Pruthi's (1946) reference to *B. hilaris* depositing eggs in the soil, and opines that perhaps in the rainy season, the bugs may prefer to oviposit on leaves instead. Parasitoid specimens from this study were not found in museums in India (Keloth Rajmohana, personal communication).

Samuel (1942) reported two species of parasitoids emerging from *B. hilaris* (as *B. picta*) around New Delhi, *Telenomus samueli* (as *Liophanurus* sp.) and *Typhodytes* sp., and made some ecological observations on each. *T. samueli* parasitized 15–25% of the eggs during the season and 30% of exposed eggs in the laboratory; only 20% were parasitized when in the soil. Freshly deposited eggs were less preferred. *Typhodytes* appeared later in the season, the adults emerging with host nymphs from the soil at a rate of 18–20% parasitism. It should be noted that *Typhodytes* may not be a valid taxon (Walker A. Jones, personal communication). However, members of the platygastrid genus *Tiphodytes* Bradley are egg parasitoids of water striders (Gerridae) rather than stink bugs.

Subba Rao and Chacko (1961) described *Trissolcus hyalinipennis* (as *Allophanurus indicus*) from *Bagrada hilaris* (as *B. cruciferarum*) in India and recorded several details on its biology including oviposition behavior, adult and larval morphology, and development and reproduction. Ghosal (2006) noted a range of 19–32 % egg parasitism in eastern India by a *Trissolcus* sp. The data apparently were collected from eggs deposited on mustard plants.

3.6.3.2 Predators

Like most other free-living insects, pentatomids are reported to be attacked by a wide array of generalist predators (McPherson and McPherson 2000). The earliest report of predation of *Bagrada hilaris* was by domestic "fowls" in South Africa (Howard 1906). Also in South Africa, Gunn (1918) reported the reduviid *Rhynocoris segmentarius* (Germar) (as *Harpactor segmentarius* Germar) preying on nymphs and adults and identified certain birds feeding on bugs infesting wheat. The only other report of predation on *B. hilaris* was by a phoretic mite (*Bochartia* sp.) in India (Thaker et al. 1969).

3.6.3.3 Other Natural Enemies

There are numerous records of entomopathogens attacking pentatomids, and certain fungi have been used in augmentation against pest species (e.g., Sosa-Gomez and Moscardi 1998, Rombach et al. 1986). Gunn (1918) reported that *Bagrada hilaris* was killed by the fungus *Isaria* sp. during the South African rainy season. Nematodes also regularly attack stink bugs (e.g., Esquivel 2011 and Stubbins et al. (2015) as do parasitic flagellates (Gibbs 1957), but there are no reports of these natural enemies attacking *B. hilaris*.

3.6.3.4 Native Natural Enemies

In the United States, four native natural enemies so far have been found attacking *Bagrada hilaris* in New Mexico: a hymenopteran egg parasitoid, *Ooencyrtus* sp. (Tessa R. Grasswitz, personal observation); the spined soldier bug, *Podisus maculiventris* (Say); the soft-winged flower beetle, *Collops vittatus* (Say); and the spined assassin bug, *Sinea diadema* (F.). The three predatory species are widely distributed in North America (Aldrich and Cantelo 1999, Voss and McPherson 2003, Eaton and Kaufman 2007, Anonymous 2015), and, therefore, may be associated with *B. hilaris* in other parts of the United States in addition to New Mexico.

Ooencyrtus sp. was recovered in New Mexico in large numbers in August 2013 when sheets of *Bagrada hilaris* eggs from a laboratory colony were placed among infested host plants in the field for 7 days. Successful parasitism of these eggs indicates that there were no intrinsic physiological barriers to parasitism by this species, but it remains to be determined if it will search for, and parasitize, eggs deposited in the soil (which appear to be the preferred oviposition site under New Mexico conditions) (Taylor, et al. 2014; Tessa R. Grasswitz, personal observations). In this respect, Narayanan et al. (1959) noted that, in India, different species of parasitoids attacked *Bagrada* eggs laid on host plant leaves versus those deposited in the soil.

To help anticipate whether key egg parasitoids of native Pentatomidae might adapt to the eggs of *Bagrada hilaris*, five species of common Platygastridae were colonized to determine if any would parasitize and develop in *B. hilaris* eggs (Walker A. Jones, unpublished data). Parasitoids used in no-choice tests were: *Trissolcus basalis* (Wollaston), *Trissolcus brochymenae* (Ashmead), *Trissolcus euschisti* (Ashmead), *Trissolcus hullensis* (Harrington), and *Telenomus podisi* Ashmead. *T. brochymenae* is the most important parasitoid of the harlequin bug, *Murgantia histrionica*, the only other common North American pentatomid in the same tribe (Strachiini) as *B. hilaris*. No native *Trissolcus* spp. accepted *B. hilaris* eggs. However, a small proportion of exposed eggs were attacked successfully by *T. podisi* females. *T. podisi*, along with all other studied platygastrid parasitoids of pentatomid eggs, have evolved host finding cues that culminate in the location and selection of host eggs glued together in masses on

FIGURE 3.9 Adult spined soldier bugs (*Podisus maculiventris*) feeding on *Bagrada hilaris*, Los Lunas, New Mexico. (Courtesy of Tessa R. Grasswitz).

plant parts (Colazza et al. 2010). No native species likely has evolved cues leading to attacking single eggs deposited in the soil.

Podisus maculiventris (**Figure 3.9**) has been reported to feed on a wide variety of prey, including various lepidopteran larvae, several species of Coleoptera, Hemiptera, Hymenoptera, and Orthoptera (McPherson 1982). It has been reported to attack other pest pentatomids such as the southern green stink bug, *Nezara viridula* (L.) (De Clercq et al. 2002). In laboratory experiments, De Clercq et al. (2002) reported that the highest daily predation rates by adult *P. maculiventris* were for first instars of *N. viridula* and the lowest rates for adults (3.4–7.6 and 0.3–0.6 prey items, respectively). Predation by adult *P. maculiventris* on adult *Bagrada hilaris* has been observed in New Mexico (Tessa R. Grasswitz, unpublished data). Female prey were consumed at a significantly higher rate than males when individual predators were presented with three mating pairs of *B. hilaris* per day. Predation on third and fifth instars of *B. hilaris* was low (an average of 0.6 and 0.8 per day, respectively), likely due to the difficulty of capturing such highly mobile stages; mating adults are probably much easier targets because coupled males and females are orientated in opposite directions, making it more difficult for them to escape from potential predators.

Collops vittatus is an important generalist predator in alfalfa, cotton, pastures, and gardens (Walker 1957). It has been reported to prey on various aphids, mites, whiteflies, Lepidoptera, and Hemiptera (including *Lygus* species) (Knowlton 1944, Walker 1957, Nielson and Henderson 1959, Ellsworth et al. 2011). In laboratory studies with adults of this beetle predator in New Mexico, individuals provisioned daily with 20 first instars of *Bagrada hilaris* killed an average of five and nine each day (male and female beetles, respectively), with only approximately half that number being fully consumed. Incomplete (partial) consumption of prey also was observed when individual *C. vittatus* were given third instars of *B. hilaris* with average predation rates of three nymphs per day for both sexes of beetles (Tessa R. Grasswitz, unpublished data). Partial consumption of captured prey also was reported by Knowlton (1944).

Sinea diadema has been reported to feed on a variety of prey, including lepidopteran larvae, several species of Coleoptera, Diptera, Hemiptera, and Hymenoptera (Voss and McPherson 2003). Reports of its feeding on *Bagrada hilaris* are limited to a single observation in New Mexico in October, 2014, when an adult *Sinea diadema* was seen feeding on fourth or fifth instars of *B. hilaris* nymphs (Tessa R. Grasswitz, personal observation).

In summary, none of the native natural enemies found attacking *B. hilaris* in the United States seems likely to inflict high enough levels of mortality to achieve adequate control of this pest under field conditions.

3.6.3.5 Classical Biological Control

Prospects for classical biological control are good. Although our knowledge of key, co-evolved natural enemies (egg parasitoids in the case of most Pentatomidae) are very poor, the apparent native geographic range is large, encompassing all or nearly all climatic and ecological subregions to which *Bagrada hilaris* may be anticipated to spread in the New World. This bug is a sporadic pest from India and Pakistan through Kenya to southern Africa. Throughout this broad geographic distribution there are certainly regional guilds of parasitoids that keep populations in check most years in most places. The fact that most eggs are uniquely deposited singly and in the soil must mean that host-specific species of parasitoids have developed unique host finding cues. These will be quite different from those species evolved to locate groups of eggs glued together on above-ground plant tissues. Such parasitoids also are likely to possess different mating systems than the typical protandrous, sib-mated parasitoid species that attack egg masses. The latter species are quasi-gregarious, have local mate competition, and female-biased sex ratios.

Selection of biocontrol agents that are adapted to the climates in areas of intended release requires a thorough analysis of the climates of the source and release sites. Currently, *Bagrada hilaris* still is spreading south and east into new areas in North America. It has been intercepted many times in commercial vehicles at the Florida border (LeVeen and Hodges 2015). Thus, it is likely that it will eventually become established in the southeastern states, which possess different climate types from those in the Southwest. However, because most commercial cole crop production occurs in the desert valleys of California and Arizona, these locations probably should be targeted first. An initial attempt to match climates using the program CLIMEX (Sutherst and Maywald 1985, Sutherst et al. 2004) compared the default settings for Bakersfield, California, with that of Africa. The closest matches within the African distribution of *B. hilaris* was South Africa and immediately adjacent areas (W. A. Jones, unpublished data). Preparations to make explorations in that region are underway.

An opportunity to explore Pakistan for *Bagrada hilaris* parasitoids appeared in 2013, which quickly yielded three egg parasitoids that were successfully collected and shipped to the United States. The United States Department of Agriculture – Agricultural Research Service (USDA-ARS) maintains the European Biological Control Laboratory (EBCL) near Montpellier, France (Jones and Sforza 2007), continuing classical biological explorations and research on invasive species begun in France in 1919. The EBCL has maintained many collaborations on biological control projects with the Commonwealth Agricultural Bureau International (CABI) South Asia laboratory, Rawalpindi, Pakistan. In 2013, EBCL received permission to release some funds for an initial attempt at collecting egg parasitoids in Pakistan. A contract was developed through the American Embassy, Paris. During 2014, personnel at CABI Pakistan established a *B. hilaris* culture on cabbage, obtained eggs that were glued to paper cards and placed in the field, both on the ground and on plants at two locations. Parasitoids were recovered and parasitized host eggs shipped to the USDA-ARS Stoneville Research Quarantine Facility, Stoneville, MS (Jones et al. 1985). Three species were colonized successfully on eggs from bugs shipped regularly (under Animal and Plant Health Inspection Service permit) from California and New Mexico (Mahmood et al. 2015). The three species of egg parasitoids tentatively were identified by E. Talamas (USDA-ARS, Systematic Entomology Laboratory, Washington, D.C., USA) and K. Rahjmohana (Zoological Survey of India, Calicut, India) as *Gryon gonikopalense* Sharma (**Figure 3.10**) and *Trissolcus hyalinipennis*

FIGURE 3.10 *Gryon gonikopalense*, egg parasitoid of *Bagrada hilaris* (lateral view). (Courtesy of Elijah J. Talamas and Ashton B. Smith, USDA-ARS).

(Platygastridae), and *Ooencyrtus* sp. (Encyrtidae). There is no biological information on *G. goniko-palense* or *Ooencyrtus* sp., but the adult and larval morphology, biology, and behavior of *T. hyalinipennis* on *B. hilaris* in India were published by Subba Rao and Chacko (1961) (as *Allophanurus indicus* n. sp.). This is the first record of an *Ooencyrtus* sp. attacking *B. hilaris*. Host specificity testing for all three parasitoids was initiated in 2015. Further exploration is planned for southern Africa.

3.7 Future Outlook

As one of our most recent invasive pentatomoids in North America, the full ecological and economic impacts of *Bagrada hilaris* are difficult to predict. Based on its rapid establishment and geographic expansion, this stink bug clearly has adjusted well to conditions in western North America; it quickly has become an economic pest of brassicaceous crops in all regions it has invaded. With the broad geographic distribution of brassicaceous hosts in North America, the limiting factors to the spread of *B. hilaris* likely will be environmental (e.g., temperature, moisture, and humidity). Currently, studies are underway to use these parameters and host availability to predict the likely distribution and risk of this bug moving into new areas (M. Papes and T.M. Perring, unpublished data). Of course, various life history parameters and thermal requirements of this stink bug may change as it adapts to new biotic and abiotic conditions.

Much work remains in developing effective conventional and organic IPM strategies for this pest. Although the literature on its host plants clearly shows that the bugs prefer members of the Brassicaceae, non-related crops such as corn, melon, pepper, potato, and wheat are feeding hosts that have experienced economic losses. A better understanding of the interactions of *Bagrada hilaris* with other species, particularly weeds and native plants, will be important considerations for management. These landscape level interactions among potential hosts need to be evaluated. Conventional insecticide use likely will continue to remain high on cole crops under attack by the *B. hilaris* until other management methods become more effective at suppressing its populations. Organic methods of management may prove to be particularly problematic due to limited management options. Also, more research is needed on effective biological control agents for this insect. The unique ovipositional behavior of *B. hilaris* effectively reduces the success of native pentatomid egg parasitoids and predators that exist in North America. Thus, of critical importance is the search and importation of natural enemies that have adapted to this bug in its native environment and the evaluation of native predators targeting prey in the soil.

3.8 Acknowledgments

We thank J. E. McPherson (Department of Zoology, Southern Illinois University, Carbondale) for his tireless efforts in the final editing of this chapter, David. A. Rider (Department of Entomology, North Dakota State University, Fargo) for developing the synonymy list used in this chapter, Sergio R. Sanchez-Peña and Ivonne Torres (Universidad Autónoma Agraria Antonio Narro, Mexico), for providing data on current distribution of *Bagrada hilaris* in Mexico, Gevork Arakelian (Department of Weights and Measures, South Gate, CA) for information on the early interception of *B. hilaris* in California, Frederic J. Klicka (Southwest Ag Service, Inc., Brawley, CA) for providing information on the efficacy of row covers for *B. hilaris* management, Timothy R. Lewis (Department of Entomology, University of California, Riverside) for generating the North American distribution map for *B. hilaris*, Jocelyn G. Millar (Department of Entomology, University of California, Riverside) for providing data on potential pheromones for *B. hilaris*, Darcy E. Oishi (Hawaii Department of Agriculture, Aiea) for providing information on the first report of *B. hilaris* in Hawaii, Elijah J. Talamas (USDA-ARS, Systematic Entomology Laboratory, National Museum of Natural History, Washington, D.C.) for identification of parasitoids, Raul T. Villanueva (Texas Agrilife Extension, Weslaco) for verification of central Texas records of *B. hilaris*, and Keloth Rajmohana (Zoological Survey of India, Calicut) for identification of parasitoids and for checking Indian parasitoid records.

3.9 References Cited

Aalbersberg, Y. K., M. C. van der Westhuizen, and P. H. Hewitt. 1989. Japanese radish as a reservoir for the natural enemies of the Russian wheat aphid *Diuraphis noxia* (Hemiptera: Aphididae). Phytophylactica 21: 241–245.

Abrol, D. P. 2009. Plant-Insect Interaction, 129–150 pp. *In* S. K. Gupta (Ed.), Biology and breeding of crucifers. CRC Press LLC, Boca Raton, FL. 405 pp.

Abrol, D. P., and A. Gupta. 2010. Insect pests attacking cauliflower (*Brassica oleracea* var. botrytis L.): Population dynamics in relation to weather factors. Green Farming 1: 167–170.

Ahuja, B., R. K. Kalyan, U. R. Ahuja, S. K. Singh, M. M. Sundria, A. Dhandapani. 2008. Integrated management strategy for painted bug, *Bagrada hilaris* (Burm.) inflicting injury at seedling stage of mustard (*Brassica juncea*) in arid Western Rajasthan. Pesticide Research Journal 20: 48–51.

Ahuja, D. B., and M. L. Joshi. 1995. Evaluation of insecticides for control of painted bug on taramira. Madras Agricultural Journal 82(12): 627–629.

Aldrich, J. R. 1986. Seasonal variation of black pigmentation under the wings in a true bug (Hemiptera: Pentatomidae): a laboratory and field study. Proceedings of the Entomological Society of Washington 88: 409–421.

Aldrich, J. R., and W. W. Cantelo. 1999. Suppression of Colorado beetle infestation by pheromone-mediated augmentation of the predatory spined soldier bug, *Podisus maculiventris* (Say) (Heteroptera: Pentatomidae). Agricultural and Forest Entomology 1: 209–217.

Ameta, O.P., A. K. Srivastava, and K. C. Sharma. 2005. Efficacy of imidacloprid seed treatment against sawfly, *Athalia lugens proxima* (Klug) and painted bug, *Bagrada cruciferarum* (Kirk) in mustard, *Brassica campestris* (L). Pestology 29: 9–13.

Anderson, T. J. 1917. Notes on insects injurious to coffee. Department of Agriculture, British East Africa, Nairobi, Agricultural Bulletin 2: 20–43.

Anonymous. 2012. Bagrada bug. http://www.infonet-biovision.org/default/ct/103/pests (accessed 1 May 2015).

Anonymous. 2015. Distribution map of *Sinea diadema* in North America. http://www.discoverlife.org/mp/20 m?kind=Sinea+diadema&mobile=iPhone (accessed 20 April 2015).

Arakelian, G. 2010. Bagrada bug. Center for Invasive Species Research. http://cisr.ucr.edu/bagrada_bug.html (accessed 14 April 2015).

Arakelian, G. 2011. Bagrada bug *Bagrada hilaris*, pp. 25–26. *In* S. Gaimari and M. O'Donnell (Eds.), California Plant Pest Disease Report, January 2008–December 2009 25: 1–108. http://www.cdfa.ca.gov/plant/ppd /PDF/CPPDR_2011_25.pdf (accessed 1 May 2015).

Atwal, A. S. 1959. The oviposition behaviour of *Bagrada cruciferarum* Kirk. (Pentatomidae: Heteroptera) and the influence of temperature and humidity on the speed of development of eggs. Proceedings National Institute Science India 25: 65–67.

Azim, M. N., and S. A. Shafee. 1986. The life cycle of *Bagrada picta* (Fabricius). Articulata 11: 261–65.

Bakhetia, D. R. C. 1986. Control of insect pests of toria, sarson and rai. Indian Farming 36(4): 41, 43–44.

Bakhetia, D. R. C., and K. S. Labana. 1978. Insect resistance in brassica crops. Crop Improvement 5: 95–103.

Bakhetia, D. R. C., and B. S. Sekhan. 1989. Insect pests and their management in rapeseed-mustard. Journal of Oilseeds Research 6: 269–99.

Banuelos, G. S., K. S. Dhillon, S. S. Banga. 2013. Oilseed Brassicas, pp 339–368. *In* B. P. Singh (Ed.), Biofuel Crops, production physiology and genetics. CAB International, Boston, MA. 525 pp.

Bath, D. S., and D. Singh. 1989. Studies on the economic threshold level of mustard aphid *Lipaphis erysimi* (Kaltenbach) on the radish seed crop in India. Tropical Pest Management 35: 154–156.

Batra, H. N. 1958. Bionomics of *Bagrada cruciferarum* Kirkaldy (Heteroptera: Pentatomidae) and its occurrence as a pest of mustard seeds. Indian Journal of Entomology 20: 130–135.

Batra, H. N., and S. Sarup. 1962. Technique of mass-breeding of the painted bug, *Bagrada cruciferarum* Kirk. (Heteroptera: Pentatomidae). Indian Oilseeds Journal 6: 135–143.

Bealmear, S., P. Warren, and K. Young. 2012. Bagrada bug: a new pest for Arizona gardeners. http://extension .arizona.edu/sites/extension.arizona.edu/files/pubs/az1588.pdf (accessed 1 May 2015).

Bhai, B. D., and S. Singh. 1961. On the biology of the painted bug, *Bagrada cruciferarum* Kirk. in the Punjab. Proceedings of the 48th Indian Science Congress, Part 3, p. 494.

Bhati, R., R. C. Sharma, and R. Singh. 2015. Studies on occurrence of insect-pests of different *Brassica* species. International Journal of Current Science 14: 125–132.

Bok, I., M. Madisa, D. Machacha, M. Moamogwe, and M., More. 2006. Manual for vegetable production in Botswana. Botswana Ministry of Agriculture, Department of Agriculture, Gaborone, Botswana. http://www.dar.gov.bw/manual1_veg_prod_botswana.pdf (accessed 12 March 2015).

Boopathi, T., and K. A. Pathak. 2012. Seasonal abundance of insect pests of broccoli in north eastern Hill Region of India. Madras Agricultural Journal 99(1–3): 125–127.

Brennan, E. B. 2013. Agronomic aspects of strip intercropping lettuce with alyssum for biological control of aphids. Biological Control 65: 302–311.

Bundy, C. S., T. R. Grasswitz, and C. Sutherland. 2012. First report of the invasive stink bug *Bagrada hilaris* (Burmeister) (Heteroptera: Pentatomidae) from New Mexico, with notes on its biology. Southwestern Entomologist 37: 411–14.

Burmeister, H. C. C. 1835. Handbuch der Entomologie, vol 2. Rhynchota. G. Reimer, Berlin. i–iv + 400 pp.

Chandel, B. S., R. Trivedi, S. Bajpai, and A. Singh. 2011. Toxicity of azadirachtin, beta-asarone, acorenone, *Acorus calamus* Linn. and *Azadirachta indica* A. Juss against painted bug, *Bagrada cruciferarum* Kirk. (Hemiptera: Pentatomidae). Life Science Bulletin. 8: 194–98.

Charleston, D. S., R. Kfir, M. Dicke, and L. Vet. 2006. Impact of botanical extracts derived from *Melia azedarach* and *Azadirachta indica* on populations of *Plutella xylostella* and its natural enemies: a field test of laboratory findings. Biological Control 39: 105–114.

Chattopadhyay, C., B. K. Bhattacharya, V. Kumar, A. Kumar, and P. D. Meena. 2011. Impact of climate change on pests and diseases of oilseeds *Brassica* – the scenario unfolding in India. Journal of Oilseed Brassica 2: 48–55.

Cheema, M. A., M. Irshad, M. Murtaza, and M. A. Ghani. 1973. Pentatomids associated with Gramineae and their natural enemies in Pakistan. Technical Bulletin of the Commonwealth Institute of Biological Control 16: 47–67.

Colazza, S., S. Guarino, and E. Peri. 2004. *Bagrada hilaris* (Burmeister) (Heteroptera: Pentatomidae) a pest of caper in the island of Pantelleria. Informatore Fitopatologico 54: 30–34.

Colazza, S., E. Peri, G. Salerno, and E. Conti. 2010. Parasitoids: exploitation of host chemical cues, pp. 97–138. *In* F .L. Consoli, J. R. P. Parra, and R. A. Zucchi [Eds.], Egg parasitoids in agroecosystems with emphasis on *Trichogramma*. Series: Progress in Biological Control, Volume 9. Springer, New York, NY. 482 pp.

Coombs, M., and D. Sands. 2000. Establishment in Australia of *Trichopoda giacomellii* (Blanchard) (Diptera: Tachinidae), a biological control agent for *Nezara viridula* (L.) (Hemiptera: Pentatomidae). Australian Journal of Entomology 39: 219–222.

Crosskey, R. W. 1976. A taxonomic conspectus of the Tachinidae (Diptera) of the Oriental Regional. Bulletin of the British Museum (Natural History), Entomological Supplement 26: 1–357.

Daiber, K. C. 1991. Insects and nematodes that attack cole crops in Southern Africa, a research review. Journal of Plant Diseases and Protection 99: 430–440.

Dallas, W. S. 1851. List of the specimens of hemipterous insects in the collection of the British Museum. Part 1. Trustees of the British Museum, London. 368 pp., plates 1–11.

Dara, S. 2012. Bagrada bug in Santa Barbara County. http://cesantabarbara.ucanr.edu/Strawberry_Production/Bagrada_Bug_Distribution (accessed 1 May 2015).

Dara, S. 2013. Entomopathogenic fungi for managing the invasive bagrada bug, *Bagrada hilaris*. http://cesantabarbara.ucanr.edu/files/170432.pdf (accessed 18 December 2014).

Dara, S. 2014. Bagrada bug distribution in California. University of California Cooperative Extension, Santa Barbara County. http://cesantabarbara.ucanr.edu/Strawberry_Production/Bagrada_Bug_Distribution (accessed 1 May 2015).

Davis C. J., and N. L. H. Krauss. 1963. Recent introductions for biological control in Hawaii–VIII. Proceedings of the Hawaiian Entomological Society. 38: 245–248.

De Clercq, P., K. Wyckhuys, H. N. De Oliveira, and J. Klapwijk. 2002. Predation by *Podisus maculiventris* on different life stages of *Nezara viridula*. Florida Entomologist 85: 197–202.

Deep, K., A. Kumar, D. P. Singh, and P. R. Yadav. 2014a. Study of the ecological factors affecting on [sic] food preference experiment and population dynamics of *Bagrada cruciferarum* Krik. [sic]. Journal of Experimental Zoology India 17. 257–260.

Deep, K., A. Kumar, D. P. Singh, and P. R. Yadav. 2014b. Study of the ecological parameter, site of oviposition, population dynamics and seasonal cycle of *Bagrada cruciferarum* on *Brassica campestris*. Journal of Experimental Zoology India 17: 331–336.

Derjanschi, V., and J. Péricart. 2005. Hémiptères Pentatomoidea Euro-Méditerranéens. Volume 1. Gènéralités systématique: première partie. Faune de France 90: 1–494.

Dharpure, S. R. 2002a. Changing scenario of insect pests of potato in Satpura Plateau of Madhya Pradesh. Journal of the Indian Potato Association 29: 135–138.

Dharpure, S. R. 2002b. New record of a painted bug, *Bagrada cruciferarum* Kirkaldy (Pentatomidae :Homoptera) [sic] on potato in Satpura Plateau of Madhya Pradesh. Journal of the Indian Potato Association 29: 155–156.

Dhiman, A., and J. R. Gandhi. 1988. Aggregation, arrival responses and growth of *Bagrada hilaris* (Heteroptera: Pentatomidae). Annals of Entomology. (Dehra Dun) 6: 19–26.

Dhuri, A. V., and K. M. Singh. 1983. Pest complex and succession of insect pests in black gram *Vigna mungo* (L.) Hepper. Indian Journal of Entomology 45(4): 396–401.

Dufour, R. 2001. Biointensive integrated pest management (IPM). Appropriate Technology Transfer for Rural Areas (ATTRA). https://attra.ncat.org/attra-pub/summaries/summary.php?pub=146 (accessed 1 May 2015).

Eaton, E. R., and K. Kaufman. 2007. Kaufman field guide to insects of North America. Houghton Mifflin Harcourt, Boston, MA. 392 pp.

Ellsworth, P., A. Mostafa, and L. Brown. 2011. *Collops* beetle: natural predator in field crops. Western Farm Press, July 8th 2011. http://westernfarmpress.com/management/collops-beetle-natural-predator-field -crops (accessed 10 March 2014).

Esquivel, J. F. 2011. *Euschistus servus* (Say) – a new host record for Mermithidae (Mermithida). Southwestern Entomologist 36(2): 207–211.

Esselbaugh, C. O. 1946. A study of the eggs of the Pentatomidae (Hemiptera). Annals of the Entomological Society of America 34: 667–691.

Fabricius, J. C. 1775. Systema entomologiae sistens insectorum classes, ordines, genera, species; adjectis synonymis, locis, descriptionibus et observationibus. Flensburgi et Lipsiae, Kortii xxxii + 832 pp.

Faúndez, E. I., A. Lüer, Á. G. Cuevas, D. A. Rider, and P. Valdebenito. 2016. First record of the painted bug *Bagrada hilaris* (Burmeister, 1835) (Heteroptera: Pentatomidae) in South America. Arquivos Entomolóxicos 16: 175–179.

Fergusson, N. D. M. 1983. The status of the genus *Allophanurus* Kieffer (Hym., Proctotrupoidea, Scelionidae). Entomologist Monthly Magazine 119: 207–210.

Fletcher, T. B. 1921. Report of the Imperial Entomologist. Sd. Reports of the Agricultural Research Institute, Pusa, 1920–21, pp. 41–59.

Gahukart, R. T., and M. G. Jotwani. 1980. Present status of field pests of sorghum and millets in India. Tropical Pest Management 26: 138–151.

Gami, J. M., J. G. Bapodra, and R. R. Rathod. 2004. Estimation of avoidable yield loss due to pest complex in mustard. Agricultural Science Digest 24(4): 309–310.

Garrison, R. 2011. Bagrada bug, Pentatomidae: *Bagrada hilaris* (Burmeister), pp. 14–15. *In* S. Gaimari and M. O'Donnell (Eds.), California Plant Pest Disease Report, January 2009–December 2009 25: 1–108.

Germar, E. F. 1838. Hemiptera Heteroptera promontorii Bonae Spei nondum descripta, quae collegit C. F. Drége. Silbermann's Revue Entomologique 5[1837]: 121–192.

Ghosal, T. K. 2006. Field observation of bio-control potential of *Trissolcus* sp. (Hymenoptera; Scelionidae). Insect Environment 12(1): 30.

Ghosal, T. K., J. Ghosh, S. K. Senapati, and D. C. Deb. 2006. Biology, seasonal incidence and impact of some insecticides on painted bug, *Bagrada hilaris* (Burm.). Journal of Applied Zoological Researches 17: 9–12.

Gibbs, A. J. 1957. *Leptomonas serpens* n. sp. parasitic in the digestive tract and salivary glands of *Nezara viridula* (Pentatomidae) and in the sap of *Solanum lycopersicum* (tomato) and other plants. Parasitology 47: 297–303.

Grasswitz, T. R. 2013a. Greenhouse evaluation of two organically acceptable foliar insecticides for control of bagrada bug, 2011. Arthropod Management Tests 38. http://amt.oxfordjournals.org/content/38/1/E11 (accessed 20 March 2015).

Grasswitz T. R. 2013b. Greenhouse evaluation of two botanical insecticides for control of brassica-feeding stink bugs, 2012. Arthropod Management Tests 38: E23. http://dx.doi.org/10.4182/amt.2013.E23. http:// amt.oxfordjournals.org/content/38/1/E23 (accessed 1 May 2015).

Grasswitz, T. R. 2014. Field evaluation of organically acceptable foliar insecticides for control of bagrada bug. Arthropod Management Tests 39: E8. http://amt.oxfordjournals.org/cgi/reprint/amt.2014.E8 (accessed 12 June 2015).

Guarino S, E. Peri, P. Lo Bue, A. La Pillo, and S. Colazza. 2007. Use of botanical insecticides to control *Bagrada hilaris* populations on caper crops on Pantelleria Island. Informatore Fitopatologico 57(7/8): 53–58.

Guarino, S., C. De Pasquale, E. Peri, G. Alonzo, and S. Colazza. 2008. Role of volatile and contact pheromones in the mating behaviour of *Bagrada hilaris* (Heteroptera: Pentatomidae). European Journal of Entomology 105: 613–617.

Gunn, D. 1918. The bagrada bug (*Bagrada hilaris*). Bulletin of the Department of Agriculture in South Africa 9: 1–19.

Gupta, J. C., and D. S. Gupta. 1970. Note on some new hosts of the painted bug (*Bagrada cruciferarum* Kirk.: Pentatomidae, Hemiptera). Indian Journal of Agricultural Science 40: 645–646.

Halbert, S. E., and J. E. Eger. 2010. Bagrada bug (*Bagrada hilaris*) (Hemiptera: Pentatomidae) an exotic pest of Cruciferae established in the western USA. Pest Alert, Fl. Dept. Agr. Consumer Services, Division of Plant Industry. http://www.freshfromflorida.com/pi/pest-alerts/pdf/bagrada-bug-pest-alert.pdf (accessed 14 April 2015).

Harris W. V. 1937. Annotated list of insects injurious to native food crops in Tanganyika. Bulletin of Entomological Research 28: 483–488.

HDOA (Hawaii Department of Agriculture). 2014. New pest advisory. Bagrada bug. Insect feeds on wide range of vegetables. NR 14–27. http://hdoa.hawaii.gov/blog/main/bagrada (accessed 26 March 2015).

Herrich-Schäffer, G. A. W. 1844. Die Wanzenartigen Insekten 7(5): 81–103. C. H. Zeh'schen Buchhandlung, Nürnburg.

Herrich-Schäffer, G. A. W. 1853. Die Wanzenartigen Insekten Index. C. H. Zeh'schen Buchhandlung, Nürnburg. 210 pp.

Herting, B., and F. J. Simmonds. 1971. A catalogue of parasites and predators of terrestrial arthropods. Section A, Volume 1. Slough U.K. (Commonwealth Agricultural Bureaux). 129 pp.

Hill, D. S. 1975. Agricultural insect pests of the tropics and their control. Cambridge University Press, New York, NY. 516 pp.

Hill, D. S. 2008. Pests of crops in warmer climates and their control. Springer, London. 704 pp.

Hinton, H. E. 1981. Biology of insect eggs, vols. 1–3. Pergamon, New York, NY. 1125 pp.

Horvath, G. 1909. Adnotationes synonymicae de Hemipteris nonnullis extraeuropaeis. Annales Musei Nationalis Hungarici 7: 631–632.

Horvath, G. 1936. Monographia Pentatomidarum genero *Bagrada*. Annales Musei Nationalis Hungarici 30: 22–47.

Howard, C. W. 1906. The bagrada bug (*Bagrada hilaris*.). Transvaal Agricultural Journal 5: 168–173.

Huang, T-I., D. A. Reed, T. M. Perring, and J. C. Palumbo. 2013. Diel activity and behavior of *Bagrada hilaris* (Hemiptera: Pentatomidae) on desert cole crops. Journal of Economic Entomology 106: 1726–1738. http://dx.doi.org/10.1603/EC13048.

Huang, T-I., D. A. Reed, T. M. Perring, and J. C. Palumbo. 2014a. Feeding damage by *Bagrada hilaris* (Hemiptera: Pentatomidae) and impact on growth and chlorophyll content of brassicaceous plant species. Arthropod-Plant Interaction 8: 89–100.

Huang T-I, D. A. Reed, T. M. Perring, and J. C. Palumbo. 2014b. Host selection behavior of *Bagrada hilaris* (Hemiptera: Pentatomidae) on commercial cruciferous host plants. Crop Protection 59: 7–13.

Husain, M. A. 1925. Annual report of the entomologist to government, Punjab, Lyallpur for the year ending 30th June, 1924. Report of the Department of Agriculture Punjab 1(2): 55–90.

Hutson, J. C. 1935. The painted bagrada bug (*Bagrada picta*). Tropical Agriculturalist 85: 191–193

Infantino A., L. Tomassoli, E. Peri, and S. Colazza. 2007. Viruses, fungi and insect pests affecting caper. European Journal of Plant Science and Biotechnology 1: 170–179.

Isaac, P. V. 1946. Report of the Imperial Entomologist. Scientific Reports of the Agricultural Research Institute, New Delhi, pp. 73–79.

Janisch. E. 1931. Experimental studies on the influence of environmental factors on insects. II. On the mortality of insects in Ceylon, and the extent of its variation, with general notes on the environmental influences and the biological optimum. Zeitschrift für Morphologie und Ökologie der Tiere 20(2–3): 287–348.

Johri, P. K., R. Maurya, D. Singh, D. Tiwari, and R. Johri. 2004. Comparative toxicity of seven indigenous botanical extracts against the infestative stage of three insect pests of agricultural importance. Journal of Applied Zoological Researches 15: 202–204.

Jones, W. A., and R. Sforza 2007. The European Biological Control Laboratory: an existing infrastructure for biological control of weeds in Europe. Bulletin OEPP/EPPO 37: 163–165.

Jones, W. A., J. E. Powell, and E. G. King. 1985. Stoneville Research Quarantine Facility: A national center for support of research on biological control of arthropod and weed pests. Bulletin of the Entomological Society of America 31: 21–26.

Joseph, S. V. 2014. Effect of trap color on captures of bagrada bug, *Bagrada hilaris* (Hemiptera: Pentatomidae). Journal of Entomological Science 49: 318–321.

Joshi, M. L., D. B. Ahuja, and B. N. Mathur. 1989. Loss in seed yield by insect pests and their occurrence on different dates of sowing in Indian mustard (*Brassica juncea* sbsp. *juncea*). Indian Journal of Agricultural Sciences 59: 166–168.

Kirkaldy, G. W. 1909. Catalogue of the Hemiptera (Heteroptera) with biological and anatomical references, lists of foodplants and parasites, etc. Volume I. Cimicidae. Berlin. xl + 392 pp.

Klug, J. C. F. 1845. Symbolae physicae, seu icones et descriptiones insectorum, quae ex itinere per African borealem et Asiam F. G. Hemprich et C. H. Ehrenberg studio novae aut illustratae redierunt. Berolini. Mittler, Berlin. Part 5, 50 plates.

Knowlton, G. F. 1944. *Collops* feeding. Journal of Economic Entomology 37: 443.

Lal, O. P., and B. Singh. 1993. Outbreak of the painted bug, *Bagrada hilaris* (Burm.) (Hemiptera: Pentatomidae) on mustard in northern India. Journal of Entomological Researches 17: 155–57.

Lambert, A. M., and T. L. Dudley. 2014. Exotic wildland weeds serve as reservoirs for a newly introduced cole crop pest, *Bagrada hilaris* (Hemiptera: Pentatomidae). Journal of Applied Entomology 138: 795–799.

Lawrence, C. 2012. Bagrada bug population explodes at organic farms. *Ventura Star.* http://www.vcstar.com /business/bagrada-bug-population-explodes-at-organic-farms (accessed 12 March 2015).

LeVeen, E., and A. C. Hodges. 2015. Bagrada bug, painted bug, *Bagrada hilaris* (Burmeister) (Insecta: Hemiptera: Pentatomidae). Univ. of Florida, IFAS Pub. EENY596, http://edis.ifas.ufl.edu/in1041 (accessed 26 March 2015).

Linnavuori, R. E. 1975. Hemiptera: Heteroptera of the Sudan with remarks on some species of the adjacent countries. Part 5. Pentatomidae. Boletim da Sociedade Portuguesa de Ciências Naturais 15: 5–128.

Lounsbury, C. P. 1898. The bagrada bug of cabbage and allied plants. Agricultural Journal of the Cape of Good Hope 13: 101–105.

Mahar, M. M. M. 1974. Carry over and host plants of painted bug, *Bagrada picta* Fabr., (Pentatomidae: Heteroptera): a pest of rabi oilseed crops. Agriculture Pakistan 24(1)[1973]: 9–10.

Mahmood, R., W. A. Jones, B. E. Bajwa, and K. Rashid. 2015. Egg parasitoids from Pakistan as possible classical biological control agents of the invasive pest Bagrada hilaris (Heteroptera: Pentatomidae). Journal of Entomological Science 50: 147–149.

Makwali, J. A., F. M. E. Wanjala, and B. M. Khaemba. 2002. *Raphanus raphanistrum* L as a diversionary host of *Brevicoryne brassicae* L and *Bagrada cruciferarum* Kirk. *In* J. M. Wesonga, T. Losenge, C. K. Ndung'u, K. Ngamau, F. K. Ombwara, S. G. Agong, A. Fricke, B. Hau and H. Stlitzel (Eds.), Proceedings of the Horticultural Seminar on Sustainable Horticultural Production in the Tropics, 3rd–6th October 2001. Jomo Kenyatta University of Agriculture and Technology, Juja, Kenya.

Malik S, T. Jabeen, B. K. Solangi, and N. A. Qureshi. 2012. Insect pests and predators associated with different mustard varieties at Tandojam. Sindh University Research Journal (Sci. Ser.) 44: 221–226.

Mani, M. S, 1941. Studies on Indian parasitic Hymenoptera II. Indian Journal of Entomology 3: 153–162.

Manzar, A., M. N. Lal, and S. S. Singh. 1999. Population dynamics of mustard sawfly (*Athalia proxima lugens* Klug.), leafminer (*Chromatomyia horticola* Gaureau) and painted bug (*Bagrada cruciferarum* Kirkaldy) on different varieties of *Brassica* oilseed crop. Journal of Entomological Research 23: 359–364.

Mari, J. M., and M. K. Lohar. 2010. Pests and predators recorded in *Brassica* ecosystem. Pakistan Journal of Agriculture, Agricultural Engineering, and Veterinary Science 26: 58–65.

Matsunaga, J. N. 2014. Bagrada bug, *Bagrada hilaris* (Burmeister) (Hemiptera: Pentatomidae). State of Hawaii Department of Agriculture, New pest advisory. 14-02, December 2014. http://hdoa.hawaii.gov /pi/files/2013/01/Bagrada-hilaris-NPA12-9-14.pdf (accessed 15 April 2015).

McPherson, J. E. 1982. The Pentatomoidea (Hemiptera) of northeastern North America with emphasis on the fauna of Illinois. Southern Illinois University Press, Carbondale, Il. 240 pp.

McPherson, J. E., and R. M. McPherson. 2000. Stink bugs of economic importance in America North of Mexico. CRC Press, Boca Raton, FL. 253 pp.

Mukerji, G. P. 1958. Some aspects of the biology of *Bagrada picta* Fabr. Agra University Journal of Research Science 7: 143–157.

Nadda, G., S. C. Eswara Reddy, and A. Shanker. 2013. Insect and mite pests on tea and their management, 317–333 pp. *In* A Gulati, R. D. Singh, R. K. Sud, and R. C. Boruah (Eds.), Science of tea technology. Scientific Publishers, Jodhpur, India. 476 pp.

Nagar, R., Y. P. Singh, R. Singh, and S. P. Singh. 2011. Biology, seasonal abundance and management of painted bug (*Bagrada hilaris* Burmeister) in eastern Rajasthan. Indian J. Entomology 73: 291–295.

Narayanan, E. S. 1954. Seasonal pests of crops. The painted bug, *Bagrada cruciferarum* Kirkaldy. Indian Farming 3(10): 8–9.

Narayanan, E. S. 1958. Insect pests of rape and mustard and methods of their control, pp. 87–97. In D. Singh (Ed.), Rape and mustard. Indian Council of Agriculture Research, New Delhi, India. 102 pp.

Narayanan, E. S., B. R. Subba Rao, and R. N. Katiyar. 1959. Population studies on *Hadrophanurus* species (Scelionidae: Hymenoptera), egg parasite of *Bagrada cruciferarum* Kirk. on maize (*Zea mays*) at Karnal. Proceedings of the National Institute of Science, India. Section B.: Biological Sciences 25: 315–320.

Nielson, M. W., and J. A. Henderson. 1959. Biology of *Collops vittatus* (Say) in Arizona, and feeding habits of seven predators of the spotted alfalfa aphid. Journal of Economic Entomology 52: 159–162.

Nyabuga, F. 2008. Sustainable management of cabbage aphid and bagrada bugs: case study from Kenya, p. 70. VDM Verlag, Saarbrücken, Germany.

Obopile M., D. C. Munthali, and B. Matilo. 2008. Farmers' knowledge, perceptions and management of vegetable pests and diseases in Botswana. Crop Protection 27: 1220–1224.

Palumbo, J. C. 2012. Control of *Bagrada hilaris* with conventional and experimental insecticides on broccoli. Arthropod Management Tests 37: E9. http://dx.doi.org/10.4182/amt.2012.E9.

Palumbo, J. C. 2014. Bagrada bug management on desert cole crops (September 17, 2014). Vegetable IPM Update, Univ. of Arizona, Coop Ext, Tucson, Volume 5, No. 20. http://ag.arizona.edu/crops/vegetables/advisories/more/insect116.html (accessed 13 June 2015).

Palumbo J. C. 2015a. Impact of bagrada bug on desert cole crops from 2010–2014, Vegetable IPM Update, Univ. of Arizona, Coop Ext, Tucson, Volume 5, No. 11. http://ag.arizona.edu/crops/vegetables/advisories/more/insect132.html (accessed 13 June 2015).

Palumbo, J. C. 2015b. Soil-surface-applied insecticides for control of *Bagrada hilaris* (Hemiptera: Pentatomidae) in broccoli, 2014. Arthropod Management Tests 40: 1–2.

Palumbo, J. C., and Y. Carrière. 2015. Association between *Bagrada hilaris* density and feeding damage in broccoli: implications for pest management. Plant Health Progress 16: 158–162. http://dx.doi.org/10.1094/PHP-RS-15-0024.

Palumbo, J. C., and T-I. Huang. 2014a. Control of *Bagrada hilaris* with conventional insecticides on broccoli, 2013. Arthropod Management Tests 39: E40. http://dx.doi.org/10.4182/amt.2014.E40.

Palumbo, J. C., and T-I. Huang. 2014b. Control of *Bagrada hilaris* with experimental insecticides on broccoli, 2013. Arthropod Management Tests 39: E41, http://dx.doi.org/10.4182/amt.2014.E41.

Palumbo, J. C., and E. T. Natwick. 2010. The bagrada bug (Hemiptera: Pentatomidae): a new invasive pest of cole crops in Arizona and California. Plant Health Progress, on-line. http://dx.doi.org/10.1094/PHP-2010-0621-01-BR. http://www.plantmanagementnetwork.org/pub/php/brief/2010/bagrada (accessed 25 March 2015).

Palumbo, J. C., T-I. Huang, T. M. Perring, D. A. Reed, and N. Prabhaker. 2013a. Control of *Bagrada hilaris* on broccoli with organically-approved insecticides. Arthropod Management Tests 38: E7. http://dx.doi.org/10.4182/amt.2013.E7.

Palumbo, J. C., T-I. Huang, T. M. Perring, D. A. Reed, and N. Prabhaker. 2013b. Evaluation of conventional insecticides for control of *Bagrada hilaris* on broccoli, 2012. Arthropod Management Tests 38: E4. http://dx.doi.org/10.4182/amt.2013.E4.

Palumbo, J. C, T-I. Huang, T. M. Perring, D. A. Reed, and N. Prabhaker. 2013c. Evaluation of experimental insecticides for control of *Bagrada hilaris* on broccoli, 2012. Arthropod Management Tests 38: E5. http://dx.doi.org/10.4182/amt.2013.E5.

Palumbo, J. C., T-I. Huang, T. M. Perring, D. A. Reed, and N. Prabhaker. 2013d. Evaluation of foliar insecticides for control of *Bagrada hilaris* and sweetpotato whitefly in broccoli, 2012. Arthropod Management Tests 38: E9. http://dx.doi.org/10.4182/amt.2013.E9.

Palumbo, J. C., T-I. Huang, T. M. Perring, D. A. Reed, and N. Prabhaker. 2013e. Evaluation of pyrethroid insecticides for control of *Bagrada hilaris* on broccoli, 2012. Arthropod Management Tests 38: E8. http://dx.doi.org/10.4182/amt.2013.E8.

Palumbo, J. C., N. Prabhaker, D. A. Reed, T. M. Perring, S. J. Castle, and T-I. Huang. 2015. Susceptibility of *Bagrada hilaris* (Hemiptera: Pentatomidae) to insecticides in laboratory and greenhouse bioassays. Journal of Economic Entomology 108: 672–682. ; http://dx.doi.org/10.1093/jee/tov010.

Pandey, U. K., M. Pandey, and S. P. S. Chuahan. 1981. Insecticidal properties of some plant material extracts against painted bug *Bagrada cruciferarum*. Indian Journal of Entomology 43: 404–407.

Parsana, G. J., H. J. Vyas, and R. K. Bharodia. 2001. Effect of irrigation, sowing date and nitrogen on the incidence of painted bug, *Bagrada hilaris* Burm. in mustard. Journal of Oilseed Research 18(1): 89–90.

Patidar, J., R. K. Patidar, R. C. Shakyawar, and M. Pathak. 2013a. Bio-efficacy of plant extracts against *Bagrada hilaris*. Annals of Plant Protection Sciences 21(2): 425–426.

Patidar, J., R. K. Patidar, R. C. Shakyawar, and M. Pathak. 2013b. Host preference and survivability of *Bagrada hilaris* (Burmeister, 1835) on off season crops. Annals of Plant Protection Sciences 21(2): 273–275.

Perring, T. M., D. A. Reed, J. C. Palumbo, T. Grasswitz, C. S. Bundy, W. Jones, and T. Royer. 2013. National pest alert: bagrada bug *Bagrada hilaris* (Burmeister) Family Pentatomidae. USDA-NIFA Regional IPM Centers. http://www.ncipmc.org/alerts/bagradabug.pdf (accessed 14 April 2015).

Pruthi, H. S. 1946. Report of the imperial entomologist. Scientific reports of the imperial agricultural research institute, New Delhi, 1941–42 to 1943–44. pp. 64–71.

Puchkova, L. V. 1961. The eggs of Hemiptera-Heteroptera: VI. Pentatomoidea, 2, Pentatomidae and Plataspidae. Entomological Review 40: 63–69.

Rajmohana, K. 2006. A checklist of the Scelionidae (Hymenoptera: Platygastroidea) of India. Zoos' Print Journal 21(12): 2506–2613.

Rajmohana, K., and T. C. Narendran. 2007. A systematic note on the genus *Trissolcus* Ashmead (Hymenoptera: Scelionidae) with a key to species from India. Records of the Zoological Survey of India 107(3): 101–103.

Rajpoot, S., R. P. Singh, and V. Pandey. 1996. Estimation of free amino acids in different developmental stages of painted bug (*Bagrada cruciferarum*, Kirkaldy). National Academy of Sciences Letters (India) 19(11–12): 214–218.

Rakshpal, R. 1949. Notes on the biology of *Bagrada cruciferarum* Kirk. Indian Journal of Entomology 11: 11–16.

Rakshpal, R. 1954. Effect of the sex of the host (*Bagrada cruciferarum*) on the sex of its parasite, *Alophora* sp. (Tachinidae). Indian Journal of Entomology 16: 80.

Rawat, R. R., and O. P. Singh. 1980. India, painted bug on wheat. FAO Plant Protection Bulletin 28(2): 77–78.

Reddy, P. P. 2013. Recent Advances in Crop Protection. Springer, New York, NY. 261 pp. http://dx.doi .org/10.1007/978-81-322-0723-8.

Reed, D. A., J. P. Newman, T. M. Perring, J. A. Bethke, and J. N. Kabashima. 2013a. Management of the bagrada bug in nurseries. https://cisr.ucr.edu/pdf/uc_ipm_bagrada_bug_nurseries.pdf (accessed 9 May 2014).

Reed, D. A., J. C. Palumbo, T. M. Perring, and C. May. 2013b. *Bagrada hilaris* (Hemiptera: Pentatomidae), an invasive stink bug attacking cole crops in the southwestern United States. Journal of Integrated Pest Management 4: 2013. http://dx.doi.org/10.1603/IPM13007.

Reed, D. A., T. M. Perring, J. P. Newman, J. A. Bethke, and J. N. Kabashima. 2014. Pest notes: Bagrada bug. University of California Agriculture and Natural Resources Publication 74166. http://www.ipm.ucdavis .edu/PMG/PESTNOTES/pn74166.html (accessed 3 March 2015).

Rider, D. A. 2006. Family Pentatomidae, pp. 233–402. *In* B. Aukema, and C. Rieger (Eds.), Catalogue of the Heteroptera of the Palaearctic Region. Volume 5. Netherlands Entomological Society, Amsterdam. 550 pp.

Rider, D. A., L. Y. Zheng, and I. M. Kerzhner. 2002. Checklist and nomenclatural notes on the Chinese Pentatomidae (Heteroptera). II. Pentatominae. Zoosystematica Rossica 11: 135–153.

Rizvi, S. M. A., R. Singh, H. M. Singh, and S. P. Singh. 1986. Painted Bug, *Bagrada hilaris* (Burm.) appears in epidemic form on maize. Narendra Deva Journal of Agricultural Research 1: 173.

Rohilla, H.R., H. Singh, H., H. Singh, and B. S. Chhillar. 2004. Biology of the painted bug, *Bagrada hilaris* (Burm.) on rapeseed mustard. Journal of Oilseeds Research 21: 303–306.

Rombach, M. C., R. M. Aguda, B. M. Shepard, and D. W. Roberts. 1986. Entomopathogenic fungi (Deuteromycotina) in the control of the black bug of rice, *Scotinophara coarctata* (Hemiptera; Pentatomidae). Journal of Invertebrate Pathology 48: 174–179.

Sabyasachi, P., T. B. Maji, and P. Mondal. 2013. Incidence of insect pest on okra, *Abelmoschus esculentus* (L) Moench in red lateritic zone of West Bengal. Journal of Plant Protection Sciences 5(1): 59–64.

Sachan, G. C., and J. P. Purwar. 2007. Integrated pest management in rapeseed and mustard, pp. 399–423. *In* P. C. Jain, and M. C. Bhargava (Eds.) Entomology: novel approaches. New India Publishing, New Delhi. 533 pp.

Sahito, H. A., A. G. Lanjar, and B. Mal. 2010. Studies on population dynamics of sucking insect pests of mustard crop (*Brassica campestris*). Pakistan Journal of Agriculture, Agricultural Engineering and Veterinary Science 26: 66–74.

Samuel, K. C. 1942. Biological notes on two new egg-parasites of *Bagrada picta* Fabr (*sic*) Pentatomidae. Indian Journal of Entomology 4: 92–93.

Sánchez-Peña, S. R. 2014. First record in Mexico of the invasive stink bug *Bagrada hilaris* on cultivated crucivers in Saltillo. Southwestern Entomologist 39: 375–377. http://dx.doi.org/10.3958/059.039.0219.

Sandhu, G. S. 1975. Seasonal movement of the painted bug, *Bagrada cruciferarum* Kirkaldy from cruciferous plants to graminaceous plants and its occurrence as a serious pest of maize, sorghum and pearl millet during spring in Punjab. Indian Journal of Entomology 37: 215–217.

Sandhu, G. S., and G. S. Deol. 1976. New records of pest on wheat. Indian Journal of Entomology 37 [1975]: 85–56.

Sandhu, G. S., B. Singh, and J. S. Bhalla. 1974. Relative efficacy of different insecticides for control of painted bug, *Bagrada cruciferarum* Kirk. (Hemiptera-Pentatomidae), on pearl millet. Indian Journal of Agricultural Sciences 44(3): 165–166.

Santa Ana, R. 2015. New insect threatens south Texas vegetable crops. The Monitor, 4 June 2015. http://www.themonitor.com/news/local/new-insect-threatens-south-texas-vegetable-crops/article_56156f68-0af3-11e5-974f-77bdfc0170de.html (accessed 5 June 2015).

Sarup, P., D. S. Singh, P. Sircar, S. A. Marpuri, R. Lal, V. S. Saxena, and V. S. Srivastava. 1972. Relative toxicity of some important pesticides to the adults of *Bagrada cruciferarum* (Pentatomidae: Hemiptera). Indian Journal of Entomology 33: 452–456.

Sharma, A. 2013. Insect pests of some important fodder trees grown under agroforestry conditions in Solan district of Himachal Pradesh. Journal of Entomological Research 37(2): 181–186.

Sharma, A., A. Sood, and T. D. Verma. 2008. Insect-pests associated with *Robinia pseudoacacia* in the agroforestry system in mid-hill regions of Himachal Pradesh. Indian Forester 134(1): 120–124.

Shivalingaswamy, T. M., and S. Satpathy. 2007. Integrated pest management in vegetable crops pp. 353–375. *In* P. C. Jain and M. C. Bhargava (Eds.), Entomology: Novel Approaches. New India Publishing, New Delhi, India. 533 pp.

Singh, A., and S. Gandhi. 2012. Agricultural insect pest: occurrence and infestation level in agricultural fields of Vadodara, Gujarat. International Journal of Science and Research Publications 2: 1–5.

Singh, A. P., P. P. Singh, and Y. P. Singh. 2007. Pest complex of Indian mustard, *Brassica juncea* (L.) Czern & Cosson, in Eastern Rajasthan. Journal of Experimental Zoology 10: 415–416.

Singh C., M. N. Lal, and P. S. Singh. 2012. Occurrence of insect pests on rapeseed, *Brassica rapa* (Var. YST-151) and its association with weather variables in eastern Uttar Pradesh. Indian Journal of Entomology 74: 183–185.

Singh, D., S. Vennila, H. R. Sardana, M. N. Bhat, and O. M. Bambawale. 2011. *Bagrada hilaris*: a new pest of tomato. Annals of Plant Protection Sciences 19(2): 472–473.

Singh, H., and V. S. Malik. 1993. Biology of (the) painted bug (*Bagrada cruciferarum*). Indian Journal of Agricultural Sciences 63: 672–674.

Singh, P. K., A. K. Singh, H. M. Singh, P. Kumar, and C. B. Yadav. 2009. Insect-pests of spine gourd (*Momordica dioica* Roxb.) and efficacy of some insecticides against the epilachnid beetle, *Henosepilachna vigintioctopunctatu* (F.), a serious pest of this crop. Pest Management and Economic Zoology. 17: 85–91.

Singh, R., and A. K. Joshi. 2004. Pests of okra (*Abelmoschus esculentus* Moench.) in Paonta Valley, Himachal Pradesh. Insect Environment 9(4): 173–174.

Singh, S. P., Y. P. Singh, and A. Kumar. 2011. Bioefficacy evaluation of chemical insecticides against painted bug, *Bagrada hilaris* (Durm.) in mustard. Pesticide Research Journal 23: 150–153.

Sooby, J. 2014. Bagrada bug update. California Certified Organic Farmers. www.ccof.org/blog/bagrada-bug-update (accessed 2 March 2015).

Sosa-Gomez, D. R., and F. Moscardi. 1998. Laboratory and field studies on the infection of stink bugs, *Nezara viridula, Piezodorus guildinii,* and *Euschistus heros* (Hemiptera: Pentatomidae) with *Metarhizium anisopliae* and *Beauveria bassiana* in Brazil. Journal of Invertebrate Pathology 71: 115–120.

Sozzi, G. O. and A. R. Vicente. 2006. Capers and caperberries, 230–256 pp. *In* K. V. Peter (Ed.), Handbook of herbs and spices, vol 3. CRC Press, Boca Raton, FL. 537 pp.

Srinivasan, K., and P. N. Krishna Moorthy. 1992. Development and adoption of integrated pest management for major pests of cabbage using Indian mustard as a trap crop, pp. 511–521. *In* N. S. Talekar (Ed.), Management of diamondback moth and other crucifer pests. Proceedings of the Second International Workshop. Asian Vegetable Research and Development Center. Shanhua, Taiwan.

Srivastava, S., and M. Srivastava. 2012. Entomo-fauna associated with bajra crop as observed in an agro-ecosystem in Rajasthan, India. International Journal of Theoretical and Applied Science 4(2): 109–121.

Stål, C. 1862. Hemiptera Mexicana enumeravit speciesque novas descripsit. Stettiner Entomologische Zeitung 23(1-2): 81–118.

Stål, C. 1865. Hemiptera Africana. Volume 1. Norstedtiana, Stockholm. iv + 256 pp.

Stebbing, E. P. 1903. Insect pest of indigo. Indian Museum Notes 5: 144.

Stichel, W. 1961. Illustrierte Bestimmungstabellen der Wanzen. II. Europa (Hemiptera-Heteroptera Europae). Volume 4. Martin-Luther, Berlin-Hermsdorf. 838 pp.

Stubbins, F., P. Agudelo, F. Reay-Jones, and J. K. Greene. 2015. First report of a mermithid nematode infecting the invasive *Megacopta cribraria* (Hemiptera: Plataspidae) in the United States. Journal of Invertebrate Pathology 127: 35–37.

Suazo, A. A., J. E. Spencer, E. C. Engel, and S. R. Abella. 2012. Responses of native and non-native Mojave Desert winter annuals to soil disturbance and water additions. Biological Invasions 14: 215–227. http://dx.doi.org/10.1007/s10530-011-9998-6.

Subba Rao, B. R., and M. J. Chacko. 1961. Studies on *Allophanurus indicus* n. sp., an egg parasite of *Bagrada cruciferarum* Kirkaldy. Beitrage zur Entomologie 11(7/8): 812–824.

Sutherst, R. W., and G. E. Maywald. 1985. A computerized system for matching climates in ecology. Agriculture, Ecosystems and Environment 13: 281–299.

Sutherst, R. W., G. E. Maywald, W. Bottomley, and A. Bourne. 2004. CLIMEX Version 2. Hearne Scientific Software, Melbourne, Australia.

Swaran Dhingra, S. 1998. Relative toxicity of some important insecticides with particular reference to change in susceptibility level of *Bagrada cruciferarum* Kirk. during the last quarter century. Journal of Entomological Research 22: 307–311.

Tandon, P. L., and B. Lal. 1977. New records of insect pests of mango from India. Indian Journal of Horticulture 34: 193–195.

Taylor, M. E. 2015. Life history and laboratory rearing of *Bagrada hilaris* (Burmeister) (Hemiptera: Heteroptera: Pentatomidae) with descriptions of immature stages and ovipositional behavior. MS Thesis, New Mexico State University. 80 pp.

Taylor, M. E., C. S. Bundy, and J. E. McPherson. 2014. Unusual ovipositional behavior of the stink bug *Bagrada hilaris* (Hemiptera: Heteroptera: Pentatomidae). Annals of the Entomological Society of America 107: 872–877. http://dx.doi.org/10.1603/AN14029.

Taylor, M. E., C. S. Bundy, and J. E. McPherson. 2015. Life history and laboratory rearing of *Bagrada hilaris* (Hemiptera: Heteroptera: Pentatomidae) with descriptions of immature stages. Annals of the Entomological Society of America 108: 536–551. http://dx.doi.org/10.1093/aesa/sav048.

Thakar, A. V., U. S. Misra, and R. R. Rawat, and S. V. Dhamdhere. 1969. A record of predatory mite, *Bochartia* sp. on *Bagrada cruciferarum* Kirkaldy. Indian Journal of Entomology 31: 86.

Thompson, W. R. 1950. A catalogue of the parasites and predators of insect pests. Section 1. Parasite host catalogue. Part 3. Parasites of the Hemiptera. Imperial Agricultural Bureau, Institute of Entomology, Parasite Service. Belleville, Ontario, Canada. 149 pp.

Tiwari, S. K., and L. Sarvanan. 2009. Life history and seasonal activity of painted bug, *Bagrada hilaris* (Burm.) infesting mustard in Allahabad, Uttar Pradesh. Pestology 33: 21–23.

Tomar, S. P. S., O. P. Dubey, and R. Tomar. 2004. Succession of insect pests on green pea. JNKVV Research Journal 38(1): 82–85.

Torres-Acosta, R. I., and S. Sanchez-Peña. 2016. Updated geographical distribution of *Bagrada hilaris* (Hemiptera: Pentatomidae) in Mexico. Journal of Entomological Science 51: 165–167.

Vekarta, M. V., and G. M. Patel. 1999. Succession of important pests of mustard in North Gujrat. Indian Journal of Entomology 61: 356–361.

Verma, G. S., and U. K. Pandey. 1981. Studies on the effect of *Acorus calamus*, *Cimicifuga foetida* and *Gynandropsis gynandra* extract against insect pests of cruciferous vegetables, painted bug *Bagrada cruciferarum* Kirk. (Hemiptera: Pentatomidae). Zeitschrift für Angewandte Zoologie 68: 109–113.

Verma, A. K., S. K. Patyal, O. P. Bhalla, and K. C. Sharma. 1993. Bioecology of painted bug (*Bagrada cruciferarum*) (Hemiptera: Pentatomidae) on seed crop of cauliflower (*Brassica oleracea* var *botrytis* subvar *cauliflora*). Indian Journal of Agricultural Sciences 63: 676–678.

Verma, S. K. 1980. Field pests of pearl millet in India. Tropical Pest Management 26: 13–20.

Vitanza, S. 2012. Texas Agrilife Extension El Paso County, IPM Program Newsletter 37(6). http://elp.tamu .edu/files/2014/04/120830.pdf (accessed 1 May 2015).

Voss, S. C., and J. E. McPherson. 2003. Life history and laboratory rearing of *Sinea diadema* (Heteroptera: Reduviidae) with descriptions of immature stages. Annals of the Entomological Society of America 96: 776–792.

Walker, J. K. 1957. A biological study of *Collops balteatus* Lec. and *Collops vittatus* (Say). Journal of Economic Entomology 50: 395–399.

Westwood, J. O. 1837. *In* F. W. Hope, A Catalogue of Hemiptera in the collection of the Rev. F. W. Hope, M. A. with short Latin descriptions of the new species. J. C. Bridgewater, London. 1837, Part 1: 1–46; 1842, Part 2: 1–26 (stating that descriptions are by J. O. Westwood).

4

Halyomorpha halys (Stål)

George C. Hamilton, Jeong Joon Ahn, Wenjun Bu, Tracy C. Leskey,
Anne L. Nielsen, Yong-Lak Park, Wolfgang Rabitsch, and Kim A. Hoelmer

Halyomorpha halys (Stål)[1]

1855	*Pentatoma halys* Stål, K. Svens. Vet.-Akad. Handl. 12(4): 181–192.
1860	*Poecilometis mistus* Uhler, Proc. Acad. Nat. Sci. Philadelphia 12: 223. [Synonymized by Josifov and Kerzhner, 1978, Frag. Faun. 23(9): 172].
1865	*Cappaea halys*: Stål, Ann. Soc. Ent. Fr. 1865(4)5: 170.
1867	*Dalpada brevis* Walker, Cat. Hem.-Het. Brit. Mus. Lond.: 226–227. [Synonymized by Josifov and Kerzhner, 1978, Frag. Faun. 23(9): 172].
1867	*Dalpada remota* Walker, Cat. Hem.-Het. Brit. Mus. Lond.: 227. [Synonymized by Josifov and Kerzhner 1978, Frag. Faun. 23(9): 172].
1874	*Halyomorpha picus*: Scott, Ann. Mag. Nat. Hist. (4)14(82): 290. (misidentification).
1881	*Halyomorpha timorensis*: Signoret, Bull. Soc. Ent. de Fr. 1881(5): 46. (misidentification).
1893	*Dalpada brevis* var. *remota*: Lethierry and Severin, Cat. Hem. 1: 98.
1914	*Halyomorpha brevis*: Bergroth, Ent. Zeit. 33(5–6): 182.
1965	*Halyomorpha mista*: Hasegawa, Japan Plant Prot. Assoc., Tokyo: 167, 171, 175, 176, 267.
1978	*Halyomorpha halys*: Josifov and Kerzhner, Frag. Faun. 23(9): 172.

CONTENTS

4.1 Introduction 244
4.2 Taxonomy and Identification 246
4.3 Worldwide Distribution 247
    4.3.1 Asia 247
    4.3.2 Europe 247
    4.3.3 North America 248
    4.3.4 Other Areas 249
4.4 Biology 249
    4.4.1 Life History 249
    4.4.2 Symbiotic Relationships 256
    4.4.3 Mate Selection 256
    4.4.4 Overwintering Behavior/Diapause 257
    4.4.5 Host Plants 258
    4.4.6 Movement between Host Plants 258
    4.4.7 Spatial and Nutritional Attributes in Relation to *Halyomorpha halys* Movement 259
        4.4.7.1 General Movement Patterns of *Halyomorpha halys* in Agricultural Landscapes 259
        4.4.7.2 Spatial Attributes Associated with Movement from Overwintering Sites to Surrounding Vegetation 259

---

[1] Synonomy adapted from David A. Rider (personal communication; see Acknowledgments).

4.4.7.3   Nutritional Attributes Associated with Host Plants and Cropping
           and Management Practices affecting *Halyomorpha halys* Movement........... 260
           4.4.7.3.1   Nutritional Value of Plants 260
           4.4.7.3.2   Intercropping 260
           4.4.7.3.3   Chemical Control 260
4.4.7.4   Movement from Crops and Surrounding Vegetation to Overwintering Sites... 260
4.4.8   Potential Geographic Distribution of *Halyomorpha halys:* Ecological Niche Modeling....261
4.5   Current Impacts.. 264
    4.5.1   *Halyomorpha halys* in Asia... 264
           4.5.1.1   Ecology of *Halyomorpha halys* and Its Economic Damage Reported in Asia... 264
                   4.5.1.1.1   Fruit 264
                   4.5.1.1.2   Field Crops and Vegetables 265
                   4.5.1.1.3   Specialty Crops: Tea, Paulownia, Orchid, and Matrimony Vine... 265
           4.5.1.2   Management of *Halyomorpha halys* in Asia....................... 266
                   4.5.1.2.1   Chemical Control Including Semiochemicals............. 266
                   4.5.1.2.2   Biological Control 266
                   4.5.1.2.3   Cultural and Mechanical Control................... 267
    4.5.2   *Halyomorpha halys* in Europe 267
           4.5.2.1   Phenology in Europe.. 268
           4.5.2.2   Host Range in Europe 268
    4.5.3   *Halyomorpha halys* in North America 268
           4.5.3.1   Urban Nuisance Pest....................................... 268
           4.5.3.2   Conventional Agricultural Pest............................. 269
           4.5.3.3   Organic Agricultural Pest 271
4.6   Current Management Tactics 271
    4.6.1   Monitoring Options 272
    4.6.2   Biofix Models 273
    4.6.3   Cultural Control... 274
    4.6.4   Biological Control... 274
    4.6.5   Chemical Control 277
4.7   What Does the Future Hold?... 278
4.8   Acknowledgments... 278
4.9   References Cited.. 279

4.1 Introduction

Halyomorpha halys (Stål), the brown marmorated stink bug (**Figure 4.1D**), is a pentatomid species native to Asia, specifically China, Japan, Korea, Myanmar, Taiwan, and Vietnam (**Figure 4.2**). In its native range, it can go through from one to three or more generations per year and, potentially, can build up in high numbers. It is considered a nuisance pest in urban areas because of its overwintering behavior and a periodic pest of agricultural crops such as apples, Asian pears, soybeans and various Asian vegetables. With the increased movement of people and cargo, and the increase in trade, between Asia and other parts of the world, *H. halys* has been found hitchhiking in air freight, shipping cargo, and passenger luggage. It is believed to have entered the United States in the mid-1990s via either the port of Manhattan, Elizabeth, or Philadelphia and subsequently was transported to Allentown, Pennsylvania (Hoebeke and Carter 2003, Hamilton 2009). Once there, populations slowly grew until they reached the point where the bugs became urban nuisance pests because of their presence in homes and buildings during the late fall and winter. As the Allentown population continued to grow, damage began to occur in surrounding agricultural areas and across the Delaware River in Pittstown, New Jersey, as early as 2006 (Nielsen and Hamilton 2009a). In 2009 and 2010, *H. halys* became a severe urban nuisance and agricultural problem in the mid-Atlantic areas of the United States. Since then, it has been detected and/or established in over 41 states and

FIGURE 4.1   **(See color insert.)** *Halyomorpha halys.* A, first instar clustering atop hatched eggs; B, fourth instar; C, fifth instar; D, adult (A–D from laboratory culture); E, adults feeding on nectarine fruit (Kearneysville, WV); F, adults congregating around exterior security light (Spring Mills, WV).

the District of Columbia in the United States, has been detected in British Columbia and Ontario, Canada, and has established populations in Ontario. Outside North America, it has been found in shipments from the United States and Asia to New Zealand and Australia, has been detected or become established in several European countries and in Santiago, Chile, and has become an agricultural pest in northern Italy.

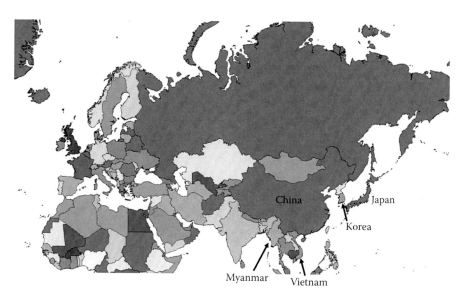

FIGURE 4.2   Distribution of *Halyomorpha halys* in Asia.

4.2 Taxonomy and Identification

There has been considerable confusion surrounding the systematics of *Halyomorpha halys* (Rider et al. 2002). First, Distant (1880, 1893, 1899) considered this taxon to be a junior synonym of *H. picus* (F.). Later, it was determined that this species was different from *H. picus*, but various authors used either *H. brevis* (Walker) or *H. remota* (Walker) as the proper name. Josifov and Kerzher (1978) determined that there was only one species of *Halyomorpha* occurring in Japan, Korea, and eastern China, and that *H. halys* (Stål) actually had priority. All citations pertaining to *Halyomorpha* species from these localities should be referred to *H. halys* (David A. Rider, personal communication). Besides its common name, the brown marmorated stink bug, other names such as the yellow-brown stink bug, the stinkwood stink bug, and the pear bug have been used in Asia.

Halyomorpha halys is relatively large as adults, ranging from 12 to 17 millimeters (mm) in length and 7 to 10 mm in width (Stål 1855) with females being larger than males (**Figure 4.1D**). They are variable in color with the hemelytra mottled (marmorated) brown in color and the antennae and legs marked with alternating light and dark banding (Hoebeke and Carter 2003). Adults also have exposed alternating light and dark bands along the connexival (outer) margin of the abdomen; the sternal surface is light brown in color and can have a reddish to pinkish hue depending on food source.

Eggs are laid in masses of 20 to 30 eggs (Hoebeke and Carter 2003, Nielsen et al. 2008a) on the undersides of host leaves. Following emergence of the nymphs from the eggs, t-shaped, blackish egg bursters are visible.

Halyomorpha halys has five nymphal instars. First instars are ≈2.4 mm in length, red-orange and black in color, and typically found on or around the egg mass (Hoebeke and Carter 2003, Nielsen et al. 2008a) (**Figure 4.1A**). Second instars are ≈3.7 mm in length, mostly black with light and dark banding on the antennae and legs. As individuals proceed through subsequent instars they increase in size with more pronounced light and dark banding on the antennae and legs (third instar, **Figure 4.1B**). This banding is most pronounced in the fifth instar (**Figure 4.1C**). Wing pads are easily visible in the third, fourth and fifth instars. In addition, in the fifth instar, the presence of a medial suture on the ventral portion of ninth abdominal segment can be used to distinguish females.

4.3 Worldwide Distribution

4.3.1 Asia

Halyomorpha halys is native to Asian countries including China (except Xinjiang and Qinghai provinces), Japan (except Hokkaido), Korea, Taiwan, Vietnam, and Myanmar (Wang and Liu 2005) (**Figure 4.2**). The presence of *H. halys* in North Korea was confirmed in 2010 when *H. halys* specimens collected from North Korea were provided to South Korea by a Hungarian natural history museum (Kim 2010).

4.3.2 Europe

Halyomorpha halys was reported in Europe for the first time in 2007 from Zurich, Switzerland (Wermelinger et al. 2008, Leskey and Nielsen 2018), but apparently had been present at least since 2004 in this area (Tim Haye, unpublished data; see Acknowledgments; Gariepy et al. 2014a) and in adjacent Liechtenstein (Balzers) (Arnold 2009). Numbers have increased steadily over the years, and the species now has spread to other cantons in Switzerland (Aargau, Basel-Stadt, Basel-Landschaft, Bern, Genève, Solothurn, St. Gallen, Schaffhausen, Ticino, Thurgau) (Wyniger and Kment 2010, Haye et al. 2014a), reaching the southern parts of Germany (Konstanz) in November 2011 (Heckmann 2012) and the adjoining eastern border of France (Strasbourg, region Alsace) in 2012 (Callot and Brua 2013). In October and December 2013, it was detected in the region of Île-de-France (Paris, Essonne) some 400 kilometers (km) further to the west (Garrouste et al. 2014). In late 2012, it was detected in Modena (region Emilia Romagna) in northern Italy (Maistrello et al. 2013) and in August 2013 in San Benigno (region Piemonte) (Pansa et al. 2013). A citizen science project uncovered approximately 200 specimens in northern Italy (70% in Emilia Romagna, 14% in Lombardia, 1% in Piemonte) and in the Swiss canton Ticino (15%) (Maistrello et al. 2014, 2016). In late 2013, nymphs and adults were found in the vicinity of Budapest, Hungary (Vétek et al. 2014) at more than several hundred kilometers from the closest known records in Italy. The first report of *H. halys* in England occurred in 2010 when two adults were found in passenger luggage originating in the US (Malumphy 2014). The second occurred in 2013 when an adult was intercepted in Teesport, North Yorhshire in a shipment of stone from China. The first records for Greece were published (Milonas and Partsinevelos 2014) based on individuals becoming a nuisance in houses in the center of Athens in autumn 2011, more than 1,000 km from the closest European populations. It has yet to be determined if the distant populations in Athens, Budapest, and Paris resulted from independent introductions from Asia and North America or were the consequence of a 'bridgehead-effect' (i.e., a translocation from one of the other European populations). The spread within Switzerland is confirmed by genetic data (mitochondrial COI) (Gariepy et al. 2014b) and could have been fostered either by natural dispersal or short-distance human translocation of infested goods or materials or other means of transportation. Barcoding data of the Italian populations revealed that the Lombardia population is similar to individuals from adjoining Ticino, indicating natural spread or short-distance translocation, whereas the population in Emilia Romagna has a low genetic diversity indicating that although their origin still could not be resolved satisfactorily, a separate invasion event is likely (Cesari et al. 2015). Recently, analyses of haplotype diversity indicated secondary invasion movements within Europe as well as the occurrence of multiple invasions from Asia (Gariepy et al. 2015). The first specimen found in Romania was collected on September 15, 2014 from a botanical garden in Bucharest (Macavei et al. 2015). Additional breeding populations were also found up to 5 km away from the original site suggesting that the bug may have been present in the city for 1-2 years. In August 2015, the first specimens were observed in the west (Vorarlberg) and east (Vienna) of Austria (Rabitsch and Friebe 2015). It seems most likely that the western population has reached the country by natural spread from nearby Swiss populations. The eastern population could have arrived from adjacent Hungary or from an independent introduction. In Serbia, *H. halys* was first found in October of 2015 via a Facebook post to Insects of Serbia and a posting to the Forum on Biological Diversity (Šeat 2015). In 2017, the first record of *H. halys* in Slovakia occurred in the southern town Sturrovo when a fifth instar was collected (Hemala and Kment

2017). The first individual found in Russia occurred in 2014 when three nymphs were collected on *Rosa* spp. and *Pittosporum tobira* (Thunb.) in Sochi which is part of the Krasnodar region (Mityushev 2016). Since then brown marmorated stink bug has become established in Sochi and has spread to Abkhazia and Georgia (Gapon 2016). The first record in Switzerland, in the heart of the continent and not at a trans-shipment center or within a shipping container, indicates airplanes and air cargo as the first introduction pathway. The vectors to Europe most likely are ornamental plants or fruits (Wermelinger et al. 2008) and the species, therefore, was introduced as a contaminant with these vectors or as a stowaway with packing materials. In the latter case, the autumnal shelter seeking behavior appears as a predetermined ultimate catalyst of such translocations. The species also was intercepted during phytosanitary border controls (e.g., 2011 in Bremerhaven, Germany, in shipments of machine parts from the United States [Freers 2012]). The utilization of these pathways may support additional influx of the species into Europe from different origins.

The spread of distance records of *Halyomorpha halys* across wide geographical areas within Switzerland, and even more so within Europe, exceeds the expected rather low natural dispersal capacity and indicates that secondary translocations through human activities (movement of ornamentals or goods; translocation along roads or railway networks) or repeated independent introductions are likely. Currently available genetic data suggest a mixed origin of the European populations and indicate a need for more sampling and haplotype identification within the native and introduced (North America) ranges (Cesari et al. 2015; Gariepy et al. 2014b, 2015).

4.3.3 North America

It is believed that the first sighting of *Halyomorpha halys* in North America occurred in fall 1996 on Adams Island (Allentown), Pennsylvania, but was not confirmed until 2001 when two specimens were sent to Richard Hoebeke for identification (Hoebeke and Carter 2003, Leskey and Nielsen 2018). An additional specimen was collected from a blacklight trap maintain by the Rutgers Cooperative Extension Vegetable IPM program at a farm in Milltown, New Jersey, in 1999 (Hamilton 2009). It initially was misidentified as *Euschistus servus* (Say). Following this, *H. halys* rapidly spread to other parts of the United States and by 2004 had been found in large areas of Pennsylvania and New Jersey, northern Delaware, and parts of Maryland and West Virginia (**Figure 4.3**) (Leskey and Hamilton 2010a,b). In 2009, populations in the east exploded and received a great deal of press interest and popular pressure. By this time, the bug had been detected and/or become established in 17 additional states including California, Oregon, North Carolina and Virginia (Leskey and Hamilton 2011). Two years later it had been detected in 17 more states (Leskey and Hamilton 2011, Leskey et al. 2012a, Rice et al. 2014). Today, *H. halys* has been detected in over 41 states and the District of Columbia (Leskey and Hamilton 2015).

The occurrence of *Halyomorpha halys* in North America is not limited to the United States. Specimens have been intercepted at Canadian ports since 1993 with the first official report coming

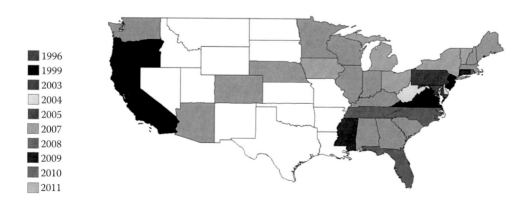

FIGURE 4.3   **(See color insert.)** Distribution of *Halyomorpha halys* in the United States, 1996-2011.

from Balzac, Alberta (Bercha 2008, Fogain and Graff 2011). Additional reports include interceptions in British Columbia in 2010 from lumber shipped from Virginia and Montreal, Quebec, in 2010 near a skid that came from the United States. The first domestic report from a residence was in Hamilton, Ontario, in 2010. Since then, this bug has become established in southern Ontario with multiple findings in Hamilton, Mississauga (two building material shipments from the United States), Burlington, Toronto, Milton, and Newboro in 2012; and Windsor, Cedar Springs, Vaughn, Niagara on the Lake and London in 2013 (Gariepy et al. 2014a).

The occurrence of *Halyomorpha halys* in North America also can be examined based on the impact it has in a geographic area (Anonymous 2015b). This shows that populations of this bug in the mid-Atlantic United States (i.e., Delaware, New Jersey, Pennsylvania, Virginia, and West Virginia) have caused severe agricultural problems and nuisance issues for the general public. Eight states (i.e., Connecticut, Indiana, Kentucky, New York, Ohio, Oregon, Tennessee, and Washington) report agricultural and nuisance issues. Ten states (i.e., Alabama, California, Georgia, Illinois, Michigan, Missouri, South Carolina, Massachusetts, New Hampshire, and Vermont) and the province of Ontario in Canada report nuisance issues only, whereas 16 states (i.e., Arkansas, Arizona, Colorado, Florida, Hawaii, Iowa, Indiana, Kansas, Maine, Minnesota, Mississippi, Nebraska, New Mexico, Texas, Wisconsin, and Utah) and Quebec report that this bug has been detected but has not become established.

The origins of *Halyomorpha halys* in North America have been investigated using DNA analysis. Xu et al. (2014) traced the origins of the United States populations using three mitochondrial DNA genes (COII, 12S Ribosomal DNA and 12S/CR) from specimens collected in 2004 and 2008 from China, South Korea, Japan, and the United States. These analyses suggest that as a single *H. halys* introduction from the Beijing area of China may be responsible for the initial dispersing United States population. Using the same genes, additional haplotype diversity was found in more recent collections of specimens from the west coast of the United States, which suggests that multiple introductions from different parts of *H. halys*'s range in Asia have occurred (M-C. Bon, unpublished data). Gariepy et al. (2014b) also examined the genetic diversity between Asia, Canada, Switzerland, and the United States using the Cytochrome Oxidase I, 28S and Cytochrome b genes from field-collected specimens from these countries and others intercepted prior to establishment in Canada. They concluded that the Canadian populations originated from the movement of established United States populations, and that the North American populations were the result of introductions from the Beijing/Hebei regions in China.

4.3.4 Other Areas

Due to *Halyomorpha halys*' proven propensity to overwinter in shipping materials, etc., areas outside the currently known distribution (Asia, Europe, and North America) are stepping up their efforts to prevent the introduction of this bug (Leskey and Hamilton 2015). Two countries that especially are concerned because of their unique ecological and agricultural systems are Australia and New Zealand (Duthie 2012, Leskey and Hamilton 2015, Lo et al. 2015). Currently, because of the potential threat and increase in interceptions, these two countries now require that inspections of shipping cargo be conducted at the port of origin to prevent these bugs from entering their countries (Leskey and Hamilton 2015).

The first report of *Halyomorpha halys* in South America comes from Chile where it was intercepted by government officals in Iquique in 2011 but was not considered to be established. Today it has become established in three areas in the Santiago metropolitan area (Faúndez and Rider 2017).

4.4 Biology

4.4.1 Life History

Halyomorpha halys is a multivoltine species with as many as five generations reported in southern China (Hoffman 1931). In the mid-Atlantic United States, an area that experiences high population densities, it is primarily bivoltine (Nielsen et al. 2008a). Adults overwinter in artificial and natural shelters and gradually emerge from these sites beginning in April. There are few host plant resources available at this time, and individual bugs are difficult to find. Mating and oviposition by the "spring adults" begin in mid-late May and may continue throughout a female's lifespan, which European studies indicate can last past August (Haye et al. 2014b). In temperate regions, females enter and leave overwintering reproductively immature and unmated (Anne L. Nielsen, unpublished data). After eclosion and presumably diapause termination, females require an additional 68 $DD_{12.7}$ – 148 DD_{25} ($DD_{12.7}$ – degree days above 12.7°C; DD_{25} – degree days above 25°C (Yanagi and Hagihara 1980, Nielsen et al. 2008a) to become reproductively mature. At 30°C, this takes 32–35 days. Adults mate and eggs are oviposited in clusters on the undersides of leaves in groups of 28 (Kawada and Kimura 1983b). Parthenogenisis has also been reported in *H. halys* (Fengjie et al. 1997). The temperature range of the egg incubation period, T_o and T_m (T_o – minimum temperature threshold, T_m – maximum temperature threshold) is similar to those for nymphs and adults at 13.94 and 37.73°C, respectively (Nielsen et al. 2008a). During oviposition, females transfer gut symbionts to the egg chorion surface. After hatching, first instars aggregate on the choria, presumably feeding on the symbionts. Disruption of this behavior significantly reduces survivorship (Nielsen et al. 2008a). Further, surface sterilization of the chorion to remove symbionts results in significant reduction in survival (Taylor et al. 2014). *H. halys* requires 538 DD_{14} to develop from the egg through the five instars to adult eclosion. The minimum (T_o) and maximum (T_m) temperature thresholds are 14.14 and 35.76°C, respectively. These thresholds are similar to those determined by Baek et al. (2017) using a nonlinear Briére model. Temperature also has been shown to impact *H. halys* regional patterns. Venugopal et al. (2016) found using multivariate analysis that average June temperatures and distance from source population negatively impacted *H. halys* populations. They also showed that *H. halys* was not present in fields when average June temperatures were above 23.5°C. When they were present, high population levels were related positively with the percent-developed open area and percent forest cover at a 250 m scale.

Halyomorpha halys is a highly mobile, polyphagous species. The exact number of reported host plants varies between authors, but over 175 have been identified throughout the United States in a collaborative factsheet by researchers (Anonymous 2015b; see **Table 4.1**).

Classification of host plants for polyphagous insects first requires a definition of 'host plant'. Here we define a host plant as a species where multiple life stages occur in multiple years (Nielsen and Hamilton 2009b). Surveys conducted at a semi-landscaped horticultural garden containing plantings of both native and non-native plantings identified host plants during the establishment phase of *Halyomorpha halys* in Allentown, Pennsylvania. All stages of *H. halys* including first instars were found on *Paulownia tomentosa* (Thunb.) Steud. (2005-2007), *Fraxinus americana* (2005–2007), *Liquidambar* spp. (2006), *Viburnum opulus* var *americanum* (2007), *V. prunifolium* (2007), *Pyrus* sp. (2006), and *Pyrus pyrifolia* (2006–2007). The identification of early-season host plants by Nielsen and Hamilton (2009b) through visual and beat surveys on ornamental hosts did not identify significant trends for adult populations. Nymphs however were significantly greater on *P. tomentosa* (Nielsen and Hamilton 2009b). Martinson et al. (2016) also examined naïve plants based on their geographis origin (Asian versus non-Asian) in commercial nurseries. These authors found that *H. halys* was more abundant on non-Asian plants when compared to Asian varieties. Late instar nymphs were observed on host plants where egg masses or young instars had not been observed previously. At the time, it was hypothesized that the nymphs changed host plants depending on nutritional needs and changing host plant phenology. For nymphs, although the first instars tend to remain aggregated around the egg mass, later instars show a strong capacity to disperse in the laboratory and field. In the laboratory, the later instars were capable of climbing 6-8 meters (m) in 15 minutes (Lee et al. 2014b). In the field, the third and fifth instars walked on average 1.3 m and 2.6 m, respectively, over 30 min on a grassy surface. *H. halys* nymphs will disperse to more attractive host plants based on plant phenology with a positive relationship observed as 'fruits' mature (Brett R. Blaauw, unpublished data; see Acknowledgments).

The importance of host plant quality for *Halyomorpha halys* nymphal and adult development has been of considerable interest to researchers. Funayama (2004) showed that females laid significantly fewer eggs when reared on apple than on a high protein diet of peanut and soybean, suggesting that apple did not meet the nutritional requirements. Funayama did not report on the survivability of these offspring on apples. However, laboratory trials by other investigators suggest that apple fruits do not support nymphal development (Acebes-Doria et al. 2016a). Of the other host plants evaluated, *H. halys* nymphs

TABLE 4.1

Known Asian, European, and North American *Halyomorpha halys* Host Plants

Scientific Name[1]	Common Name[2,3,4]	Habitat[5]	Continent Found[6]
Abelia x *grandiflora* (André) Rehd.	glossy abelia[2]	O, L	NA
Abelmoschus esculentus Moench	okra[2]	A	NA
Acacia spp.	acacia[2]	O, L, W, Wo	NA
Acer buergerianum Miq.	trident maple[2]	O, L	NA
Acer campestre L.	hedge maple[2]	O, L	NA
Acer circinatum Pursh	vine maple[2]	O, L	NA
Acer griseum (Franch.) Pax	paperbark maple[2]	O, L	NA
Acer japonicum Thunb.	Amur (Japanese Downy) maple[2]	O, L	NA
Acer macrophyllum Pursh	bigleaf maple[2]	O, L	NA
Acer negundo L.	boxelder[2]	O, L	NA
Acer palmatum Thunb.	Japanese maple[2]	O, L	NA
Acer pensylvanicum L.	striped maple[2]	O, L	NA
Acer platanoides L.	Norway maple[2]	O, L	NA
Acer rubrum L.	red maple[2]	O, L	NA
Acer saccharinum L.	silver maple[2]	O, L	NA
Acer saccharum Marshall	sugar maple[2]	O, L, W	NA
Acer tegmentosum Maxim	Manchurian snakebark maple[2]	O, L	NA
Acer x *freemanii*	Freeman maple[2]	O, L	NA
Acer spp.	maple[2]	O, L, W, Wo	NA
Aesculus x *carnea*	red horse-chestnut[3]	O, L	NA
Aesculus glabra Willd.	Ohio buckeye[2]	O, L	NA
Ailanthus altissima (Mill.) Swingle	tree of heaven[2]	W	Asia, NA
Amaranthus caudatus L.	love-lies-bleeding (amaranth)[2]	A, O, L	NA
Amelanchier laevis Wiegand (syn. *grandiflora*)	Allegheny (apple) serviceberry[2]	O, L	NA
Amelanchier spp.	shadbush[2]	O, L, W, Wo	NA
Antirrhinum majus L.	garden snapdragon[2]	A, O, L	NA
Arctium minus (Hill) Bernh.	lesser burdock[2]	W	NA
Armoracia rusticana G. Gaertn., B. Mey. & Scherb.	horseradish[2]	A	NA
Asimina triloba (L.) Dunal	pawpaw[2]	Wild	NA
Asparagus officinalis L.	asparagus[2]	A	Asia, NA
Astragalus spp.	Chinese milk vetch[4]	W	Asia
Baptisia australis Hort. ex Lehm.	blue wild indigo[2]	O, L	NA
Beta vulgaris spp. *cicla* L.	Swiss chard[2]	A	NA
Betula nigra L.	river birch[2]	O, L, W, Wo	NA
Betula papyrifera Marshall	paper birch[2]	O, L	NA
Betula pendula Roth	European white birch[2]	O, L	NA
Brassica juncea (L.) Vassiliĭ Matveievitch Czernajew	wild mustard[2]	A	NA
Brassica oleracea L.	cabbage, collards[2]	A	NA
Buddleja spp.	butterflybush[2]	O, L	NA
Camellia sinensis (L.) Kuntze	Chinese tea[2]	A	Asia
Cannabis sativa L.	hemp[2]	A	NA
Capsicum annuum L.	bell pepper, cayenne pepper[2]	A	Europe, NA
Caragana arborescens Lam.	Siberian pea shrub[2]	O, L	NA
Carpinus betulus L.	European hornbeam[2]	O, L	NA
Carya illinoinensis (Wangenh)	pecan[2]	A, O, L	NA

(Continued)

TABLE 4.1 (CONTINUED)

Known Asian, European, and North American *Halyomorpha halys* Host Plants

Scientific Name[1]	Common Name[2,3,4]	Habitat[5]	Continent Found[6]
Carya ovata (Mill.) K. Koch	shagbark hickory[2]	W, Wo	NA
Catalpa bignonioides Walter	catalpa[2]	O, L	Europe, NA
Catalpa spp.	catalpa[2]	O, L	NA
Catharanthus roseus (L.) G. Don	periwinkle[4]	O, L	Asia
Cedrus spp.	Cedar[3]	O, L, W, Wo	NA
Celastrus orbiculatus Thunb.	Oriental bittersweet[2]	O, L	NA
Celosia spp.	cock's comb[2]	O, L	NA
Celtis koraiensis Nakai	Korean hackberry[2]	O, L	NA
Celtis occidentalis (L.)	common hackberry[2]	W, Wo	NA
Cephalanthus occidentalis L.	common buttonbush[2]	O, L	NA
Cercidiphyllum japonicum Siebold & Zucc.	katsura tree[2]	O, L	
Cercis canadensis L.	eastern redbud[2]	O, L, W, Wo	NA
Cercis canadensis L. var. *texensis* (S. Watson) M.Hopkins	Texas redbud[2]	O, L	NA
Chenopodium berlandieri Moq.	pitseed goosefoot[2]	W, Wo	NA
Chionanthus retusus Lindley & Paxton	Chinese fringetree[2]	O, L	NA
Chionanthus virginicus L.	white fringetree[2]	O, L	NA
Citrus junos Sieb. ex Tanaka	Yuzu[2]	A	Asia
Citrus unshiu unshiu (Swingle) Marcow	citrus[2]	A	Asia
Cladrastis kentukea (Dum.Cours.) Rudd (syn. *lutea*)	Kentucky (American) yellowwood[2]	O, L	NA
Cornus florida L.	flowering dogwood[2]	O, L, W, Wo	NA
Cornus kousa F. Buerger ex Hance	kousa dogwood[2]	O, L	NA
Cornus macrophylla Wall.	(large-leaf) dogwood[2]	O, L	NA
Cornus officinalis Torr. Ex Dur.	Asiatic (Japanese cornel) dogwood[2]	O, L	NA
Cornus racemose Lam.	gray dogwood[2]	O, L	NA
Cornus sanguinea L.	common dogwood[2]	O, L	Europe, NA
Cornus sericea L.	redosier dogwood[2]	O, L	NA
Cornus Stellar series	dogwood[2]	O, L	NA
Corylus colurna L.	filbert, hazelnut[2]	A	NA
Crataegus laevigata (Poir.) DC	smooth (English) hawthorn[2]	O, L	NA
Crataegus monogyna Jacq.	oneseed hawthorn[2]	W, Wo	NA
Crataegus pinnatifida Bunge	haw[2]	A, W, Wo	Asia
Crataegus viridis L.	green hawthorn[2]	O, L	NA
Cucumis sativus L.	garden cucumber[2]	A	Asia, NA
Cucurbita pepo L.	field pumpkin (summer squash)[2]	A	NA
Cymbidium spp.	orchid[2]	O	Asia
Diospyros kaki Thunb.	Japanese (sweet) persimmon[3]	O, L	Asia
Elaeagnus angustifolia L.	Russian olive[2]	O, L, W, Wo	NA
Elaeagnus umbellate Thunb.	autumn olive[2]	W, Wo	NA
Ficus carica L.	edible fig[2]	A	NA
Forsythia suspense (Thunb.) Vahl	weeping forsythia[2]	O, L	NA
Fraxinus Americana L.	white (American) ash[2]	W, Wo	NA
Fraxinus excelsior L.	European ash[2]	W, Wo	Europe
Fraxinus pennsylvanica Marshall	green ash[2]	W, Wo	NA
Ginkgo biloba L.	maidenhair tree (ginkgo)[2]	O, L	NA
Gleditsia triacanthos var. *inermis* Willd.	thornless common honeylocust[2]	O, L	NA

(Continued)

TABLE 4.1 (CONTINUED)

Known Asian, European, and North American *Halyomorpha halys* Host Plants

Scientific Name[1]	Common Name[2,3,4]	Habitat[5]	Continent Found[6]
Glycine max (L.) Merr.	soybean[2]	A	Asia, NA
Halesia tetraptera L.	mountain (Carolina) silverbell[2]	O, L	NA
Hamamelis japonica Siebold & Zucc.	invasive witchhazel[2]	W, Wo	NA
Hamamelis virginiana L.	American witchhazel[2]	W, Wo	NA
Helianthus spp.	sunflower[3]	A	NA
Heptacodium miconioides Rehder	seven sons flower[2]	O, L	NA
Hibiscus moscheutos L.	crimsoneyed rosemallow[2]	O, L	NA
Hibiscus syriacus L.	rose of Sharon (hibiscus)[2]	O, L	NA
Humulus lupulus L.	common hop[2]	A	NA
Ilex aquifolium L.	English holly[2]	O, L	NA
Juglans nigra L.	black walnut[2]	W, Wo	NA
Juniperus virginiana L.	eastern redcedar[2]	O, L	NA
Koelreuteria paniculata Laxm.	goldenrain tree[2]	O, L	NA
Lagerstroemia indica (L.) Pers.	crapemyrtle[2]	O, L	NA
Larix kaempferi (Lamb.) Carr. (syn. *leptolepis*)	Japanese larch[2]	O, L	NA
Ligustrum japonicum Thunb.	Japanese or wax-leaf privet[2]	O, L, W, Wo	NA
Ligustrum sinense Lour.	Chinese privet[2]	O, L, W, Wo	NA
Liquidambar styraciflua L.	sweetgum[2]	O, L, W, Wo	NA
Liquidambar spp.	sweetgum[2]	O, L, W, Wo	NA
Liriodendron tulipifera L.	tuliptree[2]	W, Wo	NA
Lonicera tatarica L.	Tatarian honeysuckle[2]	W, Wo	NA
Lonicera spp.	honeysuckle[2]	W, Wo	NA
Lycium chinense Mill.	matrimony vine[3]	W, Wo	Asia
Lythrum salicaria L.	purple loosestrife[2]	W, Wo	NA
Magnolia grandiflora L.	southern magnolia[2]	O, L	NA
Magnolia stellata (Siebold & Zocc.) Maxim.	star magnolia[2]	O, L	NA
Mahonia aquifolium (Pursh) Nutt.	hollyleaved barberry[2] (Oregon grape)	W, Wo	NA
Malus baccata (L.) Borkh.	Siberian crab apple[2]	O, L	NA
Malus domestica Borkh.	apple[2]	O, L, A	Asia, NA
Malus pumila Mill. (syn. *domestica*)	paradise apple[2]	O, L, A	NA
Malus sargentii Rehder	Sargent's crab apple[2]	O, L	NA
Malus zumi	crab apple[2]	O, L, W, Wo	NA
Malus spp.	apple[2]	A	NA
Metasequoia glyptostroboides Hu and W.C. Cheng	dawn redwood[2]	O, L	NA
Mimosa spp.	sensitive plant (mimosa)[2]	W, Wo	NA
Morus alba L.	white mulberry[3]	W, Wo	Asia
Morus spp.	mulberry[3]	W, Wo	Asia, NA
Musineon divaricatum (pursh) Raf.	leafy wildparsley[2]	W, Wo	NA
Nyssa sylvatica Marsh.	blackgum (tupelo)[2]	O, L	NA
Panicum miliaceum L.	common millet[2]	A	Asia
Paulownia tomentosa (Thunb.) Steud.	princess tree (paulownia)[2]	O, L, W, Wo	Asia, NA
Phalaenopsis spp.	moth orchid[2]	O, L	Asia
Phaseolus lunatus L.	lima beans[2]	A	NA
Phaseolus spp.	bean[2]	A	NA

(Continued)

TABLE 4.1 (CONTINUED)

Known Asian, European, and North American *Halyomorpha halys* Host Plants

Scientific Name[1]	Common Name[2,3,4]	Habitat[5]	Continent Found[6]
Photinia (syn. *Aronia*) spp.	chokeberry[2]	W, Wo	NA
Phytolacca americana L.	American pokeweed[2]	W, Wo	NA
Pistacia chinensis Bunge	Chinese pistache[2]	O, L	NA
Pisum sativum L.	pea[3]	A	Asia, NA
Platanus occidentalis L.	American sycamore[2]	O, L	NA
Platanus x *acerifolia* spp.	London planetree[2]	O, L	NA
Platycladus orientalis (L.) Franco	Chinese arborvitae[4]	O, L	Asia
Prunus armeniaca L.	apricot[2]	A	NA
Prunus avium (L.) L.	sweet cherry[2]	O, L	NA
Prunus cerasifera Ehrh.	cherry plum[2]	O, L	NA
Prunus incisa Thunb.	Fuji cherry[2]	O, L	NA
Prunus laurocerasus L.	cherry laurel[2]	O, L	NA
Prunus persica (L.) Batsch	peach, nectarine[2]	A	Asia, Europe, NA
Prunus serotina Ehrh.	black cherry[2]	W, Wo, A	NA
Prunus serrulata Lindl.	Japanese flowering cherry[2]	O, L	NA
Prunus subhirtella Miq.	winter-flowering (Higan) cherry[2]	O, L	NA
Prunus x *incam*	okame flowering cherry[2]	O, L, A	NA
Pseudocydonia sinensis C.K. Schneid.	Chinese-quince[2]	O, L	NA
Punica granatum granatum L.	pomegranate[4]	A	Asia
Pyracantha spp.	firethorn[2]	O, L	NA
Pyrus calleryana Decne.	Callery (Bradford) pear[2]	O, L	NA
Pyrus fauriei Schneid. 'Westwood'	Korean sun pear[2]	O, L	Asia
Pyrus pyrifolia (Burm.) Nak.	Chinese (Asian) pear[2]	O, L, A	Asia, NA
Pyrus spp.	pear[2]	O, L, A	Asia, Europe, NA
Quercus alba L.	white oak[2]	O, L	NA
Quercus coccinea Muenchh.	scarlet oak[2]	O, L	NA
Quercus robur L.	English oak[2]	O, L	NA
Quercus rubra L.	northern red oak[2]	O, L	NA
Quercus spp.	oak[2]	O, L, W, Wo	NA
Rhamnus cathartica L.	common buckthorn[2]	W, Wo	NA
Robinia pseudoacacia L.	black locust[2]	W, Wo	NA
Rosa canina L.	dog (native) rose[2]	O, L	NA
Rosa multiflora Thurb.	mulitflora rose[2]	W, Wo	NA
Rosa rugose Thurb.	rugosa rose[2]	O, L	NA
Rubus phoenicolasius Maxim	wine raspberry (wineberry)[2]	A	NA
Rubus spp.	raspberry, blackberry[2]	A, W, Wo	NA
Salix spp.	willow[2]	O, L, W, Wo	Asia, NA
Sassafras albidum (Nutt.) Nees	sassafras[2]	W, Wo	NA
Secale cereale L.	cereal rye[2]	A	NA
Setaria italica (L.) P. Beauvois	pearl millet[2]	A	Asia
Solanum lycopersicum L.	garden tomato[2]	A	Asia, NA
Solanum melongena L.	eggplant[2]	A	Asia, NA
Sophora japonica (L.) Schott	Japanese pagoda tree[2]	O, L	NA
Sorbus airia Crantz	winterbeam[2]	O, L	NA
Sorbus americana Marshall	American mountain ash[2]	W, Wo	NA
Sorbus aucuparia L.	mountain ash[2]	W, Wo	Europe, NA
Sorghum bicolor (L.) Moench	sorghum[2]	A	Asia, NA
Spiraea spp.	spirea[2]	O, L	NA

(Continued)

TABLE 4.1 (CONTINUED)

Known Asian, European, and North American *Halyomorpha halys* Host Plants

Scientific Name[1]	Common Name[2,3,4]	Habitat[5]	Continent Found[6]
Stewartia pseudocamellia Maxim.	Japanese stewartia[2]	O, L	NA
Stewartia pseudocamellia var. *koreana* Sealy	Korean stewartia[2]	O, L	NA
Styphnolobium japonicum (L.) Schott	Chinese scholar tree[4]	O, L	Asia
Styrax japonicas Siebold & Zucc.	Japanese snowbell[2]	O, L	NA
Symphytum spp.	comfrey[2]	O, L	NA
Syringa pekinensis Rupr.	Peking (Chinese) tree lilac[2]	O, L	NA
Taxus cuspidata Siebold & Zucc.	Japanese yew[4]	O, L	Asia
Tetradium (syn. *Euodia*) *daniellii* (syn. *hupehensis*) (Benn.)	bee-bee tree (Korean euodia)[2]	O, L	NA
Tilia americana L.	American basswood[2]	O, L	NA
Tilia cordata Mill.	littleleaf linden[2]	O, L	NA
Tilia tomentosa Moench	silver linden[2]	O, L	NA
Tsuga canadensis (L.) Carrière	eastern hemlock[4]	W, Wo	NA
Ulmus americana L.	American elm[2]	O, L	NA
Ulmus parvifolia Jacq.	Chinese elm[2]	O, L	Asia
Ulmus procera (syn. *minor*) Salisb.	English (smoothleaf) elm[2]	O, L	NA
Ulmus spp.	elm[2]	O, L	NA
Vaccinium corymbosum L.	highbush blueberry[2]	A	NA
Viburnum dilatatum Thunb.	linden arrowwood[2]	O, L	NA
Viburnum edule (Michx.)	high bush cranberry[2]	O, L	NA
Viburnum opulus L. var *americanum* Aiton	highbush cranberry[2]	O, L	NA
Viburnum prunifolium L.	Blackhaw cranberry[2]	O, L	NA
Viburnum prunifolium L.	viburnum (blackhaw)[2]	O, L	NA
Viburnum x *burkwoodii*	viburnum[2]	O, L	NA
Vicia spp.	vetch[3]	W, Wo	Asia
Vitis riparia Michx.	riverbank wild grape[2]	W, Wo	NA
Vitis vinifera L.	wine grape[2]	A	Europe, NA
Zea mays L.	field corn, sweet corn[2]	A	Asia, NA

[1] Modified from stopBMSB.org – Host Plants of the brown marmorated stink bug in the U.S.
[2] All stages found on host (eggs, nymphs, and adults).
[3] Nymphs and adults only found on host.
[4] Adults only found on host.
[5] A – agriculture, L – Landscape, O – Ornamentals, W – Wild, Wo – Woodland.
[6] NA – North America.

only completed development on peach (*Prunus persica*) (Rosales: Rosaceae) and *Ailanthus altissima* (Mill.) Swingle (Saphindales: Simaroubaceae). Thus, although the host plant range is large, few host plants appear to be able to completely support development. It is possible that the wide host breadth is an evolutionary adaptation that helps support population growth when preferred hosts are unavailable, specifically early in the season.

Although *Halyomorpha halys* feeds primarily on the reproductive structures of plants (i.e., seeds, fruits, pods,) (Nielsen and Hamilton 2009b, Hoebeke and Carter 2003), feeding has been observed on leaves and even on the bark. Adults will feed on the trunks of woody trees, specifically *Acer rubrum*, *Malus* spp., *Platanus* x *acerifolia*, and *Ulmus americana*, which results in wounding damage of the trees (Martinson et al. 2013). This wounding increases available carbohydrates for foraging Hymenoptera, specifically Formicidae and Vespidae. Thus, *H. halys* may benefit the nutritional requirements of these hymenopteran species (Martinson et al. 2013).

Across all plant species, there appears to be a strong edge-effect at the field level on the amount of damage by *Halyomorpha halys*. In orchard crops, this has been observed in peaches and apples where *H. halys* adults are concentrated in border rows on the outer margins of orchards, coinciding with higher damage (Joseph et al. 2014, Blaauw et al. 2016). In peaches, this edge-effect has been exploited for management purposes (Blaauw et al. 2014). A geospatial analysis of field corn and soybeans found that *H. halys* was the predominant pentatomid species in these crops, and both populations and damage demonstrated strong edge effects; fields with adjacent wooded habitats had significantly higher stink bug densities than fields without adjacent wooded habitats (Venugopal et al. 2014). In addition, the plant species composition in nonmanaged woodland borders adjacent to soybean fields has been shown to impact abundance patterns in those fields with clear buildups of *H. halys* populations in borders dominated by tree of heaven in July followed by steady movement into field edges (Aigner et al. 2017).

Halyomorpha halys' polyphagous behavior may be aided by its strong capacity to disperse at landscape levels throughout most periods of its lifetime. In laboratory studies where *H. halys* adults were tethered to a flight mill, wild populations could be divided into two categories – short distance (≤5 kilometers [km]) fliers and long distance fliers (>5 km). Short distance fliers comprised 85% of the population and flew an average of 2 km in 24 hours. Long-distance fliers travelled an average of 66.54 and 75.12 km for males and females, respectively (Wiman et al. 2014). Where free flight of *H. halys* was directly observed and tracked in field studies, the mean flight speed was 3 meters/second along a straight line from take-off to landing (Lee et al. 2013b).

Adult flight activity occurs primarily at night (Wiman et al. 2014) as adults search for mates or alternate food resources. Because of this, blacklight traps are a useful monitoring tool for landscape-level movement of *H. halys*. This method has demonstrated a 75% annual increase in *H. halys'* population size in New Jersey from 2004 to 2011. Although activity changes throughout the year, a large peak in flight activity occurs at 685 $DD_{14.17}$ (Nielsen et al. 2013). Analyses of catches in blacklight traps allowed important geographic parameters to be elucidated that may be important during population spread. Early in the invasion stage, there is a strong association with urban environments. Railroads and wetlands appear to facilitate population spread. And, finally, association with agriculture has a significant relationship with population density (Wallner et al. 2014). These analyses support the hypothesis that species associated with human environments are more likely to be successful invaders (Hufbauer et al. 2012). This may have allowed *H. halys* to overcome any Allee effects associated with the genetic bottleneck (i.e., small number of initial individuals) that occurred during the introduction into the eastern United States (Wallner et al. 2014, Xu et al. 2014).

4.4.2 Symbiotic Relationships

The presence of symbiosis in the order Hemiptera is well known in the Auchenorryncha and Sternorrhyncha and is found in the diets of various pentatomorphans (Buchner 1965). These relationships are necessary and provide missing dietary components such as amino acids and vitamins not present in phloem fluids (Baumann and Moran 1997, Moran and Telang 1998, Douglas 2006). Kenyon et al. (2015) found that the endosymbiont *Candidatus* Pantoea carbekii, the primary symbiont in the gastric caeca lumina of *Halyomorpha halys*, is obtained when newly hatched first instars feed on maternal extrachorion secretions present on the surfaces of eggs. The nymphs congregate around their egg mass and apparently feed on the egg choria to obtain this endosymbiont, which is necessary for development (Bansal et al. 2014, Taylor et al. 2014). This was demonstrated by Taylor et al. (2014), who showed, through a series of experiments involving surface-sterilized and untreated eggs, that nymphal development and survival was negatively affected if this endosymbiont could not be obtained. They also noted that nymphs that were negatively affected resulted in adults with reduced fecundity. Studies conducted to determine if antimicrobials could be used to sterilize *H. halys* eggs to prevent acquisition showed that doing so significantly reduced egg hatch, nymphal survival and symbiont acquisition (Taylor et al. 2016).

4.4.3 Mate Selection

Prolonged copulation is common among Pentatomidae, ranging from 1 to 165 hours (Kawada and Kitamura 1983b, Ueno and Ito 1992). This behavior could have evolved as a mate-guarding tactic to reduce the time a female is available to mate prior to oviposition. By comparison, *Halyomorpha halys* exhibits short copulation periods, averaging 10 minutes in the laboratory. However, females are polyandrous, mating with at least five males per day (Kawada and Kitamura 1983b). Field observations of mating pairs are common, but the pairs easily disengage if disturbed. Multiple matings in *H. halys* increase female fitness, and there is an increase in the hatch rate and duration of oviposition with increased matings (Kawada and Kitamura 1983b). In the absence of mate guarding, it is hypothesized that males are assured paternity through sperm mixing and/or the high rate of available females (Sillen-Tullberg 1981, Roderick et al. 2003).

4.4.4 Overwintering Behavior/Diapause

Halyomorpha halys is well-known for causing nuisance problems, as large numbers of adults often invade human-made structures to overwinter inside protected environments (Inkley 2012). A detailed reporting in one house counted >26,000 individuals in a single year. *H. halys* will overwinter in almost any structure, between cardboard boxes, vehicles, underneath tarps, wood piles, etc., but are frequently found in attics and crawlspaces in aggregations (Inkley 2012). Survivorship is improved within aggregations when antennal contact is made, although the physiological reasoning is unknown (Toyama et al. 2006).

Anecdotal evidence that *Halyomorpha halys* will disperse from neighboring woodlots into agricultural crops, such as peach, led to an investigation of natural overwintering habitats. Lee et al. (2014b) found that adults of *H. halys* will, in fact, overwinter in the natural landscape. Overwintering adults were recovered from dry crevices in dead, standing trees with thick bark, particularly oak (*Quercus* spp.) and locust (*Robinia* spp.) that were >19.0 centimeters (cm) diameter at breast height. Of randomly sampled trees, 5.6% had overwintering *H. halys* adults. However, when the characteristics of the overwintering habitat mentioned above were applied, ≈32% of trees surveyed harbored *H. halys*, albeit at low numbers with ≈six adults/tree observed (Lee et al. 2014a). Large populations also have been observed in rock outcroppings in both Maryland and Pennsylvania (T.C. Leskey and Galen P. Dively, personal communication; see Acknowledgments).

Little is known about the effects on *Halyomorpha halys* of exposure to low winter temperatures typically found in North America. Cira et al. (2016) examined the impacts of low and freezing temperatures on mortality of this bug. They reported that *H. halys* is chill intolerant and dies before freezing and that cold tolerance varies by season, sex, and acclimation location. Acclimation site also affected the bugs supercooling point (i.e., -17.06°C ± 0.13 in Minnesota, -13.90°C ± 0.09 in Virginia).

Adult *Halyomorpha halys* overwintering in temperate regions are believed to be in facultative reproductive diapause (see **Chapter 11**). Continual observations of adults in California throughout mild winters (Charles Pickett, personal communication; see Acknowledgments) suggest that it is not. Diapause initiation and termination cues of most insect species are influenced by the interaction of intrinsic factors such as genetics as well as extrinsic factors such as temperature and photoperiod (Tauber et al. 1986). For those species that overwinter as adults, photoperiod is frequently the primarily critical cue (Tauber et al. 1986). Based on the available literature, we cautiously assume that the cues for diapause termination and induction in *H. halys* are symmetrical and determined primarily by photoperiod – thus are considered independent of temperature. Watanabe (1979) reported a 13.5–14h critical photoperiod (daylength) to terminate diapause for overwintering adults, whereas Yanagi and Hagihara (1980) suggested a longer 14.75h critical photoperiod. However, termination of diapause as indicated by the initiation of reproduction does not occur immediately upon leaving overwintering sites as females require an additional developmental period related to heat accumulation. Although emergence from overwintering sites is gradual, a peak in activity occurs in West Virginia and Virginia in April around 13.5h daylength (T.C. Leskey and J. Christopher Bergh, unpublished data; see Acknowledgements). A larger peak in adult activity (measured by leaving overwintering shelters) occurs in early-mid May, which suggests at that time emergence is the interaction of increasing daylength and temperature (Wiman et al. 2014). To further examine spring emergence, Bergh et al. (2017) placed marked adults in overwintering cages encircled by pheromone-baited and unbaited pyramid traps and followed their emergence from the cages in 2013 and 2014. The observed a small peak in activity in mid-April followed by a larger, more prolonged emergence from mid-May to early June. They also found that emerging adults did not respond to the pheromone-baited traps suggesting that a dispersal flight may be necessary before they respond to the pheromone. Current laboratory studies, reproductive activity in the field, and a predictive population model in the United States strongly point towards the diapause cue for *H. halys* being photoperiod-driven with a critical photoperiod of 13.5h (Anne L. Nielsen, unpublished data).

The impact of abiotic conditions on population dynamics of *Halyomorpha halys* during the active growing season and the overwintering period is poorly understood. In the United States, populations have fluctuated dramatically from year to year in areas in the mid-Atlantic with well-established populations since 2010, but key factors promoting or reducing survivorship as well as those influencing range expansion remain unknown.

4.4.5 Host Plants

Halyomorpha halys is a highly polyphagous, primarily arboreal species (Hoffman 1931, Hoebeke and Carter 2003, Bernon 2004, Nielsen and Hamilton 2009b), having been collected from numerous plant species in Asia, Europe, and North America. These collection records include both wild hosts and cultivated species and, often, are documented with notes on the presence of eggs, nymphs, and/or adults, and inclusion of feeding records (Hoebeke and Carter 2003, Bakken et al. 2015). In its native range (i.e., Asia), it has been reported to feed on at least 106 host plants such the Princess tree, Chinese arborvitae, Chinese milk vetch, mulberry, elm, willow, and Chinese scholar tree (Wang and Wang 1988, Bae et al. 2009, Lee et al. 2009). Agricultural hosts include apple, citrus, peach, pear, sweet persimmon, and Yuzu (Qin 1990, Fengjie et al. 1997, Funayama 2003, Kang et al. 2003, Lee et al. 2007). It also feeds on soybean, sorghum, corn, and various vegetables (Oda et al. 1980, Kawada and Kitamura 1983a, Fukuoka et al. 2002, Kang et al. 2003, Bae et al. 2009).

Outside of Asia, several authors have documented that *Halyomorpha halys* adults and/or nymphs can feed on a much wider range of hosts. In Europe, *H. halys* has been found for the first time feeding on rice grown in northern Italy (Lupi 2017). In North America, this bug has been collected from honeysuckle, walnut, shadbush, butterfly-bush, paulownia, persimmon and maple in the spring in Allenton, PA (Hoebeke and Carter 2003). Nymphs also have been collected on many of the same hosts and from basswood and catalpa in the same area in July. Nielsen and Hamilton (2009b) observed adults and nymphs on several cultivated fruit species and on rugosa rose, white ash, high bush cranberry, Blackhaw cranberry, Russian olive and Siberian pea shrub. Hedstrom et al. (2014) report feeding by this bug on hazelnuts grown in New Jersey. *H. halys* has been documented as feeding on eggplant, lima beans, okra, peppers, sweet corn, and tomatoes (Leskey et al. 2012a, Rice et. al. 2014). Zobel et al. (2016) evaluated bell pepper, eggplant, green beans, okra, pepper, sweet corn, and tomato in terms of seasonal densities, host suitability, and feeding injury. They found that overwintering adults and F1 juveniles were present during the growing season, and that *H. halys* preferred plants with fruiting structures, exhibited higher numbers, and could reproduce on crops with extended fruiting periods. Significantly higher populations were found on sweet corn, okra and bell pepper compared to eggplant, green beans, and tomato. Phillips et al. (2016) examined seasonal abundance and phenology of *H. halys* in different pepper varieties (sweet bell, hot chili and sweet banana) in Delaware, Maryland, New Jersey and Virginia. They found no differences in abundance of *H. halys* life stages or damage caused between varieties. Bakken et al. (2015) surveyed wild hosts in North Carolina and Virginia and observed nymphs and adults feeding on over 50 different species. Additional known North American host species are presented in **Table 4.1**. The utilization of ornamental hosts in commercial hosts also has been examined (Bergman 2015, Bergman et al. 2016). These authors showed that of 131 species sampled, 88 were hosts, angiosperms supported larger *H. halys* populations, and Asian cultivars supported fewer individuals than non-Asian cultivars. These results were most apparent in *Acer*, *Ulmus* and *Pyrus* species.

4.4.6 Movement between Host Plants

Adults and nymphs of *Halyomorpha halys* are highly mobile and readily move between hosts plants during the growing season (Nielsen and Hamilton 2009b, Venugopal et al. 2015a,b). In the laboratory, the vertical and horizontal dispersal capacities of nymphs were shown to vary between instars with third instars walking significantly greater horizontal distances than other life stages and third and fourth instars climbing significantly longer distances than second instars and adults (Lee et al. 2014b). In the same study, Lee et al. (2014b) demonstrated under field conditions that fifth instars walked across turf nearly twice the distance traveled by third instars. Acebes-Doria et al. (2017) demonstrated that nymphs moved up and down hosts and that movement varied between different hosts. They also showed that movement of early and older instars was dependent on the time of the year.

Movement of *Halyomorpha halys* between host plants has been shown by Nielsen and Hamilton (2009b) to be related to time of year, host maturity, and distance between hosts. They followed the abundance of *H. halys* on eleven fruit and ornamental hosts and showed that there were significant differences in abundance of nymphs and adults on each host depending on whether or not samples were taken in early, mid-, and late season and related these differences to the presence of fruit. The impact of the presence of fruit in ornamental nurseries on populations of *H. halys* has been investigated by Martinson et al. (2015). In their study of 3,844 trees across 223 cultivars, they demonstrated the presence of fruit was a good predictor of adult and nymphal seasonal abundance and within-tree distribution. They also showed that the removal of fruit suppressed the presence of *H. halys* adults and nymphs. Venugopal et al. (2015a,b) showed plant maturity to be an important indicator of *H. halys* presence within corn and soybean fields and between adjacent fields of corn and soybean. They also showed in woody plant nurseries that the distance from a bordering field of corn or soybeans or from a residential area influenced *H. halys* abundance as they moved from the exterior to the interior of the nursery planting.

4.4.7 Spatial and Nutritional Attributes in Relation to *Halyomorpha halys* Movement

Spatial attributes affecting the large-scale (i.e., state, nation, and world) movement, distribution, and potential range of *Halyomorpha halys* include annual mean temperature, maximum temperature of the warmest month, minimum temperature of the coldest month, annual precipitation, annual mean radiation, wetland rights-of-way, urban developments, railroads, and elevation (Zhu et al. 2012, Wallner et al. 2014); more detailed information can be found in **Section 4.4.8** of this chapter by Wenjun Bu. This section delineates spatial attributes in relation to movement of *H. halys* in agricultural landscapes including overwintering sites, surrounding vegetation, host plants, agricultural practices such as pest management and intercropping, and land-use types.

4.4.7.1 General Movement Patterns of *Halyomorpha halys* in Agricultural Landscapes

Halyomorpha halys is highly mobile, has a wide host range, and occupies a large geographic area. Therefore, it can utilize various resources in and around cropping areas throughout the year by movement to find the best or temporally available resources such as food, shelter, and oviposition sites. Movement of *H. halys* begins when adults emerge from overwintering sites in early spring. Then, they feed on various plants including crops and reproduce throughout the crop-growing season. *H. halys* also has been shown to move within a crop during the growing season. Hahn et al. (2017) showed that early in the season, *H. halys* did not exhibit clustering early in the season. However, as the crop matured, *H. halys* clustered in areas with ripe fruit. Rice et al. (2016) examined the influence of landscape features surrounding processing tomato fields on *H. halys* damage and demonstrated that the size and shape of the forests bordering processing tomato fields and geographic location impacted damage. They found that large areas of woods bordering the fields in southern areas of the mid-Atlantic had higher levels of damage. When temperatures decrease below 15°C in autumn (Watanabe et al. 1994), they return to overwintering sites. Therefore, knowledge of spatial arrangement of crops, surrounding vegetation, and overwintering sites is key to developing management tactics to effectively control *H. halys*.

4.4.7.2 Spatial Attributes Associated with Movement from Overwintering Sites to Surrounding Vegetation

Halyomorpha halys adults overwinter in the wild (e.g., under bark, in crevices of rocks) or in human-made structures (e.g., sheds, barns, houses). In early spring, they emerge from overwintering sites.

Because fruit trees or vegetables are generally not available as a food source at this time, they utilize surrounding uncultivated vegetation as the first food source (Adachi 1998). *Halyomorpha halys* adults, after emergence from overwintering sites, generally do not enter wheat fields and vegetable plots but disperse to woody plants such as white mulberry, elm species, willow, Chinese scholar tree, Japanese yew, and cedar trees (Yanagi and Hagihara 1980, Qin 1990). The number of adults at this time, in any particular location, is dependent upon the abundance and proximity of trees and pine cones that serve as a main food source (Ohira 2003).

When temperatures rise above 20°C in spring, the adults begin to migrate to apple, peach, or pear orchards. They also can utilize other host plants such as acacia, oriental arborvitae, and elm trees if orchards are not nearby or tree fruits are not ready for feeding (Qin 1990). Although this bug was observed in pines, it cannot survive when reared only on pine (Yanagi and Hagihara 1980). This indicates, of course, that pine is a temporal host plant early in the season before the adults move to other host plants, which generally are fruit trees and vegetables.

4.4.7.3 Nutritional Attributes Associated with Host Plants and Cropping and Management Practices affecting *Halyomorpha halys* Movement

4.4.7.3.1 Nutritional Value of Plants

Adults of *Halyomorpha halys* move into orchards or other crops when the sugar content of fruit increases (Lee et al. 2009). A high density of adults frequently is observed in apple or pear orchards at this time. However, the number of eggs laid within the orchards sometimes is lower than on other host plants around the orchard (Zhang et al. 2007) even though apple and pear can satisfy the nutritional condition of growth and development of this bug. When fruits or vegetables are not available or begin to senesce, *H. halys* disperses to utilize other plants nearby such as haw, pomegranate, elm, white mulberry, paulownia, and groundnuts (Yanagi and Hagihara 1980, Zhang et al. 2007).

Other chemical compounds in plants such as tannin can affect movement of *Halyomorpha halys* adults. In persimmon, concentration of soluble tannin decreases as fruits mature. In June and July, when persimmon is not mature, the concentration of tannin is high and the fruit is not suitable food for the bugs (Lee et al. 2009). However, in autumn, when the tannin content drops, *H. halys* moves to persimmon and begins to damage fruits.

4.4.7.3.2 Intercropping

The complicated environment in and around the orchard can be a key factor for movement and dispersal of *Halyomorpha halys* among commodities. For example, in Korea, sometimes beans or pepper are interplanted in persimmon orchards, and such intercropping provides an excellent opportunity for *H. halys* to fully utilize both bean and persimmon for their survival and reproduction with minimum movement.

4.4.7.3.3 Chemical Control

Because *Halyomorpha halys* frequently moves to utilize a variety of surrounding vegetation, Zhang et al. (2007) suggested a control strategy utilizing spraying of surrounding forest habitats using insecticides, use of attract and kill systems, bagging of fruit, and the use of natural enemies to reduce *H. halys* populations. However, Watanabe et al. (1994) discussed a similar strategy using insecticide applications to surrounding environments, stating that this might be difficult to utilize because of concerns about the impact on forest ecosystem that *H. halys* inhabits. In addition, although *H. halys* is susceptible to various insecticides, the adults have strong flying capability so they easily can escape and seek other hosts when pesticides are applied.

4.4.7.4 Movement from Crops and Surrounding Vegetation to Overwintering Sites

When available food sources become scarce and temperatures decrease in autumn, adults of *Halyomorpha halys* move to overwintering sites. Sometimes, a large number of adults aggregate and then move to overwintering sites along mountain foothills (Yanagi and Hagihara 1980). The choice of overwintering sites by *H. halys* in open areas is related to humidity and temperature conditions (Wang and Wang 1988).

Because the northern slopes of mountains receive less solar radiation and stronger winds, adults tend to gather on the sunnier southern slopes with sedimentary rock outcroppings prior to entering overwintering sites (Wang and Wang 1988).

4.4.8 Potential Geographic Distribution of *Halyomorpha halys*: Ecological Niche Modeling

Biological invasion represents a major threat to biodiversity and ecosystem functioning, resulting in great economic losses. To mitigate the often dramatic consequences, a fundamental goal in conservation biology is to predict which species will invade, as well as which areas are most vulnerable to their invasion. Recently, considerable progress has been made in identifying regions at risk of species invasion and understanding the multiple factors that influence the spread of invasive species at local and global scales. Predicting species' distributions using an ecological niche modeling (ENM) approach has emerged as a powerful tool in studying biological invasions (Zhu et al. 2012). The general idea behind correlative ENMs is to compute a model of a species' realized niche based on information on the species' occurrence and environmental data commonly stored as geographic information system (GIS) layers. The output is a map showing relative suitability of each grid cell for the species, which can be used to identify and prioritize sites for protection against invasion.

The classic applications of ENM to biological invasions involve the calibration of niche modeling in the native range and the subsequent transfer of the calibrated models to other regions to predict areas of potential invasion. Based only on niche conservatism, the calibrated model can be projected to identify areas of potential invasion. *Halyomorpha halys*, native to east Asia, has become an invasive species in the United States and Europe. In east Asia, the species spans from temperate to subtropical zones, feeding on a wide variety of fruit and ornamental trees. In the United States, this bug was imported accidentally in the late 1990s and now is emerging as a key pest in agriculture, creating major nuisance problems especially in the mid-Atlantic region. In Europe, it first was discovered in Switzerland in 2008 and later was found in Liechtenstein, Germany, and France. The northern China populations have proven to be the original sources of this bug's invasion in the United States (Xu et al. 2014) and likely also is the source of the European invasion in Switzerland.

Climate spaces occupied by the geographic separated population were compared in the ecological dimensions. The introduced occurrence clusters fell well into that of the native occurrence. This suggests that the climate niche was conserved during the invasion of *Halyomorpha halys* (Zhu et al. 2012) (**Figure 4.4**) providing the fundamental support of using ENM to predict the potential establishment of

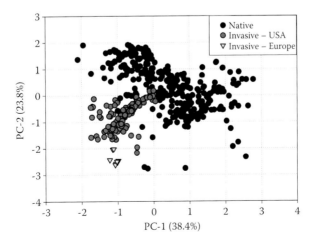

FIGURE 4.4 (**See color insert.**) Principal component analysis of 10 variables associated with occurrences of *Halyomorpha halys*. Symbols represent occurrences of the bug in native areas in Asia and introduced areas in the United States and Europe (modified from Figure 2 in Zhu et al. 2012).

this bug in new areas worldwide. In addition, the realized climate niches of the introduced United States and European populations were different from that of the native Asia population, with the introduced populations only occupying partial climate space of the native Asia population (Zhu et al. 2012) (**Figure 4.5**). These results suggest the climate space was unfilled in the introduced populations; there might be much suitable climate space available for this bug in the North America and Europe, which contributes to the continued population expansion in new areas.

Niche models were calibrated on the native Asia areas and transferred onto the global extent using both Maxent and GARP. Maxent is a machine-learning technique that predicts species' distribution by integrating detailed environmental variables with species occurrence data; it follows the principle of maximum entropy and spreads out probability as uniformly as possible but is subject to the caveat that it must match empirical information such as known presence; the Maxent models were developed using the linear, quadratic, and hinge functions to avoid the problem of over-fitting. Although Maxent has appeared superior to GARP in some previous studies, careful assessments of model quality showed no significant differences between the two. Recent studies suggested using multiple algorithms to infer a consensus estimate of niche dimensions.

At the global scale, the high climate suitable areas at risk of invasion by *Halyomorpha halys* include latitudes between 30°–50° including northern Europe, northeastern North America, southern Australia, and the North Island of New Zealand. Angola in Africa and Uruguay in South America also showed high climate suitability (Zhu et al. 2012) (**Figure 4.6**). Attention should be paid to the high-suitability areas

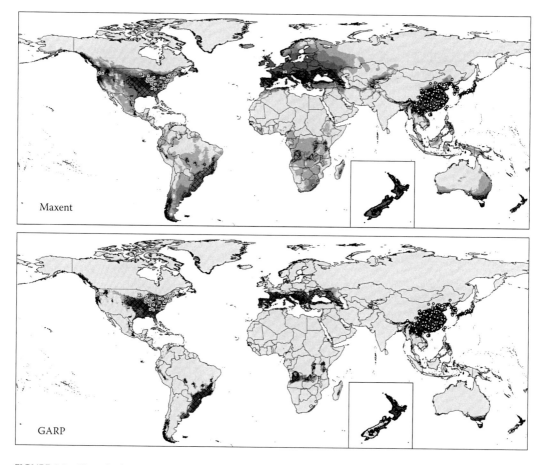

FIGURE 4.5   (See color insert.) Potential distribution of *Halyomorpha halys* worldwide based on ecological niche modeling. Niche models were calibrated on the native Asia and transferred onto the global using Maxent and GARP, dark green suggests high suitability, light green indicates low suitability (modified from Figure 6 and Supplemental Materials 1 in Zhu et al. 2012).

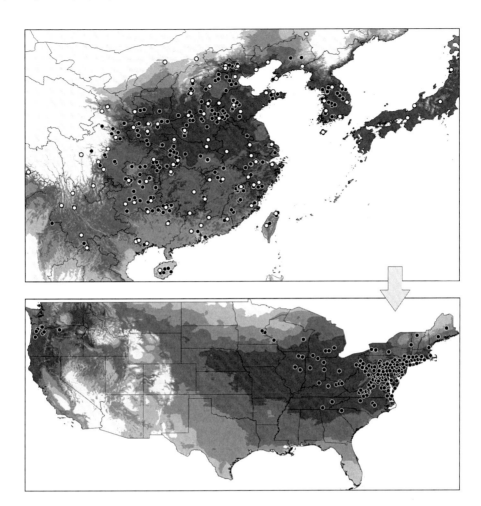

FIGURE 4.6 **(See color insert.)** Niche models for *Halyomorpha halys* based on native Asia extent and transferred onto the United States using Maxent. Dark green represents high suitability, light green indicates low suitability. Model was built using 6 variables, white and black dots represent the 95 occurrences for model calibration and the remaining for model evaluation (modified from Figure 3 in Zhu et al. 2012).

around the world, especially in developed areas with intensive trade activity with Japan, Korea or China. Strict quarantine inspection must be taken in these areas because commercial interchange might facilitate new invasions. In the United States, the models successfully identified the current disjunct distribution pattern of this bug, including successful anticipation of specific counties in Oregon, California (not shown), and Washington (Zhu et al. 2012) (**Figure 4.6**). Because some central states also showed high suitability for this bug, populations may be able to form a continuous distribution in the United States. This prediction was consistent with the observed population of *H. halys* in the United States. Recently, adults were intercepted continuously in Florida from the highly suitable area in the central Atlantic States. However, nymphs and eggs never have been found, suggesting this bug has not established populations in Florida (Julieta Brambila, personal communication; see Acknowledgments). This, most likely, is due to the unsuitable climate in southeastern United States. Nevertheless, one must keep in mind that the ENMs seek to identify suitable climate space for species but without consideration of biotic interactions or dispersal ability. Many factors influence successful establishment of non-indigenous species into a community and depends on existing species composition and richness, competitors, predators, food availability, human footprint, and climatic similarity, compared with the source areas.

In 2016, Zhu et al. (2016) used our understanding of niche filling and current data on the distribution of *Halyomorpha halys* to look at intraspecific variation in realized niche filling and unfilling and the impact of that variation on current niche-filling methods using an ordination method, ENMs, and occurrence records. They found that niche filling and expansion occurred in both North America and Europe (42.7% niche unfilling and 0.0% expansion in North America; 80.5% niche unfilling and 28.0% expansion in Europe), and that *H. halys* populations have invaded climatically novel regions in central Europe. They also estimate that there are regions of North America and Europe that are climatically similar to those of *H. halys* in its native range in Asia that are yet to be colonized.

These findings agree with those of Kriticos et al. (2017) who used CLIMEX bioclimatic niche model to estimate the potential distribution of *Halyomorpha halys*. Their model also predicted the potential for further spread in North America, especially in the the central and southern United States. In Europe, the model predicted that *H. halys* could spread further but not in the Ireland, Estonia, Latvia, Lithuania, Scandinavia, or the United Kingdom. It also predicts that in the Southern Hemisphere, areas with Mediterreanean, moist tropical and subtropical and warm-temperate climates could be at substantial risk for invasion.

4.5 Current Impacts

4.5.1 *Halyomorpha halys* in Asia

Halyomorpha halys has one to two generations per year in most regions of China, Korea, and Japan (Watanabe 1980, Bae et al. 2007, Yu and Zhang 2007) although four to six generations per year seem to be possible in southern China (Hoffman 1931). *H. halys* adults start overwintering in crevices under bark or human-made structures such as sheds, barns, and houses from late September to November (Ueno and Shoji 1978, Funayama 2012). Sometimes, high numbers of adults migrate into houses for overwintering (Zhang et al. 1993, Funayama 2012). In mountainous areas in Japan, their migration can end as early as mid-October (Saito et al. 1964, Watanabe et al. 1994).

After overwintering, adults leave overwintering sites during late March to mid-May (Watanabe et al. 1978, Wang and Wang 1988, Qin 1990, Zhang et al. 1993, Funayama 2012) when temperatures outside exceed 10°C (Qin 1990). The first host plants the adults utilize after overwintering include Chinese arborvitae (Lee et al. 2009), Chinese milk vetch (Bae et al. 2009), mulberry, elm, willow, and Chinese scholar tree (Wang and Wang 1988). During late spring or early summer, they enter orchards or crop areas to feed and reproduce on crops and surrounding vegetation.

Halyomorpha halys has a wide host range in Asia; a total of 106 host plants in 45 families have been reported in a review by Lee et al. (2013a). Among these host plants, major economic losses from these bugs occur on crops in the Fabaceae and Rosaceae. Minor, but frequent, damage by *H. halys* also has been reported on vegetables, field crops, ornamentals, and medicinal plants. In addition, this bug can serve as a vector of witches'-broom disease on paulownia, *Paulownia tomentosa* (Song et al. 2008), causing extensive economic damage to the timber industry.

4.5.1.1 Ecology of *Halyomorpha halys* and Its Economic Damage Reported in Asia

4.5.1.1.1 Fruit

Halyomorpha halys was considered a secondary pest in Asia until the 1970s when outbreaks of this bug began to be noticed (Qin 1990). Starting with a nationwide outbreak occurring in Japan in 1973 (Yanagi and Hagihara 1980), *H. halys* became a major pest of fruit trees with repeated outbreaks in China, Japan, and Korea (Ohira 2003, Zhang et al. 2007). So far, considerable feeding damage has been reported on apple, peach, pear, persimmon and yuzu (Qin 1990, Fengjie et al. 1997, Funayama 2003, Kang et al. 2003, Lee et al. 2007). Other fruit crops attacked in Asia include plum, mandarin, and grape (Oda et al. 1980). Adults do not damage new leaves but do cause severe damage to young and ripening fruits. Damaged fruits generally turn brown, collapse, and abscise during the ripening period (Watanabe 1996).

One of the reasons that *Halyomorpha halys* is a key orchard pest in Asia is the complex and diverse environment around and in the orchard. For example, intercropping with legumes and vegetables within an apple, pear, and peach orchard is not rare, especially in Korea, and is used to maximize the use of expensive agricultural land. In addition, most orchards in Asia are located in or near forested or mountainous areas, providing additional niches for this bug to utilize trees and shrubs as temporal or supplemental food sources. Because it has a wide host range and survives in various habitats, such diversified orchard environments in Asia can provide suboptimal conditions for *H. halys* to survive and reproduce throughout the growing season.

Beginning in the 2000s, development of organic orchards has become popular in China, Korea, and Japan. Consequently, *Halyomorpha halys* has become one of the key problems (Zhang et al. 2007) because of the lack of alternative control measures other than insecticides in organic farming systems. Approximately 25–30% of apples damaged by this bug often has been reported (Zhang et al. 2007) and, sometimes, more than 99% of pears injured in pear orchards has been observed (Fengjie et al. 1997).

Stink bugs including *Halyomorpha halys* are the most destructive pests of persimmon in Korea and Japan where sweet persimmon, *Diospryus kaki* (Ericales: Ebenaceae), is widely cultivated (D. Lee et al. 2001, K. Lee et al. 2002). In Japan, this bug is one of the major species damaging persimmon (Kawada and Kitamura 1983a, Adachi 1998). Feeding by *H. halys* causes blemished and corky fruits, decreasing the quality of persimmon. Also, fruit drop rate and feeding are higher with *H. halys* infestations (Lee et al. 2009).

Citrus, *Citrus unshiu* Marcow (Rosids: Rutaceae), is produced mainly in southern China, Korea, and Japan. *Halyomorpha halys* is the main pest damaging citrus fruits, producing blemish spots from September to October (Kim et al. 2000). Specifically, extensive economic losses of yuzu, *Citrus junos* Siebold ex Tanaka (Sapindales: Rutaceae), have been reported in Korea (Choi et al. 2000). Although other stink bugs such as *Glaucias subpunctatus* (Walker) and *Physopelta gutta* (Burmeister) are active at night, *H. halys* is diurnal as well as nocturnal in yuzu orchards.

4.5.1.1.2 Field Crops and Vegetables

Halyomorpha halys causes serious economic damage to soybean and grains in Asia (Kang et al. 2003). It reduces yield and quality of soybean by feeding on pods and the milky contents of the bean (Son et al. 2000, Kang et al. 2003, Bae et al. 2007, Paik et al. 2007). Although the first occurrence of *H. halys* in soybean fields generally is observed in late July, it does not cause significant feeding damage to soybeans that are in the full-flowering stage of soybean (i.e. R2 stage). Rahman and Lim (2017) showed that when a diet consisting of just pods or seeds only was provided to developing nymphs, feeding on just the pods negatively impacted development time and mortality but not fecundity or egg viability and that adults preferred to feed on seeds. However, the bug populations increase continuously as the plants develop from full seed (R6) to full maturity (R8) stages, and their feeding causes significant economic losses. *H. halys* is known to prefer indeterminate to determinate varieties of soybean and to prefer pod to seed (Son et al. 2000, Bae et al. 2007, Paik et al. 2007).

Corn damage by *Halyomorpha halys* has been reported throughout Asian countries (Fukuoka et al. 2002, Yu and Zhang 2007), but the extent of the damage is not well documented. Pearl millet (*Setaria italica* L.), sorghum (*Sorghum bicolor* L.), and common millet (*Panicum miliaceum* L.) are minor grains in Asia. *H. halys* mainly occurs in sorghum or millet fields from August to October. It prefers sorghum among the grains, and a large number of these bugs can be a serious problem in organic seed production (Kim et al. 2010).

In Asia, relatively few economic losses by *Halyomorpha halys* on vegetables have been reported. Major vegetables attacked include asparagus (Fukuoka et al. 2002), cucumber (Kawada and Kitamura 1983a, Fukuoka et al. 2002), pea, eggplant (Fukuoka et al. 2002), tomato (Oda et al. 1980), and vetch (Bae et al. 2009).

4.5.1.1.3 Specialty Crops: Tea, Paulownia, Orchid, and Matrimony Vine

Stink bugs are emerging as pests in tea production in Asia by decreasing tea quality through sucking on young leaves. They damage tea by decreasing the content of polyphenol (Murthy and Chandrasekaran 1979), water, total nitrogen, caffeine, and amino acid (Song et al. 2008). Specifically, in Korea, stink bugs such as *Plautia stali* Scott and *Halyomorpha halys* are the major pests of tea. Tea leaves damaged by these two species also show a decrease in the amount of catechin, a type of natural phenol causing stringent taste in tea; significant reduction of epigallocatechin gallate also was found by Song et al. (2008).

Witches'-broom is one of the most problematic diseases in paulownia (*Paulownia tomentosa*) plantations in Asian countries such as Korea, Japan, China, and Taiwan. Specifically, in Korea, paulownia
plantations in the 1970s and 1980s were devastated by the witches'-broom disease. The causal agent is
a phytoplasma that can be transmitted to periwinkle [*Catharanthus roseus* (L.)] by *Halyomorpha halys*
(Bak et al. 1993). The phytoplasma overwinters in xylem and spreads throughout the plant in spring. In
general, infected paulownia trees die within 2 years; big trees can live longer but eventually die (Bak et
al. 1993). Populations of *H. halys* peak two times on paulownia in May–July and August–October (Fujiie
1985). The rate of the pathogen transmission to paulownia seedlings by the nymphs is 41–62%, which is
higher than that by adults (14%) (Yu and Zhang 2007).

Damage to *Cymbidium* (Orchidaceae) and matrimony vine, *Lycium chinense* Mill (Solanales:
Solanaceae) by *Halyomorpha halys* has been reported in Asia. The bug significantly reduces the quality
of *Cymbidium*, a major export orchid in Korea, by feeding on leaves (Cho et al. 2013). Matrimony vine
is cultivated for medicinal herbs in Asia, and both *Plautia stali* and *H. halys* damage the vine by feeding
on fruit in late July. Serious economic damage has been reported on farms under environment-friendly
management (Ryu et al. 2013).

4.5.1.2 Management of *Halyomorpha halys* in Asia

4.5.1.2.1 Chemical Control Including Semiochemicals

Asian countries heavily rely on chemical control to manage *Halyomorpha halys*. For example, frequent
spraying with organophosphate insecticides is recommended in persimmon with two to three sprays per
week when outbreaks occur (Adachi 1998). Consequently, susceptibility of *H. halys* to pesticides also
has been tested extensively. These pesticides include carbamates, organophosphates, organochlorines,
pyrethroids, and neonicotinoids (see Lee et al. 2013a for a detailed list of the insecticides used against
H. halys in Asia). Qin (1990) suggested controlling nymphs of *H. halys* with insecticides in orchards
because they cannot fly during or after insecticide application, whereas the adults readily can fly and,
thus, re-invade the orchards. In addition, insecticide application on the periphery of the control area or
trees outside the orchard is suggested as a preventative management strategy for this bug (Qin 1990).

Semiochemicals also have been developed and used for monitoring and mass-trapping *Halyomorpha
halys* in Asia. Seasonal occurrence of *H. halys* generally is monitored by using mercury light traps and
an aggregation pheromone of *Plautia stali,* methyl (E, E, Z)-2, 4, 6-decatrienoate (Lee et al. 2002).
H. halys adults are caught in mercury light traps during late July, whereas the aggregation pheromone
attracts this bug during mid-May. In addition, Zhang et al. (2007) suggested a recipe for poisoned bait
as a stink bug lure: one part of 20% fenpropathrin fenpropanate butter, 20 parts of honey, and 19 parts
of water. They also suggested applying the bait to branches of old paulownia or Chinese scholar trees,
which will be effective for 10 days to attract and kill *H. halys*.

4.5.1.2.2 Biological Control

Many natural enemies of *Halyomorpha halys* are recorded from Asia. Qiu (2007) reported six egg parasitoids in the genera *Trissolcus, Telenomus, Anastatus, Ooencyrtus*, and *Acroclisoides* (the latter is probably a hyperparasitoid). In addition, three predators killing *H. halys* were reported by Qiu (2007): *Orius*
sp. (Hemiptera: Anthocoridae), *Arma chinensis* (Fallou) (Hemiptera: Pentatomidae), and *Misumena
tricuspidata* (F.) (Araneida: Thomisidae). Among these natural enemies, the predominant species was
Trissolcus halyomorphae Yang (Yang et al. 2009) [now synonymized as *Trissolcus japonicus* (Ashmead)
by Talamas et al. 2013]: its rate of parasitism ranged from 20 –70% with an average of 50% (Qiu 2007).
Li (2002) subsequently reported up to 80% egg parasitism by *T. japonicus*, a figure consistent with recent
surveys (T. Haye 2014; K. Hoelmer, unpublished).

Qiu (2007) presented a detailed life history of *T. japonicus* (as *T. halyomorphae*), a solitary egg parasitoid that typically attacks 100% of the eggs in the host egg mass. Male wasps emerge earlier than
females and wait for females to emerge in order to mate. Females preferred newly laid host eggs and laid
an average of 40.6 eggs. The developmental period from egg to adult could be as short as 7 days at 30°C,
and peak activity of the adults was from mid-June to early August in the field; thus, several generations
can occur in the field.

Other parasitoids reported in China include *Trissolcus flavipes* Thomson (now identified as *Trissolcus cultratus* (Mayr) by Talamas et al. [2015b]) with a parasitism rate of up to 63.3% (Zhang et al. 1993) and *Telenomus mitsukurii* (Ashmead) [now synonymized as *Trissolcus mitsukurii* (Ashmead), but probably a misidentification of *Trissolcus japonicus*] with an egg parasitism rate up to 84.7% (Fengjie et al. 1997). Five egg parasitoids, *T. mitsukurii*, *Trissolcus plautiae* (Watanabe), *Trissolcus itoi* Ryu, *Anastatus gastropachae* Ashmead, and *Ooencyrtus nezarae* Ishii (Arakawa and Namura 2002) were reported in Japan. All of these parasitoids probably utilize other pentatomids as hosts and are known to attack several other species of pest stink bugs (Ryu and Hirashima 1984, Qiu 2007, Matsuo et al. 2016) and the predatory stink bug *Arma chinensis* in laboratory exposures (Yang et al. 2009). A parasitic tachinid fly, *Bogosia* sp., is known to attack adult stink bugs in Japan with parasitism rates of up to 10% (Kawada and Kitamura 1992).

Following successful augmentation in China of populations of *Anastatus* sp. against the lychee bug, *Tessaratoma papillosa* Drury, Hou et al. (2009) experimented with augmentative releases of mass reared *Anastatus* against *H. halys* and reported elevated levels of parasitism in test peach orchards.

Few reports exist for pathogens of *Halyomorpha halys*. A fungal pathogen, *Ophiocordyceps nutans* (Pat.) (Hypocreales: Phiocordycipitaceae) has been reported (Sasaki et al. 2012), and an intestinal virus of *Plautia stali* (Hemiptera: Pentatomidae) was found to infect *H. halys* in the laboratory and may contribute to *H. halys* suppression (Nakashima et al. 1998).

4.5.1.2.3 Cultural and Mechanical Control

Bagging fruit is a common practice to protect tree fruits from stink bugs in Korea and Japan. The injury by *Halyomorpha halys* has been greatly reduced by bagging (Funayama 2002). Yu and Zhang (2007) suggested the following as a part of IPM in tree fruit orchards: bagging fruits 1 week earlier than normal time, mass trapping *H. halys* adults in the fall and early spring, and protecting orchard biodiversity and enhancing natural control by using parasitoids.

For the management of paulownia witches'-broom disease, it is recommended to avoid large scale plantations and to use new phytoplasma-free seedlings for new plantations. In addition, intercropping with legumes in orchards needs to be avoided because *Halyomorpha halys* can fully utilize both tree fruits and legumes to maximize their survivorship and reproduction (Adachi 1998).

In late autumn, *Halyomorpha halys* adults infest houses, buildings, and structures for overwintering. They begin to arrive on a clear day when the lowest temperature drops below 15°C, and the highest temperature is about 25°C (Watanabe et al. 1994). Peak populations moving to overwintering sites are observed when the lowest temperatures are below 10°C. Watanabe et al. (1994) suggested methods to prevent adults from entering human-made structures by using a slit trap consisting of a 30 cm x 180 cm collapsible saw-horse like frame with three plywood pieces separated by 3 mm spacers attached to one vertical face on rooftops as a tool to attract and kill them, applying pesticides such as cyphenothrin to entry points such as window frames, covering windows with pesticide (e.g., cyphenothrin), using coated polyethylene sheets, and covering the building with cyphenothrin-soaked netting.

4.5.2 *Halyomorpha halys* in Europe

Beginning with *Halyomorpha halys*' initial introduction into Europe in Switzerland in 2007 the bug now has been found in nine additional countries (Wermelinger et al. 2008, Arnold 2009, Heckman 2012, Callot and Brua 2013, Pansa et al. 2013, Garrouste et al. 2014, Maistrello et al. 2014, Milonas and Parsinevelos 2014, Vétek et al. 2014, Cesari et al. 2015, Macavei et al. 2015, Rabitsch and Friebe 2015) and with the exception of Italy is considered a nuisance pest (Sauer 2012, Pansa et al. 2013, Cesari et al. 2015, Haye et al. 2015b). Records in Switzerland still are concentrated in urban habitats and the bug generally is considered a household pest; however, infestations of ornamentals and local damage to fruit-bearing trees and peppers have been reported (Müller et al. 2011). The first detection in Germany consisted of a single female collected in southern Germany (Konstanz) and is not considered to represent an established population (Heckman 2012). In Italy, in addition to becoming an established urban pest in the Emilia Romagna, Lombardy, and Piedmont areas (Maistrello et al. 2013, 2014; Cesari et al. 2015), Pansa et al. (2013) reported that *H. halys* was found in Piedmont nectarine orchards and was causing damage to fruit. By 2014, *H. halys* was damaging pears in Modena; during summer 2015, it had spread to the Po Valley region of Moderna, Reggio Emilia, and Bologna where it caused damage to apples, apricots, peaches, pears, persimmons, plums, and tomatoes (Bariselli et al. 2016). In Hungary, *H. halys* has been found in and around Budapest (Vétek et al. 2014). The initial finding was inside a building at the Buda Arboretum. Subsequent findings were made near an apple orchard in Péterimajor. In Greece, this stink bug first was collected in Athens in the autumn of 2011 and, to date, has not been considered an agricultural pest (Milonas and Parsinevelos 2014). Recently, *H. halys* has been found in Vorarlberg and Vienna, Austria; Bucharest, Romania; and Sofia, Bulgaria (Macavei et al. 2015, Rabitsch and Friebe 2015, Simov 2016). For each Austrian detection, specimens were collected from buildings. In Romania, initial collections were made from the Botanical Garden of Bucharest. Further collections were made in urban areas in Bucharest, which were several kilometers away from the initial collection site. In 2016, *H. halys* was found infesting corn, goji, and field and edible roses with 100% crop loss occurring in goji (Roxana et al. 2016). Also in 2016, *H. halys* was reported from Girona, Spain, and Sardina, Italy (Dioli et al. 2016). Given this bug's ability to stow away in luggage, cars, and shipping containers, it is probable that it will be found in additional countries in Europe (Haye et al. 2015b).

4.5.2.1 Phenology in Europe

The life-cycle of *Halyomorpha halys* has been investigated in Swiss populations by Haye et al. (2014b). They reported that adults hibernate and appear around April with oviposition starting in early July and lasting until the end of September. However, latecomers did not struggle to survive. Development from egg to maturity required between 60 and 131 days. The new generation appeared from mid-August onwards. One generation per year developed in Switzerland, but two generations per year are to be expected in the Mediterranean area in southern Europe (Haye et al. 2014b).

4.5.2.2 Host Range in Europe

This polyphagous species has been recorded from 51 host plant species in 32 families with highest populations on the native *Sorbus aucuparia*, *Cornus sanguinea*, and *Fraxinus excelsior*; and the non-native North American *Catalpa bignonioides* and *Parthenocissus quinquefolia* (Haye et al. 2014a). There is no doubt that the list will increase with the continued spread of the species in Europe.

4.5.3 Halyomorpha halys in North America

4.5.3.1 Urban Nuisance Pest

Initially, upon its introduction into the United States, *Halyomorpha halys* was considered to be a nuisance pest because of its overwintering preference for man-made structures such as houses and commercial buildings (Hoebeke and Carter 2003). This also has been the case in Canada where it currently is considered only a nuisance pest (Gariepy et al. 2014a).

The overwintering behavior normally is the first contact humans have with *Halyomorpha halys* and frequently results in the presence of large overwintering populations (Inkley 2012). In one documented case, Inkley (2012) collected 26,205 adults during a 6-month period from January to June inside the first, second and third floors of a house. This behavior and the large numbers that can come in contact with humans within a structure has led to concern that *H. halys* could become a new aeroallergen (Merz et al. 2012), adding to the list of insects such as cockroaches and ladybugs that cause rhinitis/conjunctivitus responses upon contact (Kagen 1990, Yarbough et al. 1999, Ray and Pence 2004, Albright et al. 2006, Sharma et al. 2006). Mertz et al. (2012) investigated this potential and demonstrated that out of 15 patients tested with known prior exposure to *H. halys*, 11 exhibited reactions to this bug in skin tests.

What attracts *Halyomorpha halys* to structures and where they prefer to settle have been ongoing questions. To answer the question about attraction to buildings, The Great Stinkbug Count was conducted in 2013 (Torri J. Hancock, unpublished data; see Acknowledgments; in Leskey and Hamilton 2015). This study used a citizen science survey to evaluate several structural characteristics including color, compass orientation, and surrounding landscape and found that residences with brown exteriors were more attractive to overwintering adults than those with white, red, green, gray or tan exteriors; that stone or wood exterior-sided homes were more attractive than metal, brick, stucco, vinyl or cement exteriors; and that structures in rural areas attracted significantly more *H. halys* than those in urban areas.

To answer the question regarding settling preference, Inkley (2012) recorded the location of the over 26,000 adult *Halyomorpha halys* found in a residence (see above) between January to June. He found that within the structure, unfinished attic crawlways had the highest populations (50.2%) compared to the first and second floors of the building (40.4%) and the basement (9.4%).

In populated areas, as local populations of *Halyomorpha halys* begin to increase, the general public also can be impacted because of feeding by these bugs on ornamentals and backyard vegetable and tree-fruit plantings (Bernon 2004, Hamilton 2009, Nielsen and Hamilton 2009a, Martinson et al. 2013, Sargent et al. 2014). In response to public concerns about the impact, traps containing aggregation pheromones attractive to nymphal and adult *H. halys* have been developed and tested (Khriminan et al. 2008, 2014; Nielsen et al. 2011; Leskey et al. 2012d, 2015a; Sargent et al. 2014). Sargent et al. (2014) evaluated the use of these pheromone traps in tomatoes grown in homeowner gardens and, surprisingly, found that gardens without traps were no more likely to have *H. halys* feeding on tomato fruit than those with traps. However, gardens containing traps sustained significantly more damage to tomato fruit than gardens without traps, and individual tomato plants near the traps had more bugs present on the fruit. This phenomenon (i.e., plants near pheromone traps having higher numbers of *H. halys* and higher damage levels), has been observed in other crop systems such as peaches and apples (Anne L. Nielsen, unpublished data; Morrison et al. 2015).

Traps developed to catch adults of *Halyomorpha halys* also have been developed for indoor use. Several utilize light as an attractant. Aigner and Kuhar (2014) evaluated the efficacy of several traps using citizen scientists in Virginia during the winter and early spring of 2012 and 2013. They compared the use of two commercially available traps (Strube's Stink Bug Trap – 2012 only, Rescue® Stink Bug Trap) and two homemade traps (aluminum foil water pan trap; 2-liter bottle trap with LED light). Their results indicated that the water pan trap caught significantly more adults than the Strube, Rescue® and 2-liter bottle trap in 2012 and the 2-liter bottle trap and Rescue® trap in 2013. (The Strube trap was excluded from the study in 2013 because the company ceased production of trap in 2012).

Recently, the use of pheromone traps inside and outside man-made structures during the overwintering period has been investigated (Morrison et al. 2017a). In this study, the authors examined when *Halyomorpha halys* responds to pheromone during late winter and early spring, whether or not pheromone traps can be used to monitor for *H. halys* indoors, and whether or not pheromone traps can be used as an indoor management tool. Over a 2-year period, their results showed that adults begin responding to the pheromone in the spring following a critical photoperiod of 13.5 h, that trap captures could not be correlated with visual counts indicating their use may not be an effective indoor monitoring tool, and because only 8-20% of the known adult population responded to the pheromone-baited traps, these traps would not be an effective indoor management tool.

Halyomorpha halys' overwintering behavior also has become an issue for international commerce with countries imposing restrictions such as fumigation or heat treatments to kill specimens present in shipments prior to importation (Thompson 2014, Australian Government 2015). With heat treatments, the primary question has been what temperature should be used. Aigner and Kuhar (2016) attempted to answer this question using a series of different temperatures and periods of exposure under controlled conditions. They found that 15 minutes of exposure to 50°C was adequate to provide 100% mortality. After one hour of exposure, 45°C provided 100% mortality; after 4 hours, 42°C resulted in 92.3% mortality. The authors also examined mortality under field conditions using vehicles waiting for export using 15-minute exposure to 40°C, 50°C, and 60°C. They found that 50°C was sufficient to kill 100% of the adults placed either under the driver's seat or in the engine compartment.

Today, as populations have increased in urban areas and subsequently spread into agricultural areas, *Halyomorpha halys* has become a severe agricultural pest (Leskey et al. 2012a, Rice et al. 2014).

4.5.3.2 Conventional Agricultural Pest

The first documented agricultural damage in North America by *Halyomopha halys* was reported by Nielsen and Hamilton (2009a) but actually occurred during 2006–2007. They noted the presence of nymphs and adults and resulting damage in pear and apple orchards in Allentown, Pennsylvania, and Pittstown, New Jersey. They found that nymphs occurred at low densities throughout the season in apples, and that peak densities in pear occurred in early July and mid-August when pit hardening and swelling occurred, respectively. In addition, they found that more than 25% of the fruit in each crop was damaged by *H. halys* feeding. In apples, damage was characterized by discoloration of the outer surface and discolored dots with depressions due to the insertion of the bug's stylets into the fruit (Nielsen and Hamilton 2009b, Brown and Short 2010). Similar damage occurs in pears and peaches (Nielsen and Hamilton 2009b).

Populations of *Halyomorpha halys* continued to increase and spread throughout the mid-Atlantic United States with few reports of agricultural damage occurring. However, in 2008 and 2009, damage to apples and peaches by *H. halys* was observed for the first time, but the bug was not considered a widely distributed pest geographically until 2010 (Leskey and Hamilton 2010a, Leskey and Nielsen 2018) (**Figure 4.1E**). During the 2010 season, feeding by this bug in apples alone in the mid-Atlantic United States was estimated to have caused a loss of $37 million in apples and >90% losses by some growers in peach orchards (Leskey and Hamilton 2010b, Anonymous 2011). Feeding, at least in apples, appears to be related spatially to where fruit occurs in the tree canopy and within an orchard. Joseph et al. (2014), using a zero-inflated negative binomial model, found that apples in the upper canopy of border trees were more likely to be damaged. Acebes-Dori et al. (2016b) examined the injury caused to apples and peaches at harvest due to feeding by nymphs. Their study showed that feeding by young nymphs early in the season resulted in the least amount of damage to each crop. Feeding by adults on apples late in the season caused the most damage. However, feeding by adults on peaches early in the season resulted in the greatest amount of damage.

Since 2010, *Halyomorpha halys* has been observed damaging several vegetable crops including beans, eggplant, peppers, okra, sweet corn, and tomatoes (Leskey et al. 2012a, Rice et al. 2014). In beans and okra, damage appears as scarred or faded depressions. In tomatoes and peppers, its feeding causes white or yellow surface spotting with underlying tissue damage. Feeding in sweet corn causes discolored and sunken kernels.

In addition to peach, apple, and pear, *Halyomorpha halys* also feeds on blackberries, blueberries, raspberries, and wine grapes (Leskey and Hamilton 2010b, Leskey et al. 2012a, Rice et al. 2014). In blueberries, Wiman et al. (2015) found that feeding in early and late ripening varieties caused discoloration and internal damage to fruit and reduced fruit weight and soluble solids. Feeding by *H. halys* also has been shown to modify the chemistry and attractiveness of the fruit (Zhou et al. 2016). In the field, feeding caused lower levels of anthocyanin, phenolics and soluable solids (Brix). In blackberries and raspberries, early season feeding caused death of blossoms whereas late season feeding caused discoloration and collapse of drupes due to the bugs' inserting their stylets between individual drupes (Basnet et al. 2014, Rice et al. 2014). In wine grapes, *H. halys* adults enter vineyards in May in Virginia, lay eggs in June and July followed by the presence of all developmental stages until September (Basnet et al. 2015). Feeding by nymphs and adults on grape clusters causes necrosis and deformation of individual fruit (Pfieffer et al. 2012, Rice et al. 2014). There also are concerns that the inclusion of the bugs with harvested grape during the crush process may taint the final wine product (Tomasino et al. 2013, Mohekar et al. 2014, 2016). Mohekar et al. (2017) examined this concern and found that the release of tridecane and (E)-2-decanal by *H. halys* during processing negatively affected the final quality of the wine.

Halyomorpha halys feeds on a variety of field crops including corn, cotton, hops, sorghum, soybeans, sunflower, and wheat; however, corn and soybeans appear to be the main crops attacked thus far (Rice et al. 2014). In corn, bug populations peak mainly during milk (R3) and dough (R4) stages of ear formation. As noted above, feeding through the husk discolors and shrinks individual kernels. There is some evidence from Pennsylvania that feeding on the plants prior to the R1 stage may cause ear abortion. When bugs enter soybeans field, they remain primarily in border rows during the R4 to R6 stages of plant development and are able to feed on developing seeds through the pods (Rice et al. 2014). Late season feeding in soybeans in border rows causes a "stay green" effect characterized by plants that senesce later than the rest of the field. The effect makes it difficult to harvest the entire field thereby causing yield losses. *H. halys* recently has invaded urban areas in Minnesota. However, a 2-year survey of soybeans in Minnesota found no bugs despite its continued detection in urban areas (Koch and Pahs

2015). These results suggest that the impact of *H. halys* in soybeans may be limited to the mid-Atlantic and southern United States. However, in another study, Koch and Rich (2015) confirmed that caged *H. halys* can affect yield components such as maturity and quality of early maturing soybean varieties in Minnesota. Rich and Koch (2016) also showed that Rag1 aphid-resistant soybeans increased *H. halys* mortality in no-choice tests. In choice tests, *H. halys* preferred Rag1 aphid-resistant soybean after 4 hours but not after 24 hours.

Adults of *Halyomorpha halys* have been observed feeding on wheat during the milk and soft dough stages, which corresponds to the time when native stink bugs feed on wheat (Rice et al. 2014). So far their impact on this crop has not been determined. In a study conducted in Minnesota wheat fields using sweep net surveys during 2011 and 2012, 14 stink bug species were recovered. None was identified as *H. halys* (Koch et al. 2016).

Halyomorpha halys attacks a wide variety of ornamentals (Hoebeke and Carter 2003, Bernon 2004, Nielsen and Hamilton 2009b). Currently, it is known to feed and/or develop on at least 120 different ornamental hosts (**Table 4.1**) (Anonymous 2015b) and is capable of causing both indirect and direct damage to plants (Hoebeke and Carter 2003, Leskey et al. 2012a). Indirect damage is characterized by stippling of leaves followed by a scab-like appearance (Hoebeke and Carter 2003). Direct damage caused by this bug feeding through the bark of woody ornamentals includes large amounts of sap flow, fluxes, and discolored bark around feeding sites (Martinson et al. 2013). This type of feeding damage in ornamentals has not been reported for other stink bugs (Panizzi 1997, Martinson et al. 2013). Feeding also attracts ants and wasps to feeding sites in search of high levels of sugar available in the sap (Martinson et al. 2013).

4.5.3.3 Organic Agricultural Pest

There is no documentation of the economic impacts of *Halyomorpha halys* to the organic community. However, organic growers cannot utilize synthetic pesticides and many Organic Materials Review Institute (OMRI) approved materials, if effective, have limited residual activity; however, Cira et al. (2017) showed that a combination of azadirachtin and pyrethrins significantly reduced egg hatch and caused significant mortality of first and second instars. The authors also showed that azadirachtin and pyrethrins and sulfloxaflor significantly reduced the percentage of first instars that molted. Morehead and Kuhar (2017) further evaluated the use several OMRI approved material to control *H. halys* under laboratory and field conditions. They found that only pyrethrins resulted in moderate toxicity using a bean dip bioassay. Also, under field conditions, with one exception, none of the test materials, when applied weekly, effectively reduced damage in tomatoes or peppers. The exception involved one harvest date in peppers where the combination of pyrethrins and azadirachtin provided less damage that that seen in the unsprayed control. *H. halys* has been difficult to control for organic growers, and current research is investigating the use of trap crops, exclusion fabrics, and conservation biological control. One such study conducted by Soergel et al. (2015) investigated the use of sunflower in pepper plantings as a potential trap crop. They showed that significantly more nymphs and adults were present in peppers surrounded by sunflowers than in those not surrounded by sunflowers. Despite this, no significant difference in damage was found between peppers planted with or without sunflowers suggesting that their use might not be a suitable control tactic. Nielsen et al. (2016b) evaluated the attractiveness of sorghum, Admiral pea, millet, okra, and sunflower to *H. halys* in four Mid-Atlantic States. Sorghum was the most attractive followed by sunflower and okra. In the same study, the use of flaming, pheromone-baited traps and the application of OMRI-approved insecticides were evaluated as management techniques. The use of flaming was more effective than either of the other two techniques. Different types of physical exclusion barriers as a protective measure also have been evaluated with mixed results depending on pest pressure, barrier color, and screen size (Dobson et al. 2016).

4.6 Current Management Tactics

Management of *Halyomorpha halys* presents significant challenges to commercial growers for several reasons. First, the broad documented host range of this bug shows that over 150 wild plants and crops, if untreated, allow populations to increase and persist (Anonymous 2015b, Lee et al. 2013a); these populations serve as a nearly constant reservoir for re-infestation of susceptible crops. Second, mobile life stages of *H. halys* have strong dispersal capacities. Adults have been reported to fly over 100 m in less than one minute in the field (Lee et al. 2013b) and greater than 2 km in less than a day under laboratory conditions (Wiman et al 2014, Lee et al. 2014a), whereas nymphs have been recorded to walk over 20 m in less than 6 hours (Lee et al. 2014a). This indicates that vulnerable crops located considerable distances away from populations of this bug on wild unmanaged hosts can be at risk of damage. Finally, overwintering populations are not only present in large numbers in human-made structures such as homes and sheds (Inkley 2012), they also are present in the natural landscape in dead-standing trees (Lee et al. 2014b). These well-concealed and unchecked populations are dispersed widely throughout the landscape, likely emerging over several months in the spring (Tracy C. Leskey and J. Christopher Bergh, unpublished data; see Acknowledgements). Therefore, the need for effective tools to manage this bug is paramount in the face of the threat posed by this invasive species.

4.6.1 Monitoring Options

Because *Halyomorpha halys* is attracted to lights, landscape-level monitoring using blacklight traps is invaluable for detecting its presence and flight activity (**Figure 4.1F**). Blacklight traps have documented peaks in flight activity of this bug near tree fruit and soybean crops (Nielsen and Hamilton 2009a; Nielsen et al. 2011, 2013) as well as its spread across New Jersey, with exponential increases in populations detected annually based on data collected through 2011 (Nielsen et al. 2013). Changes in the seasonal abundance of adult populations have been mapped effectively using blacklight trap captures to provide growers with relative pest pressure of local populations (Nielsen et al. 2013). These same data have been used in conjunction with additional biological information to establish factors associated with spread of this bug across the landscape (Wallner et al. 2014). These data also have been used in combination with crowd source reports that examine the relationship between urban and agricultural situations (Hahn et al. 2016).

A variety of pheromone-based monitoring traps for *Halyomorpha halys* and other pentatomids have been developed (Adachi et al. 2007, Khrimian et al. 2008, Leskey et al. 2012d, Bae et al. 2017). On-farm monitoring can be conducted effectively using "trunk-mimicking" baited black pyramid traps (Leskey et al. 2012d). Indeed, the two-component male-produced aggregation pheromone, in a 3.5: 1 ratio of 3*S*,6*S*,7*R*,10*S*)-10,11-epoxy-1-bisabolen-3-ol and (3*R*,6*S*,7*R*,10*S*)-10,11-epoxy-1-bisabolen-3-ol (Khrimian et al. 2014), deployed in combination with the synergistic compound methyl (*E*,*E*,*Z*)-2,4,6-decatrienoate (the pheromone of the Asian stink bug species *Plauti stali*), serves as an attractive and sensitive olfactory stimulus for inclusion in traps for season-long monitoring (Weber et al. 2014, Morrison et al. 2017b). Methyl (*E*,*E*,*Z*)-2,4,6-decatrienoate alone, however, though attractive to nymphs whenever present in the field, is not attractive to adults until the late-season (Leskey et al. 2012d), thereby limiting the utility of this attractant as a stand-alone stimulus throughout much of the growing season.

In trials conducted across the United States with varying population densities, black pyramid traps baited with the aggregation pheromone and synergist (mentioned above) combination were deployed between agricultural production and nearby wild, host habitats and served as sensitive monitoring tools for the presence, abundance, and seasonal activity of nymphs and adults of *Halyomorpha halys* throughout the year. Overwintered adults first were captured in traps in early April with captures peaking in mid- to late-May. Nymphs routinely were detected in traps beginning in June though some captures were reported earlier. The smallest and largest populations of both adults and nymphs were detected in the early- and late-season, respectively (Leskey et al. 2015a). Trap sensitivity can be increased markedly by increasing the dose of the aggregation pheromone deployed alone or in combination with methyl (*E*,*E*,*Z*)-2,4,6-decatrienoate (Leskey et al. 2015b).

The purity of the aggregation pheromone itself is not critical. It is comprised of two stereoisomers of a natural sesquiterpene with a bisabolane skeleton, existing in 16 stereoisomeric forms. *Halyomorpha halys* is not only strongly attracted to pheromonal stereoisomers but also is moderately attracted to some non-pheromonal stereoisomers. This signifies that these "unnatural" stereoisomers are sufficiently similar to the true pheromone in structure to trigger behavioral responses. Importantly, mixtures of stereoisomers containing pheromone components are highly attractive to this bug even in the presence of multiple "unnatural" stereoisomers, indicating that attraction to pheromonal and non-pheromonal stereoisomers and a lack of inhibition from non-pheromonal stereoisomers increases the flexibility of developing pheromone-based products for *H. halys* (Leskey et al. 2015b). The use of plant volatiles to enhance pheromone attraction also has been examined. Morrison et al. (2017b) determined that plant volatile extract mixtures derived from either apple, peach or a green leaf volatile mixture combined with pheromone caused a small increase in *H. halys* retention on plants, and that plant species could modulate *H. halys*' retention. However, the addition of plant volatiles did not increase *H. halys* attractiveness to baited pyramid traps. For a complete review of *H. halys*' chemical ecology, see Weber et al. (2017). Ultimately, it is hoped that biological information from captures generated from baited pyramid traps can be used as decision-based support tools (i.e. thresholds) for growers. The first of these tools has been developed for apples. Short et al. (2016) showed that with the use of baited pyramid traps, a cumulative threshold of 20 adults per trap resulted in significantly higher damage levels when compared to cumulative thresholds of 1 or 10 adults per trap. They also demonstrated that a cumulative threshold of 10 adults per trap reduced insecticide applications by 40%.

Trunk traps have been developed to examine the movement of *Halyomorpha halys* nymphs up and down host trees by comparing the use of modified 'Circle', 'Hanula', and 'M&M' traps in the laboratory and in the field (Acebes-Doria et al. 2015). In the laboratory using nymphs released at the top and bottom of Tree of Heaven, *Ailanthus altissima*, logs, Circle and M&M traps caught significantly more nymphs moving up the logs. These two traps also caught the highest number of nymphs moving up and down *A. altissima* trees in the field.

Finally, real-time presence of populations of *Halyomorpha halys* in crops has been conducted using timed visual counts. Timed counts have been conducted in peaches (Blaauw et al. 2014) and apples (Leskey et al. 2012c), although no thresholds exist at present for triggering management decisions in peaches. Timed visual counts also have been used in row crops and have revealed that monitoring efforts should be focused on crops bordering wooded habitats, as these are the adjacent habitats most likely to yield significant populations of this bug infesting nearby crops (Venugopal et al. 2014, Aigner et al. 2017). Other methods have included sweep nets (Nielsen et al. 2011, Leskey et al. 2012c, Aigner et al. 2017) and beat samples (Nielsen and Hamilton 2009a,b) to document this bug's presence in various host plants and cropping systems, although they are not recommended currently to monitor its populations in most crops with the exception of soybean.

4.6.2 Biofix Models

A biofix is a biological occurrence or indicator of a particular aspect of a pest insect's development that initiates the calculation of growing-degree-days in order to time a management tactic (Anonymous 2015a). To use a biofix event in management models, the development of an insect pest must be characterized. Developmental rates and life table analyses for *Halyomorpha halys* in North America has shown that females require ≈148 DD_{25} to complete the preoviposition period and ≈538 DD_{14} to complete development from egg to adult (Nielsen et al. 2008a) with similar results reported for European populations (Haye et al. 2014b). Application of the DD requirements has been used in New Jersey to predict the peak flight period of the F1 adults as they eclose and presumably search for mates or other resources (Nielsen et al. 2013). However, there are still some misalignments in the phenology model. Nielsen et al. (2016a) addressed this through the development of an agent-based stochastic model, using temperature and photoperiod, that accurately predicts population growth and voltinism at several locations in the United States. Currently, refinements to the preoviposition period as well as the appropriate biofix threshold currently are being made by Anne L. Nielsen (unpublished data). Management programs for *H. halys* in peach orchards have used a biofix of 148 DD accumulation (or "accumulating" as written before) beginning on 1 January for initiating insecticidal sprays aimed at adult populations (Anne L. Nielsen, unpublished data; Blaauw et al. 2014) with good success. Further refinements of the model are underway and applications to other cropping systems are being pursued.

4.6.3 Cultural Control

Cultural control tactics for *Halyomorpha halys* are being investigated but have not been implemented in the United States to date. However, strategies have been reported from Asia although not widely adopted. Lee et al. (2013a and references therein) reported that mechanical removal of eggs and nymphs from crops has been considered. Other strategies proposed include destruction of nearby alternate hosts and construction of overwintering traps. Bagging fruit also has been proposed in Asia (Funayama 2002) although results were mixed. In the United States, bagging fruit also has been pursued in small-scale studies against *H. halys* (Daniel L. Frank, unpublished data; see Acknowledgements). Trap cropping also has been considered in Asia and the United States. To date, this tactic has only been evaluated in limited experimental trials. However, progress has been reported by Nielsen et al. (2016b) in a multi-state comparison trial that evaluated sunflower, grain sorghum, eggplant, pearl millet, and Admiral pea as trap crops to protect bell peppers from *H. halys* and native stink bugs. At all sites, both sunflower and sorghum hosted significantly higher seasonal populations of this bug. Importantly, there was a difference in timing of attractiveness with sunflower having higher densities of *H. halys* than sorghum because the sunflower variety flowered before the sorghum variety produced seed heads. Due to the long life cycle of *H. halys*, and the need to protect crops, sunflower and sorghum biculture currently are being evaluated as potential trap crops for this bug in organic systems to protect green peppers on organic farms. Olusola (2016) examined the use of several potential crop plants and found cowpea (Early Scarlet) to be a suitable trap crop. Mathews et al. (2017) has evaluated a polyculture system composed of sorghum and sunflowers as a combined trap crop in organic peppers. Their 2-year study showed that the trap crop was more attractive to *H. halys* than peppers for a period of 8 weeks. However, the trap crop did not prevent adults from entering the pepper crop at most sites, and the resulting reduction in damage would not make this tactic economically viable. In a separate study involving the the same plant species, Blaauw et al. (2017) used harmonic radar, immunomarking and visual sampling to evaluate movement of *H. halys* within and between a polyculture planting of peppers, sorghum, and sunflowers. They demonstrated that immuno-marking followed by dislodgement of individuals revealed higher densites in the trap crop compared to visual sampling, and that very little movement of adults between the peppers and the trap crop occurred. Harmonic tagging of adults also revealed that adults stayed longer in the trap crops and reduced their movement when compared to individuals released into the peppers.

4.6.4 Biological Control

In the invaded ranges of the United States and Europe, populations of *Halyomorpha halys* are thought to have been able to establish and increase, in some cases, to outbreak levels due in part to the enemy-release hypothesis (i.e., *H. halys* individuals escaping natural enemies from their native range). However, biological control is considered likely to be the long-term solution for this bug; the impact of predators, pathogens, and parasites is self-sustaining and can occur at landscape levels (Leskey et al. 2012a) where it is likely to be the only management option available for stink bug populations. Indeed, a number of natural enemies present within the invaded areas have been documented in different habitats and crops.

Chewing and sucking predators reported to feed on the eggs, nymphs, and adults of *Halyomorpha halys* in the United States include (but are not limited to) members of the Anthocoridae, Asilidae, Cantharidae, Carabidae, Chrysopidae, Coccinellidae, Forficulidae, Geocoridae, Gryllidae, Mantidae, Melyridae, Pentatomidae, Reduviidae and Tettigoniidae (Biddinger et al. 2012, Pote et al. 2015, Lara et al. 2016, Morrison et al. 2016). Other predators include spiders, nest-provisioning sand wasps (Crabronidae), and some gastropods (slugs) (Pote et al. 2015, Lara et al. 2016, Morrison et al. 2016, 2017c) and bats (Masslo et al. 2017). Among predatory arthropod groups, differences in overall predation rates have been observed in different years, geographic locations, and cropping systems (Rice et al. 2014). Chewing predators were thought to be more important than piercing-sucking egg predators in a comparative multistate study in organic crops (Ogburn et al. 2016). Combining laboratory observations of a range of potential predators with direct field observations and sentinel egg masses (i.e., either live or previously frozen egg masses obtained from laboratory colonies that are placed in the field to evaluate natural enemy activity) in orchards and vegetables, Morrison et al. (2016) determined that the most effective predators of eggs in orchards were various species of tettigoniids and carabid beetles followed by earwigs, salticid spiders, and crickets (gryllids). In contrast, generalist predators often found in vegetable and field crops, such as coccinellids, anthocorids, and predatory pentatomids, were the most important predators of *H. halys* eggs in those habitats. In the case of *Orius insidiosus* Say, Fraga et al. (2016) found in greenhouse, laboratory, and field experiments that adults were attracted to tridecane, a plant volatile released in response to feeding by *H. halys*. However this attraction did not result in increased predation indicating that it might be acting as an arrestant. In a separate study involving parasitoids, Rondoni et al. (2017) showed that native egg parasitoids can exploit volatiles emitted by *H. halys* and by *Vicia faba* L. in response to feeding by *H. halys*. Clearly, native predators of many types may play an important role in helping to reduce stink bug populations, especially in conservation biological control where agricultural systems may be manipulated in various ways to enhance predator populations.

Limited work has been published to date regarding pathogens. One strain (GHA) of the entomopathogenic fungus *Beauvaria bassiana* (Bals.-Criv) Vuill., which is the active ingredient in Botanigard®, and several other tested strains showed strong efficacy against adults of *Halyomorpha halys* in laboratory trials (Gouli et al. 2012). Strains of *Metarhizium anisopliae* (Metchsnikoff) Sorokin were less efficacious (Gouli et al. 2012, Pike 2014), and there is evidence that defensive compounds released by this bug, including trans-2-octenal and trans-2-decenal, may inhibit fungal growth and prevent spore germination at low concentrations (Pike 2014). Evidence of microsporidian infections of *H. halys* has also recently been found in assays conducted to determine the cause of laboratory colony die-offs. Following the discovery of laboratory infections, microsporidians also have been detected in some field populations, and further research is needed to determine their frequency and impact (Ann Hajek, unpublished data; see Acknowledgments).

Several species of tachinid flies in North America, which are parasitoids of other pentatomid species, have been reported to respond to their host pheromones and were attracted to field lures containing the pheromones (Aldrich et al. 2006); these authors reared several specimens of *Trichopoda pennipes* (F.) from field-collected *Halyomorpha halys*. The native tachinid fly *Euclytia flava* (Townsend) also was found in traps baited with commercial *H. halys* pheromone lures in California surveys (Lara et al. 2016). The flies deposit their eggs on dorsal and/or ventral surfaces of *H. halys* adults which then hatch and penetrate into the stink bug hosts; however, successful development and emergence rates have been low (Rice et al. 2014), which suggest that although the flies recognize the stink bugs as a potential hosts, the bugs are somehow physiologically unsuitable.

Because egg masses can be difficult to locate in the field, sentinel egg masses also have been used in many studies to document the levels of parasitism of stink bug eggs and species diversity of indigenous egg parasitoids in various habitats and crops in the United States and in Europe. Species of *Anastatus* (Eupelmidae), *Trissolcus*, and *Telenomus* (Scelionidae) have been the most commonly encountered hymenopterous parasitoids reared from wild (naturally laid) and sentinel *Halyomorpha halys* eggs, with occasional recoveries of *Ooencyrtus* (Encyrtidae) (e.g., *Ooencyrtus telenomicida* Vassiliex in Europe) and *Gryon* (Scelionidae in the United States). Parasitism rates by *Anastatus reduvii* (Howard), *Anastatus pearsalli* Ashmead, and *Anastatus mirabilis* (Walsh and Riley) in the United States and *Anastatus bifasciatus* (Geoffroy) in Europe; *Trissolcus brochymenae* (Ashmead), *Trissolcus edessae* Fouts, *Trissolcus euschisti* (Ashmead), *Trissolcus hullensis* (Harrington), *Trissolcus thyantae* Ashmead, *Trissolcus utahensis* (Ashmead) (as *Telenomus utahensis* Ashmead) in the United States, and *Telenomus podisi* Ashmead and *Telenomus persimilis* Ashmead in North America generally are low – well under 10% (Rice et al. 2014, Cornelius et al. 2016, Ogburn et al. 2016, Roversi et al. 2016, Dieckhoff et al. 2017). Furthermore, notable variation appears to exist in the proportion of successful development of some geographic populations of native parasitoids on *H. halys*, with some populations having a complete inability to develop to emergence (e.g., *Telenomus podisi* in Ontario, Canada [Abram et al. 2014]). Although the native parasitoids often fail to develop, they may nevertheless attack and deposit eggs, causing elevated levels of mortality to the host eggs (Abram et al. 2014, Haye et al. 2015a, Cornelius et al. 2017, Dieckhoff et al. 2017). However, when sentinel egg masses were frozen prior to field placement, the successful development of native parasitoids was increased, sometimes considerably (Haye et al. 2015a, Herlihy et al. 2016). In some cases, the use of frozen eggs permitted the successful development of parasitoid species that were not able to develop in fresh *H. halys* eggs (Haye et al. 2015a). Presumably the freezing process disables whatever physiological defensive barriers exist in fresh eggs. Indirect evidence of this has been shown by Tognon et al. (2016).

These authors examined the response of *Trissolcus erugatus* Johnson and *T. podisi* egg volatiles emitted by the eggs of *Euschistus conspersus* Uhler and *H. halys*. In the case of *H. halys*, hexadecanal, octadectanal, and eicosanal were detected. Female *T. eugatus* and *T. podisi* were also repelled by egg extracts and aldehyde blends identified from *H. halys* eggs. It is not yet known whether populations of native parasitoids will adapt over time to overcome the egg defense, but this is a question worth studying.

When compared with wild egg masses, sentinel eggs may underestimate parasitism rates because they lack kairomone cues left on leaf surfaces by the movement of parent stink bugs that help lead parasitoids to egg masses. In ornamental tree nurseries, wild egg masses had significantly greater levels of parasitism by native *Anastatus* species, which typically have very broad host ranges (Jones et al. 2014). In other habitats, differences between sentinel and wild egg parasitism rates have been less pronounced (Kim Hoelmer, unpublished data). As with predation, variability among years, geographic locations, and cropping systems regarding the most common parasitoids encountered has been reported.

Parasitism of *Halyomorpha halys* by indigenous parasitoids in the United States is substantially lower than parasitism levels of indigenous stink bugs (e.g., Koppel et al. 2009, Tillman 2010). However, because parasitism of *H. halys* by indigenous parasitoids is high in its native Asian range, a classical biological control program has been considered for suppression of *H. halys*. Exploratory surveys were conducted in China, South Korea, and Japan to collect natural enemies of *H. halys*, and the candidate agents were returned to United States quarantine facilities for evaluation as potential biological control agents for field release. Collected material was identified using extracted DNA in conjunction with morphological characters (Talamas et al. 2015a; Marie-Claude Bon and Kim Hoelmer, unpublished data). These surveys and one conducted by Zhang et al. (2017) in northern China confirmed that *Trissolcus japonicus* was the most common and effective parasitoid of *H. halys* throughout its Asian range, and *Trissolcus cultratus* was the next most abundant species. Several geographic populations of *Trissolcus* obtained from these collections (*Trissolcus japonicus, Trissolcus cultratus, Trissolcus mitsukurii*, and *Trissolcus itoi*) have been evaluated for their host specificity. To date, the species considered to be most promising is *T. japonicus*, especially because of its high parasitism rates in China, Korea, and Japan (Yang et al. 2009; Tim Haye and Kim Hoelmer, unpublished data). An unexpected impact of *H. halys* on parasitoids is its interference with plant tritrophic signaling. In a study examining the impact of *H. halys* feeding on *V. faba* on the parasitism of *Nezara viridula* (L.) eggs by *Trissolcus basalis* (Wollaston), Martorana et al. (2017) demonstrated that it interfered with the attraction of *T. basalis* to its host's eggs.

Laboratory host range evaluations in the United States (no-choice and choice assays) of nearly 60 species of phytophagous and predatory pentatomoids have shown that *Trissolcus japonicus* and *Trissolcus cultratus* are oligophagous in their host acceptance; although many of the tested species were not attacked at all, or attacked only at low rates, other species in a variety of host subfamilies were attacked, including some predatory stink bugs such as *Podisus maculiventris* (Say) (Christine Dieckhoff and Kim Hoelmer, unpublished data). Laboratory choice/no choice assays are by design highly conservative and are predictive of a natural enemy's physiological host range, but behavioral and ecological factors often reduce the realized host range under natural conditions (Van Lentern et al. 2006). Further studies are now needed to examine these factors.

Recently, an adventive population of *Trissolcus japonicus* was discovered in the wild during the conduct of sentinel egg mass surveys for native parasitoids in Beltsville, Maryland, during 2014 (Talamas et al. 2015a). Expanded surveys during 2015 found the parasitoid to be present at several other sites in Maryland, Washington, D.C., and adjacent states of Virginia and Delaware. The Asian parasitoid was found in both wild and sentinel egg masses in arboreal habitats, but not in nearby crop fields (Herlihy et al. 2016). A second adventive population of *T. japonicus* also was discovered during sentinel surveys in 2015 in Vancouver, Washington, on the opposite coast of the United States (Milnes et al. 2016). It is unknown how and when *T. japonicus* arrived in North America, but it is presumed to have been accidental (Talamas et al. 2015a). Genetic analyses have shown that both east and west coast adventive populations are distinct from *T. japonicus* populations held in North American quarantine culture and from each other (Marie-Claude Bon, unpublished data). Continued surveys are planned to document the distribution and spread of *T. japonicus* in the United States and to determine its impact on stink bug populations. The unexpected introduction of *T. japonicus* creates questions about the impact it may have not only on *Halyomorpha halys* populations but on those of native parasitoids and whether or not they can coexist with *T. japonicus*.

Konopka et al. (2017) recently investigated the question of potential coexistence using *T. japonicus* and *Anastatus bifasciatus*. They found *T. japonicus* to be a superior egg guarding and more aggressiveness and *A. bifasciatus* to be more successful at developing from multiparasitized eggs. Their finding suggest that at least these two parasitoids could complement each other in terms of *H. halys* biological control.

4.6.5 Chemical Control

Chemical control is a critical tool for managing *Halyomorpha halys* as programs that lacked effective insecticides for this bug led to devastating levels of injury in susceptible crops such as tree fruit during the 2010 outbreak in the mid-Atlantic region (Leskey et al. 2012b,c). However, the materials identified as being effective against it include broad-spectrum materials, primarily a number of pyrethroids, neonicotinoids, and carbamates (Nielsen et al. 2008b, Leskey et al. 2012b, Lee et al. 2013a, Kuhar and Kamminga 2017). Among some of the most effective materials, such as bifenthrin, dinotefuran, clothianidin and permethrin, residual activity is short, often less than 3 days in apple and peach systems (Leskey et al. 2013). Moreover, when monitoring tools were unavailable, growers responded with increased insecticide applications; in some cases, these increased four-fold higher than were applied previously before the outbreak of this bug (Leskey et al. 2012b). Thus, the use of decision-based tools for timing insecticide applications is critical not only in terms of establishing the best times to spray but also to reduce the number of applications.

Recently, several tools have emerged to improve the efficacy of management efforts using insecticides and to reduce overall inputs (Leskey et al. 2012d, 2015a,b; Blaauw et al. 2014; Lee 2015). First, traps baited with the aggregation pheromone and synergist mentioned above for *Halyomorpha halys* were used to trigger insecticide applications in apple orchards. When traps reached a cumulative threshold of 10 adults, this triggered an alternate-row-middle spray application and a second application 7 days later. This approach resulted in statistically identical levels of injury at harvest and a 40% reduction in spray applications compared with weekly spray applications. A second approach based on a border spray regime has been evaluated in peach orchards. In this case, insecticides applied against this bug were applied only to perimeter trees and the first full row of the block. Compared with frequent full-block applications, this IPM-CPR (crop perimeter restructuring) resulted in significant reductions in insecticide use and significantly less injury from *H. halys* (Blaauw et al. 2014).

Halyomorpha halys has the potential to be managed with behaviorally based approaches. One such approach is termed "attract and kill" and relies on olfactory stimuli to attract and retain mobile life stages to a limited geographic area where they can be killed, limiting the amount of, and the area requiring, insecticide sprays. In this case, the aggregation pheromone and synergist (Khrimian et al. 2014, Weber et al. 2014) have been used to attract and aggregate the nymphs and adults in border row apple trees (Morrison et al. 2015). Three important components of any "attract and kill" strategy appear to have been met based on the results to date. First, adults and nymphs must be attracted and aggregated to a geographically limited area. Using increasing doses of the pheromone and synergist with baited traps and apple trees Morrison et al. (2015) demonstrated that this bug will aggregate in an area of less than 2.5 m radius around the stimulus source regardless of dose. Second, target insects must be retained long enough to be killed. With this bug, when the pheromone and synergist (mentioned above) are deployed in association with a host plant (i.e., an apple tree), adults will remain for >20h compared with non-host substrates. Finally, an effective killing agent must be in place. With this bug, efficacious insecticides were demonstrated to provide season-long mortality in "attract and kill" apple trees located at the border of orchard blocks. This approach appears to be promising and likely will be evaluated in other cropping systems.

Another system to kill *Halyomorpha halys* nymphs and adults once they enter pheromone traps may be the use of insectide treated netting. Kuhar et al. (2017) using deltramethrin treated nets against *H. halys* in the laboratory showed that exposure to the nets for 10 seconds was enough to cause >90% and >40% mortality of nymphs and adults, respectively. They also showed that when the treated netting was placed inside a stink bug trap it provided kill rates similar or better than a standard diclorvos kill strip.

RNA interference or the use of double-stranded RNA (dsRNA)-mediated gene silencing is another potential option for the management of *Halyomorpha halys*. The technique uses gene regulatory mechanisms to introduce exogenous dsRNA into cells with the goal of silencing host cells involved in the production of critical proteins or causing modulations in the protein level produced (Ambrose 2004, Mello and Conte 2004). Ghosh et al. (2017) have taken the first step toward the use of this technique by developing a system to deliver dsRNA into *H. halys* using oral ingestion that significantly reduces the expression of genes such as JHAMT and Vg.

4.7 What Does the Future Hold?

Going forward, *Halyomorpha halys* has the potential to continue to invade new areas of the world following a pattern of population buildup in urban areas first followed by expansion into agricultural areas. In North America, *H. halys* continues to spread across the southern and midwestern United States and throughout California and the Pacific Northwest. In Canada, its spread may occur around the Great Lakes area, the Maritime Provinces, and British Columbia. In Europe, it is continuing to invade new areas and may increase its range into more European countries. Other potential expansions into areas such as New Zealand and Australia, where *H. halys* has been intercepted previously at ports but not become established, will continue to be a possibility because of international movement of cargo and people from invaded areas. It currently is unknown whether or not *H. halys* will invade Central America and the Caribbean, or more of South America; however, the potential exists. As this bug invades areas outside of North America and Asia, it can be expected to expand the list of host plants and resultant damage to agricultural commodities because of the unique crops grown in those areas.

One question that needs to be examined is the current reduction that is occurring in the mid-Atlantic United States (Leskey and Hamilton 2015). Following the appearance of peak populations throughout the area, populations, although continuing to cause agricultural damage, have declined markedly. Hypotheses including the heavy use of insecticides by growers, and changes in weather patterns (summer and winter maximum and minimum temperatures and hurricane patterns), have been proposed but, at present, have not been investigated.

Management of *Halyomorpha halys* should continue to improve in invaded areas. The development of conservation biological control options such as companion plantings (i.e., the practice of planting flowering plants in conjunction with crops to attract and retain natural enemies in infested fields) and trap crops, and the identification, release, and manipulation of biological agents, should reduce the use of current broad-spectrum insecticides to control this bug, which are disrupting historical integrated pest management (IPM) programs in agricultural systems such as tree fruit. The establishment of Asian egg parasitoids such as *Trissolcus japonicus*, whether due to adventive or deliberate introductions following regulatory review, should help suppress *H. halys* populations in habitats that cannot be actively managed. The identification of more effective, selective insecticides also will improve management and reduce current environmental issues.

4.8 Acknowledgments

The authors thank J.E. McPherson for his dedication to this book and for his keen eye for detail and his proofreading skill without which this chapter would have been a lesser contribution. The authors also thank the following people for supplying information referenced in this chapter. We wish to thank Angelita Acebes (Department of Entomology, Virginia Tech University, Winchester), J. Christopher Bergh (Department of Entomology, Virginia Tech University, Winchester), Brett R. Blaauw (Department of Entomology, Rutgers University, Upper Deerfield, New Jersey), Marie-Claude Bon (USDA ARS, Montferrier, France), Christine Dieckhoff (USDA ARS, Newark, DE), Daniel L. Frank (Agriculture & Natural Resources, West Virginia University, Morgantown), Ann Hajek (Cornell University, Ithaca, New York), Torri J. Hancock (Department of Entomology, Virginia Tech University, Blacksburg), and Tim Haye (CABI, Delémont, Switzerland) for allowing their unpublished data to be cited. We also are grateful to Julieta Brambila (USDA APHIS, Gainesville, Florida), Galen P. Dively (Department of Entomology, University of Maryland, College Park), Charles Pickett (California Department of Food and Agriculture, Sacramento), and David A. Rider (Department of Entomology, North Dakota State University, Fargo) for providing personal communications that we cited.

4.9 References Cited

Abram, P. K., T. D. Gariepy, G. Boivin, and J. Brodeur. 2014. An invasive stink bug as an evolutionary trap for an indigenous egg parasitoid. Biological Invasions 16: 1387–1395.

Acebes-Doria, A. L., T. C. Leskey, and J. C. Bergh. 2015. Development and comparison of trunk traps to monitor movement of *Halyomorpha halys* nymphs on host trees. Entomologia Experimentalis et Applicata 158: 44–53.

Acebes-Doria, A. L., T. C. Leskey, and J. C. Bergh. 2016a. Host plant effects on *Halyomorpha halys* (Hemiptera: Pentatomidae) nymphal development and survivorship. Environmental Entomology 45: 663–670.

Acebes-Doria, A. L., T. C. Leskey, and J. C. Bergh. 2016b. Injury to apples and peaches at harvest by *Halyomorpha halys* (Stål) (Hemiptera: Pentatomidae) nymphs early and late in the season. Crop Protection 89: 58–65.

Acebes-Doria, A. L., T. C. Leskey, and J. C. Bergh. 2017. Temporal and directional patterns of nymphal *Halyomorpha halys* (Hemiptera: Pentatomidae) movement on the trunk of selected wild and fruit tree hosts in the mid-Atlantic region. Environmental Entomology. doi: 10.1093/ee/nvw164.

Adachi, I. 1998. Utilization of an aggregation pheromone for forecasting population trends of the stink bugs injuring tree fruits. Plant Protection 52: 515–518.

Adachi, I., K. Uchino, and F. Mochizuki. 2007. Development of a pyramidal trap for monitoring fruit-piercing stink bugs baited with *Plautia crossota stali* (Hemiptera: Pentatomidae) aggregation pheromone. Applied Entomology and Zoology 42: 425–431.

Aigner, J. D., and T. P. Kuhar. 2014. Using citizen science to evaluate light traps for catching brown marmorated stink bugs in homes in Virginia. Journal of Extension 52: 1–9.

Aigner, J. D., and T. P. Kuhar. 2016. Lethal high temperature extremes of the brown marmorated stink bug (Hemiptera: Pentatomidae) and efficacy of commercial heat treat for control in export shipping cargo. Journal of Agricultural and Urban Entomology 32:1–6.

Aigner, B. L., D. A. Herbert, G. P. Dively, D. Venugopal, J. Whalen, B. Cissel, T. P. Kuhar, C. C. Brewster, J. W. Hogue, and E. Seymore. 2016. Comparison of two sampling methods for assessing *Halyomorpha halys* (Hemiptera: Pentatomidae) numbers in soybean fields. Journal of Economic Entomology. doi: 10.1093/jee/tow230.

Aigner, B. L., T. P. Kuhar, D. A. Herbert, C. C. Brewster, J. W. Hogue, and J. D. Aigner. 2017. Brown marmorated stink bug (Hemiptera: Pentatomidae) infestations in tree borders and subsequent patterns of abundance in soybean fields. Journal of Economic Entomology 110: 487–490.

Albright, D. D., D. Jordan-Wagner, D. C. Napili, A. L. Parker, F. Quance-Fitch, B. Whisman, J. W. Collins, and L. L. Hagan. 2006. Multicolored Asian lady beetle hypersensitivity: a case series and allergist survey. Annals of Allergy, Asthma and Immunology 97: 521–527.

Aldrich, J. R., A. Khrimian, A. Zhang, and P. W. Shearer. 2006. Bug pheromones (Hemiptera, Heteroptera) and tachanid fly host-finding. Denisia 19, zugleich Kataloge der OÖ. Landesmuseen Neue Serie 50: 1015–1031.

Ambrose, V. 2004. The functions of animal microRNSa. Nature 431: 350–355.

Anonymous. 2011. Brown marmorated stink bug causes $37 million in losses to mid Atlantic apple growers. American/Western Fruit Grower. http://growingproduce.com/article/21057/brown-marmorated-stink -bug-causes-37-million-in-losses-to-mid-atlantic-apple-growers (accessed 17 September 2015).

Anonymous. 2015a. biofix. https://en.wiktionary.org/wiki/biofix (accessed 27 November 2015).

Anonymous. 2015b. Where is BMSB? http://www.stopbmsb.org/where-is-bmsb/ (accessed 17 September 2015).

Arakawa, R., and Y. Namura. 2002. Effects of temperature on development of three *Trissolcus* spp. (Hymenoptera: Scelionidae), egg parasitoids of the brown marmorated stink bug, *Halyomorpha halys* (Stähl) (Heteroptera: Pentatomidae). Entomological Science 5(2): 215–218.

Arnold, K. 2009. *Halyomorpha halys* (Stål, 1855), eine für die europäische Fauna neu nachgewiesene Wanzenart (Insecta: Heteroptera, Pentatomidae, Pentatominae, Cappaeini). Mitt Thüringer Entomologenverbandes 16: 19.

Australian Government. 2015. Brown marmorated stink bug: emergency measures for break bulk and containerised vehicles, machinery, automotive parts and tyres. Australian Government, Department of Agriculture, Canberry City, Australia. Available at: http//www.agriculture.gov.au/SiteCollectionDocuments /biosecurity/import/general-info/ian/15/notice-to-industry-04-2015.pdf (accessed 27 July 2016).

Bae, S. -D., H. -J. Kim, G. -H. Lee, and S. -T. Park. 2007. Development of observation methods for density of stink bugs in soybean field. Korean Journal of Applied Entomology 46: 153–158.

Bae, S. -D., H. -J. Kim, Y. -N. Yoon, S. -T. Park, B. -R. Choi, and J. -K. Jung. 2009. Effects of mungbean cultivar, Jangannodgu on nymphal development, adult longevity and oviposition of soybean stink bugs. Korean Journal of Applied Entomology 48: 311–318.

Bae, S., Y. Yoon, Y. Jang, H. Kang, and R. Maharjan. 2017. Evaluation of an improved rocket traps, and baits combination for its attractiveness to hemipertan bugs in grass and soybean fields. Journal of Asia-Pacific Entomology. http://dx.doi.org/10.1016/j.aspen.2017.03.014.

Baek, S., A. Hwang, H. Kim, H. Lee, and J. -H. Lee. 2017. Temperature-dependent development and oviposition models of *Halyomorpha halys* (Hemiptera: Pentatomidae). Journal of Asia-Pacific Entomology. http://dx.doi.org/10.10106/j.aspen.2017.02.009.

Bak, W. C., W. H. Yeo, and Y. J. La. 1993. Little-leaf symptom development in the periwinkle infected with *Paulownia* witch's-broom mycoplasma-like organism by the yellow-brown stink-bug, *Halyomorpha halys* Stål (=*H. mista* Uhler). Korean Journal of Plant Pathology 9: 236–238.

Bakken, A. J., S. C. Schoof, M. Bickerton, K. L. Kamminga, J. C. Jenrette, S. Malone, M. A. Abney, D. A. Herbert, D. Reisig, T. P. Kuhar, and J. F. Walgenbach. 2015. Occurrence of brown marmorated stink bug (Hemiptera: Pentatomidae) on wild hosts in nonmanaged woodlands and soybean fields in North Carolina and Virginia. Environmental Entomology 44: 1011–1021.

Bansal, R., A. P. Mitchel, and Z. L. Sabree. 2014. The crypt-dwelling primary bacterial symbiont of the polyphagous pentatomid pest *Halyomorpha halys* (Hemiptera: Pentatomidae). Environmental Entomology 43: 617–625.

Bariselli, M., R. Bugiani, and L. Maistrello. 2016. Distribution and damage caused by *Halyomorpha halys* in Italy. Organisation Européene et Méditerranéene pour la Protection des Plantes/European and Mediterranean Plant Protection Organization Bulletin 46: 332–334. doi:10.1111/epp.12289.

Basnet, S., L. M. Maxey, C. A. Laub, T. P. Kuhar, and D. G. Pfeiffer. 2014. Stink bugs (Hemiptera: Pentatomidae) in primocane-bearing raspberries in Southwestern Virginia. Journal of Entomological Science 49: 304–312.

Basnet, S., T. P. Kuhar, C. A. Laub, and D. G. Pfeiffer. 2015. Seasonality and distribution pattern of brown mamorated stink bug (Hemiptera: Pentatomidae) in Virginia vineyards. Journal of Economic Entomology 108: 1902–1909. doi:10.1093/jee/tov124.

Baumann, P., and N. A. Moran. 1997. Non-cultivatable micro-organisms from symbiotic associations of insects and other hosts. Antonie van Leeuwenhoek 72: 39–47.

Bercha, R. 2008. Insects of Alberta [online]. www.insectsofalberta.com (accessed 18 September 2015).

Bergman, E. J. 2015. Patterns of host use by brown marmorated stink bug, *Halyomorpha halys* (Hemiptera: Pentatomidae) in woody ornamental trees and shrubs. MS Thesis. University of Maryland.

Bergman, E. J., P. D. Venugopal, H. M. Martinson, M. J. Raupp, and P. M. Shrewsbury. 2016. Host plant use by the invasive *Halyomorpha halys* (Stål) on woody ornamental trees and shrubs. PLoS ONE 11 (2): e0149975. doi:10.1371/journal.pone.0149975

Bergroth, E. 1914. Zwei neue paläarktische Hemipteren, nebst synonymischen Mitteilungen. Wiener Entomologische Zeitung 33(5–6): 177–184.

Bernon, G. 2004. Biology of *Halyomorpha halys*, the brown marmorated stink bug (BMSB). U.S. Department of Agriculture APHIS CPHST, 17. Final Report – United States Department of Agriculture APHIS CPHST 2004 T3P01.

Biddinger, D. J., J. F. Tooker, A. Surcica, and G. Krawczyk. 2012. Survey of native biocontrol agents of the brown marmorated stink bug in Pennsylvania fruit orchards and adjacent habitat. Pennsylvania Fruit News 41: 47–54.

Blaauw, B. R., D. Polk, and A. L. Nielsen. 2014. IPM-CPR for peaches: incorporating behaviorally-based methods to manage *Halyomorpha halys* and key pests in peach. Pest Management Science (doi: 10.1002/ps.3955).

Blaauw, B. R., V. P. Jones, and A. L. Nielsen. 2016. Utilizing immunomarking techniques to track *Halyomorpha halys* (Hemiptera:Pentatomidae) movement and distribution within a peach orchard. PeerJ 4: e1997; doi:10.7717/peerj.1997

Blaauw, B. R., W. R. Morrison, C. Matthews, T. C. Leskey, and A. L. Nielsen. 2017. Measuring host plant selection and retention of *Halyomorpha halys* by a trap crop. Entomologia Experimentalis et Applicata. COI:10.1111/eea.12571.

Brown, M. W., and B. D. Short. 2010. Factors affecting appearance of stink bug (Hemiptera: Pentatomidae) injury in apple. Environmental Entomology 39: 134–139.

Buchner, P. 1965. Endosymbiosis of animals with plant microorganisms. John Wiley & Sons, Chichester, United Kingdom. 909 pp.

Callot, H., and C. Brua. 2013. *Halyomorpha halys* (Stål, 1855) la Punaise diabolique, nouvelle espèce pour la faune de France (Heteroptera Pentatomidae). L'Entomologiste 69: 69–71.

Cesari, M., L. Maistrello, F. Ganzerli, P. Dioli, L. Rebecchi, and R. Guidetti. 2015. A pest alien invasion in progress: potential pathways of origin of the brown marmorated stink bug *Halyomorpha halys* populations in Italy. Journal of Pest Science 88: 1–7. doi 10.1007/s10340-014-0634-y.

Cho, M. R., S. -W. Jeon, T. J. Kang, H. H. Kim, S. -J. Ahn, and C. Y. Yang. 2013. Pests occurring on *Cymbidium*. Korean Journal of Applied Entomology 52: 403C408.

Choi, D. S., K. C. Kim, and K. C. Lim. 2000. The status of spot damage and fruit piercing pests on yuzu (*Citrus junos*) fruit. Korean Journal of Applied Entomology 39: 259–266.

Ciceoi, R. E., Mardare, T. Eliza, and I. Dobrin. 2016. The status of brown marmorated stink bug, *Halyomorpha halys*, in Bucharest area. Journal of Horticulture, Forestry and Biotechnology 20: 18–25.

Cira, T. M., R. C. Venette, J. Aigner, T. Kuhar, D. E. Mullins, S. E. Gabbert, and W. D. Hutchinson. 2016. Cold tolerance of *Halyomorpha halys* (Hemiptera: Pentatomidae) across geographic and temporal scales. Environmental Entomology 1–8. doi: 10.1093/ee/nvv220.

Cira, T. M., E. C. Burkness, R. L. Koch, and W. D. Hutchinson. 2017. *Halyomorpha halys* mortality and sublethal effects following insecticide exposure. Journal of Pest Science. doi 10.1007/s10340-017-0871-y.

Cornelius, M. L., C. Dieckhoff, B. T. Vinyard, and K. Hoelmer. 2016. Parasitism and predation on sentinel egg masses of the brown marmorated stink bug (Hemiptera: Pentatomidae) in three vegetable crops: importance of dissections for evaluating the impact of native parasitoids on an exotic pest. Environmental Entomology. doi: 10.109/ee/nvw134.

Dieckhoff, C., K. M. Tatman, and K. Hoelmer. 2017. Natural biological control of *Halyomorpha halys* by native egg parasitoids: a multi-year study in northern Delaware. Journal of Pest Science 1–16. doi: 10.1007/s10340-017-0868-6.

Dioli, P., P. Leon, and L. Maistrello. 2016. Prime segnalazioni in Spagna e in Sardegna della specie aliena *Halyomorpha halys* (Stål, 1855) e note sulla sua distribuzione in Europa (Hemiptera: Pentatomidae). Revista gaditana de entomologia 7: 539–548.

Distant, W. L. 1880. Notes on some exotic Hemiptera, with descriptions of new species. Entomologist's Monthly Magazine, 16: 201–203.

Distant, W. L. 1893. On some allied Pentatomidae, with synonymical notes. Annals and Magazine of Natural History 6(11): 389–394.

Distant, W. L. 1899. Rhynchotal notes. III Heteroptera: Discocephalinae and Pentatominae (part). Annals and Magazine of Natural History, 7(4): 421–445.

Dobson, R. C., M. Rodgers, J. L. C. Moore, and R. T. Bessin. 2016. Exclusion of the brown marmorated stink bug from organically grown peppers using barrier screens. Hort Technology 26: 191–198.

Douglas, A. E. 2006. Phloem feeding by animals: problems and solutions. Journal of Experimental Biology 57: 747–754.

Duthie, C. 2012. Risk analysis of *Halyomorpha halys* (brown marmorated stink bug) on all pathways. Ministry for Primary Industries, Wellington, New Zealand. 57 pp.

Faúndez, E. I., and D. A. Rider. 2017. The brown marmorated stink bug *Halyomorpha halys* (Stål, 1855) (Heteroptera: Pentatomidae) in Chile. Arquivos Entomoloxicos 17: 305–307.

Fengjie, C., Z. Zhifang, R. Li, and X. Liu. 1997. Study on control and observation of the bionomics characteristics of *Halyomorpha picus* Fabricius. Journal of the Hebei Agricultural University 2: 12–17.

Fogain, R., and S. Graff. 2011. First records of the invasive pest, Halyomorpha halys (Hemiptera: Pentatomidae), in Ontario and Quebec. Journal of the Entomological Society of Ontario 142: 45–48.

Fraga, D. F., J. Parker, A. C. Busoli, G. C. Hamilton, A. L. Nielsen, and C. Rodriguez-Saona. 2016. Behavioral responses of predaceous minute pirate bugs to tridecane, a volatile emitted by brown marmorated stink bug. Journal of Pest Science. doi 10.1007/s10340-016-0825-9.

Freers, A. 2012. Blinde Passagiere: Stinkwanzen, Marmorierte Baumwanze – *Halyomorpha halys*. https://ssl.bremen.de/lmtvet/sixcms/media.php/13/Blinde_Passagiere_Stinkwanze_3_2012.pdf (accessed 20 February 2015).

Fujiie, A. 1985. Seasonal life cycle of *Halyomorpha mista*. Bulletin of the Chiba Agricultural Experiment Station 26: 87–93.

Fukuoka, T., H. Yamakage, and T. Niiyama. 2002. Sucking injury to vegetables by the brown marmorated stink bug, *Halyomorpha halys* (Stål). Annual Report of the Society of Plant Protection North Japan 53: 229–231.

Funayama, K. 2002. Comparison of the susceptibility to injury of apple cultivars by stink bugs. Japanese Journal of Applied Entomology and Zoology 46: 37–40.

Funayama, K. 2003. Outbreak and control of stink bugs in apple orchards. Journal of Japanese Agricultural Technology 47: 35–39.

Funayama, K. 2004. Importance of apple fruits as food for the brown-marmorated stink bug, *Halyomorpha halys* (Stål) (Hemiptera: Pentatomidae). Applied Entomology and Zoology 39: 617–623.

Funayama, K. 2012. Nutritional states of post-overwintering adult brown-marmorated stink bugs, *Halyomorpha halys* (Stål) (Heteroptera: Pentatomidae). Japanese Journal of Applied Entomology and Zoology 56: 12–15.

Gapon, D. A. 2016. First records of the brown marmorated stink bug *Halyomorpha halys* (Stål, 1855) (Heteroptera: Pentatomidae) in Russa, Abkhazia, and Georgia. Entomological Review 96: 1086–1088.

Gariepy, T. D., H. Fraser, and C. D. Scott-Dupree. 2014a. Brown marmorated stink bug (Hemiptera: Pentatomidae) in Canada: recent establishment, occurrence, and pest status in southern Ontario. The Canadian Entomologist 146: 579–582. doi: 10.3752/cjai.2013.24.

Gariepy, T. D., T. Haye, H. Fraser, and J. Zhang, 2014b. Occurrence, genetic diversity, and potential pathways of entry of *Halyomorpha halys* in newly invaded areas of Canada and Switzerland. Journal of Pest Science 87: 17–28.

Gariepy, T. D., A. Bruin, T. Haye, P. Milonas, and G. Vétek. 2015. Occurrence and genetic diversity of new populations of *Halyomorpha halys* in Europe. Journal of Pest Science 88: 451–460.

Garrouste, R., P. Nel, A. Nel, A. Horellou, and D. Pluot-Sigwalt. 2014. *Halyomorpha halys* (Stål, 1855) en Île de France (Hemiptera: Pentatomidae: Pentatominae): surveillons la punaise diabolique. Annals of the Society of Entomology in France (Hors Serie) 50: 257–259.

Gosh, S. K. B., W. B. Hunter, A. L. Park, and D. E. Gundersen-Rindal. 2017. Double stranded RNA delivery system for plant-sap-feeding insects. PLOS ONE 12(2): e0171861. doi:10.1371/journal.pone.0171861.

Gouli, V., S. Gouli, M. Skinner, G. Hamilton, J. S. Kim, and B. L. Parker. 2012. Virulence of select entomopathogenic fungi to the brown marmorated stink bug, *Halyomorpha halys* (Stål), (Heteroptera: Pentatomidae). Pest Management Science 68: 155–157.

Hahn, N. G., A. J. Kaufman, C. Rodriguez-Saona, A. L. Nielsen, J. LaForest, and G. C. Hamilton. 2016. Exploring the spread of brown marmorated stink bug in New Jersey through the use of crowd sourced reports. American Entomologist 62: 36–45.

Hahn, N. G., C. Rodriguez-Saona, and G. C. Hamilton. 2017. Characterizing the spatial distribution of brown marmorated stink bug, *Halyomorpha halys* Stål (Hemiptera: Pentatomidae), populations in peach orchards. PLOS ONE 12(3): e0170889. https://doi.org/10.1371/journal.one.0170889.

Hamilton, G. C. 2009. Brown marmorated stink bug. American Entomologist 55: 19–20.

Hasegawa, H. 1965. Major pests of economic plants in Japan. Japan Plant Protection Association, Tokyo. 412 pp.

Haye, T. 2014. Seasonal field parasitism of *Halyomorpha halys* and co-occurring non-target species in China. Presentation at: Brown Marmorated Stink Bug Working Group Meeting, June 16, 2014, Carvel Research & Education Center Georgetown, DE. http://www.northeastipm.org/neipm/assets/File/BMSB%20Resources/BMSB-IWG-Jun-2014/Seasonal-Field-Parasitism-of-Halyomorpha-halys-in-China.pdf (accessed 2 July 2016).

Haye, T., S. Abdallah, and T. Gariepy. 2014a. Phenology, life table analysis and temperature requirements of the invasive brown marmorated stink bug, *Halyomorpha halys*, in Europe. Journal of Pest Science 87: 407–418.

Haye, T., D. Wyniger, and T. Gariepy, T. 2014b. Recent range expansion of brown marmorated stink bug in Europe, 309–314. *In* G. Müller, R. Pospischil, and W. H. Robinson (Eds.), Proceedings of the 8th International Conference on Urban Pests. Zurich, Switzerland. 469 pp.

Haye, T., S. Fischer, J. Zhang, and T. Gariepy. 2015a. Can native egg parasitoids adopt the invasive brown marmorated stink bug, *Halyomorpha halys* (Heteroptera: Pentatomidae), in Europe? Journal of Pest Science 88: 693–705.

Haye, T., T. Gariepy, K. Hoelmer, J. -P. Rossi, J. -C. Streito, X. Tassus, and N. Desneux. 2015b. Range expansion of the invasive brown marmorated stink bug, *Halyomorpha halys*: an increasing threat to field, fruit and vegetable crops worldwide. Journal of Pest Science 88: 665–673.

Heckmann, R. 2012. Erster Nachweis von *Halyomorpha halys* (Stål, 1855) (Heteroptera: Pentatomidae) für Deutschland. Heteropteron 36: 17–18.

Hedstrom, C. S., P. W. Shearer, J. C. Miller, and V. M. Walton. 2014. The effects of kernel feeding by *Halyomorpha halys* (Hemiptera: Pentatomidae) on commercial hazelnuts. Journal of Economic Entomology 107: 1858–1865.

Hemala, V., and P. Kment. 2017. First record of *Halyomorpha halys* and mass occurrence of *Nezara viridula* in Slovakia (Hemiptera: Heteroptera: Pentatomidae). Plant Protection Science 53: 1–7. doi: 10.17221/166/2016-PPS.

Herlihy, M. V., E. J. Talamas, and D. W. Weber. 2016. Attack and success of native and exotic parasitoids on eggs of *Halyomorpha halys* in three Maryland habitats. PLoS One 11(3): e0150275. doi:10.1371/journal.pone.0150275.

Hoebeke, R. E., and M. E. Carter. 2003. *Halyomorpha halys* (Stål) (Heteroptera: Pentatomidae): a polyphagous plant pest from Asia newly detected in North America. Proceedings of the Entomological Society of Washington 105: 225–237.

Hoffman, W. E. 1931. A pentatomid pest of growing beans in south China. Peking Natural History Bulletin 5: 25–27.

Hou, Z. R., H. Z. Liang, Q. Chen, Y. Hu, and H. P. Tian. 2009. Application of *Anastatus* sp. against *Halyomorpha halys*. Forest Pest and Disease 28(4): 39–43.

Hufbauer R. A., B. Facon, V. Ravigné, J. Turgeon, J. Foucaud, C. E. Lee, O. Rey, and A. Estoup. 2012. Anthropogenically induced adaptation to invade (AIAI): contemporary adaptation to human-altered habitats within the native range can promote invasions. Evolutionary Applications 5: 89–101. doi: 10.1111/j.1752-4571.2011.00211.x.

Inkley, D. B. 2012. Characteristics of home invasion by the brown marmorated stink bug (Hemiptera: Pentatomidae). Journal of Entomological Science 47: 125–130.

Jones, A. L., D. E. Jennings, C. R. R. Hooks, and P. M. Shrewsbury. 2014. Sentinel eggs underestimate rates of parasitism of the exotic brown marmorated stink bug, *Halyomorpha halys*. Biological Control 78: 61–66.

Joseph, S. V., J. W. Stallings, T. C. Leskey, G. Krawczyk, D. Polk, B. Butler, and J. C. Bergh. 2014. Spatial distribution of brown marmorated stink bug (Hemiptera: Pentatomidae) injury at harvest in mid-Atlantic apple orchards. Journal of Economic Entomology 107: 1839–1848.

Josifov, M. V., and I. M. Kerzhner. 1978. Heteroptera aus Korea. II. Teil (Aradidae, Berytidae, Lygaeidae, Pyrrhocoridae, Rhopalidae, Alydidae, Coreidae, Urostylidae, Acanthosomatidae, Scutelleridae, Pentatomidae, Cydnidae, Plataspidae). Fragmenta Faunistica 23(9): 137–196.

Kagen, S. L. 1990. Inhalant allergy to arthropods, arachnids and crustaceans. Clinical Review of Allergy 8: 99–125.

Kang, C. H., H. S. Huh, and C. -G. Park. 2003. Review on true bugs infesting tree fruits, upland crops, and weeds in Korea. Korean Journal of Applied Entomology 42: 269–277.

Kawada, H., and C. Kitamura. 1983a. Bionomics of the brown marmorated stink bug *Halyomorpha mista* Uhler. Japanese Journal of Applied Entomology and Zoology 27: 304–306.

Kawada, H., and C. Kitamura. 1983b. The reproductive behavior of the brown marmorated stink bug, *Halyomorpha mista* Uhler (Heteroptera: Pentatomidae): I. Observation of mating behavior and copulation. Applied Entomology and Zoology 18: 234–242.

Kawada, H., and C. Kitamura. 1992. The tachinid fly, *Bogosia* sp. (Diptera: Tachinidae), as a parasitoid of the brown marmorated stink bug, *Halyomorpha mista* Uhler (Heteroptera: Pentatomidae). Japanese Journal of Environmental Entomology and Zoology 4: 65–70.

Kenyon, L. J., T. Meilia, and Z. L. Sabree. 2015. Habitat visualization and genomic analysis of "*Candidatus* Pantoea carbekii," the primary symbiont of the brown marmorated stink bug. Genome Biology and Evolution 6: 763–775. doi: 10.1093/gbe/evv006.

Khrimian, A., P. W. Shearer, A. Zhang, G. C. Hamilton, and J. R. Aldrich. 2008. Field trapping of the invasive brown marmorated stink bug, *Halyomorpha halys*, with geometric isomers of methyl 2,4,6-decatrienoate. Journal of Agricultural Food Chemistry 56: 197–203.

Khrimian, A., A. Zhang, D. C. Weber, H. -Y. Ho, J .R. Aldrich, K. E. Vermillion, M. A. Siegler, S. Shirali, F. Guzman, and T. C. Leskey. 2014. Discovery of the aggregation pheromone of the brown marmorated stink bug (*Halyomorpha halys*) through the creation of stereoisomeric libraries of 1-bisabolen-3-ols. Journal of Natural Products 77: 1708–1717.

Kim, D. -W., H. -M. Kwon, and K. -S. Kim. 2000. Current status of the occurrence of the insect pests in the citrus orchards in Cheju Island. Korean Journal of Applied Entomology 39: 267–274.

Kim, J. -S., H. -C. Ko, S. -T. Yoon, Y. -H. Cho, J. -G. Kim, and C. -K. Shim. 2010. Occurrence of insect pest from organic seed producing field of minor grain germplasms. Korean Journal Crop Science 55: 58–61.

Kim, N. K. 2010. Receiving 500 North Korean insect specimens including grasshoppers and stink bugs. Yonhap News, South Korea. Retrieved from http://www.yonhapnews.co.kr/society/2010/11/09/0711000 000AKR20101109212300004.HTML.

Koch, R. L., and T. Pahs. 2015. Species composition and abundance of stink bugs (Hemiptera: Pentatomidae) in Minnesota field corn. Environmental Entomology 44: 233–238. doi: 10.1093/ee/nvv005.

Koch, R. L., and W. A. Rich. 2015. Stink bug (Hemiptera: Heteroptera: Pentatomidae) feeding and phenology on early-maturing soybean in Minnesota. Journal of Economic Entomology 108: 2335–2342. doi: 10.1093/jee/tov218.

Koch, R. L., W. A. Rich, and T. Pahs. 2016. Statewide and season-long surveys of petatomidae (Hemiptera: Heteroptera) of Minnesota wheat. Annals of the Entomological Society of America 109: 396–404.

Konopka, J. K., T. Haye, T. D. Gariepy, and J. N. McNeil. 2017. Possible coexistence of native and exotic parasitoids and their impact on control *Halyomorpha halys*. Journal of Pest Science doi 10.1007 /s10340-017-0851-2.

Koppel, A. L., D. A. Herbert, Jr., T. P. Kuhar, and K. Kamminga. 2009. Survey of stink bug (Hemiptera: Pentatomidae) egg parasitoids in wheat, soybean, and vegetable crops in Southeast Virginia. Environmental Entomology 38: 375–379.

Kriticos, D. J., J. M. Kean, C. B. Phillips, S. D. Senay, H. Acosta, and T., Haye. 2017. The potential global sistribution of the brown marmorated stink bug, *Halyomorpha halys*, a critical threat to plant biosecurity. Journal of Pest Science. doi 10.1007/s10340-017-0869-s.

Kuhar, T. P., and K. Kamminga. 2017. Review of the chemical control research on *Halyomorpha halys* in the USA. Journal of Pest Science. doi 10.1007/s10340-017-0859-7.

Kuhar, T. P., B. D. Short, G. Krawczyk, and T. C. Leskey. 2017. Deltamethrin-incorporated nets as an integrated pest management tool for the invasive *Halyomorpha halys* (Hemiptera: Pentatomidae). Journal of Economic Entomology 110: 543–545.

Lara, J., C. Pickett, C. Ingels, D. R. Haviland, E. Grafton-Cardwell, D. Doll, J. Bethke, B. Faber, S. K. Dara, and M. Hoddle. 2016. Biological control program is being developed for brown marmorated stink bug. California Agriculture 70: 15–23.

Lee, D. -H. 2015. Current status of research progress on the biology and management of *Halyomorpha halys* (Hemiptera: Pentatomidae) as an invasive species. Applied Entomology and Zoology 1–14. doi 10.1007 /s13355-015-0350-y.

Lee, D. -H., B. D. Short, S. V. Joseph, J. C. Bergh, and T. C. Leskey. 2013a. Review of the biology, ecology and management of *Halyomorpha halys* (Hemiptera: Pentatomidae) in China, Japan and the Republic of Korea. Environmental Entomology 42: 627–641.

Lee, D. -H., S. E. Wright, G. Boiteau, C. Vincent, and T. C. Leskey. 2013b. Effectiveness of glues for harmonic radar tag attachment on *Halyomorpha halys* (Hemiptera: Pentatomidae) and their impact on adult survivorship and mobility. Environmental Entomology 42: 515–523.

Lee, D. -H., J. P. Cullum, J. L. Anderson, J. L. Daugherty, L. M. Beckett, and T. C. Leskey. 2014a. Characterization of overwintering sites of the invasive brown marmorated stink bug in natural landscapes using human surveyors and detector canines. PLOS ONE. doi: 10.1371/journal.pone.0091575.

Lee, D. -H., A. L. Nielsen, and T. C. Leskey. 2014b. Dispersal capacity of nymphal stages of *Halyomorpha halys* (Hemiptera: Pentatomidae) evaluated under laboratory and field conditions. Journal of Insect Behavior 27: 639–651.

Lee, D. W., G. C. Lee, S. W. Lee, C. -G. Park, H. Y. Choo, and C. -H. Shin. 2001. Survey on pest management practice and scheme of increasing income in sweet persimmon farms in Korea. The Korean Journal of Pesticide Science 4: 45–49.

Lee, H. S., B. K. Chung, T. S. Kim, J. H. Kwon, W. D. Song, and C. W. Rho. 2009. Damage of sweet persimmon fruit by the inoculation date and number of stink bugs, *Riptortus clavatus, Halyomorpha halys*, and *Plautia stali*. Korean Journal of Applied Entomology 48: 485–493.

Lee, K. C., C. H. Kang, D. W. Lee, S. M. Lee, C. G. Park, and H. Y. Choo. 2002. Seasonal occurrence trends of hemipteran bug pests monitored by mercury light and aggregation pheromone traps in sweet persimmon orchards. Korean Journal of Applied Entomology 41: 233–238.

Lee, S. W., D. H. Lee, K. H. Choi, and D. A. Kim. 2007. A report on current management of major apple pests based on census data from farmers. Korean Journal Horticultural Science and Technology 25: 196–203.

Leskey, T. C., and G. C. Hamilton. 2010a. Brown marmorated stink bug working group meeting report, June 2010. http://projects.ipmcenters.org/Northeastern/FundedProjects/ReportFiles/Pship2010/Pship2010-Leskey -ProgressReport-237195.pdf (accessed 17 September 2015).

Leskey, T. C., and G. C. Hamilton. 2010b. Brown marmorated stink bug working group meeting report, November 2010. http://projects.ipmcenters.org/Northeastern/FundedProjects/ReportFiles/Pship2010 /Pship2010-Leskey-ProgressReport-237195-Meeting-2010_11_17.pdf.

Leskey, T. C., and G. C. Hamilton. 2011. Brown marmorated stink bug working group meeting report, November 2011. http://projects.ipmcenters.org/Northeastern/FundedProjects/ReportFiles/Pship2011 /Pship2011-Leskey-ProgressReport-4681655.pdf (accessed 17 September 2015).

Leskey, T. C., and G. C. Hamilton. 2015. Brown marmorated stink bug working group meeting report, June 2015. http://www.northeastipm.org/neipm/assets/File/BMSB-Working-Group-Meeting-Report-Jun-2015.pdf (accessed 17 September 2015).

Leskey, T. C., and A. L. Nielsen 2018. The impact of the invasive brown marmorated stink bug in North America and Europe: History, biology, ecology, and management. Annual Review of Entomology. (in press).

Leskey, T. C., G. C. Hamilton, A. L. Nielsen, D. F. Polk, C. Rodriguez-Saona, J. C. Bergh, D. A. Herbert, T. P. Kuhar, D. Pfeiffer, G. P. Dively, C. R. R. Hooks, M. J. Raupp, P. M. Shrewsbury, G. Krawczyk, P. W. Shearer, J. Whalen, C. Koplinka-Loehr, E. Myers, D. Inkley, K. A. Hoelmer, D. -H. Lee, and S. E. Wright. 2012a. Pest status of the brown marmorated stink bug, *Halyomorpha halys* (Stål), in the USA. Outlooks on Pest Management 23: 218–226.

Leskey, T. C., D. -H. Lee, B. D. Short, and S. E. Wright. 2012b. Impact of insecticides on the invasive *Halyomorpha halys* (Hemiptera: Pentatomidae): Analysis on the insecticide lethality. Journal of Economic Entomology 105: 1726–1735.

Leskey T. C., B. D. Short., B. B. Butler, and S. E. Wright. 2012c. Impact of the invasive brown marmorated stink bug, *Halyomorpha halys* (Stål) in mid-Atlantic tree fruit orchards in the United States: case studies of commercial management. Psyche. doi:10.1155/2012/535062.

Leskey T. C., S. E. Wright., B. D. Short, and A. Khrimian. 2012d. Development of behaviorally based monitoring tools for the brown marmorated stink bug, *Halyomorpha halys* (Stål) (Heteroptera: Pentatomidae) in commercial tree fruit orchards. Journal of Entomological Science 47: 76–85.

Leskey, T. C., B. D. Short, and D. -H. Lee. 2013. Efficacy of insecticide residues on adult *Halyomorpha halys* (Stål) (Hemiptera: Pentatomidae) mortality and injury in apple and peach orchards. Pest Management Science. doi: 10.1002/ps.3653.

Leskey, T. C., A. Agnello, J. C. Bergh, G. Dively, G. Hamilton, P. Jentsch, A. Khrimian, G. Krawczyk, T. Kuhar, D. -H. Lee, W. Morrison, D. Polk, C. Rodriguez-Saona, P. Shearer, B. D. Short, P. Shrewsbury, J. Walgenbach, C. Welty, J. Whalen, N. Wiman, and F. Zaman. 2015a. Attraction of the invasive *Halyomorpha halys* (Hemiptera: Pentatomidae) to traps baited with semiochemical stimuli across the United States. Environmental. Entomology 44: 746–756. doi: 10.1093/ee/nvv049.

Leskey, T. C., A. Khrimian, D. C. Weber, J. C. Aldrich, B. D. Short, D. -H. Lee, and W. R. Morrison. 2015b. Behavioral responses of the invasive *Halyomorpha halys* (Stål) (Hemiptera: Pentatomidae) to traps baited with stereoisomeric mixtures of 10,11-epoxy-1-bisabolen-3-ol. Journal of Chemical Ecology 41: 418–429.

Lethierry, L., and G. Severin. 1893. Catalogue général des Hémiptères. Bruxelles, Pentatomidae, 1: x + 286 pp.

Li, Z. X. 2002. Preliminary studies on the *Trissolcus japonicus* Ashmead (Hymenoptera: Scelionidae), a parasitoid of *Halyomorpha picus* Fabricius (Heteroptera: Pentatomidae) eggs. M.Sc. Thesis, Shandong Agricultural University.

Lo, P. L., J. T. S. Walker, and D. J. Rodgers. 2015. Risks to pest management in New Zealand's pipfruit Integrated Fruit Production programme. New Zealand Plant Protection 68: 306–312.

Lupi, D., P. Dioli, and L. Limonta. 2017. First evidence of *Halyomorpha halys* (Stå) (Hemiptera, Heteroptera Pentatomidae) feeding on rice (*Oryza sativa* L.). Journal of Entomological and Acarological Research 49: 66–71.

Macavei, R., Bâetan, I. Oltean, T. Florian, M. Varga, E. Costi, and L. Maistrello. 2015. First detection of *Halyomorpha halys* Stål, a new invasive species with a high potential of damage on agricultural crops in Romania. Lucrări Stiintifice 58: 105–108.

Maistrello, L., P. Dioli, and M. Bariselli. 2013. Trovata una cimice esotica dannosa per i frutteti. Agricoltura 6: 67–68.

Maistrello, L., P. Dioli, G. Vaccari, R. Nannini, P. Bortolotti, S. Caruso, E. Costi, A. Montermini, L. Casoli, and M. Bariselli. 2014. Primi rinvenimenti in Italia della cimice esotica *Halyomorpha halys*, una nuova minaccia per la frutticoltura. ATTI Giornate Fitopatologiche 2014(1): 283–288.

Maistrello, L., P. Dioli, M. Bariselli, G. L. Mazzoli, and I. Giacalone-Fortini. 2016. Citizen science and early detection of invasive species: phenology of first occurrences of *Halyomorpha halys* in Southern Europe. Biological Invasions pp. 1–8. doi 10.1007/s10530-016-1217-z.

Malumphy, C. 2014. Second interception of *Halyomorpha halys* (Stål) (Hemiptera: Pentatomidae) in Britain. pp. 4–5. In T. Bantock (Ed.) Het News, 3rd Series, 21. 10 pp.

Martinson H. M., M. J. Raupp, and P.M. Shrewsbury. 2013. Invasive stink bug wounds trees, liberates sugars, and facilitates native Hymenoptera. Annals of the Entomological Society of America 106: 47–52. doi: http://dx.doi.org/10.1603/AN12088.

Martinson H. M., P. D. Venugopal, E. J. Bergmann, P. M. Shrewsbury, and M. J. Raupp. 2015. Fruit availability influences the seasonal abundance of invasive stink bugs in ornamental trees. Journal of Pest Science 106: 1–8; doi 10.1007/s10340-015-0677-8.

Martinson, H. M., E. J. Bergmann, D. Venugopal, C. B. Riley, P. M. Shrewsbury, and M. J. Raupp. 2016. Invasive stink bug favors naïve plants: testing the role of plant geographic origin in diverse managed environments. Scientific Reports. doi: 10.1038/srep32646.

Martorana, L., M. C. Foti, G. Rondoni, E. Conti, S. Colazza, and E. Peri. 2017. An invasive insect herbivore disrupts plant volatile-mediated tritrophic signaling. Journal of Pest Science. doi 10.1007/s10340-017-0877-5.

Masslo, B., R. Valentin, K. Leu, K. Kerwin, G. C. Hamilton, A. Bevan, N. H. Fefferman, and D. Fonseca. 2017. Chirosurveillance: the use of native bats to detect invasive agricultural pests. PLOS ONE 12(3): e0173321. http://doi.org/10.1371/journal.pone.0173321.

Mathews, C. R., B. Blaauw, G. Dively, J. Koton, J. Moore, E. Ogburn, D. G. Pfeiffer, T. Trope, J. F. Walenbach, C. Welty, G. Zinati, and A.L. Nielsen. 2017. Evaluating a polyculture trap crop for organic management of *Halyomorpha halys* and native stink bugs in peppers. Journal of Pest Science. doi 10.1007/s10340-017-0838-z.

Matsuo, K., T. Honda, K. Itoyama, M. Toyama, and Y. Hirose. 2016. Discovery of three egg parasitoid species attacking the shield bug *Glaucias subpunctatus* (Hemiptera: Pentatomidae). Japanese Journal of Applied Entomology and Zoology 60: 43–45.

Mello, C. C., and D. Conte. 2004. Revealing the world of RNA interference. Nature 431: 338–342.

Mertz, T. L., S. B. Jacobs, T. J. Craig, and F. T. Ishmael. 2012. The brown marmorated stink bug as a new aeroallergen. Journal of Allergy Clinical Immunology 130: 1–3.

Milnes, J. M., N. G. Wiman, E. J. Talamas, J. F. Brunner, K. A. Hoelmer, M. L. Buffington, and E. H. Beers. 2016. Discovery of an exotic egg parasitoid of the brown marmorated stink bug, *Halyomorpha halys* (Stål) in the Pacific Northwest. Proceedings of the Entomological Society of Washington 118: 466–470.

Milonas, P. G., and G. K. Partsinevelos. 2014. First report of brown marmorated stink bug *Halyomorpha halys* Stål (Hemiptera: Pentatomidae) in Greece. Bulletin OEPP/EPPO Bulletin 44(2): 183–186.

Mityushev, I. M. 2016. First report of the brown marmorated stink bug, *Halyomorpha halys* Stål, in the Russian Federation. Monitoring and biological control methods of woody plan pests and pathogens: from theory to practice, Proceedings of International Conference, 18–22 April, Moscow, Russia. pps. 147–148.

Mohekar, P., N. G. Wiman, J. Osbourne, C.H. Hedstrom, V. M Walton, and E. Tomasino. 2014. Post harvest impact of brown marmorated stink bug in wine. 88th Annual Orchard Pest and Disease Management Conference, 9 January, Portland, OR. p. 20.

Mohekar, P., T. L. Lapis, N. G. Wiman, J. Lim, and E. Tomasino. 2016. Brown marmorated stink bug taint in Pinot Noir: detection and consumer rejection of thresholds of trans-2-decenal. American Journal of Enology and Viticulture 68: 120–126. doi: 10.5344/ajev.2016.15096.

Mohekar, P., J. Osborne, N. G. Wiman, V. Walton, and E. Tomasino. 2017. Influence of winemaking processing steps on the amounts of (E)-2-decanal and tridecane as off-oderants caused by brown marmorated stink bug (*Halyomorpha halys*). Journal of Agricultural and Food Chemistry. doi: 10.1021/acs.jafc.6b04268.

Moran, N. A., and A. Telang. 1998. Bacteriocyte-associated symbionts of insects. BioScience 48: 295–304.

Morehead, J. A., and T. P. Kuhar. 2017. Efficay of organically approved insecticides against brown marmorated stink bug, *Halyomorpha halys* and other stink bugs. Journal of Pest Science. doi 10.1007 /s10340-017-0879-3.

Morrison, W. R., D. -H. Lee, B. D. Short, A. Khrimian, and T. C. Leskey. 2015. Establishing the behavioral basis for an attract-and-kill strategy to manage the invasive *Halyomorpha halys* (Stål) (Hemiptera: Pentatomidae) in apple orchards. Journal of Pest Science 89: 81–96. doi 10.1007/s10340-015-0679-6.

Morrison, W. R., C. Mathews, and T. C. Leskey. 2016. Frequency, intensity, and physical characteristics of predation by generalist predators of the brown marmorated stink bug (Hemiptera: Pentatomidae) eggs. Biological Control 97: 120–130.

Morrison, W. R., A. Acebes-Doria, E. Ogburn, T. P. Kuhar, J. F. Walgenbach, J. C. Bergh, L. Nottingham, A. Dimeglio, P. Hipkins, and T. C. Leskey. 2017a. Behavioral response of the brown marmorated stink bug (Hemiptera: Pentatomide) to semiochemicals deployed inside and outside anthropogenic structures during the overwintering period. Journal of Economic Entomology. doi: 10.1093/jee/tox097.

Morrison, W. R., M. Allen, and T. C. Leskey. 2017b. Behavioural responses of the invasive *Halyomorpha halys* (Hemiptera: Pentatomidae) to host plant stimuli augmented with semiochemicals in the field. Agricultural and Forest Entomology. doi 10.1111/afe.12229.

Morrison, W. R., A. N. Bryant, B. Poling, N. F. Quinn, and T. C. Leskey. 2017c. Predation of *Halyomorpha halys* (Hemiptera: Pentatomidae) from web-building spiders associated with anthropogenic dwellings. Journal of Insect Behavior 30: 70-85. doi 10.1007/s10905-017-9599-z.

Müller, G., I. Landau Lüscher, and M. Schmidt. 2011. Data on the incidence of household arthropod pests and new invasive pests in Zurich (Switzerland), pp. 99–104. *In* W. H. Robinson and A. E. de Carvalho Campos (Eds.), Proceedings of the 7th International Conference on Urban Pests. Ouro Preto, Brazil. 417 pp.

Murthy, R. L. N., and R. Chandrasekaran. 1979. Effect of pest damage on quality of made tea, pp. 275–283. *In* C.S. Venkataram (Ed.), Proceedings of the Second Plantation Crops Symposium. Kerala, India. 555 pp.

Nakashima, N., J. Sasaki, K. Tsuda, C. Yasunaga, and H. Noda. 1998. Properties of a new picorna-like virus of the brown-winged green bug, *Plautia stali*. Journal of Invertebrate Pathology 71: 151–158.

Nielsen, A. L., and G. C. Hamilton. 2009a. Seasonal occurrence and impact of *Halyomorpha halys* (Hemiptera: Pentatomidae) in tree fruit. Journal of Economic Entomology 102: 1133–1140.

Nielsen, A. L., and G. C. Hamilton. 2009b. Life history of the invasive species *Halyomorpha halys* (Hemiptera: Pentatomidae) in Northeastern United States. Annals of the Entomological Society of America 102: 608–616.

Nielsen, A. L., G. C. Hamilton, and D. Matadha. 2008a. Developmental rate estimation and life table analysis for *Halyomorpha halys* (Hemiptera: Pentatomidae). Environmental Entomology 37: 348–355.

Nielsen, A. L., P. W. Shearer, and G. C. Hamilton. 2008b. Toxicity of insecticides to *Halyomorpha halys* (Hemiptera: Pentatomidae) using glass-vial bioassays. Journal of Economic Entomology 101: 1439–1442.

Nielsen, A. L., G. C. Hamilton, and P. W. Shearer. 2011. Seasonal phenology and monitoring of the non-native *Halyomorpha halys* (Hemiptera: Pentatomidae) in soybean. Environmental Entomology 40: 231–238.

Nielsen, A. L., K. Holmstrom, G. C. Hamilton, J. Cambridge, and J. Ingerson- Mahar. 2013. Use of blacklight traps to monitor the abundance and spread of the brown marmorated stink bug. Journal of Economic Entomology 106: 1495–1502.

Nielsen, A. L., S. Chen, and S. J. Fleischer. 2016a. Coupling developmental physiology, photoperiod, and temperature to model phenology and dynamics of an invasive heteropteran, *Halyomorpha halys*. Frontiers in Physiology 7: 1–12.

Nielsen, A. L., G. Dively, J. M. Pote, G. Zinati, and C. Mathews. 2016b. Identifying a potential trap crop for a novel insect pest, *Halyomorpha halys* (Hemiptera: Pentatomidae), in organic farms. Environmental Entomology 45: 472–478.

Oda, M., T. Sugiura, Y. Nakanishi, and Y. Ueaumi. 1980. Ecological studies of stink bugs attacking fruit trees. Report 1: the prevalence of seasonal observations by light trap, and the ecology on the occurrence of fruit trees and mulberry under field observations. Bulletin of the Nara Agricultural Experiment Station 11: 53– 62.

Ogburn E. C., R. Bessin, C. Dieckhoff, R. Dobson, M. Grieshop, K. A. Hoelmer, C. Mathews, J. Moore, A.L. Nielsen, K. Poley, J. Pote, M. Rogers, C. Welty, and J. Walgenbach. 2016. Natural enemy impact on the invasive brown marmorated stink bug, *Halyomorpha halys* (Stål) (Hemiptera: Pentatomidae), in organic agroecosystems: a regional assessment. Biological Control 101: 39–51.

Ohira, Y. 2003. Outbreak of the stink bugs attacking fruit trees in 2002. Plant Protection 57: 164–68.

Olusola, J. 2016. Trap cropping – development and deployment for control of the brown marmorated stink bug (*Halyomorpha halys* [Stål]) on cowpea (*Vigna unguiculata* L. Walp.). MS Thesis. North Carolina Agricultural and Technical State University. 108 pp.

Paik, C. -H., G. -H. Lee, M. -Y. Choi, H. -Y. Seo, D. -H. Kim, C. -Y. Hwang, and S. -S. Kim. 2007. Status of the occurrence of insect pests and their natural enemies in soybean fields in Honam province. Korean Journal of Applied Entomology 46: 275–280.

Panizzi, A. R., 1997. Wild hosts of pentatomids: ecological significance and role in their pest status on crops. Annual Review of Entomology 42: 92–122.

Pansa, M. G., L. Asteggiano, C. Costamagna, G. Vittone, and L. Tavella. 2013. First discovery of *Halyomorpha halys* in peach orchards in Piedmont. L' Informatore Agrario 69: 60–61.

Pfeiffer, D. G., T. C. Leskey, and H. J. Burrack. 2012. Threatening the harvest: the threat from three invasive insects in late season vineyards. pp. 449–474. *In* N. J. Bostanian, C. Vincent, and R. Isaacs (Eds.), Arthropod management of in vineyards: pests, approaches and future directions. Springer Science and Business Media. 504 pp. doi 10.1007/978-94-007-4032-7_19.

Pike, T. J. 2014. Interactions between the invasive brown marmorated stink bug, *Halyomorpha halys* (Hemiptera: Pentatomidae) and entomophatogenic fungi. MS Thesis. University of Maryland. 58 pp.

Phillips, C. R., T. P. Kuhar, G. P. Dively, G. Hamilton, J. Whalon, and K. Kamminga. 2016. Seasonal abundance and phenology of *Halyomorpha halys* (Hemiptera: Pentatomidae) on different pepper cultivars in the mid-Atlantic (United States). Journal of Economic Entomology. doi: 10.1093/jee/tow256.

Pote, J., A. Nielsen, K. Poley, M. Grieshop, C. Welty, J. Walgenbach, R. Bessin, E. Ogburn, C. Matthews, J. Moore, and W. Morrison. 2015. Biological control of brown marmorated stink bug. http://eorganic.info /brown-marmorated-stink-bug-organic/resources (accessed 18 November 2015).

Qin, W. 1990. Occurrence rule and control techniques of *Halyomorpha picus*. Plant Protection 16: 22–23.

Qiu, L. -F. 2007. Studies on biology of the brown marmorated stink bug *Halyomorpha halys* (Stål) (Hemiptera: Pentatomidae), an important pest for pome trees in China and its biological control. Ph.D. Dissertation, Chinese Academy of Forestry, Beijing, China. 126 pp.

Rabitsch, W., and G. J. Friebe. 2015. From the west and from the east? First records of *Halyomorpha halys* (Stål, 1855) (Hemiptera: Heteroptera: Pentatomidae) in Vorarlberg and Vienna (Austria). Beiträge zur Entomofaunistik 16: 126–129.

Rahman, M. M., and U. T. Lim. 2017. Evaluation of mature soybean pods as a food source for two pod-sucking bugs, *Riptortus pedestris* (Hemiptera: Alydidae) and *Halyomorpha halys* (Hemipera: Prntatomidae). PLoS ONE 12: 1–16 e0176187. doi.org/10.371/journal.pone.0176187.

Ray, J. N., and H. L. Pence. 2004. Ladybug hypersensitivity: report of a case and review of literature. Allergy and Asthma Proceedings 25: 133–138.

Rice, K. B, J. C. Bergh, E. J. Bergmann, D. Biddinger, C. Dieckhoff, G. Dively, H. Fraser, T. Gariepy, G. Hamilton, T. Haye, A. Herbert, K. Hoelmer, C. Hooks, A. Jones, G. Krawczyk, T. Kuhar, W. Mitchell, A. Nielsen, D. Pfeiffer, M. Raupp, C. Rodriguez-Saona, P. Shearer, P. Shrewsbury, D. Venugopal, J. Whalen, N. Wiman, T.C. Leskey, and J. Tooker. 2014. Biology, ecology, and management of brown marmorated stink bug (*Halyomorpha halys*). Journal of Integrated Pest Management 5: 1–13.

Rice, K. B., R. R. Troyer, K. M. Watrous, J. F. Tooker, and S. J. Fleischer. 2016. Landscape factors influencing stink bug injury in mid-Atlantic tomato fields. Journal of Economic Entomology. doi: 10.1093/jee/tow252.

Rich, W. A., and R. L. Koch. 2016. Effects of Rag1 aphid-resistant soybean on mortality, development, and preference of brown marmorated stink bug. Entomologia Experimentalis et Applicata 158: 109–117.

Rider, D. A., L. Y. Zheng, and I. M. Kerzhner. 2002. Check-list and nomenclatural notes on the Chinese Pentatomidae (Heteroptera). II. Pentatominae. Zoosystematica Rossica 11: 135–153.

Roderick, G., L. de Mendoza, G. Dively, and P. Folletti. 2003. Sperm precedence in Colorado potato beetle, *Leptinotarsa decemlineata* (Coleoptera: Chrysomelidae): temporal variation assessed by neutral markers. Annals of the Entomological Society of America 96(5): 631–636.

Rondonia, G., V. Bertoldi, R. Malek, M. C. Foti, E. Peri, L. Maistrello, T. Haye, and E. Conti. 2017. Native egg parasitoids recorded from the invasive *Halyomorpha halys* successfully exploit volaties emitted by the plant-herbivore complex. Journal of Pest Science. doi 10.1007/s10340-0861-0.

Roversi, P. F., F. Binazzi, L. Marianelli, E. Costi, L. Maistrello, and G. Sabbatini Peverieri. 2016. Searching for native egg-parasitoids of the invasive alien species *Halyomorpha halys* Stål (Heteroptera: Pentaomidae) in southern Europe. Redia 99: 63–70.

Ryu, J., and Y. Hirashima. 1984. Taxonomic studies on the genus *Trissolcus* Ashmead of Japan and Korea (Hymenoptera, Scelionidae). Journal of the Faculty of Agriculture, Kyushu University, 29: 35–58.

Ryu, T. -H., S. -E. Park, N. -Y. Ko, J. -G. Kim, H. -S. Shin, H. -R. Kwon, Y. -G. Kim, B. -H. Lee, M. -J. Seo, Y. -M. Yu, and Y. -N. Youn. 2013. Seasonal occurrence of insect pests and control effects of eco-friendly agricultural materials (EFAMs) in the field of *Lycium chinense* under environment-friendly management. Korean Journal of Pesticide Science 17: 402–410.

Saito, Y., S. Saito, Y. Omori, and K. Yamada. 1964. Studies on the bionomics of the bean bugs occurring in mountain regions, with particular reference to that of *Halyomorpha picus*, and to the insecticidal tests in laboratory and field. Japanese Journal of Sanitary Zoology 15: 7–16.

Sanatacruz, E. N., R. Venette, C. Dieckhoff, K. Hoelmer, and R. Koch. 2017. Cold tolerance of *Trissolcus japonicus* and *T. cultratus*, potential biological control agents of *Halyomorpha halys*, the brown marmorated stink bug. Biological Control. http://dx.doi.org/10.1016/j.biocontrol.2017.01.004.

Sargent, C., H. M. Martinson, and M. J. Raupp. 2014. Traps and trap placement may affect location of brown marmorated stink bug (Hemiptera: Pentatomidae) and increase injury to tomato fruits in home gardens. Environmental Entomology 43: 432–438.

Sasaki F., T. Miyamoto, A. Yamamoto, Y. Tamai, and T. Yajima. 2012. Relationship between intraspecific variations and host insects of *Ophiocordyceps nutans* collected in Japan. Mycoscience 53: 85–91.

Sauer, C. 2012. Die Marmorierte Baumwanze tritt neu im Deutschschweizer Gemusebau auf. Extension Gemusebau, Forschungsanstalt Agroscope Changins-Wadenswil. Gemusebau Info 28: 4–5.

Scott, J. 1874. On a collection of Hemiptera Heteroptera from Japan. Descriptions of various new genera and species. Annals & Magazine of Natural History (4)14(82): 289–304, 360–365, 426–452.

Šeat, J. 2015. *Halyomorpha halys* (Stål), 1855) (Heteroptera: Pentatomidae) a new invasive species in Serbia. Acta Entomologica Serbica 20: 167–171.

Sharma, K., S. B. Muldoon, M. F. Potter, and H. I. Pence. 2006. Ladybug hypersensitivity among homes infested with ladybugs in Kentucky. Annals of Allergy, Asthma and Immunology 97: 528–531.

Short, B. D., A. Khrimian, and T. C. Leskey. 2016. Pheromone-based decision support tools for management of *Halyomorpha halys* in apple orchards: development of a trap based treatment threshold. Journal of Pest Science. doi 10.1007/s10340-016-0812-1.

Signoret, V. 1881. Liste des Hémiptères reçueillis en Chine par M. Collin de Plancy. Bulletin de la Société Entomologique de France 1881(5): 45–47.

Sillen-Tullberg, B. 1981. Prolonged copulation: A male 'postcopulatory' strategy in a promiscuous species, *Lygaeus equestris* (Heteroptera: Lygaeidae). Behavoral Ecology and Sociobiology 9: 283–289.

Simov, N. 2016. The invasive brown marmorated stink bug *Halyomorpha halys* (Stål, 1855) (Heteroptera: Pentatomidae) already in Bulgaria. Ecologica Montenegrina 9: 51–53.

Soergel, D. C., N. Ostiguy, J. Fleischer, R. R. Troyer, E. G. Rajotte, and G. Krawczyk. 2015. Sunflower as a potential trap crop of *Halyomorpha halys* (Hemiptera: Pentatomidae) in pepper fields. Environmental Entomology 44: 1581–1589. doi: 10.1093/ee/nvv136.

Son, C. -K., S. -G. Park, Y. -H. Hwang, and B. -S. Choi. 2000. Field occurrence of stink bug and its damage in soybean. Korean Journal of Crop Science 45: 405–410.

Song, Y. -S., S. -K. Han, Y. -H. Moon, B. -C. Jeong, and J. -K. Bang. 2008. Effect on catechin and amino acid content of tea (*Camellia sinensis* L.) leaves by sucking of true bugs. Journal of the Korean Tea Society 14: 31–42.

Stål, C. 1855. Nya Hemiptera. Öfversigt af Kongliga Vetenskaps-Akademiens Förhandlingar 12(4): 181–192.

Stål, C. 1865. Hemiptera nova vel minus cognita. Annales de la Société Entomologique de France (4)5: 163–188.

Talamas, E. J., M. Buffington, and K. Hoelmer. 2013. New synonymy of *Trissolcus halyomorphae* Yang. Journal of Hymenoptera Research 33: 113–117. doi: 10.3897/JHR.33.5627.

Talamas, E. J., M. V. Herlihy, C. Dieckhoff, K. A. Hoelmer, M. L. Buffington, M. -C. Bon, and D. C. Weber. 2015a. *Trissolcus japonicus* (Ashmead) (Hymenoptera, Scelionidae) emerges in North America. Journal of Hymenoptera Research 43: 119–128. doi: 10.3897/JHR.43.4661.

Talamas, E. J., N. F. Johnson, and M. L. Buffington. 2015b. Key to Nearctic species of *Trissolcus* Ashmead (Hymenoptera, Scelionidae), natural enemies of native and invasive stink bugs (Hemiptera, Pentatomidae). Journal of Hymenoptera Research 43: 45–110.

Tauber, M. J., Tauber, C. A., and S. Masaki. 1986. Seasonal adaptation of insects. New York: Oxford University Press. 411 pp.

Taylor C. M., P. L. Coffey, D. B. DeLay, and G. P. Dively. 2014. The Importance of gut symbionts in the development of the brown marmorated stink bug, *Halyomorpha halys* (Stål). PLoS ONE 9(3): e90312. doi:10.1371/journal.pone.0090312.

Taylor, C., V. Johnson, and G. Dively. 2016. Assessing the use of antimicrobials to sterilize brown marmorated stink bug egg masses and prevent symbiont acquisition. Journal of Pest Science. doi 10.1007 /s10340-016-0814-Z.

Thompson, P. 2014. Import health standards for vehicles, machinery and tyres. Vehicle-A11-V1.1. Ministry for Primary Industries, Wellington, NJ. Available at: http://mpi.govt.nz/document-vault/1189 (accessed 28 July 2016).

Tillman, P. G. 2010. Parasitism and predation of stink bug (Heteroptera: Pentatomidae) eggs in Georgia corn fields. Environmental Entomology 39: 1184–1194. doi: 10.1603/EN09323.

Togon, R., J. Sant Ana, Q. -H., J. G. Miller, J. R. Aldrich, and F. G. Zalom. 2016. Volatiles mediating parasitism if *Euschistus conspersus* and *Halyomorpha halys* eggs by *Telenomus podisi* and *Trissolcus erugatus*. Journal of Chemical Ecology. doi: 10.1007/s10886-016-0754-3.

Tomasino E, P. Mohekar, T. Lapis, N. Wiman, V. Walton, and J. Lim. 2013. Effect of brown marmorated stink bug on wine - Impact to Pinot Noir quality and threshold determination of taint compound trans-2-decenal. In: The 15th Australian Wine Industry Technical Conference, July 13–18, Sydney, Australia., Australia: Australian Wine Industry Technical Conference.

Toyama, M., F. Ihara, and K. Yaginuma. 2006. Formation of aggregations in adults of the brown marmorated stink bug, *Halyomorpha halys* (Stål) (Heteroptera: Pentatomidae): the role of antennae in short-range locations. Applied Entomology and Zoology 41: 309–315.

Ueno, H., and Y. Ito. 1992. Sperm precedence in *Eysarcoris lewisi* Distant (Heteroptera: Pentatomidae) in relation to duration between oviposition and the last copulation. Applied Entomology and Zoology. 27: 421–426.

Ueno, W., and T. Shoji. 1978. Ecology and control of fruit pest stink bugs. Part I: overwintering locations. Northern Japanese Pest Research Bulletin 29: 16.

Uhler, P. R. 1860. Hemiptera of the North Pacific exploring expedition under Com'rs Rodgers and Ringgold. Proceedings of the Academy of Natural Sciences of Philadelphia 12: 221–231.

Van Lenteren J. C, M. J. W. Cock, T. S. Hoffmeister, and D. P. A. Sands. 2006. Host specificity in arthropod biological control, methods for testing and interpretation of the data, pp. 38–63. *In* F. Bigler, D. Babendrier, and U. Kuhlmann (Eds.), Environmental impact of invertebrates for biological control of arthropods: methods and risk assessment. CABI Publishing, Wallingford, Oxon, U.K. 229 pp.

Venugopal P. D., P. L. Coffey, G. P. Dively, and W. O. Lamp. 2014. Adjacent habitat influence on stink bug (Hemiptera: Pentatomidae) densities and the associated damage at field corn and soybean edges. PLoS ONE 9(10): e109917. doi:10.1371/journal.pone.0109917

Venugopal. P. D., G. P. Dively, and W. O. Lamp. 2015a. Spatiotemporal dynamics of the invasive *Halyomorpha halys* (Hemiptera: Pentatomidae) in and between adjacent corn and soybean fields. Journal of Economic Entomology 108: 2231–2241. doi: 10.1093/jee/tov188.

Venugopal P. D., H. M. Martinson, E. J. Bergmann, P. M. Shrewsbury, and M. J. Raupp. 2015b. Edge effects influence the abundance of the invasive *Halyomorpha halys* (Hemiptera: Pentatomidae) in woody plant nurseries. Environmental Entomology 44: 474–479. doi: 10.1093/ee/nvv061.

Venugopal P. D., G. P. Dively, A. Herbert, S. Malone, J. Whalen, and W. G. Lamp. 2016. Contrasting role of temperature in structuring regional patterns of invasive and native pestilential stink bugs. PLoS ONE 11: e0150649. doi:10.1371/journal.pone.0150649.

Vétek, G., V. Papp, A. Haltrich, and D. Rédei. 2014. First record of the brown marmorated stink bug, *Halyomorpha halys* (Hemiptera: Heteroptera: Pentatomidae), in Hungary, with description of the genitalia of both sexes. Zootaxa 3780(1): 194–200.

Walker, F. 1867. Catalogue of the specimens of Hemiptera Heteroptera in the collection of the British Museum. Part I. E. Newman, London, pp. 1–240.

Wallner, A. M., G. C. Hamilton, A. L. Nielsen, N. Hahn, E. J. Green, and C. R. Rodriguez-Saona. 2014. Landscape factors facilitating the invasive dynamics and distribution of the brown marmorated stink bug, *Halyomorpha halys* (Hemiptera: Pentatomidae), after arrival in the United States. PLoS ONE 9: e95691.

Wang, H. -J., and G. -Q. Liu. 2005. Hemiptera: Scutelleridae, Tessaratomidae, Dinindoridae and Pentatomidae, pp. 279–292. *In* K. –K. Tang, (Ed.), Insect fauna of middle-west Quinling Range and South Mountains of Gansu Province. Science Press, Beijing, China. 1055 pp.

Wang, Y., and Y. Wang. 1988. Studies on the pear bug, *Halyomorpha picus* Fab. Acta Agriculturae Boreali-Sinica 3: 96–101.

Watanabe, K. 1996. Characteristic of damages of *Lygocoris* (*Apolygus*) *lucorum* (Meyer-Dür) (Heteroptera: Miridae) and *Halyomorpha halys* (Stål) (Heteroptera: Pentatomidae) on cherry. Annual Report of the Plant Protection Society of Northern Japan 47: 143–144.

Watanabe, M. 1979. Ecology and extermination of *Halyomorpha halys*. 4. The relationship between day length and ovarian development. Annual Report Toyama Institute Health 3: 33–37.

Watanabe, M. 1980. Study of the life cycle of the brown marmorated stink bug, *Halyomorpha mista* by observing ovary development. Insectarium 17: 168–173.

Watanabe, M., K. Kamimura, and Y. Koizumi. 1978. The annual life cycle of *Halyomorpha mista* and ovarian development process. Toyama Journal of Rural Medicine 9: 95–99.

Watanabe, M., R. Arakawa, Y. Shinagawa, and T. Okazawa. 1994. Overwintering fight of brown marmorated stink bug, *Halyomorpha mista* to the buildings. Animal Health 45: 25–31.

Weber, D. C., T. C. Leskey, G. C. Walsh, and A. Khrimian. 2014. Synergy of aggregation pheromone with methyl (*E,E,Z*)-2,4,6-decatrienoate in attraction of *Halyomorpha halys* (Hemiptera: Pentatomidae). Journal of Economic Entomology 107: 1061–1068.

Weber, D. C., W. R. Morrison, A. Khrimian, K. B. Rice, T. C. Leskey, C. Rodriguez-Saona, A. L. Nielsen, and B. R. Blaauw. 2017. Chemical ecology of *Halyomorpha halys*: discoveries and applications. Journal of Pest Science. doi 10.1007/s10340-017-0876-6.

Wermelinger, B., D. Wyniger, and B. Forster. 2008. First records of an invasive bug in Europe: *Halyomorpha halys* Stål (Heteroptera: Pentatomidae), a new pest on woody ornamentals and fruit trees? Mitteilungen der Schweizerischen Entomologischen Gesellschaft 81: 1–8.

Wiman N. G., V. M. Walton, P. W. Shearer, S. I. Rondon, and J. C. Lee. 2014. Factors affecting flight capacity of brown marmorated stink bug, *Halyomorpha halys* (Hemiptera: Pentatomidae). Journal of Pest Science 88: 37–47. doi:10.1007/s10340-014-0582-6.

Wiman, N. G., J. E. Parker, C. Rodriguez-Saona, and V. M. Walton. 2015. Characterizing damage of brown marmorated stink bug (Hemiptera: Pentatomidae) in blueberries. Journal of Economic Entomology 108: 1156–1163. doi: 10.1093/jee/tov036.

Wyniger, D., and P. Kment. 2010. Key for the separation of *Halyomorpha halys* (Stål) from similar-appearing pentatomids (Insecta: Heteroptera: Pentatomidae) occurring in Central Europe, with new Swiss records. Mitteilungen der Schweizerischen Entomologischen Gesellschaft 83: 261–270.

Xu, J., D. M. Fonseca, G. C. Hamilton, K. A. Hoelmer, and A. L. Nielsen. 2014. Tracing the origin of US brown marmorated stink bugs, *Halyomorpha halys*. Biological Invasions 6: 153–166.

Yanagi, T., and Y. Hagihara. 1980. Ecology of the brown marmorated stink bug. Plant Protection 34: 315–321.

Yang, Z. Q., Y. X. Yao, L. F. Qiu, and Z. X. Li. 2009. A new species of *Trissolcus* (Hymenoptera: Scelionidae) parasitizing eggs of *Halyomorpha halys* (Heteroptera: Pentatomidae) in China with comments on its biology. Annals of the Entomological Society of America 102: 39–47.

Yarbough, J. A., J. L. Armstrong, M. Z. Blumber, A. E. Phillips, E. McGahee, and W. K. Dolan. 1999. Allergic rhinoconjunctivitus caused by *Harmonia axyridis* (Asian lady beetle, Japanese lady beetle, or lady bug). Journal of Allergy and Clinical Immunology 104: 704–705.

Yu, G. -Y., and J. -M. Zhang. 2007. The brown marmorated stink bug, *Halyomorpha halys* (Heteroptera: Pentatomidae) in P. R. China. International Workshop Biological Control Invasive Species of Forests, pp. 58–62.

Zhang, C. -T., D. -L. Li, H. -F. Su, and G. -L. Xu. 1993. A study on the biological characteristics of *Halyomorpha picus* and *Erthesina fullo*. Forest Research 6: 271–275.

Zhang, J. -M., H. Wang, L. -X. Zhao, F. Zhang, and G. -Y. Yu. 2007. The brown-marmorated stink bug, *Halyomorpha halys* (Stal) (Heteroptera: Pentatomidae) damages to an ecological apple orchard and its control strategy. Chinese Bulletin of Entomology 44: 898–901.

Zhang, J., F. Zhang, T. Gariepy, P. Mason, D. Gillespie, E. Takamas, and T. Haye. 2017. Seasonal parasitism and host specificity of *Trissolcus japonicus* in northern China. Journal of Pest Science. doi 10.1007 /s10340-017-0863-y.

Zhou, Y., M. M. Giusti, J. Parker, J. Salamanca, and C. Rodriguez-Saona. 2016. Frugivory by brown marmorated stink bug (Hemiptera: Pentatomidae) alters blueberry fruit chemistry and preference by conspecifics. Environmental Entomology 45: 1227–1234.

Zhu, G., W. Bu, Y. Gao, and G. Liu. 2012. Potential geographic distribution of brown marmorated stink bug invasion. PLoS ONE 7: 1–10.

Zhu, G., T. D. Gariepy, T. Haye, and Wenjun Bu. Patterns of niche filling and expansion across the invaded ranges of *Halyomorpha halys* in North America and Europe. 2016. Journal of Pest Science 88: 461–468. doi: 10.1007/s10340-016-0786-z.

Zobel, E. S., C. R. R. Hooks, and G. P. Dively. 2016. Seasonal abundance, host suitability, and feeding injury of the brown marmorated stink bug, *Halyomorpha halys* (Heteroptera: Pentatomidae), in selected vegetables. Journal of Economic Entomology 109: 1289–1302.

5

Megacopta cribraria (F.)

Joe E. Eger, Wayne A. Gardner, Jeremy K. Greene,
Tracie M. Jenkins, Phillip M. Roberts, and Dan R. Suiter

Megacopta cribraria (F.)[1]

1798	*Cimex cribrarius* Fabricius, Suppl. Ent. Syst.: 531 (India).
1803	*Tetyra cribraria*: Fabricius, Syst. Rhyn.: 71.
1835	*Thyreocoris cribarius* (sic): Burmeister, Handb. Ent. 2(1): 384.
1837	*Platycephala cribraria*: Westwood, Cat. Hem. Coll. Hope 1: 5.
1839	*Thyreocoris (Coptosoma) cribrarius*: Germar, Zeitschr. Ent. 1(1): 26.
1843	*Coptosoma cribrarium*: Amyot and Serville, Hist. Nat. Ins., Hém.: 66, pl. 2, fig. 4.
1867	*Coptosoma xanthochlora* Walker, Cat. Het. 1: 87 [Synonymized by Distant 1899, Ann. Mag. Nat. Hist. (7)4: 215, 226].
1896	*Coptosoma punctatissimum* Montandon, Ann. Soc. Ent. Belg. 40: 105–106 [Tentatively synonymized by Montandon 1897, Ann. Soc. Ent. Fr. 65: 457–458, Synonymized by Yang, 1934, Bull. Fan Mem. Inst. Biol. 5(3): 156, 161–164].
1977	*Megacopta cribraria*: Hsiao and Ren, Handb. Chinese Hem. Het. 1: 21–22, 293, pl. 1, fig. 13.

CONTENTS

5.1   Introduction 294
5.2   Taxonomy 295
5.3   Distribution and Spread 297
    5.3.1   Old World Distribution 297
    5.3.2   Distribution and Spread in the New World 297
5.4   General Biology 299
    5.4.1   Life Cycle 299
    5.4.2   Host Plants 300
    5.4.3   Natural Enemies 307
        5.4.3.1   Parasitoids 307
            5.4.3.1.1   Parasitoids in the Old World 307
            5.4.3.1.2   Parasitoids in the New World 308
        5.4.3.2   Parasitic Nematodes 308
        5.4.3.3   Predators 308
        5.4.3.4   Pathogens 309
    5.4.4   Endosymbionts 309
5.5   Population Genetics 310
    5.5.1   Mitochondrial DNA: Initial Studies 310
    5.5.2   Nuclear Markers 311
    5.5.3   Endosymbionts 311
    5.5.4   Origin of *Megacopta cribraria* 311

---

[1] Synonymy adapted from David A. Rider (personal communication).

5.6   Economic Impact 312
       5.6.1   Urban/Nuisance Pest 312
                5.6.1.1   Pest Status in Urban Areas 312
                5.6.1.2   Control in Urban Areas...313
       5.6.2   Crop Pest..316
                5.6.2.1   Pest Status on Crops..316
                5.6.2.2   Scouting and Control on Crops..317
       5.6.3   Biological Control Agent for Kudzu..318
       5.6.4   International Trade 319
5.7   Why Has *Megacopta cribraria* Been so Successful? 320
5.8   What Does the Future Hold?..320
5.9   Acknowledgments..321
5.10   References Cited..321

5.1 Introduction

In October 2009, a number of homes in northeastern Georgia (GA), United States, were invaded by a small round insect that previously had not been reported as a home invader. This insect was not only a nuisance because it was aggregating on and entering houses but was noxious because it released an unpleasant odor when disturbed. Homeowners, county agents, and pest control companies submitted specimens to the University of Georgia (UGA) Homeowner Insect and Weed Diagnostics Laboratory and to the UGA Urban Entomology Laboratory, both located on the UGA Griffin Campus, Griffin, GA. The insects were true bugs (Heteroptera) and resembled native members of the family Scutelleridae or related families and, as such, did not seem to warrant immediate concern. They subsequently were identified as *Megacopta cribraria* in the family Plataspidae, a family not previously known from the continental United States or the New World in general (Eger et al. 2010). In fact, the known distribution for *M. cribraria* was India and much of Asia. One of the homes that had been invaded in Hoschton, GA (Jackson County), was visited and thousands of bugs were active on the outside of the structure and on surrounding vegetation. A search of the surrounding area revealed a field of kudzu, *Pueraria montana* var. *lobata* (Willd.) Ohwi (Fabaceae), about 30 meters from the home. Adult bugs were abundant on kudzu as were large numbers of nymphs, suggesting that the kudzu field was the source of the infestation (Eger et al. 2010, Suiter et al. 2010).

A search of counties where these bugs were reported and neighboring counties was immediately initiated by UGA Griffin scientists and revealed that the bugs were present in nine counties in north Georgia (Suiter et al. 2010, Gardner et al. 2013a). *Megacopta cribraria* was found only on kudzu during this search but was known from the literature to be a pest of a number of legume crops (Eger et al. 2010) and a potential threat to soybean [*Glycine max* (L.) Merrill] (Fabaceae) production. In addition to being a home invader and potential pest of legume crops, it also has been investigated as a potential biological control agent for the invasive pest kudzu (Tayutivutikul and Yano 1990, Tayutivutikul and Kusigemati 1992a, Sun et al. 2006, Imai et al. 2011). There were no records of the species intentionally being introduced for biological control of kudzu, however. Thus, this apparently accidentally-introduced invasive bug has the unusual potential to be a pest or a beneficial depending on where it occurs. It also has the potential to interrupt trade with other countries. Gardner et al. (2013a) reported that cotton/polyester yarn shipments to Honduras contained three live and three dead *M. cribraria*, resulting in a remediation program to avoid trading sanctions. Gardner et al. (2013a) and Ruberson et al. (2013) also reported that this species, especially the nymphs, can cause a rash and stain when crushed against the skin. Although the skin rash and stain usually are not a problem, they can become a problem for people working in heavily infested soybean or kudzu fields.

Common names for *Megacopta cribraria* in India and Asia are 'bean plataspid', 'globular stink bug', and 'lablab bug'. In the United States, the accepted common name is 'kudzu bug' (Entomological Society of America 2015). How this bug arrived in the United States is unknown, but the initial infestations were located some distance from sea ports and relatively close to the Hartsfield-Jackson Atlanta International

Airport, which has many international flights. This suggests that it was introduced by travelers or trading goods arriving at that airport. Dobbs and Brodel (2004) reported that quarantine-significant organisms arrived regularly at the Miami International Airport, being found on about 10% of flights. Because this bug tends to move to overwintering sites, and the original infestation was close to a major airport, the most likely scenario is that a fertile female entered a plane in cargo or luggage bound for the United States and escaped upon arrival.

Although the North American record was the first report of the occurrence of a member of the family Plataspidae in North, Central, or South America, another species, *Coptosoma xanthogramma* (White), was found on legumes in Hawaii in 1965 (Beardsley and Fluker 1967). Froeschner (1984) reported finding three specimens of *Paracopta duodecimpunctata* (Germar) (as *Coptosoma duodecimpunctatum*) in the collection of the National Museum of Natural History with labels indicating Alaska as the collection locality. The distribution of this species is India to the Malay Peninsula and its occurrence in Alaska has not been confirmed and seems doubtful. Aiello et al. (2016) reported that a second plataspid species, *Brachyplatys vahlii* (F.), was found in Panama in 2012 and is apparently established there (Aiello et al. 2016). They based their identification on gene sequences in GenBank, but Rédei (2016) stated that, based on morphology, the species found in Panama was actually *Brachyplatys subaeneus* (Westwood).

5.2 Taxonomy

Specimens from the initial collections in Georgia were identified as *Megacopta cribraria* based on morphological characters and this determination was confirmed by David A. Rider (contributor for **Chapter 2**) and Thomas. J. Henry (personal communication). Voucher specimens from this initial collection were deposited in the National Museum of Natural History, Washington, DC, and the Florida State Collection of Arthropods, Gainesville, FL. Jenkins et al. (2010) initially sequenced mitochondrial DNA (mtDNA) genes from three *M. cribraria* adults randomly chosen from 2009 collections used to identify the species morphologically. The genetic sequence markers were specific to *M. cribraria*, and, thus, the protocol for a one-to-one correlation between genetic sequence markers and morphological taxonomy (Eger et al. 2010) was established as in other studies (Lee et al. 2005) (see **Section 5.5**, Population Genetics, for further discussion).

Megacopta cribraria originally was described as *Cimex cribrarius* by Fabricius (1798) based on specimens collected in India. It later was transferred to several genera until being assigned to *Coptosoma* Laporte by Amyot and Serville (1843) where it remained for more than a century. Montandon (1896) described a new species, *Coptosoma punctatissimum*, from Japan, indicating that it was close to *C. cribraria* but was larger and generally darker in color. The following year, Montandon (1897) indicated that he had seen specimens that were intermediate between the two species, but he did not formally designate *C. punctatissimum* as a junior synonym of *C. cribraria*. Yang (1934) revised the Chinese Plataspidae and considered *C. punctatissimum* to be a variety of *C. cribraria*, essentially synonymizing the two species. He was followed by Davidová-Vilímová (2006) who considered the two species to be conspecific. Hsiao and Ren (1977) described a new genus, *Megacopta*, designating *C. cribraria* as the type species of the genus where it resides today.

Most authors outside of Japan have treated the two names (*Megacopta cribraria* and *M. punctatissima*) as synonymous, but Japanese authors have continued to consider *M. punctatissima* as a species distinct from *M. cribraria* (e.g., Ishihara 1950, Hasegawa 1965, Hibino and Ito 1983, Hirashima 1989, Imura 2003, Himuro et al. 2006, and others). Hosokawa et al. (2007b) summarized the differences in the two species, stating that *M. cribraria* is found in the southwestern islands of Japan, is primarily a kudzu feeder, and rarely causes damage to agricultural crops. On the other hand, *M. punctatissima* occurs in mainland Japan and is a pest of soybeans. However, the two species are able to interbreed and produce viable offspring. Although there appear to be some differences in size and/or coloration between different populations of *M. cribraria* across the original distribution of the species, the same range in size and color is found in specimens from the southeastern United States. Examination of specimens from India, China, Japan and the southeastern United States did not reveal any consistent differences in external structure or in internal and external genitalia. Thus, there does not appear to be any morphological basis on which to separate populations of *M. cribraria* into different species or even subspecies. Hosokawa et al. (2014) analyzed mtDNA sequences for populations of this species from China, the southeastern United States, and Japanese islands. Using a Bayesian phylogenetic analysis, they found that neither '*M. cribraria*' nor '*M. punctatissima*' was monophyletic, being represented by eight different clades. Their conclusion that "neither *M. cribraria* nor *M. punctatissima* constitutes a monophyletic group, favoring the idea that *M. cribraria* and *M. punctatissima* do not constitute distinct species but rather represent local populations of the same species with considerable genetic and phenotypic diversity," supports the morphological conclusion that a single variable species, *M. cribraria*, is involved rather than two species.

Adults of *Megacopta cribraria* are small round bugs, 3.5 to 6.0 mm long, and light brown or olive to dark brown in color with darker punctation (**Figure 5.1E**). Eger et al. (2010) provided a key to families of Pentatomoidea from America north of Mexico. The combination of an enlarged scutellum covering most of the abdomen and two-segmented tarsi separate this bug from all others in this geography. Among North American bugs with the scutellum enlarged and covering most of the abdomen, *M. cribraria* can be recognized easily by the broadly rounded to truncate posterior of the scutellum, which is narrowly rounded in native species. Nymphs are yellow-brown, or greenish-yellow to olive-brown, relatively flattened, and moderately to strongly setose (**Figure 5.1C-D**).

FIGURE 5.1   (**See color insert.**) *Megacopta cribraria*. A, egg mass; B, underside of egg mass, arrows indicate symbiont capsules; C, second instar; D, fourth instar; E, adult female (Courtesy of Joe E. Eger); F, adults on gutter downspout; G adults on side of house, Hoschton, GA (Courtesy of Dan R. Suiter); H, infestation of adults on young soybean plants (Courtesy of Jeremy K. Greene). Dimensional lines equal 1 mm.

5.3 Distribution and Spread

5.3.1 Old World Distribution

Megacopta cribraria occurs throughout much of Asia and the Indian subcontinent. It has been reported from India, Pakistan, Sri Lanka, China, Japan, Korea, Macao, Taiwan, Indonesia, Malaysia, Myanmar, Thailand, and Vietnam (Eger et al. 2010). It also has been reported from Australia and New Caledonia by Eger et al. (2010), but those records probably are in error.

5.3.2 Distribution and Spread in the New World

Megacopta cribraria was initially discovered and identified in North America in nine northeastern Georgia counties in fall 2009 (Eger et al. 2010, Suiter et al. 2010). Homeowner complaints of aggregations of adult bugs on exterior surfaces of homes and other buildings (**Figure 5.1F-G**) prompted on-site visits by entomologists who determined the source of the aggregations as nearby patches of kudzu, *Pueraria montana* var. *lobata* (Suiter et al. 2010). The invasion of *M. cribraria* into the southeastern United States likely resulted from an inadvertent introduction on trade goods or with travelers originating from the insect's native Asian range.

Increased globalization since the 1980s has stimulated worldwide economic growth, but the associated global trade also has provided increased opportunities for accidental introductions of exotic species that become pests in their expanded ranges (McNeely 2001). Indeed, *Megacopta cribraria* adults are adept hitchhikers and have demonstrated a penchant for landing in or on trucks and automobiles, passenger and cargo compartments of airplanes, and shipping containers, especially during periods of elevated adult activity when the bugs are seeking overwintering sites in the fall and after emerging from those sites in the spring.

The port and mode of entry of this accidental introduction into the United States will likely never be identified or confirmed, but the proximity of Hartsfield-Jackson Atlanta International Airport to the site of discovery implicates arrival on either a cargo or passenger flight. However, introduction via surface transportation cannot be discounted because adults survived transit from north Georgia to Central America in 2011 and 2012 within containerized shipments of cotton yarn via ground and sea shipping through the port of Miami.

Following the initial discovery of *Megacopta cribraria* in several northeastern Georgia counties in October 2009, kudzu growing in adjacent areas and counties was sampled by sweep net for the insect. That sampling effort confirmed the occurrence of adults and late-instar nymphs in nine counties (Suiter et al. 2010, Gardner et al. 2013a), whereas sampling in 24 additional counties to the north, west, and south of those nine counties yielded no *M. cribraria*. Furthermore, scientists reported that no bugs had been observed in their sampling of kudzu for Asian soybean rust, caused by *Phakopsora pachyrhizi* Syd. and P. Syd. (Phakopsoraceae), in 40 other counties in central and south Georgia in 2009 (Robert C. Kemerait, personal communication).

Roberts (2010) reported significant infestations (≈60 insects per 20 sweeps) of *Megacopta cribraria* in soybean in ten counties in Georgia and two counties in South Carolina in July 2010. Four of those ten counties in Georgia were among the nine counties from which *M. cribraria* was reported in 2009. Based upon reports submitted by cooperators from a variety of universities and agencies in the region, *M. cribraria* spread rapidly from those nine Georgia counties into 12 additional states and the District of Columbia from 2010 through 2016 (**Figure 5.2**) (Gardner et al. 2013a, Medal et al. 2013a, Leslie et al. 2014). By the end of 2011, *M. cribraria* had been reported from all of South Carolina, most of Georgia and North Carolina, and some localities in Alabama and Virginia. During the next two years, reports confirmed the insect in all of Alabama, most of Mississippi, Tennessee, and Virginia, and in Arkansas, Delaware, the District of Columbia, Florida, Kentucky, Louisiana, and Maryland. Such rapid spread often is characteristic of invasive species that have adapted easily to the ecological conditions of their expanded ranges with a relative lack of natural enemies and interspecific competition (Sakai et al. 2001).

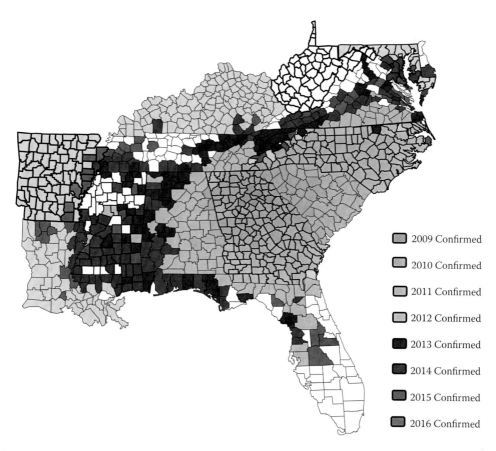

FIGURE 5.2   (See color insert.) Distribution of *Megacopta cribraria* in the southeastern United States, 2009–2016 (Courtesy of Wayne A. Gardner).

The rate of spread as reflected in the distribution map (**Figure 5.2**) is a function of several factors including, but not restricted to, the natural spread of the insect, intensity of monitoring activities, impact of natural enemies, awareness of the insect and its identity in otherwise serendipitous discoveries, and involvement of modes of human transportation and commerce in moving the insect. Efforts were made to confirm establishment of the insect in newly reported areas to avoid reporting incidental interceptions. Most reports were of established infestations in kudzu and soybean, the developmental hosts of the insect. Yet, Gardner et al. (2013a) noted that the insect also had been collected from 33 species of plants representing 15 taxonomic families. They surmised that the majority of those plants either served as resting sites or supplied some nourishment to the adults but did not serve as reproductive or developmental hosts.

The hitchhiking ability of *Megacopta cribraria* was apparent in the discovery of established populations in kudzu in Warren, Carroll, and Montgomery counties in Mississippi in 2012 adjacent to rest areas and truck stops along major east-west highway routes and distant from the then-established western edge of the distribution of the bug. Furthermore, spread might be correlated with other major highway routes including Interstates 10, 20, 26, 65, 75, 77, 85, and 95. Takano and Takasu (2016) evaluated the ability of *M. cribraria* to cling to cloth, metal, and glass surfaces at different wind velocities and found that at least some individuals were able to cling to metal and cloth surfaces for 1 minute when exposed to wind speeds of 100 km/hr, thus supporting the hitchhiking theory of spread. Gardner et al. (2013a) also postulated that adults might be moved on weather fronts, citing numerous reports of adults on windows, balconies, and rooftop gardens 100 or more meters above ground level in metropolitan condominiums, apartments, and other high rises. They pointed to the substantial northeastward spread of the insect in the spring of 2011 when a series of violent and active weather fronts moved through Alabama and Georgia into South Carolina and North Carolina.

Only a few new confirmations of *Megacopta cribraria* were reported in 2014. Those were primarily along the Gulf Coast and in the Mississippi River Valley. Leslie et al. (2014) reported that adults overwintering under leaf litter and in crevices under pine tree bark at several locations in Maryland did not survive two severe freezing events of temperatures as low as -12 to -15 °C in January 2014. Even in central Georgia, January 2014 temperatures dipped to -10 °C six times, -12 °C two times, and -14 °C one time (Georgia Automated Environmental Monitoring Network 2015). Winter mortality, therefore, might account for the lower numbers of bugs in 2014 than had been observed in previous years in many areas of the insect's distribution. It also might account for the fewer numbers of new confirmations reported in 2014. There were again only a few new counties added to the known distribution in 2015 and 2016; these were primarily west and south of the previous distribution records.

5.4 General Biology

5.4.1 Life Cycle

Zhang et al. (2012) provided basic information on the life cycle of *Megacopta cribraria* in the southeastern United States. There were two peaks in oviposition, April to early June and July to August, suggesting two generations per year. This appears to be the case across the southeastern United States although the appearance in spring may be later further north and earlier in the south. In Florida, we have found few eggs deposited on kudzu after August even though the plant may be active through December. Two generations per year also were reported by Li et al. (2004) in China, and Tayutivutikul and Kusigemati (1992b) in Japan, whereas Wang et al. (1996) and Wu et al. (2006) stated there are three generations in Zhejiang and Fujian, China, respectively.

Reported longevity of adults has been quite variable. Ramakrishna Aiyar (1913) and Srinivasaperumal et al. (1992) reported that longevity is relatively short at 2–5 days and 7 days, respectively. On the other hand, Thippeswamy and Rajagopal (2005a,b) and Zhang et al. (2012) reported ranges in adult longevity of 23–77, 25–43, and 6–25 days, respectively, which was influenced by host plant. Ruberson et al. (2013) observed longevity of 21–43 days when the bugs were held in a laboratory on pods of snap beans, *Phaseolus vulgaris* L. (Fabaceae).

As with most other pentatomoids, *Megacopta cribraria* deposits its eggs in masses, but the eggs are typically laid in two or, rarely, three rows (**Figure 5.1A**). The eggs are elongate, about 0.86 mm long, with longitudinal ridges and an operculum that is surrounded by a number of spinose projections (Zhang et al. 2012). They have been described or figured by Kershaw (1910), Moizuddin and Ahmad (1979), Ren (1984, 1992), Zhang et al. (2012), and Leslie et al. (2014). They are not deposited upright as in most other pentatomoids but are arranged on their sides with the operculum facing out. This may be an adaptation to facilitate access to the symbiont capsules, which are deposited underneath the two rows of eggs (**Figure 5.1B**) (see **Section 5.4.4**, Endosymbionts, for further discussion). Eggs usually are deposited on new leaf sheaths but may also be found on the undersides of older leaves or on stems. Tayutivutikul and Kusigemati (1992b) found that most of the eggs were found in the upper strata of the plant, whereas adults and nymphs were found in the middle to upper strata. Egg masses were reported to contain an average of 15.64 eggs in the southeastern United States (Zhang et al. 2012), 12–38 eggs in Pakistan (Ahmad and Moizuddin 1977), and an average of 31.7 eggs in Japan (Tayutivutikul and Kusigemati 1992b). Females deposited an average total of 96.5 eggs on soybean and 102.6 on lablab or field bean [*Lablab purpureus* (L.) Sweet] (Fabaceae) in India (Thippeswamy and Rajagopal 2005a), 78–274 on lablab bean in India and an average of 102 on lablab bean in Pakistan (Ahmad and Moizuddin 1977), but only 26–29 on kudzu in Japan (Tayutivutikul and Yano 1990). Shi et al. (2014) reported a range of 49 to 160 eggs depending on temperature. Ramakrishna Aiyar (1913) reported 10–40 eggs per female, but it is not clear if this is the number per egg mass or the total number of eggs deposited. Duration of the egg stage was about 6–7 days in India (Ramakrishna Aiyar 1913, Thippeswamy and Rajagopal 2005b) and 4–12 days in Pakistan (Ahmad and Moizuddin 1977).

Shi et al. (2014) studied development at different temperatures and found that eggs developed in 3.8–18.0 days at temperatures of 33–17°C.

There are five nymphal instars, which have been described or illustrated by Kershaw (1910), Moizuddin and Ahmad (1979), Zhang et al. (2012), and Leslie et al. (2014). Nymphs tend to be yellow-brown, or greenish-yellow to olive-brown, somewhat flattened, and covered in long setae (**Figure 5.1C-D**). Most native pentatomoid nymphs are relatively free of setae or have a few short setae. At different temperatures, Shi et al. (2014) found that nymphal development required 34–97 days, numbers that are close to the 32–48 days reported by Thippeswamy and Rajagopal (2005a,b). Ahmad and Moizuddin (1977) provided the following stadia for the first to fifth (1–5) instars: 1, 7.3 days; 2, 7 days; 3, 7 days; 4, 6.4 days; and 5, 8.3 days for males and 10 days for females. Ramakrishna Aiyar (1913) reported similar times of 6, 6, 9, 9, and 10 days, respectively, for instars 1-5. Tayutivutikul and Yano (1990) determined the developmental threshold for eggs and nymphs of males and females to be 14.9, 9.6, and 9.5°C, respectively, with a thermal constant of 1190–1299 degree-days for males and females. A similar study by Shi et al. (2014) found that the developmental threshold was 14.25°C with a thermal constant of 850 degree-days.

Megacopta cribraria overwinters as an adult under stones and grasses or in crevices in its natural habitat (Ren 1984). Lahiri et al. (2015) evaluated overwintering sites in woodlands adjacent to soybean fields in South Carolina, United States. They found the bugs preferred leaf litter over crevices in tree bark, and leaf litter at the base of the south-facing side of trees over leaf litter at the base of the north-facing side of trees or between trees. Movement away from feeding sites to overwintering sites resulted in the tendency for this pest to invade homes in the fall. Golec and Hu (2015) found that about 15% of overwintering females mated prior to overwintering, and these females were able to store live sperm throughout the winter. They also saw a decrease in the ratio of males to females as the winter progressed, suggesting that females are better able to survive cold temperatures than males. The fact that overwintering females may not be dependent on finding mates for reproduction when leaving overwintering sites may explain how a single female was able to establish an invasive population in the southeastern United States (see **Section 5.5**, Population Genetics, for further discussion).

5.4.2 Host Plants

Megacopta cribraria has been associated with a large number of plant species in its native and introduced ranges (**Table 5.1**). This is an active and mobile bug that migrates away from its host in preparation for overwintering in the fall and migrates out of overwintering sites in the spring. Therefore, it is not surprising that it has been observed on a number of different plants. It is considered to be primarily a legume feeder with kudzu being the preferred or natural host plant. It has been reported as a pest of several species of legumes (Fabaceae) but primarily soybean in Asia and lablab bean in China, India, and Pakistan. Eger et al. (2010) suggested that reports from plants other than legumes may be simply incidental sightings. However, Srinivasaperumal et al. (1992) reported that *M. cribraria* was able to complete development on firecracker plant, *Crossandra infundibuliformis* (L.) Nees (Acanthaceae) and cotton, *Gossypium hirsutum* L. (Malvaceae), but development was delayed and adult weight and fecundity were lower than for bugs reared on agathi, *Sesbania grandiflora* (L.) Pers. (Fabaceae). This is the only report of rearing this bug on plants other than legumes. Because the interaction with endosymbionts may affect host specificity of this bug (see **Section 5.4.4**, Endosymbionts, for further discussion), host range may differ for different populations. Yamazake and Sugiura (2005) reported *M. cribraria* feeding on a gall formed by an aphid, *Schlectendalia chinensis* Bell (Aphididae), rather than feeding directly on the plant, Chinese sumac (*Rhus chinensis* Miller) (Anacardiaceae).

In the southeastern United States, Zhang et al. (2012) studied the development of *Megacopta cribraria* on twelve potential legume host plants including lablab bean. Eggs were laid on many of these plants, but survival from egg to adult was observed only on kudzu and soybean. The lack of survival on lablab bean is interesting in that a number of studies in Pakistan and India report rearing this bug on the same host plant (Ahmad and Moizuddin 1977; Thippeswamy and Rajagopal 2005a,b). Medal et al. (2013b) conducted a similar study with 11 legumes and orange [*Citrus sinensis* (L.) Osbeck] (Rutaceae). *M. cribraria* completed development on kudzu, soybean, pigeon pea [*Cajanus cajan* (L.) Millsp.], black-eyed pea [*Vigna unguiculata* (L.) as *Vigna sinensis* (L.)], lima bean (*Phaseolus lunatus* L.), and pinto

TABLE 5.1

Plants Associated with *Megacopta cribraria*

Scientific Name	Common Name	Reference	Location
		Acanthaceae	
Crossandra infundibuliformis (L.) Nees (as *Crossandra undulaefolia* Salisb.)	Firecracker plant	Srinivasaperumal et al. 1992[1].	India
		Amaranthaceae	
Alternanthera philoxeroides (Mart.) Griseb.	Alligatorweed	Gardner et al. 2013a[2].	Southeastern United States
		Anacardiaceae	
Mangifera indica L.	Mango	Sung et al. 2006[2].	Taiwan
Rhus chinensis Mill. (as *Rhus javonica* L.)	Chinese sumac	Yamazaki and Sugiura 2005[2].	Japan
		Asteraceae	
Xanthium strumarium L.	Cocklebur	Gardner et al. 2013a[2].	Southeastern United States
Asteraceae	'Compositae'	Fletcher 1921[3], Hoffmann 1932[3].	China
		Convolvulaceae	
Ipomoea batatas (L.) Lam.	Sweet potato	Hoffmann 1932[3].	China
		Fabaceae (Leguminosae)	
Aeschynomene americana L.	American joint vetch	Medal et al. 2016a[1].	Southeastern United States
Albizia julibrissin Durazz.	Mimosa	Zhang et al. 2012[2,4].	Southeastern United States
Amphicarpaea bracteata L. Fernald	Hog peanut	Zhang et al. 2012[2].	Southeastern United States
Arachis glabrata Benth.	Perennial peanut	Medal et al. 2016a[1].	Southeastern United States
Arachis hypogaea L.	Peanut	Blount et al. 2015[4], Gardner et al. 2013a[2], Medal et al. 2013b[4].	Southeastern United States
Astragalus sinicus L.	Chinese milk vetch	Tayutivutikul and Kusigemati 1992a[3].	China, Japan
Baptisia australis (L.) R. Br.	Wild indigo	Zhang et al. 2012[2,4].	Southeastern United States
Butea monosperma (Lamark) Taubert	Flame of the forest, parrot tree	Kumar and Bharti 2015[5].	India
Cajanus cajan (L.) Millsp.	Pigeon pea, red gram	Blount et al. 2015[1]; Borah and Dutta 1999[2], 2002[5]; Borah and Sarma 2009a[5], b[3]; Borah et al. 2002[3]; Chandra and Kushwaha 2013[3]; Fletcher 1921[3]; Gardner et al. 2013a[2]; Hoffmann 1932[3]; Medal et al. 2013b[1]; Moizuddin and Ahmad 1979[3]; Ramakrishna Aiyar 1913[3]; Shroff 1920[3]; Thippeswamy and Rajagopal 2005a[3].	China, India, Myanmar, Pakistan, Southeastern United States
Cercis canadensis L.	Redbud	Zhang et al. 2012[2].	Southeastern United States
Chamaecrista fasciculata (Michx.) Greene	Partridge Pea	Medal et al. 2016a[1].	Southeastern United States

(Continued)

TABLE 5.1 (CONTINUED)

Plants Associated with *Megacopta cribraria*

Scientific Name	Common Name	Reference	Location
Cicer arietinum (L.)	Chickpea	Blount et al. 2015[4], Medal et al. 2013b[4].	Southeastern United States
Cladrastis kentuckea (Dum. Cours.) Rudd	American yellowwood	Gardner et al. 2013a[5], Zhang et al. 2012[4,5].	Southeastern United States
Crotalaria spectabilis Roth	Rattlepod	Medal et al. 2016a[4].	Southeastern United States
Cyamopsis tetragonoloba (L.) Taub. (as *Cyamopsis psoraloides* [sic])	Cluster bean	Fletcher 1921[3], Hoffmann 1932[3], Moizuddin and Ahmad 1979[3], Ramakrishna Aiyar 1913[3].	China, India
Desmodium tortuosum (Schwartz) de Candolle	Florida beggarweed	Medal et al. 2016a[4].	Southeastern United States
Glycine max (L.) Merrill	Soybean	Ali 1988[5]; Blount et al. 2015[1]; Chandra and Kushwaha 2013[3]; Del Pozo-Valdivia and Reisig 2013[1]; Gardner et al. 2013a[5]; Golec et al. 2013[2]; Golec et al. 2015[1], Greenstone et al. 2014[2]; Hirose et al. 1996[5]; Hosokawa et al. 2007b[1]; Kikuchi and Kobayashi 2010[2]; Kobayashi 1981[3]; Kono 1990[5]; Medal et al. 2013b[1], 2016a[1]; Ren 1984[3]; Seiter et al. 2013a[2], c[5], 2014a[5], b[5]; Shi et al. 2014[1]; Stubbins et al. 2014[5]; Tayutivutikul and Kusigemati 1992a[3]; Thippeswamy and Rajagopal 2005a[3]; Wang et al. 1996[3]; Xing et al. 2006[3], 2008[3]; Zhang et al. 2012[1].	Bangladesh, China, Japan, Southeastern United States
Indigofera hirsuta L.	Hairy indigo	Medal et al. 2016a[4].	Southeastern United States
Indigofera sp.	Indigo	Moizuddin and Ahmad 1979[3], Ramakrishna Aiyar 1913[2].	India
Lablab purpureus (L.) Sweet (as *Dolichos lablab* L. and *L. purpureus* var. *lignosus*)	Bean, lablab bean, field bean.	Ahmad and Moizuddin 1975a[3], b[3], 1976[5], 1977[1]; Blount et al. 2015[4]; Chandra and Kushwaha 2013[5]; Fletcher 1921[3]; Gardner et al. 2013a[2]; Hoffmann 1932[3]; Kitamura et al. 1984[2]; Moizuddin and Ahmad 1979[1]; Rajmohan and Narendran 2001[5]; Ramakrishna Aiyar 1913[1]; Rekha and Mallapur 2007[2]; Shroff 1920[3]; Thejaswi et al. 2008[2]; Thippeswamy and Rajagopal 1998a[2], b[2], 2005a[1], b[1]; Zhang et al. 2012[4,5].	China, India, Myanmar, Pakistan, Southeastern United States
Lens culinaris Medikus	Lentil	Medal et al. 2013b[4].	Southeastern United States
Lespedeza cuneata (Dum. Cours.) G. Don	Chinese lespedeza, bush clover	Zhang et al. 2012[4,5], Zheng et al. 2006[3].	Southeastern United States
Lespedeza cyrtobotrya Miq. as *Lespedeza cyrtobotria* [sic])	Lespedeza	Hibino 1985[2], 1986[2]; Hibino and Ito 1983[2].	Japan
Lespedeza hirta (L.) Hornem.	Hairy lespedeza	Zhang et al. 2012[4,5].	Southeastern United States

(Continued)

TABLE 5.1 (CONTINUED)

Plants Associated with *Megacopta cribraria*

Scientific Name	Common Name	Reference	Location
Lespedeza sp./spp.	Lespedeza	Gardner et al. 2013a[5], Tayutivutikul and Kusigemati 1992a[3].	Japan, Southeastern United States
Medicago sativa L.	Alfalfa	Gardner et al. 2013a[2], Medal et al. 2016a[1].	Southeastern United States
Melilotus alba Medikus	White sweet clover	Medal et al. 2016a[4].	Southeastern United States
Millettia sp.	none	Ren 1984[3].	China
Mucuna pruriens (L.) DC.	Velvet bean, Cowitch	Rani and Sridhar 2004[3].	India
Pachyrhyzus erosus (L.) Urb.	Jicama	Medal et al. 2013b[4].	Southeastern United States
Phaseolus lunatus L.	Lima bean	Blount et al. 2015[4], Gardner et al. 2013a[5], Golec et al. 2015[1], Hoffmann 1931[3], Medal et al. 2013b[1].	China, Southeastern United States
Phaseolus vulgaris L.	Kidney bean, green bean, pinto bean, pole bean, snap bean, string bean	Blount et al. 2015[4], Del Pozo-Valdivia and Reisig 2013[4], Easton and Pun 1997[3], Gardner et al. 2013a[2], Ishihara 1950[3], Medal et al. 2013b[1], Ren 1984[3].	Japan, Macau, Southeastern United States
Phaseolus spp.	Bean	Chandra and Kushwaha 2013[3], Hoffmann 1932[3].	China
Pisum sativum L.	Pea, winter pea, garden pea	Blount et al. 2015[4], Medal et al. 2016a[4].	Southeastern United States
Pongamia pinnata (L.) Pierre (as *P. glabra* Vent.)	Indian beech tree	Hoffmann 1932[3].	China
Pueraria montana var. *lobata* (Willd.) Ohwi (also as *P. lobata* Willd. and *P. thungergiana* Bentham)	Kudzu	Chandra and Kushwaha 2013[3]; Gardner et al. 2013a[5]; Golec et al. 2015[1], Hibino and Ito 1983[5]; Himuro et al. 2006[5]; Hosokawa and Suzuki 2000[2], 2001[5]; Hosokawa et al. 2005[5], 2007a[5], b[1], 2008[5]; Huskisson et al. 2015[5]; Imai et al. 2011[3]; Kershaw 1910[5]; Medal et al. 2013a[2], b[1], 2016a[1]; Ren 1984[1]; Seiter et al. 2014b[5]; Sun et al. 2006[5]; Takagi and Murakami 1997[5]; Takasu and Hirose 1991a[2], b[2]; Tayutivutikul and Kusigemati 1992a[3], b[5]; Tayutivutikul and Yano 1990[1]; Zhang et al. 2012[5]; Zheng et al. 2005[3].	China, Japan, Southeastern United States
Pueraria phaseoloides (Roxb.) Benth.	Tropical kudzu	Blount et al. 2015[4].	Southeastern United States
Robinia pseudoacacia L.	Black locust	Gardner et al. 2013a[5], Ren 1984[3], Zhang et al. 2012[2].	Southeastern United States
Senna obtusifolia L.	Sicklepod	Gardner et al. 2013a[2].	Southeastern United States
Sesbania bispinosa (Jacq.) W. Wight [as *S. aculeata* (Willd.) Pers.]	Dhaincha	Thippeswamy and Rajagopal 2005b[3].	China, India

(*Continued*)

TABLE 5.1 (CONTINUED)

Plants Associated with *Megacopta cribraria*

Scientific Name	Common Name	Reference	Location
Sesbania grandiflora (L.) Pers.	Agathi	Fletcher 1921[3]; Hoffmann 1932[3]; Moizuddin and Ahmad 1979[3]; Ramakrishna Aiyar 1913[3], 1920[3]; Srinivasaperumal et al. 1992[1].	China, India
Stylosanthes guianensis (Aubl.) SW	Stylo	Medal et al. 2016a[4].	Southeastern United States
Trifolium incarnatum L.	Crimson clover	Gardner et al. 2013a[5], Medal et al. 2016a[4].	Southeastern United States
Trifolium pratense L.	Red clover	Medal et al. 2016a[1].	Southeastern United States
Trifolium repens L.	White clover	Medal et al. 2016a[1].	Southeastern United States
Trifolium sp.	Clover	Gardner et al. 2013a[2].	Southeastern United States
Vachellia farnesiana (L.) Wight and Am.	Needle bush	Banerjee 1958[3].	India
Vicia angustifolia L.	Vetch	Hibino and Ito 1983[2].	Japan
Vicia faba L.	Broad bean, fava bean	Blount et al. 2015[1], Chandra and Kushwaha 2013[3].	Japan, Southeastern United States
Vicia sativa L.	Garden vetch	Zhang et al. 2012[5].	Japan
Vigna angularis (Willd.) Ohwi & Ohashi	Azuki bean	Blount et al. 2015[1], Tayutivutikul and Kusigemati 1992a[3].	Japan, Southeastern United States
Vigna mungo (L.) Hepper (also as *Phaseolus mungo* L.)	Black gram, black lentil, urd bean	Fletcher 1921[3], Thippeswamy and Rajagopal 2005b[3].	India
Vigna radiata (L.) Wilzcek	Mung bean	Blount et al. 2015[4], Easton and Pun 1997[3], Golec et al. 2015[1], Lal 1985[3], Medal et al. 2013b[4], Shroff 1920[3], Thippeswamy and Rajagopal 2005b[3].	India, Macau, Myanmar, Southeastern United States
Vigna unguiculata (L.) Walp [also as *Vigna sinensis* (L.)]	Black-eyed pea	Blount et al. 2015[4], Golec et al. 2015[4], Medal et al. 2013b[1], Zhang et al. 2012[2,4].	Southeastern United States
Wisteria brachybotrys Sieb. et Zucc.	Wisteria	Tayutivutikul and Kusigemati 1992a[3].	China, Japan
Wisteria floribunda (Willd.) DC	Japanese wisteria	Gardner et al. 2013a[5].	Southeastern United States
Wisteria frutescens (L.) Poir.	American wisteria	Gardner et al. 2013a[5], Zhang et al. 2012[4,5].	Southeastern United States
Wisteria sinensis (Sims) DC	Chinese wisteria	Gardner et al. 2013a[2], Seiter et al. 2014b[5].	Southeastern United States
Hydrangeaceae			
Deutzia crenata Siebold & Zucc.	Slender deutzia	Tayutivutikul and Kusigemati 1992a[3].	China, Japan
Juglandaceae			
Carya illinoinensis (Wagenh.) K. Koch	Pecan	Gardner et al. 2013a[2].	Southeastern United States
Malvaceae			
Corchorus capsularis L.	Jute	Hoffmann 1932[3].	China
Gossypium hirsutum L.	Cotton	Gardner et al. 2013a[2], Seiter et al. 2013b[5], Srinivasaperumal et al. 1992[1].	India, Southeastern United States

(Continued)

TABLE 5.1 (CONTINUED)

Plants Associated with *Megacopta cribraria*

Scientific Name	Common Name	Reference	Location
		Moraceae	
Ficus carica (L.)	Fig	Gardner et al. 2013a[5].	Southeastern United States
Morus alba L.	White mulberry	Zhang 1985[3], Zheng et al. 2006[3].	China
		Musaceae	
Musa acuminata Colla	Banana	Gardner et al. 2013a[2].	Southeastern United States
		Myricaceae	
Morella cerifera (L.)	Wax myrtle	Gardner et al 2013a[5].	Southeastern United States
		Oleaceae	
Ligustrum sinense Lour.	Chinese privet	Zhang et al. 2008[2].	China
		Pinaceae	
Pinus spp.	Pine trees	Gardner et al. 2013a[2].	Southeastern United States
		Poaceae	
Lolium perenne L.	Perennial ryegrass	Medal et al. 2016a[4].	Southeastern United States
Oryza sativa L.	Rice	Hoffmann 1932[3], Tayutivutikul and Kusigemati 1992a[3].	China, Japan
Saccharum officinarum L.	Sugar cane	Hoffmann 1932[3].	China
Sorghum bicolor (L.) Moench	Sorghum	Medal et al. 2016a[4].	Southeastern United States
Sorghum halepense (L.) Pers.	Johnson grass	Medal et al. 2016a[4].	Southeastern United States
Triticum aestivum L.	Wheat	Gardner et al. 2013a[2], Tayutivutikul and Kusigemati 1992a[3].	China, Japan, Southeastern United States
Zea mays L.	Corn	Medal et al. 2016a[4].	Southeastern United States
		Rosaceae	
Eriobotrya japonica Lindl.	Loquat	Gardner et al. 2013a[2].	Southeastern United States
Rubus fruticosus L.	Wild blackberry	Gardner et al. 2013a[2].	Southeastern United States
		Rutaceae	
Citrus reticulata Blanco	Tangerine	Gardner et al. 2013a[5].	Southeastern United States
Citrus × sinensis (L.) Osbeck.	Orange	Medal et al. 2013b[4].	Southeastern United States
Citrus unshiu (Swingle) Marcowicz	Satsuma mandarin	Gardner et al. 2013a[5].	Southeastern United States
Citrus spp.	Citrus	Tayutivutikul and Kusigemati 1992a[3].	China, Japan

(Continued)

TABLE 5.1 (CONTINUED)

Plants Associated with *Megacopta cribraria*

Scientific Name	Common Name	Reference	Location
		Salicaceae	
Salix nigra Marsh	Black willow	Gardner et al. 2013a[2].	Southeastern United States
		Solanaceae	
Capsicum annuum L.	Bell pepper	Medal et al. 2016a[4].	Southeastern United States
Solanum carolinense L.	Horsenettle	Imura 2003[2].	Japan
Solanum lycopersicum L.	Tomato	Medal et al. 2016a[4].	Southeastern United States
Solanum tuberosum L.	Potato	Gardner et al. 2013a[2], Hoffmann 1932[3].	China, Southeastern United States
		Vitaceae	
Vitus rotundiflora Michx.	Muscadine	Gardner et al. 2013a[2].	Southeastern United States
		Zingiberaceae	
Curcuma longa L.	Turmeric	Kotikal and Kulkarni 2000[3].	India

[1] Rearing attempted on this plant and at least some survived from egg to adult.
[2] Only adults reported from this plant.
[3] Listed plants as 'hosts' or bugs as 'pests' without further information on stages found or host suitability.
[4] Rearing attempted on this plant but no survival from egg to adult.
[5] Eggs and/or nymphs observed on this plant, usually in addition to adults.

bean (*Phaseolus vulgaris* L.) (Fabaceae), although survival was significantly lower on black-eye pea, lima bean, and pinto bean than on kudzu or soybean. Survival on pigeon pea was not significantly different than that on kudzu or soybean, which contradicts a report by Thippeswamy and Rajagopal (2005a) who stated that *M. cribraria* did not lay eggs on pigeon pea and nymphs that were placed on the plant did not survive. Results from host preference studies conducted by Blount et al. (2015) also found pigeon pea to be a suitable developmental host, but black-eyed pea, lima bean, or pinto bean were not suitable hosts. Using overwintered *M. cribraria*, Golec et al. (2015) obtained oviposition and development to adult on mung bean [*Vigna radiata* (L.) Wilzcek (Fabaceae)], lima bean, soybean and kudzu, but not on black-eyed pea. In a large study (Medal et al. 2016a), 21 plant species were evaluated to determine their suitability for development of *M. cribraria*. The plant species that were not suitable are included in **Table 5.1**. Eight species of Fabaceae were suitable for development and were statistically grouped as follows: soybean = kudzu > white sweet clover (*Melilotus alba* Medickus) = white clover (*Trifolium repens* L. > red clover (*Trifolium pratense* L.) = alfalfa (*Medicago sativa* L.) > perennial peanut (*Arachis glabrata* Benth.) = American joint vetch (*Aeschynomene americana* L.). Although there is agreement between most of the studies on the suitability of different plants for development, some of the host plants that were suitable in one study were not in others. These plants generally had fairly low survival in the trials where development was completed so differences in observed survival were not great. However, it would appear that some leguminous plants other than soybean and kudzu are suitable hosts for development in the southeastern U.S.

Huskisson et al. (2015) evaluated the density of bugs on kudzu, soybean, and lima bean in greenhouse paired choice studies. They found significantly more bugs on soybean than on lima bean. Kudzu had consistently higher numbers of bugs than soybean, but differences were significant only on one date; *M. cribraria* fed on soybeans, even in the presence of kudzu.

In the spring in Japan, *Megacopta cribraria* forms mating aggregations on *Lespedeza cyrtobotrya* Miq. (reported as *L. cyrtobotria*) and *Vicia angustifolia* L. (Fabaceae), and may feed on these plants, but rarely lays eggs on them, moving to kudzu after mating (Hibino and Ito 1983; Hibino 1985, 1986).

Gardner et al. (2013a) also reported finding adults on several plants, but few eggs and nymphs. Some of the plant associations in the United States may be mating aggregations as was observed in Japan, but the adults may derive some nutrition from feeding on these plants even if the plants are not suitable developmental hosts. In fact, Lovejoy and Johnson (2014) conducted molecular analyses of chloroplasts in the guts of wild-collected *M. cribraria* and identified chloroplasts from a diverse variety of plants as follows: pine (*Pinus* sp.) (Pinaceae), red oak (*Quercus rubra* L.) (Fagaceae), peanut (*Arachis hypogaea* L.), white sweet clover, black medic (*Medicago lupulina* L.), lespedeza (*Lespedeza* sp.) (Fabaceae), tomato (*Solanum lycopersicum* L.) (Solanaceae), lettuce (*Lactuca sativa* L.) (Asteraceae), sweet gum (*Liquidambar styraciflua* L.) (Altingiaceae), walnut (*Juglans* sp.) (Juglandaceae), and sorghum [*Sorghum bicolor* (L.)] (Poaceae). This suggests that these bugs at least feed on, and possibly acquire nutrients from, a much wider variety of plants than those that serve as developmental hosts. As the bug and its symbionts adapt to the New World, they may be able to expand the range of suitable native host plants.

5.4.3 Natural Enemies

5.4.3.1 Parasitoids

5.4.3.1.1 Parasitoids in the Old World

Heteroptera are most commonly parasitized by hymenopterous egg parasitoids and flies of the family Tachinidae, although the latter have not been reported from *Megacopta cribraria* in the Old World. Egg parasitoids reported from *M. cribraria* in the Old World are *Paratelenomus saccharalis* (Dodd), *P. tetartus* (Crawford), *Trissolcus latisulcus* (Crawford), and *Trissolcus* sp. (Scelionidae), *Ooencyrtus nezarae* Ishi (Encyrtidae), and *Ablerus* sp. and *Encarsiella boswelli* (Girault) (Aphelinidae) (Watanabe 1954, Ahmad and Moizuddin 1976; Mani and Sharma 1982; Bin and Colazza 1988; Polaszek and Hayat 1990, 1992; Hirose et al. 1996; Huang and Polaszek 1996; Johnson 1996; Rajmohan and Narendran 2001; Schmidt and Polaszek 2007). Ruberson et al. (2013) provided a detailed review of the biology of these parasitoids.

Paratelenomus saccharalis is a widespread parasitoid of the family Plataspidae and is reported from the same geographic area as that occupied by plataspids (Africa, Southern Europe, Tropical Asia, and Northern Australia) (Johnson 1996). This parasitoid has not been reported from hosts other than plataspids and is being investigated as a potential biological control agent for *Megacopta cribraria* in the United States where it has not been found to parasitize native species (Ruberson et al. 2013). In Japan, Takagi and Murakami (1997) reported that *M. cribraria* and *P. saccharalis* colonize soybean fields in early July and reproduce until the end of August, and that the parasitoid requires 12–25 days to complete development at temperatures of 20–30°C. Also in Japan, Yamagishi (1990) stated that this parasitoid was multivoltine and recorded high rates of parasitism from late May into early September. In China, Wall (1928) reported 51% total parasitism of *M. cribraria* eggs by this species. *P. saccharalis* appears to be a relatively specific and efficient parasitoid of plataspid eggs and may be suitable for biological control programs.

Other scelionids reported from *Megacopta cribraria* are *Paratelenomus tetartus* and *Trissolcus latisulcus* (Bin and Colazza 1988, Mani and Sharma 1982), although *M. cribraria* was not mentioned as a host for either in Johnson's (1991, 1996) revisions of these genera. Indeed, Wall (1931) misidentified *P. saccharalis* as *P. tetartus* (reported as *Dissolcus tetartus*). *T. latisulcus* was reported to parasitize Pentatomidae, Scutelleridae, and Tessaratomidae by Johnson (1996).

In addition to *Megacopta cribraria*, the encyrtid *Ooencyrtus nezarae* attacks eggs of a range of hosts, including the families Alydidae, Coreidae, and Pentatomidae (Takasu and Hirose 1986, Takasu et al. 2004). Parasitism of eggs of *M. cribraria* in China was reported to be 22.4–76.9% by Wu et al. (2006), and 61.4% by Zhang et al. (2003). Studies by Takasu and Hirose (1991a,b) and Takasu et al. (2004) indicated that *O. nezarae* requires a sugar source prior to oviposition, prefers to oviposit in eggs previously parasitized, and is attracted to the aggregation pheromone of *Riptortus pedestris* (Fabricius) [as *R. clavatus* (Thunberg)] (Alydidae). Although *O. nezarae* appears to be an efficient parasitoid, the broad host range would seem to preclude its use in classical biological control programs.

Aphelinids typically are parasitoids of whiteflies and armored scale insects (Schmidt and Polaszek 2007), but two species have been reported from *Megacopta cribraria*. Rajmon and Narendran (2001) reported that *Ablerus* sp. emerged from 60% of eggs sampled in India. However, they indicated that species of *Ablerus* are reported to be hyperparasitoids, so they may not be direct parasitoids of *M. cribraria*. *Encarsiella boswelli* has been reported from eggs of *M. cribraria* in India (Polaszek and Hyat 1990) but appears to be uncommon and has not been reported from Asia (Ruberson et al. 2013).

5.4.3.1.2 Parasitoids in the New World

Ruberson et al. (2013) reported that eggs of *Megacopta cribraria* were not attacked by native parasitoids in the United States but did indicate that a single tachinid fly, *Phasia robertsonii* (Townsend), emerged from an adult bug in Georgia. Golec et al. (2013) also reported parasitism of adults by a tachinid fly, *Strongygaster triangulifera* (Loew). Parasitism rates for the latter species were 5.14% overall, but females appeared to be preferred (9.34% of females and 0.93% of males were parasitized).

Eggs of *Megacopta cribraria* collected in Georgia, Alabama, and Mississippi in 2013, and Florida in 2014, were parasitized by *Paratelenomus saccharalis* (Gardner et al. 2013b, Medal et al. 2015). This parasitoid was not previously known from the New World and appears to be a separate unintentional introduction. Although the species was being evaluated in quarantine, it had not been approved for release and the genetics of the quarantined individuals and those appearing suddenly in the southeastern United States apparently are different (Wayne A. Gardner, personal communication). How this parasitoid was introduced and how it dispersed over a large geographic area before first being found are unknown. A subsequent study in organic soybeans in Georgia (Tillman et al 2016) found that oviposition and percent parasitism tended to be higher in conventional tillage soybean (58.4 %) than in no-till soybean (44.9 %). Late in the season parasitism was 82-95 %. These data suggest that this parasitoid has become a significant factor in reducing the number of *M. cribraria* in Georgia and perhaps the rest of the southeastern United States.

Other than *Paratelenomus saccharalis*, other egg parasitoids have not been reported in the US with one exception. In North Carolina and Virginia, Dhammi et al. (2016) recovered *Gonatocerus* sp. (Mymaridae), *Eretmocerus* sp. (Aphelinidae), *Encarsia* sp. (Aphelinidae), and *Ooencyrtus* sp. (Encyrtidae) from eggs of *Megacopta cribraria* but did not indicate levels of parasitism. Medal et al. (2016b) found that eggs of *M. cribraria* were not parasitized by *Trissolcus japonicus* (Ashmead) (Scelionidae), a potential biological control agent for *Halyomorpha halys* Stål (Pentatomidae), in the laboratory.

5.4.3.2 Parasitic Nematodes

There are no reports of parasitic nematodes in *Megacopta cribraria* in the Old World, but Stubbins et al. (2015) reported that a mermithid nematode was found parasitizing adults of *M. cribraria* in soybeans in South Carolina. The nematode was found in a single field (out of five sampled) and about 4.7% of females (only females were dissected) were found to be parasitized. Males and females also were held for nematode emergence and nematodes emerged from 7 (14%) of the males and 5 (10%) of the females. Most emerged from the caudal end of the bug, but one emerged from near the head. In a subsequent study (Stubbins et al. 2016b), *M. cribraria* again was found to be infected by this nematode with as many as 15% of females and 9% of males infected. Nematodes were reported from nymphs for the first time and as many as 20% of fifth instar nymphs were found to be infected. The nematode was identified as *Agamermis* sp.

5.4.3.3 Predators

There are relatively few reports identifying predators of *Megacopta cribraria* in the Old World. Ahmad and Moizuddin (1976) reported *Reduvius* sp. (as *Reduviius* sp.) (Reduviidae) preying on this bug, and Borah and Sarma (2009a) recorded a spider, *Oxyopes shweta* Tikader (Oxyopidae), and a bug, *Antilochus coquebertii* (F.) (Pyrrhocoridae), as predators of nymphs and adults of *M. cribraria*. Finally, Poorani et al.

(2008) stated that data labels on specimens of the genus *Synona* Pope (Coccinellidae) indicated the involved species were predators of *Coptosoma* spp. and *M. cribraria*.

Megacopta cribraria appears to be attacked by a number of predators in its invasive range. Ruberson et al. (2013) listed the following species that were observed feeding on *M. cribraria* in the laboratory or field: *Euthyrhynchus floridanus* (L.) (Pentatomidae); *Nabis roseipennis* (Reuter) (Nabidae); *Geocoris uliginosus* (Say), *G. punctipes* (Say) (Geocoridae); *Zelus* sp., *Sinea* sp. (Reduviidae); *Hippodamia convergens* Guérin-Méneville (Coccinellidae); and *Chrysoperla rufilabris* (Burmeister) (Chrysopidae). Greenstone et al. (2014) used molecular analysis of gut content to determine predators of *M. cribraria* in soybean and cotton fields. Nine predators tested positive for DNA of the bug as follows: *G. punctipes*, *G. uliginosus*; *Orius insidiosus* (Say) (Anthocoridae); *Podisus maculiventris* (Say) (Pentatomidae); *H. convergens*; *Zelus renardii* (Kolenati) (Reduviidae); *Oxyopes salticus* Hentz, *Peucetia viridans* (Hentz) (Oxyopidae); and *Solenopsis invicta* Buren (Formicidae). Leslie et al. (2014) provided a photograph of *Xysticus* sp. (Thomisidae) feeding on an adult of *M. cribraria*.

5.4.3.4 Pathogens

Only one pathogen, the fungus *Beauveria bassiana* (Balsamo) Vuillemin (Clavicipitaceae), has been reported from *Megacopta cribraria*. Borah and Dutta (2002) first reported this pathogen from *M. cribraria* on pigeon pea in India with 31.5% of bugs infected in 1997 and 19.3% in 1998. Pathogenicity was confirmed using Koch's postulates (Borah and Dutta 2002, Borah and Sarma 2009b) by culturing the fungus and reinfecting healthy individuals. They obtained 60–72% mortality of adults and 63–83% of nymphs treated with aqueous spore suspensions in the laboratory.

Ruberson et al. (2013) reported the occurrence of *Beauveria bassiana* in Georgia in single individuals of *Megacopta cribraria* found in kudzu and soybean. In South Carolina, Seiter et al. (2014a) also found these bugs infected with *B. bassiana* and confirmed identity and pathogenicity with Koch's postulates. Initial infestation levels in field plots averaged about 7% at peak density and ranged from 9.1 to 78.5% in plots treated with conidial suspensions, depending on density of the bugs. During the 2014 and 2015 growing seasons, widespread and significant epizootics of *B. bassiana* have been reported in soybeans in Georgia and South Carolina (Wayne A. Gardner, personal communication). In 2015, *B. bassiana* infected an average of 75.5 % of immature *M. cribraria* and 33.4 % of adults in two counties in eastern Tennessee (Britt et al. 2016). Late in the season mortality of immatures was as high as 100 % while the highest mortality of adults was a little over 60 %. In a laboratory assay (Portilla et al. 2016), strains of *B. bassiana* isolated from *M. cribraria* and *Lygus lineolaris* (Palisot de Beauvois) (Miridae) were highly effective against *M. cribraria*, with an LC_{50} of 1.98-4.98 viable spores per mm^2. These results suggest that *B. bassiana* has become an important biological control agent for this bug.

5.4.4 Endosymbionts

Megacopta species are phloem feeders, which is a carbohydrate-rich, but amino acid and vitamin-poor, food. Yet, these phloem feeders thrived, adapted, and dispersed rapidly into new environments (Prentis et al. 2008). This has been shown, in part, to be due to an intimate association with vertically transmitted bacterial endosymbionts in which insect and endosymbiont are interdependent or obligate partners (Moran 2001; Hosokawa et al. 2006, 2007b). Data are accumulating that show such insect-bacteria symbiosis serve to enhance the invasive adaptive potential of an insect (Moran 2007) by providing insects with the ability to adapt not only to nutrient-poor food sources (Gruwell et al. 2010) but to changing environments (Moran 2001).

Ovipositing females of *Megacopta cribraria* lay eggs in two rows and deposit small dark capsules under the eggs, generally between the two rows (**Figure 5.1B**). Fukatsu and Hosokawa (2002) found that the capsules contained a gammaproteobacterial endosymbiont that was not found in the eggs, demonstrating that it was transmitted orally rather than transovarially. First instar nymphs feed directly on these capsules to obtain the symbionts. Removal of the symbiont capsules results in bugs that show retarded development, arrested growth, and abnormal body coloration, suggesting that the symbionts are necessary for normal growth and development (Hosokawa et al. 2006). The symbionts also appear to affect behavior of the first instars; those acquiring the symbionts rested on the egg mass whereas those deprived of symbionts tended to wander off the mass (Hosokawa et al. 2008). Hosokawa et al. (2005) determined the components of the symbiont capsules and where they are made, stored, and excreted. They subsequently (Hosokawa et al. 2006) demonstrated that the symbiont was necessary for normal development and proposed the name Candidatus Ishikawaella capsulata for it. Hosokawa et al. (2007a) found that females produced one endosymbiont capsule for about 3.6 eggs irrespective of clutch size. They manipulated symbiont titer in first instars and estimated that the titer necessary for successful vertical transmission of the symbionts was 1.9×10^6, which was only 1/10 of the actual titer detected in newborn nymphs. Based on the actual titer detected in newborn nymphs, one capsule was sufficient for symbiont transmission to six nymphs and females produce 1.7 times more symbiont capsules than needed.

Hosokawa et al. (2007b) detailed differences between two populations of *Megacopta* in Japan, indicating that '*M. punctatissima*' is found in an area from mainland Japan to Yakushima Island and feeds primarily on kudzu but can invade soybean fields and become a pest. Bugs from the southwestern Japanese islands ('*M. cribraria*') feed mainly on 'Taiwan-kudzu' and rarely reach pest status. Exchanging symbiont capsules resulted in the non-pest developing normally on soybean and the pest not developing normally. Thus, the strain of endosymbiont appears to determine the developmental success or lack thereof on various host plants. Nikoh et al. (2011) provided a complete genome sequence for Candidatus Ishikawaella capsulata and stated that the genome suggests this symbiont provides essential amino acids for the plant-feeding host.

A second endosymbiont, the alphaproteobacterial *Wolbachia*, also occurs commonly in populations of *Megacopta cribraria* both in the United States and Japan (see **Section 5.5.3** under Endosymbionts for further discussion). Kikuchi and Fukatsu (2003) studied the diversity of these organisms and concluded *M. cribraria* from different areas of Japan each had two strains of *Wolbachia*. One was shared by both populations whereas each population had one unique strain.

5.5 Population Genetics

5.5.1 Mitochondrial DNA: Initial Studies

There were two objectives for the initial genetic studies conducted during the first two months following the discovery and identification of *Megacopta cribraria* in Georgia: (1) provide a genetic marker that would afford a one-to-one correlation between morphological taxonomy (Eger et al. 2010) and molecular taxonomy (Jenkins et al. 2010), and (2) standardize protocols for field collections obtained for DNA extraction, sequencing, and statistical analyses (Jenkins et al. 2002, 2007, 2009). Because taxonomic verification, genetic diversity, and insect origin were priorities, mtDNA was chosen because it is a non-recombining haploid molecule that is inherited maternally, evolves four times faster than nuclear DNA (nuDNA), and is relatively easy to amplify due to multiple copies in each cell (Zink and Barrowclough 2008). A 2336 bp mtDNA fragment, which included the 5'-tyrosine (Y) tRNA, cytochrome oxidase subunit I (COI), leucine tRNA (L), and partial cytochrome oxidase subunit II (COII) genes (COI-L-COII) (**Table 5.2**), was sequenced from all samples. By October, 2010, *M. cribraria* was confirmed in 80 counties in Georgia and collections processed for the 2336 bp mtDNA fragment showed only a single haplotype, designated GA1 (GenBank No. HQ444175) (**Table 5.2**).

The mitochondrial genome (15,647 bp) of five adult bugs randomly collected from four Georgia counties where *Megacopta cribraria* was found initially in 2009 (Suiter et al. 2010, Jenkins and Eaton 2011) showed consensus (GenBank No. JF288758). The COI-L-COII gene fragment within each of these genomes was GA1 (Tracie M. Jenkins, unpublished data), a conclusion independently verified by Amanda M. V. Brown. It appears, therefore, from these data that only one female lineage was introduced into Georgia and was rapidly and predictably (Zhu et al. 2012) dispersing throughout the southeast.

TABLE 5.2

Sequences and GenBank Numbers for *Megacopta cribraria*

Gene Fragment or Genome	Sequence Length (bp)	GenBank Number
MtDNA Complete Genome (GA1)	15,647	JF288758
COI-L-COII (GA1-southeastern U.S)	2336	HQ444175
COI-L-COII (GA1-Honduras)	2336	KF186211-KF186226
COI-L-COII (HN1, Honduras)	2336	KF186227
COI-L-COII (Asia)	2336	KF186196-KF186210, KC153656-KC153704
18S rRNA	1445	JQ254888
EF-1σ (Elongation Factor 1- alpha)	559	KF356283-KF356297
16S rRNA (C.I. capsulata (Cic-1)	1363	JF732916
groEL protein (C. I. capsulata (Cic-1)	1547	JF736508
wsp (outer surface protein-*Wolbachia* sp)	548	JQ266093

5.5.2 Nuclear Markers

The 18S rRNA and EF1-α nuclear gene fragments (**Table 5.2**) also were amplified and sequenced from randomly chosen adult bugs. As with the mitochondrial gene sequences, they also showed consensus among all collections studied. This lack of variation among these gene and mtDNA gene sequences strongly indicates a founding event exacerbated by a severe genetic bottleneck such as the introduction of a single female.

5.5.3 Endosymbionts

Using primers 16SA1 and 16SB1 specific for Candidatus Ishikawaella capsulata (Hosokawa et al. 2006), the 16S rRNA gene was amplified and sequenced (Jenkins et al. 2009) in three of the originally collected *Megacopta cribraria* individuals. A GenBank Basic Local Alignment Search Tool (BLAST) search (Altschul et al. 1990) matched for Candidatus Ishikawaella capsulata (AB067723) and all three specimens contained the endosymbiont (Jenkins et al. 2010). Furthermore, all individuals randomly collected in the United States and those intercepted in Honduras in 2012 had the same Candidatus Ishikawaella capsulata sequence as did random samples collected in Fukuoka Province, Japan, and Gyeongnam and Jeonnam, South Korea in 2012. Other 16S rRNA sequences from *M. cribraria* in China matched more closely gram-negative β-proteobacterium, genus *Achromobacter* (GU086442) (Tracie M. Jenkins, unpublished data). Brown et al. (2013) confirmed the identity of the symbiont and suggested that *M. cribraria* likely arrived in the United States with the capability of being a soybean pest.

A secondary endosymbiont, the α-proteobacterium *Wolbachia*, also was discovered in *Megacopta cribraria* collected in Georgia (**Table 5.2**). This endosymbiont may function to increase insect fecundity through increased growth and development (Dobson et al. 2002, Hosokawa et al. 2007b). A 548 bp fragment of the outer surface gene *wsp* was sequenced from *M. cribraria* collected in Georgia (GenBank #JQ266093) (Jenkins and Eaton 2011). A BLAST search then revealed a consensus with the *wsp* gene fragment identified as strain 1 (GenBank # AB109596), which was collected from *M. cribraria* (as *M. punctatissima*) in Japan (Kikuchi and Fukatsu 2003). Thus, the primary and secondary endosymbionts found in Georgia populations of *M. cribraria* were the same as those found in Japan.

5.5.4 Origin of *Megacopta cribraria*

Hosokawa et al. (2014) extracted DNA from 50 adult insects identified as *Megacopta cribraria* or *M. punctatissima* from native populations in Japan, South Korea, China, and Vietnam. Additionally, DNA was extracted from three *M. cribraria* adults collected in the southeastern United States. Using the *M. cribraria* mitochondrial genome sequence (GenBank # JF288758) as a reference (Jenkins and Eaton 2011), primers were designed to amplify and sequence the ND2-ND4 mtDNA region (8,687 bp),

FIGURE 5.3   Geographic origin for *Megacopta cribraria* based on COI-L-COII gene sequence from 318 samples (Tracie M. Jenkins, unpublished data).

which encompassed the COI-L-COII gene region used in previous studies (Jenkins and Eaton 2011) from each of the 53 samples (GenBank #AB872048-AB872100). The North American sequence (GenBank # AB872098-AB872100) showed consensus with GA1 (GenBank #HQ444175) (Jenkins and Eaton 2011). Phylogenetic analysis then grouped *M. cribraria* from North America into a single clade (E) with individuals collected from the island of Kyūshū, Japan. Based on this evidence, it was concluded that *M. cribraria* in North America likely came from a population in the Kyūshū region of Japan (Hosokawa et al. 2014).

A recently completed phylogenetic study of 318 adult *Megacopta cribraria* indicated the geographic area along the Korea Strait in northeast Asia, which includes the Kyūshū region, was a probable origin for the North American introductions (Tracie M. Jenkins unpublished data). The dataset included COI-L-COII gene sequence from all 2009 Georgia sampling sites, random collections across dispersal sites in the southeastern United States through 2013, adults intercepted in Honduras in 2012, and collections across Asia (China, Taiwan, South Korea, Japan, India). All but a single adult *M. cribraria* intercepted in Honduras were haplotype GA1 (Jenkins and Eaton 2011) as duplicated in other studies (Hosokawa et al. 2014). The GA1 haplotype clustered with individuals collected in the Kyūshū region of southwest Japan and Gyeongnam and Jeonnam provinces in southeast South Korea. There was little genetic differentiation among samples collected in these areas, which seems to indicate historical gene flow across the Korean Strait (**Figure 5.3**).

5.6 Economic Impact

5.6.1 Urban/Nuisance Pest

5.6.1.1 Pest Status in Urban Areas

As noted earlier, *Megacopta cribraria* was discovered in northeast Georgia in late October 2009 after its presence was noted by numerous property owners (Suiter et al. 2010, Ruberson et al. 2013). Bugs were observed on vehicles and the outside walls of homes and other buildings, prompting property owners to contact county extension agents and pest management companies for additional information. On warm, fall afternoons, the insects flew from nearby, dying kudzu patches onto houses where their presence was noted (**Figure 5.1F-G**). They were not reported as pests of agriculture (soybean) until the following summer.

The nuisance aspect of the bug has several origins, including its flying, sometimes in large numbers, around the outside of homes and aggregating on walls, doors, entryways, on and behind downspouts, fascia boards, and in cracks and crevices of walls. The excessively large numbers of these flying and aggregating bugs bring about the primary nuisance aspect of the bug. Such large numbers have altered outdoor recreational activities in the fall. The distinctive odor produced by the bug also adds to its nuisance status. More seriously, the compound(s) responsible for the bug's odor turns exposed human skin yellow, and some people have complained of adverse reactions from direct skin exposure to the chemical and a burning sensation when exposure is excessive (Ruberson et al. 2013). Finally, bugs often are so numerous that aggregations are found resting on all forms of non-host vegetation, leaving homeowners with the sense that the bug's host range is exhaustive. Incidental, large aggregations of *Megacopta cribraria* are found on azalea, pine, and tomato, and numerous other non-host trees and plants, prompting homeowners to believe, incorrectly, that the bugs have a wide host range that includes popular ornamental and food plants and even trees.

As with other seasonal, nuisance pests, such as the brown marmorated stink bug, *Halyomorpha halys*, and the multicolored Asian lady beetle, *Harmonia axyridis* (Pallas) (Coleoptera: Coccinellidae), the nuisance period of *Megacopta cribraria* is bimodal (Zhang et al. 2012). In the fall, some nuisance pests, triggered by declining temperature and day length, seek secluded sites where they spend the winter months protected from low temperatures. This behavior also apparently occurs with *M. cribraria*. The adults of the second generation occur in the fall and are numerous on kudzu at the beginning of its senescence. Concurrently, day length and temperature decrease, and the adults begin to search for overwintering sites such as cracks and crevices under tree bark, in leaf litter, and similar secluded locations. Potential overwintering sites include those on and around structures, resulting in problematic nuisance pest status for *M. cribraria*. This bug's presence was detected for the first time in north Georgia when large numbers exhibited this fall overwintering behavior. The following spring, as temperatures increased, overwintered adults became active and began to search for host plants, primarily kudzu. Because bugs often are active in Georgia (February and March) well-before kudzu begins to grow (April), they are a nuisance in the spring just as they were in the fall. As food sources become available, nuisance bug problems tend to decline as bugs find and infest kudzu in April and May and soybeans in June and July.

Megacopta cribraria especially gravitates towards light-colored surfaces. Horn and Hanula (2011) caught significantly more adults in bucket traps with white boards than in bucket traps "baited" with any other color. Anecdotally, in the fall when bugs begin to fly from kudzu patches to homes, they are known to gravitate and collect non-randomly on white surfaces, often metal downspouts and fascia boards at high points on structures. Additionally, bugs have been found on the 30th and 40th floor of high rises in downtown Atlanta, Georgia (Kyle Jordan, personal communication; Wayne A. Gardner, personal observation) and can be so numerous that they threaten the normal function of intake fans on large commercial buildings (Keith Daniels, personal communication). Clearly, this bug is a nuisance pest to homeowners, but in certain commercial situations may become a non-agricultural economic pest.

5.6.1.2 Control in Urban Areas

Megacopta cribraria is a classic seasonal, bi-modal nuisance pest, much like *Halyomorpha halys* and *Harmonia axyridis*. It is a pest in the spring and fall. Control of this bug on and around structures has proven frustrating for both homeowners and pest management professionals. Several factors combine to make it a major nuisance pest and one that is difficult to control, given the pest's biology and a pest management professional's limitations on product use and regulatory policy.

In January 2013, the United States Environmental Protection Agency, in response to mounting evidence of water contamination from use of pyrethroid insecticides in the urban environment (Domagalski et al. 2010, Jiang et al. 2012), attempted "to reduce ecological exposure from residential uses of pyrethroid and pyrethrin products" by proposing (2009) and then adopting (January 2013) an initiative to revise guidelines for pyrethroid- and pyrethrin-based pesticide products used in non-agricultural outdoor settings (United States Environmental Protection Agency 2013). The label changes specifically reduced non-agricultural outdoor use patterns for pyrethroids and pyrethrins to minimize contamination of water. For example, treatments to structural areas above impervious surfaces, such as driveways, where rain might wash insecticides into streams are prohibited, thus limiting the potential for run-off or drift to urban surface water. In addition to expanding restrictions on pyrethroid and pyrethrin use, the United States Environmental Protection Agency in 2013 implemented a series of changes to restrict the use of neonicotinoid insecticides to limit exposure to pollinators (United States Environmental Protection Agency 2015). This effort included changing pesticide labels to limit applications to protect bees, including limiting application of neonicotinoids while bees are foraging in non-agricultural settings.

The proximity of source(s) of adult *Megacopta cribraria* to structures has made sustained control difficult. Although removal of kudzu, usually the source of infestations, is an option, the source kudzu is often on private property or nearby property owned and managed by a local government municipality or other branch of government (**Figure 5.4A**). These source areas cannot be treated without permission. And, due to labor cost and other complications that might come with replacement vegetation associated with remediation, kudzu removal is typically not an option. Additionally, some state regulatory bodies may require additional licensing to treat those areas away from the structure or vegetation with any chemical, such as an insecticide or herbicide. Homeowners, though, desperate for relief from fall populations of *M. cribraria*, have been known to remove source patches of kudzu (**Figure 5.4B**). Ultimately, the elimination of *M. cribraria* from homes will require source reduction, which, in turn, may alter the pattern of bug movement towards nearby homes.

FIGURE 5.4 Association of *Megacopta cribraria*, kudzu, and residences. A, residence in Gainesville, GA, that suffered nuisance invasions of *Megacopta cribraria* from kudzu (foreground) on government-maintained property [reproduced with permission from Suiter et al. (2010)]; B, homeowners in Barnesville, GA, desperate for relief from fall invasions, removed patches of kudzu from adjacent property (Courtesy of Dan R. Suiter).

Preventing *Megacopta cribraria* from entering structures may be important, although reports of it doing so are not common; the insect primarily is a nuisance pest outside. Property owners should ensure that screen is placed over possible routes of entry into the home; that screens on windows are well-seated; and that soffit, ridge, and gable vents are screened properly. In locations where screening cannot be used, steel wool can be inserted into these openings to prevent the entry of the bugs. Lastly, doors should have a tight seal when closed, and doorsweeps should be installed.

The nuisance aspect of pests in the suburban environment often drives client complaints, especially with seasonal pests, such as *Megacopta cribraria*, where pest numbers can be overwhelming. A pest management company's most common response to these complaints is the use of insecticide sprays, typically those that result in rapid knockdown, to quickly eliminate the source of the nuisance when applied directly (topical treatment) to resting insects on structures. It also is desirable, from both the pest management company's perspective (to reduce the number of return visits) and the client's perspective (to eliminate the nuisance), that a product remain active over time and that it continue to provide rapid knockdown of subsequently invading insects that may be attracted to and land on previously-treated surfaces. Products that provide quick knockdown of *M. cribraria* are more desirable than those that allow insects to maintain their visibility, and thus nuisance status, even though they might eventually die after an extended exposure. Repeat applications may be necessary, however, because kudzu and other surrounding vegetation remain as a source of re-infestation. To reduce the frequency of re-application, longer-lasting formulations such as microencapsulated and wettable powder products should be used, if possible. Any formulation used should first be tested in a small area to be sure it does not stain, leave unsightly residues, or negatively impact the treated surface. Rather than use insecticides indoors for control of *M. cribraria*, insects that have entered the home should be vacuumed and the bagged insects then placed in hot, soapy water.

In a large laboratory study, Seiter et al. (2013b) screened nine professional pest control insecticide products (highest labeled rate for perimeter pest control) containing indoxacarb, pyrethroid, neonicotinoid, or pyrethroid + neonicotinoid mixtures, on five surfaces typically found on and around homes (metal, vinyl, painted and unpainted wood, and brick) for their residual efficacy (1, 7, 16, and 30-day-old residual) against adult *Megacopta cribraria*. In general, pyrethroid, neonicotinoid, and pyrethroid + neonicotinoid mixtures were effective at killing *M. cribraria* adults in 2-hour long forced-exposure experiments; products were faster acting on metal and vinyl than on pine plywood (latex-painted or unpainted) and brick.

In the laboratory (Dan R. Suiter, unpublished data), wild adults of *Megacopta cribraria* collected on kudzu in the fall were confined for 1-, 15-, or 60-minutes on small lengths of metal downspout treated with labeled rates of dinotefuran or cyfluthrin, and aged for one day. The bugs were then placed in clean, dry Petri dishes (no food or water). Mortality was recorded 1, 2, 24, and 48 hours following initiation of the assay. After exposure for 1 minute, mortality of bugs exposed to dinotefuran ranged from 33 to 90% after 1 and 2 hours, whereas mortality after 24 hours was 95 to 100%. After 1 and 2 hours, cyfluthrin had killed 0% and 10% of bugs, with just 15% and 20% mortality after 24 and 48 hours. One hour following initiation of a 15- or 60-minute exposure, mortality of bugs exposed to either rate of dinotefuran was 100%. Following a 15-minute exposure, cyfluthrin resulted in no mortality after 1 hour and just 20% after 48 hours. Mortality following a 60-minute exposure to cyfluthrin was 41.7, 98.3, and 100%; after 1, 2, and 24 hours, respectively.

These laboratory results were supported by an October 2013 field study (Dan R. Suiter, unpublished data) designed to evaluate the effectiveness of dinotefuran (0.10% or 0.30%) or a mixture containing 0.075% imidacloprid and beta-cyfluthrin against *Megacopta cribraria* when applied under simulated field conditions to white, metal downspouts. Ninety-one-cm lengths of metal downspout were treated to the point of runoff, allowed to air dry, and then suspended vertically above a 19 liter bucket containing soapy water.

For all 16 sampling dates, the mean number of these bugs collected from buckets placed under imidacloprid/cyfluthrin treated gutters was not significantly different ($P \geq .05$) from the mean number collected from buckets under untreated gutters. On 13 of 16 sampling dates, the mean number of bugs collected from buckets placed under a dinotefuran-treated gutter was significantly greater than the mean number of bugs collected from buckets placed under untreated or imidacloprid/cyfluthrin-treated gutters, whereas the two rates (0.10 and 0.30%) of dinotefuran were statistically ($P \geq .05$) similar on 15 of 16 sampling dates.

The rapid knockdown of *Megacopta cribraria* following exposure to residuals of the neonicotinoid dinotefuran suggests high sensitivity to this insecticide. In their study (Seiter et al. 2013b) the rapid knockdown of *M. cribraria* by dinotefuran probably was masked by the design of their assay. Adults were under constant exposure to residual deposits of each product. The earliest data presented from their study is from 2 hours of constant exposure. Exposure of *M. cribraria* to a 1-day-old residual of dinotefuran on metal resulted in 100% mortality after 2 hours. In a similar study (Dan R. Suiter, unpublished data), dinotefuran killed 100% of exposed *M. cribraria* after 1 hour of forced exposure.

Clearly, there are several insecticides that are effective against *Megacopta cribraria*, but they are not equal in initial knockdown or in residual efficacy. Although efficacy is important, regulatory issues and product labeling may limit insecticide availability and determine which insecticides are available for use outdoors on structures or nearby host plants.

5.6.2 Crop Pest

5.6.2.1 Pest Status on Crops

In the Old World, *Megacopta cribraria* is a pest of numerous legume crops. In India and Pakistan, it has been reported to be an important pest of lablab bean, hence the common name 'lablab bug' (Ahmad and Moizuddin 1975a,b; Chandra and Kushwaha 2013; Thippeswamy and Rajagopal 1998a,b). However, trials with the introduced population in the southeastern United States found that this bug did not survive to adulthood on lablab bean (Zhang et al. 2012, Medal et al. 2013b, Blount et al. 2015). In India, it has been listed as a pest of agathi, leguminous vegetable crops including mung bean , pigeon pea, urd bean [*Vigna mungo* (L.) Hepper], and velvet bean [*Mucuna pruriens* (L.) DC.] (Ramakrisha Ayyar 1920; Shroff 1920; Fletcher 1921; Lal 1980, 1985; Borah and Dutta 2002; Rani and Sridhar 2004; Thippeswamy and Rajagopal 2005b; Borah and Sarma 2009a,b). It is also a reported pest of lima bean in China (Hoffmann 1931) and kidney bean (*Phaseolus vulgaris* L.) in Japan (Ishihara 1950, Ren 1984).

Megacopta cribraria is an agriculturally important pest of soybeans in both its native range (Wang et al. 1996, Zhixing et al. 1996, Wu et al. 2006, Hosokawa et al. 2007b, Sujithra et al. 2008) and the United States (Greene et al. 2012; Seiter et al. 2013a,c; Roberts et al. 2014). Thippeswamy and Rajagopal (2005a) reported that feeding on leaves, shoots, and young pods led to "crinkled" leaves, although the effects on yield were not quantified. Likewise, Kikuchi and Kobayashi (2010) showed a reduction in growth of the main stem and in the number of trifoliates produced, as well as delayed growth, when soybeans were subjected to feeding by adults exceeding 40 individuals per plant; however, they did not quantify yield losses. Zhixing et al. (1996) showed yield losses exceeding 50% after exposure to *M. cribraria* at high densities (reported as 80 per "bunch"). Similar results were observed in soybeans in China, where yield losses up to 50% were reported (Wang et al. 1996). Yield loss in soybean has been documented in the United States when large numbers of bugs were left untreated. Seiter et al. (2013a) reported yield losses of about 60% in cage trials with initial infestation densities of 25 adults per plant; the yield components of seeds per pod and individual seed weight were reduced as initial densities of *M. cribraria* (0, 5, and 25 per plant) increased in this study. Roberts et al. (2014) reported an average yield loss of 19%, with a range from 0 to 60%, from untreated plots compared with protected plots in 37 trials conducted in Georgia from 2010 to 2013. In North Carolina, plots protected with bifenthrin had 56% fewer bugs and 6% higher yields compared to unsprayed plots (Del Pozo-Valdivia et al. 2017). Blount et al. (2016) reported that early planted soybeans (April, May, June) are more at risk for yield loss due to this insect than are later planted soybeans. In the southern United States, 6,070 hectares were infested with *M. cribraria* in 2011 and the economic impact (crop loss + cost of control) of this pest in soybeans was estimated to be US \$76,000.00 (Musser et al. 2012). The number of hectares infested increased to 148,358 in 2012, 233,434 in 2013, and 359,835 in 2014 (Musser et al. 2013, 2014, 2015) resulting in an economic impact of US \$988,034.00, \$4,286,107.00, and \$1,383,221.00, respectively. The reduction in economic impact in 2014 probably was due to a reduced number of bugs following a particularly cold winter of 2013–2014.

Despite a few reports of feeding on all plant structures in soybeans by *Megacopta cribraria*, feeding injury has been confined primarily to stems and petioles in the United States (**Figure 5.1H**) with no reports of feeding on pods. Because this bug is a stem feeder and likely a phloem feeder, yield losses in soybeans undoubtedly are due to indirect injury to the plant, probably resulting in reduced capacity to produce and transport photosynthate throughout the plant.

Although *Megacopta cribraria* has been reported from numerous plant species, many of these likely are incidental occurrences (Eger et al. 2010, Gardner et al. 2013a); however, data are limited relative to pest status across hosts (see **Section 5.4.2**, Host Plants, for further discussion). Being a host on which these bugs can develop may be a requirement for crops to suffer economic loss because field observations seem to indicate that the presence of large numbers of immatures often is associated with economic losses. In no-choice studies (Blount et al 2015; Medal et al. 2013b, 2016a; and Golec et al. 2015) black-eyed pea, lima bean, mung bean, pigeon pea, pinto bean and soybean, were reported as developmental hosts. Soybean and pigeon pea were the best developmental hosts, whereas only a few individuals of *M. cribraria* developed to adulthood on most of these plants. Sporadic occurrence of adult *M. cribraria* on cotton and other crops has been noted. Survival to adulthood on cotton was reported by Srinivasaperumal et al. (1992); however, in these no-choice assays, *M. cribraria* had reduced fecundity and size and took longer to develop than on a legume host. Occasionally, adults may be observed in cotton during periods of adult movement; however, reproducing populations have not been observed on cotton in Georgia or South Carolina, and this bug is not considered an economic pest of cotton.

5.6.2.2 Scouting and Control on Crops

Scouting and management programs for *Megacopta cribraria* are best understood in soybeans, but sampling techniques likely would be similar across other crop hosts. These bugs infest soybean during vegetative and/or reproductive stages (Del Pozo-Valdivia and Reisig 2013), feed on plant sap primarily from stems and petioles, and tend to congregate at nodes (Tayutivutikul and Yano 1990, Thippeswamy and Rajagopal 2005a, Suiter et al. 2010). The species appears to be bivoltine in soybeans in the United States (Greene et al. 2012, Zhang et al. 2012, Gardner et al. 2013a). Seiter et al. (2013c) reported that adults and nymphs of *M. cribraria* exhibit a generally aggregated spatial distribution and often are more numerous on field edges. A potential management tactic that could exploit this colonization behavior would be an in-field border application of insecticide to mitigate populations that develop around field perimeters initially. This could be a cost-saving approach to controlling these bugs, at least for the initial spray application, and the effectiveness of border sprays is being researched.

Stubbins et al. (2014) found that sweep nets and drop cloths are effective methods to estimate numbers of these bugs in soybeans. They reported that at all adult and nymph densities, fewer sweep-net samples were required for population estimations compared with the number of beat-cloth samples, but the beat-cloth method was more cost reliable than the sweep-net method at all but very low densities. Stubbins et al. (2016) compared cross-vane traps (Horn and Hanula 2011) with sweep net sampling in soybean fields. Cross-vane traps detected adults before they were found in sweep net samples, and there was a strong positive association between trap and sweep net samples at different distances from the field borders. Greene et al. (2012) noted that visual observation of the bugs on main stems and petioles in soybeans also appears to be an important means of assessing developing populations. Control of *Megacopta cribraria* with insecticides in soybeans is not difficult, as the species appears to be susceptible to a variety of insecticides (Wu et al. 1992, Li et al. 2001, Zhang and Yu 2005, Brown et al. 2015, Seiter et al. 2015b, Del Pozo-Valdivia et al. 2017) although commercial seed treated with insecticides did not offer protection from this pest (All et al. 2016). Even sublethal and low lethal doses of bifenthrin, imidicloprid, and acephate resulted in increased nymphal development time and decreased adult emergence Miao et al. 2016). Many of the insecticides currently labeled and recommended for use on other insect pests of soybeans (e.g., stink bugs), primarily the pyrethroids, are effective in controlling these bugs. However, these broad-spectrum insecticides can disrupt natural controls and increase the risk of outbreaks of other pests that are not susceptible to pyrethroids, such as soybean looper, *Chrysodeixis includens* (Walker) (Noctuidae), so scouting information throughout the growing season is vital.

TABLE 5.3

Current Sampling Methods and Treatment Thresholds for *Megacopta cribraria*
in Soybeans in Georgia and South Carolina

Sampling Method	Minimum Observations	Threshold
Sweep net (15-inch diameter)	At least ten 10-sweep samples representing entire field[1]	1 nymph/sweep
Canopy observation (visual)	At least ten observation spots representing entire field[1]	Nymphs easily found on petioles, main stems, or leaves

[1] Samples should be taken from multiple stops across the field and not just field borders.

Research efforts to determine when to treat soybeans with insecticides for *Megacopta cribraria* are ongoing and represent critical and needed information for managing this invasive species in the crop. Current recommendations are to interrupt development of each generation in soybeans with insecticide applications targeting the immature stages of this bug (Greene 2015, Roberts et al. 2015). Insecticide use is recommended when immatures are detected in sweep-net samples at approximately one nymph per sweep, regardless of instar. The critical time to protect soybeans from losses due to *M. cribraria* appears to be the time of pod and seed development, suggesting that producers should probably consider soybean growth stage as well as presence of nymphs when making control decisions (Seiter et al. 2016). Seiter et al. (2015a) and Blount et al. (2017) found that a single insecticide application timed to coincide with the presence of nymphs, particularly during the critical growth stages, prevented soybean yield losses. Sweep-net samples should be taken from all areas of the field (edges and middle) to represent the entire field, taking care not to bias sampling along border rows where populations build up initially. As an alternative to sweep-net sampling, visual inspections of insect density down in the canopy may suffice (Greene et al. 2012). If immature bugs are easily and repeatedly found on petioles and main stems using canopy observations, treatment likely is warranted (**Table 5.3**). This type of observational/visual sampling for these bugs also could help define and exploit differential colonization of border and interior areas of fields by the pest.

In addition to insecticidal control, management of *Megacopta cribraria* is aided by several other factors. A study by Del Pozo-Valdivia et al. (2016) in North Carolina and South Carolina found that early planted soybeans tended to have more bugs in untreated plots than did later planted soybeans, so planting later may help manage this bug. The authors also noted that later planted soybeans tend to experience more drought conditions and larger numbers of defoliators, so later plantings may not be a management option. The type of tillage impacts numbers of bugs with increased oviposition and up to four times as many bugs in conventional till versus no till or reduced till plots (Tillman et al. 2016, Del Pozo-Valdivia et al. 2017). Host plant resistance also has some potential. Bray et al (2016) identified six soybean lines as sources of resistance and they characterized the type of resistance as antibiosis. Fritz et al. (2016) evaluated 44 genotypes of soybean were evaluated over two growing seasons in NC and a number of these exhibited lower numbers of *M. cribraria* and reduced yield loss.

5.6.3 Biological Control Agent for Kudzu

Kudzu originally was introduced into the United States as an ornamental vine in the late 1800s and widely planted for forage and erosion control in the early to mid-1900s (Britton et al. 2002). It is a fast-growing vine that smothers all other vegetation and limits the production of pine trees, reduces plant biodiversity, damages power transmission lines, and can cause derailments along rail lines (Lindgren et al. 2013). The loss of productive forest land has been estimated at 48 to 84 dollars per acre/year, with lost production costs of up to US $588 million/year and control costs for power companies estimated at US $1.5 million/year (Britton et al. 2002, Hanula et al. 2012). Clearly the economic impact of kudzu on the United States economy is significant.

There has been only one study evaluating the impact of *Megacopta cribraria* on kudzu in the southeastern United States. Zhang et al. (2012) compared biomass of kudzu in plots that were exposed to the invasive bugs with plots from which bugs were excluded by biweekly insecticide applications. The plots that were protected from *M. cribraria* had 32.8% more biomass over one growing season than did plots that were not protected. Thus, this bug can have a significant impact on this weed and can suppress its growth. The cumulative effect of this bug over multiple growing seasons has not been evaluated, but anecdotal observations suggest that coverage of native vegetation by the kudzu plant has been reduced since the arrival of *M. cribraria*.

5.6.4 International Trade

Once *Megacopta cribraria* was recognized as a significant pest of soybeans in its expanded range in the United States, scientists and pest management practitioners focused on developing strategies and tactics for managing pest populations in crops and in urban habitats. At that time, there was little to no awareness of the potential impact that this small invasive insect might have on international trade. Yet, vigilant Honduran inspectors with the Servico de Proteccion Agropecuaria (SEPA) discovered two adult bugs in a shipment of fertile poultry eggs originating from a northeastern Georgia facility in December 2011. Inspectors subsequently discovered seven dead adults in a shipping container of frozen poultry meat originating from the same facility on 11 February 2012. Citing the latter interception, Honduran phytosanitary officials halted all agricultural imports originating from Georgia, Alabama, South Carolina, and North Carolina on 27 February 2012. Several days later, Honduran officials eased those restrictions to begin inspecting and unloading individual sea containers with cotton products to support local textile industry operations. Eventually, those self-imposed guidelines of inspecting each individual container were eased so that a smaller percentage (10%) of containers was actually inspected (Wayne A. Gardner, personal observation).

The Honduran interceptions and trade interruption raised concerns among other Central American trading partners that this invasive bug might be introduced accidentally from the southeastern United States and become established in their countries. Once established, agricultural scientists and other officials feared that it might become a significant pest of various bean crops grown by the region's subsistence farmers. Any trade restrictions likely would have dire consequences for those industries exporting goods and commodities to Central American partners as well as impacting various aspects of the economic health of the importing countries. Excluding Mexico, the United States exports annually an average of US $6.4 billion in goods and private services trade to each trading partner in Central America (United States Trade Representative 2014). An estimated average of US $624 million is in agricultural exports to each country, and the United States annually receives imports from that region averaging US $5.7 billion per country.

Devorshak (2007) explained that under provisions of the International Plant Protection Convention (IPPC), trading partners are encouraged to cooperate in minimizing the potential negative impacts of introducing new pests through trade. For example, the exporting partner should implement standards certifying that their exports will not introduce invasive pests. On the other hand, the importing partner should work to implement inspection and other phytosanitation measures that are scientifically and technically justified.

Additional interceptions of live and dead adult bugs occurred at Honduran and Guatemalan ports of entry in container shipments and in passenger jetliners originating from areas of the southeastern United States infested with the pest. Representatives of the Organismo Internacional Regional de Sanidad Agropecuaria (OIRSA) and scientists from the University of Georgia, in concert with trade specialists with the United States Department of Agriculture, Animal and Plant Health Inspection Service, Plant Protection Quarantine (USDA-APHIS-PPQ), officials with the United States Department of State, and representatives of the National Cotton Council of America, worked to remediate issues involving *Megacopta cribraria* and its impact on international trade between the United States and its Central American trading partners (Wayne A. Gardner, personal observation).

Initially, 16 officials from the OIRSA member nations participated in a 3-day educational event on the University of Georgia Griffin Campus 27–29 March 2012, including an on-site visit to a yarn spinning facility that supplies cotton products to the Latin American region, a day of scientific presentations on *Megacopta cribraria*, and round-table discussions. From these discussions, USDA-APHIS-PPQ and the National Cotton Council of America developed a Standard Operating Protocol for Loading of Containers to Ensure Freedom from *Megacopta cribraria*. The protocol outlined steps for inspecting for and eliminating bugs from the packing and loading areas and from the shipping containers prior to loading and standard certification by an inspector. The protocol was implemented by United States cotton product exporters and shippers in 2012 (National Cotton Council of America 2012).

University of Georgia scientists also assisted with various aspects of risk assessment for the import partners. Choice, no-choice, and field preference testing demonstrated that *Megacopta cribraria* did not establish successfully or develop on several varieties of pinto beans and winter pea (*Pisum sativum* L.) (Fabaceae) supplied by Guatemalan and Honduran sources (Blount et al. 2015). They concluded that if *M. cribraria* were accidentally introduced into Central America, it would not become a serious pest of pinto bean and winter pea grown in that region. Furthermore, they suggested that tropical kudzu [*Pueraria phaseoloides* (Roxburgh) Bentham], which is grown as a cover crop for nitrogen fixation in orchards in that region, is not a suitable host for *M. cribraria*. Nymphs developed to late second instar on that kudzu host, but survival rapidly decreased thereafter.

As a further risk assessment, lower lethal temperatures were established for *Megacopta cribraria* adults using a programmable water bath. No adults survived at -18 °C, the temperature at which frozen products are shipped, and there appeared to be no supercooling point in adult bugs (Wayne A. Gardner, unpublished data). Thus, there should be no concerns of adult *M. cribraria* surviving in shipments of frozen food products to import nations.

5.7 Why Has *Megacopta cribraria* Been so Successful?

The location of the introduction of this species was quite fortuitous as its preferred host, kudzu, is abundant in north Georgia. Zhu et al. (2012) used environmental parameters from the native range of the bug and determined that the southeastern United States is an area that is highly suitable for invasion by this bug. In addition to kudzu, soybeans commonly are grown in Georgia and South Carolina in areas not too far from the initial infestation. The ability of females to store sperm over the winter (Golec and Hu 2015) and the tendency of overwintering bugs to seek shelter may have enabled a single female to find its way to Georgia and start this infestation. The bug is a strong flier and has been found as high as 32 stories above ground level (Gardner et al. 2013a), enabling it to take advantage of weather patterns that may have pushed the population north and east of the originally infested area. Because kudzu, a reproductive host plant, is an invasive weed occurring in much of the area where *Megacopta cribraria* has spread, the bug has a source of food that is abundant and readily available. Shelter-seeking behavior also may have allowed this bug to spread more rapidly because its attraction to light colors (Horn and Hanula 2011) probably has resulted in its hitchhiking to new locations in vehicles. Indeed, the first findings of the bug in new locations tended to be along interstate highways or major roadways.

Because *Megacopta cribraria* belongs to a family, Plataspidae, that previously had not been found in its invasive range, the ability of native natural enemies to adapt to this species may be less than that for members of families that naturally occur in the New World (Strauss et al. 2006). The specificity of *Paratelenomus saccharalis* for plataspids suggests that these bugs are different enough to avoid most of the native parasitoids. The rapid increase in numbers of these bugs certainly suggests that there are few native natural enemies providing control of this species in its invasive range. Ruberson et al. (2013) noted that few natural enemies were found in their surveys in Georgia.

5.8 What Does the Future Hold?

Zhu et al. (2012) used niche models to predict areas suitable for colonization by *Megacopta cribraria*. Much of the eastern United States appears to be suitable with the southeastern United States being the most suitable for development. Other areas where invasion potential for this bug is high include southeastern South America, southwestern Europe, southern Africa, and eastern coastal Australia. The Pacific Northwest in the United States, where kudzu has established, shows moderate potential for invasion and establishment. Kudzu is found in much of the area in the United States where the model predicts moderate to high invasion potential, suggesting that the bug may spread to most of the area where kudzu is found. The effect of weather patterns on potential spread is unclear. Climate change as a result of global warming may play a role in the spread of this species to the north, but recent cold winters in the United States appear to have slowed its spread. Leslie et al. (2014) reported that hard freezes in January 2014 resulted in mortality of all bugs found at the base of an oak tree and only one of four overwintering bugs at another site was alive. Movement into structures or deeper into cover may have enabled these bugs to survive. The northernmost spread will probably be limited by freezing temperatures whereas western spread in the United States may be impacted by cold weather, arid conditions, or lack of reproductive hosts. Clearly, there is potential for this bug to spread to soybean producing areas of the midwestern United States, and care must be taken to prevent introduction of the bug into southeastern South America where large acreages of soybeans are grown.

Blount et al. (2017) sampled kudzu in two areas near Griffin, GA, from 2012 to 2015 and found that numbers of all life stages of *Megacopta cribraria* were significantly lower in 2014 and 2015 than in 2012 and 2013. Gardner and Olson (2016) subsequently sampled multiple kudzu patches in north and south Georgia from 2013 to 2016. Numbers of adult and immature *M. cribraria* per 20 sweeps were 115 and 155, respectively, in south Georgia in 2013 and these numbers dropped to 7 and 2 in 2016. In north Georgia, numbers decreased from 33 and 40 to 0.1 and 0.0 for adults and immatures over the same years. Authors of both papers postulated that the presence of the parasitoid *Paratelenomus saccharalis* beginning in 2013 and 2014 (Gardner et al. 2013b, Medal et al. 2015), coupled with increased infestations of *Beauveria bassiana* (Ruberson et al. 2013, Seiter et al. 2014a, Britt et al. 2016) may be responsible for the decline in numbers of *M. cribraria*. These data support anecdotal observations in the southeastern United States that the numbers of *M. cribraria* were lower in 2014 and 2015 than in previous years. So, although this bug may continue to spread, the accompanying natural enemies may prevent the large numbers we observed immediately following its introduction. In addition, areas already invaded by this bug will probably continue to see lower numbers than were previously observed.

5.9 Acknowledgments

We thank David. A. Rider (North Dakota State University, Fargo) for the synonymy of *M. cribraria*, for help with literature, and for confirming the identity of this bug. Thomas J. Henry (National Museum of Natural History, Washington, DC) also confirmed the identification of *M. cribraria*. Thanks are due to Amanda M. V. Brown (University of Montana, Bozeman) for verifying our conclusions on the GA1 genome, to Kyle Jordan (BASF Corporation, Atlanta, GA), Keith Daniels (Four Seasons Environmental, Atlanta, GA) for observations on the nuisance status of this bug in urban areas, Robert C. Kemerait (University of Georgia, Tifton) for observations of *M. cribraria* on kudzu from the soybean rust survey, and Wenjun Bu (Nankai University, Tianjin, China) for providing the English translation of the Hsiao and Ren (1977) reference. Finally, Jay McPherson's tireless efforts as editor have resulted in a much improved manuscript, and we greatly appreciate his efforts.

5.10 References Cited

Ahmad, I., and M. Moizuddin. 1975a. Scent apparatus morphology of bean plataspid *Coptosoma cribrarium* (Fabricius) (Pentatomoidea: Plataspidae) with reference to phylogeny. Pakistan Journal of Zoology 7(1): 45–49.

Ahmad, I., and M. Moizuddin. 1975b. Some aspects of internal anatomy of *Coptosoma cribrarium* (Fabr.) (Pentatomoidea: Plataspidae) with reference to phylogeny. Folia Biologica 23(1): 3–61.

Ahmad, I., and M. Moizuddin. 1976. Biological control measures of bean plataspids (Heteroptera: Pentatomoidea) in Pakistan. Proceedings of the Entomological Society of Karachi 6: 85–86.

Ahmad, I., and M. Moizuddin. 1977. Quantitative life-history of bean plataspid; *Coptosoma cribrarium* (Fabr.) (Heteroptera: Pentatomoidea). Pakistan Journal of Scientific and Industrial Research 20(6): 366–370.

Aiello, A., K. Saltonstall, and V. Young. 2016. *Brachyplatys vahlii* (Fabricius, 1787), an introduced bug from Asia: first report in the Western Hemisphere (Hemiptera: Plataspidae: Brachyplatidinae). BioInvasions Records 5(1): 7–12.

Ali, M. I. 1988. A survey of the insect pests of soybean in northern Bangladesh, their damage and occurrence. Tropical Pest Management 34(3): 328–330.

All, J., P. Roberts, and D. Kemp. 2016. Seed treatment and surface application of insecticides for supression of *Megacopta cribraria* in soybean, 2013. Arthropod Management Tests 2016 (F4), pp. 1–2.

Altschul, S. F, W. Gish, W. Miller, E. W. Myers, D. J. Lipman. 1990. Basic local alignment search tool. Journal of Molecular Biology 215(3): 403–410.

Amyot, C. J. B., and J. G. A. Serville. 1843. Histoire naturelle des insectes Hémiptères. Fain et Thunot, Paris. LXXVI + 675 + 8 pp. (Atlas), 12 plates.

Banerjee, M. R. 1958. A study of the chromosomes during meiosis in twenty-eight species of Hemiptera (Heteroptera, Homoptera). Proceedings of the Zoological Society of Calcutta 2: 9–31.

Beardsley, J. W., Jr., and S. Fluker. 1967. *Coptosoma xanthogramma* (White), (Hemiptera: Plataspidae) a new pest of legumes in Hawaii. Proceedings of the Hawaiian Entomological Society 19(3): 367–372.

Bin, F., and S. Colazza. 1988. Egg parasitoids, Hymenoptera Scelionidae and Encyrtidae, associated with Hemiptera Plataspidae, pp. 33–34. *In* J. Voegele, J. Waage, and J. Van Lenteren (Eds.), *Trichogramma* and other egg parasites. Colloques de l'INRA No. 43. Paris. 644 pp.

Blount, J. L., G. D. Buntin, and A. N. Sparks. 2015. Host preference of *Megacopta cribraria* (Hemiptera: Plataspidae) on selected edible beans and soybean. Journal of Economic Entomology 108(3): 1094–1105.

Blount, J. L., G. D. Buntin, and P. M. Roberts. 2016. Effect of planting date and maturity group on soybean yield response to injury by *Megacopta cribraria* (Hemiptera: Plataspidae). Journal of Economic Entomology 109(1):207–212.

Blount, J. L., P. M. Roberts, M. D. Toews, W. A. Gardner, G. D. Buntin, and J. N. All. 2017. Seasonal population dynamics of *Megacopta cribraria* in kudzu and soybean, and implication for insecticidal control in soybean. Journal of Economic Entomology 110(1):157–167.

Borah, B. K., and S. K. Dutta. 1999. Spatial distribution of *Megacopta cribrarium* (Fab.) adults on pigeonpea. Journal of Applied Systems Studies 12(2): 178–184.

Borah, B. K., and S. K. Dutta. 2002. Entomogenous fungus, *Beauveria bassiana* (Balsamo) Vuillemin: a natural biocontrol agent against *Megacopta cribrarium* (Fab.). Insect Environment 8(1): 7–8.

Borah, B. K., and K. K. Sarma. 2009a. Seasonal incidence of negro bug, *Megacopta cribrarium* (Fab.): on pigeonpea. Insect Environment 14(4): 147–149.

Borah, B. K., and K. K. Sarma. 2009b. Pathogenicity of entomopathogenous fungus, *Beauveria bassiana* (Balsamo) Vuillemin on *Megacopta cribrarium* (Fab.): a sucking pest of pigeonpea. Insect Environment 14(4): 159–160.

Borah, B. K., S. K. Dutta, and K. K. Sarmah. 2002. Dispersion pattern of *Megacopta cribrarium* (Fab.) nymphs on pigeonpea. Research on Crops 3(2): 441–445.

Bray, A. L., L. A. Lail, J. N. All, L. Zenglu, and W. A. Parrott. 2016. Phenotyping techniques and identification of soybean resistance to the kudzu bug. Crop Science 56(4): 1807–1816.

Britt, K., J. F. Grant, G. J. Wiggins, and S. D. Stewart. 2016. Prevalence and localized infection of the entomopathogenic fungus *Beauveria bassiana* on kudzu bug (Hemiptera: Plataspidae) in eastern Tennessee. Journal of Entomological Science 51(4): 321–324.

Britton, K., D. Orr, and J. Sun. 2002. Kudzu, pp. 325–330. *In* R. Van Driesche, B. Blossey, M. Hoddle, S. Lyon, and R. Reardon (Eds.), Biological control of invasive plants in the eastern United States. United States Department of Agriculture Forest Service Publication FHTET-2002-04, Morgantown, WV, USA. 413 pp.

Brown, A. M. V., L. Y. Huynh, C. M. Bolender, K. G. Nelson, and J. P. McCutcheon. 2013. Population genomics of a symbiont in the early stages of a pest invasion. Molecular Ecology 23(6): 1516–1530.

Brown, S. A., D. L. Kerns, T. S. Williams, K. Emfinger, and N. Jones. 2015. Evaluation of foliar insecticides for kudzu bug control in soybeans, 2014. Arthropod Management Tests 40(1): F14.

Burmeister, H. C. C. 1835. Handbuch der Entomologie. Rhyngota. T. Eslin, Berlin, Volume 2, Abt. I, ii and 400 + 4 pp., 2 plates.

Chandra, K., and S. Kushwaha. 2013. Record of hemipteran insect pest diversity on *Lablab purpureus* L: An economically important plant from Jabalpur, Madhya Pradesh. Research Journal of Agricultural Sciences 4(1): 66–69.

Davidová-Vilímová, J. 2006. Plataspidae, pp. 150–165. *In* B. Aukema and C. Rieger (Eds.), Catalogue of the Heteroptera of the Palaearctic region. Volume 5: Pentatomomorpha II. Netherlands Entomological Society, Wageningen. xiii + 550 pp.

Del Pozo-Valdivia, A. I., and D. D. Reisig. 2013. First-generation *Megacopta cribraria* (Hemiptera: Plataspidae) can develop on soybeans. Journal of Economic Entomology 106(2): 533–535.

Del Pozo-Valdivia, A. I., N. J. Seiter, D. D. Reisig, J. K. Greene, F. P. F. Reay-Jones, and J. S. Bachelor. 2016. *Megacopta cribraria* (Hemiptera: Plataspidae) population dynamics in soybeans as influenced by planting date, maturity group, and insecticide use. Journal of Economic Entomology 109(3): 1141–1155.

Del Pozo-Valdivia, A. I., D. D. Reisig, and J. S. Bacheler. 2017. Impacts of tillage, maturity group, and insecticide use on *Megacopta cribraria* (Hemiptera: Plataspidae) populations in double-cropped soybean. Journal of Economic Entomology 110(1):168–176.

Devorshak, C. 2007. Area-wide integrated pest management programmes and agricultural trade: challenges and opportunities for regulatory plant protection, Pages 407–415, *In* M. J. B. Vreysen, A. S. Robinson, and J. Hendrichs (Eds.), Area-wide insect pests. Springer Science and Business Media, Netherlands. 744 pp.

Dhammi, A., J. B. van Krestchmar, L. Ponnusamy, J. S. Bacheler, D. D. Reisig, A. Hebert, A. I. Del Pozo-Valdivia, and R. M. Roe. 2016. Biology, pest status, microbiome and control of kudzu bug (Hemiptera: Heteroptera: Plataspidae): a new invasive pest in the U. S. International Journal of Molecular Science 17, 1570, 21 pp.

Distant, W. L. 1899. Rhynchotal notes. Heteroptera: Plataspinae, Thyreocorinae, and Cydninae. Annals and Magazine of Natural History (7)4: 213–227.

Dobbs, T. T., and C. F. Brodel. 2004. Cargo aircraft as a pathway for the entry of nonindigenous pests into south Florida. Florida Entomologist 87(1): 65–78.

Dobson, S. L., E. J. Marsland, Z. Veneti, K. Bourtzis, and S. L. O'Neill. 2002. Characterization of *Wolbachia* host cell range via the in vitro establishment of infections. Applied and Environmental Microbiology 68(2): 656–660.

Domagalski, J. L., D. P. Weston, M. Zhang, and M. Hladik. 2010. Pyrethroid insecticide concentrations and toxicity in streambed sediments and loads in surface waters of the San Joaquin Valley, California, USA. Environmental Toxicology Chemistry 29(4): 813–823.

Easton, E. R., and W.-W. Pun. 1997. Observations on some Hemiptera/Heteroptera of Macau, Southeast Asia. Proceedings of the Entomological Society of Washington 99(3): 574–582.

Eger, J. E., Jr., L. M. Ames, D. R. Suiter, T. M. Jenkins, D. A. Rider, and S. E. Halbert. 2010. Occurrence of the Old World bug *Megacopta cribraria* (Fabricius) (Heteroptera: Plataspidae) in Georgia: a serious home invader and potential legume pest. Insecta Mundi 0121: 1–11.

Entomological Society of America. 2015. Common names of insects and related organisms. http://www.entsoc .org/pubs/common_names.

Fabricius, J. C. 1798. Entomologia systematica emendata et auct, secundum classes, ordines, genera, species, adjectis synonymis, locis, observationibus. Supplementum. Proft et Storch, Hafniae. ii + 572 pp.

Fabricius, J. C. 1803. Systema Rhyngotorum secundum ordines, genera, species adjectis, synonymis, locis, observationibus, descriptionibus. Carolum Reichard, Brunsvigae. x + 314 + 1 pp.

Fletcher, T. B. 1921. Annotated list of Indian crop-pests. Bulletin of the Agricultural Research Institute, Pusa 100: 1–246.

Fritz, B. J., D. D. Reisig, C. E. sorenson, A. I. Del Pozo-Valdivia, and T. E. Carter, Jr. 2016. Host plant resistance to *Megacopta cribraria* (Hemiptera: Plataspidae) in diverse soybean germplasm maturity groups V through VIII. Journal of Economic Entomology 109(3): 1438–1449.

Froeschner, R. C. 1984. Does the Old World family Plataspidae (Hemiptera) occur in North America? Entomological News 95(1): 36.

Fukatsu, T., and T. Hosokawa. 2002. Capsule transmitted gut symbiotic bacterium of the Japanese common plataspid stinkbug, *Megacopta punctatissima*. Applied and Environmental Microbiology 68(1): 389–396.

Gardner, W., and D. M. Olson. 2016. Population census of *Megacopta cribraria* (Hemiptera: Plataspidae) in kudzu in Georgia, USA, 2014-2016. Journal of Entomological Science 51(4): 325–328.

Gardner, W. A., H. B. Peeler, J. LaForest, P. M. Roberts, A. N. Sparks, Jr., J. K. Greene, D. Reisig, D. R. Suiter, J, S. Bacheler, K. Kidd, C. H. Ray, X. P. Hu, R. Kemerait, E. A. Scocco, J. E. Eger, Jr., R. R. Ruberson, E. J. Sikora, D. A. Herbert, Jr., C. Campana, S Halbert, S. D. Stewart, G. D. Buntin, M. D. Toews, and C. T. Bargeron. 2013a. Confirmed distribution and occurrence of *Megacopta cribraria* (F.) (Hemiptera: Heteroptera: Plataspidae) in the southeastern United States. Journal of Entomological Science 48(2): 118–127.

Gardner, W. A., J. L. Blount, J. R. Golec, W. A. Jones, X. P. Hu, E. J. Talamas, R. M. Evans, X. Dong, C. H. Ray, Jr., G. D. Buntin, N. M. Gerardo, and J. Couret. 2013b. Discovery of *Paratelenomus sacchara-lis* (Dodd) (Hymenoptera: Platygastridae), an egg parasitoid of *Megacopta cribraria* F. (Hemiptera: Plataspidae) in its expanded North American range. Journal of Entomological Science 48(4): 355–359.

Gardner, W., and D. M. Olson. 2016. Population census of *Megacopta cribraria* (Hemiptera: Plataspidae) in kudzu in Georgia, USA, 2014-2016. Journal of Entomological Science 51(4): 325–328.

Georgia Automated Environmental Monitoring Network. 2015. Historical data. www.GeorgiaWeather.com (accessed 18 April 2015).

Germar, E. F. 1839. Beiträge zu einer Monographie der Schildwanzen. Zeitschrift für die Entomologie 1(1): 1–146, pl. 1.

Golec, J. R., and X. P. Hu. 2015. Preoverwintering copulation and female ratio bias: life history characteristics contributing to the invasiveness and rapid spread of *Megacopta cribraria* (Heteroptera: Plataspidae). Environmental Entomology 44(2): 411–417.

Golec, J. R., X. P. Hu, C. Ray, and N. E. Woodley. 2013. *Strongygaster triangulifera* (Diptera: Tachinidae) as a parasitoid of adults of the invasive *Megacopta cribraria* (Heteroptera: Plataspidae) in Alabama. Journal of Entomological Science 48(4): 352–354.

Golec, J. R., X. P. Hu, L. Yang, and J. E. Eger. 2015. Kudzu-deprived first-generation *Megacopta cribra-ria* (F.) (Heteroptera: Plataspidae) are capable of developing on alternative legume species. Journal of Agricultural and Urban Entomology 31(1): 52–61.

Greene, J. K. 2015. Soybean Insect Control, pp. 248–258. *In* South Carolina pest management handbook. Clemson University, Clemson, SC. 312 pp.

Greene, J. K., P. M. Roberts, W. A. Gardner, F. Reay-Jones, and N. J. Seiter. 2012. Kudzu bug identification and control in soybeans. United Soybean Board Technology Transfer Publication, Clemson, SC. 10 pp.

Greenstone, M. H., P. G. Tillman, and J. S. Hu. 2014. Predation of the newly invasive pest *Megacopta cribraria* (Hemiptera: Plataspidae) in soybean habitats adjacent to cotton by a complex of predators. Journal of Economic Entomology 107(3): 947–954.

Gruwell, M. E., N. B. Hardy, P. J. Gullan, and K. Dittmar. 2010. Evoutionary relationships among primary endosybionts of the mealybug subfamily Phenacoccinae (Hemiptera: Coccoidea: Pseudococcidae). Applied and Environmental Microbiology 76(2): 7521–7525.

Hanula, J., Y. Zhang, and S. Horn. 2012. The bean plataspid, *Megacopta cribraria*, feeding on kudzu: an accidental introduction with beneficial effects, pp. 27–28. *In* K. McManus and K. W. Gottschalk (Eds.), Proceedings 23rd U.S. Department of Agriculture Interagency Research Forum on Invasive Species 2012. U. S. Forest Service, Newtown Square, PA, USA. 126 pp.

Hasegawa, H. 1965. Major pests of economic plants in Japan. Japan Plant Protection Association, Tokyo. 412 pp.

Hibino, Y. 1985. Formation and maintenance of mating aggregations in a stink bug, *Megacopta punctissimum* [sic] (Montandon) (Heteroptera, Plataspidae). Journal of Ethology 3(2): 123–129.

Hibino, Y. 1986. Female choice for male gregariousness in a stink bug, *Megacopta punctissimum* [sic] (Montandon) (Heteroptera, Plataspidae). Journal of Ethology 4(2): 91–95.

Hibino, Y., and Y. Ito. 1983. Mating aggregation of a stink bug, *Megacopta punctissimum* [sic] (Montandon) (Heteroptera: Plataspidae). Researches on Population Ecology 25(1): 180–188.

Himuro, C., T. Hosokawa, and N. Suzuki. 2006. Alternative mating strategy of small male *Megacopta punctatissima* (Hemiptera: Plataspidae) in the presence of large intraspecific males. Annals of the Entomological Society of America 99(5): 974–977.

Hirashima, Y. 1989. A check list of Japanese insects. Faculty of Agriculture, Kyushu University and Center for the Study of Japanese Field Life, Volume I. Kyushu University, Fukuoka, Japan. 540 pp.

Hirose, Y., K. Takasu, and M. Takagi. 1996. Egg parasitoids of phytophagous bugs in soybean: mobile natu-ral enemies as naturally occurring biological control agents of mobile pests. Biological Control 7(1): 84–94.

Hoffmann, W. E. 1931. Notes on Hemiptera and Homoptera at Canton, Kwangtung Province, Southern China 1924-1929. USDA Insect Pest Survey Bulletin 11(3): 138–151.

Hoffmann, W. E. 1932. Notes on the bionomics of some Oriental Pentatomidae (Hemiptera). Archivo Zoologico Italiano 16(3/4): 1010–1027.

Horn, S., and J. L. Hanula. 2011. Influence of trap color on collection of recently-introduced bean plataspid, *Megacopta cribraria* (Hemiptera: Plataspidae). Journal of Entomological Science 46(1): 85–87.

Hosokawa, T., and N. Suzuki. 2000. Mating aggregation and copulatory success by males of the stink bug, *Megacopta punctatissima* (Heteroptera: Plataspidae). Applied Entomology and Zoology 35(1): 93–99.

Hosokawa, T., and N. Suzuki. 2001. Significance of prolonged copulation under the restriction of daily reproductive time in the stink bug *Megacopta punctatissima* (Heteroptera: Plataspidae). Annals of the Entomological Society of America 94(5): 750–754.

Hosokawa, T., Y. Kikuchi, Y. M. Xien, and T. Fukatsu. 2005. The making of symbiont capsule in the plataspid stinkbug *Megacopta punctatissima*. FEMS Microbiology Ecology 54(3): 471–477.

Hosokawa, T., Y. Kikuchi, N. Nikoh, M. Shimada, and T. Fukatsu. 2006. Strict host-symbiont cospeciation and reductive genome evolution in insect gut bacteria. PLoS Biology 4(10): e337.

Hosokawa, T., Y. Kikuchi, and T. Fukatsu. 2007a. How many symbionts are provided by mothers, acquired by offspring, and needed for successful vertical transmission in an obligate insect bacterium mutualism? Molecular Ecology 16(24): 5316–5325.

Hosokawa, T., Y. Kikuchi, M. Shimada, and T. Fukatsu. 2007b. Obligate symbiont involved in pest status of host insect. Proceedings of the Royal Society Biological Sciences Series B 274(1621): 1979–1984.

Hosokawa, T., Y. Kikuchi, M. Shimada, and T. Fukatsu. 2008. Symbiont acquisition alters behaviour of stink-bug nymphs. Biology Letters 4(1): 45–48.

Hosokawa, T., N. Nikoh, and T. Fukatsu. 2014. Fine-scale geographical origin of an insect pest invading North America. PLoS ONE 9(2): e89107, 5 pp.

Hsiao, T. Y., and S. Z. Ren (as S. C. Jen). 1977. Plataspidae. pp. 14–38, 292–295, *In* T. Y. Hsiao, S. Z. Ren (as S. C. Jen), L. Y. Zheng, (as L. I. Cheng), H. L. Jing (as H. L. Ching), and S. L. Liu, 1977. A handbook for the determination of the Chinese Hemiptera-Heteroptera. Volume 1. Science Press, Beijing. 330 pp., 52 plates.

Huang, J., and A. Polaszek. 1996. The species of *Encarsiella* (Hymenoptera: Aphelinidae) from China. Journal of Natural History 30(11): 1649–1659.

Huskisson, S. M., K. L. Fogg, T. L. Upole, and C. B. Zehnder. 2015. Season dynamics and plant preferences of *Megacopta cribraria*, an exotic invasive insect species in the southeast. Southeastern Naturalist 14(1): 57–65.

Imai, K., K. Miura, H. Iida, and K. Fujisaki. 2011. Herbivorous insect fauna of invasive vine, kudzu, *Pueraria montana* (Lour.) Merr. var. *lobata* (Willd.) Maesen et S. Almeida (Leguminosae), in its native range. Japanese Journal of Applied Entomology and Zoology 55(3): 147–154.

Imura, O. 2003. Herbivorous arthropod community of an alien weed *Solanum carolinense* L. Applied Entomology and Zoology 38(3): 293–300.

Ishihara, T. 1950. The developmental stages of some bugs injurious to the kidney bean (Hemiptera). Transactions of the Shikoku Entomological Society 1: 17–31, 2 plates.

Jenkins, T. M., and T. D. Eaton. 2011. Population genetic baseline of the first plataspid stink bug symbiosis (Hemiptera: Heteroptera: Plataspidae) reported in North America. Insects 2(3): 264–272.

Jenkins, T. M., R. E. Dean, and B. T. Forschler. 2002. DNA technology, interstate commerce, and the likely origin of Formosan subterranean termite (Isoptera: Rhinotermitidae) infestation in Atlanta, Georgia. Journal of Economic Entomology 95(2): 381–389.

Jenkins, T. M., S. C. Jones, C. Y. Lee, B. T. Forschler, Z. Chen, G. Lopez-Martinez, N. T. Gallagher, G. Brown, M. Neal, B. Thistleton, and S. Kleinschmidt. 2007. Phylogeography illuminates maternal origins of exotic *Coptotermes gestroi* (Isoptera: Rhinotermitidae). Molecular Phylogenetics and Evolution. 42(3): 612–621.

Jenkins, T. M., S. K. Braman, Z. Chen, T. D. Eaton, G. V. Pettis, and D. W. Boyd. 2009. Insights into flea beetle (Coleoptera: Chrysomelidae: Galerucinae) host specificity from concordant mitochondrial and nuclear DNA phylogenies. Annals of the Entomological Society of America 102(3): 386–395.

Jenkins, T. M., T. D. Eaton, D. R. Suiter, J. E. Eger, Jr., L. M. Ames, and G. D. Buntin. 2010. Preliminary genetic analysis of a recently-discovered invasive true bug (Hemiptera: Heteroptera: Plataspidae) and its bacterial endosymbiont in Georgia, USA. Journal of Entomological Science 45(1): 62–63.

Jiang, W., D. Haver, M. Rust, and J. Gan. 2012. Runoff of pyrethroid insecticides from concrete surfaces following simulated and natural rainfalls. Water Research 46(3): 645–652.

Johnson, N. F. 1991. Revision of the Australasian *Trissolcus* species (Hymenoptera: Scelionidae). Invertebrate Taxonomy 5(1): 211–239.

Johnson, N. F. 1996. Revision of world species of *Paratelenomus* Dodd (Hymenoptera: Scelionidae). The Canadian Entomologist 128(2): 273–291.

Kershaw, J. C. W. 1910. On the metamorphoses of two coptosomine Hemiptera from Macao. (with notes by G. W. Kirkaldy). Annales de la Société Entomologique de Belgique 54: 69–73.

Kikuchi, Y., and T. Fukatsu. 2003. Diversity of *Wolbachia* endosymbionts in heteropteran bugs. Applied and Environmental Microbiology 69(10): 6082–6090.

Kikuchi, A., and H. Kobayashi. 2010. Effect of injury by adult *Megacopta punctatissima* (Montandon) (Hemiptera: Plataspidae) on the growth of soybean during the vegetative stage of growth. Japanese Journal of Applied Entomology and Zoology 54(1): 37–43.

Kitamura, C., S. Wakamura, and S. Takahashi. 1984. Identification and functions of ventral glands secretion of some Heteroptera. Applied Entomology and Zoology 19(1): 33–41.

Kobayashi, T. 1981. Insect pests of soybeans in Japan. Miscellaneous Publications of the Tohoku National Agricultural Experiment Station 2: 1–39.

Kono, S. 1990. Spatial distribution of three species of stink bugs attacking soybean seeds. Japanese Journal of Applied Entomology and Zoology 34(2): 89–96.

Kotikal, Y. K., and K. A. Kulkarni. 2000. Insect pests infesting tumeric in Northern Karnataka. Karnataka Journal of Agricultural Sciences 13(4): 858–866.

Kumar, D., and U. Bharti. 2015. Tropic niche specialization on *Bhutea monosperma* from Chandigarh. Journal of Entomology and Zoology Studies 3(4): 83–85.

Lahiri, S., D. Orr, C. Sorenson, and Y. Cardoza. 2015. Overwintering refuge sites for *Megacopta cribraria* (Hemiptera: Plataspidae). Journal of Entomological Science 50(1): 69–73.

Lal, O. P. 1980. A compendium of insect pests of vegetables in India. Bulletin of the Entomological Society of India 16(1–2) (1975): 52–88.

Lal, S. S. 1985. A review of insect pests of mungbean and their control in India. Tropical Pest Management 31(2): 105–114.

Lee, C. Y., B. T. Forschler, and T. M. Jenkins. 2005. Taxonomic questions on Malaysian termites (Isoptera: Termitidae) answered with morphology and DNA biotechnology, pp. 205–211. *In* C.-Y. Lee and W. H. Robinson (Eds.), Proceedings of the Fifth International Conference on Urban Pests, Singapore. P&Y Design Network, Penang Island, Malaysia. 571 pp.

Leslie, A. W., C. Sargent, W. E. Steiner, Jr., W. O. Lamp, J. M. Swearingen, B. B. Pagac, Jr., G. L. Williams, D. C. Weber, and M. J. Raupp. 2014. A new invasive species in Maryland: the biology and distribution of the kudzu bug, *Megacopta cribraria* (Fabricius) (Hemiptera: Plataspidae). The Maryland Entomologist 6(2): 2–23.

Li, J. F., M. Y. Tian, D. J. Gu, and J. H. Sun. 2004. Composition of arthropods and population dynamics of main insect groups on *Pueraria lobata*. Journal of South China Agricultural University 25(1): 56–58.

Li, Y. H., Z. S. Pan, J. P. Zhang, and W. S. Li. 2001. Observation of biology and behavior of *Megacopta cribraria* (Fabricius). Plant Protection Technology and Extension 21(7): 11–12.

Lindgren, C. J., K. L. Castro, H. A. Coiner, R. E. Nurse, and S. J. Darbyshire. 2013. The biology of invasive alien plants in Canada. 12. *Pueraria montana* var. *lobata* (Willd.) Sanjappa & Predeep. Canadian Journal of Plant Science 93(1): 71–95.

Lovejoy, R. T., and D. A. Johnson. 2014. A molecular analysis of herbivory in adults of the invasive bean plataspid, *Megacopta cribraria*. Southeastern Naturalist 13(4): 663–672.

Mani, M. S., and S. K. Sharma. 1982. Proctotrupoidea (Hymenoptera) from India. A review. Oriental Insects 16(2): 135–258.

McNeely, J. 2001. Invasive species: a costly catastrophe for native biodiversity. Land Use and Water Resources Research 1(2): 1–10.

Medal, J., S. Halbert, and A. Santa Cruz. 2013a. The bean plataspid, *Megacopta cribraria* (Hemiptera: Plataspidae), a new invader in Florida. Florida Entomologist 96(1): 258–260.

Medal, J., S. Halbert, T. Smith, and A. Santa Cruz. 2013b. Suitability of selected plants to the bean plataspid, *Megacopta cribraria* (Hemiptera: Plataspidae) in no-choice tests. Florida Entomologist 96(2): 631–633.

Medal, J., A. Santa Cruz, K. Williams, S. Fraser, D. Wolaver, T. Smith, and J. Davis. 2015. First record of *Paratelenomus saccharalis* (Hymenoptera: Platygastridae) on bean plataspid, *Megacopta cribraria* (Heteroptera: Plataspidae) in Florida. Florida Entomologist 98(4): 1250–1251.

Medal, J., S. Halbert, A. Santa Cruz, T. Smith, and B. J. Davis. 2016a. Greenhouse study to determine the host range of the kudzu bug, *Megacopta cribraria* (Heteroptera: Plataspidae). Florida Entomologist 99(2): 303–305.

Medal, J., G. Lotz, T. Smith, A. Santa-Cruz, E. Rohrig, K. Hoelmer, C. Dieckhoff and K. Tatman. 2016b. Pruebas de especificidad con el parasitoide de huevos *Trissolcus japonicus* (Ashmead) (Hymenoptera: Platygastridae) para el control biológico de la chinche hedionda *Halyomorpha halys* Yang (Heteroptera: Pentatomidae) en los Estados Unidos de América. Entomología Mexicana 3: 191–196.

Miao, J., D. D. Reisig, G. Li, and Y. Wu. 2016. Sublethal effects of insecticide exposure on *Megacopta cribraria* (Fabricius) nymphs: key biological traits and acetylcholinesterase activity. Journal of Insect Science 16(1): 1–6.

Moizuddin, M., and I. Ahmad. 1979. Eggs and nymphal systematics of *Coptosoma cribrarium* (Fabr.) (Pentatomoidea: Plataspidae) with a note on other plataspids and their phylogeny. Records of the Zoological Survey of Pakistan 7(1–2)(1975): 93–100.

Montandon, A. L. 1896. Plataspidinae. Nouvelle série d'études et descriptions. Annales de la Société Entomologique de Belgique 40: 86–134.

Montandon, A. L. 1897. Les Plataspidines du Muséum d'histoire naturelle de Paris. Annales de la Société Entomologique de France 1896: 436–464.

Moran, N. S. 2001. The coevolution of bacterial endosymbionts and phloem-feeding insects. Annals of the Missouri Botanical Garden 88(1): 35–44.

Moran, N. S. 2007. Symbiosis as an adaptive process and source of phenotypic complexity. Proceedings of the National Academy of Sciences of the United States of America. 104(Suppl. 1): 8627–8633.

Musser, F. R., A. L. Catchot, Jr., J. A. Davis, D. A. Herbert, Jr., B. R. Leonard, G. M. Lorenz, T. Reed, D. D. Reisig, and S. D. Stewart. 2012. 2011 soybean insect losses in the southern US. Midsouth Entomologist 6: 11–22.

Musser, F. R., A. L. Catchot, Jr., J. A. Davis, D. A. Herbert, Jr., G. M. Lorenz, T. Reed, D. D. Reisig, and S. D. Stewart. 2013. 2012 soybean insect losses in the southern US. Midsouth Entomologist 6: 12–24.

Musser, F. R., A. L. Catchot, Jr., J. A. Davis, D. A. Herbert, Jr., G. M. Lorenz, T. Reed, D. D. Reisig, and S. D. Stewart. 2014. 2013 soybean insect losses in the southern US. Midsouth Entomologist 7: 12–28.

Musser, F. R., A. L. Catchot, Jr., J. A. Davis, D. A. Herbert, Jr., G. M. Lorenz, T. Reed, D. D. Reisig, and S. D. Stewart. 2015. 2014 soybean insect losses in the southern US. Midsouth Entomologist 8: 35–48.

National Cotton Council of America. 2012. Kudzu bug remediation. 16 November 2012. www.cotton.org /tech/flow/upload/2Protocol-For-Loading-Containers-kudzu-bug.pdf (accessed 18 April 2015).

Nikoh, N., T. Hosokawa, K. Oshima, M. Hattori, and T. Fukatsu. 2011. Reductive evolution of bacterial genome in insect gut environment. Genome Biology and Evolution 3: 702–714.

Polaszek, A., and M. Hayat. 1990. *Dirphys boswelli* (Hymenoptera: Aphelinidae) an egg-parasitoid of Plataspidae (Heteroptera). Journal of Natural History 24(1): 1–5.

Polaszek, A., and M. Hayat. 1992. A revision of the genera *Dirphys* Howard and *Encarsiella* Hayat (Hymenoptera: Aphelinidae). Systematic Entomology 17(2): 181–197.

Poorani, J., A. Slipinski, and R. G. Booth. 2008. A revision of the genus *Synona* Pope, 1989 (Coleoptera: Coccinellidae: Coccinellini). Annales Zoologici 58(3): 579–594.

Portilla, M., W. Jones, O. Perera, N. Seiter, J. Greene, and R. Luttrell. 2016. Estimation of median lethal concentration of three isolates of *Beauveria bassiana* for control of *Megacopta cribraria* (Heteroptera: Plataspidac) bioassayed on solid *Lygus* spp. diet. Insects 7(3), 31, 13 pp.

Prentis, P. J., J. R. U. Wilson, E. E. Dormontt, D. M. A. Richardson, and A. J. Lowe. 2008. Adaptive evolution in invasive species. Trends in Plant Science 13(6): 288–294.

Rajmohan, K. and T. C. Narendran. 2001. Parasitoid complex of *Coptosoma cribrarium* (Fabricius) (Plataspididae: Hemiptera). Insect Environment 6(4): 163.

Ramakrishna Aiyar, T. V. 1913. On the life history of *Coptosoma cribraria* Fabr. Journal of the Bombay Natural History Society 22: 412–414.

Ramakrishna Ayyar, T. V. 1920. Some insects recently noted as injurious in South India. Report of the Indian Agricultural Research Institute, Pusa 31: 314–327.

Rani, B. J., and V. Sridhar. 2004. Record of arthropod pests on velvet bean, *Mucuna pruriens* var. *utilis* under Bangalore conditions. Journal of Medicinal and Aromatic Plant Sciences 26(3): 505–506.

Rédei, D. 2016. The identity of the *Brachyplatys* species recently introduced to Panama, with a review of bionomics (Hemiptera: Heteroptera: Plataspidae). Zootaxa 4136(1): 141–154.

Rekha, S., and C. P. Mallapur. 2007. Abundance and seasonability of sucking pests of dolichos bean. Karnataka Journal of Agricultural Sciences 20(2): 397–398.

Ren, S. 1984. Studies on the fine structure of egg-shells and the biology of *Megacopta* Hsiao et Jen from China (Hemiptera: Plataspidae). Entomotaxonomia 6(4): 327–332.

Ren, S.-Z. 1992. An Iconography of Hemiptera-Heteroptera Eggs in China. Science Press, Ltd., Beijing. 118 pp. + 80 plates.

Roberts, P. 2010. Bean plataspid, *Megacopta cribraria*, infesting soybeans in NE Georgia. University of Georgia, College of Agriculture and Environmental Sciences, Agent Update, July 14, 2010. 4 pp.

Roberts, P., M. Toews, and D. Buntin. 2014. Insect management, pp. 68–80. *In* J. Whitaker (Ed.), 2014 Georgia soybean production guide. University of Georgia, College of Agricultural and Environmental Sciences, Athens, GA. 99 pp. http://www.caes.uga.edu/commodities/fieldcrops/soybeans/prod_guide.html.

Roberts, P., M. Toews, and D. Buntin. 2015. Soybean Insect Control, pp. 559–566. *In* J. Whitaker (Ed.), 2014 Georgia pest management handbook, commercial edition. The University of Georgia Cooperative Extension Special Bulletin 28. University of Georgia, College of Agricultural and Environmental Sciences, Athens, GA. 935 pp. http://www.ent.uga.edu/pest-management.

Ruberson, J. R., K. Takasu, G. D. Buntin, J. E. Eger, Jr., W. A. Gardner, J. K. Greene, T. M. Jenkins, W. A. Jones, D. M. Olson, P. M. Roberts, D. R. Suiter, and M. D. Toews. 2013. From Asian curiosity to eruptive American pest: *Megacopta cribraria* (Hemiptera: Plataspidae) and prospects for its biological control. Applied Entomology and Zoology 48(1): 3–13.

Sakai, A. K., F. W. Allendorf, J. S. Holt, D. M. Lodge, J. Molofsky, K. A. With, S. Baughman, R. J. Cabin, J. E. Cohen, N. C. Ellstrand, D. E. McCauley, P. O'Neil, I. M. Parker, J. N. Thompson, and S. G. Weller. 2001. The population biology of invasive species. Annual Review of Ecological Systems 32: 305–332.

Schmidt, S., and A. Polaszek. 2007. The Australian species of *Encarsia* Förster (Hymenoptera, Chalcidoidea: Aphelinidae), parasitoids of whiteflies (Hemiptera, Sternorrhyncha, Aleyrodidae) and armoured scale insects (Hemiptera, Coccoidea: Diaspididae). Journal of Natural History 41(33–36): 2099–2265.

Seiter, N. J., J. K. Greene, and F. P. F. Reay-Jones. 2013a. Reduction of soybean yield components by *Megacopta cribraria* (Hemiptera: Plataspidae). Journal of Economic Entomology 106(4): 1676–1683.

Seiter, N. J., E. P. Benson, F. P. F. Reay-Jones, J. K. Greene, and P. A. Zungoli. 2013b. Residual efficacy of insecticides applied to exterior building material surfaces for control of nuisance infestations of *Megacopta cribraria* (Hemiptera: Plataspidae). Journal of Economic Entomology 106(6): 2448–2456.

Seiter, N. J., F. P. F. Reay-Jones, and J. K. Greene. 2013c. Within-field spatial distribution of *Megacopta cribraria* (Hemiptera: Plataspidae) in soybean (Fabales: Fabaceae). Environmental Entomology 42(6): 1363–1374.

Seiter, N. J., A. Grabke, J. K. Greene, J. L. Kerrigan, and F. P. F. Reay-Jones. 2014a. *Beauveria bassiana* is a pathogen of *Megacopta cribraria* (Hemiptera: Plataspidae) in South Carolina. Journal of Entomological Science 49(3): 326–330.

Seiter, N. J., J. K. Greene, and F. P. F. Reay-Jones. 2014b. Aggregation and oviposition preferences of *Megacopta cribraria* (Hemiptera: Plataspidae) in laboratory assays. Journal of Entomological Science 49(3): 331–335.

Seiter, N. J., A. I. Del Pozo-Valdiva, J. K. Greene, F. P. F Reay-Jones, P. M. Roberts, and D. R. Reisig. 2015a. Action thresholds for managing *Megacopta cribraria* (Hemiptera: Plataspidae) in soybean based on sweep-net sampling. Journal of Economic Entomology 108(4): 1818–1829.

Seiter, N. J., J. K. Greene, F. P. F. Reay-Jones, P. M. Roberts, and J. N. All. 2015b. Insecticidal control of *Megacopta cribraria* (Hemiptera: Plataspidae) in soybean. Journal of Entomological Science 50(4): 263–283.

Seiter, N. J., A. I. Del Pozo-Valdivia, J. K. Greene, F. P. F. Reay-Jones, P. M. Roberts, and D. D. Reisig. 2016. Management of *Megacopta cribraria* (Hemiptera: Plataspidae) at different stages of soybean (Fabales: Fabaceae) development. Journal of Economic Entomology 109(3): 1167–1176.

Shi, S.-S., J. Cui, and L.-S. Zang. 2014. Development, survival, and reproduction of *Megacopta cribraria* (Heteroptera: Plataspidae) at different constant temperatures. Journal of Economic Entomology 107(6): 2061–2066.

Shroff, K. D. 1920. A list of the pests of pulses in Burma, pp. 343–346. *In* T. B. Fletcher (Ed.), Proceedings of the Third Entomological Meeting, Pusa 1919, Volume 1. Superintendent of Government Printing, Calcutta, India. xi + 417 pp.

Srinivasaperumal, S., P. Samuthiravelu, and J. Muthukrishnan. 1992. Host plant preference and life table of *Megacopta cribraria* (Fab.) (Hemiptera: Plataspidae). Proceedings of the Indian National Sciences Academy, Part B (Biological Sciences) 58(6): 333–340.

Strauss, S. Y., C. O. Webb, and N. Salamin. 2006. Exotic taxa less related to native species are more invasive. PANS 103(15): 5841–5845.

Stubbins, F. L., N. J. Seiter, J. K. Greene, and F. P. F. Reay-Jones. 2014. Developing sampling plans for the invasive *Megacopta cribraria* in soybean. Journal of Economic Entomology 107(6): 2213–2221.

Stubbins, F. L., P. Agudelo, F. P. F. Reay-Jones, and J. K. Greene. 2015. First report of a mermithid nematode infecting the invasive *Megacopta cribraria* (Hemiptera: Plataspidae) in the United States. Journal of Invertebrate Pathology 127: 35–37.

Stubbins, F. L., J. K. Greene, M. D. Toews, and F. P. F. Reay-Jones. 2016a. Assessment of a cross-vane trap as a tool for sampling the invasive *Megacopta cribraria* (Hemiptera: Plataspidae) in soybean with associated evaluations of female reproductive status. Environmental Entomology 45(5): 1262–1270.

Stubbins, F. L., P. Agudelo, F. P. F. Reay-Jones, and J. K. Greene. 2016b. *Agamermis* (Nematoda: Mermithidae) infection in South Carolina agricultural pests. Journal of Nematology 48(4): 290–296.

Suiter, D. R., J. E. Eger, Jr., W. A. Gardner, R. C. Kemerait, J. N. All, P. M. Roberts, J. K. Greene, L. M. Ames, G. D. Buntin, T. M. Jenkins, and G. K. Douce. 2010. Discovery and distribution of *Megacopta cribraria* (Hemiptera: Heteroptera: Plataspidae) in northeast Georgia. Journal of Integrated Pest Management 1(1): F1–F4.

Sujithra, M., S. Srinivasan, and K. V. Hariprasad. 2008. Outbreak of lablab bug, (*Coptosoma cribraria* Fab.) on field bean, *Lablab purpureus* var. *lignosus* Medikus. Insect Environment 14(2): 77–78.

Sun, J.-H., Z.-D. Liu, K. O. Britton, P. Cai, D. Orr, and J. Hough-Goldstein. 2006. Survey of phytophagous insects and foliar pathogens in China for a biocontrol perspective on kudzu, *Pueraria montana* var. *lobata* (Willd.) Maesen and S. Almeida (Fabaceae). Biological Control 36(1): 22–31.

Sung, I-H., M.-Y. Lin, C.-H. Chang, A.-S. Cheng, and W.-S. Chen. 2006. Pollinators and their behaviors on mango flowers in southern Taiwan. Formosan Entomologist 26: 161–170.

Takagi, M., and K. Murakami. 1997. Effect of temperature on development of *Paratelenomus saccharalis* (Hymenoptera: Scelionidae), an egg parasitoid of *Megacopta punctatissimum* (Hemiptera: Plataspidae). Applied Entomology and Zoology 32(4): 659–660.

Takano, S.-i., and K. Takasu. 2016. Ability of *Megacopta cribraria* (Hemiptera: Plataspidae) to cling to different surfaces against extreme wind. Journal of Insect Behavior 29(3): 1–6.

Takasu, K., and Y. Hirose. 1986. Kudzu-vine community as a breeding site of *Ooencyrtus nezarae* Ishi (Hymenoptera: Encyrticae), an egg parasitoid of bugs attacking soybean. Japanese Journal of Applied Entomology and Zoology 30(4): 302–304.

Takasu, K., and Y. Hirose. 1991a. Host searching behavior in the parasitoid *Ooencyrtus nezarae* Ishi (Hymenoptera: Encyrtidae) as influenced by non-host food deprivation. Applied Entomology and Zoology 26(3): 415–417.

Takasu, K., and Y. Hirose. 1991b. The parasitoid *Ooencyrtus nezarae* (Hymenoptera: Encyrtidae) prefers hosts parasitized by conspecifics over unparasitized hosts. Oecologia 87(3): 319–323.

Takasu, K., S. I. Takano, N. Mizutani, and T. Wada. 2004. Flight orientation behavior of *Ooencyrtus nezarae* (Hymenoptera: Encyrtidae), an egg parasitoid of phytophagous bugs in soybean. Entomological Science, 7(3): 201–206.

Tayutivutikul, J., and K. Kusigemati. 1992a. Biological studies of insects feeding on the kudzu plant, *Pueraria lobata* (Leguminosae) I. List of feeding species. Memoirs of the Faculty of Agriculture, Kagoshima University 28(37): 89–124.

Tayutivutikul, J., and K. Kusigemati. 1992b. Biological studies of insects feeding on the kudzu plant, *Pueraria lobata* (Leguminosae). II. Seasonal abundance, habitat and development. South Pacific Study 13(1): 37–88, 1 pl.

Tayutivutikul, J., and K. Yano. 1990. Biology of insects associated with the kudzu plant, *Pueraria lobata* (Leguminosae). 2. *Megacopta punctissimum* (Hemiptera, Plataspidae). Japanese Journal of Entomology 58(3): 533–539.

Thejaswi, L., M. I. Naik, and M. Manjunatha. 2008. Studies on population dynamics of pest complex of field bean (*Lablab purpureus* L.) and natural enemies of pod borers. Karnataka Journal of Agricultural Science 21(3): 399–402.

Thippeswamy, C., and B. K. Rajagopal. 1998a. Incidence of heteropteran bugs on field bean (*Lablab purpureus* var. *liqnosus* Medikus) in Karnataka. Karnataka Journal of Agricultural Sciences 11(4): 1085–1087.

Thippeswamy, C., and B. K. Rajagopal. 1998b. Assessment of losses caused by the lablab bug, *Coptosoma cribraria* (Fabricius) to the field bean, *Lablab purpureus* var. *lignosus* Medikus. Karnataka Journal of Agricultural Sciences 11(4): 941–946.

Thippeswamy, C., and B. K. Rajagopal. 2005a. Comparative biology of *Coptosoma cribraria* Fabricius on field bean, soybean and redgram. Karnataka Journal of Agricultural Sciences 18(1): 138–140.

Thippeswamy, C., and B. K. Rajagopal. 2005b. Life history of lablab bug, *Coptosoma cribraria* Fabricius (Heteroptera: Plataspidae) on field bean, *Lablab purpureus* var. *Lignosus medikus*. Karnataka Journal of Agricultural Sciences 18(1): 39–43.

Tillman, G., J. Gaskin, D. Endale, C. Johnson, and H. Schomberg. 2016. Parasitism of *Megacopta cribraria* (Hemiptera: Plataspidae) by *Paratelenomus saccharalis* (Hymenoptera: Platygastridae) in organic soybean plots in Georgia, USA. Florida Entomologist 99(2): 300–302.

United States Environmental Protection Agency. 2013. Environmental hazard and general labeling for pyrethroid and synergized pyrethrins non-agricultural outdoor products. February 2013. http://www.epa .gov/oppsrrd1/reevaluation/environmental-hazard-statment.html.

United States Environmental Protection Agency. 2015. Protecting bees and other pollinators from pesticides. April 2015. http://www2.epa.gov/pollinator-protection.

United States Trade Representative. 2014. U.S.-Western Hemisphere trade facts. https://ustr.gov/countries -regions/americas (accessed 18 April 2015).

Walker, F. 1867. Catalogue of the Specimens of Hemiptera: Heteroptera in the Collection of the British Museum. British Museum, London, Part 1. pp. 1–240.

Wall, R. E. 1928. A comparative study of a chalcid egg parasite in three species of Plataspidinae. Lingnan Science Journal 6(3): 231–239.

Wall, R. E. 1931. *Dissolcus tetartus* Crawford, a scelionid egg parasite of Plataspidinae in China. Lingnan Science Journal 9(4): 381–382.

Wang, Z.-X., H.-D. Wang, G.-H. Chen Z.-G. Zi and C.-W. Tong. 1996. Occurrence and control of *Megacopta cribraria* (Fabricius) on soybean. Plant Protection 22(3): 7–9.

Watanabe, C. 1954. Discovery of four new species of Telenominae, egg-parasites of pentatomid and plataspid bugs, in Shikoku, Japan (Hymenoptera: Proctotrupoidea). Transactions of the Shikoku Entomological Society 4: 17–22.

Westwood, J. O. 1837, 1842. *In* F. W. Hope, A catalogue of Hemiptera in the collection of the Rev. F. W. Hope, M. A. with short Latin diagnoses of the new species. J. Bridgewater, London. Part 1: 1–46; 1842, Part 2: 1–26 (stating that descriptions are by J. O. Westwood).

Wu, M. X., Z. Q. Wu, and S. M. Hua. 2006. A preliminary study on some biological characters of globular stink bug, *Megacopta cribraria* and its two egg parasitoids. Journal of Fujian Agriculture and Forestry University Natural Science Edition 35(2): 147–150.

Wu, Y. Q., X. M. Zhang, A. Q. Hua, and J. F. Chen. 1992. Observation of the biological characteristics and control of *Megacopta cribraria* (Fabricius). Insect Knowledge 29: 272–274.

Xing, G. N., T. J. Zhao, and J. Y. Gai. 2006. Evaluation of soybean germplasm in resistance to globular stink bug. Acta Agronomica Sinica 32(4): 491–496.

Xing, G.-N., B. Zhou, T.-J. Zhao, D.-Y. Yu, H. Xing, S.-Y. Chen, and J.-Y. Gai. 2008. Mapping QTLs of resistance to *Megacopta cribraria* (Fabricius) in soybean. Acta Agronomica Sinica 34(3): 361–368.

Yamagishi, K. 1990. Notes on *Archiphanurus minor* (Watanabe) (Hymenoptera, Scelionidae). Esakia 1990(1): 193–196.

Yamazaki, K., and S. Sugiura. 2005. Hemiptera as cecidophages. Entomological News 116(3): 121–126.

Yang, W.-I. 1934. Revision of Chinese Plataspidae. Bulletin of the Fan Memorial Institute of Biology 5(3): 137–236.

Zhang, C. S., and D. P. Yu. 2005. Occurrence and control of *Megacopta cribraria* (Fabricius). China Countryside Well-off Technology 1: 35.

Zhang, S.-M. 1985. Economic insect fauna of China. Fasc. 31 Hemiptera (1). Science Press, Beijing, China x + 242 pp.

Zhang, Y. T., X. G. Du, M. Dong, and W. Shao. 2003. A preliminary investigation of egg parasitoids of *Megacopta cribraria* in soybean fields. Entomological Knowledge 40(5): 443–445.

Zhang, Y.-Z., J. L. Hanula, and J.-H. Sun. 2008. Survey for potential insect biological control agents of *Ligustrum sinense* (Scrophulariales: Oleaceae) in China. Florida Entomologist 91(3): 372–382.

Zhang, Y.-Z., J. L. Hanula, and S. Horn. 2012. The biology and preliminary host range of *Megacopta cribraria* (Heteroptera: Plataspidae) and its impact on kudzu growth. Environmental Entomology 41(1): 40–50.

Zheng, H., Y. Wu, J. Ding, D. Binion, W. Fu, and R. Reardon. 2005. Invasive plants established in the United States that are found in Asia and their associated natural enemies. Volume 2. Forest Health Technology Enterprise Team 2004–05, 2nd. Edit., Chinese Academy of Agricultural Sciences, Beijing, China and USDA Forest Service, Morgantown, WV, USA. viii + 175 pp.

Zheng, H., Y. Wu, J. Ding, D. Binion, W. Fu, and R. Reardon. 2006. Invasive plants of Asian origin established in the United States and their natural enemies. Volume 1. Forest Health Technology Enterprise Team 2004–05, 2nd. Edit., Chinese Academy of Agricultural Sciences, Beijing, China and USDA Forest Service, Morgantown, WV, USA. vii + 147 pp.

Zhixing, W., W. Haudi, C. Guihua, Z. Zi, and T. Caiwen. 1996. Occurrence and control of *Megacopta cribraria* (Fabricius) on soybean. (English Abstract). Plant Protection 22: 7–9.

Zhu, G., M. J. Petersen, and W. Bu. 2012. Selecting biological meaningful environmental dimensions of low discrepancy among ranges to predict potential distribution of bean plataspid invasion. PLoS One 7(9): e46247, 9 pp.

Zink, R M., and G. F. Barrowclough. 2008. Mitochondrial DNA under siege in avian phylogeography. Molecular Ecology 17(9): 2107–2121.

6

Murgantia histrionica (Hahn)

J. E. McPherson, C. Scott Bundy, and Thomas P. Kuhar

Murgantia histrionica (Hahn)[1]

1834	*Strachia histrionica* Hahn, Wanz. Ins. 2: 116, pl. 65, fig. 196 (Mexico).
1835	*Cimex histrionicus*: Burmeister, Handb. Ent., 2: 368.
1853	*Eurydema histrionica*: Herrich-Schaeffer, Wanz. Ins. 9: 93.
1862	*Murgantia histrionica*: Stål, Stett. Ent. Zeit. 23(1–3): 106.
1868	*Strachia histrionicha* [sic]: Glover, U. S. Dept. Agr. Rept., p. 71.
1872	*Murgantia histrionica*: Stål, K. Svens., Vet.-Akad. Handl., 10(4): 37.
1903	*Murgantia histrionica* form *nigricans* Cockerell, Bull. S. Cal. Acad. Sci., 2: 85. (Synonymized by Kirkaldy, 1909, Cat. Hem., 1: 106).

CONTENTS

6.1 Introduction 334
6.2 Host Plants 334
6.3 Life History 335
6.4 Damage 337
6.5 Control 338
    6.5.1 History 338
        6.5.1.1 B. J. Walsh and C. V. Riley 338
        6.5.1.2 J. B. Smith 339
        6.5.1.3 F. H. Chittenden 339
        6.5.1.4 W. A. Thomas 339
            6.5.1.4.1 Remedial Measures 340
            6.5.1.4.2 Trap Crops 340
            6.5.1.4.3 Burning 340
        6.5.1.5 F. B. Paddock 340
            6.5.1.5.1 Artificial Control 340
            6.5.1.5.2 Remedial Measures 340
            6.5.1.5.3 Natural Control 341
        6.5.1.6 B. B. Fulton 341
            6.5.1.6.1 Soap Solutions as an Insecticide 341
        6.5.1.7 J. G. Walker and L. D. Anderson 341
            6.5.1.7.1 Contact Insecticides 341
            6.5.1.7.2 Natural Control 341
        6.5.1.8 W. H. White and L. W. Brannon 342
            6.5.1.8.1 Cultural Methods 342

---

[1] Synonymy adapted from David A. Rider (personal communication).

	6.5.1.8.2	Contact Insecticides	342
	6.5.1.8.3	Natural Enemies	342
6.5.2	Age of Synthetic Insecticides		342
	6.5.2.1	Chlorinated Hydrocarbons	342
	6.5.2.2	Organophosphates and Carbamates	343
	6.5.2.3	Pyrethroids	343
	6.5.2.4	Neonicotinoids	343
	6.5.2.5	Other Insecticides	343
6.6	Current Management Techniques		344
6.7	Why Has *Murgantia histrionica* Been So Successful?		345
6.8	What Does the Future Hold?		345
6.9	Acknowledgments		345
6.10	References Cited		345

6.1 Introduction

Murgantia histrionica (Hahn), the harlequin bug, is an important pest of crucifers. This attractive red to yellow and black species (see **Figure 6.1D**) is a native of Central America and Mexico and, apparently, the oldest invasive stink bug in the continental United States. It first was reported for the United States by Walsh (1866) from specimens collected in Washington Co., Texas, in 1864, a record accepted by several subsequent authors (e.g., Chittenden 1908, 1920; Paddock 1915, 1918; White and Brannon 1939). However, Paddock (1918) noted that the 1864 record was as a pest and that it probably "crossed the border several years previous to 1864." It, then, spread up the Mississippi River Valley and eastward through the remaining Gulf States and up the Atlantic Coast. It reached Missouri and Kansas by 1870 (Riley 1872, 1884), North Carolina by 1867 (Riley 1870; Chittenden 1908, 1920), Tennessee by 1870 (Chittenden 1908, 1920), and Delaware by 1876 (Riley 1884; Chittenden 1908, 1920). Subsequently, it was reported from Virginia and Maryland in 1880, Indiana in 1890, Ohio in 1891 (Chittenden 1908), New Jersey in 1892, and New York in 1894 (Chittenden 1908, 1920). Westward, it was recorded during the same period from Colorado, Arizona, Nevada, and California by Uhler (1876). Today, it ranges in the continental United States from New Hampshire and New York south to Florida and west to Minnesota, South Dakota, Nebraska, and California (Froeschner 1988) and, recently, has been reported from North Dakota (Rider 2012). Thus, in about 50 years, the bug essentially had reached its current distribution. Although it may appear to inhabit much of the southern and northern states, it primarily is a southern species (e.g., Osborn 1894, Hodson and Cook 1960), occurring south of latitude 40° N. (Hodson and Cook 1960). Hodson and Cook (1960) felt that its appearance in Minnesota was due to unusual wind currents; this may also explain its discovery in other northern states (e.g., North Dakota, South Dakota). It now has been introduced into Hawaii (Froeschner 1988).

The migration of *Murgantia histrionica* was monitored closely during the late 1800s and early 1900s because of its proven pest status (e.g., Walsh 1866; Uhler 1876; Smith 1897; Chittenden 1908, 1920; Paddock 1915, 1918; Stoner 1920; Thomas 1915; White and Brannon 1939). As a result, much biological information was published during this time including host plants, life history, descriptions of the immature stages, damage, and control.

6.2 Host Plants

Murgantia histrionica feeds on a wide variety of plants but prefers crucifers (e.g., Brussels sprouts, cabbage, collard, mustard, turnip, rutabaga, radish, bitter cress, broccoli, cauliflower, kale, kohlrabi, peppergrass, shepherd's purse). It also attacks many noncruciferous plants including corn, bean, cotton, potato and others (see McPherson 1982, McPherson and McPherson 2000). In fact, Radcliffe et al. (1991) reported that it can be a minor pest of potatoes.

FIGURE 6.1   **(See color insert.)** *Murgantia histrionica* immatures, adults, and damage. A, egg cluster; B, second instars; C, fifth instars on cabbage; D, mating pair; E, damage on collards, Painter, Virginia. (Images A, C, and D: courtesy of Thomas P. Kuhar; B: courtesy of Sam E. Droege, USGS, Beltsville, Maryland; and E: courtesy of Anthony S. DiMeglio, Virginia Tech, Blacksburg).

6.3 Life History

The life cycle of *Murgantia histrionica* has been studied by several investigators (e.g., Riley 1872, 1884; Howard 1895; Sanderson 1903; Chittenden 1908, 1920; Smith 1909; Paddock 1915, 1918; Brett and Sullivan 1974). It is multivoltine, with estimates of the number of complete generations ranging from three to eight in the South (Howard 1895; Paddock 1915, 1918; Chittenden 1908, 1920) and two to five in the North (Howard 1895; Sanderson 1903; Chittenden 1908, 1920) with adults overwintering, usually under field litter and debris. However, the bugs can become active during the winter if temperatures are mild (Thomas 1915; Paddock 1915, 1918; White and Brannon 1939) with feeding, copulation, and oviposition possible (Brett and Sullivan 1974). Thus, it is possible to find all stages, including eggs, throughout the winter if temperatures are mild and hardy host plants are present. This activity, combined with overlapping generations during the year, undoubtedly is responsible for the confusion about the number of generations per year. However, it probably is bivoltine in the North and bi- or trivoltine in the South with an additional generation possible under favorable conditions. Recently, relatively speaking, Ludwig and Kok (1998a) reported that this bug is bivoltine with a partial third generation per year in southwestern Virginia, supporting an earlier report by White and Brannon (1939). Mild winter temperatures may play a major role in larger than normal populations the following season (Walker and Anderson 1933, Wallingford et al. 2011). Conversely, exposure to extreme cold temperatures (i.e., lower than 15 °C) can result in very high mortality of bug populations (DiMeglio et al. 2016).

Overwintered individuals emerge during spring, begin feeding and mating shortly thereafter. McClain (1981) found that adults preferred to colonize and oviposit within larger clusters of the host plant, at least with turnips, *Brassica napus* L., and felt that this preference enhanced nymphal survival.

English-Loeb and Collier (1987) examined nonmigratory movement of this bug on *Isomeris arborea* Nuttall in southern California. They examined the importance of age and sex of the bugs and abundance of the capsules and racemes of the host plant to movement of the adults within stands of the plants. They found that males left the release bushes more quickly than females but, subsequently, changed locations at a slower rate and less frequently than females. The authors found no sexual difference in distance moved between recaptures. Also, there was no age effect on time spent on the release bush, distance moved, or frequency of movement. Finally, females spent more time on the release bushes with more capsules and racemes.

This bug has been reared under confined conditions, and the eggs and various instars have been illustrated, photographed, and/or described (McPherson 1982) (see **Figure 6.1A-C**). Paddock (1918) reported that this insect has six instars rather than five as other authors have reported, but he undoubtedly was in error.

Precopulatory and copulatory behavior have been studied in this insect by Lanigan and Barrows (1977). The male approaches the female from the front or rear. If from the rear, he antennates the posterior part of her abdomen before moving to her head. In either the rear or front approach, he eventually begins to antennate her antennae. Subsequently, he moves posteriorly, antennating her side and then the posterior of her abdomen. If she is receptive, she will raise the tip of her abdomen approximately 30° and the male, with aedeagus extended, will rotate 180°, back into her, and insert his aedeagus. If a female is unreceptive during precopulation, she simply will walk away.

Zahn et al. (2008a) provided recent information on the rearing of this bug under controlled conditions and its reproductive behavior. Total developmental time from egg to adult at 26°C and 45% relative humidity was ≈48 days followed by a maturation period of ≈7 days. Females produced several egg masses of 12 eggs in two rows of six, thus agreeing with the reports of many earlier investigators (e.g., Smith 1897; Chittenden 1908, 1920; Smith 1909; White and Brannon 1939; Streams and Pimentel 1963; Brett and Sullivan 1974). They lived about ≈ 41 days, laying egg masses every 3 days. Males lived ≈ 25 days. Detailed information on interactions between virgin and previously mated male and females during courtship was provided.

Helmey-Hartman and Miller (2014) studied mate selection in relation to host availability and phenology. They found that both the natal host plant and the host plant where potential mates were encountered significantly affected mating success. Males and females reared on broccoli were more likely to mate than those reared on mustard, no matter the natal rearing environment or the host plant on which the opposite sex was found. Also, they found females preferred the odors of males that were the same as those the females had encountered during nymphal development. Communication is also aided by use of vibratory signals transmitted through the host plant (Cokl et al. 2004, 2007).

As with other stink bugs, the chemical ecology of *Murgantia histrionica* has been well studied (see Aldrich 1988, Aldrich et al. 1996). Zahn et al. (2008b) reported that mature males apparently produce an aggregation pheromone that is attractive to both males and females. More recently, Weber et al. (2014) showed that the two-component pheromone called murgantiol, (3S,6S,7R,10S)- and

(3S,6S,7R,10R)-10,11-epoxy-1-bisabolen-3-24 ol, is most attractive in the field to adults and nymphs in the naturally-occurring ratio of 1.4:1. Each of the two individual synthetic stereoisomers is highly attractive to male and female adults and nymphs, but they are more attractive in combination and when deployed with a host plant of *M. histrionica*.

Murgantia histrionica produces several warning secretions emitted from either the metathoracic or prothoracic gland when disturbed (Aldrich 1988, Aldrich et al. 1996). It also sequesters glucosinolates from its host plants and uses them in defense (Aliabadi et al. 2002). These black and orange bugs also sequester aglucones of glucosinolates (Aldrich et al. 1996); hence, the aposematic coloration.

This bug is attacked by several insect species. Eggs are attacked by hymenopteran parasitoids, primarily species of *Trissolcus*; nymphs by the assassin bug, *Arilus cristatus* (L.), and the nyssonid wasp *Bicyrtes quadrifasciata* (Say); and adults by the leaf-footed bug *Leptoglossus phyllopus* (L.) (unlikely record). It also is parasitized by the sarcophagid fly *Sarcodexia sternodontis* Townsend and the tachinid fly *Trichopoda pennipes* (F.) and preyed upon by the fire ant, *Solenopsis geminata* (F.) (McPherson 1982).

6.4 Damage

Adults and nymphs feed on the leaves and stems of the plants. If the damage is severe, the plants will wilt and turn brown. Small plants eventually will die; larger plants can survive but their growth may be stunted (White and Brannon 1939). Entire fields, if untreated, can be destroyed (Paddock 1915, 1918; Ludwig and Kok 2001). Although feeding may not kill the plants, it can leave white blotches at the feeding sites, making such plants as collard and kale unmarketable (Wallingford et al. 2011) (**Figures 6.1E and 6.2**).

Historically, prior to the introduction of broad-spectrum synthetic organic insecticides (e.g., dichloro-diphenyltrichloroethane [DDT]), the numbers of these bugs attacking crops seem almost unbelievable. A case in point is the report by Walker and Anderson (1933) of a severe outbreak that occurred in the summer of 1932 near Norfolk, VA. The outbreak was due to a mild winter during 1931–1932 resulting in a large population in the spring. This population reproduced on cruciferous crops in fields that had been abandoned or left standing for kale seeds. The feeding of large numbers of the resulting nymphs and the dry weather in August resulted in the nymphs migrating to new plants, "seriously damaging early fall cabbage and completely killing fields of young kale." They also fed on noncruciferous plants including corn, tomatoes, soybeans, and others. "Over 5,100 nymphs were collection on one corn plant and fully as many more fell to the ground and were not counted. This corn plant was at least a quarter of a mile from

FIGURE 6.2 Feeding injury of *Murgantia histrionica* on collards, Blacksburg, VA. (Courtesy of Anthony S. DiMeglio, Virginia Tech, Blacksburg).

the source of the infestation." "Later, a severe outbreak of the adults occurred. The host plants in some of the abandoned or neglected fields remained green longer than in others. In these fields, large numbers of the nymphs matured before their host plants died. Millions of these adults then flew into surrounding fields of kale, collards, cabbage, Chinese cabbage, broccoli, rutabagas, turnips, mustard, corn, tomatoes, and soybeans, and others collected on trees, shrubs, and grasses, and weeds of various kinds. Many of the jimson weeds, pigweeds, and ragweeds were so heavily loaded with these insects in the infested fields that the plants bent over with their weight, the under sides of the leaves often being completely covered with them. One man collected over five gallons of these bugs from tall weeds and soybeans in a kale field in less than five hours. The adults were flying over several young kale fields examined on different warm days in such great numbers that they appeared like a swarm of bees in flight. These adults completely destroyed a large number of fields of kale, collards, and cabbage, and seriously injured many others. One 40 acre field of young, vigorously growing kale was entirely killed out within three weeks after the adults began flying into it from an abandoned field. These insects were observed flying across a small lake in the face of a strong wind and landing in a field of collards on the other side, completely destroying it in a few days."

As devastating as the attacks by these bugs could be to cruciferous crops, Riley (1870) felt that even these insects had some positive qualities: "It is said that no criminal among the human race is so vile and depraved, that not one single redeeming feature can be discovered in his character. It is just so with this insect. Unlike the great majority of the extensive group (*Scutellera* Family, Order of Half-winged Bugs) to which it belongs, it has no unsavory bedbuggy smell, but on the contrary exhales a faint odor which is rather pleasant than otherwise. We have already referred to the beauty of its coloring. As offsets, therefore, to its greediness and its thievery, we have, first the fact of its being agreeable to the nose, and secondly the fact of its being agreeable to the eye. Are there not certain demons in the garb of angels, occasionally to be met with among the human species, in favor of whom no stronger arguments than the above can possibly be urged?"

6.5 Control

6.5.1 History

The pest status of this bug has changed dramatically over the years. In the late 1800s and early 1900s, it was considered a major pest because the controls available at that time were not sufficient. Control methods were discussed in several outstanding contributions, but only a few will be discussed here. But, note how similar they are to each other, several of which are still used today.[2]

6.5.1.1 B. J. Walsh and C. V. Riley

The earliest control recommendation for this bug was handpicking by Walsh (1866) who stated, "I have as yet, found no way to get clear of them, but to pick them off by hand." However, by 1884, Riley listed several additional methods, all cultural, including "hot water," and "trapping the bugs under turnip or cabbage leaves laid on the ground, between the rows." He also recommended "clean cultivation and burning of weeds and rubbish piles in winter." "We may also insist upon the point mentioned by Mr. Lintner, and often brought up by us in treating of other insects, of the great desirability of destroying, as far as possible, the early broods." "The gardener should keep a constant watch upon his cabbages, and upon the first appearance of the young bugs, should either commence careful handpicking at once or should begin the use of some one of the remedies just mentioned." He noted that "The ordinary poisonous applications have little effect upon this bug." Lastly, he mentioned the use of kerosene in passing, stating "Finally, though we have had no opportunity of testing its value in this particular case, we have little doubt but that the kerosene emulsion will here also prove most satisfactory, as it has been found so effectual against other destructive species of the same sub-order."

---

[2] See **Chapter 16** for current terminology used in control practices.

By the end of the 1800s and early 1900s, control practices still were mainly cultural. As will become apparently as we progress through the early 1900s, the three primary control practices for stink bugs were clean culture, trap cropping, and handpicking, the first two of which are still recommended today. Use of insecticides as an effective control was not strongly recommended until the 1940s with the advent of DDT.

6.5.1.2 J. B. Smith

Smith (1897) recommended "clean culture and cleanliness about the farm." However, one should keep "a heap or two of loose rubbish to attract insects seeking winter quarters." Then, during the winter, the sites can be destroyed by burning. If the remaining land is kept as free of weeds as possible, particularly crucifers, then the insects will leave the area when they emerge in spring.

Smith also discussed trap crops, recommending mustard or radishes, with mustard being preferred by the insects. He noted that diluted insecticides had proven to be largely ineffective. In higher concentrations, the insecticides were effective against the bugs but usually also killed the plants. However, this was unimportant if the plants were not important to the farmer (i.e., trap crop). Pure kerosene was effective, but no diluted emulsion was effective. Therefore, insects feeding on mustards or radishes could be collected early in the day by brushing or jarring the plants upon which they were feeding, causing them to fall into pans held beneath them containing a "scum of kerosene." Later in the season when the insects attacked cabbage, even diluted kerosene could not be used because it would be injurious to the plant.

6.5.1.3 F. H. Chittenden

Chittenden (1908; slightly revised, 1920) agreed with much of what Smith (1897) said. He (1908) stated, "The experience of years has shown that in order to obtain the best results in the treatment of the harlequin cabbage bug preventives are necessary, as there is great difficulty in obtaining insecticides with are effective and which do not at the same time injure or kill the plants." He, then, discussed clean cultural methods, trap crops, and hand methods (= handpicking) in detail.

As with Riley (1884), Smith (1897), and others, Chittenden (1908) discussed the use of kerosene as an insecticide as well as whale-oil soap. He stated that "Prof. A. F. Conradi has found that a 10-per-cent kerosene emulsion is effective in killing the nymphs, as is also whale-oil soap, at the rate of 2 pounds to 4 gallons of water. If the insects are sprayed just after they have molted these insecticides almost invariably kill them."

Chittenden (1908) also discussed "other remedies." He noted that "since the harlequin cabbage bug feed exclusively by suction and does not chew its food, the arsenicals, hellebore, and such remedies as are useful against cabbage worms are absolutely valueless against the present species." Pyrethrum was not effective and too expensive. And, "the value of hand torches for insecticidal purposes is extremely limited." However, by 1920, Chittenden had changed his mind. He stated, "The value of the hand torch for the control of this insect has been proved by experimenters in Texas, as well as by the experience of the writer and investigators working under his direction" (see additional discussion of the torch by Thomas [1915]).

Chittenden (1908) discussed natural enemies of *Murgantia histrionica* (not mentioned in the 1920 paper). He listed the hymenopteran egg parasitoids *Trissolcus brochymenae* (Ashmead) (as *Trissolcus murgantiae* Ashmead), *Trissolcus euschisti* (Ashmead) (as *Trissolcus podisi* Ashmead), and *Ooencyrtus johnsoni* (Howard) but noted that *T. euschisti* had parasitized the eggs "artificially." But, he realized the control potential of these wasps. He stated (p. 9), "It is possible that some of the natural enemies of this species, especially southern egg parasites, might be utilized in its control; i.e., by shipping parasitized eggs from localities where they are abundant to northern regions in which do not occur."

6.5.1.4 W. A. Thomas

Thomas (1915) discussed cultural control of this bug under three primary practices: (1) remedial measures (= clean culture), (2) trap crops, and (3) burning.

6.5.1.4.1 Remedial Measures

Thomas recommended destroying winter quarters and winter food plants. First, the area around the garden should be cleaned up in late fall or early winter so that no grass or other rubbish is available for winter protection. Never plant young collard or cabbage plants intended for late spring or summer near old infested plants if this can be avoided.

6.5.1.4.2 Trap Crops

Thomas recommended planting several rows of early radish, kale, turnips, mustard, or rape at intervals in the field or garden. These plants are preferred by the bugs over cabbage and should be destroyed by spraying thoroughly with kerosene or a 25% emulsion "of this material" just before the cabbage begins "to form for heading."

6.5.1.4.3 Burning

Burning was recommended for control of this bug on cabbage, collards, and kolhrabi during late summer on small plots such as gardens. Two good torches are necessary about 18 inches long made of fat pine, rags, or cotton wrapped on a stick secured with wire and saturated with kerosene. The torches are lit and brought together just below the bottom leaves for no more than 1–2 seconds. The bugs either will be destroyed by the flame or burned enough that they can no longer injure the plants. Burning should be conducted only at night so that the plants will be able to "cool off and recover before the hot sun comes out (the) next day."

6.5.1.5 F. B. Paddock

Paddock (1915, 1918) divided his control recommendations into two categories: (1) preventive measures and (2) remedial measures, and in 1918 into (1) artificial control, (2) remedial measures, and (3) natural control. The 1918 publication is, by far, the most comprehensive treatment of *Murgantia histrionica* in the older literature, although Paddock did erroneously claim that the bug had six instars rather than five.

6.5.1.5.1 Artificial Control

Here, Paddock (1918) discussed "fall destruction," "winter treatment," "spring treatment," "clean culture," and "trap crops." Fall destruction involved killing the bugs by handpicking, spraying, or burning as they concentrated in and around the remains of crops and weeds in preparation for overwintering. Winter treatment involved destroying excess plant growth in and around the field, including the remains of the last crop, by plowing under or burning, thereby destroying the overwintering sites of the bugs. Spring treatment involved destroying the bugs as they are leaving their overwintering sites but before they could reproduce. Clean culture emphasized destroying common weeds, particularly in the spring, upon which the bugs reproduce. This applied not only to the fields in which cabbage was to be grown but to the areas around the fields. Trap crops were those that were planted and were attractive to the pests before and after the main crop. Mustard was, perhaps, the best trap crop for this bug, but turnip, kale, or cabbage also could be used. These crops should be planted in the spring when they are attractive to the insects (i.e., from the time they leave overwintering sites until after oviposition). When populations build up, the trap crops could be destroyed by "spraying with pure kerosene, burning, or destroying the trap crop." In the fall, the trap crop should be planted at a time when they it is attractive because the main crop has been harvested, but the insects have not entered overwinter sites.

6.5.1.5.2 Remedial Measures

Here, Paddock (1918) included "hand picking," and "spraying." He noted that if the preventive measures (i.e., artificial control) had been followed carefully, the number of bugs present in the summer should be noticeably reduced. However, because some bugs would be missed, handpicking would probably be the most satisfactory method of control at this point. Although the process might appear expensive and tedious, it could be effective. Spraying was not particularly satisfactory. Any material that he was aware of that could kill the bugs also killed the plants. Kerosene was the most used material for killing the bugs when used in the undiluted state but, of course, also killed the plants. Therefore, it was best used against the bugs when they were on trap crops or crop residues. He also noted that a mixture of nicotine sulfate (40%), fish oil soap, and water was reported to kill 65–75% of the nymphs and 45–50% of the adults. Finally, Paddock stated that arsenical sprays such as Paris green, London purple, and arsenate of lead could not control this bug because it is a sucking insect. Contact sprays must be used, which meant that if the insects were not hit by the material, the bugs would not be injured. Finally, he, as well as some other investigators (e.g., Thomas 1915) stated that the plumber's torch could be effective in destroying these bugs on trap crops or trap remnants. "Under these conditions such treatment may be advisable, but the use of the torch is somewhat limited."

6.5.1.5.3 Natural Control

Under natural control, Paddock (1918) discussed the effects of weather on the survival of *Murgantia histrionica*. He stated that adults could survive cold much better than nymphs and, therefore, only adults were able to successfully overwinter (p. 52). In addition, excessive rainfall kills both the nymphs and adults.

Paddock (1918) saw no evidence of parasitoids for any stages of the bug during the 3 years of the study. However, "In 1892 Morgan found a parasite of the egg of the harlequin bug to be very common in Louisiana and Mississippi, and this was determined as *Trissoleus* [sic] *murgantiae* Ashm." (*Trissolcus murgantiae*; current name, *Trissolcus brochymenae*). "Morgan also records the parasite, *T. podisi* (current name, *Trissolcus euschisti*) reared from the eggs of the harlequin bug." Paddock reported the fire ant, *Solenopsis geminata*, carrying off nymphs of this bug and the leaffooted bug, *Leptoglossus phyllopus* (as *phyclopus*) (L.), destroying the adults.

6.5.1.6 B. B. Fulton

6.5.1.6.1 Soap Solutions as an Insecticide

Fulton (1930) reported on the use of soap solutions as an insecticide. He noted that soap killed the bugs "by penetrating deeply into the tracheal system through part or all of the spiracles." The spiracles are equipped with closing devices, which are better developed in adults than nymphs. Therefore, the soap can enter the tracheal system more easily in nymphs than in adults resulting in higher mortality in nymphs. Also, he found that "the efficiency of soap solution is indirectly proportional to the rate of evaporation."

6.5.1.7 J. G. Walker and L. D. Anderson

6.5.1.7.1 Contact Insecticides

Walker and Anderson (1933) reported on a severe outbreak of *Murgantia histrionica* during the summer of 1932 near Norfolk, VA, and included the results of tests of various materials in the control of this pest during this time. Among the materials tested was the 2% soap solution recommended by Fulton (1930). As Fulton had mentioned earlier, they found the adults more difficult to control than nymphs. Because the ideal conditions needed for satisfactory control, including low temperature, high humidity, and no wind, were seldom present, they tested a large series of contact sprays and wetting agents in an insectary; those showing promise then were tested in the field. The best control was with sprays containing rotenone (contained in a product known as Serrid, a derris extract) in combination with a 1% soap solution. In general, they found that nicotine, pyrethrum, and oil emulsion sprays were not effective except at high concentrations. They later reported (1939) that based on subsequent tests, thoroughly spraying infested plants with a "mixture containing 4 pounds of derris or cubé powder (rotenone content of 5 or 6%) to 50 gallons of water" plus a good wetting agent gave good control as did the concentrated derris extract plus soap they previous had recommended in 1933.

6.5.1.7.2 Natural Control

Walker and Anderson (1933) noted that swallows were observed feeding on the adults of this bug. Also, 35–55% of the eggs collected during August and September had been parasitized by *Ooencyrtus johnsoni*.

6.5.1.8 W. H. White and L. W. Brannon

The last paper we have selected for our brief survey of the early control practices for *Murgantia histrionica* is that of White and Brannon (1939). These authors discussed "Methods of Control," in which they began with the following statement, "For the best results in the treatment of the harlequin bug, preventive measures are necessary, as this insect is exceedingly difficult to combat after it has become numerous on its host plants." Following, they divided these measures into cultural methods and contact insecticides and included information on natural enemies.

6.5.1.8.1 Cultural Methods

Under cultural methods, they discussed clean culture, trap crops, handpicking, and the hand torch. These methods, by now, were standard procedure, including the use of a hand torch.

6.5.1.8.2 Contact Insecticides

Under contact insecticides, they repeated what earlier investigators has said about the difficulty in killing this bug with these insecticides because these chemicals had to come in contact with sucking insects to be effective. This was difficult because "the plants upon which the insects feed usually attain such a dense growth that it is practically impossible to reach the insects on all parts of the plant." They noted that several insecticides had been tested and that derris extract spray and derris dust was recommended for control of this bug. They also noted that more recent work by Walker and Anderson (1939) had shown that "derris or cube root powder (containing 5 or 6 percent of rotenone and used at the rate of 4 pounds in 50 gallons of water with wetting agent) is effective for control of the harlequin bug." In conclusion, they stated "It should be borne in mind that only those insects actually hit by the insecticide are killed. Thoroughness of application is therefore of prime importance."

6.5.1.8.3 Natural Enemies

White and Brannon (1939) discussed natural enemies of *Murgantia histrionica* but noted that the bug "is remarkably free from parasites and predacious insects. No internal parasite of the insect has been recorded." They stated that there were three known egg parasites, namely *Trissolcus brochymenae* (as *Trissolcus murgantiae*), *Trissolcus euschisti* (as *Trissolcus podisi*), and *Ooencyrtus johnsoni*, "but, so far as has been observed, they have never become abundant enough throughout the infested territory to serve as a natural means of control of this pest." However, they noted that *O. johnsoni* "appeared to be of considerable value in eastern Virginia during the unusually severe harlequin-bug outbreak of 1932." "During the latter part of August, approximately 50 percent of the eggs in some fields were parasitized," suggesting that "under favorable conditions this parasite may prove to be highly beneficial in some parts of the South." They also noted that "the wheel bug [*Arilus cristatus* (L.)], has been recorded as feeding on young harlequin-bug nymphs, and another, the leaf-foot bug [*Leptoglossus phyllopus* (L.)], is recorded as destroying the adult, but these are of slight importance as a means of natural control."

6.5.2 Age of Synthetic Insecticides

An "Age of Pesticides" began in the 1940s with Paul Müller's discovery of the insecticidal properties of DDT (Metcalf 1980) and, since that time, chemical control has persisted as the predominant tactic used for pest management in most agricultural systems including control of *Murgantia histrionica* on brassica vegetables. However, there have been major shifts in the classes of insecticides used over the years.

6.5.2.1 Chlorinated Hydrocarbons

In the 1940s, effective control of *Murgantia histrionica* was achieved with dust applications of DDT or other chlorinated hydrocarbons including chlorinated camphene (toxaphene), chlordane, and benzene hexachloride (BHC) (Brooks and Anderson 1947, Gaines and Deane 1948). The latter, in particular, was considered very toxic to this bug even at concentrations as low as 0.5% active γ isomer; chlordane and toxaphene were more toxic than DDT (Gaines and Dean 1948). Later, lindane (almost pure γ BHC) also would be shown to be highly efficacious on this bug (Brett and Campbell 1956) as would endosulfan and endrin (Hofmaster 1959). However, most agricultural uses of chlorinated hydrocarbon insecticides eventually would be canceled because of the persistence of these compounds in the environment, resistance that developed in several insect pests, and biomagnification of the toxicants in some wildlife food chains (Ware and Whitacre 2004). Only one chlorinated hydrocarbon insecticide, endosulfan, would maintain federal registration on vegetable crops in the United States into the 21st Century, but as of 2015, this chemical no longer is registered for use in the United States and many other countries.

6.5.2.2 Organophosphates and Carbamates

Carbamates and organophosphates gradually would replace the chlorinated hydrocarbons in vegetable production in the United States in the 1960s and 1970s. These cholinesterase-inhibiting insecticides have broad-spectrum activity against many insect pests, and *Murgantia histrionica* is no exception. Most of the compounds in these two insecticide classes including parathion, carbaryl, acephate, and diazinon, and others provided effective control of this bug (Hofmaster 1959, Rogers and Howell 1972). However, many of the organophosphates later were determined to be too toxic for safe use on vegetables and either were not registered for use or later had registrations removed particularly after full enactment of the United States Environmental Protection Agency's Food Quality Protection Act of 1996.

6.5.2.3 Pyrethroids

Throughout the 1970s and into the 1990s, a large number of synthetic pyrethroid insecticides would be developed and registered for use on vegetables throughout the world. These insecticides, which included permethrin, bifenthrin, lambda-cyhalothrin, zeta-cypermethrin, cyfluthrin, and many others were shown to be highly efficacious at extremely low application rates for control of *Murgantia histrionica* (Edelson 2004a, 2004b; Edelson and Mackey 2006a). Because of their low cost and high efficacy, pyrethroids have been the predominant insecticide class used for control of this bug and other stink bugs for the past few decades. However, because these broad-spectrum insecticides also are detrimental to important natural enemies in the crucifer crop agroecosystem (Xu et al. 2001, 2004; Cordero et al. 2007), and because of the development of insecticide resistance to them in other pest species including diamondback moth, *Plutella xylostella* (L.) (Shelton et al. 1993), green peach aphid, *Myzus persicae* (Sulzer) (Castañeda et al. 2011) and beet armyworm, *Spodoptera exigua* (Hübner) (Brewer and Trumble 1994), the use of pyrethroids for control of this bug can be problematic.

6.5.2.4 Neonicotinoids

The neonicotinoids, which target the nicotinic acetylcholine receptors in insects, were introduced in the 1990s and have become the most widely used insecticides in the world. Virtually all of the registered neonicotinoids including acetamiprid, clothianidin, dinotefuran, imidacloprid, thiacloprid, and thiamethoxam have been shown to be effective at controlling this bug when used as a foliar spray (Edelson 2004a, Edelson and Mackey 2005a,b,c, 2006b; Walgenbach and Schoof 2005). Moreover, because neonicotinoids are water soluble and can be taken up by plants through the roots and translocated through the xylem vessels to plant tissues, exposing herbivores to the toxin only when they feed (Sur and Stork 2003, Tomizawa and Casida 2005), they offer a potentially less disruptive alternative for controlling hemipteran insects. Wallingford et al. (2012) evaluated soil-applied neonicotinoids and showed that labeled rates of either imidacloprid, thiamethoxam, clothianidin, or dinotefuran provided significant control of this bug for at least two weeks after application.

6.5.2.5 Other Insecticides

Over the past couple of decades, a wide range of other insecticides (often with reduced risks to nontarget organisms) has been tested for activity on *Murgantia histrionica*. Edelson and Mackey (2006b) showed that the hemipteran feeding inhibitors pymetrozine and flonicamid were not active on this bug. These same authors also showed that emamectin benzoate, spinosad, azadirachtins, and the insect growth regulator, methoxyfenozide also were not or only slightly effective against this bug (Edelson and Mackey 2006a). Among the new diamide class of insecticides, flubendiamide and chlorantraniliprole are not effective, but cyantraniliprole and the experimental compound cyclaniliprole have shown efficacy against this bug (Kuhar and Doughty 2009, T. Kuhar, unpublished data).

For organic growers, there are no insecticides that are highly efficacious against *Murgantia histrionica*. However, suppression of this bug, especially nymphs, can be achieved with various compounds including spinosad, pyrethrins, sabadilla, and azadirachtins (Overall et al. 2007). Overall et al. (2008) evaluated the toxicity levels of organic insecticides on this bug and showed that spinosad was about 10-fold more toxic to fourth instars than pyrethrins, which were about 10-fold more toxic than azadirachtins.

6.6 Current Management Techniques

In the southern United States, in particular, populations of *Murgantia histrionica* still can reach damaging levels if proper management tactics are not used. As was demonstrated in the early 1930s, the presence of a brassica crop remaining in fields throughout the winter, coupled with warm winter, can result in serious outbreaks of this pest (Walker and Anderson 1933). Thus, as was recommended a century ago by Thomas (1915), removing or destroying overwintering crop residue is a critical first step to reducing the overwintering populations of bugs that will invade crops early the next spring.

Trap cropping is another age-old preventative control strategy that is still used today. It was recognized early as a control tactic of this bug, using radish (*Raphanus sativus* L.), turnips (*Brassica rapa* L.), mustard (*B. juncea* L.), rapeseed (*B. napus* L.), or kale (*B. oleracea* L. *acephala* group) to draw pressure away from cabbage (*B. oleracea* L. capitata group) (Thomas 1915, Chittenden 1920, Fulton 1930). Ludwig and Kok (1998b) demonstrated that a small early planting of a crop such as broccoli could aggregate the bugs and keep them off of the later-planted main crop. Destruction of the bugs is necessary to prevent dispersal from the trap crop. Because these bugs prefer certain plant species such as mustard, turnip, and Chinese cabbage over crops such as cabbage, cauliflower, broccoli, collards, and radish (Sullivan and Brett 1974), there is great potential for utilizing multiple cropping or intercropping approaches as a trap crop management strategy, particularly in the interest of reducing chemical sprays. Bender et al. (1999) found that intercropping cabbage (*B. oleracea*) and Indian mustard (*B. juncea*) reduced the need for two insecticide sprays in a heavy infestation of *Murgantia histrionica*. More recently, using field cage choice tests and small plot field experiments, Wallingford et al. (2013) determined that mustard (*B. juncea* 'Southern Giant Curled') was the most consistently selected host plant by this bug over collard in choice tests and, when planted as a double row border, was found to be an effective trap crop for reducing feeding injury on collard. Augmentation of the mustard trap crop with a systemic neonicotinoid insecticide provided no added control of this bug for the 10-week duration of the experiment.

Nonetheless, insecticides remain the most widely used rescue control strategy when these bugs reach damaging levels. Pyrethroids often are used because they are cheap and effective. However, because they are broad-spectrum toxicants, pyrethroids can be quite disruptive to natural enemies that are essential to integrated pest management (IPM) programs for other pests such as lepidopteran larvae and aphids.

Use of IPM for *Murgantia histrionica* could be improved with the development of effective sampling tools and the use of action thresholds. However, currently, there is no monitoring device or proven trapping system for this pest. Research now is being conducted in the United States on the development of an effective trap for *M. histrionica*. In laboratory and field color choice experiments, adults and large nymphs responded positively to darker colors such as green and black compared to lighter colors such as yellow; thus, future trapping devices for this pest should be green or black in color (DiMeglio et al. 2017).

6.7 Why Has *Murgantia histrionica* Been So Successful?

At the time of the invasion of this bug into Texas in the early 1860s, control practices were limited. As noted in this chapter, recommendations consisted primarily of various types of cultural control tactics, namely handpicking, clean culture, and trap cropping. Biological control was not considered important although parasites and predators had been reported. Early insecticides largely were ineffective and, often, dangerous to humans or to the crop plants. Therefore, the bugs were able to spread rapidly, but largely were confined to the southern states because of the inability of the adults to survive harsh winters. Although they were able to reach the more northern states, populations often were eliminated by subfreezing overwintering temperatures. Thus, the bugs' present and relatively permanent distribution in North America was reached by the early 1900s.

Nonetheless, despite being well established in the southern United States, and having at least three species of native hymenopteran wasps that parasitize the eggs including *Ooencyrtus johnsoni*, *Trissolcus murgantiae* (now *Trissolcus brochymenae*), and *Trissolcus podisi* (now *Trissolcus euschisti*) (White and Brannon 1939, Huffaker 1941, Ludwig and Kok 1998a, Koppel et al. 2009), natural enemies do not provide adequate control of this pest.

6.8 What Does the Future Hold?

Although the bug largely is confined to the southern states, the advent of climate change may allow it to extend its general distribution to regions that historically have not had problems with this pest. In general, the pest seems to be increasing in importance, perhaps because of a switch from broad-spectrum insecticides to more IPM-friendly insecticides for control of lepidopteran pests and aphids. Coupled with this, there has been a rapid increase in organic vegetable acres, which often maintain reservoirs of populations of this bug. Thus, the pest does not appear to be going away anytime soon and, unfortunately, there still is a need for an alternative management strategy for this bug that does not rely solely on the use of insecticides and can be integrated into current management strategies. Perhaps, transgenic (GMO) resistant crops for sucking pests could be future control strategy for this pest, but we are a long way from seeing that, especially in relatively small acreage vegetable crops. In addition, because semiochemicals are so critical to the ecology of this pest (Aldrich et al. 1996), and our understanding of aggregation pheromones and the importance of plant kairomones (Khrimian et al. 2014, Weber et al. 2014) is increasing, perhaps management strategies such as traps and toxic baits can be developed in the future.

6.9 Acknowledgments

We thank Norman F. Johnson (Department of Entomology, Ohio State University, Columbus) for his help with the synonymy of the *Telenomus* and *Trissolcus* parasitoids. We also thank Anthony S. DiMeglio (Department of Entomology, Virginia Tech, Blacksburg) and Sam E. Droege (United States Geological Service, Beltsville, MD) for providing photographs of *Murgantia histrionica*.

6.10 References Cited

Aldrich, J. R. 1988. Chemical ecology of the Heteroptera. Annual Review of Entomology 33: 211–238.

Aldrich, J. R., J. W. Avery, C. J. Lee, J. C. Graf, D. J. Harrison, and F. Bin. 1996. Semiochemistry of cabbage bugs (Heteroptera: Pentatomidae: *Eurydema* and *Murgantia*). Journal of Entomological Science 31: 172 182.

Aliabadi, A., J. A. Renwick, and D. W. Whitman. 2002. Sequestration of glucosinolates by harlequin bug, *Murgantia histrionica*. Journal of Chemical Ecology 28: 1749–1762.

Bender, D. A., W. P. Morrison, and R. E. Frisbe. 1999. Intercropping cabbage and Indian mustard for potential control of lepidopterous and other insects. Hortscience 34: 275–279.

Brett, C. H., and W. V. Campbell. 1956. Eggplant lace bug and harlequin bug susceptibility to some standard insecticide dust formulations. Journal of Economic Entomology 49: 422–423.

Brett, C. H., and M. J. Sullivan. 1974. The use of resistance varieties and other cultural practices for control of insects on crucifers in North Carolina. North Carolina Agricultural Experiment Station Bulletin 449: 1–31.

Brewer, M. J., and J. T. Trumble. 1994. Beet armyworm resistance to fenvalerate and methomyl: resistance variation and insecticide synergism. Journal of Agricultural Entomology 11: 291–300.

Brooks, W., and L. D. Anderson. 1947. Toxicity tests of some new insecticides. Journal of Economic Entomology 40: 220–228.

Burmeister, H. C. C. 1835. Handbuch der Entomologie. Tome 2. T. Enslin, Berlin. Abtheil 1: i–xii, 1–400.

Castañeda, L. E., K. Barrientos, P. A. Cortes, C. C. Figueroa, E. Fuentes-Contreras, M. Luna-Rudloff, A. X. Silva, and L. D. Bacigalupe. 2011. Evaluating reproductive fitness and metabolic costs for insecticide resistance in *Myzus persicae* from Chile. Physiological Entomology 36: 253–260.

Chittenden, F. H. 1908. The harlequin cabbage bug (*Murgantia histrionica* Hahn.). United States Department of Agriculture Bureau of Entomology Circular 103: 1–10.

Chittenden, F. H. 1920. Harlequin cabbage bug and its control. United States Department of Agriculture Farmers' Bulletin 1061: 1–13.

Cockerell, T. D. A. 1903. New bees from southern California and other records. Bulletin of the Southern California Academy of Sciences 2: 84–85.

Cokl, A., J. Presern, M. Virant-Doberlet, G. J. Bagwell, and J. G. Millar. 2004. Vibratory signals of the harlequin bug and their transmission through plants. Physiological Entomology 29: 372–803.

Cokl, A., M. Zorovic, and J. G. Millar. 2007. Vibrational communication along plants by the stink bugs *Nezara viridula* and *Murgantia histrionica*. Behavioral Proceedings 75: 40–54.

Cordero, R. J., J. R. Bloomquist, and T. P. Kuhar. 2007. Susceptibility of two diamondback moth parasitoids, *Diadegma insulare* (Cresson) (Hymenoptera; Ichneumonidae) and *Ooomyzus sokolowskii* (Kurdjumov) (Hymenoptera; Eulophidae), to selected commercial insecticides. Biological Control 42: 48–54.

DiMeglio, A. S., A. K. Wallingford, D. C. Weber, T. P. Kuhar, and D. Mullins. 2016. Supercooling points of *Murgantia histrionica* (Hemiptera: Pentatomidae) and field mortality in the Mid-Atlantic United States following lethal low temperatures. Environmental Entomology 45: 1294–1299. doi: 10.1093/ee/nvw091.

DiMeglio, A. S., T. P. Kuhar, and D. C. Weber. 2017. Color preference of harlequin bug (Heteroptera: Pentatomidae). Journal of Economic Entomology 1–3. doi: 10.1093/jee/tox179.

Edelson, J. V. 2004a. Comparison of nicotinoid insecticides for controlling harlequin bug, 2003. Arthropod Management Tests 29: E22.

Edelson, J. V. 2004b. Harlequin bug control on collard, 2003. Arthropod Management Tests 29: E24.

Edelson, J. V., and C. Mackey. 2005a. Comparison of nicotinoid insecticides for controlling harlequin bug, 2004. Arthropod Management Tests 30: E23.

Edelson, J. V., and C. Mackey. 2005b. Comparison of nicotinoid insecticides for controlling harlequin bug and thrips 2004. Arthropod Management Tests 30: E24.

Edelson, J. V., and C. Mackey. 2005c. Comparison of nicotinoid insecticides for controlling harlequin bug and thrips, 2004. Arthropod Management Tests 30: E94.

Edelson, J. V., and C. Mackey. 2006a. Comparison of nicotinoid insecticides for controlling harlequin bug, 2005. Arthropod Management Tests 31: E17.

Edelson, J. V., and C. Mackey. 2006b. Comparison of pyrethroid insecticides for controlling harlequin bug, 2005. Arthropod Management Tests 31: E18.

English-Loeb, G. M., and B. D. Collier. 1987. Nonmigratory movement of adult harlequin bugs *Murgantia histrionica* (Hemiptera: Pentatomidae) as affected by sex, age and host plant quality. The American Midland Naturalist 118: 189–197.

Froeschner, R. C. 1988. Family Pentatomidae Leach, 1815. The stink bugs, pp. 544–597. *In* T. J. Henry and R. C. Froeschner (Eds.), Catalog of the Heteroptera, or true bugs, of Canada and the continental United States. E. J. Brill, New York, New York. 958 pp.

Fulton, B. B. 1930. The relation of evaporation to killing efficiency of soap solutions on the harlequin bug and other insects. Journal of Economic Entomology 23: 625–630.

Gaines, J. C., and H. A. Dean. 1948. Comparison of insecticides for control of harlequin bugs. Journal of Economic Entomology 41: 808–809.

Glover, T. 1868. Report of the entomologist. *In* United States Department of Agriculture, Report of the Commissioner of Agriculture for the year 1867, pp. 58–76.

Hahn, C. W. 1834. Die Wanzenartigen insecten. C. H. Zeh'schen Buchhandlung, Nürnberg. 2: 33–120.

Helmey-Hartman, W. L., and C. W. Miller. 2014. Context-dependent mating success in *Murgantia histrionica* (Hemiptera: Pentatomidae). Annals of the Entomological Society of America 107: 264–273.

Herrich-Schaeffer, G. A. W. 1853. Die Wanzenartigen Insecten. C. H. Zeh'schen Buchhandlung, Nürnberg. 9 ["historischer übersicht" and "index"]: 1–210. [Dates and pagination follow Bergroth, 1919, Ent. Mitt., 8: 188–189].

Hodson, A. C., and E. F. Cook. 1960. Long-range aerial transport of the harlequin bug and the greenbug into Minnesota. Journal of Economic Entomology 53: 604–608.

Hofmaster, R. N. 1959. Effectiveness of new insecticides against the harlequin cabbage bug on collards. Journal of Economic Entomology 52: 777–778.

Howard, L. O. 1895. The harlequin cabbage bug, or calico back (*Murgantia histrionica* Hahn.). United States Department of Agriculture, Division of Entomology, Circular (Second Series) 10: 1–2.

Huffaker, C. A. 1941. Egg parasites of the harlequin bug in North Carolina. Journal of Economic Entomology 34: 117–118.

Khrimian, A., S. Shirali, K. E. Vermillion, M. A. Siegler, F. Guzman, K. Chauhan, J. R. Aldrich, and D. C. Weber. 2014. Determination of the stereochemistry of the aggregation pheromone of harlequin bug, *Murgantia histrionica*. Journal of Chemical Ecology 40: 1260–1268.

Kirkaldy, G. W. 1909. Catalogue of the Hemiptera (Heteroptera) with biological and anatomical references, lists of foodplants and parasites, etc. prefaced by a discussion on nomenclature, and an analytical table of families. Volume I: Cimicidae. Felix L. Dames, Berlin. I–XL + 392 pp.

Koppel, A. L., D. A. Herbert, Jr., T. P. Kuhar, and K. Kamminga. 2009. Survey of stink bug (Hemiptera: Pentatomidae) egg parasitoids in wheat, soybean, and vegetable crops in southeast Virginia. Environmental Entomology 38: 375–379.

Kuhar, T. P., and H. Doughty. 2009. Evaluation of soil and foliar insecticide treatments for the control of foliar insect pests in cabbage in Virginia, 2008. Arthropod Management Tests 34: E7.

Lanigan, P. J., and E. M. Barrows. 1977. Sexual behavior of *Murgantia histrionica* (Hemiptera: Pentatomidae). Psyche 84: 191–197.

Ludwig, S. W., and L. T. Kok. 1998a. Phenology and parasitism of harlequin bugs, *Murgantia histrionica* (Hahn) (Hemiptera: Pentatomidae), in southwest Virginia. Journal of Entomological Science 33: 33–39.

Ludwig, S. W. and L. T. Kok. 1998b. Evaluation of trap crops to manage harlequin bugs, *Murgantia histrionica* (Hahn) (Hemiptera: Pentatomidae) on broccoli. Crop Protection 17: 123–128.

Ludwig, S. W., and L. T. Kok. 2001. Harlequin bug, *Murgantia histrionica* (Hahn) (Heteroptera: Pentatomidae) development on three crucifers and feeding damage on broccoli. Crop Protection 20: 247–251.

McClain, D. L. 1981. Numerical responses of *Murgantia histrionica* to concentrations of its host plant. Journal of the Georgia Entomological Society 16: 257–260.

McPherson, J. E. 1982. The Pentatomoidea (Hemiptera) of northeastern North America with emphasis on the fauna of Illinois. Southern Illinois University Press, Carbondale and Edwardsville. 240 pp.

McPherson, J. E., and R. M. McPherson. 2000. Stink bugs of economic importance in America north of Mexico. CRC Press, Boca Raton, FL. 253 pp.

Metcalf, R. L. 1980. Changing role of insecticides in crop protection. Annual Review of Entomology 25: 219–256.

Osborn, H. 1894. Notes on the distribution of Hemiptera. Proceedings of the Iowa Academy of Sciences for 1893. Volume 1 (Part 4): 120–123.

Overall, L. M., J. V. Edelson, and C. Mackey. 2007. Organic insecticides to control the harlequin bug on collards and turnips, 2006. Arthropod Management Tests 32: E10.

Overall, L. M. J. V. Edelson, and C. Mackey. 2008. Effect of insecticides on harlequin bug and yellowmargined leaf beetle, 2007. Arthropod Management Tests 33: L4.

Paddock, F. B. 1915. The harlequin cabbage-bug. Texas Agricultural Experiment Station Bulletin 179: 1–9.

Paddock, F. B. 1918. Studies on the harlequin bug. Texas Agricultural Experiment Station Bulletin 227: 1–65.

Radcliffe, E. B., K. L. Flanders, D. W. Ragsdale, and D. M. Noetzel. 1991. Pest management systems for potato insects, pp. 587–621. *In* D. Pimentel (Ed.), CRC handbook of pest management in agriculture (2nd edit.) Volume III. CRC Press, Inc., Boca Raton, FL. 749 pp.

Rider, D. R. 2012. The Heteroptera (Hemiptera) of North Dakota 1: Pentatomomorpha: Pentatomoidea. The Great Lakes Entomologist 45: 312–380.

Riley, C. V. 1870. The harlequin cabbage-bug, pp. 79–80. *In* C. V. Riley and G. Vasey (Eds.), The American entomologist and botanist: an illustrated magazine of popular and practical entomology and botany, Volume II: 79–80. R. P. Studley & Co., Publishers, St. Louis, MO.

Riley, C. V. 1872. The harlequin cabbage-bug.–*Strachia* [*Murgantia*] *histrionica*, Hahn, pp. 35–38. Fourth annual report on the noxious, beneficial and other insects of the State of Missouri, made to the State Board of Agriculture, pursuant to an appropriation for this purpose from the legislature of the State. Regan & Edwards, Public Printers, Jefferson City, MO. 145 pp. + 1 p. (errata), 6 pp. (index), 8 plates.

Riley, C. V. 1884. The harlequin cabbage-bug. (*Murgantia histrionica*, Hahn.). Order Heteroptera; Family Scutelleridae, pp. 309–312. Report of the entomologist, pp. 285–418 + 2 pp. (explanation to plates), 10 plates. (Authors' edition. 1885. Annual report of Department of Agriculture for 1884). *In* Annual report of commissioner for the year 1884. United States Governing Printing Office, Washington, D. C. 581 pp.

Rogers, C. E., and G. R. Howell. 1972. Toxicity of acephate and diazinon to harlequin bugs on cabbage. Journal of Economic Entomology 66: 827–828.

Sanderson, E. D. 1903. Report of the entomologist. Fourteenth Annual Report of the Delaware College Agricultural Experiment Station, pp. 109–151.

Shelton, A. M., J. A. Wyman, N. L. Cushing, K. Apfelbeck, T. J. Denney, S. E. R. Mahr, and S. D. Eigenbrode. 1993. Insecticide resistance of diamondback moth *Plutella xylostella* (Lepidoptera: Plutellidae), in North America. Journal of Economic Entomology 86: 11–19.

Smith, J. B. 1897. The harlequin cabbage bug and the melon plant louse. New Jersey Agricultural Experiment Stations Bulletin 121: 3–14.

Smith, R. I. 1909. Biological notes on *Murgantia histrionica* Hahn. Journal of Economic Entomology 2: 108–114.

Stål, C. 1862. Hemiptera Mexicana enumeravit speciesque novas descripsit. Entomologische Zeitung Herausgegeben von dem Entomologischen Vereine zu Stettin 23(1–3): 81–118.

Stål, C. 1872. Enumeratio Hemipterorum: Bidrag till en förteckning öfver alla hittills kända Hemiptera, jemte systematiska meddelanden. Parts 1–5. Kongliga Svenska Vetenskaps-Akademiens Handlingar, 1872, part 2, 10(4): 1–159.

Stoner, D. 1920. The Scutelleroidea of Iowa. University of Iowa Studies in Natural History 8: 1–140 + 7 plates.

Streams, F. A., and D. Pimentel. 1963. Biology of the harlequin bug, *Murgantia histrionica*. Journal of Economic Entomology 56: 108–109.

Sullivan, M. J., and C. H. Brett. 1974. Resistance of commercial crucifers to the harlequin bug in the coastal plain of North Carolina. Journal of Economic Entomology 67: 262–264.

Sur, R, and A. Stork. 2003. Uptake, translocation and metabolism of imidacloprid in plants. Bulletin of Insectology 56: 35–40.

Thomas, W. A. 1915. The cabbage harlequin bug or calico bug. (*Murgantia histrionica*, Hahn). South Carolina Agricultural Experiment Station Bulletin 28: 1–3.

Tomizawa, M., and J. E. Casida. 2005. Neonicotinoid insecticide toxicology: mechanisms of selective action. Annual Review of Pharmacology and Toxicololgy 45: 247–268.

Uhler, P. R. 1876. List of Hemiptera of the region west of the Mississippi River, including those collected during the Hayden Explorations of 1873. Bulletin of the United States Geological and Geographical Survey of the Territories 2: 269–361 + plates 19–21.

Walgenbach, J. F., and S. C. Schoof. 2005. Insect control on cabbage, 2004. Arthropod Management Tests 30: E14.

Walker, H. G., and L. D. Anderson. 1933. Report on the control of the harlequin bug, *Murgantia histrionica* Hahn, with notes on the severity of an outbreak of this insect in 1932. Journal of Economic Entomology 26: 129–135.

Walker, H. G., and L. D. Anderson. 1939. Further notes on the control of the harlequin bug. Journal of Economic Entomology 32: 225–228.

Wallingford, A. K., T. P. Kuhar, P. B. Schultz, and J. H. Freeman. 2011. Harlequin bug biology and pest management in brassicaceous crops. Journal of Integrated Pest Management 2(1): 1–4.

Wallingford, A. K., T. P. Kuhar, and P. B. Schultz. 2012. Toxicity and field efficacy of four neonicotinoids on harlequin bug (Hemiptera: Pentatomidae). Florida Entomologist 95: 1123–1126.

Wallingford, A. K., T. P. Kuhar, D. G. Pfeiffer, D. B. Tholl, J. H. Freeman, H. B. Doughty, and P. B. Schultz. 2013. Host plant preference of harlequin bug (Hemiptera: Pentatomidae), and evaluation of a trap cropping strategy for its control in collard. Journal of Economic Entomology 106: 283–288.

Walsh, B. J. 1866. The Texas cabbage bug. (*Strachia histrionica* Hahn.). Practical Entomologist 1: 110.

Ware, G. W., and D. M. Whitacre. 2004. The Pesticide book (6th edit.). Meister Media Worldwide, Willoughby, Ohio. 496 pp.

Weber, D. C., G. C. Walsh, A. S. DiMeglio, M. M. Athanas, T. C. Leskey, and A. Khrimian. 2014. Attractiveness of harlequin bug, *Murgantia histrionica* (Hemiptera: Pentatomidae), aggregation pheromone: Field response to isomers, ratios and dose. Journal of Chemical Ecology 40: 1251–1259.

White, W. H., and L. W. Brannon. 1939. The harlequin bug and its control. United States Department of Agriculture Farmers' Bulletin 1712: 1–10.

Xu, J., A. M. Shelton, and X. Cheng. 2001. Variation in susceptibility of *Diadegma insulare* (Hymenoptera: Ichneumonidae) to permethrin. Journal of Economic Entomology 94: 541–546.

Xu, Y. Y., T. X. Liu, G. L. Leiby, and W. A. Jones. 2004. Effects of selected insecticides on *Diadegma insulare* (Hymenoptera: Ichneumonidae), a parasitoid of *Plutella xylostella* (Lepidoptera: Plutellidae). Biocontrol Science and Techology 14: 713–723.

Zahn, D. K., R. D. Girling, J. S. McElfresh, R. T. Carde, and J. G. Millar. 2008a. Biology and reproductive behavior of *Murgantia histrionica*. (Heteroptera: Pentatomidae). Annals of the Entomological Society of America 101: 215–228.

Zahn, D. K., J. A. Moreira, and J. G. Millar. 2008b. Identification, synthesis, and bioassay of a male-specific aggregation pheromone from the harlequin bug, *Murgantia histrionica*. Journal of Chemical Ecology 34: 238–251.

7

Nezara viridula (L.)

Jesus F. Esquivel, Dmitry L. Musolin, Walker A. Jones, Wolfgang Rabitsch, Jeremy K. Greene, Michael D. Toews, Cristiano F. Schwertner, Jocélia Grazia, and Robert M. McPherson

Nezara viridula (L.)[1]

1758	*Cimex viridulus* Linnaeus, Syst. Nat.: 444 ("Indiis" = India).
1775	*Cimex torquatus* Fabricius, Syst. Ent.: 710 (India). (Synonymized [with *smaragdulus*] by Illiger, 1807, Fauna Etrusca 2: 30).
1775	*Cimex smaragdulus* Fabricius, Syst. Ent.: 711 (Madeira). (Synonymized by Stål, 1865, Hemipt. Afr.: 193).
1783	*Cimex transversus* Thunberg, Dissert. Ent.: 40 (South Africa). (Synonymized by Stål, 1855, Öfver. Kongl. Vet.-Akad. För. 12: 345).
1789	*Cimex variabilis* Villers, Car. Linn. Ent.: 505 ("Occitania" = southern France). (Synonymized by Reuter, 1888, Rev. Synonym. Het.: 126).
1798	*Cimex spirans* Fabricius, Suppl. Ent. Syst.: 533 ("America Insulis" [= St. Kitts, Antigua, Jamaica, Barbados, Bermudas]). (Synonymized by Stål, 1868, Hem. Fabr.: 31).
1801	*Cimex viridissimus* Wolff, Icones Cimicum: 55 (India orientali). (Junior primary homonym of *Cimex viridissimus* Poda [= *Palomena viridissima* (Poda, 1761)]). (Synonymized [with *smaragdula*] by Fieber, 1861, Eur. Hem.: 330).
1818	*Pentatoma? flavicollis* Palisot de Beauvois, Ins. Rec. Afr. Am.: 185 (Haiti). (Synonymized [with *smaragdula*] by Amyot and Serville, 1843, Hist. Nat. Ins.: 144).
1837	*Pentatoma oblonga* Westwood, Cat. Hem.: 37 (Java). (Synonymized by Distant, 1901, Proc. Zool. Soc. Lond. 1900: 813).
1837	*Pentatoma unicolor* Westwood, Cat. Hem.: 38 (Java). (Synonymized by Stål, 1865, Hemipt. Afr.: 193).
1837	*Pentatoma berylina* Westwood, Cat. Hem.: 38 (India orientali). (Synonymized by Distant, 1901, Proc. Zool. Soc. Lond. 1900: 813).
1837	*Pentatoma subsericea* Westwood, Cat. Hem.: 38 (India orientali). (Synonymized by Stål, 1865, Hemipt. Afr.: 193).
1837	*Pentatoma leii* Westwood, Cat. Hem.: 38 (South Africa). (Synonymized by Stål, 1865, Hemipt. Afr.: 193).
1837	*Pentatoma 3-punctigera* Westwood, Cat. Hem.: 38 ("Insula Sti. Vincentii" = Saint Vincent, Lesser Antilles). (Synonymized by Stål, 1865, Hemipt. Afr.: 194).
1837	*Pentatoma chinensis* Westwood, Cat. Hem.: 38 (China). (Synonymized by Stål, 1865, Hemipt. Afr.: 194).
1837	*Pentatoma proxima* Westwood, Cat. Hem.: 9 (Canary Isl.). (Synonymized [with *tripunctigera*] by Westwood, Cat. Hem.: 38. [nomen nudum]).
1837	*Pentatoma chloris* Westwood, Cat. Hem.: 38 (Sierra Leone). (Synonymized by Stål, 1865, Hemipt. Afr.: 194).
1837	*Pentatoma chlorocephala* Westwood, Cat. Hem.: 38 (Brasil?). (Synonymized by Lethierry and Severin, 1893, Cat. Gen. Hem.: 167).
1837	*Pentatoma propinqua* Westwood, Cat. Hem.: 9 (Java). (Synonymized [with *unicolor*] by Westwood, 1837, Cat. Hem.: 39. [nomen nudum]).
1838	*Cimex hemichloris* Germar, Silb. Rev. Entomol. 5: 166 (South Africa). (Synonymized by Stål, 1865, Hemipt. Afr.: 194).
1843	*Pentatoma smaragdula* var. *minor* Costa, Cimicum Neapol.: 57 ("Neapolitania" = Naples env., Italy) (nomen oblitum).

---

[1] Additional references are provided in regional catalogues (e.g., Froeschner 1988, Cassis and Gross 2002, Rider 2006a).

1848   *Nezara approximata* Reiche and Fairmaire, Voy. Abyss. 3: 443 ("Abyssinie" = Ethiopia). (Synonymized by Stål, 1865, Hemipt. Afr.: 194).

1849   *Pentatoma plicaticollis* Lucas, Hist. Nat. Insectes 3: 87 (Algeria). (Synonymized [with *smaragdulus*] by Mulsant and Rey, 1866, Hist. Nat. Pun. Fr.: 296).

1851   *Rhaphigaster prasinus*: Dallas, Cat. Hem.: 274. Misidentification according to Stål, 1865, Hemipt. Afr.: 193.

1854   *Rhaphigaster orbus* Stål, Öfvers. Vet.-Akad. Förh. 10 (1853): 221 (South Africa). (Synonymized by Stål, 1865, Hemipt. Afr.: 194).

1867   *Pentatoma vicaria* Walker, Cat. Het. Hem. II: 303 ("Hindostan" = India). (Synonymized by Distant, 1900, Ann. Mag. Nat. Hist. 5: 392).

1884   *Nezara viridula* var. *aurantiaca* Costa, Atti R. Accad. Sci. Napoli 1: 37, 58 (Sardinia). (Synonymized by Reuter, 1907, Bull. Soc. Entomol. Fr. 1907: 210).

1889   *Nezara antennata* var. *icterica* Horváth, Ana. Cogn. Het. Himalay.: 31 (Japan and India). (Synonymized by Ferrari et al., 2010, Zootaxa: 17).

1903   *Nezara viridula* var. *hepatica* Horváth, Ann. Hist.-Nat. Mus. Natl. Hung. 1903: 406 (Algeria). (Synonymized by Reuter, 1907, Bull. Soc. Entomol. Fr. 1907: 210).

CONTENTS

7.1   Introduction 353
7.2   Taxonomy 354
7.3   Distribution 356
    7.3.1   Asia 356
        7.3.1.1   Range in Asia: Overview 356
        7.3.1.2   Range Expansion in Asia: A Case Study in Central Japan............................. 357
    7.3.2   Europe..361
        7.3.2.1   The Invasion History in Europe..361
    7.3.3   North America 362
    7.3.4   Caribbean (West Indies), Central America, and South America................................... 365
7.4   Biology 367
    7.4.1   Life History 367
        7.4.1.1   Eggs 368
        7.4.1.2   Nymphs 368
        7.4.1.3   Adults 369
    7.4.2   Symbiotic Relationships 369
    7.4.3   Mate Selection 370
    7.4.4   Winter and Summer Diapauses...371
        7.4.4.1   Diapause Induction in the Laboratory 371
        7.4.4.2   Diapause-Associated Reversible Color Change in Adults............................ 373
        7.4.4.3   Diapause Induction and Adult Color Change in the Field...........................375
        7.4.4.4   Diapause Maintenance in the Laboratory................................... 377
        7.4.4.5   Diapause Maintenance in the Field: Dynamics of the Physiological Condition of Adults during the Overwintering Period................................... 378
        7.4.4.6   Diapause Termination in the Laboratory................................... 378
        7.4.4.7   Diapause Termination in the Field.. 379
        7.4.4.8   Geographic Differences in Diapause Pattern 379
        7.4.4.9   Diapause Syndrome and Overwintering in the Field: Importance of the Timing of Emergence, and the Coloration and Size of Adults............. 380
    7.4.5   Host Plant Associations 381
    7.4.6   Movement among Hosts at Farmscape and Landscape Levels................................... 396
7.5   Current Impacts in Agriculture... 398
    7.5.1   Asia.. 398
    7.5.2   Europe.. 399
    7.5.3   North America 399

7.6   Impacts on General Public 400
7.7   Management 400
        7.7.1   Monitoring Options 400
        7.7.2   Cultural Control...401
        7.7.3   Biological Control..401
                  7.7.3.1   General..401
                  7.7.3.2   Classical Biological Control 402
                  7.7.3.3   Augmentation Biological Control 402
                  7.7.3.4   Conservation Biological Control..402
        7.7.4   Attract-and-Kill Strategies 403
        7.7.5   Chemical Control 403
7.8   Future Outlook 403
7.9   Acknowledgments..404
7.10   References Cited...404

7.1 Introduction

Nezara viridula (L.) is a worldwide pest of numerous crops and cropping systems. The species also is adept at exploiting a wide range of noncultivated plant species in the absence of preferred food resources and during the overwintering period, and this may contribute to its broad distribution. In fact, *N. viridula* may possess the broadest distribution range of the pentatomids (Panizzi and Slansky 1985, CABI/EPPO 1998). McPherson and McPherson (2000) is the most recent authority on the biology, ecology, distribution, and management of *N. viridula*. This chapter is intended to build upon McPherson and McPherson (2000), presenting pertinent research findings to date.

Nezara viridula is known commonly as the southern green stink bug. This current common name is only one of many that has been used for the species, including: "chinche verde" (Peña M. and Sifuentes 1972, Rodríguez Vélez 1974), "cosmopolitan stink bug" (Hokkanen 1986), "cotton green bug" (Kamal 1937), "green bug" (Hubbard 1885), "green-bug" (Drake 1920), "green-bug of India" (Atkinson 1889), "green plant-bug" (Drake 1920), "green soldier bug" (Riley and Howard 1893a, Turner 1918, Watson 1918), "green soldier-bug" (Hubbard 1885), "green stink bug" (CABI Invasive Species Compendium 2015), "green vegetable bug" (Gu and Walter 1989), "plant-bugs" (Riley and Howard 1893b), "pumpkin bug" (Watson 1918, 1919a), "southern green plant-bug" (Jones 1918), "southern green stink-bug" (Drake 1920), "southern green stinkbug" (Wolfenbarger 1947), "southern stinkbug" (Demaree 1922), "southern stink bug" (Nishida 1966), "stink-bugs" (Jones 1918), and "tomato and bean bug" (Froggatt 1916). Drake (1920) originally recommended "southern green stink-bug" as a common name; apparently, the hyphen was omitted at some point after his report.

The exact, or even approximate, date of the invasion of *Nezara viridula* into the United States is unknown, but the species was recorded in Texas in 1880 (Distant 1880). Subsequently, Hubbard (1885) first reported *N. viridula* as a pest of citrus in Florida following correspondence from West Apopka, FL, in 1883. The population dynamics of the insect in citrus and "pea-vines" influenced the pest status of *N. viridula* in Florida (Hubbard 1885). Specifically, when "pea-vines" reached maturity, *N. viridula* abandoned the vines and infested surrounding citrus trees. In reviewing the description of nymphs provided by Hubbard (1885), however, he incorrectly identified the scientific name of the insect as *Raphigaster hilaris* Fitch [sic]. He described the nymphs as: "...young are black, with white spots, which color they retain until nearly fully grown, ...". This description more closely resembles nymphs of *N. viridula* (Morrill 1910, Jones 1918) than those of *Chinavia* (as *Nezara*) *hilaris* (Say) (green stink bug), which do not have white spots (Morrill 1910, Jones 1918). Drake (1920) also noted that Hubbard (1885) was incorrect in his identification of *C. hilaris*, stating he (referring to Hubbard) "undoubtedly refers to *N. viridula* Linn. as *hilaris* Say (the northern green soldier-bug) has never been known to occur in great numbers in Florida." Additionally, Riley and Howard (1893b) identified specimens of *N. viridula* from citrus, making specific reference to the description by Hubbard (1885). Similar to the Texas invasion, Drake (1920) noted that "Owing to the paucity of data it is impossible to conjecture how, when, or even where the insect was introduced into Florida."

7.2 Taxonomy

Amyot and Serville (1843) proposed the genus *Nezara* in the group 'Rhaphigastrides' to include *Cimex smaragdulus* F. (= *viridulus* L.) and *Pentatoma marginata* Palisot de Beauvois. Several taxa were described and included in the genus over the subsequent decades (Freeman 1940), but most of them have been transferred to other genera or are recognized currently as synonyms. In the last world catalogue of the family Pentatomidae, Kirkaldy (1909) recognized six subgenera in *Nezara*, all of which are now considered as separate genera (Bergroth 1914, Freeman 1940, Day 1964, Orian 1965, Linnavuori 1972, Thomas and Yonke 1981, Rolston 1983, Grazia and Fortes 1995, Schwertner and Grazia 2007). The revision by Freeman (1940) firmly established the limits of the genus, recognizing 11 species and 21 varieties. Currently, 12 species are considered valid (Ferrari et al. 2010).

A hypothesis of phylogenetic relationship within *Nezara* was proposed by Ferrari et al. (2010). A basal dichotomy splits the genus into two clades, with *N. viridula* belonging to the clade that also includes *N. antennata* and *N. yunnana*. A close relationship between *viridula* and *antennata* also was shown by morphological similarity and interspecific mating behavior (which, actually, does not result in viable progeny) (Freeman 1940; Kiritani et al. 1963; Kon et al. 1988, 1993, 1994). The species *N. antennata* and *N. yunnana* are restricted geographically to the Oriental and southern Palearctic regions and associated with tropical and subtropical climates. *N. yunnana* occurs from northern India to southern China, and the type locality is Duanli, Yunnan, China. *N. antennata* is more widespread, having been recorded from southern China, India, Japan, the Philippines, South Korea, and Sri Lanka (Rider 2006a).

Because of its color variability and wide geographical distribution, *Nezara viridula* has been the subject of many taxonomic studies, resulting in several synonyms. The species has several 'color types' (i.e., genetic morphs) of adults but two main morphs exist (i.e., numerically dominant in most regions): var. *smaragdula* F. (G-type, completely green coloration) and var. *torquata* F. (O-type, predominantly green body with anterior yellowish coloration) (**Figure 7.1E**). Mating studies between var. *smaragdula* and var. *torquata* yielded offspring with "an orange body like Y-type (f. *aurantica* [sic] Costa) but a yellow band on the pronotum" (Ohno and Alam 1992) (**Figure 7.1F**). The so-called var. *aurantiaca* Costa (Y-type) is all orange and rare (**Figure 7.1G**, at right) (e.g., less than 1% in Brazil [Vivan and Panizzi 2002] and even less in other regions [Hokkanen 1986, Golden and Follett 2006]). Several other color morphs have been recorded from different parts of the world, including a blue form (**Figure 7.1G**, at left; Antônio R. Panizzi, personal communication). Additionally, Chen (1980, and references therein) reported the occurrence of two other forms (*N. viridula* form *duyuna* Chen, n. f., and *N. viridula* form *typica*) in China. More recently, Esquivel et al. (2015) reported the occurrence of a black morph (**Figure 7.1H**).

Different color morphs are found among established populations and do not justify separation at the species level. However, genetic evidence has demonstrated spatially structured patterns of population distribution and differentiation. Meglič et al. (2001) determined a certain amount of genetic difference among geographically isolated populations between continents and Sosa-Gómez et al. (2005) and Vivan and Panizzi (2006) within Brazilian populations. Pavlovčič et al. (2008) analyzed genetic variation between European, North and South American, Asian, and African populations and found a widely distributed haplotype, but they also found an ancient lineage in Africa and a possible hybridization lineage in Japan. Variation in genetic color polymorphism also appears to be spatially structured, with Japan hosting the greatest diversity of variants; other regions often have only a single color morph prevailing (Yukawa and Kiritani 1965). In any case, these color morphs should not be confused with environmentally determined seasonal color variants that may occur before, during, and shortly after hibernation (**Figure 7.1D**; also see **Section 7.4.4 and Chapter 11**).

Based on the distribution and frequencies of the color morphs, Yukawa and Kiritani (1965) proposed the origin of *Nezara viridula* to have been in southwest Asia. Hokkanen (1986) and Jones (1988), based on a more complete data set including parasitoids and ecological characteristics, considered the origin of *N. viridula* to have been in the Ethiopian biogeographical region of Africa. More recently, Kavar et al. (2006) sequenced 16S and 28S rDNA, cytochrome B and cytochrome C oxidase subunit I gene fragments and random amplified polymorphic DNA (RAPD) from geographically separated locations

FIGURE 7.1 **(See color insert.)** *Nezara viridula* eggs, nymphs, and adults. A, egg mass at ≈2 days old (from laboratory culture); B, fifth instars and adults feeding on fruit of *Cucurbita foetidissima*; C, mating adults on *Rapistrum rugosum* (B–C in Central Texas); D, adults expressing varying degree of overwintering russet coloration in Kyoto, Japan; E, (l. to r.) – representative adult color morphs – form *smaragdula* (common all green form, form G-type), form *torquata* (O-type), form *viridula* (R-type), and form "OR-type" resulting from cross of O-type and R-type; F, form "OY-type" resulting from cross of O-type female and either a G-type or O type male; G–H, several other color morphs include blue, orange, and black forms. (Images A–C, Courtesy of Jesus F. Esquivel, USDA, Agricultural Research Service, College Station, TX; D, Courtesy of Dmitry L. Musolin, Saint Petersburg, Russia; E–F, Adapted from K. Ohno and Md. Z. Alam, Applied Entomology and Zoology 27: 133–139, 1992, with permission; G, Blue form and orange form *aurantiaca*, courtesy of Antônio R. Panizzi, Passo Fundo, RS, Brazil; H, Black form, modified from J. F. Esquivel, V. A. Brown, R. B. Harvey, and R. E. Droleskey, Southwestern Entomologist 40: 649–652, 2015, with permission.)

(Slovenia, France, Greece, Italy, Japan, Guadeloupe, Galápagos, California, Brazil, and Botswana) and supported the 'out-of-Africa' hypothesis for the origin of *N. viridula* followed by dispersal to Asia and a more recent expansion to Europe, the Americas, and Oceania. They also suggested that the African and non-African gene pools have been separated for 3.7–4.0 million years according to the standard insect molecular clock of 2.3% pairwise divergence per million years. However, to better understand the origin of *N. viridula* and its related species, a more comprehensive study, including integrative data sources (e.g., molecular, pheromones, reproduction, bioacoustics) of all species of the Oriental clade (including *N. antennata* and *N. yunnana*) and some of the African species is required (e.g., Aldrich et al. 1987, Čokl et al. 2000, Miklas et al. 2003). The possibility of existing cryptic species was demonstrated by the discovery of behavioral differences within some populations (e.g., Ryan et al. 1996, Jeraj and Walter 1998) and indicates further research is needed.

7.3 Distribution

Distribution of *Nezara viridula* often is referred to as worldwide or cosmopolitan as the species is known to occur throughout tropical, subtropical, and warm temperate regions of Eurasia, Africa, Australia, and the Americas approximately between latitudes 45°N and 45°S (Yukawa and Kiritani 1965; Todd 1989; McPherson and McPherson 2000; Panizzi et al. 2000; Yukawa et al. 2007, 2009; Musolin 2007, 2012; Tougou et al. 2009, Panizzi and Lucini 2016). The species is constantly expanding its range (Yukawa et al. 2007, 2009; Musolin 2007, 2012; Rabitsch 2008, 2010), both in the Northern and Southern Hemispheres, by natural dispersal (e.g., Aldrich 1990, Tougou et al. 2009) and human-assisted translocation, which complicate a strict and comprehensive determination of its boundaries.

7.3.1 Asia

7.3.1.1 Range in Asia: Overview

Before describing the range of *Nezara viridula* in Asia, it should be noted that there is no agreed definition of the biogeographical boundaries between Asia and Europe, Africa, or Australia. Whereas in most cases the situation is clear, in others, the geographic, economic, and political points of view might differ in attributing a particular country or entity to a particular part of the world. This is especially true for the Transcaucasian region, with Georgia or Armenia sometimes attributed to Europe or Asia. Another problem arises when different parts of a country lie on different continents, which, as examples, are the cases with Russia, Kazakhstan, Turkey (Europe and Asia), and Egypt (Africa and Asia). Finally, there are a few countries/entities with currently uncertain official status, including the Republic of Cyprus, Hong Kong, Abkhazia, and possibly others.

Keeping these difficulties in mind, to date, *Nezara viridula* has been reported in the following countries/entities in Asia: Abkhazia, Afghanistan, Azerbaijan, Bahrain, Bangladesh, Brunei, Cambodia, China (including Hong Kong and Macau), Christmas Island, Cocos Island, Cyprus, Egypt, Georgia, India, Indonesia, Iran, Iraq, Israel, Japan, Jordan, Korea (North and South), Kuwait, Laos, Lebanon, Malaysia, Myanmar, Nepal, Pakistan, the Philippines, Qatar, Saudi Arabia, Singapore, Sri Lanka, Syria, Taiwan, Thailand, Turkey, Vietnam, and Yemen (Kirichenko 1951, 1955; Putshkov 1972; Singh 1973; Abdu and Shaumar 1985; Todd 1989; Waterhouse 1998; McPherson and McPherson 2000; Panizzi et al. 2000; Rider 2006a,b; Musolin 2012; CABI Invasive Species Compendium 2015; Abdul Aziz Mohamed, personal communication [for Bahrain]). In some of these countries (e.g., Afghanistan, Bangladesh, India, Japan, Myanmar, Taiwan, Turkey, and Vietnam), *N. viridula* is widespread whereas in others, it is present only as a local or restricted distribution (e.g., Laos, Thailand, and the Philippines) (Waterhouse 1998, Musolin 2012, CABI Invasive Species Compendium 2015).

In a few countries/entities in Asia, *Nezara viridula* has not been reported as yet. In Russia, the species is known in the south of the European part of the country (Kirichenko 1951, 1955; Rider 2006a,b; Neimorovets 2010) but not in the Asian part (Vinokurov et al. 2010). However, again, it is important which definition of a boundary between Asia and Europe is used, as the species is known in several neighbouring Transcaucasian countries/entities such as Abkhazia, Azerbaijan, and Georgia. The Asian part of Russia, east of the Ural Mountains, is too cold for *N. viridula*. It is important to note that for the species' overwintering success, the mean temperature during the coldest month of winter is believed to be critical; it should not be lower than 5°C (Musolin 2012; see below and **Section 7.4.4**).

The absence of *Nezara viridula* in some other Asian countries/entities may be explained by the climate characteristic of these regions. For example, Armenia also is located in the Transcaucasian region (i.e., partly in a subtropical climate), but we are not aware of any reliable records of *N. viridula* from this country. The species might be expected to arrive and survive (at least temporarily) in the comparatively small southern lowland part of the country located close to the boundary with Turkey where *N. viridula* is present. The remaining part of Armenia is mountainous and probably not suitable for this species.

Similarly, Kazakhstan, Kirgizia, Tajikistan, and Mongolia mostly have a continental climate and a mountainous topography and, thus, seem to be too cold for successful and permanent occurrence of *Nezara viridula*.

The territory of Turkmenistan also is mostly mountainous, but its climate is subtropical in the lowland regions and the occurrence of *Nezara viridula* is considered probable (Putshkov 1965).

Most of the territory of Uzbekistan is covered with deserts, steppes, and mountains and has a continental climate, which altogether are not suitable for *Nezara viridula* in terms of temperature and food resources.

Although we found no published records of *Nezara viridula* in Bhutan, Maldives, United Arab Emirates, and Oman, the species probably occurs in some of these countries in the regions with a milder climate.

7.3.1.2 Range Expansion in Asia: A Case Study in Central Japan

The range of *Nezara viridula* is expanding constantly, both in the Northern and Southern Hemispheres, with examples of recent invasions including California and Washington in the United States, Paraguay, southern Argentina, Hungary, Moldova, the United Kingdom, northern Switzerland, and southwest Germany (Panizzi et al. 2000; Rédei and Torma 2003; Barclay 2004; Werner 2005; Rabitsch 2008, 2010; Stewart and Kirby 2010; Grozea et al. 2012; Musolin 2012; Derjanschi and Chimişliu 2014; Looney and Murray 2016).

The phenomenon of rapid range expansion of *Nezara viridula* was documented and studied in detail in central Japan (**Figure 7.2**). The northward range expansion is believed to have been influenced strongly by changes in agricultural practices and the current climate warming (Musolin 2007, 2012; Yukawa et al. 2007, 2009; Tougou et al. 2009; Kiritani 2011; Musolin and Saulich 2012; Geshi and Fujisaki 2013). On the other hand, it is considered a part of a complex interaction with the congeneric *N. antennata*: under favorable conditions, such as a few consecutive mild winters, available food, and high population density, *N. viridula* often will replace *N. antennata* almost completely (Kiritani 1963, 1971, 2011; Kiritani and Hokyo 1970; Musolin 2007, 2012).

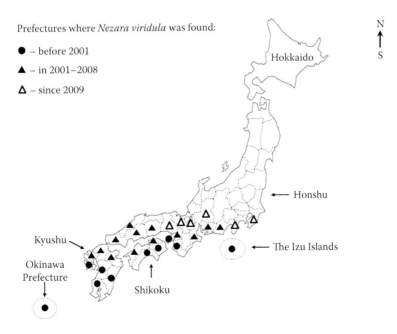

FIGURE 7.2 Changes in distribution of *Nezara viridula* in Japan. (Data from T. Koide, K. Yamaguchi, N. Ohno, and K. Morimoto, Annual Reports of the Kansai Plant Protection Society 52: 163–165, 2010; K. Kiritani, Journal of Asia-Pacific Entomology 14: 221–226, 2011; K. Suzuki, M. Nishino, and S. Shimo, Annual Reports of the Kansai Plant Protection Society 53: 133–134, 2011; D. L. Musolin, Physiological Entomology 37: 309–322, 2012; and N. Mizutani, Shokubutsu Boeki [Plant Protection] 67: 595–601, 2013, with permission.)

Japan is at the northern margin of the Asian distribution of *Nezara viridula*. In the southern portion of the country, *N. viridula* was first recorded in 1874 (Nagasaki Prefecture, Kyushu [Hasegawa 1954]), although it is difficult to conclude with certainty whether the species is native to Kyushu or was introduced to the island. Recent genetic research suggests multiple colonization of Japan by *N. viridula* (Kavar et al. 2006). In 1952, occurrence of the species was reconfirmed in the southern coastal areas of Kyushu and recorded on Shikoku and Honshu (Wakayama Prefecture [Kiritani 2011]). Since then, the range of *N. viridula* has been expanding further northward (**Figure 7.2**).

Outbreaks of *Nezara viridula* in the 1950s were promoted by cultivation of early-, mid-, and late-season planted rice. Emergence of adults of three summer generations generally coincided with the heading stages of consecutive generations of cultivated rice, an excellent food source for the bugs (Kiritani 2011; Keizi Kiritani, personal communication).

In the early 1960s, a wide-scale field survey in central Honshu mapped the northern limit of the species in the region (**Figure 7.3A**). The northern limit was shown to lie in Wakayama Prefecture (approximately 34.1°N) and to coincide roughly with the 5°C isothermal for the mean air temperature of the coldest winter month, which usually is January (Kiritani et al. 1963, Kiritani and Hokyo 1970).

In 2006–2007, a new wide-scale field survey demonstrated that, in the 45 years since the first field survey, the northern limit of the range had shifted northwards by ≈85 km (i.e., at a mean rate of 19 km per decade) (**Figure 7.3B,D**; Tougou et al. 2009, Musolin 2012). Within the next five years the northern limit moved further northward by 25 km (**Figure 7.3C,E**; Geshi and Fujisaki 2013). This northward range expansion of *Nezara viridula* in Japan raises some questions: what factors have favored this range expansion, is the species able to overwinter successfully in the recently colonized areas, and how well has the seasonal development of *N. viridula* adapted to the new environment?

Several abiotic and biotic factors are known to limit distribution ranges in insects (Uvarov 1931, Cammell and Knight 1992). Climate (mostly, thermal conditions), food, and habitat availability are among the most important constraints of a species' range. For *Nezara viridula*, food and habitat availability are unlikely to be the principal limiting factors given that the species, although exhibiting a preference for leguminous plants, is highly polyphagous (Oho and Kiritani 1960, Panizzi et al. 2000; see **Section 7.4.5**). Also, the species' range does not seem to be limited by seasonal warmth. More than one generation is produced each year (up to three generations, with a partial fourth [Kiritani et al. 1963; Musolin and Numata 2003a,b; Musolin 2007, 2012; Musolin et al. 2011; Saulich and Musolin 2014]) at locations closer to the northern margin of *N. viridula* distribution in Japan (e.g., in Wakayama, Osaka, and Kyoto Prefectures). This is unusual because, in the Northern Hemisphere, most heteropteran populations and many other insects are univoltine (or even semivoltine) towards the northern limits of their ranges, even if they are bi- or multivoltine in the more southern regions of their respective distribution ranges (Danks 1987; Saulich and Musolin 1996, 2014).

An assessment of winter survival of adult *Nezara viridula* in 16 different habitats over six winter seasons (1961–1967) shows that survival rates differ between sexes and types of hibernacula (i.e., *N. viridula* can aggregate on different overwintering host plants) and are affected by adult size and coloration (Kiritani et al. 1962, 1966; Musolin 2012). Winter temperature appears to be the principal factor that determines adult mortality during the hibernation period (**Figure 7.4**). Only 1.5% of males and 3.5% of females managed to survive the severe winter of 1962/1963 when the mean temperature in January fell to 2.9°C. Survival during moderately cold winters is much higher (40–65%) (**Figure 7.4**; Kiritani et al. 1966, Kiritani 1971, Musolin 2012). Overwintering mortality correlates negatively with the mean temperature of the coldest month (i.e., usually January), and a decrease of 1°C results in approximately a 15% increase in mean overwintering mortality. Thus, the mean January temperature was proposed to be the principal factor that determined the northern limit of the distribution of *N. viridula* in Japan (Kiritani et al. 1963). These early field data and conclusions are supported by a series of outdoor rearing experiments (**Figure 7.4**; Musolin and Numata 2003b, 2004; Musolin 2007, 2012; Tougou et al. 2009). Such a threshold may also well explain establishment of this species in Great Britain but does not explain well some other local occurrences in Central Europe (see **Section 7.3.2**).

Thus, *Nezara viridula* represents a comparatively rare case in which the northern limit of distribution of an ectothermic species is determined not by the thermal resources (i.e., available seasonal heat) for growth and development, or the availability of food resources or habitat, but by the winter temperature conditions (Musolin 2007, 2012).

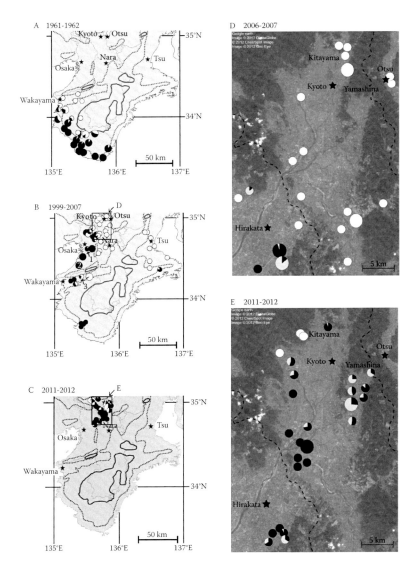

FIGURE 7.3   The shift of the northern limit of distribution of *Nezara viridula* in central Japan from the early 1960s (A, Data from K. Kiritani, N. Hokyo, and J. Yukawa, Researches on Population Ecology 5: 11–22, 1963, and K. Kiritani, Proceedings of the Symposium on Rice Insects, Tropical Agricultural Research Center, Tokyo, Japan, pp. 235–248, 1971) to 1999–2007 (B, Data from D. Tougou, D. L. Musolin, and K. Fujisaki, Entomologia Experimentalis et Applicata 30: 249–258, 2009), and 2011–2012 (C, Data from J. Geshi and K. Fujisaki, Japanese Journal of Applied Entomology and Zoology 57: 151–157, 2013). Symbols: black sections, relative abundance of *N. viridula*; white sections, relative abundance of sympatrically distributed congeneric *N. antennata*. Sample size: small circles, 1–50 specimens; large circles, >50 specimens. Six prefectures in central Japan are outlined and their capital cities are indicated as stars (in A, B, and C). Elevation (simplified): dotted line, 500 m above sea level (a.s.l.); solid black line, 1,000 m a.s.l. (in A, B, and C). Maps D and E are enlarged satellite photographs of rectangular boxes from maps B and C, respectively, showing elevation (dark areas are higher in elevation than pale areas). In addition to the data from the 2006–2007 field survey, the following data points are included (numbered on map B): (1) data from D. L. Musolin and H. Numata, Physiological Entomology 28: 65–74, 2003a, and D. L. Musolin, Global Change Biology, 13: 1565–1585, 2007 (collected in 1999); (2) data collected by H. Tanaka in 2003; see J. Yukawa, K. Kiritani, N. Gyoutoku, N. Uechi, D. Yamaguchi, and S. Kamitani, Applied Entomology and Zoology, 42: 205–215, 2007); and (3) data collected by M. Morishita in 2002; see J. Yukawa, K. Kiritani, N. Gyoutoku, N. Uechi, D. Yamaguchi, and S. Kamitani, Applied Entomology and Zoology, 42: 205–215, 2007). See text for details. Maps A and B from D. Tougou, D. L. Musolin, and K. Fujisaki, Entomologia Experimentalis et Applicata 30: 249–258, 2009, with permission; map C from J. Geshi and K. Fujisaki, Japanese Journal of Applied Entomology and Zoology 57: 151–157, 2013, with permission.

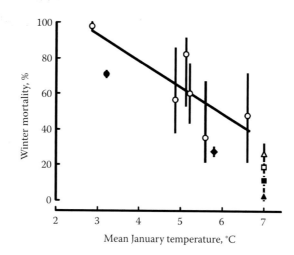

FIGURE 7.4 The effect of mean January temperature on winter mortality of *Nezara viridula* adults in central Japan. Field experiments in Asso in 1961–1967 (open circles): mean (and range of) mortality (in different types of hibernacula) in adults of both sexes (*n* = 284–1197 per winter) (Data from K. Kiritani, N. Hokyo, and K. Kimura, Annales de la Société Entomologique de France, Nouvelle Série 2: 199–207, 1966, and K. Kiritani, Proceedings of the Symposium on Rice Insects, Tropical Agricultural Research Center, Tokyo, Japan, pp. 235–248, 1971); a linear regression trend line refers to the mean mortality ($F_{1,5} = 6.81$, $P = 0.06$). Additional outdoor experiments in Osaka in 1999–2000 and in Kyoto in 2006–2007 and 2007–2008 (all other symbols): mean mortality in adults of both sexes and range (mortality in two sexes); *n* = 23–138 per cohort (Data from D. L. Musolin and H. Numata, Ecological Entomology 28: 694–703, 2003b; D. L. Musolin and H. Numata, Entomologia Experimentalis et Applicata 11: 1–6, 2004; D. L. Musolin, D. Tougou, and K. Fujisaki, Global Change Biology 16: 73–87, 2010; and K. Takeda, D. L. Musolin, and K. Fujisaki, Physiological Entomology 35: 343–353, 2010). Note that Kiritani et al. (1966) and Kiritani (1971) measured mortality during the hibernation only and did so in the wild, whereas in all other experiments we also included prewinter mortality and reared the insects in containers, thus providing protection from predators and parasites. (From D. L. Musolin, Physiological Entomology 37: 309–322, 2012, with permission.)

An analysis of historical climatic data (Tougou et al. 2009) suggests that the shift in distribution of *Nezara viridula* in Japan most likely has been promoted by the milder overwintering conditions in the region during the last few decades. The mean temperatures in the region from January to February were 1.03–1.91°C higher during 1998–2007 than during 1960–1969. The number of cold days in January and February (with mean daily temperatures below 5°C) also has decreased significantly, whereas the annual lowest temperature has risen significantly from 1960–1969 to 1998–2007 (Tougou et al. 2009). In central Japan, *N. viridula* is found predominantly near locations where the mean January temperature exceeds 5°C, the mean cumulative number of cold days (with daily mean below 5°C) does not exceed 26 during January and February, and the mean annual lowest temperature does not drop below –3.0°C. This analysis also shows that the mean January temperature and the number of cold days are the most critical factors that determine the northern distribution limit of *N. viridula*. All of the climatic data suggest that over the last 45 years, environmental conditions have become more favorable for overwintering of *N. viridula* at many locations in central Japan. This likely has promoted the northward spread of the species and represents a direct response of this species to climate warming (Tougou et al. 2009). Subsequent field collections in colder and more elevated locations documented a more rapid range shift towards the north (about 5 km per year; **Figure 7.3C,E**), thus suggesting that not only temperature but also other abiotic and, perhaps, biotic and anthropogenic factors are involved (Geshi and Fujisaki 2013).

Adults of *Nezara viridula* are strong fliers and able to disperse over long distances in the wild, presumably before and after diapause (Gu and Walter 1989). Adults of the species also are known to be carried over long distances by hurricanes (up to 750 km; Aldrich 1990) and probably also by typhoons. It is highly probable that every year, the population range limits in the northern temperate zone move further north and reproduction occurs in new areas in summer. The trend of winter warming, with both natural and anthropogenic components (e.g., in a form of the "heat islands"; see Tougou et al. 2009), improves the overwintering conditions for the species and enhances its survival. This, in turn, allows invaders to reproduce in the spring, have the advantage of reproducing in early-season on early agricultural crops, and to become established in newly colonized areas (Musolin 2012).

7.3.2 Europe

The earliest record of *Nezara viridula* within Europe apparently is the description of *Cimex smaragdulus* by Fabricius (1775) from the Macaronesian island of Madeira. The species subsequently was mentioned in most of the southern European countries with the first continental record from southern France by Villers (1789). The global distribution map suggested by DeWitt and Godfrey (1972) drew the northern limit of occurrence in Europe along southern France, northern Italy (south of the Alps), Slovenia, southern Hungary, and along the north coast of the Black Sea, which generally coincides with the Mediterranean climate or biogeographic region, particularly in the eastern European realm. However, according to these records, it is suggested that records north of the Alps at the approximate latitude of 46°N should be considered recent immigrations (Rabitsch 2008).

The reasons for the current northward expansion in Europe are unclear. Climate change favors population growth via shortened developmental cycle and extended environmentally favorable period in the summer. These latter factors result in higher population density and increased numbers of annual generations but also reduced winter mortality due to milder winters (e.g., Musolin and Numata 2003b; Musolin 2007, 2012). A similar pattern was demonstrated for Japan (Yukawa et al. 2007, Tougou et al. 2009). The suggested threshold for overwintering survival is approximately 5°C mean air temperature of the coldest winter month, 26 cold days during January and February with daily mean temperature below 5°C, and mean annual lowest temperature below –3°C (see the case study in **Section 7.3.1.2**), however, this survival threshold falls short in most western and central European climates. The average January air temperature in Budapest is ≈–1°C, but, nevertheless, the species has become established. The "heat island" effect within cities and urban habitats may play a crucial role in the establishment process of *Nezara viridula* in large parts of Europe and elsewhere (Tougou et al. 2009). It further has to be demonstrated that populations can survive winters and be sustained for longer periods of time.

Besides active dispersal, the role of passive, short-distance, human-assisted translocations within a (more or less) contiguous continent without border inspections for goods, particularly plants for transplanting (e.g., greenhouse and ornamentals) or human consumption (e.g., field crops, vegetables, and fruits), and any other transport vehicles may be equally relevant for the northward expansion. *Nezara viridula* regularly is intercepted by plant inspection authorities on a wide range of imported plants and products of different origin (Reid 2005, Malumphy and Reid 2007). The isolated records in Great Britain can only be explained from such an independent translocation event from continental Europe or from elsewhere.

7.3.2.1 The Invasion History in Europe

Assuming that records from regions with a Mediterranean climate in southern Europe fall within the native range of *Nezara viridula*, the first documented findings outside this area were during the 1920s from Germany, the 1930s from Great Britain, and the 1950–1960s from Belgium, the European part of Russia (Crimea), Finland, and Austria, with subsequent, but scattered records in Bulgaria, Hungary, Switzerland, Romania, Slovakia, and the Netherlands (**Table 7.1**). Apparently, at least some of these occurrences did not lead to the immediate establishment of permanent reproducing and overwintering populations. Some time lag (i.e., the delay between first introduction and establishment in the field) is a well-known feature of the biological invasion process (e.g., Crooks 2005). The status in Finland, Austria, Romania, Slovakia, and the Netherlands needs to be monitored further. Most recently, observations of nymphs have been made in Austria and the Netherlands in house gardens and glass houses, which indicate at least temporarily reproducing populations (Aukema 2016, Rabitsch 2016).

The leading edge distribution range dynamics of *Nezara viridula* in the eastern European realm are difficult to disentangle. According to DeWitt and Godfrey (1972), all of Bulgaria and southern Romania falls within the native range. The first mention for southern Bulgaria was by Strawinski (1959), and we here consider this geographical area as part of this bug's native range. However, *N. viridula* has been

TABLE 7.1

First Records (Interception Data and Records from the Field) and Additional Reports of *Nezara viridula* in Its Nonnative Range in Europe

Country	Year of First Record	Reference	Current Status	Additional Reports Confirming Presence
Germany	1922	Reichensperger 1922	Established	Hoffmann 1992, Rieger 1994, Schuster 1986, Voigt 1998, Werner 2005
Great Britain	1930	Salisbury et al. 2009	Established	Barclay 2004, Lansbury 1954, Reid 2006, Shardlow and Taylor 2004, Southgate and Woodroffe 1952
Belgium	1950	Schmitz 1986	Established	Dethier and Chérot 2014, Dethier and Gallant 1998, Dethier and Steckx 2010, Gallant 1996
European part of Russia	1951	Kirichenko 1951	Established	Kirichenko 1955, Neimorovets 2010
Finland	1956	Kontkanen 1956	Casual	–
Austria	1962	Dethier 1989	Unknown	Heiss 1977, Rabitsch 2016
Bulgaria north	2001	Simov et al. 2012	Established	–
Hungary north	2002	Rédei and Torma 2003	Established	Rédei and Vétek 2005
Switzerland north	2005	Werner 2005	Established	–
Romania west	2010	Grozea et al. 2012	Unknown	–
Slovakia	2014	Vétek and Rédei 2014	Unknown	–
Netherlands	2014	Aukema 2016	Unknown	–

found only recently in the north of Bulgaria (Black Sea Coast in 2001, Sofia in 2011) (Simov et al. 2012) and in western Romania (Grozea et al. 2012). These more recent records are from artificial habitats (i.e., city center, gardens, crop field), which indicate a human-assisted translocation rather than a range expansion due to natural dispersal. The invasion history within the Ukraine, Crimea, and on the northern coast of the Black Sea (i.e., Krasnodarskiy Kray in the European part of Russia [Kirichenko 1951, 1955; Rider 2006a,b; Neimorovets 2010]) is not well documented and the status remains unclear. Further north, records in Bryansk Region of Russia (Novozybkov town, 52°32'N, close to Belarus; Giterman 1931) are considered erroneous (Putshkov 1961).

There is little doubt that *Nezara viridula* will increase further in abundance and become established in new regions in Europe outside its natural range. It is likely that such records will be in cities and urban habitats and along the climatically milder western European coastline. This is supported by the number of documented interceptions of *N. viridula* by the Plant Health Authorities in the United Kingdom (23 interceptions between 1930 and 2007), most of which were associated with imported plant materials from Italy. It remains to be seen, however, if *N. viridula* is capable of adapting to the still harsher continental winter climate and thriving well in Central European territories.

7.3.3 North America

For the purposes of this section, North America includes Canada, Mexico, and the continental United States of America plus Hawaii. Dependent countries/territories of North America are discussed in **Section 7.3.4**.

DeWitt and Armbrust (1978) stated that the first record of *Nezara viridula* (form *smaragdula*) in the New World, as reported by Fabricius (1798), was in the West Indies. The first collection in the United States was from Texas and reported by Distant (1880). The distribution of *N. viridula* throughout the continental United States has been reported previously (McPherson 1982, Todd 1989, McPherson and McPherson 2000) and mapped (Commonwealth Institute of Entomology 1953, 1970; Todd and Herzog 1980; CABI/ EPPO 1998; Panizzi and Lucini 2016) but, given the species' potential for range expansion (Musolin and Saulich 2012), it is not surprising that its range now extends beyond these earlier distribution maps.

Distribution of *Nezara viridula* in the continental United States and Hawaii is presented in **Table 7.2** and depicted in **Figure 7.5**. **Table 7.2** is a compilation of representative published reports, current electronic databases on pest distribution, and personal communication with university and federal researchers. These latter communications allowed additional insight into unpublished information on the presence of this species, thus enabling a more complete assessment of its current distribution (**Figure 7.5**).

TABLE 7.2

Distribution of *Nezara viridula* in the United States of America

State	Reference
Alabama	CABI Invasive Species Compendium 2015, Discover Life 2015, Herbert et al. 1988, National Agricultural Pest Information System 2015
Arkansas	CABI Invasive Species Compendium 2015, Smith et al. 2009, Snodgrass et al. 2005
Arizona	Jones 1918, Donald B. Thomas (U. S. Department of Agriculture, Agricultural Research Service, Edinburg, TX; personal communication)
California	CABI Invasive Species Compendium 2015; Discover Life 2015; Hoffmann et al. 1987a,b; National Agricultural Pest Information System 2015
Delaware	Discover Life 2015
Florida	Blatchley 1926, Buschman and Whitcomb 1980, CABI Invasive Species Compendium 2015, Discover Life 2015, Drake 1920, Hubbard 1885, Morrill 1910
Georgia	CABI Invasive Species Compendium 2015, Discover Life 2015, McPherson et al. 2001, Turner 1918
Hawaii	CABI Invasive Species Compendium 2015, Discover Life 2015, National Agricultural Pest Information System 2015, Nishida 1966
Illinois	CABI Invasive Species Compendium 2015, McPherson and Cuda 1974
Kentucky	Weeks et al. 2012 – anecdotal, no data presented
Louisiana	Blatchley 1926, CABI Invasive Species Compendium 2015, Discover Life 2015, Morrill 1910, National Agricultural Pest Information System 2015, Orr et al. 1986
Maryland	Aldrich 1990, CABI Invasive Species Compendium 2015, National Agricultural Pest Information System 2015
Mississippi	CABI Invasive Species Compendium 2015, Discover Life 2015, Snodgrass et al. 2005
Missouri	Bailey 2009
New Mexico	Bundy 2012, Jones 1918, Donald B. Thomas (U. S. Department of Agriculture, Agricultural Research Service, Edinburg, TX; personal communication)
New Jersey	Discover Life 2015
New York	CABI Invasive Species Compendium 2015, Froeschner 1988 – adventitious, Torre-Bueno 1912, Donald B. Thomas (U. S. Department of Agriculture, Agricultural Research Service, Edinburg, TX; personal communication)
North Carolina	Discover Life 2015, Musser et al. 2014
Ohio	CABI Invasive Species Compendium 2015, Capinera 2001 – anecdotal but Balduf (1923) collected *Nezara hilaris*, Froeschner 1988 – adventitious
Oklahoma	CABI Invasive Species Compendium 2015, Mulder 2015, National Agricultural Pest Information System 2015, Richard Grantham (Oklahoma State University, Stillwater; personal communication)
Pennsylvania	Donald B. Thomas (U. S. Department of Agriculture, Agricultural Research Service, Edinburg, TX; personal communication)
South Carolina	CABI Invasive Species Compendium 2015, Discover Life 2015, Jones and Sullivan 1982, Reay-Jones 2010
Tennessee	CABI Invasive Species Compendium 2015, Musser et al. 2014, Scott Stewart (University of Tennessee, Jackson; personal communication)
Texas	Blatchley 1926, CABI Invasive Species Compendium 2015, Discover Life 2015, Eger and Ables 1981, Flitters 1963, McPherson and Sites 1989, Morrill 1910, Reinhard 1922, Vyavhare et al. 2014
Virginia	Batra et al. 1981; Blatchley 1926; Capinera 2001 – anecdotal, no data presented; CABI Invasive Species Compendium 2015; Froeschner 1988 – adventitious; Hoffman 1971 – citing Van Duzee 1904, anecdotal; Kirkaldy 1909; Van Duzee 1904 – anecdotal based on "writings of Dr. P. R. Uhler"
Washington	Anonymous 2016, Looney and Murray 2016
West Virginia	Weeks et al. 2012 – anecdotal, no data presented

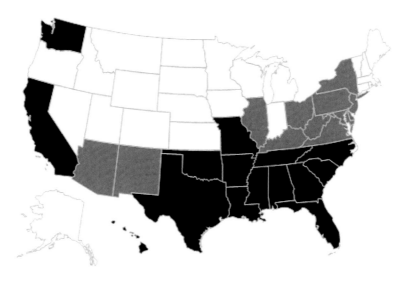

FIGURE 7.5   Distribution of *Nezara viridula* in the United States of America: black represents confirmed field populations; dark gray represents occurrence based on adventitious individuals, museum specimens, and anecdotal records; and
white represents absence of species. (Map based on data presented in Table 7.2).

Overall, the electronic databases were in agreement with the presence of *Nezara viridula* in the southeastern United States, but all varied on the occurrence of the species in more northern and southwestern
states. Predominant occurrences were in states along the Gulf of Mexico and bordering Mexico as well
as virtually the entire Atlantic seaboard (**Figure 7.5**). New York is the northernmost occurrence of
N. viridula along the Atlantic seaboard; however, this was an insect detected in a greenhouse environment (Torre-Bueno 1912). Earlier field surveys did not indicate the presence of *N. viridula* in New York
(Torre-Bueno 1903, 1908). The westernmost occurrences of *N. viridula* are California and Washington
(Hoffmann et al. 1987a,b; Discover Life 2015; Anonymous 2016; Looney and Murray 2016), but the
species also was discovered in Hawaii during 1961 (Davis 1964, Nishida 1966, Jones and Caprio 1992).
     More recent reports indicate *Nezara viridula* has been detected in and around Seattle, Washington
(Anonymous 2016, Looney and Murray 2016). Reported occurrences were at locations ranging from
47.20° to 47.62°N (Chris Looney, personal communication). These observations account for the northernmost occurrence in the western United States and lie outside the historically accepted most northern latitudinal boundary of occurrence (i.e., 45°N). The species now has been detected in Seattle for
three consecutive years (beginning in 2014), but the exact origin of the initial infestation is unknown
(Looney and Murray 2016; Chris Looney, personal communication). *N. viridula* has not been detected
in Oregon, suggesting that the infestation may have resulted from imported plant material (C. Looney,
personal communication) instead of gradual dispersal from California populations. When one considers all reported occurrences, *N. viridula* has been recorded from 27 states, over half of the contiguous
United States.
     The distribution map does not guarantee that *Nezara viridula* will be collected every year or through-
out each state (**Figure 7.5**). In fact, some state records shown in the map may be based on adventitious
specimens, particularly in the northeast. Froeschner (1988) indicated that records from more northern
states "probably are adventitious occurrences." This is supported by the fact that during the course of a
6-year trapping study for stink bugs in New Jersey, no specimens of *N. viridula* were collected (Nielsen
et al. 2013). Also supporting this statement is Aldrich (1990) who described a Maryland occurrence as
"adventitious" and likely the result of insect displacement by a hurricane (similar hurricane effects on
N. viridula populations were reported in Texas [Flitters 1963]). A third example of a probable adventitious occurrence was the collection of a third instar by sweeping roadside vegetation in southern Illinois
(McPherson 1982). In addition, a museum specimen was reported that supposedly had been collected in
central Illinois, but there is some question about the accuracy of the label information (McPherson 1982).

In several instances, determinations of occurrence were based on identification of a single insect for each location (New York, Torre-Bueno 1912; Maryland, Aldrich 1990; New Mexico, Bundy 2012; Delaware and New Jersey, Discover Life 2015; Pennsylvania [Donald B. Thomas, personal communication]), or a single insect in 2–3 localities within the state (Arizona, Donald B. Thomas, personal communication; Oklahoma, Richard A. Grantham, personal communication; NAPIS 2015); the Oklahoma occurrence was reported as an invasive in 1974 (Mulder 2015).

Confirmation of reproductive field populations throughout the distribution range is difficult because of anecdotal records (i.e., reports without, or with minimal, supporting data) (the southwestern United States and Mexico, Morrill 1910; Arizona and New Mexico, Jones 1918; Kentucky, Weeks et al. 2012; Ohio, Capinera 2001; Virginia, Van Duzee 1904, Kirkaldy 1909, Hoffman 1971, Froeschner 1988; West Virginia, Weeks et al. 2012) or conflicting reports regarding the species' presence or absence in a given locale (e.g., presence or absence in Virginia [Batra et al. 1981, Froeschner 1988, Kamminga et al. 2009]).

In addition to conflicting reports, the confined regional occurrences and annual population fluctuations can affect detection of the species' presence. Missouri and Tennessee are included in the distribution map, but these are states where species occurrence can be sporadic, in smaller population densities, and in more limited agricultural ecosystems (Bailey 2009; Scott Stewart, personal communication). In Missouri, Marston et al. (1979) reported that *Nezara viridula* had not been collected during 1972 and 1974. However, 30 years later, Bailey (2009) noted that it occurred in southern and southeastern counties. When present in Tennessee, the species occurs primarily in the southwestern region of the state (Scott Stewart, personal communication). Finally, during a 2-year field survey of stink bug species in the Brazos River Bottom of Central Texas, Suh et al. (2013) did not collect *N. viridula* despite the species having been present in previous consecutive years (Lopez et al. 2014). Within this same production region, *N. viridula* adults also were collected from native vegetation and in light traps during 2015–2017 (Jesus F. Esquivel, unpublished data). Similarly, Jones and Sullivan (1983) observed a complete absence of *N. viridula* field populations following a severe winter freeze in South Carolina.

Distribution of *Nezara viridula* spans Mexico (Discover Life 2015), from the western state of Sonora (Gibson and Carrillo S. 1959) to the Yucatán Peninsula (Rodríguez Rivera 1975). This species has been collected in the Mexican states of Chiapas, Chihuahua, Michoacán, Morelos, Nayarit, Nuevo León, Oaxaca, Puebla, Querétaro, San Luis Potosí, Sinaloa, Sonora, State of México, Tamaulipas, Veracruz, and Yucatán (Gibson and Carrillo S. 1959, Rodríguez Vélez 1974, Cancino and Blanco 2002, Ruiz et al. 2006, Discover Life 2015). Thus, in total, *N. viridula* occurs in at least 16 (of the 31) Mexican states. Similar to the United States, *N. viridula* occurs in slightly more than half of the Mexican States.

To date, *Nezara viridula* has not been reported in Alaska and, because of the climate, any occurrence would presumably be due to human intervention, adventitious occurrence, or as transients in shipping as observed in Canada (Maw et al. 2000, Paiero et al. 2013). In Canada, Maw et al. (2000) indicated detection of *N. viridula* in Ontario and Quebec but these were transient occurrences through detection at ports of entry, not detection of reproducing field populations.

7.3.4 Caribbean (West Indies), Central America, and South America

The region is comprised of about 32 countries and 20 overseas or dependent countries/territories/departments. The geographical limits of these countries/territories are enclosed between latitude 27°12'N (the Bahamas) and 56°32'S (Chile) and longitude 109°26'W (Easter Island, Chile) and 28°50'W (Martim Vaz Archipelago, Brazil).

Nezara viridula is considered widely distributed in this region (e.g., Todd 1989, Panizzi 1997) but has not been reported for all localities (**Figure 7.6**). It has been recorded from 19 countries and two territories/departments with Cuba as the northernmost limit and Argentina as the southernmost limit (**Table 7.3**). The range limit is restricted only in the south, with records occurring above latitude 40°S; the northern limit is continuous with the North American distribution. Central and South American localities not cited in the distribution range by DeWitt and Godfrey (1972) are Belize, Ecuador, and Guyana (Discover Life 2015), including a recent colonization of the Gálapagos Islands (Henry and Wilson 2004), and Central Argentina south to Buenos Aires (**Figure 7.6**). Although these latter

FIGURE 7.6   (See color insert.) Distribution of *Nezara viridula* in the Caribbean (West Indies), Central America, and South America. Solid yellow circles and yellow circles with black center dot indicate, respectively, exact localities recorded in literature and inferred localities from country or province/state/county records. The white line delimits the Amazon Basin.

observations indicate expansion in South America, the pest population has declined dramatically over the past 15 years in this continent (Panizzi and Lucini 2016). This decline has been attributed to increased use of herbicides, change in cultivation systems, inter-specific competition among stink bug species infesting major crops, increased impact of egg parasitoids, and climate change (Panizzi and Lucini 2016).

Continental countries without any records for *Nezara viridula* include El Salvador and Guatemala in Central America, and Bolivia, Peru, and Suriname in South America. For South America, DeWitt and Godfrey (1972) showed the distribution range reaching the state of Amazonas and wide distribution in the state of Pará. However, we were unable to locate records confirming the presence in the Amazonas and, contrary to the reported wide distribution in Pará, records exist for only one locality in Pará. Although the absence in the Central America countries probably is due to failed collections or observations, the absence in the Amazonian basin (**Figure 7.6**) may be due to climate or some other factors limiting the distribution of *N. viridula*. There are no studies concerning the range expansion of this species in Central and South America; most studies relate to the increase of soybean cultivation in the region (Panizzi and Slansky 1985, Panizzi et al. 2000, Vivan and Panizzi 2006).

Kavar et al. (2006) found evidence for at least four different haplotypes of *Nezara viridula* in Central and South America with two of the haplotypes unique to Brazil. The other two haplotypes showed distinct relationships: one haplotype shared by Central and northern South America, eastern Mediterranean, and North America populations; and the other shared by southeastern South America and western Mediterranean populations. The authors discussed these results as a consequence of different colonization routes of *N. viridula* in the Western Hemisphere. Interestingly, the distribution of this species in Central and South America supports a separation of at least two different populations (**Figure 7.6**).

The first record of *Nezara viridula* in this region came from Fabricius (1798) with the description of *Cimex spirans*, a junior synonym of *N. viridula* (Stål 1868), from the West Indies. Since then,

TABLE 7.3

List of Countries in the Caribbean (West Indies), Central America, and South America with records for *Nezara viridula*

Caribbean
1. Cuba
2. Dominican Republic
3. Haiti
4. Jamaica
5. Puerto Rico [USA commonwealth]
6. Saint Vincent and Grenadines

Central America
1. Belize
2. Costa Rica
3. Honduras
4. Nicaragua
5. Panama

South America
1. Argentina
2. Brazil
3. Chile
4. Colombia
5. Ecuador
6. French Guiana [French overseas department]
7. Guyana
8. Paraguay
9. Uruguay
10. Venezuela

several other records have included Central and South America (Palisot de Beauvois 1818, Amyot and Serville 1843, Pennington 1919, Costa Lima 1940, Rufinelli and Pirán 1959). It now has been reported as a vector of plant disease organisms in Puerto Rico (Kaiser and Vakili 1978).

7.4 Biology

General life-history biology of *Nezara viridula* has not deviated significantly from earlier reports (e.g., Jones 1918, Drake 1920). However, improvements in available scientific equipment and recent changes in abiotic factors have enabled researchers to more clearly elucidate behaviors related to oviposition, feeding, endosymbionts, pathogen transmission, overwintering survival frequency, and mechanisms allowing overwintering survival. This section expands on these findings.

7.4.1 Life History

The developmental life stages of *Nezara viridula* have been described by Jones (1918) and Drake (1920). Characteristics and descriptions of the eggs, nymphs of different stadia/stages, and adults have changed minimally since those original descriptions. Further, Todd (1989) and McPherson and McPherson (2000) provided an exhaustive literature review of biotic and abiotic factors affecting the biology and ecology of the species. Details on each of the life stages are summarized below, with a focus on more recent findings pertaining to the biology and ecology.

7.4.1.1 Eggs

Female *Nezara viridula* oviposit clusters of eggs (= egg masses; McLain and Mallard 1991). The numbers of eggs per cluster usually range from 60 to 90 (extreme recorded values are 1 to 184) (Musolin and Numata 2003a, Musolin et al. 2007). As eggs are oviposited, each is covered (or "smeared") with a viscous liquid or glue-like material that adheres the eggs to the oviposition substrate (e.g., leaf surface) and each other (Drake 1920) (**Figure 7.1A**). Prado et al. (2006) indicate this "smearing" process also deposits beneficial endosymbionts on the surface of the egg cluster (see **Section 7.4.2**). Ovipositing females have been observed using their hind tarsi to position eggs within these clusters (McLain and Mallard 1991, Panizzi 2006). Drake (1920) described changes in the appearance of the egg before hatch as well as the method used by the first instar to exit the egg through the operculum, and will be briefly addressed here.

Freshly deposited eggs are cream in color, darkening slightly after one day. The eyes of the developing nymph are red in color and at some point become visible through the operculum (Todd 1989). After ≈3 days, the eggs hatch and the newly eclosed first instars cluster atop the egg choria. Bundy and McPherson (2000a) reported on the morphological differences between *Nezara viridula* eggs and the eggs of several other stink bug species.

7.4.1.2 Nymphs

First instars are red in color immediately after hatching and turn black by the second day of the stadium. They remain clustered atop the egg choria, probing the egg surfaces to obtain beneficial endosymbionts (Prado et al. 2006). Dispersal from the hatched eggs to feeding sites typically begins at or near emergence of the second instars and continues through subsequent instars.

Coloration in second through fourth instars of the common green adult form (i.e., form *smaragdula*) is fairly uniform – black dorsal coloration with white spots (Rojas and Morales-Ramos 2014). However, the fourth and fifth instars can also be in a "pale form" or "dark form" (Jones 1918), represented by "green" and "[d]ark brown, nearly black" dorsal coloration, respectively. Morrill (1910) also noted the occurrence of "light" and "dark" forms for fifth instars. Light and dark forms also are present in the orange adult form of *Nezara viridula* (form *aurantiaca*) (Follett et al. 2007). The fifth instar possesses the distinct and characteristic white spots on the dorsal surface (**Figure 7.1B**) that allow differentiation between *N. viridula* and *Chinavia hilaris*; the latter possesses stripes dorsally across the abdomen (Morrill 1910). Rojas and Morales-Ramos (2014) examined the life history of black and green fifth instars and determined that green fifth instars exhibited higher survival to adulthood and adults were more fit (i.e., significantly higher adult weight and fecundity). The characters for determining sex are easily visible at the fifth instar (Esquivel and Ward 2014), becoming more pronounced as the stadium nears eclosion to adulthood.

Behaviorally, newly emerged first instars were long thought to not feed (Morrill 1910, Turner 1918). However, first instars of *Nezara viridula* provisioned with a green bean were observed inserting their stylets into the food source (Jesus F. Esquivel, personal observation). Upon microscopic inspection of this behavior, the protraction and retraction of stylets were visible through the outer surface of the green bean. These observations raised the question of whether the commonly accepted belief of nonfeeding by first instars was accurate. Using sections of green bean infected with a marker pathogen, Esquivel and Medrano (2014) demonstrated that first instars indeed feed as defined by the ingestion of the marker pathogen that, subsequently, was found within the insects. Skeptics of these findings argued that the first instars were not "feeding" but "drinking" instead. However, "feeding" by *N. viridula* is accomplished through uptake of liquefied food resources (i.e., "drinking"). Thus, "drinking" would also be defined as an aspect of "feeding," regardless.

It is common for laboratory studies to not provision first instars with a food source because of earlier reports indicating that these nymphs do not feed. The findings of Esquivel and Medrano (2014) broach the issue of whether fitness of first instars is being compromised when these insects are not provided a food source. Although these authors re-defined the feeding behavior of these nymphs, questions still remain regarding the amount of resources ingested during this stadium and the potential impact of these ingested resources on subsequent life stages. This is further emphasized because cohorts of these nymphs were allowed to molt and the pathogen also was detected in second instars, demonstrating retention of the marked pathogen through the molting process (Esquivel and Medrano 2014). The importance of ingestion and retention of pathogens by first instars is amplified further when one considers that adult *Nezara viridula* can transmit *Pantoea agglomerans* (Ewing) Fife, a pathogen that causes seed- and boll-rot disease in cotton (Medrano and Bell 2007; Medrano et al. 2007, 2009a,b; Esquivel et al. 2010; also see **Chapter 13**). Ingestion of *P. agglomerans* by early instars and retention of this pathogen through later instars and into adulthood could have implications for management of *N. viridula* (e.g., through more stringent thresholds).

7.4.1.3 Adults

Developmental time from egg to adulthood is approximately 30 days but will greatly vary based on rearing temperatures and food source. Nondiapausing adults began mating as early as 5 days for females and 6 days for males, indicating females reach sexual maturity earlier than males (**Figure 7.1C**; Mitchell and Mau 1969, Musolin and Numata 2003a). Mated females deposited viable eggs within 7–8 days after mating. This premating period (= precopulation period) differs from that reported by Fortes (2010). However, the time frame for oviposition corresponds with Fortes (2010) and Fortes et al. (2011) who observed chorionated eggs ready for fertilization at 10 days of age. Detailed descriptions of changes to the developing and presumably senescing (i.e., late-season) reproductive systems can be found in Esquivel (2009, 2016b).

Longevity of adults is variable and greatly depends on whether a particular adult reproduces directly or enters winter diapause and reproduces after overwintering. Thus, under summer field conditions in central Japan, nondiapausing (i.e., actively reproducing) adults lived on average less than 50 days, whereas those adults that emerged later in the season (e.g., in September) entered diapause and survived until the next summer (most of this time in diapause), began reproducing in April–May, and died in July–August (Musolin et al. 2010).

In the laboratory, at 25°C, postdiapause females lived up to 351 days (including a long diapause period) (Musolin et al. 2007). Fecundity of females also is variable. Under laboratory conditions at 25°C, fecundity of reproductive females ranged from 18 to 1,496 eggs, and 1 to 19 egg clusters, per female (Musolin and Numata 2003a, Musolin et al. 2007).

As noted earlier, adults exhibit several color morphs (see **Section 7.2**) including gold (or orange), blue, green, and black forms (**Figure 7.1E–H**). Although the green and orange forms have been known for some time (Yukawa and Kiritani 1965, Kiritani 1970, Vivan and Panizzi 2002, Golden and Follett 2006), a blue form was observed in a field collection in Brazil in 2015 (Antônio R. Panizzi, personal communication) and, most recently, a black form was reported in a laboratory colony in Texas (Esquivel et al. 2015). In all likelihood, polymorphism is under genetic control (Ohno and Alam 1992, Follett et al. 2007, Musolin 2012) and seasonal color polymorphism is under abiotic influences (**Figure 7.1C,D**; see **Section 7.4.4.2 and Chapter 11**; Musolin and Numata 2003a, Musolin 2012).

7.4.2 Symbiotic Relationships

Heteropterans possess endosymbiotic organisms in their gut and gastric caeca (Glasgow 1914, Malouf 1933). Glasgow (1914) observed bacteria in embryos of *Murgantia* sp., suggesting that endosymbiotic bacteria are transferred from the gut (or gastric caeca) through the germaria walls into the egg; how this transfer occurs remains unknown. Similarly, Nan et al. (2016, and references therein) observed movement of yeast-like symbionts from the fat bodies to the developing oocytes of brown planthopper,

Nilaparvata lugens Stål (Hemiptera: Delphacidae). Interestingly, Malouf (1933) also reported rod-like bacteria in the ectodermal sac of males, hypothesizing that males transferred these bacteria to females during mating. Fortes (2010) similarly "verified a rich diversity of bacteria associated with the male reproductive system of *Nezara viridula*, with predominance of the Enterobacteriaceae *Klebsiella* sp., which was previously reported associated with the gut of *N. viridula*." Prado et al. (2006) also identified a Enterobacteriaceae bacterium in the gut of *N. viridula*, adding that localization of the bacterium within the gut varied between adults and nymphs. Although Glasgow (1914) suggested transovarial transmission, Prado et al. (2006) indicated, instead, that oral transmission was responsible for transfer of symbionts to newly eclosed insects.

Some species transfer endosymbionts via bacteria-filled capsules laid underneath eggs and subsequently utilized by newly eclosed nymphs (Fukatsu and Hosokawa 2002). Ovipositing *Nezara viridula* females, however, smear the surfaces of eggs with viscous material (Drake 1920), depositing beneficial bacteria on the egg surface (Prado et al. 2006). However, it is unclear whether the viscous substance has been studied closely to determine whether the endosymbionts are, in fact, included in the viscous substance. That is, anatomically, the ovipositor and the rectum are in close proximity, and bacteria have been shown to be present in bug feces. Is it possible that the eggs are passively contaminated with bacteria due to this proximity? Alternatively, if the smeared material is the sole source of egg contamination, should we see a gland that provides a constant supply of viscous material infected with bacteria? Regardless, the egg surfaces subsequently are probed by newly hatched instars that acquire these endosymbionts for their own development. Sterilization of egg surfaces or gut resulted in compromised fitness (Hirose et al. 2006, Prado et al. 2009, Tada et al. 2011), thereby demonstrating the crucial role of endosymbionts in the biology of *N. viridula*.

Other microorganisms also are present in *Nezara viridula* (Ragsdale et al. 1979, Hirose et al. 2006). Esquivel and Medrano (2012) demonstrated that ingested bacteria seemingly colonize selective regions of the digestive system preferentially, resulting in transmission of those bacteria colonizing the rostrum and head (salivary canals). That is, although *Pantoea agglomerans*, *Pantoea ananatis* (Serrano), *Klebsiella pneumoniae* (Schroeter), and *Nematospora coryli* Peglion were ingested, only *P. agglomerans* and *N. coryli* were transmitted to cotton bolls. These latter two species also were the only microbes detected in the rostrum, head, and the alimentary canal. *K. pneumoniae* and *P. ananatis* were detected in the alimentary canal but not in the rostrum and head. The latter organism (i.e., *P. ananatis*) also occurs in, and is readily transmitted by, the cotton fleahopper [*Pseudatomoscelis seriatus* (Reuter) (Miridae)]. This raises the question whether the microenvironment (e.g., pH in salivary glands/system, temperature) of *N. viridula* is affecting colonization and transmission of *P. ananatis*. Further, if this is the case, and if we can elucidate the mechanism, can we affect the mechanism such that we can prevent *N. viridula* from transmitting other pathogenic organisms? On the related issue of affecting mechanisms, identification of neuropeptides to affect water balance to disrupt physiological processes may hold promise for managing populations (Predel et al. 2008).

7.4.3 Mate Selection

Larger females of *Nezara viridula* prefer to mate with larger males (McLain 1980). Mitchell and Mau (1969) describe the courtship process. Briefly, males initiate mating beginning with attempts (using their head) to raise the female's terminal abdominal segments so that the male can move into position to engage their respective genitalia. Once engaged (*in copula*), the insects face away from each other and can remain *in copula* "for a few minutes to several days," in some cases 10 days (**Figure 7.1C**; Mitchell and Mau 1969). Similarly, Harris and Todd (1980) reported "large variation in copulatory durations (1–165 h), number of copulations per individual (1–9), and time *in copula* prior to oviposition (1–176 h)". Borges et al. (1987) added more detailed descriptions related to long-range mate location (including pheromone extracts from males) and short-range courtship (including the indication that visual and acoustic stimuli affect mate location). Pheromone extracts contain n-dodecane, n-tridecane, sesquiterpene (hydrocarbon), sesquiterpene (mono-oxygenated), and n-nonadecane and play a key role in long-range mate location (later discussed in **Section 7.4.6**). Males are polygamous and females are polyandrous.

7.4.4 Winter and Summer Diapauses[2]

Insect diapause is "a profound, endogenously and centrally mediated interruption that routes the developmental programme away from direct morphogenesis into an alternative diapause programme of succession of physiological events; the start of diapause usually precedes the advent of adverse conditions and the end of diapause need not coincide with the end of adversity" (Koštál 2006, p. 115). *Nezara viridula* is, perhaps, one of the few true bug species in which the control of seasonal development and diapause are comparatively well understood.

Whereas diapause usually is defined as a physiological state, it also can be seen as a "syndrome of physiological and behavioural changes that adapt insects to approaching seasonal changes" (Tauber et al. 1986). This concept not only emphasizes the dynamic nature of diapause (with all its sequential stages) but also includes various forms of seasonal migration, seasonal polyphenism (both reversible and irreversible), and behavioral and physiological adaptations, all of which are related to dormancy and, in different ways, maximize survival and fitness of the diapausing individuals (Tauber et al. 1986; Saulich and Musolin 2007, 2011; Saunders 2010; Musolin 2012; also see **Chapter 11**).

In general, *Nezara viridula* is a multivoltine species producing, for example, up to three to four generations per year in central Japan (Kiritani et al. 1963, Kiritani 2011). In North America, the species can produce up to five generations at more southern latitudes (e.g., Florida [Drake 1920]) but only one or two generations at more northern latitudes (i.e., South Carolina [Jones and Sullivan 1981, 1982]). Colder temperatures at more northern latitudes play a key role in regulating the number of generations (Jones and Sullivan 1981, 1982), although the number of generations in a particular region also may be limited by the availability and phenology of preferred food plants (Velasco et al. 1995, Panizzi et al. 2000).

As with many other pentatomid species in the temperate zone, adults of *Nezara viridula* overwinter in a state of reproductive diapause under litter, bark of trees, inside dense crowns of conifers and other evergreens (e.g., cryptomeria), Spanish moss, piles of paddy straw, beneath the roofs of buildings, or in other suitable shelters (Rosenfeld 1911, Kiritani et al. 1966, Newsom et al. 1980, Jones and Sullivan 1981, Saulich and Musolin 2014). In India (23°N) and southern Brazil (23°S), however, it is thought that the species does not enter winter diapause but switches to alternative host plants and, eventually, reproduces during the mild winter (Singh 1973, Panizzi and Hirose 1995, Antônio R. Panizzi, personal communication).

Summer adult diapause (estivation), for survival during hot dry months, is reported for a population in India (23°N; Singh 1973), although there is no evidence that it occurs in the populations in central Japan (Musolin and Numata 2003a; Musolin et al. 2010, 2011). As summer diapause is poorly studied in *Nezara viridula*, the remainder of **Section 7.4.4** is devoted to facultative winter adult diapause only and mostly based on the laboratory and field observations conducted in central Japan.

7.4.4.1 Diapause Induction in the Laboratory

Diapause in *Nezara viridula*, as in many other insects with an adult (= reproductive) diapause, is manifested primarily by the degree of development of the reproductive organs and fat bodies (also see **Chapter 11 and glossary in that chapter**). Reproductively active females have mature (= chorionated) eggs or vitellogenic oocytes in their ovarioles and weakly developed or loose fat bodies (see **Figure 11.11**). In contrast, in diapausing females of a similar age, differentiation and development of the oocytes is interrupted in the early stages. In these females, the ovarioles are clear, there are no oocytes in the germaria, and the fat bodies are massive and dense (Esquivel 2009, 2011). Whereas winter diapause is studied mostly in adult females, there are diapause signs evident in males as well (see **Figure 11.11**). Reproductively active males have expanded ectodermal sacs containing milky white secretion and lightly developed or loose fat bodies (Esquivel 2011). By contrast, diapausing males have transparent,

---

[2] This section was modified and updated (in part) from "Surviving winter: diapause syndrome in the southern green stink bug *Nezara viridula* in the laboratory, in the field, and under climate change conditions" by D. L. Musolin (Physiological Entomology 37: 309–322. Copyright 2012 by the author and The Royal Entomological Society).

empty, and collapsed ectodermal sacs and massive and dense fat bodies (Musolin and Numata 2003a; Esquivel 2009, 2011; Takeda et al. 2010).

The induction of winter adult diapause in the temperate populations of *Nezara viridula* (e.g., Cotton Belt of the United States) is controlled by photoperiod (Harris et al. 1984, Seymour and Bowman 1994). Similar observations are reported for the Osaka (34.7°N, 135.5°E) population, in which induction of winter adult diapause has been tested under several constant photoperiods at 20 and 25°C (Musolin and Numata 2003a) and in other regions (Ali and Ewiess 1977, Musolin et al. 2011). A long-day photoperiodic response is found: almost all specimens are reproductive under long-day conditions, whereas those under the short-day conditions are in diapause when examined 60 days after adult ecdysis (**Figures 7.7 and 7.8**; Musolin and Numata 2003a; Musolin et al. 2007; also see **Chapter 11**). At both temperatures, the photoperiodic response curves are similar and the critical day length for diapause induction falls into a narrow range close to 12.5 h, suggesting that day length is the dominant factor in diapause induction (**Figure 7.8**; Musolin and Numata 2003a). The thermostability of the photoperiodic response within the tested range of temperatures is similar to that of several other heteropterans, namely *Riptortus pedestris* (F.) (= *clavatus*; Alydidae; Kobayashi and Numata 1995), *Arma custos* (F.) (Pentatomidae; Volkovich and Saulich 1995), and *Orius minutus* (L.) (Anthocoridae; Musolin and Ito 2008; also see **Chapter 11**). The opposite situation (i.e., when temperature influences the photoperiodic response strongly) is reported for many other insect species (Danilevsky 1961, Danks 1987, Saulich and Musolin 2011; also see **Chapter 11**).

In North American populations of *Nezara viridula*, the fifth (final) instar is reported to be the most sensitive to photoperiodic signals for diapause induction (Pitts 1977), whereas the fourth instar is suggested to be more important in a population from Egypt (Ali and Ewiess 1977). Diapausing adults of *N. viridula* are sensitive to diapause-terminating long-day signals, whereas reproductive adults can neither stop their reproductive processes nor switch to diapause in response to short-day stimuli (Musolin and Numata 2003b, Musolin et al. 2007).

In the Osaka population of *Nezara viridula*, a marked variation in the incidence of diapause is recorded under the near-critical photoperiods (L:D 12:12 and 13:11; **Figure 7.8**; Musolin and Numata 2003a). Both sexes show photoperiodic response curves of a similar shape and, in most regimes, females exhibit a higher incidence of diapause than males (**Figure 7.8**; Musolin and Numata 2003a).

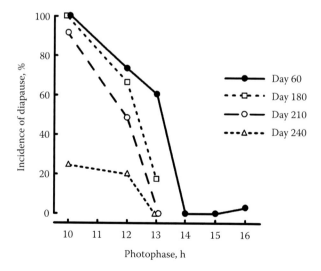

FIGURE 7.7   Effect of day length on diapause status and diapause maintenance in female *Nezara viridula* at 25°C. Incidence of diapause status was judged by dissection on day 60 after adult emergence. Incidence of diapause maintenance was judged by coloration of females on days 180, 210, and 240 after adult emergence (nonovipositing deep russet females were considered to be in diapause). (From D. L. Musolin, *Physiological Entomology* 37: 309–322, 2012, with permission.)

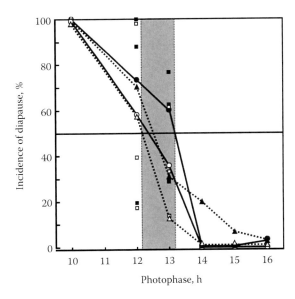

FIGURE 7.8 Photoperiodic response curves for diapause induction in females and males of *Nezara viridula*. Solid line and closed circles, females at 25°C; solid line and open circles, males at 25°C; dashed line and closed triangles, females at 20°C; dashed line and open triangles, males at 20°C (for L:D 12:12 and L:D 13:11 at 25°C, the results of three replications are also shown as closed squares, females; open squares, males). Shaded area shows a range of critical day length inducing diapause in 50% of individuals. (From D. L. Musolin and H. Numata, Physiological Entomology 28: 65–74, 2003a, with permission.)

7.4.4.2 Diapause-Associated Reversible Color Change in Adults

Nezara viridula has conspicuous body color polymorphism: more than ten, genetically determined, color morphs of adults are recognized, with the most common being a completely green morph (G-type, morph *smaragdula* F.) (**Figure 7.1E–H**; Yukawa and Kiritani 1965; Kiritani 1970, 1971; Hokkanen 1986; Ohno and Alam 1992). In addition to this genetically controlled adult body color polymorphism, *N. viridula* shows an interesting example of a reversible seasonal adult color change (seasonal polyphenism; also see **Figure 11.17 and Chapters 11 and 12**). With the possible exceptions of a recently observed blue morph in Brazil (**Figure 7.1G**; Antônio R. Panizzi, personal communication) and a black morph in North America (**Figure 7.1H**; Esquivel et al. 2015), adults of all other genetic color morphs turn gradually russet or reddish-brown when they experience diapause-inducing conditions (**Figure 7.1D and Figure 11.17**; Harris et al. 1984, Seymour and Bowman 1994, Musolin and Numata 2003a, Musolin 2012). This russet color is physiologically controlled and reversible (Yukawa and Kiritani 1965, Musolin and Numata 2003a, Musolin et al. 2007). Adult diapause termination, the resumption of active development, and the beginning of postdiapause reproduction are associated with a gradual reversion to the original body color (see below). This color change is conditioned by processes not in the cuticle but in the underlying epidermal cells and involves the pigment erythropterin (Gogala and Michieli 1962, 1967). Although erythropterin occurs in both nondiapausing (green, yellow, and others) and diapausing (reddish-brown or russet) adults, in the former it exists in aqueous solution, whereas in the latter it is predominantly crystalline and deep red (Harris et al. 1984). It has been noted that the intensity of the russet color is somewhat stronger in females than males (Kiritani and Hokyo 1970). Oxygen consumption also is much lower in russet-colored adults than in green adults (Michieli and Žener 1968). The reddish-brown winter coloration of adults apparently functions as a camouflage in the hibernacula. It seems likely that the cryptic role of the seasonal color change has never been studied experimentally in this species, whereas this type of adaptation is well known in several other species of true bugs and other insect taxa (Fuzeau-Braesch 1985, Saulich and Musolin 2012). In addition to camouflage, this seasonal adult body color change might also be important for thermoregulation during the period of diapause preparation in autumn, and/or during the overwintering period, as suggested for some beetles (Fuzeau-Braesch 1985, De Jong et al. 1996, Gross et al. 2004).

   Laboratory experiments with the green morph show that the adult body color change is under photoperiodic control (**Figure 7.9**). This also is the case in other color morphs of the species, as shown in numerous outdoor rearing experiments (Dmitry L. Musolin, unpublished data). For the following experiments, we used only adults that emerged as the green morph. Under long-day conditions, they remain green their entire life (see **Figure 7.1C**); but under short-day conditions, virtually all adults gradually start to change body color to intermediate and then russet brown after 12 days (median) at 25°C or 24 days at 20°C (**Figure 7.1D**; Musolin and Numata 2003a). Under near-critical photoperiodic conditions, adults of the intermediate color always are present and some adults change color more than once (e.g., from green to russet and then back to intermediate or green). The dynamics of color change are similar in both sexes and comparable proportions of adults of both sexes reach deep russet color. Unexpectedly, a lower temperature (20°C) fails to accelerate the change of the body color to the overwintering coloration, whereas the short-day length does: the shorter the photophase, the faster the rate of color change (**Figure 7.9**; Musolin and Numata 2003a). These observations provide strong

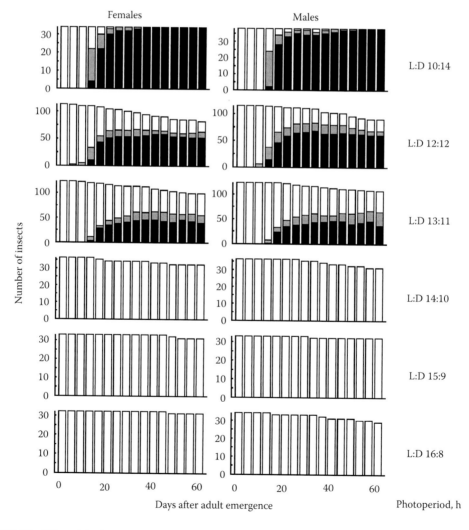

FIGURE 7.9   Effect of day length on adult body color in *Nezara viridula* at 25°C. White, gray, and black sections of the bars indicate the proportion of females and males with green, intermediately colored, and russet bodies, respectively. (From D. L. Musolin and H. Numata, *Physiological Entomology* 28: 65–74, 2003a, with permission.)

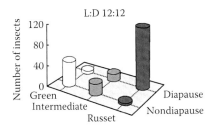
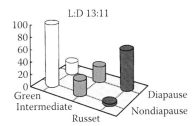

FIGURE 7.10 Diapause status and color of adult *Nezara viridula* at day 60 after emergence at L:D 12:12 and L:D 13:11 and 25°C (the sexes are combined). (From D. L. Musolin and H. Numata, Physiological Entomology 28: 65–74, 2003a, with permission.)

evidence that adult body color change is controlled by day length, not only qualitatively (russet versus green coloration), but also quantitatively (different rates of color change under different photoperiodic conditions).

The color change in adult *Nezara viridula* is associated strongly with gonadal development. Two months after adult emergence, under the short-day conditions (L:D 10:14) at 20 and 25°C, the gonads of all individuals are in a diapause state, and the adults are deeply russet brown; conversely, almost all adults are reproductively active and green under the long-day conditions at the same temperatures (compare **Figures 7.8 and 7.9**). Under the near-critical conditions of photoperiods L:D 12:12 and L:D 13:11, all color grades (i.e., green, intermediate, and russet) and both diapausing and nondiapausing adults were present (**Figure 7.10**). It is important to note that under such near-critical photoperiodic conditions, on the day of dissection, diapausing individuals can have different body colors (i.e., green, intermediate, and russet). The same is true for nondiapausing adults – they can have different body colors. However, even though most of the green adults are reproductive (i.e., nondiapausing) under such conditions, some of them are considered to be in diapause (judged by the state of their reproductive organs). The same is true for intermediately colored or russet adults – some of them might be in diapause whereas others under the same conditions might be in nondiapause (**Figure 7.10**). These results show that winter diapause induction and adult body color change are complex and dynamic processes (Musolin and Numata 2003a; also see **Chapter 11**).

The strong link between winter diapause and body coloration in *Nezara viridula* adults suggested that body color could serve as a reliable indicator of diapause in this species (Harris et al. 1984); however, the reliability of coloration as an indicator of diapause is contradicted (Seymour and Bowman 1994), and the issue is discussed in detail below.

7.4.4.3 Diapause Induction and Adult Color Change in the Field

In central Japan, the nonreproductive adults of the final annual generation of *Nezara viridula* typically start to change body color from green/yellow to intermediate/russet in mid-October. The percentage of russet adults in the population reaches a plateau of 85–100% between late November and mid-January and continues to April (**Figures 7.11 and 7.14**; Musolin and Numata 2003b, Musolin et al. 2010, Takeda et al. 2010). Beginning in October, adults minimize their walking and probing activities, as well as their consumption of food and water. They aggregate under large leaves or other shelters (**Figure 7.1D and Figure 11.17**). Their reproductive organs are in a diapause state (**Figures 11.11 and 11.12**). In females, the ovaries are clear and contain no oocytes in the germaria. The spermathecae are also small and empty. In males, the ectodermal sacs are transparent, empty, collapsed, and not readily visible. Diapausing adults of both sexes accumulate fat reserves and their fat bodies are extended and dense (Esquivel 2009, 2011; Takeda et al. 2010).

Field experiments have clarified the importance of the two leading environmental factors (namely, day length [= photoperiod] and temperature) in preparation for overwintering in *Nezara viridula*. Even though the shortening autumn days accelerate color change in the last annual (= overwintering) generation of the species, relatively high temperatures are necessary for the successful preparation for diapause

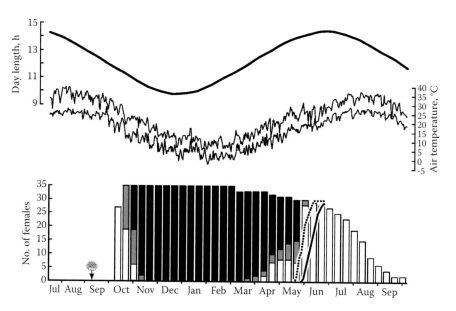

FIGURE 7.11 Seasonal body color changes in female *Nezara viridula* under natural conditions in Osaka, Japan. The experimental series corresponds to the late-season diapausing generation. Arrow marks the moment when the egg clusters were transferred into outdoor conditions. The nymphs and males are not shown. The histogram shows the number and color of females: dotted line, the total number of mating females; solid line, number of ovipositing females; light sections of bars, green females; black sections of bars, brown/russet females; and shaded sections of bars, females with intermediate body color. The temperature is shown as the minimum and maximum daily values. (Modified from D. L. Musolin and H. Numata, Ecological Entomology 28: 694–703, 2003b, with permission.)

and complete color change. If nymphs reach the adult stage too late in the season when the temperature is already too low for their development (e.g., in late October or during November), adults may fail to change body color and prepare properly for diapause, resulting in higher winter and/or spring mortality (Musolin and Numata 2004, Musolin et al. 2010).

It is likely that only nonreproductive adults of the final summer generation that have not started reproduction can successfully overwinter. Eggs can hatch and nymphs are able to survive for several weeks in the autumn or early winter if temperatures do not drop too much, but the nymphs cannot become adults or survive until the spring. The majority of adults that begin reproduction in the summer or autumn will die before or during overwintering, apparently being unable to switch from oviposition to diapause (Musolin and Numata 2003b, Musolin et al. 2010; Dmitry L. Musolin and Daisuke Tougou, unpublished data). However, in Japan, Kiritani (1963) observed a "black-spotted condition" in "females near to the end of oviposition period" and was uncertain as to the origin of these black spots. Esquivel (2009) observed what appeared to be a chorionated egg turning necrotic (or black-colored) within ovarioles of reproductive females during early fall in Central Texas (Esquivel 2009) and presumably this could be the "black-spotted condition" reported by Kiritani (1963). Observations of the "black-spotted condition" in Texas were noted at different localities within the production region and as recently as 2016 (Esquivel 2016b), indicating that the condition can be widespread and suggest that reproductive females may be resorbing eggs in preparation for overwintering. Additionally, Esquivel (2016b) expanded upon the composition and potential cause(s) of the "black-spotted condition".

Nezara viridula is a strong flier, and massive migrations pre- and postdiapause to and from the overwintering sites, respectively, comprise important phases of its seasonal cycle, although this has not been studied thoroughly (Kiritani et al. 1966, Kiritani and Hokyo 1970, Gu and Walter 1989). Adults are capable of sustained flight for as long as 12 hours under laboratory conditions (Kester and Smith 1984). They can fly over long distances in the wild and have been caught on ships stationed up to 500 km from the coasts (Hayashi et al. 1978, Gu and Walter 1989, Kiritani 2011). At least in the case of the prediapause flight, experiments suggest that females fly as virgins (Gu and Walter 1989).

7.4.4.4 Diapause Maintenance in the Laboratory

When adults enter winter diapause under constant short-day conditions in the laboratory, they remain dormant (i.e., dark-colored and mostly motionless) for varying periods of time before their diapause is terminated spontaneously. Adults that experience shorter day-lengths exhibit russet coloration for longer periods (**Figures 7.7 and 7.12**). For example, at 25°C, the median period during which females exhibited stable russet body coloration varied considerably depending on photoperiod (i.e., 84 days under L:D 13:11, 126 days under L:D 12:12, and 154 days under L:D 10:14; **Figure 7.12**; Musolin et al. 2007).

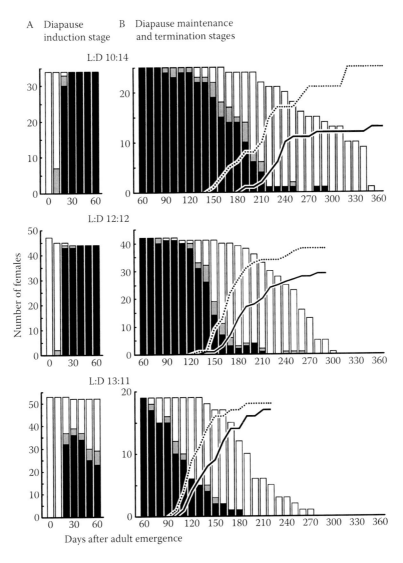

FIGURE 7.12 Effect of day length on body color change and diapause induction (A) and diapause maintenance and termination (B) in female *Nezara viridula* at 25°C under laboratory conditions. On day 60 after adult emergence, green and intermediately colored individuals were removed and only russet ones were left for further exposure (B). Histograms denote survival and relative abundance of color grades: white bars, green; gray bars, intermediate; and black bars, russet. The dotted line and solid line overlaid on the histograms denote cumulative copulation and cumulative oviposition, respectively. (From D. L. Musolin, K. Fujisaki, and H. Numata, Physiological Entomology 32: 64–72, 2007, with permission.)

7.4.4.5 Diapause Maintenance in the Field: Dynamics of the Physiological Condition of Adults during the Overwintering Period

In central Japan, adults of *Nezara viridula* essentially cease all movement by early December, although they will slowly respond if disturbed or if the temperature increases sharply. At this time, the fat bodies have the highest degree of development/accumulation. Adults remain in such a state until February, when depletion of the fat bodies becomes noticeable (Takeda et al. 2010; also see **Figure 11.12**). During most of the overwintering period, the majority of adults of both sexes remain dark-colored; the percentage of green adults varies, and increases as spring approaches (**Figure 7.11**; Takeda et al. 2010).

Previous field records indicate that many or all *Nezara viridula* males collected from hibernacula have reproductive organs that are fully developed (van Heerden 1934) or contain active sperm in their testes (Kiritani 1963). When collected in the field in January and February, females do not mate under long-day conditions in the laboratory, whereas males will mate when paired with nondiapausing laboratory females (Kiritani and Hokyo 1970). Such observations suggest that, in contrast to females who spend winter in diapause, males overwinter in a state of quiescence. This finding is used to explain the higher winter survival of females (Kiritani et al. 1966). However, regular dissections during the winter demonstrate that all adults of both sexes have reproductive organs and fat bodies in a state characteristic of true diapause until late March (Takeda et al. 2010; also see **Figure 11.12**). Thus, it is likely that both females and males overwinter in a state of true diapause, not quiescence. It appears that more research is needed to better understand the differences in diapause maintenance between the sexes.

7.4.4.6 Diapause Termination in the Laboratory

When diapausing adults are kept for an extended period under intermediate (L:D 13:11) or short-day conditions (L:D 10:14 and 12:12) at 25°C, they remain russet for varying periods of time but then start to change color spontaneously to the intermediate grade and then to green. A few adults, mostly those kept under L:D 13:11, go through a cycle of color change (i.e., from green to intermediate and russet, and then back to intermediate and green) more than once. The shorter the day length, the longer the adults remain russet, and the longer they need to reach the final stable green color (**Figure 7.12**; Musolin et al. 2007). Completion of the final body color change from russet and intermediate to green might be considered a marker of diapause termination because the attainment of the final green color almost always precedes the start of reproduction. In laboratory experiments, almost 94% of females already are green on the day of the first copulation (all three photoperiodic conditions combined), and only one russet female was found in the laboratory experiment to have copulated with a green male. Males show a similar trend. All females are green on the first day of oviposition (Musolin et al. 2007).

These data show that adult diapause in *Nezara viridula* can end spontaneously under laboratory constant short-day conditions without any preceding low temperature treatment, although in the absence of chilling and change of photoperiodic conditions from short-day to long-day, the termination of diapause and start of reproduction are poorly synchronized (Musolin et al. 2007). The time to first oviposition ranges from 106 to 158 days under different short-day conditions at 25°C (Musolin et al. 2007). At the same temperature, this timing is only 43 days under nondiapausing long-day conditions, and even less (23 days) in adults that have overwintered and started reproduction outdoors (Musolin and Numata 2003b, 2004).

From a physiological standpoint, the timing of the end of insect diapause (= the end of the termination phase, according to Koštál 2006; also see **Chapter 11**) is difficult to detect precisely, perhaps not possible at all, and the indirect markers visible externally vary between different species (Hodek 1996). The onset of copulation and oviposition are indicative that diapause has terminated and postdiapause development has commenced. The observation that almost all adults of *Nezara viridula* show a reversion of body color to green prior to copulation, and that all females are green on the day of first oviposition, suggests that the change of body color from russet to intermediate and then to green is the closest and most easily observed event indicating the end of diapause.

There has been discussion on whether the russet coloration might serve as a reliable indicator of diapause in *Nezara viridula* (Harris et al. 1984, Seymour and Bowman 1994). Results of several experiments suggest this is true in most cases: a great majority of adults that emerge in autumn are in diapause (and then in postdiapause quiescence) and have russet coloration from late autumn to early spring (**Figure 7.11**; Musolin and Numata 2003b, Musolin et al. 2010; also see **Chapters 11 and 12**). However, the value of coloration is not so apparent during the comparatively short transition periods of diapause induction in early autumn or diapause termination in spring, or when diapause is induced under inappropriate conditions (Musolin and Numata 2003a,b, 2004; Musolin et al. 2007, 2010; Takeda et al. 2010).

Photoperiodic conditions have a strong effect on the timing of diapause termination in *Nezara viridula*, even when all the photoperiods tested are short enough to induce diapause in many (L:D 13:11) or all females (L:D 10:14 and 12:12) (Musolin and Numata 2003a). The shorter the photophase, the longer adults remain russet and the later they start reproduction, thus the longer the duration of diapause. Under short-day and near-critical conditions, the preoviposition period is 10–15 times longer than under long-day conditions (L:D 14:10) at the same temperature (**Figure 7.12**; Musolin and Numata 2003a).

Comparison of the diapause induction response curve (i.e., diapause incidence recorded 60 days after adult emergence) with diapause incidence recorded on days 180, 210, and 240 after adult emergence, shows that the short-day and near-critical photoperiods differ in the effect on maintaining diapause: the shorter the photoperiod, the longer the diapause is maintained (**Figure 7.7**). It remains to be shown, however, whether the duration of diapause in *Nezara viridula* is determined during the diapause induction phase as a result of day-lengths experienced (Danks 1987, p. 136; Nakamura and Numata 2000), or if the critical photoperiod for termination of diapause decreases during the further progression of diapause development (= physiogenesis; Numata and Hidaka 1984; Danks 1987, p. 154). It is also possible that both of these processes are involved.

7.4.4.7 Diapause Termination in the Field

In the spring, the percentage of dark-colored *Nezara viridula* adults in the population gradually decreases. By mid-May in Kyoto (Takeda et al. 2010) or late May in Osaka (**Figure 7.11**; Musolin and Numata 2003b), all adults revert to the green coloration. At the same time, the behavior of adults changes dramatically: they start to walk, bask, probe, and consume food and water (Takeda et al. 2010). In Kyoto, under the quasi-natural conditions in 2008, the first copulation was recorded on 22 April and 50% of females and males were active by 28 April and 2 May, respectively (Takeda et al. 2010; see **Figure 11.12**). By mid-May, all adults were active. The state of reproductive organs also dramatically changes in the spring. In 2008, ovaries of females started to show clear signs of reproductive maturation on 28 April; the spermatheca index (used as a measure of mating status) also increased in the late April (Takeda et al. 2010). In males, the size and state of the ectodermal sacs changed considerably. Additionally, the size of the fat bodies decreased further in both sexes by this time. The incidence of diapause decreased sharply from 100% in early April to 0% by mid-May and did not differ between males and females (Takeda et al. 2010). A field survey suggested that color change in spring precedes the exit from hibernacula (Kiritani and Hokyo 1970).

7.4.4.8 Geographic Differences in Diapause Pattern

Nezara viridula is believed to be of southern origin, but occurs throughout tropical, subtropical, and warm temperate regions (see **Section 7.3**). In the areas where winters are mild, and long-term cold does not occur regularly, it may be advantageous for an insect population to be able to terminate winter diapause within a few months without having a requirement for low temperature exposure. Even though the overwintering physiology in the more southern populations of *N. viridula* has not been studied in detail, it is known that diapause in this species lasts only about two months in Australia (29°S; Coombs 2004). Nearer to the equator, populations of *N. viridula* may have no diapause as such, and low levels of reproduction are observed during the colder weeks or months in India (23°N; Singh 1973) and southern Brazil (23°S; Panizzi and Hirose 1995; Antônio R. Panizzi, personal communication). The expansion of *N. viridula* into both northern and southern temperate zones probably has been associated with the evolution of a more intensive and stable diapause, which seems to be necessary for survival in more severe winters, although termination of diapause does not, as yet, require low temperature (Musolin et al. 2007, 2011).

7.4.4.9 Diapause Syndrome and Overwintering in the Field: Importance of the Timing of Emergence, and the Coloration and Size of Adults

Whereas only nonreproductive adults of *Nezara viridula* can successfully survive winter in the temperate zone, the full picture is more complex. Many factors appear to be involved, and several conditions need to be met to ensure successful overwintering.

Body coloration of adults seems a good indicator of diapause (although not absolute) and, thus, a good predictor of winter survival (Harris et al. 1984). An experiment to evaluate winter survival in females, which included both outdoor and simulated warming conditions, showed that winter survival correlated strongly with body color. Thus, when the proportion of females that survived winter in each series was plotted against the proportion of dark-colored individuals just before the winter in these series, a significant positive relationship was found: a higher proportion of dark-colored individuals resulted in a higher rate of winter survival (**Figure 7.13A**). Females that emerge as adults later than others (15 September) do not appear to have enough time for color change, they remain green during the winter, and suffer the highest winter mortality (62.0%), even though all are nonreproductive (Musolin et al. 2010). Mortality is highest (100%) in the series that consists entirely of reproductively active green females (**Figure 7.13A**). When the winter survival is analyzed separately for each color grade in nonreproductive females, results show that winter survival differs significantly between green and russet females; females of intermediate color have an intermediate level of winter survival (**Figure 7.13B**; Musolin et al. 2010). The dark coloration likely protects adults from predators from autumn to spring (a camouflage function) but also serves as an indicator of fully formed and deep diapause.

In turn, the coloration of *Nezara viridula* adults (or more precisely, the success of seasonal color change) depends greatly on the timing of adult emergence and diapause induction. If adults emerge early in the year, they start reproducing and, thus, lose the ability to undergo true diapause, which is necessary for successful overwintering (Musolin and Numata 2003b). If they emerge very late in the autumn, they do not reproduce but may still fail to change body color properly from green to russet, probably because of the low or unstable ambient temperature, which is insufficient to allow successful and complete preparation for diapause (Musolin and Numata 2004). Support for this hypothesis is provided by experiments in which nymphs and subsequent adults of *N. viridula* are reared from the same egg masses under outdoor conditions and conditions of simulated climate warming (2.5°C; **Figure 7.14**). In the outdoor series, nymphs reached adulthood during November. None of the adults that emerged changed from green to russet coloration, and more than 50% died by the end of the winter. In contrast, in the simulated warming series, the siblings reached adulthood in late October. Most of these adults changed color to russet brown and overwintered successfully (**Figure 7.14**). The 2.5°C difference in temperature between these two treatments in the late autumn allows the siblings from the simulated warming series to prepare for diapause properly and, thus, to survive the winter (Musolin et al. 2010). This emphasizes the importance of the timing of nymphal development and of adult emergence in relation to diapause induction and successful overwintering.

Adult size also affects winter survival in *Nezara viridula*. Field surveys suggest that larger adults survive winter better than smaller adults (Kiritani and Hokyo 1970). When nymphs and subsequent adults of *N. viridula* are reared from the same egg masses, either under outdoor or simulated climate warming conditions, larger adults show significantly higher survival through the winter under both sets of conditions (Musolin et al. 2010). Also, as in other insect species, the size of adults at emergence changes during the active season: adults that emerge in September are larger than those that emerge in August (Musolin et al. 2010). Thus, a longer development time of nymphs in September under naturally decreasing temperatures results in a later emergence of larger adults, that, due to their larger size, have a greater likelihood of overwintering successfully and producing progeny during the next season. However, as discussed previously, the delayed emergence of adults can be detrimental as they are likely to not enter diapause properly and, thus, will suffer high mortality during overwintering (**Figure 7.14**; Musolin and Numata 2004).

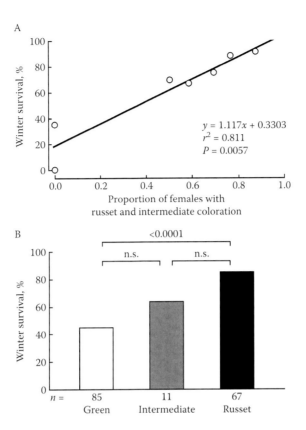

FIGURE 7.13   Effect of coloration on winter survival (from 1 December 2006 to 1 April 2007) of female *Nezara viridula* in a combined outdoor and simulated warming experiment in Kyoto, Japan. A, The relationship between winter survival and proportion of russet and intermediately colored females. The proportion of russet and intermediately colored females in each series (= experimental cohort) (shown as open circles) was calculated on 1 December as the number of russet and intermediately colored females in each series divided by the total number of all nonreproductive females (irrespective of their color) in each series. All series and treatments in which females survived the winter are included. The linear regression line and statistics after arcsine transformation are shown (all series and treatments combined; *P* of χ^2 test is shown). B, Winter survival in different color groups of nonreproductive females. Total numbers of these females of each color (*n*) are shown below the axis. See text for details. (From D. L. Musolin, Physiological Entomology 37: 309–322, 2012, with permission.)

7.4.5 Host Plant Associations

Nezara viridula is highly polyphagous (**Tables 7.4 and 7.5**). Although this species prefers leguminous plant taxa (i.e., members of the Fabaceae; **Table 7.5**), noncultivated herbaceous plants, other cultivated crops (including fruit and nut trees), ornamentals, and noncultivated trees can be exploited, or utilized, throughout the year. However, all exploited plant species may not meet the definition of a "host plant" (Smaniotto and Panizzi 2015). That is, Smaniotto and Panizzi (2015) proposed the term "associated plant" in lieu of "host plant" to recognize plant species utilized by *N. viridula*, rationalizing that "host plant" *sensu stricto* is defined as a host where an insect species can feed, reproduce, and seek shelter within. They concluded that "…the ideal host plant, the one to fulfill the three features [i.e., feeding, reproduction, and shelter within], is seldom encountered by pentatomids." Thus, as further clarified below, "associated plant" (or appropriate variant) is a more accurate term, and the term will be incorporated henceforth.

Our approach in identifying plant species associated with *Nezara viridula* is similar to Smaniotto and Panizzi (2015) because, in addition to the well-documented reproductive hosts, we include transient hosts (i.e., plants used in the absence of key hosts) and hosts used as overwintering habitat. This approach is

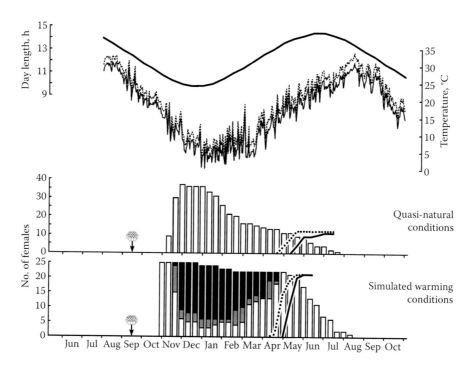

FIGURE 7.14   Seasonal development of female *Nezara viridula* under quasi-natural and simulated warming (2.5°C) conditions in Kyoto, Japan, 2006–2007. The egg mass symbol indicates the dates when the series started (15 September). Histograms denote adult female emergence, survival, and relative abundance of color grades: white bars, green; gray bars, intermediate; and black bars, russet. The dotted lines and solid lines overlaid on the histograms denote cumulative copulation and cumulative oviposition, respectively. Environmental conditions (at the top of the figure): thick line, natural day length; solid line, ambient outdoor temperature; broken line, temperature in the simulated warming incubator. See text for details. (From D. L. Musolin, Physiological Entomology 37: 309–322, 2012, with permission.)

further supported by Panizzi and Slansky (1991) who reported that host-switching from a cultivated food source to a noncultivated food source, or feeding site (i.e., fruiting structures vs. stems/leaves), can influence development time of nymphal stages, fecundity, and survival of *N. viridula* (Panizzi 1997, 2000). These latter categories enable the *N. viridula* population to subsist. In essence, if a plant species favors the continued survival of the species, we identify it as an "associated plant," or "plant associated with" *N. viridula*.

Kirkaldy (1909) and Hoffmann (1935) presented the first structured listing of "food-plants" and "food-plants," respectively, of *Nezara viridula*, suggesting that these were only plants upon which *N. viridula* fed. Because Hoffmann (1935) indicated that he "…undertook a survey of the literature…," the presumption is that he presented a world-wide listing of "food-plants" known at that time to be associated with *N. viridula*. For this reason, and because of similarities with plant species identified in North America, Hoffmann's 1935 list is included within **Table 7.4**. Likewise, tables identifying plants "…where stink bug feeding and development were observed or implied…" were presented by McPherson and McPherson (2000). However, these tables (McPherson and McPherson 2000) did not specify which plant taxa were specific to *N. viridula*, instead implying that the tables reflected plant and insect interactions for all included stink bug species. In addition to Kirkaldy (1909) and Hoffmann (1935), other previous reports specific to *N. viridula* are included here (see **Table 7.4**).

Botanical nomenclature is usually in a state of flux given synonyms for scientific names and differing common names according to location. The names reported originally are included here. Also, earlier reports presented general names of plant genera as well as more specific identification within these plant genera (e.g., Amaranthaceae in **Table 7.4**). General correspondence between the general public (i.e., growers/home gardeners) and entomologists also provided insight to plants associated with *Nezara viridula* (e.g., Riley and

TABLE 7.4

Cultivated and Wild Plants (by Family) Associated with *Nezara viridula* in North America[1]

Family	Plant Species	Common Name(s)[2]	Source[2]	Country-State[2,3]
Acanthaceae	*Asystasia gangetica* (L.) T. Anderson (syn. *Asystasia coromandeliana* Nees)	Chinese violet	Davis 1964	US-HI
Amaranthaceae	*Amaranthus hybridus* L.	Quelito morado Slim amaranth	Rodríguez Vélez 1974 Esquivel 2016a	MX-SIN US-TX
Amaranthaceae	*Amaranthus hypochondriacus* L. (syn. *Amaranthus leucocarpus* S. Watson)	Cereal plant	Drake 1920 Hoffmann 1935	US-FL WW
Amaranthaceae	*Amaranthus palmeri* S. Watson (*Amaranthus palmeri* Watts [sic])*	Quelito bledo Palmer amaranth	Rodríguez Vélez 1974*	MX-SIN
Amaranthaceae	*Amaranthus spinosus* L.	Quelito espinoso Spiny amaranth	Drake 1920 Hoffmann 1935 Jones et al. 2001 Rodríguez Vélez 1974 Todd 1976	US-FL WW US-HI MX-SIN US-GA
Amaranthaceae	*Amaranthus viridis* L.	Slender amaranth	Jones et al. 2001	US-HI
Amaranthaceae	*Amaranthus* spp.	Pigweed	Davis 1964 Hoffmann 1935	US-HI WW
Anacardiaceae	*Mangifera indica* L.	Mango	Chia et al. 1997	US-HI
Apiaceae[4] (Umbelliferae)	*Coriandrum sativum* L.	Coriander	Hoffmann 1935	WW
Asteraceae[4] (Compositae)	*Ambrosia trifida* L.	Giant ragweed	Esquivel 2016a	US-TX
Asteraceae[4] (Compositae)	*Calendula* spp.	Marigold	Anonymous 2016	US-WA
Asteraceae[4] (Compositae)	*Carduus nutans* L.	Nodding plumeless thistle	Batra et al. 1981	US-VA
Asteraceae[4] (Compositae)	*Carduus spinosissimus* Walter	Yellow thistle	Rosewall 1922	US-LA
Asteraceae[4] (Compositae)	*Cirsium texanum* Buckley	Texas thistle	Hall and Teetes 1981	US-TX
Asteraceae[4] (Compositae)	*Cirsium* spp.	"[T]histles"	Hoffmann 1935	US
Asteraceae[4] (Compositae)	*Conyza canadensis* (L.) Cronquist var. *canadensis*	Canadian horseweed	Esquivel 2016a	US-TX
Asteraceae[4] (Compositae)	*Cynara scolymus* L.	Globe artichoke	Hoffmann 1935 Jones 1918	US US-LA
Asteraceae[4] (Compositae)	*Dracopis amplexicaulis* (Vahl) Cass.	Clasping coneflower	Esquivel 2016a	US-TX
Asteraceae[4] (Compositae)	*Helianthus annuus* L.	Sunflower	Jones 1918 Hoffmann 1935 Watson 1919b	US-LA US US-FL
Asteraceae[4] (Compositae)	*Helianthus tuberosus* L.	Girasol Jerusalem artichoke	Hoffmann 1935	WW
Asteraceae[4] (Compositae)	*Lactuca sativa* L.	Lettuce	Hoffmann 1935	WW
Asteraceae[4] (Compositae)	*Parthenium hysterophorus* L.	Santa Maria feverfew	Esquivel 2016a	US-TX

(Continued)

TABLE 7.4 (CONTINUED)

Cultivated and Wild Plants (by Family) Associated with *Nezara viridula* in North America[1]

Family	Plant Species	Common Name(s)[2]	Source[2]	Country-State[2,3]
Asteraceae[4] (Compositae)	*Ratibida columnifera* (Nutt.) Wooton & Standl.	Mexican hat Upright prairie coneflower	Esquivel 2016a	US-TX
Asteraceae[4] (Compositae)	*Silybum marianum* (L.) Gaertn.	Milk thistle	Hall and Teetes 1981	US-TX
Asteraceae[4] (Compositae)	*Solidago altissima* L. [syn. *Solidago canadensis* var. *scabra* (Muhl)]	Canada goldenrod	Fontes et al. 1994	US-FL
Asteraceae[4] (Compositae)	*Solidago gigantea* Aiton	Giant goldenrod	Fontes et al. 1994	US-FL
Asteraceae[4] (Compositae)	*Solidago leavenworthii* Torr. & A. Gray	Leavenworth's goldenrod	Fontes et al. 1994	US-FL
Asteraceae[4] (Compositae)	*Sonchus oleraceus* L.	Common sowthistle	Jones et al. 2001	US-HI
Asteraceae[4] (Compositae)	*Xanthium strumarium* L.	Rough cocklebur	Esquivel 2016a	US-TX
Asteraceae[4] (Compositae)	*Xanthium* sp.	Xanthium	Batra et al. 1981	US
Boraginaceae	*Heliotropium* sp.	Heliotrope	Hoffmann 1935	WW
Brassicaceae[4] (Cruciferae)	*Bidens pilosa* L.	Spanish needle	Jones et al. 2001	US-HI
Brassicaceae[4] (Cruciferae)	*Brassica napus* L.	Dwarf essex rape Rape	Drake 1920 Hoffmann 1935	US-FL WW
Brassicaceae[4] (Cruciferae)	*Brassica nigra* (L.) W. D. J. Koch (*Brassica nigra* Koch. [sic])*	Black mustard	Ehler 2002 Hoffmann 1935*	US-CA WW
Brassicaceae[4] (Cruciferae)	*Brassica oleracea* L.	Cabbage Col	Gibson and Carrillo S. 1959 Hoffmann 1935 Jones 1918	MX-VER WW US-LA, TX
Brassicaceae[4] (Cruciferae)	*Brassica oleracea* L. var. *acephala* DC.	Collards	Drake 1920 Hoffmann 1935 Jones 1918	US-FL WW US-LA
Brassicaceae[4] (Cruciferae)	*Brassica oleracea* L. var. *botrytis* L.	Cauliflower	Hoffmann 1935 Jones 1918	WW US-LA
Brassicaceae[4] (Cruciferae)	*Brassica oleracea* L. var. *gemmifera* DC. (*Brassica oleracea* L. var. *gemmifera* Zenker [sic])*	Brussels sprouts	Hoffmann 1935* Jones 1918	WW US-LA
Brassicaceae[4] (Cruciferae)	*Brassica rapa* L. var. *rapa* (syn. *Brassica campestris* L.)	Nabo Turnip	Gibson and Carrillo S. 1959 Hoffmann 1935 Hubbard 1885 Jones 1918 Jones and Sullivan 1981 Kirkaldy 1909 Morrill 1910 Newsom et al. 1980	MX-MIC, VER WW US-FL US-LA, TX US-SC WW US-TX US-LA
Brassicaceae[4] (Cruciferae)	*Brassica* spp.	Mustard Rutabaga	Drake 1920 Hubbard 1885 Jones 1918 Newsom et al. 1980	US-FL US-FL US-LA, TX US-LA

(Continued)

TABLE 7.4 (CONTINUED)

Cultivated and Wild Plants (by Family) Associated with *Nezara viridula* in North America[1]

Family	Plant Species	Common Name(s)[2]	Source[2]	Country-State[2,3]
Brassicaceae[4] (Cruciferae)	*Lepidium virginicum* L.	Peppergrass Pepperweed	Jones and Sullivan 1982	US-SC
Brassicaceae[4] (Cruciferae)	*Nasturtium officinale* W. T. Aiton	Watercress	McHugh et al. 2003	US-HI
Brassicaceae[4] (Cruciferae)	*Raphanus raphanistrum* L.	Wild radish	Jones and Sullivan 1981, 1982	US-SC
Brassicaceae[4] (Cruciferae)	*Raphanus sativus* L.	Radish	Drake 1920 Ehler 2002 Hoffmann 1935 Jones 1918	US-FL US-CA WW US-LA
Brassicaceae[4] (Cruciferae)	*Rapistrum rugosum* (L.) All.	Turnipweed	Esquivel 2016a	US-TX
Brassicaceae[4] (Cruciferae)	*Sisymbrium irio* L.	London rocket	Esquivel 2016a	US-TX
Bromeliaceae	*Tillandsia usneoides* (L.) L.	Spanish moss (overwintering)	Rosenfeld 1911 Todd 1976	US-LA US-LA
Cannabaceae (as Moraceae)*	*Cannabis sativa* L.	Hemp	Hoffmann 1935*	WW
Cannabaceae	*Cannabis* sp.	Cannabis	Batra et al. 1981	US
Capparaceae (Capparidaceae; Cappardidaceae [sic]*)	*Cleome gynandra* L. [syn. *Gynandropsis gynandra* (L.) Briq., *Gynandropsis pentaphylla* DC.*]	Spiderwisp Wild spider flower	Davis 1964 Hoffmann 1935* Jones 1918 Jones and Sullivan 1982 Kirkaldy 1909	US-HI WW US-LA US-SC WW
Capparaceae (Capparidaceae; Cappardidaceae [sic]*)	*Cleome spinosa* Jacq. (syn. *Cleome pungens* Willd.; *Cleome spinosa* L. [sic]*)	Spiny spiderflower	Hoffmann 1935*	WW
Capparaceae (Capparidaceae; Cappardidaceae [sic]*)	*Cleome viscosa* L. [syn. *Polanisia viscosa* (L.) DC.]	Asian spiderflower	Hoffmann 1935*	WW
Capparaceae (Capparidaceae; Cappardidaceae [sic]*)	*Polanisia dodecandra* (L.) DC. (syn. *Cleome graveolens* Rafin.)*	Redwhisker clammyweed	Hoffmann 1935*	WW
Chenopodiaceae	*Beta vulgaris* L. (syn. *Beta cicla* L.)	Beet (overwintering)	Newsom et al. 1980	US-LA
Chenopodiaceae	*Beta vulgaris* L. ssp. *cicla* (L.) W. D. J. Koch (syn. *Beta cicla* L.)*	Leaf beet	Hoffmann 1935*	WW
Chenopodiaceae	*Chenopodium album* L.	Chual	Rodríguez Vélez 1974	MX-SIN
Chenopodiaceae	*Chenopodium album* L. var. *acuminatum* (Willd.) Kuntze (syn. *Chenopodium acuminatum* [sic]*)	Lamb's quarters	Drake 1920* Hoffmann 1935*	US-FL WW
Chenopodiaceae	*Chenopodium* sp.	Lamb's quarters	Hoffmann 1935 Todd 1976	WW US-GA
Chenopodiaceae	*Spinacia oleracea* L.	Spinach	Hoffmann 1935	WW

(Continued)

TABLE 7.4 (CONTINUED)

Cultivated and Wild Plants (by Family) Associated with *Nezara viridula* in North America[1]

Family	Plant Species	Common Name(s)[2]	Source[2]	Country-State[2,3]
Convolvulaceae	*Ipomoea batatas* (L.) Lam. (*Ipomoea batatas* Lam. [sic])*	Sweet potato	Hoffmann 1935*	WW
			Jones 1918	US-LA
			Kirkaldy 1909	WW
			Riley and Howard 1893a	US-LA
Convolvulaceae	*Ipomoea purpurea* (L.) Roth (*Ipomoea purpurea* (L.) Lam. [sic])*	Morning glory	Hoffmann 1935*	WW
Cucurbitaceae	*Citrullus lanatus* (Thunb.) Matsum. & Nakai var. *lanatus* (syn. *Citrullus vulgaris* Schrad.)	Sandía (=watermelon)	Rodríguez Vélez 1974	MX-SIN
Cucurbitaceae	*Cucumis melo* L.	Melón (=cantaloupe)	Rodríguez Vélez 1974	MX-SIN
Cucurbitaceae	*Cucumis sativus* L.	Cucumber	Drake 1920	US-FL
		Pepino	Hoffmann 1935	WW
			Rodríguez Vélez 1974	MX-SIN
Cucurbitaceae	*Cucurbita foetidissima* Kunth	Buffalo gourd	Esquivel 2016a	US-TX
Cucurbitaceae	*Cucurbita maxima* Duchesne (*Cucurbita maxima* Duch.)*	Japanese pumpkin Squash*	Drake 1920	US-FL
			Hoffmann 1935*	WW
Cucurbitaceae	*Cucurbita moschata* Duchesne (*Cucurbita moschata* Duch.)*	Crook-neck squash	Drake 1920	US-FL
			Hoffmann 1935*	WW
Cucurbitaceae	*Cucurbita* spp.	Squash	Hoffmann 1935	WW
			Jones 1918	US-AL
Cucurbitaceae	*Momordica charantia* L.	Balsam pear	Hoffmann 1935	China
			Jones et al. 2001	US-HI
Cucurbitaceae	*Sechium edule* (Jacq.) Sw. (*Sechium edule* Sw. [sic])*	Chayote	Drake 1920	US-FL
			Hoffmann 1935*	WW
Cupressaceae	*Juniperus* sp.	Ornamental junipers (overwintering)	Newsom et al. 1980	US-LA
Cupressaceae	*Thuja* sp.	Arborvitae (overwintering)	Newsom et al. 1980	US-LA
Cyperaceae	*Cyperus esculentus* L.	Nutgrass	Drake 1920	US-FL
		Nutsedge	Hoffmann 1935	WW
			Todd 1976	US-GA
Euphorbiaceae	*Ricinus communis* L.	Castor bean	Davis 1964	US-HI
		Castor oil bean	Drake 1920	US-FL
		Higuerilla	Hoffmann 1935	WW
			Jones et al. 2001	US-HI
			Rodríguez Vélez 1974	MX-SIN
			Shearer and Jones 1998	US-HI
			Todd 1976	US-GA
			Esquivel 2016a	US-TX
Fabaceae[4] (Leguminosae)	*Arachis hypogaea* L.	Peanut	Drake 1920	US-FL
			Hoffmann 1935	WW
			Jones 1918	US-AL
			McPherson and McPherson 2000	US

(*Continued*)

TABLE 7.4 (CONTINUED)

Cultivated and Wild Plants (by Family) Associated with *Nezara viridula* in North America[1]

Family	Plant Species	Common Name(s)[2]	Source[2]	Country-State[2,3]
Fabaceae[4] (Leguminosae)	*Cajanus cajan* (L.) Millsp. (syn. *Cajanus indicus* Spreng.; *Cajamus indicus* Spreng. [sic]*)	Pigeon pea	Hall and Teetes 1981 Hoffmann 1935*	US-TX WW
Fabaceae[4] (Leguminosae)	*Canavalia cathartica* Thouars (syn. *Canavalia microcarpa* (DC.) Piper)	Mauna loa	Nishida 1966*	US-HI
Fabaceae[4] (Leguminosae)	*Canavalia ensiformis* (L.) DC. (syn. *Canavalia gladiata* DC. [sic]*; *Canavalia ensiformis* DC. [sic]*)	Sword bean	Hoffmann 1935*	WW
Fabaceae[4] (Leguminosae)	*Cassia* spp.	Cassia Wild legume	Hoffmann 1935 Newsom et al. 1980	WW US-LA
Fabaceae[4] (Leguminosae)	*Chamaecrista fasciculata* (Michx.) Greene var. *fasciculata* (syn. *Cassia fasciculata* Michx.)	Partridge pea	Hall and Teetes 1981	US-TX
Fabaceae[4] (Leguminosae)	*Cicer arietinum* L.	Garbanzo	Rodríguez Vélez 1974	MX-SIN
Fabaceae[4] (Leguminosae)	*Clitoria* sp.	Butterfly pea	Drake 1920 Hoffmann 1935	US-FL WW
Fabaceae[4] (Leguminosae)	*Crotalaria incana* L.	Fuzzy rattlepod	Jones et al. 2001	US-HI
Fabaceae[4] (Leguminosae)	*Crotalaria juncea* L.	Bombay hemp Madras hemp Sunn-hemp	Hoffmann 1935	WW
Fabaceae[4] (Leguminosae)	*Crotalaria lanceolata* E. Mey.	Lanceleaf crotalaria	Panizzi and Slansky 1991	US-FL
Fabaceae[4] (Leguminosae)	*Crotalaria spectabilis* Roth	Showy crotalaria	Jones and Sullivan 1982	US-SC
Fabaceae[4] (Leguminosae)	*Crotalaria trichotoma* Bojer (syn. *Crotalaria usaramoensis* E. G. Baker [sic]*)	West Indian rattlebox	Drake 1920 Hoffmann 1935*	US-FL WW
Fabaceae[4] (Leguminosae)	*Crotalaria* spp.	Rattlebox	Buschman and Whitcomb 1980 Davis 1964 Hoffmann 1935 Todd 1976	US-FL US-HI WW US-GA
Fabaceae[4] (Leguminosae)	*Desmodium adscendens* (Sw.) DC. [syn. *Desmodium ovalifolium* (Prain) Wallich (sic)]	Zarzabacoa galana	Shearer and Jones 1998	US-HI
Fabaceae[4] (Leguminosae)	*Desmodium intortum* (Mill.) Urb. (*Desmodium intortum* Urban [sic]*)	Tick trefoil	Jones et al. 2001*	US-HI
Fabaceae[4] (Leguminosae)	*Desmodium tortuosum* (Sw.) DC.	Dixie ticktrefoil	Hoffmann 1935 Panizzi and Slansky 1991 Shearer and Jones 1998	WW US-FL US-HI
Fabaceae[4] (Leguminosae)	*Desmodium triflorum* (L.) DC. (*Desmodium triflorum* de Candolle)*	Threeflower ticktrefoil	Jones et al. 2001*	US-HI

(Continued)

TABLE 7.4 (CONTINUED)

Cultivated and Wild Plants (by Family) Associated with *Nezara viridula* in North America[1]

Family	Plant Species	Common Name(s)[2]	Source[2]	Country-State[2,3]
Fabaceae[4] (Leguminosae)	*Desmodium* spp.	Beggarweed	Drake 1920	US-FL
			Todd 1976	US-GA
			Watson 1918	US-FL
Fabaceae[4] (Leguminosae)	*Glycine max* (L.) Merr. (syn. *Glycine hispida* Maxim. [sic]*)	Soy bean Soya Soybean	Davis 1964	US-HI
			Hall and Teetes 1981	US-TX
			Hoffmann 1935*	WW
			Jones and Sullivan 1981, 1982	US-SC
			McPherson and McPherson 2000	US
			Musser et al. 2014	US-NC
			Newsom et al. 1980	US-LA
			Rodríguez Vélez 1974	MX-SIN
			Smith et al. 2009	US-AR
Fabaceae[4] (Leguminosae)	*Indigofera* sp.	Indigo	Buschman and Whitcomb 1980	US-FL
			Hoffmann 1935	WW
Fabaceae[4] (Leguminosae)	*Lathyrus latifolius* L.	Ever-lasting pea	Drake 1920	US-FL
Fabaceae[4] (Leguminosae)	*Lens culinaris* Medik. (syn. *Lens esculenta* Moench)	Lentil	Hoffmann 1935	WW
Fabaceae[4] (Leguminosae)	*Macroptilium lathyroides* (L.) Urb. (*Macroptilium lathyroides* Urban [sic]*)	Wild bushbean	Jones et al. 2001*	US-HI
Fabaceae[4] (Leguminosae)	*Medicago polymorpha* L.	Burclover	Esquivel 2016a	US-TX
Fabaceae[4] (Leguminosae)	*Medicago sativa* L.	Alfalfa Lucerne	Buschman and Whitcomb 1980	US-FL
			Hall and Teetes 1981	US-TX
			Hoffmann 1935	WW
			Mitchell et al. 1965	US-HI
			Rodríguez Vélez 1974	MX-SIN
Fabaceae[4] (Leguminosae)	*Mucuna pruriens* (L.) DC.	Velvet bean	Jones 1918	US-AL
Fabaceae[4] (Leguminosae)	*Neonotonia wightii* (Wight & Arn.) Lackey var. *wightii* (syn. *Glycine wightii* Verdcourt*)	Wight's neonotonia	Jones et al. 2001*	US-HI
Fabaceae[4] (Leguminosae)	*Phaseolus lunatus* L. (syn. *Phaseolus limensis* L. [sic]*)	Butter bean Butterbean Lima bean	Hoffmann 1935*	China
			Jones 1918	US-LA
			Jones and Sullivan 1982*	US-SC
			McPherson and McPherson 2000	US
Fabaceae[4] (Leguminosae)	*Phaseolus vulgaris* L.	French bean Frijol Green bean Kidney bean Snap bean String bean	Gibson and Carrillo S. 1959	MX-MIC
			Hoffmann 1935	WW
			Jones 1918	US-LA
			Newsom et al. 1980	US-LA
			Rodríguez Vélez 1974	MX-SIN
Fabaceae[4] (Leguminosae)	*Phaseolus* spp.	Bean	Davis 1964	US-HI
			Ehler 2002	US-CA
			Jones 1918	US-LA
			Mitchell et al. 1965	US-HI

(Continued)

TABLE 7.4 (CONTINUED)

Cultivated and Wild Plants (by Family) Associated with *Nezara viridula* in North America[1]

Family	Plant Species	Common Name(s)[2]	Source[2]	Country-State[2,3]
Fabaceae[4] (Leguminosae)	*Pisum sativum* L. (syn. *Pisum arvense* L.)	Peas	Drake 1920 Hoffmann 1935 Hubbard 1885 Jones 1918	US-FL WW US-FL US-AL
Fabaceae[4] (Leguminosae)	*Pisum* sp.	Peas	Hall and Teetes 1981	US-TX
Fabaceae[4] (Leguminosae)	*Pueraria montana* (Lour.) Merr. var. *lobata* (Willd.) Maesen & S.M. Almeida ex Sanjappa & Predeep [syn. *Pueraria lobata* (Willd.) Ohwi]	Kudzu	Jones and Sullivan 1982	US-SC
Fabaceae[4] (Leguminosae)	*Pueraria phaseoloides* (Roxb.) Benth. (*Pueraria phaseoloides* Bentham [sic]*)	Tropical kudzu	Jones et al. 2001*	US-HI
Fabaceae[4] (Leguminosae)	*Senna occidentalis* (L.) Link (syn. *Cassia occidentalis* L.)	Coffee senna	Jones and Sullivan 1981, 1982	US-SC
Fabaceae[4] (Leguminosae)	*Sesbania vesicaria* (Jacq.) Elliott	Bagpod	Panizzi and Slansky 1991	US-FL
Fabaceae[4] (Leguminosae)	*Sesbania* sp.	Wild legume	Newsom et al. 1980	US-LA
Fabaceae[4] (Leguminosae)	*Sophora chrysophylla* (Salisb.) Seem.	Mamani	Discover Life 2015	US-HI
Fabaceae[4] (Leguminosae)	*Tephrosia* spp.	Hoary pea	Hoffmann 1935	WW
Fabaceae[4] (Leguminosae)	*Trifolium alexandrinum* L.	Egyptian clover	Hoffmann 1935	WW
Fabaceae[4] (Leguminosae)	*Trifolium incarnatum* L.	Crimson clover	Jones and Brewer 1987 Esquivel 2016a	US-MS US-TX
Fabaceae[4] (Leguminosae)	*Trifolium pratense* L.	Red clover	Newsom et al. 1980	US-LA
Fabaceae[4] (Leguminosae)	*Trifolium repens* L.	White clover	Esquivel 2016a	US-TX
Fabaceae[4] (Leguminosae)	*Vicia faba* L.	Faba bean Fava bean	Nuessly et al. 2004	US-FL
Fabaceae[4] (Leguminosae)	*Vigna mungo* (L.) Hepper (syn. *Phaseolus mungo* L.)	Black gram	Hoffmann 1935	WW
Fabaceae[4] (Leguminosae)	*Vigna unguiculata* (L.) Walp. [syn. *Dolichos unguiculatus* L., *Vigna sinensis* (Endl.) (sic)*, *Vigna sinensis* Savi (sic)*]	Chícharo de vaca Cowpea Xpelón	Discover Life 2015 Drake 1920 Hoffmann 1935* Jones 1918 Jones and Sullivan 1982 McPherson and McPherson 2000 Newsom et al. 1980 Rodríguez Rivera 1975 Rodríguez Vélez 1974* Turner 1918 Watson 1918	US-MS US-FL US US-AL, FL, LA, TX US-SC US US-LA MX-YUC MX-SIN US-GA US, FL

(Continued)

TABLE 7.4 (CONTINUED)

Cultivated and Wild Plants (by Family) Associated with *Nezara viridula* in North America[1]

Family	Plant Species	Common Name(s)[2]	Source[2]	Country-State[2,3]
Fabaceae[4] (Leguminosae)	*Vigna unguiculata* (L.) Walp. ssp. *cylindrica* (L.) Verdc. (syn. *Dolichos biflorus* L.)	Horse gram	Hoffmann 1935	WW
Fabaceae[4] (Leguminosae)	*Vigna unguiculata* (L.) Walp. ssp. *sesquipedalis* (L.) Verdc. (syn. *Vigna sesquipedalis* L. [sic])*	Chinese long bean* Yardlong bean	Hoffmann 1935* Mau and Kessing 1991	WW US-HI
Juglandaceae	*Carya illinoinensis* (Wangenh.) K. Koch [syn. *Carya pecan* (Marshall) Engl. & Graebn.]	Pecan	Demaree 1922 Hoffmann 1935 McPherson and McPherson 2000 Turner 1918	US-GA WW US US-GA
Lamiaceae[4] (Labiatae)	*Lamium amplexicaule* L.	Henbit	Esquivel 2016a	US-TX
Lamiaceae[4] (Labiatae)	*Ocimum gratissimum* L. (syn. *Ocimum suave* Willd.)	African basil	Hoffmann 1935	WW
Lamiaceae[4] (Labiatae)	*Salvia* sp.	Salvia	Batra et al. 1981	US
Liliaceae	*Agapanthus* sp.	Agapanthus	Discover Life 2015	US-CA
Malvaceae	*Abelmoschus esculentus* (L.) Moench (syn. *Hibiscus esculentus* L.)	Okra	Drake 1920 Hoffmann 1935 Jones 1918 Jones and Sullivan 1982 McPherson and McPherson 2000	US-FL WW US-AL, LA, TX US-SC US
Malvaceae	*Abutilon grandifolium* (Willd.) Sweet (*Abutilon grandifolium* Sweet [sic]*)	Hairy abutilon	Jones et al. 2001*	US-HI
Malvaceae	*Alcea rosea* L. (syn. *Althaea rosea* Cav. [sic]*)	Hollyhock	Hoffmann 1935*	WW
Malvaceae	*Gossypium barbadense* L.	Creole cotton	Hoffmann 1935	WW
Malvaceae	*Gossypium hirsutum* L. var. *hirsutum* (syn. *Gossypium herbaceum* L.*)	Algodonero Upland cotton	Hall and Teetes 1981 Herbert et al. 2009 Hoffmann 1935* Jones 1918 Riley and Howard 1893a, b Rodríguez Vélez 1974	US-TX US-NC WW US US-FL, LA MX-SIN
Malvaceae	*Gossypium* spp.	Algodón Cotton	Drake 1920 Gibson and Carrillo S. 1959 Kirkaldy 1909 McPherson and McPherson 2000 Morrill 1910	US-FL MX-SIN, SON WW US US-FL, LA, TX
Malvaceae	*Hibiscus cannabinus* L.	Til	Hoffmann 1935	WW
Malvaceae	*Malva parviflora* L.	Cheese weed	Davis 1964 Ehler 2002	US-HI US-CA
Malvaceae	*Malvastrum coromandelianum* (L.) Garcke (*Malvastrum coromandelianum* Garcke [sic]*)	False mallow	Jones et al. 2001*	US-HI

(Continued)

TABLE 7.4 (CONTINUED)

Cultivated and Wild Plants (by Family) Associated with *Nezara viridula* in North America[1]

Family	Plant Species	Common Name(s)[2]	Source[2]	Country-State[2,3]
Moraceae	*Morus alba* L.	White mulberry	Hoffmann 1935	WW
Moraceae	*Morus nigra* L.	Black mulberry	Hoffmann 1935	WW
Moraceae	*Morus* spp.	Mulberry	Jones 1918	US
			Kirkaldy 1909	WW
Myrtaceae	*Metrosideros polymorpha* Gaudich. var. *polymorpha* [syn. *Metrosideros collina* (J. R. Forst. & G. Forst.) A. Gray]	'Ohi'a lehua	Discover Life 2015	US-HI
Nyctaginaceae	*Boerhavia coccinea* Mill.	Scarlet spiderling	Jones et al. 2001	US-HI
Onagraceae	*Oenothera speciosa* Nutt.	Pinkladies	Esquivel 2016a	US-TX
Orchidaceae	*Dendrobium* spp.	Orchids	Fleming et al. 1999	US-HI
Passifloraceae	*Passiflora incarnata* L. (*Passiflora incarinata* [sic]*)	Passion flower	Drake 1920	US-FL
			Hoffmann 1935*	WW
			Todd 1976	US-GA
Pedaliaceae	*Sesamum orientale* L. (syn. *Sesamum indicum* L.)	Ajonjoli	Hoffmann 1935	WW
		Sesame	Rodríguez Vélez 1974	MX-SIN
Phytolaccaceae	*Phytolacca americana* L. var. *americana* (syn. *Phytolacca decandra* L.)	Pokeweed	Drake 1920	US-FL
			Hoffmann 1935	WW
			Todd 1976	US-GA
Pinaceae	*Pinus taeda* L.	Loblolly pine (overwintering)	Jones and Sullivan 1981	US-SC
Poaceae[4] (Gramineae)	*Avena sativa* L.	Oats	Newsom et al. 1980	US-LA
Poaceae[4] (Gramineae)	*Cenchrus* sp.	Bur grass	Drake 1920	US-FL
		Sandbur	Hoffmann 1935	WW
Poaceae[4] (Gramineae)	*Coix lacryma-jobi* L.	Job's tears	Jones et al. 2001	US-HI
Poaceae[4] (Gramineae)	*Echinochloa* sp.	Cockspur weed	Drake 1920	US-FL
		Zacate pinto	Rodríguez Vélez 1974	MX-SIN
Poaceae[4] (Gramineae)	*Eleusine coracana* (L.) Gaertn. (*Elusine coracana* (L.) Gaert. [sic]*)	Finger millet	Hoffmann 1935*	WW
Poaceae[4] (Gramineae)	*Oryza sativa* L.	Arroz	Drake 1920	US
		Rice	Gibson and Carrillo S. 1959	MX-VER
			Hoffmann 1935	WW
			Jones 1918	US-LA
			Kirkaldy 1909	WW
			McPherson and McPherson 2000	US
			Newsom et al. 1980	US-LA
			Rodríguez Vélez 1974	MX-SIN
Poaceae[4] (Gramineae)	*Panicum* spp.	Zacate carricillo (=panicgrass)	Rodríguez Vélez 1974	MX-SIN
Poaceae[4] (Gramineae)	*Pennisetum glaucum* (L.) R. Br. [syn. *Pennisetum spicatum* (L.) Koern., *Pennisetum typhoideum* Rich.*]	Pearl millet	Hoffmann 1935*	WW
			Esquivel 2016a	US-TX
Poaceae[4] (Gramineae)	*Pennisetum setaceum* (Forssk.) Chiov.	Crimson fountaingrass	Discover Life 2015	US-HI

(Continued)

TABLE 7.4 (CONTINUED)

Cultivated and Wild Plants (by Family) Associated with *Nezara viridula* in North America[1]

Family	Plant Species	Common Name(s)[2]	Source[2]	Country-State[2,3]
Poaceae[4] (Gramineae)	*Saccharum officinarum* L.	Sugar cane Sugarcane	Hoffmann 1935 Jones 1918 Kirkaldy 1909 Jones et al. 2001	WW US-AL WW US-HI
Poaceae[4] (Gramineae)	*Secale* sp.	Rye	Anonymous 1974, cited in McPherson and McPherson 2000	US
Poaceae[4] (Gramineae)	*Sorghum bicolor* L. Moench ssp. *bicolor* (syn. *Sorghum vulgare* Pers.)	Sorghum Sorgo	Hall and Teetes 1981 Hoffmann 1935 McPherson and McPherson 2000 Rodríguez Vélez 1974	US-TX WW US MX-SIN
Poaceae[4] (Gramineae)	*Triticum aestivum* L. (syn. *Triticum vulgare* Villars, *Triticum vulgare* L. [sic]*)	Trigo Wheat	Hall and Teetes 1981 Hoffmann 1935 Jones and Sullivan 1982 McPherson and McPherson 2000 Newsom et al. 1980 Rodríguez Vélez 1974*	US-TX WW US-SC US US-LA MX-SIN
Poaceae[4] (Gramineae)	*Zea mays* L.	Corn Maiz	Hoffmann 1935 Jones 1918 Kirkaldy 1909 McPherson and McPherson 2000 Mitchell et al. 1965 Newsom et al. 1980 Rodríguez Vélez 1974	WW US-TX WW US US-HI US-LA MX-SIN
Polygonaceae	*Polygonum* sp.	Knotweed	Hoffmann 1935 Jones and Sullivan 1982	China US-SC
Polygonaceae	*Rheum rhabarbarum* L. (syn. *Rheum rhaponticum* L.)	Rhubarb	Hoffmann 1935	WW
Polygonaceae	*Rumex* spp.	Sorrel	Drake 1920 Hoffmann 1935 Todd 1976	US-GA WW US-GA
Portulacaceae	*Portulaca oleracea* L.	Verdolaga	Rodríguez Vélez 1974	MX-SIN
Proteaceae	*Macadamia integrifolia* Maiden & Betche (syn. *Macadamia ternifolia* F. Muell.)	Macadamia nut	Jones and Caprio 1992 McPherson and McPherson 2000 Mitchell et al. 1965	US-HI US US-HI
Rosaceae	*Prunus persica* (L.) Batsch	Peach	Hoffmann 1935 McPherson and McPherson 2000	WW US
Rosaceae	*Prunus serotina* Ehrh.	Black cherry	Jones and Sullivan 1982	US-SC
Rosaceae	*Prunus* spp.	Wild plum	Drake 1920 Hoffmann 1935 Todd 1976	US-FL WW US-SC
Rosaceae	*Pyrus communis* L.	Pear	Hoffman 1935	WW
Rosaceae	*Rubus* spp.	Wild blackberry	Drake 1920 Hoffmann 1935 Todd 1976	US-FL WW US-SC
Rubiaceae	*Coffea* sp.	Coffee	Hoffmann 1935	WW

(Continued)

TABLE 7.4 (CONTINUED)

Cultivated and Wild Plants (by Family) Associated with *Nezara viridula* in North America[1]

Family	Plant Species	Common Name(s)[2]	Source[2]	Country-State[2,3]
Rubiaceae	*Richardia brasiliensis* Gomes (*Richardia brasiliensis* Gomez [sic]*)	Tropical Mexican clover	Jones et al. 2001*	US-HI
Rubiaceae	*Richardia scabra* L. (syn. *Richardsonia pilosa* H. B. & K. [sic]*, *Richardsonia scabra* A. [sic]*)	Mexican clover	Drake 1920 Hoffmann 1935*	US-FL WW
Rutaceae	*Citrus × aurantiifolia* (Christm.) Swingle (pro sp.) [*medica × sp.*] (syn. *Citrus acida* Roxb., *Citrus aurantifolium* (Christm.) Swingle [sic])*	Key lime	Hoffmann 1935*	WW
Rutaceae	*Citrus × limonia* Osbeck (pro sp.) [*limon × reticulata*] (syn. *Citrus limonum* Osbeck [sic])*	Lemon*	Hoffmann 1935*	WW
Rutaceae	*Citrus × paradisi* Macfad. (pro sp.) [*maxima × sinensis*] (syn. *Citrus maxima* var. *uvacarpa* Merrill & Lee [sic])*	Grapefruit	Hoffmann 1935*	WW
Rutaceae	*Citrus × sinensis* (L.) Osbeck (pro sp.) [*maxima × reticulata*] (syn. *Citrus aurantium* L.)*	Orange	Drake 1920 Hoffmann 1935* Hubbard 1885 Kirkaldy 1909 McPherson and McPherson 2000 Morrill 1910 Riley and Howard 1893a,b Watson 1919a	US-FL WW US-FL WW US US-FL US-FL, LA US-FL
Rutaceae	*Citrus reticulata* Blanco (syn. *Citrus nobilis* Lour.)*	Tangerine	Hoffmann 1935* McPherson and McPherson 2000	WW US
Rutaceae	*Citrus* spp.	Citrus Kumquat Satsuma orange	Jones 1918 Ruiz et al. 2006 Watson 1918	US-FL MX-TMP US-FL
Scrophulariaceae	*Castilleja indivisa* Engelm.	Indian paintbrush	Hall and Teetes 1981	US-TX
Solanaceae	*Capsicum annuum* L.	Chile morrón Red pepper	Hoffmann 1935 Rodríguez Vélez 1974	WW MX-SIN
Solanaceae	*Capsicum* spp.	Pepper	Jones 1918	US-LA
Solanaceae	*Nicotiana tabacum* L.	Tobacco	Hoffmann 1935 Jones 1918 McPherson and McPherson 2000	WW US-FL US
Solanaceae	*Nicotiana* sp.	Tobacco	Discover Life 2015	US-LA
Solanaceae	*Solanum elaeagnifolium* Cav.	Silverleaf nightshade	Hall and Teetes 1981	US-TX

(Continued)

TABLE 7.4 (CONTINUED)

Cultivated and Wild Plants (by Family) Associated with *Nezara viridula* in North America[1]

Family	Plant Species	Common Name(s)[2]	Source[2]	Country-State[2,3]
Solanaceae	*Solanum lycopersicum* L. var. *lycopersicum* [syn. *Lycopersicon esculentum* Miller, *Lycopersicon lycopersicum* (L.) Karst. ex Farw., *Lycopersicum esculentum* (L.) Mill. [sic]*]	Jitomate Tomate Tomato	Drake 1920 Ehler 2002 Gibson and Carrillo S. 1959 Hoffmann 1935* Hubbard 1885 Jones 1918	US-FL US-CA MX-VER WW US-FL US-LA, SC, TX
			Jones and Sullivan 1982 McPherson and McPherson 2000 Mitchell et al. 1965 Rodríguez Vélez 1974	US-SC US US-HI MX-SIN
Solanaceae	*Solanum madrense* Fernald [*Solanum madrense* (Fenald) (sic)*]	Berenjena silvestre	Rodríguez Vélez 1974*	MX-SIN
Solanaceae	*Solanum melongena* L. (syn. *Lycopersicum melongena* L.)*	Berenjena Egg-plants Eggplant	Hoffmann 1935* Hubbard 1885 Jones 1918 Rodríguez Vélez 1974	WW US-FL US-LA, TX MX-SIN
Solanaceae	*Solanum nigrum* L. (syn. *Lycopersicum nigrum* L.)*	Black nightshade Popolo	Davis 1964 Hoffmann 1935*	US-HI WW
Solanaceae	*Solanum tuberosum* L. (syn. *Lycopersicum tuberosum* L.)*	Potato	Drake 1920 Hoffmann 1935* Jones 1918 Kirkaldy 1909 McPherson and McPherson 2000 Morrill 1910	US-FL WW US-LA WW US US-FL
Sterculiaceae	*Theobroma cacao* L.	Cacao Cocoa	Hoffmann 1935	WW
Ulmaceae	*Celtis australis* L.	Hackberry	Hoffmann 1935 Jones 1918	WW US-LA
Verbenaceae	*Lantana* spp.	Lantana	Discover Life 2015	US-HI
Vitaceae	*Vitis vinifera* L.	Cultivated grape	Hoffmann 1935	WW
Vitaceae	*Vitis* spp.	Wild grape	Drake 1920 Hoffmann 1935 Todd 1976	US-FL WW US-GA

*Asterisk denotes scientific names (and authority) and common names as originally presented by corresponding source.

[1] Includes North American taxa reflected in worldwide distribution records of Kirkaldy (1909) and Hoffmann (1935).

[2] Where multiple common names and sources are presented for a plant species, the common names and sources are sorted alphabetically and a given common name may not occur in all sources presented. The country and state designations, however, do correspond with the respective alphabetically listed sources.

[3] Country designations of Mexico (MX) and the United States (US) are followed by state abbreviations when provided within indicated source (e.g., in US – CA, California; FL, Florida; GA, Georgia; HI = Hawaii; LA, Louisiana; SC, South Carolina; TX, Texas; WA, Washington; and, in Mexico – MIC, Michoacán; SIN, Sinaloa; SON, Sonora; TMP, Tamaulipas; VER, Veracruz; YUC, Yucatán); references presenting general listing and lacking specific country localities are designated as worldwide (WW) occurrence.

[4] Current family name (original name reported in parentheses).

TABLE 7.5

Summary of Plant Families Associated with *Nezara viridula* in North America

Plant Family (*n* = 43)	No. Total Taxa	Percent of Total	No. of Plant Taxa Identified to:		Cumulative Percent of Total
			Genus[1]	Species[1]	
Fabaceae	49	24.87%	9	40	24.87%
Asteraceae	21	10.66%	2	19	35.53%
Brassicaceae	15	7.61%	2	13	43.15%
Poaceae	14	7.11%	4	10	50.25%
Solanaceae	10	5.08%	2	8	55.33%
Cucurbitaceae	9	4.57%	1	8	59.90%
Malvaceae	9	4.57%	1	8	64.47%
Amaranthaceae	6	3.05%	1	5	67.51%
Chenopodiaceae	6	3.05%	1	5	70.56%
Rutaceae	6	3.05%	1	5	73.60%
Rosaceae	5	2.54%	2	3	76.14%
Capparaceae	4	2.03%	0	4	78.17%
Lamiaceae	3	1.52%	1	2	79.70%
Moraceae	3	1.52%	1	2	81.22%
Polygonaceae	3	1.52%	2	1	82.74%
Rubiaceae	3	1.52%	1	2	84.26%
Cannabaceae	2	1.02%	1	1	85.28%
Convolvulaceae	2	1.02%	0	2	86.29%
Cupressaceae	2	1.02%	2	0	87.31%
Vitaceae	2	1.02%	1	1	88.32%
Acanthaceae	1	0.51%	0	1	88.83%
Anacardiaceae	1	0.51%	0	1	89.34%
Apiaceae	1	0.51%	0	1	89.85%
Boraginaceae	1	0.51%	1	0	90.36%
Bromeliaceae	1	0.51%	0	1	90.86%
Cyperaceae	1	0.51%	0	1	91.37%
Euphorbiaceae	1	0.51%	0	1	91.88%
Juglandaceae	1	0.51%	0	1	92.39%
Liliaceae	1	0.51%	1	0	92.89%
Myrtaceae	1	0.51%	0	1	93.40%
Nyctaginaceae	1	0.51%	0	1	93.91%
Onagraceae	1	0.51%	0	1	94.42%
Orchidaceae	1	0.51%	1	0	94.92%
Passifloraceae	1	0.51%	0	1	95.43%
Pedaliaceae	1	0.51%	0	1	95.94%
Phytolaccaceae	1	0.51%	0	1	96.45%
Pinaceae	1	0.51%	0	1	96.95%
Portulacaceae	1	0.51%	0	1	97.46%
Proteaceae	1	0.51%	0	1	97.97%
Scrophulariaceae	1	0.51%	0	1	98.48%
Sterculiaceae	1	0.51%	0	1	98.98%
Ulmaceae	1	0.51%	0	1	99.49%
Verbenaceae	1	0.51%	1	0	100.00%
Total taxa	197	100.00%	39	138	

[1] Number of taxa identified to genus or species within respective plant family.

Howard 1893a,b; Jones 1918; Drake 1920). In these instances, identification of plants was more general (e.g., "mustard" [Hubbard 1885]; "bean" [Jones 1918]; "cockspur weed" [Drake 1920]; and "thistles" [Hoffmann 1935]), and, thus, only the genus is presented to reflect these reported occurrences.

The numbers of plant taxa associated with *Nezara viridula* in North America are presented in **Tables 7.4 and 7.5**. These lists include 12 newly reported species upon which *N. viridula* was detected in Texas (Esquivel 2016a). Forty-three families have been reported overall in North America (**Tables 7.4 and 7.5**). Six plant families comprise 59.90% of plant taxa (*n* = 197) associated with *N. viridula* (**Table 7.5**). Of these overall plant taxa, the Fabaceae are preferred, representing 24.87% of associated plants, followed by the Asteraceae, Brassicaceae, Poaceae, Solanaceae, and Cucurbitaceae. Thirty-seven additional families represent the remaining 40.10% of families reported (**Table 7.5**).

Within the identified families, 158 associated plants were identified to species level in their original reports for regions of North America (**Table 7.5**). Additionally, 39 plants were identified only to genus; these included cases where only the common names of the plants were reported, and we assigned these common names to appropriate genera. Thus, in total, 197 plant taxa have been identified to date. The numbers of identified families and species in North America are similar to those previously reported in Japan where Kiritani et al. (1965), citing Oho and Kiritani (1960), indicated "as many as 145 species belonging to 32 families serve as the host plants of this insect." These data clearly reflect the polyphagous nature of *N. viridula*. Further emphasizing the opportunistic polyphagous nature, a fifth instar was observed for approximately 10 minutes with stylets inserted, and actively moving, in the stem of a banana (Musaceae; *Musa* sp. – and not included within **Tables 7.4 and 7.5**), the insect only withdrawing stylets after repeated probing with forceps by the observer (Jesus F. Esquivel, personal observation).

Identification of associated plants does not guarantee that the plants will be utilized throughout its entire distribution (Panizzi and Slansky 1991, Panizzi 1997). Panizzi and Slansky (1991) observed *Nezara viridula* on *Desmodium tortuosum* (Sw. DC.) in Florida, yet they indicated that Jones (1979) "never found them on *D. tortuosum*" in South Carolina. This could be attributed to asynchrony of pest and host plant within and between geographic locations and other biotic/abiotic factors (Panizzi 1997).

In addition to known foraging and reproductive resources, **Tables 7.4 and 7.5** include plant taxa that *Nezara viridula* has used as overwintering resources [i.e., *Beta vulgaris* L., *Juniperus* sp., *Pinus taeda* L., *Thuja* sp., *Tillandsia usneoides* (L.) L.]. Although Kirkaldy (1909) and Hoffmann (1935) appear to be worldwide listings of "foodplants," their reports typically included plant species that also occurred in North America [e.g., *Amaranthus hypochondriacus* (L.) and *Phaseolus lunatus* L. in **Table 7.4**] and were denoted as such. Therefore, these plant taxa were included in this report. Finally, McPherson and McPherson (2000) reviewed the extant literature and summarized major crops and wild plants attacked by *N. viridula* and other stink bug species, in America north of Mexico. Descriptions of damage by *N. viridula* in each crop also were provided. DeWitt and Godfrey (1972) and Panizzi (1997) present exhaustive listings of host plant-related literature worldwide, in some instances reiterating the host plant species from earlier reports. Thus, references cited in **Table 7.4** are only representative reports for each of the associated plant species in North America. For a more comprehensive summary of publications addressing plants associated with *N. viridula* outside of North America, the reader is encouraged to review references within Dewitt and Godfrey (1972), Panizzi and Slansky (1985), Velasco and Walter (1992), Panizzi (1997), McPherson and McPherson (2000), and Smaniotto and Panizzi (2015).

7.4.6 Movement among Hosts at Farmscape and Landscape Levels

Stink bugs, including *Nezara viridula*, feed on a wide range of host plants as described in **Section 7.4.5**, but prefer to utilize hosts that are actively producing seeds or fruit. Because stink bugs are active throughout the warmer parts of the year, they must rely on a variety of overlapping host plants that occur in a dynamic ecosystem. As soon as the host begins to senesce or is mowed, the bugs will disperse to a new host that is just starting to produce seeds or fruit. Bundy and McPherson (2000b) demonstrated this host plant movement for stink bugs, primarily *N. viridula*, in a soybean-cotton ecosystem. The bugs were first attracted to the early-maturing soybean cultivar and remained there until the plants began to mature. Then, they moved into the later-maturing cultivar as pods were filling with seeds and finally into conventional and transgenic cotton (i.e., Bt-cotton = cotton containing the toxin *Bacillus thuringiensis* [Bt] *kurstaki*) when flowers and bolls were forming. Additionally, the planting date of a particular agronomic crop has a profound impact on its host suitability to stink bugs. For example, Tillman (2010) observed greater populations of *Euschistus servus* (Say) (brown stink bug) and *N. viridula* in late-planted compared to early-planted corn. Understanding these movements and being able to predict when the population is likely to disperse and colonize new habitats is important for formulating management strategies.

Direct empirical evidence on stink bug dispersal is limited. However, stink bugs are widely regarded as strong fliers based on their ability to find and colonize susceptible agronomic crops that are isolated spatially in the landscape. Tillman et al. (2009) showed that an individual stink bug could be recaptured as far as 120 meters from the release site. Huang (2012) showed that marked stink bugs were much more likely to be recovered only a few meters from where they were marked in agronomic crops. A tethered stink bug, *Graphosoma rubrolineatum* (Westwood), was estimated to fly a distance of over 27 kilometers in a single 7-hour flight (Cui and Cai 2008), indicating that stink bugs are capable of long-distance flights. However, focusing on the individual as opposed to populations may oversimplify the importance of local dispersal, largely because increased dispersal distance tends to dilute the population in a particular space.

A mosaic of ecosystems comprises the broad concept embodied by the term landscape. In comparison, the spatial arrangement of resource patches utilized by a species comprises what Ehler (2000) termed the farmscape. Movement between hosts can be described at the farm (proximate) or landscape (distant) level, and the scale at which stink bugs function in those environments depends upon numerous factors related to the temporal and spatial availability of suitable hosts (Kennedy and Storer 2000, Reeves et al. 2010). At the farmscape level, stink bugs move from a field of one host to an adjacent field when the adjacent host become equally or more attractive through time (Pilkay et al. 2015). Stink bugs are extremely polyphagous (see **Tables 7.4 and 7.5**) and prefer plants with developing seeds as hosts (Jones and Sullivan 1982). Consequently, they must move, and those movements can be proximate (Kareiva 1983) if the local environment supports suitable hosts, or they can be distant into the landscape if acceptable hosts are not present locally. Stink bugs are capable of short or long flights, but tendency for longer flights changes over the season, peaking near the end of the summer months for *Halyomorpha halys* (Stål) (brown marmorated stink bug) (Wiman et al. 2015). Populations of stink bugs have been observed in mass flights (Wilbur 1939), and multiple species have been trapped in seasonal synchronous flights (Cherry and Wilson 2011).

Movement and subsequent settling of *Nezara viridula* is associated with location and quality of hosts (Velasco and Walter 1993, Herbert and Toews 2012), selection of mates (Borges et al. 1987), diapause, and overwintering (Musolin and Numata 2003b, Musolin 2012), and various abiotic factors drive these activities (Gu and Walter 1989, Todd 1989). Research indicates that stink bugs tend to aggregate near the edges of fields including cotton (Toews and Shurley 2009), corn, wheat (Reisig 2011), soybean (Venugopal et al. 2014), and peanut (Tillman et al. 2009). Soybean is an extremely suitable host plant for pentatomids, especially *N. viridula* (Velasco and Walter 1992, McPherson et al. 1993, Smith et al. 2008, Olson et al. 2011, Pilkay et al. 2015), so the species can be observed readily on the host within its worldwide distribution (Todd 1989, Clarke 1992, Vivan and Panizzi 2006). Movement at the farm level within soybeans has been studied across production systems (i.e., maturity groups, host phenology, and planting dates) (Schumann and Todd 1982, Smith et al. 2009), and limited movement by nymphs of *N. viridula* and *Piezodorus guildinii* (Westwood) has been measured in the crop (Panizzi et al. 1980). Wild hosts of *N. viridula* and other stink bugs also are important (**Tables 7.4 and 7.5**; Jones and Sullivan 1982, Panizzi 1997, Olson et al. 2012) and support movement of species at the farm and landscape level through temporal and spatial availability.

In addition to host quality, stink bug movement and intraspecific communication is influenced by pheromones and substrate-borne vibrational cues. Male-produced stink bug pheromone components have been identified from multiple sexually mature stink bug species that attract conspecific females, males, and even a few late instars (Mitchell and Mau 1971, Harris and Todd 1980, Tillman et al. 2010) (see **Chapter 15**). Although the exact ratio differs by species, both *Chinavia hilaris* and *Nezara viridula* males produce *cis*-and *trans*-(Z)-bisabolene epoxide blends (Aldrich 1988; Aldrich et al. 1987, 1989, 1993). Synthetic pheromone lures can be used to attract stink bugs to traps (Tillman et al. 2010), but stink bugs are relatively weakly attracted to pheromone-baited traps compared to the strong attraction exhibited by lepidopteran species to similar traps. This is likely because stink bugs are long-lived and can oviposit for several weeks as opposed to only a few days in Lepidoptera. Muted (= reduced) capture in pheromone traps placed in the field have limited, thus far, the usefulness of this method for predicting damage to a particular crop.

Vibrational pulses, also termed songs, are used by both sexes at medium range for calling (i.e., attracting the opposite sex) and at short range for courtship of mates, repelling, and rivalry. These songs are short in duration, lasting only a few hundred milliseconds to a second by males and several seconds in the case of the female-repelling song (Čokl et al. 2000). Distinct songs by *Nezara viridula* have been identified for location of females, courtship, a copulatory song, and a male rival song (Čokl et al. 2001). Studies show that the songs are species-specific and the frequency ranges are highly appropriate for transmission in the host plant (Kon et al. 1988, Čokl et al. 2000). *N. viridula* populations maintain similar temporal patterns and frequency characteristics in their songs despite collection of populations from different continents (Čokl et al. 2000).

7.5 Current Impacts in Agriculture

7.5.1 Asia

Nezara viridula is a highly polyphagous species, attacking both monocot and dicots. Oho and Kiritani (1960) listed as many as 145 plant species belonging to 32 families as its host plants. Todd (1989) describes the *N. viridula* host range as "over 30 families of dicotyledonous plants and a number of monocots." In North America, 43 families with a total of 197 plant taxa (i.e., 158 identified to species and 39 identified to genera) have been associated with *N. viridula* (see **Section 7.4.5 and Tables 7.4 and 7.5**). Taking into consideration that the species can breed, or at least feed, during some stages of its development on numerous cultivated and wild subspecies, cultivars or forms, it seems impossible to compile a comprehensive list of its host plants. However, it always is mentioned that the species demonstrates a strong preference for leguminous plants (**Table 7.5**; Todd and Herzog 1980, Todd 1989, Panizzi et al. 2000).

It often is mentioned that during the vegetative stages of host plants throughout the season, different generations of *Nezara viridula* breed upon or utilize different plant species (Singh 1973). Most of the damage comes from the feeding of nymphs on developing pods, seeds, or fruits and results in significant reductions in yield, quality, and germination of seeds.

In Japan, *Nezara viridula* is a pest of rice, soybean, fruits, and many other agricultural crops. For some of them (such as rice, soybean, and tomatoes), it is a major pest; for others, it is minor, one of a large complex of heteropteran pests. This variable pest status is the reason why statistics estimating damage to agriculture caused by this pest are lacking. Also, damage to rice caused by stink bugs is mostly not the yield loss but loss of grain quality (i.e., pecky grains) as a result of feeding by the bugs. If pecky grains are present at a rate over 1/1000 grains, then the quality grade of rice goes down to second class with a substantial reduction in price.

The outbreaks of *Nezara viridula* in the 1950s in Miyazaki Prefecture (Kyushu) was induced by the cultivation of early planted rice. The emergence of the first generation adults was synchronized with the heading of early planted rice, which provided suitable food for the bugs. The following generations of the pest utilized middle-season and late-planted rice. Another factor that increased damage by *N. viridula* was the cultivation of soybeans. Due to the overproduction of rice in Japan, the government encouraged and subsidized farmers to plant alternative crops such as soybeans. The soybean cultivation provided a good alternative host plant for the bug, and the pest actively utilized the crop particularly when producing its third and/or fourth generations in late autumn.

Also starting from 1950s, temperatures (and particularly winter temperatures) in central Japan demonstrated an increasing trend (Musolin 2007) and, as previously mentioned, milder overwintering conditions are beneficial for building up populations of *Nezara viridula*. Thus, it seems likely that the combination of three factors (i.e., cultivation of early-season rice, introduction of soybean, and increasing winter temperatures) have encouraged expansion of *N. viridula* in Japan.

In general, since the 1950s, the composition of heteropteran pests of rice has changed in Japan from relatively monophagous plant-sucking stink bugs with a univoltine seasonal cycle [e.g., *Scotinophara lurida* Burmeister and *Niphe elongata* (Dallas) [as *Lagynotomus elongatus* (Dallas)]] to ear-sucking stink bugs with wide host ranges and multivoltine seasonal cycles [e.g., *Nezara viridula*, *Cletus punctiger*

(Dallas), *Leptocorisa chinensis* Dallas, *Stenotus rubrovittatus* (Matsumura), and *Trigonotylus caelestialium* (Kirkaldy)] that reproduce on weeds and subsequently invade rice fields to feed on rice grains, causing pecky rice. These pests are polyphagous and multivoltine with high dispersal activity.

7.5.2 Europe

Damage caused by *Nezara viridula* to cultivated crops and ornamentals is documented in the native Mediterranean range in Europe, although damage levels usually are considerably lower than in subtropical and tropical regions. Soybeans are attacked in central Italy, with mean seed weight loss (18%), altered seed composition (reduced oil content), and reduced germination rate (up to 95% in susceptible strains) as the most important effects (Colazza et al. 1986). Damage has also been reported in France (Le Page 1996). In field experiments with castor (*Ricinus communis* L.) in France and Italy, reduced seed yield was observed (Conti et al. 1997).

Reports of damage in the introduced European range are scarce and often anecdotal. Simov et al. (2012) mention preliminary evidence that, in 2007–2008, damage (chlorosis) was observed in Bulgaria on tomato fruits, and Grozea et al. (2012) observed damage to tomatoes in Romania. Salisbury et al. (2009) listed 26 observed host plants in the United Kingdom, with runner bean (*Phaseolus coccineus* L.) as the most important. Most records were from late summer and autumn when harvesting was more or less completed, so there was little if any damage. They concluded that, for the time being, *Nezara viridula* is not a major pest in the country, but it could become problematic if populations increase and start feeding earlier in the season. This statement can be extended to the European nonnative range.

7.5.3 North America

In America north of Mexico, *Nezara viridula* ranges from Virginia south to Florida and west to Texas, California, and Washington (**Figure 7.5**; see **Section 7.3.3**); in the southern range, the species is an economic pest of numerous crops, including, but not limited to, soybean, cowpea, southern pea, lima bean, pecan, wheat, grain sorghum, corn, tomato, tobacco, and cotton (**Table 7.4**; McPherson and McPherson 2000). Major row crops, such as soybean, corn, and cotton comprise a large majority of the cultivated land area in the southern United States planted to annual crops, and these hosts support enormous populations of *N. viridula* each year. This bug has a high reproductive capacity and more generations per year (Todd 1989) than *Chinavia hilaris* and *Euschistus servus*, the two other major species observed in the complex of stink bugs important in the southern United States.

Due to this innate reproductive capacity and increased voltinism, *Nezara viridula* can make a large negative impact in agriculture in the region. Stink bugs always have been present in southern crop production systems, but the need for management has become much more apparent recently. This change is thought to be driven by cultural changes in cotton production and evolving pest pressures. Prior to boll weevil [*Anthonomus grandis* Boheman (Coleoptera: Curculionidae)] eradication and widespread adoption of Bt-cotton, growers typically made 12–14 broad spectrum insecticide applications per year to control this weevil and lepidopteran pests such as *Heliothis virescens* (F.) and *Helicoverpa zea* (Boddie) (King et al. 1987, Roof 1994). These numerous broad spectrum insecticide applications likely provided coincidental control of stink bugs. At present, the boll weevil has been eradicated from all but a few counties in Texas, and stacked gene transgenic Bt-cotton varieties provide outstanding control of virtually all serious caterpillar pests. Further, growers have developed an appreciation for the benefits of preserving natural enemies and, therefore, make less than three insecticide applications per year. The vast reduction in insecticide use has led to an increase in stink bug populations.

In field crops of the southern United States, stink bugs are considered primary pests requiring monitoring and control tactics annually. Nationwide crop losses plus management costs attributed to stink bugs in cotton exceeded $106 million (M) in 2005, but have moderated to $82.5 M in 2013 and $67.9 M in 2014 (Williams 2006, 2014, 2015). In corn and soybean, stink bugs are considered major pests, with *Nezara viridula* leading in importance, particularly in the southernmost areas of the region. When Bt soybeans are commercialized in North America, stink bugs will become the most prominent insect pests of the crop requiring foliar insecticide control in the crop, nearly identical to the situation with cotton in the southeastern United States. Stink bugs are considered major pests of corn, but there are limitations to the effectiveness of the insecticide approach for the pest group because of crop structure interfering with effective delivery of foliar-applied active ingredients and a lack of any labelled organophosphate insecticides.

In addition to physical damage caused by stink bug feeding, these pests also transmit common plant pathogens that further decimate cash crop value. Turner (1918) discovered that when he caged *Nezara viridula* on pecan nuts, it resulted in kernel spot and black pit; both conditions are reflections of reduced nut quality. More recent work shows that pathogens causing boll rot in cotton are physically present on stink bug mouthparts, and the pathogens can be transmitted when feeding. For example, feeding on developing cotton bolls can result in the loss of individual locks or the entire boll (Medrano et al. 2009a,b). Young bolls (1–2 weeks postanthesis) inoculated with these pathogens will be consumed completely by the pathogen, whereas older bolls (≈3 weeks postanthesis) likely will show only a localized infection. *N. viridula* can acquire and transmit at least two boll rot pathogen species that have been isolated from diseased cotton bolls in the Southeast. These pathogens include the bacterium *Pantoea agglomerans* and the fungal pathogen *Nematospora coryli* (Medrano et al. 2009a). Quick field assays to detect boll rot pathogens in bolls or stink bugs currently are not available.

7.6 Impacts on General Public

Aside from economic impacts (e.g., monitoring, insect control costs) on cultivated crops (see **Section 7.5**) directly affecting producers, *Nezara viridula* does not noticeably affect the general public as do some other invasive pest pentatomids such as *Halyomorpha halys* (see **Chapter 4**).

7.7 Management

7.7.1 Monitoring Options

Stink bugs are highly aggregated in the field, which complicates scouting and monitoring. It generally is easier to assess damage from stink bugs as opposed to estimating population density. Although a pheromone blend for *Nezara viridula* has been identified (e.g., Aldrich et al 1993), trapping for population estimation is not particularly effective for this insect. Mizell and Tedders (1995) modified a pyramid trap that makes it suitable for trapping stink bugs including *N. viridula*, but few captures are observed in the field. Further, research shows that more insects will be captured if the collection reservoir is baited with pheromone and provisioned with a method to anesthetize or kill the captured specimens (Cottrell 2001). Trap placement appears to be critical as Tillman et al. (2010) reported capture of *N. viridula* in pheromone-baited traps increased when the traps were placed at the interface of two crops. Pheromone traps have been used to indicate when stink bugs are dispersing into tomato fields (McPherson and McPherson 2000). Mizell et al. (1996, 1997) reported that three to five pheromone traps on a border row in the interior of a pecan orchard could provide useful information for making pest management decision support. However, additional research and breakthroughs are necessary to make traps the primary means of monitoring this pest. More recently, Endo (2016) indicated light traps could be used in Japan to predict populations of *N. viridula* and aid in assessing seasonal changes in reproductive development of collected female adults.

Reliable sampling methods for estimating stink bug abundance in the field include the sweep net, ground cloth, beat bucket, and visual counts. The sweep net is commonly used for scouting stink bugs in row crops. Todd and Herzog (1980) describe sampling methods for use in soybean utilizing 25 sweeps from a single row, and Reay-Jones (2010) used a sweep net to estimate abundance in wheat. Todd (1981), using a beat sheet to estimate stink bug density, indicated that control would be required when stink bug population exceeded 1.1 or 3.3 bugs per row meter depending on soybean maturity. Sane et al. (1999) compared sampling efficiency among the sweep net, ground cloth, vertical beat sheet, and absolute sampling methods for all stink bug species in soybean. Reay-Jones et al. (2009) compared the sweep net and beat cloth for sampling stink bugs in cotton and found evidence that a drop cloth biases the sample toward nymphs whereas the sweep net creates biases toward adults. A beat bucket sampling technique proved to be effective for estimating the number of *Oebalus pugnax* (F.) in grain sorghum (Merchant and Teetes 1992). Corn is difficult to sample for stink bugs due to its height and the fact that stink bugs tend to reside in the collar where the leaf connects to the stalk. Therefore, researchers typically use whole plant visual searches to estimate population density in corn (Tillman 2010, Reisig 2011). Similarly, stink bugs in vegetable crops are typically scouted based on visual observation (Lye and Story 1989).

Pest managers commonly utilize stink bug-induced damage when making treatment decisions. Feeding-induced injury symptoms are permanent whereas the presence of the actual bug is ephemeral or may not be present in the canopy where a sweep net will reach it. For example, Toews et al. (2008) showed that sampling internal boll injury on developing cotton bolls was 10-fold more sensitive at detecting stink bug infestations compared to use of a sweep net or beat cloth. The spatial dynamics of internal boll injury relative to captures in a sweep net have been investigated in cotton production (Reay-Jones et al. 2010). Although considerably less precise, symptoms of external feeding also can be used for estimating stink bug injury to cotton (Toews et al. 2009, Medrano et al. 2015). Stink bugs form a stylet sheath when feeding and internal feeding damage is correlated with stylet sheath counts. Apriyanto et al. (1989) clearly showed that increased numbers of stylet sheaths correlated with observed damage in early growth corn. Bundy et al. (2000) suggested that a sampling program based on stylet sheath counts could be developed as a nondestructive sampling tool. Stylet sheaths have been observed and enumerated after stink bugs have fed on pecan, rice, soybean, and tomato.

7.7.2 Cultural Control

Cultural insect control in agriculture involves the manipulation of standard production practices to reduce insect populations or make conditions less favorable for establishment and colonization. Many nonchemical approaches to controlling *Nezara viridula* have shown promise for reducing populations of or damage from the species, particularly when used in an integrated pest management program (Knight and Gurr 2007). When used alone, cultural control methods may be insufficient, but they can be effective when used in combination or paired with additional strategies such as chemical control, biological control, or attract-and-kill. Common examples of cultural control include, but are not limited to, manipulation of planting dates, planting density, sanitation, tillage operations, water, nutrients, crop residues, adjacent habitats, crop rotation, spatial isolation, and harvest timing.

The most studied example of cultural control of stink bugs in the United States is planting date manipulation, but there also are a few publications demonstrating other approaches. In Georgia, Pulakkatu-Thodi et al. (2014) demonstrated that mean lint yield and economic returns were greater in early-planted cotton as a result of escaping late summer stink bug pressure. Similarly, in Georgia, Tillman (2010) observed fewer *Nezara viridula* in early planted corn. Conversely, most authors find that early-planted soybean is subject to increased stink bug pressure in the mid-South (Baur et al. 2000) and Southeast (McPherson et al. 2001). There is some evidence that populations of immature *N. viridula* may increase under elevated phosphorus, but no evidence that either potassium or magnesium affects population density (Funderburk et al. 1991). McPherson and Bondari (1991) observed that there were more stink bugs in narrow-row planted soybean compared to wide-row plantings. Finally, Buntin et al. (1995) observed more stink bugs in no-till treatments compared with plow-tillage treatments in soybean.

7.7.3 Biological Control

7.7.3.1 General

There are many known natural enemies of *Nezara viridula*. Parasitoids are the most important (**Chapter 17**). Jones (1988) listed 57 species among two families of Diptera and five families of Hymenoptera, with 41 of these species parasitizing the egg stage. Only a few species appeared to be well-adapted or of significant importance. Although no hyperparasitoids were reported, Clarke and Seymour (1992) subsequently reported two species of *Acroclisoides* (Hymenoptera: Pteromalidae) attacking *N. viridula* eggs previously attacked by *Trissolcus basalis* (Wollaston) (Hymenoptera: Platygastridae) in Australia. Recent studies showed that *Ooencyrtus telenomicida* Vassiliev (Hymenoptera: Encyrtidae) is a facultative hyperparasitod of *T. basalis* attacking *N. viridula* in Europe (Cusumano et al. 2013). Waterhouse (1998) presented another comprehensive list of parasitoids, adding a few more species found attacking *N. viridula*.

7.7.3.2 Classical Biological Control

Nezara viridula has been the target of many classical biological control programs around the world. Parasitoids have been imported and released against this pest nearly everywhere it is found. Targeted areas include New Zealand (Cumber 1951), Australia (Wilson 1960, Coombs and Sands 2000), southern Africa (Greathead 1971, van den Berg and Greenland 1996, Farinelli et al. 1994), the continental United States (Jones et al. 1983, 1995), Taiwan (Su and Tseng 1984), Brazil (Kobayashi and Cosenza 1987), and several Pacific islands (Davis 1964, Rao et al. 1971). The egg parasitoid *Trissolcus basalis* has been the most often imported species to control *N. viridula* followed by the New World tachinids, *Trichopoda* spp. (Cumber 1951, Wilson 1960, Davis 1964, Greathead 1971, Rao et al. 1971, Jones et al. 1983, Su and Tseng 1984, Kobayashi and Cosenza 1987, Farinelli et al. 1994, Jones et al. 1995, van den Berg and Greenland 1996).

The egg parasitoid *Telenomus turesis* Walker[3] [as *Telenomus chloropus* (Thomson)] from Japan was released in Mississippi and Louisiana in the United States during 1981–1982 (Jones et al. 1983). However, it did not become established, apparently due to its inability to successfully emerge at lower humidities (Orr et al. 1985). Three species of Japanese egg parasitoids were released in Brazil, but there have been no reports of their establishment (Cosenza and Kobayashi 1986, Kobayashi and Cosenza 1987). One of the Japanese species released, *Trissolcus mitsukurii* (Ashmead) (Hymenoptera: Platygastridae), has been less able to successfully parasitize and emerge from *N. viridula* eggs than the well-established *Trissolcus basalis* (Corrêa-Ferreira and Zamataro 1989). *T. basalis* was absent from the Asian range of *N. viridula* until it was released in Taiwan in 1983 (Su and Tseng 1984). It recently was discovered in Central Honshu and Kyushu, Japan (Mita et al. 2015).

Trichopoda pennipes L. (Diptera: Tachinidae), a native of the New World, was discovered attacking *Nezara viridula* in Italy in 1988 (Colazza et al. 1996) and has since spread through much of the western Palearctic (Tschorsnig et al. 2012) following the establishment and spread of its primary host (Rabitsch 2010).

7.7.3.3 Augmentation Biological Control

Research on developing mass production and augmentation and inoculation of natural enemies to manage pentatomoids generally has shown that the process is too expensive to carry out on a commercial scale. However, a government-supported development and implementation program was executed in Brazil for many years (Corrêa-Ferreira 1993, Corrêa-Ferreira and Moscardi 1996, Corrêa-Ferreira and Panizzi 1999). That program mass-produced the egg parasitoid *Trissolcus basalis* for inoculation to manage *Nezara viridula*, *Piezodorus guildinii*, and *Euschistus heros* (F.) on 20,000 hectares of soybean per year. However, changing farming practices caused the program to be discontinued (Panizzi 2013). The mandated reduction in the use of several pesticides, coupled with the increasing pest status of *N. viridula* in greenhouse crops in southern Europe, has stimulated research on mass production of *T. basalis* for release in greenhouse crops in Spain (Canton-Ramos and Callejon-Ferre 2010).

Entomopathogens, especially fungi, have been investigated for their potential in managing *Nezara viridula* and *Piezodorus guildinii*, particularly in Brazil (e.g., Tonet and Reis 1979, Sosa-Gomez and Moscardi 1998, Sosa-Gomez and Alves 2000). Both trypanosomatids and viruses also have been reported from *N. viridula* (Williamson and von Wechmar 1995, Sosa-Gomez et al. 2005).

7.7.3.4 Conservation Biological Control

There has been some research on the use of trap cropping to enhance the utility of parasitoids of *Nezara viridula* (e.g., Rea et al. 2002). Several researchers have demonstrated that certain flowering plants attract and enhance the action of parasitoids (Tillman et al. 2015, and references therein).

---

[3] *Telenomus nakagawai* Watanabe may be a synonym of *Telenomus turesis* Walker (Norman F. Johnson, personal communication).

7.7.4 Attract-and-Kill Strategies

It is well established that stink bugs prefer to colonize seed- and fruit-producing hosts. The attract-and-kill strategy involves taking advantage of this behavior by recruiting *Nezara viridula* through use of a highly attractive crop (or lure) and then, subsequently, controlling the pests through site-specific insecticide applications and/or beneficial insects. The attractive crop (often termed a trap crop) typically is planted early on a small amount of land for the sole purpose of luring the pest away from the principle cash crop. Trap crops for *N. viridula* are well studied in many agronomic production systems including, but not limited to, cotton (Tillman 2006, Tillman et al. 2015), pecan (Smith et al. 1996), sweet corn (Rea et al. 2002), and soybean (McPherson and Newsom 1984, Todd 1989, Smith et al. 2009). Results are variable, but this practice generally is successful at reducing stink bug populations as long as the bugs are eliminated before the trap crop matures or the nymphs complete development.

In addition to recruiting populations of pest insects, trap crops can aid in recruitment of beneficial insects and lead to reduced number of insecticide applications to the main commercial crop (Tillman 2006, Tillman and Carpenter 2014). Also, trap crops such as sorghum can act as a physical barrier in the landscape for pest insects moving into commercial cotton production (Tillman 2014).

7.7.5 Chemical Control

Chemical control of insects in agriculture involves the use of synthetic or naturally derived insecticides to reduce populations before they cause economic damage equal to insect management costs. Chemical control is the most successful and widely used control strategy for stink bugs including *Nezara viridula*. Unfortunately, there are currently no selective insecticide chemistries for managing stink bugs, so growers rely on broad spectrum materials such as organophosphates (IRAC [Insecticide Resistance Action Committee] group 1B), pyrethroids (IRAC group 3A), and neonicotinoids (IRAC group 4A). Numerous studies have evaluated the efficacy of various chemical control options for stink bugs in agricultural settings (e.g., Greene et al. 2001a,b, 2005; Tillman and Mullinix 2004; Willrich et al. 2004a,b), and some research trials have provided information regarding insecticide performance specific to *N. viridula* (Greene et al. 2001a,b, 2005; Willrich et al. 2004a,b; Tillman 2006). Generally speaking, *N. viridula* is more susceptible than *Euschistus* spp. to pyrethroid insecticides including cyfluthrin, cypermethrin, and lamda-cyhalothrin (Willrich et al. 2003). More specific information on stink bug management is presented in **Chapter 16**.

7.8 Future Outlook

Management of *Nezara viridula* and other phytophagous pentatomids will require that we look outside the realm of conventional control measures (i.e., insecticide applications) for effective pest control. For example, genetically modified cotton plants (Bt-cotton) have been extremely successful in controlling heliothine pests. Developing a similar approach to deliver a control agent or toxin(s) deleterious to phytophagous pentatomids would be paramount in reducing our reliance on conventional control measures, assuming approval by governmental regulatory agencies and acceptance by the consumers.

The area of plant-insect-pathogen interactions should continue receiving attention. Although bacteria long have been associated with *Nezara viridula*, ingestion of *Pantoea agglomerans* by first instars and retention through second instars could have implications for management of this stink bug if pathogens are retained throughout adulthood (and preliminary data indicates that such is the case; Jesus F. Esquivel, unpublished data). For example, when the pathogen is detected in early instars collected from the field, perhaps more conservative treatment thresholds and/or earlier monitoring could be implemented. This approach remains to be verified, but research to date suggests a need to re-assess management tactics.

Recent simulated climate warming outdoor experiments, with *Nezara viridula* as a model species, demonstrates that responses of this pentatomid will be different for different life-history traits and seasons (Musolin et al. 2010, Musolin 2012). Thus, warming is expected to affect nymphal development negatively during the hot season (Musolin et al. 2010, Tada et al. 2011), accelerate development in the spring and autumn, and/or enhance survival of adults in the winter (Musolin et al. 2010, Takeda et al. 2010).

Elevated temperatures, in combination with short-day autumnal conditions, might be beneficial for diapause induction. The milder winters will increase the likelihood of adult survival until spring. Consequently, these adults will have a greater chance to pass their genes to the next generation. In a cumulatively and complex way, these factors will affect the population dynamics of *N. viridula*, its relationship with other members of the biotic community and, likely, its pest status. In general, it is known that insect species have the potential to respond to climate change through phenotypic flexibility or rapid evolutionary (genetic) responses to strong selection (Bale et al. 2002). As noted by Thomas et al. (2001), "Improving environmental conditions at existing margins ... are likely to initiate range extensions purely on the basis of ecological, physiological and population-dynamic processes–requiring no evolutionary change" (p. 579). This is likely what is being observed in the case of *N. viridula* – the improved overwintering conditions in central Japan have stimulated northward range expansion of the species (Musolin 2007; Yukawa et al. 2007, 2009; Tougou et al. 2009). However, Bradshaw and Holzapfel (2008) contend that all known responses to climate warming involve genetic changes related to seasonality and diapause syndrome and none of the responses involves an increase in thermal optimum or in heat tolerance. Careful monitoring of the performance of *N. viridula* in its recently colonized areas, as well as a detailed examination of the species' plant-insect-symbionts-competitors-natural enemies complex during all seasons will be essential for understanding its adaptation to continued climate change.

Nezara viridula will further spread within Europe. As a stowaway and contaminant, it will make use of the increasing movements of vegetables, crops, and goods with any vehicle. It also will profit from climate change, responding with increasing population growth that may lead to subsequent spread from points-of-entry to natural habitats. It is to be expected, although with considerable uncertainty, that economic damage will increase in Europe in the future. Negative impacts on native biodiversity are less likely but might be worth investigating in situations of mass occurrences of these bugs.

7.9 Acknowledgments

We thank the following individuals for their assistance and support during this endeavor: Tana B. Luna, Christopher T. Parker, Lauren A. Ward, and Kendall Wolff. We also thank Norman F. Johnson (Department of Entomology, Ohio State University, Columbus) for his help with the synonymy of the *Telenomus* parasitoids. Research of Dmitry L. Musolin on diapause in *Nezara viridula* in Japan was aided by the fruitful collaboration with Professor Hideharu Numata (Graduate School of Science, Kyoto University, Japan) and Professor Keiji Fujisaki (Kyoto University, Japan), and supported by the Ministry of Education, Culture, Science, Sports and Technology of Japan (Grants-in-Aid for JSPS Fellows No. 98116 and L-4562 and STA Fellow No. 200141 and via The 21st Century COE Program at Kyoto University). The following individuals contributed via personal communications: Berend Aukema, Naturalis Biodiversity Center, Leiden, Netherlands; Richard Grantham, Oklahoma State University, Stillwater; Keizi Kiritani, Shizuoka, Japan; Chris Looney, Washington State Department of Agriculture, Olympia; Abdul Aziz Mohamed, Arabian Gulf University, Manama, Kingdom of Bahrain; Antônio R. Panizzi, EMBRAPA Trigo, Passo Fundo, Brasil; Scott Stewart, University of Tennessee, Jackson; Donald B. Thomas, U. S. Dept. Agriculture, Agricultural Research Service, Edinburg, TX; and Daisuke Tougou, Kyoto University, Japan.

7.10 References Cited

Abdu, R. M., and N. F. Shaumar. 1985. A preliminary list of the insect fauna of Qatar. Qatar University Science Bulletin 5: 215–232.

Ali, M., and M. A. Ewiess. 1977. Photoperiodic and temperature effects on rate of development and diapause in the green stink bug, *Nezara viridula* L. (Heteroptera: Pentatomidae). Zeitschrift für Angewandte Entomologie 84: 256–264.

Aldrich, J. R. 1988. Chemical ecology of the Heteroptera. Annual Review of Entomology 33: 211–238.

Aldrich, J. R. 1990. Dispersal of the southern green stink bug, *Nezara viridula* (L.) (Heteroptera: Pentatomidae), by hurricane Hugo. Proceedings of the Entomological Society of Washington 92: 757–759.

Aldrich, J. R., W. R. Lusby, J. P. Kochansky, and J. A. Lockwood. 1987. Pheromone strains of the cosmopolitan pest, *Nezara viridula* (Heteroptera: Pentatomidae). Journal of Experimental Zoology 244: 171–175.

Aldrich, J. R., W. R. Lusby, B. E. Marron, K. C. Nicolaou, M. P. Hoffmann, and L. T. Wilson. 1989. Pheromone blends of green stink bugs and possible parasitoid selection. Naturwissenschaften 76: 173–175.

Aldrich, J. R., H. Numata, M. Borges, F. Bin, G. K. Waite, and W. R. Lusby. 1993. Artifacts and pheromone blends from *Nezara* spp. and other stink bugs (Heteroptera: Pentatomidae). Zeitschrift für Naturforschung C. 48: 73–79.

Amyot, C. J. B., and J. G. A. Serville. 1843. Histoire naturelle des insects. Hémiptéres. Paris: Fain et Thunot. 675 pp.

Anonymous. 1974. Small grains. Cooperative Economic Insect Report 24(15): 223–224.

Anonymous. 2016. *Nezara viridula*. P-Patch Community Gardening Program, Seattle Department of Neighborhoods, Seattle, WA. P-Patch Tipsheet PP410. 2 pp.

Apriyanto, D., J. D. Sedlacek, and L. H. Townsend. 1989. Feeding activity of *Euschistus servus* and *E. variolarius* (Heteroptera: Pentatomidae) and damage to an early growth stage of corn. Journal of the Kansas Entomological Society 62: 392–399.

Atkinson, E. T. 1889. Notes on Indian insect pests. Rhyncota. Indian Museum Notes 1: 1–8.

Aukema, B. 2016. Nieuwe en interessante nederlandse Wantsen VI (Hemiptera: Heteroptera). Nederlandse Faunistische Mededelingen 46: 57–86.

Bailey, W. C. 2009. New stink bug found in Missouri soybean. University of Missouri, Division of Plant Sciences, Integrated Pest & Crop Management; http://ipm.missouri.edu/ipcm/2009/12/New-Stink-Bug -Found-in-Missouri-Soybean/ (accessed 16 September 2015).

Balduf, W. V. 1923. The insects of the soybean in Ohio. Ohio Agricultural Experiment Station Bulletin 366: 147–181.

Bale, J. S., G. J. Masters, I. D. Hodkinson, C. Awmack, T. M. Bezemer, V. K. Brown, J. Butterfield, A. Buse, J. C. Coulson, J. Farrar, J. E. G. Good, R. Harrington, S. Hartley, T. H. Jones, R. L. Lindroth, M. C. Press, I. Symrnioudis, A. D. Watt, and J. B. Whittaker. 2002. Herbivory in global climate change research: direct effects of rising temperature on insect herbivores. Global Change Biology 8: 1–16.

Barclay, M. 2004. The green vegetable bug *Nezara viridula* (L., 1758) (Hem.: Pentatomidae), new to Britain. Entomologist's Record 116: 55–58.

Batra, S. W. T., J. R. Coulson, P. H. Dunn, and P. E. Boldt. 1981. Insects and fungi associated with *Carduus* thistles (Compositae). United States Department of Agriculture Technical Bulletin No. 1616: 1–100.

Baur, M. E., D. J. Boethel, M. L. Boyd, G. R. Bowers, M. O. Way, L. G. Heatherly, J. Rabb, and L. Ashlock. 2000. Arthropod populations in early soybean production systems in the mid-South. Environmental Entomology 29: 312–328.

Bergroth, E. 1914. Notes on some genera of Heteroptera. Annales de la Société Entomologique de Belgique 58: 23–28.

Blatchley, W. S. 1926. Heteroptera or true bugs of eastern North America with especial reference to the faunas of Indiana and Florida. Nature Publishing Company, Indianapolis, IN. 1116 pp.

Borges, M., P. C. Jepson, and P. E. Howse. 1987. Long-range mate location and close-range courtship behavior of the green stink bug, *Nezara viridula* and its mediation by sex pheromones. Entomologia Experimentalis et Applicata 44: 205–212.

Bradshaw, W. E., and C. M. Holzapfel. 2008. Genetic response to rapid climate change: it's seasonal timing that matters. Molecular Ecology 17: 157–166.

Bundy, C. S. 2012. An annotated checklist of the stink bugs (Heteroptera: Pentatomidae) of New Mexico. The Great Lakes Entomologist 45: 196–209.

Bundy, C. S., and R. M. McPherson. 2000a. A morphological examination of stink bug (Heteroptera: Pentatomidae) eggs on cotton and soybeans, with a key to genera. Annals of the Entomological Society of America 93: 616–624.

Bundy, C. S., and R. M. McPherson. 2000b. Dynamics and seasonal abundance of stink bugs (Heteroptera: Pentatomidae) in a cotton/soybean ecosystem. Journal of Economic Entomology 93: 697–706.

Bundy, C. S., R. M. McPherson, and G. A. Herzog. 2000. An examination of the external and internal signs of cotton boll damage by stink bugs (Heteroptera: Pentatomidae). Journal of Entomological Science 35: 402–410.

Buntin, G. D., D. V. McCracken, and W. L. Hargrove. 1995. Populations of foliage-inhabiting arthropods on soybean with reduced tillage and herbicide use. Agronomy Journal 87: 789–794.

Buschman, L. L., and W. H. Whitcomb. 1980. Parasites of *Nezara viridula* (Hemiptera: Pentatomidae) and other Hemiptera in Florida. Florida Entomologist 63: 154–167.

CABI/EPPO. 1998. *Nezara viridula* (Linnaeus), Heteroptera: Pentatomidae. *In* Distribution maps of plant pests, Map 27 (2nd revision). CAB International, Wallingford, United Kingdom. 7 pp.

CABI Invasive Species Compendium. 2015. *Nezara viridula* (green stink bug): http://www.cabi.org/isc /datasheet/36282 (accessed 21 January 2015).

Cammell, M. E., and J. D. Knight. 1992. Effect of climate change on the population dynamics of crop pests. Advances of Ecological Research 22: 117–162.

Cancino, E. R., and J. M. C. Blanco. 2002. Orden Hemiptera, pp. 147–160. *In* Artrópodos terrestres de los estados de Tamaulipas y Nuevo León, México. Universidad Autónoma de Tamaulipas, Serie Publicaciones Científicas No. 4, Ciudad Victoria, Tamaulipas, Mexico. 377 pp.

Canton-Ramos, J., and A. J. Callejon-Ferre. 2010. Raising *Trissolcus basalis* for the biological control of *Nezara viridula* in greenhouses of Almeria (Spain). African Journal of Agricultural Research 5: 3207–3212. doi10.5897/AJAR10.651.

Capinera, J. L. 2001. Southern green stink bug, pp. 272–275. *In* Handbook of vegetable pests. Academic Press, San Diego, CA. 729 pp.

Cassis, G., and G. F. Gross. 2002. Hemiptera: Heteroptera (Pentatomomorpha). *In* W. W. K. Houston and A. A. Wells (Eds.), Zoological catalogue of Australia. Volume 27.3B. CSIRO Publishing, Melbourne, Australia. xiv + 737 pp.

Chen, F. 1980. A new form of *Nezara viridula* (Linnaeus). Entomotaxonomia 2: 31–32.

Cherry, R., and A. Wilson. 2011. Flight activity of stink bug (Hemiptera: Pentatomidae) pests of Florida rice. Florida Entomologist 94: 359–360.

Chia, C. L., R. A. Hamilton, and D. O. Evans. 1997. Mango. College of Tropical Agriculture & Human Resources, University of Hawaii at Manoa. CTAHR Fact Sheet HC-2, 2 pp.

Clarke, A. R. 1992. Current distribution and pest status of *Nezara viridula* (L.) (Hemiptera: Pentatomidae) in Australia. Journal of the Australian Entomological Society 31: 289–297.

Clarke, A. R., and J. E. Seymour. 1992. Two species of *Acroclisoides* (Girault & Dodd) (Hymenoptera: Pteromalidae) parasitic on *Trissolcus basalis* (Wollaston), a parasitoid of *Nezara viridula* (L.) (Hemiptera: Pentatomidae). Journal of the Australian Entomological Society 31: 299–300.

Čokl, A., M. Virant-Doberlet, and N. Stritih. 2000. The structure and function of songs emitted by southern green stink bugs from Brazil, Florida, Italy and Slovenia. Physiological Entomology 25: 196–205.

Čokl, A., H. L. McBrien, and J. G. Millar. 2001. Comparison of substrate-borne vibrational signals of two stink bug species, *Acrosternum hilare* and *Nezara viridula* (Heteroptera: Pentatomidae). Annals of the Entomological Society of America 94: 471–479.

Colazza, S., E. Ciriciofolo, and F. Bin. 1986. *Nezara viridula* (L.) injurious to soyabean in central Italy (Rhynchota Pentatomidae). Informatore Agrario 42: 79–84.

Colazza, S., G. Giangiuliani, and F. Bin. 1996. Fortuitous introduction and successful establishment of *Trichopoda pennipes* F.: adult parasitoid of *Nezara viridula* (L.). Biological Control 6: 409–411.

Commonwealth Institute of Entomology. 1953. Distribution maps of insect pests, Ser. A, Map 27, *Nezara viridula*. London: Commonwealth Institute of Entomology. 4 pp.

Commonwealth Institute of Entomology. 1970. Distribution maps of insect pests, Ser. A, Map 27, *Nezara viridula*. London: Commonwealth Institute of Entomology. 4 pp. (Revised)

Conti, E., F. Bin, A. Estragnat, and F. Bardy. 1997. The green stink bug, *Nezara viridula* L., injurious to castor, *Ricinus communis* L., in southern France. International conference on pests in agriculture, 6–8 January 1997, at le Corum, Montpellier, France, Volume 3: 1045–1052.

Coombs, M. 2004. Overwintering survival, starvation resistance, and post-diapause reproductive performance of *Nezara viridula* (L.) (Hemiptera: Pentatomidae) and its parasitoid *Trichopoda giacomellii* Blanchard (Diptera: Tachinidae). Biological Control 30: 141–148.

Coombs, M., and D. P. A. Sands. 2000. Establishment in Australia of *Trichopoda giacomellii* (Blanchard) (Diptera: Tachinidae), a biological control agent for *Nezara viridula* (L.) (Hemiptera: Pentatomidae). Australian Journal of Entomology 39: 219–222.

Corrêa-Ferreira, B. S. 1993. Utilização do parasitóide de ovos *Trissolcus basalis* (Wollaston) no controle de percevejos da soja. Empresa Brasileira de Pesquisa Agropecuária, Centro Nacional de Pesquisa de Soja, Circular Técnica 11: 40 pp.

Corrêa-Ferreira, B. S., and F. Moscardi. 1996. Biological control of soybean stink bugs by inoculative release of *Trissolcus basalis*. Entomologia Experimentalis et Applicata 79: 1–7.

Corrêa-Ferreira, B. S., and A. R. Panizzi. 1999. Percevejos da soja e seu manejo. Empresa Brasileira de Pesquisa Agropecuária, Centro Nacional de Pesquisa de Soja, Circular Técnica 24: 7–45.

Corrêa-Ferreira, B. S., and C. E. O. Zamataro. 1989. Capacidade reprodutiva e longevidade dos parasitóides de ovos *Trissolcus basalis* (Wollaston) e *Trissolcus mitsukurii* Ashmead (Hymenoptera: Scelionidae). Revista Brasileira de Biologia 49: 621–626.

Cosenza, G. W., and T. Kobayashi. 1986. Introdução de *Trissolcus mitsukurii* para o controle de percevejos da soja, pg. 180. *In* Resumos X Congresso Brasileiro de Entomologia, Rio de Janeiro, Brasil. 451 pp.

Costa, A. 1843. Cimicum regni Neapolitani. Centuria, Napoli. 76 pp.

Costa, A. 1884. Notizie ed osservazioni sulla geo-fauna Sarda. Memoria Terza. Risultamento delle ricerche fatte in Sardegna nella estate del 1883. Atti della Reale Accademia delle Scienze Fisiche e Matematiche di Napoli (2)1 (Fasc. 9): 1–64.

Costa Lima, A. M. 1940. Insetos do Brasil. Hemipteros. Tomo II. Rio de Janeiro: Escola Nacional de Agronomia. 351 pp.

Cottrell, T. E. 2001. Improved trap capture of *Euschistus servus* and *Euschistus tristigmus* (Hemiptera: Pentatomidae) in pecan orchards. Florida Entomologist 84: 731–732.

Crooks, J. A. 2005. Lag times and exotic species: the ecology and management of biological invasions in slow-motion. Ecoscience 12: 316–329.

Cui, J., and W. Cai. 2008. Flight performance of a tethered stink bug, *Graposoma* [sic] *rubrolineata* (Westwood) (Hemiptera: Pentatomidae). Journal of Life Sciences 2: 5–11.

Cumber, R. A. 1951. The introduction into New Zealand of *Microphanurus basalis* Woll. (Scelionidae: Hym.), egg-parasite of the green vegetable bug, *Nezara viridula* L. (Pentatomidae). New Zealand Journal of Science and Technology 32 (B): 30–37.

Cusumano, A., E. Peri, V. Amodeo, J. N. McNeil, and S. Colazza. 2013. Intraguild interactions between egg parasitoids: window of opportunity and fitness costs for a facultative hyperparasitoid. http://dx.doi.org/10.1371/journal.pone.0064768.

Dallas, W. S. 1851. List of the specimens of hemipterous insects in the collection of the British Museum. Part I: 1–592, plates 1–15. Richard Taylor, London, UK.

Danilevsky, A. S. 1961. Photoperiodism and seasonal development of insects. Leningrad State University Press, Leningrad, the USSR. 243 pp. (in Russian) [Translated into English by J. Johnson and N. Waloff, Oliver & Boyd, London, UK, 1965, 284 pp.].

Danks, H. V. 1987. Insect dormancy: an ecological perspective. Biological Survey of Canada Monograph Series No. 1, Ottawa, Canada. x + 439 pp.

Davis, C. J. 1964. The introduction, propagation, liberation, and establishment of parasites to control *Nezara viridula* variety *smaragdula* (Fabricius) in Hawaii (Heteroptera: Pentatomidae). Proceedings of the Hawaiian Entomological Society 18: 369–375.

Day, G. M. 1964. Revision of *Acrosternum* auctt. nec Fieber from Madagascar. Annals and Magazine of Natural History 7: 559–565.

De Jong, P. W., S. W. S. Gussekloo, and P. M. Brakefield. 1996. Differences in thermal balance, body temperature and activity between non-melanic and melanic two-spot ladybird beetles (*Adalia bipunctata*) under controlled conditions. Journal of Experimental Biology 199: 2655–2666.

Demaree, J. B. 1922. Kernel-spot of the pecan and its cause. United States Department of Agriculture, Division of Entomology 1102: 1–15.

Derjanschi, V., and C. Chimişliu. 2014. *Nezara viridula* (Linnaeus 1758) (Heteroptera, Pentatomidae): nouă semnalare pe teritoriul României, pp. 140–141. *In* T. Ion (Ed.), Sustainable use and protection of animal world diversity (International Symposium dedicated to 75th anniversary of Professor Andrei Munteanu). Tipografia Academiei de Ştiinţe a Moldovei, Chişinău. 248 pp. (in Romanian)

Dethier, M. 1989. Les Pentatomoidea de la collection Kappeller. Archives des Sciences Genéve 42: 553–568.

Dethier, M., and F. Chérot. 2014. Alien Heteroptera in Belgium: a threat for our biodiversity or agroforestry? Andrias 20: 51–55.

Dethier, M., and J. B. Gallant. 1998. Hétéroptères remarquables pour la faune belge. Natura Mosana 51: 75–86.

Dethier, M., and E. Steckx. 2010. Note sur quelques Hétéroptères intéressants pour la faune de Belgique. Natura Mosana 63: 9–22.

DeWitt, J. R., and E. J. Armbrust. 1978. Feeding preference studies of adult *Nezara viridula* (Hemiptera: Pentatomidae) morphs from India and the United States. The Great Lakes Entomologist 11: 67–69.

DeWitt, N. B., and G. L. Godfrey. 1972. The literature of arthropods associated with soybeans. II. A bibliography of the southern green stink bug, *Nezara viridula* (Linnaeus) (Hemiptera: Pentatomidae). Illinois Natural History Survey 78: 1–23.

Discover Life. 2015. http://www.discoverlife.org/mp/20m?act=make_map&r=.2&la=%2036&lo=-90&kind =Nezara+viridula (accessed 11 September 2015).

Distant, W. L. 1880. Insecta. Rhyncota. Hemiptera-Heteroptera. *In* Biologia Centrali-Americana. Volume 1. R. H. Porter, London, UK. xx + 462 pp.

Distant, W. L. 1900. Rhynchotal notes IV. Heteroptera: Pentatominae (part). Annals and Magazine of Natural History (7)5: 386–397.

Distant, W. L. 1901. Revision of the Rhynchota belonging to the family Pentatomidae in the Hope collection at Oxford. Proceedings of the Zoological Society of London 1900: 807–824.

Drake, C. J. 1920. The southern green stink-bug in Florida. Florida State Plant Board Quarterly Bulletin 4: 41–94.

Eger, J. E., Jr., and J. R. Ables. 1981. Parasitism of Pentatomidae by Tachinidae in South Carolina and Texas. Southwestern Entomologist 6: 28–33.

Ehler, L. E. 2000. Farmscape ecology of stink bugs in northern California. Memoirs Thomas Say Publications in Entomology. Entomological Society of America Press, Lanham, MD. 59 pp.

Ehler, L. E. 2002. An evaluation of some natural enemies of *Nezara viridula* in northern California. BioControl 47: 309–325.

Endo, N. 2016. Effective monitoring of the population dynamics of *Nezara viridula* and *Nezara antennata* (Heteroptera: Pentatomidae) using a light trap in Japan. Applied Entomology and Zoology 51: 341–346.

Esquivel, J. F. 2009. Stages of gonadal development of the southern green stink bug (Hemiptera: Pentatomidae): improved visualization. Annals of the Entomological Society of America 102: 303–309.

Esquivel, J. F. 2011. Improved visualization of fat body cell conditions and abundance in the southern green stink bug (Hemiptera: Pentatomidae). Journal of Entomological Science 46: 52–61.

Esquivel, J. F. 2016a. *Nezara viridula* (L.) in Central Texas: I. New host plant associations and reproductive status of adults encountered within. Southwestern Entomologist 41: 895–904.

Esquivel, J. F. 2016b. *Nezara viridula* (L.) in Central Texas: II. Seasonal occurrence of black-spotted condition in adult females. Southwestern Entomologist 41: 905–911.

Esquivel, J. F., and E. G. Medrano. 2012. Localization of selected pathogens of cotton within the southern green stink bug. Entomologia Experimentalis et Applicata 142: 114–120.

Esquivel, J. F., and E. G. Medrano. 2014. Ingestion of a marked bacterial pathogen of cotton conclusively demonstrates feeding by first instar southern green stink bug (Hemiptera: Pentatomidae). Environmental Entomology 43: 110–115.

Esquivel, J. F., and L. A. Ward. 2014. Characteristics for determining sex of late-instar *Nezara viridula* (L.). Southwestern Entomologist 39: 187–189.

Esquivel, J. F., E. G. Medrano, and A. A. Bell. 2010. Southern green stink bugs (Hemiptera: Pentatomidae) as vectors of pathogens affecting cotton bolls – a brief review. Southwestern Entomologist 35: 457–461.

Esquivel, J. F., V. A. Brown, R. B. Harvey, and R. E. Droleskey. 2015. A black color morph of adult *Nezara viridula* (L.). Southwestern Entomologist 40: 649–652.

Fabricius, J. C. 1775. Systema entomologiae sistens insectorum classes, ordines, genera, species; adjectis synonymis, locis, descriptionibus et observationibus. Flensburgi et Lipsiae: Kortii. 832 pp.

Fabricius, J. C. 1798. Supplementum Entomologiae Systematicae. Proft et Storch, Hafniae. 572 pp.

Farinelli, D., M. A. van den Berg, and M. Maritz. 1994. Report of work on *Trichopoda pennipes* F. (Diptera: Tachinidae) an adult parasitoid of the green stink bug, *Nezara viridula* (L.) (Hemiptera: Pentatomidae). The Southern African Macadamia Growers Association Yearbook 2: 22–23.

Ferrari, A., C. F. Schwertner, and J. Grazia. 2010. Review, cladistic analysis and biogeography of *Nezara* Amyot & Serville (Hemiptera: Pentatomidae). Zootaxa 2424: 1–41.

Fieber, F. X. 1861. Die europäischen Hemiptera. Halbflügler (Rhynchota Heteroptera), pp. 113–444. *In* Nach der analytischen Methode bearbeitet. Wien: Gerold. 452 pp.

Fleming, K. J. Halloran, A. Hara, T. Hara, K. Leonhardt, E. Mersino, K. Sewake, and J. Uchida. 1999. Plant bug, seed bug, and stink bug, p. 38. *In* K. Leonhardt and K. Sewake (Eds.), Growing dendrobium orchids in Hawaii: a production and pest management guide. College of Tropical Agriculture & Human Resources, University of Hawaii at Manoa. 96 pp.

Flitters, N. E. 1963. Observations on the effect of Hurricane "Carla" on insect activity. International Journal of Biometeorology 6: 85–90.

Follett, P. A., F. Calvert, and M. Golden. 2007. Genetic studies using the orange body color type of *Nezara viridula* (Hemiptera: Pentatomidae): inheritance, sperm precedence, and disassortative mating. Annals of the Entomological Society of America 100: 433–438.

Fontes, E. M. G., D. H. Habeck, and F. Slansky, Jr. 1994. Phytophagous insects associated with goldenrods (*Solidago* spp.) in Gainesville, Florida. Florida Entomologist 77: 209–221.

Fortes, P. 2010. Fisiologia reprodutiva de *Nezara viridula* (Linnaeus, 1758) (Hemiptera: Pentatomidae). Doctoral Thesis, Escola Superior de Agricultura Luiz de Queiroz, Universidade de São Paulo. 123 pp. http://www.teses.usp.br/teses/disponiveis/11/11146/tde-20042010-110715/en.php.

Fortes, P., G. Salvador, and F. L. Cônsoli. 2011. Ovary development and maturation in *Nezara viridula* (L.) (Hemiptera: Pentatomidae). Neotropical Entomology 40: 89–96.

Freeman, P. 1940. A contribution to the study of the genus *Nezara* Amyot & Serville (Hemiptera, Pentatomidae). Transactions of the Royal Entomological Society of London 90: 351–374.

Froeschner, R. C. 1988. Family Pentatomidae Leach, 1815. The stink bugs, pp. 544–597. *In* T. J. Henry and R. C. Froeschner (Eds.), Catalogue of the Heteroptera, or true bugs, of Canada and the continental United States. E. J. Brill, New York, New York. 958 pp.

Froggatt, W. W. 1916. The tomato and bean bug (*Nezara viridula*, Linn.). Agricultural Gazette of New South Wales 27: 649–650.

Fukatsu, T., and T. Hosokawa. 2002. Capsule-transmitted gut symbiotic bacterium of the Japanese common plataspid stinkbug, *Megacopta punctatissima*. Applied and Environmental Microbiology 68: 389–396.

Funderburk, J. E., F. M. Rhoads, and I. D. Teare. 1991. Population dynamics of soybean insect pests vs. soil nutrient levels. Crop Science 31: 1629–1633.

Fuzeau-Braesch, S. 1985. Colour changes, pp. 549–589. *In* G. A. Kerkut and L. I. Gilbert (Eds.), Comprehensive insect physiology and pharmacology. Volume 9. Behaviour. Pergamon Press, Oxford, UK. 734 pp.

Gallant, J. B. 1996. Note hémiptèrologique. *Nezara viridula* (L.) (Heteroptera, Pentatomidae), espèce en progression sur notre territoire? Bulletin de la Société royale belge d'Entomologie 132: 405–406.

Germar, E. F. 1838. Hemiptera Heteroptera promontorii Bonae Spei nondum descripta, quae collegit C. F. Drege. Silbermann's Revue Entomologique 5: 121–192.

Geshi, J., and K. Fujisaki. 2013. Northward range expansion of *Nezara viridula* in Kinki District, Japan: expansion speed. Japanese Journal of Applied Entomology and Zoology 57: 151–157.

Gibson, W. W., and J. L. Carrillo S. 1959. *Nezara viridula* (Linn.) [= *smaragdula* (Fabr.)], p. 32. *In* Lista de insectos en la collección entomológica de la Oficina de Estudios Especiales, S. A. G. Folleto Miscelaneo, No. 9. 254 pp.

Giterman, G. E. 1931. Materials to the Hemiptera fauna of the BSSR, pp. 77–104. *In* Materials to the studies of flora and fauna of Belorussia. Volume 6. Nauka, Minsk. 200 pp. (in Russian)

Glasgow, H. 1914. The gastric cæca and the cæcal bacteria of the Heteroptera. Biological Bulletin 26: 101–171.

Gogala, M., and Š. Michieli. 1962. Sezonsko prebarvanje pri nekaterih vrstah stenic (Heteroptera). Biološki Vestnik (Ljubljana) 10: 33–44. (in Slovenian with German summary)

Gogala, M., and Š. Michieli. 1967. Berichtigung zu unseren Veröffentlichungen über Farbstoffe bei Heteropteren. Bulletin of Science, Conseil Academy RSF Yugoslavie (Section A) (Zagreb) 12: 6.

Golden, M., and P. A. Follett. 2006. First report of *Nezara viridula* f. *aurantiaca* (Hemiptera: Pentatomidae) in Hawaii. Proceedings of the Hawaiian Entomological Society 38: 131–132.

Grazia, J., and N. D. F. Fortes. 1995. Revisão do gênero *Rio* Kirkaldy, 1909 (Heteroptera, Pentatomidae). Revista Brasileira de Entomologia 39: 409–430.

Greathead, D. J. 1971. A review of biological control in the Ethiopian region. Commonwealth Institute of Biological Control, Commonwealth Agricultural Bureau, Farnham Royal, Slough, UK. 162 pp.

Greene, J. K., S. G. Turnipseed, M. J. Sullivan, and O. L. May. 2001a. Treatment thresholds for stink bugs (Hemiptera: Pentatomidae) in cotton. Journal of Economic Entomology 94: 403–409.

Greene, J. K., G. A. Herzog, and P. M. Roberts. 2001b. Management decisions for stink bugs, pp. 913–917. *In* D. D. Hardee and E. Burris (Eds.), Proceedings of the Beltwide Cotton Conference, National Cotton Council of America, Memphis, TN. Volume 2: 741–1454. (CD-ROM)

Greene, J., C. Capps, G. Lorenz, G. Studebaker, J. Smith, R. Luttrell, S. Kelley, and W. Kirkpatrick. 2005. Management considerations for stink bugs – 2004, pp. 1527–1538. *In* P. Dugger and D. Richter (Eds.), Proceedings of the Beltwide Cotton Production Conferences, National Cotton Council of America, Memphis, TN. 3164 pp. (CD-ROM)

Gross, J., E. Schmolz, and M. Hilker. 2004. Thermal adaptations of the leaf beetle *Chrysomela lapponica* (Coleoptera: Chrysomelidae) to different climes of central and northern Europe. Environmental Entomology 33: 799–806.

Grozea I., R. Stef, A. M. Virteiu, A. Cărăbeţ, and L. Molnar. 2012. Southern green stink bugs (*Nezara viridula* L.) a new pest of tomato crops in western Romania. Research Journal of Agricultural Science 44: 24–27.

Gu, H., and G. H. Walter. 1989. Flight of green vegetable bugs *Nezara viridula* (L.) in relation to environmental variables. Journal of Applied Entomology 108: 347–354.

Hall, D. G., IV, and G. L. Teetes. 1981. Alternate host plants of sorghum panicle-feeding bugs in southeast central Texas. Southwestern Entomologist 6: 220–228.

Harris, V. E., and J. W. Todd. 1980. Temporal and numerical patterns of reproductive behaviour in the southern green stink bug, *Nezara viridula* (Hemiptera: Pentatomidae). Entomologia Experimentalis et Applicata 27: 105–116.

Harris, V. E., J. W. Todd, and B. G. Mullinix. 1984. Color change as an indicator of adult diapause in the southern green stink bug, *Nezara viridula*. Journal of Agricultural Entomology 1: 82–91.

Hasegawa, H. 1954. Notes on *Nezara viridula* (Linné) and its allied green stink-bugs in Japan. Bulletin of the National Institute of Agricultural Sciences. Ser. C 4: 215–228. (in Japanese, with English summary)

Hayashi, K., H. Suzuki, and S. Asahina. 1978. Note on the transoceanic insects captured on East China Sea in 1977. Tropical Medicine (Nagasaki) 20: 131–142. (in Japanese, with English summary)

Heiss, E. 1977. Zur Heteropterenfauna Nordtirols (Insecta: Heteroptera) VI: Pentatomoidea. Veröffentlichungen des Tiroler Landesmuseums Ferdinandeum 57: 53–77.

Henry, T. J., and M. R. Wilson. 2004. First records of eleven true bugs (Hemiptera: Heteroptera) from the Galápagos Islands, with miscellaneous notes and corrections to published reports. Journal of the New York Entomological Society 112: 75–86.

Herbert, A., E. Blinka, J. Bacheler, J. Van Duyn, J. Greene, M. Toews, P. Roberts, and R. H. Smith. 2009. Managing stink bugs in cotton: research in the southeast region. Virginia Cooperative Extension, Virginia Polytechnic Institute and State University. Publication 444–390: 1–16. http://pubs.ext.vt .edu/444/444-390/444-390.pdf.

Herbert, D. A., T. P. Mack, R. B. Reed, and R. Getz. 1988. Degree-day maps for management of soybean insect pests in Alabama. Circular, Alabama Agricultural Experiment Station, Auburn University, 591: 1–19.

Herbert, J. J., and M. D. Toews. 2012. Seasonal abundance and population structure of *Chinavia hilaris* and *Nezara viridula* (Hemiptera: Pentatomidae) in Georgia farmscapes containing corn, cotton, peanut, and soybean. Annals of the Entomological Society of America 105: 582–591.

Hirose, E., A. R. Panizzi, J. T. De Souza, A. J. Cattelan, and J. A. Aldrich. 2006. Bacteria in the gut of southern green stink bug (Heteroptera: Pentatomidae). Annals of the Entomological Society of America 99: 91–95.

Hodek, I. 1996. Diapause development, diapause termination and the end of diapause. European Journal of Entomology 93: 475–487.

Hoffman, R. L. 1971. Genus *Nezara* Amyot & Serville, p. 47. *In* The insects of Virginia: No. 4. Shield bugs (Hemiptera: Scutelleroidea: Scutelleridae, Corimelaenidae, Cydnidae, Pentatomidae). Virginia Polytechnic Institute and State University, Research Division Bulletin Number 67: 1–61.

Hoffmann, H.-J. 1992. Zur Wanzenfauna (Hemiptera-Heteroptera) von Köln. Decheniana, Beiheft 31: 115–164.

Hoffmann, M. P., L. T. Wilson, and F. G. Zalom. 1987a. The southern green stink bug, *Nezara viridula* Linnaeus (Heteroptera: Pentatomidae); new location. Pan-Pacific Entomologist 63: 333.

Hoffmann, M. P., L. T. Wilson, and F. G. Zalom. 1987b. Control of stink bugs in tomatoes. California Agriculture 41(5/6): 4–6.

Hoffmann, W. E. 1935. The foodplants of *Nezara viridula* Linn. (Hem. Pent.), pp. 811–816. Proceedings of the VI International Congress of Entomology. Madrid, Spain. (total proceedings page numbers not readily available)

Hokkanen, H. 1986. Polymorphism, parasites, and the native area of *Nezara viridula* (Hemiptera, Pentatomidae). Annales Entomologici Fennici 52: 28–31.

Horváth, G. 1889. Analecta ad cognitioonem Heteropterorum Himalayensium conscripsit. Természetrajzi Füzetek 12: 29–40.

Horváth, G. 1903. Pentatomidae novae Extraeuropaeae. Annales Musei Nationalis Hungarici 1: 400–409.

Huang, Ta-I. 2012. Local dispersal of stink bugs (Hemiptera: Pentatomidae) in mixed agricultural landscapes of the Coastal Plain. Ph.D. Dissertation, University of Georgia, Athens, GA. 129 pp.

Hubbard, H. G. 1885. The green soldier bug (*Raphigaster* [sic] *hilaris*, Fitch), pp. 159–162. *In* Part II – Miscellaneous insects affecting the orange, Chapter X – Insects affecting the twigs and leaves, pp. 132–163. Insects affecting the orange, report on the insects affecting the culture of the orange and other plants of the citrus family, with practical suggestions for their control or extermination, made, under direction of the entomologist. Division of Entomology, United States Department of Agriculture, Washington, DC. 227 pp.

Illiger, C. 1807. Fauna Etrusca sistens Insecta quae in Provinciis Florentina et Pisana praesertim collegit Petrus Rossius. Volume 2. C. G. Fleckeisen, Helmstadii. 511 pp.

Jeraj, M., and G. H. Walter. 1998. Vibrational communication in *Nezara viridula*: response of Slovenian and Australian bugs to one another. Behavioural Processes 44: 51–58.

Jones, T. H. 1918. The southern green plant-bug. United States Department of Agriculture Bulletin 689: 1–27.

Jones, V. P., and L. C. Caprio. 1992. Damage estimates and population trends of insects attacking seven macadamia cultivars in Hawaii. Journal of Economic Entomology 85: 1884–1890.

Jones, V. P., D. M. Westcott, N. Finson, and R. K. Nishimoto. 2001. Relationship between community structure and southern green stink bug (Heteroptera: Pentatomidae) damage in macadamia nuts. Environmental Entomology 30: 1028–1035.

Jones, W. A. 1979. The distribution and ecology of pentatomid pests of soybeans in South Carolina. Ph.D. Dissertation, Clemson University, Clemson, SC. 114 pp.

Jones, W. A. 1988. World review of the parasitoids of the southern green stink bug, *Nezara viridula* (L.) (Heteroptera: Pentatomidae). Annals of the Entomological Society of America 81: 263–273.

Jones, W. A., Jr., and F. D. Brewer. 1987. Suitability of various host plant seeds and artificial diets for rearing *Nezara viridula* (L.). Journal of Agricultural Entomology 4: 223–232.

Jones, W. A., Jr., and M. J. Sullivan. 1981. Overwintering habitats, spring emergence patterns, and winter mortality of some South Carolina Hemiptera. Environmental Entomology 10: 409–414.

Jones, W. A., Jr., and M. J. Sullivan. 1983. Seasonal abundance and relative importance of stink bugs in soybean. South Carolina Agricultural Experiment Station Technical Bulletin 1087. South Carolina Agricultural Experiment Station, Clemson University, Clemson, South Carolina. 6 pp.

Jones, W. A., and M. J. Sullivan. 1982. Role of host plants in population dynamics of stink bug pests in South Carolina. Environmental Entomology 11: 867–875.

Jones, W. A., Jr., S. Y. Young, M. Shepard, and W. H. Whitcomb. 1983. Use of imported natural enemies against insect pests of soybean, pp. 63–76. *In* H. N. Pitre (Ed.), Natural enemies of arthropod pests in soybean. Southern Cooperative Series Bulletin 285. South Carolina Agricultural Experiment Station, Clemson University, Clemson, South Carolina. 90 pp.

Jones, W. A., L. E. Ehler, M. P. Hoffmann, N. A. Davidson, L. T. Wilson, and J. W. Beardsley. 1995. Southern green stink bug, pp. 81–83. *In* J. R. Nechols, L. A. Andres, J. W. Beardsley, J. D. Goeden, and C. G. Jackson (Eds.), Biological control in the western United States. Accomplishments and Benefits of Regional Research Project W-84, 1964-1989. University of California, Division of Agriculture and Natural Resources, Oakland, CA. Publication 3361. 336 pp.

Kaiser, W. J., and N. G. Vakili. 1978. Insect transmission of pathogenic xanthomonads to bean and cowpea in Puerto Rico. Phytopathology 68: 1057–1063.

Kamal, M. 1937. The cotton green bug, *Nezara viridula* L., and its important egg parasite *Microphanus megacephalus* (Ashmead). Bulletin de la Societe Entomologique d'Egypte 21: 175–207.

Kamminga, K. L., D. A. Herbert, Jr., T. P. Kuhar, S. Malone, and A. Koppel. 2009. Efficacy of insecticides against *Acrosternum hilare* and *Euschistus servus* (Hemiptera: Pentatomidae) in Virginia and North Carolina. Journal of Entomological Science 44: 1–10.

Kareiva, P. M. 1983. Local movement in herbivorous insects: applying a passive diffusion model to mark-recapture field experiments. Oecologia 57: 322–327.

Kavar, T., P. Pavlovčič, S. Sušnik, V. Meglič, and M. Virant-Doberlet. 2006. Genetic differentiation of geographically separated populations of the southern green stink bug *Nezara viridula* (Hemiptera: Pentatomidae). Bulletin of Entomological Research 96: 117–128.

Kennedy, G. G., and N. P. Storer. 2000. Life systems of polyphagous arthropod pests in temporally unstable cropping systems. Annual Review of Entomology 45: 467–493.

Kester, K. M., and C. M. Smith. 1984. Effects of diet on growth, fecundity and duration of tethered flight of *Nezara viridula*. Entomologia Experimentalis et Applicata 35: 75–81.

King, E. G., J. R. Phillips, and R. B. Head. 1987. 40th annual conference report on cotton insect research and control, pp. 170–192. *In* E. G. King, J. R. Phillips, and R. B. Head (Eds.), 1987 Proceedings: Beltwide Cotton Production Research Conferences, National Cotton Council of America, Memphis, TN. 559 pp. (CD-ROM)

Kirichenko, A. N. 1951. True bugs of European part of the USSR (Hemiptera): keys and bibliography (Keys for the fauna of the USSR published by the Zoological Institute of The Academy of Sciences of The USSR, Volume 42). Publishing House of the Academy of Sciences of the USSR, Moscow–Leningrad. 424 pp. (in Russian)

Kirichenko, A. N. 1955. Order Hemiptera – True bugs, pp. 737–757. *In* E. N. Pavlovskiy and A. A. Shtakelberg (Eds.), Pest of forests. A reference book. Volume 2. Publishing House of the Academy of Sciences of the USSR, Moscow–Leningrad. 1098 pp. (in Russian)

Kiritani, K. 1963. The change in reproductive system of the southern green stink bug, *Nezara viridula*, and its application to forecasting of the seasonal history. Japanese Journal of Applied Entomology and Zoology 7: 327–337.

Kiritani, K. 1970. Studies on the adult polymorphism in the southern green stink bug, *Nezara viridula* (Hemiptera: Pentatomidae). Researches on Population Ecology 12: 19–34.

Kiritani, K. 1971. Distribution and abundance of the southern green stink bug, *Nezara viridula*, pp. 235–248. *In* Anonymous (Ed.), Proceedings of the Symposium on Rice Insects. Tropical Agricultural Research Center, Tokyo, Japan. (total proceedings page numbers not readily available)

Kiritani, K. 2011. Impacts of global warming on *Nezara viridula* and its native congeneric species. Journal of Asia-Pacific Entomology 14: 221–226.

Kiritani, K., and N. Hokyo. 1970. Studies on the population ecology of the southern green stink bug, *Nezara viridula* L. (Heteroptera: Pentatomidae). Agriculture, Forestry and Fisheries Research Council, Ministry of Agriculture, Forestry and Fisheries of Japan, Tokyo. 260 pp. (Shitei Shiken [Insect Pests and Diseases Series], Volume 9). (in Japanese)

Kiritani, K., N. Hokyo, and K. Kimura. 1962. Differential winter mortality relative to sex in the population of the southern green stink bug, *Nezara viridula* (Pentatomidae, Hemiptera). Japanese Journal of Applied Entomology and Zoology 6: 242–246.

Kiritani, K., N. Hokyo, and J. Yukawa. 1963. Co-existence of the two related stink bugs *Nezara viridula* and *N. antennata* under natural conditions. Researches on Population Ecology 5: 11–22.

Kiritani, K., N. Hokyo, K. Kimura, and F. Nakasuji. 1965. Imaginal dispersal of the southern green stink bug, *Nezara viridula* L., in relation to feeding and oviposition. Japanese Journal of Applied Entomology and Zoology 9: 291–296.

Kiritani, K., N. Hokyo, and K. Kimura. 1966. Factors affecting the winter mortality in the southern green stinsk bug, *Nezara viridula* L. Annales de la Société Entomologique de France, Nouvelle Série 2: 199–207 (Sunn Pest Memoirs 9).

Kirkaldy, G. W. 1909. Catalogue of the Hemiptera (Heteroptera) with biological and anatomical references, lists of foodplants and parasites, etc. Volume I: Cimicidae. Felix L. Dames, Berlin. 392 pp.

Knight, K. M. M., and G. M. Gurr. 2007. Review of *Nezara viridula* (L.) management strategies and potential for IPM in field crops with emphasis on Australia. Crop Protection 26: 1–10.

Kobayashi, S., and H. Numata. 1995. Effects of temperature and photoperiod on the induction of diapause and the determination of body coloration in the bean bug, *Riptortus clavatus*. Zoological Science 12: 343–348.

Kobayashi, T., and G. W. Cosenza. 1987. Integrated control of soybean stink bugs in the Cerrados. Japan Agricultural Research Quarterly 20: 229–236.

Koide, T., K. Yamaguchi, N. Ohno, and K. Morimoto. 2010. The situation of distribution of the southern green stink bug, *Nezara viridula* (Hemiptera: Pentatomidae), and its damage on soybean in Aichi prefecture. Annual Reports of the Kansai Plant Protection Society 52: 163–165. (in Japanese)

Kon, M., A. Oe, H. Numata, and T. Hidaka. 1988. Comparison of the mating behaviour between two sympatric species, *Nezara viridula* and *N. antennata* (Heteroptera: Pentatomidae), with special reference to sound emission. Journal of Ethology 6: 91–98.

Kon, M., A. Oe, and H. Numata. 1993. Intra- and interspecific copulations in the two congeneric green stink bugs, *Nezara antennata* and *N. viridula* (Heteroptera, Pentatomidae), with reference to postcopulatory changes in the spermatheca. Journal of Ethology 11: 83–89.

Kon, M., A. Oe, and H. Numata. 1994. Ethological isolation between two congeneric green stink bugs, *Nezara antennata* and *N. viridula* (Heteroptera, Pentatomidae). Journal of Ethology 12: 67–71.

Kontkanen, P. 1956. Kokousselostuksia – Sitzungsberichte 29.XI.1956. Annales Entomologici Fennici 22: 185–188.

Koštál, V. 2006. Eco-physiological phases of insect diapause. Journal of Insect Physiology 52: 113–127.

Lansbury, I. 1954. Two additional records of *Nezara viridula* (L.) (Hem., Pentatomidae) in Britain. The Entomologist's Monthly Magazine 90: 168.

Le Page, R. 1996. Soja. Punaises: a surveiller pour traiter a temps. Oleoscope 33: 23–26.

Lethierry, L., and G. Severin. 1893. Catalogue Général des Hemiptères. Pentatomidae I: i–x, 1–286. Bruxelles: Museé Royal Historique Naturel de Belgique.

Linnaeus, C. 1758. Systema naturæ per regna tria naturæ secundum classes, ordines, genera, species, cum characteribus, differentiis, synonymis, locis. Editio decima, reformata. Laurentii Salvii, Holmiae. Volume 1: 823 + 1 pp.

Linnavuori, R. E. 1972. Studies on African Pentatomoidea. Arquivos do Museu Bocage 3: 395–434.

Looney, C., and T. Murray. 2016. *Nezara viridula* – settler or sightseer, p. 17. *In* Research Reports of the 75th Annual Pacific Northwest Insect Management Conference, 11-12 January 2016, Portland, OR. 81 pp.

Lopez, J. D., M. A. Latheef, and W. C. Hoffmann. 2014. A multiyear study on seasonal flight activity based on captures of southern green stink bug (Hemiptera: Pentatomidae) in blacklight traps in Central Texas. Journal of Cotton Science 18: 153–165.

Lucas, H. 1849. Onzième Famille. Les Scutellériens. Genus *Pentatoma*, Oliv. *Cimex*, Linn. Fabr. *Edessa* et Cydnus, ejusd. *Raphigaster*, H. Schaff. *Nezara*, Am. et Serv., pp. 86–102. *In* Histoire naturelle des animaux articulés. Cinquième classe, Insectes. Exploration scientifique de l'Algérie pendant les années 1840, 1841, 1842. Sciences Physiques. Zoologie III. Imprimerie Nationale, Bertrand, Paris. 527 pp.

Lye, B.-H., and R. N. Story. 1989. Spatial dispersion and sequential sampling plan of the southern green stink bug (Hemiptera: Pentatomidae) on fresh market tomatoes. Environmental Entomology 18: 139–144.

Malouf, N. S. R. 1933. Studies on the internal anatomy of the stink bug, *Nezara viridula* L. Bulletin de la Société Royale Entomologique d'Egypte. Séance du 20 Juin: 96–119. With plates I–VII, and 1 text figure.

Malumphy, C., and S. Reid. 2007. Non-native Heteroptera associated with imported plant material in England during 2006 & 2007. HetNews 10: 2–3.

Marston, N. L., G. D. Thomas, C. M. Ignoffo, M. R. Gebhardt, D. L. Hostetter, and W. A. Dickerson. 1979. Seasonal cycles of soybean arthropods in Missouri: effect of pesticidal and cultural practices. Environmental Entomology 8: 165–173.

Mau, R. F. L., and J. L. M. Kessing. 1991. *Nezara viridula* (Linnaeus). http://www.extento.hawaii.edu/kbase /crop/type/nezara.htm (site last updated April 2007; accessed 21 March 2016).

Maw, H. E. L., R. G. Foottit, K. G. A. Hamilton, and G. G. E. Scudder. 2000. Checklist of the Hemiptera of Canada and Alaska. NRC Research Press, Ottawa, Ontario, Canada. 220 pp.

McHugh, J. J., Jr., L. N. Constantinides, and C. Tarutani-Weissman. 2003. Crop profile for watercress in Hawaii. http://citeseerx.ist.psu.edu/viewdoc/download?doi=10.1.1.670.5363&rep=rep1&type=pdf (accessed 21 March 2016).

McLain, D. K. 1980. Female choice and the adaptive significance of prolonged copulation in *Nezara viridula* (Hemiptera: Pentatomidae). Psyche 87: 325–336.

McLain, D. K., and S. D. Mallard. 1991. Sources and adaptive consequences of egg size variation in *Nezara viridula* (Hemiptera: Pentatomidae). Psyche 98: 135–164.

McPherson, J. E. 1982. The Pentatomoidea (Hemiptera) of northeastern North America with emphasis on the fauna of Illinois. Southern Illinois University Press, Carbondale, IL. 240 pp.

McPherson, J. E., and J. P. Cuda. 1974. The first record in Illinois of *Nezara viridula* (Hemiptera: Pentatomidae). Transactions of the Illinois State Academy of Science 67: 461–462.

McPherson, J. E., and R. M. McPherson. 2000. Stink bugs of economic importance in America north of Mexico. CRC Press, Boca Raton, FL. 253 pp.

McPherson, J. E., and R. W. Sites. 1989. Annotated records of species of Pentatomoidea (Hemiptera) collected at lights. The Great Lakes Entomologist 22: 95–98.

McPherson, R. M., and K. Bondari. 1991. Influence of planting date and row width on abundance of velvet-bean caterpillars (Lepidoptera: Noctuidae) and southern green stink bugs (Heteroptera: Pentatomidae) in soybean. Journal of Economic Entomology 84: 311–316.

McPherson, R. M., and L. D. Newsom. 1984. Trap crops for control of stink bugs in soybean. Journal of the Georgia Entomological Society 19: 470–480.

McPherson, R. M., G. K. Douce, and R. D. Hudson. 1993. Annual variation in stink bug (Heteroptera: Pentatomidae) seasonal abundance and species composition in Georgia soybean and its impact on yield and quality. Journal of Entomological Science 28: 61–72.

McPherson, R. M., C. S. Bundy, and M. L. Wells. 2001. Impact of early soybean production system on arthropod pest populations in Georgia. Environmental Entomology 30: 76–81.

Medrano, E. G., and A. A. Bell. 2007. Role of *Pantoea agglomerans* in opportunistic bacterial seed and boll rot of cotton (*Gossypium hirsutum*) grown in the field. Journal of Applied Microbiology 102: 134–143.

Medrano, E. G., J. F. Esquivel, and A. A. Bell. 2007. Transmission of cotton seed and boll rotting bacteria by the southern green stink bug (*Nezara viridula* L.). Journal of Applied Microbiology 103: 436–444.

Medrano, E. G., J. F. Esquivel, A. A. Bell, J. Greene, P. Roberts, J. Bacheler, J. J. Marois, D. L. Wright, R. L. Nichols, and J. Lopez. 2009a. Potential for *Nezara viridula* (Hemiptera: Pentatomidae) to transmit bacterial and fungal pathogens into cotton bolls. Current Microbiology 59: 405–412.

Medrano, E. G., J. F. Esquivel, R. L. Nichols, and A. A. Bell. 2009b. Temporal analysis of cotton boll symptoms resulting from southern green stink bug (*Nezara viridula* L.) feeding and transmission of a bacterial pathogen. Journal of Economic Entomology 102: 36–42.

Medrano, E. G., A. A. Bell, J. K. Greene, P. M. Roberts, J. S. Bacheler, J. J. Marois, D. L. Wright, J. F. Esquivel, R. L. Nichols, and S. Duke. 2015. Relationship between piercing-sucking insect control and internal lint and seed rot in Southeastern cotton (*Gossypium hirsutum* L.). Journal of Economic Entomology 108: 1540–1544.

Meglič, V., M. Virant-Doberlet, J. Šuštar-Vozlič, S. Sušnik, A. Čokl, N. Mikla, and M. Renou. 2001. Diversity of the southern green stink bug *Nezara viridula* (L.) (Heteroptera: Pentatomidae). Journal of Central European Agriculture 2: 241–249.

Merchant, M. E., and G. L. Teetes. 1992. Evaluation of selected sampling methods for panicle-infesting insect pests of sorghum. Journal of Economic Entomology 85: 2418–2424.

Michieli, Š., and B. Žener. 1968. Der Sauerstoffverbrauch verschiedener Farbstadien bei der Wanze *Nezara viridula* (L.). Zeitschrift für vergleichende Physiologie 58: 223–224.

Miklas, N., A. Čokl, M. Renou, and M. Virant-Doberlet. 2003. Variability of vibratory signals and mate choice selectivity in the southern green stink bug. Behavioural Processes 61: 131–142.

Mita, T., H. Nishimoto, N. Shimizu, and N. Mizutani. 2015. Occurrence of *Trissolcus basalis* (Hymenoptera, Platygastridae), an egg parasitoid of *Nezara viridula* (Hemiptera, Pentatomidae), in Japan. Applied Entomology and Zoology 50: 27–31. doi: 10.1007/s13355-014-0298-3.

Mitchell, W. C., and R. F. L. Mau. 1969. Sexual activity and longevity of the southern green stink bug, *Nezara viridula*. Annals of the Entomological Society of America 62: 1246–1247.

Mitchell, W. C., and R. F. L. Mau. 1971. Response of the female southern green stink bug and its parasite, *Trichopoda pennipes*, to male stink bug pheromones. Journal of Economic Entomology 64: 856–859.

Mitchell, W. C., R. M. Warner, and E. T. Fukunaga. 1965. Southern green stink bug, *Nezara viridula* (L.), injury to macadamia nut. Proceedings of the Hawaiian Entomological Society 19: 103–109.

Mizell, R. F., and W. L. Tedders. 1995. A new monitoring method for detection of the stinkbug complex in pecan orchards. Proceedings of the Southeastern Pecan Growers Association 88: 36–40.

Mizell, R. F., III, H. C. Ellis, and W. L. Tedders. 1996. Traps to monitor stink bugs and pecan weevils. The Pecan Grower 8: 17–20.

Mizell, R. F., III, W. L. Tedders, C. E. Yonce, and J. A. Aldrich. 1997. Stink bug monitoring – an update. Proceedings of the Southeastern Pecan Growers Association 90: 50–52.

Mizutani, N. 2013. Recent distribution and occurrences of the southern green stink bug, *Nezara viridula* in Japan. Shokubutsu Boeki [Plant Protection] 67: 595–601. (in Japanese)

Morrill, A. W. 1910. Pentatomid bugs of the genus *Nezara*, pp. 78–83. *In* Plant-bugs injurious to cotton bolls, United States Department of Agriculture, Division of Entomology, Bulletin Number 86. Washington: Government Printing Office. 110 pp.

Mulder, P. 2015. Oklahoma State University Insect Survey and Detection Report submitted to United States Department of Agriculture Current Research Information System. http://portal.nifa.usda.gov/web/crisprojectpages/0010303-insect-survey-and-detection.html (accessed 18 September 2015).

Mulsant, E., and C. Rey. 1866. Histoire naturelle des Punaises de France. Annales de la Société Linnéenne de Lyon 13: 293–267.

Musolin, D. L. 2007. Insects in a warmer world: ecological, physiological and life-history responses of true bugs (Heteroptera) to climate change. Global Change Biology 13: 1565–1585.

Musolin, D. L. 2012. Surviving winter: diapause syndrome in the southern green stink bug *Nezara viridula* in the laboratory, in the field, and under climate change conditions. Physiological Entomology 37: 309–322.

Musolin, D. L., and K. Ito. 2008. Photoperiodic and temperature control of nymphal development and induction of reproductive diapause in two predatory *Orius* bugs: interspecific and geographic differences. Physiological Entomology 33: 291–301.

Musolin, D. L., and H. Numata. 2003a. Photoperiodic and temperature control of diapause induction and colour change in the southern green stink bug *Nezara viridula*. Physiological Entomology 28: 65–74.

Musolin, D. L., and H. Numata. 2003b. Timing of diapause induction and its life-history consequences in *Nezara viridula*: is it costly to expand the distribution range? Ecological Entomology 28: 694–703.

Musolin, D. L., and H. Numata. 2004. Late-season induction of diapause in *Nezara viridula* and its effect on adult coloration and post-diapause reproductive performance. Entomologia Experimentalis et Applicata 11: 1–6.

Musolin, D. L., and A. Kh. Saulich. 2012. Responses of insects to the current climate change: from physiology and behaviour to range shifts. Entomological Review 92: 715–740.

Musolin, D. L., K. Fujisaki, and H. Numata. 2007. Photoperiodic control of diapause termination, colour change and postdiapause reproduction in the southern green stink bug, *Nezara viridula*. Physiological Entomology 32: 64–72.

Musolin, D. L., D. Tougou, and K. Fujisaki. 2010. Too hot to handle? Phenological and life-history responses to simulated climate change of the southern green stink bug *Nezara viridula* (Heteroptera: Pentatomidae). Global Change Biology 16: 73–87.

Musolin, D. L., D. Tougou, and K. Fujisaki. 2011. Photoperiodic response in the subtropical and warm-temperate zone populations of the southern green stink bug *Nezara viridula*: why does it not fit the common latitudinal trend? Physiological Entomology 36: 379–384.

Musser, F. R., A. L. Catchot, Jr., J. A. Davis, D. A. Herbert, Jr., G. M. Lorenz, T. Reed, D. D. Reisig, and S. D. Stewart. 2014. 2013 Soybean insect losses in the southern US. Midsouth Entomologist 7: 15–28.

Nakamura, K., and H. Numata. 2000. Photoperiodic control of the intensity of diapause development in the bean bug, *Riptortus clavatus* (Heteroptera: Alydidae). European Journal of Entomology 97: 19–23.

Nan, G.-H., Y.-P. Xu, Y.-W. Yu, C.-X. Zhao, C.-X. Zhang, and X.-P. Yu. 2016. Oocyte vitellogenesis triggers the entry of yeast-like symbionts into the oocyte of brown planthopper (Hemiptera: Delphacidae). Annals of the Entomological Society of America 109: 753–758.

National Agricultural Pest Information System (NAPIS). 2015. Reported status of southern green stink bug/ *Nezara viridula*. https://napis.ceris.purdue.edu/generate (accessed 15 September 2015).

Neimorovets, V. V. 2010. True bugs (Heteroptera) of the Krasnodar Territory and the Republic of Adygea: checklist. Saint Petersburg – Pushkin: All-Russian Institute of Plant Protection, Russian Academy of Agricultural Sciences. 103 pp. (Plant Protection News, Supplement) (in Russian, with English summary)

Newsom, L. D., M. Kogan, F. D. Miner, R. L. Rabb, S. G. Turnipseed, and W. H. Whitcomb. 1980. General accomplishments toward better pest control in soybean, pp. 51–98. *In* C. B. Huffaker (Ed.), New technology of pest control. Wiley-Interscience Publishing, John Wiley and Sons, Inc. New York, NY. 500 pp.

Nielsen, A. L., K. Holmstrom, G. C. Hamilton, J. Cambridge, and J. Ingerson-Mahar. 2013. Use of black light traps to monitor the abundance, spread, and flight behavior of *Halyomorpha halys* (Hemiptera: Pentatomidae). Journal of Economic Entomology 106: 1495–1502.

Nishida, T. 1966. Behavior and mortality of the southern stink bug *Nezara viridula* in Hawaii. Researches on Population Ecology 8: 78–88.

Nuessly, G. S., M. G. Hentz, R. Beiriger, and B. T. Scully. 2004. Insects associated with faba bean, *Vicia faba* (Fabales: Fabaceae), in southern Florida. Florida Entomologist 87: 204–211.

Numata, H., and T. Hidaka. 1984. Photoperiodic control of adult diapause in the bean bug, *Riptortus clavatus* Thunberg (Heteroptera: Coreidae): III. Diapause development and temperature. Applied Entomology and Zoology 19: 356–360.

Oho, N., and K. Kiritani. 1960. Bionomics and control of the southern green stink bug. Shokubutsu Boeki [Plant Protection] 14: 237–241. (in Japanese)

Ohno, K., and Md. Z. Alam. 1992. Hereditary basis of adult color polymorphism in the southern green stink bug, *Nezara viridula* Linné (Heteroptera: Pentatomidae). Applied Entomology and Zoology 27: 133–139.

Olson, D. M., J. R. Ruberson, A. R. Zeilinger, and D. A. Andow. 2011. Colonization preference of *Euschistus servus* and *Nezara viridula* in transgenic cotton varieties, peanut, and soybean. Entomologia Experimentalis et Applicata 139: 161–169.

Olson, D. M., J. R. Ruberson, and D. A. Andow. 2012. Effects on stink bugs of field edges adjacent to woodland. Agriculture, Ecosystems & Environment 156: 94–98.

Orian, A. J. E. 1965. A new genus of Pentatomidae from Africa, Madagascar and Mauritius (Hemiptera). Proceedings of the Royal Entomological Society of London 34: 25–29.

Orr, D. B., D. J. Boethel, and W. A. Jones. 1985. Development and emergence of *Telenomus chloropus* and *Trissolcus basalis* (Hymenoptera: Scelionidae) at various temperatures and relative humidities. Annals of the Entomological Society of America 75: 615–619.

Orr, D. B., J. S. Russin, D. J. Boethel, and W. A. Jones. 1986. Stink bug (Hemiptera: Pentatomidae) egg parasitism in Louisiana soybeans. Environmental Entomology 15: 1250–1254.

Paiero, S. M., S. A. Marshall, J. E. McPherson, and M.-S. Ma. 2013. Stink bugs (Pentatomidae) and parent bugs (Acanthosomatidae) of Ontario and adjacent areas: a key to species and a review of the fauna. Canadian Journal of Arthropod Identification Number 24. 183 pp. doi:10.3752/cjai.2013.24 (accessed 25 September 2015).

Palisot de Beauvois, A. M. F. J. 1818. Insectes recueillis en Afrique et en Amérique, dans les royaumes d'Oware et de Benin, a Saint-Domingue et dans les États-Unis pendant les années 1786–1797. Fain et Compagnie (Parts 11–12): 173–208.

Panizzi, A. R. 1997. Wild hosts of pentatomids: ecological significance and role in their pest status on crops. Annual Review of Entomology 42: 99–122.

Panizzi, A. R. 2000. Suboptimal nutrition and feeding behavior of hemipterans on less preferred plant food sources. Anais da Sociedade Entomológica do Brasil 29: 1–12.

Panizzi, A. R. 2006. Possible egg positioning and gluing behavior by ovipositing southern green stink bug, *Nezara viridula* (L.) (Heteroptera: Pentatomidae). Neotropical Entomology 35: 149–151.

Panizzi, A. R. 2013. History and contemporary perspectives of the integrated pest management of soybean in Brazil. Neotropical Entomology 42: 119–127.

Panizzi, A. R., and E. Hirose. 1995. Seasonal body weight, lipid content, and impact of starvation and water stress on adult survivorship and longevity of *Nezara viridula* and *Euschistus heros*. Entomologia Experimentalis et Applicata 76: 247–253.

Panizzi, A. R., and T. Lucini. 2016. What happened to *Nezara viridula* (L.) in the Americas? Possible reasons to explain populations decline. Neotropical Entomology 45: 619–628.

Panizzi, A. R., and F. Slansky, Jr. 1985. Review of phytophagous pentatomids (Hemiptera: Pentatomidae) associated with soybean in the Americas. Florida Entomologist 68: 184–214.

Panizzi, A. R., and F. Slansky, Jr. 1991. Suitability of selected legumes and the effect of nymphal and adult nutrition in the southern green stink bug (Hemiptera: Heteroptera: Pentatomidae). Journal of Economic Entomology 84: 103–113.

Panizzi, A. R., M. H. M. Galileo, H. A. O. Gastal, J. F. F. Toledo, and C. H. Wild. 1980. Dispersal of *Nezara viridula* and *Piezodorus guildinii* nymphs in soybeans. Environmental Entomology 9: 293–297.

Panizzi, A. R., J. E. McPherson, D. G. James, M. Javahery, and R. M. McPherson. 2000. Chapter 13. Stink bugs (Pentatomidae), pp. 421–474. *In* C. W. Schaefer and A. R. Panizzi (Eds.), Heteroptera of economic importance. CRC Press LLC, Boca Raton, FL. 828 pp.

Pavlovčič, P., T. Kavar, V. Meglič, and M. V. Doberlet. 2008. Genetic population structure and range colonisation of *Nezara viridula*. Bulletin of Insectology 61: 191–192.

Peña M., R., and J. A. Sifuentes. 1972. Lista de nombres científicos y comunes de plagas agrícolas en México, 1972. Agricultura Técnica en México 3: 132–144.

Pennington, M. S. 1919. Notas sobre las especies Argentinas del género *Nezara* A. & S. Physis 4: 527–530.

Pilkay, G. L., F. P. F. Reay-Jones, M. D. Toews, J. K. Greene, and W. C. Bridges. 2015. Spatial and temporal dynamics of stink bugs in southeastern farmscapes. Journal of Cotton Science 15: 1–13. doi: 093/jisesa /iev006.

Pitts, J. R. 1977. Effect of temperature and photoperiod on *Nezara viridula* L. M.S. Thesis. Louisiana State University, Baton Rouge, LA. ix + 91 pp.

Poda, N. 1761. Insecta Musei Graecensis, quae in ordines genera et species juxta Systema Naturae Linnaei digessit. Graecii: Widmanstad. 127 pp.

Prado, S. S., D. Rubinoff, and R. P. P. Almeida. 2006. Vertical transmission of a pentatomid caeca-associated symbiont. Annals of the Entomological Society of America 99: 577–585.

Prado, S. S., M. Golden, P. A. Follett, M. P. Daugherty, and R. P. P. Almeida. 2009. Demography of gut symbiotic and aposymbiotic *Nezara viridula* L. (Hemiptera: Pentatomidae). Environmental Entomology 38: 103–109.

Predel, R., W. K. Russell, D. H. Russell, J. Lopez, J. Esquivel, and R. J. Nachman. 2008. Comparative peptidomics of four related hemipteran species: pyrokinins, myosuppresin, corazonin, adipokinetic hormone, sNPF, and periviscerokinins. Peptides 29: 162–167.

Pulakkatu-Thodi, I., D. Shurley, and M. D. Toews. 2014. Influence of planting date on stink bug injury, yield, fiber quality, and economic returns in Georgia cotton. Journal of Economic Entomology 107: 646–653.

Putshkov, V. G. 1961. Pentatomoidea. Fauna Ukrainy 21(1): 1–338. (in Ukrainian)

Putshkov, V. G. 1965. Shield-bugs of Middle Asia (Hemiptera, Pentatomoidea). Publishing House Ilim (The Academy of Sciences of Kyrgyz SSR, Institute of Biology), Frunze (Kyrgyz SSR, the USSR). 332 pp. (in Russian)

Putshkov, V. G. 1972. Order Hemiptera (Heteroptera) – True Bugs, pp. 222–262. *In* O. L. Kryzhanovskiy and E. M. Danzig (Eds.), Insects and mites – pest of agricultural plants. Volume I. Insects with incomplete development. Publishing House Nauka, Leningrad. 324 pp. (in Russian)

Rabitsch, W. 2008. Alien true bugs of Europe (Insecta: Hemiptera: Heteroptera). Zootaxa 1827: 1–44.

Rabitsch, W. 2010. True bugs (Hemiptera, Heteroptera), pp. 407–433. *In* A. Roques, M. Kenis, D. Lees, C. Lopez-Vaamonde, W. Rabitsch, J.-Y. Rasplus, and D. Roy (Eds.), Alien terrestrial arthropods of Europe. BioRisk 4(1): 407–433.

Rabitsch, W. 2016. Notes on the true bug fauna of Vienna with five first records for Austria (Hemiptera: Heteroptera). Beiträge zur Entomofaunistik 17: 39–54.

Ragsdale, D. W., A. D. Larson, and L. D. Newsom. 1979. Microorganisms associated with feeding and from various organs of *Nezara viridula*. Journal of Economic Entomology 72: 725–727.

Rao, V. P., M. A. Ghani, T. Sankaran, and K. C. Mathur. 1971. A review of the biological control of insects and other pests in South-East Asia and the Pacific region. Technical Communication No. 4. Commonwealth Agricultural Bureaux. Bangalore Press, Bangalore, India. 149 pp.

Rea, J. H., S. D. Wratten, R. Sedcole, P. J. Cameron, and S. I. Davis. 2002. Trap cropping to manage green vegetable bug *Nezara viridula* (L.) (Heteroptera: Pentatomidae) in sweet corn in New Zealand. Agriculture and Forestry Entomology 4: 101–107.

Reay-Jones, F. P. F. 2010. Spatial and temporal patterns of stink bugs (Hemiptera: Pentatomidae) in wheat. Environmental Entomology 39: 944–955.

Reay-Jones, F. P. F., J. K. Greene, M. D. Toews, and R. B. Reeves. 2009. Sampling stink bugs (Hemiptera: Pentatomidae) for population estimation and pest management in southeastern cotton production. Journal of Economic Entomology 102: 2360–2370.

Reay-Jones, F. P. F., M. D. Toews, J. K. Greene, and R. B. Reeves. 2010. Spatial dynamics of stink bugs (Hemiptera: Pentatomidae) and associated boll injury in southeastern cotton fields. Environmental Entomology 39: 956–969.

Rédei, D., and A. Torma. 2003. Occurrence of the southern green stink bug, *Nezara viridula* (Heteroptera: Pentatomidae) in Hungary. Acta Phytopathologica et Entomologica Hungarica 38: 365–367.

Rédei, D., and G. Vétek. 2005. Mass occurrence of and damage caused by the southern green stink bug in Budapest. Kertészet és Szőlészet 54: 10. (in Hungarian)

Reeves, R. B., J. K. Greene, F. P. F. Reay-Jones, M. D. Toews, and P. D. Gerard. 2010. Effects of adjacent habitat on populations of stink bugs (Heteroptera: Pentatomidae) in cotton as part of a variable agricultural landscape in South Carolina. Environmental Entomology 39: 1420–1427.

Reiche, L., and L. Fairmaire. 1848. Ordre des Hémiptères, pp. 433–550. *In* P. V. Ferret and J. G. Galinier (Eds.), Voyage en Abyssinie dans les Provinces du Tigre, du Samen et de l'Ahmara. Volume 3. Paulin, Libraire-Éditeur, Paris. 536 pp.

Reichensperger, A. 1922. Rheinlands Hemiptera heteroptera. I. Verhandlungen des Naturhistorischen Vereins der preußischen Rheinlande und Westfalens 77: 35–77.

Reid, S. 2005. Heteroptera recently intercepted in England & Wales on foreign plant material. HetNews 6: 11.

Reid, S. 2006. A significant interception of the green vegetable bug, *Nezara viridula* (Linnaeus) (Hemiptera: Pentatomidae) in the UK. The Entomologist's Record and Journal of Variation 118: 123–125.

Reinhard, H. J. 1922. Host records of some Texas Tachinidae (Diptera). Entomological News 33: 72–73.

Reisig, D. D. 2011. Insecticidal management and movement of brown stink bug, *Euschistus servus*, in corn. Journal of Insect Science 11 (Article 168): 1–11. http://jinsectscience.oxfordjournals.org/content/11/1/168 (accessed 17 July 2016).

Reuter, O. M. 1888. Synonymische Revision der von den älteren Autoren (Linné 1758–Latreille 1806) beschriebenen Palaearktischen Heteropteren. I. Acta Societatis Scientiarum Fennicae 15(1): 241–315; 15(2): 443–812.

Reuter, O. M. 1907. Sur quelques variétés prétendues des genres Palomena Muls. et Rey, Nezara Am. et Serv. [Hémipt. Hétéropt]. Bulletin de la Société Entomologique de France 1907: 209–210.

Rider, D. A. 2006a. Family Pentatomidae, pp. 233–402. *In* B. Aukema and C. Rieger (Eds.), Catalogue of the Heteroptera of the Palaearctic Region. Volume 5. The Netherlands Entomological Society, Amsterdam. xiii + 550 pp.

Rider, D. A. 2006b. *Nezara viridula* (Linnaeus, 1758): http://www.ndsu.nodak.edu/ndsu/rider/Pentatomoidea /Species_Nezarini/Nezara_viridula.htm (accessed 21 January 2015).

Rieger, C. 1994. Ein Fund von *Nezara viridula* (Linnaeus, 1758) in Süddeutschland (Heteroptera: Pentatomidae). Entomolgische Zeitung 104: 469–472.

Riley, C. V., and L. O. Howard. 1893a. Extracts from correspondence – The sweet-potato root weevil, p. 261. *In* C. V. Riley and L. O. Howard (Eds.), Insect Life 5: 213–288.

Riley, C. V., and L. O. Howard. 1893b. Extracts from correspondence – Plant-bugs injuring oranges in Florida, pp. 264–265. *In* C. V. Riley and L. O. Howard (Eds.), Insect Life 5: 213–288.

Rodríguez Rivera, R. 1975. Estimación de las pérdidas ocasionadas por chinches y plaga de almacén al xpelón (*Vigna sinensis*) en Yucatán. Folia Entomológica Mexicana 33: 30.

Rodríguez Vélez, J. 1974. Observaciones sobre la biologia de la chinche verde, *Nezara viridula* (L [sic]), en el Valle del Fuerte Sin. Folia Entomológica Mexicana 28: 5–12.

Rojas, M. G., and J. A. Morales-Ramos. 2014. Juvenile coloration as a predictor of health in *Nezara viridula* (Heteroptera: Pentatomidae) rearing. Journal of Entomological Science 49: 166–175.

Rolston, L. H. 1983. A revision of the genus *Acrosternum* Fieber, subgenus *Chinavia* Orian, in the Western Hemisphere (Hemiptera: Pentatomidae). Journal of the New York Entomological Society 9: 97–176.

Roof, M. E. 1994. Control of bollworm/budworm and other insects before and after boll weevil eradication in South Carolina, pp. 798-799. *In* Proceedings of the 1994 Beltwide Cotton Conferences. National Cotton Council of America, Memphis, TN. 1751 pp. (CD-ROM)

Rosenfeld, A. H. 1911. Insects and spiders in Spanish moss. Journal of Economic Entomology 4: 398–409.

Rosewall, O. W. 1922. Insects of the yellow thistle (Hem., Col., Lepid., Dip., Hym.). Entomological News and Proceedings of the Entomological Section of the Academy of Natural Sciences of Philadelphia 33: 176–180.

Ruffinelli, A., and A. A. Pirán. 1959. Hemipteros heteropteros del Uruguay. Montevideo: Facultad de Agronomia de Montevideo, Boletín no. 51: 1–60.

Ruiz, E. C., J. M. A. Coronado B., and S. N. Myartseva. 2006. Situación actual del manejo de las plagas do los cítricos en Tamaulipas, México. Manejo Integrado de Plagas y Agroecología (Costa Rica) 78: 94–100.

Ryan, M. A., A. Čokl, and G. H. Walter. 1996. Differences in vibratory sound communication between a Slovenian and an Australian population of *Nezara viridula* (Heteroptera: Pentatomidae). Behavioural Processes 36: 183–193.

Salisbury, A., M. V. Barclay, S. Reid, and A. Halstead. 2009. The current status of the southern green shield bug, *Nezara viridula* (Hemiptera: Pentatomidae), an introduced pest species recently established in south-east England. British Journal of Entomology and Natural History 22: 189–194.

Sane, I., D. R. Alverson, and J. W. Chapin. 1999. Efficiency of conventional sampling methods for determining arthropod densities in close-row soybeans. Journal of Agricultural and Urban Entomology 16: 65–84.

Saulich, A. Kh., and D. L. Musolin. 1996. Univoltinism and its regulation in some temperate true bugs (Heteroptera). European Journal of Entomology 93: 507–518.

Saulich, A. Kh., and D. L. Musolin. 2007. Times of the year: the diversity of seasonal adaptations and ecological mechanisms controlling seasonal development of true bugs (Heteroptera) in the temperate climate, pp. 25–106. *In* A. A. Stekolnikov (Ed.), Adaptive strategies of terrestrial arthropods to unfavorable environmental conditions: a collection of papers in memory of Professor Viktor Petrovich Tyshchenko (Proceedings of the Biological Institute of Saint Petersburg State University 53). Saint Petersburg State University, Saint Petersburg, Russia. 387 pp. (in Russian, with English summary)

Saulich, A. Kh., and D. L. Musolin. 2011. Biology and ecology of the predatory bug *Podisus maculiventris* (Say) (Heteroptera, Pentatomidae) and possibility of its application against the Colorado potato beetle, *Leptinotarsa decemlineata* Say (Coleoptera, Chrysomelidae). A textbook for the course Seasonal Cycles of Insects. Saint Petersburg State University, Saint Petersburg, Russia. 84 pp. (in Russian)

Saulich, A. Kh., and D. L. Musolin. 2012. Diapause in the seasonal cycle of stink bugs (Heteroptera, Pentatomidae) from the temperate zone. Entomological Review 92: 1–26.

Saulich, A. Kh., and D. L. Musolin. 2014. Seasonal cycles in stink bugs (Heteroptera, Pentatomidae) from the temperate zone: diversity and control. Entomological Review 94: 785–814.

Saunders, D. S. 2010. Photoperiodism in insects: migration and diapause responses, pp. 218–257. *In* R. J. Nelson, D. L. Denlinger, and D. E. Somers (Eds.), Photoperiodism: the biological calendar. Oxford University Press, New York. xiv + 600 pp.

Schmitz, G. 1986. Captures "insolites" d'hétéroptères. Bulletin and Annales de la Société royale belge d'Entomologie 122: 33–38.

Schumann, F. W., and J. W. Todd. 1982. Population dynamics of the southern green stink bug (Heteroptera: Pentatomidae) in relation to soybean phenology. Journal of Economic Entomology 75: 748–753.

Schuster, G. 1986. Zur Wanzenfauna Schwabens und der Schwäbischen Alb. Bericht der Naturforschenden Gesellschaft Augsburg 42: 1–36.

Schwertner, C. F., and J. Grazia. 2007. O gênero *Chinavia* Orian (Hemiptera, Pentatomidae, Pentatominae) no Brasil, com chave pictórica para os adultos. Revista Brasileira de Entomologia 51: 416–435.

Seymour, J. E., and G. J. Bowman. 1994. Russet coloration in *Nezara viridula* (Hemiptera: Pentatomidae): an unreliable indicator of diapause. Environmental Entomology 23: 860–863.

Shardlow, M. E. A., and R. Taylor. 2004. Is the southern green shield bug, *Nezara viridula* (L.) (Hemiptera Pentatomidae) another species colonizing Britain due to climate change? British Journal of Entomology and Natural History 17: 143–146.

Shearer, P. W., and V. P. Jones. 1998. Suitability of selected weeds and ground covers as host plants of *Nezara viridula* (L.) (Hemiptera: Pentatomidae). Proceedings of the Hawaiian Entomological Society 33: 75–82.

Simov, N., M. Langourov, S. Grozeva, and D. Gradinarov. 2012. New and interesting records of alien and native true bugs (Hemiptera: Heteroptera) from Bulgaria. Acta Zoologica Bulgarica 64: 241–252.

Singh, Z. 1973. Southern green stink bug and its relationship to soybeans: bionomics of the southern green stink bug *Nezara viridula* (Hemiptera: Pentatomidae) in Central India. Metropolitan Book Co. (PVT) Ltd., Delhi, India. 106 pp.

Smaniotto, L. F., and A. R. Panizzi. 2015. Interactions of selected species of stink bugs (Hemiptera: Heteroptera: Pentatomidae) from leguminous crops with plants in the Neotropics. Florida Entomologist 98: 7–17.

Smith, J. F., R. G. Luttrell, and J. K. Greene. 2008. Seasonal abundance of stink bugs (Heteroptera: Pentatomidae) and other polyphagous species in a multi-crop environment in south Arkansas. Journal of Entomological Science 43: 1–12.

Smith, J. F., R. G. Luttrell, and J. K. Greene. 2009. Seasonal abundance, species composition, and population dynamics of stink bugs in production fields of early and late soybean in South Arkansas. Journal of Economic Entomology 102: 229–236.

Smith, M. W., D. C. Arnold, R. D. Eikenbary, N. R. Rice, A. Shiferaw, B. S. Cheary, and B. L. Carroll. 1996. Influence of ground cover on beneficial arthropods in pecan. Biological Control 6: 164–176.

Snodgrass, G. L., J. J. Adamczyk, and J. Gore. 2005. Toxicity of insecticides in a glass-vial bioassay to adult brown, green, and southern green stink bugs (Heteroptera: Pentatomidae). Journal of Economic Entomology 98: 177–181.

Sosa-Gómez, D. R., and S. B. Alves. 2000. Temperature and relative humidity requirements for conidiogenesis of *Beauveria bassiana* (Deuteromycetes: Moniliaceae). Anais da Sociedade Entomologica do Brasil 29: 515–521.

Sosa-Gómez, D. R., and F. Moscardi. 1998. Laboratory and field studies on the infection of stink bugs, *Nezara viridula*, *Piezodorus guildinii*, and *Euschistus heros* (Hemiptera: Pentatomidae) with *Metarhizium anisopliae* and *Beauveria bassiana* in Brazil. Journal of Invertebrate Pathology 71: 115–120.

Sosa-Gómez, D. R., J. J. Silva, F. Costa, E. Binneck, S. R. R. Marin, and L. Nepomuceno. 2005. Population structure of the Brazilian southern green stink bug, *Nezara viridula*. Journal of Insect Science 23: 1–10.

Southgate, B. J., and G. E. Woodroffe. 1952. *Nezara viridula* L. (Hem., Pentatomidae) in Britain. The Entomologist's Monthly Magazine 88: 19.

Stål, C. 1854. Nya Hemiptera från Cafferlandet. Öfversigt af Kongliga Vetenskaps-Akademiens Förhandlingar 10: 209–227.

Stål, C. 1855. Entomologiska notiser: om Thumbergska hemipterarter. Öfversigt af Kongliga Vetenskaps-Akademiens Förhandlingar 12: 343–347.

Stål, C. 1865. Hemiptera Africana I. Stockholm: Norstedtiana. iv + 256 pp.

Stål, C. 1868. Hemiptera Fabriciana. Fabricianska Hemipterarter, efter de i Köpenhavn och Kiel Förvarade typexemplaren granskade och beskrifne. Kongliga Svenska Vetenskaps-Akademiens Handligar 7: 1–148.

Stewart, A. J. A., and P. Kirby. 2010. Hemiptera, pp. 512–530. *In* N. Maclean (Ed.), Silent summer: the state of wildlife in Britain and Ireland. Cambridge University Press, Cambridge. 768 pp.

Strawinski, K. 1959. Heteroptera for the fauna of Bulgaria. Bulletin de l'institute de Zoologie et Musée 8: 77–82.

Su, T. H., and H. K. Tseng. 1984. The introduction of an egg parasite, *Trissolcus basalis* (Wollaston), for control of the southern green stink bug, *Nezara viridula* (L.) in Taiwan. Journal of Agriculture and Forestry 33: 49–54.

Suh, C. P., J. K. Westbrook, and J. F. Esquivel. 2013. Species composition of stink bugs in cotton and other major row crops in the Brazos River Bottom production area of Texas. Southwestern Entomologist 38: 561–570.

Suzuki, K., M. Nishino, and S. Shimo. 2011. Expanded distribution of *Nezara viridula* (Hemiptera: Pentatomidae) in Mie prefecture. Annual Report of the Kansai Plant Protection Society 53: 133–134. (in Japanese)

Tada, A., Y. Kikuchi, T. Hosokawa, D. L. Musolin, K. Fujisaki, and T. Fukatsu. 2011. Obligate association with gut bacterial symbiont in Japanese populations of the southern green stinkbug *Nezara viridula* (Heteroptera: Pentatomidae). Applied Entomology and Zoology 46: 483–488.

Takeda, K., D. L. Musolin, and K. Fujisaki. 2010. Dissecting insect responses to climate warming: overwintering and post-diapause performance in the southern green stink bug, *Nezara viridula*, under simulated climate-change conditions. Physiological Entomology 35: 343–353.

Tauber, M. J., C. A. Tauber, and S. Masaki. 1986. Seasonal adaptations of insects. Oxford University Press, New York. xvi + 412 pp.

Thomas, C. D., E. J. Bodsworth, R. J. Wilson, A. D. Simmons, Z. G. Davies, M. Musche, and L. Conradt. 2001. Ecological and evolutionary processes at expanding range margins. Nature 411: 577–581.

Thomas, D. B., and T. R. Yonke. 1981. A review of the Nearctic species of the genus *Banasa* Stål (Hemiptera: Pentatomidae). Journal of the Kansas Entomological Society 54: 233–248.

Thunberg, C. P. 1783. Dissertatio entomologica novas insectorum species, sistens, cujus partem secundam, cons. exper. facult. med. upsal., publice ventilandam exhibent. Part 2: 29–52, plate 2. J. Edman, Upsaliae.

Tillman, P. G. 2006. Susceptibility of pest *Nezara viridula* (Heteroptera: Pentatomidae) and parasitoid *Trichopoda pennipes* (Diptera: Tachinidae) to selected insecticides. Journal of Economic Entomology 99: 648–657.

Tillman, P. G. 2010. Composition and abundance of stink bugs (Heteroptera: Pentatomidae) in corn. Environmental Entomology 39: 1765–1774.

Tillman, P. G. 2014. Physical barriers for suppression of movement of adult stink bugs into cotton. Journal of Pest Science 87: 419–427.

Tillman, P. G., and J. E. Carpenter. 2014. Milkweed (Gentianales: Apocynaceae): a farmscape resource for increasing parasitism of stink bugs (Hemiptera: Pentatomidae) and providing nectar to insect pollinators and monarch butterflies. Environmental Entomology 43: 370–376.

Tillman, P. G., and B. G. Mullinix, Jr. 2004. Comparison of susceptibility of pest *Euschistus servus* and predator *Podisus maculiventris* (Heteroptera: Pentatomidae) to selected insecticides. Journal of Economic Entomology 97: 800–806.

Tillman, P. G., T. D. Northfield, R. F. Mizell, and T. C. Riddle. 2009. Spatiotemporal patterns and dispersal of stink bugs (Heteroptera: Pentatomidae) in peanut-cotton farmscapes. Environmental Entomology 38: 1038–1052.

Tillman, P. G., J. R. Aldrich, A. Khrimian, and T. E. Cottrell. 2010. Pheromone attraction and cross-attraction of *Nezara*, *Acrosternum*, and *Euschistus* spp. stink bugs (Hemiptera: Pentatomidae) in the field. Environmental Entomology 39: 610–617.

Tillman, P. G., A. Khrimian, T. E. Cottrell, R. F. Mizell, and W. C. Johnson. 2015. Trap cropping systems and a physical barrier for suppression of stink bugs (Heteroptera: Pentatomidae) in cotton. Journal of Economic Entomology 108: 2324–2334.

Todd, J. W. 1976. Effects of stink bug feeding on soybean seed quality, pp. 611–618. *In* L. D. Hill (Ed.), World Soybean Research: Proceedings of the World Soybean Research Conference I, Interstate Printers and Publishers, Danville, IL. 1073 pp.

Todd, J. W. 1981. Effects of stinkbug damage on soybean quality. *In* Proceedings of the International Congress on Soybean Seed Quality and Stand Establishment 22: 46–51.

Todd, J. W. 1989. Ecology and behavior of *Nezara viridula*. Annual Review of Entomology 34: 273–292.

Todd, J. W., and D. C. Herzog. 1980. Sampling phytophagous Pentatomidae on soybean, pp. 438–478. *In* M. Kogan and D.C. Herzog (Eds.), Sampling methods in soybean entomology. Springer, New York. 587 pp.

Toews, M. D., and W. D. Shurley. 2009. Crop juxaposition affects cotton fiber quality in Georgia farmscapes. Journal of Economic Entomology 102: 1515–1522.

Toews, M. D., J. Greene, F. P. F. Reay-Jones, and R. B. Reeves. 2008. A comparison of sampling techniques for stink bugs in cotton, pp. 1193–1203. *In* S. Boyd, M. Huffman, D. Richter, and B. Robertson (Eds.), Proceedings of the Beltwide Cotton Conferences, National Cotton Council of America, Memphis, TN. 1934 pp. (CD-ROM)

Toews, M. D., E. L. Blinka, J. W. Van Duyn, D. A. Herbert, Jr., J. S. Bacheler, P. M. Roberts, and J. K. Greene. 2009. Fidelity of external boll feeding lesions to internal damage for assessing stink bug damage in cotton. Journal of Economic Entomology 102: 1344–1351.

Tonet, G. L., and E. M. Reis. 1979. Pathogenicity of *Beauveria bassiana* to soybean pest insects. Pesquisa Agropecuaria Brasileira 14: 89–95.

Torre-Bueno, J. R. de la. 1903. A preliminary list of the Pentatomidæ within fifty miles of New York. Journal of the New York Entomological Society 11: 128–129.

Torre-Bueno, J. R. de la. 1908. Hemiptera Heteroptera of Westchester County, N. Y. Journal of the New York Entomological Society 16: 223–228.

Torre-Bueno, J. R. de la. 1912. *Nezara viridula* Linné, an Hemipteron new to the northeastern United States. Entomological News 23: 316–318.

Tougou, D., D. L. Musolin, and K. Fujisaki. 2009. Some like it hot! Rapid climate change promotes shifts in distribution ranges of *Nezara viridula* and *N. antennata* in Japan. Entomologia Experimentalis et Applicata 30: 249–258.

Tschorsnig, H.-P., P. Cerretti, and T. Zeegers. 2012. Eight "alien" tachinids in Europe?, pp. 11–13. *In* J. O'Hara (Ed.), *The Tachinid Times* 25th Anniversary issue. 28 pp.

Turner, W. F. 1918. *Nezara viridula* and kernel spot of pecan, pp. 490–491. *In* Science New Series, 47, No. 1220: 471–494.

Uvarov, B. P. 1931. Insects and climate. Transactions of the Entomological Society of London 79: 1–247.

van den Berg, M. A., and J. Greenland. 1996. Further releases of *Trichopoda pennipes*, parasitoid of the green stinkbug, *Nezara viridula*, in South Africa. Tachinid Times 9: 2.

Van Duzee, E. P. 1904. Annotated list of the Pentatomidæ recorded from America north of Mexico, with descriptions of some new species. Transactions of the American Entomological Society 30: 1–80.

van Heerden, P. W. 1934. The green stink-bug (*Nezara viridula* Linn.). Annals of the University of Stellenbosch 11A: 1–24.

Velasco, L. R. I., and G. H. Walter. 1992. Availability of different host plant species and changing abundance of the polyphagous bug *Nezara viridula* (Hemiptera: Pentatomidae). Environmental Entomology 21: 751–759.

Velasco, L. R. I., and G. H. Walter. 1993. Potential of host-switching in *Nezara viridula* (Hemiptera: Pentatomidae) to enhance survival and reproduction. Journal of Economic Entomology 22: 326–333.

Velasco, L. R. I., G. H. Walter, and V. E. Harris. 1995. Voltinism and host plant use by *Nezara viridula* (L.) (Hemiptera: Pentatomidae) in Southeastern Queensland. Journal of the Australian Entomological Society 34: 193–203.

Venugopal, P. D., P. L. Coffey, G. P. Dively, and W. O. Lamp. 2014. Adjacent habitat influence on stink bug (Hemiptera: Pentatomidae) densities and the associated damage at field corn and soybean edges. PloS ONE 9 (10): e109917. doi:10.1371/journal.pone.0109917.

Vétek, G., and D. Rédei. 2014. First record of the southern green stink bug, *Nezara viridula*, from Slovakia (Hemiptera: Heteroptera: Pentatomidae). Klapalekiana 50: 241–245.

Villers, C. J. de. 1789. Caroli Linnæi entomologia, faunæ Sueciæ descriptionibus aucta: D. D. Scopoli, Geoffroy, De Geer, Fabricii, Schrank, etc., speciebus vel in systemate non enumeratis, vel nuperime detectis, vel speciebus Galliæ Australis locupletata, generum specierumque rariorum iconibus ornata, curante et augente Carolo de Villers. Lugduni (Lyon) 1: 1–765.

Vinokurov, N. N., E. V. Kanyukova, and V. B. Golub. 2010. Catalogue of the Heteroptera of Asian part of Russia. Nauka Publishing House, Novosibirsk (Russia). 320 pp. (in Russian)

Vivan, L. M., and A. R. Panizzi. 2002. Two new morphs of the southern green stink bug, *Nezara viridula* (L.) (Heteroptera: Pentatomidae), in Brazil. Neotropical Entomology 31: 475–476.

Vivan, L. M., and A. R. Panizzi. 2006. Geographical distribution of genetically determined types of *Nezara viridula* (L.) (Heteroptera: Pentatomidae) in Brazil. Neotropical Entomology 35: 175–181.

Voigt, K. 1998. *Nezara viridula* erneut in Süddeutschland gefunden! (Heteroptera, Pentatomidae). Carolinea 56: 121–122.

Volkovich, T. A., and A. Kh. Saulich. 1995. The predatory bug *Arma custos*: photoperiodic and temperature control of diapause and colouration. Entomological Review 74: 151–162.

Vyavhare, S. S., M. O. Way, and R. F. Medina. 2014. Stink bug species composition and relative abundance of the redbanded stink bug (Hemiptera: Pentatomidae) in soybean in the Upper Gulf Coast Texas. Environmental Entomology 43: 1621–1627.

Walker, F. 1867. Catalogue of the specimens of heteropterous Hemiptera in the collection of the British Museum. Part II: Scutata, pp. 241–417. E. Newman, London, UK.

Waterhouse, D. F. 1998. Biological control of insect pests: Southeast Asian prospects. Australian Centre for International Agricultural Research, Canberra. viii + 548 pp. (ACIAR Monograph No. 51).

Watson, J. R. 1918. Green soldier bug or pumpkin bug (*Nezara viridula* Linn.), pp. 231–235. *In* Insects of a citrus grove. University of Florida Agricultural Experiment Station Bulletin 148: 165–267.

Watson, J. R. 1919a. Pumpkin bug (*Nezara viridula*), p. 161. *In* Florida truck and garden insects. University of Florida Agricultural Experiment Station Bulletin 151 (a revision of Bulletin 134): 113–211.

Watson, J. R. 1919b. Sunflowers, pp. 185–186. *In* Florida truck and garden insects. University of Florida Agricultural Experiment Station Bulletin 151 (a revision of Bulletin 134): 113–211.

Weeks, J. A., A. C. Hodges, and N.C. Leppla. 2012. Citrus pests – southern green stink bug. University of Florida Fact Sheet. http://idtools.org/id/citrus/pests/factsheet.php?name=Southern+green+stink+bug (accessed 29 November 2015).

Werner, D. J. 2005. *Nezara viridula* (Linnaeus, 1758) in Köln und in Deutschland (Heteroptera, Pentatomidae). Heteropteron 21: 29–31.

Westwood, J. O. 1837, 1842. *In* F. W. Hope (Ed.), A catalogue of Hemiptera in the collection of the Rev. F. W. Hope, M. A. with short Latin descriptions of the new species. J. Bridgewater, London, UK. 1837, Part 1: 1–46; 1842, Part 2: 1–26 (stating that descriptions are by J. O. Westwood).

Wilbur, D. A. 1939. Mass flights of the pentatomid, *Thyanta custator* (Fabr.), in Kansas. Journal of the Kansas Entomological Society 12: 77–80.

Williams, M. R. 2006. Cotton insect losses - 2005, pp. 1151–1204. *In* M. Huffman and D. Richter (Eds.), Proceedings of the Beltwide Cotton Conferences. National Cotton Council of America, Memphis, TN. 2534 pp. (CD-ROM)

Williams, M. R. 2014. Cotton insect loss estimates - 2013, pp. 798–812. *In* S. Boyd, M. Huffman, and B. Robertson (Eds.), Proceedings of the Beltwide Cotton Conferences. National Cotton Council of America, Memphis, TN. 1071 pp. (CD-ROM)

Williams, M. R. 2015. Cotton insect loss estimates - 2014, pp. 494–506. *In* S. Boyd and M. Huffman (Eds.), Proceedings of the Beltwide Cotton Conferences. National Cotton Council of America, Memphis, TN. 1019 pp. (CD-ROM)

Williamson, C., and M. B. von Wechmar. 1995. The effects of two viruses on the metamorphosis, fecundity, and longevity of the green stinkbug, *Nezara viridula*. Journal of Invertebrate Pathology 65: 174–178.

Willrich, M. M., D. R. Cook, and B. R. Leonard. 2003. Laboratory and field evaluations of insecticide toxicity to stink bugs (Heteroptera: Pentatomidae). Journal of Cotton Science 7: 156–163.

Willrich, M. M., B. R. Leonard, D. R. Cook, and R. H. Gable. 2004a. Evaluation of insecticides for control of stink bugs on cotton, 2003. Arthropod Management Tests 29: F61.

Willrich, M. M., M. E. Bau, B. R. Leonard, D. J. Boethel, and R.H. Gable. 2004b. Evaluation of insecticides for control of *Thyanta custator*, brown stink bug, and southern green stink bug on cotton, 2003. Arthropod Management Tests 29: F62.

Wilson, F. 1960. A review of the biological control of insects and weeds in Australia and Australian New Guinea. Technical Communication No. 1, Commonwealth Institute of Biological Control, Commonwealth Agricultural Bureaux, Ottawa, Canada. 101 pp.

Wiman, N. G., V. M. Walton, P. W. Shearer, S. I. Rondon, and J. C. Lee. 2015. Factors affecting flight capacity of brown marmorated stink bug, *Halyomorpha halys* (Hemiptera: Pentatomidae). Journal of Pest Science 88: 37–47.

Wolfenbarger, D. O. 1947. Tests of some newer insecticides for control of subtropical fruit and truck crop pests. Florida Entomologist 29: 37–44.

Wolff, J. F. 1801. Icones Cimicum descriptionibus illustratae. J. J. Palm, Erlangen. Volume 2: 41–84, plates 5–8.

Yukawa, J., and K. Kiritani. 1965. Polymorphism in the southern green stink bug. Pacific Insects 7: 639–642.

Yukawa, J., K. Kiritani, N. Gyoutoku, N. Uechi, D. Yamaguchi, and S. Kamitani. 2007. Distribution range shift of two allied species, *Nezara viridula* and *N. antennata* (Hemiptera: Pentatomidae), in Japan, possibly due to global warming. Applied Entomology and Zoology 42: 205–215.

Yukawa, J., K. Kiritani, T. Kawasawa, Y. Higashiura, N. Sawamura, K. Nakada, N. Gyotoku, A. Tanaka, S. Kamitani, K. Matsuo, S. Yamauchi, and Y. Takematsu. 2009. Northward range expansion by *Nezara viridula* (Hemiptera: Pentatomidae) in Shikoku and Chugoku Districts, Japan, possibly due to global warming. Applied Entomology and Zoology 44: 429–437.

8

Piezodorus guildinii (Westwood)

C. Scott Bundy, Jesus F. Esquivel, Antônio R. Panizzi,
Joe E. Eger, Jeffrey A. Davis, and Walker A. Jones

Piezodorus guildinii (Westwood)[1]

1837	*Raphagaster* [sic] *guildinii* Westwood, Cat. Hope 7: 31.
1872	*Piezodorus guildinii*: Stål, K. Svens., Vet.-Akad. Handl. 10(4): 45.
1891	*Nezara hebes* Bergroth, Rev. d'Ent. 10: 227. (Synonymized with *P. guildinii* by Rolston, 1983, J. New York Ent. Soc. 91: 101).
1894	*Piezodorus guildingi* [sic]: Uhler, Proc. Zool. Soc. London: 175.
1904	*Piezodorus guildingi* [sic]: Van Duzee, Trans. Am. Ent. Soc., 30: 61.
1916	*Piezodorus guildinii*: Van Duzee, Check List Hem. Am.: 8.

CONTENTS

8.1   Introduction 426
8.2   Taxonomy and Identification 426
    8.2.1   Taxonomy 426
    8.2.2   Identification and Comparison with Other Stink Bugs 426
8.3   Distribution 427
    8.3.1   Range Expansion in South America 428
    8.3.2   Range Expansion in North America 428
8.4   Biology 429
    8.4.1   Life History 429
    8.4.2   Overwintering 431
8.5   Host Plants 431
8.6   Economic Impact 436
    8.6.1   Feeding and Injury 436
    8.6.2   Impact of Other *Piezodorus* spp. 437
8.7   Management 437
    8.7.1   Monitoring 437
    8.7.2   Cultural Control 438
    8.7.3   Biological Control 438
    8.7.4   Chemical Control 439
8.8   Future Outlook 440
8.9   Acknowledgments 440
8.10 References Cited 440

---

[1] Synonymy adapted from David A. Rider (personal communication) and Froeschner (1988).

8.1 Introduction

Piezodorus guildinii (Westwood), commonly known as the redbanded stink bug (United States), the small green stink bug (Brazil), and the alfalfa stink bug (Argentina), is an important pest of soybean [*Glycine max* (L.) Merr.] wherever the crop is grown in warmer regions of the New World (Panizzi and Slansky 1985a, McPherson and McPherson 2000, Panizzi et al. 2000b), including, more recently, Cuba (Artabe and Martínez 2003). It also is a pest of other leguminous crops in South America including alfalfa, *Medicago sativa* L., and various forage legumes (Quintanilla et al. 1967–1968, Ochoa 1969, Fraga and Ochoa 1972, Alzugaray and Ribeiro 2000, Ribeiro et al. 2009, Zerbino and Alzugaray 2010, Zerbino et al. 2015).

Piezodorus guildinii, originally described from material collected on the island of St. Vincent in the Lesser Antilles in the Caribbean Sea (Westwood 1837, Froeschner 1988), currently ranges from Argentina and Brazil through Central America to the southern United States (Panizzi and Slansky 1985b, Panizzi et al. 2000b, Grazia et al. 2015, Panizzi 2015). Its range increased within Brazil in response to expanded soybean production (Kogan and Turnipseed 1987, Panizzi et al. 2012).

In South America, *Piezodorus guildinii* first was reported in soybean in Brazil in the early 1970s. Its distribution followed the great expansion of soybean production during that period, eventually becoming the most important pest of soybean in Brazil (Panizzi and Smith 1976a, Panizzi and Slansky 1985b, Panizzi et al. 2000b). It also is the most abundant stink bug pest of soybean in Uruguay and Argentina (Zerbino et al. 2016).

In North America, *Piezodorus guildinii* first was mentioned as a minor pest of soybean in the Florida Everglades in the early 1960s (Genung and Green 1962, Genung et al. 1964). Subsequently, Jones and Sullivan (1982, 1983) and Panizzi and Slansky (1985a) reported it in low numbers in southern South Carolina and in north central Florida, respectively, in the 1980s. In recent years, it appears to have suddenly expanded its range in the United States. However, unlike Brazil, this expansion cannot be attributed to expanded soybean production (see **Section 8.3.2**).

8.2 Taxonomy and Identification

8.2.1 Taxonomy

Piezodorus is a predominantly Old World genus with about a dozen valid species worldwide, some of which are difficult to separate (Leston 1953, Stadden and Ahmad 1995). Nearly half of these species impact crops (see discussion in **Section 8.6.2**). These include *P. guildinii*, the only New World member; *P. hybneri* (Gmelin) (= *P. rubrofasciatus* F.) in Asia and Africa; *P. lituratus* (F.) in Europe and western Asia; *P. oceanicus* (Montrouzier) in Australia; *P. punctiventris* (Dallas) in West Africa; and *P. purus* Stål in sub-Saharan Africa.

Piezodorus guildinii is a member of the *hybneri* group, which also includes *P. hybneri*, *P. oceanicus*, and *P. purus*; there have been questions about whether these actually compose different species (Distant 1902, Leston 1953, Stadden and Ahmad 1995). *P. guildinii* apparently is closely related to *P. hybneri* (Staddon and Ahmad 1995) and both share at least one pheromone component (Leal et al. 1998, Borges et al. 2007, Moraes et al. 2008, Endo et al. 2012). *P. oceanicus*, the redbanded shield bug, long thought to be a form of *P. hybneri*, was described as new (twice), as *P. grossi* Ahmad (Ahmad 1995) and *P. grossi* Staddon (Staddon 1997), before finally being recognized under an earlier described name, *P. oceanicus* (Cassis and Gross 2002).

8.2.2 Identification and Comparison with Other Stink Bugs

Piezodorus guildinii (**Figure 8.1B,C**) is a small pentatomid that often is green with a red band across the pronotum. It often is confused with other small green-colored stink bugs in North America, especially the redshouldered stink bug, *Thyanta custator accerra* McAtee, and other *Thyanta* spp. These species

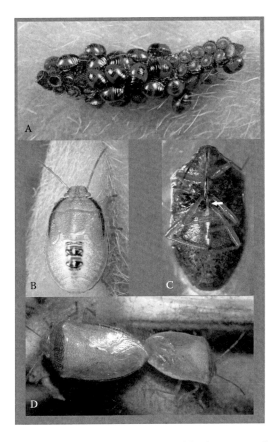

FIGURE 8.1 **(See color insert.)** *Piezodorus guildinii.* A, egg mass and first instars on soybean; B, fifth instar on soybean; C, ventral view of adult showing abdominal spine; D, mating pair. (Images A and B: courtesy of Ted C. MacRae; C: courtesy of Jeffrey A. Davis; D: courtesy of J. J. da Silva).

often occur together with *P. guildinii* on soybean in the southern United States. The initial appearance of *P. guildinii* in new areas in the United States was overlooked because of its resemblance to *Thyanta* spp. In soybeans in South America, it may be confused with other green-colored stink bugs such as *Thyanta* spp. and *Chinavia* spp. Despite similarities in color, *P. guildinii* may easily be separated from *Thyanta* spp. by the presence of a ventral abdominal spine, which reaches the coxae of the middle legs (**Figure 8.1C**); this spine is absent in *Thyanta* spp. (McPherson and McPherson 2000, Akin et al. 2007). *Chinavia* spp. also have a ventral abdominal spine, which usually is shorter than that of *P. guildinii.* Also, *Chinavia* spp. generally are larger than *P. guildinii* and lack the red band across the pronotum.

8.3 Distribution

Piezodorus guildinii is the only species in the genus that occurs in the New World. Distant's (1880–1893) treatise on Central American Heteroptera mentioned its occurrence in Guatemala, Cuba, and South America; Van Duzee (1907) in Jamaica; Kirkaldy (1909) in Grenada, Hispaniola, St. Vincent, and Mexico; and Wolcott (1941) in Puerto Rico. Its current range, excluding the Caribbean Islands listed above, is Argentina north through Mexico into Texas and then east through Missouri, Arkansas, and Louisiana to South Carolina and Florida (Barber 1914, Blatchley 1926, Panizzi and Slansky 1985b, Froeschner 1988, Baldwin 2004, Temple et al. 2013a). The original report of this species' occurrence in New Mexico by Van Duzee (1904) was not confirmed by Bundy (2012) in his list of Pentatomidae from the state.

8.3.1 Range Expansion in South America

The growth of agribusiness in the Neotropics in recent years has been dramatic. For example, in Brazil alone, soybean production increased 10-fold from 1960 to 1973 (Turnipseed and Kogan 1976). Production increased from 15.5 million tons in 1980 to about 97 million tons in 2016 (Anonymous 2016). A similar situation occurred in other countries such as Argentina and Paraguay. Under such dramatic ecosystem reformations via displacement of other crop and non-crop areas, changes in pest status of insects can be dramatic (Turnipseed 1973, Panizzi et al. 2012).

Much research has been conducted on pest pentatomids in South America because of their long history as soybean pests (Panizzi and Slansky 1985b; Panizzi et al. 2000b, 2012). *Piezodorus guildinii* seldom was seen in soybean in Brazil until the early 1970s (Panizzi et al. 2000b). As soybean cultivation expanded to the central, western, and northeastern parts of Brazil, this insect eventually became the most important soybean pest in that country.

During this great expansion of soybean cultivation in South America, *Nezara viridula* (L.) and *Piezodorus guildinii* were the dominant phytophagous stink bugs in soybean in Argentina (Vicentini and Jimenez 1977) and in southern Brazil (Kogan 1977). In Uruguay, *P. guildinii* first was observed on soybean in 1969 but was not considered a pest until 1981 (Bourokhovitch and Morey 1981, Gonnet 2007). It is now one of the primary pests of agricultural-pastoral production systems in Uruguay (Castiglioni 2004, Zerbino et al. 2016).

Beginning in the late 1970s, *Piezodorus guildinii* began to replace *Nezara viridula* as the dominant pest of soybean in some areas of Brazil (Panizzi et al. 1977, Panizzi 1985a, Panizzi and Slansky 1985b). As a smaller and more mobile pentatomid, *P. guildinii* apparently is better adapted to warmer climates than *N. viridula* and also seems to be more capable of colonizing soybean in Brazil early in the season (Panizzi and Smith 1976a, Heinrichs 1976, Panizzi 1985a). However, the reasons for this apparent species shift are not completely clear. In addition to mobility and early colonization of the crop, such factors as lower rates of parasitism and differential susceptibility to insecticides may be involved (Kogan and Turnipseed 1987). Direct seeding, changes in planting dates, and the use of early maturing varieties also may have contributed to the changing importance of pests.

8.3.2 Range Expansion in North America

Soybean production in the United States doubled from 1960 to 1973, with the greatest rate of increase in the southeastern states (Turnipseed and Kogan 1976). In 10 southern states, soybean production tripled to 6.5 million hectares while in the original production areas in Corn Belt states, production increased by 47% to 13.3 million hectares. Following the expansion of soybean cultivation in the southern states, a complex of stink bugs soon became a major factor in limiting yields from South Carolina to East Texas. They accounted for "more than $68 million in losses some years in the southern states from insecticide costs and crop damage" (McPherson et al. 1994).

The most common species in the stink bug complex for soybean in the United States typically have been the southern green stink bug, *Nezara viridula*; green stink bug, *Chinavia hilaris* (Say) [previously reported as *Acrosternum hilare* (Say)]; and brown stink bug, *Euschistus servus* (Say) (Miner 1966; Turnipseed 1972, 1973; Carner et al. 1974; Jones and Sullivan 1983; McPherson et al. 1993, 1994; Funderburk et al. 1999; Baur et al. 2000; Gore et al. 2006). Other minor species of phytophagous stink bugs often mentioned in soybean include other *Euschistus* spp. and *Thyanta* spp., plus several more for which soybean probably is not a developmental host. The minor species vary with geography. Jones and Sullivan (1983) is the only one of the earlier reports mentioning *Piezodorus guildinii* (in small numbers) in southern South Carolina in the early 1980s.

Although first mentioned as a pest in south Florida (Genung and Green 1962), *Piezodorus guildinii* probably has existed in peninsular Florida and Mexico as a long-term resident, breeding on native legumes such as indigo (*Indigofera* spp.) and clovers (*Trifolium* spp.) (Panizzi and Slansky 1985a,b). As with *Nezara viridula*, cold winters likely limit its northern boundaries in the United States (Jones and Sullivan 1981), but it is capable of expanding northward again following mild winters (Musolin 2007). In 2008, in Mississippi, yield loss and control costs due to stink bugs was estimated to be $29.8 million with *P. guildinii* being the most abundant species (Musser et al. 2009).

The reasons for the permanent appearance of *Piezodorus guildinii* as a major pest of soybean north and west of Florida in the United States are not clear. Published records fail to provide a picture of the gradual movement of this insect across the southern states. However, by the early 2000s, it was being reported as a soybean pest in several southern states where it had never been reported previously (Baur et al. 2010).

Menezes (1981) found a single *Piezodorus guildinii* female near Quincy in northwestern Florida ("panhandle") in 1977, and several more in 1978. By 1979, about 300 adults and nymphs were collected during the season. Subsequently, specimens were collected in several southwestern Georgia counties adjoining Florida with Menezes (1981) reporting that "This species is rapidly increasing in abundance" and warned that it had almost completely displaced *Nezara viridula* in Brazil. McPherson et al. (1993) demonstrated that *P. guildinii* was among four stink bug that composed 98% of recorded species from soybean during a 1987–1991 study in southeastern Georgia. However, a report of insect losses due to stink bugs in Georgia soybean for 1996 mentioned *N. viridula*, *Chinavia hilaris* (as *Acrosternum hilare*), and *Euschistus servus* but not *P. guildinii* (Riley et al. 1997). A survey of stink bugs in soybean in southeastern Texas by Drees and Rice (1990) during 1981–1983 did not yield *P. guildinii*, but a second survey by Vyavhare et al. (2014) during 2011–2013 found that *P. guildinii* now composed 65% of the total population of major species sampled. In 2007, in Mississippi, the reported stink bug complex included *N. viridula*, *C. hilaris* (as *A. hilare*), *E. servus*, *Thyanta* spp., and *P. guildinii* as the major pests of soybean, infesting every acre sampled and causing both yield and quality losses (Musser and Catchot 2008). In 2008, *P. guildinii* again was listed among the same species that were the most economically important insects attacking soybean in Mississippi and Tennessee (Musser et al. 2009).

Prior to 2000, *Piezodorus guildinii* had not been reported in Louisiana soybean, and specimens from Louisiana were not present in the Louisiana State Arthropod Museum (Davis et al. 2011, Temple et al. 2013a). It initially was identified in southern Louisiana during 2000 by crop consultants and Louisiana Cooperative Extension personnel (Baldwin 2004). Within two years, it exceeded the action threshold for stink bugs (9 per 25 sweeps), requiring insecticide applications on much of the soybean crop in southern Louisiana (Baldwin 2004). It quickly became the most devastating stink bug pest in Louisiana soybean production (Temple et al. 2013a). The sudden appearance and subsequent expansion across the state of Louisiana appears to be unique and is well documented, having been reported in all soybean producing regions in Louisiana by 2006 (Davis et al. 2011, Temple et al. 2013a).

Subsequently, *Piezodorus guildinii* migrated north from Louisiana to Arkansas and Missouri (Temple et al. 2013a). It was observed as a soybean pest in Arkansas for the first time in 2005, and breeding populations were seen in 2006 and 2007 (Smith et al. 2009). It was reported for the first time in Missouri in 2007 (Tindall and Fothergill 2011). By 2008, stink bugs had become the most economically important pest group attacking soybeans across the southern states from South Carolina to Texas, with *P. guildinii* the most important species (Musser et al. 2010).

8.4 Biology

8.4.1 Life History

Piezodorus guildinii has been a serious pest of soybean and leguminous pasture crops in South America for many years and, thus, there have been several studies on its biology and ecology (see references in Panizzi and Slansky 1985b, Panizzi et al. 2000b).

Most pentatomids, including *Piezodorus guildinii*, mate end-to-end (**Figure 8.1D**). Eggs of pentatomids usually are deposited in clusters (masses) composed of many rows (see **Chapter 1**). However, eggs of *P. guildinii* typically are deposited in characteristic double rows (e.g., Fraga and Ochoa 1972, Panizzi and Smith 1977, Grazia et al. 1980, Bundy and McPherson 2000, Galileo et al. 2007). Stink bugs that deposit eggs in two rows usually prefer longitudinal substrates rather than leaf surfaces (Silva and Panizzi 2008; Walker A. Jones and J. E. McPherson, personal observations). Panizzi and Smith (1977) reported that 60% of 500 egg masses were deposited on soybean pods, and that the mean number of eggs per mass was 15.1. Link et al. (1980) recorded that the number of eggs per mass for 542 masses ranged between 4 and 39 and averaged 17.5; they also reported that oviposition mainly was on pods and, additionally, that eggs were deposited over 30 cm above ground. Temple et al. (2016) reported that the number of eggs per mass ranged from 2 to 55 and averaged 16.6. Fraga and Ochoa (1972) reported 13–17 eggs per mass.

Recently, oviposition has been studied in relationship to soybean maturity group (MG), plant structure, and vertical distribution within the canopy (Temple et al. 2016). Plant structure by maturity group interactions resulted in more egg masses deposited on leaves in MG IV (79.4%) and more on pods in MG V (72.7%) (Temple et al. 2016). Only 29.9% of egg clusters in MG IV and 18.3% of egg clusters in MG V were deposited in the upper 35 cm of the soybean canopy.

Eggs typically are barrel-shaped and black with a light band around the middle and have an overall wooly or prickly appearance under low magnification (**Figure 8.1A**) (Bundy and McPherson 2000, Akin et al. 2007, Silva and Panizzi 2008, La Porta et al. 2013). The eggs of certain other *Piezodorus* spp. are similar (Hinton 1981, Candan 1998). As with most stink bug species, females of *P. guildinii* glue each egg to the substrate and to adjacent eggs, probably using the last tarsomere of the hind leg to help position each egg (Panizzi 2006). In *Nezara viridula*, the glue is produced by the follicular cells of the ovarioles, and a component of this secretion acts as a host recognition kairomone for the egg parasitoid, *Trissolcus basalis* (Wollaston) (Hymenoptera: Platygastridae) (Bin et al. 1993). *T. basalis* also has been reported as a parasitoid of *P. guildinii*, and the relationship between the glue and the parasitoid probably is similar in *P. guildinii*.

Panizzi and Smith (1977) were the first to detail the basic biological characters of *Piezodorus guildinii* in Brazil. Under controlled laboratory conditions (24°C; 80% relative humidity), the minimum preoviposition period was 22.6 days, with the total mean number of eggs deposited per female being 31.1 eggs with a maximum of 114. The egg incubation period averaged 7.5 days, and the first instars clustered atop the egg choria during the stadium (**Figure 8.1A**). The total nymphal stadia (hatching to adult) averaged 31.5 days.

Serra and La Porta (2001) studied development, longevity, and fecundity of the bugs when fed green beans, *Phaseolus vulgaris* L. (Fabaceae), in the laboratory. Nymphal mortality was approximately 60%, adult longevity ranged from 60 to 79 days, and the mean fecundity was 86.3 eggs per female.

Oliveira and Panizzi (2003) compared nymphal development and adult reproduction when the bugs were provided different growth stages of soybean [see description of soybean development in Fehr et al. (1971)]. Nymphs fed pods in the pod-filling stages (R5–R6) showed greater survivorship and shorter developmental time compared to those fed pods in earlier (R3–R4) or later (R7–R8) stages. Similarly, adults lived longer and had higher fecundity when fed soybean pods at R5–R6 compared to those fed R3-R4 or R7-R8 soybean pods.

Though bacterial endosymbionts are known to impact stink bugs (both positively and negatively), little information is available for symbionts of *Piezodorus guildinii*. Recent research indicates that *Enterobacter hormaechei* may be the primary symbiont of *P. guildinii* (Husseneder et al. 2017).

Piezodorus guildinii is multivoltine. It has been reported to complete five generations per year (on soybean and other crops) in Argentina (Fraga and Ochoa 1972, Panizzi and Slansky 1985b) and Brazil (Panizzi 1997) and at least three generations per year in Louisiana (one to two on clover, two on soybean) (Jeffrey A. Davis, unpublished data).

Abiotic factors can significantly affect the physiology and morphology of *Piezodorus guildinii*. Temperature and day length affect the development rate, longevity, fecundity, and egg viability (Zerbino et al. 2013). Generally, a long photophase with warmer temperatures positively increases biological responses. Zerbino (2014), in a laboratory study, found that under a short photophase and low temperature, *P. guildinii* adults accumulate lipid reserves, have undeveloped reproductive organs, are smaller in size, exhibit darker colored pronotal bands and connexiva, feed less frequently, and enter reproductive diapause. Zerbino et al. (2013) found that, in Colonia, Uruguay, *P. guildinii* did not reproduce unless soybean was available. Cividanes and Parra (1994) estimated that the number of generations of *P. guildinii* in Brazil during the soybean season was higher than that of *Nezara viridula* (i.e., 5.0 vs. 3.7 generations). Finally, Panizzi et al. (1980) and Costa and Link (1982) found that nymphs and adults of *P. guildinii*, respectively, were more mobile than those of *N. viridula*.

8.4.2 Overwintering

Few studies have been conducted to determine the overwintering habits of *Piezodorus guildinii* in the Americas. The most detailed studies have been conducted in South America, specifically Brazil and Uruguay. In Brazil, studies were conducted in northern Paraná state in Londrina (S 23° - W 50°) where five generations are completed per year. Three of the generations are completed on the soybean crop during spring/summer months. The pest then moves to other legume plants such as lanceleaf crotalaria, *Crotalaria lanceolata* E. Mey.; and pigeon pea, *Cajanus cajan* (L.), where it completes another generation. Finally, with the start of autumn and during the mild winter, the stink bug moves to indigo plants where it may complete a fifth generation, depending on the severity of low temperatures (Panizzi 1997). *P. guildinii* returns to soybean the following spring (Panizzi 1997). In cooler areas of southern Brazil, in Rio Grande do Sul state, during fall/winter it is found on (but not reproducing on) alternate host plants such as chickling pea, *Vicia sativa* L.; wild radish, *Raphanus sativus* L.; and white lupine, *Lupinus albus* L. (Silva et al. 2006). In Passo Fundo (S 28°- W 52°), *P. guildinii* also might be found in a state of reduced activity underneath crop residues and on the soil in areas with undisturbed vegetation (Antônio R. Panizzi, unpublished data).

Further south in Uruguay at Colonia (S 34°- W 57°), *Piezodorus guildinii* has three generations per year (Antônio R. Panizzi, personal communication). It reproduces on soybean and forage-cultivated legumes such as alfalfa, *Medicago sativa* L.; red clover, *Trifolium pratense* L.; and bird's-foot trefoil, *Lotus corniculatus* L., during spring-summer. As temperatures drop, they move to trees [e.g., *Pittosporum undulatum* Ventenat (Pittosporaceae) and *Ligustrum lucidum* Aiton (Oleaceae)] and bamboo, *Phyllostachys* sp. (Poaceae), looking for shelter. With the additional decrease in winter temperatures, adults move to eucalyptus leaf litter, *Eucalyptus* spp., where they stay protected until the following spring (Zerbino et al. 2016).

Zerbino et al. (2015) studied the temporal morphological and physiological changes of *Piezodorus guildinii* that occurred in the field on different host plants through the year. As with laboratory studies discussed above (Zerbino et al. 2014), they demonstrated that during autumn/winter adults accumulate lipid reserves, both females and males showed undeveloped reproductive organs and smaller body size; at this time, females showed darker coloration of the pronotum band and of the connexiva; these traits indicated reproductive diapause (see **Chapter 11** for discussion of diapause).

8.5 Host Plants

The plants associated with *Piezodorus guildinii* have been documented by Panizzi and Slansky (1985b) for the Americas and Smaniotto and Panizzi (2015) for the Neotropics. This list, updated to include all records found in the literature for the Americas, is summarized in **Table 8.1** and includes 75 plant species. The range of known hosts is smaller than that for *Nezara viridula* (see **Table 7.5** in **Chapter 7**), although, as with *N. viridula*, most hosts for *P. guildinii* also are in the legume family (Fabaceae). *Indigofera* spp. apparently are the most suitable hosts for *P. guildinii* (Panizzi and Slansky 1985a, Panizzi 1992). Panizzi and Smith (1977) referred to Monte (1937) as originally reporting nymphs and adults on *Crotalaria* spp. in Brazil. Quintanilla et al. (1967–1968) described *P. guildinii* morphology, damage, and host plants in Argentina. Panizzi (1985b) identified and described the effects of non-crop hosts on the biology of *P. guildinii*.

As discussed above, soybean is an important host for *Piezodorus guildinii* throughout the insect's range. In Argentina, Quintanilla et al. (1967–1968) also mentioned soybean as a host and referred to other early reports on this bug's distribution and host plants. Panizzi and Smith (1976a) and Heinrichs (1976) reported that *P. guildinii* was abundant in Brazilian soybean, entering the crop in low numbers from germination to flowering. When pods began to form, these insects began to increase in numbers and populations peaked when the pods were nearing maturity.

Although *Piezodorus guildinii* is a serious pest of soybean, the crop apparently is a poor reproductive host compared to preferred non-crop hosts. Females deposit as many as 37 egg masses on *Indigo* spp. compared to an average of three masses on soybean (Panizzi et al. 2000b). Fraga and Ochoa (1972),

TABLE 8.1

Plants Associated with *Piezodorus guildinii* (Westwood) in the Americas[1]

Scientific Name	Type of Association	Location	Reference
Amaranthaceae			
Hebanthe erianthos (Poiret) Pederson [syn. *Pfaffia paniculata* (Martius) Kuntz]	Incidental	Brazil	Ferreira and Panizzi 1982, Smaniotto and Panizzi 2015
Anacardiaceae			
Schinus terebinthifolius Raddi	Incidental	USA	Cassani 1986
Spondias mombin L.	Incidental	Costa Rica	Ballou 1937
Apiaceae (Umbelliferae)[2]			
Foeniculum vulgare Miller	Incidental	Brazil	Lopes et al. 1974, Link and Grazia 1987
Aquifoliaceae			
Ilex paraguariensis A. Saint-Hilaire	Incidental	Argentina	Quintanilla et al. 1981
	Incidental	Brazil	Chiaradia 2010
Asteraceae			
Bidens pilosa L.	Incidental	Brazil	Ferreira and Panizzi 1982
Gochnatia polymorpha (Lessing) Cabrera	Incidental	Brazil	Garlet et al. 2010
Helianthus annuus L.	Incidental	Brazil	Malaguido and Panizzi 1998
Bignoniaceae			
Adenocalymma comosum (Chamisso) de Candolle	Incidental	Brazil	Ferreira and Panizzi 1982
Pyrostegia venusta Miers	Incidental	Brazil	Ferreira and Panizzi 1982
Brassicaceae			
Brassica napus L.	Incidental	Brazil	Link and Grazia 1987
Raphanus sativus L.	Incidental	Brazil	Silva et al. 2006
Cactaceae			
Peireskia aculeata Miller	Incidental	Brazil	Link and Grazia 1987
Cucurbitaceae			
Sechium edule (Jacquin) Swartz	Incidental	Brazil	Lopes et al. 1974
Ericaceae			
Vaccinium corymbosum L.	Incidental	Argentina	Rocco and Greco 2011
Euphorbiaceae			
Ricinus communis L.	Incidental	Brazil	Ferreira and Panizzi 1982
Fabaceae			
Cajanus cajan (L.) Millspaugh	Reproductive Host	Brazil	Panizzi et al. 2000b
	Reproductive Host	Central America	Saunders et al. 1983
Chamaecrista nictitans (L.) Moench [syn. *Chamaecrista aeschinomene* (de Candolle ex Colladon) Greene]	Incidental	Puerto Rico	Wolcott 1936
Chamaecrista fasciculata (Michaux) Greene	Reproductive Host	USA	J. Eger, personal observation
Crotalaria brevidens Benth	Incidental	USA	Panizzi and Slansky 1985a

(Continued)

TABLE 8.1 (CONTINUED)

Plants Associated with *Piezodorus guildinii* (Westwood) in the Americas[1]

Scientific Name	Type of Association	Location	Reference
Crotalaria lanceolata E. Mey.	Reproductive Host	Brazil	Panizzi et al. 2002
	Reproductive Host	USA	Panizzi and Slansky 1985a
Crotalaria pallida Aiton	Incidental	Colombia	Hallman 1979
Crotalaria sp.	Reproductive Host	Brazil	Monte 1937
Desmodium intortum (Miller) Urban	Incidental	Brazil	Ferreira and Panizzi 1982
Desmodium tortuosum (Swartz) de Candolle	Reproductive Host	USA	J. Eger, personal observation
	Incidental	Colombia	Hallman 1979
Desmodium uncinatum (Jacquin) de Candolle	Incidental	Brazil	Link and Grazia 1987
Glycine max (L.) Merrill	Reproductive Host	Argentina	Quintanilla et al. 1967-68
	Reproductive Host	Bolivia	Ventura 1988
	Reproductive Host	Brazil	Silva et al. 1968
	Reproductive Host	Colombia	Hallman 1979
	Reproductive Host	Cuba	Artabe and Martínez 2003
	Reproductive Host	Ecuador	Stansly and Orellana M. 1990
	Reproductive Host	Uruguay	Bourokhovitch and Morey 1981
	Reproductive Host	USA	Genung and Green 1962
Indigofera endecaphylla Jacquin	Reproductive Host	Brazil	Panizzi 1992
Indigofera hirsuta L.	Reproductive Host	Brazil	Panizzi 1992
	Reproductive Host	Colombia	Hallman 1979
	Reproductive Host	USA	Panizzi and Slansky 1985a
Indigofera suffruticosa Miller	Reproductive Host	Brazil	Panizzi 1992
Indigofera truxillensis Kunth	Reproductive Host	Brazil	Panizzi 1992
Lablab purpureus (L.) Sweet (syn. *Dolichos lablab* L.)	Reproductive Host	Argentina	Fraga and Ochoa 1972
Lens culinaris Medikus	Reproductive Host	Brazil	Lopes et al. 1974
Lotononis bainesii Baker	Reproductive Host	Brazil	Lopes et al. 1974
Lotus corniculatus L.	Reproductive Host	Brazil	Lopes et al. 1974
	Reproductive Host	Uruguay	Alzugaray 1995
Lupinus albus L.	Reproductive Host	Brazil	Link and Grazia 1987
Lupinus angustifolius L.	Reproductive Host	Brazil	Link and Grazia 1987
Lupinus luteus L.	Reproductive Host	Brazil	Link and Grazia 1987
Macroptilium lathyroides (L.) Urban (syn. *Phaseolus lathyroides* L.)	Reproductive Host	Colombia	Hallman 1979
Medicago polymorpha L. [syn. *Medicago hispida* (Gaertner)]	Reproductive Host	Brazil	Lopes et al. 1974, Link and Grazia 1987
Medicago sativa L.	Reproductive Host	Argentina	Quintanilla et al. 1967-68
	Reproductive Host	Brazil	Zerbino et al. 2015
	Reproductive Host	Paraguay	Gomez et al. 2013
	Reproductive Host	Uruguay	Alzugaray 1995
Phaseolus lunatus L.	Reproductive Host	USA	Genung 1959
Phaseolus vulgaris L.	Reproductive Host	Brazil	Silva et al. 1968
	Reproductive Host	Cuba	Gonzalez et al. 2011
	Reproductive Host	USA	Genung 1959
Pisum sativum L.	Reproductive Host	Brazil	Lopes et al. 1974

(Continued)

TABLE 8.1 (CONTINUED)

Plants Associated with *Piezodorus guildinii* (Westwood) in the Americas[1]

Scientific Name	Type of Association	Location	Reference
Sesbania bispinosa (Jacquin) W. Wight [syn. *Sesbania aculeata* (Willdenow) Persoon]	Reproductive Host	Brazil	Panizzi 1985b
Trifolium pratense L.	Reproductive Host	Uruguay	Alzugaray 1995
	Reproductive Host	Argentina	Cingolani et al. 2014
	Reproductive Host	Brazil	Zerbino et al. 2015
Trifolium repens L.	Reproductive Host	Brazil	Lopes et al. 1974
	Reproductive Host	Uruguay	Alzugaray 1995
Vicia sp.	Reproductive Host	Brazil	Lopes et al. 1974
Vigna unguiculata (L.) Walpers	Reproductive Host	Brazil	Jackai and Daoust 1986
	Reproductive Host	Peru	Jackai and Daoust 1986
	Reproductive Host	USA	Genung 1959
Lauraceae			
Nectandra sp.	Incidental	Brazil	Ferreira and Panizzi 1982
Linaceae			
Linum usitatissimum L.	Incidental	Brazil	Link and Grazia 1987
Malpighiaceae			
Malpighia glabra L.	Incidental	Brazil	Albuquerque et al. 2002
Malvaceae			
Gossypium hirsutum L.	Incidental	Argentina	Quintanilla et al. 1967-68
	Incidental	Brazil	Silva et al. 1968
	Incidental	Puerto Rico	Wolcott 1941
Melastomataceae			
Miconea cinerascens Miquel	Incidental	Brazil	Garlet et al. 2010
Myrtaceae			
Eugenia uniflora L.	Incidental	Brazil	Costa et al. 1995
Myrciaria tenella (DC.) O. Berg	Incidental	Brazil	Costa et al. 1995
Psidium guajava L.	Incidental	Costa Rica	Ballou 1937
Nictaginaceae			
Bougainvillea glabra Choisy	Incidental	Brazil	Ferreira and Panizzi 1982
Oleaceae			
Ligustrum lucidum W. T. Aiton	Incidental	Brazil	Panizzi and Grazia 2001
	Incidental	Uruguay	Zerbino et al. 2015
Phytolaccaceae			
Phytolacca dioica L.	Incidental	Brazil	Lopes et al. 1974
Piperaceae			
Piper sp.	Incidental	Puerto Rico	Wolcott 1936
Pittosporaceae			
Pittosporum undulatum Ventenat	Incidental	Uruguay	Zerbino et al. 2015
Poaceae			
Oryza sativa L.	Incidental	Central America	Saunders et al. 1983
Phyllostachys sp.	Incidental	Uruguay	Zerbino et al. 2015

(Continued)

TABLE 8.1 (CONTINUED)

Plants Associated with *Piezodorus guildinii* (Westwood) in the Americas[1]

Scientific Name	Type of Association	Location	Reference
Schizachyrium condensatum (Kunth) Nees	Incidental	Argentina	Cánepa et al. 2015
Triticum aestivum L.	Incidental	Brazil	Cividanes et al. 1987
Rosaceae			
Fragaria × *ananassa* Duchesne	Incidental	Brazil	Link and Grazia 1987
Rubus spp.	Incidental	Brazil	Pasini and Lúcio 2014
Rubiaceae			
Coffea arabica L.	Incidental	Brazil	Silva et al. 1968
Sapindaceae			
Serjania fuscifolia Radlkofer	Incidental	Brazil	Ferreira and Panizzi 1982
Solanaceae			
Capsicum spp.	Incidental	Central America	Saunders et al. 1983
	Incidental	Puerto Rico	Wolcott 1936
Solanum mauritianum Scopoli	Incidental	Brazil	Garlet et al. 2010
Violaceae			
Anchietea pyrifolia A. Saint-Hilaire (syn. *Anchietea salutaris* A. Saint-Hilaire)	Incidental	Brazil	Ferreira and Panizzi 1982
Hybanthus atropurpureus (A. Saint-Hilaire) Taubert	Incidental	Brazil	Ferreira and Panizzi 1982

[1] Reproductive hosts are plants on which bugs complete development. Incidental indicates a collection or observation on a plant that probably does not support development.

[2] Current family name (original name reported in parentheses).

working in alfalfa, and Panizzi and Smith (1977), working in soybean, suggested that egg and nymphal development times generally were similar to those reported for other pentatomids (e.g., McPherson 1971, 1974; Owusu-Manu 1974; Munyaneza and McPherson 1994; Bharathimeena and Sudharma 2008; Parveen et al. 2015).

Panizzi et al. (2002) compared nymphal and adult performance between the uncultivated weed host lanceleaf crotalaria and soybean and found that the weed was more suitable for nymphal development, but both plants were equally suitable for adult reproduction. Zerbino et al. (2016) compared feeding performance of *Piezodorus guildinii* on soybean, alfalfa, bird's-foot trefoil, and red clover, the main cultivated host plants in Uruguay. Fitness was best when fed immature seeds of alfalfa and soybean and lowest on immature pods and seeds of bird's-foot trefoil and red clover, respectively. Saluso et al. (2010) reported that the life cycle of *P. guildinii* fed alfalfa was longer and adults weighed less than those fed soybean.

Piezodorus guildinii also feeds on several pasture legumes in parts of South America. In Argentina, producers of seed alfalfa in the Saldo River area have reported that this insect can completely destroy this crop (Fraga and Ochoa 1972). Fraga and Ochoa (1972) referred to earlier reports by Ochoa (1969) and Quintanilla et al. (1967–1968) that *P. guildinii* is a pest of alfalfa. Finally, in Argentina, *P. guildinii* and *Nezara viridula* together can destroy up to 90% of a seed alfalfa crop (Moschetti et al. 2007). In Uruguay, *P. guildinii* is a pest of other leguminous forage crops such as bird's-foot trefoil, red clover, and white clover, *Trifolium repens* L. (Alzugaray and Ribeiro 2000, Zerbino and Alzugaray 2010, Ribeiro et al. 2009, Zerbino et al. 2015).

In Puerto Rico, *Piezodorus guildinii* has been recorded on cotton, *Gossypium hirsutum* L. (Wolcott 1941), but this is an incidental record because it is not considered a pest of cotton.

8.6 Economic Impact

8.6.1 Feeding and Injury

Stink bug feeding on soybean pods reduces yield and quality, decreases seed weight, delays crop maturity, reduces seed oil content, and reduces germination of harvested seed (Miner 1966, Duncan and Walker 1968, Jensen and Newsom 1972, Todd and Turnipseed 1974, Todd 1976, Miller et al. 1977, McPherson et al. 1979, Panizzi et al. 1979, Russin et al. 1987, Brier and Rogers 1991). Soybean seed with severe stink bug injury does not have oil or meal value (Todd 1976, 1982), resulting in price dockage or sale rejection (McPherson et al. 1994).

Stink bugs, including *Piezodorus guildinii*, also are capable of transmitting bacteria (some of which are plant pathogens) to their hosts when feeding (Husseneder et al. 2017). The full impact of this transmission on plant injury and the range of pathogen species involved currently is unknown for *P. guildinii*.

Delayed soybean maturity, defined as soybeans that retain leaves and have green stems and/or green pods long after normal senescence (Boethel et al. 2000), is associated with stink bug injury. This syndrome results in delayed harvest and decreased quality of harvested soybeans. In some cases, this problem has prevented seed harvest and resulted in crop destruction. Harvesters are unable to efficiently process green stems as they become entangled in the threshing apparatus. This decreases threshing efficiency while increasing wear on machinery, reducing fuel efficiency, and limiting yields. Furthermore, green stems and pods retain moisture, which can then be transferred to harvested grain. High moisture will create postharvest spoilage, and producers are penalized when selling grain with high moisture content.

Historically, in the United States, delayed maturity has been reported in response to infestations of *Nezara viridula* and *Euschistus servus* in Arkansas, Georgia, and Louisiana (Daugherty et al. 1964, Duncan and Walker 1968, Todd and Turnipseed 1974, Boethel et al. 2000). Boethel et al. (2000) showed that *N. viridula* infestations at a density of six stink bugs per 0.3 meter of row for 7–14 d between R3 and R5.5 growth stages resulted in delayed maturity. More recently, Vyavhare et al. (2015) demonstrated that *Piezodorus guildinii* at densities of eight adults per 0.3 meter caused delayed maturity in United States soybeans.

In Brazil, delayed maturity has been associated with *Piezodorus guildinii* infestations in several reports. Panizzi et al. (1979) reported that soybeans retained leaves when *P. guildinii* infested soybean during pod development and pod fill stages (R3–R6). Populations of two to five *P. guildinii* adults per plant caused excessive green foliage retention when infested at the R4 stage (Costa and Link 1977), whereas continuous infestations of six to ten *P. guildinii* per meter during seed development (R4–R6) caused green foliage retention, but foliage retention did not occur in soybean infested prior to seed development (R1–R3) (Galileo and Heinrichs 1978). Sosa-Gómez and Moscardi (1995) investigated differences in leaf retention among different stink bug species and found that *P. guildinii* caused greater leaf retention than *N. viridula* and *Euschistus heros* (F.).

In the United States, *Nezara viridula* and *Chinavia hilaris* historically were considered the most damaging species in soybean compared to other species (Miner 1961, McPherson et al. 1979). However, this has changed with the arrival of *Piezodorus guildinii*. Preliminary data from caged experiments indicates *P. guildinii* is an aggressive soybean feeder, damaging 94% of pods, 79% of seeds, and reducing seed weight by 78% within 72 hr (Parker 2012). However, literature currently is not available comparing *P. guildinii* damage potential to *N. viridula* and *C. hilaris*.

Current action thresholds for most stink bugs in the United States range from 20 to 36 per 100 sweeps (Greene and Davis 2015). However, because *Piezodorus guildinii* damages soybeans quickly, is more tolerant of insecticides, recolonizes soybean fields faster, and appears to reproduce faster than other stink bug species, the current threshold for *P. guildinii* in Louisiana is 16 insects per 100 sweeps (Ring et al. 2015).

In South America, *Piezodorus guildinii* has the greatest potential to damage soybean (Vicentini and Jimenez 1977). In Brazil, Corrêa-Ferreira and Azevedo (2002) directly compared soybean yield loss and seed damage by *P. guildinii*, *Nezara viridula*, and *Euschistus heros* in the field and greenhouse. In field trials at four stink bugs per meter, yield differences were not detected among the species but *P. guildinii* damaged more seed. In greenhouse infestations at two stink bugs per plant, lower yields occurred in plants infested with *P. guildinii*. Feeding by *P. guildinii* resulted in significantly more empty pods than the other stink bug species (Corrêa-Ferreira and Azevedo 2002). *P. guildinii* has been shown to be more damaging than *N. viridula* and *Euschistus heros* due to deeper stylet penetration into seeds, greater enzymatic activity of saliva, and larger food and salivary canals (Depieri and Panizzi 2011).

8.6.2 Impact of Other *Piezodorus* spp.

In addition to *Piezodorus guildinii*, at least five species are crop pests: *P. hybneri* (Gmelin) (= *P. rubro-fasciatus* F.) in Asia and Africa, *P. lituratus* (F.) in Europe and western Asia, *P. oceanicus* (Montrouzier) in Australia, *P. punctiventris* (Dallas) in West Africa, and *P. purus* Stål in sub-Saharan Africa. The one-banded stink bug, *P. hybneri*, is a major pest of soybean and other leguminous crops in Asia and Africa (Ishihara 1950, Joseph 1953, Gentry 1965, Kobayashi et al. 1972, Gill 1987, Singh et al. 1989, Tengkano et al. 1991, Higuchi 1993, Dwomoh et al. 2008, Srivastava and Srivastava 2012, Bayu and Tengkano 2014). It is difficult to control because of its frequent movement (Ishikura et al. 1955). This situation has led to many studies on alternative management strategies in Japan (Higuchi 1993, Kikuchi et al. 1995), Indonesia (Tengkano et al. 1991, Bayu and Tengkano 2014), and Malaysia (van den Berg et al. 1995).

Piezodorus lituratus, the gorse shield bug, a common insect in Europe and western Asia, is a pest of soybean in Italy (Zandigiacomo 1990, 1992) and a minor pest of hazelnut in Turkey (Tuncer et al. 2014). Blatchley (1926) reported it from Florida, but Van Duzee (1904) doubted that this species occurred in North America. Froeschner (1988) agreed with this latter opinion. *P. oceanicus*, the redbanded shield bug, is a major pest of soybean and other legume crops in Australia (Gross 1976; Evans 1985; Bailey 2007; Brier 2007, 2010). *P. punctiventris* is a pest of soybean; okra, *Abelmoschus esculentus* (L.) Moench; and other vegetables in Nigeria (Ezueh and Dina 1979, Akinlosotu 1983, Jackai et al. 1998). *P. purus* has been implicated in transmitting a bacterial disease in cotton in southern Africa (Pearson 1934). It also has been reported as an occasional serious nuisance in South Africa, with hordes of bugs migrating from the field and descending upon homes, resulting in occupants having to vacate (Haines 1935, Sithole and Oelofose 2006).

8.7 Management

8.7.1 Monitoring

In North America, the most common protocol for monitoring stink bugs in soybean relies on the sweep net (38 cm diam) that measures infestations in the upper canopy. The number of stink bugs per 100 sweeps is then recorded. Once thresholds have been met, control measures are initiated. Though alternative methods such as fumigation and ground cloth exist to sample populations, the sweep net is the most efficient (Kogan and Pitre 1980). Stink bugs disperse in clumped patterns resulting in diverse densities within a field (Kogan and Pitre 1980). To maintain precision, the sampling distribution should include sampling across the entire field, which the sweep net can accomplish in a short period of time. In addition, Russin et al. (1987) demonstrated that a complex of stink bugs (*Nezara viridula*, *Chinavia hilaris*, and *Euschistus servus*) preferred to feed on pods in the upper half of the plant canopy, increasing the chance of detection by sweep net.

For *Piezodorus guildinii* in the United States, damaging populations in soybean will occur from growth stage R4 to R7 (Temple et al. 2013a). Furthermore, the earlier the variety matures, the more likely that heavier population densities will increase (Temple et al. 2013a). With limited alternative hosts available during the summer months, populations remain concentrated in soybean fields where the bugs are capable of quickly building to economically important levels. Thus, monitoring should begin before R3 (pod initiation) in order to track population growth. Finally, because *P. guildinii* has the propensity to develop large numbers (3 to 5 times action threshold) quickly in soybean (Temple et al. 2011), recommendations are to monitor every 5 days.

In South America, the ground cloth or beat cloth method is the most popular method of sampling stink bugs on soybean. Originally developed by Boyer and Dumas (1963), it was improved by Shepard et al. (1974) and, later, by Drees and Rice (1985) who introduced the vertical beat sheet (VBS). Today, use of the VBS is widespread in South America to sample stink bugs, including *Piezodorus guildinii*, in soybean fields (see review by Corrêa-Ferreira 2012). In general, 10 randomly allocated samples are taken per hectare to estimate the stink bug population; when populations reach two adults or late instar nymphs per meter, control measures should be taken to avoid economic grain yield reduction. In the case of soybean fields for seed production, one adult or late instar nymph per meter justifies control measures (Panizzi et al. 2012).

8.7.2 Cultural Control

Producers can manipulate agronomic practices to avoid stink bug injury in soybean. By managing crop maturity through planting of early maturing soybean varieties, producers can manipulate crop phenology to have pod production and pod fill occurring when stink bugs are at their lowest numbers. Earlier in the season, stink bugs are fewer in number and widely dispersed in the landscape. Conversely, during the latter part of the production season, late-maturing soybeans concentrate stink bug populations as available host acreage diminishes. In the southern United States, soybean producers adopted an early soybean production system, planting maturity group IV and V instead of VI and VII to allay late season drought stress and stink bug pressure (Baur et al. 2000). This worked well until the arrival of *Piezodorus guildinii*. Because *P. guildinii* reproduces primarily on legumes, earlier maturing soybeans concentrate populations of this insect and, thus, insecticide applications for stink bugs have increased (Temple et al. 2013b).

Another cultural control tactic available to producers is trap crops. Trap crops concentrate stink bugs in a confined area, reducing their movement and total acreage to be sprayed (Hokkanen 1991, Todd et al. 1994, Shelton and Badenes-Perez 2006). McPherson and Newsom (1984) confined 70 to 85% of the *Nezara viridula* population to soybean trap crops using only 10% of the total acreage, thus reducing stink bugs in the main crop to 0.1 per meter.

In South American soybean, the trap crop concept developed by Newsom and Herzog (1977) in the United States for soybean pests was introduced and tested and results were positive for managing stink bugs (Panizzi 1980). Although trap crops are effective, grower acceptance is low because trap crops usually are insect species-specific, often are planted at times different from those of the main crop, and are not harvestable (Shelton and Badenes-Perez 2006).

A final cultural tactic combines the understanding of stink bug movement and the use of insecticides. Field colonization behavior of many stink bugs is known to be aggregated (e.g., Todd and Herzog 1980, Venugopal et al. 2014). Therefore, site-specific targeting of insecticide applications is possible, particularly for aggregations of stink bugs within field margins. This could reduce pesticide applications, saving growers money while conserving natural enemies. Davis et al. (2011) conducted small plot (0.5 acre) field experiments to test this hypothesis. Treatments included plots that never were sprayed, plots that had the entire area treated (100% of acreage treated), and plots that had only the four rows on either side treated (25% of acreage treated). Three insecticide applications occurred from R5 (pod filling) to R7 (mature pods). Perimeter insecticide applications reduced *Piezodorus guildinii* plot colonization by two weeks compared to the untreated plots. Treating only the perimeter also kept *P. guildinii* below the action threshold (16 stink bugs per 100 sweeps) (Davis et al. 2011).

8.7.3 Biological Control

Egg parasitoids are important biological control agents that cause natural mortality of stink bug eggs in soybean (Yeargan 1979, Orr et al. 1986, Koppel et al. 2009). Most egg parasitoids of stink bugs belong to the genera *Trissolcus* and *Telenomus* (Hymenoptera: Platygastridae) (Johnson 1984). In Brazil, *Telenomus podisi* Ashmead (previously reported as *T. mormideae* Costa Lima) parasitized 27% of *Piezodorus guildinii* egg masses (Panizzi and Smith 1976b); more recently, over 50% of *P. guildinii* and *Nezara viridula* egg masses were parasitized by *T. podisi* (Corrêa-Ferreira and Moscardi 1994). In Uruguay, Castiglioni et al. (2010) found the following species of egg parasitoids of *P. guildinii* on soybean and on forage legumes: *Telenomus podisi*, *Trissolcus basalis* (Wollaston), *Trissolcus brochymenae* (Ashmead), *Trissolcus urichi* (Crawford), and *Trissolcus teretis* (Johnson), with *T. podisi* most common. In Argentina, Cingolani et al. (2014) reported *T. podisi*, *T. basalis*, and *T. urichi* as egg parasitoids of *P. guildinii* on soybean and alfalfa fields, but they did not occur on egg masses collected from red clover. In the United States on soybean, *T. podisi* also is the most common egg parasitoid of stink bugs (Yeargan 1979, Orr et al. 1986, Koppel et al. 2009). These studies were conducted before *P. guildinii* became an important pest of soybeans in the United States (Temple et al. 2013a). However, at this time, *T. podisi*, *T. basalis*, and *Gryon* sp. have been shown to parasitize eggs of *P. guildinii* in Florida (Buschman and Whitcomb 1980, Temerak and Whitcomb 1984).

Tachinid flies also have been reported as parasitoids of *Piezodorus guildinii*, though they do not appear to have the impact of egg parasitoids (Panizzi and Smith 1976b). In Brazil, Panizzi and Smith (1976b) recovered low levels of *Eutrichopodopsis nitens* (Blanchard) from the bugs on soybean. In Florida, also on soybean, *Trichopoda pennipes* (F.) and *Euthera tentatrix* Loew have been reported from *P. guildinii* (Buschman and Whitcomb 1980, Panizzi and Slansky 1985c, McPherson and McPherson 2000).

Entomogenous nematodes have been utilized as insect biological control agents (Smart 1995), and some research has examined the susceptibility of stink bugs to these natural enemies. *Nezara viridula* has been reported as being susceptible to Steinernematidae under laboratory conditions (Wassink and Poinar 1984) and has been found infested with *Pentatomimermis* spp. in Russia and India (Rubtsov 1977, Bhatnagar et al. 1985). Field-level nematode infections of *Hexamermis* spp. have been found in *Chinavia hilaris* in Louisiana (Kamminga et al. 2012) and *Hexamermis* or *Mermis* spp. have been found in *Piezodorus guildinii* in Louisiana (Kamminga et al. 2012) and Uruguay (Ribeiro and Castiglioni 2009). Although nematode populations can be found infesting stink bugs (Esquivel 2011), infections are rare and do not appear to reduce field populations of stink bugs. A survey in Louisiana of parasitoids infesting *N. viridula* in soybean and clover indicated that only about 2% were infected with nematodes (Fuxa et al. 2000).

Predators have received little attention in biological control efforts, but native predatory species may be important in various cropping systems. Tillman et al. (2015) analyzed the gut contents of various predators collected from soybean and cotton fields in Georgia for the presence of DNA from *Piezodorus guildinii*. The following species tested positive: *Geocoris* spp. (Geocoridae); *Orius insidiosus* (Say) (Anthocoridae); *Hippodamia convergens* Guérin-Méneville, *Harmonia axyridis* (Pallas), *Scymnus* sp. (Coccinellidae); *Solenopsis invicta* Buren (Formicidae); *Mecaphesa asperata* (Hentz) (Thomisidae); and *Oxyopes salticus* (Hentz) (Oxyopidae).

8.7.4 Chemical Control

Chemical control is the primary management tactic used to manage *Piezodorus guildinii* in soybean (Temple et al. 2013b). Prior to *P. guildinii* becoming a soybean pest in the United States, it was relatively easy to manage stink bugs with low to medium rates of pyrethroids or organophosphates (Willrich et al. 2000). Unfortunately, *P. guildinii* is less sensitive to commonly used insecticides in comparison to other stink bug species (Temple et al. 2013b). Therefore, farmers spray more frequently to manage *P. guildinii* infestations (Davis et al. 2011). The average number of insecticide applications in Louisiana soybean has increased from one to two per season to three to five per season (Temple et al. 2013b). Likewise, increased insecticide applications to soybean have been reported in Texas due to an increase in *P. guildinii* abundance (Vyavhare et al. 2014).

Baur et al. (2010) reported baseline toxicity data for Louisiana populations of *Piezodorus guildinii* to several insecticides including acephate, cypermethrin, and methamidophos in glass vial bioassays. This work also demonstrated initial field efficacy data for control of *P. guildinii* in soybean, but made no direct comparisons to other stink bugs. Control of *P. guildinii* with all insecticides was estimated at approximately 50 to 80% in those field tested. Temple et al. (2013b) conducted insecticide field efficacy experiments comparing *Nezara viridula* and *P. guildinii* control. They found that labeled rates of pyrethroids provided 94% and 75% control, organophosphates 90% and 85% control, and neonicotinoids 78% and 63% control, for *N. viridula* and *P. guildinii*, respectively (Temple et al. 2013b). *P. guildinii* was four to eight-fold less susceptible to pyrethroids and two to eight-fold less susceptible to organophosphates than *N. viridula* (Temple et al. 2013b).

In South American soybean, chemical control of stink bugs also has increased in recent years. The frequency of application has increased from one to two applications per season to four to five applications. In most cases, insecticides are applied in combination with preplanting and postplanting herbicides used to eliminate weeds and also in combination with fungicides primarily used to control soybean rust (*Phakopsora* spp.) (Bueno et al. 2015). The most common insecticides used to control *Piezodorus guildinii* are neonicotinoids combined with pyrethroids (Baur et al. 2010). In Brazil, Farias et al. (2006) reported greater than 80% control with thiamethoxam + lamda-cyhalothrin at the rates of 21.1+15.9 and 28.2+21.2 g of active ingredient/ hectare. Similar results were reported in Argentina for several species of stink bugs on soybean (Gamundi et al. 2007). Other insecticides such as acephate and endosulfan, commonly used in the past, have been abandoned. Stadler et al. (2006) in Argentina reported resistant population of *P. guildinii* to endosulfan.

8.8 Future Outlook

Stink bugs in general are becoming a growing concern worldwide as pests. The following factors have been identified to explain their success: they are highly polyphagous; a reduction in insecticide applications in cotton for boll weevil control, which had secondarily helped suppress stink bug populations; development of species-specific insecticides for heliothine control in cotton; lack of transgenic products to control insects with piercing-sucking mode of feeding; greater ability to survive under unfavorable conditions; and innate ability to adapt easily to new environments. Additionally, stink bugs are taking advantage of rising temperatures worldwide, new cultivation systems (no tillage), early soybean production systems, multiple cropping systems, and increased international trade of agricultural commodities among countries (Panizzi 2015). For *Piezodorus guildinii*, in particular, these and other factors, such as interspecific competition (Tuelher et al. 2016), surely are favoring the recent growth in its populations in several areas of the Neotropical Region and the southern United States.

Through the years, several studies have indicated that *Piezodorus guildinii* is the most damaging among the many species of pentatomids known to colonize soybean in the Americas (Vicentini and Jimenez 1977, Panizzi et al. 1979, Sosa-Gómez and Moscardi 1995, Corrêa-Ferreira and Azevedo 2002, Depieri and Panizzi 2011, Tuelher et al. 2016). Despite these many studies, there still is need of detailed work to better explain the feeding and resulting damage of *P. guildinii* to soybean. Worthy of note is the recently published study conducted by Lucini et al. (2016) on electronically monitoring the feeding behavior of *P. guildinii* on soybean using the EPG (Electrical Penetration Graph). They found that adult bugs use the cell rupture strategy to feed from the seed endosperm in the soybean pod and switch to salivary sheath feeding for xylem in leaves and stems. These feeding behaviors explain symptoms of damage to soybean. Moreover, the elucidation of the specific feeding sites opens up opportunities to develop transgenic soybean varieties expressing toxins or blocking proteins that might inhibit the action of the salivary enzymes in these sites to avoid or mitigate *P. guildinii* damage.

In conclusion, considering the growing importance of *Piezodorus guildinii* as a pest of soybean in the Americas, additional field and laboratory studies are needed to advance our knowledge about this insect to help design more effective management programs to mitigate its impact as a pest.

8.9 Acknowledgments

We thank J. E. McPherson (Department of Zoology, Southern Illinois University, Carbondale) for his efforts in the final editing of this chapter, and J. J. da Silva (Embrapa Soja, Brazil) and Ted C. MacRae (Monsanto, Chesterfield, Missouri) for the images of *Piezodorus guildinii* used in this chapter. We also thank Tiago Lucini (Embrapa Trigo, Brazil) for help in tracking down some obscure references.

8.10 References Cited

Ahmad, I. 1995. A review of pentatomine legume bug genus *Piezodorus* Fieber (Hemiptera: Pentatomidae; Pentatominae) with its cladistic analysis. Proceedings of the Pakistan Congress of Zoology 15: 329–358.

Akin, S., J. Phillips, and D. T. Johnson. 2007. Biology, identification and management of the redbanded stink bug. University of Arkansas Cooperative Extension Service Printing Services. FSA7078-PD-8-11N. 4 pp.

Akinlosotu, T. A. 1983. Destructive and beneficial insects associated with vegetables in south-western Nigeria. Acta Horticulturae 123: 217–130.

Albuquerque, F. A. de, F. C. Pattaro, L. M. Borges, R. S. Lima, and A. V. Zabini. 2002. Insetos associados a cultura da aceroleira (*Malpighia glabra* L.) na regiao de Maringa, Estado do Parana. Acta Scientiarum 24: 1245–124.

Alzugaray, R. 1995. Seguimiento de poblaciones de insectos en semilleros de leguminosas forrajeras, pp. 57–75. *In* D. F. Risso, E. J. Berretta, and A. Morón (Eds.), Producción y manejo de pasturas, Instituto Nacional de Investigación Agropecuaria, Tacuarembó, Uruguay. 246 pp.

Alzugaray, R., and A. Ribeiro. 2000. Insectos en pasturas, pp. 13–30. *In* M. S. Zerbino and A. Ribeiro (Eds.), Manejo de plagas en pasturas y cultivos. Instituto Nacional de Investigación Agropecuaria, Serie Técnica 112: 1–106.

Anonymous. 2016. Conab companhia nacional de abastecimento. http://www.conab.gov.br (accessed 10 June 2016).

Artabe, L. M., and M. A. Martínez. 2003. Ocurrencia de Heteropteros en agroecosistemas Cubanos de soya (*Glycine max* (L.) Merril) [sic]. Revista Protección Vegetale 18: 98–103.

Bailey, P. T. 2007. Pests of field crops and pastures: identification and control. Commonwealth Science and Industrial Research Organisation Publishing. Victoria, Australia. 520 pp.

Baldwin, J. 2004. Stubborn new stink bug threatens Louisiana soybean. Louisiana Agriculture 47: 4.

Ballou, C. V. 1937. Insect notes from Costa Rica in 1936. Insect Pest Survey Bulletin 17: 483–529.

Barber. H. G. 1914. Insects of Florida. II. Hemiptera. Bulletin of the American Museum of Natural History 33: 495–535.

Baur, M. E., D. J. Boethel, M. L. Boyd, G. R. Bowers, M. O. Way, L. G. Heatherly, J. Rabb, and L. Ashlock. 2000. Arthropod populations in early soybean production systems in the Mid-South. Environmental Entomology 29: 312–328.

Baur, M. E., D. R. Sosa-Gomez, J. Ottea, B. R. Leonard, I. C. Corso, J. J. da Silva, J. Temple, and D. J. Boethel. 2010. Susceptibility to insecticides used for control of *Piezodorus guildinii* (Heteroptera: Pentatomidae) in the United States and Brazil. Journal of Economic Entomology 103: 869–876.

Bayu, M. S. Y. I., and W. Tengkano. 2014. Endemik kepik hijau pucat, *Piezodorus hybneri* Gmelin (Hemiptera: Pentatomidae) dan pengendaliannya. Buletin Palawija 28: 73–83.

Bergroth, E. 1891. Contributions a l'etude des Pentatomides. Revue d'Entomologie 10: 200–235.

Bharathimeena, T., and K. Sudharma. 2008. Biological studies on the southern green stink bug, *Nezara viridula* (L.) and the smaller stink bug, *Piezodoms rubrofasdatus* [sic] (F.) (Pentatomidae: Hemiptera) infesting vegetable cowpea. Pest Management in Horticultural Ecosystems 14: 30–36.

Bhatnagar, V. C., C. S. Pawar, D. R. Jadhav, and J. C. Davies. 1985. Mermithid nematodes as parasites of *Heliothis* spp. and other crop pests in Andhra Pradesh, India. Proceedings of the Indian Academy of Sciences, Animal Sciences 94: 509–515.

Bin, F., S. B. Vinson, M. R. Strand, S. Colazza, and W. A. Jones. 1993. Source of an egg kairomone for *Trissolcus basalis*, a parasitoid of *Nezara viridula*. Physiological Entomology 18: 7–15.

Blatchley, W. S. 1926. Heteroptera or true bugs of Eastern North America with especial reference to the faunas of Indiana and Florida. Nature Publishing Company, Indianapolis. 1116 pp.

Boethel, D. J., J. S. Russin, A. T. Wier, M. B. Layton, J. S. Mink, and M. L. Boyde. 2000. Delayed maturity associated with southern green stink bug (Heteroptera: Pentatomidae) injury at various soybean phenological stages. Journal of Economic Entomology 93: 707–712.

Borges, M., J. G. Millar, R. A. Laumann, and M. C. B. Moraes. 2007. A male-produced sex pheromone from the Neotropical redbanded stink bug, *Piezodorus guildinii* (W.). Journal of Chemical Ecology 33: 1235–1248.

Bourokhovitch, M., and C. Morey. 1981. Aspectos sanitarios del cultivo de la soja. Revista de la Asociación de Ingenieros Agrónomos, Uruguay, 20: 9–18.

Boyer, W. P., and W. A. Dumas. 1963. Soybean insect survey as used in Arkansas. Cooperative Economic Insect Report 13: 91–92.

Brier, H. 2007. Pulses — summer (including peanuts), pp. 169–258. *In* P. T. Bailey (Ed.), Pests of field crops and pastures: identification and control. Commonwealth Science and Industrial Research Organisation Publishing. Victoria, Australia. 528 pp.

Brier, H. 2010. Integrated pest management in Australian mungbeans and soybeans — swings, roundabouts and conundrums, pp. 1–10. *In* Proceedings of the 1st Australian Summer Grains Conference, 21st–24th June 2010, Gold Coast, Australia. The State of Queensland, Department of Employment, Economic Development and Innovation.

Brier, H. B., and D. J. Rogers. 1991. Susceptibility of soybeans to damage by *Nezara viridula* (L.) (Hemiptera: Pentatomidae) and *Riptortus serripes* (F.) (Hemiptera: Alydidae) during three stages of pod development. Journal of the Australian Entomological Society 30: 123–128.

Bueno, A. F., B. S. Corrêa-Ferreira, S. Roggia, and R. Bianco. 2015. Silenciosos e daninhos. Cultivar, Pelotas, RS, Brazil 196: 25–27.

Bundy, C. S. 2012. An annotated checklist of the stink bugs (Heteroptera: Pentatomidae) of New Mexico. The Great Lakes Entomologist 45: 196–209.

Bundy, C. S., and R. M. McPherson. 2000. Morphological examination of stink bug (Heteroptera: Pentatomidae) eggs on cotton and soybeans, with a key to genera. Annals of the Entomological Society of America 93: 616–624.

Buschman, L. L., and W. H. Whitcomb. 1980. Parasites of *Nezara viridula* (Hemiptera: Pentatomidae) and other Hemiptera in Florida. Florida Entomologist 63: 154–162.

Candan, S. 1998. External morphology of the eggs of *Piezodorus lituratus* (L.). Türkiye Entomoloji Dergisi 22: 307–313.

Cánepa, M. E., G. A. Montero, and I. M. Barberis. 2015. Matas de gramíneas como refugios de artrópodos invernantes en agriecosistemas pampeanos: efectos del tamaño, del agrupamiento y de la arquitectura de las plantas. Ecología Austral 25: 119–127.

Carner, G. R., M. Shepard, and S. G. Turnipseed. 1974. Seasonal abundance of insect pests of soybeans. Journal of Economic Entomology 67: 487–493.

Cassani, J. R. 1986. Arthropods on Brazilian peppertree, *Schinus terebinthifolious* (Anacardiaceae), in South Florida. Florida Entomologist 69: 184–196.

Cassis, G., and G. G. Gross. 2002. Zoological catalogue of Australia. Commonwealth Science and Industrial Research Organisation Publishing. Victoria, Australia. 737 pp.

Castiglioni, E. 2004. La soja avanza sobre el paisaje y la chinche avanza sobre la soja. Changué 26: 2–6.

Castiglioni, E., A. Ribeiro, R. Alzugaray, H. Silva, I. Ávila, and M. Loiácono. 2010. Prospección de parasitoides de huevos de *Piezodorus guildinii* (Westwood) (Hemiptera: Pentatomidae) em el litoral oeste de Uruguay. Agrociencia Uruguay 14: 22–25.

Chiaradia, L. A. 2010. Artropodofauna associada a erva-mate em Chapeco, SC. Revista de Ciências Agroveterinárias 9: 134–142.

Cingolani, M. F., N. M. Greco, and G. G. Liljesthröm. 2014. Egg parasitism of *Piezodorus guildinii* and *Nezara viridula* (Hemiptera: Pentatomidae) in soybean, alfalfa and red clover. Revista de la Facultad de Ciencias Agrarias 46: 15–27.

Cividanes, F. J., and J. R. P. Parra. 1994. Zoneamento ecológico de *Nezara viridula* (L.), *Piezodorus guildinii* (West.) e *Euschistus heros* (Fabr.) (Heteroptera: Pentatomidae) em quatro estados produtores de soja do Brasil. Anais da Sociedade Entomológica do Brasil 23: 219–226.

Cividanes, F. J., L. H. Silvestre, and M. J. Thomazini. 1987. Levantamento populacional de insetos na cultura de trigo. [Population survey of insects on wheat crop]. Semina Londrina 8: 14–16.

Corrêa-Ferreira, B. S. 2012. Amostragem de pragas da soja, pp. 631–672. *In* C. B. Hoffmann-Campo, B. S. Corrêa-Ferreira, and F. Moscardi. (Eds.), Soja – Manejo integrado de insetos e outros artrópodes-praga. Embrapa, Brasília, DF, Brazil. 859 pp.

Corrêa-Ferreira, B. S., and J. Azevedo. 2002. Soybean seed damage by different species of stink bugs. Agricultural and Forestry Entomology 4: 145–150.

Corrêa-Ferreira, B. S., and F. Moscardi. 1994. Temperature effect on the biology and reproductive performance of the egg parasitoid *Trissolcus basalis* (Woll.). Anais da Sociedade Entomológica do Brasil 23: 399–408.

Costa, E. C., and D. Link. 1977. Danos causados por algumas espécies de Pentatomidae em duas variedades de soja. Revista do Centro de Ciências Rurais 7: 199–206.

Costa, E. C., and D. Link. 1982. Dispersão de adultos de *Piezodorus guildinii* e *Nezara viridula* (Hemiptera: Pentatomidae) em soja. Revista do Centro de Ciências Rurais 12: 51–57.

Costa, E. C., P. C. Bogorni, and V. H. Bellomo. 1995. Percevejos coletados em copas de diferentes espécies florestais. Pentatomidae-1. Revista Ciências Florestais 5: 123–128.

Daugherty, D. M., M. H. Neustadt, C. W. Gehrke, L. E. Cavanah, L. F. Williams, and D. E. Green. 1964. An evaluation of damage to soybeans by brown and green stink bugs. Journal of Economic Entomology 57: 719–722.

Davis, J. A., K. L. Kamminga, A. R. Richter, and B. R. Leonard. 2011. New integrated pest management strategies for stink bug control in Louisiana soybean. Louisiana Agriculture, Spring 2011: 26–27.

Depieri, R. A., and A. R. Panizzi. 2011. Duration of feeding and superficial and in-depth damage to soybean seed by selected species of stink bugs (Heteroptera: Pentatomidae). Neotropical Entomology 40: 197–203.

Distant, W. L. 1880–1893. Insecta. Rhynchota. Hemiptera-Heteroptera. Volume I. *In* F. D. Goodman and O. Salvin (Eds.), Biologia Centrali-Americana. London. i–xx + 462 pp., 39 plates.

Distant, W. L. 1902. Fauna of British India, including Ceylon and Burma. Rhynchota. Volume I (Heteroptera). Taylor and Francis, London. 193 pp.

Drees, B. M., and M. E. Rice. 1985. Vertical beet sheet: a new device for sampling soybean insects. Journal of Economic Entomology 78: 1507–1510.

Drees, B. M., and M. E. Rice. 1990. Population dynamics and seasonal occurrence of soybean insect pests in southeastern Texas. Southwestern Entomologist 15: 49–56.

Duncan, R. G., and J. R. Walker. 1968. Some effects of the southern green stink bug on soybeans. Louisiana Agriculture 12: 10–11.

Dwomoh, E. A., J. B. Ackonor, and J. V. K. Afun. 2008. Survey of insect species associated with cashew (*Anacardium occidentale* Linn.) and their distribution in Ghana. African Journal of Agricultural Research 3: 205–214.

Endo, N., T. Yasuda, T. Wada, and S. Muto. 2012. Age-related and individual variation in male *Piezodorus hybneri* (Heteroptera: Pentatomidae) pheromones. Psyche 2012, 4 pp. http://dx.doi.org/10.1155/2012/609572.

Esquivel, J. F. 2011. *Euschistus servus* (Say) – a new host record for Mermithidae (Mermithida). Southwestern Entomologist 36: 207–211.

Evans, M. L. 1985. Arthropod species in soybeans in southeast Queensland. Journal of the Australian Entomological Society 24: 169–177.

Ezueh, M. I., and S. O. Dina. 1979. Pest problems of soybeans and control in Nigeria, pp. 275–283. *In* World Soybean Research Conference II, Westview Press, Boulder, Colorado. 897 pp.

Farias, J. R., J. A. S. França, F. Sulzbach, M. Bigolin, R. A. Fiorin, H. Maziero, and J. V. C. Guedes. 2006. Eficiência de thiamethoxan + lambda-cyhalothrin no controle do percevejo-verde-pequeno, *Piezodorus guildini* (Westwood, 1837) (Hemiptera: Pentatomidae) e seletividade para predadores na cultura da soja. Revista da Faculdade de Zootecia, Veterinária e Agronomia, Uruguaiana, RS, Brazil 13: 10–19. (in Portuguese, with English abstract)

Fehr, W. R., C. E. Caviness, D. T. Burmood, and J. S. Pennington. 1971. Stage of development descriptions for soybeans, *Glycine max* (L.) Merrill. Crop Science 11: 929–930.

Ferreira, B. S. C., and A. R. Panizzi. 1982. Percevejos pragas da soja no Norte do Paraná: abundância em relação à fenologia da planta e hospedeiros intermediários. Anais do II Seminário Nacional de Pesquisa de Soja, Brasília, DF, Brazil, Volume 2: 140–151.

Fraga, C. P., and L. H. Ochoa. 1972. Aspectos morfológicos y bioecológicos de *Piezodorus guildinii* (West.) (Hemiptera, Pent.). Informativo de Investigaciones Agropecuarias (Argentina) Supplemento 28: 103–117.

Froeschner, R. C. 1988. Family Pentatomidae Leach, 1815. The stink bugs, pp. 544–597. *In* T. J. Henry and R. C. Froeschner (Eds.), Catalog of the Heteroptera, or true bugs, of Canada and the continental United States. E. J. Brill, New York, New York. 958 pp.

Funderburk, J., R. McPherson, and D. Buntin. 1999. Soybean insect management, pp. 273–290. *In* L. G. Heatherly and H. F. Hodges (Eds.), Soybean production in the Midsouth. CRC Press, Boca Raton, FL. 404 pp.

Fuxa, J. E., J. R. Fuxa, A. R. Richter, and E. H. Weidner. 2000. Prevalence of a trypanosomatid in the southern green stink bug, *Nezara viridula*. Journal of Eukaryotic Microbiology 47: 388–394.

Galileo, M. H. M., and E. A. Heinrichs. 1978. Efeito dos danos causados por *Piezodorus guildinii* (Westwood, 1837) (Hemiptera, Pentatomidae), em diferentes níveis e épocas de infestação, no rendimento de grãos de soja [*Glycine max* (L.) Merrill]. Anais da Sociedade Entomológica do Brasil 7: 20–25.

Galileo, M. H. M., H. A. de O. Gastal, and J. Grazia. 2007. Levantamento populacional de Pentatomidae (Hemiptera) em cultura de soja (*Glycine max* (L.) Merr.) no Município de Guaíba, Rio Grande do Sul. Revista Brasileira de Biologia 37: 111–120.

Gamundi, J. C., E. Perotti, and A. Molinari. 2007. Evaluación de insecticidas para el control de chinches en cultivos de soja. Instituto Nacional de Tecnología Agropecuaria Estación Experimental Agropecuaria (Oliveros, Santa Fe, Argentina) 36: 112–114.

Garlet, J., M. Roman, and E. C. Costa. 2010. Pentatomídeos (Hemiptera) associados a espécies nativas em Itaara, RS, Brasil. Biotemas 23: 91–96.

Gentry, J. W. 1965. Crop insects of Northeast Africa-Southwest Asia. United States Department of Agriculture, Agriculture Research Service, Agriculture Handbook 273. 210 pp.

Genung, W. G. 1959. Investigations for control of insects attacking the pods of table legumes. Proceedings of the Florida State Horticultural Society 71: 25–29.

Genung, W. G., and V. E. Green, Jr. 1962. Insects attacking soybeans with emphasis on varietal susceptibility. Florida Agricultural Experiment Station Journal, Series Number 1602: 138–141.

Genung, W. G., V. E. Green, and C. Wehlburg. 1964. Inter-relationship of stinkbug and diseases to everglades soybean production. Proceedings of the Soil Crop Science Society of Florida 24: 131–137.

Gill, K. S. 1987. Insect pests of linseed, pp. 342–355. *In* A. M. Wadhwani (Ed.), Linseed. Indian Council of Agricultural Research (ICAR), New Delhi, India. 386 pp.

Gomez, V. A., E. F. Gaona, O. R. Arias, M. B. de Lopez, and O. E. Ocampos. 2013. Aspectos biológicos de *Piezodorus guildinii* (Westwood) (Hemiptera: Pentatomidae) criados com diferentes dietas em condiciones de laboratório. Revista de la Sociedad Entomolologica Argentina 72: 27–34.

Gonnet, A. F. R. 2007. Fluctuaciones poblacionales de *Piezodorus guildinii* (Westwood) (Hemiptera: Pentatomidae) y caracterización de sus enemigos naturales en soja y alfalfa. Tesis, Magíster en Ciencias, Universidad de la República Oriental del Uruguay, Agrarias, Montevideo. 64 pp.

González, Y. R., J. R. G. Sousa, R. E. Ruíz, E. M. Ferrer, and C. A. Caturla. 2011. Afectaciones directas producidas por el complejo de chinches (Hemiptera: Pentatomidae) en granos de frijol común (*Phaseolus vulgaris* L.) y determinación de *Nematospora* sp. Fitosanidad 15: 179–183.

Gore, J., C. A. Abel, J. J. Adamczyk, and G. Snodgrass. 2006. Influence of soybean planting date and maturity group on stink bug (Heteroptera: Pentatomidae) populations. Environmental Entomology 35: 531–536.

Grazia, J., M. C. Del Vechio, E. M. P. Balestieri, and Z. A. Ramiro. 1980. Estudo das ninfas de pentatomídeos (Heteroptera) que vivem sobre soja (*Glycine max* (L.) Merrill): 1 - *Euschistus heros* (Fabricius, 1798) e *Piezodorus guildinii* (Westwood, 1837). Anais da Sociedade Entomológica do Brasil 9: 39–51.

Grazia, J., A. R. Panizzi, C. Greve, C. F. Schwertner, L. A. Campos, T. de A. Garbelotto, and J. A. M. Fernandes. 2015. Stink bugs (Pentatomidae), pp. 681–756. *In* A. R. Panizzi and J. Grazia (Eds.), True bugs (Heteroptera) of the Neotropics. Springer Science+Business Media, Dordrecht. 901 pp.

Greene, J., and J. A. Davis. 2015. Stink bugs, pp. 146–149. *In* G. L. Hartman, J. C. Rupe, E. F. Sikora, L. L. Domier, J. A. Davis, and K. L. Steffey (Eds.), Compendium of soybean diseases and pests, 5th ed. American Phytopathological Society, St. Paul, MN, USA. 201 pp.

Gross, G. F. 1976. Plant-feeding and other bugs (Hemiptera) of South Australia. Heteroptera-Part II: 251–501. Handbook of the flora and fauna of South Australia. A. B. James, Government Printer, Adelaide, South Australia.

Haines, G. C. 1935. Cluster bugs. Farming South Africa 10: 182–188.

Hallman, G. 1979. Importancia de algumas relaciones naturales plantas-artropodos en la agricultura de la zona calida del Tolima central. Revista Colombiana de Entomologia 5: 19–26.

Heinrichs, E. A. 1976. Stink bug complex in soybeans, pp. 173–177. *In* R. M. Goodman (Ed.), Expanding the use of soybeans. Proceedings of a Conference for Asia and Oceania. International Soy Series Number 10, Urbana, IL, USA.

Higuchi, H. 1993. Seasonal prevalence of egg parasitoids attacking *Piezodorus hybneri* (Heteroptera: Pentatomidae). Applied Entomology and Zoology 28: 347–352.

Hinton, H. E. 1981. Biology of insect eggs in three volumes. Volume II, v–xviii + 475–778. Pergamon Press, Oxford, England.

Hokkanen, H. M. T. 1991. Trap cropping in pest management. Annual Review of Entomology 36: 119–138.

Husseneder, C., J.-S. Park, A. Howells, C. V. Tikhe, and J. A. Davis. 2017. Bacteria associated with *Piezodorus guildinii* (Hemiptera: Pentatomidae), with special reference to those transmitted by feeding. Environmental Entomology 46: 159–166.

Ishihara, P. 1950. The developmental stages of some bugs injurious to the kidney bean (Hemiptera). Transactions of the Shikoku Entomological Society 1: 2.

Ishikura, H., N. Nagano, T. Kobayashi, and I. Tamura. 1955. Studies on insect pests of soybean crop, III. On the damages by stink bugs, their ecology and control measures against them. Bulletin of the Shikoku National Agricultural Experiment Station 2: 147–195. (in Japanese, with English summary)

Jackai, L. E. N., and R. A. Daoust. 1986. Insect pests of cowpeas. Annual Review of Entomology 31: 95-119.

Jackai, L. E. N., K. E. Dashiell, and L. L. Bello. 1998. Evaluation of soybean genotypes for field resistance to stink bugs in Nigeria. Crop Protection 7: 48–54.

Jensen, R. L., and L. D. Newsom. 1972. Effect of stink bug-damaged soybean seeds on germination, emergence, and yield. Journal of Economic Entomology 65: 261–264.

Johnson, N. F. 1984. Systematics of Nearctic *Telenomus*: classification and revisions of the *podisi* and *phymatae* species groups (Hymenoptera: Scelionidae). Bulletin of the Ohio Biological Survey (n. ser.) 6(3): x + 113 pp.

Jones, W. A., Jr., and M. J. Sullivan. 1981. Overwintering habitats, spring emergence patterns, and winter mortality of some South Carolina Hemiptera. Environmental Entomology 10: 409–414.

Jones, W. A., and M. J. Sullivan. 1982. Role of host plants in population dynamics of stink bug pests of soybean in South Carolina. Environmental Entomology 11: 867–875.

Jones, W. A., Jr., and M. J. Sullivan. 1983. Seasonal abundance and relative importance of stink bugs in soybean. South Carolina Agricultural Experiment Station Technical Bulletin 1087. 6 pp.

Joseph, T. 1953. On the biology, bionomics, seasonal incidence and control of *Piezodorus rubrofasciatus* Fabr., a pest of linseed and lucerne at Delhi. Indian Journal of Entomology 15: 33–37.

Kamminga, K. L., J. A. Davis, S. P. Stock, and A. R. Richter. 2012. First report of a mermithid nematode infecting *Piezodorus guildinii* and *Acrosternum hilare* (Hemiptera: Pentatomidae) in the United States. Florida Entomologist 95: 214–217.

Kikuchi, A., A. Naito, and H. Matsuura. 1995. Attempts to increase percent parasitism of three hymenopterous species on the eggs of two soybean (*Glycine max*) stink bugs by host eggs added artificially in an open field. Bulletin of the National Agriculture Research Center 24: 61–66.

Kirkaldy, G. W. 1909. Catalogue of the Hemiptera (Heteroptera) with biological and anatomical references, lists of foodplants and parasites, etc. Volume I: Cimicidae. Felix L. Dames, Berlin. I–XL + 392 pp.

Kobayashi, T., T. Hasegawa, and K. Kegasawa. 1972. Major insect pests of leguminous crops in Japan. Tropical Agriculture Research Series 6: 109–126.

Kogan, M. 1977. Soybean Entomology Program. Consultant's Report. Londrina, Empresa Brasileira de Pesquisa Agropecuária, Centro Nacional de Pesquisa de Soja. Paraná, Brazil. 10 pp.

Kogan, M., and H. N. Pitre, Jr. 1980. General sampling methods for above-ground populations of soybean arthropods, pp. 30–60. *In* M. Kogan and D. C. Herzog (Eds.), Sampling methods in soybean entomology. Springer-Verlag, New York, USA. 587 pp.

Kogan, M., and S. G. Turnipseed. 1987. Ecology and management of soybean arthropods. Annual Review of Entomology 32: 507–538.

Koppel, A. L., D. A. Herbert, Jr., t. P. Kuhar, and K. Kamminga. 2009. Survey of stink bug (Hemiptera: Pentatomidae) egg parasitoids in wheat, soybean, and vegetable crops in southeast Virginia. Environmental Entomology 38: 375–379.

La Porta, N., M. Loiacono, and C. B. Margaria. 2013. Platigástridos (Hymenoptera: Platygastridae) parasitoides de Pentatomidae en Córdoba. Caracterización de las masas de huevos parasitoidizadas y aspectos biológicos. Revista de la Sociedad Entomológica Argentina 72: 179–194.

Leal, W. S., S. Kuwahara, and X. Shi. 1998. Male-released sex pheromone of the stink bug *Piezodorus hybneri*. Journal of Chemical Ecology 24: 1817–1829.

Leston, D. 1953. Notes on the Ethiopian Pentatomoidea. 10. Some specimens from southern Africa in the South African Museum, with a note on the remarkable pygophore of *Elvisura irrorata* Spinola and description of a new species of *Piezodorus* Fieber. Annals of the South African Museum 41: 48–60.

Link, D., and J. Grazia. 1987. Pentatomídeos da região central do Rio Grande do Sul. Anais da Sociedade Entomológica do Brasil 16: 115–129.

Link, D., J. A. V. Ranichi, and L. C. Concatto. 1980. Oviposition of the stink bug *Piezodorus guildinii* (Westwood, 1837) on bean plant. Revista do Centro de Ciências Rurais 10: 271–276.

Lopes, O. J., D. Link, and L. V. Basso. 1974. Pentatomídeos de Santa Maria – lista preliminar de plantas hospedeiras. Revista do Centro de Ciências Rurais 4: 317–322.

Lucini, T., A. R. Panizzi, and E. A. Backus. 2016. Characterization of an EPG waveform library for red-banded stink bug, *Piezodorus guildinii* (Hemiptera: Pentatomidae), on soybean plants. Annals of the Entomological Society of America 109: 198–210.

Malaguido, A. B., and A. R. Panizzi. 1998. Pentatomofauna associated with sunflower in Northern Paraná State, Brazil. Anais da Sociedade Entomológica do Brasil 27: 473–475.

McPherson, J. E. 1971. Laboratory rearing of *Euschistus tristigmus*. Journal of Economic Entomology 64: 1339–1340.

McPherson, J. E. 1974. Notes on the biology of *Mormidea lugens* and *Euschistus politus* (Hemiptera: Pentatomidae) in southern Illinois. Annals of the Entomological Society of America 67: 940–942.

McPherson, J. E., and R. M. McPherson. 2000. Stink bugs of economic importance in America north of Mexico. CRC Press, Boca Raton, USA. 254 pp.

McPherson, R. M., and L. D. Newsom. 1984. Trap crops for control of stink bugs in soybean. Journal of the Georgia Entomological Society 19: 470–480.

McPherson, R. M., L. D. Newsom, and B. F. Farthing. 1979. Evaluation of four stink bug species from three genera affecting soybean yield and quality in Louisiana. Journal of Economic Entomology 72: 188–194.

McPherson, R. M., G. K. Douce, and R. D. Hudson. 1993. Annual variation in stink bug (Heteroptera: Pentatomidae) seasonal abundance and species composition in Georgia soybean and its impact on yield and quality. Journal of Entomological Science 28: 61–72.

McPherson, R. M., J. W. Todd, and K. V. Yeargan. 1994. Stink bugs, pp. 87–90. *In* L. G. Higley and D. J. Boethel (Eds.), Handbook of soybean insect pests. Entomological Society of America, Lanham, MD. 136 pp.

Menezes, E. B. 1981. Population dynamics of the stinkbug (Hemiptera: Pentatomidae) complex on soybean and comparison of two relative methods of sampling. Ph.D. Dissertation, University of Florida, Gainesville. 259 pp.

Miller, L. A., H. A. Rose, and F. J. D. McDonald. 1977. The effects of damage by the green vegetable bug, *Nezara viridula* (L.) on yield and quality of soybeans. Journal of the Australian Entomological Society 16: 421–426.

Miner, F. D. 1961. Stink bug damage to soybeans. Arkansas Farm Research 10: 12.

Miner, F. D. 1966. Biology and control of stink bugs on soybeans. Arkansas Experiment Station Bulletin 708. 40 pp.

Monte, O. 1937. Notas hemipterológicas. O Campo 8: 70–71.

Moraes, M. C. B., M. Pareja, R. A. Laumann, and M. Borges. 2008. The chemical volatiles (semiochemicals) produced by Neotropical stink bugs (Hemiptera: Pentatomidae). Neotropical Entomology 37: 489–505.

Moschetti, E. M., E. M. Echeverría, and L. M. Ávalos. 2007. Producción de semilla de alfalfa, pp. 405–448. *In* D. H. Basigalup (Ed.), El cultivo de la alfalfa en la Argentina. Ediciones Instituto Nacional de Tecnologia Agropecuaria, Buenos Aires, Argentina. 476 pp.

Munyaneza, J., and J. E. McPherson. 1994. Comparative study of life histories, laboratory rearing, and immature stages of *Euschistus servus* and *Euschistus variolarius* (Hemiptera: Pentatomidae). The Great Lakes Entomologist 26: 263–274.

Musolin, D. L. 2007. Insects in a warmer world: ecological, physiological and life-history responses of true bugs (Heteroptera) to climate change. Global Change Biology 13: 1565–1585.

Musser, F. R., and A. L. Catchot. 2008. Mississippi soybean insect losses. Midsouth Entomologist 1: 29–36.

Musser, F. R., S. D. Stewart, and A. L. Catchot, Jr. 2009. 2008 Soybean insect losses for Mississippi and Tennessee. Midsouth Entomologist 2: 42–46.

Musser, F. R., G. M. Lorenz, S. D. Stewart, and A. L. Catchot, Jr. 2010. 2009 Soybean insect losses for Mississippi, Tennessee, and Arkansas. Midsouth Entomologist 3: 48–54.

Newsom, L. D., and D. C. Herzog. 1977. Trap crops for control of soybean pests. Louisiana Agriculture 20: 14–15.

Ochoa, L. H. 1969. La chinche verde de los alfalfares. Desarollo 2: 6.

Oliveira, E. D. M., and A. R. Panizzi. 2003. Performance of nymphs and adults of *Piezodorus guildinii* (Westwood) (Hemiptera: Pentatomidae) on soybean pods at different developmental stages. Brazilian Archives of Biology and Technology 46: 187–192.

Orr, D. B., J. S. Russin, D. J. Boethel, and W. A. Jones. 1986. Stink bug (Hemiptera: Pentatomidae) egg parasitism in Louisiana soybeans. Environmental Entomology 15: 1250–1254.

Owusu-Manu, E. 1974. Biology of *Bathycoelia thalassina* (H.-S.) (Heteroptera: Pentatomidae), in Ghana, pp. 3–9. *In* R. Kumar (Ed.), Proceedings of the 4th Conference of West African Cocoa Entomologists, Legon, Ghana. Zoology Department, University of Ghana.

Panizzi, A. R. 1980. Uso de cultivar armadilha no controle de percevejos. Trigo e Soja 47: 11–14.

Panizzi, A. R. 1985a. Dynamics of phytophagous pentatomids associated with soybean in Brazil, pp. 674–680. *In* R. Shibles (Ed.), Proceedings of the World Soybean Research Conference III. Boulder, Colorado. Westview Press. 1262 pp.

Panizzi, A. R. 1985b. *Sesbania aculeata*: nova planta hospedeira de *Piezodorus guildinii* (Hemiptera: Pentatomidae) no Paraná. Pesquisa Agropecuária Brasileira 20: 1237–1238.

Panizzi, A. R. 1992. Performance of *Piezodorus guildinii* on four species of *Indigofera*. Entomologia Experimentalis et Applicata 63: 221–228.

Panizzi, A. R. 1997. Wild hosts of pentatomids: ecological significance and role in their pest status on crops. Annual Review of Entomology 42: 99–122.

Panizzi, A. R. 2006. Possible egg positioning and gluing behavior by ovipositing southern green stink bug, *Nezara viridula* (L.) (Heteroptera: Pentatomidae). Neotropical Entomology 35: 149–151.

Panizzi, A. R. 2015. Growing problems with stink bugs (Hemiptera: Heteroptera: Pentatomidae): species invasive to the U.S. and potential Neotropical invaders. American Entomologist 61: 223–233.

Panizzi, A. R., and J. Grazia. 2001. Stink bugs (Heteroptera, Pentatomidae) and a unique host plant in the Brazilian subtropics. Iheringia Série Zoologia 90: 21–35.

Panizzi, A. R., and F. Slansky, Jr. 1985a. Legume host impact on performance of adult *Piezodorus guildinii* (Westwood) (Hemiptera: Pentatomidae). Environmental Entomology 14: 237–242.

Panizzi, A. R., and F. Slansky, Jr. 1985b. Review of phytophagous pentatomids associated with soybean in the Americas. Florida Entomologist 68: 184–214.

Panizzi, A. R., and F. Slansky, Jr. 1985c. *Piezodorus guildinii* (Hemiptera: Pentatomidae): an unusual host of the tachinid *Trichopoda pennipes*. Florida Entomologist 68: 485–486.

Panizzi, A. R., and J. G. Smith. 1976a. Ocorrência de Pentatomidae em soja no Paraná 1973/74. O Biológico 42: 173–176.

Panizzi, A. R., and J. G. Smith. 1976b. Observações sobre inimigos natuais de *Piezodorus guildinii* (Westwood, 1837) (Hemiptera, Pentatomidae) em soja. Anais da Sociedade Entomológica do Brasil 5: 11–17.

Panizzi, A. R., and J. G. Smith. 1977. Biology of *Piezodorus guildinii*: oviposition, development time, adult sex ratio, and longevity. Annals of the Entomological Society of America 70: 35–39.

Panizzi, A. R., B. S. Corrêa-Ferreira, D. L. Gazzoni, E. B. de Oliveira, G. G. Newman, and S. G. Turnipseed. 1977. Insetos da soja no Brasil. Empresa Brasileira de Pesquisa Agropecuária. Centro Nacional de Pesquisa de Soja. Boletim Técnico 1. Londrina, Paraná, Brazil. 20 pp.

Panizzi, A. R., J. G. Smith, L. A. G. Pereira, and J. Yamashita. 1979. Efeitos dos danos de *Piezodorus guildinii* no rendimento e qualidade da soja. Anais 1 Seminário Nacional de Pesquisa de Soja, Londrina, PR 2: 59–78.

Panizzi, A. R., M. H. M. Galileo, H. A. O. Gastal, J. F. F. Toledo, and C. H. Wild. 1980. Dispersal of *Nezara viridula* and *Piezodorus guildinii* nymphs in soybeans. Environmental Entomology 9: 293–297.

Panizzi, A. R., S. R. Cardoso, and E. D. M. Oliveira. 2000a. Status of pigeonpea as an alternative host of *Piezodorus guildinii* (Hemiptera: Pentatomidae), a pest of soybean. Florida Entomologist 83: 335–342.

Panizzi, A. R., J. E. McPherson, D. G. James, M. Javahery, and R. M. McPherson. 2000b. Stink bugs (Pentatomidae), pp. 421–474. *In* C. W. Schaefer and A. R. Panizzi (Eds.), Heteroptera of economic importance. CRC Press, Boca Raton, USA. 828 pp.

Panizzi, A. R., S. R. Cardoso, and V. R. Chocorosqui. 2002. Nymph and adult performance of the small green stink bug, *Piezodorus guildinii* (Westwood) on lanceleaf crotalaria and soybean. Brazilian Archives of Biology and Technology 45: 53–58.

Panizzi, A. R., A. F. Bueno, and F. A. C. Silva. 2012. Insetos que atacam vagens e grãos, pp. 335-420. *In* C. B. Hoffmann-Campo, B. S. Corrêa-Ferreira, and F. Moscardi (Eds.), Soja manejo integrado de insetos e outros artrópodes-praga. Embrapa Soja, Londrina, PR, Brazil. 859 pp.

Parker, J. L. 2012. Assessment of stink bug feeding damage in Louisiana soybean: use of a no-choice feeding field protocol. M.S. Thesis, Louisiana State University, Baton Rouge. 82 pp.

Parveen, S., J. S. Choudhary, A. Thomas, and V. V. Ramamurthy. 2015. Biology, morphology and DNA barcodes of *Tessaratoma javanica* (Thunberg) (Hemiptera: Tessaratomidae). Zootaxa 3939: 261–271.

Pasini, M. P. B., and A. Dal'Col Lúcio. 2014. Pentatomids associated with blackberry. Ciência e Agrotecnologia 38: 256–261.

Pearson, E. O. 1934. Investigations on cotton stainers and internal boll disease, pp. 146–152. *In* Report of the 2nd Congress on Cotton Growing Problems. Empire Cotton Growing Corporation, London. 340 pp.

Quintanilla, R. H., A. Margheritis, and H. F. Rizzo. 1967–1968. Catalogo de hemipteros hallados en la Provincia de Entre Ríos. Revista de la Facultad de Agronomía y Veterinaria de Buenos Aires 16: 29–38.

Quintanilla, R. H., H. F. Rizzo, A.S. de Núñez. 1981. Catalogo preliminar de hemípteros hallados em la Provincia de Misiones (Argentina). Revista de la Facultad de Agronomía 2: 145–161.

Ribeiro, A., and E. Castiglioni. 2009. Fluctuaciones de poblacionales de *Piezodorus guildinii* (Westwood) (Hemiptera: Pentatomidae) en soja y alfalfa em Paysandú, Uruguay. Agrociencia Uruguay 23: 32–36.

Ribeiro, A, E. Castiglioni, H. Silva, and S. Bartaburu. 2009. Fluctuaciones poblacionales de pentatómidos (Hemiptera: Pentatomidae) en soja (*Glycine max*) y lotus (*Lotus corniculatus*). Boletín de Sanidad Vegetal Plagas 35: 429–438.

Riley, D. G., G. K. Douce, and R. M. McPherson. 1997. Summary of losses from insect damage and costs of control in Georgia 1996. Georgia Agricultural Experiment Station, Special Publication 91. 59 pp.

Ring, D. R., A. L. Morgan, D. P. Reed, F. Huang, L. D. Foil, M. J. Stout, T. D. Schowalter, T. Smith, J. Beuzelin, J. A. Davis, S. Brown, D. L. Kerns, K. Healy, T. E. Reagan, and S. J. Johnson. 2015. Insect pest management guide. Louisiana State University AgCenter Publication 1838. 241 pp.

Rocca, M. and N. Greco. 2011. Diversity of herbivorous communities in blueberry crops of different regions of Argentina. Environmental Entomology 40: 247–259.

Rolston, L. H. 1983. A revision of the genus *Acrosternum* Fieber, subgenus *Chinavia* Orian, in the Western Hemisphere (Hemiptera: Pentatomidae). Journal of the New York Entomological Society 91: 97–176.

Rubtsov, I. A. 1977. A new genus and two new species of mermithids from bugs. Parazitologiya 11: 541–544. (In Russian, with English abstract)

Russin, J. S., M. B. Layton, D. B. Orr, and D. J. Boethel. 1987. Within-plant distribution of, and partial compensation for, stink bug (Heteroptera: Pentatomidae) damage to soybean seeds. Journal of Economic Entomology 80: 215–220.

Saluso, A., L. Xavier, F. A. C. Silva, and A. R. Panizzi. 2010. Situação atual de hemípteros fitófagos associados aos sistemas agrícolas da Região Pampeana Argentina. *In* Resumos XXIII Congresso Brasileiro de Entomologia. Natal, Rio Grande do Norte. http://www.seb.org.br/eventos/CBE/XXIIICBE/resumos/resumos XXIIICBE (accessed 19 January 2016).

Saunders, J. L., A. B. S. King, and C. L. Vargas S. 1983. Plagas de cultivos en América Central. Una lista de referencia. Centro Agronomico Tropical de Investigacion Y Enseñanze, CATIE, Boletin Tecnico No. 9. 90 pp.

Serra, G. V., and N. C. La Porta. 2001. Aspectos biológicos y reproductivos de *Piezodorus guildinii* (West.) (Hemiptera: Pentatomidae) en condiciones de laboratorio. Agriscientia 18: 51–57.

Shelton, A. M., and F. R. Badenes-Perez. 2006. Concepts and applications of trap cropping in pest management. Annual Review of Entomology 51: 285–308.

Shepard, B. M., G. R. Carner, and S. G. Turnipseed. 1974. A comparison of three sampling methods for arthropods in soybean. Environmental Entomology 3: 227–232.

Silva, A. G. A., C. R. Gonçalves, D. M. Galvão, A. J. L. Gonçalves, J. Gomes, M. N. Silva, and L. Simoni. 1968. Quarto catálogo dos insetos que vivem nas plantas do Brasil: seus parasitos e predadores: insetos, hospedeiros e inimigos naturais. Parte II, Tomo 1 Ministério da Agricultura Rio de Janeiro, Rio de Janeiro, RJ, Brasil. 622 pp.

Silva, F. A. C., and A. R. Panizzi. 2008. The adequacy of artificial oviposition substrates for laboratory rearing of *Piezodorus guildinii* (Westwood) (Heteroptera, Pentatomidae). Revista Brasileira de Entomologia 52: 131–134.

Silva, M. T. B., B. S. C. Ferreira, and D. R. Sosa-Gómez. 2006. Controle de percevejos em soja, pp. 109–123. *In* L. D. Borges (Ed.), Tecnologia de aplicação de defensivos agrícolas. Plantio Direto Eventos, Passo Fundo, RS, Brazil. 146 pp.

Singh, O. P., K. J. Singh, and R. D. Thakur. 1989. Studies on the bionomics and chemical control of stink bug, *Piezodorus rubrofasciatus* Fabricius, a new pest of soybean in Madhya Pradesh. Indian Journal of Plant Protection 18: 81–83.

Sithole, H., and J. Oelofose. 2006. Insects getting more attention in Kruger. https://www.sanparks.org/docs /parks_kruger/conservation/scientific/interesting_facts/mooiplaas_insect_oubreaks_2006.pdf (accessed 19 January 2016).

Smaniotto, L. F., and A. R. Panizzi. 2015. Interactions of selected species of stink bugs (Hemiptera: Heteroptera: Pentatomidae) from leguminous crops with plants in the Neotropics. Florida Entomologist 98: 7–17.

Smart, G. C. 1995. Entomopathogenic nematodes for the biological control of insects. Supplement to the Journal of Nematology 27(4S): 529–534.

Smith, J. F., R. G. Luttrell, and J. K. Greene. 2009. Seasonal abundance, species composition, and population dynamics of stink bugs in production fields of early and late soybean in south Arkansas. Journal of Economic Entomology 102: 229–236.

Sosa-Gómez, D. R., and F. Moscardi. 1995. Retenção foliar diferencial em soja provocada por percevejos (Heteroptera: Pentatomidae). Anais da Sociedade Entomológica do Brasil 24: 401–404.

Srivastava, S., and M. Srivastava. 2012. Entomo-fauna associated with bajra crop as observed in an agroecosystem in Rajasthan, India. International Journal of Theoretical and Applied Sciences 4: 109–121.

Staddon, B. W. 1997. A new shield-bug species *Piezodorus grossi* (Heteroptera: Pentatomidae) from the Australasian region previously confused with *P. hybneri* (Gmelin). Journal of Natural History 31: 1859–1863.

Staddon, B. W., and I. Ahmad. 1995. Species problems and species groups in the genus *Piezodorus* Fieber (Hemiptera: Pentatomidae). Journal of Natural History 29: 787–802.

Stadler, T., M. Buteler, and A. A. Ferrero. 2006. Susceptibilidad a endosulfan y monitoreo de resistencia en poblaciones de *Piezodorus guildinii* (Insecta: Heteroptera: Pentatomidae), en cultivos de soja de Argentina. Revista de la Sociedad de Entomologia de Argentina 65: 109–119.

Stål, C. 1872. Enumeratio Hemipterorum: Bidrag till en förteckning öfver alla hittills kända Hemiptera, jemte systematiska meddelanden. Parts 1–5. Kongliga Svenska Vetenskaps-Akademiens Handlingar, 1872, part 2, 10(4): 1–159.

Stansly, P. A., and G. J. Orellana M. 1990. Field manipulation of *Nomuraea rileyi* (Moniliales: Moniliaceae): effects on soybean defoliators in Coastal Ecuador. Journal of Economic Entomology 83: 2193–2195.

Temerak, S. A., and W. H. Whitcomb. 1984. Parasitoids of predaceous and phytophagous pentatomid bugs in soybean fields at two sites of Alachua County, Florida. Zeitschrift für Angewandte Entomologie 97: 279–282.

Temple, J. H., J. A. Davis, J. T. Hardke, P. Price, S. Micinski, C. Cookson, A. R. Richter, and B. R. Leonard. 2011. Seasonal abundance and occurrence of the redbanded stink bug in Louisiana soybeans. Louisiana Agriculture, Spring 2011: 20–21.

Temple, J. H., J. A. Davis, S. Micinski, J. T. Hardke, P. Price, and B. R. Leonard. 2013a. Species composition and seasonal abundance of stink bugs (Hemiptera: Pentatomidae) in Louisiana soybean. Environmental Entomology 42: 648–657.

Temple, J. H., J. A. Davis, J. Hardke, J. Moore, and B. R. Leonard. 2013b. Susceptibility of southern green stink bug and redbanded stink bug to insecticides in soybean field experiments and laboratory bioassays. Southwestern Entomologist 38: 393–406.

Temple, J. H., J. A. Davis, J. T. Hardke, P. P. Price, and B. R. Leonard. 2016. Oviposition and sex ratio of the redbanded stink bug, *Piezodorus guildinii*, in soybean. Insects 7, 27. http://dx.doi.org/10.3390/insects 7020027.

Tengkano, W., A. Naito, and A. M. Tahir. 1991. The effect of trap crop *Sesbania rostrata* for control of sucking bugs of soybean, pp. 92–113. Proceeding of Final Seminar of the Strengthening of Pioneering Research for Palawija Crops Production (ATA-378). AARD, CRIFC, BORIF and JICA. Bogor, Indonesia (4–5 March 1991).

Tillman, P. G., M. H. Greenstone, and J. S. Hu. 2015. Predation of stink bugs (Hemiptera: Pentatomidae) by a complex of predators in cotton and adjoining soybean habitats in Georgia, USA. Florida Entomologist 98: 1114–1126.

Tindall, K. V., and K. Fothergill. 2011. First records of *Piezodorus guildinii* in Missouri. Southwestern Entomologist 36: 203–205.

Todd, J. W. 1976. Effects of stink bug feeding on soybean seed quality, pp. 611–618. *In* L. D. Hill (Ed.), World soybean research: proceedings of the world soybean research conference. The Interstate Printers and Publishers, Inc., Danville, IL, USA. 1073 pp.

Todd, J. W. 1982. Effects of stink bug damage on soybean quality, pp. 46–50. *In* J. B. Sinclair and J. A. Jackobs (Eds.), Soybean seed quality and stand establishment. Proceedings of a conference for scientists of Asia, Colombo, Sri Lanka. International Soybean Program. University of Illinois, INTSOY series number 22. 206 pp.

Todd, J. W., and D. C. Herzog. 1980. Sampling phytophagous Pentatomidae in soybean, pp. 438–478. *In* M. Kogan, and D. C. Herzog (Eds.), Sampling methods in soybean entomology. Springer-Verlag, New York, USA. 587 pp.

Todd, J. W., and S. G. Turnipseed. 1974. Effects of southern green stink bug damage on yield and quality of soybeans. Journal of Economic Entomology 67: 421–426.

Todd, J. W., R. M. McPherson, and D. J. Boethel. 1994. Management tactics for soybean insects, pp. 115–117. *In* L. G. Higley and D. J. Boethel (Eds.), Handbook of soybean insect pests. Entomological Society of America, Lanham, MD. 136 pp.

Tuelher, E. S., E. H. Silva, E. Hirose, R. N. C. Guedes, and E. E. Oliveira. 2016. Competition between the phytophagous stink bugs *Euschistus heros* and *Piezodorus guildinii* in soybeans. Pest Management Science. https://www.researchgate.net/publication/301763913_Competition_between_the_phytophagous _stink_bugs_Euschistus_heros_and_Piezodorus_guildinii_in_soybeans. (accessed 15 June 2016).

Tuncer, C., İ. Saruhan, and İ. Akça. 2014. Seasonal occurrence and species composition of true bugs in hazelnut orchards. Acta Horticulturae 1052: 263–268.

Turnipseed, S. G. 1972. Management of insect pests of soybeans. Proceedings, Annual Tall Timbers Conference on Ecology of Animal Control by Habitat Management 4: 189–203.

Turnipseed, S. G. 1973. Insects, pp. 545–572. *In* B. E. Caldwell (Ed.), Soybeans: improvement, production, and uses. American Society of Agronomy. Madison, WI. 681 pp.

Turnipseed, S. G., and M. Kogan. 1976. Soybean entomology. Annual Review of Entomology 21: 247–282.

Uhler, P. R. 1894. On the Hemiptera-Heteroptera of the island of Granada, West Indies. Proceedings of the Zoological Society of London. Part I: 167–224.

van den Berg, H., A. Bagus, K. Hassan, A. Muhammad, and S. Zega. 1995. Predation and parasitism on eggs of two pod-sucking bugs, *Nezara viridula* and *Piezodorus hybneri* in soybean. International Journal of Pest Management 41: 134–142.

Van Duzee, E. P. 1904. Annotated list of the Pentatomidae recorded from America north of Mexico, with descriptions of some new species. Transactions of the American Entomological Society 30: 1–80.

Van Duzee, E. P. 1907. Notes on Jamaican Hemiptera: a report on a collection of Hemiptera made on the island of Jamaica in the spring of 1906. Bulletin of the Buffalo Society of Natural Sciences 8: 1–79.

Van Duzee, E. P. 1916. Check list of the Hemiptera (excepting the Aphididae, Aleurodidae and Coccidae) of America, north of Mexico. New York Entomological Society, New York. i–xi + 111 pp.

Ventura, A. Q. 1988. El cultivo y la investigacion de la soja em Bolivia, pp. 110–119. *In* B. Ramakrishna (Ed.), V Seminario manejo de suelos em sistemas de producción de soya. Quito, Ecuador. 290 pp.

Venugopal, P. D., P. L. Coffey, G. P. Dively, and W. O. Lamp. 2014. Adjacent habitat influence on stink bug (Hemiptera: Pentatomidae) densities and the associated damage at field corn and soybean edges. PloS One 9: e109917.

Vicentini, R., and H. A. Jimenez. 1977. El vaneo de los frutos en soja. Instituto Nacional de Tecnología Agropecuaria, Serie Tecnica 47. 30 pp.

Vyavhare, S. S., M. O. Way, and R. F. Medina. 2014. Stink bug species composition and relative abundance of the redbanded stink bug (Hemiptera: Pentatomidae) in soybean in the upper gulf coast Texas. Environmental Entomology 43: 1621–1627.

Vyavhare, S. S., M. O. Way, and R. F. Medina. 2015. Redbanded stink bug (Hemiptera: Pentatomidae) infestation and occurrence of delayed maturity in soybean. Journal of Economic Entomology 108: 1516–1525.

Wassink, H., and Poinar, Jr., G. O. 1984. Nematological reviews – resenas nematologicas use of the entomogenous nematode, *Neoaplectana carpocarpsae* Weiser (Steinernematidae: Rhabditida), in Latin America. Nematropica 14: 97–109.

Westwood, J. O. 1837, 1842. *In* F. W. Hope, A catalogue of Hemiptera in the collection of the Rev. F. W. Hope, M. A. with short Latin descriptions of the new species. J. C. Bridgewater, London, UK. 1837, Part 1: 1–46; 1842, Part 2: 1–26 (stating that descriptions are by J. O. Westwood).

Willrich, M. M., C. G. Clemens, M. E. Baur, B. J. Fitzpatrick, and D. J. Boethel. 2000. Evaluation of insecticides for southern green and brown stink bug control on soybean, 1999. Arthropod Management Tests 25(1): F152.

Wolcott, G. N. 1936. "Insectae Borinquenses". A revision of " 'Insectae Portoricensis'. A preliminary annotated check-list of the insects of Porto Rico, with descriptions of some news [sic] species" and "first supplement to Insectae Portoricensis". The Journal of Agriculture of the University of Puerto Rico 20: 1–600.

Wolcott, G. N. 1941. A supplement to "Insectae Borinquenses". The Journal of Agriculture of the University of Puerto Rico 25: 33–158.

Yeargan, K. 1979. Parasitism and predation of stink bug eggs in soybean and alfalfa fields. Environmental Entomology 8: 715–719.

Zandigiacomo, P. 1990. I prinicipali fitofagi della soia nel 1989 nell'Italia nord-orientale. Informatore Fitopatologico 40: 55–58.

Zandigiacomo, P. 1992. Pest found in soybean crops in north-eastern Italy. Informatore Agrario 48: 57–59.

Zerbino, M. S. 2014. Adaptaciones y respuestas inducidas de *Piezodorus guildinii* (Westwood, 1837) (Hemiptera: Pentatomidae) a las variaciones estacionales del ambiente - effecto del fotoperíodo, de la temperatura y del alimento en la biologia, fisiología y fenología. Tesis Doctora en Ciencias Agrárias, Universidad de la Republica, Montevideo, Uruguay. 196 pp.

Zerbino, S., and R. Alzugaray. 2010. *Piezodorus guildinii* (Westwood), pp 241–242. *In* C. M. Bentancourt and I. B. Scatoni (Eds.), Guía de insectos y ácaros de importancia agrícola y forestal em el Uruguay. Facultad de Agronomía, Universidad de la República Oriental del Uruguay, Montevideo, Uruguay. 582 pp.

Zerbino, M. S., N. A. Altier, and A. R. Panizzi. 2013. Effect of photoperiod and temperature on nymphal development and adult reproduction of *Piezodorus guildinii* (Westwood) (Heteroptera: Pentatomidae). Florida Entomologist 96: 572–582.

Zerbino, M. S., N. A. Altier, and A. R. Panizzi. 2014. Phenological and physiological changes in adult *Piezodorus guildinii* (Hemiptera: Pentatomidae) due to variation in photoperiod and temperature. Florida Entomologist 97: 734–742.

Zerbino, M. S., N. A. Altier, and A. R. Panizzi. 2015. Seasonal occurrence of *Piezodorus guildinii* on different plants including morphological and physiological changes. Journal of Pest Science 88: 495–505.

Zerbino, M. S., N. A. Altier, and A. R. Panizzi. 2016. Performance of nymph and adult of *Piezodorus guildinii* (Westwood) (Hemiptera: Pentatomidae) feeding on cultivated legumes. Neotropical Entomology 45: 114–122.

Section IV

Potentially Invasive Pentatomoidea

9

Oebalus spp. and *Arvelius albopunctatus* (De Geer)

J. E. McPherson and C. Scott Bundy

Oebalus Stål, 1862[1]

1862   *Oebalus* Stål. Stett. Ent. Zeit., 23(1–3): 102. Type-species: *Cimex typhoeus* Fabricius, 1803, Syst. Rhyn., p. 162; junior synonym of *Cimex pugnax* Fabricius, 1775, Syst. Ent., p. 704. (Designated by Kirkaldy, 1909, Cat. Hem., 1: 61).

1891   *Solubea* Bergroth, Rev.d'Ent. 10: 235. (Unnecessary new name for *Oebalus* Stål; Sailer, 1957, Proc. Ent. Soc. Wash. 59: 41).

Arvelius Spinola, 1840[1]

1840   *Arvelius* Spinola, Essai Hem., p. 344. Type species: *Cimex gladiator* Fabricius, 1775; junior synonym of *Cimex albopunctatus* De Geer, 1773, Mem. Ins. 3, 331, p. 34, fig. 6. (Designated by Kirkaldy, 1909, Cat. Hem., 1: 150).

CONTENTS

9.1   General Introduction (*Oebalus* and *Arvelius*) 455
9.2   Overview of *Oebalus* spp. 455
9.3   Key to Species of *Oebalus* in America North of Mexico 457
9.4   *Oebalus insularis* Stål 458
9.5   *Oebalus ypsilongriseus* (De Geer) 458
9.6   *Arvelius albopunctatus* (De Geer) 459
9.7   Overview of *Arvelius albopunctatus* 459
9.8   Acknowledgments 460
9.9   References Cited 460

9.1 General Introduction (*Oebalus* and *Arvelius*)

We herein include species that occur in the United States but do not meet our definition of invasive species discussed in **Chapter 1** (i.e., *"invasive pentatomoids are non-native species whose populations become established and spread, sometimes exhibiting uncontrolled population growth and displacing native populations, and cause adverse socioeconomic, environmental, or human-health effects."*). These species have become established but do not fully meet the rest of the definition.

9.2 Overview of *Oebalus* spp.

The genus *Oebalus* Stål (1862) is listed in much of the older literature as *Solubea* Bergroth (1891). Sailer (1944) reviewed the genus as *Solubea* but later (1957) indicated that the name was a junior synonym of *Oebalus*. It contains seven species (*O. grisescens* is invalid, see below), which are primarily Neotropical

---

[1] Synonymy from Froeschner (1988).

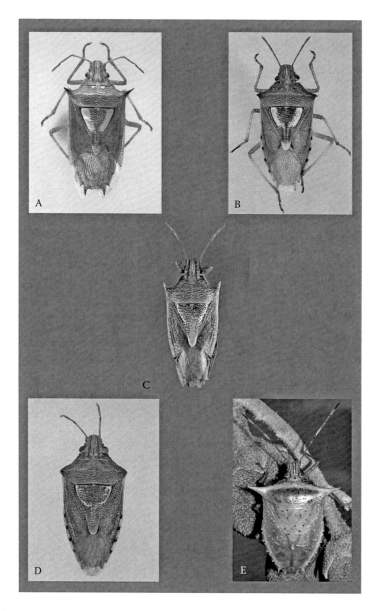

FIGURE 9.1 (**See color insert.**) *Oebalus* spp. and *Arvelius albopunctatus*, Dorsal habitus. A, *O. ypsilongriseus* with humeral spines; B, *O. ypsilongriseus* lacking humeral spines; C, *O. pugnax*; D, *O. insularis*; E, *A. albopunctatus*. (Images A, B, and D: courtesy of J. E. Eger; Image C: courtesy of C. Scott Bundy; Image E: courtesy of J. J. da Silva).

in distribution (Sailer 1944). Three of the species occur in America north of Mexico, *O. pugnax* (F.), *O. insularis* Stål, and *O. ypsilongriseus* (De Geer) (**Figure 9.1A–D**).

Oebalus pugnax (the rice stink bug) originally was described from "America" by Fabricius (1775) and, today, consists of two subspecies, *O. p. pugnax* and *O. p. torrida* (Sailer) (Sailer 1944). It is found throughout much of the United States south to Mexico and Colombia and in the West Indies (Froeschner 1988). *O. p. pugnax* has the same range as the species except in Colombia, which is where *O. p. torrida* occurs (Sailer 1944, Froeschner 1988). *O. pugnax* is a well known pest of rice, hence its common name.

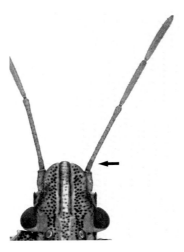

FIGURE 9.2   Antenna of *Oebalus ypsilongriseus*. (Courtesy of J. E. Eger).

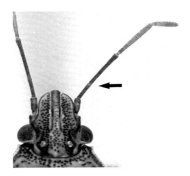

FIGURE 9.3   Antenna of *Oebalus insularis*. (Courtesy of J. E. Eger).

Oebalus insularis Stål and *O. ypsilongriseus* recently have been discovered in the United States from rice fields in Florida (Cherry et al. 1998, Cherry and Nuessly 2010). Both are noted pests of rice in South and Central America. Although they have increased in numbers in Florida, they essentially have not expanded their ranges beyond the state. Therefore, they do not easily fit our definition of an "invasive species" but certainly have the potential to become invasive.

9.3 Key to Species of *Oebalus* in America North of Mexico[2]

1. Antennal segment 2 about 0.5 X length of 1 (**Figure 9.2**)...............................*O. ypsilongriseus*
1'. Antennal segment 2 about 1.2 X length of 1 (**Figure 9.3**)...2
2. Humeri with well developed spines directed anterolaterally; scutellum yellowish, including apex (**Figure 9.1C**) .... *O. pugnax*
2'. Humeri acute or extended as short spines laterally; scutellum reddish brown, apex yellowish, lateral margins, including basal angles, with yellowish calloused areas in basal ½, these areas often approximate distally, forming U-shaped pattern (**Figure 9.1D**).....................*O. insularis*

---

[2] See Sailer (1944) for full descriptions of species.

9.4 *Oebalus insularis* Stål[3]

1857   *Pentatoma geographica*: Guėrin-Méneville. Ord. Hém. Het. Latr. Paris 7: 369–370. Misidentification.

1872   *Oebalus insularis* Stål, K. Svens. Vet.-Akad. Handl. 10(4): 22 (Cuba); Sailer, 1957, Proc. Ent. Soc. Wash. 59: 41.

1893   *Mormidea Guerini* Lethierry and Severin, Cat. Gen. Hém. 1: 123. New name for *Pentatoma geographica*: Guérin-Méneville, 1857, not Fabricius, 1803, Syst. Rhyn., pp. 159–160. (Synonymized by Sailer, 1944, Proc. Ent. Soc. Wash. 46: 119).

1909   *Solubea insularis*: Kirkaldy, Cat. Hem. 1:61.

1914   *Mormidea guerini*: Barber, Bull. Am. Mus. Nat. Hist. 33: 522.

1932   *Solubea insularis*: Barber and Bruner, J. Dept. Agr. Puerto Rico 16: 252.

This species was described by Stål in 1872 from Cuba (p. 22). Prior to its recent discovery in large numbers in Florida (Cherry and Nuessly 2010), it also was listed from Mexico and various locations in South and Central America and the Caribbean (Barber and Bruner 1932, Barber 1939, Sailer 1944, King and Saunders 1984). Sailer (1944) listed it from Florida, as reported by Barber (1914, 1939) and Barber and Bruner (1932), but we have not been able to confirm this earlier record. It is primarily a rice pest (e.g., Essig 1928, King and Saunders 1984, Pantoja et al. 1999, Vivas and Notz 2010) but will attack other plants including sorghum (King and Saunders 1984, Cherry et al. 2013, Nuessly et al. 2013) and wild Gramineae (King and Saunders 1984).

Oebalus insularis was reported from Florida rice fields by Cherry and Nuessly (2010) who noted that it first was observed in 2007. During 2008 and 2009, they conducted an extensive survey of rice fields in southern Florida and found that it occurred in all fields sampled and was the second most abundant species comprising 20% of all stink bugs collected; it also was widespread and well established. As far as they knew, this species had not been reported in commercial rice fields in other states. Cherry and Wilson (2011) studied flight activity of this bug throughout the year from 1 January 2008 through 1 January 2010.

The life history of this species is similar to that of *Oebalus pugnax* (McPherson and McPherson 2000). The bugs become active early in the year feeding on wild grasses in and around the rice fields and invade the rice when the crop begins to flower. They feed on the developing grains during the milk and dough stages and can cause empty or sterile grains (King and Saunders 1984) with the net effect being lower panicle weight (Pantoja et al. 1993). Feeding also can cause discoloration of the grains (pecky rice) (King and Saunders 1984). Economic injury levels and economic threshold for rice have been determined for this bug (Vivas and Notz 1010).

This species has been reared on rice in the laboratory under controlled conditions from egg to adult, and the incubation period and nymphal stadia have been determined; also determined have been the preovipositional and ovipositional periods and fecundity (Pantoja et al. 1999).

9.5 *Oebalus ypsilongriseus* (De Geer)[3]

1773   *Cimex ypsilon-griseus* De Geer, Mem. Hist. Ins. 3: 333–334, pl. 34, fig. 9. (Surinam).

1790   *Cimex litteratus* Gmelin, Systs. Nat. 1(4): 2148. Unnecessary new name for *Cimex ypsilongriseus* De Geer, 1773.

1803   *Cimex inscriptus* Fabricius, Syst. Rhyn., p. 159. [Synonymized By Stål, 1868, K. Svens. Vet.-Akad. Handl. 7(11): 28].

1862   *Oebalus Ypsilon griseus*: Stål, Stett. Ent. Zeit. 23(1–3): 102.

1878   *Oebalus ypsilonoides* Berg, Hem. Argentina, An. Soc. Cient. Arg. 5(6): 302–303. (Synonymized by Sailer, 1944, Proc. Ent. Soc. Wash. 46: 116).

1909   *Solubea ypsilongriseus*: Kirkaldy, Cat. Hem. 1: 62.

1909   *Solubea ypsilonoides*: Kirkaldy, Cat. Hem. 1: 62. (Synonymized by Sailer, 1957, Proc. Ent. Soc. Wash. 59: 41).

1944   *Solubea grisescens* Sailer, Proc. Ent. Soc. Wash. 46: 118–119, pl. 10, figs. 3, 12. (Synonymized by Vecchio et al., 1994, Rev. Bras. Ent. 38: 101–108).

1957   *Oebalus ypsilongriseus*: Sailer, Proc. Ent. Soc. Wash. 59: 41.

---

[3] Synonomy from Sailer (1944).

This species was described by De Geer as *Cimex ypsilon-griseus* in 1773 from Surinam (p. 333), then known as Dutch Guiana. It occurs throughout much of South and Central America (Pantoja et al. 1995, Panizzi et al. 2000) and now is reported from Florida (e.g., Mead 1983, Cherry et al. 1998). It is an important pest of rice (Vecchio and Grazia 1992a, Pantoja et al. 1995, Panizzi et al. 2000) but will attack other plants such as sorghum (Cherry et al. 2013, Nuessly et al. 2013).

This species has had a somewhat complicated taxonomic history. Sailer (1944), in his review of the genus, gave detailed descriptions of the morphology, biology, and distribution of the genus and species, including *Oebalus ypsilongriseus*. He also described a new species, *O. grisescens*, which he stated was similar to *S. ypsilongriseus*. In 1983, Mead reported the first record of *Solubea grisescens* in the United States from Homestead, Dade County, Florida. However, *O. grisescens* now has been shown to be a junior synonym of *O. ypsilongriseus*, based on the study of Vecchio et al. (1994) who showed that *O. grisescens* is the hibernating morph of *O. ypsilongriseus*.

Cherry et al. (1998) reported that *Oebalus ypsilongriseus* first was observed in Florida rice fields in 1994. They conducted an extensive survey of rice fields in the Everglades Agricultural Area during 1995 and 1996 and found this species occurred in all fields sampled and was the second most abundant species comprising 10.4% of all stink bugs collected; it also was widespread and well established. As noted for *O. insularis*, Cherry and Nuessly (2010) stated that *O. ypsilongriseus* had not been reported in commercial rice fields in other states. Cherry and Deren (2000) used sweep nets to determine the effect of time of day on vertical migration of this stink bug in Florida rice fields and the effects of temperature and wind speed on size of samples. Time of day did not affect sample sizes and temperature and wind speed had little or no effect. Finally, Cherry and Wilson (2011) studied flight activity of this bug throughout the year from 1 January 2008 through 1 January 2010.

This species has been reared in the laboratory under controlled conditions from egg to adult (Vecchio and Grazia 1993a), its fecundity investigated (Vecchio and Grazia 1992a), and the eggs (including embryonic development) (Vecchio and Grazia 1992b) and first to fifth instars (Vecchio and Grazia 1993b) have been described.

Some information is available on the effects of feeding by this insect on rice. Silva et al. (2002) studied the effects of low levels of infestations (number of bugs) by *Oebalus ypsilongriseus* and *O. poecilus* (Dallas) on grain yield and quality of isolated panicles and found no significant effects on percentages of weight loss and empty spikelets. However, based on the presence of feeding sheaths in the spikelets, *O. ypsilongriseus* was more active than *O. poecilus*. Finally, infestation by these species reduced total grain milling and increased the number of spikelets that were damaged but did not change the percentage of whole kernels.

9.6 *Arvelius albopunctatus* (De Geer)[1]

1773   *Cimex albopunctatus* De Geer, Mem. Ins. 3, 331, p. 34, fig. 6. (Surinam).
1876   *Arvelius albopunctatus*: Uhler, Bull. U. S. Geol. Geogr. Surv. Terr., 1: 290.

9.7 Overview of *Arvelius albopunctatus*

Arvelius albopunctatus (De Geer) (the tomato stink bug) (**Figure 9.1E**) originally was described as *Cimex albopunctatus* from Surinam by De Geer (1773). It is found throughout most of South America north through Baja California (Mexico) to Arizona, Texas, and Florida in the United States; it also has been reported from the West Indies (Uhler 1876, Froeschner 1988, Rider 2015).

This species appears to feed preferentially on wild and cultivated members of the Solanaceae. It is a pest of tomato, eggplant, peppers, potato, and sweet potato in Brazil and is an important pest of cherry tomatoes in Baja California, Mexico (Panizzi 2015). *Arvelius albopunctatus* also feeds on solanaceous weeds such as tropical soda apple in Florida (Diaz et al. 2012) and silverleaf nightshade in Texas (Goeden 1971); it was evaluated (and subsequently rejected) as a potential biological control agent of the latter

---

[1] Synonymy from Froeschner (1988).

weed species in South Africa (Olckers and Zimmerman 1991). In addition to members of the night-shade family, however, this stink bug also feeds on soybeans, green beans, okra, and sunflower in Brazil (Panizzi and Slansky 1985, Panizzi 2015) and soybeans in Argentina (Rizzo 1976, Panizzi et al. 2000). Not much information is available on the damage caused by feeding of this bug; however, it has been reported to cause wilting of potato leaves (Dias et al. 1985, Panizzi et al. 2000) and presumably, causes injury to fruiting structures (Olckers and Zimmermann 1991).

The immature life stages of *Arvelius albopunctatus* have been studied in Brazil by Grazia et al. (1984), who reported that eggs are deposited in clusters of 7 to 65 on host plants. Campos et al. (2007), also in Brazil, described three color morphs for fourth and fifth instars. Nymphs feed on fruit, leaves, and stems of their host plants (Panizzi et al. 2000).

9.8 Acknowledgments

We thank J. E. Eger (Dow AgroSciences, Tampa, FL) and J. J. da Silva (Embrapa Soja, Brazil) for the images of *Oebalus* spp. and *Arvelius albopunctatus*, respectively, used in this chapter.

9.9 References Cited

Barber, H. G. 1914. Insects of Florida. II. Hemiptera. Bulletin of the American Museum of Natural History 33: 495–535.

Barber, H. G. 1939. Insects of Porto Rico and the Virgin Islands–Hemiptera-Heteroptera (excepting the Miridae and Corixidae). Scientific Survey of Porto Rico and the Virgin Islands. New York Academy of Sciences 14: 263–441.

Barber, H. G., and S. C. Bruner. 1932. The Cydnidae and Pentatomidae of Cuba. Journal of the Department of Agriculture of Puerto Rico 16: 231–284, plates 24–26.

Berg, C. 1878. Hemiptera Argentina. Anales de la Sociedad Científica Argentina 5(6): 297–314.

Bergroth, E. 1891. Contributions à l'étude des Pentatomides. Revue d'Entomologie 10: 200–235.

Campos, L. A., R. A. Teixeira, and F. S. Martins. 2007. Three new color patterns of nymphs of *Arvelius albopunctatus* (De Geer) (Hemiptera: Pentatomidae). Neotropical Entomology 36: 972–975.

Cherry, R., and C. Deren. 2000. Sweep net catches of stink bugs (Hemiptera: Pentatomidae) in Florida rice fields at different times of day. Journal of Entomological Science 35: 490–493.

Cherry, R., and G. Nuessly. 2010. Establishment of a new stink bug pest, *Oebalus insularis* (Hemiptera: Pentatomidae), in Florida rice. Florida Entomologist 93: 291–293.

Cherry, R., and A. Wilson. 2011. Flight activity of stink bug (Hemiptera: Pentatomidae) pests of Florida rice. Florida Entomologist 94: 359–360.

Cherry, R., D. Jones, and C. Deren. 1998. Establishment of a new stink bug pest, *Oebalus ypsilongriseus* (Hemiptera: Pentatomidae) in Florida rice. Florida Entomologist 81: 216–220.

Cherry, R., Y. Wang, G. Nuessly, and R. Raid. 2013. Effect of planting date and density on insect pests of sweet sorghum grown for biofuel in southern Florida. Journal of Entomological Science 48: 52–60.

De Geer, C. 1773. Mémoires pour servir à l'histoire des insectes. Volume III. Hesselberg, Stockholm. 696 pp. + 1 (errata), 44 plates.

Dias, J. A. C. de S., A. S. Costa, V. A. Yuki, and N. P Granja. 1985. Murcha transitória do ponteiro da batata associado a alimentação do percevejo-do-tomateiro. Anais da Sociedade Entomológica do Brasil 14: 335–339.

Diaz, R., K. Hibbard, A. Samayoa, and W. A. Overholt. 2012. Arthropod community associated with tropical soda apple and natural enemies of *Gratiana boliviana* (Coleoptera: Chrysomelidae) in Florida. Florida Entomologist 95: 228–232.

Essig, E. O. 1928. Rice bugs. The Pan-Pacific Entomologist 4: 128.

Fabricius, J. C. 1775. Systema entomologiae, sistens insectorum classes, ordines, genera, species, adjectis synonymis, locis, descriptionibus, observationibus. Flensburgi et Lipsiae, Kortii. xxvii + 832 pp.

Fabricius, J. C. 1803. Systema Rhyngotorum secundum ordines, genera, species adjectis synonymis, locis, observationibus, descriptionibus. Carolum Reichard, Brunsvigae. x + 314 pp. + 1 [emendanda].

FIGURE 3.1 A, Comparison of adult *Murgantia histrionica* and *Bagrada hilaris* (dorsal view); B, *B. hilaris* egg; C, *B. hilaris* egg covered in soil; D, "Blind" cabbage injury from *B. hilaris* feeding, Yuma, Arizona; E, 2nd and 3rd instar nymphs of *B. hilaris*; F, large cluster of *B. hilaris* nymphs feeding on broccoli seed heads. (Images A-C and D-F: courtesy of C. Scott Bundy and John C. Palumbo, respectively).

FIGURE 4.1 *Halyomorpha halys.* A, first instar clustering atop hatched eggs; B, fourth instar; C, fifth instar; D, adult (A–D from laboratory culture); E, adults feeding on nectarine fruit (Kearneysville, WV); F, adults congregating around exterior security light (Spring Mills, WV).

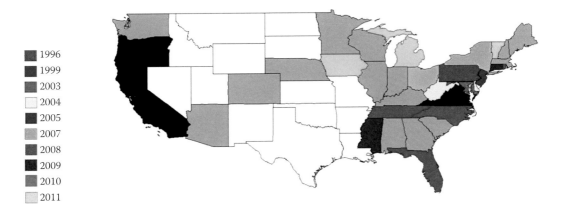

FIGURE 4.3 Distribution of *Halyomorpha halys* in the United States, 1996-2011.

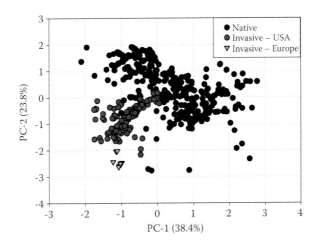

FIGURE 4.4 Principal component analysis of 10 variables associated with occurrences of *Halyomorpha halys*. Symbols represent occurrences of the bug in native areas in Asia and introduced areas in the United States and Europe (modified from Figure 2 in Zhu et al. 2012).

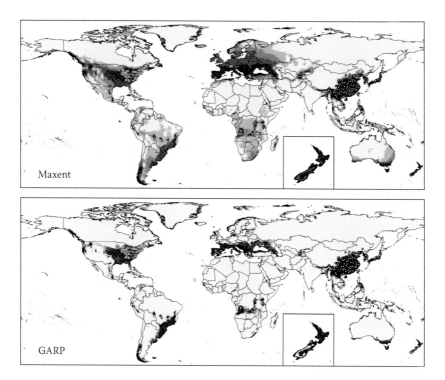

FIGURE 4.5 Potential distribution of *Halyomorpha halys* worldwide based on ecological niche modeling. Niche models were calibrated on the native Asia and transferred onto the global using Maxent and GARP, dark green suggests high suitability, light green indicates low suitability (modified from Figure 6 and Supplemental Materials 1 in Zhu et al. 2012).

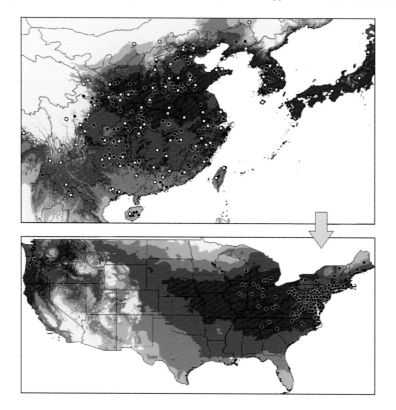

FIGURE 4.6 Niche models for *Halyomorpha halys* based on native Asia extent and transferred onto the United States using Maxent. Dark green represents high suitability, light green indicates low suitability. Model was built using 6 variables, white and black dots represent the 95 occurrences for model calibration and the remaining for model evaluation (modified from Figure 3 in Zhu et al. 2012).

FIGURE 5.1 *Megacopta cribraria.* A, egg mass; B, underside of egg mass, arrows indicate symbiont capsules; C, second instar; D, fourth instar; E, adult female (Courtesy of Joe E. Eger); F, adults on gutter downspout; G adults on side of house, Hoschton, GA (Courtesy of Dan R. Suiter); H, infestation of adults on young soybean plants (Courtesy of Jeremy K. Greene). Dimensional lines equal 1 mm.

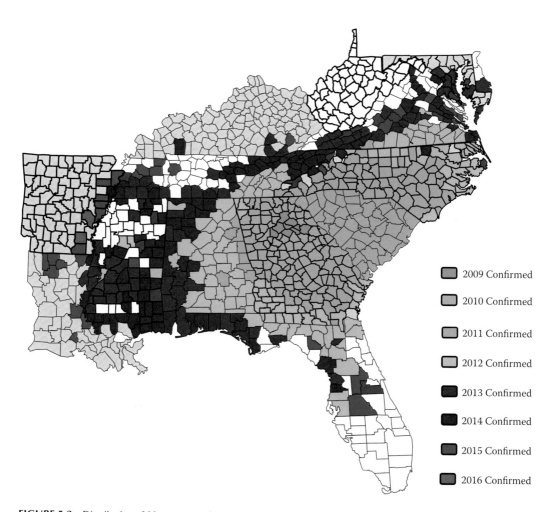

2009 Confirmed

2010 Confirmed

2011 Confirmed

2012 Confirmed

2013 Confirmed

2014 Confirmed

2015 Confirmed

2016 Confirmed

FIGURE 5.2 Distribution of *Megacopta cribraria* in the southeastern United States, 2009–2016 (Courtesy of Wayne A. Gardner).

FIGURE 6.1 *Murgantia histrionica* immatures, adults, and damage. A, egg cluster; B, second instars; C, fifth instars on cabbage; D, mating pair; E, damage on collards, Painter, Virginia. (Images A, C, and D: courtesy of Thomas P. Kuhar; B: courtesy of Sam E. Droege, USGS, Beltsville, Maryland; and E: courtesy of Anthony S. DiMeglio, Virginia Tech, Blacksburg).

FIGURE 7.1 *Nezara viridula* eggs, nymphs, and adults. A, egg mass at ≈2 days old (from laboratory culture); B, fifth instars and adults feeding on fruit of *Cucurbita foetidissima*; C, mating adults on *Rapistrum rugosum* (B–C in Central Texas); D, adults expressing varying degree of overwintering russet coloration in Kyoto, Japan; E, (l. to r.) – representative adult color morphs – form *smaragdula* (common all green form, form G-type), form *torquata* (O-type), form *viridula* (R-type), and form "OR-type" resulting from cross of O-type and R-type; F, form "OY-type" resulting from cross of O-type female and either a G-type or O-type male; G–H, several other color morphs include blue, orange, and black forms. (Images A–C, Courtesy of Jesus F. Esquivel, USDA, Agricultural Research Service, College Station, TX; D, Courtesy of Dmitry L. Musolin, Saint Petersburg, Russia; E–F, Adapted from K. Ohno and Md. Z. Alam, Applied Entomology and Zoology 27: 133–139, 1992, with permission; G, Blue form and orange form *aurantiaca*, courtesy of Antônio R. Panizzi, Passo Fundo, RS, Brazil; H, Black form, modified from J. F. Esquivel, V. A. Brown, R. B. Harvey, and R. E. Droleskey, Southwestern Entomologist 40: 649–652, 2015, with permission.)

FIGURE 7.6 Distribution of *Nezara viridula* in the Caribbean (West Indies), Central America, and South America. Solid yellow circles and yellow circles with black center dot indicate, respectively, exact localities recorded in literature and inferred localities from country or province/state/county records. The white line delimits the Amazon Basin.

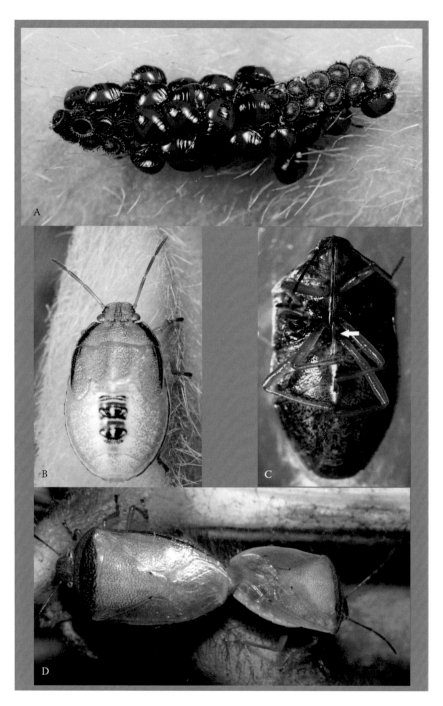

FIGURE 8.1 *Piezodorus guildinii.* A, egg mass and first instars on soybean; B, fifth instar on soybean; C, ventral view of adult showing abdominal spine; D, mating pair. (Images A and B: courtesy of Ted C. MacRae; C: courtesy of Jeffrey A. Davis; D: courtesy of J. J. da Silva).

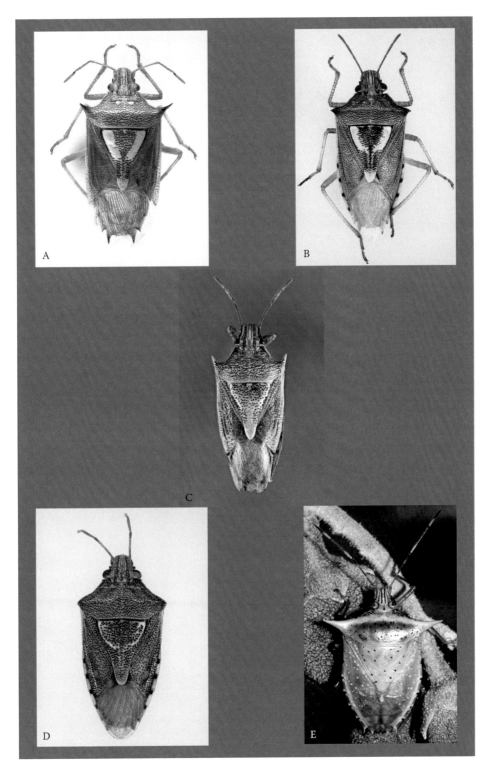

FIGURE 9.1 *Oebalus* spp. and *Arvelius albopunctatus*, Dorsal habitus. A, *O. ypsilongriseus* with humeral spines; B, *O. ypsilongriseus* lacking humeral spines; C, *O. pugnax*; D, *O. insularis*; E, *A. albopunctatus*. (Images A, B, and D: courtesy of J. E. Eger; Image C: courtesy of C. Scott Bundy; Image E: courtesy of J. J. da Silva).

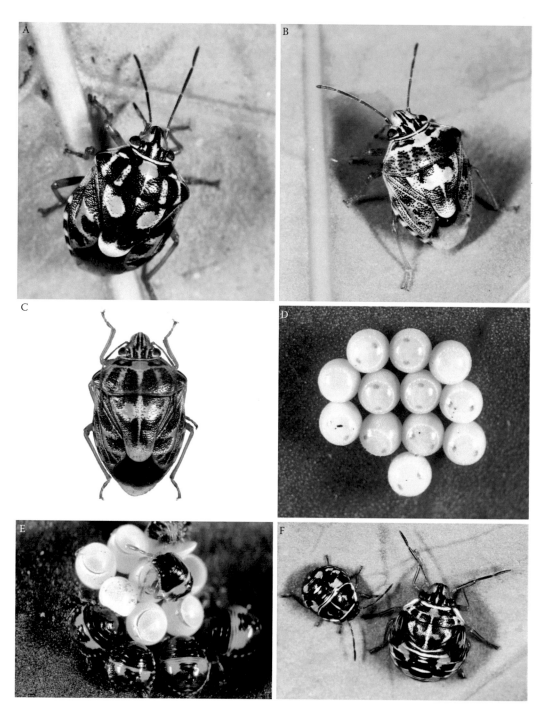

FIGURE 10.1 *Antestiopsis* spp. adults and immatures. A–C, adults: A, *Antestiopsis thunbergii bechuana* (Kiambu *Coffea arabica* plantation, Kenya, 2015); B, *A. facetoides* (Kiambu *Coffea arabica* plantation, Kenya, 2015); C, *A. intricata* (Foumbot *Coffea arabica* plantation, Cameroon, 1985). D–F, *A. thunbergii bechuana* immatures: D, eggs; E, first instars on egg choria; F, third and fifth instars. Immature stages are from a laboratory colony, icipe, Kenya, 2015. (Images A–B, D–F: Courtesy of Robert S. Copeland, Biosystematic Unit, International Centre of Insect Physiology and Ecology, Nairobi, Kenya; C: Courtesy of Régis Babin.)

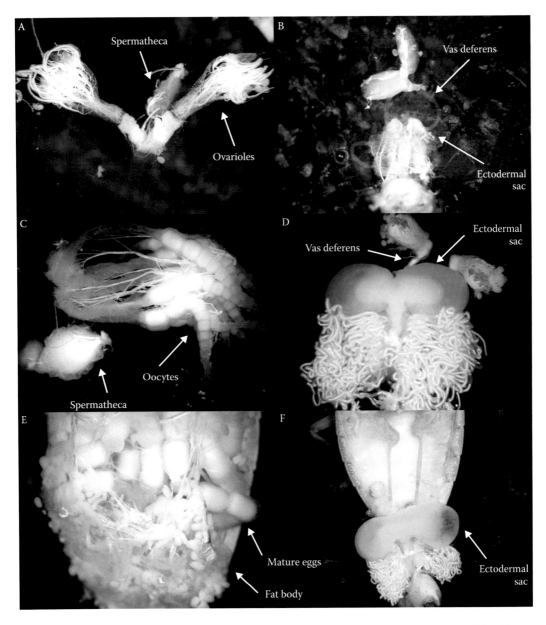

FIGURE 11.11 State of gonadal development in diapause and nondiapause adults of the southern green stink bug, *Nezara viridula*. Nonreproductive female (virgin, prereproductive or in diapause): no oocytes in germaria, clear ovarioles, and empty spermatheca (A). Nonreproductive male (virgin, prereproductive or in diapause): clear vasa deferentia and collapsed ectodermal sac (B). Reproductive female (nondiapause): developing oocytes with yolk and expanded spermatheca (C); chorionated eggs in ovarioles and loose fat body (E). Reproductive male (nondiapause): yellow vasa deferentia and expanded ectodermal sac containing milky white secretion (D, F). For details on morphology and description of stages of gonadal development in *N. viridula* see Esquivel (2009). Note that in the original publication (Esquivel 2009) images (A) and (B) refer to the gonads without development in virgin adults. In diapause adults, the gonads remain basically in the same state until diapause termination. (A, B, C, and F are from J. F. Esquivel, Annals of the Entomological Society of America 102: 303–208, 2009, with permission; D and E are courtesy of Dr. Jesus F. Esquivel.)

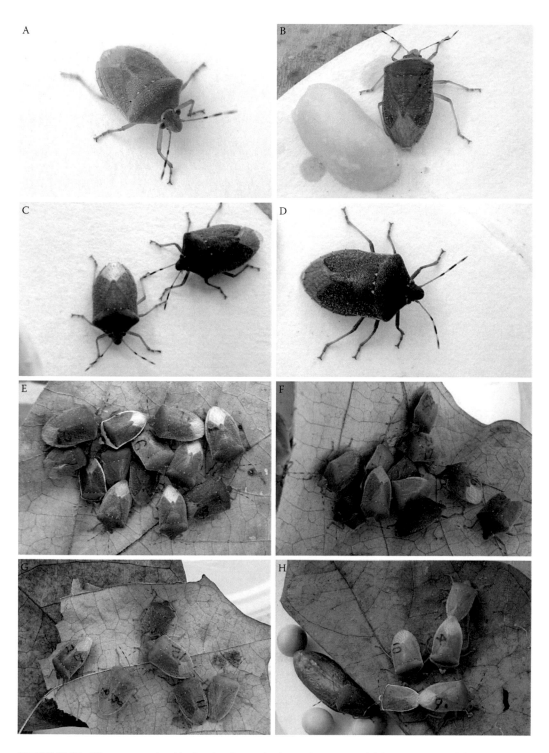

FIGURE 11.17 Diapause-associated body color changes in the southern green stink bug, *Nezara viridula*, reared under laboratory or quasi-natural outdoor conditions. Laboratory rearing in Kyoto (Japan): reproductive (i.e., nondiapause) green adult (A). Adults of the intermediate body color grade (B, C left). Diapausing russet (brown) adults (C right, D). Outdoor rearing in Kyoto: mostly intermediate colored and russet overwintering adults at the early stages of diapause (November–December; E, F); adults of different body color grades at the later stage of diapause (March; G); diapause termination, body color change and beginning of postdiapause reproduction (April; H). (From D. L. Musolin, Physiological Entomology 37: 309–322, 2012, with permission.)

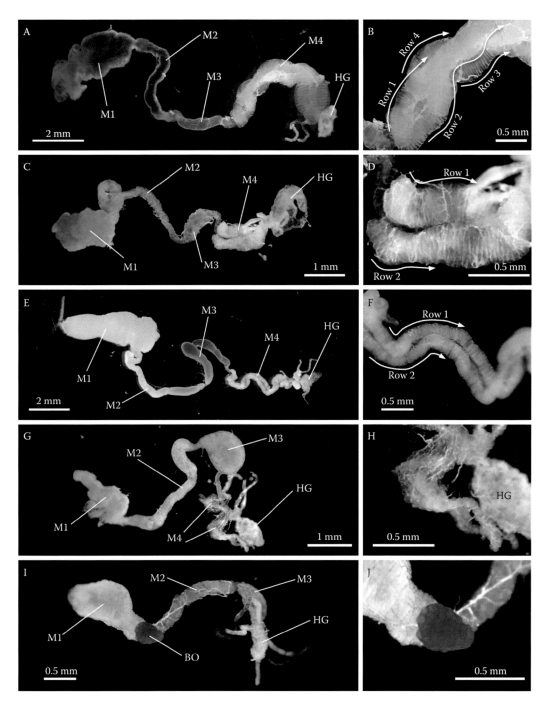

FIGURE 14.1 Symbiotic organ in the Pentatomomorpha. Dissected midgut and symbiotic organ of (A-B) *Menida scotti* (family Pentatomidae), (C-D) *Elasmucha putoni* (Acanthosomatidae), (E-F) *Riptortus pedestris* (Alydidae), (G-H) *Togo hemipterus* (Rhyparochromidae), and (I-J) *Kleidocerys resedae* (Lygaeidae). (B) Midgut crypts arranged in 4 rows. (D) Midgut crypts arranged in 2 rows, whose luminal entrances are completely sealed. (F) Midgut crypts arranged in 2 rows. (H) Tubular outgrowths. (J) Bacteriome. Abbreviations: M1, midgut first section; M2, midgut second section; M3, midgut third section; M4, midgut fourth section with crypts (symbiotic organ); HG, hindgut; BO, Bacteriome.

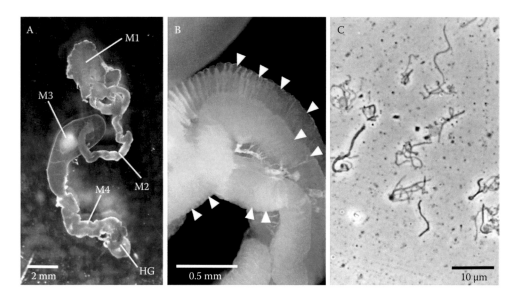

FIGURE 14.3 Microbial symbiosis in *Nezara viridula*. (A) Dissected midgut. Abbreviations: M1, midgut first section; M2, midgut second section; M3, midgut third section; M4, midgut fourth section with crypts (symbiotic organ); HG, hindgut. (B) Enlarged image of midgut crypts. Arrowheads indicate crypts. (C) Symbiotic bacteria of *N. viridula* (phase contrast microscopy).

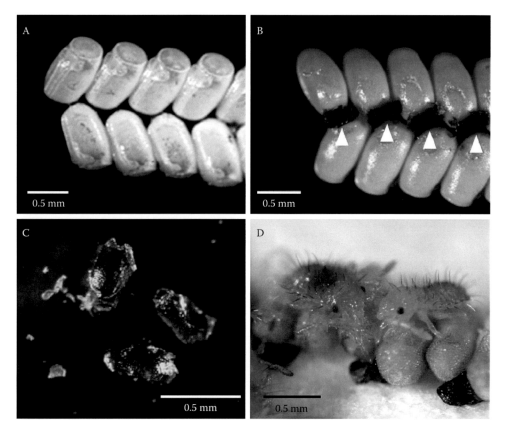

FIGURE 14.4 Capsule symbiosis in plataspid insects. (A-B) Egg cluster of *Brachyplatus subaeneus*: A. Dorsolateral view; B, Posterolateral view. Arrows indicate symbiont capsules. (C) Symbiont capsules removed from egg mass. (D) First instars of *Megacopta cribraria*. Note: nymphs feeding on capsule contents.

Froeschner, R. C. 1988. Family Pentatomidae Leach, 1815. The stink bugs, pp. 544–597. *In* T. J. Henry and R. C. Froeschner (Eds.), Catalog of the Heteroptera, or true bugs, of Canada and the continental United States. E. J. Brill, New York, NY. 958 pp.

Gmelin, J. F. 1790. Hemiptera. *In* Caroli a Linné Systema Naturae (13th edit.). J. F. Gmelin (Ed.) 1(4): 1520, 1523, 2041–2224. For an explanation of the 1790 date, see Henry and Froeschner, 1988.

Goeden, R. D. 1971. Insect ecology of silverleaf nightshade. Weed Science 19: 45–51.

Grazia, J., R. Hildebrand, and A. Mohr. 1984. Estudo das ninfas de *Arvelius albopunctatus* (De Geer, 1773) (Heteroptera, Pentatomidae). Anais da Sociedade Entomológica do Brasil 13: 141–150.

Guérin-Méneville, F. E. 1857. Ordre des Hémiptères, Latr. Première section. Hétéroptères, Latr., pp. 359–424, plate 13. *In* M. R. de la Sagra's Historie Physique, Politique et Naturelle de l'Ile de Cuba. Arthus Bertrand, Paris 7. 868 pp.

Henry, T. J., and R. C. Froeschner (Eds.). 1988. Catalog of the Heteroptera, or true bugs, of Canada and the continental United States. E. J. Brill, New York, NY. 958 pp.

King, A. B. S., and J. L. Saunders. 1984. The invertebrate pests of annual food crops in Central America. A guide to their recognition and control. Tropical Development and Research Institute (TDRI), Overseas Development Adminstration, London. 166 pp.

Kirkaldy, G. W. 1909. Catalogue of the Hemiptera (Heteroptera) with biological and anatomical references, lists of foodplants and parasites, etc. prefaced by a discussion on nomenclature, and an analytical table of families. Volume I: Cimicidae. Felix L. Dames, Berlin. I–XL + 392 pp.

Lethierry, L. F., and G. Severin. 1893. Catalogue Général des Hémiptères. Hétéroptères. Pentatomidae, 1: i–x, 1–275. R. Friedländer and Fils, Bruxelles and Berlin.

McPherson, J. E., and R. M. McPherson. 2000. Stink bugs of economic importance in America north of Mexico. CRC Press LLC, Boca Raton, FL. 253 pp.

Mead, F. W. 1983. Insect detection: a stink bug, *Oebalus grisescens* (Sailer). Tri-ology 22(11): 4.

Nuessly, G. S., Y. Wang, H. Sandhu, N. Larsen, and R. H. Cherry. 2013. Entomologic and agronomic evaluations of 18 sweet sorghum cultivars for biofuel in Florida. Florida Entomologist 96: 512–528.

Olckers, T., and H. G. Zimmermann. 1991. Biological control of silverleaf nightshade, *Solanum elaeagnifolium*, and bugweed, *Solanum mauritianum*, (Solanaceae) in South Africa. Agriculture, Ecosystems, and Environment 37: 137–155.

Panizzi, A. R. 2015. Growing problems with stink bugs (Hemiptera: Heteroptera: Pentatomidae): species invasive to the U. S. and potential Neotropical invaders. American Entomologist 61: 223–233.

Panizzi, A. R., and F. Slansky, Jr. 1985. Review of phytophagous pentatomids (Hemiptera: Pentatomidae) associated with soybean in the Americas. Florida Entomologist 68: 184–214.

Panizzi, A. R., J. E. McPherson, D. G. James, M. Javahery, and R. M. McPherson. 2000. Chapter 13. Stink bugs (Pentatomidae), pp. 421–474. *In* C. W. Schaefer and A. R. Panizzi (Eds.), Heteroptera of economic importance. CRC Press LLC, Boca Raton, FL. 828 pp.

Pantoja, A., E. Daza, and M. C. Duque. 1993. Effecto de *Oebalus ornatus* (Sailer) y *Oebalus insularis* Stal (sic) (Hemiptera: Pentatomidae) sobre el arroz: una comparacion entre especies. Manejo Integrado de Plagas (Costa Rica) 26: 31–33.

Pantoja, A., E. Daza, C. A. Garcia, O. I. Mejía, and D. A. Rider. 1995. Relative abundance of stink bugs (Hemiptera: Pentatomidae) in southwestern Colombia rice fields. Journal of Entomological Science 30: 463–467.

Pantoja, A., E. Daza, O. I. Mejía, C. A. Garcia, M. C. Duque, and L. E. Escalona. 1999. Development of *Oebalus ornatus* (Sailer) and *Oebalus insularis* (Stal) [sic] (Hemiptera: Pentatomidae) on rice. Journal of Entomological Science 34: 335–338.

Rider, D. A. 2015. *Arvelius* Spinola, 1837. Pentatomoidea home page. North Dakota State University. https://www.ndsu.edu/ndsu/rider/Pentatomoidea/Genus_Chlorocorini/Arvelius.htm (accessed 9 February 2016).

Rizzo, H. F. E. 1972. Enemigos animales del cultivo de la soja. Revista Institucional de la Bolsa de Cereales 2851: 6 pp.

Sailer, R. I. 1944. The genus *Solubea* (Heteroptera: Pentatomidae). Proceedings of the Entomological Society of Washington 46: 105–127.

Sailer, R. I. 1957. *Solubea* Bergroth, 1891, a synonym of *Oebalus* Stål, 1862, and a note concerning the distribution of *O. ornatus* (Sailer) (Hemiptera, Pentatomidae). Proceedings of the Entomological Society of Washington 59: 41–42.

Silva, D. R., E. Ferreira, and N. R. de A. Vieira. 2002. Avaliação de perdas causadas por *Oebalus* spp. (Hemiptera: Pentatomidae) em arroz de terras altas. Pesquisa Agropecuária Tropical 32: 39–45.

Stål, C. 1862. Hemiptera Mexicana enumeravit speciesque novas descripsit. Stettin Entomologische Zeitung (Entomologische Zeitung Herausgegeben von dem Entomologischen Vereine zu Stettin) 23(1–3): 81–118.

Stål, C. 1868. Hemiptera Fabriciana. Fabricianska Hemipterarter, efter de i Köpenhamn och Kiel förvarade typexemplaren granskade och beskrifne. 1. Kongliga Svenska Vetenskaps-Akademiens Handlingar 7(11): 1–148.

Stål, C. 1872. Enumeratio Hemipterorum. Bidrag till en förteckning öfver alla hittills kända Hemiptera, jemte systematiska meddelanden. Parts 1–5. Konglilga Svenska Vetenskaps-Akademiens Handlingar, 1872, part 2, 10(4): 1–159.

Uhler, P. R. 1876. List of Hemiptera of the region west of the Mississippi River, including those collected during the Hayden explorations of 1873. Bulletin of the United States Geological and Geographical Survey of the Territories 1: 267–361, plates 1–21.

Vecchio, M. C. Del, and J. Grazia. 1992a. Obtenção de posturas de *Oebalus ypsilongriseus* (De Geer, 1773) em laboratório (Heteroptera: Pentatomidae). Anais da Sociedade Entomológica do Brasil 21: 367–373.

Vecchio, M. C. Del, and J. Grazia. 1992b. Estudo dos imaturos de *Oebalus ypsilogriseus* (De Geer, 1773): I–Descrição do ovo e desenvolvimento embrionário (Heteroptera: Pentatomidae). Anais da Sociedade Entomológica do Brasil 21: 374–382.

Vecchio, M. C. Del, and J. Grazia. 1993a. Estudo dos imatuaros de *Oebalus ypsilongriseus* (De Geer, 1773): II–Descrição das ninfas (Heteroptera: Pentatomidae). Anais da Sociedade Entomológica do Brasil 22: 109–120.

Vecchio, M. C. Del, and J. Grazia. 1993b. Estudo dos imaturos de *Oebalus ypsilongriseus* (De Geer, 1773): III–Duração e mortalidade dos estágios de ovo e ninfa (Heteroptera: Pentatomidae). Anais da Sociedade Entomológica do Brasil 22: 121–129.

Vecchio, M. C. Del, J. Grazia, and G. S. Albuquereque. 1994. Dimorfismo sazonal em *Oebalus ypsilongriseus* (De Geer, 1773) (Hemiptera, Pentatomidae) e uma nova sinonímia. Revista Brasileira de Entomologia 38: 101–108.

Vivas, L. E. and A. Notz. 2010. Determination of damage threshold and level of economic vaneadora rice bug on the variety Cimarrón in Calabozo, Guarico State, Venezuela. Agronomía Tropical 60: 271–281.

Section V

A Noninvasive Group
(Antestia Complex)

10

The Antestia Bug Complex in Africa and Asia

Régis Babin, Pierre Mbondji Mbondji, Esayas Mendesil, Harrison M. Mugo, Joon-Ho Lee, Mario Serracin, N. D. T. M. Rukazambuga, and Thomas A. Miller

CONTENTS

10.1 Introduction 466
10.2 *Antestiopsis* Species and Their Distribution on Coffee in Africa and Asia 467
    10.2.1 *Antestiopsis* spp. 467
    10.2.2 *Antestiopsis thunbergii* (Gmelin) 467
    10.2.3 *Antestiopsis intricata* (Ghesquière and Carayon) 468
    10.2.4 *Antestiopsis facetoides* Greathead 469
    10.2.5 *Antestiopsis clymeneis* (Kirkaldy) 469
    10.2.6 *Antestiopsis cruciata* (F.) 469
    10.2.7 Antestia Species Closely Related to the Group but Not Found on Coffee 470
10.3 Life History 470
    10.3.1 Ecological Preferences 470
        10.3.1.1 Relationships with Elevation 470
        10.3.1.2 Relationships with Shade and Microclimate 470
        10.3.1.3 Ecological Preferences at Tree Level 471
    10.3.2 Feeding Habits 471
        10.3.2.1 Host Plants 471
        10.3.2.2 Feeding on Coffee 472
    10.3.3 Phenology with Descriptions of Developmental Stages 472
        10.3.3.1 *Antestiopsis thunbergii* 472
            10.3.3.1.1 Egg Stage 472
            10.3.3.1.2 Nymphal Instars 473
            10.3.3.1.3 Adult Stage 473
        10.3.3.2 *Antestiopsis intricata* 474
            10.3.3.2.1 Egg Stage 474
            10.3.3.2.2 Nymphal Instars 474
            10.3.3.2.3 Adult Stage 475
    10.3.4 Reproduction on Coffee 475
        10.3.4.1 *Antestiopsis thunbergii* 475
        10.3.4.2 *Antestiopsis intricata* 475
    10.3.5 Population Dynamics on Coffee 476
        10.3.5.1 Seasonal Variation 476
        10.3.5.2 Spatial Distribution in Plantations 477
        10.3.5.3 Factors Involved in Population Dynamics 477
10.4 Predators and Parasites 478
    10.4.1 Predators 478
    10.4.2 Egg Parasitoids 478
    10.4.3 Nymph and Adult Parasitoids 478

10.5   Damage to Coffee 479
     10.5.1   Direct Damage on Flower Buds, Berries, and Shoots 479
     10.5.2   Microbe Transmission 479
     10.5.3   Coffee Potato Taste Defect and Possible Causing Mechanisms 480
10.6   Management and Control 480
     10.6.1   Economic Impact 480
     10.6.2   Control Methods 481
         10.6.2.1   Cultural Methods 481
             10.6.2.1.1   Coffee Pruning and Shade Management 481
             10.6.2.1.2   Handpicking 481
         10.6.2.2   Control with Insecticides 481
             10.6.2.2.1   Poison-Bait Sprays 481
             10.6.2.2.2   Botanical Insecticides 481
             10.6.2.2.3   Chemical Insecticides 482
             10.6.2.2.4   Organization of Chemical Control 483
         10.6.2.3   Biological Control 483
             10.6.2.3.1   Parasitism Rate and Parasitoid Value 483
             10.6.2.3.2   Parasitism Seasonality 483
             10.6.2.3.3   Parasitoid Rearing 484
             10.6.2.3.4   Biological Control Attempts 486
     10.6.3   Future Control Strategies 486
         10.6.3.1   Conservation Biological Control through Plant Diversification 486
         10.6.3.2   Semiochemical Control 486
10.7   Why Are Antestia Bugs Not Invasive Species? 487
     10.7.1   A Limited Distribution Compared with Coffee Berry Borer 487
     10.7.2   Limited Opportunities of Spreading by Human Transport 487
     10.7.3   Relatively Slow Growing Species 487
     10.7.4   A Strong Pressure of Natural Enemies 488
     10.7.5   Constraining Ecological Preferences 488
     10.7.6   A Close Relationship with Host Plants of the Family Rubiaceae 488
10.8   Acknowledgments 488
10.9   References Cited 489

10.1 Introduction

Amongst the Pentatomidae, a group of species commonly known as Antestia bugs or, less frequently, "variegated coffee bugs", are important pests of coffee Arabica in Africa. This group of a dozen species and subspecies initially was part of the genus *Antestia* but is now placed in the genus *Antestiopsis*, although it keeps the common name of Antestia bugs, which has been widely used by African coffee farmers for decades. Because of their economic importance, Antestia bugs have been the object of many studies covering the pest biology and ecology as well as the damage to the plant and control measures.

     Most of basic knowledge on Antestia bugs is due to important research work conducted by entomologists who lived in Eastern Africa in the first half of 20th century, including Kenya (Anderson 1919, Le Pelley 1932), Uganda (Gowdey 1918, Wilkinson 1924, Hargreaves 1936), and Tanzania (Kirkpatrick 1937). In Western Africa, further studies were conducted later by Nanta (1950) and Lavabre (1952) in the Ivory Coast and Carayon (1954a,b) and Bruneau de Miré (1969) in Cameroon and the Central African Republic. In Rwanda and Burundi, Foucart and Brion (1959) conducted extensive research on Antestia bugs in the second half of the 1950s. In Uganda, Greathead (1966a) proposed a review of the taxonomy of the group, shedding light on the classification and distribution of the various *Antestiopsis* species. Most recent research works on Antestia bugs are credited to Abebe (1987, 1999), Mendesil and Abebe (2004), and Mendesil et al. (2012) in Ethiopia; Bouyjou and Cilas (1992), Cilas et al. (1998), and Bouyjou et al. (1999) in Burundi; Mbondji Mbondji (1997, 1999, 2001) in Cameroon; and Mugo (1994), Mugo and Ndoiru (1997) and Mugo et al. (2013) in Kenya, to name a few.

The present chapter reviews our knowledge of the Antestia bugs' distribution, life history, natural enemies as well as their damage, role in coffee potato taste defect, economic impact, and control. We discuss the ecological range and distribution on coffee of the Antestia species, which currently are limited to Africa and Asia. Interestingly, other major coffee pests originating in Africa, such as the coffee berry borer, have spread globally to coffee plantations everywhere.

10.2 *Antestiopsis* Species and Their Distribution on Coffee in Africa and Asia

10.2.1 *Antestiopsis* spp.

1952	*Antestiopsis* Leston. Rev. Zool. Bot. Afr., 45: 269. Orthotype, *Cimex anchora*. Thunberg, 1783, Dissert. Ent. Nov. Ins. Sp., Sistens, J. 2:47 (Upsala).

In 1952, Leston proposed *Antestiopsis* as a new genus for coffee bugs formerly classified in the genus *Antestia*. The genus *Antestiopsis* currently includes a dozen species, most of which have an African distribution although one species is reported on coffee in India. African species are present everywhere Arabica coffee is grown on the continent. The appearance of all of these African *Antestiopsis* species is similar in shape, size, and color (**Figure 10.1A–C, E–F).** However, any given species can show variation in size, intensity, and disposition of color patterns on the body. Given that the distributions of *Antestiopsis* species often overlap, the morphological characteristics of these species created confusion in the early taxonomy of these bugs. Most species have been described several times and have been given different names, and some authors still use inappropriate names today.

There have been several attempts to clarify the classification of the group in the first half of 20th century (Kirkpatrick 1937, Leston 1952, Carayon 1954a), but the most complete work is that of Greathead (1966a, 1968) who proposed for African species an in-depth study of the taxonomy of the group including identification keys, distribution maps, and morphometric and biological data with crossbreeding trials. Information presented here is largely taken from the Greathead studies, with the exception of *Antestiopsis cruciata*, which has an Asian distribution. Taxonomy of this Asian species is based on the study of Rider et al. (2002). The taxonomy of one of the most important African species of the group, *A. thunbergii*, has been revised recently by Rider (1998), and this study also will be included.

10.2.2 *Antestiopsis thunbergii* (Gmelin)

1783	*Cimex variegatus* Thunberg, Dissert. Ent. Nov. Ins. Sp., Sistens, 2: 48–49. (preoccupied).
1790	*Cimex thunbergii* Gmelin, Car. Lin. Syst. Nat., 1: 2158. (new name).
1822	*Cimex olivaceus* Thunberg, Dissert. Ent. Hem. Rostr. Capens., Acad. Typo. 2–3. (preoccupied; synonym).
1837	*Pentatoma orbitalis* Westwood, Cat. Hem. Coll. Rev. F.W. Hope, M.A., 8, 35 (London). (synonym).
1838	*Cimex facetus* Germar, Rev. Ent., 5: 172. (synonym).
1854	*Pentatoma lineaticollis* Stål, Ofvers. Vet.-Akad. Förh., 10[1853]: 220 (Stockholm). (synonym).
1867	*Strachia pentatomoides* Walker, Cat. Sp. Het. Hem. Coll. Brit. Mus., 2, 325 (London). (synonym).
1952	*Antestiopsis faceta*: Leston, Rev. Zool. Bot. Afr., 45: 269.
1953	*Antestiopsis orbitalis*: Leston, Ann. S. Afr. Mus., 51: 58.
1959	*Antestiopsis lineaticollis*: Le Pelley, East Afr. High Com., 54, 185 (Nairobi).
1966a	*Antestiopsis orbitalis orbitalis*: Greathead, Bull. Ent. Res., 58: 520, 521–525.
1998	*Antestiopsis thunbergii*: Rider, Proc. Ent. Soc. Wash., 100(3): 450–451, new combination for *Cimex thunbergii* Gmelin, 1790.

Antestiopsis thunbergii occurs from the southernmost part of Cape Province in South Africa to the highlands of Kenya and Ethiopia and usually is the most common Antestia bug on coffee where it is grown

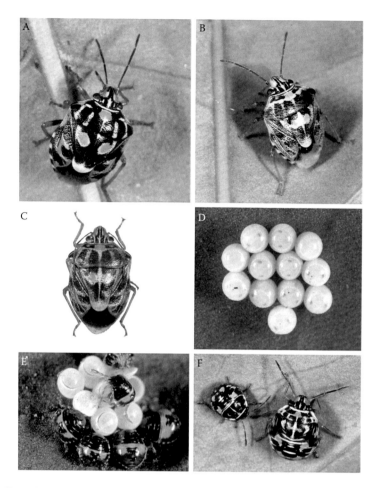

FIGURE 10.1   (See color insert.) *Antestiopsis* spp. adults and immatures. A–C, adults: A, *Antestiopsis thunbergii bechuana* (Kiambu *Coffea arabica* plantation, Kenya, 2015); B, *A. facetoides* (Kiambu *Coffea arabica* plantation, Kenya, 2015); C, *A. intricata* (Foumbot *Coffea arabica* plantation, Cameroon, 1985). D–F, *A. thunbergii bechuana* immatures: D, eggs; E, first instars on egg choria; F, third and fifth instars. Immature stages are from a laboratory colony, icipe, Kenya, 2015. (Images A–B, D–F: Courtesy of Robert S. Copeland, Biosystematic Unit, International Centre of Insect Physiology and Ecology, Nairobi, Kenya; C: Courtesy of Régis Babin.)

in this region. The western limit of the distribution of Antestia bugs on coffee seems to be the Katanga region in the Democratic Republic of Congo (DRC). The species includes several forms, collected from different provinces of South Africa, as well as two well defined subspecies, *A. thunbergii bechuana* (Kirkaldy) and *A. thunbergii ghesquierei* Carayon. *A. thunbergii bechuana* is the main pest of coffee from Zimbabwe to Central and Eastern Kenya with collections from Zimbabwe, Malawi, Zambia, Tanzania, and Kenya. *A. thunbergii ghesquierei* has a more western distribution from northeastern DRC to Ethiopia, with specimens collected from these two countries as well as from Burundi, Rwanda, Uganda, and Western Tanzania (Greathead 1966a).

10.2.3 *Antestiopsis intricata* (Ghesquière and Carayon)

1936	*Antestia faceta*: Hargreaves, East Afr. Agr. J., 1: 448–452. (misidentification).
1948	*Antestia intricata* Ghesquière and Carayon, Rev. Zool. Bot. Afr., 41: 58–59.
1954a	*Antestiopsis lineaticollis intricata*: Carayon, Bull. Sci. Sect. Tech. Agr. Trop., 5: 365–368.
1959	*Antestiopsis intricata*: Le Pelley, East Afr. High Com., 54 (Nairobi).

Antestiopsis intricata has a broad west-east distribution in tropical Africa, starting from the western coast to the eastern horn of the continent. It is the only species of Antestia bugs reported on coffee in West Africa. Carayon (1954a) reported that *A. intricata* was collected for the first time in the Grand-Bassam area of the Ivory Coast in 1855. The species, at first, was not considered distinct from *A. faceta* but because of the study of Ghesquière and Carayon (1948), the species was later named correctly as *A. intricata*. In 1942, the species (as *A. faceta*) was reported in large numbers on *Coffea canephora* (Robusta coffee) and *Coffea liberica*, in the southern Ivory Coast, especially in the Man area (Carayon 1954a). Some specimens were collected in the early 19th century from neighboring West African countries such as Guinea, Ghana, and Benin. Other specimens from Nigeria, Cameroon, Central African Republic, DRC, Uganda, Western Kenya, Ethiopia and Sudan were addressed by Carayon (1954b) and Greathead (1966a).

10.2.4 *Antestiopsis facetoides* Greathead

1876	*Antestia variegata*: Stål, K. Svens. Vet.-Akad. Handl., 14: 96 (Zanzibar). (part, misidentification).
1937	*Antestia faceta*: Kirkpatrick, Trans. R. Ent. Soc. Lond., 86: 250–251, 339–340. (misidentification).
1959	*Antestiopsis* sp. indescr.: Le Pelley, East Afr. High Com., 54: 54–55, 126, 234. (Nairobi).
1966a	*Antestiopsis facetoides* Greathead, Bull. Ent. Res., 58: 529–531, (new name for *variegata*: Stål, 1876).

The distribution of *Antestiopsis facetoides* Greathead is limited to Eastern Kenya and Eastern Tanzania, including Zanzibar, where it often is found mixed with *A. thunbergii bechuana* on coffee, sometimes displacing it in plantations at lower elevations.

10.2.5 *Antestiopsis clymeneis* (Kirkaldy)

1861	*Pentatoma confusa* Signoret, Faune Hém. Madagascar, Deux. Part. Hét., 3: 932. (preoccupied).
1865	*Antestia confusa*: Stål, Hem. Afr., Tom. Prim. 4: 201–202 (Stockholm).
1909	*Antestia clymeneis* Kirkaldy, Cat. Hem., 1: 128. (new name).
1934	*Antestia clymeneis* var. *flaviventris* Frappa, Bull. Econ. Madagascar, 89: 371. (synonym).
1952	*Antestia confusa*: Cachan, Mém. Inst. Scient. Madagascar, 1: 429.
1969	*Antestiopsis clymeneis*: Greathead, Bull. Ent. Res., 59: 307–310, new combination for *Antestia clymeneis* Kirkaldy, 1909.

Antestiopsis clymeneis (Kirkaldy) is known only from Madagascar, where it is reported on Arabica coffee and wild Rubiaceae. Greathead (1969) addressed the classification of the species and distinguished two subspecies: *A. clymeneis galtiei* (Frappa), which is the most common on coffee, and *A. clymeneis frappai*, a new subspecies reported on wild Rubiaceae of the genera *Gaertnera*, *Saldinia*, and *Mapouria*.

10.2.6 *Antestiopsis cruciata* (F.)

1775	*Cimex cruciatus* Fabricius, Syst. Ent. : 714–715.
1837	*Pentatoma pantherina* Westwood, Cat. Hem. Coll. Rev. F.W. Hope, M.A., 8: 34 (London). (synonym).
1843	*Pentatoma cruciata*: Amyot and Serville, Hist. Nat. Ins. Hem., 132–133.
1861	*Strachia geometrica* Nietner, Obs. enemies coffee tree Ceylon, 18. (synonym).
1867	*Strachia velata* Walker, Cat. Spec. Hem. Het. Coll. Brit. Mus., 329. (synonym).
1902	*Antestia cruciata*: Distant, Fauna Br. India, incl. Ceylon Burma, 1: 185. (new combination).
1918	*Plautia picturata* Distant, Fauna Br. India, incl. Ceylon Burma, 7: 136 (synonym).
1952	*Antestiopsis cruciata* Leston, Rev. Zool. Bot. Afr., 45: 269, new combination for *Antestia cruciata* Distant, 1902.

Although Leston (1952) included this Asian species in the genus *Antestiopsis*, some authors from Asia still use the name *Antestia cruciata* (e.g., Chandra et al. 2012). *Antestiopsis cruciata* is known from

Pakistan, India, Myanmar, Sri Lanka, Southern China, and Southeast Asia (Rider et al. 2002, Azim 2011). It is a well-known coffee pest in Asia (e.g., Nietner 1861). The species also is known as a pest of Jasmine (Baliga 1967).

10.2.7 Antestia Species Closely Related to the Group but Not Found on Coffee

In his taxonomic study of the group, Greathead (1966a) reported at least four additional *Antestiopsis* species, morphologically related to the group but not found on coffee and sometimes with questionable classification. *A. falsa* Schouteden was reported to occur in South Africa, Mozambique and Tanzania but was not found on coffee. *A. crypta* (Greathead) [= *A. transvaalia* (Distant)] was reported from the Katanga region of the DRC. *A. lepelleyi* (Greathead) and *A. littoralis* were reported from Central Kenya and the Kenyan Coast, respectively, but also have never been found on coffee. The Asian species *A. anchora* (Thunberg) has a wide distribution and was reported as a pest of rice (Barrion and Litsinger 1994).

10.3 Life History

Our knowledge of Antestia bug life history is mainly based on studies of the two economically important coffee pests of the group, *Antestiopsis thunbergii* and *A. intricata*. The biology of the other Antestia species rarely has been addressed because these species are considered of secondary importance due to their limited distribution and low damage on coffee.

10.3.1 Ecological Preferences

10.3.1.1 Relationships with Elevation

Antestia bugs are found mostly on Arabica coffee between 1000 and 2100 meters above sea level (m asl). However, their distributions suggest that ecological preferences differ between species. In Eastern Africa, *Antestiopsis thunbergii bechuana* is associated with the climate of the highest elevations, whereas *A. intricata* is more common in the warmer climate of forest areas and low to mid-altitude coffee growing areas (Greathead 1966a, Abebe 1987, Wrigley 1988). In Central Kenya, *A. facetoides* prefers coffee grown at lower elevations, although the species can be found mixed with *A. thunbergii bechuana* in plantations at mid-elevations (Greathead 1966a). In Rwanda, *A. thunbergii ghesquierei* is found in all coffee growing zones, from 1000 to above 2000 m asl (Foucart and Brion 1959). In West Africa, *A. intricata* has been collected on Robusta coffee at low elevations in the southern Ivory Coast (Carayon 1954a). In Cameroon, however, this species has not been collected below 1000 m asl and was absent totally from Robusta coffee (Mbondji Mbondji 1997).

10.3.1.2 Relationships with Shade and Microclimate

There are conflicting reports regarding the influence of shade on Antestia bugs. In Ethiopia where coffee is normally grown under shade trees, *Antestiopsis intricata* is a more serious pest in shaded coffee at lower altitudes (Abebe 1987). In Cameroon, a study by Mbondji Mbondji (1999) reported high infestations (more than nine bugs per tree) by *A. intricata* in shaded coffee plantations located at ≈1000 m asl whereas unshaded plantations were less infested in the same locality. In line with this, a recent work conducted in Western Kenya by Mugo et al. (2013) reported higher infestations by *A. thunbergii* in shaded compared to unshaded coffee grown between 1300 and 1500 m asl. Other studies have shown that at high elevations (≈2000 m asl) Antestia bugs were more numerous in warmer unshaded plantations (Le Pelley 1968). Most of the above authors agreed that shade did not directly affect Antestia populations; instead, the resulting microclimate was responsible. Kirkpatrick (1937) suggested that Antestia bug infestation levels in coffee plantations were affected by interactions of the pest with temperature and moisture.

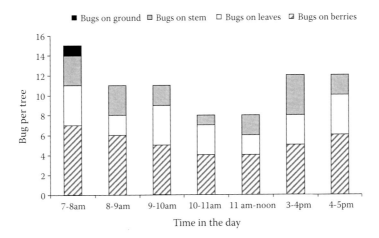

FIGURE 10.2 Mean number of Antestia bugs per tree observed on coffee berries, leaves, and stems and on the ground at various times of the day, from a sample of three coffee trees sheltering at least 12 Antestia bugs, in Rwanda (N. D. T. M. Rukazambuga, unpublished data).

10.3.1.3 Ecological Preferences at Tree Level

Kirkpatrick (1937) reported that *Antestiopsis thunbergii* adults and nymphs avoided direct sunlight during the hottest hours of the day when they usually hid within the coffee bush in the leaf cover. The author also wrote that Antestia bugs usually were more visible in the morning and evening or during cloudy days. Foucart and Brion (1959) reported that feeding activity of *A. thunbergii ghesquierei* usually occurred between 10 a.m. and noon and between 7 p.m. and 8 p.m. in Rwanda and Burundi. By contrast, the long flights of adults, involved in the process of new plantation colonization, usually occurred in the warmer hours of the day.

A recent study in Rwanda described the distribution of Antestia bug adults on different parts of the coffee tree during the day. The results presented in **Figure 10.2** showed that the bugs spent more time on green berries than on other parts of the tree, and that they declined in number from early morning to mid-day, then increased again in the afternoon. The numbers of bugs on leaves and stems were more variable, although they varied with a similar trend over the day. The number of bugs on the ground was negligible (N. D. T. M. Rukazambuga, unpublished data).

Most authors agreed that bushy coffee trees sheltered more Antestia bugs than open ones (e.g., Le Pelley 1968). In line with this observation, coffee pruning is a common recommendation for Antestia bug control. Although Antestia bug adults and nymphs are mobile insects, their movements between or within coffee trees are limited and presumably allow them to avoid large variations in temperature and moisture.

10.3.2 Feeding Habits

10.3.2.1 Host Plants

Antestia bugs have been found on a wide range of plants from different botanic families. However, feeding and/or breeding have been observed for a limited number of them, mainly from the Rubiaceae. According to old insect collections in Eastern Africa, Antestia bugs, most likely *Antestiopsis thunbergii*, were found on Poaceae such as finger millet (*Eleusine coracana* Gaertn.), sorghum (*Sorghum* spp.), and maize (*Zea mays* L.); on Musaceae such as banana (*Musa*); on Euphorbiaceae, such as the castor oil plant (*Ricinus communis* L.); and on Malvaceae such as okra (*Abelmoschus esculentus* (L.) Moench.); Antestia bugs also were found on fruits of Annonaceae, and on flowers of *Gliricidia maculata* Kunth (Fabaceae), *Symphonia globulifera* L. (Clusiaceae) and *Bridelia micrantha* (Hochst) Baill. (Phyllanthaceae) (Anderson 1919, Wilkinson 1924). In addition, eggs of Antestia bugs have been found on unidentified herbaceous weeds around coffee stems as well as on castor oil trees and filaos, *Casuarina equisetifolia* L. (Casuarinaceae), which are often associated with Arabica coffee as windbreaks (Wilkinson 1924). *A. thunbergii* also has been reported on *Chrysanthemoides monilifera* (L.) Norlindh (Asteraceae) and *Leucosidea sericea* Eckl. and Zeyh. (Rosaceae) (Le Pelley 1942, Greathead 1966b). In the DRC, Lefèvre and Hendrickx (1942) collected Antestia bugs on forest jasmine, *Jasminum abyssinicum* Hochst. ex DC. (Oleaceae) and on *Trema orientalis* (L.) Blume (Cannabaceae) in the Kivu region. In Cameroon, *A. intricata* was collected on Asteraceae such as *Vernonia amygdalina* Delile, *Ageratum conyzoides* L., *Bidens pilosa* L., and Amaranthaceae (*Achyranthes* sp.) (Mbondji Mbondji 1999). In Ethiopia, *A. intricata* once was recorded on Mauritius thorn, *Caesalpinia decapetala* (Roth) Alston (Fabaceae), during the time when green berries were not available on coffee trees (Crowe and Gebremedhin 1984), and on orange flowers, *Citrus sinensis* (L.) Osbeck (Rutaceae) (Bayissa and Tadesse 1981).

Rearing of Antestia bugs on some of these plants has been attempted in the past, with limited success. Best results were obtained in the DRC, where Antestia bugs fed and developed until the fourth instar on *Galinsoga parviflora* Cav. (Asteraceae) (Lefèvre and Hendrickx 1942), whereas in Cameroon, some adults were obtained from nymphs reared on *Vernonia amygdalina* (Mbondji Mbondji 1999). In the Ivory Coast, the full development of *Antestiopsis intricata* was obtained on *Solanum anomalum* Thonn. (Solanaceae), which abounds as a weed in coffee plantations (Lavabre 1952). However, because most of these plants are common within or in close association with coffee plantations, they usually are considered to be temporary hosts of Antestia bugs.

Authors uniformly agree that coffee, and especially *Coffea arabica*, is the favorite host plant of Antestia bugs. In the Ivory Coast and the DRC, *Antestiopsis intricata* has been found on two additional cultivated coffee species, *C. canephora* Pierre ex A. Froehner and *C. liberica* W. Bull ex Hiern, but these two hosts are considered occasional because no significant and long-term infestation has been reported for them. Five wild species in the family Rubiaceae (i.e., *Pavetta elliottii* K. Schum. and K. Krause, *Psychotria nairobiensis* Bremek, *Galiniera coffeoides* Delile, *Vangueria apiculata* (Verdc.) Lantz, and *Canthium* sp.) are known to be host plants for *A. thunbergii* in Eastern Africa (Hargreaves 1936, Le Pelley 1942). These plants allow the development of large populations of Antestia bugs and, because they are common in environments surrounding coffee, they evidently form reservoirs for reinfestation of coffee plants.

10.3.2.2 Feeding on Coffee

Antestia bug species feeding on coffee plants have similar habits. They feed on flower buds, berries at different stages of development and maturation, and green shoots and leaves. The feeding preferences for these different organs are controversial. Most authors agree that green berries, flower buds, and green twigs are consumed first. Mature leaves and berries with solid beans inside are chosen only if no other feeding sources are available (Kirkpatrick 1937). Le Pelley (1942, 1968) reported that large green berries and red berries are preferred, followed by small berries, and then shoots. Foucart and Brion (1959) wrote that feeding habits of *Antestiopsis thunbergii ghesquierei* were a crucial parameter of the pest development, more important that climate conditions. However, they reported that food preferences varied over the bug's development as well as over the year.

10.3.3 Phenology with Descriptions of Developmental Stages

10.3.3.1 Antestiopsis thunbergii

10.3.3.1.1 Egg Stage

Eggs of *Antestiopsis thunbergii* are 1.2 mm long and 1.0 mm large, white and attached to the substrate by a secretion of adhesive material. They usually are laid in clusters of twelve (**Figure 10.1D**). The incubation period varies greatly with temperature. Because most past rearing attempts were conducted in insectaries at room temperature, the locality, elevation, and period of the year are crucial factors affecting development duration. In Kenya, at ≈1800 m asl, Anderson (1919) reported an egg incubation period of 7 to 9 days from January to April and 13 to 15 days from May to December, whereas Le Pelley (1968) reported an incubation period of 6 to 15 days at mean temperatures from 22.5 to 16.9°C. In Uganda, at ≈1200 m asl, the incubation period was reported as 4 to 9 days by Hargreaves (1930); in Tanzania, at ≈900 m asl, from 5 to 6 days in February and March and from 8 to 10 days from June to August (Kirkpatrick 1937); and in Rwanda and Burundi, 9 to 13 days, with a significant influence of humidity (Foucart and Brion 1959). A recent study conducted with individuals of *A. thunbergii* collected on Kilimanjaro, Tanzania, and maintained in climatic chambers as a colony in Kenya reported mean egg stage durations of 8.9, 5.0 and 4.0 days at constant temperatures of 20, 25, and 30°C, respectively (Ahmed et al. 2016). Kirkpatrick (1937) reported that, when they were ready to hatch, all of the eggs in a cluster hatched in 2 or 3 hours. He also stated that in the absence of parasitoids, egg viability was high, with ≈100% of the eggs hatching at outside temperatures. In Kenya, egg viability of 94, 57, and 15% was reported at constant temperatures of 20, 25, 30°C, respectively, with the colony from Kilimanjaro (Ahmed et al. 2016).

10.3.3.1.2 Nymphal Instars

Antestiopsis thunbergii has five instars. Newly hatched first instars usually stay together on the egg choria for 1–3 days (**Figure 10.1E**) before beginning to feed. Sometimes, they can molt to the second instar without feeding, but, in this case, mortality is high (Kirkpatrick 1937).

Reports from Anderson (1919) and Kirkpatrick (1937) showed high variations in the durations of the nymphal stadia, depending on elevation and time of the year, with mean durations between 5.8 and 15.1 for the first stadium, 9.4 and 33 days for the second stadium, 10.4 and 22.5 days for the third stadium, 10.5 and 22.5 days for the fourth stadium, and 13.0 and 22.3 days for the fifth stadium. Foucart and Brion (1959) reported that both the duration of nymphal development and nymphal survival varied during the year, even in rearing conditions with standard feeding conditions. They assumed that nymphal development time was linked to evaporation, with high evaporation leading to shorter development time.

Table 10.1 shows results obtained by Le Pelley (1968) with a population from Kabete, Kenya, reared in an open-air shelter and those of a recent study with a population from Kilimanjaro, Tanzania, in incubators at constant temperatures (Ahmed et al. 2016). Clearly, as is usual for insects, development of nymphal instars was faster at higher temperatures. The above authors agreed that there was no difference in the duration of development between males and females.

10.3.3.1.3 Adult Stage

Morphology of *Antestiopsis thunbergii* adults is quite variable in size and coloration patterns in such a way that it is not easy to give an accurate description of the species. The bugs are between 7.5 and 9.5 mm, with females usually somewhat larger than males. Irrespective of the sex, the pronotum and scutellum have striking variable patterns with black, orange and white as colors (e.g., **Figure 10.1A**). The cuticle of the wing has orange and white marks, whereas the membrane and hind-wing are dark brown (Le Pelley 1968).

TABLE 10.1

Development Duration of the Five Nymphal Instars of *Antestiopsis thunbergii* Reared at Four Average Temperatures in an Open-Air Shelter in Kenya (Le Pelley 1968)[1] and at 3 Constant Temperatures in Incubators with Individuals from Kilimanjaro, Tanzania (Ahmed et al. 2016)[2]

	Development Stage Duration (in Days)					
	Stage 1	**Stage 2**	**Stage 3**	**Stage 4**	**Stage 5**	**Egg-Adult**
Average temp. (°C)[1]						
16-17	12-15	28-35	15-26	21-30	25-40	101-146
17-18	10-13	18-28	15-19	14-24	17-32	74-116
18-19	7-12	17-29	11-18	9-24	14-30	58-113
19-20	7-9	17-26	10-19	6-23	14-31	54-108
Constant temp. (°C)[2]						
20	8.5	25.4	15.6	13.6	19.4	89.6
25	5.0	11.3	10.7	12.7	18.9	63.1
30	3.3	9.4	9.6	10.1	18.9	55.8

In Tanzania, monitoring of adults at outdoor temperatures showed that the female lifespan was longer than the male, with averages between 121 and 131 days for females and 98 and 106 days for males (Kirkpatrick 1937). In Kenya, a mean lifespan of 93 days for females and 86 days for males was obtained for adults of a population from Kilimanjaro, Tanzania, reared in incubators at 20°C (Ahmed et al. 2016). In Rwanda and Burundi, Foucart and Brion (1959) reported an average female lifespan of 76.5 days, with a maximum value of 245 days, whereas males survived 50.2 days on average with a maximum of 96 days. The complete life cycle was estimated to be between 3 and 6 months depending on temperature (Kirkpatrick 1937, Ahmed et al. 2016). In Rwanda and Burundi, duration of development from egg to adult was reported between 72 and 82 days, whereas the duration from egg to mature female (i.e., ready to oviposit) was between 90 and 99 days (Foucart and Brion 1959).

10.3.3.2 *Antestiopsis intricata*

10.3.3.2.1 *Egg Stage*

Eggs are almost spherical and are 1.05 mm in length and 0.90 mm in width (Mbondji Mbondji 2001). They are dull white and develop two reddish lines just before hatching (Le Pelley 1968, Abebe 1987). Eggs usually are laid on the underside of leaves in clusters of 12, but sometimes they are laid on other plant parts such as berries, branches, and the trunk (Le Pelley 1968). They are bonded to the support by glue excreted by the female during oviposition.

Under laboratory conditions in Ethiopia, the incubation period of eggs ranged from 3 to 5 days with an average of 3.6 days at a mean temperature of 25.1°C (Abebe 1987). In another study in Ethiopia, incubation periods ranged from 6 to 8 days, with an average of 7.4 days at a mean temperature of 27°C (Mendesil and Abebe 2004). In outside conditions in Western Cameroon (1100 m asl), incubation periods were 6 to 7 days between June and October, at a mean temperature of 25.8°C and 4 to 5 days between November and March, at a mean temperature of 27.4°C (Mbondji Mbondji 2001).

10.3.3.2.2 *Nymphal Instars*

Antestiopsis intricata has five instars, which are circular in shape and similar in color to the adults. Molts from one instar to another usually occur during the night. The first instar is ≈1.1 mm in length, circular in shape, and slightly flattened dorsoventrally. The thoracic segments are similar, brown in color as is the head. The second instar is similar but longer, 2.0 mm on average. The third instar is longer, 2.8 mm, and the prothorax begins to extend posteriorly. The fourth instar is 4.2 mm on average and has well-developed wing pads. The fifth instar has wing pads that overlap the first abdominal segments. Global shape and colors, as noted above, are similar to those of adults. At this stage, male and female nymphs begin to differentiate in size with average lengths for males and females of 5.70 and 6.45 mm, respectively. The mean durations of the various nymphal stadia from two studies, one in Western Cameroon (Mbondji Mbondji 2001) and one in Ethiopia (Mendesil and Abebe 2004), are given in **Table 10.2**.

TABLE 10.2

Development Duration of the Five Nymphal Instars of *Antestiopsis intricata* Reared at Two Average Temperatures in Outside Conditions Calculated in Western Cameroon (1100 m) for Two Distinct Periods, June to October and November to March (Mbondji Mbondji 2001),[1] and at One Average Temperature Under Laboratory Conditions in Ethiopia (Mendesil and Abebe 2004)[2]

	Development Stage Duration (in Days)					
	Stage 1	Stage 2	Stage 3	Stage 4	Stage 5	Egg-Adult
Average temp. (°C)[1]						
25.8	6.0	11.0	9.5	9.5	13.5	56.0
27.4	6.0	10.6	8.6	8.6	12.5	50.8
Average temp. (°C)[2]						
27.0 ± 0.5	6.6	10.3	7.5	8.8	9.7	50.3
	(6–8)	(9–13)	(7–10)	(6–14)	(7-14)	(41-67)

10.3.3.2.3 Adult Stage

Morphology of *Antestiopsis intricata* adults is less variable compared with that of *A. thunbergii* (**Figure 10.1C**) and relatively constant throughout its distribution area. Abebe (1987) described specimens collected in Ethiopia as follows: "The adult is black to brownish or greyish black with white to Ivory longitudinal stripes. There are orange spots on the dorsal surface. It is about 8 mm in length…" He also reported a mean life span of 95 days at 24.5°C, with 94 days and 104 days as minimum and maximum lifespans, respectively.

10.3.4 Reproduction on Coffee

10.3.4.1 *Antestiopsis thunbergii*

In coffee plantations, *Antestiopsis thunbergii* females were more numerous than males, with a sex ratio of 0.61:0.39 (female:male) in Tanzania (Kirkpatrick 1937). In the laboratory in Kenya, a sex ratio of 0.53:0.47 (female:male) was reported at a constant rearing temperature of 20°C, with a population from Kilimanjaro, Tanzania (Ahmed et al. 2016). Mating occurred at dusk and during the night. A pre-oviposition period was recorded as 13 to 36 days, with a mean of 19 days by Anderson (1919), whereas the study in Kenya reported a mean pre-oviposition period of ≈24 days for females reared at 20°C and 26 days for those reared at 25°C (Ahmed et al. 2016).

Fecundity data are quite variable. In Kenya, Anderson (1919) reported a mean fecundity of 126 eggs per female, within a period of 89 days, with a maximum fecundity of 485 eggs for a laying period of 256 days. In Tanzania, Kirkpatrick (1937) reported a mean fecundity of ≈150 eggs per female and not more than 240 eggs. In Rwanda and Burundi, Foucart and Brion (1959) reported a mean fecundity of 158 eggs per female. A recent study reported a mean fecundity of 133 eggs for a population from Kilimanjaro, Tanzania, reared at 20°C in Kenya and a drastic drop of the fecundity at 25°C and 30°C (Ahmed et al. 2016).

Fertilized *Antestiopsis thunbergii* females laid eggs in batches of about 12, ranging between 9 and 15 eggs. The last batch or the last two batches usually were incomplete, containing only 4 to 6 eggs (Anderson 1919). The eggs were laid on the underside of a coffee leaf, sometimes on a berry or the trunk, and rarely on dried leaves. Approximately 90% of the eggs were laid on the underside of the leaf (Anderson 1919).

10.3.4.2 *Antestiopsis intricata*

Under field conditions, *Antestiopsis intricata* females were more numerous than males. A sex ratio of 0.67:0.33 (female:male) was reported from a sample of 500 individuals collected in Western Cameroon (Carayon 1954a). A reared population gave a sex ratio of 0.56:0.44 (female:male) in the same locality (Mbondji Mbondji 2001).

Also under field conditions in Western Cameroon, females became sexually mature 5 to 6 days after emergence (Mbondji Mbondji 2001). In Ethiopia, females maintained on Arabica green berries under laboratory conditions became sexually mature 7.3 days after emergence (Mendesil and Abebe 2004). In Western Cameroon, mating occurred during the cooler parts of the day, mainly in the late afternoon, or during the night and lasted 3 hours on average (Mbondji Mbondji 1999, 2001). In the same conditions, females started laying eggs 4 to 7 days after mating. In Ethiopia, oviposition started 4.7 days after mating and lasted up to 227 days, with an average oviposition period of 134.8 days (Mendesil and Abebe 2004).

As with *Antestiopsis thunbergii*, eggs of *A. intricata* usually were laid in batches of twelve eggs under the coffee leaves or, sometimes, on berries or shoots. The study conducted in Ethiopia reported a daily fecundity of 12 eggs on average and a total of 61 to 457 eggs per female with an average of 324 eggs. In this study, the peak oviposition period was recorded at 25 days after adult emergence, and the number of eggs laid declined with increasing age. There was an inverse relationship between age and the number of eggs laid per female (r= −0.91, P<0.01) (Mendesil and Abebe 2004). In Western Cameroon, females laid a maximum of 13 batches, or 135 eggs, with an oviposition period of 68 days (Mbondji Mbondji 1999, 2001).

10.3.5 Population Dynamics on Coffee

10.3.5.1 Seasonal Variation

A study of seasonal variation in population size of *Antestiopsis intricata* was conducted for ≈10 years in southwestern Ethiopia. It showed that the pest populations increased from February to March, peaked in May to June, and declined thereafter (Abebe 1987). Similar trends were reported by Chichaybelu (2008) for seasonal variations of the same species in the same area.

Another study conducted in Cameroon at two different elevations revealed different trends (Mbondji Mbondji 2001); at 1100 m asl, *Antestiopsis intricata* population density remained at ≈2 bugs per tree between January and May, then decreased between June and October, and then increased quickly to reach a peak in December, with ≈9 bugs per tree (**Figure 10.3A**). In this study, pest infestations at 1100 m asl reached or were above the economic threshold of 2 bugs per tree for half of the year, which was used to trigger control measures. By contrast in this study, *A. intricata* populations never reached the density of 1 bug per tree at 1800 m asl (**Figure 10.3B**) but showed similar seasonal variation as populations in southwestern Ethiopia.

In Rwanda, a study conducted in the middle of 1950s described seasonal variations in populations of *Antestiopsis thunbergii ghesquierei* (Foucart and Brion 1959). Densities were assessed using the "pyrethrum test," a knock-down technique based on pyrethrum spraying. Population densities were high with not less than 5 bugs per coffee tree during the period, and the peak density was 45 bugs per tree in February (**Figure 10.4**). Populations were between 10 and 18 bugs per tree from July to September, with a small peak in August, then quickly increased from November to reach a peak in February and steadily

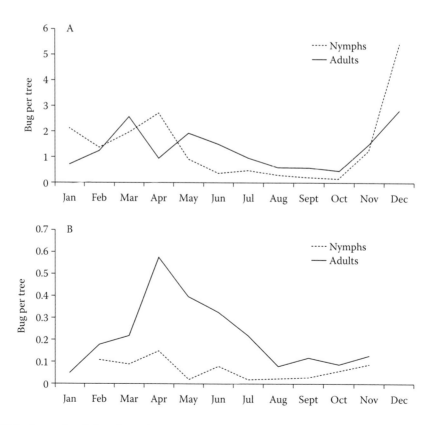

FIGURE 10.3 Seasonal variation of two *Antestiopsis intricata* populations at 1100 m asl (A) and 1800 m asl (B) in Western Cameroon (from Mbondji Mbondji 2001). Densities were obtained using the "knock-down" technique by spraying 100 coffee trees sampled in four plantations with insecticide every month during 2 years.

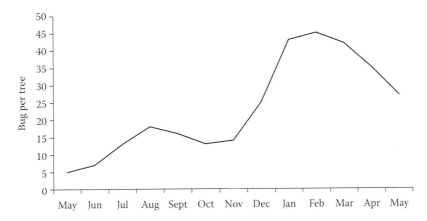

FIGURE 10.4   Seasonal variation of *Antestiopsis thunbergii ghesquierei* populations in Rwanda-Burundi from May 1954 to May 1955 (from Foucart and Brion 1959). Densities were obtained using the "pyrethrum test," a knock-down technique using pyrethrum insecticide.

dropped from March to May. Recent studies in Burundi reported lower densities, although using the same assessment technique (Leblanc 1993, Bouyjou et al. 1999).

10.3.5.2 Spatial Distribution in Plantations

Some studies have addressed the spatial distribution of Antestia bug populations in coffee plantations in order to recommend reliable sampling procedures for accurate infestation estimations. For instance in Burundi, Cilas et al. (1998) showed that *Antestiopsis thunbergii ghesquierei* populations were aggregated in coffee plantations, meaning that some coffee trees sheltered several insects whereas others were totally free from insects. Moreover, the study showed that *A. thunbergii* adults usually were more aggregated than nymphs and the authors suggested that this may have been due to reproductive behavior and may have involved communication with semiochemicals. The study also showed that population distributions were not structured spatially; in other words, infested coffee trees were not grouped but randomly distributed in the plantation. The above authors suggested that, while laying eggs, females could move through the plantation and detect more attractive coffee trees, ultimately more favorable for their offspring.

However, a recent study conducted in a plantation in Rwanda found that the population of *Antestiopsis thunbergii ghesquierei* was distributed mainly on the coffee plants in the border of the plantation (85.5%), whereas a small percentage of the population (14.5%) was on coffee plants in the middle of the field. Moreover, the study showed that 63% of the population was on coffee trees with berries, whereas only 37% was on trees without berries (N. D. T. M. Rukazambuga, unpublished data).

10.3.5.3 Factors Involved in Population Dynamics

Most authors agree that, in the absence of control practices, Arabica coffee plantations offer conditions for the survival of Antestia bug populations throughout the year (Le Pelley 1968). However, as shown above, populations varied significantly in time and space in coffee plantations, and this may be linked to different agroecological factors affecting pest behaviour and development. Although the feeding habits of the bugs (see **Section 10.3.2**, Feeding Habits) suggest that wild Rubiaceae present in close association with plantations may encourage infestation, coffee tree phenology also is an important factor because these bugs show a preference for some parts of the tree over others, particularly green berries and shoots. Thus, Foucart and Brion (1959) suggested that the low number of berries on coffee trees as well as the dry weather were factors affecting the annual population decrease in Rwanda and Burundi. Microclimate conditions (see **Section 10.3.2.1**, Host Plants) induced by elevation, shade, weather, and coffee tree morphology also are crucial factors impacting population dynamics in plantation. Another factor that will be addressed in the next section is natural enemies, which are numerous for Antestia bugs.

10.4 Predators and Parasites

10.4.1 Predators

Carayon (1954a), in his study of *Antestiopsis* in West Africa, recorded some generalist predators of *A. intricata* including assassin bugs (Heteroptera: Reduviidae), specifically *Pseudophonoctonus formosus* (Distant), *Hediocoris fasciatus* Reuter, *Nagusta punctaticollis* Stål, and *Rhynocoris albopunctatus* Stål. A recent study in Cameroon confirmed the presence of these species in Arabica coffee and showed the dominance of *P. formosus*, which was both the most common and the most active predator (Mbondji Mbondji 2001). Some mantid species (Mantodea: Mantidae) also were recorded on Arabica coffee preying on nymphs and adults of *A. intricata* in West Africa (Carayon 1954a) and in Ethiopia (IAR 1984), and on nymphs and adults of *A. thunbergii ghesquierei* in Rwanda (Sylver Habumugisha, unpublished data).

10.4.2 Egg Parasitoids

Most reported Antestia egg parasitoids are from the family Platygastridae: the Scelioninae *Gryon fulviventre* (Crawford) (=*Hadronotus antestiae* Dodd) has been collected from eggs of *Antestiopsis intricata, A. thunbergii bechuana, A. thunbergii ghesquierei* and *A. facetoides*, with collections on coffee from Kenya, Tanzania, Uganda, Rwanda, Burundi, the DRC, and Cameroon (Brion 1963, Greathead 1966b, Mbondji Mbondji 2001). The telenomine *Telenomus* (=*Asolcus*) *sechellensis* (Kieffer) is an egg parasitoid of *A. intricata, A. facetoides* and of different *A. thunbergii* forms and subspecies, with a distribution on coffee in Kenya, Tanzania, Ethiopia, Uganda, Rwanda, Burundi, Seychelles, South Africa, and Cameroon (Brion 1963, Greathead 1966b, Mugo and Ndoiru 1997, Mbondji Mbondji 2001). Two telenomines, *Trissolcus* (=*Asolcus*) *mopsus* (Nixon) and *Trissolcus* (=*Asolcus*) *suranus* (Nixon), were recorded from eggs of *A. thunbergii ghesquierei* and *A. intricata*. *T. mopsus* is distributed on coffee in Uganda, Rwanda, Burundi, Cameroon, Ethiopia, and South Africa, whereas *T. suranus* is present only in Uganda.

The family Eupelmidae includes one egg parasitoid, *Anastatus antestiae* Ferrière, collected from *Antestiopsis intricata, A. thunbergii bechuana,* and *A. thunbergii ghesquierei*. This parasitoid is distributed on coffee in Kenya, Tanzania, Uganda, Rwanda, Burundi, the DRC, and Cameroon. The family Pteromalidae also includes an egg parasitoid of Antestia bugs, *Acroclisoides africanus* Ferrière, with similar hosts and distribution as the Eupelmidae. A genus of Encyrtidae, *Ooencyrtus*, was noted as an Antestia egg parasitoid in Kenya (Le Pelley 1968).

Pediobius and *Aprostocetus*, hyperparasitoids in the family Eulophidae, were collected from Antestia eggs in Uganda, and Uganda and Rwanda, respectively. Another hyperparasitoid, from the subfamily Scelioninae, was identified as *Baryconus* sp. in Kenya (Le Pelley 1968).

10.4.3 Nymph and Adult Parasitoids

Three species in the family Braconidae have been reported as nymphal parasitoids of Antestia bugs. The most common species, *Aridelus coffeae* Brues, was collected from nymphs of *Antestiopsis thunbergii bechuana, A. thunbergii ghesquierei, A. facetoides,* and *A. intricata* with a distribution on coffee in Kenya, Tanzania, and Uganda; *Aridelus taylori* Nixon has been reared from *A. thunbergii ghesquierei* and *A. intricata* in Uganda, and *Aridelus rufus* Cameron was obtained from *A. thunbergii bechuana* in Tanzania. These three species preferred to attack young nymphs such as the second and third instars. Parasitized first instars usually die thus depriving the parasite a chance to develop (Kirkpatrick 1937, Greathead 1966b, Le Pelley 1968).

The fly family Tachinidae includes various species of the genus *Bogosia* that are known to be parasitoids of Antestia bug adults. The best known and most widely distributed is the species *B. rubens* Villeneuve, which has been reared from *Antestiopsis thunbergii ghesquierei* and *A. intricata* collected on coffee in Tanzania, Kenya, Uganda, DRC, Rwanda and Ethiopia.

The strepsipteran *Corioxenos antestiae* Blair (Corioxenidae) also is a parasitoid of Antestia bugs in the adult stage. The host range includes most of the *Antestiopsis* species of economic importance. For example, *C. antestiae* has been collected on coffee in Kenya, Tanzania, Uganda, Ethiopia, and Cameroon (Kirkpatrick 1937; Greathead 1966b, 1971; Abebe 1999; Mbondji Mbondji 2001).

10.5 Damage to Coffee

10.5.1 Direct Damage on Flower Buds, Berries, and Shoots

Antestia bugs cause a considerable amount of feeding damage. Feeding on flower buds results in browning or blackening of buds causing failure to set fruit (Le Pelley 1968, Mugo 1994). Loss of flowers is significant when high infestations of Antestia bugs occur during the onset of rains (Waller et al. 2007). Severe infestations may even prevent the tree from flowering (Wrigley 1988). Feeding on young green berries results in premature fruit drop. The berries are shed by the production of an abscission layer on the stock leading to a significant crop loss, which is difficult to assess (Foucart and Brion 1959).

Feeding on well-developed berries results in different types of damage, all of them affecting strongly the gustative quality of coffee beans (Ribeyre and Avelino 2012). This damage is often difficult to detect on berries and is visible only after the coffee washing process when beans are stripped and cleaned of parchment. Berries damaged by these bugs usually results in beans with crater-like depressions (Leblanc 1993). If the feeding lesions are infected by *Eremothecium* fungi, the beans become brown or black and sometimes show the damage known as "brown zebra coffee beans" (Ribeyre and Avelino 2012).

Developed leaves in the field also are attacked by Antestia bugs; however, in the complete absence of flowers or berries, the shoots are preferred over leaves. When young leaves at the growing points are damaged, they become scarred and distorted. Multiple branching of shoots always is associated with Antestia damage where in severe cases this leads to "witches-broom" effect. The first effect of the feeding of *Antestiopsis* on the shoot is marked by a reduction in growth that is accompanied by a shortening of internodes. This is followed by a duplication or multiplication of branches as a result of the feeding on the growing point, producing a typical bunchy or matted growth (Le Pelley 1968, Wrigley 1988).

10.5.2 Microbe Transmission

Entry of pathogenic bacteria or fungi into the plant tissue through insect feeding wounds is common (e.g., Mitchell 2004). In this same manner, Antestia bugs may provide fungi with access to berries during feeding (Le Pelley 1942). Fungal species involved are *Eremothecium coryli* and *Ashbya* (=*Eremothecium*) *gossypii* (Wallace 1931, 1932; cited in Le Pelley 1942). However, the exact transmission mechanism, direct or passive introduction, has yet to be elucidated. The characteristic injury symptom of these fungal infections is bean rot. These fungi have been observed in berries with feeding lesions of *Antestiopsis thunbergii* and *A. intricata* (Mitchell 2004).

Recent research by Matsuura et al. (2014) identified bacterial symbionts associated with *Antestiopsis thunbergii*. They characterized a gammaproteobacterial gut symbiont as the primary symbiotic associate of an obligate nature for *A. thunbergii*, and three facultative symbionts in the genera *Sodalis*, *Spiroplasma*, and *Rickettsia*. J. Carayon (personal communication) suggested that symbionts might play a crucial part in Antestia bug alimentation. He hypothesized that the behavior of newly hatched first instars, which usually stay together on the egg choria for 1 to 3 days before dispersing and beginning to feed, may facilitate the acquisition of these symbionts from the female adult to her progeny.

10.5.3 Coffee Potato Taste Defect and Possible Causing Mechanisms

Potato taste defect (PTD), also known as "peasy off-flavour," is a flavor defect characterized as a potato-like taste in coffee, which diminishes coffee quality and, thus, makes coffee undesirable (Jackels et al. 2014). Current research is designed to elucidate the role of Antestia bugs in PTD transmission with the hypotheses that PTD is caused by Antestia-derived chemicals originating from the bugs feeding on berries or from Antestia-vectored pathogens or Antestia-borne microorganisms such as symbiotic bacteria (Matsuura et al. 2014).

Bouyjou et al. (1993, 1999) reported that PTD flavor was produced by an Enterobacteriaceae bacterium, and they hypothesized the linkage of Antestia bugs, bacteria, and PTD. The bacterium recently was identified as *Pantoea coffeiphila* (Gueule et al. 2015). The compounds responsible for PTD first were identified as 3-isopropyl-2-methoxypyrazine (= 2-isopropyl-3-methoxypyrazine, IPMP) and 2-isobutyl-3-methoxypyrazine (IBMP) (Becker et al. 1988). The recent study of Jackels et al. (2014) revealed that IPMP is produced inside the beans but not deposited on the bean surface by a bacterial growth because they found IPMP only in interior volatiles of PTD coffee, not in surface volatiles. Also, in the PTD coffee surface volatile, as well as on desiccated *Antestiopsis thunbergii*, they found three major alkanes including tridecane, using Gas Chromatography–Mass Spectrometry. Their results supported the hypothesis of Bouyjou et al. (1993). In addition, links between bacterial symbionts of Antestia bugs in the development of potato taste defect in coffee may be worth investigating (Matsuura et al. 2014).

10.6 Management and Control

10.6.1 Economic Impact

Antestia bugs are major pests of Arabica coffee in Africa everywhere it is grown on the continent. They cause significant losses both in yield and quality of coffee. Globally, production loss due to all the pests of the group is estimated between 20 and 35%, but in East Africa crop losses up to 45% are reported. In Ethiopia, a strong correlation between berry drop and the presence of *Antestiopsis intricata* in coffee plantations was noted and the crop loss due to berry drop was assessed at 9% (Tadesse et al. 1993). Studies in the same country also showed that Antestia bugs caused ≈48% darkened coffee beans (Abebe 1988; IAR 1996a, b). A study by Chichaybelu (2008) using artificial infestation with Antestia bugs showed that four pairs of the pest on a branch caused 54% berry drop and 90% damaged berries.

In Ethiopia, Antestia bugs, in general, are a more serious problem in large plantations compared to coffee home gardens and coffee forest production systems. In Kenya, a study showed that 2 to 4 Antestia bugs per tree caused a crop loss between 15 and 27% in total bean weight (Wanjala 1980), whereas in Uganda, 20, 36, and 51% yield losses were attributed to densities of 0.5, 1, and 2 bugs per coffee bush, respectively (McNutt 1979). In Rwanda, density of 5 bugs per tree was shown to cause ≈20% damaged berries, and ≈7% total crop loss. The same study showed that with high density at 30 bugs per tree, crop loss was more than 30% (Foucart and Brion 1959). Following a large sampling effort in washing stations of various coffee growing areas in Rwanda in the 1960s and 1970s, Leblanc (1993) reported rates of damaged berries between 2 and 40% and total crop losses between 1 and 32%.

The Great Lakes region in Africa also is strongly affected by PTD, with a presence centered in the western Rift Valley areas of Rwanda, Burundi, Eastern DRC, and Western Uganda. In Rwanda, for instance, the presence of PTD with other quality defects disqualifies up to 80% of the high quality green coffee that initially was qualified to be specialty coffee grade as opposed to standard grade. This effect has great negative impacts on revenue from coffee as the standard grade has a much lower price than specialty coffee (N. D. T. M. Rukazambuga, unpublished data).

10.6.2 Control Methods

10.6.2.1 Cultural Methods

10.6.2.1.1 Coffee Pruning and Shade Management

Shade-tree regulation and pruning of coffee trees are the common cultural practices recommended to reduce Antestia bug outbreaks. Because the pests prefer dense foliage, pruning is recommended to keep the coffee bush open. Shade and pruning measures modify the microclimate and create unfavorable conditions for Antestia bugs, reducing infestations (Crowe and Gebremedhin 1984, Wrigley 1988, IAR 1996b). In Ethiopia, studies on the effect of pruning indicated a significant reduction in the amount of coffee bean damage and berry loss due to Antestia bugs. These measures also increased insecticide efficiency towards Antestia bugs (Chichaybelu 1993).

10.6.2.1.2 Handpicking

In the 1930s, handpicking was used to control Antestia bugs in East Africa. Prior to collection, bugs were forced down on the coffee stem by using smoke obtained from smoldering cow dung. Le Pelley (1968) reported that many millions of Antestia bugs were collected this way in Kenya in 1933 and 1934 but with little benefit compared to the costs.

10.6.2.2 Control with Insecticides

10.6.2.2.1 Poison-Bait Sprays

This control practice was used in the past in East Africa, but although it no longer is used, it gave good results in Tanzania and Uganda (Le Pelley 1968). The method is based on the fact that Antestia bugs usually drink water from dew on coffee leaves and are attracted by sugar baits. The method was to spray a water solution of arsenic and sugar on coffee bushes (Le Pelley 1968).

10.6.2.2.2 Botanical Insecticides

Botanical insecticides such as pyrethrum and neem developed for use on tropical crops were notable because they were produced locally and cheaper than commercial synthetic insecticides. For decades they have been considered good alternatives to commercial insecticides because of their harmlessness for human health and the environment due mainly to selectivity of action.

Pyrethrum extracted in kerosene has been used in Kenya for the control of Antestia bugs since the early 1930s. The method consisted of covering coffee bushes with cotton sheets and spraying them with the insecticide using a hand-atomizer (Le Pelley 1932). The high toxicity of pyrethrum for Antestia bugs has led researchers to improve the method, and pyrethrum water emulsions and dusts have been developed and used with success in East Africa (Crowe et al. 1961).

In Rwanda, pyrethrum products are currently approved for the control of Antestia bugs in organic certified coffee. Because of the short persistence of photounstable pyrethrum (24 hours only), spraying is recommended when the bugs are active (i.e., in the morning between 6 and 9 and in mid-afternoon between 3 and 5 [N. D. T. M. Rukazambuga, unpublished data]).

Recently, in Ethiopia, different plant species present in coffee growing areas were tested under laboratory conditions for the control of *Antestiopsis intricata*. Extracts of *Millettia ferruginea* (Hochst.) Baker, a natural product, resulted in 84% adult and 78% nymphal mortality, 24 hours after treatment, whereas *Chrysanthemum cinerariifolium* L. (pyrethrum) extracts caused 79% adult and 80% nymphal mortality under the same laboratory conditions (Mendesil and Abdeta 2007). The study also showed toxicity of some plant extracts against eggs of Antestia bugs. In another laboratory-based experiment, essential oils of the three plants *Dysphania* (*Chenopodium*) *ambrosioides* (L.) Mosyakin & Clemants, *Thymus vulgaris* L., and *Ruta chalepensis* L. resulted in 92.5, 90, and 87.5% mortality of Antestia bugs, respectively (Mendesil et al. 2012).

10.6.2.2.3 Chemical Insecticides

In Cameroon and Kenya, first attempts to control Antestia bugs with synthetic insecticides were carried out in the middle of the 1940s, by using broad spectrum dichlorodiphenyltrichloroethane (DDT) (Thelu 1946, Le Pelley 1968). This insecticide quickly proved to be very efficient, but after a few years, some problems emerged, particularly secondary population outbreaks of other coffee pests, such as leaf-miners and mealybugs, which had been, until then, limited by high levels of parasitism. In Cameroon, DDT was quickly abandoned in favor of hexachlorocyclohexane (HCH) and lindane, its gamma isomer. Lindane was not used in Kenya because it was thought to taint coffee (Le Pelley 1968). From the middle of the 1950s, organochlorine insecticides were replaced progressively by organophosphates. Those organophosphates that were allegedly moderately toxic to humans were recommended widely in Africa for decades for the control of Antestia bugs (McNutt 1979, Crowe and Gebremedhin 1984). Some organophosphates, such as chlorpyrifos-ethyl, malathion, fenthion, and fenitrothion, still are recommended and/or in use today for Antestia bug control. For example, a survey conducted in the early 2010s in Kenya showed that the four most commonly used insecticides in coffee farms were organophosphates, namely chlorpyrifos-ethyl, fenitrothion, diazinon, and dimethoate (Mugo et al. 2011). The toxicity of organophosphates to humans and adverse effects on the environment are well known, and the use of most of them largely has been restricted, if not totally prohibited, in temperate regions.

New molecules have been tested and recommended since the early 1990s. In Ethiopia, Chichaybelu (1993) screened different insecticides for Antestia bug control and reported that two synthetic pyrethroids, cypermethrin and lambda-cyhalothrin gave good results. Today, these two molecules also are considered to be highly toxic for the environment, primarily because of toxicity to fishes. In Cameroon, Mbondji Mbondji and Ngollo Dina (1992) conducted large field experiments to compare efficacy of various insecticides. This study showed good ability of pyrethroids to control *Antestiopsis intricata* on coffee (**Table 10.3**).

New neonicotinoid insecticides were recommended recently for pest control on coffee. In Rwanda, for example, imidacloprid, a systemic insecticide was recommended for the control of Antestia bugs starting in 2013 and showed a 3-month persistence and effectiveness in the field. An optimal timing of spraying well ahead of harvest is recommended to avoid pesticide residues in coffee berries (N. D. T. M. Rukazambuga, unpublished data).

TABLE 10.3

Comparison of Field Efficacy (in % of the Total Antestia Bug Population Killed) of Different Insecticides for the Control of *Antestiopsis intricata* on Coffee in Cameroon (from Mbondji Mbondji and Ngollo Dina 1992)

Active Ingredient	Group	Dose (g/ha)	Field Efficacy (%)
Lindane	Organochlorine	320.0	95.3
Isoprocarb	Carbamate	184.5	67.5
Propoxur	Carbamate	184.5	94.8
Diazinon	Organophosphate	455.0	41.3
Fenthion	Organophosphate	1000.0	40.9
Chlorpyriphos-ethyl	Organophosphate	750.0	94.3
Deltamethrin	Pyrethroid	8.0	94.5
Alpha-cypermethrin	Pyrethroid	24.0	97.4
Fenobucarb + Fenvalerate	Carbamate + Pyrethroid	375.0 + 25.0	92.5
Fenitrothion + Cypermethrin	Organophosphate + Pyrethroid	360.0 + 6.0	92.2
Fenitrothion + Lambda-cyhalothrin	Organophosphate + Pyrethroid	360.0 + 5.0	98.0

10.6.2.2.4 Organization of Chemical Control

Insecticide sprayings may be the only efficient control measure to stop an Antestia bug outbreak that could lead to high crop loss. However, chemical control has some disadvantages, besides cost and human and environment health issues. Ideal insecticide use requires sophistication in rational use that may be absent is some locations. Economic thresholds have been fixed for Antestia bug infestations by research institutions to help coffee farmers and private companies organize chemical sprayings. Economic threshold is that pest density (given in number of insects per tree) that likely will cause crop loss that exceeds the cost of chemical control. The bugs can cause considerable damage even when present in relatively small numbers, and economic thresholds usually are low for Antestia bugs. They vary from a country to country and sometimes between regions of the same country. For example, in Kenya, in East of Rift Valley, a mean of 2 bugs per coffee tree is considered to have attained the economic injury level. In West of Rift, which is a wetter area, economic thresholds are exceeded when more than 1 bug per tree is present (Mugo et al. 2011). In Uganda, Rwanda and Burundi, sprayings are recommended if mean density exceeds 1 bug per tree (McNutt 1979, Leblanc 1993). In Ethiopia, 5 bugs per tree is the current economic threshold for chemical control (Crowe and Gebremedhin 1984).

A method for a quick and cheap monitoring of Antestia bug populations on coffee has been available since the 1930s and used for decades, with some improvements. The method, commonly referred to as "pyrethrum test," consists of spraying a determined number of coffee trees with pyrethrum and counting the Antestia bugs that fall on plastic sheets that were spread on the ground around the trees before spraying (Le Pelley 1968). The number of trees to be sprayed and the spraying frequency for a good population assessment were controversial. A study conducted in the 1990s in Burundi statistically confirmed that 10 trees per plantation was the good sampling size for optimal results (Bouyjou and Cilas 1992).

10.6.2.3 Biological Control

10.6.2.3.1 Parasitism Rate and Parasitoid Value

As discussed in **Section 10.4**, Antestia bugs are hosts for a wide range of parasitoids, especially egg parasitoids. Good parasitism rates indicate these species are promising agents for biological control. Greathead (1966b) reviewed the literature on Antestia bug parasitoids in Africa and obtained information on parasitoid life history and their potential value in biological control. His main points are summarized as follows: (1) egg parasitism was always high regardless of the area studied, with rates ranging from 40 to 95%; the predominant egg parasitoid species was *Telenomus* (= *Asolcus*) *sechellensis*; (2) nymphal parasitism reached 50% and was due to *Aridelus* spp., with promising perspectives in some areas; and (3) adult parasitism due to *Bogosia* spp. (Tachinidae) and *Corioxenos antestiae* (Strepsiptera) usually did not exceed 10%, and, therefore, these parasitoids were not considered to be good candidates for biological control.

In Rwanda and Burundi, parasitoids have been found to reduce the Antestia population by 68% (Foucart and Brion 1959, Brion 1963). In these countries, egg parasitism was due mainly to *Telenomus sechellensis*, *Gryon fulviventre*, and *Trissolcus mopsus*, and, to a lesser extent *Anastatus antestiae* and *Acroclisoides africanus*. The adult tachinid parasitoid *Bogosia rubens* also was present, but had a limited impact (11.5% parasitism rate) (Foucart and Brion 1959, Brion 1963).

A recent study conducted in Ethiopia by Abebe (1999) reported 45-50% egg parasitism by three species, namely *Trissolcus* (= *Asolcus*) *suranus*, *Gryon fulviventre* (= *Hadronotus antestiae*) and *Anastatus antestiae*, with *T. suranus* being the most abundant. On the other hand, *Corioxenos antestiae* and *Bogosia rubens* contributed only 5% of Antestia bug parasitism in this study.

10.6.2.3.2 Parasitism Seasonality

In Cameroon, a study conducted in the main area of Arabica coffee production showed high rates of egg parasitism of *Antestiopsis intricata* due to five parasitoid species *Telenomus sechellensis*, *Trissolcus mopsus*, *Gryon fulviventre*, *Pediobius* sp., and *Anastatus antestiae* (Mbondji Mbondji 2001). In the Foumbot location in the West region of Cameroon, a large monthly collection and laboratory rearing of

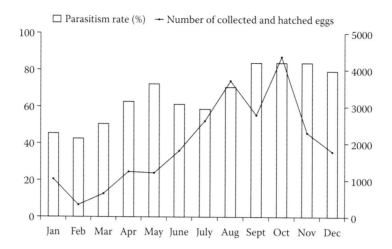

FIGURE 10.5   Seasonal variation of parasites emerging from eggs of *Antestiopsis intricata* (white bars) and number of sampled eggs (black line) in Foumbot in the West region in Cameroon (from Mbondji Mbondji 2001).

eggs gave parasitism rates between ≈40 and 80% (**Figure 10.5**). The parasitism rate had its lowest value in February, increased from March to August, decreased in September, and peaked in October.

Mbondji Mbondji (2001) explained that the relatively low level of egg parasitism in January and February was due to farmer practices before harvesting, especially that of plantation weeding and chemical spraying. The study also showed that in Foumbot (**Figure 10.6A**), as well as in Santa (**Figure 10.6B**), another location in North-West region of Cameroon, *Telenomus sechellensis* was the most common parasitoid species, followed by *Trissolcus mopsus* and *Pediobius* sp. Parasitoid abundance peaked from July to October which is the time of rainy season in this area. The species *Anastatus antestiae* was not collected in Santa, and the author suggested that this species might not be adapted to coffee growing at high elevation (1900 m asl). The strepsipteran *Corioxenos antestiae* was found locally on *Antestiopsis intricata* adults in Western Cameroon, but usually with low stylopization rates (< 1% in Mbondji Mbondji 2001).

A similar study conducted in Rwanda-Burundi reported the seasonal variations of the three most important egg parasitoid species of *Antestiopsis thunbergii ghesquierei* in the area (see below) (Brion 1963). High levels of egg parasitism ranging between ≈35 and 70% were recorded over the year with the exception of October, which showed a parasitism rate of ≈15% (**Figure 10.7**). The results also showed different variations depending on the parasitoid species: *Trissolcus* sp. was mainly present from January to June, with a peak in May, whereas *Telenomus sechellensis* and *Gryon fulviventre* were present all through the year although parasitism rates varied. The same study recorded relatively high levels of adult parasitism by *Bogosia rubens*. The parasitism rate was variable over the year and reached a maximum of 50% (Brion 1963).

10.6.2.3.3 Parasitoid Rearing

Mass rearing for production of biological control agents has been attempted on other pentatomid species. For example, *Telenomus sechellensis* was reared from eggs of the predatory bug *Macrorhaphis acuta* Dallas; and *Gryon fulviventre* from eggs of *M. acuta*; the Sudan millet bug, *Agonoscelis versicoloratus* (Turton); and the southern green stink bug, *Nezara viridula* (L.) (Le Pelley 1979). Moreover, these two parasitoids have shown the ability to develop in unfertilized eggs of Antestia bugs, which can be preserved for a longer time than fertilized eggs. This characteristic is sometimes suggested by researchers to explain the fact that egg parasitoid populations can remain in coffee plantations even outside Antestia bug reproduction periods (Le Pelley 1979).

In Cameroon, *Telenomus sechellensis* was the only species reared successfully on other pentatomid eggs. The species used was *Aspavia hastator* (F.), which is common on weeds from Compositae and Amaranthaceae around coffee plantations, notably on *Achyranthes aspera* L. (Mbondji Mbondji 2001).

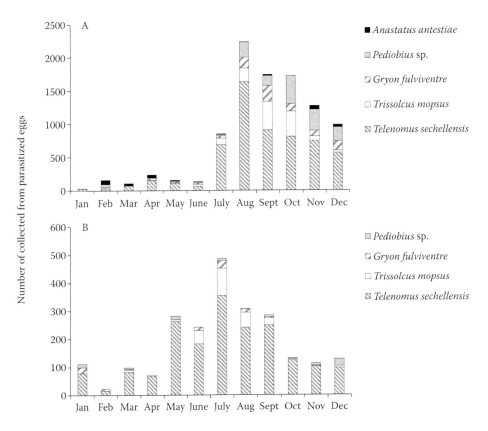

FIGURE 10.6 Seasonal variation of the number of parasitoids that emerged from parasitized *Antestiopsis intricata* eggs collected on coffee in two locations in Cameroon, (A). Foumbot (1100 m asl) in the West region, (B). Santa (1900 m asl) in the North-West region (from Mbondji Mbondji 2001).

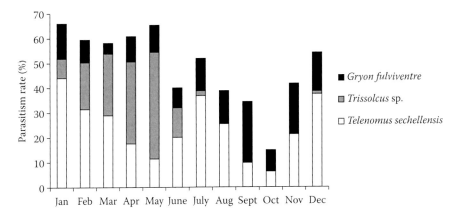

FIGURE 10.7 Seasonal variation of *Antestiopsis thunbergii ghesquierei* parasitism rate for the three most important parasitoid species of the study area in Rwanda-Burundi, *Gryon fulviventre*, *Telenomus sechellensis* and *Trissolcus* sp. (from Brion 1963).

10.6.2.3.4 Biological Control Attempts

True attempts of biological control of Antestia bugs are scarce. Because egg parasitism is naturally high in coffee plantations, main release attempts have employed parasitoids of nymphs and adults. From the 1940s to the 1960s, the strepsipteran *Corioxenos antestiae* and the tachinid *Bogosia* sp. were introduced into some localities of Kenya, Tanzania, and Uganda where they had been absent previously, but the introductions met with limited success (Greathead 1971).

10.6.3 Future Control Strategies

10.6.3.1 Conservation Biological Control through Plant Diversification

Antestia parasitoids already are present in coffee plantations with satisfactory parasitism potential, especially egg parasitoids. Thus, there is no need to introduce new species, which could compete with local species and do more harm than good. A good strategy might be to enhance and maintain indigenous parasitoid populations at such a level that they could keep Antestia bug populations under economic thresholds. To this end, some research studies have been conducted to better understand parasitoid ecology. For instance, the tachinid *Bogosia rubens* pupates in the soil, and pupal development is possible only with good moisture. Therefore, shade, maintenance of plants covering the ground, or mulching are practices favorable for this parasitoid (e.g., Taylor 1945).

In a more recent study conducted in Western Cameroon, flowers of weeds such as *Ageratum conyzoides* L., *Galinsoga parviflora* Cav. and *Bidens pilosa* L. (Asteraceae), and some crops associated with coffee such as banana (*Musa* sp.), have been examined for parasitoids with promising results. These plants also have been tested for their ability to maintain adult parasitoids in the laboratory. Banana flowers gave good results with a maximum parasitoid longevity of 26 days (Mbondji Mbondji 2001).

Plant diversification in crops is regarded as one of the main tenets of agroecology. There are many examples where adding plants to the crop helps by encouraging pest regulation by natural enemies (e.g., Ratnadass et al. 2012). Taking into account high levels of indigenous parasitism for Antestia bugs, a promising control strategy may be to add plants within coffee cropping systems, which could provide shelter and food for adult parasitoids. Much work needs to be done to determine which plant species and cropping design would best achieve these goals.

10.6.3.2 Semiochemical Control

As with most true bugs, semiochemicals play a crucial role in Antestia bug communication. For these bugs, scent glands are present as early as the first instar, where orifices are visible on the abdomen. For adults, scent glands are well developed on the metathorax and fully functional. Gland secretions are repellent and irritant compounds involved in the defense of the bugs against enemies, and it is likely that the brightly colored body of Antestia bugs is a visual signal alerting predators to the unpleasant taste of the bugs. A recent study by Jackels et al. (2014) showed that the three main compounds present in desiccated adults of *Antestiopsis thunbergii ghesquierei* from Rwanda were tridecane, dodecane, and tetradecane. These compounds are known to be involved in true bug defense (Millar 2005).

In their study in Burundi, Cilas et al. (1998) assumed that the aggregative distribution of Antestia bug nymphs in coffee plantations was due to semiochemical communication, and it is likely that reproduction behavior also involves pheromones. An olfactometry study recently conducted at icipe, Nairobi, Kenya, showed that *Antestiopsis thunbergii bechuana* nymphs are attracted to Arabica coffee by the fully developed green berries but not by the mature red berries (Teresiah N. Njihia, unpublished data). This result confirms the feeding preference of Antestia bug nymphs for immature berries (see **Section 10.3.2.2**, Feeding on Coffee). Moreover, the study showed that 30-day-old adults of both sexes are attracted to adult males of same age (Teresiah N. Njihia, unpublished data). This suggests the presence of a male-aggregative pheromone involved in Antestia bug reproduction, as has been demonstrated for other stink bugs (Millar 2005). If the results of this study are promising, then future research in the field of Antestia bug chemical ecology should be encouraged to help develop monitoring or mass trapping systems, as well as habitat management (push-pull strategies), for controlling these bugs.

10.7 Why Are Antestia Bugs Not Invasive Species?

10.7.1 A Limited Distribution Compared with Coffee Berry Borer

Worldwide, coffee is grown over 10 million hectares in humid regions of more than 70 tropical countries (FAOSTAT 2013). With ≈2 million hectares grown with coffee, Africa accounts for 20% of the global coffee area (FAOSTAT 2013). Originating in Africa in co-evolution with coffee plants, Antestia bugs have not followed the host coffee plant as it was dispersed abroad unlike so many of the other coffee pests. Today, the dozen species and subspecies of *Antestiopsis* have a distribution on coffee limited to Africa. *Antestiopsis cruciata* is the only Asian species known on coffee, and it probably originated in Asia. Antestia bugs are completely absent from Latin America, although that region accounts for more than 50% of the world coffee-growing area.

By contrast, another coffee pest that originated in Africa, the coffee berry borer, *Hypothenemus hampei* Ferrari (Coleoptera: Scolytinae), exhibits a different fate. Today, this tiny beetle that feeds upon coffee beans inside the berries is present virtually everywhere coffee is grown, with the exception of high elevation Arabica coffees, causing considerable production losses (Jaramillo et al. 2006, Vega et al. 2009). With this present section of the chapter, we will consider the reasons why Antestia bugs have not been as successful as coffee berry borer in the colonization process of coffee plantations. So, why have the Antestia bugs not become invasive?

10.7.2 Limited Opportunities of Spreading by Human Transport

Interregional and international travels of people and goods are known to be one of the main factors supporting biological invasions (Mack et al. 2000). The coffee berry borer is a good example of an invasive pest that reached new coffee growing areas largely by escaping detection during human transport. This pest can survive for several months in green coffee beans that are often imported for commercial coffee production or, to a lesser extent, as plant material for coffee growing (Chapman et al 2015). By contrast, feeding habits of Antestia bugs do not allow these pests to survive on coffee beans alone (see **Section 10.3.2**, Feeding Habits). The only way they could be transported over long distances is through coffee plants (seedlings) or plant parts, such as berries or grafting materials (cuttings and rootstocks). Transport of coffee plant material usually is limited and subject to phytosanitary inspection and quarantine restriction. Antestia bugs, which are both easily detected (eggs are big, white, and grouped in clusters [see **Section 10.3.3**]) and sensitive to insecticides, would be unlikely to survive these measures.

10.7.3 Relatively Slow Growing Species

Another characteristic of invasive insects is a short life cycle and high fecundity leading to quick growth of populations. Prolific invading species are more likely to compete successfully with indigenous species and expand their range (Mack et al. 2000). With a life cycle of ≈1 month and an average fecundity of 200 eggs per female at 25°C (Jaramillo et al. 2009), the coffee berry borer is considered a prolific species, with potentially eight or nine generations per year in favorable conditions (Damon 2000). By comparison, *Antestiopsis thunbergii* has a life cycle of between 3 and 6 months, depending on temperature. *A. intricata* has a somewhat shorter life cycle than *A. thunbergii* and seems to be adapted to higher temperatures (see **Section 10.3.3**). Mean fecundities of 150 and 200 eggs per female have been reported for *A. thunbergii* and *A. intricata*, respectively, with maximum values of more than 400 eggs per females for both species. However, these species can be considered slow growing as demonstrated by a recent study on *A. thunbergii*, which reported an intrinsic rate of increase (r) of 0.013 at 20°C and 0.006 at 25°C and a mean generation time of 3 and 4 months, depending on temperature (Ahmed et al. 2016).

10.7.4 A Strong Pressure of Natural Enemies

If Antestia bug fecundity can be considered as relatively high, the same is not true for egg viability in nature. In fact, as shown in **Section 10.6.2.3** (Biological Control), egg parasitism, mainly due to platygastrids, can reach rates as high as 95% in coffee plantations, with rates usually exceeding 50%. Antestia bug nymphs and adults also are parasitized by Hymenoptera. Thus, the growing capacity of Antestia bug populations may be slowed by strong pressure from parasitoids.

Moreover, most of the parasitoid species listed in **Section 10.6.2.3** have been collected from different species and subspecies of Antestia bugs over wide geographical areas from Western to Eastern and Southern African coffee-producing countries. This suggests that these parasitoids are adapted to a wide range of habitats, at least wider than that of their hosts. Moreover, this suggests that most of these parasitoids may be opportunistic rather than specific to Antestia bugs and may not depend on these bugs alone to develop. Therefore, egg parasitoid ubiquity and density in coffee plantations may play a crucial role in the limitation of Antestia bug expansion to new areas.

10.7.5 Constraining Ecological Preferences

The distribution of Antestia bug species over Africa as well as the biological data reported in the literature suggest that expansion of these pests is limited by constraining ecological preferences, which differ from one species to another. *Antestiopsis thunbergii*, including its subspecies and different geographic forms, is more adapted to climatic conditions of Eastern and Southern African highlands. The optimal temperature for *A. thunbergii bechuana* was found to be 23°C in a recent study (Ahmed et al. 2016). The other species of major economic importance, *A. intricata*, is present on coffee in West and Central African highlands, from the Atlantic Ocean to Uganda and Ethiopia, and seems to be adapted to a wider range of temperatures because it has been found at low elevations on Robusta coffee. The overlapping area of these two species is limited to Ethiopia and the countries of the Great Lakes region (Uganda, Rwanda, and Burundi), where both *A. thunbergii ghesquierei* and *A. intricata* are present on Arabica coffee. This region is considered to be a transition zone between different biogeographical regions in Africa (Linder et al. 2012), and the different Antestia bug species apparently find here suitable living conditions. However, globally, Antestia bugs are distributed according to the main biogeographical regions of Africa, as considered by Linder et al. (2012). For example, *A. thunbergii* mainly occurs in the Somalian, Zambezian and South African biogeographical regions, whereas *A. intricata* mainly occurs in the Guinean and Congolian biogeographical regions. Clearly, Antestia bug species have not expanded to all Arabica coffee-growing areas of Africa, showing that they have constraining ecological preferences.

10.7.6 A Close Relationship with Host Plants of the Family Rubiaceae

Antestia bugs have been found on a large range of host plants (**Section 10.3.2**, Feeding Habits). However, coffee and related species of the Rubiaceae are the main host plants of the group (Hargreaves 1936, Le Pelley 1942). Some of the wild plants in this family are present in close proximity to coffee plantations. These alternative host plants may play a role in the Antestia life cycle and in the coffee colonization process in such a manner that their presence is crucial for optimal development of Antestia populations. Thus, this close relationship with the Rubiaceae may limit the expansion of Antestia bugs on coffee where these alternative host plants are absent.

10.8 Acknowledgments

The authors thank the following individuals for providing information referenced in this chapter. We wish to thank Teresiah N. Njihia (Plant Health Department, International Centre of Insect Physiology and Ecology, Nairobi, Kenya) and Sylver Habumugisha (School of Agriculture, Rural Development and Agricultural Economic, University of Rwanda, Butare) for allowing their unpublished data to be cited. We also are grateful to the late Jacques Carayon (Entomologie Agricole et Tropicale, Museum National d'Histoire

Naturelle, Paris) for providing personal communication that we cited. Also, we would like to thank Robert S. Copeland (Biosystematic Unit, International Centre of Insect Physiology and Ecology, Nairobi, Kenya) for providing pictures of the montage of **Figure 10.1**, Gérard Delvare (Biological System Department, Cirad/ UMR CBGP, Montpellier, France) for giving us access and guiding us in the Cirad/CBGP insect collection, Abdelmutalab Gesmalla Ahmed, Dickson M. Munyasya and Ephantus Guandaru (Plant Health Department, International Centre of Insect Physiology and Ecology, Nairobi, Kenya), for providing the Antestia bugs from the field and the laboratory colony photographed for the montage, and Sisay Tesfaye (Jimma Agricultural Research Center, Jimma, Ethiopia) for help in finding some of the references that we cited. Finally, we would like to thank J. E. McPherson and Dmitry Musolin for their assistance in writing this chapter.

10.9 References Cited

Abebe, M. 1987. Insect pests of coffee with special emphasis on Antestia, *Antestiopsis intricata* in Ethiopia. International Journal of Tropical Insect Science 8: 977–980.

Abebe, M. 1988. Coffee bean darkening (discoloration), a new and unidentified problem on coffee. Institute of Agricultural Research Newsletter 3: 4–5, Ethiopia.

Abebe, M. 1999. The role of parasites in the natural control of Antestia bug, *Antestiopsis intricata* (Ghesquiere and Carayon). Possibilities for further control using exotic parasites, pp. 492–496. *In* Proceedings of the 18th International Conference on Coffee Science, 2–6 August 1999, Helsinki, Finland. Association for Science and Information on Coffee (ASIC), Paris. 559 pp.

Ahmed, A. G., L. K. Murungi, and R. Babin. 2016. Developmental biology and demographic parameters of antestia bug *Antestiopsis thunbergii* (Hemiptera: Pentatomidae), on *Coffea arabica* (Rubiaceae) at different constant temperatures. International Journal of Tropical Insect Science 36: 119–127.

Amyot, C. J. B., and Audinet Serville. 1843. Histoire naturelle des insectes. Hémiptères. Atlas. La Librairie Encyclopédique de Roret, Paris. 795 pp.

Anderson, T. J. 1919. The coffee bug, *Antestia lineaticollis* Stål, British East Africa Department of Agriculture, Bulletin of the Division of Entomology, Nairobi. 53 pp.

Azim, M. N. 2011. Taxonomic survey of stink bugs (Heteroptera: Pentatomidae) of India. Halteres 3: 1–10.

Baliga, H. 1967. A new species of *Corioxenos* (Stylopoidea) parasitizing *Antestiopsis cruciata* (F.) (Homoptera, Pentatomidae) in India. Bulletin of Entomological Research 57: 387–393.

Barrion, A. T., and J. A. Litsinger. 1994. Taxonomy of rice insect pests and their arthropod parasites and predators, pp. 13–359. *In* E. A. Heinrich (Ed.), Biology and management of rice insects. International Rice Research Institute, New Delhi. 794 pp.

Bayissa, M., and G. M. Tadesse. 1981. Oranges as alternative host plants of Antestia bug of coffee. Committee of Ethiopian Entomologists' Newsletter 1: 9.

Becker, R., B. Döhla, S. Nitz, and O.G. Vitzthum. 1988. Identification of the "peasy" off-flavour note in Central African coffees, pp. 203–215. *In* Proceedings of the 12th International Scientific Colloquium on Coffee, 29 June – 3 July 1987, Montreux, Suisse. Association for Science and Information on Coffee (ASIC), Paris. 960 pp.

Bouyjou, B., and C. Cilas. 1992. Etude de l'échantillonnage par le "test pyrèthre" pour évaluer la population d'*Antestiopsis orbitalis* dans la caféière burundaise, pp. 80–83. *In* Deuxième symposium sur la recherche caféière au Burundi, 20–21 October 1992, Bujumbura, Burundi, ISABU, Bujumbura. 168 pp.

Bouyjou, B., G. Fourny, and D. Perreaux. 1993. Le goût de pomme de terre du café arabica au Burundi, pp. 357–369. *In* Proceedings of the 15th International Scientific Colloquium on Coffee, 6–11 June 1993, Montpellier, France. Association for Science and Information on Coffee (ASIC), Paris. 900 pp.

Bouyjou, B., B. Decazy, and G. Fourny. 1999. L'élimination du "goût de pomme de terre" dans le café Arabica du Burundi. Plantations, Recherche, Développement 6: 107–115.

Brion, L. 1963. Le parasitisme naturel de la punaise du caféier Arabica au Rwanda. Bulletin Agricole du Rwanda 4: 115–130.

Bruneau de Miré, P, 1969. Etude sur l'*Antestiopsis lineaticollis*, pp. 130–135. *In* Annual report of the French Institute for Coffee and Cacao (IFCC), 1969, Cameroun.

Cachan, P. 1952. Les Pentatomidae de Madagascar (Hémiptères Hétéroptères). Mémoire de l'Institut Scientifique de Madagascar. Série E : Entomologie, 1952, 1: 231–462.

Carayon, J. 1954a. Les *Antestiopsis* (Hemipt. Pentatomidae) du caféier en Afrique tropicale française. Bulletin des Sciences, Section Techniques Agricoles Tropicales 5: 363–373.

Carayon, J. 1954b. A propos d'une récente attaque du Caféier Robusta par les *Antestiopsis* (Hemiptera, Pentatomidae) dans l'Oubangui (A. E. F.). Journal d'Agriculture Tropicale et de Botanique Appliquée 1: 204–209.

Chandra, K., S. Kushwaha, S. Sambath, and B. Biswas. 2012. Distribution and diversity of Hemiptera fauna of Veerangana Durgavati Wildlife Sanctuary, Damoh, Madhya Pradesh (India). Biological Forum-An International Journal 4: 68–74.

Chapman, E. G., R. H. Messing, and J. D. Harwood. 2015. Determining the origin of the coffee berry borer invasion of Hawaii. Annals of the Entomological Society of America 108: 585–592.

Chichaybelu, M. 1993. Importance and control of Antestia, *Antestiopsis intricata* (Ghesquiere and Carayon) on *Coffea arabica* L. at Bebeka Coffee Plantation Development Project in south west Ethiopia. M.Sc. Thesis. Alemaya University of Agriculture, Alemaya, Ethiopia. 63 pp.

Chichaybelu, M. 2008. Seasonal abundance and importance of Antestia bug (*Antestiopsis intricata*) in Southwest Ethiopia, pp. 291–295. *In* G. Adugna, B. Belachew, T. -Shimber, E. Taye and T. Kufa (Eds.), Coffee diversity and knowledge. Ethiopian Institute of Agricultural Research (EIAR), Addis Ababa, Ethiopia. 510 pp.

Cilas, C., B. Bouyjou, and B. Decazy. 1998. Frequency and distribution of *Antestiopsis orbitalis* Westwood (Hem., Pentatomidae) in coffee plantations in Burundi: implications for sampling techniques. Journal of Applied Entomology 122: 601–606.

Crowe, T. J., and T. Gebremedhin. 1984. Coffee pests in Ethiopia: Their biology and control. Institute of Agriculture Research, Addis Ababa. 45 pp.

Crowe, T. J., G. D. G. Jones, and R. Williamson. 1961. The use of pyrethrum formulations to control *Antestiopsis* on coffee in East Africa. Bulletin of Entomological Research 52: 31–41.

Damon, A. 2000. A review of the biology and control of the coffee berry borer, *Hypothenemus hampei* (Coleoptera: Scolytidae). Bulletin of Entomological Research 90: 453–465.

Distant, W. L. 1902. The Fauna of British India, including Ceylon and Burma. Volume 1. Heteroptera, London. xxxviii + 438 pp.

Distant, W. L. 1918. The Fauna of British India, including Ceylon and Burma. Volume 7. Homoptera: Appendix (cont.), Heteroptera: Addenda, London. viii + 210 pp.

Fabricius, J. C. 1775. Systema entomologiae sistens insectorum classes, ordines, genera, species; adjectis synonymis, locis, descriptionibus et observationibus, Flensburgi et Lipsiae. xxxii + 832 pp.

FAOSTAT. 2013. Food and Agriculture Organization of the United Nations, Statistic Division. http://faostat3.fao.org/home (accessed 1 January 2015).

Foucart, G., and L. Brion. 1959. Contribution à l'étude de la punaise du caféier Arabica au Rwanda-Urundi. Rwanda Agricultural Research Institute (ISAR), Butare, Rwanda. 334 pp.

Frappa, C. 1934. Les insectes nuisibles au caféier à Madagascar. IV : Insectes nuisibles aux fruits et aux fèves. Bulletin Economique de Madagascar 89: 370–379.

Germar, E. F. 1838. Hemiptera Heteroptera promontorii Bonae Spei nondum descripta, quae collegit C. F. Drége. Revue Entomologique (Silbermann) 5: 121–192.

Ghesquière, J., and J. Carayon. 1948. A propos de quelques *Antestia* et *Helopeltis* de l'Afrique tropicale (Hemiptera Pentatomidae et Miridae). Revue de Zoologie et de Botanique Africaines 41: 55–65.

Gmelin, J. F. 1790. Caroli a Linné Systema Naturae; Edition 13 aucta, reformata. Tome I, Pars IV, pp. 1517–2224, Beer, Lippsiae.

Gowdey, C. C. 1918. Report of the Government Entomologist, pp. 42–51. *In* Annual report of the government entomologist. Uganda Department of Agriculture, annual report for the year ending 31st March 1917, Kampala.

Greathead, D. J. 1966a. A taxomomic study of the species of *Antestiopsis* associated with *Coffea arabica* in Africa. Bulletin of Entomological Research 56: 515–554.

Greathead, D. J. 1966b. The parasites of *Antestiopsis* spp. (Hem. Pentatomidae) in East Africa and a discussion of the possibilities of biological control. Technical Bulletin of the Commonwealth Institute of Biological Control 7: 113–137.

Greathead, D. J. 1968. Supplementary notes on the *orbitalis* (Westw.) group of *Antestiopsis* (Hemiptera, Pentatomidae). Bulletin of Entomological Research 58: 227–232.

Greathead, D. J. 1969. On the taxonomy of *Antestiopsis* spp. (Hem., Pentatomidae) of Madagascar, with notes on their biology. Bulletin of Entomological Research 59: 307–315.

Greathead, D. J. 1971. A review of biological control in the Ethiopian region. Technical communication n°5, Commonwealth Institute of Biological Control, Commonwealth Agricultural Bureaux, England. 162 pp.

Gueule, D., G. Fourny, E. Ageron, A. Le Flèche-Matéos, M. Vandenbogaert, P. A. D. Grimont, and C. Cilas. 2015. *Pantoea coffeiphila* sp. nov. cause of 'potato taste' of Arabica coffee from African Great Lakes region. International Journal of Systematic and Evolutionary Microbiology 65: 23–29.

Hargreaves, H. 1930. Variegated coffee bug (*Antestia* spp.). Department of Agriculture of Uganda, Circular n°22, Entebbe, Uganda.

Hargreaves, H. 1936. Variegated coffee bug in Uganda. East African Agricultural Journal 1: 448–452.

IAR. 1984. Coffee research team progress report for the period 1983–1984, Institute of Agricultural Research, Addis Ababa, Ethiopia. 149 pp.

IAR. 1996a. Jimma research center progress report for the period 1986–1991, Institute of Agricultural Research, Jimma, Ethiopia. 139 pp.

IAR. 1996b. Jimma research center progress report for the period 1992–1993, Institute of Agricultural Research, Jimma, Ethiopia. 123 pp.

Jackels, S. C., E. E. Marshall, A. G. Omaiye, R. L. Gianan, F. T. Lee, and C. F. Jackels. 2014. GCMS Investigation of volatile compounds in green coffee affected by potato taste defect and the Antestia bug. Journal of Agricultural and Food Chemistry 62: 10222–10229.

Jaramillo, J., C. Borgemeister, and P. Baker. 2006. Coffee berry borer *Hypothenemus hampei* (Coleoptera: Curculionidae): Searching for sustainable control strategies. Bulletin of Entomological Research 96: 223–233.

Jaramillo, J., A. Chabi-Olaye, C. Kamonjo, A. Jaramillo, F. E. Vega, H. M. Poehling, and C. Borgemeister. 2009. Thermal tolerance of the coffee berry borer *Hypothenemus hampei*: Predictions of climate change impact on a tropical insect pest. Plos One 4: 1–11.

Kirkaldy, G. W. 1909. Catalogue of the Hemiptera (Heteroptera) with biological and anatomical references, lists of foodplants and parasites, etc. prefaced by a discussion on nomenclature, and an analytical table of families. Volume I: Cimicidae. Felix L. Dames, Berlin. I–XL + 392 pp.

Kirkpatrick, T. W. 1937. Studies on the ecology of coffee plantations in East Africa. II. The autecology of *Antestia* spp. (Pentatomoidea) with a particular account of a Strepsipterous parasite. Transactions of the Royal Entomological Society 86: 247–343.

Lavabre, E. M. 1952. Sur une plante pouvant héberger la punaise du caféier *Antestia lineaticollis* s. sp. *intricata* Ghesq. and Carayon. Agronomie Tropicale 7: 150–151.

Leblanc, L. 1993. I. Méthodes d'estimation des populations et des dégâts causés par les principaux ennemis du caféier au Rwanda (punaises, scolytes des baies, rouille, anthracnose). II. Liste annotée des insectes et maladies du caféier au Rwanda, avec références bibliographiques sur les acquis de la recherche depuis 1954. Document de travail n°19, ISAR, Rubona, Rwanda. 67 pp.

Lefèvre, P. C., and F. L. Hendrickx. 1942. Observations récentes sur les Antestia. Institut National pour l'Etude Agronomique du Congo Belge 26: 23–33.

Le Pelley, R. H. 1932. On the control of *Antestia lineaticollis*, Stål (Hem., Pentatom.) on coffee in Kenya colony. Bulletin of Entomological Research 23: 217–228.

Le Pelley, R. H. 1942. The food and feeding habits of Antestia in Kenya. Bulletin of Entomological Research 33: 71–89.

Le Pelley, R. H. 1959. Agricultural insects of East Africa. The East African High Commission, Nairobi, Kenya. 307 pp.

Le Pelley, R. H. 1968. Pests of coffee. Longmans Green and Co Ltd, London. 590 pp.

Le Pelley, R. H. 1979. Some scelionid egg-parasites reared from coffee bugs (*Antestiopsis* spp.) and from some unusual pentatomid hosts. Entomophaga 24: 255–258.

Leston, D. 1952. Notes on the Ethiopian Pentatomidae (Hem.) I. The genotype of *Antestia* Stål. Revue de Zoologie et de Botanique Africaines 45: 268–270.

Leston, D. 1953. Notes on the Ethiopian Pentatomoidea (Hem.). X. Some specimens from southern Africa in the South African Museum, with a note on the remarkable pygophore of *Elvisura irrorata* Spinola and description of a new species of *Piezodorus* Fieber. Annals of the South African Museum 51: 48–60.

Linder, H. P., H. M. de Klerk, J. Born, N. D. Burgess, J. Fjeldså, and C. Rahbek. 2012. The partitioning of Africa: statistically defined biogeographical regions in sub-Saharan Africa. Journal of Biogeography 39: 1189–1205.

Mack, R. N., D. Simberloff, W. Mark Lonsdale, H. Evans, M. Clout, and F. A. Bazzaz. 2000. Biotic invasions: causes, epidemiology, global consequences, and control. Ecological Applications 10: 689–710.

Matsuura, Y., T. Hosokawa, M. Serracin, G.M. Tulgetske, T.A. Miller, and T. Fukatsu. 2014. Bacterial symbionts of a devastating coffee plant pest, the stinkbug *Antestiopsis thunbergii* (Hemiptera: Pentatomidae). Applied and Environmental Microbiology 80: 3769–3775.

Mbondji Mbondji, P. 1997. Abondance, diversité et distribution géographique des Hémiptères nuisibles ou associés aux caféiers au Cameroun, pp. 744-751. *In* Proceedings of the 17th International Scientific Colloquium on Coffee, 20–25 July 1997, Nairobi, Kenya. Association for Science and Information on Coffee (ASIC), Paris. 828 pp.

Mbondji Mbondji, P. 1999. Observations éco-biologiques sur *Antestiopsis lineaticollis intricata* au Cameroun (Heteroptera : Pentatomidae). Annales de la Société Entomologique de France 35: 77–81.

Mbondji Mbondji, P. 2001. Recherches fauniques et écobiologiques sur les Hémiptères nuisibles ou associés aux caféiers au Cameroun. Thèse de Doctorat es-sciences naturelles. Université de Yaoundé I et Museum National d'Histoire Naturelle de Paris. 180 pp.

Mbondji Mbondji, P., and E. Ngollo Dina. 1992. Recherche d'efficacité de différents insecticides contre *Antestiopsis lineaticollis* Stål. (Het. Pentatomidae) et *Epicampoptera marantica* Tams. (Lep. Drepanidae), ravageurs des caféiers au Cameroun. Mémoires de la Société Royale Belge d'Entomologie 35: 423–428.

McNutt, D. N. 1979. Control of *Antestiopsis* spp. on coffee in Uganda. Proceedings of the Academy of Natural Sciences 25: 5–15.

Mendesil, E., and M. Abebe. 2004. Biology of Antestia bug, *Antestiopsis intricata* (Ghesquière & Carayon) (Hemiptera: Pentatomidae) on *Coffea arabica* L. Journal of Coffee Research 32: 30–39.

Mendesil, E., and C. Abdeta. 2007. Toxicity and ovicidal activity of some botanicals against Antestia bug, *Antestiopsis intricata* (Ghesquière & Carayon), pp. 1399–1403. *In* 21st International Conference on Coffee Science, 11–15 September 2006, Montpellier, France, Association for Science and Information on Coffee (ASIC), Lausanne.[CD-ROM].

Mendesil, E., M. Tadesse, and M. Negash. 2012. Efficacy of plant essential oils against two major insect pests of coffee (Coffee berry borer, *Hypothenemus hampei*, and Antestia bug, *Antestiopsis intricata*) and maize weevil, *Sitophilus zeamais*. Archives of Phytopathology and Plant Protection 45: 366–372.

Millar, J. G. 2005. Pheromones of true bugs, pp. 37-84. *In* S. Schulz (Ed.) The chemistry of pheromones and other semiochemicals II, Volume 240. Springer Berlin Heidelberg, New York. XII + 333 pp.

Mitchell, P. L. 2004. Heteroptera as vectors of plant pathogens. Neotropical Entomology 33: 519–545.

Mugo, H. M. 1994. Coffee insect pests attacking flowers and berries in Kenya: a review. Kenya Coffee 59: 1777–1783.

Mugo, H. M., and S. K. Ndoiru. 1997. Use of *Telenomus* (*Asolcus*) *seychellensis* (Hymenoptera: Scelionidae) in biological control of Antestia bugs, *Antestiopsis* spp. (Hemiptera: Pentatomidae) in coffee. Kenya Coffee 62: 2455–2459.

Mugo, H. M., L. W. Irungu, and P. N. Ndegwa. 2011. The insect pests of coffee and their distribution in Kenya. International Journal of Science and Nature 2: 564–569.

Mugo, H. M., J. K. Kimemia, and J. M. Mwangi. 2013. Severity of Antestia bugs, *Antestiopsis* spp. and other key insect pests under shaded coffee in Kenya. International Journal of Science and Nature 4: 324–327.

Nanta, J. 1950. Quelques observations sur les stades larvaires *d'A. occidentalis*, punaise des caféiers en Côte d'Ivoire. Bulletin du Centre de Recherche Agronomique de Bingerville 1: 34–35.

Nietner, J. 1861. Observations on the enemies of the coffee-tree in Ceylon, Ceylon Times Office, Colombo. 31 pp.

Ratnadass, A., P. Fernandes, J. Avelino, and R. Habib. 2012. Plant species diversity for sustainable management of crop pests and diseases in agroecosystems: a review. Agronomy for Sustainable Development 32: 273–303.

Ribeyre, F., and J. Avelino. 2012. Impact of field pests and diseases on coffee quality, pp. 151–176. *In* T. Oberthür, P. Läderach, H. A. J. Pohlan and J. H. Cock (Eds.), Specialty coffee: managing coffee. International Plant Nutrition Institute, Penang, Malaysia. X + 347 pp.

Rider, D. A. 1998. Nomenclatural changes in the Pentatomoidea (Hemiptera-Heteroptera: Cydnidae, Pentatomidae). II. Species level changes. Proceedings of the Entomological Society of Washington 100: 449–457.

Rider, D. A., L. Y. Zheng, and I. M. Kerzhner. 2002. Checklist and nomenclatural notes on the Chinese Pentatomidae (Heteroptera). II. Pentatominae. Zoosystematica Rossica 11: 135–153.

Signoret, V. 1861. Faune des Hémiptères de Madagascar. Deuxième Partie. Hétéroptères. Annales de la Société Entomologique de France 3: 917–972.

Stål, C. 1853. Nya Hemiptera från Cafferlandet. Öfversigt af Kongliga Vetenskaps-Akademiens Förhandlingar 10: 209–227.

Stål, C. 1864. Hemiptera Africana. Tomus Primus. IV + 256 pp. Holmiae.

Stål, C. 1876. Enumeratio Hemipterorum: Bidrag till en förteckning öfver alla hittills kända Hemiptera, jemte systematiska meddelanden. Kongliga Svenska Vetenskaps-Akademiens Handlingar 1876, part 5, 14(4): 1–162.

Tadesse, M., M. Abebe, and T. Erge. 1993. Antestia as a possible cause of coffee berry fall at Tepi state farm, p. 24. *In* Proceedings of the Crop Protection Society of Ethiopia, 5–6 March 1992, Addis Ababa, CPSE, Ethiopia. 67 pp.

Taylor, T. C. H. 1945. Recent investigations of *Antestia* species in Uganda. East African Agricultural Journal 10: 223–33 ; 11: 47–55.

Thelu, P. 1946. Utilisation du D.D.T. dans la lutte contre les punaises (*Antestia lineaticollis* Stål) du caféier au Cameroun. Comptes Rendus de l'Académie Agricole de France 32: 157.

Thunberg, C. P. 1783. Dissertatio entomologica novas insectorum species, sistens, Johan Edman 2: 29–70, Upsala.

Thunberg, C. P. 1822. Dissertatio entomologica de Hemipteris rostratis Capensibus, part. 2, Academiæ Typographi, Upsala. 8 pp.

Vega, F. E., F. Infante, A. Castillo, and J. Jaramillo. 2009. The coffee berry borer, *Hypothenemus hampei* (Ferrari) (Coleoptera: Curculionidae): a short review, with recent findings and future research directions. Terrestrial arthropod reviews 2: 129–147.

Walker, F. 1867. Catalogue of the specimens of heteropterous Hemiptera in the collection of the British Museum. Part II. Scutata. E. Newman, London, pp. 241–417.

Wallace, G. B. 1931. A coffee bean disease: a parasitic disease of coffee beans. Tropical Agriculture (Trinidad) 8: 14–17.

Wallace, G. B. 1932. Coffee bean disease: relation of *Nematospora gossypii*. Tropical Agriculture (Trinidad) 8: 127.

Waller, J. M., M. Bigger, and R. J. Hillocks. 2007. Coffee pests, diseases and their management. CAB International, Wallingford, UK. 434 pp.

Wanjala, F. M. E. 1980. Effect of infesting green coffee berries with different population levels of *Antestiopsis lineaticollis* Stål (Heteroptera: Pentatomidae) in Kenya. Turrialba 30: 109–111.

Westwood, J. O. 1837. *In* F. W. Hope (Ed.), A catalogue of Hemiptera in the collection of the Rev. F. W. Hope, M. A. with short Latin descriptions of the new species. J. Bridgewater, London, Part 1: 1–46.

Wilkinson, H. 1924. The coffee bug. Department of Agriculture of Uganda, Circular n°13, Entebbe.

Wrigley, G. 1988. Coffee. Longman Scientific and Technical, Harlow, UK. 639 pp.

Section VI

Diapause and Seasonal Cycles of Pentatomoidea

11

Diapause in Pentatomoidea[1]

Dmitry L. Musolin and Aida Kh. Saulich

CONTENTS

11.1 Introduction...498
11.2 Diapause as a Form of Dormancy in Pentatomoidea 499
    11.2.1 Phases of Diapause...499
    11.2.2 Three Types of Diapause in Pentatomoidea: Embryonic (Egg), Nymphal,
        and Adult Diapause...500
    11.2.3 Two Forms of Diapause: Obligate and Facultative Diapause 501
    11.2.4 Two Seasonal Classes of Diapause: Winter and Summer Diapause................502
    11.2.5 Diversity of Winter Diapause Patterns in Pentatomoidea...............................502
11.3 Environmental Factors Controlling Induction of Winter Diapause 509
    11.3.1 Day Length..509
        11.3.1.1 Photoperiodic Response of Diapause Induction................................509
        11.3.1.2 Developmental Stage(s) Sensitive to Day Length.............................511
        11.3.1.3 Required Day Number..511
    11.3.2 Temperature...513
        11.3.2.1 Effect of Temperature on the Photoperiodic Response Curve during
            Induction of Winter Diapause...514
        11.3.2.2 Temperature Optimum of Photoperiodic Response 515
    11.3.3 Food...515
11.4 Winter Diapause *Per Se* 518
    11.4.1 Peculiarities of Diapause in Females and Males 519
    11.4.2 Cold Hardiness..520
11.5 Diapause Development and Termination of Winter Diapause 521
    11.5.1 Spontaneous Termination of Winter Diapause..522
    11.5.2 Cold Termination of Winter Diapause 523
    11.5.3 Photoperiodic Termination of Winter Diapause 523
11.6 Environmental Factors Controlling Postdiapause Development in Spring 525
    11.6.1 Day Length..526
    11.6.2 Temperature...526
    11.6.3 Food...526
11.7 Seasonal Adaptations Associated with Winter Diapause..527
    11.7.1 Migrations to and from Overwintering Sites 527
    11.7.2 Formation of Aggregations 528
    11.7.3 Photoperiodic Control of Nymphal Growth Rate 529
    11.7.4 Seasonal Polyphenism...530

---

[1] This chapter was modified, expanded, and updated from "Diapause in the seasonal cycle of stink bugs (Heteroptera, Pentatomidae) from the Temperate Zone" by A. Kh. Saulich and D. L. Musolin (2012) (Copyright 2012 authors and Pleiades Publishing, Ltd.). Full Latin and common names along with authorities of pentatomoids mentioned in the text are given in **Table 11.2**. All specific eco-physiological terms that are **boldfaced** when mentioned the first time in the text are explained in the Glossary at the end of this chapter.

11.8   Summer Diapause (Estivation) 537
11.9   Other Seasonal Adaptations..538
11.10 Conclusions 539
11.11 Acknowledgments..539
11.12 References Cited 540
11.13 Glossary 555

11.1 Introduction

In most habitats over the globe, both terrestrial and aquatic, environmental conditions constantly change. The nature and magnitude of these changes are different – they are rhythmic or unique, long-term or short, severe or mild, and involve only the physical environment or also include the biota. Accordingly, the effects of these environmental changes on individuals, populations, and species vary greatly (Danilevsky 1961, Tauber et al. 1986, Danks 1987).

To survive under conditions of the annual rhythm of climate and to cope with seasonal changes of the environment, insects, as with other living organisms, need special **seasonal adaptations**. Some of these adaptations are behavioral such as migration or burrowing. However, others often are represented by sequential changes of seasonal physiological states and involve periods of dormancy.

Dormancy is defined as *a state of suppressed development (developmental arrest), which is adaptive (that is ecologically or evolutionarily meaningful and not just artificially induced) and usually accompanied with metabolic suppression* (Koštál 2006). Insects have different forms of dormancy, which vary in their intensity (sometimes called deepness), but all of them are associated with increased nonspecific resistance to unfavorable conditions. Any form of dormancy in insects is a complex phenomenon that is aimed first at solving different seasonal ecological problems such as survival of cold winters, hot and dry periods of summers, rainy seasons, or times of years when food is scarce or of low quality. Another extremely important ecological function of insect dormancy is synchronization of intra- and interspecific relationships with other organisms of the ecosystem (Danilevsky 1961, Saunders 1976, Tauber et al. 1986, Saulich and Volkovich 2004). A widespread example of dormancy is diapause, which can be subdivided into winter diapause and summer diapause (see **Section 11.2.4**).

According to current views, **diapause** is *a profound, endogenously and centrally mediated interruption that routes the developmental program away from direct morphogenesis into an alternative diapause program of succession of physiological events; the start of diapause usually precedes the advent of adverse conditions and the end of diapause need not coincide with the end of adversity* (Koštál 2006). This state of an organism is characterized by a complex of morphological, physiological, and behavioral traits known as the **diapause syndrome** (Tauber et al. 1986). As a rule, insects in diapause have a lowered water content, decreased oxygen consumption, and an increased ability to survive suboptimal low and/or high temperatures as well as other environmental and antropogenic stresses (e.g., radiation or pesticides).

Another related term that needs to be defined is **quiescence**, which is *an immediate response (without central regulation) to a decline of any limiting environmental factor(s) below the physiological thresholds with immediate resumption of the processes if the factor(s) rise above them* (Koštál 2006). Quiescence is in many ways related to diapause but is definitely less ecologically important.

Different species of insects can survive adverse season (e.g., cold winters, hot peaks of summers, dry or rainy seasons) in different physiological states. Thus, some species can **overwinter** (i.e., simply survive winter) in a physiological state of deep and profound winter diapause or in quiescence. Similar diversity of patterns can be observed in summer; although some species are active throughout the summer, others have profound summer diapause or short quiescence.

Phenomena analogous to insect diapause have been observed in various groups of animals, plants, and fungi. All of these phenomena are aimed at solving the same ecological problem, namely adaptation to the rhythmicity of climatic conditions. This eco-physiological adaptation has been more thoroughly studied in insects than in other organisms, as evidenced by the vast literature devoted to the problem itself, and to the specific traits of diapause in different groups of insects and other arthropods. However, the level of knowledge varies strongly between different insect taxa. The true bugs (Heteroptera) have received much attention in studies of diapause, although the number of species studied and the depth of the research have been noticeably less than in the Lepidoptera and Diptera.

The Pentatomoidea is one of the largest superfamilies of Heteroptera (see **Chapter 2**). Species in this superfamily have a wide range of seasonal adaptations, and many are economically important (Schuh and Slater 1995; Musolin and Saulich 1996a; McPherson and McPherson 2000; Panizzi et al. 2000; Saulich and Musolin 2007a,b, 2012, 2014; Henry 2009; Musolin 2012; Panizzi and Grazia 2015). Despite the relatively large size of pentatomoid bugs and the availability of specimens, only a few species that are economically important have been studied in detail in terms of seasonal cycles and seasonal development. The Pentatomidae is the third largest family of the Heteroptera. It comprises about 4,840–4,950 species (over 10% of the entire order) that are grouped into 900–940 genera and 8–11 subfamilies (Gapon 2008, Henry 2009; Vinokurov et al. 2010, see **Chapter 2**). Within this family, representatives of only three subfamilies (i.e., Asopinae [= Stiretrinae], Podopinae, Pentatominae) have been studied in terms of diapause, seasonal cycles, and seasonal development.

In this chapter, we review and discuss specific traits of diapause and associated phenomena in pentatomids and include rare examples from other (less well studied in this respect) families of Pentatomoidea.

11.2 Diapause as a Form of Dormancy in Pentatomoidea

Diapause is one of the most widely spread forms of insect dormancy. Vast literature is devoted to insect diapause, including its physiology, ecological functions, and, recently, genetics (Danilevsky 1961, Tauber et al. 1986, Danks 1987, Saulich and Volkovich 2004, Denlinger 2008, Denlinger and Lee 2010). In this section, we provide a brief overview of what is known about diapause in Pentatomoidea.

Diapause now is understood as a complex and dynamic process. In the less common cases, diapause is **obligate** (or **obligatory**), and the initiation of such diapause needs no external (i.e., **exogenous**) signals or cues because it represents a fixed and genetically strongly controlled component of the ontogenetic program, which is realized regardless of the environmental conditions in each **generation**. In more widespread cases, however, diapause is **facultative** and external (i.e., **exogenous**) token stimuli are necessary to induce the diapause state and, thus, individuals can switch between two ontogenetic alternatives, i.e., **direct** (or **active**) **development** or diapause (Koštál 2006; see **Section 11.2.3** for more details).

11.2.1 Phases of Diapause

Many different views on the stages or phases of diapause have been presented in the literature (see Danilevsky 1961; Saunders 1976; Tauber et al. 1986, Hodek 1996, 2002; Saulich and Volkovich 2004 for reviews). Recently Koštál (2006) suggested a simplified model of diapause which consisted of three major phases: prediapause, diapause, and postdiapause (**Figure 11.1**).

During the **prediapause phase**, direct ontogenetic development (morphogenesis) continues. This phase has two subphases in the species with facultative diapause and most likely only one in the species with obligate diapause. In the first group (i.e., the species with facultative diapause), diapause needs to be induced by an environmental cue and, thus, this cue needs to be perceived, transmitted, and interpreted by the neurohormonal system of the individual. This cue switches the ontogenetic pathway from direct development to diapause, and this period is called the **induction subphase**. At the same time, in the species with obligate diapause, there is no induction subphase because diapause is not induced. In other words, in the species with obligate diapause, this diapause is a necessary step (i.e., arrest) of development in each generation. Another part of the prediapause phase is a **preparation subphase**, during which individuals undergo behavioral and/or physiological change (e.g., acquire energy resources such as lipids or in some cases – starch, etc.), void the digestive system (gut), migrate and/or simply look for protective microhabitats (often called **hibernaculum** or **hibernacula**), sometimes change body color and so on.

During the next and more prolonged **diapause phase**, direct active development is **endogenously** (i.e., internally) arrested, and an alternative program of still mostly unknown physiological events proceeds. The diapause phase can be divided into three subphases. During the **initiation subphase**, direct development ceases, deep physiological preparations take place, and intensity (or deepness) of diapause may

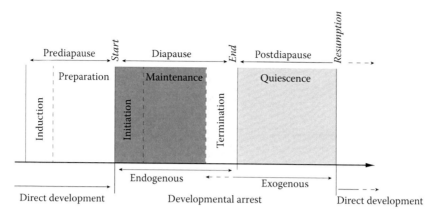

FIGURE 11.1 Schematic depiction of sequential phases of insect diapause. Thick line with arrowhead in the lower part of the figure indicates the passage of life of a hypothetical insect individual. Three major phases, namely prediapause, diapause, and postdiapause are named (on the top). Prediapause can be divided into the induction and preparation subphases and diapause into the initiation, maintenance, and termination subphases; postdiapause can be represented by quiescence. Developmental arrest may be endogenously (i.e., internally) or exogenously (i.e., externally) controlled (see at the bottom of the figure). More details are given in **Section 11.2.1**. Note that not all the (sub)phases must necessarily be found in all species and situations and this depiction might be applied to any type of diapause (i.e., embryonic, nymphal or adult; see **Section 11.2.2** for details). (Modified from V. Koštál, *Journal of Insect Physiology* 52: 113–127, 2006, with permission.)

increase. In some cases, individuals may continue accepting food or water, building energy reserves, and seeking suitable microhabitat during this subphase. These physiological processes are followed by the **maintenance subphase**, during which the endogenous developmental arrest persists regardless of environmental conditions. Specific token stimuli may help maintain diapause (or, in other words, prevent its termination). During this subphase, metabolic rate is relatively low and constant. Largely unknown physiological process(es) lead to more or less gradual decrease of diapause intensity and increase of sensitivity to diapause-terminating environmental conditions. With time, changes in environmental conditions can stimulate the decrease of diapause intensity to its minimum level and, thus, lead to the **termination subphase**. During this subphase, the intensity of diapause further decreases. By the end of the termination subphase, a usual active physiological state is mostly reached. Then, if conditions (primarily temperature) permit, direct development may overtly resume and the insects can begin moving, feeding, molting (in the case of nymphal/larval diapause) or copulating (in the case of adult diapause). However, if conditions are not yet permissive (usually, temperature is still too low and/or food is not available), the covert potentiality for direct development is restored but not realized and, as a result, the insects do not leave the diapause microhabitat and do not become fully active.

The diapause phase is followed by the **postdiapause phase**. Under field conditions, the diapause phase often ends as early as mid- or late winter or early spring, but temperature or other environmental conditions do not allow active development. In this case, insects experience **postdiapause quiescence**, an exogenously (i.e., externally) imposed inhibition of development and metabolism. When environmental conditions permit, the quiescence is followed by the full resumption of active development.

Our understanding of the nature and sequence of phases of insect diapause is still incomplete. The scheme outlined above and graphically represented in **Figure 11.1** was suggested by V. Koštál (2006), but a few other views and models have been suggested in the literature (Danilevsky 1961; Tauber et al. 1986; Danks 1987; Zaslavski 1988; Hodek 1996, 2002; Saulich and Volkovich 2004; Belozerov 2009). Below, we shall consider these sequential phases of diapause, using pentatomoids as examples.

11.2.2 Three Types of Diapause in Pentatomoidea: Embryonic (Egg), Nymphal, and Adult Diapause

Diapause in different insect species can be confined to any developmental stadium from embryo to adult. In the Heteroptera, three types of diapause are recognized including **embryonic** (or **egg**), **nymphal**, and

TABLE 11.1

Different Forms and Types of Winter Diapause in Pentatomoidea

Family	Total Number of Species Studied	Associations between Two Forms (Obligate and Facultative) and Three Types of Winter Diapause[1]		
		Embryonic	Nymphal	Adult
Acanthosomatidae	5			O – 3; F – 2
Cydnidae	6			O – 4; F – 2
Parastrachiidae	1			O – 1
Pentatomidae[2]	48	O – 2; F – 1	O – 1; F – 1	O – 3; F – 38
Plataspidae	4		O – 2	F – 2
Scutelleridae[2]	11		O – 3; F – 1	O – 4; F – 2
Tessaratomidae	2		O – 1	O – 1
Thaumastocoridae	1			F – 1
Thyreocoridae	4			O – 4
Subtotal		O – 2; F – 1	O – 7; F – 2	O – 20; F – 47
Total[2]	82	3	9	67

[1] Forms of diapause: F – facultative, O – obligate.

[2] In the rows Pentatomidae, Scutelleridae, and Total, the total number of species studied differs from the total number of diapause cases because some species do not have winter diapause (they have homodynamic seasonal development; see **Section 11.2.3**).

adult (sometimes called **reproductive**, or **imaginal**) **diapause** (**Table 11.1**). However, each species of insects, as a rule, can form diapause only at one particular developmental stage. Often (if not always), even within one stadium, there is a strict association of diapause with a particular ontogenetic stage. Cobben (1968) gave 16 different species-specific examples of timing of diapause only within the embryonic diapause in Heteroptera.

In the literature, however, there are examples of overwintering of insects from the same species or population at different stages of their life cycle. For example, nymphs and adults have been reported to overwinter together in *Ischnodemus sabuleti* (Fallen) (Lygaeidae; Tischler 1960) and *Chiloxanthus pilosus* (Fallen) (Saldidae; Cobben 1968). Finally, some species of true bugs have life cycles that last two or even more years and, thus, these species form diapause more than once in their life cycle; it usually happens during different developmental stadia (e.g., in Aradidae, Aphelocheiridae, and Reduviidae).

11.2.3 Two Forms of Diapause: Obligate and Facultative Diapause

As briefly mentioned above, diapause in insects may be of two forms. In some species, diapause is **obligate** (or **obligatory**) in which case, it does not need to be induced; it is determined hereditarily and always present in each generation regardless of external conditions. Obligate diapause in a particular species strictly determines a **univoltine seasonal cycle** (i.e., a pattern with one generation per year) over the entire range of the species because, in this case, active development is interrupted by obligate diapause in each generation (see **Chapter 12**). In other cases, diapause is **facultative**, and, therefore, it is induced by external (i.e., **exogenous**) factors and does not occur necessarily in each generation (**Table 11.1**). The facultative nature of diapause makes it possible to produce two or more generations during a year, with individuals of the last generation in the season entering facultative winter diapause.

Due to the facultative nature of diapause, different geographic populations of some species can produce different numbers of annual generations. **Voltinism** may also differ between particularly cold and warm years even in the same location.

In Pentatomoidea, both forms of diapause (i.e., facultative and obligate) have been documented. Among 79 experimentally studied species of stink bugs and their allies that have winter diapause, 29 species (about 37%) had an obligate winter diapause, whereas others (about 63%) had a facultative winter diapause (**Table 11.1**).

At the same time, no pronounced relationship between the form of diapause (i.e., facultative or obligate) and its type (i.e., association with a particular developmental stage – embryonic, nymphal or adult) or taxonomic position of a species has been detected. In other words, if sufficiently studied, any combination of type and form of diapause can be found in any taxon within the Pentatomoidea.

The form of diapause (i.e., obligate or facultative) usually is considered a species-specific character in Heteroptera. However, biological diversity virtually is unlimited and, in some cases, when different populations of a species were studied, examples of intraspecific variation of diapause form and mechanisms of voltinism control have been reported (Hodek 1977). Thus, within one species with a wide range in the Northern Hemisphere, a tendency to enter diapause was much stronger in the northern populations with all bugs of these populations sometimes demonstrating obligate diapause. Closer to the range core in the Temperate Zone, populations were heterogeneous with some individuals having obligate and others having facultative diapause. Further south, most populations have facultative diapause or show a strong tendency towards **homodynamic seasonal development** (nondiapause development in any season and nonstop sequence of generations without any pronounced period of seasonal dormancy). It is likely that **day length** is less important in winter diapause induction as well as winter diapause, itself, because it is less ecologically important in the (sub)tropic populations of true bugs than in populations in the more temperate or cold climates. These tendencies have been demonstrated in the pentatomids *Euthyrhynchus floridanus* (Mead 1976, Richman and Whitcomb 1978) and *Podisus maculiventris* (De Clercq and Degheele 1993) in Florida (USA).

11.2.4 Two Seasonal Classes of Diapause: Winter and Summer Diapause

Insect diapause often is associated with winter and, in these cases, is called **winter diapause**, or **hibernation**. Such examples have been documented in thousands of species from all major insect orders (Danilevsky 1961, Saunders 1976, Tauber et al. 1986, Danks 1987). However, diapause as a special physiological state can take place during other times of the year too. Frequently, insects can enter facultative or obligate diapause in summer; in this case, such diapause is called **summer diapause**, or **estivation**. It is important to understand that in some species, even one individual can have two different diapauses at two different ontogenetic stages. This is the case, for example, in the pentatomid *Picromerus bidens*: this predatory species passes winter in the state of obligate winter embryonic diapause and then adults may enter facultative summer adult diapause (Musolin and Saulich 2000; see **Chapter 12**).

The ecological importance of dormancy during summer might be related to survival of extremely high temperatures and dry conditions. In some other cases, summer diapause also is important for survival during a rainy season or a period when food is too scarce. Finally, similarly to winter diapause, summer diapause may be important for fine synchronization of the species' seasonal cycle with local environmental conditions (see **Chapter 12**). Summer diapause in Heteroptera has been studied much less than winter diapause (Saulich and Musolin 2007b), and some interesting cases will be discussed in **Section 11.8**. However, in this chapter, we will concentrate mostly on winter diapause.

11.2.5 Diversity of Winter Diapause Patterns in Pentatomoidea

Among 82 species of Pentatomoidea in which seasonal development and/or winter dormancy have been studied at least to some extent, most species (67, or ≈82%) overwinter as adults (**Tables 11.1 and 11.2**). In five families (i.e., Acanthosomatidae, Cydnidae, Parastrachiidae, Thaumastocoridae, and Thyreocoridae) only adult diapause has been reported so far. Three species (i.e., *Picromerus bidens*, *Apoecilus* [= *Apateticus*] *cynicus*, and *Trochiscocoris hemipterus*) overwinter in the embryonic stage and all belong to Pentatomidae. In nine species, nymphs overwinter; among them, two species belong to Pentatomidae (*Carbula humerigera* and *Pentatoma rufipes*), two to Plataspidae (*Coptosoma mucronatum* and *Coptosoma scutellatum*), four to Scutelleridae (*Odontoscelis dorsalis*, *Odontoscelis fuliginosa*, *Odontoscelis lineola*, and *Poecilocoris lewisi*) and one to Tessaratomidae (*Musgraveia sulciventris*). Among all these species, only in *P. lewisi* and *M. sulciventris* does diapause appear to be strongly linked to a particular nymphal instar (fifth instar in *P. lewisi* [Tanaka et al. 2002] and second instar, before the commencement of feeding, in *M. sulciventris* [Cant et al. 1996]). In all other cases, nymphs of different instars can overwinter: second through third/fourth instars do this more often than fifth instars; there are no records so far of winter diapause in the first instar in Pentatomoidea (see **Table 11.2** for references).

TABLE 11.2

Diapause and Associated Seasonal Adaptations Facilitating Synchronization of the Seasonal Cycle with Environmental Conditions in Species of the Superfamily Pentatomoidea[1]

Species (Common Name)	Form of Diapause[2]	Factors (Cues) Inducing Diapause and Associated Seasonal Adaptations[3]	Type of Diapause	References
Family ACANTHOSOMATIDAE				
Subfamily Acanthosomatinae				
Acanthosoma denticaudum Jakovlev (= *A. denticauda japonica* Jensen-Haarup)	O	–	Adult	Hori et al. 1993
Acanthosoma haemorrhoidale angulatum Jakovlev (hawthorn shield bug)	O	–	Adult	Hori et al. 1993
Elasmostethus atricornis (Van Duzee)	O	–	Adult	Carter and Hoebeke 2003, Jones and McPherson 1980
Elasmucha grisea grisea (L.) (parent bug)	F	–	Adult	Melber et al. 1981, Ogorzałek and Trochimczuk 2009
Elasmucha lateralis (Say) (birch bug)	F	–	Adult	Jones and McPherson 1980
Family CYDNIDAE				
Subfamily Cydninae				
Cydnus aterrimus (Forster)	F	–	Adult	Putshkov 1961
Subfamily Sehirinae				
Adomerus biguttatus (L.)	O	–	Adult	Putshkov 1961
Sehirus cinctus cinctus (Palisot de Beauvois) (white-margined burrower bug)	O	–	Adult	Sites and McPherson 1982
Sehirus luctuosus Mulsant & Rey (forget-me-not shield bug)	O	–	Adult	Putshkov 1961, Southwood and Leston 1959
Tritomegas bicolor (L.) (pied shield bug)	F (?)	–	Adult	Putshkov 1961 (1 generation), Southwood and Leston 1959 (2 generations)
Subfamily Amnestinae				
Amnestus pusillus Uhler	O	–	Adult	Froeschner 1941
Family PARASTRACHIIDAE				
Parastrachia japonensis (Scott)	O	–	Adult	Filippi et al. 2000a,b; Tachikawa and Schaefer 1985; Tojo et al. 2005a,b
Family PENTATOMIDAE				
Subfamily Podopinae				
Tribe Graphosomatini				
Graphosoma lineatum (L.) (Italian striped bug)	F	PhP, T, PhP control of body color in adults	Adult	Gamberale-Stille et al. 2010; Johansen et al. 2010; Musolin and Saulich 1996a, 2001; Nakamura et al. 1996; Tullberg et al. 2008
Graphosoma rubrolineatum (Westwood)	F	North: PhP, T; South: PhP, Fd	Adult	Nakamura and Numata 1999

(Continued)

TABLE 11.2 (CONTINUED)

Diapause and Associated Seasonal Adaptations Facilitating Synchronization of the Seasonal Cycle with Environmental Conditions in Species of the Superfamily Pentatomoidea[1]

Species (Common Name)	Form of Diapause[2]	Factors (Cues) Inducing Diapause and Associated Seasonal Adaptations[3]	Type of Diapause	References
		Tribe Scotinopharini		
Scotinophara lurida (Burmeister) (black rice bug)	F	PhP, T	Adult	Cho et al. 2007, 2008; Fernando 1960; Lee et al. 2001
Dybowskyia reticulata (Dallas)	F	PhP, T	Adult	Nakamura and Numata 1997a, 1998; Numata 2004
		Subfamily Asopinae (= Stiretrinae)		
		Tribe Amyotini		
Apoecilus (= *Apateticus*) *cynicus* (Say)	O	–	Embryonic	Javahery 1994, Jones and Coppel 1963, Whitmarsh 1916
Arma custos (F.)	F	PhP, T, PhP control of body color in nymphs	Adult	Volkovich and Saulich 1995
Euthyrhynchus floridanus (L.)	F	–	Adult	Ables 1975, Mead 1976, Oetting and Yonke 1975
Podisus maculiventris (Say) (spined soldier bug)	F	PhP, T	Adult	Goryshin et al. 1988a, Volkovich et al. 1991b
		Tribe Jallini		
Perillus bioculatus F. (twospotted stink bug)	F	PhP, T, Fd	Adult	Horton et al. 1998; Izhevskii and Ziskind 1981; Jasič 1967, 1975; Shagov 1977; Volkovich et al. 1991a
Zicrona caerulea (L.) (blue shield bug)	F	PhP, T	Adult	McPherson 1982, Saulich (unpublished data)
		Tribe Platinopini		
Andrallus spinidens (F.)	F	T	Adult	Shintani et al. 2010
Picromerus bidens (L.) (spined stink bug)	O (winter) F (summer)	– PhP	Embryonic Adult	Larivière and Larochelle 1989, Leston 1955, Musolin 1996, Musolin and Saulich 2000
		Subfamily Pentatominae		
		Tribe Aeliini		
Aelia acuminata (L.) (bishop's mitre shield bug)	F	PhP, T	Adult	Hodek 1971a, 1977; Honěk 1969
Aelia fieberi Scott	F	PhP, T	Adult	Nakamura and Numata 1995, 1997b
Aelia rostrata Boheman (wheat stink bug)	F (winter) O (summer)	– –	Adult Adult	Dikyar 1981, Cakmak et al. 2008
Aelia sibirica Reuter	F	PhP, T	Adult	Burov 1962
		Tribe Antestiini		
Plautia stali Scott	F	PhP, PhP control of body color in nymphs and adults	Adult	Kotaki 1998a,b; Kotaki and Yagi 1987; Numata and Kobayashi 1994

(Continued)

TABLE 11.2 (CONTINUED)

Diapause and Associated Seasonal Adaptations Facilitating Synchronization of the Seasonal Cycle with Environmental Conditions in Species of the Superfamily Pentatomoidea[1]

Species (Common Name)	Form of Diapause[2]	Factors (Cues) Inducing Diapause and Associated Seasonal Adaptations[3]	Type of Diapause	References
		Tribe Carpocorini		
Dichelops melacanthus (Dallas)	F	PhP, PhP control of growth rates in nymphs and body color in adults	Adult	Chocorosqui and Panizzi 2003
Dolycoris baccarum (L.) (sloe bug, berry bug, hairy shield bug)	F	PhP, PhP control of nymphal growth	Adult	Conradi-Larsen and Sømme 1973, Hodek and Hodková 1993, Hodková et al. 1989, Nakamura 2003, Perepelitsa 1971
Palomena angulosa (Motschulsky)	O	PhP control of growth rates in nymphs	Adult	Hori 1986, Hori and Kimura 1993
Palomena prasina (L.) (green shield bug)	O	PhP control of growth rates in nymphs, seasonal change of body color in adults	Adult	Saulich and Musolin 1996, 2007b; Southwood and Leston 1959
		Tribe Eysarcorini		
Carbula humerigera (Uhler)	F (winter)	PhP, PhP control of growth rates in nymphs	Nymphal (2nd–5th instars)	Kiritani 1985a,b
	F (summer)	PhP	Adult	
Eysarcoris aeneus (Scopoli)	F	PhP	Adult	Yao 2002
Eysarcoris lewisi (Distant)	F	PhP, PhP control of growth rates in nymphs	Adult	Hori and Inamura 1991, Hori and Kimura 1993
Eysarcoris ventralis (Westwood)	F	PhP	Adult (females)	Nakazawa and Hayashi 1983, Noda and Ishii 1981
		Tribe Menidini		
Menida disjecta (Uhler) (= *M. scotti* Puton)	O	–	Adult	Koshiyama et al. 1993, 1994, 1997
		Tribe Pentatomini		
Chinavia hilaris (Say) (green stink bug)	F	PhP, T	Adult	Javahery 1990 (as *Acrosternum hilare* [Say]), McPherson and Tecic 1997, Wilde 1969
Euschistus conspersus Uhler (consperse stink bug)	F	PhP, PhP control of body color in adults	Adult	Cullen and Zalom 2000, 2006, Toscano and Stern 1980
Euschistus heros F. (Neotropical brown stink bug)	F	PhP, PhP control of body color in adults	Adult	Mourao and Panizzi 2000, 2002; Panizzi and Niva 1994
Euschistus ictericus (L.)	F	PhP, PhP control of body color in adults	Adult	McPherson and Paskewitz 1984
Euschistus servus (Say) (brown stink bug)	F	PhP, PhP control of body color in adults	Adult	Borges et al. 2001, McPherson 1982
Euschistus tristigmus tristigmus (Say) (dusky stink bug)	F	PhP, PhP control of body color in adults	Adult	McPherson 1974, 1975a,b, 1979, 1982

(Continued)

TABLE 11.2 (CONTINUED)

Diapause and Associated Seasonal Adaptations Facilitating Synchronization of the Seasonal Cycle with Environmental Conditions in Species of the Superfamily Pentatomoidea[1]

Species (Common Name)	Form of Diapause[2]	Factors (Cues) Inducing Diapause and Associated Seasonal Adaptations[3]	Type of Diapause	References
Nezara antennata Scott (Oriental green stink bug)	F (winter)	PhP, PhP control of body color in adults (winter)	Adult	Musolin (unpublished data), Noda 1984
	F (summer)	PhP, no color change in adults (summer)	Adult	
Nezara viridula (L.) (southern green stink bug)	F	PhP, PhP control of growth rates in nymphs and body color in adults	Adult	Ali and Ewiess, 1977; Musolin 2012; Musolin and Numata 2003a,b; Musolin et al. 2007
Oebalus poecilus (Dallas) (small green stink bug)	F	PhP, PhP control of body color in adults	Adult	Albuquerque 1993, Greve et al. 2003, Santos et al. 2003
Oebalus pugnax pugnax (F.) (rice stink bug)	F	–	Adult	McPherson 1982, McPherson and Mohlenbrock 1976
Oebalus ypsilongriseus (De Geer) (rice stink bug)	F	PhP, PhP control of body color in adults	Adult	Vecchio et al. 1994
Pentatoma rufipes (L.) (forest bug)	O	–	Nymphal (2nd–3rd instars)	Putshkov 1961, Saulich (unpublished data), Southwood and Leston 1959
Thyanta calceata (Say)	F	PhP, PhP control of body color in adults	Adult	McPherson 1977, 1978, 1982; Oetting and Yonke 1971
Tribe Piezodorini				
Piezodorus guildinii (Westwood) (redbanded stink bug)	F	PhP	Adult	Zerbino et al. 2013, 2014, 2015
Piezodorus hybneri (Gmelin)	F	PhP	Adult	Endo et al. 2007, Higuchi 1994
Tribe Strachiini				
Bagrada hilaris (Burmeister) (= *B. cruciferarum* Kirkaldy) (bagrada bug, painted bug)	Homodynamic development (i.e., no dormancy period), at least in South			Panizzi 1997, Siddiqui 2000, Singh and Malik 1993, Taylor et al. 2015
Murgantia histrionica (Hahn) (harlequin bug)	Homodynamic development (i.e., no dormancy period), at least in South			McPherson and McPherson 2000, Siddiqui 2000
Trochiscocoris hemipterus (Jakovlev)	F	–	Embryonic	Akramowskaya 1959, Asanova and Kerzhner 1969
Tribe Eurydemini				
Eurydema oleracea (L.) (brassica bug, green cabbage bug)	F	PhP	Adult	Fasulati 1979
Eurydema rugosa (= *E. rugosum*) Motschulsky (cabbage bug)	F	PhP, Fd	Adult	Ikeda-Kikue and Numata 1994, 2001; Numata and Yamamoto 1990
Tribe Rhynchocorini				
Biprorulus bibax Breddin	F	PhP	Adult	James 1990a,b, 1991, 1993
Tribe Cappaeini				
Halyomorpha halys (Stål) (= *Dalpada brevis* Walker) (brown marmorated stink bug)	F	PhP, T, PhP control of growth rates in nymphs and body color in nymphs and adults	Adult	Hoebeke and Carter 2003; Lee et al. 2013; Niva and Takeda 2002, 2003; Toyama et al. 2006; Watanabe 1980

(Continued)

TABLE 11.2 (CONTINUED)

Diapause and Associated Seasonal Adaptations Facilitating Synchronization of the Seasonal Cycle with Environmental Conditions in Species of the Superfamily Pentatomoidea[1]

Species (Common Name)	Form of Diapause[2]	Factors (Cues) Inducing Diapause and Associated Seasonal Adaptations[3]	Type of Diapause	References
Family PLATASPIDAE				
Coptosoma mucronatum Seedenstücker	O	–	Nymphal (2nd–4th instars)	Davidová-Vilímová and Štys 1982
Coptosoma scutellatum (Geoffroy)	O	–	Nymphal (2nd–4th instars)	Davidová-Vilímová 2006, Davidová-Vilímová and Štys 1982, Putshkov 1961, Saulich and Musolin 1996, Werner 2005
Megacopta cribraria (F.) (kudzu bug)	F	–	Adult	Chen et al. 2009, Eger et al. 2010, Gardner et al. 2013, Hosokawa et al. 2014, Zhang and Yu 2005, Zhang et al. 2012
Megacopta punctatissima (Montandon)	F	–	Adult	Tayutivutikul and Kusigemati 1992, Tayutivutikul and Yano 1990
Family SCUTELLERIDAE				
Subfamily Eurygastrinae				
Eurygaster integriceps Puton (sunn pest)	O	–	Adult	Brown 1962, Javahery 1995, Shinyaeva 1980, Viktorov 1967
Subfamily Scutellerinae				
Chrysocoris purpureus (Westwood)	F (summer)	T, H (?)	Adult	Roychoudhury 1998, 1999
Poecilocoris lewisi (Distant) (clown stink bug)	F (winter)	PhP	Nymphal (5th instar)	Tanaka et al. 2002, Tomokuni et al. 1993
	F (summer)	PhP	Adult	
Subfamily Odontotarsinae				
Phimodera flori Fieber	O	–	Adult	Davidová-Vilímová and Král 2003
Phimodera humeralis (Dalman) (= *Podops nodicollis* Burmeister)	O	–	Adult	Putshkov 1961
Subfamily Pachycorinae				
Pachycoris klugii Burmeister	F	–	Adult	Peredo 2002
Pachycoris stallii Uhler	F	–	Adult	Williams et al. 2005
Tetyra bipunctata (Herrich-Schäffer)	O	–	Adult	Gilbert et al. 1967, McPherson 1982
Subfamily Odontoscelinae				
Odontoscelis dorsalis (F.)	O	–	Nymphal (3rd–4th instars)	Putshkov 1961
Odontoscelis fuliginosa (L.)	O	–	Nymphal (3rd–4th instars)	Putshkov 1961
Odontoscelis lineola Rambur (lesser streaked shield bug)	O	–	Nymphal (3rd–4th instars)	Hawkins 2003, Putshkov 1961

(Continued)

TABLE 11.2 (CONTINUED)

Diapause and Associated Seasonal Adaptations Facilitating Synchronization of the Seasonal Cycle with Environmental Conditions in Species of the Superfamily Pentatomoidea[1]

Species (Common Name)	Form of Diapause[2]	Factors (Cues) Inducing Diapause and Associated Seasonal Adaptations[3]	Type of Diapause	References
Family TESSARATOMIDAE				
Subfamily Natalicolinae				
Encosternum delegorguei Spinola	O	–	Adult	Dzerefos et al. 2009
Subfamily Oncomerinae				
Musgraveia sulciventris (Stål) (bronze orange bug)	O (winter)	–	Nymphal (2nd instar)	Cant et al. 1996, Hely 1964, McDonald 1969
	F (summer)	Fd	Adult	
Family THAUMASTOCORIDAE				
Subfamily Xylastodorinae				
Discocoris drakei Slater & Ashlock	F	–	Adult	Couturier et al. 2002
Family THYREOCORIDAE (= Family CORIMELAENIDAE)				
Corimelaena lateralis lateralis (F.)	O	–	Adult	McPherson 1972
Corimelaena obscura McPherson & Sailer	O	–	Adult	Bundy and McPherson 1997
Galgupha ovalis Hussey	O	–	Adult	Biehler and McPherson 1982
Thyreocoris scarabaeoides (L.) (scarab shield bug)	O	–	Adult	Putshkov 1961

[1] Only a few examples, particularly experimental studies that have been more thorough, are given for each family. Only families with several examples are divided into subfamilies and tribes. Species are listed alphabetically within each family/tribe/subfamily.

[2] Forms of diapause: F – facultative, O – obligate. If not indicated otherwise, diapause is winter diapause.

[3] Factors (cues) inducing diapause: PhP – photoperiod, T – temperature, Fd – food, H – humidity or precipitation; symbol "–" means *not applicable* for obligate diapause and *unknown* for facultative diapause.

In each particular insect species, usually only one ontogenetic stage can enter winter diapause. Within one species, the ability to enter diapause at different stages has been assumed to occur among pentatomoid species only in *Pentatoma rufipes*: Southwood and Leston (1959) and then Putshkov (1961) suggested that not only nymphs but also adults of this species can overwinter, although such an unusual pattern has not been proven and rare records of adults of this species early in spring might be misleading and related to parasitism (see **Chapter 12** for more details).

In several pentatomid species (e.g., *Perillus bioculatus* [Jasič 1975]; *Graphosoma lineatum* [Putshkov 1961, Nakamura et al. 1996]), some individuals likely can live longer than 1 year and, thus, overwinter twice (i.e., enter winter diapause in 2 consecutive years). This possibility also cannot be excluded in *Sinopla perpunctatus* Signoret (Faúndez and Osorio 2010).

The pentatomids *Bagrada hilaris* and *Murgantia histrionica* can be physiologically active throughout the whole year (i.e., apparently develop without entering pronounced diapause at any season), at least in southern areas (e.g., India [Singh and Malik 1993, Siddiqui 2000] or the United States of America [McPherson and McPherson 2000, Taylor et al. 2015]).

Based on the data from **Table 11.2**, we can conclude that winter adult diapause is the most widespread diapause in Pentatomoidea. This conclusion supports the earlier but less representative estimations that winter adult diapause is most characteristic of the whole Heteroptera (Hertzel 1982; Ruberson et al. 1998; Saulich and Musolin 2007b, 2012; Esenbekova et al. 2015), although exceptions occur. For example, most species of plant bugs (Miridae) overwinter in the egg stage (Wheeler 2001).

11.3 Environmental Factors Controlling Induction of Winter Diapause

Ecological factors often have a dual mode of action on living organisms. They determine environmental conditions under which these organisms live; thus, they have a **vital function**. At the same time, many ecological factors might be used as reliable predictors of environmental changes that the ecosystem is going to face in the future; thus, they have a **signal function** and act as cues (Tyshchenko 1980). For instance, a vital mode of action of temperature determines a range within which a particular species can live, whereas the daily rhythm of temperature (**thermorhythm**) has a signal function and predicts the coming seasonal environmental changes.

Several abiotic and biotic factors that have regular rhythmicity in nature can be used by insects as signals (or cues) for synchronization of their seasonal development with environmental conditions. In many cases, insects use more than one cue, and, thus, the mode of action of different factors can be complex.

11.3.1 Day Length

Day length has an astronomic preciseness and no environmental factor can affect it. Thus, natural day length is the most reliable environmental cue available. Many species of insects use day length as a reliable cue for structuring their seasonal cycle and synchronizing their seasonal development with local environmental conditions. Precise seasonal dynamics of day length (and **night length**) are of critical importance here, but not the changes in energy or intensity of optical radiation.

In eco-physiological laboratory experiments when day length is set artificially, it is called **photoperiod** and determined as a ratio between the duration of the light period (i.e., **photophase**) and the dark period (i.e., **scotophase**). Graphically, it can be shown in the following way: L:D 16:8, meaning 16 hours of light followed by 8 hours of darkness every day in the laboratory.

Many insect species respond in different ways to day length and such physiological responses are called **photoperiodic responses (PhPR)**. These responses have been found in many species, including those in the Pentatomoidea (**Table 11.2**). In the Northern Hemisphere, for species with winter diapause, short (or shortening) day length becomes a signal of the approaching autumnal decrease of temperature and serves as a trigger for a hormonal cascade leading to a pause in active metamorphosis (i.e., diapause).

11.3.1.1 Photoperiodic Response of Diapause Induction

In the Northern Hemisphere, formation of facultative winter diapause is controlled by the **PhPR of a long-day type** (i.e., one that allows active development under long-day conditions and induces diapause under short-day conditions). Under long-day conditions (i.e., simulating early summer), individuals develop directly (so called active development) and produce the next generation; under short-day conditions (i.e., simulating autum or winter), individuals enter winter diapause.

Critical photoperiod (i.e., **critical day length**, or **photoperiodic threshold**), is one of the most important ecological characteristics of the PhPR. This parameter corresponds to the day length at which 50% of individuals of a particular population enter diapause. Typical PhPR of a long-day type has been found in many pentatomoid species. **Figure 11.2** demonstrates the PhPRs of diapause induction in two pentatomids, *Aelia fieberi* and *Plautia stali*. These insects had been reared and then maintained under particular constant photoperiodic conditions at 25°C from the day when they hatched from eggs. All females that experienced long-day conditions (i.e., with photophases 15 or 16 hours) became reproductive, whereas most females reared and then maintained under **short-day** conditions (scotophase of 12 hours or more) entered adult diapause. Thus, at 25°C, the critical photoperiod was approximately 13.5 hours for *P. stali* and about 14.5 hours for *A. fieberi* (Numata and Nakamura 2002). These examples clearly

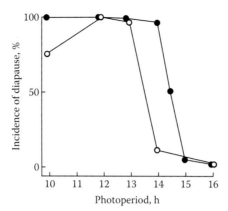

FIGURE 11.2 Photoperiodic responses of winter adult diapause induction in females of pentatomids *Plautia stali* (open circles) and *Aelia fieberi* (close circles) from Osaka area, Japan (34.7°N). Adults were reared from eggs and then maintained at 25°C under photoperiod indicated on the horizontal axis. (Modified from H. Numata and K. Nakamura, European Journal of Entomology 99: 155–161, 2002, with permission.)

demonstrate the **qualitative PhPR** of insects to day length: each individual responds in an "All or None" (i.e., "Yes or No") manner by choosing one of two alternative pathways: in the case of adult diapause – diapause or direct development (i.e., reproduction).

However, there are examples when the PhPR controls quantitative parameters such as duration of a particular stage, size, or degree of body pigmentation; in such cases, it is called the **quantitative PhPR** (Tyshchenko 1977, Zaslavski 1988, Numata and Kobayashi 1994, Musolin and Saulich 1997). **Figure 11.3** demonstrates a typical example of such a response. Females of the black rice bug, *Scotinophara lurida*, were reared from eggs to adults at 25°C under a long-day photoperiod L:D 16:8. Upon emergence, they were transferred to one of five short-day and long-day photoperiods and four temperatures (15, 20, 25, 30°C) to determine the PhPR, using the Days of first oviposition (DFO) as a criterion of the response (Cho et al. 2008). Under the three higher temperatures, there generally was a decreasing DFO in response to increasing day length, clearly demonstrating the quantitative nature of the PhPR; there was little or no response at 15°C.

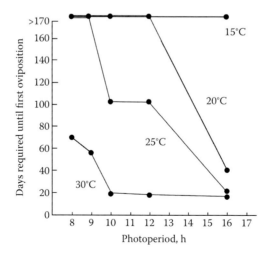

FIGURE 11.3 Effect of photoperiod and temperature on duration of the preoviposition period (i.e., days required until first oviposition) in females of the black rice bug, *Scotinophara lurida*, from Daesan, Korea (37°N). Females were reared from eggs to adults at 25°C under a photoperiod L:D 16:8 and upon emergence, adults were transferred to various experimental conditions (photoperiod is indicated on the horizontal axis, temperature is indicated next to the curves). (Data from J. R. Cho, M. Lee, H. S. Kim, and K. S. Boo, Journal of Asia-Pacific Entomology 11: 53–57, 2008, with permission.)

It also should be remembered that in the field, insects experience not constant but day-to-day **changing day-length conditions**: day length increases from winter until the day of summer solstice (June 20 to 22 in the Northern Hemisphere [i.e., **increasing day length**]) and then decreases until winter solstice (December 20 to 23 in the Northern Hemisphere [i.e., **decreasing day length**]). Many insect species can sense these small daily changes of day length and, thus, experiments with changing photoperiods often can give more information about mechanisms controlling species' seasonal development than experiments using constant photoperiods (Danilevsky 1961, Saunders 1976, Tauber et al. 1986, Saulich and Volkovich 2004).

11.3.1.2 Developmental Stage(s) Sensitive to Day Length

Diapause induction which happens during the prediapause phase (see **Section 11.2.1 and Figure 11.1**) involves the **day-length sensitive stage** and the process of accumulation of photoperiodic information. Perception of day-length signals/cues is known to occur at a certain stage of insect development, which is strictly species-specific and usually directly precedes the **diapausing stage** (Saunders 1976, Saulich and Volkovich 2004). **Sensitivity to day length** in different species may appear at different stages of development, from the egg to the adult, or extend over several stages but, again, is always strictly species-specific. The adult diapause, typical of most pentatomoids, is characterized by the greatest variation in which stage or stages are susceptible to day-length influence. Even in the same type of diapause, evaluation of the day-length information may take place at different development stages and have a different duration (**Table 11.3**; Musolin and Saulich 1999).

The duration and identity of the sensitive stage largely determine the entire pattern of seasonal development and the adaptive capabilities of the species (Saulich 1995). This is evident especially in cases of artificial displacement or transfer of insects into new geographic regions, which will be discussed in detail in **Chapter 12**.

11.3.1.3 Required Day Number

The sensitive stage is a necessary component of the PhPR. During this stadium, the daily photoperiodic signals are accumulated. Apparently, one short- or long-day signal is not enough to induce a response. The number of photoperiodic cycles triggering diapause or active development has been referred to as **a packet of photoperiodic information** (Goryshin and Tyshchenko 1972) or **the required day number** (Saunders 1976).

TABLE 11.3

Stages Sensitive to Day Length in Species of the Superfamily Pentatomoidea with Photoperiodically Induced Winter Diapause[1]

Sensitive Stages	Species (References)
Nymphal Diapause	
Nymphs	*Carbula humerigera* (Kiritani 1985a,b)
Adult Diapause	
Nymphs of 2nd instar	*Oebalus poecilus* (Albuquerque 1993)
Nymphs starting from 3rd instar and adults	*Podisus maculiventris* (Volkovich et al. 1991b)
Nymphs of last two (4th and 5th) instars, and adults or nymphs of last (5th) instar and adults	*Dolycoris baccarum* (Perepelitsa 1971)
	Halyomorpha halys (Niva and Takeda 2003)
	Eysarcoris lewisi (Hori and Kimura 1993)
Mostly adults	*Graphosoma lineatum* (Musolin and Maisov 1998)
	Arma custos (Saulich and Volkovich 1996)
	Perillus bioculatus (Jasič 1967, 1975)
	Aelia acuminata (Hodek 1971a)
	Chinavia hilaris (Wilde 1969)
	Nezara viridula (Ali and Ewiess 1977)
	Plautia stali (Kotaki and Yagi 1987)

[1] All species belong to Pentatomidae. Examples from other families are not available.

This parameter indicates how many short days are required for photoperiodic induction of winter diapause or how many long days are needed for induction of physiological activity (i.e., nondiapause state) in all individuals of a particular local population of the species. Together with the critical photoperiod (see **Section 11.3.1.1**), the required day number is an important component of the insect PhPR. These two parameters play different roles: critical photoperiod indicates when exactly in the season the diapause induction shall start, whereas the packet of photoperiodic information designates how many days after the arrival of critical photoperiod are needed for diapause induction in all members of the population.

The process of accumulation of photoperiodic signals has been studied only in few species of pentatomoids. However, in all cases, a particular state (diapause or active development) was induced only by complete packets of **short-day** or **long-day** photoperiodic information. For example, the experimentally determined packet of short-day information for female *Podisus maculiventris* was 10 or 11 days at 20°C. In other words, experience of 10 or 11 short days at 20°C was enough to induce diapause in 100% of females. At a higher temperature (24°C), the same number of days under short-day conditions proved to be insufficient for diapause induction: even larger packets of short-day signals (16 short days) induced diapause only in 30% of females. Thus, at higher temperatures, the packet of photoperiodic information (or the required day number) for diapause induction must be larger, and this might be related to the rates of nymphal growth. With increase of temperature from 20 to 24°C, nymphs grow faster and the duration of the nymphal period is reduced from 28 to 21 days, and to 16 days at 28°C. Because the nymphs become sensitive to day length starting from the third instar, only a small fraction of individuals have the time to accumulate the needed number of short-day signals at high temperatures and enter diapause; all other females fail to enter diapause under such conditions (Volkovich et al. 1991b).

The end of the diapause preparation and the start of the initiation subphases (see **Figure 11.1**) cannot always be determined. A reliable indicator of the completely formed diapause is the survival rate of diapausing individuals at low temperatures. For example, the higher survival rate of adult *Podisus maculiventris* at the favorable overwintering temperature of 8°C was observed in the individuals that were transferred into the cold 17–19 days after the emergence of adults (**Figure 11.4**). It is possible that prediapause feeding of the adults stopped, and the diapause was established completely at that particular moment. The adults transferred into the cold before or after this moment showed a lower resistance to adverse overwintering conditions and suffered higher mortality.

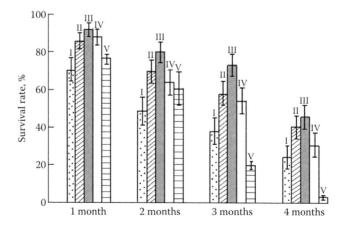

FIGURE 11.4 Effect of age on survival of low temperatures exposure of different duration (1 to 4 months at 8°C, photoperiod L:D 12:12, humidity 90–95%) in females of spined soldier bug, *Podisus maculiventris*. Nymphs were reared to adults and then maintained under constant experimental conditions: photoperiod L:D 12:12 at 20°C and then exposed to cold treatment of different durations. Age of females (days after emergence of adults): 11–13 (group I), 14–16 (group II), 17–19 (group III), 20–22 (group IV), 22–25 (group V). Horizontal line: experimental series (duration of the cold treatment). The laboratory culture originated from Missouri, the United States of America (about 38°N). (From N. I. Goryshin, T. A. Volkovich, A. Kh. Saulich and I. A. Borisenko, Manuscript deposited in the VINITI (Vsesojuzniy Institut Nauchnotehnicheskoy Informacii [All-Union Institute of Scientific and Technical Information], Moscow, No. 115-B-90, 1989, with permission.)

11.3.2 Temperature

Although temperature usually acts as a mere modifier of the photoperiodic effect, in some insects it is known to be the main cue of diapause induction. The leading role of temperature in diapause induction has been most clearly demonstrated in tropical insects (Denlinger 1986).

Among the pentatomoids in which temperature has been studied from this standpoint, winter adult diapause has been found to be controlled primarily by temperature only in the pentatomid *Andrallus spinidens*. This predaceous, polyphagous species is distributed in the tropical and subtropical regions. In southern Japan (Takanabe; 32°N), it occurs on herbaceous plants in rice and other fields where it actively feeds on larvae of the noctuid moths *Spodoptera litura* F. and *Aedia leucomelas* (L.) (Shintani et al. 2010) and usually produces three or four generations a year. To control such a multivoltine seasonal cycle and form winter adult diapause in autumn, insects usually use day length as a cue (see **Section 11.3.1 and Chapter 12**), but this species utilizes a different cue. The long-term mean temperature in this region is 26.8°C at the beginning of September and 22.3°C at the end of September. Adults enter winter diapause primarily in response to this decrease of temperature, not the change of day length. Nymphs hatching at the end of October usually are not able to complete development to adults before winter and, apparently, cannot form diapause. Only adults can enter diapause and survive until spring. Therefore, it is important for the population to enter diapause before October.

In laboratory experiments, when nymphs and then adults of *Andrallus spinidens* were reared and then maintained under constant conditions, diapause could be induced in all photoperiodic regimes tested (with duration of the photophase of 12 to 16 h) but only at temperatures below 25°C. At higher temperatures, all individuals actively developed (i.e., were nondiapausing) irrespective of day length (**Figure 11.5**). Sensitivity to temperature is present at the nymphal and adult stages in this species. Diapause is terminated during winter under the influence of low temperatures, and the adult bugs resume their activity in spring (Shintani et al. 2010). Thus, the multivoltine seasonal cycle of *A. spinidens* is controlled primarily by the temperature, whereas photoperiod seems to perform only an additive function (Saulich and Musolin 2012).

Other examples of the control of winter diapause induction by temperature as in *Andrallus spinidens* are rare. For insects in the Temperate Zone, temperature's effect as a modifier of the PhPR during diapause induction seems to be more important and widespread. Even though many environmental factors likely influence the PhPR, in most species the modifying effect of temperature is the strongest.

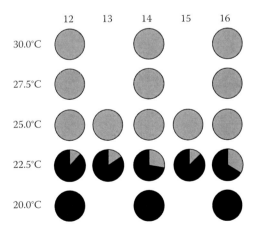

FIGURE 11.5 Effect of photoperiod and temperature on induction of winter adult diapause in the pentatomid *Andrallus spinidens* from Takanabe, Japan (32°N). Nymphs were reared to adults and then maintained under constant experimental conditions: temperature is indicated on the left; photoperiod on the top of the figure (as photophases, in hours). Light sectors: reproductive (i.e., nondiapause) females; black sectors: diapause females. (From Y. Shintani, Y. Masuzawa, Y. Hirose, R. Miyahara, F. Watanabe, and J. Tajima, Entomological Science 13: 273–279, 2010, with permission.)

11.3.2.1 Effect of Temperature on the Photoperiodic Response Curve during Induction of Winter Diapause

Low temperature usually promotes induction of winter diapause. In some species, this effect is most evident around the critical photoperiod, but, in others, it is most evident under short-day or long-day conditions. High temperatures (such as 30 to 32°C) can strongly suppress the effect of day length and even completely prevent induction of winter diapause in some species. On the other hand, low temperatures (usually 15°C or below) enhance the tendency towards winter diapause and all individuals enter diapause irrespective of the preceding day length.

Degree of sensitivity of the critical photoperiod to temperature varies in different species. It is even possible to organize different species on the basis of temperature sensitivity of their PhPRs of diapause induction from a strong dependence of the critical photoperiod on temperature to a temperature-stable PhPR with the critical photoperiod almost insensitive to temperature (Saulich and Volkovich 2004). Based on the data from many insect species, it has been shown that within a range of 20 to 27°C, a temperature shift of 5°C can cause a shift of critical photoperiod of approximately 1 hour (Danilevsky 1961). In other species, including *Podisus maculiventris*, a critical photoperiod of diapause induction basically remains stable within a temperature range of 17.5 to 25.5°C, even though a proportion of diapausing individuals generally decreases under short-day conditions and PhPR can be strongly suppressed under high temperatures (**Figure 11.6**; Goryshin et al. 1988a).

The results of experimental studies of winter adult diapause in the pentatomid *Scotinophara lurida* demonstrate that temperature can affect the quantitative PhPR in a similar way: the higher the temperature, the more diapause is suppressed even under typically diapause-inducing conditions (**Figure 11.3**; Cho et al. 2008).

The main ecological function of temperature in insect diapause onset is the optimal timing of diapause induction during the season. In warm years, winter diapause induction shifts to later dates due to the critical photoperiod decreasing under the action of high temperatures whereas in cold years, an earlier induction of winter diapause takes place. In general, an increase in temperature suppresses the tendency to enter winter diapause, whereas a drop in temperature facilitates induction of winter diapause. However, the signal function of temperature cannot always be distinguished (or separated) from the direct suppression of activity by low temperatures. In many cases, especially in southern species that overwinter as adults, the absence of oviposition in the field might often be caused by the direct suppression of maturation and/or oviposition by the low ambient temperature rather than by induction of winter adult diapause (Saulich and Musolin 2009).

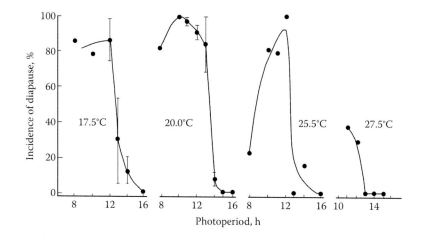

FIGURE 11.6 Effect of constant temperature on the photoperiodic response of winter adult diapause induction in females of the spined soldier bug, *Podisus maculiventris*. The laboratory culture originated from Missouri, the United States of America (about 38°N). Nymphs were reared to adults and then maintained under constant experimental conditions. Vertical lines are ranges of results in different replicates of the experiments. (From N. I. Goryshin, T. A. Volkovich, A. Kh. Saulich, M. Vagner, and I. A. Borisenko, Zoologicheskii Zhurnal [Zoological Journal] 67: 1149–1161, 1988, with permission.)

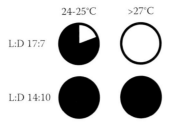

FIGURE 11.7 Effect of photoperiod and temperature in winter diapause induction in females of the predatory pentatomid *Arma custos* from Belgorod Province, Russia (50°N). Nymphs were reared to adults and then maintained under
constant experimental conditions. Light sectors: reproductive (i.e., nondiapause) females; black sectors: diapause females.
(Modified from T. A. Volkovich and A. Kh. Saulich, Entomological Review 74: 151–162, 1995, with permission.)

11.3.2.2 Temperature Optimum of Photoperiodic Response

As with other physiological responses, the PhPR has its own temperature optimum. Within this optimal
range, the response is clear and ecologically meaningful (e.g., winter diapause is normally induced under
short-day conditions), whereas outside the optimum, the PhPR does not work properly (e.g., too low or
too high temperatures do not allow diapause to be properly formed and insects suffer high mortality).
This temperature optimum is an ecologically important characteristic of PhPR because, to a high degree,
it determines the effectiveness of the PhPR under natural conditions (Danilevsky 1961). The wider the
range of the temperature optimum of the PhPR, the more important the role of day length in the control
of the species' seasonal development.

The range of temperatures under which the PhPR fully manifests itself differs significantly between
different species of insects. Temperature optimum likely evolved in intense relationships with other
critically important characteristics of the PhPR. Thus, in the predatory pentatomid *Arma custos*, the
temperature optimum of the PhPR is narrow (**Figure 11.7**) and associated with high temperatures, generally allowing nondiapause development in most females only when the temperature is higher than 27°C
(Volkovich and Saulich 1995). This peculiarity of the PhPR of *A. custos* makes it virtually impossible for
the seasonal cycle of this species to be bi- or multivoltine in the forest-steppe zone in Europe, in spite of
the facultative (i.e., nonobligate) nature of its adult diapause: it is simply too cold in the region to allow
nondiapause development and, thus, winter adult diapause is induced in each generation. Realized number of annual generations often is reduced because the time is limited when food is available, abundant,
and of good quality (Saulich and Volkovich 1996).

Somewhat similar results were obtained in experiments with the pentatomid *Dybowskyia reticulata*
in Japan (Nakamura and Numata 1998). The temperature optimum of the PhPR of this species also is
shifted into the high temperature range; adequate response to day length occurs only at 27.5°C or higher,
whereas even a slight decrease in temperature to 25°C induces diapause in all the individuals under both
long- and short-day conditions. In Osaka (Japan; 34.7°N), *D. reticulata* completes one generation in cold
years and two generations in warmer years. The relatively low summer temperature (25°C and lower)
"switches off" the physiological mechanism of response to day length so that all the adults enter diapause
regardless of the dates of their emergence. The need to limit the number of generations is related to the
fact that *D. reticulata* is a narrow oligophage feeding on the seeds of umbellates, which are only available briefly in summer and shatter before the beginning of September (Nakamura and Numata 1998).

11.3.3 Food

The interaction of photoperiod and temperature creates a reliable ecological mechanism controlling the
timely onset of diapause in a particular season. However, there are cases when another factor, namely
trophic (food, or diet), is added to this usual tandem.

The primary value of the trophic factor in regulation of the seasonal development has been studied
in great detail in the cabbage bug, *Eurydema rugosa* (= *E. rugosum*), which has a winter adult diapause. In Osaka (Japan; 34°N), the nymphs of this species feed on leaves and seeds of various crucifers.

Individuals of the first generation feed on wild crucifers (e.g., brown mustard) that die out before early summer. The resulting adults emerge in mid-June and most enter diapause. Because the cultivated crucifers (such as radish and cabbage) remain green much longer, most of the first-generation adults feeding on them are reproductively active and give rise to individuals of the second generation, which become adults at the end of summer and form the overwintering population (Ikeda-Kikue and Numata 2001). To determine the role of individual factors in control of seasonal development of this species, the nymphs were reared to adults and then maintained on two different diets (i.e., leaves and seeds of rape, *Brassica napus* L.) under experimental conditions similar to those in the natural environment (**Figure 11.8**).

The nymphs of *Eurydema rugosa* developed in June under long-day conditions synchronously on the two diets. Most adults of both sexes that molted in July and were maintained on rape leaves remained physiologically active; the females copulated and laid eggs and gave rise to the second generation. By contrast, nearly all the adults that molted at the same time but were maintained on rape seeds entered diapause. The nymphs of the second generation also were reared on rape leaves; however, most of the adults that emerged in September entered diapause (**Figure 11.8**). These experiments clearly showed that although day length played the principal role in induction of winter adult diapause, the response was modified significantly by the food in mid-summer, which thus affected the voltinism of the population.

In laboratory experiments (**Figure 11.9**), it was shown further that food (again, leaves and seeds of rape) acted as a signal in this species only under long-day conditions. Under short-day conditions, all females of *Eurydema rugosa* entered diapause regardless of the diet (Numata and Yamamoto 1990, Numata 2004).

The role of the food plant in diapause induction is known for many species of phytophagous insects. The shortage of food or a decrease in its quality usually increase the tendency to enter diapause. However, in *Eurydema rugosa* the trophic conditions facilitating the onset of diapause were no less favorable than

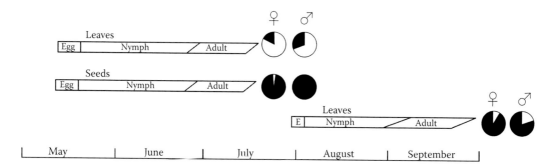

FIGURE 11.8 Effect of diet (leaves or seeds of rape) on winter adult diapause induction in the pentatomid *Eurydema rugosa* in Osaka, Japan (34.7°N) under quasi-natural conditions (nymphs and adults experienced natural day length and temperature). Light sectors: reproductive (i.e., nondiapause) adults; black sectors: diapause adults. (From K. Ikeda-Kikue and H. Numata, Acta Societatis Zoologicae Bohemoslovenicae 65: 197–205, 2001, with permission.)

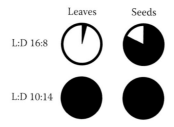

FIGURE 11.9 Effect of photoperiod and diet (leaves and seeds of rape) in winter adult diapause induction in females of the pentatomid *Eurydema rugosa* in Osaka, Japan (34.7°N). Nymphs were reared to adults and then maintained under constant experimental conditions (indicated on the left) at 25°C. Light sectors: reproductive (nondiapause) females; black sectors: diapause females. (From H. Numata, Applied Entomology and Zoology 39: 565–573, 2004, with permission.)

those promoting nondiapause development: both the survival rate and the body weight of nymphs fed on rape seeds were higher than for nymphs fed on leaves because the nutritional quality of seeds is higher than that of leaves (Numata and Yamamoto 1990). Therefore, the cue for diapause induction in this species, at least in part, was the phenological phase of development of the food plant (i.e., type of diet) rather than simply its nutritional value. The absence of leaves and the availability of seeds act as ecological signals of the approaching end of the **vegetative season** and the need for a winter diapause for *E. rugosa*.

It has been demonstrated experimentally that diapause induced by short-day conditions in insects feeding on leaves or seeds (i.e., the short-day diapause) differs in its properties from the diapause induced under long-day conditions in insects feeding on rape seeds (i.e., the **food-mediated, or trophic diapause**). The short-day diapause was terminated under the influence of low temperatures, which is typical of most species with winter diapause, after which the bugs became completely insensitive to day length. The food-mediated diapause was not terminated by low temperature, at least in the laboratory, and its properties remain to be studied (Ikeda-Kikue and Numata 1994).

The effect of diet on the PhPR of diapause induction has some specific traits in predatory bugs as well. For example, in the pentatomid *Perillus bioculatus* (a laboratory culture that originated from Canada; about 46.0°N) kept in the same photoperiodic regime (photophase 16 hours), all individuals feeding on eggs and larvae of the Colorado potato beetle *Leptinotarsa decemlineata* Say remained physiologically active, whereas those feeding on the diapausing adults of this beetle entered diapause (Shagov 1977). Similar results were obtained later in experiments with a population that originated from the United States of America (46.5°N): under the same photoperiods (photophases 14, 15, and 16 hours), the fraction of diapausing *P. bioculatus* adults was greater on the diet of older instar larvae of *L. decemlineata* than on eggs and younger instar larvae of the beetle (**Figure 11.10**). Thus, the age structure of the prey population acted as a cue for winter diapause induction in the predator: presence of mature prey (food) was interpreted by the predatory pentatomid as a signal of approaching autumn (Horton et al. 1998). However, similar to phytophagous bugs, the signal role of food manifested itself only under long-day conditions.

Unlike the oligophagous pentatomid *Perillus bioculatus*, the polyphagous predaceous stink bug *Podisus maculiventris* has a wide trophic range including no less than 75 species of insects from eight orders

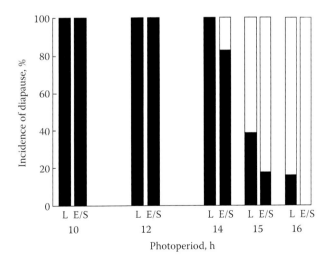

FIGURE 11.10 Effect of photoperiod and diet (eggs and larvae of different instars of the Colorado potato beetle, *Leptinotarsa decemlineata*) on diapause induction in the female pentatomid *Perillus bioculatus* from Wapato, Washington, the United States of America (46.5°N). Nymphs were reared to adults and then maintained under constant experimental conditions (photoperiod is indicated) at 23°C on two different diets. E/S – eggs and small larvae (younger instars) of the beetle; L – large larvae (older instars) of the beetle. Light sections of bars: reproductive (i.e., nondiapause) females, black sections of bars: diapause females. (Modified from D. R. Horton, T. Hinojosa, and S. R. Olson, *The Canadian Entomologist* 130: 315–320, 1998, with permission.)

(McPherson 1982). As could be expected, this predaceous stink bug revealed a much weaker influence of food on the PhPR of diapause induction. The fraction of diapausing individuals among those reared and then maintained in the laboratory on an unfavorable diet (larvae of the house fly, *Musca domestica* L.) increased only around the critical photoperiod (Goryshin et al. 1988b).

11.4 Winter Diapause *Per Se*

As noted above and evident from **Table 11.2**, the great majority of pentatomoids overwinter as adults. Winter diapause at this stage has been studied mostly in females, where it manifests itself most clearly in arrested ovarian development, suppressed oogenesis, the absence of oviposition, and presence of well-developed fat bodies. Thus, reproductively active females of *Nezara viridula* have mature (= chorionated) eggs or vitellogenic oocytes in their ovarioles, and weakly developed or loose fat bodies (**Figure 11.11C,E**).

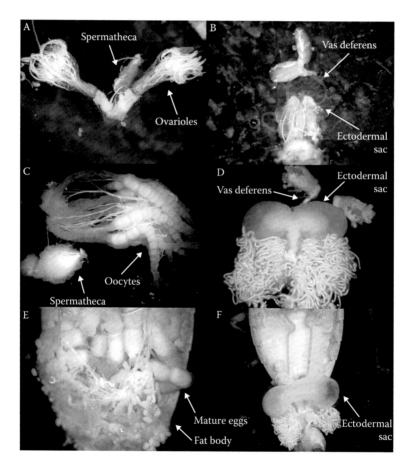

FIGURE 11.11   (See color insert.) State of gonadal development in diapause and nondiapause adults of the southern green stink bug, *Nezara viridula*. Nonreproductive female (virgin, prereproductive or in diapause): no oocytes in germaria, clear ovarioles, and empty spermatheca (A). Nonreproductive male (virgin, prereproductive or in diapause): clear vasa deferentia and collapsed ectodermal sac (B). Reproductive female (nondiapause): developing oocytes with yolk and expanded spermatheca (C); chorionated eggs in ovarioles and loose fat body (E). Reproductive male (nondiapause): yellow vasa deferentia and expanded ectodermal sac containing milky white secretion (D, F). For details on morphology and description of stages of gonadal development in *N. viridula* see Esquivel (2009). Note that in the original publication (Esquivel 2009) images (A) and (B) refer to the gonads without development in virgin adults. In diapause adults, the gonads remain basically in the same state until diapause termination. (A, B, C, and F are from J. F. Esquivel, Annals of the Entomological Society of America 102: 303–208, 2009, with permission; D and E are courtesy of Dr. Jesus F. Esquivel.)

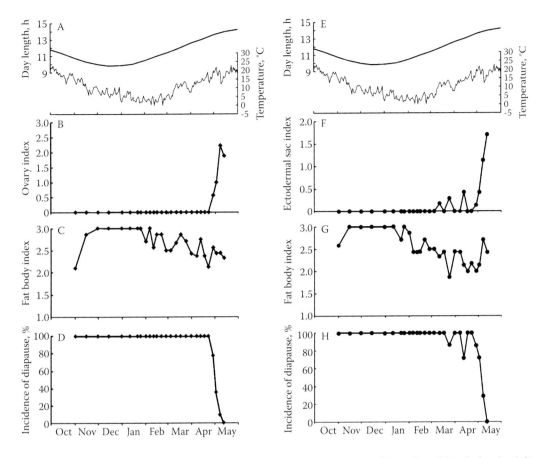

FIGURE 11.12 Dynamics of physiological indices during overwintering under quasi-natural conditions in females (left) and males (right) of the southern green stink bug, *Nezara viridula*, in Kyoto, Japan (35°N). A and E – natural day length and temperature; B – mean ovary index in females: from 0 (diapause: transparent ovarioles, no oocytes in germarium) to 3 (reproduction: semi-transparent ovarioles with mature eggs); C and G – mean fat body index: from 1 (reproduction: fat body small, loose, weakly developed) to 3 (diapause: fat body expanded, dense, well developed); D – incidence of diapause in females; F – mean ectodermal sac (accessory gland) index: from 0 (diapause: sacs transparent, empty, compact) to 3 (reproduction: sacs enlarged, filled with semi-transparent white-yellow secretion); H – incidence of diapause in males. (Modified from K. Takeda, D. L. Musolin, and K. Fujisaki, Physiological Entomology 35: 343–353, 2010, with permission.)

In contrast, in diapausing females of a similar age, differentiation and development of the oocytes is interrupted in the early stages. In these females, the ovarioles are clear, there are no oocytes in germaria, and the fat body is massive and dense (**Figure 11.11A**). In males the onset of diapause is usually (but not always!) marked with suppression of sexual activity and pheromone production, arrested or deeply suppressed development of the testes and/or accessory glands, and development of massive and dense fat bodies (compare **Figure 11.11B and Figure 11.11D,F**).

In both sexes, preparation for winter diapause is accompanied by active growth of the fat body (**Figure 11.12**), changes in the biochemical composition of tissues, in some cases – wax (or similar compounds) secretion (Dzerefos et al. 2009), migration (see **Section 11.7.1**), accumulation of specific nutrients (such as starch [Fedotov 1947]), and changes in behavior (**Figure 11.12**) and/or coloration (see **Section 11.7.4**). A period of diapause development is characterized by reduced oxygen consumption.

11.4.1 Peculiarities of Diapause in Females and Males

In many species of Pentatomoidea that exhibit winter adult diapause, those that are facultative diapausers show no significant difference between the sexes in the parameters of photoperiodic induction of diapause. This has been demonstrated in the pentatomids *Aelia fieberi* (Nakamura and Numata 1997b), *Dybowskyia reticulata* (Nakamura and Numata 1998), *Nezara viridula* (**Figure 11.12**; Musolin and Numata 2003a, Takeda et al. 2010), and many other species. In many species, the gonads of both sexes remain inactive until the end of the diapause or even postdiapause quiescence (Takeda et al. 2010).

At the same time, there are species with pronounced difference in physiological state of females and males during winter diapause. Thus, the well studied sunn pest, *Eurygaster integriceps*, has a deep obligate winter adult diapause. In diapausing females of this species, all morphogenetic processes stop (or become deeply suppressed) whereas in diapausing males, spermatogenesis continues and by the end of diapause males have mature sperm (Shinyaeva 1980).

Some other species copulate in autumn and females store sperm until the next spring. In such case, males sometimes even do not survive until spring. This strategy is known in various heteropteran families such as Nabidae (Kott et al. 2000), Anthocoridae (Kobayashi and Osakabe 2009, Saulich and Musolin 2009), and Pyrrhocoridae (Socha 2010). Among the Pentatomoidea, this strategy has been recorded in *Menida disjecta* (= *M. scotti*). The winter adult diapause is obligate in this species, but the males already have mature sperm in autumn. In the process of mating, which may occur even during winter, the males supply the females with nutrients and, thus, likely increase the females' chances of successfully overwintering (Koshiyama et al. 1993, 1994).

Males of the white-spotted stink bug, *Eysarcoris ventralis*, also have mature testes in autumn, winter, and spring and are ready to copulate after transfer to high temperature in the laboratory, whereas longday conditions are required for the start of ovarian development in females (Noda and Ishii 1981).

In the plataspid *Megacopta cribraria*, approximately 15% of overwintered females store live sperm from autumn until as late as mid-March. This trait not only allows them to oviposit without additional copulation in the spring but also likely increases the invasive potential of this species while it colonizes new areas because even one fertilized female can establish a new population of the pest (Golec and Hu 2015).

11.4.2 Cold Hardiness

Cold hardiness is an important issue in insect ecology and usually understood as the ability of an organism to survive at low temperatures (Leather et al. 1993). Diapause provides general nonspecific tolerance of insects to various adverse environmental conditions, including cold. The survival of insects at low temperatures recently has attracted considerable attention (e.g., Lee and Denlinger 1991; Bale 1993, 1996; Leather et al. 1993; Hodková and Hodek 2004; Danks 2005; Denlinger and Lee 2010), but the data for Heteroptera still are scarce.

In general, responses of insects to cold are complex and, as a rule, differ between diapausing and nondiapausing individuals, at different periods of the year, in different ontogenetic stages, and between populations. Nevertheless, based on their response to temperatures below the melting point of their body fluids, it has been suggested that insects can use three different strategies to cope with low temperatures:

(1) **freeze intolerance** (also called freeze avoidance, freeze susceptibility, or chill intolerance),

(2) **freeze tolerance** (also called freezing tolerance), and

(3) **cryoprotective dehydration**.

The freeze intolerant species cannot survive the formation of ice within their bodies and, thus, have evolved a set of biochemical, physiological, behavioral, and ecological measures/adaptations to prevent ice formation. In contrast, freeze tolerant insects can withstand ice formation, usually only in the extracellular fluids, and have a set of characteristics that enables them to survive such ice formation. Adoption of the strategy of cryoprotective dehydration allows the third group of insects to survive subzero temperatures by losing water to the surrounding environment, so resulting in an increase of the concentration of their body fluids and, thus, a decline in their melting point (to equilibration with the ambient temperature). As a result, they cannot freeze (Zachariassen 1985, Bale 2002, Sinclair et al. 2003, Chown and Nicolson 2004, Berman et al. 2013, Storey and Storey 2015).

As shown above, insects differ in their strategies to cope with cold, but most species, including all heteropterans studied thus far, follow the strategy of freeze intolerance. Even under harsh winter conditions (e.g., in Alaska), the freeze intolerant parent bug *Elasmostethus interstinctus* survives winter by **supercooling** (i.e., the physical phenomenon by which water and aqueous solutions remain unfrozen below their melting point if ice nucleating agents are absent; Barnes et al. 1996, Duman et al. 2004).

The relation between winter diapause and cold hardiness has been considered in numerous special publications (e.g., Denlinger 1991, Leather et al. 1993, Danks 2000, Bale 2002, Denlinger and Lee 2010). In general, winter diapause is thought to be necessary for increasing cold hardiness and successful overwintering of insects living in the Temperate Zone. However, there are several exceptions to this rule, where insects can survive winter without deep diapause, apparently using other specific ecophysiological strategies (Denlinger 1991, Šlachta et al. 2002).

The cold hardiness of insects under experimental conditions usually is estimated by **the supercooling point (SCP)** (i.e., the temperature at which spontaneous freezing occurs in a supercooled liquid, also referred to as the **crystallization temperature**). In several species, the SCP value is not constant throughout the year. For example, the SCP value of the Italian striped bug, *Graphosoma lineatum*, during 2000–2001 in the Czech Republic was about –7°C in May–June, decreased to –14 to –12°C in August–October, dropped to –18°C in December–January, and then increased again by spring (Šlachta et al. 2002). A similar pattern of SCP dynamics was observed in the stink bugs *Scotinophara lurida* in Korea (Cho et al. 2007) and *Halyomorpha halys* in the United States (Cira et al. 2016).

Seasonal trends are not always so distinct, however. For example, in a laboratory culture of the predaceous stink bug *Podisus maculiventris* originating from the United States of America (38°N), the SCP values of nondiapausing eggs and first instars were –34.1 ± 0.28°C and –29.0 ± 0.40°C, respectively, despite the fact that this species overwinters as adults. At the same time, the SCP values of diapausing and nondiapausing females were similar: –17.8 ± 0.46°C and – 15.0 ± 0.60°C, respectively (Borisenko 1987). The diapausing (–11.7 ± 0.7°C) and nondiapausing (–10.4 ± 0.8°C) adults of *Nezara viridula* from South Carolina (USA) also showed almost no difference in this parameter (Elsey 1993). These data testify to a weak relation or no relation at all between cold hardiness and diapause in the above species.

11.5 Diapause Development and Termination of Winter Diapause

The gradual changes that occur during the central diapause phase and finally result in its ending (i.e., termination) are usually referred to as **diapause development**. The term reflects the fact that diapause is not only a specific physiological state but also a dynamic process whose ending is followed by resumption of active development and often morphogenesis. Termination of diapause is achieved by resumption of activity of neurosecretory centers as a result of spontaneous or induced processes.

The specific features of the state of winter diapause and the processes taking place during diapause are still insufficiently studied. Based on the research of the gradual changes that occur during winter diapause development, Hodek (1983) distinguished two processes: **horotelic** (slow and spontaneous) and **tachytelic** (fast and induced; evolving at a rate faster than in the case of horotelic process).

Horotelic processes represent slow and internally regulated diapause development under more or less stable conditions (i.e., those under which diapause was induced). In this case, spontaneous diapause termination is free of external influence and does not require any stimuli. In contrast, the tachytelic processes take place when diapause development is influenced and accelerated by environmental conditions and diapause is externally and prematurely terminated by action of, for example, low temperatures (**cold termination of diapause**) or changes in the day length (**photoperiodic termination of diapause**). In other words, slow horotelic processes result in spontaneous diapause termination, whereas fast tachytelic processes accelerate diapause development and finally end up with externally induced diapause termination (Hodek 1983, 1996, 2002; Zaslavski 1988). These two processes are explained here separately and can be studied separately in the laboratory, but, in nature, stable conditions almost never exist and, thus, the tachytelic process (caused by, for example, cold in winter) is likely to override the slow horotelic process.

11.5.1 Spontaneous Termination of Winter Diapause

This type of termination of winter diapause is based on endogenous (horotelic) processes and may proceed under the same conditions under which diapause was induced. **Spontaneous diapause termination** is thought to be most important in species with a weak diapause, which generally is typical of insects of tropical and subtropical origin. Because the conditions never remain constant in most parts of the Earth, true spontaneous diapause termination can be observed only under stable laboratory conditions. The possibility of this type of winter diapause termination under constant conditions has been shown in laboratory experiments for many heteropterans, including the stink bugs *Carbula humerigera* (Kiritani 1985b), *Plautia stali* (Kotaki 1998a,b), *Nezara viridula* (Musolin et al. 2007), and others.

Spontaneous diapause termination usually follows a prolonged period of diapause development, with the timing of diapause termination varying between individuals. For example, Musolin et al. (2007) showed the timing of postdiapause oviposition between the earliest and the latest females of *Nezara viridula* in different short-day regimes at 25°C varied from 106 days (photophase 13 hours; **Figure 11.13C**) to 158 days (photophase 10 hours; **Figure 11.13A**), whereas in the nondiapausing females the range of variation was only 43 days at the same temperature (photophase 14 hours; **Figure 11.13D**). Under natural conditions, the difference in the timing of oviposition after overwintering between the earliest and the latest females was only 23 days (Musolin et al. 2007). These results demonstrate that rates of spontaneous diapause termination vary greatly between individuals.

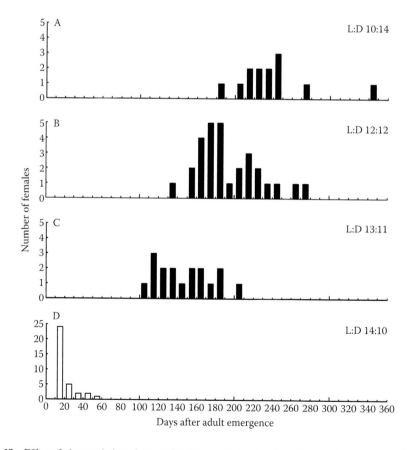

FIGURE 11.13 Effect of photoperiod on the preoviposition period in females of the southern green stink bug, *Nezara viridula,* from Osaka, Japan (34.7°N) at 25°C. Nymphs were reared to adults and then maintained under constant experimental conditions (From D. L. Musolin, K. Fujisaki, and H. Numata, Physiological Entomology 32: 64–72, 2007, with permission.)

The rate of spontaneous winter diapause termination in *Nezara viridula* depended on the photoperiodic conditions during the preceding diapause induction and the subsequent regime. In other words, the diapause that was induced and maintained under different photoperiodic conditions varied in its intensity: a shorter photophase corresponded to a stronger diapause and a later onset of the postdiapause oviposition (**Figure 11.13A–D**).

11.5.2 Cold Termination of Winter Diapause

Cold termination of diapause has been shown experimentally to be of primary significance for most insect species in the Temperate Zone although under field conditions, its effect often is difficult to separate from the spontaneous diapause termination processes (Hodek 1983, 1996, 2002).

Insect activity usually resumes after exposure of diapausing individuals to temperatures ranging from 0 to 10°C; some species have narrower ranges of temperature favorable for diapause termination. Negative temperatures usually hinder the diapause termination process, as do temperatures exceeding 15°C. The temperature requirements of diapausing stages are determined mostly by the living conditions and geographic origin of the species but are almost independent of the stage at which overwintering occurs (Saulich and Volkovich 2004).

Environmental conditions during overwintering and the diapause termination subphase (see **Figure 11.1**) affect the physiological state of the subsequent stages. For example, diapause in female *Podisus maculiventris* was most efficiently terminated by temperature from 6 to 8°C; such conditions generally facilitated the highest survival rate of the adults during diapause and highest reproductive indices after diapause (e.g., fecundity). Even slight deviations from the optimal conditions during diapause may have considerable negative consequences after diapause (e.g., low fecundity and/or survival rate; Goryshin et al. 1989).

The duration of cold exposure required for winter diapause termination varies from 1 to 6 months depending on the species. The neuroendocrine centers gradually resume activity in response to cold exposure and become capable of providing immediate stimulation when the temperature rises in spring (Tauber et al. 1986).

11.5.3 Photoperiodic Termination of Winter Diapause

After photoperiodic induction of diapause, many diapausing insect species remain sensitive to day length and diapause in such species can be terminated by changes of photoperiodic conditions. For example, if diapause was induced by short-day conditions, it can later be terminated by exposure to long-day conditions. This type of winter diapause termination is typical of species with larval (nymphal) and adult diapause.

Photoperiodic termination of winter diapause also is based on the interaction of spontaneous (i.e., horotelic) and induced (i.e., tachytelic) processes. This is indicated by the variable duration of the period required for long-day diapause termination at different stages of diapause. During the initiation subphase (see **Figure 11.1**), diapause is not intense/deep and not completely formed, but, nonetheless, the diapause termination capacity is blocked most strongly. Therefore, insects transferred in autumn from short-day to laboratory long-day conditions usually do not undergo fast photoperiodic termination of diapause (as evidenced, for example, by oviposition in adult diapause). Later, due to the progress of the horotelic process of diapause development, the blocking of morphogenesis becomes weaker, and the time required for photoperiodic termination of diapause (i.e., induced, tachytelic process) gradually shortens (Hodek 1983, 2002; Koštál 2006).

The photoperiodic responses of diapause termination sometimes show amazing similarity with those of diapause induction, and the critical photoperiod values may be nearly the same. The coinciding PhPR curves of diapause induction and termination may indicate that the terminating effect results from the same physiological mechanism that controls the onset of diapause. In other cases, for example in *Nezara viridula*, the PhPR curves may differ somewhat in shape, suggesting that more complicated mechanisms are involved (Musolin et al. 2007).

The interaction of spontaneous and induced processes during diapause termination were demonstrated in two studies using *Podisus maculiventris* that originated from Missouri, the United States of America (about 38°N). In the first experiment, diapause induction, termination, and postdiapause oviposition were studied after exposure of the bugs to different photoperiodic and temperature conditions (**Figure 11.14**; Chloridis et al. 1997). When females were reared from egg to adult and then maintained at long-day conditions (L:D 16:8) and 23°C and remained under the same conditions further, all of them were non-diapause and soon started oviposition (curve A). Basically, the same was recorded when females were reared from egg to adult and then maintained under the same long-day conditions (L:D 16:8) and 23°C but were transferred to short-day conditions (L:D 8:16) on day 13 after the final molt: these females were reproductive and did not stop oviposition at least during 27 days following the transfer from long-day to short-day conditions (curve B). On the contrary, all females reared from egg to adult and then maintained under short-day conditions (L:D 8:16) at 23°C entered diapause (curves C, D, and E). As a result of spontaneous diapause termination under constant short-day conditions (L:D 8:16) at 23°C (curve C), the females started laying eggs on day 47, which indicated that the diapause formed at the photoperiod L:D 8:16 and 23°C was not very deep and stable. Exposure to cold under the same short-day conditions (4°C for 10 days; curve E) hastened the onset of the postdiapause oviposition and increased the fraction of ovipositing females as compared to those in the trial in which the females were kept under constant short-day conditions at 23°C without any cold treatment (curve C). However, even on day 130, the fraction of ovipositing females was only slightly over 40% (curve E). A much greater diapause terminating effect was observed after consecutive action of cold (10 days at 4°C and darkness) and long-day conditions (L:D 16:8) and 23°C (curve D): oviposition started on day 25 (i.e., 10 days of low-temperature treatment plus 15 days of reproduction stimulating conditions), and all the females terminated diapause by day 70. These results show that adults of *P. maculiventris* remain sensitive to day length during winter diapause, which is a prerequisite for its photoperiodic termination. Moreover, day-length sensitivity is preserved even after exposure to cold.

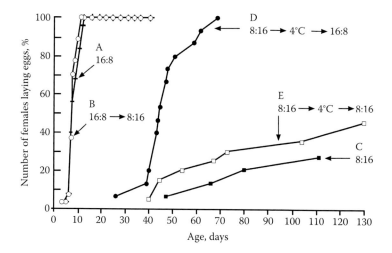

FIGURE 11.14 Oviposition dynamics in females of the spined soldier bug, *Podisus maculiventris*, under different photoperiodic conditions at 23°C (if otherwise not indicated). A – females reared from eggs to adults and then maintained at long-day conditions (L:D 16:8); B – females reared from eggs to adults and then maintained at long-day conditions (L:D 16:8); on day 13, after the final molt, females were transferred to short-day conditions (L:D 8:16); C – females reared from eggs to adults and then maintained at constant short-day conditions (L:D 8:16); D – females reared from eggs to adults and then maintained at short-day conditions (L:D 8:16); then on day 13, after the final molt, females were transferred to 4°C and darkness for 10 days, then to long-day conditions (L:D 16:8); E – females reared from eggs to adults and then maintained at short-day conditions (L:D 8:16); then on day 13, after the final molt, females were transferred to 4°C and darkness for 10 days, then to short-day conditions (L:D 8:16). The laboratory culture originated from Missouri, the United States of America (about 38°N). (From A. S. Chloridis, D. S. Koveos, and D. C. Stamopoulos, Entomophaga 42: 427–434, 1997, with permission.)

TABLE 11.4

Photoperiodic Termination of Diapause in Adult *Podisus maculiventris* (From N. I. Goryshin, T. A. Volkovich, A. Kh. Saulich, and I. A. Borisenko, Manuscript deposited in the VINITI (Vsesojuzniy Institut Nauchno-tehnicheskoy Informacii [All-Union Institute of Scientific and Technical Information]), Moscow, No. 115-B-90, 1989, with permission)[1]

Temperature, °C	Age of Adults at the Moment of Transfer to L:D 16:8	Number of Pairs	Proportion of Individuals That Terminated Diapause, %	Preoviposition Period, Days after Transfer (Mean ± S.E.)
20.0	25	23	100	26.1 ± 1.37
	27–29	11	100	25.2 ± 2.42
24.0	25–30	26	100	17.8 ± 0.93

[1] Diapause was induced under short-day conditions of L:D 12:12 at 20°C and then adults were transferred to the long-day conditions of L:D 16:8 at 20 and 24°C at different ages. The culture originated from Missouri, the United States of America (about 38°N).

The possibility of photoperiodic termination of winter adult diapause in *Podisus maculiventris* also was shown in another experiment with bugs that originated from the same Missouri population (**Table 11.4**). In that experiment, winter adult diapause was induced by short days (L:D 12:12) at 20°C, and adults on days 25 to 30 after emergence were transferred into long-day conditions (L:D 16:8) at 20 or 24°C. Under both temperatures, diapause soon terminated and females started oviposition. The higher temperature (24°C) had a stronger diapause terminating effect than the lower temperature (20°C), because preoviposition period was shorter at 24°C than at 20°C (**Table 11.4**). However, we cannot exclude that the difference in duration of the preoviposition period was caused by the effect of temperature on the postdiapause maturation rates rather than diapause termination process.

The photoperiodic sensitivity during winter diapause appears to be typical of many species overwintering as adults. In particular, this phenomenon was observed in the pentatomids *Halyomorpha halys* (Yanagi and Hagihara 1980), *Graphosoma lineatum* (Nakamura et al. 1996), and *Eysarcoris lewisi* (Hori and Kimura 1993). It is interesting that diapause induction requires exposure to short-day conditions starting from the third instar in *Podisus maculiventris* (see **Section 11.3.1.3 and Table 11.3**) and from the fifth instar in *E. lewisi* (Hori and Kimura 1993), whereas photoperiodic termination of winter diapause in both species requires exposure to long-day conditions only in the adult stage (Hori and Kimura 1993, Chloridis et al. 1997). Thus, the sensitive periods for induction and termination of winter adult diapause differ in length: the processes of diapause induction require a much longer action of the cue and likely involve more profound changes in the endocrine system than the processes leading to diapause termination. This conclusion was later supported by the results of experiments on winter diapause termination in *Scotinophara lurida* (Cho et al. 2008).

11.6 Environmental Factors Controlling Postdiapause Development in Spring

Studies of various insect species from the Temperate Zone in the Northern Hemisphere have shown that for most species, winter diapause ends before December, and the most severe part of winter is spent in a state of postdiapause quiescence (**Figure 11.1**; e.g., Danilevsky 1961; Hodek 1971b, 1996; Hodková 1982; Ushatinskaya 1990; Saulich and Volkovich 2004; Koštál 2006; Saulich and Musolin 2007b). According to a very precise definition suggested by Koštál (2006; p. 121), **postdiapause quiescence** is *an exogenously imposed inhibition of development and metabolism, which follows the termination of diapause when conditions are not favorable for resumption of direct development.* Postdiapause quiescence performs both functions of diapause: survival and synchronization of development; it complements winter diapause rather than replaces it, ensuring more precise seasonal synchronization (Veerman 1985, Belozerov 2009). Among the external factors controlling the resumption of active development in spring, the most important for pentatomoids in the temperate latitudes are day length, temperature, and presence of food (i.e., the same factors and cues that control the onset of winter diapause in autumn; see **Section 11.3**).

Resumption of active development of pentatomoids after winter dormancy (see **Figure 11.1**) might manifest itself in different ways and involve different life processes depending on the type of winter diapause. In embryonic (egg) diapause, embryogenesis comes to an end; in larval (nymphal) diapause, metamorphosis continues; and, finally, in adult diapause, the blocking of oogenesis is removed and activity of reproductive glands resumes. Eco-physiological mechanisms controlling resumption of active development in spring have been studied mostly in species with winter adult diapause.

11.6.1 Day Length

Prolonged exposure to cold usually results in temporary or permanent **photoperiodic refractoriness** (i.e., insensitivity to photoperiod, when the insects lose the ability to measure or respond to day length and, thus, they develop without entering diapause under any day-length conditions). Therefore, in spring, with the onset of warm weather, most species of pentatomoids resume activity regardless of the day length and reproduce until the end of their lives. Such a neutral response to day length after diapause first was described in the fire bug, *Pyrrhocoris apterus* L., and referred to as ***Pyrrhocoris*-like response** (Hodek 1971b, 1977).

In contrast with species that lose photoperiodic sensitivity irreversibly, in some other species this sensitivity is lost in autumn or winter but restored at the beginning of summer after a short refractory period. Such a type of response first was discovered in the bishop's mitre shield bug, *Aelia acuminata*, and referred to as ***Aelia*-like**, or **recurrent response** (Hodek 1971a). This phenomenon later was observed in other pentatomoids, such as *Dolycoris baccarum* (Hodek 1977), *Eurydema rugosa* (Ikeda-Kikue and Numata 1992), and *Graphosoma lineatum* (Nakamura et al. 1996). The resumed photoperiodic sensitivity may allow the insects to enter diapause more than once during their lifespan and, therefore, switch to a prolonged perennial, or semivoltine, life cycle (see **Chapter 12**). This type of response was suggested as a possible option in the predaceous stink bug *Perillus bioculatus* as well (Jasič 1967).

11.6.2 Temperature

It is well known that resumption of active development in spring is controlled by increasing temperatures. However, because the temperature regime in spring is highly unstable, some species overwintering as nymphs or adults, and forming close associations with particular food plants (i.e., mono- or oligophages) or their phenological phases, would benefit from using more precise external cues, in particular day length, as triggers of spring activation. Nevertheless, according to the data available, most species capable of photoperiodic diapause termination under laboratory conditions irreversibly lose their daylength sensitivity during overwintering in the field. Therefore, in spring, with the onset of warm weather, the bugs resume activity and start to reproduce regardless of the day length.

Winter adult diapause in overwintered females of *Eurydema rugosa* in central Japan was shown to be terminated completely by the beginning of April. The bugs at that time were in a state of postdiapause quiescence and did not start to reproduce due to the suppressing effect of low temperatures. This suppression could not be eliminated by either the presence of food or long-day conditions. Oviposition started only after the temperature exceeded the lower threshold of postdiapause morphogenesis (Ikeda-Kikue and Numata 1992). In a similar manner, the females of the pentatomid *Aelia fieberi* transferred into the laboratory (25°C) in late March or early April and supplied with favorable food started to oviposit much earlier than in the field, where their oviposition was suppressed by low temperature (Nakamura and Numata 1997b). Once started, oviposition continued until the end of the females' lives.

11.6.3 Food

One of the important components of the environment, essential for insect winter diapause termination and resumption of activity in spring, is the presence of adequate food resources. The role of food is particularly apparent in regulation of postdiapause development of species feeding on fruits and seeds. Food

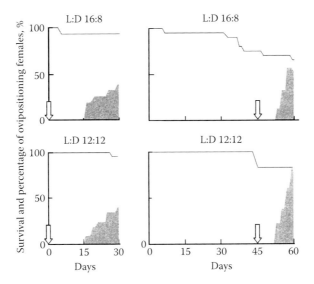

FIGURE 11.15 Survival and oviposition of females of the pentatomid *Dybowskyia reticulata* from Osaka, Japan (34.7°N) transferred in late March from the field to the laboratory short-day (L:D 12:12) and long-day (L:D 16:8) conditions at 25°C. Light areas: nonovipositing females; shaded areas: females that started laying eggs. Arrows mark the moment when food became available. (From H. Numata, Applied Entomology and Zoology 39: 565–573, 2004, with permission.)

as a trigger for spring reproduction was demonstrated in experiments with the pentatomid *Dybowskyia reticulata* (Nakamura and Numata 1997a). The females collected in the field and transferred into the laboratory in March started to lay eggs under both long-day and short-day conditions, but only in the presence of food (**Figure 11.15**).

In the absence of food, oviposition was delayed by a considerable period under both long- and short-day conditions. Availability of food stimulated reproduction (the moment of the appearance of food is marked with an arrow in **Figure 11.15**). Thus, the onset of reproduction in this species in spring is not controlled by either temperature or day length but is determined solely by the availability of food. In summer, the presence of food and its quality usually act as secondary cues whereas in spring, the absence of food becomes the main factor hindering gonad maturation.

11.7 Seasonal Adaptations Associated with Winter Diapause

In addition to winter diapause, pentatomoids have a diverse set of seasonal adaptations that allow them to synchronize their growth and reproduction with the seasons optimal for such activities and dormancy periods with the harsh periods of the year.

11.7.1 Migrations to and from Overwintering Sites

Migration is a complex phenomenon widely spread throughout the Heteroptera, including the Pentatomoidea. It is difficult to define insect migration but Dingle (1996, p. 38) listed five characteristics that distinguish migration from other forms of movements. These are: (1) it is persistent; (2) it is straightened out; (3) it is undistracted by resources that would ordinarily halt it; (4) there are distinct departing and arriving behaviors; and (5) energy is reallocated to sustain it. Dingle (1996) also stressed that not all migrants display all of these characteristics all of the time, but most will display most of them at least part of the time during which they are migrating. Migrations often happen seasonally and differ from sporadic and short-distance dispersal and other forms of movements in space aimed at search for habitats for feeding, oviposition, etc. Distances covered by migrating true bugs might differ manyfold, from hundreds of meters to hundreds of kilometers (Saulich and Musolin 2007a,b).

Migration behavior in insects now is understood as a special behavioral and physiological syndrome (Johnson 1969, Dingle 1996). Migrating individuals are characterized by enhanced motion activity and suppression of other functions, particularly reproduction and feeding. Usually, migrating individuals are in adult diapause, and migration is triggered by temperature conditions and/or movements of air.

Seasonal migrations might be linked to both winter and summer diapauses, and microhabitats chosen by diapausing individuals in winter and summer might be the same or different in different species. Seasonal migrations also are strongly linked to particular species-specific stages of ontogenesis, take place in particular periods of the year, and lead to adaptive changes of habitats.

Within the Pentatomoidea, some species are considered to be strongly migratory (e.g., *Eurygaster integriceps, Eurygaster maura, Aelia rostrata, Aelia melanota*) whereas others are semi-migratory (e.g., *Aelia furcula, Dolycoris penicillatus*) or nonmigratory (*Aelia acuminata*) (Brown 1962, Javahery 1995).

Migrations towards overwintering microhabitats (called also **hibernation quarters**) and sites occupied during summer diapause (called also **estivation quarters**) have been studied in Scutelleridae, particularly in the sunn pest, *Eurygaster integriceps*, in Eastern Europe. Even though *Eurygaster* species do not fly as well as some lepidopterans or orthopterans do (Arnoldi 1947), distance, duration, and regularity of their migrations deserve special attention (Critchley 1998).

Three regular migration events can be distinguished in the life cycle of *Eurygaster integriceps*:

(1) in spring from the hibernation quarters to the fields,

(2) in summer from the fields to the estivation quarters in mountains, and

(3) in autumn from the estivation quarters to the hibernation quarters.

Distances covered by these bugs seem to depend on geographic locations. Distant migrations (150–200 km) are typical for southern populations of *Eurygaster integriceps* that live in hot regions and migrate to overwinter at higher altitudes (e.g., Central Asia). Individuals from the more northern (and, thus, colder) regions and from lowland populations normally do not need to fly far to find cooler **overwintering quarters** and, thus, have shorter migrations (i.e., 20–50 km). Thus, in the center of the European part of Russia, *E. integriceps* overwinters in valley forest and forest belts and does not need to fly long distances. In such regions, seasonal migrations usually cover not more than 10–15 km. In special experiments with bugs labeled with radioactive isotopes in Stavropol Province (Russia), young overwintering adults were recorded up to 10 km from their feeding sites (Andrejev et al. 1958, 1964).

Eurygaster integriceps has three clear physiological/behavioral states (previously called *instincts*, Arnoldi 1947): nomadic, aggregative, and migratory. The nomadic state can be seen not only in adults but also in nymphs that often move around. The aggregative state may be observed both during periods of activity and dormancy. Often, adults are nonuniformly distributed at the hibernation and estivation sites; they are numerous at some microhabitats, whereas other similar and close microhabitats are almost unpopulated. And, finally, the migratory state clearly is aimed at active search of favorable habitats first for estivation and then for hibernation (Arnoldi 1947, Brown 1962, Critchley 1998).

11.7.2 Formation of Aggregations

Formation of large aggregations at different times of the year is characteristic of many species from various families of Pentatomoidea. Among these species, the phenomenon is more visible in the sunn pest, *Eurygaster integriceps* (Brown 1962), the parent bug *Elasmostethus humeralis* (Kobayashi and Kimura 1969), pentatomids *Menida disjecta* (Inaoka et al. 1993) and *Halyomorpha halys* (Hoebeke and Carter 2003, Nielsen and Hamilton 2009, Nielsen et al. 2011), and plataspids *Coptosoma scutellatum*, *C. mucronatum* (Davidová-Vilimová and Štys 1982), *Caternaultiella rugosa* Schouteden (Gibernau and Dejean 2001), and *Megacopta cribraria* (Eger et al. 2010, Suiter et al. 2010).

Aggregations can differ in size ranging from comparatively small groups of up to a few dozen bugs (e.g., in pentatomids *Biprorulus bibax* [James 1990a,b], *Euschistus heros* [Panizzi and Niva 1994], *Halys fabricii* [as *Halys dentatus*], and *Erthesina fullo* [Dhiman et al. 2004]) to groups of thousands (e.g., the parastrachiid *Parastrachia japonensis* can have as many as 4,000 adults that form an overwintering aggregation of up to 2 meters in size [Tachikawa and Schaefer 1985]).

In most species, aggregations are formed by adults, but cases are known where nymphs aggregate (e.g., overwintering aggregations of plataspid *Coptosoma scutellatum*; Davidová-Vilimová and Štys 1982) or both adults and nymphs do so (*Caternaultiella rugosa* Schouteden; Gibernau and Dejean 2001).

In large aggregations, individuals of different categories (hibernating, estivating, or nondiapausing) can have advantages over nonaggregated individuals including enhanced mating opportunities (Hibino 1985); shelter from adverse environmental conditions (Kiritani 2006) or parasitoids (*Caternaultiella rugosa* Schouteden; Gibernau and Dejean 2001); reduced desiccation (Lockwood and Storey 1986, Vulinec 1990); and combined chemical defense against predators (Cocroft 2001). However, there are possible negative effects too, because large aggregations are likely to attract predators and parasitoids or stimulate development of pathogens. This has been shown in the case of *Nezara viridula* and its parasitoid *Trissolcus basalis* (Wollaston) (Hymenoptera: Scelionidae) in Hawaii (Nishida 1966, Jones and Westcot 2002).

Ecological importance of seasonal aggregations during diapause has been studied in the subsocial shield bug *Parastrachia japonensis* in East Asia. This subsocial species is monophagous and feeds only on seeds of *Schoepfia jasminodora* Siebold et Zuccarini (Olacaceae). The fruits and seeds of the shrub are available only for a couple of weeks per year and to synchronize its seasonal cycle with that of the shrub, *P. japonensis* spends about 10 months annually in adult diapause and forms large aggregations (Tachikawa and Schaefer 1985; Tsukamoto and Tojo 1992; Nomakuchi et al. 1998; Filippi et al. 2000b; Tojo et al. 2005a,b). Formation of these aggregations decreases the metabolic rate in diapausing bugs which, in turn, increases their survival rate during long dormancy over hot summer or cold winter periods when food is not available. Elegant laboratory experiments demonstrated that oxygen consumption was twice as low in bugs in aggregations compared to isolated individuals. Interestingly, group size was not important if compared to the physical contacts with other individuals of the same species. It was suggested that such contacts stimulate excretion of a chemical compound functioning as an aggregation pheromone and promoting formation of aggregations (Tojo et al. 2005b). A few other eco-physiological adaptations allow the species to survive for an extended period when food resources are unreliable (Tojo et al. 2005a,b).

Pentatomoids also produce aggregation pheromones for either food or mate location or to identify overwintering habitats. Thus, males of the stink bug *Halyomorpha halys* produce a recently identified two-component aggregation pheromone (Khrimian 2005; Khrimian et al. 2008, 2014). The species also responds to a kairomone, which is an aggregation pheromone of a sympatric Asian pentatomid *Plautia stali*, although this stimulus is only attractive beginning in early August (Aldrich et al. 2009, Nielsen et al. 2011, Weber et al. 2014).

In autumn, during a period of preparation for winter diapause, many pentatomoids (e.g., *Halyomorpha halys, Menida disjecta, Urochela quadrinotata*) search for hibernation quarters and, in so doing, can enter houses and other buildings in large numbers, often becoming a serious nuisance (Kobayashi and Kimura 1969, Inaoka et al. 1993, Watanabe et al. 1995, Hoebeke and Carter 2003, Inkley 2012, Lee et al. 2014).

11.7.3 Photoperiodic Control of Nymphal Growth Rate

The growth rate of nymphs and, correspondingly, the duration of the nymphal stadia in Pentatomoidea are affected largely by ambient temperatures; an increase in temperature within the temperature optimum range hastens development, and a decrease hinders it. However, the developmental rate depends on other factors and cues as well. In particular, one of the important seasonal adaptations in insects is **photoperiodic control (i.e., regulation) of the nymphal growth rate**; nymphs may develop faster under certain photoperiodic conditions and slower under others. Such adaptation is a quantitative PhPR (see **Section 11.3.1.1**). In several species, under low to moderate temperatures, development is accelerated by short-day conditions. As day length decreases in autumn, the nymphal growth rate increases so as to reach the overwintering stage before environmental conditions get worse. Such an adaptation first was described in fire bug, *Pyrrhocoris apterus* (Pyrrhocoridae; Saunders 1983, Numata et al. 1993, Saulich et al. 1993), and later found in the predatory stink bug *Arma custos* (Volkovich and Saulich 1995), green shield bug, *Palomena prasina* (Saulich and Musolin 1996, Musolin and Saulich 1999), and many other heteropteran species (Musolin and Saulich 1997). Recently, Niva and Takeda (2003) noted that in *Halyomorpha halys*, short day accelerated nymphal development, whereas long day accelerated reproductive maturation. The two types of photoperiodic responses at different stages may help maintain the univoltinism of *H. halys* in the field, assuring the right timing for diapause and reproduction.

As with diapause induction, there must be some (preceding) stages that are sensitive to the cue inducing this response (i.e., day length). For some species, these stages already are known. Thus, in the stink bug *Eysarcoris lewisi*, acceleration or retardation of further nymphal growth is controlled by the photoperiodic conditions experienced by the nymphs during the third instar, whereas the physiological state (reproduction versus diapause) is controlled by conditions experienced only by nymphs during the fifth instar or adult stage (Hori and Kimura 1993).

However, despite its clear adaptive significance, photoperiodic control of growth rate is not a universal phenomenon in pentatomoids. Nymphs of some bugs grow faster under long-day conditions; moreover, responses to day length may be directly opposite in different populations of the same species. Such differences were found, in particular, between populations of the sloe bug, *Dolycoris baccarum*, from Norway and Japan (Conradi-Larsen and Sømme 1973, Nakamura 2003) and between populations of *Nezara viridula* from Egypt and Japan (Ali and Ewiess 1977, Musolin and Numata 2003a). These examples show that this trait can manifest itself at the population level, ensuring a high level of adaptation of the local population to specific living conditions, and that seasonal adaptations can differ between populations.

At the same time, in some species, the nymphal growth rate does not depend on photoperiodic conditions. For example, no distinct relations between the durations of nymphal stadia and photoperiodic conditions were observed in *Podisus maculiventris* (Goryshin et al. 1988b). In *Picromerus bidens* (Musolin and Saulich 1997, 2000), the effect of day length was small.

The physiological mechanism underlying this adaptation still has not been studied sufficiently. Further research will be necessary to understand the exact nature of the effect of day length (acceleration of development under certain conditions or retardation under other conditions) and the relationships between these phenomena and winter or summer diapauses.

11.7.4 Seasonal Polyphenism

The external appearance of individuals of the same species and developmental stage can change seasonally in some insect species. These changes can happen during the ontogenetic development of the same individuals or occur in representatives of different generations of the same species. Such cases often are difficult to notice and classify, and genetic and physiological mechanisms behind these changes are poorly understood. However, diversity of such forms often are described in terms of polymorphism or polyphenism.

Polymorphism usually is understood as the presence in a population of two or more distinct phenotypes (morphs, forms) at the same ontogenetic stage (discontinuous variation; Kennedy 1961; Walker 1986; Nijhout 2003; Saulich and Musolin 2007a,b; Simpson et al. 2011; Rogers 2015). Polymorphism can be divided into:

– **genetic polymorphism** (different phenotypes are produced by different genotypes), and

– **environmental**, or **ecological polymorphism**, or **polyphenism**, or **conditional polyphenism** (different phenotypes are produced by one genotype under different environmental conditions).

Several distinct genetically controlled color morphs of *Nezara viridula* (Hokkanen 1986, Ohno and Alam 1992, Musolin 2012; see **Chapter 7**), wing size polymorphism in many aquatic and semi-aquatic Heteroptera (Saulich and Musolin 2007a), and all cases of sexual dimorphism, can be considered examples of genetic polymorphism.

Wing polymorphism recently was reported in the Neotropical genus *Braunus* Distant (Pentatomidae). However, degree of manifestation, nature (i.e., genetic or environmental polymorphism), and control mechanism are not known, as yet, because most species of *Braunus* are known from only few specimens (Barão et al. 2016).

Polyphenism covers cases when environmental conditions determine which phenotype will be realized. Phenotypic plasticity can result from variation in developmental, physiological, biochemical, and behavioral processes that are sensitive to environmental variables (Nijhout and Davidowitz 2009, Simpson et al. 2011). If changes in frequencies of phenotypes are regular (annual) and controlled by environmental conditions, then such cases of polymorphism can be called **seasonal polyphenism**.

In Heteroptera, cases of **seasonal changes of wing size and degree of development of wing muscles or other organs** are good examples of seasonal polyphenism. They allow many true bug species to survive unfavorable seasons, migrate or disperse, and effectively use available resources. However, whereas wing size/wing muscle seasonal polyphenism is widely represented in some ecological and taxonomic groups of Heteroptera (e.g., aquatic and semi-aquatic bugs [Gerromorpha and Nepomorpha]; Saulich and Musolin 2007a), it remains basically unknown in Pentatomoidea. The only known exception is possibly a burrower bug, *Scaptocoris carvalhoi* Becker, distributed in Brazil (Nardi et al. 2008) and represented by two distinct wing forms. **Long-winged** (i.e., **macropterous**) individuals demonstrate a greater locomotion capacity than **short-winged** (i.e., **brachypterous**) individuals, and only long-winged adults can fly. The exact mechanism of control of this wing polyphenism is not known, but a significant increase in the frequency of long-winged adults has been noticed during the swarming season suggesting that the wing polyphenism is seasonal and highly functional. Furthermore, this seasonality may be related to the scarceness of rain during the developmental period of the nymphs. In this case, the lack of rain can result in a decrease in the moisture content of the soil and act as an inducing factor for wing development mechanisms (Nardi et al. 2008).

Another category of seasonal polyphenism is **seasonal body color polyphenism** or **seasonal body color change**, widely represented in the Pentatomoidea. There are numerous examples of this phenomenon where seasonal body color polyphenism often is linked to changes in the physiological state of individuals, namely formation of winter diapause, and often is under photoperiodic control (Musolin and Saulich 1996b, 1999; Saulich and Musolin 2007b).

There are two forms of adults of the dusky stink bug, *Euschistus tristigmus tristigmus*, which differ in morphology and body color. For many years, the two forms were considered different species or subspecies, but it was discovered they were seasonal forms of the same species (McPherson 1975a). The two forms could be produced under experimental conditions by manipulation of the rearing photoperiods (long-day versus short-day; McPherson 1974, 1975a,b, 1979). Subsequently, similar patterns of photoperiodic control of adult body coloration were reported in a few other pentatomids: *Thyanta calceata* (McPherson 1977, 1978), *Plautia stali* (Kotaki and Yagi 1987), *Oebalus ypsilongriseus* (Vecchio et al. 1994, Panizzi 2015), *Euschistus servus* (Borges et al. 2001), *Euschistus conspersus* (Cullen and Zalom 2006), *Nezara viridula* (Musolin and Numata 2003a, Musolin 2012), *Piezodorus guildinii* (Zerbino et al. 2014, 2015), and others. In even more pentatomoids, seasonal body color polyphenism has been reported, but its control mechanism remains unknown (e.g., *Halys fabricii* [as *Halys dentatus*] and *Erthesina fullo*; Dhiman et al. 2004).

In *Halyomorpha halys*, seasonal polyphenism of body color and other morphological characters manifests itself in both nymphal and adult stages (Niva and Takeda 2002). For example, the red color on the sternum of adults is more common in nondiapausing adults and may be related to reproductive maturity. The pronotum of fifth instars reared from hatching under short-day conditions (L:D 11:13) shows a darker, brown-marbled color pattern with less creamy-yellowish speckles, than that of the nymphs reared under long-day conditions (L:D 16:8). Temperature also influences body coloration of nymphs, and higher temperature enhances the long-day effect. The fifth instars reared under short-day conditions, which are destined to diapause when they become adults, have shorter white stripes on the pronotum, smaller body size, less frequent feeding, and more lipid accumulation than the long-day reared nymphs. In another experiment, cohorts of nymphs were transferred from long-day and high temperature conditions (L:D 16:8 at 25°C) to short-day and low temperature conditions (L:D 11:13 at 20°C) for the durations of the second–fifth and fourth–fifth stadia and for the fifth stadium. It was found that the longer the exposure to short-day and low temperature conditions during the nymphal stage, the greater the expression of short-day-associated characteristics observed in the fifth instars and adults (Niva and Takeda 2002, 2003).

The seasonal color polyphenism is irreversible in some species; in others, coloration changes gradually and may be reversible. For example, the freshly molted adults of *Nezara viridula* may be either green or yellow, depending on the genetic morph (a case of genetically controlled color polymorphism; see above in this **Section and in Chapter 7**). The same green or yellow body coloration is preserved during reproduction (**Figures 11.16 and 11.17A**). However, diapausing individuals of both sexes turn from green or yellow to reddish-brown soon after the adults emerge (**Figures 7.1D, and Figures 11.16 and 11.17B–F**; Musolin and Numata 2003a) and retain this coloration until the complete termination of diapause. These adult body color changes are controlled by day length and correlated with the physiological state (diapause versus nondiapause) of the individual (Harris et al. 1984, Musolin and Numata 2003a, Musolin et al. 2007, Musolin 2012).

The same pattern for *Nezara viridula* also was observed under field conditions in central Japan: reproductive adults of the summer generations were green (or yellow), although adults of the late-season generation normally did not reproduce but changed body color to reddish-brown and entered winter diapause (**Figure 7.1C,D**). After overwintering, the body color changed back to the initial green (yellow; **Figure 11.17G–H**), and the bugs began to reproduce (**Figure 11.18**; Musolin and Numata 2003b, Musolin et al. 2010, Takeda et al. 2010). A similar pattern was observed in the field in the redbanded stink bug *Piezodorus guildinii* (Zerbino et al. 2014, 2015).

Not so dramatic, but apparently an adaptive ontogenetic seasonal polyphenism was found in *Graphosoma lineatum*. In Sweden, the majority of newly eclosed adults of this species appearing in the late summer have no or very little red pigmentation. Instead, they exhibit a pale, light brownish (epidermis) and black (melanized cuticula) striation. These adults leave their host plants for overwintering in the ground. When they appear again on their flowering host plants in early summer after winter adult diapause, they show the typical red-and-black striation. Thus, the pale stripes turn red sometime before the postdiapause reproductive period. It is stressed that the two broad functions of protective coloration, camouflage and warning coloration, need not be mutually exclusive. It also is apparent that the five nymphal instars are all colored in various shades of brown and black and appear quite cryptic when feeding on seeds in the dried umbels of host plants (Tullberg et al. 2008).

In *Podisus maculiventris*, the degree of melanization also changes over the seasons, and adults are brighter in mid-summer than in spring or autumn. However, this response is mostly controlled by temperature, not day length (Aldrich 1986).

Analysis of the above examples of seasonal body color changes in pentatomoids easily reveals the dominant trend: the prevalence of brown coloration or dull texture of the integuments in overwintering insects. This makes them less conspicuous, providing passive protection from predators (**the seasonal camouflage**). Dark coloration also may give a certain adaptive advantage in **thermoregulation**, even during the winter. However, exceptions also are known (e.g., the pentatomid *Oebalus poecilus* is darker during the reproductive season than during overwintering; Albuquerque 1993).

The significance of seasonal body color changes is especially high in species forming large aggregations in winter and/or summer diapause sites; these color changes are common among pentatomoids.

It is important to note that day-length-controlled body color polyphenism has been found not only in adults where, in some cases it precedes diapause or is not directly linked to diapause, but also has been found in nymphs. In the nymphs of the pentatomid *Plautia stali* in Japan, six coloration forms (phenotypes) can be distinguished with cuticle color varying from green to dark brown (**Figure 11.19**). The incidence of these forms is controlled by photoperiod: under long-day conditions, incidence of brightly-colored nymphs is higher; under short-day conditions, more intense pigmentation is evident. Diapause in this species also is controlled by day length, but it is linked to the adult stage (**Figure 11.19**; Numata and Kobayashi 1994).

Somewhat similar was found in the predatory pentatomid *Arma custos* (Volkovich and Saulich 1995). In this species, short-day and low-temperature conditions stimulated appearance of dark-colored nymphs, whereas under long-day and high-temperature conditions incidence of nymphs with such coloration pattern was much lower and most or all nymphs were brightly colored (**Figure 11.20**). Similarly to the just described case of *Plautia stali*, induction of adult diapause and body color determination in nymphs in *A. custos* are two independently controlled processes because they are not only linked to different developmental stages (adults and nymphs) but also have different temperature optima. Thus, the

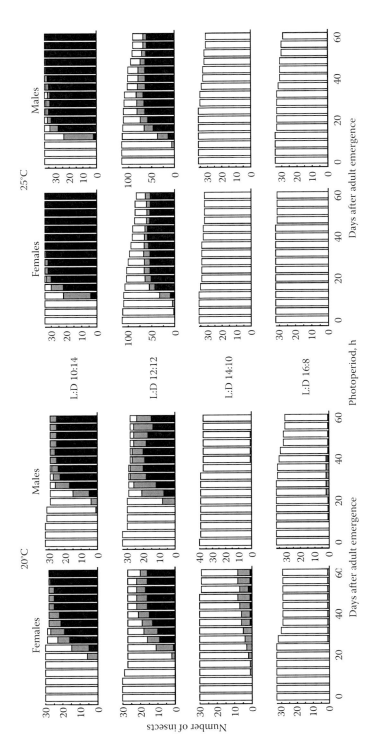

FIGURE 11.16 Effect of photoperiod and temperature on body coloration of adults of the southern green stink bug, *Nezara viridula*, from Osaka, Japan (34.7°N). Nymphs were reared to adults and then maintained under constant experimental conditions indicated in the figure: photoperiod is indicated in the center; temperature was 20°C (left) or 25°C (right). The body color of adults is shown as follows: light sections of bars: green; shaded sections of bars: intermediate coloration; black sections of bars: brown/russet. Note that normally brown/russet coloration of adults is associated with diapause (see text for details). (From D. L. Musolin and H. Numata, Physiological Entomology 28: 65–74, 2003, with permission.)

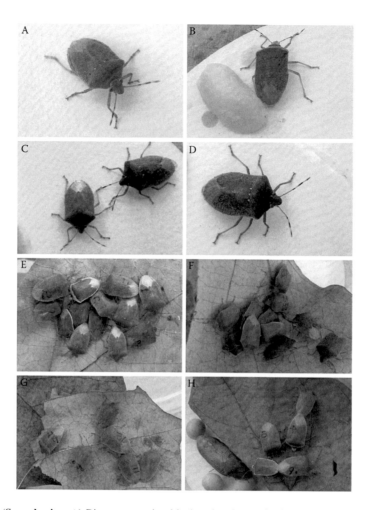

FIGURE 11.17 **(See color insert.)** Diapause-associated body color changes in the southern green stink bug, *Nezara viridula*, reared under laboratory or quasi-natural outdoor conditions. Laboratory rearing in Kyoto (Japan): reproductive (i.e., nondiapause) green adult (A). Adults of the intermediate body color grade (B, C left). Diapausing russet (brown) adults (C right, D). Outdoor rearing in Kyoto: mostly intermediate colored and russet overwintering adults at the early stages of diapause (November–December; E, F); adults of different body color grades at the later stage of diapause (March; G); diapause termination, body color change and beginning of postdiapause reproduction (April; H). (From D. L. Musolin, *Physiological Entomology* 37: 309–322, 2012, with permission.)

PhPR of adult diapause induction is most clear at high temperatures (27 to 30°C) and completely suppressed by low temperature, whereas photoperiodic control of body color in nymphs manifested itself under all tested conditions (Volkovich and Saulich 1995). Taking into consideration that incidence of dark-colored nymphs is much higher under short-day and low-temperature conditions (**Figure 11.20**), it might be speculated that appearance of pigmented nymphs is related to thermoregulation and takes place in colder periods of late spring (and early autumn in the regions where two generations can be produced per year; see **Chapter 12**). It is well documented that melanization of insect cuticle in the spring and autumn (when days are short) enhances absorption of solar insulation and, as a result, the body temperature can be up to 10 to 15°C higher than the air temperature (Hoffmann 1974, Tauber et al. 1986). Lighter color, due to low pigmentation, helps avoid overheating during the hottest days in mid-summer.

An interesting and complex case of seasonal body color change was reported in the parent bug, *Sinopla perpunctatus* (Faúndez and Osorio 2010). In southern Chile (53°S), this species has a univoltine seasonal cycle. Young adults of both sexes are dark green when they emerge in late summer as are the leaves on

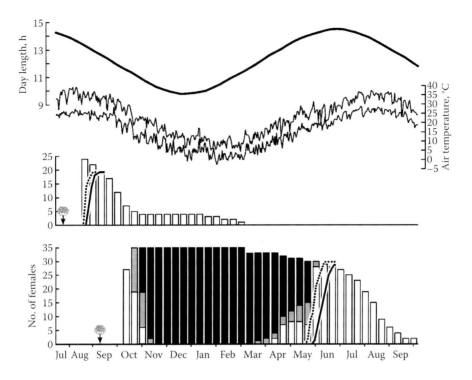

FIGURE 11.18 Seasonal body color changes in female of the southern green stink bug, *Nezara viridula*, under natural conditions in Osaka, Japan (34.7°N). The upper experimental series corresponds to the mid-summer reproductive (i.e., nondiapause and directly breeding) generation; the lower series, the late-season diapausing generation. Arrow marks the moment when the egg clusters were transferred into outdoor conditions to start each experimental series. The nymphs and males are not shown. The histogram shows the number and color of females as follows: Light sections of bars: green females; shaded sections of bars: females with intermediate body color; black sections of bars: brown/russet females. Dotted line: the total number of mating females; solid line: that of ovipositing females. The temperature is shown as the minimum and maximum daily values. (Modified from D. L. Musolin and H. Numata. Ecological Entomology 28: 694–703, 2003, with permission.)

their host plant. For overwintering, adults change body color to orange, which also is adaptive because leaves are reddish/brown in the autumn and the litter in which the adults overwinter is dark. In spring, both females and males change body color and starting at this moment the trajectories of body color begin to differ between the sexes. Males turn dark green and die after copulation. Females, on the other hand, turn light green, and their body color corresponds to the color of the lower surfaces of leaves where females guard their only egg cluster; they remain with the resulting brood until the nymphs reach the last instar. Only after this prolonged guarding, females turn dark green and then either die or probably overwinter again. It has been speculated that this seasonal polyphenism (or at least one of its phase, namely, light green to dark green transition) is not regulated by day length but, rather, linked to sexual maturation and copulation, as this transition is characteristic only of females. The control mechanism of body color change in this species still needs to be determined. Nevertheless, the whole pattern looks very adaptive and linked to parental care, a behavioral strategy well known in acanthosomatids and some other true bugs (Cobben 1968, Tallamy and Wood 1986, Kudo 2006).

In the case of **seasonal morphological polyphenism**, **seasonal forms** (sometimes called **seasonal morphs**) can differ morphologically. For example, the bugs can have different shapes of spines on the pronotum as in the pentatomids *Euschistus tristigmus tristigmus* (McPherson 1975a,b), *Oebalus ypsilongriseus* (Vecchio et al. 1994, Panizzi 2015), *Euschistus heros* (Mourão and Panizzi 2002), and *Dichelops melacanthus* (Chocorosqui and Panizzi 2003). These are the examples of irreversible seasonal polyphenism.

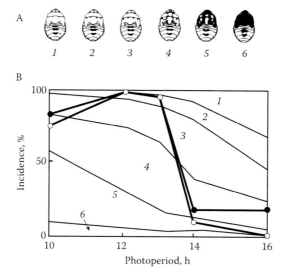

FIGURE 11.19   Effect of photoperiod on body color of the fifth instar nymphs, and photoperiodic response of adult diapause induction in the pentatomid *Plautia stali* from Tawaramoto, Japan (34.5°N). Grades of body color in fifth instars
(A) are shown above the graph (B) with comparison of the photoperiodic response curves controlling the induction of adult
diapause in females and males (thick lines) and the determination of body color of nymphs (thin lines). Close circles, males;
open circles, females. Incidence on the vertical axis refers to the incidence of diapause (thick lines) and the incidence of
color grades of nymphs (thin lines). Nymphs were reared and adults were then maintained under constant photoperiodic
conditions (indicated under the horizontal axis) at temperature 25°C. Numerals on the graph indicate grades of body color
in nymphs. (From H. Numata and S. Kobayashi, Experientia 50: 969–971, 1994, with permission.)

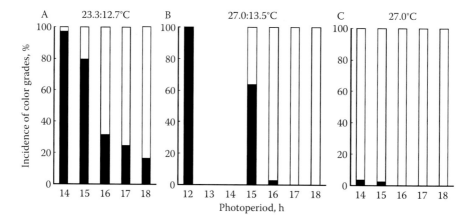

FIGURE 11.20   Effect of photoperiod and temperature on body color of nymphs of the predatory pentatomid *Arma custos*
from Belgorod Province, Russia (50°N). Nymphs were reared to adults and then maintained under constant photoperiods
(indicated under the horizontal axes) but different temperatures: A – outdoor rearing under constant photoperiods and
mean day temperature 23.3°C and mean night temperature 12.7°C (i.e., a rhythm of mean temperatures 23.3:12.7°C);
B – laboratory rearing under constant photoperiods and constant day temperature 27.0°C and constant night temperature
13.5°C (i.e., rhythm of temperature 27.0:13.5°C); C – laboratory rearing under constant photoperiods and constant temperature
27.0°C. Overall mean temperature in the treatment B was higher than in the treatment A, but lower than in the treatment
C. Light sections of bars: brightly-colored nymphs; black sections of bars: dark-colored nymphs. (Modified from T. A.
Volkovich and A. Kh. Saulich, Entomological Review 74: 151–162, 1995, with permission.)

Interestingly, if the definition of polyphenism, or environmental polymorphism, is analysed (different phenotypes are produced by one genotype under different environmental conditions), it becomes apparent that any case of photoperiodically controlled diapause induction can be considered as an example of such polyphenism: diapause and nondiapause phenotypes are produced by one genotype even though after overwintering (or the end of summer diapause), the diapause phenotype becomes undistinguishable from the nondiapause phenotype. This concept was introduced by Walker (1986), but in the literature facultative diapause and polyphenism are still usually treated separately.

11.8 Summer Diapause (Estivation)

As briefly discussed above (**Section 11.2.4**), diapause can take place not only in winter, but also in summer. In the latter case, diapause is formed under conditions of long day and high temperature and often associated with low humidity and low food availability. Such diapause not only ensures survival of insects during a season of unfavorable conditions but may also perform synchronizing functions in summer by postponing the subsequent ontogenetic stages to later periods (see **Chapter 12**). Similarly to winter diapause, summer diapause can be obligate (i.e., occurring every year and in each generation) or facultative (i.e., induced by external factors such as day length, temperature, or food; see **Section 11.2.3**; Masaki 1980; Saulich and Volkovich 1996, 2004). It also may be linked to any developmental stage from egg to adult, but this link is usually species-specific (i.e., a species can estivate only in a particular stage; see **Section 11.2.2**). In contrast to winter diapause, summer diapause does not last long and its duration is unlikely to be more than a few weeks.

Some true bug species may be able to form both winter and summer diapauses in their seasonal cycle (e.g., pentatomids *Scotinophara lurida* [Fernando 1960] and *Picromerus bidens* [Musolin and Saulich 1996b, 2000]). These diapauses can be linked to the same or (what happens more often) different ontogenetic stages and be of the same or different forms (obligate or facultative).

Even though environmental conditions (at least thermal and photoperiodic) differ greatly between summer and winter, in many respects summer diapause is similar to winter diapause. Both of them are formed well in advance of the actual deterioration of environmental conditions. Preparation for summer diapause is associated with pronounced physiological changes and increasing nonspecific resistance to unfavorable conditions similar to a set of adaptations well known for winter diapause. However, summer diapause is much less studied and likely to be less common among Pentatomoidea as well as other insects.

Although summer diapause is observed more frequently in insects from the tropical or subtropical zones (Masaki 1980), species that include summer diapause in their seasonal cycles also can be found in the Temperate Zone. Among pentatomoids, studied examples of summer adult diapause include the Oriental green stink bug, *Nezara antennata* (Noda 1984), *Carbula humerigera* (Kiritani 1985a,b), the predatory pentatomid *Picromerus bidens* (Musolin 1996; Musolin and Saulich 1996b, 2000), and a scutellerid *Poecilocoris lewisi* (Tanaka et al. 2002).

Adults of some species migrate into the mountains before summer diapause (the pentatomid *Aelia rostrata* in Turkey ascends to about 1,500 meters above sea level; Şişli 1965). Some species can migrate twice – first to the estivation quarters and later in the season – from the estivation quarters to the hibernation quarters (e.g., berry bug, *Dolycoris baccarum*; Krambias 1987).

Even these few example of summer diapause in pentatomoids demonstrate clearly the importance of this seasonal adaptation in insect life cycles and the ways in which this additional period of dormancy can optimize seasonal development of populations.

Finally, in some species or populations, diapause that is formed in summer does not end in autumn and instead lasts until the end of winter. Such a pattern might be called **summer–winter diapause**. This diapause pattern, for example, might explain what is occurring in the plataspid *Coptosoma scutellatum* (see **Chapter 12**). Every such case must be studied comprehensively because summer–winter diapause might, in fact, turn out to consist of two diapauses (i.e., summer and winter) with short and hardly detectable period between them.

11.9 Other Seasonal Adaptations

It is obvious that winter diapause is an important seasonal adaptation of pentatomoids as well as the great majority of other terrestrial and aquatic insects. It not only is crucial for surviving winters in many (if not most) regions of the globe but also important for synchronizing insects' life cycles with local conditions (Danilevsky 1961, Saunders 1976, Tauber et al. 1986, Danks 1987). However, winter diapause is actually only one (although very complex) seasonal adaptation out of many species- or population-specific physiological, biochemical, behavioral, ecological, or morphological adaptations utilized by true bugs and other insects in seasonally variable environments (Musolin and Saulich 1996b, 1997, 1999; Saulich and Musolin 1996, 2007a,b, 2009, 2012, 2014; Numata and Nakamura 2002; Numata 2004). Many of these adaptations are still more or less directly linked to overwintering (as with seasonal migrations or polyphenism; see **Section 11.7**), whereas others likely are not. A few examples of adaptations that are not directly related to overwintering, but still enhance the ability of pentatomoids to survive and take advantage in utilization of seasonally variable environmental conditions, are given below in this section.

Parental care (or **maternal care, care behavior, maternal instinct**) is an interesting peculiarity of Pentatomoidea. In most cases, this subsocial behavior is probably not linked directly to winter diapause but apparently is an important element of seasonal development and the reproductive cycle. Parental care is characteristic of many families of Pentatomoidea, such as Acanthosomatidae (Tallamy and Wood 1986, Tallamy and Schaefer 1997, Tallamy 2001, Faúndez and Osorio 2010), Cydnidae (Filippi et al. 2009), Scutelleridae (Nakahira 1994, Peredo 2002), Parastrachiidae (Filippi et al. 2000b, 2001; Gibernau and Dejean 2001), and Tessaratomidae (Gogala et al. 1998, Monteith 2006).

A comparative analysis of parental care in heteropteran taxa demonstrates that there are different levels of complexity of such behavior ranging from rather simple responses to complex patterns (Tallamy and Schaefer 1997, Hanelová and Vilímová 2013). In numerous primitive cases, females simply protect egg masses with their bodies from parasitoids and predators (including conspecific males) until nymphs hatch. In other cases (e.g., burrower bugs *Sehirus luctuosus* and *Tritomegas bicolor*), females not only guard egg clusters but also carry them from one microhabitat to another, trying to find the most favorable conditions for embryogenesis (Korinek 1940). In even more sophisticated cases of parental care, females not only guard their eggs and nymphs but also bring food for the progeny or even produce additional nonfertilized eggs as food for young nymphs after hatching (e.g., *Adomerus triguttulus* [Nakahira 1994], *Parastrachia japonensis* [Filippi et al. 2000b], *Canthophorus niveimarginatus* [Filippi et al. 2009]). The ecological importance of parental care as a seasonal adaptation is well documented in the Acanthosomatidae, Cydnidae, and Parastrachiidae but still needs to be further evaluated and properly understood.

Seasonal change in reproductive allocation varies during the reproductive period in the parent bug, *Elasmostethus interstinctus* (Mappes et al. 1996). At the beginning and middle of the reproductive period, females lay smaller eggs than at the end of the period. Cluster size and number of eggs per clutch decrease in laying sequence, earlier clusters being much larger than later clusters. Lifetime fecundity of females correlate positively with female size: large females produce more eggs and live longer than small ones. At the same time, egg size does not vary with female size. Offspring survival until adulthood increases with egg weight. In this species, individuals overwinter before reproduction, and, because the nymphs from later-laid eggs have the least time to gather resources before overwintering, it may be important for the later-laid eggs to be of higher quality. Reproductive allocation varies during the reproductive period; females allocate relatively more resources to offspring number at the beginning of the reproductive season and more to offspring quality at the end of their life (Mappes et al. 1996).

Seasonal food plant change allows insects to fully utilize the warm season even if the primary food plant is not available from early spring to late autumn. Thus, in Germany, during spring and early summer, nymphs of the parent bug *Elasmucha grisea* develop exclusively on birch (*Betula pendula* Roth) and cannot develop successfully on alder [*Alnus glutinosa* (L.) Gärtner] before early August. Only then do the adults shift to alder and utilize it until the end of the season. Alternative choice tests and oviposition experiments in the laboratory also showed a preference of adult *E. grisea* for birch in spring and alder in summer. This changing food plant preference guarantees an optimum efficiency of the reproductive potential of this species as the two generations of *E. grisea* feed on the two food plants successively (Melber et al. 1981).

A somewhat similar situation has been found in a few polyphagous pentatomids such as *Halyomorpha halys* (= *H. mista*) (Kawada and Kitamura 1983), *Nezara viridula*, *Piezodorus guildinii*, and *Euschistus heros* (Panizzi 1997, 2007). Undoubtedly, polyphagy allows many pentatomoids to use several food plants, produce more generations, and fully utilize local thermal conditions by earlier commencement of active development in spring and later formation of winter diapause (or developing without any dormancy period in some tropical regions) compared to monophagous true bugs.

Seasonal variation of microhabitat selection is somewhat different than the just described seasonal food plant change. In tropical Brazil, *Phloea subquadrata* (Phloeidae) does not need the pronounced winter diapause; its adults and nymphs are active all year round, but the egg-laying season is restricted mainly to the warm, rainy season. To deal with the annual sequence of dry and wet seasons, this species has evolved a seasonal change in microhabitat selection: under dry weather conditions, phloeids are found closer to the base of the host tree [*Plinia cauliflora* (Mart.) Kausel; Myrtaceae] trunks; in the rainy season, the bugs climb and live higher on the same tree (Salomão and Vasconcellos-Neto 2010).

11.10 Conclusions

The above review of diapause in the experimentally studied pentatomoids (82 species), mostly from the Temperate Zone, has revealed no distinct trends in the type (embryonic, nymphal, or adult), form (obligate or facultative), or seasonal class (winter or summer) of diapause in particular taxa within the superfamily.

Earlier analysis of seasonal development of Pentatomidae resulted in a similar conclusion; evolution of their seasonal adaptations does not correspond precisely to the phylogeny of the family (Saulich and Musolin 2012). However, similar complexes of seasonal adaptations might be formed within individual genera (e.g., *Palomena* with an obligate diapause and photoperiodic control of nymphal growth rates, or *Euschistus* with photoperiodic control of body color polyphenism), small tribes (e.g., Aeliini), or even subfamilies (e.g., Podopinae). At the same time, the large subfamily Pentatominae, despite being a polyphyletic taxon, is quite uniform in that most of the species have facultative winter adult diapause.

The seasonal changes of body coloration are more widespread in Pentatomidae (mostly those having winter adult diapause) than in other taxa of true bugs. However, other seasonal adaptations that are common in some heteropteran taxa such as wing size/wing muscle seasonal polyphenism (Saulich and Musolin 2007a,b) have been found so far only in one pentatomoid species (burrower bug *Scaptocoris carvalhoi*; Nardi et al. 2008).

Knowledge of ecology and seasonal development of a species may in some cases help clarify its taxonomic position. For example, *Dybowskyia reticulata* is considered a member of the tribe Tarisini by some authors (Vinokurov et al. 2010) and a member of the Graphosomatini by others (Gapon 2008). In its ecological characteristics, *D. reticulata* resembles species of Graphosomatini and this ecological trait should be taken into consideration in its classification.

In general, despite intensive studies of seasonal adaptations in pentatomoids and other heteropteran taxa in recent decades, diapause and ecological mechanisms of its control have been studied in less than 1% of the known species of true bugs. Clearly, more research is needed.

The role that individual seasonal adaptations and their complexes play in the structure of the species' seasonal cycle will be discussed in detail in **Chapter 12**.

11.11 Acknowledgments

Our research on heteropteran diapause was aided by the fruitful collaboration with many colleagues, including Nikolai Goryshin and Tatyana Volkovich (Department of Entomology, Leningrad / Saint Petersburg State University, Saint Petersburg), Hideharu Numata (Graduate School of Science, Kyoto University, Kyoto), Kiyomitsu Ito (National Agricultural Research Center for Hokkaido Region, Sapporo), and Kenji

Fujisaki (Graduate School of Agriculture, Kyoto University, Kyoto). We are very grateful to D. Kutcherov (Department of Entomology, Saint Petersburg State University, Saint Petersburg) for critical reading of the earlier version of the manuscript, Professor David A. Rider (Department of Entomology, North Dakota State University, Fargo) for consultations on taxonomical issues and Professor J. E. McPherson (Southern Illinois University, Department of Zoology, Southern Illinois University, Carbondale) for careful and patient editing of the chapter. In different years, research was partly supported by the Ministry of Education, Culture, Science, Sports and Technology of Japan (Grants-in-Aid for JSPS Fellows No. 98116 and L-4562 and STA Fellow No. 200141 and via The 21st Century COE Program at Kyoto University); the Council for Grants of the President of the Russian Federation, State Support of the Leading Scientific Schools (project No. 3332.2010.4); and the Inessa Charitable Foundation.

11.12 References Cited

Ables, J. R. 1975. Notes on the predacious pentatomid *Euthyrhynchus floridanus* (L.). Journal of the Georgia Entomological Society 10: 353–356.

Akramowskaya, E. G. 1959. True bugs (Hemiptera-Heteroptera) of the Armenian SSR, pp. 79–144. *In* Anonymous (Ed.), Zoological collection of papers, 11 (Data on the fauna of the Armenian SSR, Volume IV, Zoological Institute of the Academy of Sciences of the Armenian SSR). Zoological Institute of the Academy of Sciences of the Armenian SSR, Yerevan. 192 pp.

Albuquerque, G. S. 1993. Planting time as a tactic to manage the small rice stink bug, *Oebalus poecilus* (Hemiptera, Pentatomidae), in Rio Grande do Sul, Brazil. Crop Protection 12: 627–630. http://dx.doi .org/10.1016/0261-2194(93)90128-6.

Aldrich, J. R. 1986. Seasonal variation of black pigmentation under the wings in a true bug (Hemiptera: Pentatomidae): a laboratory and field study. Proceedings of the Entomological Society of Washington 88: 409–421.

Aldrich, J. R., A. Khrimian, X. Chen, and M. J. Camp. 2009. Semiochemically based monitoring of the invasion of the brown marmorated stink bug and unexpected attraction of the native green stink bug (Heteroptera: Pentatomidae) in Maryland. Florida Entomologist 92: 483–491. http://dx.doi.org/10.1653/024.092.0310.

Ali, M., and A. Ewiess. 1977. Photoperiodic and temperature effects on rate of development and diapause in the green stink bug, *Nezara viridula* L. (Heteroptera: Pentatomidae). Zeitschrift für Angewandte Entomologie 84: 256–264. http://dx.doi.org/10.1111/j.1439-0418.1977.tb04286.x.

Andrejev, S. B., A. V. Vsevodin, C. A. Molchanova, and A. V. Khotjanovitch. 1958. Some results of the use of tracer technique in the study of plant protection, pp. 18–26. *In* Anonymous (Ed.), Second United Nations International conference on the peaceful uses of atomic energy. Geneva (English translation is available at http://www.osti.gov/scitech/biblio/4291659; accessed on July 21, 2016)

Andrejev, S. B., C. A. Molchanova, B. A. Martens, and A. A. Rakitin. 1964. Radioactive labeling in the study of insect pests. Vestnik Sel'skokhoziajstvennoy Nauki Vsesojuznoy Akademii Sel'skokhoziajstvennih Nauk [Bulletin of Agricultural Sciences of All-Union Academy of Agricultural Sciences] 2: 122–128. (in Russian)

Arnoldi, K. V. 1947. The sunn pest (*Eurygaster integriceps*) in the Central Asian wildlife as related to the ecological and biocenological aspects of its biology, pp. 136–269. *In* D. M. Fedotov (Ed.), The sunn pest *Eurygaster integriceps* Put. Volume 1. Publishing House of The Academy of Sciences of the USSR, Moscow. 272 pp. (in Russian)

Asanova, R. B., and I. M. Kerzhner. 1969. Eine Ubersicht der Gattung *Trochiscocoris* Reuter mit Beschreibung einer neuen Unterart aus dem zentralen Kasahstan (Heteroptera: Pentatomidae). Beitrage zur Entomologie 19: 115–121. (in German with English summary)

Bale, J. S. 1993. Classes of insect cold hardiness. Functional Ecology 7: 751–753.

Bale, J. S. 1996. Insect cold hardiness: a matter of life and death. European Journal of Entomology 93: 369–382.

Bale, J. S. 2002. Insect and low temperatures: from molecular biology to distributions and abundance. Philosophical Transactions of the Royal Society of London B Biological Sciences 357: 849–862. http:// dx.doi.org/10.1098/rstb.2002.1074.

Barão, K. R., T. De Almeida Garbelotto, L. A Campos, J. Grazia. 2016. Unusual looking pentatomids: reassessing the taxonomy of *Braunus* Distant and *Lojus* McDonald (Hemiptera: Heteroptera: Pentatomidae). Zootaxa 4078(1): 168–186. http://dx.doi.org/10.11646/zootaxa.4078.1.16.

Barnes, B. M., J. L. Barger, J. Seares, P. C. Tacquard, and G. L. Zuercher. 1996. Overwintering in yellow-jacket queens (*Vespula vulgaris*) and green stinkbugs (*Elasmostethus interstinctus*) in subarctic Alaska. Physiological Zoology 69: 1469–1480.

Belozerov, V. N. 2009. New aspects in investigations of diapause and non-diapause dormancy types in insects and other arthropods. Entomological Review 89: 127–136. http://dx.doi.org/10.1134/S0013873809020018.

Berman, D. I., A. N. Leirikh, and E. P. Bessolitsina. 2013. Three strategies of cold tolerance in click beetles (Coleoptera, Elateridae). Doklady Biological Sciences 450: 168–172. http://dx.doi.org/10.1134/S0012496613030186.

Biehler, J. A., and J. E. McPherson. 1982. Life history and laboratory rearing of *Galgupha ovalis* (Hemiptera: Corimelaenidae) with descriptions of immature stages. Annals of the Entomological Society of America 75: 465–470. http://dx.doi.org/10.1093/aesa/75.4.465.

Borges, M., A. J. Zhang, M. J. Camp, and J. R. Aldrich. 2001. Adult diapause morph of the brown stink bug, *Euschistus servus* (Say) (Heteroptera: Pentatomidae). Neotropical Entomology 30: 179–182. http://dx.doi.org/10.1590/S1519-566X2001000100028.

Borisenko, I. A. 1987. Dynamics of cold tolerance during development of the spined soldier bug *Podisus maculiventris* Say. Byulleten' Vsesojuznogo Instituta Zashchiti Rasteniy [Bulletin of the All-Union Institute of Plant Protection] 69: 12–16. (in Russian)

Brown, E. S. 1962. Researches on the ecology and biology of *Eurygaster integriceps* Put. (Hemiptera, Scutelleridae) in Middle East countries, with special reference to the overwintering period. Bulletin of Entomological Research 53: 445–514. http://dx.doi.org/10.1017/S0007485300048252.

Bundy, C. S., and J. E. McPherson. 1997. Life history and laboratory rearing of *Corimelaena obscura* (Heteroptera: Thyreocoridae) with descriptions of immature stages. Annals of the Entomological Society of America 90: 20–27. http://dx.doi.org/10.1093/aesa/90.1.20.

Burov, V. N. 1962. Factors affecting population dynamics and harmfulness of wheat stink bugs of the genus *Aelia* (Heteroptera, Pentatomidae). Entomologicheskoe Obozrenie [Entomological Review] 41: 262–273. (in Russian with English summary)

Cakmak, O., M. Bashan, and E. Kocak. 2008. The influence of life-cycle on phospholipid and triacylglycerol fatty acid profiles of *Aelia rostrata* Boheman (Heteroptera: Pentatomidae). Journal of the Kansas Entomological Society 81: 261–275. http://dx.doi.org/10.2317/JKES-709.11.1.

Cant, R. G., R. N. Spooner-Hart, G. A. C. Beattie, and A. Meats. 1996. The biology and ecology of the bronze orange bug, *Musgraveia sulciventris* (Stål) – A literature review. General and Applied Entomology 27: 19–42.

Carter, M. E., and E. R. Hoebeke. 2003. Biology and seasonal history of *Elasmostethus atricornis* (Van Duzee) (Hemiptera: Acanthosomatidae), with descriptions of the immature stages and notes on Pendergrast organs. Proceedings of the Entomological Society of Washington 105: 525–534.

Chen, Q., J. L. Wang, S. J. Guo, H. X. Bai, and X. N. Zhuo. 2009. Studies on the biological characteristics of *Megacopta cribraria* (Fabricius). Journal of Henan Agricultural Science 4: 88–90.

Chloridis, A. S., D. S. Koveos, and D. C. Stamopoulos. 1997. Effect of photoperiod on the induction and maintenance of diapause and the development of the predatory bug *Podisus maculiventris* (Hemiptera: Pentatomidae). Entomophaga 42: 427–434. http://dx.doi.org/10.1007/BF02769836.

Cho, J. R., M. Lee, H. S. Kim, and K. S. Boo. 2007. Cold hardiness in the black rice bug, *Scotinophara lurida*. Physiological Entomology 32: 167–174. http://dx.doi.org/10.1111/j.1365-3032.2007.00565.x.

Cho, J. R., M. Lee, H. S. Kim, and K. S. Boo. 2008. Effect of photoperiod and temperature on reproductive diapause of *Scotinophara lurida* (Burmeister) (Heteroptera: Pentatomidae). Journal of Asia-Pacific Entomology 11: 53–57. http://dx.doi.org/10.1016/j.aspen.2008.04.003.

Chocorosqui, V. R., and A. R. Panizzi. 2003. Photoperiod influence on the biology and phenological characteristics of *Dichelops melacanthus* (Dallas, 1851) (Heteroptera: Pentatomidae). Brazilian Journal of Biology 63: 655–664. http://dx.doi.org/10.1590/S1519-69842003000400012.

Chown, S. L., and S. W. Nicolson. 2004. Insect physiological ecology. Mechanisms and patterns. Oxford University Press, Oxford. x + 244 pp.

Cira, T. M., R. C. Venette, J. Aigner, T. Kuhar, D. E. Mullins, S. E. Gabbert, and W. D. Hutchison. 2016. Cold tolerance of *Halyomorpha halys* (Hemiptera: Pentatomidae) across geographic and temporal scales. Environmental Entomology 45: 484–491. http://dx.doi.org/10.1093/ee/nvv220.

Cobben, R. H. 1968. Evolutionary trends in Heteroptera. Part I. Eggs, Architecture of the shell, gross embryology and eclosion. Centre for Agricultural Publishing and Documentation, Wageningen. 376 pp.

Cocroft, R. B. 2001. Vibrational communication and the ecology of group-living, herbivorous insects. American Zoologist 41: 1215–1221. http://az.oxfordjournals.org/content/41/5/1215.

Conradi-Larsen, E. M., and L. Sømme. 1973. Notes on the biology of *Dolycoris baccarum* L. (Het., Pentatomidae). Norsk Entomologiskl Tidsskrift 20: 245–247.

Couturier, G., M. S. Padilha De Oliveira, P. Beserra, D. Pluot-Sigwalt, and F. Kahn. 2002. Biology of *Discocoris drakei* (Hemiptera: Thaumastocoridae) on *Oenocarpus mapora* (Palmae). Florida Entomologist 85: 261–266. http://dx.doi.org/10.1653/0015-4040(2002)085[0261:BODDHT]2.0.CO;2.

Critchley, B. R. 1998. Literature review of sunn pest *Eurygaster integriceps* Put. (Hemiptera, Scutelleridae). Crop Protection 17: 271–287. http://dx.doi.org/10.1016/S0261-2194(98)00022-2.

Cullen, E. M., and F. G. Zalom. 2000. Phenology-based field monitoring for consperse stink bug (Hemiptera: Pentatomidae) in processing tomatoes. Environmental Entomology 29: 560–567. http://dx.doi.org/10.1603/0046-225X-29.3.560.

Cullen, E. M., and F. G. Zalom. 2006. *Euschistus conspersus* female morphology and attraction to methyl (2E,4Z)-decadienoate pheromone-baited traps in processing tomatoes. Entomologia Experimentalis et Applicata 119: 163–173. http://dx.doi.org/10.1111/j.1570-7458.2006.00407.x.

Danilevsky, A. S. 1961. Photoperiodism and seasonal development of insects. Leningrad University Press, Leningrad, the USSR. 244 pp. (in Russian). (Translated into English by J. Johnson assisted by N. Waloff, Oliver & Boyd, Edinburgh, UK, 1965. 284 pp.).

Danks, H. V. 1987. Insect dormancy: An ecological perspective. Biological Survey of Canada, Ottawa, Canada. x + 439 pp. (Monograph Ser. No. 1).

Danks, H. V. 2000. Dehydration in dormant insects. Journal of Insect Physiology 46: 837–852. http://dx.doi.org/10.1016/S0022-1910(99)00204-8.

Danks, H. V. 2005. Key themes in the study of seasonal adaptations in insects. I. Patterns of cold hardiness. Applied Entomology and Zoology 40: 199–211. http://doi.org/10.1303/aez.2005.199.

Davidová-Vilímová, J. 2006. Family Plataspidae Dallas, 1851, pp. 150–165. *In* B. Aukema and C. Rieger (Eds.), Catalogue of the Heteroptera of the Palaearctic Region. Volume 5. Pentatomomorpha II. Nederlandse Entomologische Vereniging, Ponsen & Looijen, Wageningen, The Netherlands. 550 pp.

Davidová-Vilímová, J., and D. Král. 2003. The occurrence of the psammophilous species *Phimodera flori* (Heteroptera: Scutelleridae) in the Czech Republic. Acta Societatis Zoologicae Bohemicae 67: 175–178.

Davidová-Vilímová, J., and P. Štys. 1982. Bionomics of European *Coptosoma* species (Heteroptera, Plataspidae). Acta Universitatis Carolinae Biologica 11: 463–484.

De Clercq, P., and D. Degheele. 1993. Cold storage of the predatory bugs *Podisus maculiventris* (Say) and *Podisus sagitta* (Fabricius) (Heteroptera: Pentatomidae). Parasitica 1–2: 27–41.

Denlinger, D. L. 1986. Dormancy in tropical insects. Annual Review of Entomology 31: 239–264. http://doi.org/10.1146/annurev.en.31.010186.001323.

Denlinger, D. L. 1991. Relationship between cold hardiness and diapause, pp. 174–198. *In* R. E. Lee, Jr., and D. L. Denlinger (Eds.), Insects at low temperature. New York, Chapman and Hall. 513 pp.

Denlinger, D. L. 2008. Why study diapause? Entomological Research 38: 1–9. http://doi.org/10.1111/j.1748-5967.2008.00139.x.

Denlinger, D. L., and R. E. Lee, Jr. (Eds.). 2010. Low temperature biology of insects. New York, Cambridge University Press. xiv + 390 pp.

Dhiman, S. C., Y. K. Yadav, and D. Sharma. 2004. Occurrence of two pentatomid bugs, *Halys dentatus* Fabr. and *Erthesina fullo* Thunb., together on some economic forest trees and their seasonal occurrence. Indian Forester 130: 821–824.

Dikyar, R. 1981. Biology and control of *Aelia rostrata* in central Anatolia. Bulletin of European Mediterranean Plant Protection Organization 11: 39–41. http://doi.org/10.1111/j.1365-2338.1981.tb01762.x.

Dingle, H. 1996. Migration: the biology of life on the move. Oxford University Press, N.Y. vi + 474 pp.

Duman, J. G., V. Bennett, T. Sformo, R. Hochstrasser, and B. M. Barnes. 2004. Antifreeze proteins in Alaskan insects and spiders. Journal of Insect Physiology 50: 259–266. http://doi.org/10.1016/j.jinsphys.2003.12.003.

Dzerefos, C. M., E. T. F. Witkowski, and R. Toms. 2009. Life-history traits of the edible stinkbug, *Encosternum delegorguei* (Hem., Tessaratomidae), a traditional food in southern Africa. Journal of Applied Entomology 133: 749–759. http://doi.org/10.1111/j.1439-0418.2009.01425.x.

Eger, J. E., Jr., L. M. Ames, D. R. Suiter, T. M. Jenkins, D. A. Rider, and S. E. Halbert. 2010. Occurrence of the Old World bug *Megacopta cribraria* (Fabricius) (Heteroptera: Plataspidae) in Georgia: a serious home invader and potential legume pest. Insecta Mundi 121: 1–11.

Elsey, K. D. 1993. Cold tolerance of the southern green stink bug (Heteroptera: Pentatomidae). Environmental Entomology 22: 567–570.

Endo, N., T. Yasuda, K. Matsukura, T. Wada, S. Muto, and R. Sasaki. 2007. Possible function of *Piezodorus hybneri* (Heteroptera: Pentatomidae) male pheromone: effects of adult age and diapause on sexual maturity and pheromone production. Applied Entomology and Zoology 42: 637–641. http://dx.doi .org/10.1303/aez.2007.637.

Esenbekova, P. A., M. Z. Nurushev, and J. Homziak. 2015. Aquatic Hemiptera (Heteroptera) of Kazakhstan, with notes on life history, ecology and distribution. Zootaxa 4013(2): 195–206. http://dx.doi.org/10.11646 /zootaxa.4013.2.2.

Esquivel, J. F. 2009. Stages of gonadal development of the southern green stink bug (Hemiptera: Pentatomidae): improved visualization. Annals of the Entomological Society of America 102: 303–309.

Fasulati, S. R. 1979. Photoperiodic response and coloration of the brassica bug *Eurydema oleracea* L. (Heteroptera, Pentatomidae). Entomologicheskoe Obozrenie [Entomological Review] 57: 15–23. (in Russian with English summary)

Faúndez, E. I., and G. A. Osorio. 2010. New data on the biology of *Sinopla perpunctatus* Signoret, 1864 (Hemiptera: Heteroptera: Acanthosomatidae). Boletin de Biodiversidad de Chile 3: 24–31.

Fedotov, D. M. 1947. Changes in the internal state of the sunn pest (*Eurygaster integriceps* Put.) during a year, pp. 35–80. *In* D. M. Fedotov (Ed.), The sunn pest *Eurygaster integriceps* Put. Volume 1. Publishing House of The Academy of Sciences of the USSR, Moscow. 272 pp. (in Russian)

Fernando, H. E. 1960. A biological and ecological study of the rice pentatomid bug, *Scotinophara lurida* (Burm.) in Ceylon. Bulletin of Entomological Research 51: 559–576. http://dx.doi.org/10.1017/S0007485300055164.

Filippi, L., S. Nomakuchi, M. Hironaka, and S. Tojo. 2000a. Insemination success discrepancy between long-term and short-term copulations in the provisioning shield bug, *Parastrachia japonensis* (Hemiptera: Cydnidae). Journal of Ethology 18: 29–36. http://dx.doi.org/10.1007/s101640070021.

Filippi, L., M. Hironaka, S. Nomakuchi, and S. Tojo. 2000b. Provisioned *Parastrachia japonensis* (Hemiptera: Cydnidae) nymphs gain access to food and protection from predations. Animal Behavior 60: 757–763. http://dx.doi.org/10.1006/anbe.2000.1526.

Filippi, L., M. Hironaka, and S. Nomakuchi. 2001. A review of the ecological parametres and implications of subsociality in *Parastrachia japonensis* (Hemiptera: Cydnidae), a semelparous species that specializes on a poor resource. Population Ecology 43: 41–50. http://dx.doi.org/10.1007/PL00012014.

Filippi, L., N. Baba, K. Inadomi, T. Yanagi, M. Hironaka, and S. Nomakuchi. 2009. Pre- and post-hatch trophic egg production in the subsocial burrow bug, *Canthophorus niveimarginatus*. Naturwissenschaften 96: 201–211. http://dx.doi.org/10.1007/s00114-008-0463-z.

Froeschner, R. C. 1941. Contributions to a synopsis of the Hemiptera of Missouri, Part 1. Scutelleridae, Podopidae, Pentatomidae, Cydnidae, Thyreocoridae. American Midland Naturalist 26: 122–146.

Gamberale-Stille, G., A. I. Johansen, and B. S. Tullberg. 2010. Change in protective coloration in the striated shieldbug *Graphosoma lineatum* (Heteroptera: Pentatomidae): predator avoidance and generalization among different life stages. Evolutionary Ecology 24: 423–432. http://dx.doi.org/10.1007 /s10682-009-9315-3.

Gapon, D. A. 2008. A taxonomic review of the world fauna of stink bugs (Heteroptera: Pentatomidae) of the subfamilies Asopinae and Podopinae. Ph.D. dissertation in Biology. St. Petersburg State University, St. Petersburg, Russia. 178 + 798 pp. (Supplement). (in Russian)

Gardner, W. A., H. B. Peeler, J. LaForest, P. M. Roberts, A. N. Sparks, Jr., J. K. Greene, D. Reisig, D. R. Suiter, J. S. Bacheler, K. Kidd, C. H. Ray, X. P. Hu, R. C. Kemerait, E. A. Scocco, J. E. Eger, Jr., J. R. Ruberson, E. J. Sikora, D. A. Herbert, Jr., C. Campana, S. Halbert, S. D. Stewart, G. D. Buntin, M. D. Toews, and C. T. Bargeron. 2013. Confirmed distribution and occurrence of *Megacopta cribraria* (F.) (Hemiptera: Heteroptera: Plataspidae) in the southeastern United States. Journal of Entomological Science 48: 118–127. http://dx.doi.org/10.18474/0749-8004-48.2.118.

Gibernau, M., and A. Dejean 2001. Ant protection of a Heteropteran trophobiont against a parasitoid wasp. Oecologia 126: 53–57. http://dx.doi.org/10.1007/s004420000479.

Gilbert, B. L., S. J. Barras, and D. M. Norris. 1967. Bionomics of *Tetyra bipunctata* (Hemiptera: Pentatomidae: Scutellerinae) as associated with *Pinus banksiana* in Wisconsin. Annals of the Entomological Society of America 60: 698–701. http://dx.doi.org/10.1093/aesa/60.3.698.

Gogala, M., H. Yong, and C. Brühl. 1998. Maternal care in *Pygoplatys* bugs (Heteroptera: Tessaratomidae). European Journal of Entomology 95: 311–315.

Golec, J. R., and X. P. Hu. 2015. Preoverwintering copulation and female ratio bias: life history characteristics contributing to the invasiveness and rapid spread of *Megacopta cribraria* (Heteroptera: Plataspidae). Environmental Entomology 44: 411–417. http://dx.doi.org/10.1093/ee/nvv014.

Goryshin, N. I., and G. F. Tyshchenko. 1972. Experimental analysis of photoperiodic induction in insects. Trudy Biologicheskogo Instituta Leningraskogo Gosudarstvennogo Universiteta [Proceedings of the Biological Institute of the Leningrad State University] 21: 68–89. (in Russian)

Goryshin, N. I., T. A. Volkovich, A. Kh. Saulich, M. Vagner, and I. A. Borisenko. 1988a. The role of temperature and photoperiod in the control over development and diapause of the carnivorous bug *Podisus maculiventris* (Hemiptera, Pentatomidae). Zoologicheskii Zhurnal [Zoological Journal] 67: 1149–1161. (in Russian with English summary)

Goryshin, N. I., A. Kh. Saulich, T. A. Volkovich, I. A. Borisenko, and N. P. Simonenko. 1988b. The effect of the trophic factor on development and photoperiodic reaction in the spined soldier bug *Podisus maculiventris* (Hemiptera, Pentatomidae). Zoologicheskii Zhurnal [Zoological Journal] 67: 1324–1332. (in Russian with English summary)

Goryshin, N. I., T. A. Volkovich, A. Kh. Saulich, and I. A. Borisenko. 1989. Ecology of the spined soldier bug *Podisus maculiventris*. Manuscript deposited in the VINITI (Vsesojuzniy Institut Nauchno-tehnicheskoy Informacii [All-Union Institute of Scientific and Technical Information], Moscow). No. 115-B-90. 49 pp. (in Russian)

Greve, C., N. D. F. Fortes, and J. Grazia. 2003. Estágios imaturos de *Oebalus poecilus* (Heteroptera, Pentatomidae). Iheringia. Seria Zoologia 3: 89–96. http://dx.doi.org/10.1590/S0073-47212003000100010. (in Portuguese with English summary)

Hanelová, J., and J. Vilímová. 2013. Behaviour of the central European Acanthosomatidae (Hemiptera: Heteroptera: Pentatomoidea) during oviposition and parental care. Acta Musei Moraviae, Scientiae Biologicae (Brno) 98: 433–457.

Harris, V. E., J. W. Todd, and B. G. Mullinix. 1984. Color change as an indicator of adult diapause in the southern green stink bug, *Nezara viridula*. Journal of Agricultural Entomology 1: 82–91.

Hawkins, R. D. 2003. Shieldbugs of Surrey. Surrey Wildlife Trust, Surrey, UK. 192 pp.

Haye, T., S. Abdallah, T. Gariepy, and D. Wyniger. 2014. Phenology, life table analysis and temperature requirements of the invasive brown marmorated stink bug, *Halyomorpha halys*, in Europe. Journal of Pest Science 87: 407–418. http://dx.doi.org/10.1007/s10340-014-0560-z.

Hely, P. C. 1964. Two insect pests of citrus. World Crops 16: 53–57.

Henry, T. J. 2009. Biodiversity of Heteroptera, pp. 223–263. *In* R. G. Foottit and P. H. Adler (Eds.), Insect biodiversity: science and society. Wiley-Blackwell, Oxford, UK. xxi + 632 pp.

Hertzel, G. 1982. Zur Phänologie und Fortpflanzungsbiologie einheimischer Pentatomiden-Arten (Heteroptera). Entomologische Nachrichten und Berichte. 26: 69–72.

Hibino, Y. 1985. Formation and maintenance of mating aggregations in a stinkbug, *Megacopta punctissimum* (Montandon) (Heteroptera: Plataspidae). Journal of Ethology 3: 123–129. http://dx.doi.org/10.1007/BF02350302.

Higuchi, H. 1994. Photoperiodic induction of diapause, hibernation and voltinism in *Piezodorus hybneri* (Heteroptera: Pentatomidae). Applied Entomology and Zoology 29: 585–592.

Hodek, I. 1971a. Sensitivity to photoperiod in *Aelia acuminata* (L.) after adult diapause. Oecologia (Berlin) 6: 152–155. http://dx.doi.org/10.1007/BF00345716.

Hodek, I. 1971b. Termination of adult diapause in *Pyrrhocoris apterus* (Heteroptera: Pyrrhocoridae) in the field. Entomologia Experimentalis et Applicata 14: 212–222. http://dx.doi.org/10.1111/j.1570-7458.1971.tb00158.x.

Hodek, I. 1977. Photoperiodic response in spring in three Pentatomidae (Heteroptera). Acta Entomologica Bohemoslovaca 74: 209–218.

Hodek, I. 1983. Role of environmental factors and endogenous mechanisms in the seasonality of reproduction in insects diapausing as adults, pp. 9–33. *In* V. K. Brown and I. Hodek (Eds.), Diapause and life cycle strategies in insects. Dr. W. Junk Publisher, The Hague (Ser. Entomologica). 273 pp.

Hodek, I. 1996. Diapause development, diapause termination and the end of diapause. European Journal of Entomology 93: 475–487.

Hodek, I. 2002. Controversial aspects of diapause development. European Journal of Entomology 99: 163–173. http://dx.doi.org/10.14411/eje.2002.024.

Hodek, I., and M. Hodková. 1993. Role of temperature and photoperiod in diapause regulation in Czech populations of *Dolycoris baccarum* (Heteroptera, Pentatomidae). European Journal of Entomology 90: 95–98.

Hodková, M. 1982. Interaction of feeding and photoperiod in regulation of the corpus allatum activity in females of *Pyrrhocoris apterus* L. (Hemiptera). Zoologische Jahrbucher Abteilung fuer Allgemeine Zoologie und Physiologie der Tiere 86: 477–488.

Hodková, M., and I. Hodek. 2004. Photoperiod, diapause and cold-hardiness. European Journal of Entomology 101: 445–458. http://dx.doi.org/10.14411/eje.2004.064.

Hodková, M., I. Hodek, and L. Sømme. 1989. Cold is not a prerequisite for the completion of photoperiodically induced diapause in *Dolycoris baccarum* from Norway. Entomologia Experimentalis et Applicata. 52: 185–188. http://dx.doi.org/10.1111/j.1570-7458.1989.tb01266.x.

Hoebeke, E. R., and M. E. Carter. 2003. *Halyomorpha halys* (Stål) (Heteroptera: Pentatomidae): a polyphagous plant pest from Asia newly detected in North America. Proceedings of the Entomological Society of Washington 105: 225–237.

Hoffmann, R. J. 1974. Environmental control of seasonal variation in the butterfly *Colias eurytheme*: effects of photoperiod and temperature on pteridine pigmentation. Journal of Insect Physiology 20: 1913–1924. http://dx.doi.org/10.1016/0022-1910(74)90098-5.

Hokkanen, H. 1986. Polymorphism, parasites, and the native area of *Nezara viridula* (Hemiptera, Pentatomidae). Annales Entomologici Fennici 52: 28–31.

Honěk, A. 1969. Induction of diapause in *Aelia acuminata* (L.) (Heteroptera, Pentatomidae). Acta Entomologica Bohemoslovaca 66: 345–351.

Hori, K. 1986. Effect of photoperiod on nymphal growth of *Palomena angulosa* Motschulsky (Hemiptera: Pentatomidae). Applied Entomology and Zoology 21: 597–605. http://doi.org/10.1303/aez.22.528.

Hori, K., and R. Inamura. 1991. Effects of stationary photoperiod on reproductive diapause, nymphal growth, feeding and digestive physiology of *Eysarcoris lewisi* Distant (Heteroptera: Pentatomidae). Applied Entomology and Zoology 26: 493–499. http://doi.org/10.1303/aez.26.493.

Hori, K., and A. Kimura. 1993. Effect of stationary photoperiod on diapause induction of *Eysarcoris lewisi* Distant (Heteroptera: Pentatomidae) and the developmental stage sensitive to stimulus for reproductive diapause. Applied Entomology and Zoology 28: 53–58. http://doi.org/10.1303/aez.28.53.

Hori, K., K. Nakamura, and K. Goto. 1993. Nymphal development of *Acanthosoma denticauda* and *A. haemorrhoidale angulata* (Heteroptera, Pentatomidae) under natural and laboratory conditions. Japanese Journal of Entomology 61: 55–63.

Horton, D. R., T. Hinojosa, and S. R. Olson. 1998. Effect of photoperiod and prey type on diapause tendency and preoviposition period in *Perillus bioculatus* (Hemiptera: Pentatomidae). The Canadian Entomologist 130: 315–320. http://dx.doi.org/10.4039/Ent130315-3.

Hosokawa, T., N. Nikoh, and T. Fukatsu. 2014. Fine-scale geographical origin of an insect pest invading North America. PLoS ONE 9: e89107. http://dx.doi.org/10.1371/journal.pone.0089107.

Ikeda-Kikue, K., and H. Numata. 1992. Effects of diet, photoperiod and temperature on the postdiapause reproduction in the cabbage bug, *Eurydema rugosa*. Entomologia Experimentalis et Applicata 64: 31–36. http://dx.doi.org/10.1111/j.1570-7458.1992.tb01591.x.

Ikeda-Kikue, K., and H. Numata. 1994. Effect of low temperature on the termination of photoperiodic and food-mediated diapause in the cabbage bug, *Eurydema rugosa* Motschulsky (Heteroptera: Pentatomidae). Applied Entomology and Zoology 29: 229–236. http://doi.org/10.1303/aez.29.229.

Ikeda-Kikue, K., and H. Numata. 2001. Timing of diapause induction in the cabbage bug *Eurydema rugosum* (Heteroptera: Pentatomidae) on different host plants. Acta Societatis Zoologicae Bohemoslovenicae 65: 197–205.

Inaoka, T., M. Watanabe, Y. Kosuge, and T. Kohama. 1993. Biology of house-invading stink bugs and some control trials against them in an elementary school in Hokkaido. 1. Species composition and seasonal fluctuations of house-invading activity. Japanese Journal of Sanitary Zoology 44: 341–347. (in Japanese with English summary)

Inkley, D. B. 2012. Characteristics of home invasion by the brown marmorated stink bug (Hemiptera: Pentatomidae). Journal of Entomological Science 47: 125–130. http://dx.doi.org/10.18474/0749-8004-47.2.125.

Izhevskii, S. S., and L. A. Ziskind. 1981. The prospects of using of introduced stink bugs *Perillus bioculatus* (Fabr.), *Podisus maculiventris* (Say), and *Oplomus nigripennis* var. *pulcher* Dull. (Pentatomidae: Hemiptera) against *Leptinotarsa decemlineata* Say (Chrysomelidae: Coleoptera), pp. 20–37. *In* Smetannik A.I. (Ed.), Biological suppression of quarantine pests and weeds. VNITIKIZR, Moscow. 105 pp. (in Russian)

James, D. G. 1990a. Energy reserves, reproductive status and population biology of overwintering *Biprorulus bibax* (Hemiptera: Pentatomidae) in southern New South Wales citrus groves. Australian Journal of Zoology 38: 415–422. http://dx.doi.org/10.1071/ZO9900415.

James, D. G. 1990b. Seasonality and population development of *Biprorulus bibax* Breddin (Hemiptera: Pentatomidae) in southwestern New South Wales. General and Applied Entomology 22: 61–66.

James, D. G. 1991. Maintenance and termination of reproductive dormancy in an Australian stink bug, *Biprorulus bibax*. Entomologia Experimentalis et Applicata 60: 1–5. http://dx.doi.org/10.1111/j.1570-7458.1991.tb01515.x.

James, D. G. 1993. Apparent overwintering of *Biprorulus bibax* Breddin (Hemiptera: Pentatomidae) on *Eremocitrus glauca* (Rutaceae). Australian Entomologist 20: 129–132.

Jasič, I. 1967. A contribution to the knowledge of the diapause in *Perillus bioculatus* (Fabr.) (Heteroptera, Pentatomidae). Acta Entomologica Bohemoslovaca 64: 333–334.

Jasič, I. 1975. On the life cycle of *Perillus bioculatus*. Acta Entomologica Bohemoslovaca 72: 383–390.

Javahery, M. 1990. Biology and ecological adaptation of the green stink bug (Hemiptera: Pentatomidae) in Quebec and Ontario. Annals of the Entomological Society of America 83: 201–206. http://dx.doi.org/10.1093/aesa/83.2.201.

Javahery, M. 1994. Development of eggs in some true bugs (Hemiptera-Heteroptera). Part I. Pentatomidae. The Canadian Entomologist 126: 401–433.

Javahery, M. 1995. A technical review of sunn pests (Heteroptera: Pentatomoidea) with special reference to *Eurygaster integriceps* Puton. FAO/RNE. Cairo. 80 pp.

Johansen, A. I., A. Exnerová, K. H. Svádová, P. Štys, G. Gamberale-Stille, and B. S. Tullberg. 2010. Adaptive change in protective coloration in adult striated shieldbugs *Graphosoma lineatum* (Heteroptera: Pentatomidae): test of detectability of two colour forms by avian predators. Ecological Entomology 35: 602–610. http://dx.doi.org/10.1111/j.1365-2311.2010.01219.x.

Johnson, C. G. 1969. Migration and dispersal of insects by flight. Methuen, London, UK. 763 pp.

Jones, P. A., and H. C. Coppel. 1963. Immature stages of *Apateticus cynicus* (Say) (Hemiptera: Pentatomidae). The Canadian Entomologist 95: 770–779.

Jones, V. P., and D. Westcot. 2002. The effect of seasonal changes on *Nezara viridula* (L.) (Hemiptera: Pentatomidae) and *Trissolcus basalis* (Wollaston) (Hymenoptera: Scelionidae) in Hawaii. Biological Control 23: 115–120. http://dx.doi.org/10.1006/bcon.2001.0996.

Jones, W. A., Jr., and J. E. McPherson. 1980. The first report of the occurrence of acanthosomatids in South Carolina. Journal of the Georgia Entomological Society 15: 286–289.

Kawada, H., and C. Kitamura. 1983. Bionomics of the brown marmorated stink bug, *Halyomorpha mista*. Japanese Journal of Applied Entomology and Zoology 27: 304–306. (in Japanese)

Kennedy, J. S. 1961. A turning point in the study of insect migration. Nature 189(4767): 785–791. http://dx.doi.org/10.1038/189785a0.

Khrimian, A. 2005. The geometric isomers of methy-2,4,6-decatrienoate, including pheromones of at least two species of stink bugs. Tetrahedron 61: 3651–3657. http://dx.doi.org/10.1016/j.tet.2005.02.032.

Khrimian, A., P. W. Shearer, A. Zhang, G. C. Hamilton, and J. R. Aldrich. 2008. Field trapping of the invasive brown marmorated stink bug, *Halyomorpha halys*, with geometric isomers of methyl 2,4,6-decatrienoate. Journal of Agricultural and Food Chemistry 56: 197–203. http://dx.doi.org/10.1021/jf072087e.

Khrimian, A., A. Zhang, D. C. Weber, H.-Y. Ho, J. R. Aldrich, K. E. Vermillion, M. A. Siegler, S. Shirali, F. Guzman, and T. C. Leskey. 2014. Discovery of the aggregation pheromone of the brown marmorated stink bug (*Halyomorpha halys*) through the creation of stereoisomeric libraries of 1-bisabolen-3-ols. Journal of Natural Products 77: 1708–1717. http://dx.doi.org/10.1021/np5003753.

Kiritani, K. 2006. Predicting impacts of global warming on population dynamics and distribution of arthropods in Japan. Population Ecology 48: 5–12. http://dx.doi.org/10.1007/s10144-005-0225-0.

Kiritani, Y. 1985a. Timing of oviposition and nymphal diapause under the natural daylengths in *Carbula humerigera* (Heteroptera: Pentatomidae). Applied Entomology and Zoology 20: 252–256. http://doi.org/10.1303/aez.20.252.

Kiritani, Y. 1985b. Effect of stationary and changing photoperiods on nymphal development in *Carbula humerigera* (Heteroptera: Pentatomidae). Applied Entomology and Zoology 20: 257–263. http://doi.org/10.1303/aez.20.257.

Kobayashi, T., and S. Kimura. 1969. The studies on the biology and control of house entering stink bugs. Part 1. The actual state of the hibernation of stink bugs in houses. Bulletin of the Tohoku National Agricultural Experiment Station (Morioka) 37: 123–138 (in Japanese).

Kobayashi, T., and M. H. Osakabe. 2009. Pre-winter copulation enhances overwintering success of *Orius females* (Heteroptera: Anthocoridae). Applied Entomology and Zoology 44: 47–52. http://doi .org/10.1303/aez.2009.47.

Korinek, V. V. 1940. Fauna of some true bugs of the Khoperskiy State Nature Reserve. Trudy Khoperskogo Gosudarstvennogo Zapovednika [Proceedings of the Khoperskiy State Nature Reserve] 1: 174–218. (in Russian)

Koshiyama, Y., H. Tsumuki, M. Muraji, K. Fujisaki, and F. Nakasuji. 1993. Transfer of male secretions to females through copulation in *Menida scotti* (Heteroptera: Pentatomidae). Applied Entomology and Zoology 28: 325–332.

Koshiyama, Y., K. Fujisaki, and F. Nakasuji. 1994. Mating and diapause in hibernating adults of *Menida scotti* Puton (Heteroptera: Pentatomidae). Researches on Population Ecology (Kyoto) 36: 87–92. http://dx.doi .org/10.1007/BF02515089.

Koshiyama, Y., K. Fujisaki, and F. Nakasuji. 1997. Effect of mating during hibernation on life history traits of female adults of *Menida scotti* (Heteroptera: Pentatomidae). Chugoku Kontyu (Japan) 11: 11–18.

Koštál, V. 2006. Eco-physiological phases of insect diapause. Journal of Insect Physiology 52: 113–127. http:// dx.doi.org/10.1016/j.jinsphys.2005.09.008.

Kotaki, T. 1998a. Age-dependent change in effects of chilling on diapause termination in the brown-winged green bug, *Plautia crossota stali* Scott (Heteroptera: Pentatomidae). Entomological Science 1: 485–489.

Kotaki, T. 1998b. Effects of low temperature on diapause termination and body colour change in adults of a stink bug, *Plautia stali*. Physiological Entomology 23: 53–61. http://dx.doi.org/10.1046/j.1365-3032 .1998.2310053.x.

Kotaki, T., and S. Yagi. 1987. Relationship between diapause development and coloration change in brown-winged green bug, *Plautia stali* Scott (Heteroptera: Pentatomidae). Japanese Journal of Applied Entomology and Zoology 31: 285–290. http://dx.doi.org/10.1303/jjaez.31.285.

Kott, P., S. Roth, and K. Reinhardt. 2000. Hibernation mortality and sperm survival during dormancy in female Nabidae (Heteroptera: Nabidae). Opuscula Zoologica Fluminensia 182: 1–6.

Krambias, A. 1987. Host plant, seasonal migration and control of the berry bug *Dolycoris baccarum* L. in Cyprus. FAO Plant Protection Bulletin. 35: 25–26.

Kudo, S. 2006. Within-clutch egg-size variation in a subsocial bug: the position effect hypothesis. Canadian Journal of Zoology 84: 1540–1544. http://dx.doi.org/10.1139/z06-163.

Larivière, M.-C., and A. Larochelle. 1989. *Picromerus bidens* (Heteroptera: Pentatomidae) in North America, with a world review of distribution and bionomics. Entomological News 100: 133–146.

Leather, S. R., K. F. A. Walker, and J. S. Bale. 1993. The ecology of insect overwintering. Cambridge University Press, Cambridge. 255 pp.

Lee, D.-H., B. D. Short, S. V. Joseph, J. C. Bergh, and T. C. Leskey. 2013. Review of the biology, ecology, and management of *Halyomorpha halys* (Hemiptera: Pentatomidae) in China, Japan, and the Republic of Korea. Environmental Entomology 42: 627–641. http://dx.doi.org/10.1603/EN13006.

Lee, D.-H., J. P. Cullum, L. J. Anderson, J. L. Daugherty, L. M. Beckett, and T. C. Leskey. 2014. Characterization of overwintering sites of the invasive brown marmorated stink bug in natural landscapes using human surveyors and detector canines. PLoS ONE 9: e91575. http://dx.doi.org/10.1371/journal.pone.0091575.

Lee, K.-Y., K.-S. Ahn, H.-J. Kang, S.-K. Park, and T.-S. Kim. 2001. Host plants and life cycle of rice black bug, *Scotinophara lurida* (Burmeister) (Heteroptera: Pentatomidae). Korean Journal of Applied Entomology 40: 309–313.

Lee, R. E., and D. L. Denlinger (Eds.). 1991. Insects at low temperature. Chapman and Hall, New York. 513 pp.

Leston, D. 1955. The life-cycle of *Picromerus bidens* (L.) (Hemiptera, Pentatomidae) in Britain. The Entomologist's Monthly Magazine 91: 109.

Lockwood, J. A., and R. N. Storey. 1986. Adaptive functions of nymphal aggregation in the southern green stinkbug, *Nezara viridula* (L.) (Hemiptera: Pentatomidae). Environmental Entomology 15: 739–749. http://dx.doi.org/10.1093/ee/15.3.739.

Mappes, J., A. Kaitala, and V. Rinne. 1996. Temporal variation in reproductive allocation in a shield bug *Elasmostethus interstinctus*. Journal of Zoology 240: 29–35. http://dx.doi.org/10.1111/j.1469-7998.1996 .tb05483.x.

Masaki, S. 1980. Summer diapause. Annual Review of Entomology 25: 1–25. http://dx.doi.org/10.1146/annurev .en.25.010180.000245.

McDonald, F. J. D. 1969. Life cycle of the bronze orange bug *Musgraveia sulciventris* (Stål) (Hemiptera: Tessaratomidae). Australian Journal of Zoology 17: 817–820.

McPherson, J. E. 1972. Life history of *Corimelaena lateralis lateralis* (Hemiptera: Thyreocoridae) with descriptions of immature stages and list of other species of Scutelleroidea found with it on wild carrot. Annals of the Entomological Society of America 65: 906–911. http://dx.doi.org/10.1093/aesa/65.4.906.

McPherson, J. E. 1974. Photoperiod effects in a southern Illinois population of the *Euschistus tristigmus* complex (Hemiptera: Pentatomidae). Annals of the Entomological Society of America 67: 943–952. http://dx.doi.org/10.1093/aesa/67.6.943.

McPherson, J. E. 1975a. Life history of *Euschistus tristigmus tristigmus* (Hemiptera: Pentatomidae) with information on adult seasonal dimorphism. Annals of the Entomological Society of America 68: 333–334. http://dx.doi.org/10.1093/aesa/68.2.333.

McPherson, J. E. 1975b. Effect of development photoperiod on adult morphology in *Euschistus tristigmus tristigmus* (Say) (Hemiptera: Pentatomidae). Annals of the Entomological Society of America 68: 1107–1110. http://dx.doi.org/10.1093/aesa/68.6.1107.

McPherson, J. E. 1977. Effect of developmental photoperiod on adult color and pubescence in *Thyanta calceata* (Hemiptera: Pentatomidae) with information on ability of adults to change color. Annals of the Entomological Society of America 70: 373–376. http://dx.doi.org/10.1093/aesa/70.3.373.

McPherson, J. E. 1978. Effects of various photoperiods on color and pubescence in *Thyanta calceata* (Hemiptera: Pentatomidae). The Great Lakes Entomologist 11: 155–158.

McPherson, J. E. 1979. Effects of various photoperiods on morphology in *Euschistus tristigmus tristigmus* (Hemiptera: Pentatomidae). The Great Lakes Entomologist 12: 23–26.

McPherson, J. E. 1982. Pentatomoidea (Hemiptera) of Northeastern North America. Southern Illinois University Press, Carbondale and Edwardsville, IL. 240 pp.

McPherson, J. E., and R. M. McPherson. 2000. Stink bugs of economic importance in America north of Mexico. CRC Press, LLC, Boca Raton, FL. 253 pp.

McPherson, J. E., and R. H. Mohlenbrock. 1976. A list of the Scutelleroidea of the La Rue-Pine Hills Ecological Area with notes on biology. The Great Lakes Entomologist 9: 125–169.

McPherson, J. E., and S. M. Paskewitz. 1984. Life history and laboratory rearing of *Euschistus ictericus* (Hemiptera: Pentatomidae), with descriptions of immature stages. Journal of the New York Entomological Society 92: 53–60.

McPherson, J. E., and D. L. Tecic. 1997. Notes on the life histories of *Acrosternum hilare* and *Cosmopepla bimaculata* (Heteroptera: Pentatomidae) in southern Illinois. The Great Lakes Entomologist 30: 79–84.

Mead, F. W. 1976. A predatory stink bug, *Euthyrhynchus floridanus* (Linnaeus) (Hemiptera: Pentatomidae). Florida Department of Agriculture and Consumer Services, Division of Plant Industry, Entomology Circulars 174: 1–2.

Melber, A., H. G. Klindworth, and G. H. Schmidt. 1981. Saisonaler Wirtspflanzenwechsel bei der baumbewohnenden Wanze *Elasmucha grisea* L. (Heteroptera: Acanthosomatidae) [Seasonal food plant change in the parent bug *Elasmucha grisea* L. (Heteroptera: Acanthosomatidae)]. Zeitschrift für Angewandte Entomologie 91: 55–62. (in German with English summary)

Monteith, G. B. 2006. Maternal care in Australian oncomerine shield bugs (Insecta, Heteroptera, Tessaratomidae), pp. 1135–1152. *In* W. Rabitsch (Ed.), Hug the bug. For love of true bugs. Festschrift zum 70. Geburtstag von Ernst Heiss. Denisia 19: 1184 pp.

Mourão, A. P. M., and A. R. Panizzi. 2000. Diapausa e diferentes formas sazonais em *Euschistus heros* (Fabr.) (Hemiptera: Pentatomidae) no Norte do Paraná [Diapause and different seasonal morphs of *Euschistus heros* (Fabr.) (Hemiptera: Pentatomidae) in Northern Paraná State]. Anais da Sociedada Entomológica do Brasil 29: 205–218. http://dx.doi.org/10.1590/S0301-80592000000200002.

Mourão, A. P. M., and A. R. Panizzi. 2002. Photophase influence on the reproductive diapause, seasonal morphs, and feeding activity of *Euschistus heros* (Fabr., 1798) (Hemiptera: Pentatomidae). Revista Brasileira de Zoologia 62: 231–238. http://dx.doi.org/10.1590/S1519-69842002000200006. (in Portuguese with English summary)

Musolin, D. L. 1996. Photoperiodic induction of aestivation in the stink bug *Picromerus bidens* (Heteroptera, Pentatomidae). A preliminary report. Entomological Review 76: 1058–1060.

Musolin, D. L. 2012. Surviving winter: diapause syndrome in the southern green stink bug *Nezara viridula* in the laboratory, in the field, and under climate change conditions. Physiological Entomology 37: 309–322. http://dx.doi.org/10.1111/j.1365-3032.2012.00846.x.

Musolin, D. L., and A. V. Maisov. 1998. Day-length sensitivity during diapause induction and termination in the shield bug *Graphosoma lineatum* L. (Heteroptera, Pentatomidae), p. 44. *In* G. S. Medvedev et al. (Eds.), Problemi entomologii v Rossii [Problems of Entomology in Russia]. Proceedings of the XI Congress of the Russian Entomological Society. Volume 2. St. Petersburg. 240 pp. (in Russian)

Musolin, D. L., and H. Numata. 2003a. Photoperiodic and temperature control of diapause induction and colour change in the southern green stink bug *Nezara viridula*. Physiological Entomology 28: 65–74. http://dx.doi.org/10.1046/j.1365-3032.2003.00307.x.

Musolin, D. L., and H. Numata. 2003b. Timing of diapause induction and its life-history consequences in *Nezara viridula*: is it costly to expand the distribution range? Ecological Entomology 28: 694–703. http://dx.doi.org/10.1111/j.1365-2311.2003.00559.x.

Musolin, D. L., and A. Kh. Saulich. 1996a. Factorial regulation of the seasonal cycle in the stink bug *Graphosoma lineatum* L. (Heteroptera, Pentatomidae). I. Temperature and photoperiodic responses. Entomological Review 75: 84–93.

Musolin, D. L., and A. Kh. Saulich. 1996b. Photoperiodic control of seasonal development in bugs (Heteroptera). Entomological Review 76: 849–864.

Musolin, D. L., and A. Kh. Saulich. 1997. Photoperiodic control of nymphal growth in true bugs (Heteroptera). Entomological Review 77: 768–780.

Musolin, D. L., and A. Kh. Saulich. 1999. Diversity of seasonal adaptations in terrestrial true bugs (Heteroptera) from the Temperate Zone. Entomological Science 2: 623–639.

Musolin, D. L., and A. Kh. Saulich. 2000. Summer dormancy ensures univoltinism in the predatory bug *Picromerus bidens* (Heteroptera, Pentatomidae). Entomologia Experimentalis et Applicata 95: 259–267. http://dx.doi.org/10.1046/j.1570-7458.2000.00665.x.

Musolin, D. L., and A. Kh. Saulich. 2001. Environmental control of voltinism of the stink bug *Graphosoma lineatum* in the forest-steppe zone (Heteroptera: Pentatomidae). Entomologia Generalis 25: 255–264. http://dx.doi.org/10.1127/entom.gen/25/2001/255.

Musolin, D. L., K. Fujisaki, and H. Numata. 2007. Photoperiodic control of diapause termination, colour change and postdiapause reproduction in the southern green stink bug, *Nezara viridula*. Physiological Entomology 32: 64–72. http://dx.doi.org/10.1111/j.1365-3032.2006.00542.x.

Musolin, D. L., D. Tougou, and K. Fujisaki. 2010. Too hot to handle? Phenological and life-history responses to simulated climate change of the southern green stink bug *Nezara viridula* (Heteroptera: Pentatomidae). Global Change Biology 16: 73–87. http://dx.doi.org/10.1111/j.1365-2486.2009.01914.x.

Nakahira, T. 1994. Production of trophic eggs in the subsocial burrower bug *Adomerus triguttulus*. Naturwissenschaften 81: 413–414.

Nakamura, K. 2003. Effect of photoperiod on development and growth in a pentatomid bug, *Dolycoris baccarum*. Entomological Science 6: 11–16. http://dx.doi.org/10.1046/j.1343-8786.2003.00006.x.

Nakamura, K., and H. Numata. 1995. Photoperiodic sensitivity in adult of *Aelia fieberi* (Heteroptera: Pentatomidae). European Journal of Entomology 92: 609–613.

Nakamura, K., and H. Numata. 1997a. Effects of environmental factors on diapause development and postdiapause oviposition in a phytophagous insect, *Dybowskyia reticulata*. Zoological Science 14: 1019–1024. http://dx.doi.org/10.2108/zsj.14.1019.

Nakamura, K., and H. Numata. 1997b. Seasonal life cycle of *Aelia fieberi* (Heteroptera: Pentatomidae) in relation to the phenology of its host plants. Annals of the Entomological Society of America 90: 625–630. http://dx.doi.org/10.1093/aesa/90.5.625.

Nakamura, K., and H. Numata. 1998. Alternative life cycles controlled by temperature and photoperiod in the oligophagous bug, *Dybowskyia reticulata*. Physiological Entomology 23: 69–74. http://dx.doi.org/10.1046/j.1365-3032.1998.2310069.x.

Nakamura, K., and H. Numata. 1999. Environmental regulation of adult diapause of *Graphosoma rubrolineatum* (Westwood) (Heteroptera: Pentatomidae) in southern and northern populations of Japan. Applied Entomology and Zoology 34: 323–326. http://dx.doi.org/10.1303/aez.34.323.

Nakamura, K., I. Hodek, and M. Hodková. 1996. Recurrent photoperiodic response in *Graphosoma lineatum* (Heteroptera: Pentatomidae). European Journal of Entomology 93: 519–523.

Nakazawa, K., and H. Hayashi. 1983. Bionomics of the stink bugs and allied bugs causing the pecky rice. 1. Development and occurrence of diapausing females in *Eysarcoris ventralis* Westwood and *Cletus punctiger* (Dallas). Bulletin of the Hiroshima Prefecture Agricultural Experiment Station 46: 21–32. (in Japanese with English summary)

Nardi, C., P. M. Fernandes, and J. M. S. Bento. 2008. Wing polymorphism and dispersal of *Scaptocoris carval-hoi* (Hemiptera: Cydnidae). Annals of the Entomological Society of America 101: 551–557. http://dx.doi .org/10.1603/0013-8746(2008)101[551:WPADOS]2.0.CO;2.

Nielsen, A. L., and G. C. Hamilton. 2009. Life history of the invasive species *Halyomorpha halys* (Hemiptera: Pentatomidae) in northeastern United States. Annals of the Entomological Society of America 102: 608–616. http://dx.doi.org/10.1603/008.102.0405.

Nielsen, A. L., G. C. Hamilton, and P. W. Shearer. 2011. Seasonal phenology and monitoring of the non-native *Halyomorpha halys* (Stål) (Hemiptera: Pentatomidae) in soybean. Environmental Entomology 40: 231–238. http://dx.doi.org/10.1603/EN10187.

Nijhout, H. F. 2003. Development and evolution of adaptive polyphenisms. Evolution & Development 5: 9–18. http://dx.doi.org/10.1046/j.1525-142X.2003.03003.x.

Nijhout, H. F., and G. Davidowitz. 2009. The developmental-physiological basis of phenotypic plasticity, pp. 589–608. *In* D. W. Whitman and T. N. Ananthakrishnan (Eds.), Phenotypic plasticity of insects: mechanisms and consequences. Science Publishers, Enfield, New Hampshire. 894 pp.

Nishida, T. 1966. Behavior and mortality of the southern stink bug *Nezara viridula* in Hawaii. Researches on Population Ecology (Kyoto) 8: 78–88. http://dx.doi.org/10.1007/BF02524749.

Niva, C. C., and M. Takeda. 2002. Color changes in *Halyomorpha brevis* (Heteroptera: Pentatomidae) corre-lated with distribution of pteridines: regulation by environmental and physiological factors. Comparative Biochemistry and Physiology. B. Comparative Biochemistry 132: 653–660. http://dx.doi.org/10.1016 /S1096-4959(02)00081-7.

Niva, C. C., and M. Takeda. 2003. Effects of photoperiod, temperature and melatonin on nymphal develop-ment, polyphenism and reproduction in *Halyomorpha halys* (Heteroptera: Pentatomidae). Zoological Science 20: 963–970. http://dx.doi.org/10.2108/zsj.20.963.

Noda, H., and T. Ishii. 1981. Effect of photoperiod and temperature on the ovarian development of the white-spotted stink bug, *Eysarcoris ventralis* (Heteroptera: Pentatomidae). Japanese Journal of Applied Entomology and Zoology 25: 33–38. http://doi.org/10.1303/jjaez.25.33. (in Japanese with English summary)

Noda, T. 1984. Short day photoperiod accelerates the oviposition in the oriental green stink bug, *Nezara antennata* Scott (Heteroptera: Pentatomidae). Applied Entomology and Zoology 19: 119–120.

Nomakuchi, S., L. Filippi, and S. Tojo. 1998. Selective foraging behavior in nest-provisioning females of *Parastrachia japonensis* (Hemiptera: Cydnidae): cues for preferred food. Journal Insect Behavior 11: 605–619. http://dx.doi.org/10.1023/A:1022305423994.

Numata, H. 2004. Environmental factors that determine the seasonal onset and termination of reproduction in seed-sucking bugs (Heteroptera) in Japan. Applied Entomology and Zoology 39: 565–573. http://doi .org/10.1303/aez.2004.565.

Numata, H., and S. Kobayashi. 1994. Threshold and quantitative photoperiodic responses exist in an insect. Experientia 50: 969–971. http://dx.doi.org/10.1007/BF01923489.

Numata, H., and K. Nakamura. 2002. Photoperiodism and seasonal adaptations in some seed-sucking bugs (Heteroptera) in central Japan. European Journal of Entomology 99: 155–161.

Numata, H., and K. Yamamoto. 1990. Feeding on seeds induces diapause in the cabbage bug, *Eurydema rugosa*. Entomologia Experimentalis et Applicata 57: 281–284. http://dx.doi.org/10.1111/j.1570-7458.1990 .tb01440.x.

Numata, H., A. Kh. Saulich, and T. A. Volkovich. 1993. Photoperiodic responses of the linden bug, *Pyrrhocoris apterus*, under conditions of constant temperature and under thermoperiodic conditions. Zoological Science 10: 521–527.

Oetting, R. D., and T. R. Yonke. 1971. Biology of some Missouri stink bugs. Journal of the Kansas Entomological Society 44: 446–459.

Oetting, R. D., and T. R. Yonke. 1975. Immature stages and biology of *Euthyrhynchus floridanus* (L.) (Hemiptera: Pentatomidae). Annals of the Entomological Society of America 68: 659–662. http://dx.doi .org/10.1093/aesa/68.4.659.

Ogorzałek, A., and A. Trochimczuk. 2009. Ovary structure in a presocial insect, *Elasmucha grisea* (Heteroptera, Acanthosomatidae). Arthropod Structure and Development 38: 509–519. http://dx.doi .org/10.1016/j.asd.2009.08.001.

Ohno, K., and M. Z. Alam. 1992. Hereditary basis of adult color polymorphism in the southern green stink bug, *Nezara viridula* Linné (Heteroptera: Pentatomidae). Applied Entomology and Zoology 27: 133–139.

Panizzi, A. R. 1997. Wild hosts of pentatomids: ecological significance and role in their pest status on crops. Annual Review of Entomology 42: 99–122. http://dx.doi.org/10.1146/annurev.ento.42.1.99.

Panizzi, A. R. 2007. Nutritional ecology of plant feeding arthropods and IPM, pp. 170–222. *In* M. Kogan and P. Jepson (Eds.), Perspectives in ecological theory and integrated pest management. Cambridge University Press, Cambridge, UK. 588 pp.

Panizzi, A. R. 2015. Growing problems with stink bugs (Hemiptera: Heteroptera: Pentatomidae): species invasive to the U.S. and potential Neotropical invaders. American Entomologist 61: 223–233. http://dx.doi.org/10.1093/ae/tmv068.

Panizzi, A. R., and J. Grazia (Eds.). 2015. True bugs (Heteroptera) of the Neotropics. Springer Science+Business Media, Dordrecht. xxii + 902 pp. (Entomology in Focus, Volume 2). http://dx.doi.org/10.1007/978-94-017-9861-7.

Panizzi, A. R., and C. C. Niva. 1994. Overwintering strategy of the brown stink bug in northern Paraná. Pesquisa Agropecuária Brasileira. 29: 509–511.

Panizzi, A., J. E. McPherson, D. G. James, M. Javahery, and R. M. McPherson. 2000. Chapter 13. Stink bugs (Pentatomidae), pp. 421–474. *In* C. W. Schaefer and A. R. Panizzi (Eds.), Heteroptera of economic importance. CRC Press, LLC, Boca Raton, FL. 828 pp.

Peredo, L. C. 2002. Description, biology, and maternal care of *Pachycoris klulgii* (Heteroptera: Scutelleridae). Florida Entomologist 85: 464–473. http://dx.doi.org/10.1653/0015-4040(2002)085[0464:DBAMCO]2.0.CO;2.

Perepelitsa, L. V. 1971. The role of photoperiod in the development of *Dolycoris baccarum*. Byulleten' Vsesojuznogo Instituta Zashchiti Rasteniy [Bulletin of the All-Union Institute of Plant Protection] 21: 11–13. (in Russian with English summary)

Putshkov, P. V. 1961. Fauna of Ukraine. Shield bugs. 21. Akademya Nauk UkrSSR [Academy of Sciences of the Ukrainian SSR], Kiev. 338 pp. (in Ukrainian)

Richman, D. B., and W. H. Whitcomb. 1978. Comparative life cycles of four species of predatory stink bugs. Florida Entomologist 61: 113–119.

Rogers, S. M. 2015. Mechanisms of polyphenism in insects, pp. 1–38. *In* K. H. Hoffmann (Ed.), Insect molecular biology and ecology. CRC Press, Boca Raton, FL. 418 pp.

Roychoudhury, N. 1998. Adult diapause in *Chrysocoris purpureus* Westwood. Insect Environment 4: 105.

Roychoudhury, N. 1999. Occurrence of reproductive diapause in *Chrysocoris purpureus* Westwood (Heteroptera: Scutelleridae). The Indian Forester 125: 637–639.

Ruberson, J. R., T. J. Kring, and N. Elkassabany. 1998. Overwintering and the diapause syndrome of predatory Heteroptera, pp. 49–69. *In* M. Coll and J. R. Ruberson (Eds.), Predatory Heteroptera: their ecology and use in biological control. Entomological Society of America, Lanham, MA. 233 pp.

Salomão, A. T., and J. Vasconcellos-Neto. 2010. Population dynamics and structure of the Neotropical bark bug *Phloea subquadrata* (Hemiptera: Phloeidae) on *Plinia cauliflora* (Myrtaceae). Environmental Entomology 39: 1724–1730. http://dx.doi.org/10.1603/EN09282.

Santos, R. S. S., L. R. Redaelli, L. M. G. Diefenbach, H. P. Romanowski, and H. F. Prando. 2003. Characterization of the imaginal reproductive diapause of *Oebalus poecilus* (Dallas) (Hemiptera: Pentatomidae). Revista Brasileira de Biologia 63: 695–703. http://dx.doi.org/10.1590/S1519-69842003000400017.

Saulich, A. Kh. 1995. Role of abiotic factors in forming secondary ranges of adventive insect species. Entomological Review 74: 1–15.

Saulich, A. Kh., and D. L. Musolin. 1996. Univoltinism and its regulation in some temperate true bugs (Heteroptera). European Journal of Entomology 93: 507–518.

Saulich, A. Kh., and D. L. Musolin. 2007a. Seasonal development of aquatic and semi-aquatic bugs (Heteroptera). St. Petersburg University Press, St. Petersburg, Russia. 205 pp. (in Russian with English summary)

Saulich, A. Kh., and D. L. Musolin. 2007b. Times of the year: the diversity of seasonal adaptations and ecological mechanisms controlling seasonal development of true bugs (Heteroptera) in the temperate climate, pp. 25–106. *In* A. A. Stekolnikov (Ed.), Adaptive strategies of terrestrial arthropods to unfavorable environmental conditions: a collection of papers in memory of Professor Viktor Petrovich Tyshchenko (Proceedings of the Biological Institute of St. Petersburg State University 53). 387 pp. (in Russian with English summary)

Saulich, A. Kh., and D. L. Musolin. 2009. Seasonal development and ecology of anthocorids (Heteroptera, Anthocoridae). Entomological Review 89: 501–528. http://dx.doi.org/10.1134/S0013873809050017.

Saulich, A. Kh., and D. L. Musolin. 2012. Diapause in the seasonal cycle of stink bugs (Heteroptera, Pentatomidae) from the Temperate Zone. Entomological Review 92: 1–26. http://dx.doi.org/10.1134 /S0013873812010010.

Saulich, A. Kh., and D. L. Musolin. 2014. Seasonal cycles in stink bugs (Heteroptera, Pentatomidae) from the Temperate Zone: diversity and control. Entomological Review 94: 785–814. http://dx.doi.org/10.1134 /S0013873814060013.

Saulich, A. Kh., and T. A. Volkovich. 1996. Monovoltinism in insects and its regulation. Entomological Review 76: 205–221.

Saulich, A. Kh., and T. A. Volkovich. 2004. Ecology of photoperiodism in insects. St. Petersburg University Press, St. Petersburg. 276 pp. (in Russian)

Saulich, A. Kh., T. A. Volkovich, and H. Numata. 1993. Temperature and photoperiodic control of Linden bug *Pyrrhocoris apterus* (Hemiptera, Pyrrhocoridae) development under natural conditions. Vestnik Sankt-Peterburgskogo Universiteta. Seria 3 (Biologija) [Proceedings of Saint Petersburg State University. Series 3 (Biology)] 4(24): 31–39. (in Russian with English summary)

Saunders, D. S. 1976. Insect clocks. Pergamon Press, Oxford, UK. 280 pp.

Saunders, D. S. 1983. A diapause induction-termination asymmetry in the photoperiodic responses of the linden bug *Pyrrhocoris apterus* and the effect of near critical photoperiods on development. Journal of Insect Physiology 29: 399–405. http://dx.doi.org/10.1016/0022-1910(83)90067-7.

Schuh, R. T., and J. A. Slater. 1995. True bugs of the world (Hemiptera: Heteroptera): classification and natural history. Cornell University Press, Ithaca, NY. xii + 336 pp.

Shagov, E. M. 1977. Photoperiodic response and its variation in *Perillus*. Ekologiya [Ecology] 4: 751–753. (in Russian with English summary)

Shintani, Y., Y. Masuzawa, Y. Hirose, R. Miyahara, F. Watanabe, and J. Tajima. 2010. Seasonal occurrence and diapause induction of a predatory bug *Andrallus spinidens* (F.) (Heteroptera: Pentatomidae). Entomological Science 13: 273–279. http://dx.doi.org/10.1111/j.1479-8298.2010.00386.x.

Shinyaeva, L. I. 1980. Spermatogenesis during prediapause and diapause formation in *Eurygaster integriceps*. Zoologicheskii Zhurnal [Zoological Journal] 69: 1025–1032. (in Russian with English summary)

Siddiqui, A. S. 2000. A revision of *Strachiine* genera (Hemiptera: Pentatomidae: Pentatominae). Ph.D. Thesis, Department of Zoology, University of Karachi, Karachi, Pakistan. 260 pp.

Simpson, S. J., G. A. Sword, and N. Lo. 2011. Polyphenism in insects. Current Biology 21: R738–R749. http:// dx.doi.org/10.1016/j.cub.2011.06.006.

Sinclair, B. J., P. Vernon, C. J. Klok, and S. L. Chown. 2003. Insects at low temperatures: an ecological perspective. Trends in Ecology and Evolution 18: 257–262. http://dx.doi.org/10.1016/S0169-5347(03)00014-4.

Singh, Z., and V. S. Malik. 1993. Biology of painted bug (*Bagrada cruciferarum*). Indian Journal of Agricultural Science 63: 672–674.

Şişli, M. N. 1965. The effect of the photoperiod on the induction and termination of the adult diapause of *Aelia rostrata* Boh. (Hemiptera: Pentatomidae). Communications Faculty of Sciences University of Ankara 10: 62–69.

Sites, R. W., and J. E. McPherson. 1982. Life history and laboratory rearing of *Sehirus cinctus cinctus* (Hemiptera: Cydnidae), with description of immature stages. Annals of the Entomological Society of America 75: 210–215. http://dx.doi.org/10.1093/aesa/75.2.210.

Šlachta, M., J. Vambera, H. Zahradníčkova, and V. Koštál. 2002. Entering diapause is a prerequisite for successful cold-acclimation in adult *Graphosoma lineatum* (Heteroptera: Pentatomidae). Journal of Insect Physiology 48: 1031–1039. http://dx.doi.org/10.1016/S0022-1910(02)00191-9.

Socha, R. 2010. Pre-diapause mating and overwintering of fertilized adult females: new aspects of the life cycle of the wing-polymorphic bug, *Pyrrhocoris apterus* (Heteroptera: Pyrrhocoridae). European Journal of Entomology 107: 521–525. http://dx.doi.org/10.14411/eje.2010.059.

Southwood, T. R. E., and D. Leston. 1959. Land and water bugs of the British Isles. Frederick Warne and Co., London, UK. 436 pp.

Storey, K. B., and J. M. Storey. 2015. Insects in winter: metabolism and regulation of cold hardiness, pp. 245–270. *In* K. H. Hoffmann (Ed.), Insect molecular biology and ecology. CRC Press, LLC, Boca Raton, FL. 418 pp.

Suiter, D. R., J. E. Eger Jr., W. A. Gardner, R. C. Kemerait, J. N. All, P. M. Roberts, J. K. Greene, L. M. Ames, G. D. Buntin, T. M. Jenkins, and G. K. Douce. 2010. Discovery and distribution of *Megacopta cribraria* (Hemiptera: Heteroptera: Plataspidae) in northeast Georgia. Journal of Integrated Pest Management 1: 1–4. http://dx.doi.org/10.1603/IPM10009.

Tachikawa, S., and C. W. Schaefer. 1985. Biology of *Parastrachia japonensis* (Hemiptera: Pentatomoidea: ?-idae). Annals of the Entomological Society of America 78: 387–397. http://dx.doi.org/10.1093/aesa/78.3.387.

Takeda, K., D. L. Musolin, and K. Fujisaki. 2010. Dissecting insect responses to climate warming: overwintering and post-diapause performance in the southern green stink bug, *Nezara viridula*, under simulated climate-change conditions. Physiological Entomology 35: 343–353. http://dx.doi.org/10.1111/j.1365-3032.2010.00748.x.

Tallamy, D. W. 2001. Evolution of exclusive paternal care in arthropods. Annual Review of Entomology 46: 139–165. http://dx.doi.org/10.1146/annurev.ento.46.1.139.

Tallamy, D. W., and C. W. Schaefer. 1997. Maternal care in the Hemiptera: ancestry, alternatives, and current adaptive value, pp. 94–115. *In* J. C. Choe and B. J. Crespi (Eds.), The evolution of social behaviour in insects and arachnids. Cambridge University Press, New York, NY. 541 pp.

Tallamy, D. W., and T. K. Wood. 1986. Convergence patterns in subsocial insects. Annual Review of Entomology 31: 369–390. http://dx.doi.org/10.1146/annurev.en.31.010186.002101.

Tanaka, S. I., C. Imai, and H. Numata. 2002. Ecological significance of adult summer diapause after nymphal winter diapause in *Poecilocoris lewisi* (Distant) (Heteroptera: Scutelleridae). Applied Entomology and Zoology 37: 469–475. http://doi.org/10.1303/aez.2002.469.

Tauber, M. J., C. A. Tauber, and S. Masaki. 1986. Seasonal adaptations of insects. Oxford University Press, New York. 412 pp.

Taylor, M. E., C. S. Bundy, and J. E. McPherson. 2015. Life history and laboratory rearing of *Bagrada hilaris* (Hemiptera: Heteroptera: Pentatomidae) with descriptions of immature stages. Annals of the Entomological Society of America 108: 536–551. http://dx.doi.org/10.1093/aesa/sav048.

Tayutivutikul, J., and K. Kusigemati. 1992. Biological studies of insects feeding on the kudzu plant, *Pueraria lobata* (Leguminosae). II. Seasonal abundance, habitat and development. South Pacific Study 13: 45–46.

Tayutivutikul, J., and K. Yano. 1990. Biology of insects associated with the kudzu plant, *Pueraria lobata* (Leguminosae). 2. *Megacopta punctissimum* (Hemiptera, Plataspidae). Japanese Journal of Entomology 58: 533–539. (in Japanese with English summary)

Tischler, W. 1960. Studien zur Bionomie und Ökologie der Schmalwanze *Ischnodemus sabuleti* Fall. (Hem., Lygaeidae). Zeitschrift für wissenschaftliche Zoologie 163: 168–209.

Tojo, S., Y. Nagase, and L. Filippi. 2005a. Reduction of respiration rates by forming aggregations in diapausing adults of the shield bug, *Parastrachia japonensis*. Journal of Insect Physiology 51: 1075–1082. http://dx.doi.org/10.1016/j.jinsphys.2005.05.006.

Tojo, S., S. Nomakuchi, M. Hironaka, and L. Filippi. 2005b. Physiological and behavioural adaptation of a subsocial shield-bug, *Parastrachia japonensis*, that allow it to survive on the drupes of its sole host plant available only two weeks a year, pp. 35–36. *In* V. E. Kipyatkov (Ed.), Proceedings of the 3rd European Congress on Social Insects (Meeting of European Section of the IUSSI). St. Petersburg University Press, St. Petersburg, Russia. 204 pp.

Tomokuni, M., T. Yasunaga, M. Takai, I. Yamashita, M. Kawamura, and T. Kawasawa. 1993. A field guide to Japanese bugs: terrestrial heteropterans. Zenkoku Noson Kyoiku Kyokai, Tokyo, Japan. 380 pp. (in Japanese)

Toscano, N. C., and V. M. Stern. 1980. Seasonal reproductive condition of *Euschistus conspersus*. Annals of the Entomological Society of America 73: 85–88. http://dx.doi.org/10.1093/aesa/73.1.85.

Toyama, M., F. Ihara, and K. Yaginuma. 2006. Formation of aggregations in adults of the brown marmorated stink bug, *Halyomorpha halys* (Stål) (Heteroptera: Pentatomidae): the role of antennae in short-range locations. Applied Entomology and Zoology 4: 309–315. http://doi.org/10.1303/aez.2006.309.

Tsukamoto, L., and S. Tojo. 1992. A report of progressive provisioning in a stink bug, *Parastrachia japonensis* (Hemiptera: Cydnidae). Journal of Ethology 10: 21–29. http://dx.doi.org/10.1007/BF02350183.

Tullberg, B., G. Gamberale-Stille, T. Bohlin, and S. Merilaita. 2008. Seasonal ontogenetic colour plasticity in the adult striated shieldbug *Graphosoma lineatum* (Heteroptera) and its effect on detectability. Behavioral Ecology and Sociobiology 62: 1389–1396. http://dx.doi.org/10.1007/s00265-008-0567-7.

Tyshchenko, V. P. 1977. Physiology of photoperiodism in insects. Trudy Vsesojuznogo Entomologicheskogo Obshchestva [Proceedings of the All Union Entomological Society]. Nauka, Leningrad, USSR. 59: pp. 1–156. (in Russian)

Tyshchenko, V. P. 1980. Signal and vital modes of action of ecological factors. Zhurnal Obshey Biologii [Journal of General Biology] 41: 655–667. (in Russian)

Ushatinskaya, R. S. 1990. Cryptic life and anabiosis. Nauka, Moscow, USSR. 182 pp. (in Russian)

Vecchio, M. C., J. Grazia, and G. S. Albuquerque. 1994. Seasonal dimorphism in *Oebalus ypsilongriseus* (De Geer, 1773) (Hemiptera, Pentatomidae) and a new synonym. Revista Brasileira de Entomologia 38: 101–108.

Veerman, A. 1985. Diapause, pp. 279–316. *In* W. Helle and M. W. Sabelis (Eds.), Spider mites: their biology, natural enemies and control. Volume 1A. Elsevier Science Publisher, Amsterdam, The Netherlands. 406 pp.

Viktorov, G. A. 1967. Problems in insect population dynamics with reference to the sunn pest. Nauka, Moscow, USSR. 271 pp. (in Russian)

Vinokurov, N. N., E. V. Kanyukova, and V. B. Golub. 2010. Catalogue of the Heteroptera of Asian part of Russia. Nauka, Novosibirsk, Russia. 320 pp. (in Russian with English summary)

Volkovich, T. A. and A. Kh. Saulich. 1995. The predatory bug *Arma custos*: photoperiodic and temperature control of diapause and colouration. Entomological Review 74: 151–162.

Volkovich, T. A., L. I. Kolesnichenko, and A. Kh. Saulich. 1991a. The role of thermal rhythms in the development of *Perillus bioculatus* (Hemiptera, Pentatomidae). Entomological Review 70(1): 68–80.

Volkovich, T. A., A. Kh. Saulich, and N. I. Goryshin. 1991b. Day-length sensitive stage and accumulation of photoperiodic information in the predatory bug *Podisus maculiventris* Say (Heteroptera: Pentatomidae). Entomological Review 70: 159–167.

Vulinec, K. 1990. Collective security: aggregation by insects as a defense, pp. 251–288. *In* D. L. Evans and J. O. Schmidt (Eds.), Insect defenses: adaptive mechanisms and strategies for prey and predators. State University of New York, New York. 482 pp.

Walker, T. J. 1986. Stochastic polyphenism: coping with uncertainty. Florida Entomologist 69: 46–62.

Watanabe, M. 1980. Study of the life cycle of the brown marmorated stink bug, *Halyomorpha mista*. Insectarium (Japan) 17: 168–173.

Watanabe, M., T. Inaoka, Y. Kosuge, and T. Kohama. 1995. Biology of house-invading stink bugs and some control trials against them in an elementary school in Hokkaido. 2. Preventive effect against invasion of stink bugs by soundproof window and application of concentrated insecticide to window frames. Japanese Journal of Sanitary Zoology 46: 349–353. (in Japanese)

Weber, D. C., T. C. Leskey, G. C. Walsh, and A. Khrimian. 2014. Synergy of aggregation pheromone with methyl (E,E,Z)-2,4,6-decatrienoate in attraction of *Halyomorpha halys* (Hemiptera: Pentatomidae). Journal of Economic Entomology 107: 1061–1068. http://dx.doi.org/10.1603/EC13502.

Werner, D. J. 2005. Biologie, Ökologie und Verbreitung der Kugelwanze *Coptosoma scutellatum* (Heteroptera, Plataspidae) in Deutschland. Entomologie Heute 17: 65–90.

Wheeler, A. G., Jr. 2001. Biology of the plant bugs (Hemiptera: Miridae): pests, predators, opportunists. Cornell University Press, Ithaca. xvi + 508 pp.

Whitmarsh, R. D. 1916. Life-history notes on *Apateticus cynicus* and *maculiventris*. Journal of Economic Entomology 9: 51–53. http://dx.doi.org/10.1093/jee/9.1.51.

Wilde, G. E. 1969. Photoperiodism in relation to development and reproduction in the green stink bug. Journal of Economic Entomology 62: 629–630. http://dx.doi.org/10.1093/jee/62.3.629.

Williams, III, L., M. C. Coscarón, P. M. Dellapé, and T. M. Roane. 2005. The shield-backed bug, *Pachycoris stallii*: description of immature stages, effect of maternal care on nymphs, and notes on life history. 13 pp. Journal of Insect Science 5: 29 <insectscience.org/5.29>.

Yanagi, T., and Y. Hagihara. 1980. Ecology of the brown marmorated stink bug. Plant Protection 34: 315–321. (in Japanese)

Yao, M. 2002. Development, number of annual generations, and the relationship of effective heat unit and to abundance of overwintered adults in the following year of the white-spotted spined bug, *Eysarcoris aeneus* (Scopoli) (Heteroptera: Pentatomidae). Japanese Journal of Applied Entomology 46: 15–21. http://doi.org/10.1303/jjaez.2002.15. (in Japanese with English summary)

Zachariassen, K. E. 1985. Physiology of cold tolerance in insects. Physiological Reviews 65: 799–832.

Zaslavski, V. A. 1988. Insect development, photoperiodic and temperature control. Springer, Berlin, Germany. 187 pp.

Zerbino, M. S., N. A. Altier, and A. R. Panizzi. 2013. Effect of photoperiod and temperature on nymphal development and adult reproduction of *Piezodorus guildinii* (Westwood) (Heteroptera: Pentatomidae). Florida Entomologist 96: 572–582. http://dx.doi.org/10.1653/024.096.0223.

Zerbino, M. S., N. A. Altier, and A. R. Panizzi. 2014. Phenological and physiological changes in adult *Piezodorus guildinii* (Hemiptera: Pentatomidae) due to variation in photoperiod and temperature. Florida Entomologist 97: 734–743. http://dx.doi.org/10.1653/024.097.0255.

Zerbino, M. S., N. A. Altier, and A. R. Panizzi. 2015. Seasonal occurrence of *Piezodorus guildinii* on different plants including morphological and physiological changes. Journal of Pest Science 88: 495–505. http://doi.org/10.1007/s10340-014-0630-2.

Zhang, C. S., and D. P. Yu. 2005. Occurrence and control of *Megacopta cribraria* (Fabricius). Chinese Countryside Well-off Technology 1: 35.

Zhang, Y., J. L. Hanula, and S. Horn. 2012. The biology and preliminary host range of *Megacopta cribraria* (Heteroptera: Plataspidae) and its impact on kudzu growth. Environmental Entomology 41: 40–50. http://doi.org/10.1603/EN11231.

11.13 Glossary

Definitions apply to usage in **Chapters 11 and 12** and do not necessarily cover all meanings of particular terms. Cross-referenced entries within a definition are **boldfaced**. In a few cases of specific terms, references are given (for full reference citations, see References Cited section of this chapter).

Active development: see **Active seasonal development** and **Active physiological state**

Active physiological state: (1) on the organism level, physiological state of an individual (i.e., growth and metamorphosis), opposite of **dormancy** (first of all **diapause**); **(2)** on the population level, corresponds to **active seasonal development**. Synonyms: **Active development**, **Nondiapause development**

Active seasonal development: development of nondiapause individuals or **generation(s)** (e.g., growth, metamorphosis, and reproduction). Synonyms: **Active development**, **Active physiological state**, **Nondiapause development**. Antonyms: **Diapause**, **Dormancy**

Adult diapause: diapause at the adult stage (both in females and males), typically manifests itself as arrested development of reproductive system (i.e., arrested maturation of gonads, activity of reproduction-related glands, blocked oogenesis, absence of oviposition, etc.) and absence of reproductive behavior. Synonyms: **Imaginal diapause**, **Reproductive diapause**

***Aelia*-like response:** restoration of sensitivity to **day length** after a short refractory period (during **winter diapause**; see Hodek 1971a). Synonym: **Recurrent response**

Aestivation: see **Summer diapause**. Alternative spelling: **Estivation**

Annual cycles: see **Seasonal cycle**

Apterous adults: see **Aptery**

Aptery: anatomical condition (i.e., morph or form) of an adult insect completely lacking any wings. Synonyms: **Apterous adults**, **Winglessness**

Bivoltinism: special case of **multivoltinism**, characterized by development of only two **generations** per year, one in spring (or early summer) and one in autumn (or late summer)

Body color polyphenism: see **Seasonal body color polyphenism**

Brachypterous adults: see **Short-winged adults**

Brachyptery: see **Short-winged adults**

Care behavior: see **Parental care**

Changing day length: (1) under field conditions: natural seasonal change of **day length**; day increases until the day of summer solstice (June 20 to 22 in the Northern Hemisphere [i.e., **increasing day length, or photoperiod**]) and then decreases until winter solstice (December 20 to 23 in the Northern Hemisphere [i.e., **decreasing day length, or photoperiod**]); **(2)** under laboratory conditions: a special experimental protocol mimicking natural seasonal change of **day length** set up as an artificial shortening or lengthening of **photophase** and **scotophase** from one day to another (or with another interval)

Changing photoperiod: see **Changing day length**

Chill intolerance: see **Freeze intolerance**

Civil twilights: the lightest parts of twilights; consist of the morning civil twilight (begins when the geometric center of the Sun is 6° below the horizon [i.e., civil dawn] and ends at sunrise or when the geometric center of the Sun is 0°50′ below the horizon) and evening civil twilight (begins at sunset or when the geometric center of the sun is 0°50′ below the horizon and ends when the geometric center of the Sun reaches 6° below the horizon [civil dusk]); many insect species perceive civil twilights as a part of **photophase**

Cold hardiness: ability of an organism to survive at low temperature; it can be achieved by three strategies: **cryoprotective dehydration, freeze intolerance**, and **freeze tolerance**. Synonym: **Cold tolerance**

Cold termination of winter diapause: termination of **winter diapause** in response to exposure to low temperatures (usually above 0°C); often considered as a result of the **tachytelic process**

Cold tolerance: see **Cold hardiness**

Conditional polyphenism: see **Polyphenism**

Critical day length: see **Critical photoperiod**

Critical photoperiod: under field conditions or in the laboratory, **photoperiod** (i.e., **day length**) at which 50% of individuals of a particular population at particular temperature (especially in the laboratory) demonstrate clearly **photoperiodic response** (e.g., enter **diapause**). Synonyms: **Critical day length, Photoperiodic threshold**

Cryophase: lower-temperature phase of the **thermorhythm** in a laboratory experiment

Cryoprotective dehydration: one of three strategies of **cold hardiness**; survival of subzero temperatures by losing osmotic water to the surrounding environment, so resulting in an increase of the concentration of their body fluids and, thus, a decline in their melting point (to equilibration with the ambient temperature); as a result, they cannot freeze (Zachariassen 1985, Bale 2002, Sinclair et al. 2003)

Crystallization temperature: see **Supercooling point**

Day length: duration (in hours and minutes) of the light part of a daily cycle (i.e., **photophase**). Synonym: **Photoperiod** (note that **day length** usually refers to the field situation, whereas **photoperiod** to laboratory experiment). Antonym: **Night length** (i.e., **Scotophase**). Laboratory photoperiod is usually indicated as, e.g., L:D 16:8, where L is light period, or **photophase** (16 hours), and D is dark period, or **scotophase** (8 hours)

Day-length sensitive stage: see **Sensitive stage**

Decreasing day length: see **Changing day length**

Decreasing photoperiod: see **Changing day length**

Degree-days: a method of estimation of thermal (i.e., temperature) requirements of organisms (populations, species) or resources of particular locations or regions; total degree-days from an appropriate starting date or temperature level are used to understand **voltinism** of species or develop pest control strategy; computed as the integral of a function of time that generally varies with temperature. See **Sum of effective temperatures**

Development(al) threshold: see **Lower development threshold**

Diapause: profound, endogenously, and centrally mediated interruption that routes the developmental program away from direct morphogenesis into an alternative diapause program of succession of physiological events; the start of diapause usually precedes the advent of adverse conditions, but the end of diapause need not coincide with the end of adversity (Koštál 2006)

Diapause, forms of: diapause can be of two forms – **facultative diapause** and **obligate diapause**

Diapause, seasonal classes of: diapause can be of two seasonal classes – **winter diapause** and **summer diapause**

Diapause, types of: diapause can be linked to four (in Heteroptera – three) ontogenetic (i.e., developmental) stages and, thus, be of four (in Heteroptera – three) types – **embryonic** (i.e., **egg**) **diapause, nymphal** (i.e., **larval**) **diapause**, pupal diapause (not in Heteroptera), and **adult** (i.e., **reproductive**, or **imaginal**) **diapause**

Diapause development: slow and dynamic changes (i.e., physiological processes) in internal state of diapausing individual leading to **diapause termination**

Diapause induction: in species with **facultative diapause**, such **diapause** needs to be induced (i.e., initiated) by an environmental cue and, thus, this cue needs to be perceived, transmitted, and interpreted by the neurohormonal system of the individual; this cue switches the ontogenetic pathway from **active physiological state** to **diapause**

Diapause phase: central phase of **diapause**; consists of three subphases: **initiation**, **maintenance**, and **termination**

Diapause syndrome: complex of morphological, physiological, and behavioral traits associated with **diapause** (Tauber et al. 1986); often used as a synonym of **diapause**

Diapause termination: end of **diapause**; gradual changes that occur during **diapause development** and result in its ending

Diapausing stage: ontogenetic (i.e., developmental) stage at which diapause occurs. In pentatomoids, it can be the egg (i.e., embryonic), nymphal, or adult stages

Direct development: individual **active physiological state** without interruption for physiological **dormancy**; opposite to **diapause** or any other form of **dormancy**. Synonym: **Nondiapause development**. Antonym: **Diapause, Diapause development**

Dispersal: general term for movement of insects; scattering or spreading of members of one population in space, with different purposes, usually resulting in increasing of mean distance among members of the population (see Dingle 1996)

Dormancy: general term covering any state of suppressed development (i.e., developmental arrest) that is adaptive (that is ecologically or evolutionarily meaningful and not just artificially induced) and usually accompanied with metabolic suppression (Koštál 2006)

Ecological polymorphism: see **Polyphenism**

Egg diapause: see **Embryonic diapause**

Embryonic diapause: diapause at the embryonic (i.e., egg) stage of ontogenesis; typically manifests itself as arrested embryogenesis (i.e., postponed hatching of nymphs). Synonym: **Egg diapause**

Endogenous processes: processes that originate from within an organism; internal. Antonym: **Exogenous processes**

Endogenous univoltinism: pattern of **univoltinism** based on **obligate diapause**; seasonal development with completion of strictly only one **generation** during the **vegetative season** or year. See **Univoltine seasonal cycle**

Environmental polymorphism: see **Polyphenism**

Estivation: see **Summer diapause.** Alternative spelling: **Aestivation**

Estivation quarters: microhabitats used by insects to survive during **summer diapause** (i.e., **estivation**). Synonyms: **Estivation sites**

Estivation sites: see **Estivation quarters**

Exogenous processes: processes that originate from outside of an organism, from its environment; external. Antonym: **Endogenous processes**

Exogenous univoltinism: pattern of **univoltinism** in species or populations that potentially can have **multivoltine seasonal cycles**, but whose **seasonal development** is limited by completion of strictly only one **generation** during the **vegetative season** or year; such pattern is controlled by external factors. See **Univoltine seasonal cycle**

Facultative: optional or discretionary, something that must be induced. See, for example, **Facultative diapause**. Antonym: **Obligate**

Facultative diapause: **diapause** that is not obligate but induced in particular **generation** by external factors (e.g., **day length**, temperature, food, humidity); it can, but does not necessarily, occur in each **generation**

Food-mediated diapause: **facultative diapause** that is induced by trophic factor (i.e., food or diet) – seasonal change of quality or availability of food. Synonym: **Trophic diapause**

Freeze avoidance: see **Freeze intolerance**

Freeze intolerance: one of three strategies of **cold hardiness**; the freeze intolerant species cannot survive the formation of ice within their bodies and, thus, have evolved a set of biochemical, physiological, behavioral, and ecological measures/adaptations aimed at prevention of ice formation. Synonyms: **Freeze avoidance, Freeze susceptibility**

Freeze susceptibility: see **Freeze intolerance**

Freeze tolerance: one of three strategies of **cold hardiness**; the freeze tolerant insects can withstand ice formation, usually only in the extracellular fluids, and have a set of characteristics that enables them to survive such ice formation

Generation: (1) all individuals of the population living and developing at the same time (i.e., started development approximately at the same time) and usually existing in the same physiological state (i.e., **active physiological state** or **diapause**); **(2)** a period of time necessary for completion of development of a full **life cycle**

Genetic polymorphism: case of **polymorphism** in which two or more different phenotypes (i.e., morphs or forms) are produced by different genotypes. Also see **Polyphenism**

Growing season: see **Vegetative season**

Heterodynamic seasonal cycle: **seasonal cycle** that corresponds to **heterodynamic seasonal development**

Heterodynamic seasonal development: type of **seasonal development** in which periods of **active seasonal development** alternate with periods of seasonal **dormancy** of varying duration and intensity (e.g., **winter diapause** or **summer diapause**). Also see **Heterodynamic seasonal cycle**

Hibernacula (singular: **Hibernaculum**): see **Hibernation quarters**

Hibernaculum (plural: **Hibernacula**): see **Hibernation quarters**

Hibernation: see **Winter diapause**

Hibernation quarters: microhabitats used by insects to survive during **winter diapause** (i.e., **hibernation**). Synonyms: **Hibernaculum** (plural: **Hibernacula**), **Hibernation sites**, **Overwintering sites**

Hibernation sites: see **Hibernation quarters**

Homodynamic seasonal cycle: seasonal cycle that corresponds to **homodynamic seasonal development**

Homodynamic seasonal development: type of **seasonal development** in which **active seasonal development** is not interrupted by periods of seasonal **dormancy**; it is typical only for very stable and warm environmental conditions (e.g., tropical and subtropical regions, caves, subterranial microhabitats, artificial constructions such as grain storage barns). See **Homodynamic seasonal cycle**

Horotelic process: slow, spontaneous, and endogenous internal physiological processes of **diapause development** that proceeds under stable environmental conditions (i.e., when there are no dramatic changes of environmental conditions) and leads to **spontaneous diapause termination**. See **Tachytelic process**

Imaginal diapause: see **Adult diapause**

Increasing day length: see **Changing day length**

Increasing photoperiod: see **Changing day length**

Induced diapause termination: diapause termination based on physiological processes induced by external conditions (i.e., exogenous, nonspontaneous); it is equivalent and the result of **tachytelic process**; can proceed only under influence of changes in environmental conditions

Induction subphase: only in the case of **facultative diapause**, subphase of **prediapause phase** during which the ontogenetic pathway is switched from **direct development** to **diapause**

Initiation subphase: subphase of **diapause phase** during which direct development ceases, deep physiological preparations take place, and intensity (or deepness) of diapause may increase

Larval diapause: see **Nymphal diapause**

Life cycle: sequence of life stages (in Heteroptera – egg, nymphal, and adult stages) that an organism undergoes from birth to reproduction and death. Compare to **Seasonal cycle**

Long day: for a particular population and conditions (first of all temperature), daily cycle with **photophase** longer than the **critical photoperiod**

Long-day conditions: day-length (i.e., **photoperiodic**) **conditions** with **photophase** longer than **critical photoperiod**; for majority of insect in the Northern Hemisphere, the conditions with **photophase** that induces **active physiological state** (i.e., **nondiapause development**). Antonym: **Short-day conditions**

Long-day diapause: facultative summer diapause induced under **long-day conditions** in early or mid-summer in the Northern Hemisphere. In laboratory, such **diapause** usually can be induced under **photoperiodic conditions** with **photophase** longer than 12 hours of light

Long-day photoperiodic response: see **Long-day type photoperiodic response of diapause induction**

Long-day type photoperiodic response of diapause induction: **photoperiodic response** that induces **active physiological state** under **long-day conditions** and **facultative diapause** under **short-day conditions**; a typical **photoperiodic response** in populations that have facultative **winter diapause** and **multivoltine seasonal development**. Synonym: **Long-day photoperiodic response**

Long-winged adults: adults with fully developed wings. Wing length may be controlled genetically (see **Wing polymorphism**) or environmentally (see **Wing polyphenism**). Synonym: **Macropterous adults, Macroptery**. Antonyms: **Short-winged adults, Brachypterous adults, Brachyptery**

LDT: see **Lower development threshold**

Lower development threshold: species- or ontogenetic-stage-specific temperature below which growth and development (i.e., morphogenesis) do not take place. Synonym: **Development(al) threshold**. Abbreviation: **LDT**

Macropterous adults: see **Long-winged adults**

Macroptery: see **Long-winged adults**

Maintenance subphase: subphase of **diapause phase** during which the endogenous developmental arrest persists regardless of environmental conditions

Maternal care: see **Parental care**

Maternal instinct: see **Parental care**

Migration: complex form of movement of individuals and populations characterized by the following parameters: (1) it is persistent; (2) it is straightened out; (3) it is undistracted by resources that would ordinarily halt it; (4) there are distinct departing and arriving behaviors; and (5) energy is reallocated to sustain it; not all migrants display all of these characteristics all of the time, but most will display most of them at least part of the time during which they are migrating; migrations often happen seasonally (Dingle 1996). See **Seasonal migrations**

Monovoltine seasonal cycle: see **Univoltine seasonal cycle**

Monovoltinism: see **Univoltinism**

Multivoltine seasonal cycle: **seasonal cycle** typical for **multivoltine seasonal development**. Synonym: **Polyvoltine seasonal cycle**

Multivoltine seasonal development: type of **seasonal development** with completion of two or more **generations** during the **vegetative season** or year; in the last seasonal **generation,** facultative **winter diapause** is formed

Multivoltinism: type of **seasonal development** with **multivoltine seasonal cycle**. See **Multivoltine seasonal development**

Night length: duration (in hours and minutes) of the night (i.e., dark) part of daily cycle (i.e., **scotophase**). Antonym: **Day length**

Nondiapause development: see **Direct development**, **Active physiological state**, and **Active seasonal development**

Nymphal diapause: diapause at the nymphal (i.e., larval) stage, typically manifests itself as arrested metamorphosis (i.e., absence of molting to the next nymphal or adult stage). Synonym: **Larval diapause**

Obligate: by necessity; genetically determined; something that should not be induced. See **Obligate diapause**. Synonym: **Obligatory**. Antonym: **Facultative**

Obligate diapause: diapause in which initiation needs no external signal or cues because it represents a fixed component of the ontogenetic program that is realized regardless of the environmental conditions in each **generation**; one of two forms of **diapause**. Antonym: **Facultative diapause**

Obligatory: see **Obligate**

Oligopause: form of **dormancy** with less intensive suppression of development than **diapause**

Overwintering: process of passing winter season with all associated unfavorable conditions (e.g., cold and/or subzero temperatures, ice, snow, and limited food availability)

Overwintering quarters: see **Hibernation quarters**

Overwintering sites: see **Hibernation quarters**

Packet of photoperiodic information: number of photoperiodic cycles (**short days** or **long days**) triggering **diapause** or **active physiological state**. Synonym: **Required day number**

Parental care: complex of behavioral traits that enhance the fitness of offspring. Synonyms: **Maternal care, Care behavior, Care behavior, Maternal instinct**

Partial generation: fairly common pattern in which, at the end of the appropriate **season**, some part of the population gives rise to the subsequent **generation** whereas the other part (usually the one that completed development somewhat later) enters **diapause**

Perennial seasonal cycle: see **Semivoltine seasonal cycle**

Phenology: (**1**) study of periodic plant and animal **seasonal cycle** events and how these are influenced by seasonal and interannual variations in climate, as well as habitat factors (such as elevation); (**2**) **seasonal development** of a population in a particular year

Phenophase(s): particular phase(s) of **seasonal development** of a local population

Photoperiod: see **Day length**

Photoperiodic conditions: important characteristic of natural (i.e., field) conditions or laboratory regime in terms of light; usually understood as a ratio between **day length** (i.e. **photophase**) and **night length** (i.e., **scotophase**), but also might refer to intensity of light, its specter, duration of cycle (in laboratory), etc.

Photoperiodic control (i.e., regulation) of the nymphal growth rate: type of **photoperiodic response** that manifests itself as different rates of nymphal growth under different **photoperiodic conditions** (e.g., acceleration of growth under **short-day conditions** and retardation of growth under **long-day conditions**); note that some species do not have such **photoperiodic response**

Photoperiodic diapause termination: diapause termination that happens in response to change in **photoperiodic conditions**; often considered as a result of the **tachytelic process**

Photoperiodic refractoriness: period of insensitivity of insects to **day length**; usually after **overwintering**

Photoperiodic response: physiological reaction of an organism to the experienced **photoperiodic conditions** (e.g., **photoperiodic response of diapause induction**). Abbreviation: **PhPR**

Photoperiodic response of diapause induction: (**1**) physiological reaction of an organism to the experienced **photoperiodic conditions** that manifests itself in induction of one of two alternative physiological states – **facultative diapause** or **active physiological state**; (**2**) total of reactions of a particular population or laboratory cohort

Photoperiodic threshold: see **Critical photoperiod**

Photophase: light part of daily cycle. Antonym: **Scotophase**

PhPR: see **Photoperiodic response**

Polymorphism: presence in a population of two or more distinct phenotypes (i.e., morphs, forms) at the same ontogenetic stage (discontinuous variation). See **Genetic polymorphism, Polyphenism**

Polyphenism: special case of **polymorphism** when changes in frequencies of phenotypes are controlled by environmental conditions. Synonyms: **Environmental polymorphism, Ecological polymorphism, Conditional polyphenism**

Polyvoltine seasonal cycle: see **Multivoltine seasonal cycle**

Polyvoltinism: see **Multivoltinism**

Postdiapause phase: final phase of **diapause**; follows **termination subphase**; during this phase, insect often experience **postdiapause quiescence**

Postdiapause quiescence: quiescence that insects often experience in late winter and/or early spring when **active development** and metabolism are exogenously (i.e., externally) inhibited

Prediapause phase: first phase of **diapause** during which **direct development** (morphogenesis) continues and **diapause** is induced (in species with **facultative diapause**) or formed (in species with **facultative diapause** as well as **obligate diapause**)

Preparation subphase: subphase of **prediapause phase** during which individuals undergo behavioral and/or physiological change (e.g., acquire energy resources such as lipids), void digestive system (gut), migrate and/or simply look for protective microhabitats (i.e., **hibernaculum**), sometimes change body color and so on

Pyrrhocoris-**like response:** insensitivity to **day length** when, in response to prolonged exposure to cold, some insects lose the ability to measure or respond to **day length** and, thus, develop without entering **diapause** under any **day-length conditions**

Qualitative photoperiodic response: photoperiodic response in which each individual responds in an "All or None" (i.e., "Yes or No") manner by choosing one of two alternative pathways (e.g., in the case of **adult diapause** – **diapause** or **direct development** [i.e., reproduction]). Antonym: **Quantitative photoperiodic response**

Qualitative PhPR: see **Qualitative photoperiodic response**

Quantitative photoperiodic response: photoperiodic response that controls quantitative parameters such as size, duration of a particular stage, or degree of pigmentation, etc. Antonym: **Qualitative photoperiodic response**

Quantitative PhPR: see **Quantitative photoperiodic response**

Quasi-natural experimental conditions: experimental set up in which insects are reared/maintained outdoors under conditions as much as possible mimicking the wild environmental conditions; for example, insect are reared in captivity in containers (e.g., Petri dishes, cages), but the containers are placed outdoors in such a way that insects can experience natural **day length**, temperature, and humidity, usually being only protected from direct sun light, rain, large predators, and parasitoids

Quiescence: an immediate response (without complex preceeding central neuroendocrine regulation) to a decline of any limiting environmental factor(s) below the physiological thresholds with immediate resumption of the processes if the factor(s) rise above them (Koštál 2006)

Recurrent response: see *Aelia*-**like response**

Reproductive diapause: see **Adult diapause**

Required day number: see **Packet of photoperiodic information**

Scotophase: night (i.e., dark) part of daily cycle. Antonym: **Photophase**

SCP: see **Supercooling point**

Seasonal adaptations: ability of organisms (in the form of physiological, biochemical or behavioral responses) to survive, take advantage in utilization of resources and form a specific and stable pattern of **seasonal development** (i.e., **seasonal cycle**) under seasonally changing local environmental conditions

Seasonal body color change: see **Seasonal body color polyphenism**

Seasonal body color polyphenism: example of **seasonal polyphenism** when during the season, two or more forms exist of the same ontogenetic stage (e.g., nymphs [larvae], pupae, or adults) with difference in body coloration. These different body color forms can be in different generations or coloration of one particular individual can change during its life (e.g., diapausing individuals change body color from green to russet and, then, upon diapause termination after overwintering, the body color changes back to green). Synonym: **Seasonal body color change**

Seasonal camouflage: ability an organism to avoid observation or detection by other organisms; passive protection from predators

Seasonal changes of degree of development of wing muscles: example of **seasonal polyphenism** when during the season, two or more forms of adults exist with difference in degree of development of wing muscles (e.g., a form with fully developed wing muscles and capable of flight and a form with weakly developed or reduced wing muscles and incapable of flight). Also see **Seasonal changes of wing size**

Seasonal changes of reproductive allocation: refers to the seasonal changes in the proportion of an organism's energy budget (or investment of resources) allocated to reproduction

Seasonal changes of wing size: example of **seasonal polyphenism** when during the season, two or more forms of adults exist with difference in degree of development of wings (in some species, can range from winglessness to fully developed wings, or **macroptery**). See **Long-winged adults**, **Short-winged adults**, and **Seasonal changes of degree of development of wing muscles**

Seasonal cycle: specific and stable pattern of realization of the **life cycle** of a species or population against the background of seasonally changing local environmental conditions; might either include or not periods of seasonal **dormancy**, **migrations** or other **seasonal adaptations**. Synonym: **Annual cycles**

Seasonal development: consecutive realization (i.e., completion) of **generation(s)** against the background of seasons in a particular location. **Active seasonal development** might alternate with periods of seasonal **dormancy** (i.e., **heterodynamic seasonal development**) or might go without such alteration (i.e., **homodynamic seasonal development**)

Seasonal food plant change: situation when during the **vegetative season** individuals of the same or different **generations** of a particular local population consequently use different plants as food plants (e.g., one plant species as a primary food plant in the early summer and then another plant species as a primary food plant in the late summer)

Seasonal forms: categories of individuals in a population that look differently as a result of **seasonal morphological polyphenism**; representation of **seasonal morphological polyphenism**. Synonym: **Seasonal morphs**

Seasonal migrations: strictly regular (seasonal) **migrations** linked to particular stages of **life cycle** and **seasonal cycle**, usually between breeding habitats and **hibernation quarters** and/or **estivation quarters**; as a rule, an essential part of the **seasonal cycle** of a population

Seasonal morphological polyphenism: example of **seasonal polyphenism** when during the season, two or more forms of the same ontogenetic stage (e.g., nymphs [larvae], pupae, or adults) exist and differ in morphology (e.g., differ in shape, size) of the whole body or particular organs

Seasonal morphs: see **Seasonal forms**

Seasonal polyphenism: special case of **polyphenism** when changes in frequencies of phenotypes are regular (i.e., annual, seasonal) and controlled by environmental conditions

Seasonal variation of microhabitat selection: situation when during the **vegetative season**, individuals of the same or different **generations** of a particular local population use different microhabitats (e.g., one microhabitat is preferred in summer and then another microhabitat is preferred in winter)

Semivoltine seasonal cycle: seasonal cycle typical for **semivoltine seasonal development**

Semivoltine seasonal development: seasonal development with completion of one **generation** over a period that is longer than one **vegetative season** or year

Semivoltinism: type of seasonal development with **semivoltine seasonal cycle**. See **Semivoltine seasonal development**

Sensitive stage: developmental (i.e., ontogenetic) stage(s) during which individuals of a particular species are sensitive to external signals (e.g., to **day length**). See **Sensitivity to day length**

Sensitivity to day length: ability of individuals to measure **day length** and discriminate **short days** from **long days**

SET: see **Sum of effective temperatures**

Short day: for a particular population and conditions (first of all temperature), daily cycle with **photophase** shorter than the **critical photoperiod**

Short-day conditions: photoperiodic conditions with **photophase** shorter than **critical photoperiod**; for majority of insect in the Northern Hemisphere, the conditions that induce **winter diapause**

Short-day diapause: facultative winter diapause induced under **short-day conditions** in late summer or autumn in the Northern Hemisphere; in laboratory, such **diapause** usually can be induced under **short-day conditions**

Short-day photoperiodic response: see **Short-day type photoperiodic response of diapause induction**

Short-day type photoperiodic response of diapause induction: photoperiodic response which induces **active physiological state** under **short-day conditions** and **facultative diapause** under **long-day conditions**; a typical **photoperiodic response** in populations that have facultative **summer diapause** and **bivoltine or univoltine seasonal development**. More common in insects living in tropical and subtropical regions; rare in the Temperate Zone. Synonym: **Short-day photoperiodic response**

Short-winged adults: adults with small (i.e., reduced in size, undeveloped) wings usually unsuitable for flight. Wing length may be controlled genetically (see **Wing polymorphism**) or

environmentally (see **Wing polyphenism**). Synonym: **Brachypterous adults, Brachyptery.**
Antonyms: **Long-winged adults, Macropterous adults, Macroptery**

Signal factor: see **Signal function of ecological factors**

Signal function of ecological factors: reflects dual mode of action of ecological factors (primarily temperature, but also light, food, etc.) on living organisms; some ecological factors (e.g., day length, thermorhythm, quality of food) can be used as signals that predict the coming seasonal environmental changes (Tyshchenko 1980). See **Vital function of ecological factors**

Spontaneous diapause termination: diapause termination based on spontaneous (i.e., not induced by external conditions, endogenous) physiological processes; equivalent and result of **horotelic process**; can proceed without changes of environmental conditions

Sum of effective temperatures: sum of **degree-days** above the **lower development threshold** required for an insect stage or the ontogenesis to complete development. Abbreviation: **SET**

Summer diapause: diapause that takes place in summer; one of two seasonal classes of **diapause**. Synonym: **Estivation**. Alternative spelling: **Aestivation**

Summer–winter diapause: diapause that is formed in summer but does not end in autumn and instead lasts until the end of winter. In some cases, it is likely to consist of two diapauses (i.e., summer and winter ones) with short and hardly detectable period between them

Supercooling: the process of lowering the temperature of a liquid (i.e., water or any body liquids) below its freezing point without it becoming a solid (i.e., ice)

Supercooling point: the temperature at which spontaneous freezing occurs in a supercooled liquid. Synonym: **Crystallization temperature**. Abbreviation: **SCP**

Tachytelic process: fast and induced physiological process that evolves at a rate faster than in the case of **horotelic process**; internal physiological process of **diapause development** that proceeds under influence of change of environmental conditions and lead to **induced diapause termination** (see Hodek 1983, 1996, 2002). See **Horotelic process**

Temperature optimum of photoperiodic response: range of temperatures under which a particular **photoperiodic response** manifests itself adequately and is not suppressed by too low or too high suboptimal temperatures (e.g., in the case of **long-day diapause**, clear **active physiological state** is observed under **long-day conditions** and apparent **diapause** is induced under **short-day conditions**)

Termination subphase: final subphase of central **diapause phase** during which the intensity of **diapause** decreases and by the end of this subphase, a usual **active physiological state** is mostly reached

Thermoperiod: one of characteristics of **thermorhythm**; ratio between duration of **thermophase** (i.e., phase with higher temperature) and duration of **cryophase** (i.e., phase with lower temperature) in daily temperature cycle

Thermophase: higher-temperature phase of the **thermorhythm** in laboratory experiment

Thermoregulation: ability of an organism to utilize special behavioral or physiological adaptations in order to keep its body temperature within certain boundaries, even when the surrounding temperature is different

Thermorhythm: (1) under field conditions, a sinusoid changes of ambient temperature during daily cycle; **(2)** under laboratory conditions, experimental regime of temperature mimicking the natural dynamics of ambient temperature during daily cycle; in a simple case, **thermorhythm** (similar to **photoperiod**) consists of two phases (i.e., parts) – **thermophase** (with higher temperature) and **cryophase** (with lower temperature), which might (or might not) coincide with **photophase** and **scotophase**, respectively

Trivoltinism: special case of **multivoltinism**, characterized by development of three **generations** per **vegetative season** or year

Trophic diapause: see **Food-mediated diapause**

Univoltine seasonal cycle: seasonal cycle typical for **univoltine seasonal development**. Synonym: **Monovoltine seasonal cycle**

Univoltine seasonal development: seasonal development with completion of strictly only one **generation** during the **vegetative season** or year

Univoltinism: type of **seasonal development** with **univoltine seasonal cycles**. Synonym: **Monovoltinism**. See **Univoltine seasonal development**

Vegetation season: see **Vegetative season**

Vegetative season: period of time in a year when the climate is prime for plants to experience the most growth. Synonyms: **Growing season**, **Vegetation season**

Vital factor: see **Vital function of ecological factors**

Vital function of ecological factors: reflects dual mode of action of ecological factors (primarily temperature, but also light, food, etc.) on living organisms; determines a range within which a particular species can live (Tyshchenko 1980). See **Signal function of ecological factors**

Voltinism: term used to indicate the number of **generations** realized/produced by a population during a year

Wing muscle seasonal polyphenism: special case of **polyphenism** when degree of development of wing muscles change during a **vegetative season** or year; strongly linked to flight ability

Wing polymorphism: special case of **polymorphism** when within a population two or more discrete genetically controlled morphs (or forms) exist with wings of different length or degree of development (e.g., with fully developed, reduced, or totally absent wings). See, for example, **Long-winged adults**, **Short-winged adults**, and **Apterous adults**

Wing polyphenism: special case of **polyphenism** when, within a population, two or more discrete phenotypes (i.e., forms) with wings of different length or degree of development (e.g., with fully developed, reduced, or totally absent wings) are produced by one genotype under different environmental conditions. See, for example, **Long-winged adults**, **Short-winged adults**, and **Apterous adults**

Winglessness: see **Aptery**

Winter diapause: diapause that takes place in winter; one of two seasonal classes of **diapause**. Synonym: **Hibernation**

12

Seasonal Cycles of Pentatomoidea[1]

Aida Kh. Saulich and Dmitry L. Musolin

CONTENTS

12.1 Introduction..565
12.2 The Univoltine Seasonal Cycle...567
    12.2.1 The Endogenously Controlled Univoltine Seasonal Cycle.........................568
        12.2.1.1 The Univoltine Seasonal Cycle Based on Obligate Embryonic Diapause 568
        12.2.1.2 The Univoltine Seasonal Cycle Based on Obligate Nymphal Diapause 571
        12.2.1.3 The Univoltine Seasonal Cycle Based on Obligate Adult Diapause 573
    12.2.2 The Exogenously Controlled Univoltine Seasonal Cycle............................574
12.3 The Multivoltine Seasonal Cycle..578
    12.3.1 The True Multivoltine Seasonal Cycle 579
    12.3.2 The Strictly Bivoltine Seasonal Cycle 584
    12.3.3 The Partially Bivoltine Seasonal Cycle 585
12.4 The Semivoltine (Perennial) Seasonal Cycle 587
12.5 The Significance of Photoperiodic and Thermal Responses for Expansion of Insects
Beyond Their Natural Distribution Ranges 587
    12.5.1 Natural or Accidental Invasions—Case Studies of the Southern Green Stink Bug,
Nezara viridula, and Brown Marmorated Stink Bug, *Halyomorpha halys*.................587
    12.5.2 Intentional Introductions—Case Studies of the Spined Soldier Bug, *Podisus
maculiventris*, and Twospotted Stink Bug, *Perillus bioculatus*................................592
12.6 Conclusions 595
12.7 Acknowledgments..598
12.8 References Cited 599

12.1 Introduction

Insects provide extensive material for studying the evolution and diversity of **seasonal adaptations** and **annual cycles** based on these adaptations. Every insect species, and in many cases every population, possesses its own specific annual cycle that differs from those of other populations or species, including those that are taxonomically close or sympatric.

A comparative study of seasonal adaptations in particular taxa combined with taxonomic analysis makes an efficient approach to solving some fundamental problems of evolution. This principle underlies the existing hypotheses of the evolution of seasonal adaptations in insects. However, studies in this direction are limited, being mostly focused on some families of Lepidoptera (Tyshchenko 1983, Masaki and Yata 1988), Coleoptera (e.g., Carabidae; Matalin 2007), Neuroptera (e.g., Chrysopidae; Volkovich 2007), Orthoptera (e.g., Gryllidae; Tauber et al. 1986, Masaki and Walter 1987), and Hymenoptera (Kipyatkov

---

[1] This chapter was modified, expanded and updated from "Seasonal cycles in stink bugs (Heteroptera, Pentatomidae) from the Temperate Zone: diversity and control" by A. Kh. Saulich and D. L. Musolin (2014b) (Copyright 2014 authors and Pleiades Publishing, Ltd.). Note: Many specific eco-physiological terms that are **boldfaced** when mentioned the first time in the text of this chapter are explained in the Glossary to **Chapter 11 (Section 11.13)**.

and Lopatina 2007) but also include some Chelicerata (Belozerov 2007, 2012) and Crustacea (Alekseev 1990, Hairston and Cáceres 1996). The study of seasonal cycles in representatives of the Pentatomoidea would expand the capacities of such an analysis.

In our previous communications (Saulich and Musolin 2007a,b, 2011, 2014b; **Chapter 11**), we proposed a system of seasonal adaptations involved in the formation of various annual cycles of insects. Among such seasonal adaptations, we considered four categories of phenomena that determine the seasonal cycles of insects:

- **active** (i.e., **nondiapause**) **physiological state** and responses controlling **active development**, mostly its rate (e.g., control of growth rate by temperature and **day length**, different behavioral responses aimed at maximizing of fitness);

- **diapause** and responses controlling the formation, development, and termination of the state of physiological **dormancy** (first of all diapause), both **facultative** and **obligate**;

- **migrations** and responses allowing the insects to actively avoid the adverse conditions by movement;

- **seasonal polyphenism** and responses controlling the morphological and physiological characters (e.g., coloration; body shape, size, and proportions; and the degree of development of wings and/or wing muscles) that often are closely associated with diapause or some other form of seasonal dormancy (see **Chapter 11** for details and examples).

Combinations of these seasonal adaptations underlie the diversity of seasonal patterns of insects, which, in turn, can be classified into several basic types. First of all, one can distinguish between the **homodynamic** and **heterodynamic** types of **seasonal development**, and the corresponding seasonal cycles. In the former case, the insects remain in the active physiological state all year-round; in the latter case, periods of active development alternate with periods of seasonal dormancy of varying duration and intensity.

Homodynamic seasonal cycles are largely characteristic of species living under relatively stable conditions: inhabitants of the tropical and subtropical zones, synanthropic and cave species, and some soil-dwelling insects. In the Northern Hemisphere, homodynamic seasonal development often can be observed in the southern geographic populations of those species, which are heterodynamic in the temperate climate. For example, populations of the spined soldier bug, *Podisus maculiventris* (Say), in Florida, the United States of America (30°N), develop without diapause (De Clercq and Degheele 1993), so that adults of this species can be found there all year-round (Richman and Mead 1980). Populations of the southern green stink bug, *Nezara viridula* (L.), living still closer to the equator also lack diapause: in India (23°N; Singh 1973) and Brazil (23°S; Panizzi and Hirose 1995), oviposition of these bugs was recorded even in winter, which is impossible in the temperate part of the species range. Homodynamic seasonal development also is likely to be characteristic of the bagrada bug (painted bug), *Bagrada hilaris* (Burmeister), and the harlequin bug, *Murgantia histrionica* (Hahn), in southern areas such as India (Singh and Malik 1993, Siddiqui 2000) and the United States of America (McPherson and McPherson 2000, Taylor et al. 2015; see also **Chapters 3 and 6**).

However, the permanent existence of insects under more severe climatic conditions with pronounced seasonality of climate in subtropic, temperate, and polar latitudes depends on invariable alternation of periods of active development and seasonal dormancy. Such a **heterodynamic seasonal cycle** can exist in the forms of **univoltinism** (i.e., a type of seasonal development with **univoltine cycles**), **multivoltinism** (i.e., a type of seasonal development with **bivoltine**, **trivoltine** or **multivoltine cycles**), and **semivoltinism** (i.e., a type of seasonal development with cycles, that are longer than one calendar year).

The main distinguishing trait of a univoltine cycle is completion of strictly only one **generation** during the **vegetative season** whereas in the multivoltine cycle, two or more generations can be formed during a season. An important trait of the multivoltine seasonal cycle is that the overwintered generation gives rise to consecutive summer generation(s), and the last annual generation, in turn, forms diapause and overwinters. The ecological control of multivoltinism has proven to be essentially similar in most taxa studied (Danilevsky 1961, Tauber et al. 1986, Danks 1987). It usually is based on **the photoperi-odic response** (**PhPR**) of facultative diapause induction of the **long-day type**; diapause is induced by a decrease in day length in autumn often combined with a decrease in temperature (**Chapter 11**). The system synchronizing the multivoltine seasonal development of insects with the periodically chang-ing external conditions is made more reliable by the ability of different species to respond simultane-ously to different external factors such as temperature (both its mean value and the daily and seasonal rhythms), seasonal dynamics of day length, and qualitative composition of diet. Therefore, due to the modifying effects of external conditions on the parameters of the PhPR of diapause induction, the tim-ing of diapause and the number of annual generations can vary depending on the weather conditions of a given year. The geographic variation of PhPR of diapause induction brings the seasonal cycle of each geographic population into strict correspondence with the specific traits of the local climatic conditions (**Chapter 11** and references cited therein).

A comparatively small proportion of insect species or populations develop slowly and fail to finish the preadult development within one calendar year. This is more typical of polar regions with limited thermal conditions or low temperature environments (such as caves or mountain streams). Under such conditions, insects may have **semivoltine seasonal cycles** that extend over two years or even longer.

In studies of insect seasonal cycles, of greatest interest is the geographic variation of **voltinism** and dia-pause. According to climatic zonality, the annual number of generations of most insects in the Northern Hemisphere increases from north to south. Usually, there is a regular transition from semi- or univolt-inism in the northern parts of the species' range to bi- and multivoltinism and, finally, to the homody-namic seasonal cycle in the south. In many cases, the parameters of diapause, such as intensity, duration, and tendency to form diapause, are modified in tropical populations but diapause is not completely lost (Denlinger 1986). In the Northern Hemisphere, most of the potentially multivoltine species of true bugs and other insects become univoltine or even semivoltine at the northern boundaries of their distribution. On the contrary, in the Southern Hemisphere, potentially multivoltine species switch to univoltine devel-opment towards the Southern Pole (Danks 1987, Saulich and Musolin 1996, Saulich 2010).

However, this rule has exceptions. Thus, stable annual development of more than one generation in the north of the species' range rarely is observed but has been recorded. For example, *Nezara viridula*, in Japan, has two or even three generations a year even close to its northern distribution boundary (Kiritani et al. 1963; Musolin and Numata 2003a,b; Musolin 2007, 2012; **Chapters 7 and 11**).

In this chapter, we analyze the most studied seasonal development patterns of the Pentatomoidea living in the temperate climate, attempt to reveal the eco-physiological mechanisms participating in the formation of a certain type of seasonal cycle, and estimate the similarities and differences in the seasonal development patterns of species of different taxa within this superfamily of true bugs. The avail-able published data and our own experimental material mostly focus on 3 out of 11 presently recognized subfamilies of Pentatomidae (Asopinae [= Stiretrinae], Pentatominae, and Podopinae) because only a few examples are available for other pentatomoid families. Unfortunately, the great majority of true bugs of the world have not been studied as yet in respect to their seasonal development and the control mech-anisms of their seasonal cycles.

12.2 The Univoltine Seasonal Cycle

In the univoltine seasonal cycle, only one generation develops during a vegetative season or year. Until recently, it had been assumed that the only cause of univoltinism was obligate diapause that formed in each development cycle (generation) regardless of the external conditions. However, it now is known that a univoltine seasonal cycle can be maintained in a variety of ways, but the exact eco-physiological mechanisms underlying it can be revealed only by special experiments (Saulich and Musolin 1996, Saulich and Volkovich 1996).

Univoltinism presently is subdivided into **endogenous univoltinism** (based on obligate diapause in each generation) and **exogenous univoltinism** (controlled by external factors that limit the number of generations to only one per year in different parts of the ranges of potentially multivoltine species).

12.2.1 The Endogenously Controlled Univoltine Seasonal Cycle

This seasonal cycle is characterized by obligate diapause starting invariably in each generation of the species or population, regardless of the external conditions. In the temperate latitudes, the formation of winter diapause occurs well before the drop of temperature in autumn. If a species has obligate diapause, only one generation can develop in all its populations across the entire distribution range. This is the distinctive feature of this type of seasonal cycle.

However, heterogeneity of voltinism pattern has been recorded within some species. In such cases, a population consists of two fractions, one of which has obligate diapause and produces only one generation per year under any conditions, whereas the second fraction of the same population has facultative diapause, induction of which is controlled by external conditions (Hodek 1977, Zaslavski 1988; **Chapter 11**). As a result, the first fraction of the population produces only one generation per year, whereas the second fraction might also produce a second generation if the conditions of a particular year permit this to occur.

Diapause can occur at any ontogenetic stage, but that stage is strictly species-specific (**Chapter 11**). In particular, among the various pentatomoids that have been studied, obligate **embryonic diapause** has been found in *Picromerus bidens* (L.) and *Apoecilus cynicus* (Say) (subfamily Asopinae); obligate **nymphal diapause** in *Pentatoma rufipes* (L.) (Pentatominae); and obligate **adult diapause** in *Menida disjecta* (Uhler) (= *Menida scotti* Puton), *Palomena angulosa* (Motschulsky), *Palomena prasina* (L.) (Pentatominae) and many species from Acanthosomatidae, Scutelleridae, and Cydnidae (see **Chapter 11**).

The apparent simplicity of the univoltine seasonal cycle often is combined with various adaptations, which not only help to maintain such a cycle but also ensure the proper timing of individual developmental stages with certain periods of the year that are most favorable for active development or dormancy. Such adaptations can be revealed only by special experiments. In the following sections, we review the most studied examples of the different causes and mechanisms of the univoltine cycle in pentatomoids as well as the role of additional seasonal adaptations in maintaining this cycle.

12.2.1.1 The Univoltine Seasonal Cycle Based on Obligate Embryonic Diapause

Picromerus bidens. Most authors consider the seasonal cycle of this predatory pentatomid as univoltine with obligate diapause at the embryonic stage (Leston 1955, Southwood and Leston 1959, Putshkov 1961, Javahery 1986, Larivière and Larochelle 1989). However, there are a few reports in the literature of living adult bugs being found in early spring (Schumacher 1910–1911, Butler 1923), which would be impossible if all the adults died soon after mating and oviposition in late summer or early autumn. To explain the findings of these adults in spring, a hypothesis has been proposed that assumes simultaneous realization of two patterns of seasonal development: a primary one, with **overwintering** in the embryonic stage; and a rarer secondary one, with overwintering in the adult stage (Larivière and Larochelle 1989).

Our analysis of the collections in the Zoological Institute (the Russian Academy of Sciences, Saint Petersburg) confirmed the early season findings of adult *Picromerus bidens* in various regions of Russia in April–May (**Figure 12.1**).

Phenological records demonstrate that in the temperate belt of Russia and in Ukraine, the great majority of *Picromerus bidens* pass through the final molt in July (Putshkov 1961); in Great Britain, the final molt occurs mostly in August–September (Southwood and Leston 1959, Hawkins 2003; **Figure 12.1**). Because the findings of adults in spring (**Figure 12.1**) do not fit into the phenological pattern with overwintering in the embryonic stage, Leston (1955) assumed that the overwintering adults were those infested in the late summer with tachinid flies of the subfamily Phasiinae (Diptera: Tachinidae). The possibility of overwintering in an unusual stage later was confirmed for other species of insects (Viktorov 1976, Tauber et al. 1986, Danks 1987). The presence of endoparasitoids greatly affects the seasonal development of their hosts by modifying their physiological state. Phasiinine flies infest adult bugs that appear in the late summer or autumn. Their larvae suppress development of the host's reproductive system. The infested bugs do not participate in reproduction but live longer and overwinter; thus, the parasitic tachinid larvae can overwinter inside the hosts and complete their development in the spring of the subsequent year.

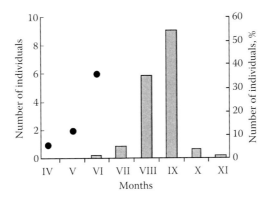

FIGURE 12.1 Findings of adults of the predatory pentatomid *Picromerus bidens* in the field in Great Britain (the right ordinate axis and columns; *n* > 300; data from D. Leston. The Entomologist's Monthly Magazine 91: 109, 1955) and early-season findings of adults in various regions of Russia (the left ordinate axis and circles; based on the records in the collections of the Zoological Institute, the Russian Academy of Sciences, Saint Petersburg; data from D. L. Musolin. Ph. D. Dissertation in Biology. Saint Petersburg, Russia, 1997; late-season records of adults in Russia are not shown). The early-season findings of adults likely represent overwintered individuals that had been parasitized by tachinid flies of the subfamily Phasiinae (Diptera: Tachinidae) in the previous late summer or autumn, did not reproduce and then die before winter, but, in fact, overwintered and survived until April–June (see more explanation in **Section 12.2.1.1**).

The seasonal development of *Picromerus bidens* under natural conditions was studied experimentally in Belgorod Province, Russia (50°N). The results allowed us to characterize the seasonal cycle of this bug in the temperate climate of the forest-steppe zone (**Figure 12.2**). The field experiment showed that nymphs of this species hatched in spring from overwintered eggs. Long days slightly retarded nymphal development. The final molt occurred in June but oviposition was delayed until the second half of August so that females laid diapausing eggs only in autumn (Musolin and Saulich 2000).

This kind of seasonal cycle is unusual for pentatomoids, and its relatively simple pattern seems to leave no place for photoperiodic control. However, it was experimentally shown that the seasonal cycle of *Picromerus bidens* included not one but two distinct dormancy periods: the obligate winter embryonic diapause and the facultative summer adult diapause (i.e., reproductive diapause, or **estivation**; see **Chapter 11** for details).

The onset of facultative summer adult diapause in *Picromerus bidens* is controlled by a **short-day type photoperiodic response of diapause induction** of the adults (**Figure 12.3**). Under photoperiodic

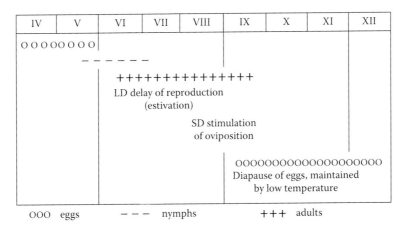

FIGURE 12.2 Seasonal development of the predatory pentatomid *Picromerus bidens* in Belgorod Province, Russia (50°N). LD is long day; SD is short day; months are indicated on the top of the figure see text for explanation. (Data from D. L. Musolin. Ph. D. Dissertation in Biology. Saint Petersburg, Russia, 1997.)

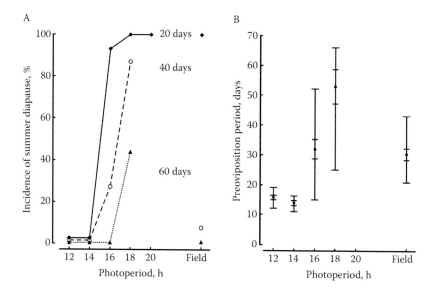

FIGURE 12.3 Photoperiodic induction of facultative summer adult diapause in females of the predatory pentatomid *Picromerus bidens*. A, Photoperiodic response of summer diapause induction under laboratory (24.5°C) and field conditions in Belgorod Province, Russia (50°N; the preoviposition period was between July 6 and August 23; the photoperiodic response was determined on days 20, 40, and 60 after the final molt); B, Duration of the preoviposition period under the same conditions (mean ± SE; min–max). (From D. L. Musolin and A. Kh. Saulich, Entomologia Experimentalis et Applicata 95: 259–267, 2000, with permission.)

regimes of 12 and 14 hours of light per day at a temperature of 24.5°C, all the females started ovipositing synchronously about day 15, on average, following the final molt, whereas under the long-day laboratory conditions and under the outdoor conditions in July (when natural day is still long), the bugs did not reproduce but entered estivation (**Figure 12.3A**), which was manifested as a significantly prolonged preoviposition period (**Figure 12.3B**; Musolin and Saulich 2000).

The adaptive significance of these long-day-stimulated delays in the nymphal growth and adults' gonad development of *Picromerus bidens* were understood only after the properties of the obligate winter embryonic diapause were studied. This diapause was not deep; at 25°C, nymphs started to hatch from the diapausing eggs 2–3 weeks after oviposition (Musolin and Saulich 2000). This means that in the field, eggs laid in the middle of summer would develop into nymphs in the summer of the same year; however, such nymphs would inevitably perish in winter because they cannot enter winter diapause and survive the cold season. Because of a 1.5–2.0-month long adult estivation period induced by the long day preceding reproduction, oviposition is shifted to the end of summer or beginning of autumn. At this time, the late-season drop of temperature suppresses embryogenesis, thus preventing the nymphs from hatching in autumn, which would be fatal for them (**Figure 12.2**).

Based on these results, we hypothesized that the weakness of the obligate winter embryonic diapause in *Picromerus bidens* (which starts irrespective of the external conditions) required the formation of an eco-physiological mechanism that would ensure a delay in oviposition until the autumn drop of temperature that, in turn, would arrest embryonic development and prevent the nymphs from hatching in autumn. This mechanism would be established in the form of facultative summer adult diapause induced by the long day. In the field, females of *P. bidens* terminate estivation and start reproducing after only 2 months of dormancy and under the conditions of naturally decreasing day length (see **Figure 12.2**; Musolin and Saulich 2000).

The univoltine seasonal cycle with winter embryonic diapause also has been described in North American species of the genus *Apoecilus* (formerly *Apateticus*) (also of the same pentatomid subfamily Asopinae): *A. cynicus* and *A. bracteatus* (**Fitch**) (as *A. crocatus* Uhler) (Whitmarsh 1916, Downes 1920, Jones and Coppel 1963, Evans and Root 1980, Javahery 1994). The winter embryonic diapause occurs in these species at the stage of the developed blastoderm and is terminated only by the action of low winter temperature (Javahery 1994). Adults of the summer generation emerge in June–July but reproduce only late in autumn. Unfortunately, the cited authors did not study the physiological state of the females in summer, so that the existence of summer adult diapause was not confirmed. At the same time, phenological observations clearly indicate the presence of obligate winter diapause at the egg stage and the development of only one generation per season.

12.2.1.2 The Univoltine Seasonal Cycle Based on Obligate Nymphal Diapause

Obligate nymphal diapause is comparatively rare in the Pentatomoidea; however, it has been recorded in a few species (see **Chapter 11 and Tables 11.1 and 11.2**).

A plataspid *Coptosoma scutellatum* (Geoffroy) occupies meadows, grasslands, and other open, well insulated and, thus, warm habitats. Its host plants are perennial legumes (Fabaceae): alfalfa, clover, broom, and many others (Putshkov 1961). The species always produces only one generation per year. Nymphs of the third and fourth instars overwinter in aggregations (Davidová-Vilimová and Štys 1982; **Figure 12.4**). For the population from Central Bohemia (The Czech Republic), it was reported that in some cases, younger nymphs (first and second instars) also can overwinter in temperature- (but not day-length-) controlled **oligopause** (i.e., a less intensive suppression of development than diapause; Davidová-Vilimová and Štys 1982). Later, however, laboratory studies of this species in the forest-steppe zone in Russia (50°N) showed that, normally, only nymphs of the third and fourth instars can overwinter in a state of true diapause (not **quiescence** or oligopause; Saulich and Musolin 1996, 2014a; Musolin 1997). Effect of environmental factors such as two temperatures (24.5 and 28.0°C) and different constant and **increasing**

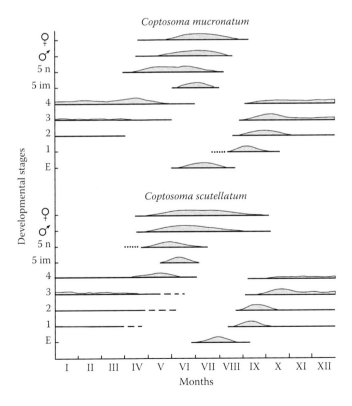

FIGURE 12.4 Seasonal presence of different life stages and instars of two plataspids – *Coptosoma scutellatum* (in Bohemia, The Czech Republic) and *C. mucronatum* (Southern Slovakia). E – eggs; 1–5 – nymphal instars; 5 im – fifth instar nymphs parasitized by the braconid *Aridelus egregius* (Schmiedeknecht); 5 n – normal (i.e., nonparasitized) fifth instar nymphs. (From J. Davidová-Vilímová and P. Štys, Acta Universitatis Carolinae Biologica 11: 463–484, 1982, with permission.)

photoperiods failed to provoke direct (i.e., nondiapause) development of nymphs: all of them formed diapause in the third and fourth instars, and none of them molted into the fifth instar before diapause (Musolin 1997). Nymphs completed metamorphosis only after overwintering in spring – they molted into the fifth instar, became adults, and continued ontogenesis. Their progeny grew slowly and entered obligate nymphal diapause as third and fourth instars.

Apparently, some nymphs reach the **diapausing stage** as early as mid-July or early August. Adaptive value of such a seasonal pattern of *Coptosoma scutellatum* is not clear. Early formation of winter diapause (in July) seems not to be efficient, as the vegetative season is not fully used for feeding and growth. Moreover, diapausing nymphs do not actively move during dormancy and, thus, are more susceptible in summer and autumn to abiotic and biotic influence than nondiapausing older nymphs and adults. Nevertheless, such dormant nymphs experience and must survive the hottest period of summer (Musolin 1997, Saulich and Musolin 2014a).

It then was suggested by Musolin (1997) that in addition to the winter nymphal diapause in the fourth instar, *Coptosoma scutellatum* might pass mid-summer in the state of summer diapause in the younger (most likely the third) instar. This hypothesis was supported indirectly by observations in the field and laboratory. Thus, under the outdoor conditions in Ukraine, first and second instars last (in total) 25–30 days and the third instar about another month (Putshkov 1961).

According to laboratory data obtained at 24.5°C, third instars took twice as long to develop under long-day L:D 18:6 conditions (on average 22.5 days) than under short-day L:D 15:9 conditions (on average 10.6 days; Musolin 1997). It is likely that this pronounced photoperiodically controlled retardation of nymphal growth during the third instar postpones molting into the fourth instar in which nymphs enter winter diapause. Moreover, in the laboratory, starting from the third instars, nymphs were reluctant to move from old food to new food, whereas younger nymphs did so quickly. In such cases, we can suppose that in the seasonal cycle of *Coptosoma scutellatum*, there are two dormancy periods: facultative summer diapause of third instars and obligate winter diapause of fourth instars. These two dormancy periods might look like a single long **summer–winter diapause** of third and fourth instars. Similar patterns are known in many insect species with prolonged diapause of an unclear nature that starts early or in the middle of summer. To better understand the nature of the dormancy and structure of the seasonal cycle of this species, additional eco-physiological research in the field and laboratory is needed.

Coptosoma mucronatum **Seedenstücker** is a second East-Palaearctic plataspid species studied in the neighboring region – southern Slovakia. Ecologically, it is similar to *C. scutellatum* and has a similar seasonal cycle (**Figure 12.4**). Older nymphs of both plataspid species can be parasitized by the braconid *Aridelus egregius* (Schmiedeknecht) (Hymenoptera: Braconidae, Euphorinae). Development of parasitized nymphs is retarded, fifth instars appear somewhat later in the season than nonparasitized nymphs (**Figure 12.4**), and these nymphs cannot become adults (Davidová-Vilimová and Štys 1982).

The forest bug, *Pentatoma rufipes*, is a typical large pentatomid that is widespread in the Palaearctic and inhabits broad-leaved forests. Detailed phenological data confirming univoltinism of this species have been obtained from southwestern England. The species apparently overwinters as young nymphs, as tiny nymphs have been seen in September and large third and fourth instar nymphs have been recorded in May. Fifth instar nymphs typically are found in June–July and adults mostly in July–September. Copulation has been recorded in August and September (Hawkins 2003). Nymphal diapause probably is obligate, which ensures the univoltine seasonal cycle of this pentatomid across its entire range (Putshkov 1961). Some cases of overwintering adults of *P. rufipes* have been recorded (Southwood and Leston 1959), but it is likely that only the adults infested with phasiinine flies (Diptera: Tachinidae) can overwinter. The bivoltine tachinid fly *Phasia hemiptera* (F.) is known to parasitize *P. rufipes* in spring and the green shield bug, *Palomena prasina*, in autumn (Sun and Marshall 2003).

Univoltine seasonal development with nymphal (and likely obligate) diapause also is recorded in the shieldbacked bugs *Odontoscelis fuliginosa* (L.), *Odontoscelis dorsalis* (F.), and *Irochrotus lanatus* (Pallas) (Putshkov 1961).

In general, seasonal cycles with overwintering nymphs are quite rare in Pentatomoidea and other true bugs. They have been recorded in no more than 1.7–11.0% of the species that have been studied, according to various sources (Saulich and Musolin 2007b; also see **Chapter 11, Table 11.2**).

12.2.1.3 The Univoltine Seasonal Cycle Based on Obligate Adult Diapause

Strictly univoltine seasonal cycle with obligate adult diapause over the entire range of a species is well documented in **the sunn pest, *Eurygaster integriceps* Puton**, a serious agricultural pest. The activity of this species is limited to only 2.5–3.0 months per year. During the remaining 9.0–9.5 months, the bugs are in dormancy, which is formed in young adults in June–July and lasts until the next spring. This prolonged dormancy of adults consists of two stages – summer diapause (i.e., estivation) and winter diapause (i.e., **hibernation**). The first stage starts after a short but intensive prediapause feeding, which is followed by migration of adults into their **estivation quarters**. In the lowland regions, forest edges, meadows, artificial tree belts, parks, and gardens can be used as estivation quarters. In mountainous regions, the bugs can fly up to 2,500–2,800 meters a.s.l. Summer diapause lasts about two months and during this time, adults lose about 20% of the reserves accumulated during prediapause preparation feeding (Ushatinskaya 1955). In autumn, as summer heat decreases, adults migrate downward and overwinter in a state of obligate winter adult diapause. Survival over a long period with such contrasting conditions is possible only because of the reserves accumulated during the prediapause feeding. In addition to the usual accumulation of fat in the fat body, *E. integriceps*, and related species, can accumulate half-digested food (such as starch) in the first section of the midgut (Fedotov 1947). These accumulated reserves provide resources not only for survival during the whole dormancy period but also for maturation of sperm during winter and the next spring (Shinyaeva 1980), a phenomenon very unusual for Heteroptera.

Maturation of males before winter, and difference in diapause development between sexes, may not be as rare as was believed in the past. The pentatomid ***Menida disjecta*** (= *M. scotti*), found in Asia, has a univoltine seasonal cycle in Japan with an obligate winter adult diapause. The physiological states of females and males, however, differ during the winter. Even though adult diapause is obligate, males already have mature sperm in autumn and copulate with females during the winter. While mating, the males transfer nutrients to females, which likely increase the females' chances of successfully overwintering. Egg maturation in females, however, starts only in spring (Koshiyama et al. 1993, 1994).

A population of the **green shield bug, *Palomena prasina*,** from the forest-steppe zone of Russia (Belgorod Province, 50°N), has been studied experimentally. In spring, after a short period of feeding, overwintered adults start ovipositing, which lasts until the end of July. The nymphs feed, grow slowly, and begin to molt to adults by mid- to late July. These adults hibernate after a period of prediapause feeding (**Figure 12.5**; Saulich and Musolin 1996).

In the laboratory, *Palomena prasina* invariably formed adult diapause at all photoperiodic and temperature regimes tested, and reproduction could not be induced by any special conditions. Based on these results, we concluded that winter adult diapause in *P. prasina* is obligate. However, day length still had a certain regulatory effect on seasonal development of the species, influencing the growth rate of the nymphs (Saulich and Musolin 1996, Musolin and Saulich 1997): the shorter the day length, the faster the nymphs grew. The difference in the growth rate became noticeable by the third instar and increased gradually, reaching the maximum in the fifth instar. This growth acceleration was recorded at all temperature regimes tested (20, 26, and 30°C), being the greatest at 20°C. At this temperature, nymphs reared under **short-day conditions** molted to adults almost 20 days earlier than those under

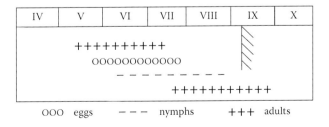

IV	V	VI	VII	VIII	IX	X

OOO eggs   − − − nymphs   +++ adults

FIGURE 12.5 Seasonal development of the green shield bug *Palomena prasina* in the forest-steppe zone (Belgorod Province, Russia; 50°N). Vertical line marks the date when the natural day length plus half the civil twilight equals 15 hours; months are indicated on the top of the figure (Data from D. L. Musolin. Ph. D. Dissertation in Biology. Saint Petersburg, Russia, 1997.)

long-day conditions. At higher temperatures, acceleration of development was less pronounced and often disappeared completely. This fact can probably be explained by the existence of a certain temperature optimum and limits of this PhPR, which is typical of any biological process (see **Chapter 11**). It is at moderate autumn temperatures that short-day acceleration of nymphal development becomes adaptive because it increases the probability that late-hatching nymphs will reach the diapausing stage (adult) and prepare for overwintering before the onset of unfavorable cold weather in autumn. The nymphs hatching in the middle of summer do not face this problem because they have enough time to reach the adult stage. Besides acceleration of development at the end of the vegetative season, photoperiodic control of the nymphal growth rate synchronizes the adult molt, shifting the appearance of the overwintering stage to the period most favorable for diapause formation (Musolin and Saulich 1997). It was fortunate that photoperiodic control of the nymphal development rate was discovered in a species in which onset of winter adult diapause was not affected by the day length.

A similar **quantitative photoperiodic control** of the nymphal growth rate was described in a closely related congeneric species, **Palomena angulosa**, in Japan (Hori 1986). The nymphs of this species, which hatch earlier in the season, have slower growth rates than those hatching later, the difference reaching 50–60%. On the one hand, such an adaptive strategy is determined by the time of fruiting of the food plants. Although the species is polyphagous and can feed not only on different species but also on different parts of plants, its nymphs cannot successfully reach the adult stage without feeding on fruits (Hori et al. 1985). On the other hand, acceleration of nymphal growth in autumn reflects the need to complete preadult development before the onset of autumn cold, since *P. angulosa*, like *P. prasina*, can overwinter only in the adult stage.

12.2.2 The Exogenously Controlled Univoltine Seasonal Cycle

Similar to multivoltinism, **exogenously controlled univoltinism** is a common type of seasonal development. Nearly all species having a potentially multivoltine seasonal cycle can (or actually do) switch to univoltine development in some parts of their ranges, for a variety of reasons.

This type of seasonal cycle is found in the predatory pentatomid **Arma custos (F.)**, which was studied in the forest-steppe zone (Belgorod Province, Russia, 50°N) where it always has only one annual generation (Saulich and Volkovich 1996). According to published data, a single generation per vegetative season was recorded not only in the northern and central regions of Europe (Chelnokova 1980) and in the forest-steppe zone of Siberia (Petrova 1975) but also in southern Ukraine (Putshkov 1961), Bulgaria (Iosifov 1981), and Kirghizia (Putshkov 1965). However, this species may have two generations in Abkhazia because its second instars were found there at the end of July (T.A. Volkovich, personal communication). In addition, a nondiapausing culture was obtained from the Krasnodar population (45°N) of *A. custos* by rearing the nymphs and adults under long-day conditions of L:D 18:6 at 28°C (Ismailov and Oleshchenko 1977). These data indicate that the onset of winter diapause in *A. custos* is externally controlled.

Indeed, experimental studies of *Arma custos* showed that induction of winter adult diapause in this species was controlled by the **PhPR of the long-day type** (Volkovich and Saulich 1995): diapause was formed under the short-day conditions (**photophase** 14 hours or shorter) at temperature of 29–30°C (**Figure 12.6, line 1**), whereas female maturation, mating, and oviposition took place under long-day conditions. However, when temperature was a few degrees lower (24–25°C), which is still relatively high for the origin of this population, incidence of diapause dramatically increased (**Figure 12.6, line 2**). In another experiment with the same species, it was shown that the range of temperatures permitting full manifestation of the PhPR of diapause induction was somewhat expanded in the case of **thermorhythm** (27°C during the photophase and 13.5°C during the **scotophase**). However, at the mean temperature of 20°C, about 60% of the females still entered diapause under long-day conditions (Volkovich and Saulich 1995).

Thus, *Arma custos* can realize the multivoltine seasonal cycle only in those regions where the temperature is high enough to prevent formation of winter adult diapause. However, in the greatest part of its range, temperature of the warmest month (July) does not reach the level at which development without diapause would be possible.

Shift to univoltinism in *Arma custos* is supported by one more specific trait that is rare among insects, namely, **sensitivity to day length** exclusively (or mostly) at the adult stage (Saulich and Volkovich 1996;

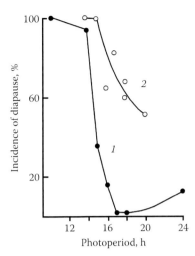

FIGURE 12.6 The photoperiodic response of adult diapause induction in the predatory pentatomid *Arma custos* from Belgorod Province, Russia (50°N). Nymphs were reared to adults and then maintained at temperature of 29–30°C (*1*) and 24–25°C (*2*) under constant photoperiodic conditions (indicated under the horizontal axis). (From T. A. Volkovich and A. Kh. Saulich, Entomological Review 74: 151–162, 1995, with permission.)

also see **Chapter 11**). Correspondingly, the physiological state of the adults in the summer generation and the possibility of diapause formation are determined by the timing of the final molt. If adults emerge when day length is below the **critical photoperiod** of diapause induction, they all enter diapause. In field experiments carried out in the forest-steppe zone, nymphs that started development on different dates always molted to adults no earlier than mid-July, when the day length was already below the critical photoperiod of that particular population. Therefore, we could not obtain nondiapause adult bugs in the same vegetative season because they all formed facultative winter adult diapause. Attempts to prevent the onset of diapause were successful in only one experimental series, when nymphal development was advanced (i.e., artificially shifted towards earlier dates) as compared with the natural **phenology** of the local population. In that case, the nymphs reared under **quasi-natural conditions** molted into adults much earlier than usually, already at the end of June. As a result, 37% of the females were nondiapausing and soon started oviposition. However, this scenario never is realized in the field. According to our observations, even in the warmest years, the first adults of *A. custos* appear in Belgorod Province no earlier than the end of July, when reproduction is already photoperiodically suppressed. Therefore, univoltinism of *A. custos* under the forest-steppe conditions is ensured not by the invariable, genetically determined onset of diapause but by the specific traits of its PhPR of diapause induction, namely its high temperature optimum and the position of the day-length sensitive stage.

The ecological reasons for reduction of the number of annual generations of *Arma custos* can be only hypothesized. On one hand, this reduction is certainly related to the temperature; the temperature conditions of the forest-steppe zone permit reliable development of only one generation of this species per year. Although the **sum of effective temperatures (SET)** in the region greatly exceeds the value needed for completion of one generation, it is not sufficient for successful development of two generations. On the other hand, this predatory bug feeds on the larvae of leaf beetles, most of which produce only one generation, at least in the forest-steppe zone where our research was carried out. It may be that the combined action of these two factors (i.e., temperature and availability of food) ultimately determine a shift of *A. custos* to the univoltine cycle in the region studied.

Another species of this small genus, ***Arma chinensis* (Fallou)**, occurs in southeast Asia, has a multivoltine seasonal cycle, and produces two generations per year in Harbin (China; 45.5°N) and three generations per year in more southern regions (Zou et al. 2012).

Among phytophagous pentatomoids with potentially multivoltine cycles, a similar transition from multi- or bivoltinism to univoltinism in response to food quality and local SET values that are insufficient for two complete generations can be observed in the well-studied **Italian striped bug, *Graphosoma*** *lineatum* (**L.**). This species also uses day length as a **signal factor** for the facultative winter adult diapause induction. Laboratory experiments showed that the winter adult diapause of *G. lineatum* was controlled by a long-day PhPR, which manifested itself at fairly high temperatures. The critical photoperiod was 17 hours 15 minutes even at a constant temperature of 24°C (Musolin and Saulich 1996). Such a high critical photoperiod induced diapause in the first generation of *G. lineatum* in the forest-steppe zone and ensured the univoltine seasonal cycle, even though the required SET value of the population was comparatively small (Musolin and Saulich 2001).

This conclusion was confirmed by an experiment under quasi-natural conditions. When, during the experiment, the timing of nymphal development corresponded to the natural phenology of *Graphosoma lineatum* in Belgorod Province of Russia (50°N), all individuals in the series entered diapause (**Figure 12.7**:

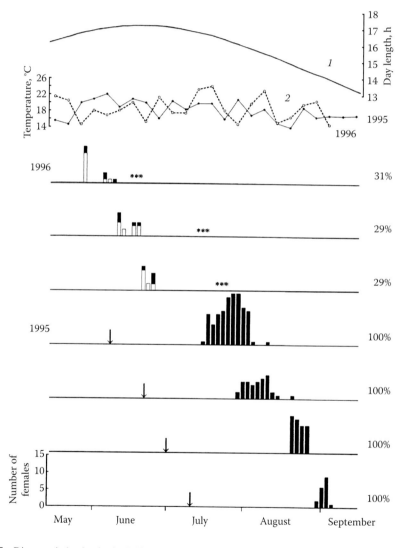

FIGURE 12.7 Diapause induction in the Italian striped bug, *Graphosoma lineatum*, under quasi-natural conditions in 1995 and 1996 in Belgorod Province, Russia (50°N). Horizontal lines: series of the experiment. Arrows: dates of nymph hatching in 1995; nymphs of second–fifth instars were transferred to quasi-natural conditions in 1996. Histograms: dates of the final molt and the physiological status of females: light columns, active (i.e., reproductive, or nondiapausing) females; shaded columns, diapausing females. Asterisks: onset of oviposition in the series of 1996. Curve: day length with half the civil twilight (*1*); broken line: temperature (*2*). (From D. L. Musolin and A. Kh. Saulich, Entomologia Generalis 25: 255–264, 2001, with permission.)

series of 1995). By contrast, in the experimental series of 1996, the dates of the final molt were artificially shifted (in the same way as in the experiments with *Arma custos* described just above) towards an early period in summer during which the natural day length exceeded the critical day length of the species. In this case, up to 70% of the females remained physiologically active and started oviposition. These results clearly confirm the role of day length in induction of facultative winter diapause in *G. lineatum* and explain the univoltine type of its seasonal development in the forest-steppe zone. The immediate cause of the shift to the univoltine seasonal cycle may be the trophic factor: the developing nymphs and prediapause adults feed on the seeds of umbellates that ripen in the second half of summer and may not be abundant later in the season.

Similar results also were obtained in experiments with the phytophagous pentatomid ***Dybowskyia reticulata* (Dallas)** in Japan (Nakamura and Numata 1998). The temperature optimum of the PhPR of adult diapause induction in this species also was shifted into the high temperature range: adequate response to day length was observed only at 27.5°C and higher temperatures, whereas a decrease in temperature even to 25°C induced diapause in all adults under long- and short-day conditions. In Osaka (Japan, 34.7°N), *D. reticulata* completes only one generation in cold years and two generations in warmer ones. The relatively low summer temperature (25°C and lower) "switches off" the physiological mechanism of response to day length, so that all the adults enter diapause regardless of the dates of their emergence. The need to limit the number of generations likely is related to the fact that *D. reticulata* is a narrow oligophage feeding on the seeds of umbellates that are available briefly in summer and shatter by the beginning of September.

Interesting results were obtained in experiments with another pentatomid ***Graphosoma rubrolineatum* (Westwood)** (Nakamura and Numata 1999), which also mostly feeds on the seeds of umbellates. Two populations of this species were studied in Japan: the northern (Hokkaido Island, 44.2°N) and the southern one (Osaka, 34.7°N). Individuals of both populations revealed a common and strong tendency to enter diapause and, as a result, their PhPRs of diapause induction look similar (**Figure 12.8**).

The difference between the populations was manifested in the action of high temperatures on the parameters of the PhPR of diapause induction: in the northern population, the fraction of reproductive females increased as the temperature grew under the long-day conditions, but no such effect was observed in the southern population. The strong influence of high temperatures on the percentage of reproductive individuals may lead to the production of a partial second generation in particularly warm years in the north of the species' range (see below), whereas in more southern regions all the first-generation adults enter diapause. It is believed that such an inversion of voltinism may be related to the availability of food. According to observations, in the Osaka region, seeds of umbellates ripen by the middle of summer and shatter by the beginning of autumn; whereas on Hokkaido Island, they remain available until late autumn so that the second-generation nymphs would not experience shortage of food.

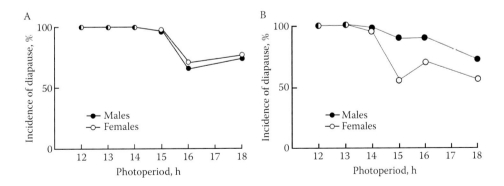

FIGURE 12.8 The photoperiodic response of facultative winter adult diapause induction in the pentatomid *Graphosoma rubrolineatum*. A, Population from Hokkaido Island, Japan (44.2°N); B, Population from Osaka, Japan (34.7°N). Nymphs were reared to adults and then maintained at 25°C under constant photoperiodic conditions (indicated under the horizontal axis). (Modified from K. Nakamura and H. Numata, Applied Entomology and Zoology 34: 323–326, 1999, with permission.)

Thus, the main factor limiting the number of generations is the absence of suitable food for the southern population in the later part of the season and the insufficient heat for the northern ones, but the **signal function** of day length is preserved in both cases (Nakamura and Numata 1999).

A somewhat more complex case of the univoltine seasonal cycle was described in the North American stink bug *Chinavia hilaris* (**Say**) [as *Acrosternum hilare* (**Say**)] in Canada (Javahery 1990). The winter adult diapause of this species is controlled by the long-day PhPR as in many other pentatomoids. Laboratory experiments showed that bugs actively reproduce at long-day conditions L:D 16:8 and form diapause at short-day conditions L:D 8:16 (Wilde 1969). Two generations of the species were observed in the southern part of its distribution in Arkansas (Miner 1966) and in southern Illinois (McPherson and Tecic 1997). In Canada, overwintered adults emerge when the temperature rises above 20°C in spring. In late May and early June, the bugs actively feed on leaves and young shoots of buckthorn, *Rhamnus cathartica* L.; white basswood, *Tilia heterophonia* L.; and clammy locust, *Robinia viscosa* Vent. (Javahery 1990). The females lay eggs from the third week of June to the beginning of August. The nymphs molt into adults from the end of August to the third week of October. Thus, in southern Canada, the species has one generation per vegetative season, and its development is shifted onto the second half of summer. The young adults feed on buckthorn fruits for 1–3 weeks and then migrate onto the leaves of deciduous trees in the forests nearby and stay there without feeding from the end of September to the end of October or beginning of November, flying over short distances in warm sunny days. The physiological nature of this late-autumn resting period was not studied experimentally, but Javahery (1990) regarded it as a summer diapause. With the onset of frost at the end of October, the bugs migrate into the litter and overwinter there until April. The shift of emergence of the new generation towards the second half of summer is probably related to the need of additional feeding of the adults on buckthorn fruits that ripen in September.

Comparatively late active seasonal development also was observed in the European representative congeneric pentatomid *Acrosternum heegeri* **Fieber** (Putshkov 1961).

Thus, in the above examples, the main factor determining the number of generations in a given population is the availability of food (i.e., a **vital factor**), whereas temperature and day length act together as signal factors. This results in a sophisticated mechanism controlling the seasonal development of each population and ensures tuned adaptation to seasonality of the local environmental conditions.

Transition from multivoltinism to univoltinism based on the presence of two facultative diapauses (summer and winter) is a rare phenomenon in the stink bug family. This seasonal pattern was discovered in *Carbula humerigera* (**Uhler**) (Kiritani 1985a,b). Adult diapause in this species is formed under the influence of **increasing day length** in late spring and early summer. Oviposition is delayed by photoperiodic conditions and it begins only after the summer solstice, when day length starts to decrease. The summer delay of oviposition (i.e., summer adult diapause) should be regarded as an adaptive mechanism that phenologically shifts the nymphal development towards a later part of the season, when the day-length conditions would stimulate the formation of a facultative winter nymphal diapause. Thus, two facultative diapauses – the winter nymphal one and the summer adult one – participate in the formation of the univoltine seasonal cycle in *C. humerigera*.

12.3 The Multivoltine Seasonal Cycle

As mentioned above, the main distinctive feature of the multivoltine seasonal cycle is that the overwintered generation gives rise to consecutive summer generations that are concluded with formation of diapause and overwintering. Depending on the external conditions (first of all, temperature and the availability of food), there may be two, three, or more such summer generations. Development of two generations per season is often designated in the literature as **bivoltinism**, and development of three generations as **trivoltinism**. In addition, there is a fairly common pattern in which at the end of the season, some part of the population gives rise to the subsequent generation whereas the other part of the population (usually the one that completes development somewhat later) enters diapause. This case is referred to as development of a **partial generation**. It should be noted that all the above examples given earlier in this paragraph are specific cases of the multivoltine seasonal cycle.

Only a few multivoltine species of pentatomoids have been studied experimentally and thoroughly analyzed in terms of the mechanisms ensuring their seasonal cycle. Detailed experimental data obtained in the laboratory are rarely supplemented with reliable records and comprehensive analysis of seasonal development in the field. In the following sections, we review only those few multivoltine pentatomoid species whose seasonal development has been analyzed based on experimental data.

12.3.1 The True Multivoltine Seasonal Cycle

As mentioned above, insects having a multivoltine seasonal cycle are characterized by the number of annual generations changing with altitude and latitude, most often under the influence of thermal conditions. Such species usually develop in one generation near the poleward boundaries of their ranges, under the conditions of heat deficit; the number of generations increases towards the equator in accordance with the growing local SET. The seasonal development and timing of winter diapause in each geographic population are adjusted to the local conditions by the specific parameters of its PhPR of diapause induction.

An example of precise agreement between the number of completed generations and the thermal and day-length conditions of a particular habitat is provided by the pentatomid *Piezodorus hybneri* **(Gmelin)**, which was studied in Japan (Higuchi 1994). This species is one of the major soybean pests in southwestern Japan where it has four annual generations. The induction of the facultative winter adult diapause of *P. hybneri* is controlled by a long-day type PhPR with the critical photoperiod between 12 and 13 hours (**Figure 12.9**). The sensitive stages are the fifth (last) instar and adult stage.

According to the laboratory data, the **lower development threshold (LDT)** of this species from egg to adult is 14.2°C, and completion of its preadult development requires a SET value of about 280 **degree-days** (283 for females and 278 for males). The LDT of adult maturation is 18.4°C, and oviposition starts upon accumulation of 70 degree-days. Thus, about 350 degree-days are needed for one complete generation of *Piezodorus hybneri* (Higuchi 1994).

Under field conditions in Kumamoto, southwestern Japan (32.9°N), oviposition by overwintered adults was observed in late April and early May; the timing of the subsequent oviposition peaks was consistent with accumulation of the SET needed for each subsequent generation (**Figure 12.10**). Thus, the thermal resources of southwestern Japan are sufficient for completion of four generations of *P. hybneri*, whereas timely induction of diapause in the last (fourth) annual generation is ensured by PhPR of diapause induction.

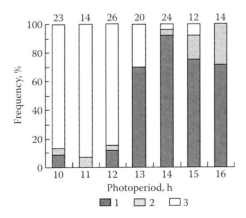

FIGURE 12.9 Effect of photoperiod on maturation of females of the pentatomid *Piezodorus hybneri* from Kumamoto, Japan (32.9°N). Nymphs were reared to adults and then maintained at 25°C under constant photoperiodic conditions (indicated under the horizontal axis). Proportions of females: 1 – mature (i.e., reproductive, or nondiapausing), 2 – with developed oocytes in ovarioles; 3 – with undeveloped oocytes (i.e., diapausing). The number of females tested is given above the columns. (From H. Higuchi, Applied Entomology and Zoology 29: 585–592, 1994, with permission.)

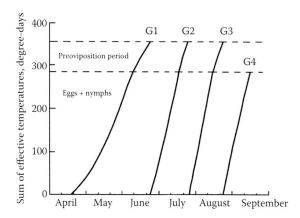

FIGURE 12.10   A scheme of seasonal development of the pentatomid *Piezodorus hybneri* in Kumamoto, southwestern Japan (32.9°N), based on the determined thermal parameters of development. G1–G4 are the consecutive generations of the species. (From H. Higuchi, Applied Entomology and Zoology 29: 585–592, 1994, with permission.)

The sloe bug, *Dolycoris baccarum* (L.), is a widespread Palaearctic pentatomid with a multivoltine seasonal cycle. Its populations from Russia (Kamenkova 1958, Perepelitsa 1971), Kazakhstan (Asanova and Iskakov 1977), the Czech Republic (Hodek 1977, Babrakzai and Hodek 1987, Hodek and Hodková 1993), Norway (Conradi-Larsen and Sømme 1973, 1978; Hodková et al. 1989), Israel (Yathom 1980), Turkey (Karsavuran 1986), and Japan (Nakamura and Numata 2006) have been studied in detail. The species produces only one generation per year in the north of its range (e.g., Norway) and two generations in the temperate latitudes (e.g., Voronezh, Russia) and in southern regions (e.g., Krasnodar in Russia; Almaty in Kazakhstan).

The local population of *Dolycoris baccarum* on Cyprus Island develops in one generation because its seasonal cycle includes summer diapause lasting from June to November. The overwintered adult bugs migrate in March from their **hibernation quarters** to the plains. In April–May, they mate, and the females lay eggs. The nymphs pass through five stadia and become adults, which migrate in June to their estivation quarters in the mountains at latitudes of about 1,300–1,500 meters a.s.l. In December, the adult bugs move to lower places (about 1,200 meters a.s.l.), where they overwinter until the end of March (Krambias 1987).

In more northern multivoltine populations of *Dolycoris baccarum*, which lack summer diapause, induction of the winter adult diapause is controlled by a long-day type PhPR: the adults actively develop at long day and enter diapause at short day. Sensitivity to day length is present in the nymphs starting from the fourth instar; however, the influence of short-day conditions only on the fifth instars and adults is sufficient for diapause induction in 100% of individuals (Perepelitsa 1971).

Geographic variation of the parameters of PhPR of diapause induction in this species was first discovered during a comparative study of its populations from Voronezh (51.7°N) and Krasnodar (45°N; Perepelitsa 1971). The difference between the values of critical photoperiod was about 1 hour (**Figure 12.11**).

In Krasnodar, emergence of adults of the first generation starts on June 20 and continues through all of July (Kamenkova 1958). Because the natural day length during this period considerably exceeds the critical day length of adult diapause induction of the local population measured in the laboratory (Perepelitsa 1971), all the adults of the first generation remain physiologically active and participate in reproduction. The critical day length including half the **civil twilights** (15 hours 30 minutes) is reached at the latitude of Krasnodar at the end of July, which is exactly when fourth instars (the day-length-sensitive stage) of the second generation emerge. The subsequent development of the second generation of *Dolycoris baccarum* proceeds under short-day conditions. By September 1, the day length decreases to 14 hours of light, which ensures the onset of diapause in all the adults.

In Voronezh, which is positioned 6.7 degrees to the north of Krasnodar, diapause formation in *Dolycoris baccarum* starts 1 hour later, at a day length of 16 hours 30 minutes, in accordance with

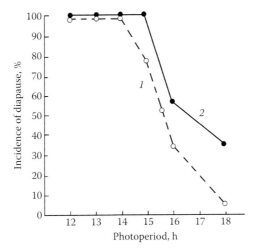

FIGURE 12.11 Geographic variation of the photoperiodic response of diapause induction in the sloe bug, *Dolycoris baccarum*. Populations tested: *1* – Krasnodar (45°N) and *2* – Voronezh (51.7°N), both in Russia. Nymphs were reared and then adults were maintained at 28°C under constant photoperiodic conditions (indicated under the horizontal axis). (Data from L. V. Perepelitsa, Byulleten' Vsesojuznogo Instituta Zashchiti Rasteniy [Bulletin of the All-Union Institute of Plant Protection] 21: 11–13, 1971.)

geographic variation of the critical day length of diapause induction (Perepelitsa 1971). Considering the civil twilights, such a day length at the latitude of Voronezh is observed on July 20; at the end of August the day length decreases to 15 hours, which induces diapause in all the bugs.

Thus, the seasonal development of *Dolycoris baccarum* populations in their natural habitats proceeds in perfect agreement with the parameters of the PhPR of diapause induction determined in laboratory experiments. The populations develop in two generations. The timing of diapause induction is determined by hereditary properties of the PhPR of diapause induction of each geographic population and corresponds to the local conditions in the best way possible. The Krasnodar and Voronezh populations of *D. baccarum* develop without summer diapause.

The existence of geographic variation of the PhPR of diapause induction in *Dolycoris baccarum* was also clearly demonstrated during a comparative study of its populations from Japan (**Figure 12.12**): those from Osaka (34.7°N) and Hokkaido Island (44.2°N). The critical photoperiod of diapause induction

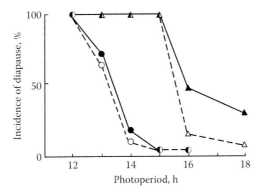

FIGURE 12.12 The photoperiodic response of facultative winter adult diapause induction in different geographic populations of the sloe bug, *Dolycoris baccarum*. Circles: population from Osaka (34.7°N), triangles: population from Hokkaido Island (44.2°N), both in Japan; black symbols: males; light symbols: females. Nymphs were reared to adults and then maintained at 25°C under constant photoperiodic conditions (indicated under the horizontal axis). (From K. Nakamura and H. Numata, Applied Entomology and Zoology 41: 105–109, 2006, with permission.)

response of the Osaka population was about 13 hours 30 minutes at 25°C. The day length (with half the civil twilights) in Osaka at the beginning of September is 13 hours; this value determined the onset of facultative winter adult diapause (Nakamura and Numata 2006).

Many species of true bugs studied in this region of Japan have similar critical photoperiods of winter diapause induction and produce three generations per season (see **Section 12.5.1** below). Based on this, the cited authors assumed that *Dolycoris baccarum* also developed in three generations in the Osaka region, although, according to other data (Kobayashi 1972), this species has a bivoltine seasonal cycle there.

The critical photoperiod for the Hokkaido population was about 16 hours, which permitted development of two complete generations (**Figure 12.12**). The difference in the critical photoperiod of diapause induction between the populations from Osaka and Hokkaido was more than 2 hours, indicating a clinal geographic variation (**Figure 12.12**). The critical photoperiod changed by about 1 hour every 5 degrees of latitude, similar to what had been determined earlier for many insect species (Danilevsky 1961, Saulich and Volkovich 2004). According to the earlier cited data of Perepelitsa (1971) for Voronezh and Krasnodar, the critical day length of European populations also varied within 1 hour per 5 degrees of latitude.

The temperature-related variation of the long-day PhPR of diapause induction usually is manifested in the following way: the proportion of diapausing individuals decreases in all the photoperiods as the temperature rises. Such variation was observed in the European populations of *Dolycoris baccarum* from Norway (Conradi-Larsen and Sømme 1973) and The Czech Republic (Hodek and Hodková 1993). By contrast, in the Japanese populations, the number of diapausing individuals decreased noticeably only under the long-day conditions L:D 16:8 whereas at short-day conditions L:D 12:12, the number of diapausing bugs did not decrease even when the temperature rose to 27.5°C and 30°C (**Figure 12.13**). The effect of the temperature was noticeable only in the duration of diapause, which was much shorter at 30°C than at 25°C in the bugs of both populations under the short-day conditions (Nakamura and Numata 2006).

Stink bugs of the genus ***Aelia*** have been well studied because of their economic importance as cereal pests. In particular, the development of two species, ***A. acuminata*** (**L.**) and ***A. sibirica*** **Reuter**, was studied in the north of Kazakhstan (Kustanai Province, 53°N) where both species can produce two generations per vegetative season (Burov 1962). Despite slight differences in the parameters of the PhPR of diapause induction (**Figure 12.14**), the seasonal cycles of the two species are noticeably different. *A. sibirica* shows a greater tendency for univoltinism, with more than 30% of adults of the first generation forming winter adult diapause even in the most favorable years.

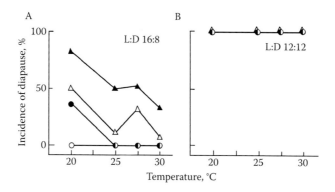

FIGURE 12.13 The effect of temperature on winter diapause induction in the populations of the sloe bug, *Dolycoris baccarum*, from Osaka (34.7°N) and Hokkaido Island (44.2°N), both in Japan, under long-day (A) and short-day (B) conditions. Nymphs were reared to adults and then maintained at four different temperatures (indicated under the horizontal axis) under two constant photoperiodic conditions – long day L:D 16:8 and short day L:D 12:12. Open circles, females of the Osaka population; closed circles, males of the Osaka population; open triangles, females of the Hokkaido population; closed triangles, males of the Hokkaido population. (From K. Nakamura and H. Numata, Applied Entomology and Zoology 41: 105–109, 2006, with permission.)

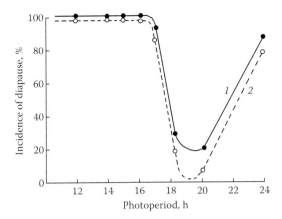

FIGURE 12.14 Photoperiodic responses of winter adult diapause induction in females of two pentatomids from the northern Kazakhstan (53°N): *Aelia sibirica* (*1*) and *Aelia acuminata* (*2*). Nymphs were reared to adults and then maintained at 25°C under constant photoperiodic conditions (indicated under the horizontal axis). (From V. N. Burov, Entomologicheskoe Obozrenie 41: 262–273, 1962, with permission.)

The SET required for completion of the developmental cycle is 480 degree-days in *Aelia sibirica* and 396 degree-days in *A. acuminata*. According to a lower SET value (by 84 degree-days), the final molt in the first generation of *A. acuminata* occurs almost a week earlier, and the great majority of adults remains reproductive (i.e., nondiapause). Under the conditions of Kustanai Province, the SET exceeding the LDT of *A. acuminata* (13°C) barely exceeds 900 degree-days; therefore, even a slight decrease in temperature will result in a situation when only some nymphs of the second generation (mostly those hatching from early egg clusters) have sufficient time to develop into adults and properly form diapause, whereas the remaining individuals cannot prepare for overwintering and gradually die out in winter (Burov 1962). Due to its greater tendency for univoltinism, *A. sibirica* is better adapted to the conditions of North Kazakhstan. Because adults of this species have enough time to feed before winter diapause, they overwinter successfully and remain permanently abundant in the study region (Burov 1962).

The population of *Aelia acuminata* occurring in the southwest of Slovakia also shows a considerable tendency towards univoltinism that is based, however, not on facultative but on obligate winter adult diapause. Individuals of this population are highly heterogeneous with respect to voltinism. In the greater part of the population, diapause occurs invariably in each generation whereas in the other part, diapause is facultative. However, in Central Europe, the species usually produces only one generation even though the critical photoperiod of the PhPR of diapause induction in the fraction with facultative diapause was found to be much lower than in the North Kazakhstan population: between 15 and 16 hours of light (Honěk 1969, Hodek and Honěk 1970). The female life span in this species is relatively long, reaching two months under laboratory condition. Hodek (1977) felt the females do not fully realize their reproductive potential within one vegetative season and probably after reproduction, they can form a repeated diapause and survive until the next year to reproduce again.

For this possibility to be realized, insects have to preserve the ability to evaluate day length during most of their life. It is well known that photoperiodic sensitivity after winter diapause varies between species, even among the phytophagous species with facultative winter adult diapause (Nakamura and Numata 1995, Saulich and Musolin 2007b; see **Chapter 11, Section 11.6.1**). Some species lose photoperiodic sensitivity irreversibly whereas others, in particular *Aelia acuminata*, lose sensitivity in autumn or winter but restore it in early summer after a short refractory period in spring (Hodek 1971). This ability allows the insects to form diapause more than once during their life and to switch to the multiyear adult stage with repeated reproduction (i.e., semivoltine cycle; see **Section 12.4**).

Unlike *Aelia acuminata*, **Aelia fieberi Scott** does not preserve photoperiodic sensitivity after winter diapause but loses it irreversibly so that after overwintering, the adult bugs continue ovipositing until the end of their lives (Nakamura and Numata 1995). In the warm climate of Japan, the period favoring the activity of *A. fieberi* is long enough for this species to produce two complete generations. Adults of the first generation reproduce actively in summer and die after oviposition. Adults of the second generation form winter diapause in autumn under the influence of short-day conditions (Nakamura and Numata 1997). With such a seasonal cycle, there seems to be no need for repeated overwintering; therefore, restoration of photoperiodic sensitivity becomes unnecessary.

According to laboratory data, the critical photoperiod of diapause induction of *Aelia fieberi* at 25°C is 14 hours 30 minutes; it determines winter diapause induction as early as the beginning of August, when the temperature conditions would still be favorable for further activity and development of the third generation. Experiments carried out under field conditions showed that the nymphs developed successfully under field temperatures until October, and the adults that emerged under such conditions could overwinter successfully. Thus, the temperature of the autumn months did not suppress development of the potential third generation. However, it was determined that voltinism of this species also was controlled by a trophic factor: the seeds that ripened late proved to be unsuitable for feeding of the nymphs. As a result, under the influence of natural selection, *A. fieberi* in central Japan produces only two generations per season, whereas the day length remains the principal control cue of its seasonal cycle (Nakamura and Numata 1997).

***Aelia rostrata* Boheman** also has a facultative winter adult diapause (Şişli 1965), but its seasonal cycle studied in Turkey clearly differs from those of other species of the genus *Aelia*. At the beginning of May, when the temperature rises above 20°C, the overwintered adults migrate from their mountain overwintering sites onto the plains, mostly to cereal fields. After 10–15 days of intense feeding, they mature and start oviposition. Preadult development of the new (first) generation takes about a month, so that young adults of the summer generation appear in the second half of June and enter summer diapause. In the study region, *A. rostrata* produces only one generation per season (Babaroğlu and Uğur 2001). As the cereals ripen and the summer temperatures rise, the young adults migrate to the mountain estivation (i.e., summer diapause) quarters positioned at 1,200–2,000 meters a.s.l., where they gradually enter hibernation (i.e., winter diapause) with the onset of cold in autumn. In other species having a similar seasonal cycle, namely *Dolycoris baccarum* (Krambias 1987) and shieldbacked bugs of the family Scutelleridae (Arnoldi 1947, Brown 1962), the summer and winter diapauses are separated by short active periods during which time the adults fly short distances and migrate from one dormancy place to another (Krambias 1987). There is no evidence yet of such migrations between summer and winter diapauses in *A. rostrata* (Cakmak et al. 2008).

12.3.2 The Strictly Bivoltine Seasonal Cycle

The strictly bivoltine seasonal cycle can be regarded as a special case of multivoltinism, characterized by development of only two generations per year, one in spring (or early summer) and one in autumn (or late summer). The two periods of activity usually are separated by summer diapause (i.e., estivation), which ensures survival during the unfavorable period in mid-summer with high temperatures and scarce food. This happens every year, even though the temperature resources would usually be sufficient for a greater number of generations. Obligate or facultative estivation occurs in summer under the conditions of high temperatures and long day. This frequently is observed in species living in the tropical and subtropical belts and is typical of the inhabitants of deserts and semi-deserts, even though it is quite common in the temperate latitudes as well. Estivation can be linked to any stage of development, from egg to adult, but the estivating stage is strictly species-specific (Masaki 1980; see **Chapter 11** for details).

This pattern of seasonal development first was studied in the cabbage moth, *Mamestra brassicae* L. (Lepidoptera, Noctuidae), which has only two generations even in the extreme south of Japan, although, theoretically, it would be able to produce up to six generations per season (Masaki and Sakai 1965).

Among the pentatomids, an example of a species with a strictly bivoltine seasonal cycle is **the Oriental green stink bug, *Nezara antennata* Scott**.

There are two species of the genus ***Nezara*** distributed in Japan: *N. antennata* and *N. viridula*. Of these, *N. antennata* is known only from Asia (Hokkanen 1986, Rider 2006; see **Chapter 7**). This species is widespread in Japan, its range covering the islands of Okinawa, Kyushu, Shikoku, Honshu, and

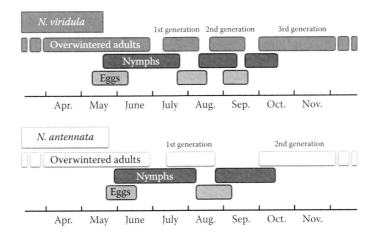

FIGURE 12.15　Seasonal development of *Nezara viridula* and *N. antennata*, in central Japan (Data from K. Kiritani, N. Hokyo and J. Yukawa, Researches on Population Ecology 5: 11–22, 1963.)

Hokkaido (Tomokuni et al. 2003; Rider 2006; Musolin 2007, 2012; Ferrari et al. 2010). *Nezara viridula* has a much wider distribution, almost cosmopolitan, and is continuously expanding its range; in Japan, it overlaps with the southern part of the range of *N. antennata* (Kiritani et al. 1963, Tougou et al. 2009, Musolin 2012; see **Chapter 7**).

Both species, but especially *Nezara viridula*, are broad polyphages, feeding on more than 197 species from more than 43 families of plants (Todd 1989, Panizzi et al. 2000; see **Tables 7.4 and 7.5**). They occupy similar ecological niches and occur sympatrically; in some cases, *N. viridula* has replaced *Nezara antennata* almost completely in a mere several years (Kiritani et al. 1963; Kiritani 1971; Kon et al. 1994; Musolin 2007, 2012; Yukawa et al. 2007, 2009; Tougou et al. 2009).

Nezara antennata, in central Japan, has only two generations per year and does not form a third generation (even in exceptionally warm years) unlike *N. viridula*, which can produce three generations per year (**Figure 12.15**). Laboratory experiments (Noda 1984) have shown that adults of *N. antennata* maintained at a long or increasing day length did not quickly start oviposition but formed summer diapause: the preoviposition period under such conditions was more than 40 days at 25°C, as compared with about 20 days in the adults transferred from long-day to short-day conditions. Field experiments also have demonstrated that induction of estivation in the middle of summer resulted in a delay of oviposition. The estivating adults did not change their color from green to brown whereas during winter diapause, the color of adults did change, both in this species and in *N. viridula* (D. L. Musolin, unpublished data). After the end of estivation, the adults started to reproduce and gave rise to the second generation, which formed facultative winter adult diapause under the short-day conditions (Numata and Nakamura 2002). Thus, *N. antennata* has one generation in spring and one in autumn; the summer adult diapause allows the species to avoid exposure to excessive heat. Such pattern might be adaptive as it has been shown that the species is not heat-tolerant and that up to 68% of the nymphs died even at 30°C (Kariya 1961).

12.3.3 The Partially Bivoltine Seasonal Cycle

A seasonal function of summer diapause is quite different in **the clown stink bug, *Poecilocoris lewisi* (Distant),** another pentatomoid species well studied in Japan (Tanaka et al. 2002). Nymphs of *P. lewisi* feed on seeds of the dogwood *Cornus controversa* Hemsley (Cornaceae), which gain their full weight and ripen only at the beginning of July.

This shieldbacked bug is **partially bivoltine** and has two facultative diapauses: winter nymphal diapause and summer adult diapause. Both of them are under photoperiodic control and both have critical photoperiods of diapause induction close to 14 hours 30 minutes (**Figure 12.16**). Short day induces winter diapause in the fifth instar and long day induces fairly short adult summer diapause manifested

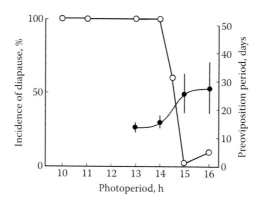

FIGURE 12.16 Two photoperiodic responses of the clown stink bug, *Poecilocoris lewisi*, from Osaka (34.7°N), Japan. Induction of facultative winter nymphal diapause: open circles, left vertical axis; induction of facultative summer adult diapause in females: closed circles and right axis (preoviposition period is shown as mean ± SD). Nymphs were reared to adults and then maintained at 25°C under constant photoperiodic conditions (indicated under the horizontal axis). (Modified from S. I. Tanaka, C. Imai, and H. Numata, Applied Entomology and Zoology 37: 469–475, 2002, with permission.)

as an about 20-day-long delay in gonads' maturation in adults in summer. Both diapauses terminate spontaneously: nymphal diapause terminates in early spring after overwintering and adult diapause terminates in mid-summer (Tanaka 2002, Tanaka et al. 2002).

Overwintered nymphs of *Poecilocoris lewisi* molt to the adult stage in May. Then, in response to the long-day conditions at that time, all females enter summer diapause and start oviposition only after termination of this estivation in July, when seeds of dogwood become available (**Figure 12.17**). Thus, summer adult diapause synchronizes emergence of nymphs of the bug's first summer generation with the period when food is available and fresh. Earlier nymphs of this first summer generation reach the

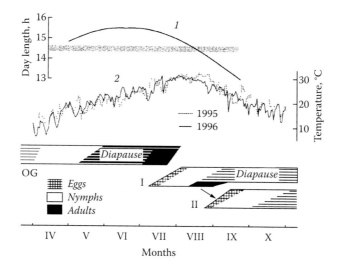

FIGURE 12.17 Seasonal development of the clown stink bug, *Poecilocoris lewisi*, in Osaka (34.7°N), Japan. The upper panel shows the day length including 1 hour of the twilight period (*1*) and daily mean temperatures recorded at an outdoor experimental cage in 1995 and 1996 (*2*). The shaded bar shows the critical day length for the induction of nymphal and adult diapauses. The lower panel shows the possible pathways of the species' life cycle in Osaka. OG, overwintered generation; I is the first generation (full); II is the second generation (partial); horizontal hatching shows periods of diapauses in nymphs and adults of different generations; an arrow shows that the second generation (partial) is produced by the non-diapausing fraction of the first generation. (Modified from S. I. Tanaka, C. Imai, and H. Numata, Applied Entomology and Zoology 37: 469–475, 2002, with permission.)

fifth instar when the day is still long (exceeding 14 hours 30 minutes) and, thus, they do not enter winter nymphal diapause but develop directly and quickly molt to the adult stage. These adults of the first summer generation also develop directly and start oviposition without entering any dormancy because the day is already too short for induction of the summer adult diapause at this time (<14 hours 30 minutes). Their progeny (i.e., eggs and newly emerged nymphs) belong to the partial second generation. In response to short day, nymphs of this generation enter diapause when they reach the fifth instar and overwinter.

In contrast to the earlier nymphs of the first summer generation, the later nymphs of the same generation reach the fifth instar after mid-August. Under already short-day conditions (<14 hours 30 minutes) these nymphs enter facultative winter nymphal diapause and overwinter together with the fifth instars of the partial second generation (**Figure 12.17**).

Thus, in *Poecilocoris lewisi*, the facultative summer adult diapause removes critical pressure of the trophic factor by postponing the beginning of reproduction until the period when food is available for the progeny. At the same time, this summer diapause does not prevent realization of the partial second generation (**Figure 12.17**; Tanaka et al. 2002).

12.4 The Semivoltine (Perennial) Seasonal Cycle

Perennial seasonal cycles are widespread in the Insecta and have been found in representatives of many orders. They are formed for different reasons and result from different modifications of univoltine seasonal cycles. Transition to this strategy can be realized in various ways, for example: (1) retardation of preadult development, (2) inclusion of a long diapause in the life cycle, and (3) extension of the adult life span with several periods of reproduction (Danks 1992, Saulich 2010). There must be other, still undescribed, models of transition to the semivoltine cycle.

So far, semivoltine seasonal cycles have not been found in pentatomoids although, potentially, they may exist. As mentioned above, many species, in particular the pentatomids *Dolycoris baccarum*, *Eurydema rugosum*, and *Graphosoma lineatum*, are known to preserve or restore photoperiodic sensitivity after the first overwintering. Because of this ability, they can, potentially, form diapause more than once during the individual life cycle and probably switch to the semivoltine seasonal cycle. Therefore, it is quite possible that, eventually, such a strategy will be discovered in some representatives of the large and ecologically diverse superfamily Pentatomoidea.

12.5 The Significance of Photoperiodic and Thermal Responses for Expansion of Insects Beyond Their Natural Distribution Ranges

The results of many experiments indicate that exact correspondence between seasonal development and local conditions creates a serious obstacle to the free movement of insects even within the species' range (Danilevsky and Kuznetsova 1968, Saulich 1999, Volkovich 2007). However, for various reasons (e.g., as the result of accidental introduction, intended introduction of biological control agents, and climate changes), insects often enter into new habitats and must adapt or perish (Musolin 2007, Musolin and Saulich 2012b). Detailed analysis of specific examples allows one to determine the cause of success or failure of invasion or introduction and, in some cases, to predict the possibility of naturalization of certain species outside their original ranges.

12.5.1 Natural or Accidental Invasions—Case Studies of the Southern Green Stink Bug, *Nezara viridula*, and Brown Marmorated Stink Bug, *Halyomorpha halys*

A convenient model species for such analysis is **the southern green stink bug, *Nezara viridula***. This pentatomid is characterized by diffuse range expansion. It must have originated in the Ethiopian region of Africa and dispersed first into Asia and relatively recently into Europe and the American supercontinent (Kavar et al. 2006). In the middle of the 19th century, *N. viridula* appeared on the southern islands of Japan, from whence it spread to the north and reached the environs of Osaka (34.7°N) by the end of the 20th century. At the beginning of the 21st century, it advanced still further northward (Tomokuni et al. 1993; Musolin 2007, 2012; Yukawa et al. 2007, 2009; Musolin and Saulich 2011; Geshi and Fujisaki 2013; see **Chapter 7** for details).

In the warm temperate climate of central Japan, *Nezara viridula* usually has three annual generations (**Figure 12.15**), although part of its population seems to be able to produce an incomplete fourth generation (Kiritani and Hokyo 1962).

In February, the overwintered adults start changing color from brown/russet to intermediate and by March–April, most of them acquire green (or yellow) coloration typical of the reproductive state. Mating starts in April, and oviposition starts in April–May (Kiritani et al. 1963, Musolin et al. 2010, Takeda et al. 2010, Musolin 2012). Adults of the first generation appear in July and soon give rise to the second generation, which reaches the adult stage in the second half of August. Adults of the third generation appear at the end of September. Most of them enter winter diapause and only a minor fraction of adults reproduces in autumn (**Figure 12.18**; Kiritani et al. 1963, Musolin 2012).

Induction of the facultative winter adult diapause in *Nezara viridula* is controlled by a long-day PhPR with a critical photoperiod of about 12 hours 30 minutes at temperatures of 20°C and 25°C (Musolin and Numata 2003a). Considering the fact that this species has reached the environs of Osaka only recently, Musolin and Numata (2003b) carried out a field experiment to determine how well its population has adapted to the local climatic conditions. Attention was focused on the dates of emergence of adults and formation of facultative winter adult diapause because these phases of the seasonal cycle often determine the possibility of naturalization of invasive insects in temperate regions.

The natural day length in Osaka (without the civil twilights) is 12 hours 53 minutes on September 1, 12 hours 27 minutes on September 15, and 11 hours 48 minutes on October 1. All females of *Nezara viridula* that emerged before September 1 were reproductively active and laid eggs. Among females that emerged in the second half of September, approximately 60% entered facultative winter diapause (**Figure 12.18; series 2 and 3**). The remaining approximately 40% of females were reproductive but up to 60% of their progeny were destined to die during winter because of low temperature (Musolin and Numata 2003b). In females that emerged in the first half of October (**Figure 12.18, series 4**), incidence of diapause reached 100%. Thus, in the outdoor experiment, diapause was induced very late in the season, but it was in a good agreement with the critical photoperiod (12 hours 30 minutes) estimated in the laboratory. If compared to several native true bug species studied earlier (**Figure 12.19**), *N. viridula* had a somewhat lower critical photoperiod and, correspondingly, the majority of females of this species formed diapause somewhat later in the season (e.g., only ≈75% at the beginning of October).

According to experimental data (Musolin and Numata 2003b), the fate of the nymphs of *Nezara viridula* hatching from the eggs laid in autumn varied. Only the nymphs that hatched before mid-September could successfully complete preadult development, whereas those that hatched after that time died during winter; the later they hatched, the earlier the stage at which they encountered the lethal low winter temperatures. From the eggs laid in mid-September, adults emerged only in November. Because they developed under late-autumn short-day conditions, these adults did not reproduce. Although these adults had some chance of surviving until spring, they could not get properly prepared for overwintering: they did not have enough time to change their body color and to accumulate sufficient energy resources (Musolin and Numata 2004). The formation of a complete diapause state in true bugs usually takes considerable time. For example, according to our observations, preparation for diapause in another pentatomid *Podisus maculiventris* requires at least 17–19 days at a moderate temperature of about 20°C (Saulich and Musolin 2011, 2012).

Adults of *Nezara viridula* entering diapause are known to turn brown or russet soon after the final molt (**Figure 12.18**; Musolin and Numata 2003a, Musolin 2012; see **Chapters 7 and 11**) and to remain brown/russet until the complete termination of winter diapause in the spring. The change of body coloration is controlled by day length and correlated with the physiological state of the individual (Harris et al. 1984, Musolin and Numata 2003a, Musolin 2012; see **Chapters 7 and 11**). However, the change of body color in the adults that emerged in another field experiment in November proceeded slowly during winter and ended as late as the end of March (Musolin and Numata 2003b; see **Chapter 11**). Moreover, 20% of individuals still remained green even at that time. The temperature during the period of emergence

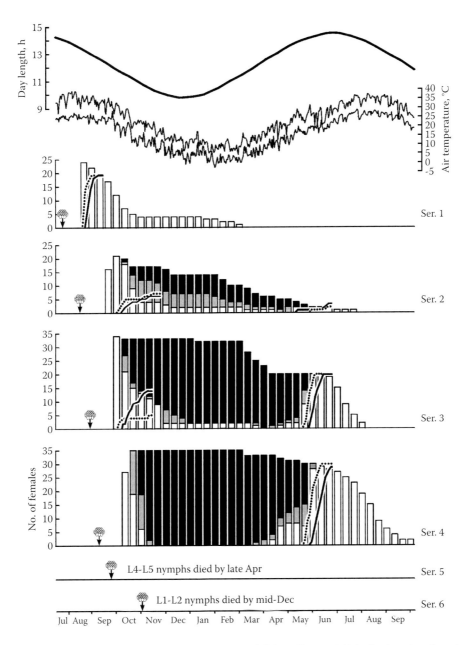

FIGURE 12.18 Seasonal development of the southern green stink bug, *Nezara viridula*, in six series of quasi-natural experiments performed in Osaka (34.7°N), Japan. Arrows: dates when egg clusters were transferred to field conditions. Histograms: number and coloration of females (nymphs and males not shown). Dotted lines: total number of mating females; solid lines: number of ovipositing females. Light columns: green adults; shaded columns: adults with intermediate coloration; filled columns: brown or russet adults. L1–L5 are nymphal instars. The natural conditions: curve: day length; broken lines: daily minima and maxima of temperature. (From D. L. Musolin and H. Numata, Ecological Entomology 28: 694–703, 2003, with permission.)

probably was sufficient for the prediapause feeding but too low for the physiological processes associated with the change of body color (Musolin and Numata 2003b).

Thus, we are dealing here only with the initial stage of expansion of *Nezara viridula* into the central part of Japan. To become completely and successfully established in this region, the population should undergo adaptive changes in the PhPR parameters that determine the dates of induction of the winter

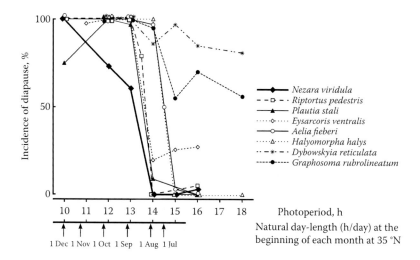

FIGURE 12.19 The photoperiodic responses of facultative winter diapause induction in females of several seed-feeding true bug species in Osaka and the nearest regions of Japan. Nymphs were reared to adults and then maintained at 25°C under constant photoperiodic conditions (indicated under the horizontal axis). Details and references: *Nezara viridula* (Osaka, 34.7°N, 135.5°E; Musolin and Numata 2003a); *Riptortus pedestris* (= *R. clavatus*) (Kyoto, 35°N, 135.8°E; Kobayashi and Numata 1993); *Plautia stali* (Tawaramoto, 34.5°N, 135.8°E; Numata and Kobayashi 1994); *Eysarcoris ventralis* (Izumo, 35.4°N, 132.8°E; Noda and Ishi 1981); *Aelia fieberi* (Osaka, 34.7°N, 135.5°E; Nakamura and Numata 1997); *Halyomorpha halys* (Kobe, 34.7°N, 135.3°E; Niva 2003); *Dybowskyia reticulata* (Osaka, 34.7°N, 135.5°E; Nakamura and Numata 1998); *Graphosoma rubrolineatum* (Osaka, 34.7°N, 135.5°E; Nakamura and Numata 1999). All of the species belong to the family Pentatomidae, except *R. pedestris* (Alydidae). (From D. L. Musolin, Global Change Biology 13: 1565–1585, 2007, with permission.)

adult diapause. These changes most likely would affect the critical photoperiod: its value would probably increase, so that diapause would be induced earlier (i.e., before mid-September). This would prevent both nonadaptive reproduction in October–November and incomplete diapause formation in the late-autumn adults. The current late diapause formation not only has direct negative effects on the population (such as high adult mortality during overwintering and low reproductive potential after diapause) but also influences the viability of the spring generation (Musolin and Numata 2003b).

An essentially similar but oppositely directed process was observed in Japan in the fall webworm, *Hyphantria cunea* Drury (Lepidoptera: Arctiidae). The population of this pest species was introduced accidentally into Japan in 1945 and established in the territory where only two generations could be completed per year. At the end of the 20th century, a population was discovered that had switched to a trivoltine seasonal cycle due to changes in its eco-physiological traits, particularly the lowering of the critical photoperiod of the winter pupal diapause induction (Gomi 1997, Musolin and Saulich 2012a).

The unique adaptive ability to disperse and naturalize outside of its natural range has been demonstrated by **the brown marmorated stink bug, *Halyomorpha halys* (Stål)**. The natural range of this pentatomid covers southeastern Asia (China, Japan, North Korea, and South Korea; see **Chapter 4**). However, in the mid-1990s, the species was found in the United States of America (Pennsylvania). Since then, it has actively spread across the continent and has been registered in 40 states and territories of the United States of America and Canada (Hoebeke and Carter 2003, Lee et al. 2013). Recently, the species also has been reported in different locations in Europe (Wermelinger et al. 2008, Haye et al. 2014, Milonas and Partsinevelos 2014). However, the invasive North American and European populations differ in their genetic composition, and it is believed that the species invaded North America from the Beijing region (China) and Europe from another (not yet determined) region of Asia (Gariepy et al. 2014, Xu et al. 2014).

Halyomorpha halys attracts serious attention as it is widely polyphagous and can feed on and, thus, damage more than 300 species of both wild and cultivated plants. Moreover, as the species apparently prefers to overwinter in large aggregations in houses and other constructions, which buffer unfavorable low temperatures, it can become a serious nuisance pest (Hoebeke and Carter 2003, Nielsen and Hamilton 2009, Nielsen et al. 2011, Inkley 2012, Cambridge 2015; see **Chapter 4**).

In different parts of its native range in Asia, *Halyomorpha halys* produces one or two generations per year (Lee et al. 2013). In laboratory experiments with the bugs from the Nagano population (Japan; 36.5°N), Yanagi and Hagihara (1980) have demonstrated that induction of the facultative winter adult diapause is controlled by day length. At 25°C under a day length of 14 hours 45 minutes or shorter, adults entered winter diapause, whereas all adults were reproductive if the day was 15 hours 30 minutes or longer (**Figure 12.20**).

When the nymphs were reared to adults and then maintained under semi-natural outdoor conditions in Nagano, it was demonstrated that adults that had emerged after July 15–25 did not reproduce but began to enter facultative winter adult diapause (**Figure 12.21**). At this time of the year, natural day length is similar to the critical photoperiod determined under laboratory conditions plus 30 minutes that usually are reserved for the civil twilights because, as had been shown experimentally, many insect species recognize the civil twilights as a part of photophase (i.e., a light part of the day cycle; Goryshin and Geispitz 1975).

Thus, even early studies of the seasonal development of *Halyomorpha halys* within its natural range clearly demonstrated that this species has the day-length-controlled facultative winter adult diapause, which allows it to produce different number of annual generations in different parts of its native range depending on the local temperature or other environmental conditions. Later, it also was demonstrated

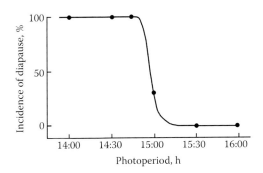

FIGURE 12.20 Effect of photoperiod on induction of the facultative winter adult diapause in the brown marmorated stink bug, *Halyomorpha halys*, from Nagano (36.5°N), Japan. Nymphs were reared to adults and then maintained at 25°C under constant photoperiodic conditions (indicated under the horizontal axis). (From T. Yanagi and Y. Hagihara, Shokubutsu-Boeki [Plant Protection] 34: 315–321, 1980, with permission.)

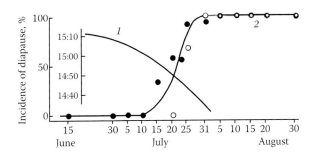

FIGURE 12.21 Diapause induction in the brown marmorated stink bug, *Halyomorpha halys*, under natural day length in Nagano (36.5°N), Japan. Curve *1* is natural day length in Nagano + 30 minutes of the civil twilight (inserted vertical axis); curve *2* is incidence of facultative winter adult diapause (black symbols are at outdoor temperature; white symbols are at room temperature). (From T. Yanagi and Y. Hagihara, Shokubutsu-Boeki [Plant Protection] 34: 315–321, 1980, with permission.)

that older nymphs and adults are more sensitive to day length than younger nymphs (Niva 2003, Niva and Takeda 2003).

In the bivoltine seasonal cycle exhibited in southern China, adults of the first generation of *Halyomorpha halys* emerge in late June or early July when local photoperiodic conditions do not limit the possibility of a second generation because day length during adult emergence is longer than the critical photoperiod (Lee et al. 2013). As a result, adults of this generation are reproductive. In contrast, adults of the second generation emerge in late August or early September and enter facultative winter adult diapause because the natural day length at that time is already short.

The seasonal cycle of *Halyomorpha halys* also is well studied within its invasive range. In Pennsylvania and New Jersey, the United States of America, only one generation per year is recorded. In these regions, the SET for a full generation is approximately 700 degree-days with the LDT of 14°C. The peak of abundance of the bug was recorded in August, when the accumulated SET is approximately 800–1,000 degree-days, thus determining univoltine seasonal development in this part of the invasive range (Nielsen and Hamilton 2009, Leskey et al. 2012). At this time of year, natural day length becomes short, strongly limiting further active development of the species and promoting induction of facultative winter adult diapause.

In another part of the invasive range – in Switzerland, Europe – *Halyomorpha halys* is strictly univoltine (Haye et al. 2014). Adults of the summer generation emerge in mid-August when natural day length can be as short as 14 hours 15 minutes. The critical photoperiod of the PhPR of diapause induction of the local population is not known but it likely is close to the one estimated for the Nagano population (15 hours). Short-day conditions apparently prevent reproduction of these adults and promote induction of the winter adult diapause. To ensure successful realization of two complete generations, adults of the first generation should emerge no later than in late June when the day is longer than 15 hours; this does not happen under natural conditions.

The available body of information on distribution and seasonal development of *Halyomorpha halys* in its invasive range does not allow one to confidently predict the further secondary range of this dangerous invader. Zhu et al. (2012) suggested that *H. halys* potentially could become established in nearly all parts of Europe, but southern Europe would be the most favorable. Recently, it has been found in northern Italy (Haye et al. 2014) and Greece (Milonas and Partsinevelos 2014). It is likely that the species will continue to spread at least throughout the Mediterranean area. Temperature conditions in southern Europe would clearly be more favorable for its development, which can result in production of two or even more annual generations in many regions.

12.5.2 Intentional Introductions—Case Studies of the Spined Soldier Bug, *Podisus maculiventris*, and Twospotted Stink Bug, *Perillus bioculatus*

Two North American predatory pentatomids, *Podisus maculiventris* and *Perillus bioculatus* (F.), can serve as examples of intentional introductions of true bugs into several European countries that were imported with the sole purpose of controlling two invasive and economically important pests, the fall webworm, *Hyphantria cunea* (Lepidoptera: Arctiidae), and the Colorado potato beetle, *Leptinotarsa decemlineata* Say (Coleoptera: Chrysomelidae), respectively. Both phytophagous pests have natural ranges more or less similar to those of the two predatory bugs, and both pests became successfully established in Europe in the 20th century. Based on the similarity of these ranges, it was expected that the predatory bugs also would become successfully established on a new continent. However, fates of these two introduced pentatomids in Europe differ substantially. Extensive research programs resulted in a large body of eco-physiological information, which improved our understanding of the possibilities of introductions and naturalization of pest and useful insects.

Availability of suitable food, a SET sufficient for at least one annual generation, and ability to survive cold season (usually in a state of diapause) are the main conditions that determine the possibility and success of any invasion or introduction to a new region. If at least these three conditions are met, then the initial invasion/introduction and survival are possible. For further naturalization, the seasonal cycle of the species should become well adapted to and coordinated with local environmental conditions.

As discussed in **Chapter 11**, day length plays a dual signal function: (1) it acts as a trigger controlling seasonal changes of eco-physiological states of an organism (e.g., induction or termination of diapause and reproduction) and (2) also is used for synchronization of seasonal development of different components of the ecosystem (e.g., between a phytophagous insect and its host plant or between a parasitoid and its insect host). Usually, both these functions are well coordinated. However, when organisms are transferred to distant locations, coordination between the seasonal rhythm of the organisms and environmental conditions could be seriously disturbed or successful establishment even prevented in the new location. In such cases, diapause might still be formed, but it could happen too late or too early in the season and, as a result, organisms might fail to form a proper diapause. In other words, day length might still play a signal role inducing physiological change but fail to do it in appropriate time, i.e. to successfully synchronize the seasonal development of the organism with its new environmental conditions (Tyshchenko 1983).

Introduction of **Podisus maculiventris** into Europe was not particularly successful. Development of one generation of this polyphagous predatory bug requires a SET of about 400 degree-days above the LDT of 11°C (Saulich 1995, De Clercq 2000). The hibernating adults survive cooling down to −15°C (Borisenko 1987). It was believed that all these traits would allow the predator to become naturalized in the greater part of Europe where it would produce from one generation in the north to four generations in the south, depending on the local temperature conditions. However, the parameters of the PhPR of adult diapause induction of the population originating from Missouri, the United States of America (38°N), could not ensure the timely formation of winter diapause in Europe and, thus, the survival of the species in the temperate climate remains doubtful.

Indeed, photoperiodic sensitivity of *Podisus maculiventris* starts in the third instar (see **Chapter 11**), which means that for proper induction of facultative winter adult diapause, the nymphs starting from the third instar should experience day length below the critical value (13 hours 30 min). Development of fourth and fifth instars takes 14 to 16 days at 20°C. In addition, according to experimental data, the prediapause feeding period of adults lasts 17 to 19 more days until complete formation of diapause. Thus, this species needs at least a month of favorable temperatures (not lower than 15 to 16°C) from the date of the critical photoperiod until the onset of diapause (Volkovich et al. 1991b, Saulich and Musolin 2011).

The critical photoperiod for diapause induction in *Podisus maculiventris* (13 hours 30 min including the civil twilights) is reached approximately on the same date (September 23 to 25) at all altitudes over Eastern Europe. Due to this fact, the zone of possible acclimation of this species, for example, in Russia can be determined easily. The regions favorable for the pentatomid should have a long and warm autumn, with the temperatures of October exceeding 15 to 16°C. In Russia, such conditions occur only in the extreme south of Krasnodar Territory. For example, in Sochi (43.5°N), the mean daily temperature drops below 12°C on October 30, whereas the SET above 11°C for the vegetative period is about 2,000 degree-days. Such a SET value would support development of four generations of *P. maculiventris*, the last of which would experience the photo-thermal conditions inducing adult diapause. The thermal requirements of diapausing adults also would be met in this region. Thus, *P. maculiventris* can acclimate successfully on the Black Sea coast of Krasnodar Territory but not further north. More northern native populations of this species, for example, those from southern Canada, probably would have a greater potential for naturalization in Europe (Saulich 1995).

The critical importance of the timing and duration of the day-length-sensitive stage clearly was demonstrated in the comparative analysis of seasonal development outside of the natural ranges of two species, *Podisus maculiventris* (population originated from North America, 38°N) and *Riptortus pedestris* Thunberg [= *R. clavatus* (F.)] (Heteroptera, Alydidae; population originated from Japan, 35°N). These two species are similar in many eco-physiological traits such as a SET of one full generation and facultative winter adult diapause controlled by a thermostable PhPR of diapause induction with the critical photoperiod of approximately 13 hours. However, in terms of the sensitive stage, induction of diapause in *P. maculiventris* is determined by the conditions experienced mostly by nymphs starting from the third instar and, to a lesser degree, by adults; whereas in *R. pedestris*, diapause induction is more flexible and faster, being controlled by the conditions experienced by fourth and fifth instars but mostly by adults (Numata 1985, 1990).

Ability of these two species to enter diapause outside their natural ranges was tested in field experiments under quasi-natural conditions in the forest-steppe zone in Russia (50°N). In accordance with the temperature conditions of the region, both species potentially can produce two generations per year with emergence of adults of the second generation in late August and early September. Natural day length at this time of the year already is shorter than the critical day length for winter adult diapause induction in both species.

In *Riptortus pedestris*, adults can measure day length and respond accordingly, even if the nymphs develop under different photoperiodic conditions. Thus, when adults emerge in late August, they respond to natural day length as short days and, after a comparatively brief period of prediapause feeding under still warm conditions, they successfully enter facultative winter adult diapause.

In *Podisus maculiventris*, for diapause to be induced, the short-day signals must be received by nymphs beginning with the third instar; it is not enough if the short-day signals are perceived only by adults. Thus, the adults emerging under short-day conditions in late August already are programmed for nondiapause development because as nymphs, they developed under long-day conditions. Photoperiodic sensitivity of the adult stage is not sufficient to override the response formed during the nymphal stage, especially because the period when it is still warm is short. The process of diapause induction in this species starts early (from the third instar) and takes longer than in *Riptortus pedestris*: adults would need a month to properly form diapause. Temperature in late August and in September is definitely low for prediapause feeding of adults of *P. maculiventris* and, thus, adults of the second generation emerging in late August fail to properly form diapause (Musolin 1997).

Seasonal development of **Perillus bioculatus**, another predatory pentatomid, was studied in field experiments in the southwest of Slovakia (about 45°N) during five seasons (Jasič 1975). The experiments were carried out with bugs originally brought from Canada (Ontario Province, 45°N) and maintained in culture for many years in various European countries. Results showed that under Central European conditions, this species could complete two or three generations per year, depending on the temperature, and form facultative winter adult diapause. Because of sufficient cold hardiness (down to −12°C), the adults can overwinter in shelters under the snow cover.

However, it was not clear whether the synchronizing role of day length during the diapause induction period was preserved after introduction of the species into Europe. As discussed above and in **Chapter 11**, an ecological factor can act as an external cue or as an inductor of a certain physiological state, both components being equally important. Some examples are known in which the inductor function works properly (i.e., diapause as a physiological state is formed correctly) but the signal function is impaired (i.e., the seasonal timing of diapause is wrong; Tyshchenko 1980).

To answer this question, parameters of the PhPR of diapause induction of this population of *Perillus bioculatus* were studied in the laboratory (Volkovich et al. 1991a). The critical photoperiod changed from 15 hours 30 minutes to 14 hours 30 minutes as the temperature rose from 24°C to 27°C (**Figure 12.22; lines 1 and 2**). Furthermore, the critical photoperiod value decreased noticeably under the thermorhythm conditions: it was approximately 14 hours 30 minutes under natural thermorhythm with amplitude 13.7°C (scotophase) to 26.4°C (photophase) and the mean temperature 19.1°C (**Figure 12.22; line 3**). Proceeding from these data, it can be assumed that the critical photoperiod of the PhPR of diapause induction (14 hours 30 min, considering the natural temperature dynamics) was reached in the study region in Slovakia by August 20 when the temperature dropped to 19°C and induction of winter adult diapause started. However, it is known that complete formation of diapause requires not only a single short-day signal but also accumulation of a so-called **packet of photoperiodic information** (Goryshin and Tyshchenko 1972, Tyshchenko 1977), or the **required day number** (Saunders 1976), which is a species-specific number of short days required to trigger diapause (also see **Chapter 11**).

In most species studied, this packet of photoperiodic information consists of 18–20 short days. The filling of the packet of photoperiodic information usually is accompanied by a long preparation for diapause. Therefore, winter diapause induction in *Perillus bioculatus* has to be finished by the end of September, which indeed was observed in the southwest of Slovakia (Jasič 1975). Thus, the multivoltine seasonal cycle realized in the relocated populations remains under the control of the same physiological responses that regulate the seasonal development of the species in its native range.

Extensive programs of acclimation of *Perillus bioculatus* also were conducted in the Soviet Union in the 1960–1970s, but the numerous attempts yielded no practical results. The main reason for the failure of acclimation of this pentatomid in the new territory was a mismatch between the emergence of the predator and its prey in spring after overwintering, which already had been observed in the experiments of Jasič (1975). In spring, the overwintered adults emerged earlier than their prey and, being narrow

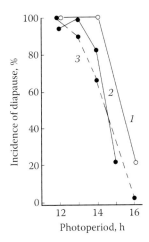

FIGURE 12.22 The effect of temperature on the photoperiodic response of diapause induction in the predatory pentato-mid *Perillus bioculatus* from the southwest of Slovakia (about 45°N). Nymphs were reared to adults and then maintained under constant photoperiodic conditions (indicated under the horizontal axis) at constant temperatures of 24°C (open circles and solid line *1*), 27°C (closed circles and solid line *2*) or under the natural thermorhythm with amplitude 26.4°C (photophase) to 13.7°C (scotophase) and the mean temperature of 19.1°C (closed circles and dashed line *3*). (From T. A. Volkovich, L. I. Kolesnichenko, and A. Kh. Saulich, Entomological Review 70: 68–80, 1991, with permission.)

oligophages, they faced a shortage of food. The research programs finally were stopped due to lack of positive applicable results.

However, examination of wild thickets of the common ragweed *Ambrosia artemisiifolia* L. (Asteraceae) in Krasnodar Territory in May 2008 revealed numerous nymphs of *Perillus bioculatus* (with field density up to 20 individuals/m²) that were feeding actively on different stages of the ragweed leaf beetle, *Zygogramma suturalis* F. (Coleoptera: Chrysomelidae) (Ismailov and Agasieva 2010). These nymphs probably originated from occasional individuals released in the course of the early research programs that had survived since that time as small inconspicuous populations in some favorable habitats. Later, when additional suitable prey species appeared (in particular, the ragweed leaf beetle), the pentatomids became naturalized and dispersed over the south of Russia. According to Ismailov and Agasieva (2010), *P. bioculatus* also fed on the larvae of the ragweed moth, *Tarachidia candefacta* Hübner (Lepidoptera: Noctuidae); this fact contradicts the previous characteristic of this species as a narrow oligophage. The ragweed moth is a species new to Europe, which also was introduced intentionally into Russia in the 20th century to control common ragweed; the prospects of its naturalization as an agent of ragweed control were studied on the Black Sea coast of the Caucasus (Nayanov 1973, Esipenko 2012). As a result of this introduction event (and possibly some other cases of invasion), the ragweed moth has become a common species in the steppe zone of the south of Russia and Ukraine (Klyuchko et al. 2004, Klyuchko 2006, Poltavsky and Artokhin 2006).

For *Perillus bioculatus*, deficiency of food resources after the end of overwintering apparently became a serious obstacle during the initial stage of introduction. Emergence of a new food source in the form of *Tarachidia candefacta* and *Zygogramma suturalis* removed the limiting action of the trophic factor. Further establishment of *P. bioculatus* proceeded successfully because of the good match of the Canadian population' parameters of the PhPR of diapause induction to the photo-thermal environmental conditions of southern Russia and Ukraine: *P. bioculatus* can successfully enter facultative winter adult diapause in a timely manner in this part of its secondary range.

12.6 Conclusions

Detailed consideration of the published data on the seasonal cycles of insects in general, and Pentatomoidea in particular, is impeded by the scarcity of experimental data that are needed for proper analysis of the seasonal development of individual species and populations. Still, the available original and published data allow us to discuss the reasons and regularities of formation of the various seasonal patterns within Pentatomidae and some other related families of true bugs and the eco-physiological responses involved in their formation.

Among the 82 pentatomoid species analyzed (see **Chapter 11, Table 11.2**, for the list), the great majority of species is characterized by the potentially multivoltine seasonal cycle with a facultative winter adult diapause regulated by the PhPR of diapause induction of the long-day type. This property is realized in the polyphagous species occurring in the regions with moderate humidity and temperatures at middle and low latitudes, given the optimal trophic conditions and the choice of host plants (for phytophages) or prey (for predators). The number of realized annual generations largely depends on local thermal conditions, whereas day length acts as a cue and provides information necessary for optimal timing of active development and dormancy in the given location.

However, under actual natural conditions in the Temperate Zone, few species of pentatomoids produce more than one annual generation. The number of realized generations often is reduced in much of the species' range. Thus, for example, populations living at high and middle latitudes are usually univoltine. The main factors limiting the number of generations are food and temperature. Transition to univoltine development often is observed in the climatic belts where thermal conditions exceed the requirements of one generation but are not sufficient for completion of two generations within one vegetation season. To stop the series of nondiapausing generations at the only stage capable of overwintering, the population requires a reliable cue to warn it of the impending changes in the environment. The day length usually acts as such a cue for cessation of active development and transition to seasonal dormancy. The exact eco-physiological mechanisms involved in each particular case can be determined only in special experiments: they may include an increase in the thermal optimum of the PhPR of diapause induction (as in *Arma custos*), a high critical photoperiod (as in pentatomids of the genus *Graphosoma*), and likely other responses ensuring the timely formation of diapause.

The univoltine seasonal cycle based on obligate diapause is quite rare in Pentatomoidea but still has been found in species with different types of winter diapause: embryonic (e.g., in *Apoecilus cynicus, A. bracteatus,* and *Picromerus bidens*), nymphal (e.g., in *Pentatoma rufipes*), and adult (e.g., in *Palomena prasina* and *P. angulosa*). The onset of diapause usually corresponds strictly to the period to which diapause is adapted; otherwise, overwintering would be unsuccessful. Precise seasonal timing of diapause induction can be ensured in different ways. Species with embryonic and nymphal diapause have acquired adaptations slowing down the preadult development, so that the only stage capable of overwintering can be shifted onto the autumn period. This is accomplished by inclusion of facultative adult estivation into the seasonal cycle (e.g., in *P. bidens*), the result of which shifts oviposition to a later period, more favorable for the overwintering eggs. In species with winter adult diapause (e.g., in the genus *Palomena*), the nymphal growth rate is controlled by day length: decelerated under long-day conditions and accelerated under short-day conditions. This seasonal adaptation solves the opposite problem, allowing the only overwintering stage (the adult) to appear before the autumn drop of temperature when preparation to diapause would be difficult. The above seasonal adaptations have the same effect: the diapausing stage is formed during a specific period of the year. In both cases, the formation of obligate winter diapause is determined hereditarily in each generation, but the exact timing of appearance of the diapausing stage is completely controlled by day length.

Adaptation of invasive species to new conditions is difficult, regardless of the type of seasonal cycle. Polyphagous species with a univoltine seasonal cycle based on obligate diapause, which is to some extent independent of the environmental conditions, adapt more easily. The species or populations with a multivoltine cycle, whose seasonal development is controlled by external conditions and strictly depends on the local climate, face the greatest difficulties when relocated. One of the main obstacles to naturalization is a mismatch between the parameters of the PhPR of diapause induction of the introduced population and the new climatic conditions. If this obstacle is overcome, the phases of the life cycle become synchronized with the periods of the year to which they are adapted.

The same seasonal adaptations participate in formation of different types of seasonal cycle, but their occurrence can vary between taxa. In particular, photoperiodic regulation of the growth rate is used both in univoltine (obligate or exogenously controlled) and in multivoltine cycles, but this adaptation seems to be more functional in the case of univoltinism. In the stink bug subfamily Pentatominae, which mainly unites the potentially multivoltine species, this adaptation was found more often in representatives of the tribes Carpocorini and Eysarcorini. The responses to day length can be totally opposite even within one species: some populations of a species can accelerate their development under short-day conditions and others, under long-day conditions. This trait is manifested at the population level and ensures a high degree of adaptation of local populations to their living conditions.

Summer diapause is seldom included in the seasonal cycle of Pentatomoidea but has been discovered and experimentally studied in several species. In the arid climate, summer diapause is more likely to be obligate (as in *Aelia rostrata*). The onset of estivation accompanied by migrations allows true bugs to survive high summer temperatures. This seasonal strategy is similar to the well-studied seasonal cycle of Scutelleridae. In the temperate latitudes, summer diapause ensures synchronization of the phases of univoltine seasonal development with local environmental conditions and usually is controlled by day length (e.g., in *Picromerus bidens* and *Carbula humerigera*).

At the present stage of research of seasonal cycles, it would be premature to make any conclusions about the occurrence of particular types of seasonal development in different taxa of stink bugs. Phylogenetic reconstruction based on morphological characters has no predictive value as concerns the pattern of seasonal development of a particular species or its populations. For example, in Pentatomidae, *Andrallus spinidens* (F.) and *Picromerus bidens* (both Asopinae, Platinopini) belong to two sister genera (**Figure 12.23**; Gapon 2008) but have clearly different seasonal cycles: a multivoltine cycle mostly regulated by the temperature in *A. spinidens* and a univoltine cycle based on obligate embryonic and facultative adult diapauses in *P. bidens*.

The species *Apoecilus cynicus* and *Podisus maculiventris* from the tribe Amyotini of the pentatomid subfamily Asopinae are morphologically similar but belong to different clades (**Figure 12.24**; Gapon 2008). They also have highly different seasonal development patterns: obligate univoltinism in

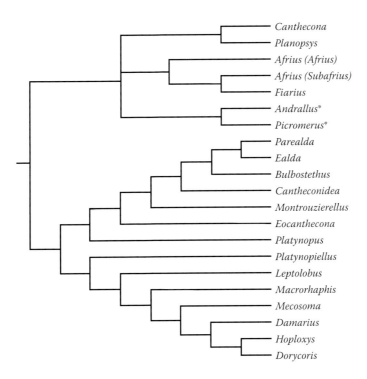

FIGURE 12.23 Phylogeny of the pentatomid tribe Platinopini Gapon 2008. The genera *Andrallus* Bergroth and *Picromerus* Amyot et Serville are indicated by asterisks. (From D. A. Gapon. A taxonomic review of the world fauna of stink bugs [Heteroptera: Pentatomidae] of the subfamilies Asopinae and Podopinae. Ph. D. Dissertation in Biology. Saint Petersburg, Russia, 2008.)

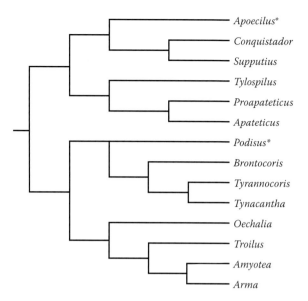

FIGURE 12.24   Phylogeny of the pentatomid tribe Amyotini Gapon 2008. The genera *Apoecilus* Stål and *Podisus* Herrich-Schäffer are indicated by asterisks. (From D. A. Gapon. A taxonomic review of the world fauna of stink bugs [Heteroptera: Pentatomidae] of the subfamilies Asopinae and Podopinae. Ph. D. Dissertation in Biology. Saint Petersburg, Russia, 2008.)

A. cynicus and photoperiodically regulated multivoltinism in *P. maculiventris*. On the other hand, the seasonal cycles of *Picromerus bidens* and *A. cynicus* are similar though these species belong to different pentatomid tribes. There seem to be more such examples, but the data on the seasonal development of most species are absent or fragmentary. Nevertheless, the type of seasonal cycle does not appear to be related to the taxonomic position of the species.

Analysis of the available material clearly shows that a certain seasonal cycle may not be typical of the whole species but may be characteristic for particular populations, being completely determined by the specific living conditions of each population. In view of this, it is not totally correct to apply the terms **multivoltinism** or **univoltinism** to a species because the same species may be represented by univoltine populations under some conditions and multivoltine populations under others, depending on the geographic position. Therefore, these terms should probably refer to the seasonal cycle of a population in a particular region. The species as a whole can be described as *potentially multivoltine*. The term **univoltine** should be used only in those cases where the species is known to complete only one generation within its entire range, or where it has been shown experimentally that nondiapause development of a series of generations cannot be induced without some special techniques.

In general, even though the study of seasonal cycles of Pentatomoidea and their control is needed both for practical reasons and for solving theoretical issues, this field of the true bugs' biology is still far from being well-explored.

12.7 Acknowledgments

We are grateful to Professor J. E. McPherson (Southern Illinois University, Department of Zoology, Southern Illinois University, Carbondale) for careful and patient editing of the chapter and Dr. Tatiana A. Volkovich (Department of Entomology, Saint Petersburg State University, Russia) for permission to cite her personal communication. The research in different years was supported by the Council for Grants of the President of the Russian Federation; State Support of the Leading Scientific Schools (project No. 3332.2010.4); the Inessa Charitable Foundation; and the Ministry of Education, Culture, Science, Sports and Technology of Japan (Grants-in-Aid for JSPS Fellows No. 98116 and L-4562 and STA Fellow No. 200141 and via The 21st Century COE Program at Kyoto University).

12.8 References Cited

Alekseev, V. R. 1990. Diapause of crustaceans: the ecophysiological aspects. Nauka, Moscow. 144 pp. (in Russian with English summary)

Arnoldi, K. V. 1947. The sunn pest (*Eurygaster integriceps*) in the Central Asian wildlife as related to the ecological and biocenological aspects of its biology, pp. 136–269. *In* D. M. Fedotov (Ed.), The sunn pest *Eurygaster integriceps* Put. Volume 1. Publishing House of The Academy of Sciences of the USSR, Moscow. 272 pp. (in Russian)

Asanova, R. B., and B. V. Iskakov. 1977. Harmful and beneficial bugs of Kazakhstan. Kainar, Alma-Ata. 204 pp. (in Russian)

Babaroğlu, N., and F. Uğur. 2001. Investigation on fecundity of cereal bug *Aelia rostrata* Boh. (Heteroptera: Pentatomidae) on some wheat and barley varieties. Plant Protection Bulletin 41: 1–16. (in Turkish with English summary)

Babrakzai, Z., and I. Hodek. 1987. Diapause induction and termination in a population of *Dolycoris baccarum* (Heteroptera, Pentatomidae) from Central Bohemia. Věstník Československé Společnosti Zoologické 51: 85–88.

Belozerov, V. N. 2007. Distribution of dormant stages in the life cycles of Acari (Chelicerata: Arachnida) in comparison with mandibulate arthropods (insects and crustaceans), pp. 193–233. *In* A. A. Stekolnikov (Ed.), Adaptive strategies of terrestrial arthropods to unfavorable environmental conditions: a collection of papers in memory of Professor Viktor Petrovich Tyshchenko (Proceedings of the Biological Institute of Saint Petersburg State University 53). 387 pp. (in Russian with English summary)

Belozerov, V. N. 2012. Dormant stages and their participation in adjustment and regulation of life cycles of harvestmen (Arachnida, Opiliones). Entomological Review 92: 688–714. http://dx.doi.org/10.1134/S0013873812060073.

Borisenko, I. A. 1987. Dynamics of cold tolerance during development of the spined soldier bug *Podisus maculiventris* Say. Byulleten' Vsesojuznogo Instituta Zashchiti Rasteniy [Bulletin of the All-Union Institute of Plant Protection] 69: 12–16. (in Russian)

Brown, E. S. 1962. Researches on the ecology and biology of *Eurygaster integriceps* Put. (Hemiptera, Scutelleridae) in Middle East countries, with special reference to the overwintering period. Bulletin of Entomological Research 53: 445–514. http://dx.doi.org/10.1017/S0007485300048252.

Burov, V. N. 1962. Factors affecting population dynamics and harmfulness of wheat stink bugs of the genus *Aelia* (Heteroptera, Pentatomidae). Entomologicheskoe Obozrenie 41: 262–273. (in Russian with English summary)

Butler, E. A. 1923. Biology of British Hemiptera-Heteroptera. H. F. and G. Witherby, London. viii + 682 pp.

Cakmak, O., M. Bashan, and E. Kocak. 2008. The influence of life-cycle on phospholipid and triacylglycerol fatty acid profiles of *Aelia rostrata* Boheman (Heteroptera: Pentatomidae). Journal of the Kansas Entomological Society 81: 261–275. http://dx.doi.org/10.2317/JKES-709.11.1.

Cambridge, J., A. Payenski, and G. C. Hamilton. 2015. The distribution of overwintering brown marmorated stink bugs (Hemiptera: Pentatomidae) in college dormitories. Florida Entomologist 98: 1257–1259. http://dx.doi.org/10.1653/024.098.0442.

Chelnokova, T. A. 1980. An ecological characteristic and biotopic distribution of predatory bugs of the family Pentatomidae (Hemiptera–Heteroptera) in the forest zone of the Middle Volga basin. Trudy Kuibyshevskogo Pedagogicheskogo Instituta [Proceedings of the Kuibyshev Teacher's Institute] 243: 82–86. (in Russian)

Conradi-Larsen, E. M., and L. Sømme. 1973. Notes on the biology of *Dolycoris baccarum* L. (Het., Pentatomidae). Norsk Entomologiskl Tidsskrift 20: 245–247.

Conradi-Larsen, E. M., and L. Sømme. 1978. The effect of photoperiod and temperature on imaginal diapause in *Dolycoris baccarum* from southern Norway. Journal of Insect Physiology 24: 243–249. http://dx.doi.org/10.1016/0022-1910(78)90042-2.

Danilevsky, A. S. 1961. Photoperiodism and seasonal development of insects. Leningrad University Press, Leningrad, the USSR. 243 pp. (in Russian). (Translated into English by J. Johnson assisted by N. Waloff, Oliver & Boyd, Edinburgh, UK, 1965. 284 pp.).

Danilevsky, A. S., and I. A. Kuznetsova. 1968. Intraspecific adaptations of insects to climatic zonality, pp. 5–51. *In* A. S. Danilevsky (Ed.), Photoperiodic adaptations in insects and acarines. Leningrad University Press, Leningrad, the USSR. 270 pp. (in Russian with English summary)

Danks, H. V. 1987. Insect dormancy: an ecological perspective. Biological Survey of Canada, Ottawa, Canada. x + 399 pp. (Monograph Ser. No. 1).

Danks, H. V. 1992. Long life cycles in insects. The Canadian Entomologist 134: 167–187. http://dx.doi .org/10.4039/Ent124167-1.

Davidová-Vilímová, J., and P. Štys. 1982. Bionomics of European *Coptosoma* species (Heteroptera, Plataspidae). Acta Universitatis Carolinae Biologica 11: 463–484.

De Clercq, P. 2000. Predaceous stinkbugs (Pentatomidae: Asopinae), pp. 737–789. *In* C. W. Schaefer and A. R. Panizzi (Eds.), Heteroptera of economic importance. CRC Press LLC, Boca Raton, FL. 828 pp.

De Clercq, P., and D. Degheele. 1993. Cold storage of the predatory bugs *Podisus maculiventris* (Say) and *Podisus sagitta* (Fabricius) (Heteroptera: Pentatomidae). Parasitica 1–2: 27–41.

Denlinger, D. L. 1986. Dormancy in tropical insects. Annual Review of Entomology 31: 239–264. http:// dx.doi.org/10.1146/annurev.en.31.010186.001323.

Downes, W. 1920. The life history of *Apateticus crocatus* Uhl. (Hemiptera). Proceedings of the Entomological Society of British Columbia 16: 21–27.

Esipenko, L. P. 2012. A new method of the ragweed (*Ambrosia artemisiifolia* L.) inhibition in southern Russia. Nauchniy Zhurnal KubGAU [Scientific Journal of the Kuban State Agrarian University] 79(05): 1–11. (in Russian with English summary)

Evans, E. W., and R. B. Root. 1980. Group molting and other lifeways of a solitary hunter, *Apateticus bracteatus* (Hemiptera: Pentatomidae). Annals of the Entomological Society of America 73: 270–274.

Fedotov, D. M. 1947. Changes in the internal state of adults of the sunn pest (*Eurygaster integriceps* Put.) during a year, pp. 35–80. *In* D. M. Fedotov (Ed.), The sunn pest *Eurygaster integriceps* Put. Volume 1. Publishing House of The Academy of Sciences of the USSR, Moscow, Leningrad. 272 pp. (in Russian)

Ferrari, A., C. F. Schwertner, and J. Grazia. 2010. Review, cladistic analysis and biogeography of *Nezara* Amyot & Serville (Hemiptera: Pentatomidae). Zootaxa 2424: 1–41. http://dx.doi.org/10.11646/%25x.

Gapon, D. A. 2008. A taxonomic review of the world fauna of stink bugs (Heteroptera: Pentatomidae) of the subfamilies Asopinae and Podopinae. Ph.D. Dissertation in Biology. Saint Petersburg State University, Saint Petersburg, Russia. 178 pp. + 798 pp. (Supplement). (in Russian)

Gariepy, T. D., T. Haye, H. Fraser, and J. Zhang. 2014. Occurrence, genetic diversity, and potential pathways of entry of *Halyomorpha halys* in newly invaded areas of Canada and Switzerland. Journal of Pest Science 87: 17–28. http://dx.doi.org/10.1007/s10340-013-0529-3.

Geshi, Y., and K. Fujisaki. 2013. Northward range expansion of *Nezara viridula* in Kinki District, Japan: expansion speed. Japanese Journal Applied Entomology and Zoology 57: 151–157. http://dx.doi.org/10 .1303/jjaez.2013.151. (in Japanese with English summary)

Gomi, T. 1997. Geographic variation in critical photoperiod for diapause induction and its temperature dependence in *Hyphantria cunea* Drury (Lepidoptera: Arctiidae). Oecologia (Berlin) 111: 160–165. http:// dx.doi.org/10.1007/s004420050220.

Goryshin, N. I., and K. F. Geispitz. 1975. Some urgent problems of phenological analysis in entomology. Zoologicheskii Zhurnal [Zoological Journal] 54: 895–912. (in Russian)

Goryshin, N. I., and G. F. Tyshchenko. 1972. Experimental analysis of photoperiodic induction in insects. Trudy Biologicheskogo Instituta Leningraskogo Gosudarstvennogo Universiteta [Proceedings of the Biological Institute of the Leningrad State University] 21: 68–89. (in Russian)

Hairston, N. G., Jr., and C. E. Cáceres. 1996. Distribution of crustacean diapause: micro- and macroevolutionary pattern and process. Hydrobiologia 320(1–3): 27–44. http://dx.doi.org/10.1007/BF00016802.

Harris, V. E., J. W. Todd, and B. G. Mullinix. 1984. Color change as an indicator of adult diapause in the southern green stink bug, *Nezara viridula*. Journal of Agricultural Entomology 1: 82–91.

Hawkins, R. D. 2003. Shieldbugs of Surrey. Surrey Wildlife Trust, Surrey, UK. 192 pp.

Haye, T., S. Abdallah, T. Gariepy, and D. Wyniger. 2014. Phenology, life table analysis and temperature requirements of the invasive brown marmorated stink bug, *Halyomorpha halys*, in Europe. Journal of Pest Science 87: 407–418. http://dx.doi.org/10.1007/s10340-014-0560-z.

Higuchi, H. 1994. Photoperiodic induction of diapause, hibernation and voltinism in *Piezodorus hybneri* (Heteroptera: Pentatomidae). Applied Entomology and Zoology 29: 585–592.

Hodek, I. 1971. Sensitivity to photoperiod in *Aelia acuminata* (L.) after adult diapause. Oecologia (Berlin) 6: 152–155. http://dx.doi.org/10.1007/BF00345716.

Hodek, I. 1977. Photoperiodic response in spring in three Pentatomidae (Heteroptera). Acta Entomologica Bohemoslovaca 74: 209–218.

Hodek, I., and M. Hodková. 1993. Role of temperature and photoperiod in diapause regulation in Czech populations of *Dolycoris baccarum* (Heteroptera, Pentatomidae). European Journal of Entomology 90: 95–98.

Hodek, I., and A. Honěk. 1970. Incidence of diapause in *Aelia acuminata* (L.) population from southwest Slovakia (Heteroptera). Věstník Československé Společnosti Zoologické 34: 170–183.

Hodková, M., I. Hodek, and L. Sømme. 1989. Cold is not a prerequisite for the completion of photoperiodically induced diapause in *Dolycoris baccarum* from Norway. Entomologia Experimentalis et Applicata 52: 185–188. http://dx.doi.org/10.1111/j.1570-7458.1989.tb01266.x.

Hoebeke, E. R., and M. E. Carter. 2003. *Halyomorpha halys* (Stål) (Heteroptera: Pentatomidae): a polyphagous plant pest from Asia newly detected in North America. Proceedings of the Entomological Society of Washington 105: 225–237.

Hokkanen, H. 1986. Polymorphism, parasites, and the native area of *Nezara viridula* (Hemiptera, Pentatomidae). Annales Entomologici Fennici 52: 28–31.

Honěk, A. 1969. Induction of diapause in *Aelia acuminata* (L.) (Heteroptera, Pentatomidae). Acta Entomologica Bohemoslovaca 66: 345–351.

Hori, K. 1986. Effect of photoperiod on nymphal growth of *Palomena angulosa* Motschulsky (Hemiptera: Pentatomidae). Applied Entomology and Zoology 21: 597–605.

Hori, K., K. Kuramochi, and S. Nakabayashi. 1985. Effect of several different food plants on nymphal development of *Palomena angulosa* Motschulsky (Hemiptera: Pentatomidae). Research Bulletin of Obihiro University (Series 1) 14: 239–246.

Inkley, D. B. 2012. Characteristics of home invasion by the brown marmorated stink bug (Hemiptera: Pentatomidae). Journal of Entomological Science 47: 125–130. http://dx.doi.org/10.18474/0749-8004 -47.2.125.

Iosifov, M. V. 1981. Fauna of Bulgaria. Volume 12: Heteroptera, Pentatomidae. Publishing House of Bulgarian Academy of Sciences, Sofia, Bulgaria. 205 pp. (in Bulgarian)

Ismailov, V. Ya., and I. S. Agasieva. 2010. The predaceous bug *Perillus bioculatus* F.: a new view of the prospects of acclimation and use. Zashchita i Karantin Rasteniy [Protection and Quarantine of Plants] 2: 30–31. (in Russian with English summary)

Ismailov, V. Ya., and I. N. Oleshchenko. 1977. Laboratory breeding of the predaceous bug *Arma custos* F. (Hemiptera: Pentatomidae) and some aspects of its biology. Nauchnye Doklady Vysshei Shkoly. Biologicheskie Nauki [Research Reports of the Higher Education. Biological Sciences] 4: 54–58. (in Russian)

Jasič, I. 1975. On the life cycle of *Perillus bioculatus*. Acta Entomologica Bohemoslovaca 72: 383–390.

Javahery, M. 1986. Biology and ecology of *Picromerus bidens* (Hemiptera: Pentatomidae) in southeastern Canada. Entomological News 97: 87–98.

Javahery, M. 1990. Biology and ecological adaptation of the green stink bug (Hemiptera: Pentatomidae) in Quebec and Ontario. Annals of the Entomological Society of America 83: 201–206. http://dx.doi .org/10.1093/aesa/83.2.201.

Javahery, M. 1994. Development of eggs in some true bugs (Hemiptera-Heteroptera). Part I. Pentatomidae. The Canadian Entomologist 126: 401–433.

Jones, P. A., and H. C. Coppel. 1963. Immature stages of *Apateticus cynicus* (Say) (Hemiptera: Pentatomidae). The Canadian Entomologist 95: 770–779. http://dx.doi.org/10.4039/Ent95770-7.

Kamenkova, K. V. 1958. Biology and ecology of the stink bug *Dolycoris baccarum* L., an accessory host of the sunn pest oviphages in Krasnodar territory. Entomologicheskoe Obozrenie [Entomological Review] 37: 654–579. (in Russian with English summary)

Kariya, H. 1961. Effect of temperature on the development and the mortality of the southern green stink bug, *Nezara viridula* and the oriental green stink bug, *N. antennata*. Japanese Journal of Applied Entomology and Zoology 5: 191–196.

Karsavuran, Y. 1986. Investigations on the biology and ecology of *Dolycoris baccarum* (L.) (Heteroptera, Pentatomidae) which attacks various plants of economic importance at Bornova (Izmir). Türkiye Bitki Koruma Dergisi 10: 213–230.

Kavar, T., P. Pavlovčič, S. Sušnik, V. Meglic, and M. Virant-Doberlet. 2006. Genetic differentiation of geographically separated populations of the southern green stink bug *Nezara viridula* (Hemiptera: Pentatomidae). Bulletin of Entomological Research 96: 117–128. http://dx.doi.org/10.1079/BER2005406.

Kipyatkov, V. E., and E. B. Lopatina. 2007. Seasonal cycles and strategies in ants: structure, diversity, and adaptive traits, pp. 107–192. *In* A. A. Stekolnikov (Ed.), Adaptive strategies of terrestrial arthropods to unfavorable environmental conditions: a collection of papers in memory of Professor Viktor Petrovich Tyshchenko (Proceedings of the Biological Institute of Saint Petersburg State University 53). 387 pp. (in Russian with English summary)

Kiritani, K. 1971. Distribution and abundance of the southern green stink bug, *Nezara viridula*, pp. 235–248. *In* Proceedings of Symposium on Rice Insects. Tropical Agricultural Research Center, Tokyo. 288 pp.

Kiritani, K., and N. Hokyo. 1962. Studies on the life table of the southern green stink bug, *Nezara viridula*. Japanese Journal of Applied Entomology and Zoology 6: 24–40.

Kiritani, K., N. Hokyo, and J. Yukawa. 1963. Co-existence of the two related stink bugs *Nezara viridula* and *N. antennata* under natural conditions. Researches on Population Ecology (Kyoto) 5: 11–22.

Kiritani, Y. 1985a. Timing of oviposition and nymphal diapause under the natural daylengths in *Carbula humerigera* (Heteroptera: Pentatomidae). Applied Entomology and Zoology 20: 252–256.

Kiritani, Y. 1985b. Effect of stationary and changing photoperiods on nymphal development in *Carbula humerigera* (Heteroptera: Pentatomidae). Applied Entomology and Zoology 20: 257–263.

Klyuchko, Z. F. 2006. Noctuid moths of Ukraine. Izdatel'stvo Raevskogo, Kiev. 248 pp. (in Ukrainian)

Klyuchko, Z. F., Yu. I. Budashkin, and V. P. Gerasimov. 2004. New and little known species of Noctuidae (Lepidoptera) of Ukraine. Vestnik Zoologii [Bulletin of Zoology] 38: 94. (in Russian with English summary)

Kobayashi, S., and H. Numata. 1993. Photoperiodic responses controlling the induction of adult diapause and the determination of seasonal form in the bean bug, *Riptortus clavatus*. Zoological Science 10: 983–990.

Kobayashi, T. 1972. Biology of insect pests of soybean and their control. Japan Agricultural Research Quarterly 6: 212–218.

Kon, M., A. Oe, and H. Numata. 1994. Ethological isolation between two congeneric green stink bugs, *Nezara antennata* and *N. viridula* (Heteroptera: Pentatomidae). Journal of Ethology 12: 67–71. http://dx.doi.org/10.1007/BF02350082.

Koshiyama, Y., H. Tsumuki, M. Muraji, K. Fujisaki, and F. Nakasuji. 1993. Transfer of male secretions to females through copulation in *Menida scotti* (Heteroptera: Pentatomidae). Applied Entomology and Zoology 28: 325–332.

Koshiyama, Y., K. Fujisaki, and F. Nakasuji. 1994. Mating and diapause in hibernating adults of *Menida scotti* Puton (Heteroptera: Pentatomidae). Researches on Population Ecology (Kyoto) 36: 87–92. http://dx.doi.org/10.1007/BF02515089.

Krambias, A. 1987. Host plant, seasonal migration and control of the berry bug *Dolycoris baccarum* L. in Cyprus. FAO Plant Protection Bulletin 35: 25–26.

Larivière, M.-C., and A. Larochelle. 1989. *Picromerus bidens* (Heteroptera: Pentatomidae) in North America, with a world review of distribution and bionomics. Entomological News 100: 133–146.

Lee, D.-H., B. D. Short, S. V. Joseph, J. C. Bergh, and T. C. Leskey. 2013. Review of the biology, ecology, and management of *Halyomorpha halys* (Hemiptera: Pentatomidae) in China, Japan, and the Republic of Korea. Environmental Entomology 42: 627–641. http://dx.doi.org/10.1603/EN13006.

Leskey, T. C., S. E. Wright, B. D. Short, and A. Khrimian. 2012. Development of behaviorally-based monitoring tools for the brown marmorated stink bug, *Halyomorpha halys* (Stål) (Heteroptera: Pentatomidae) in commercial tree fruit orchards. Journal of Entomological Science 47: 76–85. http://dx.doi.org/10.18474/0749-8004-47.1.76.

Leston, D. 1955. The life-cycle of *Picromerus bidens* (L.) (Hemiptera, Pentatomidae) in Britain. The Entomologist's Monthly Magazine 91: 109.

Masaki, S. 1980. Summer diapause. Annual Review of Entomology 25: 1–25. http://dx.doi.org/10.1146/annurev.en.25.010180.000245.

Masaki, S., and T. Sakai. 1965. Summer diapause in the seasonal life cycle of *Mamestra brassicae* L. (Lepidoptera: Noctuidae). Japanese Journal of Applied Entomology and Zoology 9: 191–205. (in Japanese with English summary)

Masaki, S., and T. J. Walter. 1987. Cricket life cycles. Evolutionary Biology 21: 349–423.

Masaki, S., and O. Yata. 1988. Seasonal adaptation and photoperiodism in butterflies. Special Bulletin of Lepidopteran Society of Japan 6: 341–383. (in Japanese with English summary)

Matalin, A. V. 2007. Typology of life cycles of ground beetles (Coleoptera, Carabidae) in Western Palaearctic. Entomological Review 87: 947–972. http://dx.doi.org/10.1134/S0013873807080027.

McPherson, J. E., and R. M. McPherson. 2000. Stink bugs of economic importance in America north of Mexico. CRC Press, Boca Raton, FL. 254 pp.

McPherson, J. E., and D. L. Tecic. 1997. Notes on the life histories of *Acrosternum hilare* and *Cosmopepla bimaculata* (Heteroptera: Pentatomidae) in southern Illinois. The Great Lakes Entomologist 30: 79–84.

Milonas, P. G., and G. K. Partsinevelos. 2014. First report of brown marmorated stink bug *Halyomorpha halys* Stål (Hemiptera: Pentatomidae) in Greece. Bulletin OEPP/EPPO Bulletin 44: 183–186. http://dx.doi .org/10.1111/epp.12129.

Miner, F. D. 1966. Biology and control of stink bugs on soybeans. Arkansas Agricultural Experiment Station Bulletin 708: 1–40.

Musolin, D. L. 1997. Seasonal cycles of true bugs (Heteroptera): diversity and environmental control. Ph.D. Dissertation in Biology. Saint Petersburg State University, Saint Petersburg, Russia. 168 pp. (in Russian)

Musolin, D. L. 2007. Insects in a warmer world: ecological, physiological and life-history responses of true bugs (Heteroptera) to climate change. Global Change Biology 13: 1565–1585. http://dx.doi.org /10.1111/j.1365-2486.2007.01395.x.

Musolin, D. L. 2012. Surviving winter: diapause syndrome in the southern green stink bug *Nezara viridula* in the laboratory, in the field, and under climate change conditions. Physiological Entomology 37: 309–322. http://dx.doi.org/10.1111/j.1365-3032.2012.00846.x.

Musolin, D. L., and H. Numata. 2003a. Photoperiodic and temperature control of diapause induction and colour change in the southern green stink bug *Nezara viridula*. Physiological Entomology 28: 65–74. http://dx.doi.org/10.1046/j.1365-3032.2003.00307.x.

Musolin, D. L., and H. Numata. 2003b. Timing of diapause induction and its life-history consequences in *Nezara viridula*: is it costly to expand the distribution range? Ecological Entomology 28: 694–703. http://dx.doi.org/10.1111/j.1365-2311.2003.00559.x.

Musolin, D. L., and H. Numata. 2004. Late-season induction of diapause in *Nezara viridula* and its effect on adult coloration and postdiapause reproductive performance. Entomologia Experimentalis et Applicata 111: 1–6. http://dx.doi.org/10.1111/j.0013-8703.2004.00137.x.

Musolin, D. L., and A. Kh. Saulich. 1996. Factorial regulation of the seasonal cycle in the stink bug *Graphosoma lineatum* L. (Heteroptera, Pentatomidae). I. Temperature and photoperiodic responses. Entomological Review 75: 84–93.

Musolin, D. L., and A. Kh. Saulich. 1997. Photoperiodic control of nymphal growth in true bugs (Heteroptera). Entomological Review 76: 768–780.

Musolin, D. L., and A. Kh. Saulich. 2000. Summer dormancy ensures univoltinism in the predatory bug *Picromerus bidens* (Heteroptera, Pentatomidae). Entomologia Experimentalis et Applicata 95: 259–267. http://dx.doi.org/10.1046/j.1570-7458.2000.00665.x.

Musolin, D. L., and A. Kh. Saulich. 2001. Environmental control of voltinism of the stink bug *Graphosoma lineatum* in the forest-steppe zone (Heteroptera: Pentatomidae). Entomologia Generalis 25: 255–264. http://dx.doi.org/10.1127/entom.gen/25/2001/255.

Musolin, D. L., and A. Kh. Saulich. 2011. Changes in the natural ranges of insects under the conditions of recent climate warming. Izvestia Sankt-Peterburgskoj Lesotehniceskoj Akademii [Proceedings of Saint Petersburg Forest Technical Academy, Saint Petersburg, Russia] 196: 246–254. (in Russian with English summary)

Musolin, D. L., and A. Kh. Saulich. 2012a. Insect voltinism under the conditions of recent climate changes. Izvestia Sankt-Peterburgskoj Lesotehniceskoj Akademii [Proceedings of Saint Petersburg Forest Technical Academy, Saint Petersburg, Russia] 200: 208–221. (in Russian with English summary)

Musolin, D. L., and A. Kh. Saulich. 2012b. Responses of insects to the current climate changes: from physiology and behavior to range shifts. Entomological Review 92: 715–740. http://dx.doi.org/10.1134/S00 13873812070019.

Musolin, D. L., D. Tougou, and K. Fujisaki. 2010. Too hot to handle? Phenological and life-history responses to simulated climate change of the southern green stink bug *Nezara viridula* (Heteroptera: Pentatomidae). Global Change Biology 16: 73–87. http://dx.doi.org/10.1111/j.1365-2486.2009.01914.x.

Nakamura, K., and H. Numata. 1995. Photoperiodic sensitivity in adult of *Aelia fieberi* (Heteroptera: Pentatomidae). European Journal of Entomology 92: 609–613.

Nakamura, K., and H. Numata. 1997. Seasonal life cycle of *Aelia fieberi* (Heteroptera: Pentatomidae) in relation to the phenology of its host plants. Annals of the Entomological Society of America 90: 625–630. http://dx.doi.org/10.1093/aesa/90.5.625.

Nakamura, K., and H. Numata. 1998. Alternative life cycles controlled by temperature and photoperiod in the oligophagous bug, *Dybowskyia reticulata*. Physiological Entomology 23: 69–74. http://dx.doi.org /10.1046/j.1365-3032.1998.2310069.x.

Nakamura, K., and H. Numata. 1999. Environmental regulation of adult diapause of *Graphosoma rubrolineatum* (Westwood) (Heteroptera: Pentatomidae) in southern and northern populations of Japan. Applied Entomology and Zoology 34: 323–326. http://dx.doi.org/10.1303/aez.34.323.

Nakamura, K., and H. Numata. 2006. Effect of photoperiod and temperature on the induction of adult diapause in *Dolycoris baccarum* (L.) (Heteroptera: Pentatomidae) from Osaka and Hokkaido, Japan. Applied Entomology and Zoology 41: 105–109. http://dx.doi.org/10.1303/aez.2006.105.

Nayanov, N. I. 1973. On acclimation of the ragweed moth *Tarachidia candefacta* Hübn. (Lepidoptera, Noctuidae) in the south of European Russia. Entomologicheskoe Obozrenie 70: 759–767. (in Russian with English summary)

Nielsen, A. L., and G. C. Hamilton. 2009. Life-history of the invasive species *Halyomorpha halys* (Hemiptera: Pentatomidae) in the Northeastern United States. Annals of Entomological Society of America 102: 608–616. http://dx.doi.org/10.1603/008.102.0405.

Nielsen, A. L., G. C. Hamilton, and P. W. Shearer. 2011. Seasonal phenology and monitoring of the non-native *Halyomorpha halys* (Stål) (Hemiptera: Pentatomidae) in soybean. Environmental Entomology 40: 231–238. http://dx.doi.org/10.1603/EN10187.

Niva, C. C. 2003. Molecular and neuroendocrine mechanisms of photoperiodism in *Halyomorpha halys* (Heteroptera: Pentatomidae). Ph.D. Dissertation. Kobe University, Kobe, Japan. 114 pp.

Niva, C. C., and M. Takeda. 2003. Effects of photoperiod, temperature and melatonin on nymphal development, polyphenism and reproduction in *Halyomorpha halys* (Heteroptera: Pentatomidae). Zoological Science 20: 963–970. http://dx.doi.org/10.2108/zsj.20.963.

Noda, H., and T. Ishii. 1981. Effect of photoperiod and temperature on the ovarian development of the white-spotted stink bug, *Eysarcoris ventralis* (Heteroptera: Pentatomidae). Japanese Journal of Applied Entomology and Zoology 25: 33–38. (in Japanese with English summary)

Noda, T. 1984. Short day photoperiod accelerates the oviposition in the oriental green stink bug, *Nezara antennata* Scott (Heteroptera: Pentatomidae). Applied Entomology and Zoology 19: 119–120.

Numata, H. 1985. Photoperiodic control of adult diapause in the bean bug *Riptortus clavatus*. Memoirs of Faculty of Science, Kyoto University, Series Biology (Kyoto, Japan) 10: 29–48.

Numata, H. 1990. Photoperiodic induction of the first and the second diapause in the bean bug, *Riptortus clavatus*: a photoperiodic history effect. Journal of Comparative Physiology A. Sensory, Neural, and Behavioral Physiology 167: 167–171. http://dx.doi.org/10.1007/BF00188108.

Numata, H., and S. Kobayashi. 1994. Threshold and quantitative photoperiodic responses exist in an insect. Experientia 50: 969–971. http://dx.doi.org/10.1007/BF01923489.

Numata, H., and K. Nakamura. 2002. Photoperiodism and seasonal adaptations in some seed-sucking bugs (Heteroptera) in central Japan. European Journal of Entomology 99: 155–161. http://dx.doi.org/10.14411 /eje.2002.023.

Panizzi, A. R., and E. Hirose. 1995. Seasonal body weight, lipid content, and impact of starvation and water stress on adult survivorship and longevity of *Nezara viridula* and *Euschistus heros*. Entomologia Experimentalis et Applicata 76: 247–253. http://dx.doi.org/10.1111/j.1570-7458.1995.tb01969.x.

Panizzi, A., J. E. McPherson, D. G. James, M. Javahery, and R. M. McPherson. 2000. Chapter 13. Stink bugs (Pentatomidae), pp. 421–474. *In* C. W. Schaefer and A. R. Panizzi (Eds.), Heteroptera of economic importance. CRC Press LLC, Boca Raton, FL. 828 pp.

Perepelitsa, L. V. 1971. The role of photoperiod in the development of *Dolycoris baccarum*. Byulleten' Vsesojuznogo Instituta Zashchiti Rasteniy [Bulletin of the All-Union Institute of Plant Protection] 21: 11–13. (in Russian with English summary)

Petrova, V. P. 1975. Stink bugs (Hemiptera, Pentatomoidea) of Western Siberia. Novosibirsk State Teacher's University, Novosibirsk, Russia. 238 pp. (in Russian)

Poltavsky, A., N. and K. S. Artokhin. 2006. *Tarachidia candefacta* (Lepidoptera: Noctuidae) in the South of European Russia. Phegea 34: 41.

Putshkov, P. V. 1961. Fauna of Ukraine. Shield bugs. 21. Akademya Nauk UkrSSR [Academy of Sciences of the Ukrainian SSR], Kiev. 338 pp. (in Ukrainian)

Putshkov, P. V. 1965. Shield-bugs of Middle Asia (Hemiptera, Pentatomoidea). Publishing House Ilim [The Academy of Sciences of Kyrgyz SSR, Institute of Biology], Frunze [Kyrgyz SSR, the USSR]. 332 pp. (in Russian)

Richman, D. V., and F. W. Mead. 1980. Stages in the life cycle of a predatory stink bug, *Podisus maculiventris* (Say) (Hemiptera: Pentatomidae). Entomology Circular (Florida Department of Agriculture and Consumer Services, Division of Plant Industry) 216: 1–2.

Rider, D. A. 2006. Family Pentatomidae Leach, 1815, pp. 233–414. *In* B. Aukema and C. Rieger (Eds.), Catalogue of the Heteroptera of the Palaearctic Region. Volume 5. Pentatomomorpha II. Nederlandse Entomologische Vereniging, Ponsen & Looijen, Wageningen, The Netherlands. 550 pp.

Saulich, A. Kh. 1995. Role of abiotic factors in forming secondary ranges of adventive insect species. Entomological Review 74: 591–605.

Saulich, A. Kh. 1999. Seasonal development and dispersal potential of insects. Saint Petersburg State University, Saint Petersburg, Russia. 248 pp. (in Russian)

Saulich, A. Kh. 2010. Long life cycles in insects. Entomological Review 90: 1127–1152. http://dx.doi.org /10.1134/S0013873810090010.

Saulich, A. Kh., and D. L. Musolin. 1996. Univoltinism and its regulation in some temperate true bugs (Heteroptera). European Journal of Entomology 93: 507–518.

Saulich, A. Kh., and D. L. Musolin. 2007a. Seasonal development of aquatic and semiaquatic true bugs (Heteroptera). Saint Petersburg State University, Saint Petersburg, Russia. 205 pp. (in Russian with English Summary)

Saulich, A. Kh., and D. L. Musolin. 2007b. Times of the year: the diversity of seasonal adaptations and ecological mechanisms controlling seasonal development of true bugs (Heteroptera) in the temperate climate, pp. 25–106. *In* A. A. Stekolnikov (Ed.), Adaptive strategies of terrestrial arthropods to unfavorable environmental conditions: a collection of papers in memory of Professor Viktor Petrovich Tyshchenko (Proceedings of the Biological Institute of Saint Petersburg State University 53). 387 pp. (in Russian with English summary)

Saulich, A. Kh., and D. L. Musolin. 2011. Biology and ecology of the predatory bug *Podisus maculiventris* (Say) (Heteroptera, Pentatomidae) and possibility of its application against the Colorado potato beetle, *Leptinotarsa decemlineata* Say (Coleoptera, Chrysomelidae). A textbook for the course Seasonal Cycles of Insects. Saint Petersburg State University, Saint Petersburg, Russia. 84 pp. (in Russian)

Saulich, A. Kh., and D. L. Musolin. 2012. Diapause in the seasonal cycle of stink bugs (Heteroptera, Pentatomidae) from the temperate zone. Entomological Review 92: 1–26. http://dx.doi.org/10.1134/S0013873812010010.

Saulich, A. Kh., and D. L. Musolin. 2014a. Seasonal development of plataspid shield bugs (Heteroptera: Pentatomoidea: Plataspidae). Vestnik Moskovskogo Gosudarstvennogo Universiteta Lesa – Lesnoi Vestnik [Moscow State Forest University Bulletin – Forest Bulletin] 18(6): 193–201. (in Russian with English summary)

Saulich, A. Kh., and D. L. Musolin. 2014b. Seasonal cycles in stink bugs (Heteroptera, Pentatomidae) from the temperate zone: diversity and control. Entomological Review 94: 785–814. http://dx.doi.org/10.1134 /S0013873814060013.

Saulich, A. Kh., and T. A. Volkovich. 1996. Monovoltinism in insects and its regulation. Entomological Review 76: 205–217.

Saulich, A. Kh., and T. A. Volkovich. 2004. Ecology of photoperiodism in insects. Saint Petersburg University Press, Saint Petersburg. 276 pp. (in Russian)

Saunders, D. S. 1976. Insect clocks. Pergamon Press, Oxford, UK. 280 pp.

Schumacher, F. 1910–1911. Beiträge zur Kenntnis der Biologie der Asopiden. Zeitschrift fur Wissensch Insectenbiol 6: 263–266, 376–383, 430–437; 7: 40–47. (in German with English summary)

Shinyaeva, L. I. 1980. Spermatogenesis during prediapause and diapause formation in *Eurygaster integriceps*. Zoologicheskii Zhurnal [Zoological Journal] 69: 1025–1032. (in Russian with English summary)

Siddiqui, A. S. 2000. A revision of *Strachiine* genera (Hemiptera: Pentatomidae: Pentatominae). Ph.D. Thesis, Department of Zoology, University of Karachi, Karachi, Pakistan. 260 pp.

Singh, Z. 1973. Southern green stink bug and its relationship to soybeans: bionomics of the southern green stink bug *Nezara viridula* (Hemiptera: Pentatomidae) in central India. Metropolitan Book Co. (PVT) Ltd, Delhi, India. 106 pp.

Singh, Z., and V. S. Malik. 1993. Biology of painted bug (*Bagrada cruciferarum*). Indian Journal of Agricultural Science 63. 672–674.

Şişli, M. N. 1965. The effect of the photoperiod on the induction and termination of the adult diapause of *Aelia rostrata* Boh. (Hemiptera: Pentatomidae). Communications Faculty of Sciences University of Ankara 10: 62–69.

Southwood, T. R. E., and D. Leston. 1959. Land and water bugs of the British Isles. Frederick Warne and Co., London, UK. xi + 436 pp.

Sun, X., and S. A. Marshall. 2003. Systematics of *Phasia* Latreille (Diptera: Tachinidae). Zootaxa 276: 1–320.

Takeda, K., D. L Musolin, and K. Fujisaki. 2010. Dissecting insect responses to climate warming: overwintering and post-diapause performance in the southern green stink bug, *Nezara viridula*, under simulated climate-change conditions. Physiological Entomology 35: 343–353. http://dx.doi.org/10.1111/j.1365-3032.2010.00748.x.

Tanaka, S. I. 2002. Experimental analysis of seasonal adaptations in *Poecilocoris lewisi*. Ph.D. Dissertation. Osaka City University, Osaka, Japan. 74 pp.

Tanaka, S. I., C. Imai, and H. Numata. 2002. Ecological significance of adult summer diapause after nymphal winter diapause in *Poecilocoris lewisi* (Distant) (Heteroptera: Scutelleridae). Applied Entomology and Zoology 37: 469–475. http://doi.org/10.1303/aez.2002.469.

Tauber, M. J., C. A. Tauber, and S. Masaki. 1986. Seasonal adaptations of insects. Oxford University Press, New York. xvi + 412 pp.

Taylor, M. E., C. S. Bundy, and J. E. McPherson. 2015. Life history and laboratory rearing of *Bagrada hilaris* (Hemiptera: Heteroptera: Pentatomidae) with descriptions of immature stages. Annals of the Entomological Society of America 108: 536–551. http://dx.doi.org/10.1093/aesa/sav048.

Todd, J. W. 1989. Ecology and behavior of *Nezara viridula*. Annual Review of Entomology 34: 273–292. http://dx.doi.org/10.1146/annurev.en.34.010189.001421.

Tomokuni, M., T. Yasunaga, M. Takai, I. Yamashita, M. Kawamura, and T. Kawasawa. 1993. A field guide to Japanese bugs: terrestrial heteropterans. Zenkoku Noson Kyoiku Kyokai, Tokyo, Japan. 380 pp. (in Japanese)

Tougou, D., D. L. Musolin, and K. Fujisaki. 2009. Some like it hot! Rapid climate change promotes changes in distribution ranges of *Nezara viridula* and *Nezara antennata* in Japan. Entomologia Experimentalis et Applicata 130: 249–258. http://dx.doi.org/10.1111/j.1570-7458.2008.00818.x.

Tyshchenko, V. P. 1977. Physiology of photoperiodism in insects. Trudy Vsesoyuznogo Entomologicheskogo Obshchestva (Proceedings of the All-Union Entomological Society). Nauka Publishing House, Leningrad, the USSR. 59: 1–156. (in Russian)

Tyshchenko, V. P. 1980. Signal and vital modes of action of ecological factors. Zhurnal Obshey Biologii [Journal of General Biology] 41: 655–667. (in Russian)

Tyshchenko, V. P. 1983. Evolution of seasonal adaptations in insects. Zhurnal Obshey Biologii [Journal of General Biology] 44: 10–22. (in Russian)

Ushatinskaya, R. S. 1955. Physiological peculiarities of the sunn pest (*Eurygaster integriceps* Put.) during a period of dormancy, overwintering in mountains and lowlands, pp. 134–170. *In* D. M. Fedotov (Ed.), Sunn pest *Eurygaster integriceps* Put. Volume 3. Publishing House of The Academy of Sciences of the USSR, Moscow. 280 pp. (in Russian)

Viktorov, G. A. 1976. Ecology of entomoparasites. Nauka Publishing House, Moscow, the USSR. 152 pp. (in Russian)

Volkovich, T. A. 2007. Diapause in the life cycles of lacewings (Neuroptera, Chrysopidae), pp. 234–304. *In* A. A. Stekolnikov (Ed.), Adaptive strategies of terrestrial arthropods to unfavorable environmental conditions: a collection of papers in memory of Professor Viktor Petrovich Tyshchenko (Proceedings of the Biological Institute of Saint Petersburg State University, 53). 387 pp. (in Russian with English summary)

Volkovich, T. A., and A. Kh. Saulich. 1995. The predatory bug *Arma custos*: photoperiodic and temperature control of diapause and colouration. Entomological Review 74: 151–162.

Volkovich, T. A., L. I. Kolesnichenko, and A. Kh. Saulich. 1991a. The role of thermal rhythms in the development of *Perillus bioculatus* (Hemiptera, Pentatomidae). Entomological Review 70(1): 68–80.

Volkovich, T. A., A. Kh. Saulich, and N. I. Goryshin. 1991b. Day-length sensitive stage and accumulation of photoperiodic information in the predatory bug *Podisus maculiventris* Say (Heteroptera; Pentatomidae). Entomological Review 70(1): 159–167.

Wermelinger, B., D. Wyniger, and B. Forster. 2008. First records of an invasive bug in Europe: *Halyomorpha halys* Stål (Heteroptera: Pentatomidae), a new pest on woody ornamentals and fruit trees? Mitteilungen der Schweizerischen Entomologischen Gesellschaft 81: 1–8.

Whitmarsh, R. D. 1916. Life-history notes on *Apateticus cynicus* and *maculiventris*. Journal of Economic Entomology 9: 51–53. http://dx.doi.org/10.1093/jee/9.1.51.

Wilde, G. E. 1969. Photoperiodism in relation to development and reproduction in the green stink bug. Journal of Economic Entomology 62: 629–630. http://dx.doi.org/10.1093/jee/62.3.629.

Xu, J., D. M. Fonseca, G. C. Hamilton, K. A. Hoelmer, and A. L. Nielsen. 2014. Tracing the origin of US brown marmorated stink bugs, *Halyomorpha halys*. Biological Invasions 16: 153–166. http://dx.doi.org/10.1007/s10530-013-0510-3.

Yanagi, T., and Y. Hagihara. 1980. Ecology of the brown marmorated stink bug. Shokubutsu-Boeki [Plant Protection] 34: 315–321. (in Japanese)

Yathom, S. 1980. An outbreak of *Dolycoris baccarum* L. (Heteroptera: Pentatomidae) on sunflower in Israel. Israel Journal of Entomology 14: 25–28.

Yukawa, J., K. Kiritani, N. Gyoutoku, N. Uechi, D. Yamaguchi, and S. Kamitani. 2007. Distribution range shift of two allied species, *Nezara viridula* and *N. antennata* (Hemiptera: Pentatomidae), in Japan, possibly due to global warming. Applied Entomology and Zoology 42: 205–215. http://dx.doi.org/10.1303/aez.2007.205.

Yukawa J., K. Kiritani, T. Kawasawa, Y. Higashiura, N. Sawamura, K. Nakada, N. Gyotoku, A. Tanaka, S. Kamitani, K. Matsuo, S. Yamauchi, and Y. Takematsu. 2009. Northward range expansion by *Nezara viridula* (Hemiptera: Pentatomidae) in Shikoku and Chugoku districts, Japan, possibly due to global warming. Applied Entomology and Zoology 44: 429–437. http://dx.doi.org/10.1303/aez.2009.429.

Zaslavski, V. A. 1988. Insect development, photoperiodic and temperature control. Springer, Berlin, Germany. 187 pp.

Zhu, G., W. Bu, Y. Gao, and G. Liu. 2012. Potential geographic distribution of brown marmorated stink bug invasion (*Halyomorpha halys*). PLoS ONE 7:e31246. http://dx.doi.org/10.1371/journal.pone.0031246.

Zou, D., M. Wang, L. Zhang, Y. Zhang, X. Zhang, and H. Chen. 2012. Taxonomic and bionomic notes on *Arma chinensis* (Fallou) (Hemiptera: Pentatomidae: Asopinae). Zootaxa 3382: 41–52.

Section VII

Vectors of Plant Pathogens

13

Pentatomoids as Vectors of Plant Pathogens

Paula Levin Mitchell, Adam R. Zeilinger, Enrique Gino Medrano, and Jesus F. Esquivel

CONTENTS

13.1 Introduction 612
    13.1.1 Modes of Transmission of Pathogens by Hemiptera...612
    13.1.2 Heteropteran Feeding Mechanisms...612
    13.1.3 Pentatomid Mouthparts..614
        13.1.3.1 Structure of Mouthparts 614
        13.1.3.2 Formation of the Stylet Bundle 614
        13.1.3.3 Rostrum and Feeding Mechanics 615
        13.1.3.4 Stylet Penetration Potential 615
    13.1.4 Overview of Pathogens Transmitted by Pentatomoidea.. 616
    13.1.5 Selective Transmission — Role of Pathogen Localization...................................... 616
    13.1.6 Objectives..617
13.2 Viruses 617
    13.2.1 General Overview of Viruses 617
    13.2.2 Longan and Lychee 617
13.3 Phytoplasmas..618
    13.3.1 General Overview of Phytoplasmas 618
    13.3.2 Empress Tree 618
13.4 Bacterial Pathogens...619
    13.4.1 General Overview of Bacterial Pathogens 619
    13.4.2 Cotton 619
    13.4.3 Legumes... 620
13.5 Fungal Pathogens.. 620
    13.5.1 General Overview of Fungal Pathogens...620
    13.5.2 Coffee...620
    13.5.3 Cotton.. 621
    13.5.4 Legumes... 621
    13.5.5 Pistachio.. 622
    13.5.6 Rice.. 624
    13.5.7 Other Crops.. 624
13.6 Trypanosomatids.. 625
    13.6.1 General Overview of Trypanosomatids...625
    13.6.2 Coconut.. 626
    13.6.3 Oil Palms 627
13.7 Epidemiology of Pentatomid-Borne Pathogens... 628
    13.7.1 Symbiotic Pathogens... 628
    13.7.2 Facultative Pathogens.. 630
13.8 Future Research Directions... 631
13.9 Acknowledgments 632
13.10 References Cited 632

13.1 Introduction

13.1.1 Modes of Transmission of Pathogens by Hemiptera

Vector-borne pathogens can be categorized functionally according to the degree of symbiosis that they acquire with their vectors. Three modes of transmission have been broadly described: non-persistent, semi-persistent, and persistent. Although Nault (1997) originally compiled these modes specifically for viruses transmitted by sternorrhynchan and auchenorrhynchan vectors, they can usefully be applied to all taxa of microbial pathogens and hemipteran vectors.

The differences in persistence among different transmission forms are directly related to which vector tissue(s) are colonized by the pathogen (Nault 1997). Pathogens transmitted in a non-persistent manner tend to colonize the stylets of hemipteran vectors and generally require only brief periods to be acquired by vectors and to inoculate subsequent hosts (on the order of minutes); infections also are lost quickly by vectors thereby requiring the vectors to move quickly from infected to healthy hosts (Uzest et al. 2010, Blanc et al. 2014). Semi-persistent pathogens colonize the foregut of their vectors; as a consequence, acquisition and inoculation access periods are longer and vectors remain infective for longer periods of time (Nault 1997). Finally, persistent pathogens colonize vector tissues more systemically, colonizing the digestive system and haemocoel. Importantly, salivary glands eventually are colonized from which inoculation of hosts occurs (Nault 1997). In addition, a subset of persistent pathogens also multiply in the tissues of vectors; these "persistent-propagative" pathogens often are transovarially transmitted as well. While transmission modes can be useful for understanding vector-pathogen relationships, they also relate to patterns of induced phenotypic changes in host plants (Mauck et al. 2012), are a conserved trait within virus taxa (Nault 1997), and define the epidemiology of associated diseases (Madden et al. 2000; see **Section 13.7**).

13.1.2 Heteropteran Feeding Mechanisms

The ability to transmit pathogens depends on feeding behaviors, which, to a great extent, are determined by the structure of the mouthparts. All Hemiptera are characterized by mouthparts composed of four piercing structures, two outer mandibular stylets and two inner maxillary stylets, enclosed (when not in use) by the 4-segmented labium, which usually functions only as a protective sheath. The labrum is reduced, and there are no maxillary or labial palps. A food canal and a salivary canal are formed by the interlocking maxillary stylets, and liquid food is sucked into the precibarium and then through the cibarium into the esophagus (**Figure 13.1**; also **Section 13.1.3.2**, Formation of the Stylet Bundle). However,

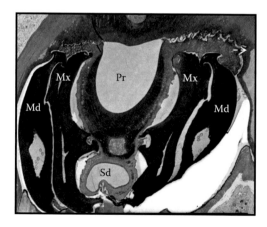

FIGURE 13.1 Convergence of the maxillary and mandibular stylets with the precibarium and salivary duct within the head of *Nezara viridula*. Md, mandibular stylet; Mx, maxillary stylet; Pr, precibarium; Sd, Salivary duct. (Courtesy of Robert E. Droleskey, USDA, College Station, Texas.)

the attachment to the head, the innervation of the stylets and labium, the movement of the stylets in relation to one another, the composition of the saliva, and the manner in which these structures are used to obtain food vary among the three suborders (i.e., Heteroptera, Auchenorrhyncha, Sternorrhyncha) (Backus 1988), affecting their behaviors and, consequently, their role as vectors.

In Heteroptera, the rostrum connects flexibly near the front of the head and the maxillary and mandibular stylets move together when inserted in food or prey, with the apices of mandibular stylets slightly ahead. One or both sets of stylets may be barbed or serrated (Cobben 1978). Watery saliva is produced by all true bugs, but only the phytophagous Pentatomomorpha are capable of producing the gelling saliva, which also is found in Auchenorrhyncha and Sternorrhyncha (Miles 1968, Cobben 1978, Backus 1988). Chemosensilla are found on the antennae and labium and within the precibarium, allowing both contact chemoreception and tasting of sampled fluid drawn up the stylets (Backus 1988). Labial dabbing (tapping the plant surface [with or without exuding watery saliva onto the plant] to sample the surface chemicals) has been reported for both Cimicomorpha and Pentatomomorpha (Cook and Neal 1999, Cline and Backus 2002). The path of stylet insertion is intracellular, in vivid contrast to the delicate meandering of some aphid stylets between the palisade cells (Pollard 1973). In addition to damage along the entry path, damage to the target tissue also may be extensive. This may occur simply because the stylet bundle is large, or ingestion may be preceded by mechanical laceration and/or enzymatic activity from watery saliva. If pathogens must be delivered to uninjured, functional plant host cells (e.g., phloem sieve tube cells [Pollard 1977]), it is difficult to understand how larger bugs can effectively inoculate phloem-limited phytoplasmas or fastidious bacteria. The diameter of the stylets, and hence the food and salivary canals, also affects the ability to acquire and inoculate infective particles. The maxillary stylet canals of some pentatomids, for example, permit entry of the ascospores of *Eremothecium* (formerly *Nematospora*) *coryli* (Peglion) Kurtzman (Ragsdale et al. 1979), likely facilitating the close association between this fungus and the larger pentatomids and coreids.

Four modes of feeding, or feeding strategies, were described for Hemiptera by Miles and Taylor (1994): stylet sheath, lacerate and flush, macerate and flush, and osmotic pump. Stylet sheath (or salivary sheath) feeding involves a complete sheath of gelling saliva extending from the plant surface to the target vascular tissue and sealing the stylets into the cell of choice (sieve tube or xylem vessel). Lacerate and flush feeders use the stylets to physically rip open plant cells. Watery saliva then flushes out the cell contents and the mix is ingested. A partial salivary sheath may be produced within plant tissue, or a deposit (flange) of gelling saliva may be secreted on the plant surface without any internal deposition. Macerate and flush, associated with plant bugs (Miridae), employs enzymes in watery saliva to chemically macerate the cells rather than through physical laceration. Enzymatic activity also characterizes osmotic-pump feeding in which salivary sucrase alters the osmotic balance in the intercellular spaces, inducing parenchyma cells to leak their contents and phloem sieve tube cells to unload. This feeding method is associated with several tribes of Coreidae. The salivary enzymes of hemipteroids recently have been reviewed by Sharma et al. (2013), with a discussion of their role in pathogen transmission.

Miles and Taylor's (1994) organizational scheme was simplified by Backus et al. (2005) who identified just two basic strategies, sheath feeding and cell-rupture feeding, with other behaviors (lacerate and flush, macerate and flush, lacerate and sip, lance and ingest) relegated to the status of "tactics" within cell-rupture feeding. Sharma et al. (2013) presented a similar arrangement, but with osmotic-pump feeding included as a strategy. However, it is essential to realize that despite the confusing nomenclature, the various strategies are not mutually exclusive. That is, sheath-producing Heteroptera often employ other strategies and in some species, a salivary sheath may accompany the stylets for some distance as they approach the target tissue to be ruptured. Obviously, the cimicomorphs (unable to produce gelling saliva) must by definition be cell-rupture feeders, but Pentatomomorpha are exceptionally plastic in their feeding behaviors. Some species alter their strategy as development proceeds (e.g., *Eurydema rugosa* Motschulsky [Hori 1968]) or may switch from cell-rupture feeding to sheath feeding as the target tissue changes. *Oncopeltus fasciatus* (Dallas) uses the salivary-sheath strategy on phloem but employs lacerate and flush tactics (i.e., cell-rupture feeding) on seeds (Miles 1959). Similarly, *Nezara viridula* (L.) produces a complete salivary sheath when hydrating from xylem but feeds via cell rupture on developing seeds of soybean (Cooke 2014). Pentatomid species vary in their preferences for plant structures as well;

ingestion may occur on a variety of plant parts, including leaves, stems, buds, fruits, developing seeds, or mature seeds (McPherson and McPherson 2000, Silva et al. 2012).

Several methods have been used to characterize and measure heteropteran feeding, but determining preferred target tissue after the stylets penetrate the plant tissue has always been a challenge. Counts of the number of surface flanges of gelling saliva (often called "stylet sheaths") have been used frequently since first introduced by Bowling (1979). In most cases, these counts have accurately reflected crop damage caused by mechanical or enzymatic injury and pathogen introduction (Zeilinger et al. 2015 and references therein), but not always (e.g., blueberries [Wiman et al. 2015]). In contrast, the relationship between counts of flanges and food consumption or preference is inconsistent and unreliable, depending on the species and stage (Zeilinger et al. 2015). Growth rate is related significantly to flange counts only for fifth instars of *Euschistus servus* (Say) but not for *Nezara viridula* nymphs nor for adults of either species (Zeilinger et al. 2015). Histological studies have been in use for over a century to follow stylets or the course of sheath material in plant tissue (e.g., Horsfall 1923 and references therein) but are time-consuming. A model for estimating potential stylet penetration by pentatomids and other heteropterans has been developed by Esquivel (2011, 2015, and see **Section 13.1.3.4**, Stylet Penetration Potential). Major advances in our understanding of pathogen transmission by hemipterans have come from electrical-penetration graphing (EPG) techniques. Fereres and Moreno (2009) reviewed the role of EPG in interpreting virus transmission by aphids; the technique has also been invaluable for studying auchenorrhynchan vectors (e.g., sharpshooters [Backus 2016]). Few electrical-monitoring studies of heteropterans have been published to date; however, interest in this technique is growing as indicated by recent publications on pentatomids (Wiman et al. 2014, Lucini and Panizzi 2016), and the potential for application to vector studies certainly exists.

13.1.3 Pentatomid Mouthparts

13.1.3.1 Structure of Mouthparts

Pentatomids possess the typical heteropteran mouthparts described previously (see **Section 13.1.2**, Heteropteran Feeding Mechanisms), composed of interlocking needle-like stylets to pierce food sources and obtain nutritional resources. This unique mode of feeding involves complex linkages among the individual stylets to form a stylet bundle, and deliberate changes in insect posture while feeding that affect stylet-penetration potential (Esquivel 2011, 2015).

13.1.3.2 Formation of the Stylet Bundle

Butt (1943), in examining *Chinavia hilaris* (Say) [as *Acrosternum hilare* (Say)] and *Oncopeltus fasciatus* as representatives of the "terrestrial plant-feeding forms" of Hemiptera, provided a detailed description of the development of the individual maxillary and mandibular stylets that form the stylet bundle. Briefly, within the insect head and on either side of the precibarium and alimentary canal, the maxillary and mandibular stylets arise from "walls of the bristle pouches." Levers are associated with these pouches to aid in protraction and retraction of stylets. Additionally, at their bases, stylets are flattened with attached protractor and retractor muscles (Butt 1943). Similar flattening of the stylets is observed in *Nezara viridula* (Jesus F. Esquivel [JFE], unpublished data). These muscles presumably interact with the levers for the protraction and retraction of the stylets from the labium (further clarified in **Section 3.1.3.3**). Dendritic nerve bundles are present in the neural canals of each stylet, and these nerve bundles innervate along the length of the stylet as suggested by the reduced number of bundles visible in cross-sections at the distal end of the maxillary stylets (JFE, unpublished data).

Smith (1926) presented a general diagram of the configuration of the mouthparts for hemipterans. As described above, the stylets originate from pouches within the head on either side of the precibarium and salivary duct. From their origin, and moving towards the apex of the head, the maxillary and mandibular stylets begin to converge upon the precibarium and salivary duct. This convergence brings together the maxillary stylets from either side of the head to form the food and salivary canals which align with the precibarium and salivary duct, respectively (**Figure 13.1**). The salivary canal is used to egest saliva and other enzymatic components to break down host tissue for ingestion via the food canal. The mandibular stylets also converge at the same location but are external to, and enclose, the maxillary stylet bundle. This configuration of an internal maxillary stylet bundle encased by mandibular stylets forms the stylet bundle.

In a cross-section of the stylet bundle, the apposing maxillary stylets interlink at three locations of the maxillary bundle (top, middle, and bottom; Brozek and Herczek 2004). The top and bottom linkages primarily couple the two maxillary stylets. The middle linkage functions similarly and, more importantly, has the added purpose of delineating and separating the food and salivary canals. Thus, the location of the middle linkage relative to the top and bottom affects food and salivary canal diameters. In hemipterans/ heteropterans, a range of configurations for these linkages occur (Cobben 1978), and the middle linkage placement can be influenced by the type of feeding. Based on the orientation of the stylet bundle axis, Butt (1943) indicated the possibility of a twisting aspect to the maxillary bundle along its length. A similar twisting – or auger-type – aspect is indeed present in the stylet bundle of *Nezara viridula* (JFE, unpublished data). Additionally, within the stylet bundle, the maxillary bundle itself displays a similar apical twisting of the interlocked stylets (JFE, unpublished data).

13.1.3.3 Rostrum and Feeding Mechanics

External to the head, the stylet bundle is housed within the rostrum. The rostrum consists of the 4-segmented labium and paired mandibular and maxillary stylets (also see **Section 13.1.2**, Heteropteran Feeding Mechanisms). Several types of sensory hairs are distributed along the labium surface and at the apex of the labium (Rani and Madhavendra 1995). These sensory hairs aid in host selection.

In resting position, the rostrum is held against the body with segment 1 lying within the bucculae on the ventral side of the head. The stylet bundle lies within a groove in the labium but extends anteriorly beyond segment 1. This extended portion of the stylet bundle, as well as the portion along the length of segment 1, is almost completely enclosed by the labrum (JFE, unpublished data), which opens narrowly by a ventral longitudinal groove. The labrum, although not part of the rostrum, still plays a vital role in the movement of the stylets during feeding (see below).

When investigating a potential food source, the insect keeps the labrum against the head venter. However, as the insect begins to insert its stylets into the host, it retains segment 1 within the bucculae but articulates segment 2 away from the insect venter, thereby forming variable angles between segments 1 and 2 when feeding (Esquivel 2011); simultaneously, the labrum (and enclosed portion of stylet bundle) also articulates away from the insect venter and the stylets are extracted from segments 1 and 2 to align the stylet bundle with segment 3. Segment 3 articulates at the juncture with segment 2, and this latter articulation maintains segments 3 and 4 perpendicular with the food source. The latter segments also aid in protecting and guiding the stylet bundle. None of the labial segments penetrates the food source. The insect not only mechanically protracts and retracts the stylet bundle into and from the food source, respectively, it also incorporates its entire body as leverage for penetration (JFE, personal observation).

13.1.3.4 Stylet Penetration Potential

Historically, staining of feeding locations has been used to estimate the depth of stylet penetration by pentatomids and other hemipterans (e.g., Smith 1926). Esquivel (2011), however, devised a practical mathematical model to calculate penetration estimates for pentatomids and mirids (Esquivel 2015) without the use of the labor-intensive staining methods. This new model primarily relies on the interaction between labial segments 1 and 2, as well as the mechanics of the labrum (described above), when an adult *Nezara viridula* assumes a feeding position. For illustrations and mathematical details of the model, the interested reader may consult Esquivel (2011).

For *Nezara viridula*, stylet penetration was significantly deeper as the angles between segment 1 and the proximal end of segment 2 became more acute (Esquivel 2011). At all angles tested (90° — 50°), males differed significantly from females, and all instars differed significantly from one another. Length of rostra and individual segments increased proportionally as the insect progressed through stadia to adulthood (Esquivel 2011). Similar measurements of other adult pentatomids indicated that longer rostra did not always equate to deeper penetration (Esquivel 2015), and the length of the second segment appeared to influence this observation (JFE, unpublished data). All penetration estimates are conservative because the model does not account for bodily behaviors when feeding, but the model should be applicable to all heteropterans feeding in a similar manner and may be used to determine both potential feeding damage and delivery of stylet-borne pathogens.

13.1.4 Overview of Pathogens Transmitted by Pentatomoidea

No heteropterans are known to transmit spiroplasmas or fastidious xylem-colonizing bacteria. The former are transmitted exclusively by leafhoppers (Mitchell 2004) and all known insect vectors of the latter are sharpshooters (Cicadellidae) or spittlebugs (Machaerotidae) (i.e., dedicated xylem-ingesting Auchenorrhyncha [Eden-Green et al. 1992, Purcell and Hopkins 1996]). No pentatomid species are yet known to transmit fastidious phloem-colonizing bacteria, although a coreid of comparable size, *Anasa tristis* De Geer, transmits cucurbit yellow-vine disease (Bruton et al. 2003). Pentatomoids are most strongly associated with the transmission of fungal diseases but also have been implicated in the transmission of a virus, a phytoplasma, non-fastidious bacteria, and trypanosomatid plant parasites (Mitchell 2004). The close association with fungi primarily revolves around the intimate relationship between Heteroptera and *Eremothecium coryli*, which causes disease on a wide variety of crops. Of 36 species of bugs associated with this yeast, 17 are pentatomids (Mitchell 2004). The economic importance of pentatomids as vectors is illustrated clearly by species in the genus *Lincus* Stål, many of which transmit potentially devastating trypanosomatid diseases of oil palm and coconut (Camargo and Wallace 1994).

13.1.5 Selective Transmission — Role of Pathogen Localization

Nezara viridula has been shown to have the capacity to transmit a strain of an opportunistic bacterium *Pantoea agglomerans* (Ewing and Fife) marked with antibiotic resistance into cotton bolls resulting in disease (Medrano et al. 2007). In that study, insects were provided a food source contaminated with the bacterial pathogen and then caged with cotton bolls for 2 days. To determine whether the insects acquired the marked pathogen, the bugs were surface sterilized and pulverized, and suspensions of the whole insect homogenate were selectively cultured on bacterial media with the antibiotic. Bolls were harvested 2 weeks following the caging period with the insects. Minimal seed and lint damage along with an absence of disease was observed in bolls caged with insects not harboring the marked *P. agglomerans* strain. Conversely, bolls fed upon by insects that carried the cotton pathogen had infection symptoms comparable to those observed from field-grown diseased bolls. Further, *P. agglomerans* with antibiotic resistance was recovered from diseased seed and lint tissue; thus, transmission by *N. viridula* was established. Using the same vector model, Medrano et al. (2009a) tested the potential of *N. viridula* to both acquire and transmit other opportunistic cotton pathogens including two bacterial pathogens [*P. ananatis* (Serrano) and *Klebsiella pneumoniae* (Schroeter)] and the fungus *Eremothecium coryli*. Both bacteria were marked with antibiotic resistance; the fungus is not part of the insect's normal microbial flora. Interestingly, all three microbes were detected in insects that were provided a food source contaminated with one of the three pathogens; however, only the fungus was transmitted into cotton bolls.

   The selective transmission of both *Pantoea agglomerans* and *Eremothecium coryli* (but not *P. ananatis* or *Klebsiella pneumoniae*) was then studied by surgically removing and analyzing components of the insect including the rostrum, head (i.e., salivary glands), and the alimentary canal (Esquivel and Medrano 2012). Although all of the microbes were detected in the alimentary canal, only those transmitted by *Nezara viridula* into cotton (i.e., *P. agglomerans* and *E. coryli*) were also found in the rostrum and head. Therefore, the vector potential of the insect is based not only on whether it is harboring a pathogen but where the microbe is residing within the insect. All three of the bacteria used in this work are bacilli (i.e., rod-shaped). However, both *P. agglomerans* and *P. ananatis* are flagellated, which is important in both movement and adhesion (Bergey et al. 1974). Conversely, *K. pneumoniae* is not motile (i.e., no flagella), yet the cells exude a thick, sticky exopolysaccharide (mucus). *Eremothecium coryli* consists of multiple cellular forms including budding yeast, spores, and mycelia (Koopmans 1977). The genomes of both opportunistic strains of *P. agglomerans* (Medrano and Bell 2012) and *P. ananatis* (Medrano and

Bell 2015) have been sequenced, and these data may assist in providing insight into the vector-microbe localization relationship. Regarding *N. viridula*, future research should focus on determining whether the insect's internal chemistry and/or morphology affect colonization by the pathogens in the rostrum and the head. Collectively, these attributes and others likely play a critical role in determining where in the body the microbes colonize this stink bug.

13.1.6 Objectives

The role of phytophagous heteropterans in pathogen transmission historically has been underappreciated. Often, vector screening focuses exclusively on leafhoppers or aphids, and the larger bugs are not even tested until all other options have been exhausted (Mitchell 2004). This review of the literature will enhance our understanding of the role of stink bugs as vectors and, perhaps, stimulate further research. The chapter is organized at the level of the pathogen (e.g., virus, phytoplasma, bacteria, fungi, and trypanosomatids) and subdivided by crop type. Pentatomid taxonomy follows Rider (2015); the sources for bacterial and fungal nomenclature are, respectively, Bergey et al. (1974) and Index Fungorum Partnership (2015).

13.2 Viruses

13.2.1 General Overview of Viruses

Virus transmission is associated primarily with aphids, leafhoppers, and planthoppers; few confirmed cases of virus transmission by Heteroptera are known. Early literature incorrectly attributed many disease symptoms to "viruses" that subsequently were associated with spiroplasmas, phytoplasmas, or simply damage by insect saliva and stylets; thus, older reports of mirids and other true bugs as virus vectors are unreliable. Even today, the term virus may be used in cases where the causative agent is not known with certainty; for example, beet savoy disease, transmitted by a piesmatid, *Piesma cinereum* (Say), is listed variously in reference books as a virus and a suspected virus (e.g., Narisu 2000, Ruppel 2003). Nonetheless, two definitive examples of heteropteran transmission of viruses have been documented: *P. quadratum* (Fieber) (Piesmatidae) transmits the rhabdovirus that causes beet leafcurl disease (Proeseler 1980) in Europe, and *Engytatus* (=*Cyrtopeltis*) *nicotianae* (Koningsberger) (Miridae) transmits velvet tobacco mottle virus in Australia (Gibb and Randles 1991). Modes of transmission in these two cases are quite different. In *P. quadratum*, transmission occurs via saliva (Eisbein 1976) whereas in *E. nicotianae*, the virus particles are not found in salivary glands but in the gut, hemolymph, and feces. An ingestion-defecation mode was proposed to explain this vector-virus relationship (Gibb and Randles 1991).

13.2.2 Longan and Lychee

Longan witches'-broom disease affects the shoots, inflorescences, leaves, and seeds of longan, *Dimocarpus longan* Lour., in China, Thailand, and Vietnam (Australian Government 2003, Nguyen et al. 2012). Leaves are deformed, inflorescences fail to expand properly, and no fruits are produced (Chen et al. 2001). A similar, possibly identical, disease has been reported in China from lychee, *Litchi sinensis* Sonn. (Koizumi 1995). The causative agent has been identified as a filamentous virus by several Chinese researchers (Chen et al. 2001), but the disease is attributed to a phytoplasma in Thailand and Vietnam (Chantrasri et al. 1999, Nguyen et al. 2012). The Australian Department of Agriculture, Fisheries and Forestry thus classifies longan witches'-broom as "aetiological agent unconfirmed" (Australian Government 2003, 2004). In China, the stink bug *Tessaratoma papillosa* Drury transmits the disease in longan and possibly between longan and lychee; however, it is not the only vector; a psyllid (*Cornegenapsylla sinica* Yang and Li) also has been implicated and the parasitic plant, dodder, *Cuscuta campestris* Yunck., also may play a role (Chen et al. 2001). Studies of *T. papillosa* (Chen et al. 2001 and references therein) showed that both nymphs and adults were capable of vectoring the disease agent, and the virus was present in the bug's salivary glands. The nymphal transmission rate was 26–45%, whereas that of adults was 18–36%. This stink bug is not associated with the disease in Thailand or Vietnam.

13.3 Phytoplasmas

13.3.1 General Overview of Phytoplasmas

Originally called mycoplasma-like organisms (MLOs), these prokaryotes are classified as Mollicutes and are related to spiroplasmas and mycoplasmas. Lacking a cell wall, they are thought to have evolved from gram-positive bacteria, and have the smallest genome among the prokaryotes. Because phytoplasmas cannot be cultured in vitro, Koch's postulates cannot be fully satisfied for phytoplasma diseases (Weintraub and Beanland 2006). Koch's postulates, a set of rules for experimental proof of pathogenicity, require that a pathogen be associated continually with the disease and that the organism be isolated and grown in pure culture. The pure culture then must be used to inoculate a healthy plant, causing disease symptoms, and the same pathogen must be isolated from experimentally infected plants (Roberts and Boothroyd 1972). Even though Koch's postulates cannot be satisfied fully for phytoplasma diseases, molecular techniques based on 16S ribosomal DNA (rDNA) polymorphisms have aided extensively in our understanding of phytoplasma diseases and vectors in recent years. Fruit and timber trees, rice, legumes, maize, and potato are among the crops affected by these obligate plant parasites. Phytoplasma diseases are characterized by "yellows," "greening," stunting, phyllody (formation of leaf-like structures from floral parts), and overproduction of axillary buds ("witches' broom" effect) (Bertaccini and Duduk 2009). All phytoplasmas are phloem-limited and vectored by piercing-sucking feeders, especially deltocephaline leafhoppers and several families of planthoppers; in addition, two psyllids are known vectors (Weintraub and Beanland 2006). Infection is acquired passively by ingestion from phloem; these organisms are small enough to pass through pores in sieve elements and are translocated with the flow of assimilate. Phytoplasmas then move into the gut and hemolymph and eventually into the salivary glands (Bertaccini and Duduk 2009).

For many years, it was thought that transmission of phloem-colonizing pathogens, presumably via hemipteran watery saliva, had to involve non-destructive, selective feeding and hence relatively small insect mouthparts, as acquisition and especially inoculation were assumed to occur only in undamaged sieve tube cells (Weintraub and Beanland 2006). However, a lace bug was shown to transmit the coconut root wilt phytoplasma (Mathen et al. 1990), and these insects typically feed by rupturing cells rather than selective feeding in the phloem. Subsequent discovery that a relatively large coreid bug was the vector for a phloem-limited bacterium causing cucurbit yellow vine disease (Pair et al. 2004), and that a pentatomid (see below) transmitted the phytoplasma of Paulownia witches'-broom (Okuda et al. 1998), has called this assumption into question. Recently, a phytoplasma associated with coconut lethal yellowing in Mozambique has been found in *Platacantha lutea* (Westwood), and this pentatomid has been proposed as a potential vector (Dollet et al. 2011). More studies clearly are needed to fully understand the transmission of phloem-limited prokaryotes by true bugs.

13.3.2 Empress Tree

The Empress tree, *Paulownia tomentosa* (Thunb.) Sieb. and Zucc. ex Steud., is native to eastern and central China where it serves as an important timber tree and for urban landscaping. Planted extensively in Japan and Korea for timber and on other continents for economic or ornamental purposes, it has escaped cultivation and become moderately to severely invasive in some regions of the United States and is considered potentially invasive in Australia (Hiruki 1999, Innes 2009). In East Asia, wood quality is severely reduced by a phytoplasma infection known as Paulownia witches'-broom disease. Infected shoots bear sterile, virescent flowers, leaves are faded and distorted, root growth is reduced, phloem necrosis is evident, cells of the wood are disarranged, and saplings are stunted and killed. Up to 70% of mature orchard trees may be infected in China and this phytoplasma is the most important factor affecting empress tree timber production in Japan.

The disease may be propagated vegetatively, but the confirmed insect vector is the polyphagous pentatomid *Halyomorpha halys* (Stål) (Hiruki 1999, Okuda et al. 1998). Some reports in the literature mention *H. mista* (Uhler) or *H. picus* (F.) as vectors, but the former is a junior synonym and the latter resulted from confusion with an Indian species, so these observations now are considered to refer to *H. halys* in

East Asia (Rider et al. 2002). Both nymphs and adults can transmit the phytoplasma from paulownia to periwinkle in Japan; presence in the plant has been confirmed by electron microscopy and in both periwinkle and bugs by molecular analyses of 16S rDNA (Okuda et al. 1998). In China, stink bugs were shown to transmit the pathogen from diseased plants to healthy empress tree seedlings. A 10-day acquisition period was followed by a 30-day incubation period and 5–7 days feeding on healthy plants with 1–3 insects per seedling. Electron microscopy of bug salivary glands and leaf veins and petioles of the seedlings revealed the presence of the phytoplasma, and disease symptoms developed in 33 of 107 inoculated plants (Shao et al. 1982; cited in Hiruki 1999).

Although both empress tree and *Halyomorpha halys* now co-occur in the United States, there is no evidence that paulownia witches'-broom disease is found currently in the United States (Rice et al. 2014) or elsewhere outside of East Asia (Hiruki 1999). Damage from *H. halys* in the United States occurs primarily on reproductive structures of fruit trees, vegetables, and row crops (Hoebeke and Carter 2003, Rice et al. 2014). Nymphs also feed on leaves and stems, but the resultant stippling damage (see photographs in Hoebeke and Carter 2003) suggests cell rupture feeding rather than ingestion from vascular tissues. However, stylet insertion through bark of common landscape trees recently has been observed, resulting in sugary wound exudates (Martinson et al. 2013, Rice et al. 2014), and interpreted as ingestion from the phloem (Martinson et al. 2013). Thus, this species may have the potential to transmit pathogens other than paulownia witches'-broom into vascular tissue.

13.4 Bacterial Pathogens

13.4.1 General Overview of Bacterial Pathogens

Bacteria are single-celled, ubiquitous microorganisms that colonize diverse environments including stink bug tissues (Kaiser and Vakili 1978, Ragsdale et al. 1979, Mitchell 2004, Hirose et al. 2006, Medrano et al. 2009b). The majority is harmless, yet a subset consists of plant pathogens. Plant pathogenic bacteria include fastidious (i.e., difficult to culture outside of the host) and non-fastidious (i.e., grow readily in bacterial culture media) species (Agrios 2005). Both types include infective bacterial strains that cause opportunistic plant infections. The basic penetrative feeding mechanism used by stink bugs wounds plant hosts and provides a physical conduit for acquisition and transmission of a bacterial pathogen. Interestingly, the number of studies focused on this vector potential is limited, and any resulting plant infection is grouped with the damage associated with insect feeding (Williams 2014). In this section, we focus on stink bugs known to be vectors of bacterial pathogens.

13.4.2 Cotton

In the late 1990s, a non-traditional cotton boll rot noticeably began to decrease cotton yields in South Carolina (Hollis 2001). Symptoms of this malady are manifested exclusively inside developing green bolls with the outside appearing normal. Thus, early diagnosis of infected fields is based on cross-sectioning of green bolls. Medrano and Bell (2007) reported that an infective bacterium identified as *Pantoea agglomerans* that was isolated from a diseased field boll was a causative agent of the disease. Infections only occurred if the boll wall was pierced mechanically using a needle allowing an entry way for the pathogen. In other work, Medrano et al. (2007) found that *Nezara viridula* was a capable vector of the same *P. agglomerans* strain used in earlier studies. Stink bugs reared in the laboratory acquired the cotton pathogen after they were provided a bacterial contaminated food source. The pathogen was transmitted into bolls upon feeding by these insects resulting in diseased seed and lint with symptoms identical to initial reports and a symptomless boll exterior. Feeding by insects not exposed to the opportunist caused negligible internal damage to the bolls. Interestingly, temporal studies subsequently showed that bolls became immune to infections following three weeks of development (Medrano et al. 2009b). Also, bolls at two weeks into development could tolerate insect feeding provided that there was no pathogen present. Collectively, these studies dissected the dynamics between *N. viridula* and a vector-borne infection.

13.4.3 Legumes

Nezara viridula has been examined as a potential vector of bacterial blight of beans (*Phaseolus vulgaris* L.) and cowpea [*Vigna unguiculata* (L.) Walp.], and various bacteria capable of causing necrotic water-soaked spots on leaves of soybean [*Glycine max* (L.) Merrill] (Kaiser and Vakili 1978, Ragsdale et al. 1979). When only the stylets were allowed to touch artificial medium during feeding, field-collected *N. viridula* adults in Louisiana transferred 31 types of bacteria representing 13 genera. Ragsdale et al. (1979) obtained many of these bacteria from dissected insect bodies, but only isolates of *Cornyebacterium* and *Pseudomonas* caused leaf vein necrosis and spots when applied to artificially damaged vegetative soybean plants (Ragsdale et al. 1979). They concluded that only a "casual relationship" existed between the stink bugs and the transferred microflora, but that *N. viridula* did have potential for significant vector activity. Subsequent studies of field-collected soybean seeds exposed to different stink bug population densities in Louisiana showed levels of seedborne bacteria to rise with increasing stink bug population levels. However, this effect was seen only in the upper portion of plants, which suffered higher levels of feeding damage, and only in 1 of the 2 years of the study. A similar but non-significant trend was observed in the second year. Bug species present were predominantly *N. viridula* plus *Chinavia hilaris*, and *Euschistus* spp.; bacterial isolates were not identified (Russin et al. 1988). In Puerto Rico, several isolates of *Xanthomonas axonopodis* (as *Xanthomonas phaseoli)*, which causes bacterial blight of beans and cowpea, were obtained by washing field-collected *N. viridula* in sterile distilled water. Of 10 insects tested, 5 carried *Xanthomonas* pathogenic to either bean or cowpea. Stem cankers formed when stink bugs were exposed to a bacterial suspension and allowed to feed on cowpea, but no transmission was observed when bugs collected from blight-infested plants fed upon caged test plants. In contrast, leaf-feeding beetles did transmit the xanthomonads in controlled feeding trials. Therefore, the chrysomelid *Cerotoma ruficornis* Olivier was suggested as the primary vector rather than *N. viridula* (Kaiser and Vakili 1978).

13.5 Fungal Pathogens

13.5.1 General Overview of Fungal Pathogens

Associations between pentatomid feeding and fungal pathogen transmission have been widely reported (Mitchell 2004). Generally, fungi comprise the majority of plant pathogens (Agrios 2005). In contrast to bacteria that typically require a host wound or other entry point, certain fungi produce structures called appressoria that physically penetrate plant tissues (Mendgen et al. 1996). Once inside the host, necrotrophic fungi absorb nutrients, killing the infected plant cells. In addition to filamentous forms, some fungi are dimorphic growing as budding yeasts depending on environmental conditions (Webster and Weber 2007). Fungal cells, with hyphae that are 55–100 μm long and 7–10 μm wide (Marasas 1971), are much larger than bacterial cells (1.0–2.5 μm long by 0.8–1.0 μm wide), but they are still small enough to be internalized by pentatomids. Here, we focus on several fungal diseases that are linked to vector-feeding activities.

13.5.2 Coffee

In coffee-growing countries of east Africa—including Uganda, Kenya, and Burundi—the pentatomid species in the genus *Antestiopsis* are associated with numerous forms of damage to coffee plants (*Coffea arabica* L.). According to Kirkpatrick (1937) and Le Pelley (1942), all *Antestiopsis* spp. are oligophagous—feeding only on hosts in the Rubiaceae, although both studies were not comprehensive. It also appears that *C. arabica* is a highly preferred host plant over numerous wild hosts as well as *Coffea canephora* Pierre ex A. Froehner (= *C. robusta*) (Kirkpatrick 1937, Le Pelley 1942). Feeding by *A. thunbergii* (Gmelin) (as *Antestia lineaticollis* Stål) on growing coffee shoots causes "die back" and reduced growth, and feeding on green and red coffee cherries results in abscission of fruit (Le Pelley 1942). The role that microbial pathogens play in shoot die-back or fruit abscission remains unclear.

Nonetheless, feeding by *Antestiopsis thunbergii* likely is associated with potato taste defect (PTD) (Jackels et al. 2014). PTD disease is named for the moldy potato-like odor associated with diseased beans when roasted. Three candidate causal pathogens have been proposed: *Eremothecium coryli*, *Nematospora gossypii* Ashby and Nowell [= *Eremothecium gossypii* (Ashby and Nowell) Kurtzman] and the bacterium *Pantoea coffeiphila* (Wallace 1931, 1932, Gueule et al. 2015). Importantly, transmission tests have not been conducted and Koch's postulates have not been satisfied for any of these candidate agents. Rather, the evidence is more circumstantial. Kirkpatrick (1937) hypothesized that all *Antestiopsis* spp. are able to transmit fungal pathogens, that *E. coryli* and *E. gossypii* are widespread in coffee agro-ecosystems, and that infection prevalence of vectors varies spatially and temporally. Based on previous work on pentatomid transmission of *E. coryli* and *E. gossypii*, we hypothesize that they are transmitted by *Antestiopsis* spp. in a semi-persistent manner (Mitchell 2004). The potential for insect transmission and plant pathogenicity of *P. coffeiphila* are currently unknown, although the newly described species is associated with PTD (Gueule et al. 2015). Finally, although more speculative, Small (1923) argued that *Antestiopsis* spp. also can infect coffee cherries with *Phoma* spp. (fungi).

PTD incidence tends to be low but can nonetheless severely reduce crop yield and quality (Jackels et al. 2014). Management of PTD primarily focuses on management of *Antestiopsis* spp. which can be effective (Le Pelley 1942). Economic threshold estimates vary between one and four individuals per coffee plant (Meulen and Schoeman 1990, Cilas et al. 1998); such low thresholds are consistent with other pentatomid-borne diseases (Greene et al. 2001; see **Section 13.5.3**, Cotton).

13.5.3 Cotton

Nowell (1918) reported a "stainer bug", *Dysdercus suturellus* (Herrich-Schaeffer) that deposited fungi into cotton bolls in the West Indies. This early work involved collecting field insects and then exposing them to cotton. Damage to seed and lint quality was labeled as a type of stigmatomycosis that involved *Eremothecium* (formerly *Nematospora*). The infection occurred strictly within the bolls and had a greater effect on young developing fruit. Symptoms included yellowing of the lint, which decreased the value of the affected bolls. It was found that control of the sucking insect consequently also decreased the disease incidence.

Marasas (1971) reported that the yeast *Eremothecium ashbyi* Guillermond [as *Crebrothecium ashbyi* (Guilliermond) Routien] caused an inner boll rot of cotton in South Africa as well as other parts of the continent. Additionally, three yeast-like species of Spermophthoraceae: *Ashbya gossypii* (Ashby and Nowell) Guilliermond (=*Eremothecium gossypii*), *Eremothecium coryli* and *E. ashbyi* were reported to cause cotton staining in association with sucking insects. The taxonomy of this group is controversial, yet they all apparently stain cotton similarly. In this study, open, stained bolls were tested for the presence of fungal structures such as hyphae and spores. It was presumed that a sucking insect had transmitted the fungi as insects were not collected.

More recently, Medrano et al. (2009a) demonstrated the ability of *Nezara viridula* to transmit *Eremothecium coryli* into cotton bolls. The yeast was isolated from field-grown diseased bolls collected in Florida and Georgia (USA). Interestingly, bolls collected from North and South Carolina did not yield the fungus. Insects reared in the laboratory were tested for fungal growth and none was detected. Stink bugs were then provided a food source that had been dipped in a yeast suspension. Finally, the insects were housed with greenhouse grown bolls. Disease did indeed occur in bolls fed upon by insects that harbored the fungus. Conversely, bolls fed upon by insects that did not possess the pathogen were neither stained nor diseased. Further, the fungus was recovered from both the contaminated insects and diseased bolls; thus, Koch's postulates were fulfilled. The infection symptomology observed was comparable to previous descriptions with yellow, matted lint. In contrast, bacterial inner boll infections caused lint and seed necrosis with a dark brown to black coloration.

13.5.4 Legumes

The association between *Eremothecium coryli* and bug feeding has been recognized for nearly a century (Nowell 1918). Although the earliest observations were of cotton stainers, Ashby and Nowell (1926) reported that in the West Indies, the abundant and polyphagous *Nezara viridula* transferred the fungus from cultivated legumes to cotton. Daugherty (1967) reported that on soybean and other legumes,

E. coryli infection causes cream-colored, sunken areas or necrotic, discolored lesions on the seed coat, hence the name yeast spot. Infection during pod formation causes abscission; infection during seed development produces spots, depressions, shriveling, and change in texture and color of the cotyledon. Preston and Ray (1943) found that lima bean, cowpea, mung bean, and soybean could be affected. The role of stink bugs in the transmission of this disease on lima beans (as *Nematospora phaseoli* Wingard) was confirmed by Wingard (1925), but research on vectors of yeast spot of soybean did not begin until the 1960s. Prior to this time, the symptoms of yeast infection often had been attributed to direct stink bug feeding damage (Daugherty 1967).

Daugherty (1967) reported that *Thyanta custator* (F.), *Euschistus variolarius* (Palisot de Beauvois), *E. tristigmus* (Say), *E. servus euschistoides* (Vollenhoven), *E. servus servus* (Say), and *Chinavia hilaris* transmitted yeast spot disease in cage studies in the United States; macerated heads of *E. servus servus* were found to contain the fungus. Two other pentatomids, *Holcostethus limbolarius* (Stål) and *Cosmopepla lintneriana* Kirkaldy (as *C. bimaculata* Thomas), also were tested by Daugherty (1967), but they failed to transmit the fungus to soybeans. Earlier studies of the disease in lima bean already had implicated *C. hilaris* (Wingard 1925, Schoene and Underhill 1933), although Leach and Clulo (1943) questioned whether the fungus could be carried internally. Subsequently, more detailed studies of *C. hilaris* conducted by Foster and Daugherty (1969) yielded viable yeast from surface-sterilized nymphs and adults, and from all three tagmata of macerated bodies of surface-sterilized adult bugs, although more frequently from heads and abdomens. Analysis of organs revealed spores to be present primarily in the stylets, salivary receptacles, and hindgut. Clark and Wilde (1970a,b) studied retention and acquisition of the fungus. Laboratory inoculation of *C. hilaris* using a yeast suspension or infected pods showed retention of *Eremothecium coryli* for 90 days and at least 60 days, respectively. Heads were more likely than other body regions to carry the fungus. Viable *E. coryli* also could be isolated from feces. Fifth instars lost the ability to transmit after the adult molt, but their exuviae were infective (Clark and Wilde 1970a). Adults could acquire the fungus by feeding on soybean or dogwood (*Cornus drummondii* Meyer) berries previously fed upon by infected bugs (Clark and Wilde 1970b). A survey of microorganisms associated with *Nezara viridula* collected from soybean in Louisiana (Ragsdale et al. 1979) showed only 2 of 118 insects tested carried *E. coryli* internally, although the dimensions of the salivary and food canals in this species are large enough to admit the ascospores. In Japan, both *E. coryli* and *E. ashbyi* cause yeast spot of soybean (Kimura et al. 2008b). Although research has focused on the primary vector, the alydid *Riptortus pedestris* (F.) [as *Riptortus clavatus* (Thunberg)], several pentatomid species also have been shown to carry and transmit *E. coryli* but at lower rates. Transmission rates for adults of *N. antennata* Scott, *Dolycoris baccarum* (L.), and *Piezodorus hybneri* (Gmelin) were 50.0, 40.0, and 16.7%, respectively, compared with 81.6% for *R. pedestris*. No *E. coryli* was isolated from nymphs of *N. antennator* or *D. baccarum*, although the carrying rate for nymphs of *R. pedestris* was 11.5% (Kimura et al. 2008a).

Fungi other than *Eremothecium coryli* also are associated with stink bugs on soybean, but the relationship is not as intimate as that for yeast spot. Two *Penicillium* spp. isolates also were included among the internal microflora of *Nezara viridula* (Ragsdale et al. 1979). Seed infection by *Fusarium* spp. increased significantly with stink bug feeding damage in controlled field plots in Louisiana (Russin et al. 1988). In the same study, rates of seedborne *Alternaria* spp., *Colletotrichum truncatum* (Schw.) Andrus & W. D. Moore, and *Phomopsis* spp. were unaffected by population levels of stink bugs, which included *N. viridula* (77.9%), *Euschistus* spp. (16.6%), and *C. hilaris* (5.5%) (Russin et al. 1988). However, Nyvall (1999) noted that infection rates for seedborne *Alternaria* spp. were associated with damage by bean leaf beetle and stink bugs. In Brazil, increased seed infection by *Fusarium* sp., *Phomopsis sojae* Lehman, and *C. truncatum* is associated with feeding by *Piezodorus guildinii* (Westwood) (Panizzi et al. 1979). These fungi occur naturally on soybean and can infect seeds in the absence of stink bugs. The role of the bugs is to alter the incidence of infection; whether this is due to the physical damage from feeding or a change in nutritional quality of the damaged seed is uncertain (Russin et al. 1988).

13.5.5 Pistachio

In California, several hemipterans cause direct damage to pistachio (*Pistacia vera* L.) fruits, producing a condition known as epicarp lesion and characterized by the appearance of necrotic regions followed by fruit abortion. Feeding through the shell later in the season produces discolored, spongy, sunken regions in the nutmeat, known as kernel necrosis (Rice et al. 1985). In addition to this direct feeding damage, coreids and pentatomids have been implicated in the transmission of two fungal diseases, pistachio stigmatomycosis and panicle and shoot blight (Mitchell 2004).

Stigmatomycosis is an older term for fungal infections caused by *Eremothecium coryli* and other yeasts associated with hemipteran feeding damage (Ashby and Nowell 1926). The term still is used commonly in pistachio for infections by *E. coryli* and *Aureobasidium pullulans* (de Bary) G. Arnaud, although similar fungal infections in other crops have different names (e.g., yeast spot, fruit rot, dry rot). The disease is found throughout the pistachio-growing areas of California and in Iran, Russia, and Greece (Michailides and Morgan 1990, 1991, and references therein). In pistachio, the disease may appear as underdeveloped kernels; wet, rancid, distorted, but full-sized kernels; or white, jelly-like kernels (Michailides 2014). Incidence of this disease in orchards ranges from 0.7 to 29.1%, depending on mode of irrigation (Michailides and Morgan 1990), and peaks in August and September, which coincides with a period of kernel feeding by large hemipterans (Michailides and Morgan 1991).

The pentatomid species associated with transmission of these fungal infections in California are *Thyanta pallidovirens* (Stål), *Chlorochroa uhleri* (Stål), and *C. ligata* (Say); a coreid, *Leptoglossus clypealis* (Heidemann), also is a known vector (Michailides and Morgan 1990). Infection was significantly enhanced when fruit clusters sprayed with a suspension of *A. pullulans* were caged in the field with *T. pallidovirens*; incidence of stigmatomycosis in this treatment reached 28.6%. When *T. pallidovirens* were allowed to feed on cultures of this yeast and then exposed in field cages to sterilized fruit clusters, significant transmission levels were also documented (Michailides and Morgan 1991). The involvement of *A. pullulans* in stigmatomycosis is unusual; the common causative agents in most crops are *Eremothecium coryli* and related yeasts. Experiments with the two *Chlorochroa* species, *L. clypealis*, and *T. pallidovirens* feeding on *E. coryli* produced similar results, with infection rates ranging from 17.7 to 29.6% for the pentatomids, compared with 43.6% for the coreid (Michailides and Morgan 1990). These bugs can, therefore, acquire fungal propagules from the plant surface or from infected pistachio fruits or other crops and transmit directly into the kernel. Sprayed fungicides have no effect on the incidence of stigmatomycosis in pistachio, suggesting that the yeasts may be carried internally (Michailides and Morgan 1990) as shown for pyrrhocorids (Frazer 1944) and other pentatomids (see **Section 13.5.4**, Legumes).

Panicle and shoot blight, first noted in California orchards in 1984, is a devastating disease considered to be a serious threat to California pistachios (Holtz 2002, Michailides and Morgan 2004). Symptoms begin as necrotic spots on shoots, rachises, and leaves (and later on fruits) that coalesce as the season progresses. Spots on petioles lead to leaf abscission and defoliation, whereas blight on fruiting structures causes the rachis to collapse, ultimately destroying whole fruit clusters. Losses may reach 100% (Rice et al. 1985, Holtz 2002). The causative agent is an ascomycete fungus, *Botryosphaeria dothidea* (Mougeot) Cesati and De Notaris, which is ubiquitous in woodlands on trees and shrubs as well as on other cultivated tree fruits (Michailides et al. 1998). Conidia are spread by rain, birds, and sprinkler irrigation as well as insects (Holtz 2002); the latter may transfer the fungus within orchards, between orchards, or perhaps even from surrounding vegetation to pistachio orchards. The large propagules (pycnidiospores measure 15–29 × 5–8 μm) are thought to be carried externally on hemipteran mouthparts and legs. Bug feeding and wounding in general facilitate entry into the rachis or fruits, although infection can occur via stomata or lenticels (Michailides and Morgan 2004, Daane et al. 2005).

Thyanta pallidovirens, *Chlorochroa uhleri* and an unidentified *Chinavia* sp. (as *Acrosternum*) have been studied to determine their role in transmission of panicle and shoot blight, along with several coreoid and mirid species. Feeding punctures correlate significantly with the incidence of infected pistachios. In a series of field experiments, a significant increase in infection rates occurred when *T. pallidovirens*, the unidentified *Chinavia* sp., or the coreoids were caged on fruit sprayed with *Botryosphaeria dothidea*; results for *C. uhleri* showed a similar trend (Michailides et al. 1998). However, subsequent trials (Daane et al. 2005) showed that mechanical wounding with a pin increased infection rates significantly more than did stink bug feeding. In addition, these bugs are poor vectors; of a large sample of hemipterans collected in infected orchards, <0.2% carried inoculum. Furthermore, spores remained attached to stink bugs for only 3 days, compared with 10 days for a leaffooted bug. The most important role of pentatomids, coreids, and mirids in the orchards may be to provide germination sites for existing spores by causing fluids to ooze from feeding punctures (Daane et al. 2005). Stink bug monitoring and control is only one part of a management plan that includes pruning, fungicide sprays, and modification of irrigation methods (Holtz 2002).

13.5.6 Rice

Pecky rice is a general term for rice of inferior quality with imperfections; the kernel is discolored and often breaks during milling. The condition is associated with stink bug feeding and is thought to result also from fungi introduced via feeding because two distinctly different types of spots are produced, a round chalky lesion and a darker discoloration. A detailed discussion of the relationship between *Oebalus pugnax pugnax* (F.), fungal infections, and pecky rice is given by McPherson and McPherson (2000), and therefore only selected papers will be summarized here.

Gelling saliva is produced by pentatomids as the stylets contact and penetrate plant tissue, forming a surface flange and a sheath lining the path of the stylets. Superficial salivary sheaths of *Oebalus pugnax* harbor fungi on their interior and exterior walls. Hull penetration (indicated by an internal sheath) results in the presence of fungi in and around puncture wounds to the kernel and underneath the hull. As stylets are removed, sheath material usually seals the flange; however, scanning electron microscopy showed some sheaths on the rice hull remained open, which may provide an additional infection route. The authors concluded that fungi may enter the kernel either via the stylets during the feeding process or through an open puncture after feeding (Hollay et al. 1987).

Five different fungi, isolated from damaged rice, caused peck symptoms when inoculated with a wire into the endosperm of rice kernels at the soft dough stage; the wire was intended to mimic stylet penetration. Inoculation attempts without the wire were unsuccessful (Lee et al. 1993). Two fungi that caused discoloration in inoculation trials also were cultured from excised stylets of *O. pugnax*: *Alternaria alternata* (Fr.) Keissler and *Curvularia lunata* (Wakker) Boedijn. Based on these studies and additional field experiments, these authors concluded that the discolored spots characteristic of pecky rice are caused by fungi introduced as *O. pugnax* feeds, but that the relationship between the bug and the fungi is a loose one (Lee et al. 1993).

13.5.7 Other Crops

Eremothecium coryli infests fruits representing at least 16 genera (Ashby and Nowell 1926); in all cases when the vector has been sought, it has been a heteropteran. In addition to the crops already discussed above, serious outbreaks of this pathogen have been reported in tomato and in citrus. Fruit rot of tomato in California in 1998 led to extensive losses where portions of fields were completely abandoned due to deterioration of ripe fruit at harvest time. The presence of *Euschistus conspersus* Uhler in the fields prompted inoculation experiments simulating bug probing. Lesions appeared only if *E. coryli* inoculum was injected to a depth of 4 mm; an ascospore suspension placed on the surface of the fruit failed to cause infection (Miyao et al. 2000).

In Cuba, *Eremothecium coryli* infests tomato fruit but also produces round yellow spots on the surface of green oranges, often causing fruit drop and rendering the oranges unmarketable (Grillo and Alvarez 1983). In the center of each yellow lesion, a salivary sheath was evident with stylets penetrating to the juice vesicles. The coreid *Leptoglossus gonagra* (F.) was responsible for most damage, but *Nezara viridula* also was present in orange groves. Fruits exposed in the laboratory to insects of both species (field-collected from citrus and cucurbit weeds) developed the typical yellow lesions, with ascospores and vegetative cells of *E. coryli* present in the vesicles. Vegetative cells also were found in the hindgut of both *L. gonagra* and *N. viridula*. Heads of *L. gonagra* were dissected and subjected to histological examination, but no *E. coryli* were observed in any form, possibly due to the small sample size. The bugs likely transfer the yeast in the field from their cucurbit hosts (e.g., *Momordica charantia* L.) to citrus (Grillo and Alvarez 1983). Two other pentatomid species are reported by Frazer (1944) to transmit stigmatomycosis of citrus: *Rhynchocoris poseidon* Kirkaldy (as *Rhynchocoris serratus* Don.) and *Cappaea taprobanensis* (Dallas). In Australia, the disease is known as dry rot, and may be associated with the spined citrus bug, *Biprorulus bibax* Breddin, which was present in large numbers when the disease first appeared (Shivas et al. 2005).

Halyomorpha halys has been found to transmit *Eremothecium coryli* in the laboratory to several vegetables and fruits. The yeast was recovered consistently from internal tissues damaged by the feeding of *H. halys* in the field. In laboratory trials, *H. halys* was able to infect tomato, green pepper, green beans, apples, pears, and nectarines (Brust and Rane 2011). Given the highly polyphagous habits of this stink bug, the potential for dissemination of *E. coryli* in the United States will be increased greatly as this invasive stink bug continues to spread.

Aflatoxins are produced by the fungus *Aspergillus flavus* Link and related species. In cornfields, the spatial patterns of stink bug-damaged kernels and aflatoxin levels are related; both variables were found to be aggregated with a strong field-edge effect, and aflatoxin levels were positively correlated with stink bug damage. The predominant ear-feeding stink bug was *Euschistus servus* (Ni et al. 2011). In a further study, results were confirmed; indices of association between these factors were significant at several sampling locations (Ni et al. 2014b). In contrast, feeding by *E. servus* had no effect on the incidence of common smut [*Ustilago maydis* DC (Corda)] infection in maize (Ni et al. 2014a). In field cage studies, smut infection percentages did not differ significantly between plants that were exposed to bugs shaken with smut spores, bugs without smut spores, or no bugs. Interestingly, post-harvest analyses of cornmeal for aflatoxin in this cage study (Ni et al. 2014a) indicated no significant relationship between bug presence and the level of aflatoxin, unlike the two previous field studies (Ni et al. 2011, 2014b).

Kernel-spot lesions on pecan, caused by feeding of *Nezara viridula* and other stink bugs, harbor a wide variety of fungi including some toxic *Penicillium* spp. (Payne and Wells 1984). Whether *N. viridula* is involved in transmission is unknown, but *Penicillium* spp. are known to be part of the internal microflora of *N. viridula* collected from soybean (Ragsdale et al. 1979).

Moniliasis disease of cacao, or frosty pod rot, caused by *Moniliophthora roreri* H.C. Evans, is a serious threat to cocoa production from Mexico to South America (Phillips-Mora et al. 2006). Losses may reach 80–90% within a few years of establishment of the fungus. The cacao bug, *Antiteuchus tripterus* (F.) was implicated as a vector at one time (Eberhard 1974; cited in Agrios 1980) but distribution and transmission currently are attributed to wind and human activity (Krauss 2014).

13.6 Trypanosomatids

13.6.1 General Overview of Trypanosomatids

Most entomologists are familiar with the insect-vectored trypanosomatids causing human illness — Chagas' disease, leishmaniasis, and African sleeping sickness— but a closely related group associated with plants and insects is less well known. These plant-associated trypanosomatids are mostly nonpathogenic, but a few cause serious disease of coconut, oil palms, and coffee. Monoxenous trypanosomatids (i.e., species that have only a single host) live within the midgut and hindgut of insects, particularly Hemiptera and Diptera (Maslov et al. 2013); these occasionally may be found in fruits (Conchon et al. 1989), but the life cycle is not considered truly digenetic. Dixenous species, in contrast, inhabit both plants and insects during part of their life cycles. Many of these were placed originally in the monoxenous genera *Leptomonas* or *Herpetomonas*, but eventually were classified together in the genus *Phytomonas* on the basis of morphology and the plant host (Hollar and Maslov 1997). Despite earlier uncertainty about this somewhat arbitrary placement, recent molecular phylogenetic studies have confirmed that, indeed, *Phytomonas* is monophyletic (Hollar and Maslov 1997, Maslov et al. 2013). Three groupings within *Phytomonas* based on the plant structure inhabited (i.e., latex, fruit, or phloem) (Vickerman 1994) also are supported by molecular analyses (Hollar and Maslov 1997, Jaskowska et al. 2015). A review of *Phytomonas* research is provided by Jaskowska et al. (2015); the hemipteran vectors are reviewed thoroughly by Camargo and Wallace (1994) and more briefly by Mitchell (2004). An extensive bibliography is provided by Solarte et al. (1995).

Latex-inhabiting forms generally are considered to be commensals rather than parasites of euphorbs, milkweeds, and other plants with laticifers. Bugs in several heteropteran families have been implicated as vectors, including Lygaeidae, Rhyparochromidae, Stenocephalidae, and Pentatomidae (Camargo and Wallace 1994). *Edessa loxdalii* Westwood was able to transmit *Phytomonas* sp. to seedlings of *Cecropia palmata* Willd. in Suriname and 51% of the wild population harbored these organisms in the salivary glands (Kastelein 1985; cited in Camargo and Wallace 1994). Only one instance of pathogenicity of a latex-inhabiting species has been reported, in cassava (Vainstein and Roitman 1986). However, the insect vector was never determined and the cassava cultivar susceptible to infection is no longer grown (Camargo 1999).

Fruit-inhabiting forms, isolated from tomato, corn, and other crops, are vectored by a number of heteropterans, including coreids (Jankevicius et al. 1989) and the pentatomids *Nezara viridula* (Gibbs 1957) and *Arvelius albopunctatus* (De Geer) (Kastelein and Camargo 1990). In both pentatomid species, the flagellates are found in the digestive tract and the salivary glands. This group of trypanosomatids is not of serious economic concern. Infection is suspected to cause fruit spotting (Camargo 1999), but infected fruits may be unblemished and there is no clear evidence of pathogenicity.

Phloem-limited trypanosomatids are by far the most economically damaging, contributing to substantial losses in coconut, oil palm, and coffee production in South America. The vector of coffee necrosis, caused by *Phytomonas leptovasorum* Stahel, has not been determined. *Ochlerus* spp. (Pentatomidae: Discocephalinae: Ochlerini) harbor the trypanosome, but transmission has not been demonstrated experimentally (Vermeulen 1968; cited in Camargo 1999). Suspicion also has fallen on two species of the pentatomid genus *Lincus* Stål, namely *L. spathuliger* Breddin (Stahel 1954; cited in Dollet 1984), and *L. styliger* Breddin (Dollet 1984), but, again, no transmission could be shown.

Two serious diseases of palms transmitted by phloem-limited trypanosomatids, hartrot of coconut and "marchitez sorpresiva" (sudden wilt) of oil palm, are caused by *Phytomonas staheli* McGhee and McGhee and were shown to be transmitted by pentatomids in the genus *Lincus* (Camargo and Wallace 1994). The palm diseases caused by trypanosomatids are restricted to Central and South America and the Caribbean. Several species of palms other than the major economic species also are affected by *Phytomonas staheli*, including açaí (Camargo et al. 1990 and references therein, Meneguetti and Trevisan 2010). A wilt of ornamental red ginger [*Alpinia purpurata* (Vieillard) K. Schumann] in the Caribbean is caused by a similar strain of this trypanosomatid (Dollet et al. 2001).

13.6.2 Coconut

Hartrot of coconut, *Cocos nucifera* L., is a lethal wilt that has been reported from Brazil, Columbia, Ecuador, Suriname, Trinidad, Costa Rica, and French Guiana (Dollet 1984, McCoy et al. 1984, Resende et al. 1986). Leaves wilt and turn brown, starting from the lowest leaves and advancing upwards. Floral buds become necrotic, unripe fruit drops, the apical shoot rots, and eventually the entire heart of palm is destroyed, including the roots. From visible symptoms to death takes 10 weeks to several months and because the spread is rapid, entire plantations may be demolished by this disease (Bezerra and Figueiredo 1982, Dollet 1984, Camargo 1999).

Hartrot first was noted in Suriname over 100 years ago, but the etiological agent remained unknown until 1976, and the relationship between the trypanosomatid and its insect vectors was not established until the 1980s (Camargo 1999). Early studies linked the disease to the genus *Lincus*, but, often, the transmission studies used a mix of species, or identification was taken only to genus. Taxonomic work by Dolling (1984) and Rolston (1983, 1989) has helped to clarify the vectors. Hartrot, most likely, is transmitted in French Guiana by *L. bipunctatus* (Spinola) (Louise et al. 1986, as *L. croupius* Rolston); this species can acquire trypanosomatids in the laboratory from diseased coconut, which then can be recovered from the salivary glands (Dollet et al. 2002). Cage studies documented transmission of hartrot from wild-collected bugs to healthy coconut palms, but two other species, *L. dentiger* Breddin and *L. apollo* Dolling, also were included initially in these experiments (Louise et al. 1986). In Suriname, *L. vandoesburgi* Rolston and *L. lamelliger* Breddin occur on diseased coconut, and one or both species are responsible for disease transmission (Asgarali and Ramkalup 1985; cited in Camargo and Wallace 1994). In Brazil, *L. lobuliger* Breddin is the vector (Resende et al. 1986).

A simulation model incorporating the plant, the trypanosomatid *Phytomonas staheli*, and the vector *Lincus lobuliger* was developed for coconut in Brazil. These bugs gather on the undersides of leaf petioles, feeding from the phloem sap of stipules. At night, adults and nymphs may be found moving about on the ground. The model predicted that chemical control would delay spread of the disease but would need to be used continually to be effective. Another option was banding trees or otherwise blocking upward colonization from the ground, but ultimately breeding programs may be the optimal solution (Sgrillo et al. 2005).

The genome of an isolate from South American coconut recently was sequenced and compared with that of a latex-inhabiting species from *Euphorbia*. Interestingly, both strains have the requisite enzyme to metabolize trehalose, presumably for nutrition in their insect vectors, but only the phloem-inhabiting pathogen (i.e., the coconut hartrot isolate, *Cocos nucifera*) possesses multiple copies of an invertase homolog, suggesting a greater ability to metabolize sugars (Porcel et al. 2014).

13.6.3 Oil Palms

"Marchitez" of oil palms, *Elaeis guineensis* Jacquin, first was reported in Colombia in 1963 shortly after African oil palms were first cultivated there (Lopez et al. 1975; cited in Dollet 1984). The disease, also called sudden wilt, is prevalent across northern South America, including Venezuela, Peru, Ecuador, Colombia, Brazil, and Suriname. Trypanosomatids were recognized quickly as the etiological agent, and the connection between hartrot and marchitez was established when *Phytomonas staheli* was described from both coconut and oil palms in Suriname (McGhee and McGhee 1979). On the basis of light and electron microscopic analysis of flagellate morphology, McCoy and Martinez-Lopez (1982) showed both coconut and oil palm pathogens in Colombia to be indistinguishable from *P. staheli*. Subsequent DNA analyses (Serrano et al. 1999) supported the similarity of these pathogens, although multiple strains exist (i.e., between hartrot and marchitez flagellates and even among hartrot isolates from different geographic areas). Hartrot strains differed from those of marchitez by a single cytosine/thymine substitution in the region of DNA examined. Using these methods, all of the palm isolate sequences could be distinguished easily from isolates of a different species from tomato fruit (Serrano et al. 1999). Transmission experiments in Brazil demonstrated that *Lincus lobuliger* transferred from coconut infected with hartrot could pass the disease to healthy oil palms (Resende et al. 1986). The progression of symptoms in marchitez also is similar to that of hartrot. Leaves turn brown and then gray, beginning with the lower leaves and moving upward, fruit rots and drops, and the root system is weakened (Dollet 1984 and references therein). Recently, a much slower progression of marchitez disease has appeared in Peru, called slow wilt as opposed to sudden wilt. Initial leaf chlorosis is evident, followed by dehydration and breakage of the leaves. The pattern of fruit decay, root rot, and leaf discoloration differs from that of sudden wilt (Trelles Di Lucca et al. 2013).

Of the 35 species of *Lincus*, 13 species have been reported from economically important palms (Maciel et al. 2015 and references therein). Confirmed vectors of marchitez (sudden wilt), in addition to *L. lobuliger* in Brazil, include *L. lethifer* Dolling (Ecuador) and *L. tumidifrons* Rolston (Colombia) (Desmier de Chenon 1984, Dolling 1984, Perthuis et al. 1985, Alvarez 1993). In a study of possible vectors of slow wilt, a high proportion of *L. spurcus* Rolston was found to carry trypanosomes in the abdomen and head, but actual transmission could not be verified, possibly because of the long incubation time of the disease (Trelles Di Lucca et al. 2013). Surprisingly little is known of the biology of *Lincus* spp., which are difficult to observe, occupying crevices between the bases of fronds and other hidden locations (Camargo and Wallace 1994). Desmier de Chenon (1984) described the biology of *L. lethifer* on oil palms in Ecuador. Diseased trees harbored hundreds of bugs in the crown between sheaths of the flower stalks and under the root bulb; apparently healthy trees supported bug colonies as well, but, within a few months, all of the these trees also had developed symptoms of marchitez. As with adults and juveniles, eggs were found in protected areas between sheaths or in cracks in the root. Bugs were most active and visible in the evening, as was observed for *L. lobuliger* in Brazil

All stages of *Lincus tumidifrons* were collected from the bases of fronds on oil palms in Colombia (Alvarez 1993) and reared in the laboratory to obtain life history data. Successful rearing required dark conditions; no adults were obtained from cages exposed to white light. Rearing techniques using continuous darkness also have been developed for the hartrot vector *L. bipunctatus* along with a method of inoculating the bugs' digestive tract with *Phytomonas* cultured in vitro (Dollet et al. 2002, as *L. croupius*).

Ground insecticidal sprays controlled the disease in oil palm plantations, presumably because the bugs moved among plants by walking rather than flying (Desmier de Chenon 1984); a similar effect was observed in coconut plantations when a ring of endrin was applied around the trees (Louise et al. 1986).

The potential for biological control by a wheel bug, *Arilus* sp., was investigated in Brazil; 32 of 50 field-collected specimens were found to harbor the *Phytomonas* in their digestive tracts, strongly suggesting predation on *L. lobuliger*, the local vector species (Meneguetti and Trevisan 2010).

Ecological observations of several *Lincus* spp. inhabiting native palms in the genus *Astrocaryum* in Peru (Couturier and Kahn 1989, 1992; Llosa et al. 1990) have added to our knowledge of the group. These bugs reside within the green leaf sheaths on the back of the petiole, hiding between the spines and the leaf sheath fibers, or they hide under litter. Of the three species observed on these native palms in Peru (i.e., *L. malevolus* Rolston, *L. hebes* Rolston, and *L. spurcus*), only *L. spurcus* has been reported from oil palm (*Elaeis guineensis*) plantations.

Lincus spp. are strictly Neoptropical, and these trypanosome diseases are not known to occur among African oil palms in their native habitat. However, with worldwide movement of plants, the possibility always exists for inadvertent colonization, and the effect of such invasive introductions outside of the Neotropics could be disastrous.

13.7 Epidemiology of Pentatomid-Borne Pathogens

As mentioned in **Section 13.1**, transmission modes largely define the epidemiology of plant pathogens. Madden et al. (2000) hypothesized that the rate of spread of a given vector-borne pathogen would be most sensitive to different epidemiological parameters depending on the pathogen's mode of transmission (see **Section 13.1.1**). They developed a series of compartmental epidemic models to explore the parameters to which each transmission mode would be most sensitive: rates of spread for non-persistently transmitted pathogens were most sensitive to vector movement rates, whereas persistently transmitted pathogens were most sensitive to vector longevity, acquisition rate, and inoculation rate. Semi-persistent pathogens were an intermediate case.

Little work has been done on the epidemiology of pentatomid-borne pathogens. Sgrillo et al. (2005) developed an epidemic model of the spread of *Phytomonas staheli* by *Lincus lobuliger* and showed that the spread of *P. staheli* was most sensitive to movement rates of the vectors. Given the paucity of epidemiological work, we developed two epidemic models specific to pentatomid-borne pathogens to explore the factors most important for pathogen spread. Plant pathogens can have either symbiotic or facultative relationships with pentatomid vectors. Symbiotic pathogens colonize the bodies of their pentatomid vectors, whereas facultative pathogens use feeding holes made by pentatomids as entry points into the tissues of their host plants. We developed separate models for the epidemiology of these two pathogen groups.

13.7.1 Symbiotic Pathogens

Pentatomid-borne pathogens occupy nearly all transmission modes—from non-persistent to persistent. As such, our model focuses on epidemiological factors that vary among these forms of transmission. Following Brauer et al. (2008), model 13.1 includes two host compartments and two vector compartments:

$$\frac{dS}{dt} = -\frac{\beta SV}{N} + aI$$

$$\frac{dI}{dt} = \frac{\beta SV}{N} - aI$$

$$\frac{dU}{dt} = b_u\left(\frac{K-U}{K}\right)U + \mu V - \frac{\alpha UI}{N} - d_u U \tag{13.1}$$

$$\frac{dV}{dt} = b_v\left(\frac{K-V}{K}\right)V + \frac{\alpha UI}{N} - \mu V - d_v V$$

The variables *S*, *I*, *U*, and *V* relate to susceptible hosts, infected hosts, non-infectious vectors, and infectious vectors, respectively. As this model consists of a system of differential equations, the values of these variables change over time and as a function of the model parameters, which do not vary with time. Susceptible hosts become infected at an inoculation rate β, and non-infectious vectors acquire the pathogen at an acquisition rate α, both in a frequency-dependent manner (Wonham et al. 2006). Infected hosts lose infection through recovery at a recovery rate a, and infectious vectors lose infection at a recovery rate μ. Host and vector recovery could be caused by different processes, including death and birth of new susceptible individuals, acquired immunity/resistance, or simple loss of infection (e.g., vectors of non-persistently transmitted pathogens). The host population is assumed to be closed (i.e., total population size does not change). The vector population dynamics, on the other hand, are governed by logistic growth—with a carrying capacity K, constant birth rate b_i, and constant death rate d_i, where $i = v$ or u to represent rates of infectious or non-infectious vectors, respectively (Gaff and Gross 2007). Thus both birth and death rates can vary according to vector infection status.

Different transmission modes should influence epidemiology through (a) vector movement, (b) vector recovery, and (c) vector population dynamics (Madden et al. 2000). Pentatomid species exhibit similar levels of movement among hosts, at least compared to Hemiptera overall, so our model ignores vector movement. Rather, we focus on parameters specific to the vector-pathogen relationship—vector recovery, μ, and birth rate of infectious vectors, b_v. Although we are unaware of any heteropteran vectors with transovarial transmission, resulting in $b_v > 0$, this process is known for numerous persistent-propagative pathogens such as Candidatus Liberibacter spp. transmitted by psyllid and triozid vectors (Haapalainen 2014). Furthermore, vertical transmission of endosymbionts occurs in several families of true bugs including pentatomids (Prado et al. 2006). Thus, some form of vertical pathogen transmission in pentatomids may be possible. Vectors of non-persistently transmitted pathogens by definition have high recovery rates, whereas those of persistently transmitted pathogens have low recovery rates. Likewise, persistent-propagative, vertically transmitted pathogens can be differentiated from other transmission modes by setting $b_v > 0$. We explored the effects of vector recovery and infectious vector birth by running numerical simulations of model 13.1 by varying μ and b_v while keeping all other parameters constant. Numerical simulations were run in R 3.2.1 with the deSolve package (Soetaert et al. 2010, R Core Team 2015).

As vector recovery rate increases, the percent of infected vectors in the population decreases linearly (**Figure 13.2**). Interestingly, the percent of infected hosts is greatest at intermediate vector recovery

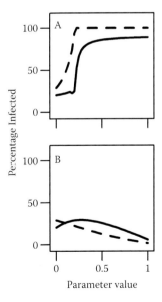

FIGURE 13.2 Percent infected hosts (*I*, solid line) and vectors (*V*, dashed line) from an epidemic model of pentatomid-borne symbiotic pathogens (13.1) at varying values of (A) infectious vector birth rate, b_v, and (B) vector recovery rate, μ. Values are taken as equilibrial (i.e., long-term) values from numerical simulations after 2,000 time steps. Preliminary simulations using 10,000 time steps verified these results. Other parameters were held constant: α = β = 0.4, a = 0.1, b_u = 0.3, $d_u = d_v = 0.2$, K = 250, $S + I$ = 100; μ = 0 when varying b_v and vice-versa. R code for models and simulations are available at https://github.com/arzeilinger/pentatomid_vector_epidemic_model.

rates, which relates to semi-persistently transmitted pathogens. Not surprisingly, infectious vector birth rates show the opposite effects from vector recovery; as b_v increases, the percent of both infected hosts and vectors increases in a sigmoidal manner (**Figure 13.2**). Although this model makes many simplifying assumptions of pentatomid ecology, these results may help explain why non-persistently transmitted pathogens are relatively rare among the Pentatomidae (Mitchell 2004).

13.7.2 Facultative Pathogens

For facultative pathogens, the vector-pathogen relationship will differ dramatically from that of symbiotic pathogens. Likewise, their epidemiology also should differ dramatically. We propose an alternative model for spread of facultative pentatomid-borne pathogens:

$$\frac{dS}{dt} = -\beta S + aE + aI$$

$$\frac{dE}{dt} = \beta S - aE - \frac{\delta EI}{N} \qquad (13.2)$$

$$\frac{dI}{dt} = \frac{\delta EI}{N} - aI$$

The facultative model 13.2 differs from the symbiotic model 13.1 most drastically in that vector-infection status is not included. Instead, we have added an additional compartment for exposed hosts, *E*. Traditionally, exposed compartments in epidemic models have been used to model latent periods in which hosts have been inoculated but are not yet infectious (e.g., Zeilinger and Daugherty 2014). Here, we propose that susceptible hosts become fed upon by pentatomids at a rate β, which is a prerequisite

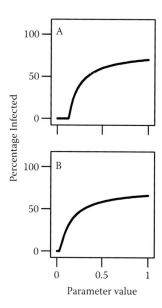

FIGURE 13.3 Percent infected hosts, *I*, from an epidemic model of pentatomid-associated facultative pathogens (13.2) at varying values of (A) transmission rate, δ, and (B) herbivory rate, β. Values are taken as equilibrial (i.e., long-term) values from numerical simulations after 2,000 time steps. Preliminary simulations using 10,000 time steps verified these equilibrial results. Other parameters were held constant: $a = 0.1$, $S + I = 100$; $\beta = 0.4$ when varying δ and vice-versa. R code for models and simulations are available at https://github.com/arzeilinger/pentatomid_vector_epidemic_model.

for becoming infected at a rate δ, in a frequency-dependent manner, as in model 13.1. We assume that infected hosts in the population serve as sources for subsequent infections, making the rate of new infections dependent on the density of infected hosts, I. Although no vector population dynamics are included in model 13.2, the parameter β could be modeled as a function of vector dynamics in future extensions of the model. Results from model 13.2 indicate that spread of facultative pathogens is affected by pentatomid herbivory rates and transmission rates in qualitatively similar ways (**Figure 13.3**). Moreover, these dynamics are similar to those of directly transmitted diseases (Brauer et al. 2008).

In summary, as little work has been done on the epidemiology of pentatomid-borne pathogens, we developed a series of models to describe the most relevant features of hemipteran-pathogen relationships and their effects on pathogen spread. We hypothesize that the epidemiology of pathogens with symbiotic or facultative relationships with pentatomids are fundamentally different. As with other groups of symbiotic pathogens (Madden et al. 2000), we predict that vector-recovery rates and transovarial transmission of pathogens are both important determinants of pathogen spread. Given this, future work should investigate if pathogens with persistent-propagative relationships with pentatomid vectors can be vertically transmitted and, if so, at what rates.

For the sake of model generality and tractability, we have ignored many important features of vector biology in our models. Nonetheless, they provide useful hypotheses that hopefully will motivate future empirical and theoretical work.

13.8 Future Research Directions

It is clear that pentatomids are important vectors of pathogens. What still is in its infancy is work focused on sources of pathogen acquisition and differentiating between damage due solely to insect feeding and vector-borne disease. These factors are significant and should be incorporated into insect thresholds to adjust appropriate detection levels accordingly. Toews et al. (2010) reported the interrelationship between cotton losses due to stink bug pressure and boll disease. Based on 10 years of data, the number of bales attributed to loss due to stink bugs did not correlate consistently with the number of stink bugs detected in a given year. In other words, some seasons reported relatively low insect numbers and higher than expected bale losses and vice versa. This disproportion was hypothesized to be due to higher losses occurring in years that the insects were harboring and transmitting pathogens. This concept was supported by a separate study that compared the number of diseased bolls collected in plots that were protected from sucking pests by repetitive insecticide applications with diseased boll numbers from unprotected plots (Medrano et al. 2015). The incidence of disease was significantly lower in insecticide-protected plots. These findings warrant future research to identify phytopathogen reservoirs (i.e., weeds, water, soil, or other crops planted in adjacent fields). Further, are there particular weather situations that are conducive to elevate pathogen prevalence? Adding the pathogen location information to a farmscape scenario may assist in developing pathogen-avoidance strategies and adjusting thresholds during seasons that pathogens are expected at higher rates in the environment.

It has been shown that a spectrum of pathogens acquired and retained by pentatomids are, in turn, transmitted, causing plant host infection. Future work should be based on aseptic or at least sanitary dissections of insect tissues for accurate detection of the targeted phytopathogens. Careful analysis will expand our knowledge and ability to recognize which insect tissues the pathogens invade and inhabit. Identification of colonization patterns also will assist in learning whether a pathogen resides transiently or persistently. Further, information of this type could provide a baseline to determine whether the insects acquire the pathogens horizontally or vertically.

Future work towards definitively demonstrating the vector potential of pentatomids should be considered. Currently, there are many more associations of vector activities in affected crops than fulfillment of Koch's postulates. One manner to address this is the formation of collaborations between entomologists and plant pathologists. Knowledge gained could be incorporated and assist in further development of predictive epidemic disease models, such as those described in the preceding section, to increase their robustness.

13.9 Acknowledgments

Paula Levin Mitchell thanks Winthrop University for a 3-hour teaching reduction in spring 2015 that provided essential time for literature research and writing. Adam R. Zeilinger thanks Tom Miller (Department of Entomology, University of California Riverside, Riverside) for fruitful discussions on the *Antestiopsis*-coffee system and Rodrigo Almeida (Department of Environmental Science, Policy, and Management, University of California Berkeley, Berkeley) and Matt Daugherty (Department of Entomology, University of California Riverside, Riverside) for helpful discussions of transmission modes and epidemic modeling. Finally, Enrique Gino Medrano thanks Alois Bell (Insect Control and Cotton Disease Research Unit, US Department of Agriculture, Agricultural Research Service, College Station, Texas) for interesting concepts regarding the infection dynamic incited by vector-pathogen interactions.

13.10 References Cited

Agrios, G. N. 1980. Insect involvement in the transmission of fungal pathogens, pp. 293–324. *In* K. F. Harris and K. Maramorosch (Eds.), Vectors of plant pathogens. Academic Press, New York, NY. 467 pp.

Agrios, G. N. 2005. Plant pathology (5th edit.). Elsevier Academic Press, Burlington, MA. 922 pp.

Alvarez, F. A. 1993. Ciclo de vida de *Lincus tumidifrons* Rolston (Hemiptera: Pentatomidae), vector de la Marchitez Sorpresiva de la palma de aceite. Revista Colombiana de Entomología 19(4): 167–174.

Asgarali, J., and P. Ramkalup. 1985. Study of *Lincus* sp. (Pentatomidae) as a possible vector of hartrot in coconut. Surinaamse Landbouw 33: 56–61.

Ashby, S. F., and W. Nowell. 1926. The fungi of stigmatomycosis. Annals of Botany (London) 40(157): 69–83.

Australian Government. 2003. Longan and lychee fruit from the People's Republic of China and Thailand. Draft Import Risk Analysis Report. Part B. Department of Agriculture, Fisheries and Forestry, Government of Australia, Canberra, Australia.

Australian Government. 2004. Longan and lychee fruit from the People's Republic of China and Thailand. Final Import Risk Analysis Report Part A. Department of Agriculture, Fisheries and Forestry, Government of Australia, Canberra, Australia.

Backus, E. A. 1988. Sensory systems and behaviours which mediate hemipteran plant-feeding: A taxonomic overview. Journal of Insect Physiology 34(3): 151–165.

Backus, E. A. 2016. Sharpshooter feeding behavior in relation to transmission of *Xylella fastidiosa*: A model for foregut-borne transmission mechanisms, Chapter 13. *In* J. Brown (Ed.), Vector-mediated transmission of plant pathogens. APS Press, St. Paul, MN. (in press)

Backus, E. A., M. S. Serrano, and C. M. Ranger. 2005. Mechanisms of hopperburn: An overview of insect taxonomy, behavior, and physiology. Annual Review of Entomology 50: 125–151.

Bergey, D. H., R. E. Buchanan, and N. E. Gibbons. 1974. Bergey's manual of determinative bacteriology. Williams & Wilkins, Baltimore, MD. 1246 pp.

Bertaccini, A., and B. Duduk. 2009. Phytoplasma and phytoplasma diseases: a review of recent research. Phytopathologia Mediterranea 48: 355–378.

Bezerra, J. L., and J. M. de Figueiredo. 1982. Ocorrência de *Phytomonas staheli* McGhee & McGhee em coqueiro (*Cocos nucifera* L.) no estado da Bahia, Brasil. Fitopatologia Brasileira 7: 139–143.

Blanc, S., M. Drucker, and M. Uzest. 2014. Localizing viruses in their insect vectors. Annual Review of Phytopathology 52: 403–425.

Bowling, C. C. 1979. The stylet sheath as an indicator of feeding activity of the rice stink bug. Journal of Economic Entomology 72: 259–260.

Brauer, F., P. van den Driessche, and J. Wu. 2008. Mathematical epidemiology. Springer Verlag, Berlin. 414 pp.

Brozek, J., and A. Herczek. 2004. Internal structure of the mouthparts of the true bugs. Polskie Pismo Entomologiczne (Polish Journal of Entomology) 73: 79–106.

Brust, G., and K. Rane. 2011. Transmission of the yeast *Eremothecium coryli* to fruits and vegetables by the brown marmorated stink bug. University of Maryland Extension, College Park, Maryland. https://extension.umd.edu/sites/default/files/_docs/articles/YeastTransmissionByBMSBinVegetables.pdf (accessed 9 June 2015).

Bruton, B. D., F. Mitchell, J. Fletcher, S. D. Pair, A. Wayadande, U. Melcher, J. Brady, B. Bextine, and T. W. Popham. 2003. *Serratia marcescens*, a phloem-colonizing, squash Bug -transmitted bacterium: Causal agent of cucurbit yellow vine disease. Plant Disease 87(8): 937–944.

Butt, F. H. 1943. Comparative study of mouth parts of representative Hemiptera-Homoptera. Cornell University Agricultural Experiment Station Memoir 254. Cornell University, Ithaca, New York. 19 pp. + 8 plates.

Camargo, E. P. 1999. *Phytomonas* and other trypanosomatid parasites of plants and fruit. Advances in Parasitology 42: 29–112.

Camargo, E. P., and F. G. Wallace. 1994. Vectors of plant parasites of the genus *Phytomonas*. Advances in Disease Vector Research. 10: 333–359.

Camargo, E. P., P. Kastelein, and I. Roitman. 1990. Trypanosomatid parasites of plants. Parasitology Today 6: 22–25.

Chantrasri, P., V. Sardsud, and W. Srichart. 1999. Transmission studies of phytoplasma, the causal agent of witches' broom disease of longan. The 25th Congress on Science and Technology of Thailand, 20–22 October 1999, Pitsanulok, Thailand (abstract). http://www.thaiscience.info/Article%20for%20 ThaiScience/Article/2/Ts-2%20transmission%20studies%20of%20phytoplasma,causal%20agent%20 of%20witches%20broom%20disease%20in%20longan.pdf (accessed 18 May 2015).

Chen, J., J. Chen, and X. Xu. 2001. Advances in research of longan witches' broom disease. Acta Horticulturae 558: 413–416.

Cilas, C., B. Bouyjou, and B. Decazy. 1998. Frequency and distribution of *Antestiopsis orbitalis* Westwood (Hem., Pentatomidae) in coffee plantations in Burundi: implications for sampling techniques. Journal of Applied Entomology – Zeitschrift Für Angewandte Entomologie 122: 601–606.

Clark, R. G., and G. E. Wilde. 1970a. Association of the green stink bug and the yeast spot disease organism of soybeans. I. Length of retention, effect of molting, isolation from feces and saliva. Journal of Economic Entomology 63: 200–204.

Clark, R. G., and G. E. Wilde. 1970b. Association of the green stink bug and the yeast spot disease organism of soybeans. II. Frequency of transmission to soybeans, transmission from insect to insect, isolation from field population. Journal of Economic Entomology 63: 355–357.

Cline, A. R., and E. A. Backus. 2002. Correlations among AC electronic monitoring waveforms, body postures, and stylet penetration behaviors of *Lygus hesperus* (Hemiptera: Miridae). Environmental Entomology 31: 538–549.

Cobben, R. H. 1978. Evolutionary trends in Heteroptera. Part II. Mouthpart-structures and feeding strategies. Mededelingen Landbouwhogeschool Wageningen 78-5: 1–407.

Conchon, I., M. Campaner, C. Sbravate, and E. P. Camargo. 1989. Trypanosomatids, other than *Phytomonas* spp., isolated and cultured from fruit. Journal of Protozoology 36: 412–414.

Cook, C. A., and J. J. Neal. 1999. Feeding behavior of larvae of *Anasa tristis* (Heteroptera: Coreidae) on pumpkin and cucumber. Environmental Entomology 28: 173–177.

Cooke, S. B. 2014. Probing behavior of southern green stink bug, *Nezara viridula* (Hemiptera: Pentatomidae), on the soybean plant, *Glycine max*. MS Thesis, Winthrop University, Rock Hill, SC. 100 pp.

Couturier, G., and F. Kahn. 1989. Bugs of *Lincus* spp. vectors of marchitez and hartrot (oil palm and coconut diseases) on *Astrocaryum* spp., Amazonian native plants. Principes 33(1): 19–20.

Couturier, G., and F. Kahn. 1992. Notes on the insect fauna on two species of *Astrocaryum* (Palmae, Cocoeae, Bactridinae) in Peruvian Amazonia, with emphasis on potential pests of cultivated palms. Bulletin de l'Institut Français d'Études Andines 21(2): 715–725.

Daane, K. M., J. Millar, R. E. Rice, P. G. daSilva, W. Bentley, R. H. Beede, and G. Weinberger. 2005. Stink bugs and leaffooted bugs, pp. 186–196. *In* L. Ferguson (Ed.), Pistachio production manual (4th edit.). University of California Davis Fruit & Nut Information Center, Davis, CA.

Daugherty, D. M. 1967. Pentatomidae as vectors of leaf spot disease of soybeans. Journal of Economic Entomology 60: 147–152.

Desmier de Chenon, R. 1984. Recherches sur le genre *Lincus* Stål, Hemiptera, Pentatomidae, Discocephalinae, et son rôle éventual dans la marchitez du palmier à huile et du hart-rot du cocotier. Oléagineux 39: 1–6.

Dollet, M. 1984. Plant diseases caused by flagellate protozoa (*Phytomonas*). Annual Review of Phytopathology 22: 115–132.

Dollet, M., N. R. Sturm, and D. A. Campbell. 2001. The spliced leader RNA gene array in phloem-restricted plant trypanosomatids (*Phytomonas*) partitions into two major groupings: Epidemiological implications. Parasitology 122: 289–297.

Dollet, M., D. Gargani, E. Muller, K. Vezian, S. Birot, and J. M. Maldes. 2002. *Lincus croupius* (Heteroptera: Pentatomidae): rearing, cycle and trials for experimental transmission of plant trypanosomatids (*Phytomonas* spp.), p. 17. *In* I. M. Kerzhner (Ed.), Second quadrennial meeting of the International Heteropterists' Society abstracts. Russian Academy of Sciences, Zoological Institute, St. Petersburg. 59 pp. (abstract).

Dollet, M., F. Macome, A. Vaz, and S. Fabre. 2011. Phytoplasmas identical to coconut lethal yellowing phytoplasmas from Zambesia (Mozambique) found in a pentatomide bug in Cabo Delgado province. Bulletin of Insectology 64 (Supplement): S139–S140.

Dolling, W. R. 1984. Pentatomid bugs (Hemiptera) that transmit a flagellate disease of cultivated palms in South America. Bulletin of Entomological Research 74(3): 473–476.

Eberhard, W. 1974. Insectos y hongos gue atacán a la chinche del cacao, *Antiteuchus tripterus*. Revista Facultad National de Agronomia Medellín 29: 65–68.

Eden-Green, S. J., R. Balfas, T. Sutarjo, and Jamalius. 1992. Characteristics of the transmission of Sumatra disease of cloves by tube-building cercopoids, *Hindola* spp. Plant Pathology 41(6): 702–712.

Eisbein, K. 1976. Untersuchungen zum elektronenmikroskopischen Nachwies des Rübenkräusel-Virus (Beta virus 3) in *Beta vulgaris* L. und *Piesma quadratum* Fieb. Archiv für Phytopathologie und Pflanzenschutz, Berlin 12: 299–313.

Esquivel, J. F. 2011. Estimating potential stylet penetration of southern green stink bug – a mathematical modeling approach. Entomologia Experimentalis et Applicata 140: 163–170.

Esquivel, J. F. 2015. Stylet penetration estimates for a suite of phytophagous hemipteran pests of row crops. Environmental Entomology 44: 619–626.

Esquivel, J. F., and E. G. Medrano. 2012. Localization of selected pathogens of cotton within the southern green stink bug. Entomologia Experimentalis et Applicata 142: 114–120.

Fereres, A., and A. Moreno. 2009. Behavioural aspects influencing plant virus transmission by homopteran insects. Virus Research 141: 158–168.

Foster, J. E., and D. M. Daugherty. 1969. Isolation of the organism causing yeast-spot disease from the salivary system of the green stink bug. Journal of Economic Entomology 62: 424–427.

Frazer, H. L. 1944. Observations on the method of transmission of internal boll disease of cotton by the cotton stainer-bug. Annals of Applied Biology 31: 271–291.

Gaff, H. D., and L. J. Gross. 2007. Modeling tick-borne disease: a metapopulation model. Bulletin of Mathematical Biology 69: 265–288.

Gibb, K. S., and J. W. Randles. 1991. Transmission of velvet tobacco mottle virus and related viruses by the mirid *Cyrtopeltis nicotianae*. Advances in Disease Vector Research 7: 1–17.

Gibbs, A. J. 1957. *Leptomonas serpens* n. sp., parasitic in the digestive tract and salivary glands of *Nezara viridula* (Pentatomidae) and in the sap of *Solanum lycopersicum* (tomato) and other plants. Journal of Parasitology 47: 297–303.

Greene, J. K., S. G. Turnipseed, M. J. Sullivan, and O. L. May. 2001. Treatment thresholds for stink bugs (Hemiptera: Pentatomidae) in cotton. Journal of Economic Entomology 94: 403–409.

Grillo, H., and M. Alvarez. 1983. *Nematospora coryli* Peglion (Nematosporaceae: Hemiascomycetidae) y sus trasmisoresen el cultivo de los cítricos. Centro Agrícola 10(2): 13–34.

Gueule, D., G. Fourny, E. Ageron, A. Le Fleche-Mateos, M. Vandenbogaert, P. A. D. Grimont, and C. Cilas. 2015. *Pantoea coffeiphila* sp. nov., cause of the "potato taste" of Arabica coffee from the African Great Lakes region. International Journal of Systematic and Evolutionary Microbiology 65: 23–29.

Haapalainen, M. 2014. Biology and epidemics of *Candidatus* Liberibacter species, psyllid-transmitted plant-pathogenic bacteria. Annals of Applied Biology 165: 172–198.

Hirose, E. P., A. R Panizzi, J. T. De Souza, A. J. Cattelan, and J. R. Aldrich. 2006. Bacteria in the gut of southern green stink bug (Heteroptera: Pentatomidae). Annals of the Entomological Society of America 99: 91–95.

Hiruki, C. 1999. Paulownia witches'-broom disease important in East Asia. Acta Horticulturae 496: 63–68.

Hoebeke, E. R., and M. E. Carter. 2003. *Halyomorpha halys* (Stål) (Heteroptera: Pentatomidae): A polyphagous plant pest from Asia newly detected in North America. Proceedings of the Entomological Society of Washington 105: 225–237.

Hollar, L., and D. A. Maslov. 1997. A phylogenetic view on the genus *Phytomonas*. Molecular and Biochemical Parasitology 89: 295–299.

Hollay, M. E., C. M. Smith, and J. F. Robinson. 1987. Structure and formation of feeding sheaths of rice stink bug (Heteroptera: Pentatomidae) on rice grains and their association with fungi. Annals of the Entomological Society of America 80: 212–216.

Hollis, P. 2001. Seed rot has lowered South Carolina cotton yields. Southeast Farm Press. http://southeastfarmpress .com/mag/farming_seed_rot_lowered/ (accessed 15 August 2015).

Holtz, B. A. 2002. Plant protection for pistachio. HortTechnology 12(4): 626–632.

Hori, K. 1968. Feeding behavior of the cabbage bug, *Eurydema rugosa* Motschulsky (Hemiptera: Pentatomidae) on the cruciferous plants. Applied Entomology and Zoology 3(1): 26–36.

Horsfall, J. L. 1923. The effects of feeding punctures of aphids on certain plant tissues. Pennsylvania State College Agricultural Experiment Station Bulletin 182: 1–22.

Index Fungorum Partnership. 2015. Index Fungorum. www.indexfungorum.org (accessed 24 September 2015).

Innes, R. J. 2009. *Paulownia tomentosa*. *In*: Fire Effects Information System, [Online]. U.S. Department of Agriculture, Forest Service, Rocky Mountain Research Station, Fire Sciences Laboratory (Producer), Missoula, Montana. http://www.fs.fed.us/database/feis/plants/tree/pautom/all.html (accessed 28 May 2015).

Jackels, S. C., E. E. Marshall, A. G. Omaiye, R. L. Gianan, F. T. Lee, and C. F. Jackels. 2014. GCMS investigation of volatile compounds in green coffee affected by potato taste defect and the Antestia bug. Journal of Agricultural and Food Chemistry 62: 10222–10229.

Jankevicius, J. V., S. I. Jankevicius, M. Campaner, I. Conchon, L. A. Maeda, M. M. G. Teixeira, E. Freymuller and E.P. Camargo. 1989. Life cycle and culturing of *Phytomonas serpens* (Gibbs), a trypanosomatid parasite of tomatoes. Journal of Protozoology 36: 265–271.

Jaskowska, E., C. Butler, G. Preston, and S. Kelly. 2015. *Phytomonas*: Trypanosomatids adapted to plant environments. PLoS Pathogens 11(1): e1004484. http://dx.doi.org/10.1371/journal.ppat.1004484 (accessed 3 June 2015).

Kaiser, W. J., and N. G. Vakili. 1978. Insect transmission of pathogenic xanthomonads to bean and cowpea in Puerto Rico. Phytopathology 68: 1057–1063.

Kastelein, P. 1985. Transmission of *Phytomonas* sp. (Trypanosomatidae) by the bug *Edessa loxdali* (Pentatomidae). Surinaamse Landbouw 33: 62–64.

Kastelein, P., and E. P. Camargo. 1990. Trypanosomatid protozoa in fruit of Solanaceae in southeastern Brazil. Memórias do Instituto Oswaldo Cruz 85: 413–417.

Kimura, S., S. Tokumaru, and A. Kikuchi. 2008a. Carrying and transmission of *Eremothecium coryli* (Peglion) Kurtzman as a causal pathogen of yeast-spot disease in soybeans by *Riptortus clavatus* (Thunberg), *Nezara antennata* Scott, *Piezodorus hybneri* (Gmelin) and *Dolycoris baccarum* (Linnaeus). Japanese Journal of Applied Entomology and Zoology 52: 13–18. (in Japanese with English abstract)

Kimura, S., S. Tokumaru, and K. Kuge. 2008b. *Eremothecium ashbyi* causes soybean yeast-spot and is associated with stink bug, *Riptortus clavatus*. Journal of General Plant Pathology 74: 275–280. http://dx.doi .org/10.1007/s10327-008-0097-1.

Kirkpatrick, T. W. 1937. Studies on the ecology of coffee plantations in East Africa. II. The autecology of *Antestia* spp. (Pentatomidae) with a particular account of a strepsipterous parasite. Transactions of the Royal Entomological Society of London 86: 247–343.

Koizumi, M. 1995. Problems of insect-borne virus diseases of fruit trees in Asia. Food and Fertilizer Technology Center, Taipei, Taiwan. en.fftc.org.tw/htmlarea_file/library/20110712175437/eb417b.pdf (accessed 24 May 2015).

Koopmans, A. 1977. A cytological study of *Nematospora coryli* Pegl. Genetica 47: 187–195.

Krauss, U. 2014. Moniliophthora roreri (frosty pod rot). Invasive Species Compendium http://www.cabi.org/isc /datasheet/34779 (accessed 26 June 2015).

Leach, J. G., and G. Clulo. 1943. Association between *Nematospora phaseoli* and the green stink bug. Phytopathology 33: 1209–1211.

Lee, F. N., N. P. Tugwell, S. J. Fanna, and G. J. Weidemann. 1993. Role of fungi vectored by rice stink bug (Heteroptera: Pentatomidae) in discoloration of rice kernels. Journal of Economic Entomology 86(2): 549–556.

Le Pelley, R. H. 1942. The food and feeding habits of *Antestia* in Kenya. Bulletin of Entomological Research 33: 71–89.

Llosa, J. F., G. Couturier, and F. Kahn. 1990. Notes on the ecology of *Lincus spurcus* and *L. malevolus* (Heteroptera: Pentatomidae: Discocephalinae) on Palmae in forests of Peruvian Amazonia. Annales de la Société Entomologique de France. (N.S.) 26(2): 249–254.

Lopez, G., P. Genty, and M. Ollagnier. 1975. Contrôle préventif de la "Marchitez sorpresiva" de l'*Elaeis guineensis* en Amérique latine. Oléagineux 30: 243–50.

Louise, C., M. Dollet, and D. Mariau. 1986. Recherches sur le hartrot du cocotier, maladie à *Phytomonas* (Trypanosomatidae) et sur son vecteur *Lincus* sp. (Pentatomidae) en Guyane. Oléagineux 41(10): 437–449.

Lucini, T., and A. R. Panizzi. 2016. Waveform characterization of the soybean stem feeder *Edessa meditabunda* (F.) (Hemiptera: Heteroptera: Pentatomidae): Overcoming the challenge of wiring pentatomids for EPG. Entomologia Experimentalis et Applicata 158(2): 118-132. http://dx.doi.org/10.1111/eea.12389.

Maciel, A. S., T. de A. Garbelotto, I. C. Winter, T. Roell, and L. A. Campos. 2015. Description of the males of *Lincus singularis* and *Lincus incisus* (Hemiptera: Pentatomidae: Discocephalinae). Zoologia (Curitiba) 32(2): 157–161. http://www.scielo.br/scielo.php?script=sci_arttext&pid=S1984-46702015000200157&lng=en&tlng=en. 10.1590/S1984-46702015000200007 (accessed 2 June 2015).

Madden, L. V., M. J. Jeger, and F. van den Bosch. 2000. A theoretical assessment of the effects of vector-virus transmission mechanism on plant virus disease epidemics. Phytopathology 90: 576–594.

Marasas, W. F. O. 1971. Cotton staining caused by *Crebrothecium ashbyi* in South Africa. Bothalia 10: 407–410.

Martinson, H. M., M. J. Raupp, and P. M. Shrewsbury. 2013. Invasive stink bug wounds trees, liberates sugars, and facilitates native Hymenoptera. Annals of the Entomological Society of America 106: 47–52.

Maslov, D. A., J. Votýpka, V. Yurchenko, and J. Lukeš. 2013. Diversity and phylogeny of insect trypanosomatids: all that is hidden shall be revealed. Trends in Parasitology 29(1): 43–52.

Mathen, K., P. Rajan, C. P. R. Nair, M. Sasikala, M. Gunasekharan, M. P. Govindankutty, and J. J. Solomon. 1990. Transmission of root (wilt) disease to coconut seedlings through *Stephanitis typica* (Distant) (Heteroptera: Tingidae). Journal of Tropical Agriculture 67: 69–73.

Mauck, K., N. A. Bosque-Pérez, S. D. Eigenbrode, C. M. De Moraes, and M. C. Mescher. 2012. Transmission mechanisms shape pathogen effects on host–vector interactions: evidence from plant viruses. Functional Ecology 26: 1162–1175.

McCoy, R. E., and G. Martinez-Lopez. 1982. *Phytomonas staheli* associated with coconut and oil palm diseases in Colombia. Plant Disease 66: 675–677.

McCoy, R. E., R. A. Rodriguez, and J. A. Guzman. 1984. *Phytomonas* flagellates associated with diseased coconut palm in Central America. Plant Disease 68: 537.

McGhee, R. B., and A. H. McGhee. 1979. Biology and structure of *Phytomonas staheli* sp. n., a trypanosomatid located in sieve tubes of coconut and oil palms. Journal of Protozoology 26: 348–351.

McPherson, J. E., and R. M. McPherson. 2000. Stink bugs of economic importance in America north of Mexico. CRC Press, Boca Raton, FL. 253 pp.

Medrano, E. G., and A. A. Bell. 2007. Role of *Pantoea agglomerans* in opportunistic bacterial seed and boll rot of cotton (*Gossypium hirsutum*) grown in the field. Journal of Applied Microbiology 102: 134–143.

Medrano, E. G., and A. A. Bell. 2012. Genome sequence of *Pantoea* sp. strain Sc 1, an opportunistic cotton pathogen. Journal of Bacteriology 194: 3019.

Medrano, E. G., and A. A. Bell. 2015. Genome sequence of *Pantoea ananatis* strain CFH 7-1, which is associated with a vector-borne cotton fruit disease. Genome Announcements 3: e01029-15. http://dx.doi.org/10.1128/genomeA.01029-15.

Medrano, E. G., J. F. Esquivel, and A. A. Bell. 2007. Transmission of cotton seed and boll rotting bacteria by the southern green stink bug (*Nezara viridula* L.). Journal of Applied Microbiology 103: 436–444.

Medrano, E. G., J. F. Esquivel, A. Bell, J. Greene, P. Roberts, J. Bacheler, J. Marois, D. Wright, R. Nichols, and J. Lopez. 2009a. Potential for *Nezara viridula* (Hemiptera: Pentatomidae) to transmit bacterial and fungal pathogens into cotton bolls. Current Microbiology 59: 405–412.

Medrano, E. G., J. F. Esquivel, R. L. Nichols, and A. A. Bell. 2009b. Temporal analysis of cotton boll symptoms resulting from southern green stink bug feeding and transmission of a bacterial pathogen. Journal of Economic Entomology 102: 36–42.

Medrano, E. G., A. A. Bell, J. K. Greene, P. M. Roberts, J. A. Bacheler, J. J. Marios, D. L. Wright, J. F. Esquivel, R. L. Nichols, and S. Duke. 2015. Relationship between piercing-sucking insect control and internal lint and seed rot in southeastern cotton (*Gossypium hirsutum* L.). Journal of Economic Entomology 108: 1540–1545.

Mendgen, K., M. Hahn, and H. Deising. 1996. Morphogenesis and mechanisms of penetration by plant pathogenic fungi. Annual Review of Phytopathology 34: 367–386.

Meneguetti, D. U. de O., and O. Trevisan. 2010. Ocorrência de protozoários morfologicamente semelhante a *Phytomonas staheli* em Reduviidae e potencial do *Arilus* sp como controlador biológico. Revista Científica da Faculdade de Educação e Meio Ambiente 1(1): 84–93.

Meulen, H. J. van der, and A. S. Schoeman. 1990. Aspects of the phenology and ecology of the antestia stink bug, *Antestiopsis orbitalis orbitalis* (Hemiptera: Pentatomidae), a pest of coffee. Phytophylactica 22(4): 423–426.

Michailides, T. J. 2014. Pistachio Stigmatomycosis. UC IPM Pest Management Guidelines: Pistachio. UC ANR Publication 3461, University of California, Davis, California. http://www.ipm.ucdavis .edu/PMG/r605100511.html#REFERENCE (accessed 19 June 2015).

Michailides, T. J., and D. P. Morgan. 1990. Etiology and transmission of stigmatomycosis disease of pistachio in California, pp. 88–95. *In* California Pistachio Industry Annual Report, Crop Year 1989-90. California Pistachio Commission, Fresno, CA. 136 pp.

Michailides, T. J., and D. P. Morgan. 1991. New findings on the stigmatomycosis disease of pistachio in California. pp. 106–110. *In* California Pistachio Industry Annual Report, Crop Year 1990-91. California Pistachio Commission, Fresno, CA. 136 pp.

Michailides, T. J., and D. P. Morgan. 2004. Panicle and shoot blight of pistachio: A major threat to the California Pistachio Industry. APSnet Features. Online.: http://dx.doi.org/10.1094/APSnetFeature-2004-0104 (accessed 17 June 2015).

Michailides, T. J., D. P. Morgan, and D. Felts. 1998. Spread of *Botryosphaeria dothidea* in central California pistachio orchards. Acta Horticulturae 470: 582–591.

Miles, P. W. 1959. The salivary secretions of a plant-sucking bug, *Oncopeltus fasciatus* (Dall.) (Heteroptera: Lygaeidae)—I The types of secretions and their roles during feeding. Journal of Insect Physiology 3: 243–255.

Miles, P. W. 1968. Insect secretions in plants. Annual Review of Phytopathology 6: 137–164.

Miles, P. W., and G. S. Taylor. 1994. 'Osmotic pump' feeding by coreids. Entomologia Experimentalis et Applicata 73: 163–173.

Mitchell, P. L. 2004. Heteroptera as vectors of plant pathogens. Neotropical Entomology 33(5): 519–545.

Miyao, G. M., R. M. Davis, and H. J. Phaff. 2000. Outbreak of *Eremothecium coryli* fruit rot of tomato in California. Plant Disease 84(5): 594.

Narisu. 2000. Ash-gray leaf bugs (Piesmatidae), pp. 265–270. *In* C. W. Schaefer and A. R. Panizzi (Eds.), Heteroptera of economic importance. CRC Press, Boca Raton, FL. 828 pp.

Nault, L. R. 1997. Arthropod transmission of plant viruses: A new synthesis. Annals of the Entomological Society of America 90: 521–541.

Nguyen, T. T. D., S. Paltrinieri, J. F. Meija, H. X. Trinh, and A. Bertaccini. 2012. Detection and identification of phytoplasmas associated with longan witches' broom in Vietnam. Phytopathogenic Mollicutes 2(1): 23–27. http://www.indianjournals.com/ijor.aspx?target=ijor:mollicutes&volume=2&issue=1&arti cle=004 (abstract) (accessed 27 May 2015).

Ni, X., J. P. Wilson, G. D. Buntin, B. Guo, M. D. Krakowsky, R. D. Lee, T. E. Cottrell, B. T. Scully, A. Huffaker, and E. A. Schmelz. 2011. Spatial patterns of aflatoxin levels in relation to ear-feeding insect damage in pre-harvest corn. Toxins 3: 920–931.

Ni, X., M. D. Toews, G. D. Buntin, J. E. Carpenter, A. Huffaker, E. A. Schmelz, T. E. Cottrell, and Z. Abdo. 2014a. Influence of brown stink bug feeding, planting date and sampling time on common smut infection of maize. Insect Science 21: 564–571. http://dx.doi.org/10.1111/1744-7917.12149.

Ni, X., J. P. Wilson, M. D. Toews, G. D. Buntin, R. D. Lee, X. Li, Z. Lei, K. He, W. Xu, X. Li, A. Huffaker, and E. A. Schmelz. 2014b. Evaluation of spatial and temporal patterns of insect damage and aflatoxin level in the pre-harvest corn fields to improve management tactics. Insect Science 21: 572–583. http://dx.doi .org/10.1111/j.1744-7917.2012.01531.x.

Nowell, W. 1918. Internal diseases of cotton bolls in the West Indies. West Indian Bulletin 17: 1–26.

Nyvall, R. F. 1999. Field crop diseases. Iowa State University Press, Ames, IA. 102 pp.

Okuda, S., Y. Nakano, T. Goto, and T. Natsuaki. 1998. 16S RDNAs of Paulownia witches' broom phytoplasma transmitted by *Halyomorpha mista*. 7th International Congress of Plant Pathology, Edinburgh. www .bspp.org.uk/icpp98/3.7/33.html (accessed 10 June 2015).

Pair, S. D., B. D. Bruton, F. Mitchell, J. Fletcher, A. Wayadande, and U. Melcher. 2004. Overwintering squash bugs harbor and transmit the causal agent of cucurbit yellow vine disease. Journal of Economic Entomology 97: 74–78.

Panizzi, A. R., J. G. Smith, L. A. G. Pereira, and J. Yamashita. 1979. Efeitos dos danos de *Piezodorus guildinii* (Westwood, 1837) no rendimento e qualidade da soja. Anais do I Seminario Nacional de Pesquisa de Soja 2: 59–78.

Payne, J. A., and J. M. Wells. 1984. Toxic penicillia isolated from lesions of kernel-spotted pecans. Environmental Entomology 13: 1609–1612.

Perthuis, B., R. Desmier de Chenon, and E. Merland. 1985. Mise en évidence du vecteur de la marchitez sorpresiva du palmier à huile, la punaise *Lincus lethifer* Dolling (Hemiptera Pentatomidae Discocephalinae). Oléagineux 40(10): 473–476.

Phillips-Mora, W., A. Coutiño, C. F. Ortiz, A. P. López, J. Hernández, and M. C. Aime. 2006. First report of *Moniliophthora roreri* causing frosty pod rot (moniliasis disease) of cocoa in Mexico. Plant Pathology 55: 584. http://dx.doi.org/10.1111/j.1365-3059.2006.01418.x.

Pollard, D. G. 1973. Plant penetration by feeding aphids (Hemiptera: Aphidoidea): A review. Bulletin of Entomological Research 62: 631–714.

Pollard, D. G. 1977. Aphid penetration of plant tissue, pp. 105–111. *In* K. F. Harris and K. Maramorosch (Eds.), Aphids as virus vectors. Academic Press, New York, NY. 559 pp.

Porcel, B. M., F. Denoeud, F. Opperdoes, B. Noel, M-A. Madoui, T. C. Hammarton, M. C. Field, C. Da Silva, A. Couloux, J. Poulain, M. Katinka, K. Jabbari, J-M. Aury, D. A. Campbell, R. Cintron, N. J. Dickens, R. Docampo, N. R. Sturm, V. L. Koumandou, S. Fabre, P. Flegontov, J. Lukeš, S. Michaeli, J. C. Mottram, B. Szöőr, D. Zilberstein, F. Bringaud, P. Wincker, M. Dollet. 2014. The streamlined genome of *Phytomonas* spp. relative to human pathogenic kinetoplastids reveals a parasite tailored for plants. PLoS Genetics 10(2): e1004007. http://dx.doi.org/10.1371/journal.pgen.1004007 (accessed 3 June 2015).

Prado, S. S., D. Rubinoff, and R. P. P. Almeida. 2006. Vertical transmission of a pentatomid caeca-associated symbiont. Annals of the Entomological Society of America 99: 577–585.

Preston, D. A., and W. W. Ray. 1943. Yeast spot disease of soybean reported from Oklahoma and North Carolina. Plant Disease Reporter 27(22): 601–602.

Proeseler, G. 1980. Piesmids, pp. 97–113. *In* K. F. Harris and K. Maramorosch, (Eds.), Vectors of plant pathogens. Academic Press, New York, NY. 467 pp.

Purcell, A. H., and D. L. Hopkins. 1996. Fastidious xylem-limited bacterial plant pathogens. Annual Review of Phytopathology 34: 131–151. http://dx.doi.org/10.1146/annurev.phyto.34.1.131.

R Core Team. 2015. R: A language and environment for statistical computing. R Foundation for Statistical Computing, Vienna, Austria.

Ragsdale, D. W., A. D. Larson, and L. D. Newsom. 1979. Microorganisms associated with feeding and from various organs of *Nezara viridula*. Journal of Economic Entomology 72: 725–727.

Rani, P. U., and S. S. Madhavendra. 1995. Morphology and distribution of antennal sense organs and diversity of mouthpart structures in *Odontopus nigricornis* (Stall [sic]) and *Nezara viridula* L. (Hemiptera). International Journal of Insect Morphology and Embryology 24: 119–132.

Resende, M. L. V., R. E. L. Borges, J. L. Bezerra, and D. P. Oliveira. 1986. Transmissão da murcha-de-Phytomonas a coqueiros e dendezeiros por *Lincus lobuliger* (Hemiptera, Pentatomidae): Resultados preliminares. Revista Theobroma 16: 149–154.

Rice, K. B., C. J. Bergh, E. J. Bergmann, D. J. Biddinger, C. Dieckhoff, G. Dively, H. Fraser, T. Gariepy, G. Hamilton, T. Haye, A. Herbert, K. Hoelmer, C. R. Hooks, A. Jones, G. Krawczyk, T. Kuhar, H. Martinson, W. Mitchell, A. L. Nielsen, D. G. Pfeiffer, M. J. Raupp, C. Rodriguez-Saona, P. Shearer, P. Shrewsbury, P. D. Venugopal, J. Whalen, N. G. Wiman, T. C. Leskey, and J. F. Tooker. 2014. Biology, ecology, and management of brown marmorated stink bug (Hemiptera: Pentatomidae). Journal of Integrated Pest Management 5(3): A1-A13. http://dx.doi.org/10.1603/IPM14002 (accessed 2 June 2015).

Rice, R. E., J. K. Uyemoto, J. M. Ogawa, and W. M. Pemberton. 1985. New findings on pistachio problems. California Agriculture 39: 15–18.

Rider, D. A. 2015. Pentatomoidea home page. North Dakota State University. www.ndsu.edu/faculty/rider/Pentatomoidea/ (accessed 24 May 2015).

Rider, D. A., L. Y. Zhang, and I. M. Kerzhner. 2002. Checklist and nomenclatural notes on the Chinese Pentatomidae (Heteroptera). II. Pentatominae. Zoosystematica Rossica 11(1): 135–153.

Roberts, D. A., and C. W. Boothroyd. 1972. Fundamentals of plant pathology. W. H. Freeman and Co., San Francisco. 402 pp.

Rolston, L. H. 1983. A revision of the genus *Lincus* Stål (Hemiptera: Pentatomidae: Discocephalinae: Ochlerini). Journal of the New York Entomological Society 91(1): 1–47.

Rolston, L. H. 1989. Three new species of *Lincus* (Hemiptera: Pentatomidae) from palms. Journal of the New York Entomological Society 97(3): 271–276.

Ruppel, E. G. 2003. Diseases of beet (*Beta vulgaris* L.). *In* Common names of plant diseases. American Phytopathological Society. http://www.apsnet.org/publications/commonnames/Pages/Beet.aspx (updated 8 April 2003; accessed 10 June 2015).

Russin, J. S., D. B. Orr, M. B. Layton, and D. J. Boethel. 1988. Incidence of microorganisms in soybean seeds damaged by stink bug feeding. Phytopathology 78(3): 306–310.

Schoene, W. J., and G. W. Underhill. 1933. Economic status of the green stinkbug with reference to the succession of its wild hosts. Journal of Agricultural Research 46: 863–866.

Serrano, M. G., M. Campaner, G. A. Buck, M. M. G. Teixeira, and E. P. Camargo. 1999. PCR amplification of the spliced leader gene for the diagnosis of trypanosomatid parasites of plants and insects in methanol-fixed smears. FEMS Microbiology Letters 176: 241–246.

Sgrillo, R. B., J. I. L. Moura, and K. R. P. A. Sgrillo. 2005. Simulation model for phytomona epidemics in coconut trees. Neotropical Entomology 34(4): 527–538.

Shao, P. X., R. F. Hong, Q. D. Tong, C. H. Peng, J. Y. Shen, and Z. Y. Chen. 1982. Studies on the insect vector of paulownia witches'-broom. I. Transmission by *Halyomorpha halys* [Stal] and electron microscopical observation. Shandong Forest Science and Technology 1: 42–45.

Sharma, A., A. N. Khan, S. Subrahmanyam, A. Raman, G. S. Taylor, and M. J. Fletcher. 2013. Salivary proteins of plant-feeding hemipteroids – implication in phytophagy. Bulletin of Entomological Research 104(2):117–36. http://dx.doi.org/10.1017/S0007485313000618.

Shivas, R. G., M. W. Smith, T. S. Marney, T. K. Newman, D. L. Hammelswang, A. W. Cooke, K. G. Pegg, and I. G. Pascoe. 2005. First record of *Nematospora coryli* in Australia and its association with dry rot of *Citrus*. Australasian Plant Pathology 34: 99–101.

Silva, F. A. C., J. J. da Silva, R. A. Depieri, and A. R. Panizzi. 2012. Feeding activity, salivary amylase activity, and superficial damage to soybean seed by adult *Edessa meditabunda* (F.) and *Euschistus heros* (F.) (Hemiptera: Pentatomidae). Neotropical Entomology 41 (5): 386–390. http://dx.doi.org/10.1007/s13744-012-0061-9.

Small, W. 1923. The diseases of Arabica coffee in Uganda. Department of Agriculture Uganda Protectorate Circular 9. Uganda Protectorate Department of Agriculture, Entebbe, Uganda. 22 pp.

Smith, K. M. 1926. A comparative study of the feeding methods of certain Hemiptera and of the resulting effects upon the plant tissue, with special reference to the potato plant. Annals of Applied Biology 13: 109–139.

Soetaert, K., T. Petzoldt, and R. W. Setzer. 2010. Solving differential equations in R: Package deSolve. Journal of Statistical Software 33: 1–25.

Solarte, R. Y., E. A. Moreno, and J. V. Scorza. 1995. Flageliasis de plantas; Comentarios sobre una revision bibliografica. Revista de Ecología Latinoamericana 3(1–3): 57–68.

Stahel, G. 1954. Die Siebröhrenkrankheit (Phloemnekrose, flagellatose) des kaffeebaumes. Netherlands Journal of Agricultural Science 4: 260–264.

Toews, M., P. Roberts, and E. G. Medrano. 2010. Interrelationships among stink bug management, cotton fiber quality and boll rot. Proceedings of the Beltwide Cotton Conferences, New Orleans, Louisiana, January 245–247.

Trelles Di Lucca, A. G., E. F. T. Chipana, M. J. T. Albújar, W. D. Peralta, Y. C. M. Piedra and J. L. A. Zelada. 2013. Slow wilt: another form of Marchitez in oil palm associated with trypanosomatids in Peru. Tropical Plant Pathology 38(6): 522–533.

Uzest, M., D. Gargani, A. Dombrovsky, C. Cazevieille, D. Cot, and S. Blanc. 2010. The "acrostyle": A newly described anatomical structure in aphid stylets. Arthropod Structure and Development 39: 221–229.

Vainstein, M. H., and I. Roitman. 1986. Cultivation of *Phytomonas françai* associated with poor development of root system of cassava. Journal of Protozoology 33(4): 511–513.

Vermeulen, H. 1968. Investigations into the cause of the phloem necrosis disease of *Coffea liberica* in Surinam, South America. Netherlands Journal of Plant Pathology 74: 202–218.

Vickerman, K. 1994. The evolutionary expansion of the trypanosomatid flagellates. International Journal for Parasitology 24(8): 1317–1331.

Wallace, G. B. 1930. A coffee-bean disease. Tropical Agriculture 7: 14–17.

Wallace, G. B. 1932. Coffee bean disease. Relation of *Nematospora gossypii* to the disease. Tropical Agriculture 9: 127.

Webster, J., and R. W. S. Weber. 2007. Introduction to Fungi. Cambridge University Press, NY. 841 pp.

Weintraub, P. G., and L. Beanland. 2006. Insect vectors of phytoplasmas. Annual Review of Entomology 51: 91–111.

Williams, M. R. 2014. Cotton insect losses – 2014. http://www.entomology.msstate.edu/resources/cottoncrop .asp (accessed 1 September 2015).

Wiman, N. G., V. M. Walton, P. W. Shearer, and S. I. Rondon. 2014. Electronically monitored labial dabbing and stylet 'probing' behaviors of brown marmorated stink bug, *Halyomorpha halys*, in simulated environments. PLoS ONE 9(12): e113514. http://dx.doi.org/10.1371/journal.pone.0113514.

Wiman, N. G., J. E. Parker, C. Rodriguez-Saona, and V. M. Walton. 2015. Characterizing damage of brown marmorated stink bug (Hemiptera: Pentatomidae) in blueberries. Journal of Economic Entomology 108(3): 1156–1163. http://dx.doi.org/10.1093/jee/tov036.

Wingard, S. A. 1925. Studies on the pathogenicity, morphology, and cytology of *Nematospora phaseoli*. Bulletin of the Torrey Botanical Club 52: 249–290.

Wonham, M. J., M. A. Lewis, J. Renclawowicz, and P. Van den Driessche. 2006. Transmission assumptions generate conflicting predictions in host–vector disease models: a case study in West Nile virus. Ecology Letters 9: 706–725.

Zeilinger, A. R., and M. P. Daugherty. 2014. Vector preference and host defense against infection interact to determine disease dynamics. Oikos 123: 613–622.

Zeilinger, A. R., D. M. Olson, T. Raygoza, and D. A. Andow. 2015. Do counts of salivary sheath flanges predict food consumption in herbivorous stink bugs (Hemiptera: Pentatomidae)? Annals of the Entomological Society of America 108: 109–116. http://dx.doi.org/10.1093/aesa/sau011.

Section VIII

Symbiotic Microorganisms

14

Symbiotic Microorganisms Associated with Pentatomoidea

Yoshitomo Kikuchi, Simone S. Prado, and Tracie M. Jenkins

CONTENTS

14.1 Microbial Symbiosis in Pentatomomorphan Insects: An Overview..644
    14.1.1 Introduction 644
    14.1.2 Microbial Symbiosis in Insects 644
    14.1.3 Microbial Symbiosis in Pentatomorphan Insects...645
        14.1.3.1 Overview...645
        14.1.3.2 Symbiotic Organ: Midgut Crypts, M3 Bulb, and Bacteriocytes 654
        14.1.3.3 Phylogenetic Diversity of Symbiotic Bacteria 654
        14.1.3.4 Transmission Mechanism of Symbiotic Bacteria 656
        14.1.3.5 Biological Functions of Symbiotic Bacteria 657
        14.1.3.6 Symbiont Genome..657
14.2 Symbiosis in the Southern Green Stink Bug, *Nezara viridula* 659
    14.2.1 Overview 659
    14.2.2 Fitness Effect of Symbiotic Bacteria 659
    14.2.3 Effect of High Temperature on Pentatomid Symbiosis..660
    14.2.4 Symbiotic Associations in Allied Pentatomids 661
14.3 From Eastern Hemisphere to Western Hemisphere: Candidatus Ishikawaella Capsulata
    and Its Plataspid Host..661
    14.3.1 Overview 661
    14.3.2 Genetic Baseline..662
        14.3.2.1 Vertical Transmission in Symbiont Capsules 662
        14.3.2.2 Ishikawaella, Insect Development and Pest Status..663
        14.3.2.3 Genome Evolution...663
    14.3.3 Invasion of *Megacopta cribraria* 664
        14.3.3.1 Overview..664
        14.3.3.2 Ishikawaella Genome in the Western Hemisphere 664
        14.3.3.3 Genomic Landscape...665
    14.3.4 Hypotheses from a Natural Laboratory Experiment...665
14.4 Facultative Symbionts in the Pentatomomorpha...666
14.5 Concluding Remarks..666
14.6 Acknowledgments..667
14.7 References Cited..667

14.1 Microbial Symbiosis in Pentatomomorphan Insects: An Overview

14.1.1 Introduction

Heinrich Anton De Barry (1879), a German plant pathologist, defined "symbiosis" as a close association between different organisms. Among diverse symbiotic associations in nature, the most cohesive form is "endosymbiosis," an association in which one partner is a "symbiont" (usually microorganisms including fungi, Archaea, and bacteria) that lives inside the body of and intimately interacts with the other partner called "host" (usually animals, plants, and some protists). Sometimes, the microbial associates are harmful or even lethal to the host organisms, and, in those cases, the microbe is regarded as a "pathogen" or "parasite." Some microorganisms that have neutral effects on the host organisms are called "commensalists." In contrast, several microbial associates benefit the host organisms owing to their versatile metabolic abilities that result in significant ecological and evolutionary advantages to the host organisms; such microbial associates are called "mutualists." For example, leguminous plants can grow under nitrogen-deficient soil conditions; this ability is conferred to the plant by *Rhizobium* symbionts that fix atmospheric dinitrogen and provide the hosts with fixed nitrogen compounds such as nitrate and amino acids (Gualtieri and Bisseling 2000). Bobtail squids harbor luminescent *Vibrio* in a specific light organ and use the bacteria to hide from predators (Nyholm and McFall-Ngai 2004). Corals are tightly associated with photosynthetic dinoflagellates (Muscatine 1973). The human gut is colonized by a large number of diverse microorganisms that play pivotal roles in the processes of metabolism, immunity, and development of gut organs (Ley et al. 2006). In addition, the evolutionary origin of mitochondria and chloroplasts is believed to be endosymbiotic microorganisms (Margulis and Fester 1991). Hence, nature is full of endosymbiosis that have accelerated organismal evolution.

Recent development of molecular experimental techniques such as polymerase chain reaction (PCR), DNA sequencing, molecular phylogenetic analysis, *in situ* hybridization, and next-generation sequencing has revolutionized the entire world of microbial ecology, which has in turn revealed the ecological and evolutionary aspects of endosymbiosis. In this chapter, we present an overview of the current knowledge concerning the diversity of endosymbioses in pentatomomorphan insects, with a special focus on the well-studied symbiotic systems of the southern green stink bug, *Nezara viridula* (L.) (family Pentatomidae) and the kudzu bug, *Megacopta cribraria* (F.) (family Plataspidae).

14.1.2 Microbial Symbiosis in Insects

Symbiotic associations with bacteria occur in many animals, plants, fungi, and protists (Margulis and Fester 1991, Ruby et al. 2004), among which insects are regarded as the largest group that takes significant advantages of bacterial symbionts (Buchner 1965, Bourtzis and Miller 2003, Kikuchi 2009). The insects that feed exclusively on restricted diets, such as plant sap, vertebrate blood, or woody materials, usually host symbiotic microorganisms in their bodies, where the symbionts are involved in the provision of essential nutrients and/or digestion of food materials for the host (Douglas 1989). Carnivorous and herbivorous insects that feed on animal or plant tissues usually do not possess specific symbionts.

Symbiotic bacteria generally show strict host tissue tropism and are localized in specialized body parts called "symbiotic organs." Locations of symbionts are diverse in insects, ranging from extracellular within the lumen of intestinal symbiotic organs called "crypts" to intracellular within specialized cells called "bacteriocytes" (Kikuchi 2009). Bacteriocytes are enlarged cells specialized for harboring symbiotic bacteria in the cytoplasm, which are sometimes clustered and form a large symbiotic organ called a "bacteriome" in many insect species that are associated with intracellular symbionts.

Endosymbiosis between aphids and their intracellular bacterium *Buchnera aphidicola* are among the most well-known examples of such nutritional mutualistic associations. Almost all aphids possess bacteriocytes for housing the intracellular symbionts (Douglas 1989, Baumann 2005). *Buchnera* provides the host aphid with essential amino acids that are almost lacking in the plant phloem sap. The symbiont is passed from the mother to offspring via transovarial transmission, where *Buchnera* directly infects the embryos from the mother's bacteriocytes (Miura et al. 2003, Koga et al. 2012). The symbiont phylogeny perfectly mirrors the aphid phylogeny, indicating the ancient origin of the symbiosis and the strict vertical transmission during evolution. Such long-term associations have resulted in greater host–symbiont dependency: the symbiont cannot survive outside the host cells (i.e., in an uncultivable state), whereas the host aphid suffers from retarded growth, high mortality, and sterility if the symbiont is eliminated by antibiotic treatments.

Such highly developed (or highly interdependent) host–symbiont associations also have been identified in diverse insect groups such as *Wigglesworthia* symbionts in tsetse flies (Aksoy 2000), *Baumannia* symbionts in sharpshooters (Moran et al. 2005, Wu et al. 2006), *Carsonella* symbionts in psyllids (Thao et al. 2000), *Tremblaya* symbionts in mealybugs (Thao et al. 2002), *Blochmannia* symbionts in carpenter ants (Sauer et al. 2000), and *Nardonella* symbionts in weevils (Lefevre et al. 2004). In addition to such nutritional metabolic roles, recent studies have unveiled more diverse functions of symbiotic microorganisms in insects. For example, antibiotic-producing actinomycete symbionts that protect the host from pathogenic/ harmful microorganisms have been discovered from leaf-cutting ants (Currie et al. 1999, Currie 2001), ambrosia beetles (Scott et al. 2008), and beewolves (Kaltenpoth et al. 2005). Staphylinid beetles possess gammmaproteobacterial symbionts that produce a toxic compound, pederine, that protects the insect against predators (Kellner 2002). Antlions employ a heat-shock protein produced by symbiotic bacteria as a venom for predation (Yoshida et al. 2001).

14.1.3 Microbial Symbiosis in Pentatomorphan Insects

14.1.3.1 Overview

The heteropteran group is divided into seven infraorders including Enicocephalomorpha, Dipsocoromorpha, Gerromorpha, Nepomorpha, Leptopodomorpha, Cimicomorpha, and Pentatomomorpha (Schuh and Slater 1995, Weirauch and Schuh 2011). Of these, symbiotic bacteria have been reported from the Cimicomorpha and Pentatomomorpha (Glasgow 1914, Buchner 1965, Dasch et al. 1984). In the Cimicomorpha, blood-sucking species representing the families Reduviidae (assassin bugs) and Cimicidae (bed bugs) harbor symbiotic bacteria in their gut cavity and in bacteriocytes, respectively (Glasgow 1914, Buchner 1965, Usinger 1966, Dasch et al. 1984, Hosokawa et al. 2010b). Almost all pentatomomorphans, except for the predatory Asopinae and the mycophagous Aladoidae, are phytophagous, most of which possess symbiotic bacteria.

Although sap-feeding hemipterans usually possess intracellular symbiosis and well-developed bacteriomes (Baumann 2005, Moran 2007), pentatomomorphan species harbor symbiotic bacteria in their gut extracellularly. The insect gut generally is divided into three regions, namely the foregut, midgut (mesenteron, ventriculus), and hindgut. The foregut and hindgut are lined with cuticle, which serves as a barrier for most digestion and absorption. The midgut, however, lacks cuticle, and, therefore, most of these two processes take place in this portion of the gut. In many insects, the midgut is simply organized and consists of only two or three different sections (Lehane and Billingsley 1996). On the other hand, the midgut of pentatomomorphan species often is divided anatomically into four distinct sections: (1) the voluminous first section (M1), (2) the tubular second section (M2), (3) the ovoid third section (M3), and (4) the fourth section (M4) with numerous sacs or tubular-outgrowths, called as crypts, which are densely populated by specific bacterial symbionts (**Figure 14.1A–H**). The M4 crypts commonly are developed in phytophagous species of the superfamilies Pentatomoidea, Coreoidea, and Lygaeoidea. The biological roles of the four sections are not known exactly, although it has been suggested that the M1 serves for transient food storage and digestion, the M2 and the M3 perform food digestion and absorption, and the M4 is specialized for harboring the symbiotic bacteria.

Although the association of microbes with the midgut crypts of Pentatomidae has been known since the late 1800s, these microbes remained unidentified. The first comprehensive work on their identification was authored by Glasgow (1914), who revealed that, although bacteria from the midgut crypts of different hosts were morphologically different, their morphology remained constant for each species, and they always were present in a monoculture. Rosenkranz (1939) suggested that these bacteria played significant roles in relation to insect nutrition. Furthermore, he discussed the possibility that mother stink bugs covered their eggs with bacteria that were acquired orally by nymphs soon after hatching. Thereafter, classic histological observations and recent molecular works have unveiled many other different types

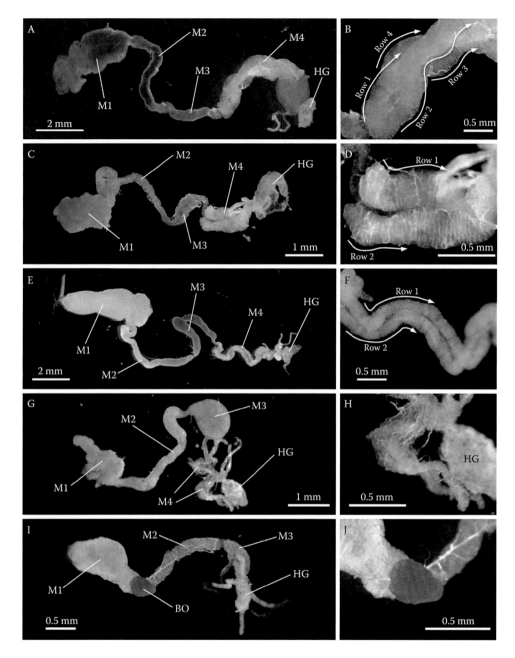

FIGURE 14.1 (**See color insert.**) Symbiotic organ in the Pentatomomorpha. Dissected midgut and symbiotic organ of (A-B) *Menida scotti* (family Pentatomidae), (C-D) *Elasmucha putoni* (Acanthosomatidae), (E-F) *Riptortus pedestris* (Alydidae), (G-H) *Togo hemipterus* (Rhyparochromidae), and (I-J) *Kleidocerys resedae* (Lygaeidae). (B) Midgut crypts arranged in 4 rows. (D) Midgut crypts arranged in 2 rows, whose luminal entrances are completely sealed. (F) Midgut crypts arranged in 2 rows. (H) Tubular outgrowths. (J) Bacteriome. Abbreviations: M1, midgut first section; M2, midgut second section; M3, midgut third section; M4, midgut fourth section with crypts (symbiotic organ); HG, hindgut; BO, Bacteriome.

of symbiotic associations in the Pentatomomorpha. Symbiotic association within the Pentatomomorpha is remarkably diverse in terms of the morphology of symbiotic organs, symbiont transmission mechanisms, symbiont phylogeny, and biological functions of symbiotic bacteria (**Table 14.1**). In the following sections, we review the diverse symbiotic systems reported up to the present in pentatomomorphan insects.

TABLE 14 1

Bacterial Symbiosis in the Pentatomomorpha: Taxonomical Affiliation of Symbiont, Symbiotic Organ, and Transmission Mechanism

Species	Symbiont	Phylum/Class	Symbiotic Organ Type	Transmission Mechanism	Reference
Superfamily PENTATOMOIDEA					
Family ACANTHOSOMATIDAE					
Acanthosoma denticaudum Jakovlev	*Rosenkranzia clausaccus*	γ-*Proteobacteria*	M4 crypts, 2 rows	Egg smearing	Kikuchi et al. 2009
Acanthosoma forficula Jakovlev	*Rosenkranzia clausaccus*	γ-*Proteobacteria*	M4 crypts, 2 rows	Egg smearing	Kikuchi et al. 2009
Acanthosoma firmatum (Walker) [= *A. gigavteum* Matsumura]	*Rosenkranzia clausaccus*	γ-*Proteobacteria*	M4 crypts, 2 rows	Egg smearing	Kikuchi et al. 2009
Acanthosoma haemorrhoidale (L.)	*Rosenkranzia clausaccus*	γ-*Proteobacteria*	M4 crypts, 2 rows	Egg smearing	Kikuchi et al. 2009
Acanthosoma labiduroides Jakovlev	*Rosenkranzia clausaccus*	γ-*Proteobacteria*	M4 crypts, 2 rows	Egg smearing	Kikuchi et al. 2009
Elasmostetus brevis Lindberg	*Rosenkranzia clausaccus*	γ-*Proteobacteria*	M4 crypts, 2 rows	Egg smearing	Kikuchi et al. 2009
Elasmostetus humeralis Jakovlev	*Rosenkranzia clausaccus*	γ-*Proteobacteria*	M4 crypts, 2 rows	Egg smearing	Kikuchi et al. 2009
Elasmostetus nubilus (Dallas)	*Rosenkranzia clausaccus*	γ-*Proteobacteria*	M4 crypts, 2 rows	Egg smearing	Kikuchi et al. 2009
Elasmucha dorsalis (Jakovlev)	*Rosenkranzia clausaccus*	γ-*Proteobacteria*	M4 crypts, 2 rows	Egg smearing	Kikuchi et al. 2009
Elasmucha putoni Scott	*Rosenkranzia clausaccus*	γ-*Proteobacteria*	M4 crypts, 2 rows	Egg smearing	Kikuchi et al. 2009
Elasmucha signoreti Scott	*Rosenkranzia clausaccus*	γ-*Proteobacteria*	M4 crypts, 2 rows	Egg smearing	Kikuchi et al. 2009
Lindbergicoris gramineus (Distant)	*Rosenkranzia clausaccus*	γ-*Proteobacteria*	M4 crypts, 2 rows	Egg smearing	Kikuchi et al. 2009
Sastragala esakii Hasegawa	*Rosenkranzia clausaccus*	γ-*Proteobacteria*	M4 crypts, 2 rows	Egg smearing	Kikuchi et al. 2009
Sastragala scutellata (Scott)	*Rosenkranzia clausaccus*	γ-*Proteobacteria*	M4 crypts, 2 rows	Egg smearing	Kikuchi et al. 2009
Family CYDNIDAE					
Adomerus rotundus (Hsiao)	Uncharacterized γ-proteobacterium	γ-*Proteobacteria*	M4 crypts, 2 rows	Egg smearing	Hosokawa et al. 2012b
Adomerus trigattulus (Motschulsky)	Uncharacterized γ-proteobacterium	γ-*Proteobacteria*	M4 crypts, 2 rows	Egg smearing	Hosokawa et al. 2012b
Adomerus variegatus (Signoret)	Uncharacterized γ-proteobacterium	γ-*Proteobacteria*	M4crypts, 2 rows	Egg smearing	Hosokawa et al. 2012b
Adrisa mazna (Uhler)	Uncharacterized γ-proteobacterium	γ-*Proteobacteria*	M4 crypts, 2 rows	Egg smearing	Hosokawa et al. 2012b
Canthophorus niveimarginatus Scott	Uncharacterized γ-proteobacterium	γ-*Proteobacteria*	M4 crypts, 2 rows	Egg smearing	Hosokawa et al. 2012b
Macroscytus japonensis Scott	Uncharacterized γ-proteobacterium	γ-*Proteobacteria*	M4 crypts, 2 rows	Egg smearing	Hosokawa et al. 2012b
Schiodtella japonica Imura and Ishikawa	Uncharacterized γ-proteobacterium	γ-*Proteobacteria*	M4 crypts, 2 rows	Egg smearing	Hosokawa et al. 2012b
Family PARASTRACHIIDAE					
Parastrachia japonensis (Scott)	*Benitsuchiphilus tojoi*	γ-*Proteobacteria*	M4 crypts, 2 rows	Egg smearing	Hosokawa et al. 2010a

(Continued)

TABLE 14.1 (CONTINUED)

Bacterial Symbiosis in the Pentatomomorpha: Taxonomical Affiliation of Symbiont, Symbiotic Organ, and Transmission Mechanism

Species	Symbiont	Phylum/Class	Symbiotic Organ Type	Transmission Mechanism	Reference
Family PENTATOMIDAE					
Antestiopsis thunbergii (Gmelin)	Uncharacterized γ-proteobacterium	γ-*Proteobacteria*	M4 crypts, 4 rows	Egg smearing	Matsuura et al. 2014
Antiteuchus costaricensis Ruckes	Uncharacterized γ-proteobacterium	γ-*Proteobacteria*	M4 crypts, 4 rows	Egg smearing	Bistolas et al. 2014
Arvelius porrectispinus Breddin	Uncharacterized γ-proteobacterium	γ-*Proteobacteria*	M4 crypts, 4 rows	Egg smearing	Bistolas et al. 2014
Chinavia hilaris (Say)	Uncharacterized γ-proteobacterium	γ-*Proteobacteria*	M4 crypts, 4 rows	Egg smearing	Prado and Almeida 2009a
Chlorochroa ligata (Say)	Uncharacterized γ-proteobacterium	γ-*Proteobacteria*	M4 crypts, 4 rows	Egg smearing	Prado and Almeida 2009a
Chlorochroa sayi (Stål)	Uncharacterized γ-proteobacterium	γ-*Proteobacteria*	M4 crypts, 4 rows	Egg smearing	Prado and Almeida 2009a
Chlorochroa uhleri (Stål)	Uncharacterized γ-proteobacterium	γ-*Proteobacteria*	M4 crypts, 4 rows	Egg smearing	Prado and Almeida 2009a
Edessa bugabensis Distant	Uncharacterized γ-proteobacterium	γ-*Proteobacteria*	M4 crypts, 4 rows	Egg smearing	Bistolas et al. 2014
Edessa eburatula Breddin	Uncharacterized γ-proteobacterium	γ-*Proteobacteria*	M4 crypts, 4 rows	Egg smearing	Bistolas et al. 2014
Edessa irrorata Dallas	Uncharacterized γ-proteobacterium	γ-*Proteobacteria*	M4 crypts, 4 rows	Egg smearing	Bistolas et al. 2014
Edessa jugata Westwood	Uncharacterized γ-proteobacterium	γ-*Proteobacteria*	M4 crypts, 4 rows	Egg smearing	Bistolas et al. 2014
Edessa junix Stål	Uncharacterized γ-proteobacterium	γ-*Proteobacteria*	M4 crypts, 4 rows	Egg smearing	Bistolas et al. 2014
Edessa sp.	Uncharacterized γ-proteobacterium	γ-*Proteobacteria*	M4 crypts, 4 rows	Egg smearing	Bistolas et al. 2014
Eurydema dominulus (Scopoli)	Uncharacterized γ-proteobacterium	γ-*Proteobacteria*	M4 crypts, 4 rows	Egg smearing	Kikuchi et al. 2012
Eurydema gebleri rugosa Motschulsky	Uncharacterized γ-proteobacterium	γ-*Proteobacteria*	M4 crypts, 4 rows	Egg smearing	Kikuchi et al. 2012
Euschistus heros (F.)	Uncharacterized γ-proteobacterium	γ-*Proteobacteria*	M4 crypts, 4 rows	Egg smearing	Prado and Almeida 2009a
Euschistus sp.	Uncharacterized γ-proteobacterium	γ-*Proteobacteria*	M4 crypts, 4 rows	Egg smearing	Bistolas et al. 2014
Halyomorpha halys (Stål)	*Pantoea carbekii*	γ-*Proteobacteria*	M4 crypts, 4 rows	Egg smearing	Bansal et al. 2014
Loxa sp.	Uncharacterized γ-proteobacterium	γ-*Proteobacteria*	M4 crypts, 4 rows	Egg smearing	Bistolas et al. 2014
Mormidea collaris Dallas	Uncharacterized γ-proteobacterium	γ-*Proteobacteria*	M4 crypts, 4 rows	Egg smearing	Bistolas et al. 2014
Mormidea ypsilon (L.)	Uncharacterized γ-proteobacterium	γ-*Proteobacteria*	M4 crypts, 4 rows	Egg smearing	Bistolas et al. 2014
Murgantia histrionica (Hahn)	Uncharacterized γ-proteobacterium	γ-*Proteobacteria*	M4 crypts, 4 rows	Egg smearing	Prado and Almeida 2009a
Nezara viridula (L.)	Uncharacterized γ-proteobacterium	γ-*Proteobacteria*	M4 crypts, 4 rows	Egg smearing	Prado et al. 2006
Plautia stali Scott	Uncharacterized γ-proteobacterium	γ-*Proteobacteria*	M4 crypts, 4 rows	Egg smearing	Hosokawa et al. 2016
Sibaria englemani Rolston	Uncharacterized γ-proteobacterium	γ-*Proteobacteria*	M4 crypts, 4 rows	Egg smearing	Bistolas et al. 2014
Thyanta pallidovirens (Stål)	Uncharacterized γ-proteobacterium	γ-*Proteobacteria*	M4 crypts, 4 rows	Egg smearing	Prado and Almeida 2009a

(Continued)

TABLE 14 1 (CONTINUED)

Bacterial Symbiosis in the Pentatomomorpha: Taxonomical Affiliation of Symbiont, Symbiotic Organ, and Transmission Mechanism

Species	Symbiont	Phylum/Class	Symbiotic Organ Type	Transmission Mechanism	Reference
Family PLATASPIDAE					
Brachyplatys subaeneus (Westwood)	*Ishikawaella capsulata*	*γ-Proteobacteria*	M4 crypts, 2 rows[1]	Capsule transmission	Hosokawa et al. 2006
Brachyplatys vahlii (F.)	*Ishikawaella capsulata*	*γ-Proteobacteria*	M4 crypts, 2 rows[1]	Capsule transmission	Hosokawa et al. 2006
Coptosoma japonicum Matsumura	*Ishikawaella capsulata*	*γ-Proteobacteria*	M4 crypts, 2 rows[1]	Capsule transmission	Hosokawa et al. 2006
Coptosoma parvipictum Montandon	*Ishikawaella capsulata*	*γ-Proteobacteria*	M4 crypts, 2 rows[1]	Capsule transmission	Hosokawa et al. 2006
Coptosoma sphaerula (Germar)	*Ishikawaella capsulata*	*γ-Proteobacteria*	M4 crypts, 2 rows[1]	Capsule transmission	Hosokawa et al. 2006
Megacopta cribraria (F.) [= *M. punctatissima* (Montandon)]	*Ishikawaella capsulata*	*γ-Proteobacteria*	M4 crypts, 2 rows[1]	Capsule transmission	Hosokawa et al. 2006
Family SCUTELLERIDAE					
Cantao ocellatus (Thunberg)	Uncharacterized γ-proteobacterium	*γ-Proteobacteria*	M4 crypts, 4 rows	Egg smearing	Kaiwa et al. 2010
Eucorysses grandis (Thunberg)	Uncharacterized γ-proteobacterium	*γ-Proteobacteria*	M4 crypts, 4 rows	Egg smearing	Kaiwa et al. 2011
Family UROSTYLIDIDAE					
Urochela luteovaria Distant	*Tachikawaea gelatinosa*	*γ-Proteobacteria*	M4 crypts, 4 rows[1]	Jelly transmission	Kaiwa et al. 2014
Urochela quadrinotata (Reuter)	*Tachikawaea gelatinosa*	*γ-Proteobacteria*	M4 crypts, 4 rows[1]	Jelly transmission	Kaiwa et al. 2014
Urostylis annulicornis Scott	*Tachikawaea gelatinosa*	*γ-Proteobacteria*	M4 crypts, 4 rows[1]	Jelly transmission	Kaiwa et al. 2014
Urostylis riicornis Scott	*Tachikawaea gelatinosa*	*γ-Proteobacteria*	M4 crypts, 4 rows[1]	Jelly transmission	Kaiwa et al. 2014
Urostylis westwoodii Scott	*Tachikawaea gelatinosa*	*γ-Proteobacteria*	M4 crypts, 4 rows[1]	Jelly transmission	Kaiwa et al. 2014
Superfamily COREOIDEA					
Family ALYDIDAE					
Alydus calcaratus (L.)	Uncharacterized *Burkholderia*	*β-Proteobacteria*	M4 crypts, 2 rows	Environmental acquisition	García et al. 2014
Alydus corspersus Montandon	Uncharacterized *Burkholderia*	*β-Proteobacteria*	M4 crypts, 2 rows	Environmental acquisition	García et al. 2014
Alydus tornentosus Fracker	Uncharacterized *Burkholderia*	*β-Proteobacteria*	M4 crypts, 2 rows	Environmental acquisition	García et al. 2014
Daclera levana Distant	Uncharacterized *Burkholderia*	*β-Proteobacteria*	M4 crypts, 2 rows	Environmental acquisition	Kikuchi et al. 2011
Leptocorisa acuta (Thunberg)	Uncharacterized *Burkholderia*	*β-Proteobacteria*	M4 crypts, 2 rows	Environmental acquisition	Kikuchi et al. 2011
Leptocorisa chinensis Dallas	Uncharacterized *Burkholderia*	*β-Proteobacteria*	M4 crypts, 2 rows	Environmental acquisition	Kikuchi et al. 2005
Leptocorisa oratoria (F.)	Uncharacterized *Burkholderia*	*β-Proteobacteria*	M4 crypts, 2 rows	Environmental acquisition	Kikuchi et al. 2011
Megalotomus costalis Stål	Uncharacterized *Burkholderia*	*β-Proteobacteria*	M4 crypts, 2 rows	Environmental acquisition	Kikuchi et al. 2011

(Continued)

TABLE 14.1 (CONTINUED)

Bacterial Symbiosis in the Pentatomomorpha: Taxonomical Affiliation of Symbiont, Symbiotic Organ, and Transmission Mechanism

Species	Symbiont	Phylum/Class	Symbiotic Organ Type	Transmission Mechanism	Reference
Megalotomus quinquespinosus (Say)	Uncharacterized *Burkholderia*	β-*Proteobacteria*	M4 crypts, 2 rows	Environmental acquisition	Garcia et al. 2014
Paraplesius unicolor Scott	Uncharacterized *Burkholderia*	β-*Proteobacteria*	M4 crypts, 2 rows	Environmental acquisition	Kikuchi et al. 2011
Riptortus linearis (F.)	Uncharacterized *Burkholderia*	β-*Proteobacteria*	M4 crypts, 2 rows	Environmental acquisition	Kikuchi et al. 2011
Riptortus pedestris (F.)	Uncharacterized *Burkholderia*	β-*Proteobacteria*	M4 crypts, 2 rows	Environmental acquisition	Kikuchi et al. 2005
Family COREIDAE					
Acanthocoris sordidus (Thunberg)	Uncharacterized *Burkholderia*	β-*Proteobacteria*	M4 crypts, 2 rows	Environmental acquisition	Kikuchi et al. 2011
Anacanthocoris striicornis (Scott)	Uncharacterized *Burkholderia*	β-*Proteobacteria*	M4 crypts, 2 rows	Environmental acquisition	Kikuchi et al. 2011
Cletus punctiger (Dallas)	Uncharacterized *Burkholderia*	β-*Proteobacteria*	M4 crypts, 2 rows	Environmental acquisition	Kikuchi et al. 2011
Cletus rusticus Stål [= *C. punctiger* (Dallas)]	Uncharacterized *Burkholderia*	β-*Proteobacteria*	M4 crypts, 2 rows	Environmental acquisition	Kikuchi et al. 2011
Cletus trigonus (Thunberg)	Uncharacterized *Burkholderia*	β-*Proteobacteria*	M4 crypts, 2 rows	Environmental acquisition	Kikuchi et al. 2011
Dasynus coccocinctus (Burmeister)	Uncharacterized *Burkholderia*	β-*Proteobacteria*	M4 crypts, 2 rows	Environmental acquisition	Kikuchi et al. 2011
Homoeocerus dilatatus Horváth	Uncharacterized *Burkholderia*	β-*Proteobacteria*	M4 crypts, 2 rows	Environmental acquisition	Kikuchi et al. 2011
Homoeocerus marginiventris Dohrn	Uncharacterized *Burkholderia*	β-*Proteobacteria*	M4 crypts, 2 rows	Environmental acquisition	Kikuchi et al. 2011
Homoeocerus unipunctatus (Thunberg)	Uncharacterized *Burkholderia*	β-*Proteobacteria*	M4 crypts, 2 rows	Environmental acquisition	Kikuchi et al. 2011
Hygia lativentris (Motschulsky)	Uncharacterized *Burkholderia*	β-*Proteobacteria*	M4 crypts, 2 rows	Environmental acquisition	Kikuchi et al. 2011
Hygia opaca (Uhler)	Uncharacterized *Burkholderia*	β-*Proteobacteria*	M4 crypts, 2 rows	Environmental acquisition	Kikuchi et al. 2011
Leptoglossus australis (F.) [= *L. gonagra* (F.)]	Uncharacterized *Burkholderia*	β-*Proteobacteria*	M4 crypts, 2 rows	Environmental acquisition	Kikuchi et al. 2011
Molipteryx fuliginosa (Uhler)	Uncharacterized *Burkholderia*	β-*Proteobacteria*	M4 crypts, 2 rows	Environmental acquisition	Kikuchi et al. 2011
Notobitus meleagris (F.)	Uncharacterized *Burkholderia*	β-*Proteobacteria*	M4 crypts, 2 rows	Environmental acquisition	Kikuchi et al. 2011
Paradasynus spinosus Hsiao	Uncharacterized *Burkholderia*	β-*Proteobacteria*	M4 crypts, 2 rows	Environmental acquisition	Kikuchi et al. 2011
Plinachtus basalis (Westwood)	Uncharacterized *Burkholderia*	β-*Proteobacteria*	M4 crypts, 2 rows	Environmental acquisition	Kikuchi et al. 2011
Plinachtus bicoloripes Scott	Uncharacterized *Burkholderia*	β-*Proteobacteria*	M4 crypts, 2 rows	Environmental acquisition	Kikuchi et al. 2011
Thasus neocalifornicus Brailovsky and Barrera	Uncharacterized *Burkholderia*	β-*Proteobacteria*	M4 crypts, 2 rows	Environmental acquisition	Olivier-Espejel et al. 2011
Family STENOCEPHALIDAE					
Dicranocephalus agilis (Scopoli)	Uncharacterized *Burkholderia*	β-*Proteobacteria*	M4 crypts, 2 rows	Environmental acquisition	Küechler et al. 2016
Dicranocephalus albipes (F.)	Uncharacterized *Burkholderia*	β-*Proteobacteria*	M4 crypts, 2 rows	Environmental acquisition	Küechler et al. 2016

(Continued)

TABLE 14.1 (CONTINUED)

Bacterial Symbiosis in the Pentatomomorpha: Taxonomical Affiliation of Symbiont, Symbiotic Organ, and Transmission Mechanism

Species	Symbiont	Phylum/Class	Symbiotic Organ Type	Transmission Mechanism	Reference
Dicranocephalus lateralis (Signoret)	Uncharacterized *Burkholderia*	β-*Proteobacteria*	M4 crypts, 2 rows	Environmental acquisition	Küechler et al. 2016
Dicranocephalus medius (Mulsant et Rey)	Uncharacterized *Burkholderia*	β-*Proteobacteria*	M4 crypts, 2 rows	Environmental acquisition	Küechler et al. 2016
Superfamily LYGAEOIDEA					
Family ARTHENEIDAE					
Chilacis typhae (Perris)	*Rohrkolberia cinguli*	γ-*Proteobacteria*	Bacteriome[2]	Transovarial transmission	Küechler et al. 2011
Family BERYTIDAE					
Metatropis rufescens (Herrich-Schäffer)	Uncharacterized *Burkholderia*	β-*Proteobacteria*	M4 crypts, Tubes	Environmental acquisition	Kikuchi et al. 2011
Yemma exilis Horváth	Uncharacterized *Burkholderia*	β-*Proteobacteria*	M4 crypts, Tubes	Environmental acquisition	Kikuchi et al. 2011
Family BLISSIDAE					
Blissus insularis Barber	Uncharacterized *Burkholderia*	β-*Proteobacteria*	M4 crypts, Tubes	Environmental acquisition	Boucias et al. 2012
Cavelerius saccharivorus (Okajima)	Uncharacterized *Burkholderia*	β-*Proteobacteria*	M4 crypts, Tubes	Environmental acquisition	Itoh et al. 2014
Dimorphopterus pallipes (Distant)	Uncharacterized *Burkholderia*	β-*Proteobacteria*	M4 crypts, Tubes	Environmental acquisition	Kikuchi et al. 2011
Ischnodemus sabuleti (Fallén)	Uncharacterized γ-proteobacterium	γ-*Proteobacteria*	Bacteriome[2]	Transovarial transmission	Küechler et al. 2012
Family LYGAEIDAE					
Arocatus longiceps Stål	Uncharacterized γ-proteobacterium	γ-*Proteobacteria*	Bacteriome[2]	Transovarial transmission	Küechler et al. 2012
Belonochilus numenius (Say)	Uncharacterized γ-proteobacterium	γ-*Proteobacteria*	Bacteriome[2]	Transovarial transmission	Küechler et al. 2012
Kleidocerys resedae (Panzer)	*Kleidoceria schneideri*	γ-*Proteobacteria*	Bacteriome[2]	Transovarial transmission	Küechler et al. 2010
Nysius expressus Distant	*Schneideria nysicola*	γ-*Proteobacteria*	Bacteriome[2]	Transovarial transmission	Matsuura et al. 2012
Nysius plezeius Distant	*Schneideria nysicola*	γ-*Proteobacteria*	Bacteriome[2]	Transovarial transmission	Matsuura et al. 2012
Nysius sp. 1	*Schneideria nysicola*	γ-*Proteobacteria*	Bacteriome[2]	Transovarial transmission	Matsuura et al. 2012
Nysius sp. 2	*Schneideria nysicola*	γ-*Proteobacteria*	Bacteriome[2]	Transovarial transmission	Matsuura et al. 2012
Orsillus depressus (Mulsant and Rey)	Uncharacterized γ-proteobacterium	γ-*Proteobacteria*	Bacteriome[2]	Transovarial transmission	Küechler et al. 2012
Ortholomus punctipennis (Herrich-Schäffer)	Uncharacterized γ-proteobacterium	γ-*Proteobacteria*	Bacteriome[2]	Transovarial transmission	Küechler et al. 2012
Family PACHYGRONTHIDAE					
Pachygrontha antennata (Uhler)	Uncharacterized *Burkholderia*	β-*Proteobacteria*	M4 crypts, 2 rows[1]	Environmental acquisition	Kikuchi et al. 2011

(Continued)

TABLE 14.1 (CONTINUED)

Bacterial Symbiosis in the Pentatomomorpha: Taxonomical Affiliation of Symbiont, Symbiotic Organ, and Transmission Mechanism

Species	Symbiont	Phylum/Class	Symbiotic Organ Type	Transmission Mechanism	Reference
Family RHYPAROCHROMIDAE					
Gyndes pallicornis (Dallas)	Uncharacterized *Burkholderia*	β-*Proteobacteria*	M4 crypts, Tubes	Environmental acquisition	Kikuchi et al. 2011
Horridipamera inconspicua (Dallas)	Uncharacterized *Burkholderia*	β-*Proteobacteria*	M4 crypts, Tubes	Environmental acquisition	Kikuchi et al. 2011
Horridipamera nietneri (Dohrn)	Uncharacterized *Burkholderia*	β-*Proteobacteria*	M4 crypts, Tubes	Environmental acquisition	Kikuchi et al. 2011
Metochus abbreviatus Scott	Uncharacterized *Burkholderia*	β-*Proteobacteria*	M4 crypts, Tubes	Environmental acquisition	Kikuchi et al. 2011
Neolethaeus assamensis (Distant)	Uncharacterized *Burkholderia*	β-*Proteobacteria*	M4 crypts, 2 rows	Environmental acquisition	Kikuchi et al. 2011
Neolethaeus dallasi (Scott)	Uncharacterized *Burkholderia*	β-*Proteobacteria*	M4 crypts, 2 rows	Environmental acquisition	Kikuchi et al. 2011
Pachybrachius luridus Hahn	Uncharacterized *Burkholderia*	β-*Proteobacteria*	M4 crypts, Tubes	Environmental acquisition	Kikuchi et al. 2011
Panaorus albomaculatus (Scott)	Uncharacterized *Burkholderia*	β-*Proteobacteria*	M4 crypts, Tubes	Environmental acquisition	Kikuchi et al. 2011
Panaorus japonicus (Stål)	Uncharacterized *Burkholderia*	β-*Proteobacteria*	M4 crypts, Tubes	Environmental acquisition	Kikuchi et al. 2011
Paromius exiguus (Distant)	Uncharacterized *Burkholderia*	β-*Proteobacteria*	M4 crypts, Tubes	Environmental acquisition	Kikuchi et al. 2011
Togo hemipterus (Scott)	Uncharacterized *Burkholderia*	β-*Proteobacteria*	M4 crypts, Tubes	Environmental acquisition	Kikuchi et al. 2011
Superfamily PYRRHOCOROIDEA					
Family LARGIDAE					
Physopelta cincticollis Stål	Uncharacterized *Burkholderia*	β-*Proteobacteria*	M4 crypts, Tubes	Environmental acquisition	Sudakaran et al. 2015
Physopelta gutta (Burmeister)	Uncharacterized *Burkholderia*	β-*Proteobacteria*	M4 crypts, Tubes	Environmental acquisition	Takeshita et al. 2015
Physopelta parviceps Blöte	Uncharacterized *Burkholderia*	β-*Proteobacteria*	M4 crypts, Tubes	Environmental acquisition	Takeshita et al. 2015
Physopelta slanbuschii (F.)	Uncharacterized *Burkholderia*	β-*Proteobacteria*	M4 crypts, Tubes	Environmental acquisition	Takeshita et al. 2015
Family PYRRHOCORIDAE					
Antilochus coquebertii (F.)	Diverse gut microbiota	Diverse[3]	Enlarged M3	Egg smearing	Sudakaran et al. 2015
Antilochus nigripes (Burmeister)	Diverse gut microbiota	Diverse[3]	Enlarged M3	Egg smearing	Sudakaran et al. 2015
Cenaeus carnifex (.F.)	Diverse gut microbiota	Diverse[3]	Enlarged M3	Egg smearing	Sudakaran et al. 2015
Dindymus lanius Stål	Diverse gut microbiota	Diverse[3]	Enlarged M3	Egg smearing	Sudakaran et al. 2015

(Continued)

TABLE 14.1 (CONTINUED)

Bacterial Symbiosis in the Pentatomomorpha: Taxonomical Affiliation of Symbiont, Symbiotic Organ, and Transmission Mechanism

Species	Symbiont	Phylum/Class	Symbiotic Organ Type	Transmission Mechanism	Reference
Dysdercus andreae (L.)	Diverse gut microbiota	Diverse[3]	Enlarged M3	Egg smearing	Sudakaran et al. 2015
Dysdercus cingulatus (F.)	Diverse gut microbiota	Diverse[3]	Enlarged M3	Egg smearing	Sudakaran et al. 2015
Dysdercus decussatus Boisduval	Diverse gut microbiota	Diverse[3]	Enlarged M3	Egg smearing	Sudakaran et al. 2015
Dysdercus fasciatus Signoret	Diverse gut microbiota	Diverse[3]	Enlarged M3	Egg smearing	Sudakaran et al. 2015
Dysdercus olivaceus (F.)	Diverse gut microbiota	Diverse[3]	Enlarged M3	Egg smearing	Sudakaran et al. 2015
Dysdercus poecilus (Herrich-Schäffer)	Diverse gut microbiota	Diverse[3]	Enlarged M3	Egg smearing	Sudakaran et al. 2015
Dysdercus suturellus (Herrich-Schäffer)	Diverse gut microbiota	Diverse[3]	Enlarged M3	Egg smearing	Sudakaran et al. 2015
Dysdercus voelkeri Schmidt	Diverse gut microbiota	Diverse[3]	Enlarged M3	Egg smearing	Sudakaran et al. 2015
Probergrothius angolensis (Distant)	Diverse gut microbiota	Diverse[3]	Enlarged M3	Egg smearing	Sudakaran et al. 2015
Probergrothius nigricornis (Stål)	Diverse gut microbiota	Diverse[3]	Enlarged M3	Egg smearing	Sudakaran et al. 2015
Probergrothius sanguinolens (Amyot and Serville)	Diverse gut microbiota	Diverse[3]	Enlarged M3	Egg smearing	Sudakaran et al. 2015
Probergrothius sexpunctatus (Laporte)	Diverse gut microbiota	Diverse[3]	Enlarged M3	Egg smearing	Sudakaran et al. 2015
Pyrrhocoris apterus (L.)	Diverse gut microbiota	Diverse[3]	Enlarged M3	Egg smearing	Sudakaran et al. 2012
Pyrrhocoris sibiricus Kuschakewitsch	Diverse gut microbiota	Diverse[3]	Enlarged M3	Egg smearing	Sudakaran et al. 2015
Scantius aegypticus (L.)	Diverse gut microbiota	Diverse[3]	Enlarged M3	Egg smearing	Sudakaran et al. 2015
Scantius obscurus Distant	Diverse gut microbiota	Diverse[3]	Enlarged M3	Egg smearing	Sudakaran et al. 2015

[1] Crypt arrangement is not necessarily clear.

[2] Intracellular symbiosis.

[3] Gut microbiota consists of *Coriobacterium*, *Gordonibacter* (both are Actinobacteria), *Clostridium* (Firmicutes), *Klebsiella* (Proteobacteria), and many others.

14.1.3.2 Symbiotic Organ: Midgut Crypts, M3 Bulb, and Bacteriocytes

As discussed above, phytophagous species of the superfamilies Pentatomoidea, Coreoidea, and Lygaeoidea commonly develop midgut crypts in the posterior region of the midgut, although the morphology and arrangement of the symbiotic organ are diverse. In the Pentatomoidea and Coreoidea, several small sacs are arranged in rows of two or four along the midgut (**Figure 14.1B,D,F**). In the Lygaeoidea, most species, except members of the family Pachygronthidae, develop tubular-type crypts in the M4 region. Pachygronthid species commonly possess small sac-type crypts, as reported in the pentatomoid and coreoid insects. Although the luminal cavity of the midgut crypts generally open into the lumen of the gut (Goodchild 1963), members of the family Acanthosomatidae develop an exclusively unique type of crypt organization, in which the crypts superficially connect to the midgut, but the entrance into the crypts remains completely closed (Rosenkranz 1939). In the family Plataspidae, the alimentary tract is completely disconnected between the M3 and M4 regions, and the two gut portions are connected only by a tiny threadlike connecting tissue in adult insects (Buchner 1965, Fukatsu and Hosokawa 2002, Hosokawa et al. 2006). However, the lumen of the two gut portions are connected in hatchlings so that symbiotic bacteria can pass through and colonize in the M4 region.

Although several species of the superfamily Lygaeoidea harbor extracellular symbionts in the midgut crypts, the members of the family Lygaeidae generally lack the symbiotic organ. On the other hand, several lygaeid species possess large, red-colored bacteriome(s) housing intracellular symbionts in their body cavities (**Figure 14.1I–J**) (Buchner 1965; Küchler et al. 2010, 2011, 2012; Matsuura et al. 2012, 2015). Such intracellular symbiosis also has been reported in the genus *Ischnodemus* of the family Blissidae, although other members of this family harbor extracellular gut symbionts in the tubular-type midgut crypts (Buchner 1965, Küechler et al. 2012). Those species that possess intracellular symbionts commonly have neither gut symbionts nor midgut crypts (**Figure 14.1I**). Although most lygaeoid species possess extracellular gut symbionts of the class Betaproteobacteria, intracellular symbionts of these same species belong to the Gammaproteobacteria (see **Section 14.1.3.3**). This strongly suggests that the gut symbionts have been replaced by the intracellular symbionts in the evolution of the Lygaeoidea (Matsuura et al. 2012). It is unclear why such a symbiotic transition has occurred in the lygaeoid lineage. One possibility is that, although their localization pattern is quit different, there is no functional difference between the extracellular and intracellular symbiotic bacteria, and the transition has occurred by chance. Alternatively, the intracellular symbionts may confer ecological and evolutionary advantages on the host insects that the gut symbionts do not, resulting in the shift of the symbionts.

The members of the family Pyrrhocoridae (superfamily Pyrrhocoroidea) lack midgut crypts. Instead, M3 of these species is remarkably enlarged and swollen, the lumen of which harbors a large number of bacterial cells (Kaltenpoth et al. 2009, Sudakaran et al. 2015).

14.1.3.3 Phylogenetic Diversity of Symbiotic Bacteria

Many pentatomomorphan symbionts are unculturable outside their hosts (Y. Kikuchi, unpublished data). Therefore, phylogenetic placements of the symbionts has been estimated based on DNA sequences, usually of the 16S rRNA gene. **Figure 14.2** shows the molecular phylogeny of the symbiotic bacteria in pentatomomorphan species. Microbial symbiosis in the Pentatomomorpha is usually of the mono-association type, wherein each host species houses a single species/strain of bacteria. Symbiotic bacteria of the superfamily Pentatomoidea (including the families Pentatomidae, Scutelleridae, Acanthosomatidae, Cydnidae, Parastrachiidae, Plataspidae, and Urostylididae) investigated so far belong to the bacterial family Enterobacteriaceae of the class Gammaproteobacteria (Hosokawa et al. 2006, 2010a, 2012b, 2016; Kikuchi et al. 2009; Prado and Almeida 2009a; Kaiwa et al. 2010, 2011; Bansal et al. 2014; Bistolas et al. 2014; Kaiwa et al. 2014; Matsuura et al. 2014). In the families Plataspidae, Acanthosomatidae, and Urostylididae, symbiotic bacteria form a monophyletic group that depends on the host families, which suggests that each family had a single evolutionary origin of symbiosis (Hosokawa et al. 2006, Kikuchi et al. 2009, Kaiwa et al. 2014). Symbionts of other families such as the Pentatomidae, Scutelleridae, and Cydnidae are paraphyletic and phylogenetically more diverse (Prado and Almeida 2009a; Kaiwa et al. 2010,

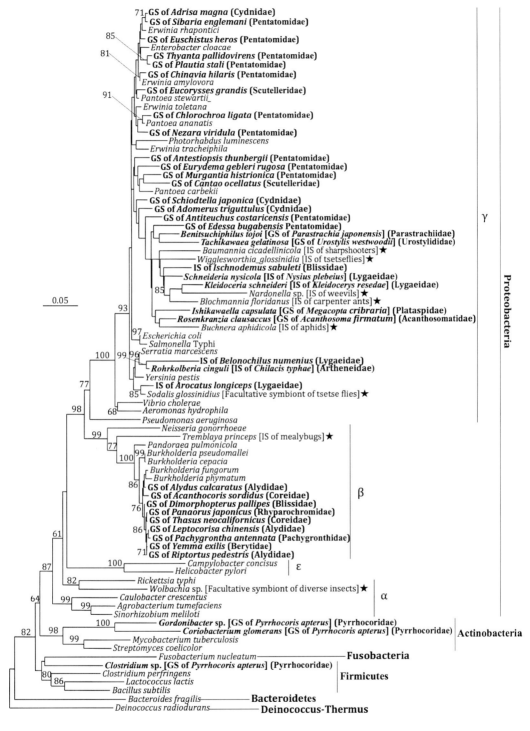

FIGURE 14.2 Phylogenetic tree based on the sequence of the 16S rRNA gene (neighbor-joining analysis). Bacterial phyla/classes are shown on the right side. Symbiotic bacteria of pentatomomorphan species are represented in boldface, and stink bug families are shown in parentheses. Abbreviations: GS, gut symbiont; IS, intracellular symbiont. Asterisks indicate symbiotic bacteria of other insect groups. Several symbionts of stink bugs and other insects have been microbiologically characterized and named, whose host insects are informed in brackets. Bootstrap supporting values higher than 60% are depicted at the nodes. The scale bar represents 0.05 changes per base.

2011; Hosokawa et al. 2012b, 2016; Bistolas et al. 2014), which indicates frequent horizontal transmission of the symbiotic bacteria and/or multiple evolutionary origins of the symbiotic associations.

Symbiotic bacteria of the superfamilies Coreoidea and Lygaeoidea belong to the genus *Burkholderia* of the class Betaproteobacteria (Kikuchi et al. 2005, 2011; Olivier-Espejel et al. 2011; Boucias et al. 2012; Itoh et al. 2014). Most of the *Burkholderia* species isolated from the coreoid and lygaeoid insects form a monophyletic group in the genus *Burkholderia*, which is called the stink bug-associated beneficial and environmental (SBE) group (Kikuchi et al. 2011, Itoh et al. 2014). In addition to the insect-associated *Burkholderia*, this bacterial clade includes several strains from plant galls and environmental soils, among which the insect-associated, plant-gall-associated, and environmentally isolated strains do not form coherent groups (Kikuchi et al. 2011); this indicates frequent horizontal transmission between them. In fact, as discussed in **Section 14.1.3.4**, it has been reported that the coreoid and lygaeoid species do not vertically transmit the *Burkholderia* symbionts but, rather, acquire the symbionts from the environment in every generation (Kikuchi et al. 2007). Such a transmission mechanism may lead to the development of the tangled phylogeny of the *Burkholderia* symbionts.

Bacteriocyte-associated symbionts of the families Lygaeidae and Blissidae belong to the Enterobacteriaceae of the Gammaproteobacteria. These intracellular symbionts do not form a monophyletic group within the bacterial group (Küechler et al. 2012), which strongly suggests their multiple evolutionary origins.

In contrast to the simple, mono-associated gut microbiota in the superfamilies Pentatomoidea, Coreoidea, and Lygaeoidea, the gut bacterial community in the family Pyrrhocoridae is relatively complex, wherein *Actinomyces*, Firmicutes, and Proteobacteria are housed in the M3 section of the midgut, which is swollen (**Table 14.1**) (Sudakaran et al. 2012). A broad survey of the gut community in diverse pyrrhocorid species suggests that the *Actinomyces*-dominated, complex microbiota has evolved as the common ancestor of this family (Sudakaran et al. 2015).

14.1.3.4 Transmission Mechanism of Symbiotic Bacteria

Diverse modes of mechanisms for symbiont transmission have been identified in several insect groups (Buchner 1965, Kikuchi 2009, Salem et al. 2015). Endocellular symbionts such as *Buchnera* typically are passed to the next generation via transovarial transmission, wherein the symbiont directly infects the embryos in the maternal body (Mira and Moran 2002, Braendle et al. 2003, Miura et al. 2003, Frydman et al. 2006, Koga et al. 2012). A similar mechanism has been reported in the intracellular symbioses of the families Lygaeidae and Blissidae (Buchner 1965, Küechler et al. 2012, Matsuura et al. 2012). On the other hand, insects harboring extracellular symbionts have evolved elaborate post-ovipositional mechanisms for symbiont transmission such as superficial bacterial contamination of eggs (egg smearing with excrement) and probing of parental bacteria-containing excrement (coprophagy). In the Pentatomomorpha, the egg-smearing manner is possibly the most common and reported mechanism in several species of the superfamilies Pentatomoidea and Pyrrhocoroidea (Abe et al. 1995, Prado et al. 2006, Kaltenpoth et al. 2009, Kaiwa et al. 2010, Tada et al. 2011, Kikuchi et al. 2012, Hosokawa et al. 2013). In the family Pentatomidae, female adults have enlarged crypts in the posterior end of M4, which are believed to be symbiotic organs involved in the egg-smearing manner of symbiont transmission (Hayashi et al. 2015). Acanthosomatid bugs bear closed crypts resulting in symbionts that never are excreted directly from the crypts; these bugs develop a female-specific symbiotic organ called a "lubricating organ" along the reproductive canal, which smears the symbiotic bacteria on the egg surfaces during oviposition (Rosenkranz 1939, Kikuchi et al. 2009). Mother stink bugs usually contaminate eggs during oviposition, except for an exceptional case reported in a subsocial species, *Parastrachia japonensis* Scott (Parastrachiidae). In *P. japonensis*, which is well known for maternal care of eggs and hatchlings, females smear their egg clusters with symbiont-containing excrement immediately before egg hatching (Hosokawa et al. 2012a), indicating they can determine when their eggs will hatch by an unrevealed sensing-mechanism.

In addition, a unique and elaborate transmission mechanism, called "capsule transmission," has been reported for the family Plataspidae, in which symbiont-filled particles, referred to as the "symbiont capsules," are deposited in association with eggs, and hatchlings probe the content of the capsules to acquire the symbiont (Schneider 1940; Müller 1956, Fukatsu and Hosokawa 2002; Hosokawa et al. 2005, 2006) (see **Section 14.3**). Furthermore, a novel type of symbiont transmission was reported recently in the Urostylididae. During egg oviposition, females cover their egg clusters with a symbiont-containing gel-like substance, called "symbiont jelly," and the hatchlings feed on the jelly to acquire the symbiont (Kaiwa et al. 2014).

Coreoid and lygaeoid species associated with *Burkholderia* symbionts do not vertically transmit the symbiont but, rather, acquire the microbes from the surrounding environment in every generation (Kikuchi et al. 2007, 2011; Olivier-Espejel et al. 2011; Boucias et al. 2012; Itoh et al. 2014). Such a transmission manner, called "environmental acquisition" (or environmental transmission), is common among symbiotic associations in marine invertebrates such as squid-*Vibrio* symbiosis (Nyholm and McFall-Ngai 2004), and in terrestrial plants such as legume-*Rhizobium* symbiosis (Gualtieri and Bisseling 2000). However, this type of symbiont transmission/acquisition rarely has been reported in insects, except for coreoid and lygaeoid species. Although most of the vertically transmitted symbionts are unculturable outside their hosts, probably because of their host-dependent lifestyles and extremely reduced genomes, the *Burkholderia* symbionts are easily culturable (Kikuchi et al. 2007, 2011). Experimental inspections of the bean bug, *Riptortus pedestris* Thunberg [= *R. clavatus* (F.)], have revealed that the host acquires the *Burkholderia* symbionts mainly during the second stadium, and that a 50% infective dose (ID_{50}: the amount of symbionts required for the colonization of 50% of the tested insects) was remarkably low; only 80 symbiont cells were required, suggesting high efficiency of the symbiont acquisition (Kikuchi and Yumoto 2013), which probably stabilizes the symbiotic association without vertical transmission.

14.1.3.5 Biological Functions of Symbiotic Bacteria

Numerous studies have demonstrated repeatedly that symbiotic bacteria play a pivotal role in the life of a pentatomomorphan host; experimental elimination of symbionts by egg-surface sterilization or by removing capsule/jelly causes retarded growth, nymphal mortality, small body size, and/or abnormal body coloration (**Table 14.2**).

Buchner (1965) noted a general pattern that insects living on nutritionally unbalanced diets, such as plant sap or vertebrate blood, tend to harbor endosymbiotic microorganisms, whereas carnivorous and herbivorous insects feeding on animal or plant tissues do not, which implies that symbiotic bacteria provide essential nutrients that are limited in species that feed on insects. For instance, on the basis on physiological and genomic analyses, *Buchnera* spp. provide their aphid hosts with essential amino acids that are mostly lacking in plant phloem sap (Sasaki and Ishikawa 1998, Shigenobu et al. 2000). The same situation also may apply to the sap-feeding groups of the Pentatomomorpha, such as the Plataspidae and Urostylididae; this is strongly supported by recent genome analyses of their symbionts (Nikoh et al. 2011, Kaiwa et al. 2014) (also see **Sections 14.1.3.6 and 14.3.2.3**). However, in phytophagous pentatomomorphans, not only sap feeders but also many seed feeders possess well-developed endosymbiotic systems. Because plant seeds seem to be more nutritionally rich and balanced than plant phloem sap, biological functions of the symbiotic bacteria in seed feeders remain unclear. The use of an artificial diet for the rearing of African cotton stainer, *Dysdercus fasciatus* Signoret (Pyrrhocoridae), elucidated this long-standing question and clearly demonstrated that symbiotic bacteria provide B-vitamins to the host insect (Salem et al. 2014). Such a nutritional supplement is probably a major biological function of the symbiotic bacteria in seed-feeding pentatomomorphans.

14.1.3.6 Symbiont Genome

The genome sequence of symbiotic bacteria has been reported so far from Candidatus Ishikawaella capsulata in *Megacopta punctatissima* (Montandon) (Plataspidae) (Nikoh et al. 2011), Candidatus Tachikawaea gelatinosa in *Urostylis westwoodii* Scott (Urostylididae) (Kaiwa et al. 2014), Candidatus Pantoea carbekii in *Halyomorpha halys* (Stål) (Pentatomidae) (Kenyon et al. 2015), and two strains of the *Burkholderia* symbiont in *Riptortus pedestris* (Alydidae) (Shibata et al. 2013, Takeshita et al. 2014). Intracellular obligate symbionts of diverse insects, including *Buchnera* in aphids and *Wigglesworthia* in tsetse flies, share unique genetic traits such as an AT (Adenine, Thymine)-biased nucleotide composition, accelerated molecular evolution, and remarkably reduced genome size. In comparison with a

TABLE 14.2

Fitness Defects Incurred by Stink Bugs because of Experimental Symbiont Elimination

Species	Defect by Symbiont Elimination	Reference
	Superfamily PENTATOMOIDEA **Family ACANTHOSOMATIDAE**	
Elasmostethus humeralis	Severe nymphal mortality, retarded growth, and abnormal (whitish) body coloration	Kikuchi et al. 2009
	Family CYDNIDAE	
Adomerus rotundus	Retarded growth and small body size	Hosokawa et al. 2013
Adomerus triguttulus	Nymphal mortality, retarded growth, small body size, and abnormal (whitish) body coloration	Hosokawa et al. 2013
	Family PARASTRACHIIDAE	
Parastrachia japonensis	Small body size and abnormal (whitish) body coloration	Hosokawa et al. 2012a
	Family PENTATOMIDAE	
Chinavia hilaris	Nymphal mortality, retarded growth, small body size, and low fecundity	Prado and Almeida 2009b
Eurydema gebleri rugosa	Retarded growth, small body size, and abnormal (whitish) body coloration	Kikuchi et al. 2012
Halyomorpha halys	Nymphal mortality, retarded growth, small body size, and low fecundity	Taylor et al. 2014
Nezara viridula	Severe nymphal mortality (reported in a Japanese population, but not in a Hawaiian population)	Tada et al. 2011
Plautia stali	Severe nymphal mortality, small body size, and Low fecundity	Hosokawa et al. 2016
Sibaria englemani	Retarded growth	Bistolas et al. 2014
	Family PLATASPIDAE	
Brachyplatys subaeneus	Severe nymphal mortality	Hosokawa et al. 2006
Coptosoma parvipictum	Severe nymphal mortality	Hosokawa et al. 2006
Megacopta cribraria [= *M. punctatissima*]	Nymphal mortality, retarded growth, small body size, and abnormal (whitish) body coloration	Hosokawa et al. 2006
	Family UROSTYLIDIDAE	
Urostylis westwoodii	Severe nymphal mortality	Kaiwa et al. 2014
	Superfamily COREOIDEA **Family ALYDIDAE**	
Riptortus pedestris	Retarded growth, small body size, and low fecundity	Kikuchi and Fukatsu 2014
	Superfamily LYGAEOIDEA **Family BLISSIDAE**	
Blissus insularis	Nymphal mortality and retarded growth	Boucias et al. 2012
	Superfamily PYRRHOCOROIDEA **Family PYRRHOCORIDAE**	
Dysdercus fasciatus	Nymphal mortality, retarded growth, and low fecundity	Salem et al. 2013
Pyrrhocoris apterus	Nymphal mortality, retarded growth, and low fecundity	Salem et al. 2013

free-living bacterium, *Escherichia coli*, with a genome size and AT contents of 4.6 megabase (Mb) and 49%, respectively, the corresponding values were 0.6 Mb and 74% in *Buchnera* (Shigenobu et al. 2000), and 0.7 Mb and 77% in *Wigglesworthia* (Akman et al. 2002). Similar genetic traits are also shared by the stink bug gut symbionts, wherein genome size and AT contents are 0.75 Mb and 70% in Ishikawaella capsulata (Nikoh et al. 2011), 0.71 Mb and 65% in Tachikawaea gelatinosa (Kaiwa et al. 2014), and 1.2 Mb and 69% in Pantoea carbekii (Kenyon et al. 2015). Such a drastic genome reduction also has been reported in Candidatus Rosenkranzia clausaccus in *Elasmostethus humeralis* Jak., *E. nubilus* Dallas, and *Sastragala esakii* Hasegawa (Acanthosomatidae) (Kikuchi et al. 2009) and in Candidatus

Benitsuchiphilus tojoi in *Parastrachia japonensis* (Parastrachiidae) (Hosokawa et al. 2010a). Although it remains to be discovered what evolutionary mechanisms have led to these peculiar genetic traits in insect symbionts, theoretical studies have suggested that the endosymbiotic lifestyle may relax the symbiont's genome against environmental selection pressures, whereas severe bottlenecks during vertical transmission and absence of recombination and/or horizontal gene transfer may accelerate molecular evolution and genome reduction (Moran 1996, Wernegreen 2002).

Gene compositions of Candidatus Ishikawaella and Candidatus Tachikawaea symbionts are basically similar to those of *Buchnera*, probably because the host stink bugs feed exclusively on plant phloem sap (Nikoh et al. 2011, Kaiwa et al. 2014). These symbionts, having host-dependent lifestyles, have lost several genes that are involved in the primary metabolic processes such as replication, transcription, and energy production (e.g., TCA cycle) and cell structures such as cell walls, flagella, and pilli. In contrast, these symbiotic bacteria retain several biosynthesis pathways for nutrients that are limited in plant sap and, thus, are demanded by the host insects such as essential amino acids and vitamins (for Ishikawaella symbiont, see **Section 14.3** for more information). The gene composition of Candidatus Pantoea carbekii symbiont in *Halyomorpha halys* is similar to those of Ishikawaella and Tachikawaea, whereas the symbiont retains intact TCA cycle and cell-wall synthesis pathways (Kenyon et al. 2015).

Probably because the *Burkholderia* symbionts possess a free-living life stage in the environmental soil, in contrast to the vertically transmitted symbionts, genome sizes of the *Burkholderia* symbionts are remarkably larger with a genome size and lower AT content of 6.96 Mb and 36.8% in strain RPE64 and 8.69 Mb and 36.6% in strain RPE67 (Shibata et al. 2013, Takeshita et al. 2014). Their genomes consist of almost a full-set of genes of primary metabolic processes and cell structures and also include a number of genes with no homologous genes.

14.2 Symbiosis in the Southern Green Stink Bug, *Nezara viridula*

14.2.1 Overview

Nezara viridula (Pentatomidae) is distributed worldwide (Todd 1989) and is an economically important pest of many crops such as soybean, cotton, and macadamia nuts (Schaefer and Panizzi 2000). *N. viridula* can cause direct damage, as with other heteropterans, by sucking the developing seeds and reducing yield and quality of various commercial crops (Schaefer and Panizzi 2000, Jones et al. 2001). In addition, the insects cause physical damage to plants by their probing behavior, which may facilitate the penetration of opportunistic bacterial and fungal pathogens into the plant tissues (Russin et al. 1988, Lee et al. 1993). Furthermore, the bugs act as vectors of plant pathogens, causing diseases such as yeast-spot disease in soybean (Clarke and Wilde 1970a, b). Because *N. viridula* is economically important, extensive research has been conducted on various aspects of its biology, and rearing and experimental systems have been established. For these reasons, *N. viridula* was chosen as a model species for studying bacterial symbionts associated with midgut crypts.

14.2.2 Fitness Effect of Symbiotic Bacteria

Hirose et al. (2006) and Prado et al. (2006) consistently found a bacterium associated with the midgut M4 region of stink bugs (**Figure 14.3**). Hirose et al. (2006) investigated symbionts inside the midgut of insects from Brazil; and Prado et al. (2006) investigated symbionts of insects from Hawaii, California, and South Carolina. Although the M4 region has a low concentration of cultivable bacteria, molecular analyses have revealed that midgut crypts consist of a large number of *Erwinia*-like bacterial symbionts (Hirose et al. 2006; Prado et al. 2006, 2009; Tada et al. 2011, Prado and Zucchi 2012).

Early studies on the effects of surface sterilization of pentatomid eggs on insect development had shown conflicting results; however, increased knowledge in this field and the development of molecular tools have confirmed that egg-surface sterilization eliminates the crypt-associated bacterium, generating aposymbiotic insects. The influence of this aposymbiotic status on insect fitness, however, varies between individuals and/or populations (Prado et al. 2006, 2009; Tada et al. 2011; Prado and Zucchi

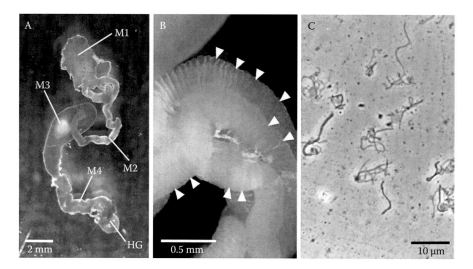

FIGURE 14.3 (**See color insert.**) Microbial symbiosis in *Nezara viridula*. (A) Dissected midgut. Abbreviations: M1, midgut first section; M2, midgut second section; M3, midgut third section; M4, midgut fourth section with crypts (symbiotic organ); HG, hindgut. (B) Enlarged image of midgut crypts. Arrowheads indicate crypts. (C) Symbiotic bacteria of *N. viridula* (phase contrast microscopy).

2012). Although no obvious fitness defects were observed after symbiont elimination in a Hawaiian population of *Nezara viridula* (Prado et al. 2006), the gut symbiont was essential in a Japanese population in which aposymbiotic individuals rarely reached adulthood (Tada et al. 2011).

14.2.3 Effect of High Temperature on Pentatomid Symbiosis

In their study on the impact of high temperatures and egg surface sterilization on *Nezara viridula*'s nymphal development rate and reproductive performance, Prado et al. (2009) showed that the symbiont's maintenance is affected by both treatments. They detected the symbiont in 100, 84, and 8.3% of the untreated control insects at 20, 25, and 30°C, respectively. In insects originating from surface sterilized egg masses, the symbionts were detected in only one of 21 insects at 25°C. Nymphs and adults from sterilized egg masses at 20°C lived longer, but the time taken to reach adulthood was longer, and the females never laid eggs. Mean generation time was remarkably longer at 20°C than at 25 and 30°C, regardless of surface sterilization. Additionally, the pre-oviposition period and number of eggs were significantly different between the surface sterilized and control treatments only at 20°C. The results of the study by Prado et al. (2009) emphasized that not only egg surface sterilization but also higher temperature (30°C) have an impact on the maintenance of symbionts and negatively affect the host's development and reproduction in *N. viridula*.

Prado and Almeida (2009b) demonstrated in an allied species, *Chinavia hilaris* (Say) (formerly known as *Acrosternum hilare*), that developmental time, survival, and reproductive parameters were negatively affected by egg surface sterilization. However, no such effects were observed in *Murgantia histrionica* (Hahn) (Prado and Almeida 2009b). Prado et al. (2010) demonstrated that *C. hilaris* and *M. histrionica* lost their gut symbionts within two generations when the insects were reared at 30°C. In addition, survival and reproductive rates of both *C. hilaris* and *M. histrionica* reared at 30°C were lower than those reared at 25°C. Based on these results, it is assumed that the decrease in host fitness is linked with, and potentially mediated by, symbiont loss at 30°C (Prado et al. 2010). In pentatomid insects, these crypt-associated symbionts probably play an important role in stink bug adaptation to climate change. Although not well documented, this unique style of transmission (i.e., egg smearing with excrement) in which symbionts live for a certain period on the egg surface, might be notable because the transmission mechanism may allow environmental factors such as climate change to interfere with symbiont transmission/acquisition, which could lead to serious fitness defects in stink bug hosts.

14.2.4 Symbiotic Associations in Allied Pentatomids

Prado and Almeida (2009a) showed that not only *Nezara viridula* but also other pentatomid species, such as *Chinavia hilaris*, *Murgantia histrionica*, *Euschistus heros* (F.), *Chlorochroa ligata* (Say), *C. sayi* (Stål), *C. uhleri* (Stål), *Plautia stali* Scott, and *Thyanta pallidovirens* (Stål), carried symbiotic bacteria in their midgut crypts. Phylogenetic placement estimated by using 16S rRNA gene sequences demonstrated that all of the crypt-associated symbionts form a clade with plant-associated *Erwinia* and *Pantoea* species. Recently, it has been observed that other pentatomid stink bugs, including *Dichelops melacanthus* Dallas, *Edessa meditabunda* F., *Loxa deducta* Walker, *Pellaea stictica* Dallas, *Piezodorus guildinii* (Westwood), *Thyanta perditor* (F.), *Eurydema rugosa*, *Eurydema dominulus*, and *Halyomorpha halys*, house symbionts associated with their midgut crypts (**Table 14.1**) (Kikuchi et al. 2012, Prado and Zucchi 2012, Kenyon et al. 2015). However, all the symbionts' gene sequences differed from each other, suggesting that most of the stink bug symbionts occasionally are replaced by other taxonomically similar bacteria over the evolutionary period, probably due to their postnatal symbiont transmission mechanism (Prado et al. 2010, Prado and Zucchi 2012, Hosokawa et al. 2016).

14.3 From Eastern Hemisphere to Western Hemisphere: Candidatus Ishikawaella Capsulata and Its Plataspid Host

14.3.1 Overview

Extensive genetic, phylogenetic, and genomic research has helped clarify bacteriocyte symbiosis or the mutually obligate symbiosis between unculturable endocellular Gammaproteobacteria and their hemipteran hosts (Wernegreen 2002, Gil et al. 2004, Baumann 2005, Moran et al. 2008). Hemipterans are the only insect group to feed throughout their life cycles on nutriently depauperate phloem sap (Douglas 1998, Spaulding and von Dohlen 1998, Baumann 2005). These bacterial endosymbionts, which are vertically transmitted across generations, have severely reduced genomes lacking in replicative genetics (Wernegreen 2002, Baumann 2005, Moran et al. 2008) that are function specific. They supply dietary needs, particularly essential amino acids, necessary for host survival and reproduction (Buchner 1965, Douglas 1998, Sasaki and Ishikawa 1998, Charles and Ishikawa 1999, Shigenobu et al. 2000, Douglas 2003) and have co-evolved with their host over millions of years (Moran et al. 1993; Baumann 2005; Moran et al. 2008; Bennett and Moran 2013, 2015). Undeterred by food source, these bacteriocyte symbioses often involve invasive pests expanding ranges across myriad environmental niches (Moran 2007, Pérez-Brocal et al. 2011, Bennett and Moran 2013). But what of vertically transmitted extracellular microorganisms found in the gut of phloem-feeding Heteroptera in the family Pentatomidae living in mutually obligate relationships with their hosts? Are their genomes reduced and, if so, for what purpose? Does the heritable symbiosis foster range expansion into diversified herbivorous ecologies? Does the obligate symbiont confer pest status on the host?

We have explored these questions by examining the obligate symbiosis between the gut microorganism, "Candidatus Ishikawaella capsulata" (Hosokawa et al. 2006), and its plataspid *Megacopta* host. Although discovered for the first time in the Western Hemisphere in 2009 (Eger et al. 2010, Jenkins et al. 2010, Suiter et al. 2010), the obligate symbiosis had been genetically and behaviorally studied in Japan since 2002 (Fukatsu and Hosokawa 2002; Hosokawa et al. 2005, 2006, 2007a, 2007b, 2008; Nikoh et al. 2011).

A seminal study by Fukatsu and Hosokawa (2002) showed that the obligate extracellular gammaproteobacteria, "Candidatus Ishikawaella capsulata" (i.e., Ishikawaella) (Hosokawa et al. 2006), is associated with the plataspid *Megacopta punctatissima*, a pest species of legumes in Japan (Hosokawa et al. 2007b). As with the well-studied intracellular Gammaproteobacteria, Ishikawaella was shown to have an acutely reduced genome, accelerated protein evolution, and AT-biased codons and to have co-speciated with its *Megacopta* host (Hosokawa et al. 2006, Nikoh et al. 2011). These studies provided an extensive genetic baseline of an orally transmitted microorganism (Fukatsu and Hosokawa 2002) indigenous to the Eastern Hemisphere that would soon become a problem in the Western Hemisphere. In October 2009,

Megacopta cribraria (Eger et al. 2010) and its obligate gut endosymbiont, Ishikawaella (Jenkins et al. 2010), made its Western Hemisphere debut in Northeast Georgia in the United States. It was discovered feeding and developing on kudzu, *Pueraria montana* Loureiro (Merrill) variety *lobata* (Willdenow) (Suiter et al. 2010). In less than a year, *M. cribraria*, commonly identified as the kudzu bug in the United States, was found feeding and developing on soybean, *Glycine max* (L.) Merrill (Suiter et al. 2010, Zhang et al. 2012, Del Pozo-Valdivia and Reisig 2013). Although kudzu is an invasive plant in the United States (Zhang et al. 2012), soybean, worth about $40 billion to the United States economy, is not (Jenkins and Eaton 2011). The Western Hemisphere has, therefore, become a natural laboratory experiment for testing two hypotheses: (1) A mutually obligate extracellular symbiosis indigenous to the Eastern Hemisphere will adapt rapidly and expand its range into novel environments in the Western Hemisphere, or (2) The symbiont genome will rapidly evolve as the symbiosis expands its range across the Western Hemisphere.

14.3.2 Genetic Baseline

14.3.2.1 Vertical Transmission in Symbiont Capsules

Fukatsu and Hosokawa (2002) were the first to genetically characterize the vertically transmitted extracellular gammaproteobacterium, Candidatus Ishikawaella capsulata (Ishikawaella) and its obligate relationship with its indigenous plataspid host, *Megacopta punctatissima* (Hemiptera: Heteroptera: Plataspidae). Studies showed that, like bacteriocyte symbioses, Ishikawaella has a drastically reduced, AT-rich genome (Hosokawa et al. 2006, Nikoh et al. 2011), is vertically transmitted, has a long history of co-evolution with its host, and provides its host with the essential nutrients lacking in phloem for insect growth and development (Fukatsu and Hosokawa 2002; Hosokawa et al. 2005, 2006, 2007a,b, 2008; Nikoh et al. 2011).

Adults of *Megacopta punctatissima* were shown to harbor the bacterium in the crypts of the midgut posterior region (Fukatsu and Hosokawa 2002, Hosokawa et al. 2006). The capsules, likely produced in the enlarged end section of the crypt-bearing midgut part (Fukatsu and Hosokawa 2002), were shown to be vertically transmitted and deposited underneath and proximal to the eggs (**Figure 14.4**) (Fukatsu and Hosokawa 2002). Newly hatched nymphs were observed to probe these capsules (**Figure 14.4**), ingest the symbiont, aggregate, and then become quiescent 1–2 days before dispersing to feed on phloem

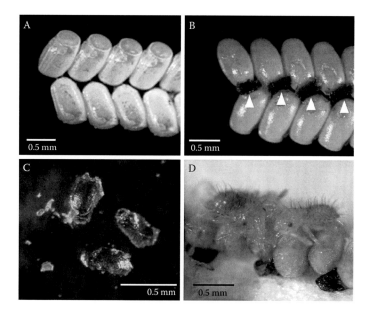

FIGURE 14.4   **(See color insert.)** Capsule symbiosis in plataspid insects. (A-B) Egg cluster of *Brachyplatus subaeneus*: A. Dorsolateral view; B, Posterolateral view. Arrows indicate symbiont capsules. (C) Symbiont capsules removed from egg mass. (D) First instars of *Megacopta cribraria*. Note: nymphs feeding on capsule contents.

(Hosokawa et al. 2007a, 2008). Nymphs that did not ingest symbionts from the capsules were easily identified by their wandering behavior and ultimate failure to flourish (Hosokawa et al. 2008).

14.3.2.2 *Ishikawaella, Insect Development and Pest Status*

Megacopta punctatissima is a legume pest of mainland Japan whereas *M. cribraria*, found across the southwestern islands of Japan, is not (Hosokawa et al. 2007b). The difference between the pest and non-pest status of these two plataspids appeared to be the gut symbiont Ishikawaella (Hosokawa et al. 2007b).[1] Host eggs were, therefore, manipulated to determine if Ishikawaella, shown to co-speciate with its plataspid host (Hosokawa et al. 2006) also was responsible for the pest status of *M. punctatissima*. *M. punctatissima* eggs were combined with symbiont capsules from *M. cribraria*, and *M. cribraria* eggs were combined with symbiont capsules from *M. punctatissima*. *M. cribraria* that received capsules from *M. punctatissima* thrived on legumes, including soybean, whereas *M. punctatissima*, that received symbiont capsules from *M. cribraria* had high mortality on soybean. Hosokawa et al. (2007b) concluded, therefore, that it was the Ishikawaella gut symbiont in capsules from *M. punctatissima* that were essential for normal development and reproduction as well as conferred pest status on the host.

14.3.2.3 *Genome Evolution*

The Ishikawaella genome, like the genomes of bacteriocyte-associated endosymbionts, had similar patterns of gene reduction (Hosokawa et al. 2006, Nikoh et al. 2011) but significant differences as well. Nikoh et al. (2011) suggested that these differences might have functional and ecological consequences for Ishikawaella. Unlike the bacteriocyte-associated endosymbionts, *Buchnera* in aphids (Shigenobu et al. 2000) or *Blochmannia* in carpenter ants (Gil et al. 2003), Ishikawaella had more genes for the synthesis of amino acids as well as some genes for the synthesis of vitamins and cofactors. Nikoh et al. (2011) speculated that Ishikawaella either had a more recent evolutionary history or could supply more nutritional needs to its host than either endosymbionts *Buchnera* or *Blochmannia*. Ishikawaella, however, had fewer genes for cofactor synthesis than bacteriocyte-associated endosymbionts *Baumannia* in leafhoppers (Wu et al. 2006) or *Wigglesworthia* in tsetse flies (Akman et al. 2002), which reflected the nutritional needs of their respective hosts (Nikoh et al. 2011). Ishikawaella compensated for the lack of essential amino acids and vitamins in phloem (Nikoh et al. 2011) and *Baumannia* and *Wigglesworthia* compensated for the lack of cofactors in the diet of their respective cicadellid and glossinid hosts (Akman et al. 2002, Wu et al. 2006). Nikoh et al. (2011) also observed that, like endocellular symbiont genomes, the Ishikawaella genome retained many genes for translation, replication, and energy production. Unlike bacteriocyte-associated genomes, however, the Ishikawaella genome had fewer genes for cell wall synthesis or lipid metabolism. The question thus arises, how does Ishikawaella survive the extracellular condition?

Nikoh et al. (2011) speculated that the reductive evolution of the symbiont genome with its lack of genes for cell wall synthesis or lipid metabolism may be possible because of the composition of the vertically transmitted symbiont capsules. Ishikawaella cells are anchored within the capsules by a secretion matrix (Hosokawa et al. 2005). When capsules are deposited upon oviposition, this matrix may be simulating the intracellular composition of the cytoplasm and, therefore, supplying Ishikawaella with metabolites that it cannot make with its reduced genome. These capsules also may serve to protect the bacteria from dehydration and radiant energy (Nikoh et al. 2011).

Insect-bacterial mutualists are examples of "complementarity and syntrophy between host and symbiont" (Shigenobu et al. 2000, International Aphid Genomics Consortium 2010, Wernegreen 2012). Ishikawaella has a long history of mutually dependent co-evolution with its plataspid hosts including *Megacopta punctatissima* and *M. cribraria* (Hosokawa et al. 2006). Genome evolution of the extracellular Ishikawaella also is comparable to the evolutionary patterns of endocellular symbiont genomes,

---

[1] For further discussion of the taxonomic status of *M. cribraria* and *M. punctatissima*, see **Chapter 5**.

especially the *Buchnera* genome (Nikoh et al. 2011). But the genome differences of Ishikawaella are notable: more genes for the synthesis of amino acids, vitamins, and cofactors; the retention of many genes specific to translation and replication; and genes specific to energy production. Finally, Ishikawaella is stably transmitted by the female plataspid in symbiont capsules (Hosokawa et al. 2005, 2006), which may function to nourish and protect the bacteria. The subsequent introduction of *M. cribraria* into the southeastern United States and the Western Hemisphere and the accompanying *Megacopta*-Ishikawaella symbiosis provided a natural laboratory to test the ability of the symbiosis to facilitate dispersal evolution through range expansion.

14.3.3 Invasion of *Megacopta cribraria*

14.3.3.1 Overview

Megacopta cribraria (Eger et al. 2010, Jenkins et al. 2010, Jenkins and Eaton 2011) debuted on kudzu vine in northeast Georgia, in the United States, in October 2009 (Suiter et al. 2010). Previous research strongly suggested that the obligate symbiosis would likely disperse from kudzu into soybean fields (Fukatsu and Hosokawa 2002; Hosokawa et al. 2005, 2006, 2007b), which, in fact, occurred within a year of its discovery (Seiter et al. 2013). Interestingly, only one female line, designated GA1, was confirmed from sequence data from initial Georgia collections (Jenkins et al. 2010). By October 2010, *M. cribraria* had expanded in Georgia from nine counties in 2009 to 80 counties. A single female line, GA1, was confirmed from consensus mitochondrial genome sequences (15,647 bp) from individuals collected in five counties and mitochondrial fragment sequences (2,336 bp) collected from all 80 counties (83 sequences) (Jenkins and Eaton 2011). The Ishikawaella endosymbiont also was confirmed in all *M. cribraria* (Jenkins and Eaton 2011). Using ecological niche modeling, it had been predicted that *M. cribraria* had the potential to expand its range across the southeastern United States and, subsesquently, invade southwestern Europe, southeastern South America, southern Africa, and the eastern coast of Australia (Zhu et al. 2012). By 2013, the end of the study, the GA1 haplotype and the Ishikawaella endosymbiont had been confirmed genetically in over 300 collections from across eight southeastern states in the United States as well as from collections in Honduras, Central America (Tracie Jenkins, unpublished data). As of March 2015, this bug has been confirmed in 12 states in the United States (http://wiki.bugwood .org/Kudzu_bug) as well as Central America. Because Ishikawaella co-evolves with its host (Hosokawa et al. 2006) as well as determines the pest status of the insect (Hosokawa et al. 2007b), has the genetic fitness landscape of the genome changed since initial invasion? If so can these changes provide insights into protein evolution relating to the pest status of the host?

14.3.3.2 Ishikawaella Genome in the Western Hemisphere

If evolution is imagined as movement of a population by natural selection across an adaptive or fitness landscape (Wright 1932), and if the genome is the landscape on which the force of natural selection acts, then the evolutionary trajectories along that landscape can be explored.

Brown et al. (2014) used population genomics to evaluate the evolutionary trajectory of the Ishikawaella genome. Twenty-four Ishikawaella genomes were sequenced from three plant species across 23 sites from 2009, during the first days of insect discovery, to 2011, at the conclusion of the study. The focus of the research, which was greatly facilitate by the reference sequence from Nikoh et al. (2011), was to compare the genomics of the pest-conferring symbiont in Asia to the genomics of the United States symbiont to determine the direction and extent of the genetic changes in the genome over the 2-year period of the study. Insect size and color are phenotypic differences in Asia indicative of pest status (i.e., the smaller and lighter insect was the nonpest or aposymbiotic insect) (Hosokawa et al. 2006, 2007b). This is significant for United States populations because color and size variation, often associated with host plant, had been observed and could be indicative of genetic variations (Brown et al. 2014).

Brown et al. (2014) did not observe genotypic changes among symbionts sampled during the study regardless of insect phenotype or whether the host fed on kudzu or soybean. Comparative analyses between the Ishikawaella genome of *Megacopta cribraria* from the United States with the complete pest-conferring Ishikawaella genome of *M. punctatissima* from Japan (Nikoh et al. 2011) showed little difference. It was therefore hypothesized that the symbiosis introduced into the United States in 2009 was capable of being a legume pest and has maintained that status from 2009 to 2011.

14.3.3.3 Genomic Landscape

Except for being 46 bp longer than the Japanese reference genome (Nikoh et al. 2011), and having a single difference in the riboflavin synthase (*rib*C) gene, the Ishikawaella genomes from all invasive *Megacopta cribraria* examined were concordant in gene order and orientation (Brown et al. 2014) with the Japanese Ishikawaella genome from the soybean pest *M. punctatissima* (Nikoh et al. 2011). No fixed differences were observed to accumulate in the genomes over the 2 years of the study. Comparative analyses, however, across the United States-collected endosymbionts showed that negative allele frequency changes including those for "lipid transport and metabolism, cell cycle control, cell division and chromosome partitioning and amino acid metabolism" were not significantly different between endosymbionts taken from plataspids feeding on kudzu and those feeding on soybean. Positive allele frequency changes for functions such as replication, recombination and repair, biogenesis, and lipid and coenzyme transport and metabolism were significantly different between Ishikawaella genomes from *M. cribraria* feeding on kudzu and *M. cribraria* feeding on soybean.

Brown et al. (2014) observed increased allele frequency changes as the population rapidly expanded its range from the end of 2010 into 2011, which is typical of an expanding population or negative (purifying) selection. The correlation of increased allele frequency change with distance from the grand zero (the point where the invasion started) also was correlated with the insect's host plants, particularly soybean (Brown et al. 2014, suggesting that the evolutionary processes that have occurred (or on-going) depend on the host plants. This phenotypic plasticity could be supporting the adaptive potential of the insect by allowing the symbiosis to respond differentially to novel environments (Hunt et al. 2011).

Except for a single difference in the riboflavin synthase (*rib*C) gene between the United States and Japanese genomes, the United States Ishikawaella genome appeared to be functionally similar to the Japanese Ishikawaella genome. Both had the same gene order and organizational landscape. Although the significance of the point mutation, which showed consensus across the United States samples, in *rib*C was unclear because there is a lack of understanding of how riboflavin availability differs across legume species, it did suggest questions and lines of research. Ishikawaella, therefore, appeared to confer pest status on its plataspid host upon arrival in northeast Georgia in 2009. Thus, the emergence of *Megacopta cribraria* on soybean during the first year in the United States was not due to a change in the endosymbiont. Jenkins and Eaton (2011) documented female lineage consensus in 83 individuals in 36 counties, which was suggestive of limited genetic diversity in the insect. The phenotypic variation observed in host-plant distribution as well as phenotypic differences in insect color and size likely was due, therefore, to the insect, not the endosymbiont (Brown et al. 2014).

14.3.4 Hypotheses from a Natural Laboratory Experiment

The Western Hemisphere has been an ongoing natural laboratory for testing two hypotheses as they relate to a plataspid pest symbiosis indigenous to Asia and recently introduced into North America. The hypotheses are as follows:

1. A mutually obligate extracellular symbiosis indigenous to the Eastern Hemisphere will rapidly adapt and expand its range into novel environments in the Western Hemisphere. This hypothesis cannot be rejected based on the research reported. The symbiosis has rapidly expanded its range across the southeastern United States and into Central America.

2. The symbiont genome will rapidly evolve as the symbiosis expands its range across the Western Hemisphere. Population genome sequence studies of the United States samples across 23 sites and in comparison to genome reference sequence from Asia (Nikoh et al. 2011) showed consistency between gene order and orientation between the two genomes. The United States genomes also showed patterns consistent with positive and negative selection, but more research

is needed to understand the processes and their effects on adaptive evolution. Single nucleotide polymorphisms observed within genes or pathways such as those involved in riboflavin biosynthesis appear promising for understanding feeding success on soybean, but much work remains to be done. Rapid genome evolution, however, was not observed over the 2-year span of the Brown et al. (2014) study.

Continuous accumulation of field data, in conjunction with genome sequencing analyses and capsule-exchange experiments, would reveal which hypothesis is more plausible.

14.4 Facultative Symbionts in the Pentatomomorpha

In addition to the specific symbionts localized in symbiotic organs, facultative bacterial endosymbionts have been reported from diverse species of the Pentatomomorpha. Facultative symbionts or secondary symbionts in insects are generally not 100% prevalent but are partially infected among host populations; they are not localized but, rather, spread to various host tissues and are not essential for the host's development and survival. The genus *Wolbachia*, famous as a sex-ratio distorter in many insects and other arthropods (O'Neill et al. 1997, Werren et al. 2008), commonly is detected in the Pentatomomorpha. Of the species that were investigated by Kikuchi and Fukatsu (2003), more than 30% were infected with the bacteria. Actinobacteria species, some of which are known to produce a diverse array of antibiotics, frequently have been detected from the midgut crypts of several pentatomid species including *Thyanta perditor* (F.), *Edessa meditabunda* F., *Loxa deducta* Walker, *Pellaea stictica* (Dallas), *Piezodorus guildinii* (Westwood), and *Nezara viridula* (Zucchi et al. 2012). *Sodalis glossinidius* first was described as a facultative symbiont of tsetse flies, and then *Sodalis*-allied bacteria were detected from different insects such as bird lice and spittle bugs (Fukatsu et al. 2007, Koga et al. 2013). A recent survey revealed that the *Sodalis* symbionts are prevalent among the Pentatomoidea; in total, 13.6% of the investigated 108 species were infected with the symbiont (Hosokawa et al. 2015). Although the biological functions of these facultative symbionts remain unclear, the prevalence of the symbionts implies their biological roles in the host stink bugs. For example, facultative symbionts in aphids, such as *Serratia symbiotica* and *Hamiltonella defensa*, confer the hosts with high-temperature tolerance, resistance against pathogens and parasitoids, and body color alteration (Oliver et al. 2003; Tsuchida et al. 2004, 2010; Scarborough et al. 2005). Future studies are required to reveal the biological importance of the facultative symbionts in the Pentatomomorpha.

14.5 Concluding Remarks

Because of our continuous efforts over this last decade, the enormous diversity of microbial symbioses in the Pentatomomorpha has been unveiled. This has highlighted the importance of stink bug symbiosis not only in elucidating several long-standing evolutionary questions, such as the effect of the manner of symbiont transmission (vertical or horizontal routes) on the evolutionary consequence of intimate interactions and the effect of intra- and extracellular lifestyles on the genome evolution of the symbiotic bacteria but, also, secondarily improving strategies to control pest stink bugs. At the present time, symbiotic associations of several key families remain to be investigated, such as those of the families Dinidoridae, Tessaratomidae, and Phloeidae (Pentatomoidea). Further effort would provide us with deeper and broader views on the evolutionary process of the microbial symbiosis in pentatomomorphan insects, which should open a new window in the world of basic and applied science, ranging from evolutionary ecology to agriculture biology.

In addition to the advanced accumulation of genomic information on symbiotic bacteria, host transcriptome recently has been analyzed in several pentatomomorphan species such as *Riptortus pedestris* and *Dysdercus fasciatus* (Futahashi et al. 2013, Bauer et al. 2014). Furthermore, in the near future, recent advanced sequencing technologies will facilitate analysis of host genome as well as symbiont genome. In fact, draft genomes of several heteropteran species, including *Halyomorpha halys*, have become open access (see http://www.ncbi.nlm.nih.gov/genome?term=txid33345[orgn]). As shown in several species, symbiotic bacteria play pivotal metabolic roles in pentatomomorphan host insects, implying that the symbiotic associations could be good targets for pest stink bug control. The growing body of information on the molecular mechanisms involved in symbiont infection, colonization, establishment, and maintenance processes may provide a novel molecular target toward pest control of the harmful pentatomomorphan species.

14.6 Acknowledgments

We thank David A. Rider (Entomology Department, North Dakota State University, Fargo) and Tom J. Henry (Systematic Entomology Laboratory, USDA-ARS, c/o National Museum of Natural History, Washington, DC) for their help in reviewing the taxonomic status of the heteropterans listed in **Table 14.1**.

14.7 References Cited

Abe, Y., K. Mishiro, and M. Takanashi. 1995. Symbiont of brown-winged green bug, *Plautia stali* Scott. Japanese Journal of Applied Entomology and Zoology 39: 109–115.

Akman, L., A. Yamashita, H. Watanabe, K. Oshima, T. Shiba, M. Hattori, and S. Aksoy. 2002. Genome sequence of the endocellular obligate symbiont of tsetse flies, *Wigglesworthia glossinidia*. Nature Genetics 32: 402–407. doi:10.1038/ng986.

Aksoy, S. 2000. Tsetse – A haven for microorganisms. Parasitology Today 16: 114–118. doi:10.1016/S0169-4758(99)01606-3.

Bansal, R., A.P. Michel, and Z.L. Sabree. 2014. The crypt-dwelling primary bacterial symbiont of the polyphagous pentatomid pest *Halyomorpha halys* (Hemiptera: Pentatomidae). Environmental Entomology 43: 617–625. doi:10.1603/EN13341.

Bauer, E., H. Salem, M. Marz, H. Vogel, and M. Kaltenpoth. 2014. Transcriptomic immune response of the cotton stainer *Dysdercus fasciatus* to experimental elimination of vitamin-supplementing intestinal symbionts. PLoS One 9: e114865. doi:10.1371/journal.pone.0114865.

Baumann, P. 2005. Biology bacteriocyte-associated endosymbionts of plant sap-sucking insects. Annual Review of Microbiology 59: 155–189. doi:10.1146/annurev.micro.59.030804.121041.

Bennett, G.M., and N.A. Moran. 2013. Small, smaller, smallest: the origins and evolution of ancient dual symbioses in a Phloem-feeding insect. Genome Biology and Evolution 5: 1675–1688. doi:10.1093/gbe/evt118.

Bennett, G.M., and N.A. Moran. 2015. Heritable symbiosis: The advantages and perilsof an evolutionary rabbit hole. Proceedings of the National Academy of Sciences of the United States of America. doi:10.1073/pnas.1421388112.

Bistolas, K.S., R.I. Sakamoto, J.A. Fernandes, and S.K. Goffredi. 2014. Symbiont polyphyly, co-evolution, and necessity in pentatomid stinkbugs from Costa Rica. Frontiers in Microbiology 5: 349. doi:10.3389/fmicb.2014.00349.

Boucias, D.G., A. Garcia-Maruniak, R. Cherry, H. Lu, J.E. Maruniak, and V.-U. Lietze. 2012. Detection and characterization of bacterial symbionts in the Heteropteran, *Blissus insularis*. FEMS Microbiology Ecology 82: 629–641. doi:10.1111/j.1574-6941.2012.01433.x.

Bourtzis, K., and T.A. Miller. 2003. Insect symbiosis. CRC press, Boca Raton, FL. 347 pp.

Braendle, C., T. Miura, R. Bickel, A.W. Shingleton, S. Kambhampati, and D.L. Stern. 2003. Developmental origin and evolution of bacteriocytes in the aphid-*Buchnera* symbiosis. PLoS Biology 1: 70–76. doi:10.1371/journal.pbio.0000021.

Brown, A.M., L.Y. Huynh, C.M. Bolender, K.G. Nelson, and J.P. McCutcheon. 2014. Population genomics of a symbiont in the early stages of a pest invasion. Moleculae Ecology 23: 1516–1530. doi:10.1111/mec.12366.

Buchner, P. 1965. Endosymbiosis of animals with plant microorganisms. Interscience Publishers, New York, NY. 909 pp.

Charles, H., and H. Ishikawa. 1999. Physical and genetic map of the genome of *Buchnera*, the primary endosymbiont of the pea aphid *Acyrthosiphon pisum*. Journal of Molecular Evolution 48: 142–150. doi:10.1007/PL00006452.

Clarke, R.G., and G.E. Wilde. 1970a. Association of the green stink bug and the yeast-spot disease organism of soybeans. I. Lenght of retention, effect of molting, isolation from feces and saliva. Journal of Economic Entomology 63: 200–204. doi:http://dx.doi.org/10.1093/jee/63.1.200.

Clarke, R.G., and G.E. Wilde. 1970b. Association of the green stink bug and the yeast-spot disease organism of soybeans. II. Frequency of transmission to soybeans, transmission from insect to insect, isolation from field population. Journal of Economic Entomology 63: 355–357. doi:http://dx.doi.org/10.1093/jee/63.2.355.

Currie, C.R. 2001. A community of ants, fungi, and bacteria: a multilateral approach to studying symbiosis. Annual Reviews in Microbiology 55: 357–380. doi:10.1146/annurev.micro.55.1.357.

Currie, C.R., J.A. Scott, R.C. Summerbell, and D. Malloch. 1999. Fungus-growing ants use antibiotic-producing bacteria to control garden parasites. Nature 398: 701–704. doi:10.1038/19519.

Dasch, G.A., E. Weiss, and K.P. Chang. 1984. Endosymbionts of insects, pp. 811-833. *In* N. R. Krieg and J. G. Holt (Eds), Bergey's manual of systematic bacteriology, 1st ed., Volume 1. Williams & Wilkins, Baltimore, MD. 964 pp.

de Bary, A. 1879. Die Erscheinung der Symbiose. Verlag Karl J. Trübner, Strasbourg. 30 pp.

Del Pozo-Valdivia, A.I., and D.D. Reisig. 2013. First-generation *Megacopta cribraria* (Hemiptera: Plataspidae) can develop on soybeans. Journal of Economic Entomology 106: 533–535. doi:http://dx.doi.org/10.1603/EC12425.

Douglas, A.E. 1989. Mycetocyte symbiosis in insects. Biological Reviews 64: 409–434. doi:10.1111/j.1469-185X.1989.tb00682.x.

Douglas, A.E. 1998. Nutritional interactions in insect-microbial symbioses: aphids and their symbiotic bacteria *Buchnera*. Annual Review of Entomology 43: 17–37. doi:10.1146/annurev.ento.43.1.17.

Douglas, A.E. 2003. The nutritional physiology of aphids. Advances in Insect Physiology 31: 73–140. doi:doi:10.1016/S0065-2806(03)31002-1.

Eger, J.E., L.M. Ames, D.R. Suiter, T.M. Jenkins, D.A. Rider, and S.E. Halbert. 2010. Occurrence of the Old World bug *Megacopta cribraria* (Fabricius) (Heteroptera: Plataspidae) in Georgia: a serious home invader and potential legume pest. Insecta Mundi 0121: 1–11.

Frydman, H.M., J.M. Li, D.N. Robson, and E. Wieschaus. 2006. Somatic stem cell niche tropism in *Wolbachia*. Nature 441: 509–512. doi:10.1038/nature04756.

Fukatsu, T., and T. Hosokawa. 2002. Capsule-transmitted gut symbiotic bacterium of the Japanese common plataspid stinkbug, *Megacopta punctatissima*. Applied and Environmental Microbiology 68: 389–396. doi:10.1128/AEM.68.1.389-396.2002.

Fukatsu, T., R. Koga, W.A. Smith, K. Tanaka, N. Nikoh, K. Sasaki-Fukatsu, K. Yoshizawa, C. Dale, and D.H. Clayton. 2007. Bacterial endosymbiont of the slender pigeon louse, *Columbicola columbae*, allied to endosymbionts of grain weevils and tsetse flies. Applied and Environmental Microbiology 73: 6660–6668. doi:10.1128/AEM.01131-07.

Futahashi, R., K. Tanaka, M. Tanahashi, N. Nikoh, Y. Kikuchi, B.L. Lee, and T. Fukatsu. 2013. Gene expression in gut symbiotic organ of stinkbug affected by extracellular bacterial symbiont. PLoS One 8: e64557. doi:10.1371/journal.pone.0064557.

Garcia, J.R., A.M. Laughton, Z. Malik, B.J. Parker, C. Trincot, S.S.L. Chiang, E. Chung, and N.M. Gerardo. 2014. Partner associations across sympatric broad-headed bug species and their environmentally acquired bacterial symbionts. Molecular Ecology 23: 1333–1347. doi: 10.1111/mec.12655.

Gil, R., A. Latorre, and A. Moya. 2004. Bacterial endosymbionts of insects: insights from comparative genomics. Environmental Microbiology 6: 1109–1122. doi:10.1111/j.1462-2920.2004.00691.x.

Gil, R., F.J. Silva, E. Zientz, F. Delmotte, F. González-Candelas, A. Latorre, C. Rausell, J. Kamerbeek, J. Gadau, B. Hölldobler, R.C.H.J. van Ham, R. Gross, and A. Moya. 2003. The genome sequence of *Blochmannia floridanus*: comparative analysis of reduced genomes. Proceedings of the National Academy of Sciences of the United States of America 100: 9388–9393. doi:10.1073/pnas.1533499100.

Glasgow, H. 1914. The gastric caeca and the caecal bacteria of the Heteroptera. The Biological Bulletin 3: 101–171.

Goodchild, A.J.P. 1963. Studies on the functional anatomy of the intestines of Heteroptera. Proceedings of the Zoological Society of London 141: 851–910. doi:10.1111/j.1469-7998.1963.tb01631.x.

Gualtieri, G., and T. Bisseling. 2000. The evolution of nodulation. Plant Molecular Biology 42: 181–194. doi:10.1023/A:1006396525292.

Hayashi, T., T. Hosokawa, X.Y. Meng, R. Koga, and T. Fukatsu. 2015. Female-specific specialization of a posterior end region of the midgut symbiotic organ in *Plautia splendens* and allied stinkbugs. Applied and Environmental Microbiology 81: 2603–2611. doi:10.1128/AEM.04057-14.

Hirose, E., A.R. Panizzi, J.T. De Souza, A.J. Cattelan, and J.R. Aldrich. 2006. Bacteria in the gut of southern green stink bug (Heteroptera: Pentatomidae). Annals of the Entomological Society of America 99: 91–95. doi:http://dx.doi.org/10.1603/0013-8746(2006)099%5B0091:BITGOS%5D2.0.CO;2.

Hosokawa, T., Y. Kikuchi, X.Y. Meng, and T. Fukatsu. 2005. The making of symbiont capsule in the plataspid stinkbug *Megacopta punctatissima*. FEMS Microbiology Ecology 54: 471–477. doi:10.1016/j.femsec.2005.06.002.

Hosokawa, T., Y. Kikuchi, N. Nikoh, M. Shimada, and T. Fukatsu. 2006. Strict host-symbiont cospeciation and reductive genome evolution in insect gut bacteria. PLoS Biology 4: e337. doi:10.1371/journal.pbio.0040337.

Hosokawa, T., Y. Kikuchi, and T. Fukatsu. 2007a. How many symbionts are provided by mothers, acquired by offspring, and needed for successful vertical transmission in an obligate insect-bacterium mutualism? Molecular Ecology 16: 5316–5325. doi:10.1111/j.1365-294X.2007.03592.x.

Hosokawa, T., Y. Kikuchi, M. Shimada, and T. Fukatsu. 2007b. Obligate symbiont involved in pest status of host insect. Proceedings of the Royal Society B 274: 1979–1984. doi:10.1098/rspb.2007.0620.

Hosokawa, T., Y. Kikuchi, M. Shimada, and T. Fukatsu. 2008. Symbiont acquisition alters behaviour of stinkbug nymphs. Biology Letters 4: 45–48. doi:10.1098/rsbl.2007.0510.

Hosokawa, T., Y. Kikuchi, N. Nikoh, X.Y. Meng, M. Hironaka, and T. Fukatsu. 2010a. Phylogenetic position and peculiar genetic traits of a midgut bacterial symbiont of the stinkbug *Parastrachia japonensis*. Applied and Environmental Microbiology 76: 4130–4135. doi:10.1128/aem.00616-10.

Hosokawa, T., R. Koga, Y. Kikuchi, X.Y. Meng, and T. Fukatsu. 2010b. *Wolbachia* as a bacteriocyte-associated nutritional mutualist. Proceedings of the National Academy of Sciences of the United States of America 107: 769–774. doi:10.1073/pnas.0911476107.

Hosokawa, T., M. Hironaka, H. Mukai, K. Inadomi, N. Suzuki, and T. Fukatsu. 2012a. Mothers never miss the moment: a fine-tuned mechanism for vertical symbiont transmission in a subsocial insect. Animal Behaviour 83: 293–300. doi:10.1016/j.anbehav.2011.11.006.

Hosokawa, T., Y. Kikuchi, N. Nikoh, and T. Fukatsu. 2012b. Polyphyly of gut symbionts in stinkbugs of the family Cydnidae. Applied and Environmental Microbiology 78: 4758–4761. doi:10.1128/AEM.00867-12.

Hosokawa, T., M. Hironaka, K. Inadomi, H. Mukai, N. Nikoh, and T. Fukatsu. 2013. Diverse strategies for vertical symbiont transmission among subsocial stinkbugs. PLoS One 8: e65081. doi:10.1371/journal.pone.0065081.

Hosokawa, T., N. Kaiwa, Y. Matsuura, Y. Kikuchi, and T. Fukatsu. 2015. Infection prevalence of *Sodalis* symbionts among stinkbugs. Zoological Letters 1: 5. doi:10.1186/s40851-014-0009-5.

Hosokawa, T., Y. Ishii, N. Nikoh, M. Fujie, N. Satoh, and T. Fukatsu. 2016. Obligate bacterial mutualists evolving from environmental bacteria in natural insect populations. Nature Microbiology 1: 15011. doi:10.1038/nmicrobiol.2015.11.

Hunt, B.G., L. Ometto, Y. Wurm, D. Shoemaker, S.V. Yi, L. Keller, and M.A. Goodisman. 2011. Relaxed selection is a precursor to the evolution of phenotypic plasticity. Proceedings of the National Academy of Sciences of the United States of America 108: 15936–15941. doi:10.1073/pnas.1104825108.

International Aphid Genomics Consortium. 2010. Genome sequence of the pea aphid *Acyrthosiphon pisum*. PLoS Biology 8: e1000313. doi:10.1371/journal.pbio.1000313.

Itoh, H., M. Aita, A. Nagayama, X.Y. Meng, Y. Kamagata, R. Navarro, T. Hori, S. Ohgiya, and Y. Kikuchi. 2014. Evidence of environmental and vertical transmission of *Burkholderia* symbionts in the oriental chinch bug, *Cavelerius saccharivorus* (Heteroptera: Blissidae). Applied and Environmental Microbiology 80: 5974–5983. doi:10.1128/AEM.01087-14.

Jenkins, T.M., and T.D. Eaton. 2011. Population genetic baseline of the first plataspid stink bug symbiosis (Hemiptera: Heteroptera: Plataspidae) reported in North America. Insects 2: 264–272. doi:10.3390/insects2030264.

Jenkins, T.M., T.D. Eaton, D.R. Suiter, J.E. Eger, L.M. Ames, and G.D. Buntin. 2010. Preliminary genetic analysis of a recently-discovered Invasive true bug (Hemiptera: Heteroptera: Plataspidae) and its bacterial endosymbiont in Georgia, USA. Journal of Entomological Science 45: 1–2.

Jones, V.P., D.M. Westcott, N.N. Finson, and R.K. Nishimoto. 2001. Relationship between community structure and southern green stink bug (Heteroptera: Pentatomidae) damage in macadamia nuts. Environmental Entomology 30: 1028–1035. doi:http://dx.doi.org/10.1603/0046-225X-30.6.1028.

Kaiwa, N., T. Hosokawa, Y. Kikuchi, N. Nikoh, X.Y. Meng, N. Kimura, M. Ito, and T. Fukatsu. 2010. Primary gut symbiont and secondary, *Sodalis*-allied symbiont of the scutellerid stinkbug *Cantao ocellatus*. Applied and Environmental Microbiology 76: 3486–3494. doi:10.1128/aem.00421-10.

Kaiwa, N., T. Hosokawa, Y. Kikuchi, N. Nikoh, X.Y. Meng, N. Kimura, M. Ito, and T. Fukatsu. 2011. Bacterial symbionts of the giant jewel stinkbug *Eucorysses grandis* (Hemiptera: Scutelleridae). Zoological Science 28: 169–174. doi:10.2108/zsj.28.169.

Kaiwa, N., T. Hosokawa, N. Nikoh, M. Tanahashi, M. Moriyama, X.Y. Meng, T. Maeda, K. Yamaguchi, S. Shigenobu, M. Ito, and T. Fukatsu. 2014. Symbiont-supplemented maternal investment underpinning host's ecological adaptation. Current Biology 24: 2465–2470. doi:10.1016/j.cub.2014.08.065.

Kaltenpoth, M., W. Gottler, G. Herzner, and E. Strohm. 2005. Symbiotic bacteria protect wasp larvae from fungal infestation. Current Biology 15: 475–479. doi:10.1016/j.cub.2004.12.084.

Kaltenpoth, M., S.A. Winter, and A. Kleinhammer. 2009. Localization and transmission route of *Coriobacterium glomerans*, the endosymbiont of pyrrhocorid bugs. FEMS Microbiology Ecology 69: 373–383. doi:10.1111/j.1574-6941.2009.00722.x.

Kellner, R.L. 2002. Molecular identification of an endosymbiotic bacterium associated with pederin biosynthesis in *Paederus sabaeus* (Coleoptera: Staphylinidae). Insect Biochemistry and Molecular Biology 32: 389–395. doi:10.1016/S0965-1748(01)00115-1.

Kenyon, L.J., T. Meulia, and Z.L. Sabree. 2015. Habitat visualization and genomic analysis of "Candidatus *Pantoea carbekii*," the primary symbiont of the brown marmorated stink bug. Genome Biology and Evolution 7: 620–635. doi:10.1093/gbe/evv006.

Kikuchi, Y. 2009. Endosymbiotic bacteria in insects: their diversity and culturability. Microbes and Environments 24: 195–204. doi:10.1264/jsme2.ME09140S.

Kikuchi, Y., and T. Fukatsu. 2003. Diversity of *Wolbachia* endosymbionts in heteropteran bugs. Applied and Environmental Microbiology 69: 6082–6090. doi: 10.1128/AEM.69.10.6082-6090.2003.

Kikuchi, Y., and T. Fukatsu. 2014. Live imaging of symbiosis: spatiotemporal infection dynamics of a GFP-labelled *Burkholderia* symbiont in the bean bug *Riptortus pedestris*. Molecular Ecology 23: 1445–1456. doi:10.1111/mec.12479.

Kikuchi, Y., and I. Yumoto. 2013. Efficient colonization of the bean bug *Riptortus pedestris* by an environmentally transmitted *Burkholderia* symbiont. Applied and Environmental Microbiology 79: 2088–2091. doi:10.1128/AEM.03299-12.

Kikuchi, Y., X.Y. Meng, and T. Fukatsu. 2005. Gut symbiotic bacteria of the genus *Burkholderia* in the broad-headed bugs *Riptortus clavatus* and *Leptocorisa chinensis* (Heteroptera: Alydidae). Applied and Environmental Microbiology 71: 4035–4043. doi:10.1128/AEM.71.7.4035-4043.2005.

Kikuchi, Y., T. Hosokawa, and T. Fukatsu. 2007. Insect-microbe mutualism without vertical transmission: a stinkbug acquires a beneficial gut symbiont from the environment every generation. Applied and Environmental Microbiology 73: 4308–4316. doi:10.1128/AEM.00067-07.

Kikuchi, Y., T. Hosokawa, N. Nikoh, X.Y. Meng, Y. Kamagata, and T. Fukatsu. 2009. Host-symbiont cospeciation and reductive genome evolution in gut symbiotic bacteria of acanthosomatid stinkbugs. BMC Biology 7: 2. doi:10.1186/1741-7007-7-2.

Kikuchi, Y., T. Hosokawa, and T. Fukatsu. 2011. An ancient but promiscuous host-symbiont association between *Burkholderia* gut symbionts and their heteropteran hosts. The ISME Journal 5: 446–460. doi:10.1038/ismej.2010.150.

Kikuchi, Y., T. Hosokawa, N. Nikoh, and T. Fukatsu. 2012. Gut symbiotic bacteria in the cabbage bugs *Eurydema rugosa* and *Eurydema dominulus* (Heteroptera: Pentatomidae). Applied Entomology and Zoology 47: 1–8. doi:10.1007/s13355-011-0081-7.

Koga, R., X.Y. Meng, T. Tsuchida, and T. Fukatsu. 2012. Cellular mechanism for selective vertical transmission of an obligate insect symbiont at the bacteriocyte-embryo interface. Proceedings of the National Academy of Sciences of the United States of America 109: E1230–1237. doi:10.1073/pnas.1119212109.

Koga, R., G.M. Bennett, J.R. Cryan, and N.A. Moran. 2013. Evolutionary replacement of obligate symbionts in an ancient and diverse insect lineage. Environmental Microbiology 15: 2073–2081. doi:10.1111/1462-2920.12121.

Küechler, S.M., K. Dettner, and S. Kehl. 2010. Molecular characterization and localization of the obligate endosymbiotic bacterium in the birch catkin bug *Kleidocerys resedae* (Heteroptera: Lygaeidae, Ischnorhynchinae). FEMS Microbiology Ecology 73: 408–418. doi:10.1111/j.1574-6941.2010.00890.x.

Küechler, S.M., K. Dettner, and S. Kehl. 2011. Characterization of an obligate intracellular bacterium in the midgut epithelium of the bulrush bug *Chilacis typhae* (Heteroptera, Lygaeidae, Artheneinae). Applied and Environmental Microbiology 77: 2869–2876. doi:10.1128/AEM.02983-10.

Küechler, S.M., P. Renz, K. Dettner, and S. Kehl. 2012. Diversity of symbiotic organs and bacterial endosymbionts of lygaeoid bugs of the families Blissidae and Lygaeidae (Hemiptera: Heteroptera: Lygaeoidea). Applied and Environmental Microbiology 78: 2648–2659. doi:10.1128/AEM.07191-11.

Küechler, S.M., Y. Matsuura, K. Dettner, and Y. Kikuchi. 2016. Phylogenetically diverse *Burkholderia* associated with midgut crypts of spurge bugs, *Dicranocephalus* spp. (Heteroptera: Stenocephalidae). Microbes and Environments 31: 145-153. doi:10.1264/jsme2.ME16042.

Lee, F.N., N.P. Tugwell, S.J. Fannah, and G.J. Weidemann. 1993. Role of fungi vectored by rice stink bug (Heteroptera: Pentatomidae) in discoloration of rice kernels. Journal of Economic Entomology 86: 549–556. doi:http://dx.doi.org/10.1093/jee/86.2.549.

Lefevre, C., H. Charles, A. Vallier, B. Delobel, B. Farrell, and A. Heddi. 2004. Endosymbiont phylogenesis in the Dryophthoridae weevils: Evidence for bacterial replacement. Molecular Biology and Evolution 21: 965–973. doi:10.1093/molbev/msh063.

Lehane, M.J., and P.F. Billingsley. 1996. Biology of the insect midgut. Chapman & Hall, London, UK. 486 pp.

Ley, R.E., D.A. Peterson, and J.I. Gordon. 2006. Ecological and evolutionary forces shaping microbial diversity in the human intestine. Cell 124: 837–848. doi:10.1016/j.cell.2006.02.017.

Margulis, L., and R. Fester. 1991. Symbiosis as a source of evolutionary innovation. MIT Press, Cambridge, MA. 470 pp.

Matsuura, Y., Y. Kikuchi, T. Hosokawa, R. Koga, X.Y. Meng, Y. Kamagata, N. Nikoh, and T. Fukatsu. 2012. Evolution of symbiotic organs and endosymbionts in lygaeid stinkbugs. The ISME Journal 6: 397–409. doi:10.1038/ismej.2011.103.

Matsuura, Y., T. Hosokawa, M. Serracin, G.M. Tulgetske, T.A. Miller, and T. Fukatsu. 2014. Bacterial symbionts of a devastating coffee plant pest, the stinkbug *Antestiopsis thunbergii* (Hemiptera: Pentatomidae). Applied and Environmental Microbiology 80: 3769–3775. doi:10.1128/AEM.00554-14.

Matsuura, Y., Y. Kikuchi, T. Miura, and T. Fukatsu. 2015. *Ultrabithorax* is essential for bacteriocyte development. Proceedings of the National Academy of Sciences of the United States of America 112: 9376–9381. doi:10.1073/pnas.1503371112.

Mira, A., and N.A. Moran. 2002. Estimating population size and transmission bottlenecks in maternally transmitted endosymbiotic bacteria. Microbial Ecology 44: 137–143. doi:10.1007/s00248-002-0012-9.

Miura, T., C. Braendle, A. Shingleton, G. Sisk, S. Kambhampati, and D.L. Stern. 2003. A comparison of parthenogenetic and sexual embryogenesis of the pea aphid *Acyrthosiphon pisum* (Hemiptera: Aphidoidea). Journal of Experimental Zoology Part B: Molecular and Developmental Evolution 295: 59–81. doi:10.1002/jez.b.3.

Moran, N.A. 1996. Accelerated evolution and Muller's rachet in endosymbiotic bacteria. Proceedings of the National Academy of Sciences of the United States of America 93: 2873–2878.

Moran, N.A. 2007. Symbiosis as an adaptive process and source of phenotypic complexity. Proceedings of the National Academy of Sciences of the United States of America 104: 8627–8633. doi:10.1073/pnas.0611659104.

Moran, N.A., M.A. Munson, P. Baumann, and H. Ishikawa. 1993. A molecular clock in endosymbiotic bacteria is calibrated using the insect hosts. Proceedings of the Royal Society B 253: 167–171. doi:10.1098/rspb.1993.0098.

Moran, N.A., P. Tran, and N.M. Gerardo. 2005. Symbiosis and insect diversification: an ancient symbiont of sap-feeding insects from the bacterial phylum *Bacteroidetes*. Applied and Environmental Microbiology 71: 8802–8810. doi:10.1128/AEM.71.12.8802-8810.2005.

Moran, N.A., J.P. McCutcheon, and A. Nakabachi. 2008. Genomics and evolution of heritable bacterial symbionts. Annual Review of Genetics 42: 165–190. doi:10.1146/annurev.genet.41.110306.130119.

Müller, H.J. 1956. Experimentelle studien an der symbiose von *Coptosoma scutellatum* Geoffr. (Hem. Heteropt.). Zeitschrift für Morphologie und Ökologie der Tiere 44: 459–482. doi:10.1007/BF00407170.

Muscatine, L. 1973. Nutrition of corals, pp. 77–115. *In* O. A. Jones and R. Endean (Eds), Biology and geology of coral reefs. Academic Press, New York.

Nikoh, N., T. Hosokawa, K. Oshima, M. Hattori, and T. Fukatsu. 2011. Reductive evolution of bacterial genome in insect gut environment. Genome Biology and Evolution 3: 702–714. doi:10.1093/gbe/evr064.

Nyholm, S.V., and M.J. McFall-Ngai. 2004. The winnowing: establishing the squid-*Vibrio* symbiosis. Nature Reviews Microbiology 2: 632–642. doi:10.1038/nrmicro957.

O'Neill, S.L., A.A. Hoffmann, and J.H. Werren. 1997. Influential Passenger: Inherited Microorganisms and Arthropod Reproduction. Oxford University Press, New York, NY. 226 pp.

Oliver, K.M., J.A. Russell, N.A. Moran, and M.S. Hunter. 2003. Facultative bacterial symbionts in aphids confer resistance to parasitic wasps. Proceedings of the National Academy of Sciences of the United States of America 100: 1803–1807. doi:10.1073/pnas.0335320100.

Olivier-Espejel, S., Z.L. Sabree, K. Noge, and J.X. Becerra. 2011. Gut microbiota in nymph and adults of the giant mesquite bug (*Thasus neocalifornicus*) (Heteroptera: Coreidae) is dominated by *Burkholderia* acquired *de novo* every generation. Environmental Entomology 40: 1102–1110. doi:10.1603/EN10309.

Pérez-Brocal, V., R. Gil, A. Moya, and A. Latorre. 2011. New insights on the evolutionary history of aphids and their primary endosymbiont *Buchnera aphidicola*. International Journal of Evolutionary Biology 2011: 250154. doi:doi: 10.4061/2011/250154.

Prado, S.S., and R.P. Almeida. 2009a. Phylogenetic placement of pentatomid stink bug gut symbionts. Current Microbiology 58: 64–69. doi:10.1007/s00284-008-9267-9.

Prado, S.S., and R.P. Almeida. 2009b. Role of symbiotic gut bacteria in the development of *Acrosternum hilare* and *Murgantia histrionica*. Entomologia Experimentalis et Applicata 132: 21–29. doi:10.1111/j.1570-7458.2009.00863.x.

Prado, S.S., and T.D. Zucchi. 2012. Host-symbiont interactions for potentially managing heteropteran pests. Psyche: A Journal of Entomology 2012: 1–9. doi:http://dx.doi.org/10.1155/2012/269473.

Prado, S.S., D. Rubinoff, and R.P. Almeida. 2006. Vertical transmission of a pentatomid caeca-associated symbiont. Annals of the Entomological Society of America 99: 577–585. doi:http://dx.doi.org/10.1603/0013-8746(2006)99%5B577:VTOAPC%5D2.0.CO;2.

Prado, S.S., M. Golden, P.A. Follett, M.P. Daugherty, and R.P. Almeida. 2009. Demography of gut symbiotic and aposymbiotic *Nezara viridula* L. (Hemiptera: Pentatomidae). Environmental Entomology 38: 103–109. doi:http://dx.doi.org/10.1603/022.038.0112.

Prado, S.S., K.Y. Hung, M.P. Daugherty, and R.P. Almeida. 2010. Indirect effects of temperature on stink bug fitness, via maintenance of gut-associated symbionts. Applied and Environmental Microbiology 76: 1261–1266. doi:10.1128/aem.02034-09.

Rosenkranz, W. 1939. Die symbiose der Pentatomiden. Zeitschrift für Morphologie und Ökologie der Tiere 36: 279–309. doi:10.1007/BF00403148.

Ruby, E., B. Henderson, and M. McFall-Ngai. 2004. We get by with a little help from our (little) friends. Science 303: 1305–1307. doi:10.1126/science.1094662.

Russin, J.S., D.B. Orr, M.B. Layton, and D.J. Boethel. 1988. Incidence of microorganisms in soybean seeds damaged by stink bug feeding. Phytopathology 78: 306–310. doi:10.1094/Phyto-78-306.

Salem, H., E. Kreutzer, S. Sudakaran, and M. Kaltenpoth. 2013. Actinobacteria as essential symbionts in firebugs and cotton stainers (Hemiptera, Pyrrhocoridae). Environmental Microbiology 15: 1956–1968. doi:10.1111/1462-2920.12001.

Salem, H., E. Bauer, A.S. Strauss, H. Vogel, M. Marz, and M. Kaltenpoth. 2014. Vitamin supplementation by gut symbionts ensures metabolic homeostasis in an insect host. Proceedings of the Royal Society B 281: 20141838. doi:10.1098/rspb.2014.1838.

Salem, H., L. Florez, N. Gerardo, and M. Kaltenpoth. 2015. An out-of-body experience: the extracellular dimension for the transmission of mutualistic bacteria in insects. Proceedings of the Royal Society B 282: 20142957. doi:10.1098/rspb.2014.2957.

Sasaki, T., and H. Ishikawa. 1998. Production of essential amino acids from glutamate by mycetocyte symbionts of the pea aphid, *Acyrthosiphon pisum*. Journal of Insect Physiology 41: 41–46. doi:doi:10.1016/0022-1910(94)00080-Z.

Sauer, C., E. Stackebrandt, J. Gadau, B. Holldobler, and R. Gross. 2000. Systematic relationships and cospeciation of bacterial endosymbionts and their carpenter ant host species: proposal of the new taxon Candidatus Blochmannia gen. nov. International Journal of Systematic and Evolutionary Microbiology 5: 1877–1886. doi:10.1099/00207713-50-5-1877.

Scarborough, C.L., J. Ferrari, and H.C. Godfray. 2005. Aphid protected from pathogen by endosymbiont. Science 310: 1781. doi:10.1126/science.1120180.

Schaefer, C.W., and A.R. Panizzi. 2000. Heteroptera of economic importance. CRC Press, Florida, FL. 856 pp.

Schneider, G. 1940. Beiträge zur Kenntnis der symbiontischen Einrichtungen der Heteropteren. Zeitschrift für Morphologie und Ökologie der Tiere 36. doi:10.1007/978-3-662-38370-4.

Schuh, R.T., and J.A. Slater. 1995. True bugs of the world (Hemiptera: Heteroptera). Cornell University Press, New York, NY. 336 pp.

Scott, J.J., D.C. Oh, M.C. Yuceer, K.D. Klepzig, J. Clardy, and C.R. Currie. 2008. Bacterial protection of beetle-fungus mutualism. Science 322: 63. doi:10.1126/science.1160423.

Seiter, N.J., F.P. Reay-Jones, and J.K. Greene. 2013. Within-field spatial distribution of *Megacopta cribraria* (Hemiptera: Plataspidae) in soybean (Fabales: Fabaceae). Environmental Entomology 42: 1363–1374. doi:10.1603/EN13199.

Shibata, T.F., T. Maeda, N. Nikoh, K. Yamaguchi, K. Oshima, M. Hattori, T. Nishiyama, M. Hasebe, T. Fukatsu, Y. Kikuchi, and S. Shigenobu. 2013. Complete genome sequence of *Burkholderia* sp. strain RPE64, bacterial symbiont of the bean bug *Riptortus pedestris*. Genome Announcements 1: e00441-00413. doi:10.1128/genomeA.00441-13.

Shigenobu, S., H. Watanabe, M. Hattori, Y. Sakaki, and H. Ishikawa. 2000. Genome sequence of the endocellular bacterial symbiont of aphids *Buchnera* sp. APS. Nature 407: 81–86. doi:10.1038/35024074.

Spaulding, A.W., and C.D. von Dohlen. 1998. Phylogenetic characterization and molecular evolution of bacterial endosymbionts in psyllids (Hemiptera: Sternorrhyncha). Molecular Biology and Evolution 15: 1506–1513.

Sudakaran, S., H. Salem, C. Kost, and M. Kaltenpoth. 2012. Geographical and ecological stability of the symbiotic mid-gut microbiota in European firebugs, *Pyrrhocoris apterus* (Hemiptera, Pyrrhocoridae). Molecular Ecology 21: 6134–6151. doi:10.1111/mec.12027.

Sudakaran, S., F. Retz, Y. Kikuchi, C. Kost, and M. Kaltenpoth. 2015. Evolutionarytransition in symbiotic syndromes enabled diversification of phytophagous insects on an imbalanced diet. The ISME Journal. 9: 2587-2604. doi:10.1038/ismej.2015.75.

Suiter, D.R., J.E. Eger, W.A. Gardner, R.C. Kemerait, J.N. All, P.M. Roberts, J.K. Greene, L.M. Ames, G.D. Buntin, T.M. Jenkins, and G.K. Douce. 2010. Discovery and distribution of *Megacopta cribraria* (Hemiptera: Heteroptera: Plataspidae) in Northeast Georgia. Journal of Integrated Pest Management 1: 1–4. doi:10.1603/IPM10009.

Tada, A., Y. Kikuchi, T. Hosokawa, D. L. Musolin, K. Fujisaki, and T. Fukatsu. 2011. Obligate association with gut bacterial symbiont in Japanese populations of the southern green stinkbug *Nezara viridula* (Heteroptera: Pentatomidae). Applied Entomology and Zoology 46: 483–488. doi:10.1007/s13355-011-0066-6.

Takeshita, K., T.F. Shibata, N. Nikoh, T. Nishiyama, M. Hasebe, T. Fukatsu, S. Shigenobu, and Y. Kikuchi. 2014. Whole-genome sequence of *Burkholderia* sp. strain RPE67, a bacterial gut symbiont of the bean bug *Riptortus pedestris*. Genome Announcements 2: e00556-00514. doi:10.1128/genomeA.00556-14.

Takeshita, K., Y. Matsuura, H. Itoh, R. Navarro, T. Hori, T. Sone, Y. Kamagata, P. Mergaert, Y. Kikuchi. 2015. *Burkholderia* of plant-beneficial group are symbiotically associated with bordered plant bugs (Heteroptera: Pyrrhocoroidea: Largidae). Microbes and Environments 30: 321–329. doi:10.1264/jsme2.ME15153.

Taylor, C.M., P.L. Coffey, B.D. DeLay, and G.P. Dively. 2014. The importance of gut symbionts in the development of the brown marmorated stink bug, *Halyomorpha halys* (Stal). PLoS One 9: e90312. doi:10.1371/journal.pone.0090312.

Thao, M.L., N.A. Moran, P. Abbot, E.B. Brennan, D.H. Burckhardt, and P. Baumann. 2000. Cospeciation of psyllids and their primary prokaryotic endosymbionts. Applied and Environmental Microbiology 66: 2898–2905. doi:10.1128/AEM.66.7.2898-2905.2000.

Thao, M.L., P.J. Gullan, and P. Baumann. 2002. Secondary (γ-*Proteobacteria*) endosymbionts infect the primary (β-*Proteobacteria*) endosymbionts of mealybugs multiple times and coevolve with their hosts. Applied and Environmental Microbiology 68: 3190–3197. doi:10.1128/AEM.68.7.3190-3197.2002.

Todd, J.W. 1989. Ecology and behavior of *Nezara viridula*. Annual Review of Entomology 34: 273–292. doi:10.1146/annurev.en.34.010189.001421.

Tsuchida, T., R. Koga, and T. Fukatsu. 2004. Host plant specialization governed by facultative symbiont. Science 303: 1989. doi:10.1126/science.1094611.

Tsuchida, T., R. Koga, M. Horikawa, T. Tsunoda, T. Maoka, S. Matsumoto, J.C. Simon, and T. Fukatsu. 2010. Symbiotic bacterium modifies aphid body color. Science 330: 1102–1104. doi:10.1126/science.1195463.

Usinger, R. 1966. Monograph of Cimicidae (Hemiptera, Heteroptera). College Park, M.D.: Entomological Society of America.

Weirauch, C., and R.T. Schuh. 2011. Systematics and evolution of Heteroptera: 25 years of progress. Annual Review of Entomology 56: 487–510. doi:10.1146/annurev-ento-120709-144833.

Wernegreen, J.J. 2002. Genome evolution in bacterial endosymbionts of insects. Nature Reviews Genetics 3: 850–861. doi:10.1038/nrg931.

Wernegreen, J.J. 2012. Endosymbiosis. Current Biology 22: R555-561. doi:10.1016/j.cub.2012.06.010.

Werren, J.H., L. Baldo, and M.E. Clark. 2008. *Wolbachia:* master manipulators of invertebrate biology. Nature Reviews Microbiology 6: 741–751. doi:10.1038/nrmicro1969.

Wright, S. (1932) The roles of mutation, inbreeding, crossbreeding, and selection in evolution, pp. 356–366. *In* D. F. Jones (Ed.), Proceedings of the Sixth International Congress on Genetics (Volume 1) Brooklyn Botanic Garden.

Wu, D., S.C. Daugherty, S.E. Van Aken, G.H. Pai, K.L. Watkins, H. Khouri, L.J. Tallon, J.M. Zaborsky, H.E. Dunbar, P.L. Tran, N.A. Moran, and J.A. Eisen. 2006. Metabolic complementarity and genomics of the dual bacterial symbiosis of sharpshooters. PLoS Biology 4: e188. doi:10.1371/journal.pbio.0040188.

Yoshida, N., K. Oeda, E. Watanabe, T. Mikami, Y. Fukita, K. Nishimura, K. Komai, and K. Matsuda. 2001. Protein function: Chaperonin turned insect toxin. Nature 411: 44. doi:10.1038/35075148.

Zhang, Y., J.L. Hanula, and S. Horn. 2012. The biology and preliminary host range of *Megacopta cribraria* (Heteroptera: Plataspidae) and its impact on kudzu growth. Environmental Entomology 41: 40–50. doi:10.1603/EN11231.

Zhu, G., M.J. Petersen, and W. Bu. 2012. Selecting biological meaningful environmental dimensions of low discrepancy among ranges to predict potential distribution of Bean plataspid invasion. PLoS One 7: e46247. doi:10.1371/journal.pone.0046247.

Zucchi, T.D., S.S. Prado, and F.L. Consoli. 2012. The gastric caeca of pentatomids as a house for actinomycetes. BMC Microbiology 12: 101. doi:10.1186/1471-2180-12-101.

Section IX

Semiochemistry

15

Semiochemistry of Pentatomoidea

Donald C. Weber, Ashot Khrimian, Maria Carolina Blassioli-Moraes, and Jocelyn G. Millar

CONTENTS

15.1 Introduction 678
    15.1.1   Scope: Families in Hemiptera: Heteroptera 678
    15.1.2   Scope: Definitions of Semiochemicals 678
    15.1.3   Justification for Research on Stink Bug Semiochemistry 679
15.2 Pheromones: Overview and Species Accounts 679
    15.2.1   *Agroecus griseus* Dallas [Pentatomidae: Pentatominae: Carpocorini] 681
    15.2.2   *Bagrada hilaris* Burmeister [Pentatomidae: Pentatominae: Strachiini] 681
    15.2.3   *Chinavia hilaris* (Say), *Chinavia ubica* (Rolston), and *Chinavia impicticornis*
            (Stål) [Pentatomidae: Pentatominae: Nezarini] 682
    15.2.4   *Chlorochroa sayi* (Stål), *Chlorochroa uhleri* (Stål), and *Chlorochroa ligata* (Say)
            [Pentatomidae: Pentatominae: Nezarini] 683
    15.2.5   *Edessa meditabunda* (F.) [Pentatomidae: Edessinae] 683
    15.2.6   *Eocanthecona furcellata* (Wolff) [Pentatomidae: Asopinae] 684
    15.2.7   *Euschistus conspersus* Uhler, *Euschistus heros* (F.), and *Euschistus servus* (Say)
            [Pentatomidae: Pentatominae: Carpocorini] 684
          15.2.7.1 *Euschistus conspersus*: Pheromone-Based Monitoring 684
          15.2.7.2 *Euschistus heros*: Effects of Food on Pheromone Emission,
                and Pheromone-Based Monitoring 685
          15.2.7.3 *Euschistus servus*: Use of Pheromone in Trap Crop 685
    15.2.8   *Eysarcoris lewisi* (Distant) [Pentatomidae: Pentatominae: Eysarcorini] 685
    15.2.9   *Halyomorpha halys* (Stål) [Pentatomidae: Pentatominae: Cappaeini] 688
          15.2.9.1 Aggregation Pheromone 689
          15.2.9.2 Attraction of *Halyomorpha halys* to the Pheromone of *Plautia stali* 691
    15.2.10 *Murgantia histrionica* (Hahn) [Pentatomidae: Pentatominae: Strachiini] 692
    15.2.11 *Nezara viridula* (L.) and *Nezara antennata* Scott [Pentatomidae: Pentatominae:
          Nezarini] 693
    15.2.12 *Oebalus poecilus* (Dallas) [Pentatomidae: Pentatominae: Carpocorini] 693
    15.2.13 *Pallantia macunaima* Grazia [Pentatomidae: Pentatominae: Pentatomini] 694
    15.2.14 *Piezodorus guildinii* (Westwood) and *Piezodorus hybneri* (Gmelin)
          [Pentatomidae: Pentatominae: Piezodorini] 695
          15.2.14.1 Sex Pheromone of *Piezodorus guildinii* 695
          15.2.14.2 Aggregation Pheromone of *Piezodorus hybneri*: Functions and Variation 696
          15.2.14.3 Attraction of *Piezodorus hybneri* to the Pheromone of *Riptortus clavatus*
                 Thunberg (Heteroptera: Coreoidea: Alydidae) 696
    15.2.15 *Plautia stali* Scott [Pentatomidae: Pentatominae: Antestiini] 696
    15.2.16 *Sehirus cinctus* (Palisot de Beauvois) [Cydnidae: Sehirinae] 699
    15.2.17 *Tessaratoma papillosa* (Drury) [Tessaratomidae: Tessaratominae] 699
    15.2.18 *Thyanta custator* (F.), *Thyanta pallidovirens* (Stål), and *Thyanta perditor* (F.)
          [Pentatomidae: Pentatominae: Antestiini] 699
    15.2.19 *Tibraca limbativentris* (Stål) [Pentatomidae: Pentatominae: Carpocorini] 700

15.3 Allomones: Defensive Chemicals 700
    15.3.1 Range of Chemistry 700
    15.3.2 Effect of MTG and DAG Secretions against Natural Enemies 702
    15.3.3 Other Behavioral Roles: Within Species 702
15.4 Kairomones for Natural Enemies 703
    15.4.1 Pentatomid Compounds as Kairomones for Egg Parasitoids 704
    15.4.2 Pentatomid Compounds as Kairomones for Diptera 706
    15.4.3 Pentatomid Compounds as Kairomones for Arthropod Predators 707
    15.4.4 Predatory Stink Bugs Exploiting Semiochemicals of Their Prey 707
    15.4.5 Tritrophic Interactions Involving Pentatomoid Bugs 707
15.5 Overview: Semiochemicals in Life History of Pentatomoids, and Practical Applications 708
    15.5.1 Patterns of Production and Response to Pheromones 708
        15.5.1.1 Cross-Species Attraction 709
        15.5.1.2 Multiple Components: Ratios and Variability 710
        15.5.1.3 Dose and Release Rates of Stink Bug Semiochemicals 711
    15.5.2 Current Role of Stink Bug Semiochemicals in Pest Management 712
    15.5.3 Future Directions in Research and Applications 713
15.6 Acknowledgments 714
15.7 References Cited 714

15.1 Introduction

15.1.1 Scope: Families in Hemiptera: Heteroptera

We review here semiochemicals identified from species in the true bug superfamily Pentatomoidea, which includes among others the families Pentatomidae, Acanthosomatidae, Plataspidae, Scutelleridae, Cydnidae, and Tessaratomidae. This does not include some well-studied groups in the infraorder Pentatomomorpha, such as Coreoidea (Alydidae, Coreidae, and Rhopalidae), Lygaeoidea (Berytidae, Blissidae, Cymidae, Geocoridae, and Lygaeidae), Aradidae, and Pyrrhocoridae. Subfamilies and tribes of Pentatomidae are classified according to Rider (2015).

15.1.2 Scope: Definitions of Semiochemicals

Semiochemicals identified from the Pentatomoidea, as with other organisms, fulfill a wide range of communication roles for senders and receivers (terminology from Nordlund and Lewis 1976). Pheromones are used for communication between members of the same species and may be classified as sex pheromones, aggregation pheromones, alarm pheromones, and other categories, according to their functional roles. We cover these in detail in **Section 15.2**. Allomones (**Section 15.3**), which the common name "stink bug" connotes, are repellent or repugnant chemicals used against members of other species such as predators and parasitoids; as such, they benefit the sender at the expense of the receiver. Conversely, kairomones benefit the receiver at the expense of the sender, and most examples discussed below involve the exploitation of pheromonal or allomonal signals by natural enemies whose hosts or prey are pentatomoid bugs, as summarized in **Section 15.4**. Plant volatiles that are released in response to pentatomoid herbivory are considered synomones when they attract natural enemies of phytophagous bugs, thus conferring a benefit to the plant which produces them, in helping to protect it from herbivory (**Section 15.4.5**). In addition, several stink bug aggregation pheromones attract other species of stink bugs, which could be considered a kairomonal or synomonal interaction, depending on whether the sender experiences fitness losses or gains, respectively (**Section 15.5.1.1**). From the above statements it is apparent that a single semiochemical, or a blend of semiochemicals released together, may have multiple functions, and, therefore, may be designated as a pheromone, allomone, kairomone, or synomone, depending on the context in which it is being used or exploited.

In spite of the fascinating complexity of interactions involving semiochemicals, historically, many investigations into presumed stink bug semiochemicals have failed to adequately demonstrate the purported functions of the various compounds. Instead, differences in the volatiles or gland contents produced by adult males and females have been used to propose functions for chemicals that are exclusive to or much more abundant in one sex. However, until rigorous and meaningful bioassays demonstrating the behavioral responses of conspecifics to specific compounds have been carried out, designation as pheromone components is not justified (Millar 2005). The same caveat applies to other classes of semiochemicals: chemistry and biology must be investigated in tandem to establish whether a given substance, or component of a multi-component emission, is indeed a semiochemical that elicits a behavioral response.

15.1.3 Justification for Research on Stink Bug Semiochemistry

Pentatomoidea are important in virtually all ecosystems of the world, and, as with other superfamilies of true bugs, many are herbivorous. Other species are predators of other insects as well as feeding on plants at least occasionally. Many stink bugs are pests whose importance has increased in recent years because of the replacement of broad-spectrum insecticides with more selective suppression tactics for control of other key pests, such as the deployment of transgenic crops that express insecticidal proteins, use of mating disruption and insect growth regulators in control of tree fruit pests, and Lepidoptera- and Homoptera-selective control tactics used in vegetable crops (Millar et al. 2010). All species of stink bugs for which pheromones have been identified have male-produced pheromones that attract females and, in many cases, also males and nymphs. These pheromones can potentially be exploited for monitoring and possibly in attract-and-kill or other tactics for control of pest populations.

Identification of active pheromone components must be performed rigorously so as to ascertain: (1) the minimum blend of compounds that are both necessary and sufficient to elicit activity equivalent to that elicited by pheromones released by live bugs (Millar 2014), and (2) the ratio and release rate of components that are optimally attractive under field conditions. It also is important to determine the functional roles of other components that are present in crude extracts, particularly compounds that might inhibit attraction of the species under study, or of congeners (i.e., to prevent cross-attraction between related species). Thus, a methodical and thorough approach to the identification of semiochemicals is required in order to exploit their full utility for integrated pest management (Millar 2014).

15.2 Pheromones: Overview and Species Accounts

Pheromones have been identified for approximately 45 species of Pentatomoidea, in 25 genera, all but two of them in the Pentatomidae, as shown in **Table 15.1**). Approximately half of these have been discovered since the thorough review of Millar (2005), and this chapter will focus primarily on advances made since the publication of that review. This review covers the literature up until October 2015, when the manuscript was submitted.

Almost all of the pheromones described to date are produced by adult male bugs. The exceptions are the aggregation pheromones identified from nymphal pentatomids of a few species (Fucarino et al. 2004) and the so-called solicitation pheromones produced by cydnid nymphs (see below). The pattern of responses elicited by the male-produced pheromones is divided roughly equally between those that appear to attract only females, and those that attract both sexes, as well as nymphs when this has been tested. Even this distinction may be variable: for example, the male-produced pheromone of *Piezodorus hybneri* attracts approximately 90% females during the summer, but equal proportions of males and females in the fall (Endo et al. 2010).

Below, we provide case studies of new discoveries since Millar's (2005) comprehensive review, some of which were also reported in Moraes et al. (2008b) for neotropical species. The chemistry is diverse and challenging from the standpoint of stereochemical specificity, frequently involving molecules such as sesquiterpenoids with multiple chiral centers. In spite of the analytical and synthetic chemistry challenges embodied in the often complex structures of pentatomid pheromones, many species will respond

TABLE 15.1

Pheromones Investigated in the Pentatomoidea. All Pheromones Are Produced by Adult Males, Except for the Final Species, *Sehirus cinctus*, in Which the Pheromone Is Produced by Nymphs

Section	Species	Responder[a]	Key Citation(s)
	Pentatomidae: Asopinae		
15.2.6	*Eocanthecona furcellata*	?	Ho et al. 2003, 2005
--	*Oplomus dichrous*	?	Aldrich et al. 1986b
--	*Oplomus ebulinus*	?	Aldrich and Lusby 1986
--	*Perillus bioculatus*	?	Aldrich et al. 1986b
--	*Perillus strigipes*	?	Aldrich and Lusby 1986
--	*Podisus fretus*	M?, F?	Aldrich et al. 1986a
--	*Podisus maculiventris*	M, F, n	Aldrich et al. 1984
--	*Stiretrus anchorago*	M, F, n	Aldrich et al. 1986b, Kochansky et al. 1989
--	*Tynacantha marginata*	?	Kuwahara et al. 2000
	Pentatomidae: Edessinae		
15.2.5	*Edessa meditabunda*	F	Zarbin et al. 2012
	Pentatomidae: Pentatominae: Antestiini		
15.2.15	*Plautia stali*	M, F, n?	Sugie et al. 1996
15.2.18	*Thyanta custator*	?	McBrien et al. 2002
15.2.18	*Thyanta pallidovirens*	F	McBrien et al. 2002, Millar 1997
15.2.18	*Thyanta perditor*	F	Moraes et al. 2005a
	Pentatomidae: Pentatominae: Cappaeini		
15.2.9	*Halyomorpha halys*	M, F, n	Khrimian et al. 2014a
	Pentatomidae: Pentatominae: Carpocorini		
15.2.1	*Agroecus griseus*	F	Fávaro et al. 2012
15.2.7	*Euschistus conspersus*	M, F, n	see Millar 2005
15.2.7	*Euschistus heros*	F	see Millar 2005
--	*Euschistus ictericus*	?	see Millar 2005
--	*Euschistus obscurus*	F	see Millar 2005
--	*Euschistus politus*	M, F, n	see Millar 2005
15.2.7	*Euschistus servus*	M, F, n	see Millar 2005
--	*Euschistus tristigmus*	M, F, n	see Millar 2005
15.2.12	*Oebalus poecilus*	F	de Oliveira et al. 2013
15.2.19	*Tibraca limbativentris*	F	Borges et al. 2006
	Pentatomidae: Pentatominae: Eysarcorini		
15.2.8	*Eysarcoris lewisi*	M, F, n	Takita et al. 2008, Mori et al. 2008
--	*Eysarcoris parvus*	F	Alizadeh et al. 2002
	Pentatomidae: Pentatominae: Nezarini		
--	*Chinavia aseada*	?	see Millar 2005
15.2.3	*Chinavia hilaris*	M, F, n	McBrien et al. 2001
15.2.3	*Chinavia impicticornis*	F	Blassioli-Moraes et al. 2012
--	*Chinavia marginata*	?	see Millar 2005
--	*Chinavia pensylvanica*	?	see Millar 2005
15.2.3	*Chinavia ubica*	F	Blassioli-Moraes et al. 2012
15.2.4	*Chlorochroa ligata*	M?, F	Ho and Millar 2001b
15.2.4	*Chlorochroa sayi*	M, F, n?	Ho and Millar 2001a, Millar et al. 2010
15.2.4	*Chlorochroa uhleri*	M, F, n?	Ho and Millar 2001b, Millar et al. 2010

(Continued)

TABLE 15.1 (CONTINUED)

Pheromones Investigated in the Pentatomoidea. All Pheromones Are Produced by Adult Males,
Except for the Final Species, *Sehirus cinctus*, in Which the Pheromone Is Produced by Nymphs

Section	Species	Responder[a]	Key Citation(s)
15.2.11	*Nezara viridula*	M, F, n	Aldrich et al. 1987, Baker et al. 1987
15.2.11	*Nezara antennata*	M, F, n?	Aldrich et al. 1989, 1993
	Pentatomidae: Pentatominae: Pentatomini		
15.2.13	*Pallantia macunaima*	F	Fávaro et al. 2013
	Pentatomidae: Pentatominae: Piezodorini		
15.2.14	*Piezodorus guildinii*	F	Borges et al. 2007
15.2.14	*Piezodorus hybneri*	M, F, n	Leal et al. 1998, Endo et al. 2010
	Pentatomidae: Pentatominae: Rhynchocorini		
--	*Biprorulus bibax*	M, F, n?	James et al. 1994, 1996
	Pentatomidae: Pentatominae: Strachiini		
15.2.2	*Bagrada hilaris*	F	Guarino et al. 2008
15.2.10	*Murgantia histrionica*	M, F, n	Zahn et al. 2008, Khrimian et al. 2014b
	Tessaratomidae: Tessaratominae		
15.2.17	*Tessaratoma papillosa*	?	Zhao et al. 2012, Wang et al. 2012
	Cydnidae: Cephalocteninae		
15.2.16	*Sehirus cinctus*	F	Kölliker et al. 2006

[a] M=males; F=females; n=nymphs.

well to mixtures containing unnatural ratios of active pheromone components and/or unnatural isomers
of the pheromone components. This offers opportunities to synthesize relatively inexpensive mixtures of
pheromone stereoisomers that may be of considerable practical value in management of stink bug pests,
whereas pure substances would be prohibitively expensive to produce in even gram quantities.

The attraction of some stink bug species to the pheromones of other species, either by partial overlap
of active components or, surprisingly, to pheromones that the responding species does not itself produce,
also has been exploited in stink bug monitoring. Natural enemies also are attracted to many of the pheromones and allomones of the Pentatomoidea (**Section 15.4**), co-opting them for use as kairomones to find
their hosts or prey. This is understood more easily than the cross-attraction among bug species, which is
discussed in **Section 15.5.1.1**.

15.2.1 *Agroecus griseus* Dallas [Pentatomidae: Pentatominae: Carpocorini]

Agroecus griseus ranges from Panama to northern Argentina and is a major pest of corn in Brazil (Rider
and Rolston 1987). This was the first *Agroecus* species in which semiochemicals were studied. Aerations
of male and female *A. griseus* showed the presence of a male-specific compound, which was identified as methyl 2,6,10-trimethyltridecanoate of unspecified stereochemistry (Fávaro et al. 2012). This
compound had been described previously as a sex pheromone of *Euschistus heros* (Aldrich et al. 1991)
and *E. obscurus* (Borges and Aldrich 1994). A synthetic mixture of all eight stereoisomers of methyl
2,6,10-trimethyltridecanoate was attractive to *A. griseus* females, but not to males, in a Y-tube olfactometer (Fávaro et al. 2012). In addition to the male-produced sex pheromone, Fávaro et al. (2012) identified
defensive compounds from *A. griseus* (See Allomones, **Section 15.3**).

15.2.2 *Bagrada hilaris* Burmeister [Pentatomidae: Pentatominae: Strachiini]

Bagrada hilaris, known as the painted bug or bagrada bug, is native to Africa but has spread to south
Asia and the Mediterranean region and, within the past decade, to the southwestern United States.

B. hilaris feeds primarily on plants in the Brassicaceae (mustard family) and Capparaceae (caper family) and is extremely damaging to cole crops, bok choi, and capers (Guarino et al. 2008, Reed et al. 2013, Biovision Foundation 2014). Guarino et al. (2008) studied possible volatile and contact pheromones that might mediate mating behaviors. In Y-tube bioassays, female *B. hilaris* were attracted to odors from males and also to a hexane extract of volatile compounds collected from males, whereas odors of females did not attract either gender. GC-MS analyses showed that both males and females produce nonanal, decanal, and (*E*)-2-octenyl acetate, the latter produced in substantially higher quantities by males than females. Thus, (*E*)-2-octenyl acetate was suggested to be a long-range mate-location pheromone, but to date, results of field bioassays have not been published. Interestingly, in the alydid bug *Leptocorisa chinensis*, a blend of (*E*)-2-octenyl acetate and octanol is also produced by both sexes but is attractive only to males (Leal et al. 1996).

In studies of possible short-range or contact pheromones using open arena bioassays, males displayed characteristic courtship behaviors in the presence of virgin females (Guarino et al. 2008). Courtship behavior also was displayed in response to females killed by freezing, but not to freeze-killed females washed with hexane, suggesting a nonpolar contact pheromone in the cuticular lipids. De Pasquale et al. (2007) identified thirteen homologous cuticular *n*-alkanes (nC_{17}-nC_{29}) from both sexes. The hydrocarbon profiles of males and females were qualitatively similar, but marked sex-specific quantitative differences were observed for some of the components, which were suggested to be involved in mate recognition (De Pasquale et al. 2007). However, bioassays with pure compounds or blends of compounds have not yet been reported.

15.2.3 *Chinavia hilaris* (Say), *Chinavia ubica* (Rolston), and *Chinavia impicticornis* (Stål) [Pentatomidae: Pentatominae: Nezarini]

Blassioli-Moraes et al. (2012) found that males of two sympatric neotropical stink bug species, *Chinavia ubica* and *C. impicticornis*, both produced epoxybisaboladienes, which attracted conspecific females in laboratory assays. Diastereomeric (2*S*,3*R*,6*S*,7*Z*)-2,3-epoxy-7,10-bisaboladiene (**1**) and (2*R*,3*S*,6*S*,7*Z*)-2,3-epoxy-7,10-bisaboladiene (**2**) had been identified previously as sex pheromone components for *C. hilaris* (= *Acrosternum hilare*) and *Nezara viridula* (L.) with different ratios of **1** and **2** providing species specificity (reviewed in Millar 2005). Microprobe NMR spectra of volatiles from male *C. ubica* showed that they produced a 9:1 mixture of **1** and **2**, whereas *C. impicticornis* emitted almost exclusively epoxide **2** (**Figure 15.1**).

The absolute configurations of epoxides **1** and **2** were established by oxidative cleavage to 4-acetyl-1,2-epoxy-1-methylcyclohexanes **3** and **4** followed by comparison of their retention times on a chiral stationary phase β-DEX GC column with those of standards synthesized from commercially available (-)- and (+)-limonenes (Blassioli-Moraes et al. 2012). Interestingly, bioassay data suggested that the absolute configurations of the epoxides appeared to be less important for conspecific recognition than the relative configurations.

Chinavia hilaris also represents an example of a species that is attracted to a compound that it does not produce but which is a pheromone of another stink bug species. Thus, Aldrich et al. (2007, 2009) demonstrated that *C. hilaris* was attracted to methyl (2*E*,4*E*,6*Z*)-2,4,6-decatrienoate, and that this attraction was even stronger than that to the 95:5 blend of epoxides **1** and **2** identified as the natural pheromone blend of this species (McBrien et al. 2001). Tillman et al. (2010) confirmed this cross-attraction although

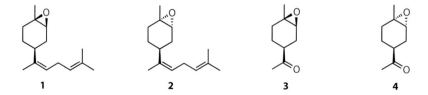

    1   **2**   **3**   **4**

FIGURE 15.1   *Chinavia ubica* and *Chinavia impicticornis* sex pheromone components **1** and **2**, and products of cleavage thereof, **3** and **4**, used to determine the absolute configurations of **1** and **2**.

also finding that the *C. hilaris* pheromone was not attractive, at least under the conditions in which it was tested. Consequently, methyl (2*E*,4*E*,6*Z*)-2,4,6-decatrienoate is suggested as the best available lure for monitoring this species.

15.2.4 *Chlorochroa sayi* (Stål), *Chlorochroa uhleri* (Stål), and *Chlorochroa ligata* (Say) [Pentatomidae: Pentatominae: Nezarini]

Chlorochroa is primarily a New World genus with about 35 known species, several of which are occasional pests of fruit and field crops in the western United States (Millar et al. 2010, Rider 2015). Ho and Millar (2001a, b) identified the pheromones for three species, *Chlorochroa ligata*, *C. sayi*, and *C. uhleri*. Millar et al. (2010) carried out field trials with reconstructed blends of the male-produced compounds using custom-made cylindrical screen traps. In each case, males as well as females were trapped, although only females had responded to these compounds in laboratory bioassays (Ho and Millar 2001a, b). The field trials confirmed the previously reported findings of Ho and Millar (2001b) that the main component of the *C. uhleri* male-specific volatiles, methyl (3*R*,6*E*)-2,3-dihydrofarnesoate (deployed as the racemic methyl (6*E*)-2,3-dihydrofarnesoate), was the only component attractive to adults in field assays. Addition of the two minor male-produced components did not affect catches nor did addition of synthetic odors of alfalfa, a host plant. When *C. uhleri* pheromone was combined with both the 3-component pheromone of *C. sayi* and the single-component pheromone of *Nezara viridula*, trap captures were reduced, indicating interference by one or both of these pheromones. For *C. sayi*, all three components of its pheromone (methyl geranate, methyl citronellate, and methyl (6*E*)-2,3-dihydrofarnesoate) were required for maximal attraction. In contrast to *C. uhleri*, captures of male and female *C. sayi* were enhanced, not reduced, by combination with *C. uhleri* and *N. viridula* pheromones, compared to its 3-component pheromone alone (Millar et al. 2010). These results highlight the asymmetry of cross-species attraction as well as the complexity in determining the attraction of males and females to different blends of pheromones and other components in laboratory versus field bioassays. Furthermore, for the *C. sayi* components, differences in volatility produced varying ratios of pheromone components as the multicomponent lure aged in the field, highlighting the challenges in obtaining stable semiochemical release ratios from lures for extended periods of time.

15.2.5 *Edessa meditabunda* (F.) [Pentatomidae: Edessinae]

The genus *Edessa* is one of the largest in the Pentatomidae, containing approximately 260 described species, but information on possible pheromonal communication in these species is sparse. Zarbin et al. (2012) identified a male-produced sex pheromone from *Edessa meditabunda*, a major pest of soybeans in Brazil, as methyl 4,8,12-trimethylpentadecanoate (**5**) (**Figure 15.2**). The absolute configuration of the pheromone has not been determined. Methyl 4,8,12-trimethyltetradecanoate also was present as a minor component in the volatiles collected from males, but it did not elicit electroantennogram responses nor did it improve attraction of females to **5** in a laboratory assay. Therefore, it does not appear to be a component of the pheromone. Ester **5** bears strong similarities to methyl 2,6,10-trimethyltridecanoate and methyl 2,6,10-trimethyldodecanoate, sex pheromone components of *Euschistus heros* (Aldrich et al.

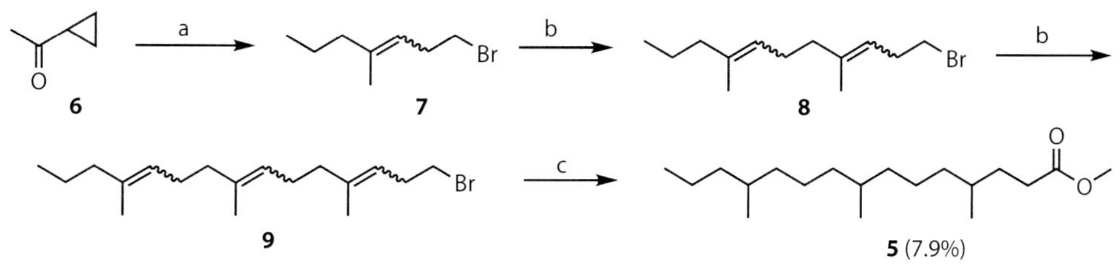

FIGURE 15.2 Non-stereoselective synthesis of male-produced sex pheromone of *Edessa meditabunda*. a) 1. C$_3$H$_7$MgBr, 2. HBr; b) 1. Mg, 2. **6**, 3. HBr; c) 1. NaOAc/HMPA, 2. H$_2$/Pd, 3. LiAlH$_4$, 4. CrO$_3$, 5. CH$_2$N$_2$.

1991) and *E. obscurus* (Borges and Aldrich 1994), respectively. A non-stereoselective synthesis of *E. meditabunda* pheromone **5** (**Figure 15.2**) started from cyclopropyl methyl ketone (**6**) and used the Julia cyclopropane cleavage in three iterative steps to place all three methyl groups (**Figure 15.2**), in a manner similar to syntheses of other methyl-branched bug pheromones (reviewed in Millar 2005). Interestingly, the mixture of all eight possible stereoisomers represented by synthetic **5** was attractive to female *E. meditabunda* in a Y-tube bioassay, demonstrating that the unnatural stereoisomers were not inhibitory (Zarbin et al. 2012).

15.2.6 *Eocanthecona furcellata* (Wolff) [Pentatomidae: Asopinae]

Ho et al. (2003, 2005) investigated volatile compounds from metathoracic glands, dorsal abdominal glands, and abdominal setae of the Asian generalist predatory stink bug *Eocanthecona furcellata*. Metathoracic glands of both male and female bugs contained (*E*)-2-decenal as the major component with lesser amounts of (2*E*,4*E*)-2,4-decadienal, (2*E*,4*Z*)-2,4-decadienal and (*E*)-2,9-decadienal. The amount of (*E*)-2,9-decadienal in gland extracts from female *E. furcellata* was about five times higher than in extracts from males, and, conversely, the amount of (2*E*,4*Z*)-2,4-decadienal was higher in males than in females. (*E*)-2,9-Decadienal has been reported as an attractant for males of the anthocorid bug *Orius insidiosus* (Say) (Aldrich and Puapoomchareon 1996). In *E. furcellata*, there was no sexual dimorphism in the morphology of the dorsal abdominal glands of adults, even though their contents were different between the sexes. Geraniol was found only in males, and limonene and α-terpineol were found only in females, whereas linalool oxide isomers linalool, neral, and geranial were found in both sexes. The dorsal abdominal gland contents of nymphs also were analyzed, with 4-oxo-(*E*)-2-hexenal, 4-oxo-(*E*)-2-octenal, tridecane, and tetradecanal being major components. Male bugs but not females possessed abdominal setae, and 6,10,13-trimethyltetradecyl isovalerate (stereochemistry not characterized) was identified in extracts of the setae of males. Ho et al. (2005) investigated production of this male-specific compound under various rearing conditions with the intriguing result that production was strongly dependent on rearing density, decreasing over 100-fold as rearing density increased. This effect was partially reversible and did not depend on the sex of the accompanying bugs. The density cues (e.g., tactile, olfactory, or visual) that regulate production of the compound are not known (Ho et al. 2005).

15.2.7 *Euschistus conspersus* Uhler, *Euschistus heros* (F.), and *Euschistus servus* (Say) [Pentatomidae: Pentatominae: Carpocorini]

The genus *Euschistus* consists of >80 species, and several are pests of tree fruit, tomato, soybean, cotton, and other crops in the New World (McPherson and McPherson 2000). Species for which pheromones are known fall into two distinct groups, one group based on methyl decadienoates, the other on trimethyltridecanoates (Millar 2005). Based on this knowledge, several practical advances have been made in these bugs' monitoring and management.

15.2.7.1 *Euschistus conspersus*: Pheromone-Based Monitoring

Aldrich et al. (1991) identified methyl (2*E*,4*Z*)-2,4-decadienoate as the major component of the aggregation pheromone of the consperse stink bug, *Euschistus conspersus*. In tomato fields in California, Cullen and Zalom (2005) discovered that responses of males and females differed by crop stage, and that the onset of female-biased pheromone trap catches could be used as a biofix, which could then be used to calculate degree-days to predict subsequent nymphal development in the field to optimally time insecticide treatments targeting this susceptible stage (Cullen and Zalom 2006). Such models give growers of processing tomatoes more flexibility to use insecticides with lower risks of remaining residues at harvest (Cullen and Zalom 2007). Krupke et al. (2001, 2006) attracted *E. conspersus* to pheromone-baited mullein plants on the borders of apple orchards in Washington state, but efforts to trap the bugs were not successful. These results, and similar results from other species, highlight the fact that over shorter ranges, phytophagous stink bugs probably use substrate-borne vibrational signals rather than volatile pheromones to locate conspecifics once they are on the same plant (Millar 2005). Thus, volatile pheromones may effectively attract bugs to the vicinity of a pheromone lure but may not induce bugs to enter traps.

15.2.7.2 *Euschistus heros*: Effects of Food on Pheromone Emission, and Pheromone-Based Monitoring

Moraes et al. (2008c) held 9-day-old *E. heros* males under 4 regimes: supplied with green beans, supplied with water only, held in humidified air, or held in dry air. During the following week, bean-fed bugs released all three pheromone components in the ratio reported by Zhang et al. (2003) for all seven days. In contrast, insects without food stopped releasing methyl 2,6,10-trimethyltridecanoate and methyl 2,6,10-trimethyldodecanoate after one day, only continuing emission of methyl (2*E*,4*Z*)-2,4-decadienoate, demonstrating the importance of diet as a strong influence on pheromone production and blend.

Borges et al. (2011) found that a transparent trap fitted with lures containing 1 mg of the *Euschistus heros* sex pheromone component methyl 2,6,10-trimethyltridecanoate (as a stereoisomeric mixture) was more effective and less time-consuming than the beat-cloth technique in monitoring soybean infestation by *E. heros* in central Brazil. Borges et al. (2011) also caught *Edessa meditabunda* and *Piezodorus guildinii* in the same traps, at >10% of total captures, but statistical comparisons to unbaited traps were not reported.

15.2.7.3 *Euschistus servus*: Use of Pheromone in Trap Crop

Tillman and Cottrell (2012) tested the effectiveness of a trap crop of grain sorghum, with and without pyramidal traps baited with methyl (2*E*,4*Z*)-2,4-decadienoate, for preventing movement of *E. servus* from corn and peanut fields into cotton crops in Georgia, United States. Although the trap crops reduced movement into cotton, the presence of the pheromone traps had no effect on the efficacy of the trap crop.

15.2.8 *Eysarcoris lewisi* (Distant) [Pentatomidae: Pentatominae: Eysarcorini]

Eysarcoris lewisi is one of the major pests of rice in northern Japan (Kiritani 2007). The possibility of using pheromone-baited traps for monitoring *E. lewisi* was examined by Takita (2007), who found that males reared under a long-day photoperiod attracted adults of both sexes and nymphs to water-pan traps in the field, in contrast to short-day males, which did not attract conspecifics. Mori (2007) demonstrated the pheromone to be either (2*Z*,6*R*,1′*S*,5′*S*)- or (2*Z*,6*R*,1′*R*,5′*R*)-2-methyl-6-(4′-methylenebicyclo[3.1.0]hexyl)hept-2-en-1-ol by synthesis and bioassay of four diastereomeric mixtures of isomers. Takita et al. (2008) isolated the pheromone and confirmed its attractiveness to both sexes and nymphs of *E. lewisi* in field trials. These workers also determined that the pheromone had the (2*Z*)-configuration from NMR spectra. Mori et al. (2008) and Tashiro and Mori (2008) then determined its absolute configuration to be (2*Z*,6*R*,1′*S*,5′*S*)-2-methyl-6-(4′-methylenebicyclo[3.1.0]hexyl)hept-2-en-1-ol (**10**) by syntheses of individual stereoisomers, using an enzyme-based kinetic resolution of an intermediate in a key step in the syntheses, shown in **Figure 15.3**.

Thus, (*R*)-citronellal was converted in several routine steps into diazo compound **14**. Reaction of **14** with copper and copper salts produced a carbene that intramolecularly inserted into the terminal alkene, producing bicyclic ketone **15**. Reduction of ketone **15** produced alcohols **16** and **17**, which were subjected to Lipase PS-D catalyzed acetylation, whereupon only **16** was acetylated to yield acetate **18**. The acetate **18** and unreacted alcohol **17** were then readily separated and converted to **19** and **20** respectively, the absolute configurations of which were determined by circular dichroism spectroscopy (Mori et al. 2008, Tashiro and Mori 2008). The conversion of ketone **20** to the final product **10** involved several steps, including Ando's Z-selective olefination (Ando 1998), producing stereoisomer **10** in 3.3% overall yield from (*R*)-citronellal. This isomer matched the compound produced by *E. lewisi* (Tashiro and Mori 2008). These authors reported that the LiB(*sec*-Bu)$_3$H (L-Selectride®) reduction of the diastereomeric mixture of ketones **15** strongly favored (>95%) delivery of the hydride from the face opposite the cyclopropane ring to avoid steric hindrance (Tashiro and Mori 2008). However, when Khrimian et al. (2011) used the same conditions

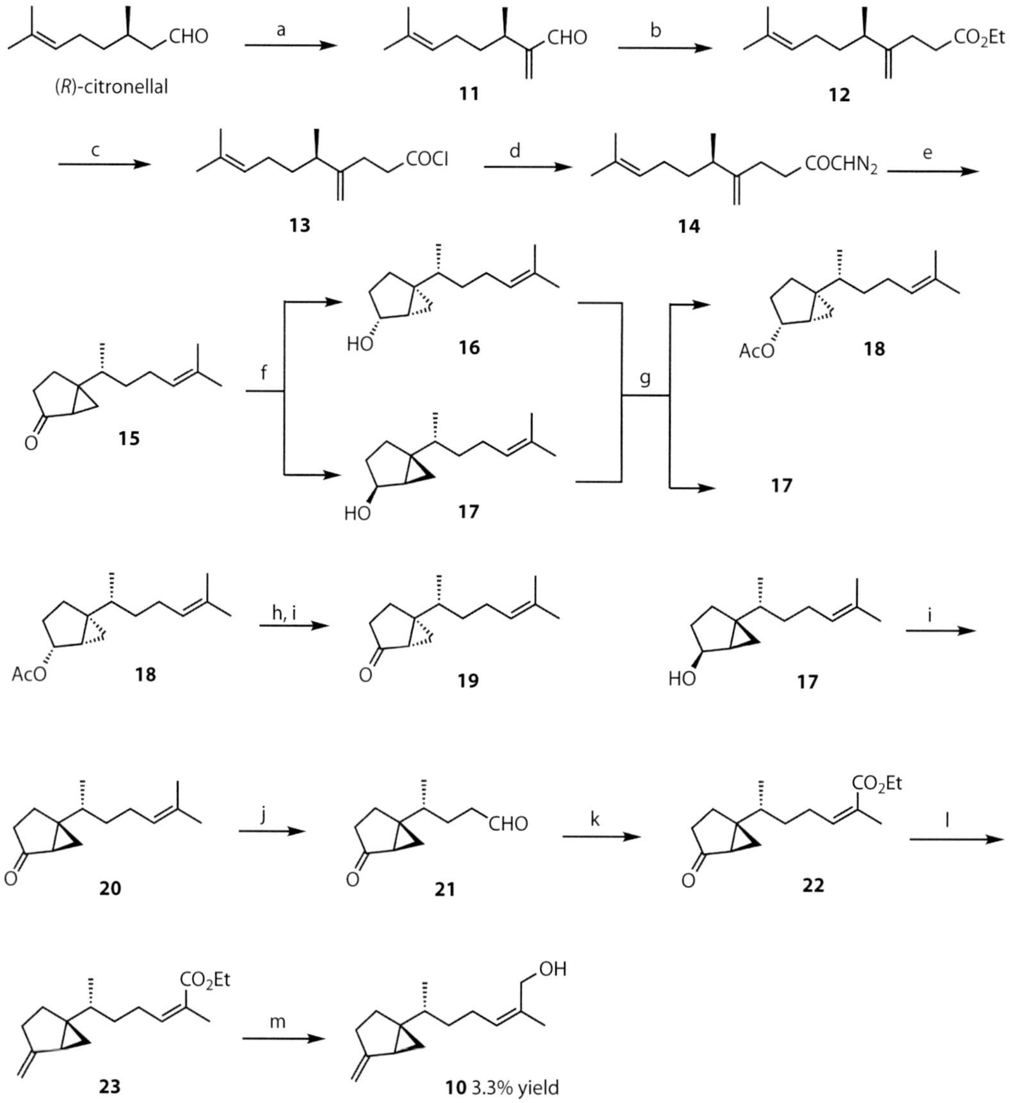

FIGURE 15.3 Synthesis of *Eysarcoris lewisi* pheromone **10** via Lipase PS-D diastereomeric resolution of bicyclic alcohols **16** and **17**. a) CH_2O, $EtCO_2H/Py$; b) 1. $LiAlH_4$, 2. $MeC(OEt)_3/EtCO_2H$; c) 1. KOH, 2. NaOEt, 3. $(COCl)_2$; d) CH_2N_2; e) Cu/$CuSO_4$; f) $LiB(sec-Bu)_3H$; g) Lipase PS-D/$CH_2=CHOAc$; h) KOH; i) TPAP/NMO; j) $OsO_4/NaIO_4$; k) $(o-MeC_6H_4O)_2P(O)CHMeCO_2Et/NaH$; l) $(C_6H_5)_3PMeBr/n-BuLi$; m) $(i-Bu)_2AlH$.

for the L-Selectride® reduction of **24,** they were not able to reproduce the >95% endo-selectivity reported by Tashiro and Mori (2008). Instead, reduction of ketone **24** with L-Selectride® (**Figure 15.4**) produced a ~77:23 mixture of the *endo*-alcohols (*ent*-**17** and *ent*-**16**) and the *exo*-alcohols **25** and **26** (Khrimian et al. 2011). The alcohols **25** and **26** were separated by silica gel chromatography and oxidized back to ketones *ent*-**20** and *ent*-**19** with (diacetoxyiodo)benzene in the presence of (2,2,6,6-tetramethyl-1-piperidinyl)oxyl (TEMPO) (**Figure 15.4**). Thus, the decreased selectivity in the L-Selectride® reduction of the diastereomeric mixture **24** fortuitously offered an expedient way of producing pure ketones *ent*-**20** and *ent*-**19**, obviating the need for the kinetic resolution step used by Tashiro and Mori (2008) (**Figure 15.3**).

The second asymmetric synthesis of the *E. lewisi* pheromone developed by Tashiro and Mori (2008) was entirely diastereoselective and employed Hodgson's intramolecular cyclopropanation (Hodgson et al. 2007). Alcohol **27**, prepared from (*R*)-citronellal in two steps (Mori 2007), was treated with

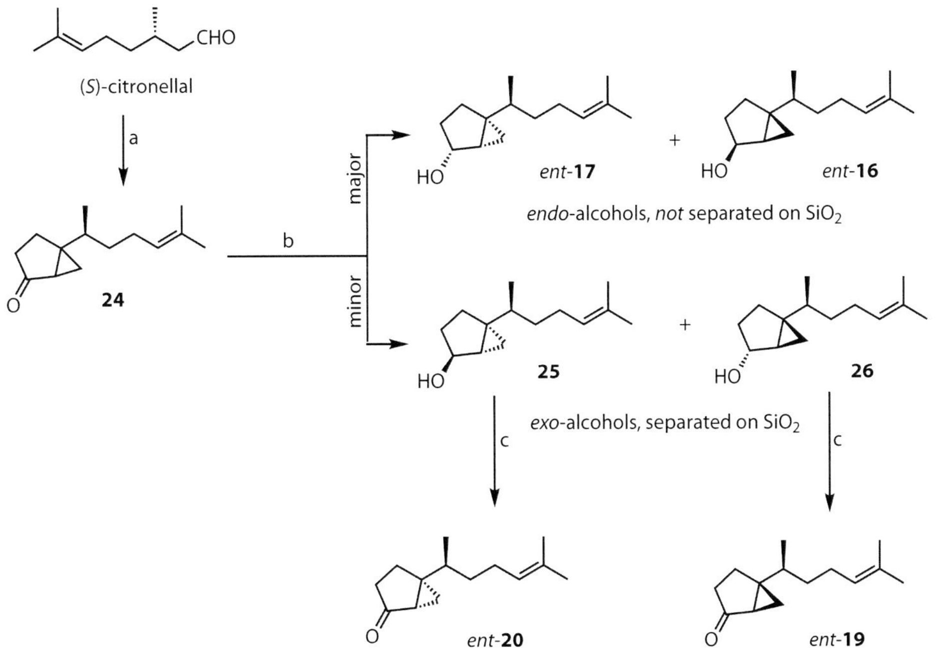

FIGURE 15.4 Diastereomeric resolution of ketone **24** via LiB(sec-Bu)₃H reduction, chromatographic separation, and oxidation. a) Mori (2007); b) LiB(sec-Bu)₃H; c) PhI(OAc)₂/TEMPO.

methanesulfonyl chloride to give chloride **28** via the intermediate mesylate (**Figure 15.5**). Reaction of the Grignard reagent from **28** with (*R*)-epichlorohydrin in the presence of CuI gave the key intermediate **29**. Hodgson's intramolecular cyclopropanation of **29** triggered by lithium 2,2,6,6-tetramethylpiperidide afforded *exo*-alcohol *ent*-**25** as a single stereoisomer. Oxidation of *ent*-**25** with TPAP and *N*-methylmorpholine *N*-oxide furnished ketone **20**, a late-stage intermediate of the *E. lewisi* pheromone in the previous synthesis (**Figure 15.3**). The yield of ketone **20** from (*R*)-citronellal in this six-step synthesis was 31% versus 11% in the previous 11-step synthesis (**Figure 15.3**).

The third synthesis of **10** was based on two key steps (**Figure 15.6**): asymmetric addition of allyl-zinc to an alkynyl carbonyl intermediate and diastereoselective, gold-catalyzed cycloisomerization of

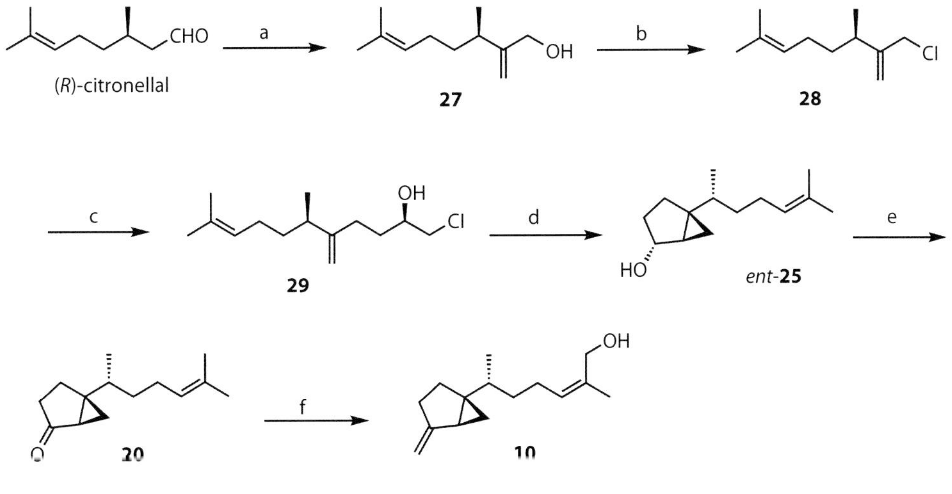

FIGURE 15.5 Asymmetric synthesis of *Eysarcoris lewisi* pheromone **10**. a) Mori (2007); b) MsCl/Py; c) 1. Mg, 2. (*R*)-epichlorohydrin/CuI; d) n-BuLi, 2,2,6,6-tetramethylpiperidine; e) TPAP/NMO; f) Tashiro and Mori (2008).

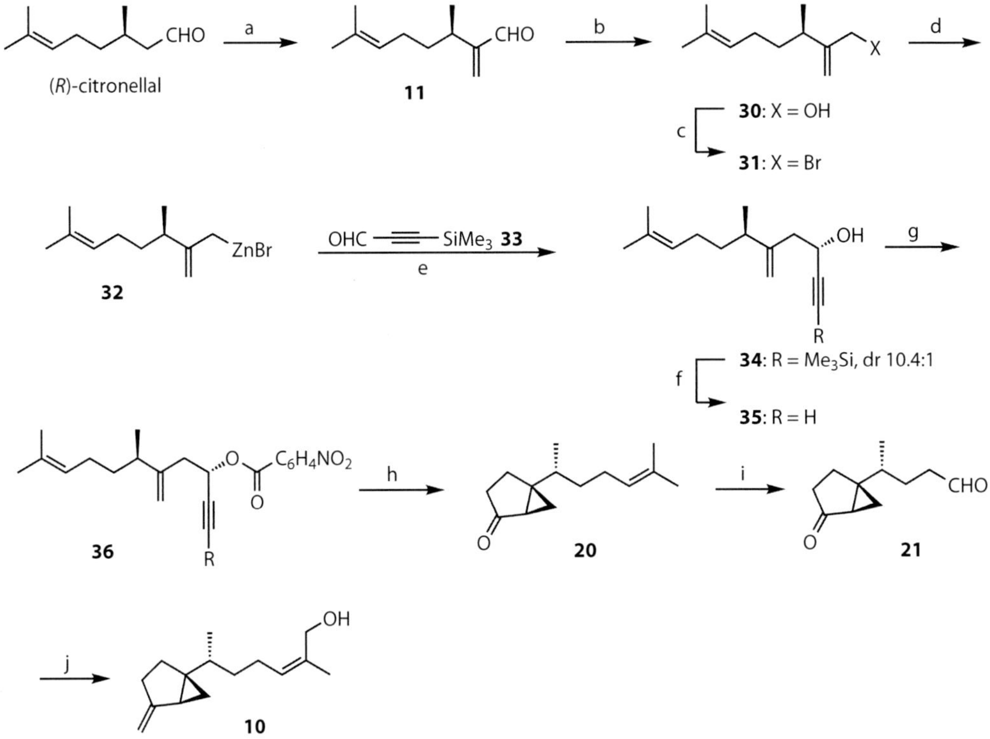

FIGURE 15.6 Stereocontrolled synthesis of *Eysarcoris lewisi* pheromone **10** through asymmetric addition of allylzinc **32** to acetylenic aldehyde **33** and gold-catalyzed cycloisomerization of chiral enynol **36**. a) CH$_2$O, EtCO$_2$H/Py; b) NaBH$_4$/CeCl$_3$; c) Br$_2$/Ph$_3$P; d) Zn; e) 2,2'-methylene-*bis*-[(4S)-4-phenyl-2-oxazoline]; f) K$_2$CO$_3$/MeOH; g) *p*-NO$_2$C$_6$H$_4$COCl/Py/DMAP; h) 1. AuCl$_3$/Py, 2. LiOH; i) O$_3$; j) Tashiro and Mori (2008).

allylpropargyl alcohols to bicyclo[3.1.1]hexanes (Fürstner and Schlecker 2008). Thus, the reaction of chiral allylzinc **32**, generated in several steps from (*R*)-citronellal, with 3-trimethylsilylpropynal (**33**) in the presence of a chiral, deprotonated bisoxazoline ligand under Nakamura reaction conditions (Nakamura et al. 1998) at -100°C was appreciably enantioselective and afforded (3*S*)-enynol **34** (R = Me$_3$Si) with ~ 10:1 dr. At the more convenient -78°C, the reaction gave a somewhat lower 8.8:1 diastereomeric ratio and without the chiral ligand, it was completely non-diastereoselective. After removal of the Me$_3$Si protecting group, propargylic alcohol **35** was converted to the *p*-nitrobenzoate **36** followed by AuCl$_3$(Py)-catalyzed cycloisomerization to produce ketone **20** (dr 19:1) after hydrolysis of the initially formed enol ester (Fürstner and Schlecker 2008). Cleavage of ketone **20** with ozone gave aldehyde **21**, which was converted to **10** in three steps as described previously (**Figure 15.3**). Availability of multiple synthetic procedures for this unique sesquiterpene stink bug pheromone with a bicyclo[3.1.1]hexane skeleton may afford sufficient material to fully assess its biological activity. The biological activity has been studied with the natural pheromone isolated from *E. lewisi* (Takita et al. 2008), and diastereomeric mixtures (Mori 2007), and with synthetic stereoisomer **10,** showing that it attracted significantly more bugs than the diastereomer with (2*Z*,6*R*,1'*R*,5'*R*) configuration, which appeared inactive. Furthermore, synthetic **10** was as attractive as natural extract or live males. Intriguingly, neither synthetic **10** nor natural extracts attracted any males in field tests. Furthermore, isomers with the (2*Z*,6*S*) configuration appeared neither to inhibit nor enhance attraction to **10** (Mori et al. 2008).

15.2.9 *Halyomorpha halys* (Stål) [Pentatomidae: Pentatominae: Cappaeini]

The brown marmorated stink bug, *Halyomorpha halys*, first identified in the United States in Pennsylvania in 2001 (Hoebeke and Carter 2003), has now been found in 41 states and Canada. It also has successfully invaded Europe, and now is established in Switzerland, France, Italy, Greece, and Hungary (Haye et al. 2015). It already has caused significant economic damage to agriculture in >12 of the United States, in addition to its nuisance status as an invader of homes and other buildings for over-wintering (Northeastern IPM Center 2015). The species is native to and widespread in northeastern Asia; in its native and introduced ranges, *H. halys* feeds on a wide variety of fruit crops (Lee et al. 2013), and, not surprisingly, it has become a significant agricultural pest in the United States (Leskey et al. 2012a,b).

15.2.9.1 Aggregation Pheromone

Khrimian et al. (2014a) identified a male-produced aggregation pheromone of *H. halys* as a 3.5:1 mixture of two epoxides, (3*S*,6*S*,7*R*,10*S*)-10,11-epoxy-1-bisabolen-3-ol (**37**) and (3*R*,6*S*,7*R*,10*S*)-10,11-epoxy-1-bisabolen-3-ol (**38**) (**Figure 15.7**). The basic skeletons of the pheromone stereoisomers first were identified by comparison to spectral data from mixtures of the stereoisomers generated during the identification of the pheromone of another stink bug, *Murgantia histrionica* (Zahn et al. 2008, *vide infra*). The specific stereoisomers produced by *H. halys* then were identified by synthesis of stereoisomeric libraries of 10,11-epoxy-1-bisabolen-3-ols and screening those on two enantioselective gas chromatography columns (Khrimian et al. 2014a,b). A novel and straightforward route to all stereoisomers of 1,10-bisaboladien-3-ol and 10,11-epoxy-1-bisabolen-3-ol was developed (Khrimian et al. 2014a; **Figure 15.7**) via the rhodium-catalyzed asymmetric addition of trimethylaluminum (Siewert et al. 2007) to diastereomeric mixtures of cyclohex-2-enones **39** and **48**. Ketones **39** and **48** were synthesized from (*R*)- and (*S*)-citronellals, respectively, using 1,4-conjugate addition to methyl vinyl ketone mediated by diethylamino(trimethyl)silane, followed by intramolecular cyclization with potassium hydroxide catalyzed by tetrabutylammonium hydroxide (Hagiwara et al. 2002). **Figure 15.7**.

Addition of trimethylaluminum to ketone **39** in the presence of chloro(1,5-cyclooctadiene)rhodium(I) dimer ([Rh(cod)Cl]$_2$) and (*R*)-BINAP (**Figure 15.7**, top) yielded stereoisomers **40** and **41** as major and minor products, respectively. The selective formation was attributed to relatively facile delivery of the methyl group to the sterically less hindered bottom face of the (6*S*,7*R*)-stereoisomer of **39** to give (3*S*,6*S*,7*R*)-1,10-bisaboladien-3-ol (**40**), whereas the reaction with sterically less favorable (6*R*,7*R*)-**39** to give (3*S*,6*R*,7*R*)-stereoisomer **41** proceeded in low yield and modest diastereoselectivity. Alcohol **40** was stereoselectively dihydroxylated with AD-mix-β, and the resulting triol **42** was converted to epoxybisabolenol **37**, the main pheromone component of *H. halys* (Khrimian et al. 2014a) and *M. histrionica* (Khrimian et al. 2014b). Dihydroxylation of **40** with AD-mix-α, followed by cyclization of intermediate triol **43**, provided epoxybisabolenol **44**, the minor aggregation pheromone component of *M. histrionica* (discussed below).

Addition of trimethylaluminum to **39** in the presence of [Rh(cod)Cl]$_2$ and (*S*)-BINAP (**Figure 15.7**, middle) led to the stereoisomers **45** and **46**, from predominant delivery of the methyl group from the top face, with (6*R*,7*R*)-**39** being sterically more favorable than (6*S*,7*R*)-**39**. The minor *trans*-stereoisomer **46** was dihydroxylated with AD-mix-β and converted as described above to epoxybisabolenol **38** that matched a minor component of the *H. halys* pheromone blend.

The stereochemistries of the addition of trimethylaluminum to the ketone **48** (**Figure 15.7**, bottom) with the (7*S*)-configuration in the presence of [Rh(cod)Cl]$_2$ and (*S*)- and (*R*)-BINAP were governed essentially by the same rules as described above for the (7*R*)-ketone **39**. One notable application in this series is the synthesis of (3*R*,6*R*,7*S*)-1,10-bisaboladien-3-ol (= zingiberenol) (**50**), which has been identified as a sex pheromone component of the rice stink bug, *Oebalus poecilus* (de Oliveira et al. 2013), but the compound's synthesis was not described.

The assignments of the relative and absolute configurations of the stereoisomeric 1,10-bisaboladien-3-ols and 10,11-epoxy-1-bisabolen-3-ols were carried out by single-crystal X-ray crystallography of underivatized (3*R*,6*S*,7*R*,10*S*)- and (3*S*,6*R*,7*R*,10*S*)- triols and correlations with stereodefined 1,3,10-bisabolatrienes (= zingiberenes) and 1,3(15),10-bisabolatrienes (= sesquiphellandrenes) (Khrimian et al. 2014a,b). Thus, 1,10-bisaboladien-3-ols and 10,11-epoxy-1-bisabolen-3-ols, having lower thin layer chromatography R$_f$ values and longer GC retention times (**38**, **41**, **46**, and **49**), were found to have the *trans* relative configuration. Conversely, the less polar, shorter retention time stereoisomers **37**, **40**, **44**, **45**, and **50** had the *cis* configuration.

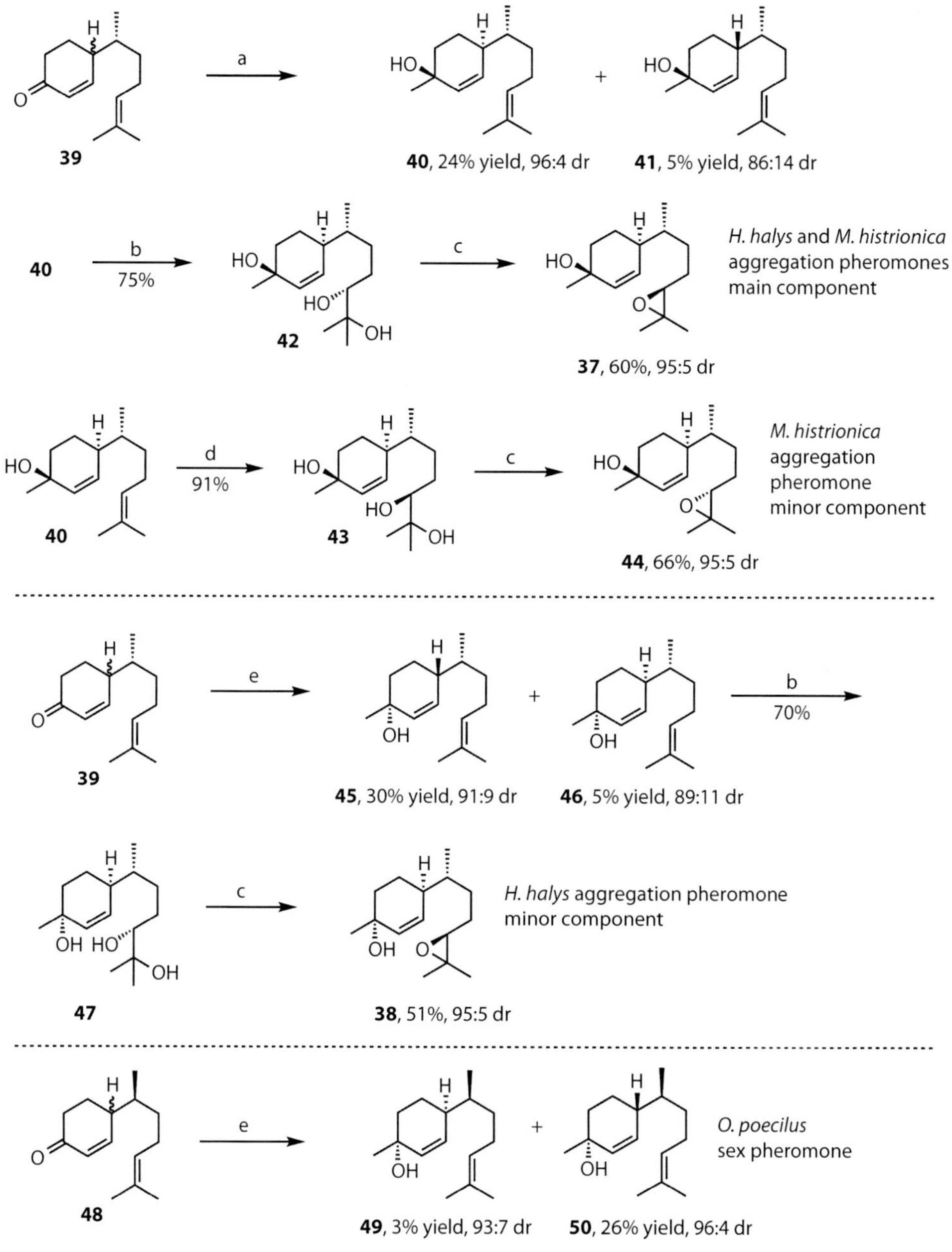

FIGURE 15.7 Syntheses of individual stereoisomers of 1,10-bisaboladien-3-ols and 10,11-epoxy-1-bisabolen-3-ols. a) Me$_3$Al/[Rh(cod)Cl]$_2$/(*R*)-BINAP/SiO$_2$ separation; b) AD-mix-b; c) 1. MsCl/Py, 2. KOH/MeOH; d) AD-mix-a; e) Me$_3$Al/[Rh(cod)Cl]$_2$/(*S*)-BINAP/SiO$_2$ separation.

The identifications of aggregation pheromone components of *H. halys* and *M. histrionica* were accomplished by careful matching of GC retention times of components in aeration extracts of male bugs with standards of known relative and absolute stereochemistries from the stereoisomeric library of 10,11-epoxy-1-bisabolen-3-ols, on two enantioselective columns (Chiraldex G-TA and Hydrodex β-6TBDM; Khrimian et al. 2014a,b).

In field bioassays, both the major (**37**) and the minor (**38**) pheromone components attracted adult *H. halys* of both sexes. Epoxide **37** was more attractive than the minor component **38**, whereas the mixture at the natural 3.5:1 ratio was more attractive than either component alone. The mixture and the major component also attracted nymphs (Khrimian et al. 2014a). Furthermore, lures prepared from mixtures of stereoisomers also were attractive (Khrimian et al. 2014a, Weber et al. 2014a, Leskey et al. 2015b), demonstrating that the presence of additional stereoisomers apparently does not hinder attraction of *H. halys*. This may make it economically feasible to produce isomer mixtures of aggregation pheromone for *H. halys* management. In an effort to simplify large-scale production and reduce the cost of the *H. halys* pheromone, Leskey et al. (2015b) reported a synthesis of "technical grade" pheromone from either (*R*)- or racemic citronellals using only one distillation and no chromatography throughout the whole synthesis. The resulting mixtures of isomers attracted bugs in field bioassays (Leskey et al. 2015b), with or without the synergist methyl (2*E*,4*E*,6*Z*)-2,4,6-decatrienoate (*vide infra*), and were used to monitor populations of *H. halys* in ten states across the United States (Leskey et al. 2015a). Although systematic studies of lure dose and formulations have not yet been conducted, rubber septa impregnated with 10.7 mg of more purified pheromone, or 31 mg of the "technical grade" mixture of eight isomers produced from the large-scale synthesis from (*R*)-citronellal (both containing ~2 mg of the most active pheromone component **37**) remained attractive for 2–3 weeks in field trials (Leskey et al. 2015a,b). Controlled-release formulations of *H. halys* pheromone are currently being developed by several semiochemical companies (A.K., pers. comm.)

15.2.9.2 Attraction of *Halyomorpha halys* to the Pheromone of *Plautia stali*

It was discovered serendipitously that *H. halys* is attracted to methyl (2*E*,4*E*,6*Z*)-2,4,6-decatrienoate (Tada et al. 2001a,b; Lee et al. 2002), the aggregation pheromone of the brown-winged green stink bug, *Plautia stali* (Sugie et al. 1996), with which it is sympatric in northeast Asia. This cross-attraction was confirmed in the United States after *H. halys* became established (Aldrich et al. 2007, Khrimian et al. 2008). Although this compound is somewhat unstable under field conditions, Khrimian et al. (2008) demonstrated that about 80% of methyl (2*E*,4*E*,6*Z*)-2,4,6-decatrienoate could be preserved on rubber septum dispensers if the dispensers were shielded from direct sunlight (see also section below for *Thyanta perditor*). However, field studies showed that preventing isomerization of methyl (2*E*,4*E*,6*Z*)-2,4,6-decatrienoate was not essential because *H. halys* adults and nymphs were still attracted to mixtures of methyl 2,4,6-decatrienoate isomers, indicating that other *cis-trans* isomers were not antagonistic. Furthermore, methyl (2*Z*,4*E*,6*Z*)-2,4,6-decatrienoate and methyl (2*E*,4*Z*,6*Z*)-2,4,6-decatrienoate (pheromone of pentatomid *Thyanta* spp.) attracted *H. halys*, possibly through isomerization to the (2*E*,4*E*,6*Z*)-isomer. Alternatively, one or more of these other isomers may be attractive in their own right. Until the actual pheromone was identified, (2*E*,4*E*,6*Z*)-2,4,6-decatrienoate was the primary lure used in traps to monitor *H. halys* after it became established in the United States (Leskey et al. 2012b). However, the lures primarily were attractive in the late season in both Japanese and American studies, limiting their value for monitoring during a major part of the growing season (Funayama 2008, Leskey et al. 2012b). Funayama (2008) assessed the nutritional status of captured *H. halys* and hypothesized that *H. halys* might be eavesdropping on *P. stali* pheromone as an indirect method of host location. Intriguingly, he found a strong response in early season to methyl (2*E*,4*E*,6*Z*)-2,4,6-decatrienoate only during a year (2001) of high populations, and scattered early-season response in years of moderate populations (2002 and 2006). Weber et al. (2014a) reported analogous results from field trials conducted in Maryland in April 2012.

The identification of the pheromone actually produced by *H. halys* males offered new opportunities for management of this species. Weber et al. (2014a) found that a combination of a mixed-isomer preparation of the *H. halys* pheromone with methyl (2*E*,4*E*,6*Z*)-2,4,6-decatrienoate acted synergistically. In season-long trials using pyramid traps, lures containing both the *H. halys* pheromone and the decatrienoate caught 1.9 to 3.2 times more adults, and 1.4 to 2.5 times more nymphs than would be expected from additive effects of the compounds deployed individually. The pattern of captures of males and females was similar. Mixed-isomer lures derived from (*R*)-citronellal, containing eight stereoisomers of 10,11-epoxy-1-bisabolen-3-ol, (and including the two pheromone components), were as effective at attracting adults and nymphs with or without methyl (2*E*,4*E*,6*Z*)-2,4,6-decatrienoate, as were lures loaded only with pure **37** and **38** in the natural ratio. Captures also increased with dose for the mixed-isomer lures with and without accompanying methyl (2*E*,4*E*,6*Z*)-2,4,6-decatrienoate (Leskey et al. 2015b). These results suggest that a combination of relatively inexpensive semiochemicals could be used for detection, monitoring, and potentially control of this polyphagous invasive pest.

However, in a study in which commercial pyramid traps baited with methyl (2*E*,4*E*,6*Z*)-2,4,6-decatrienoate lures were tested in community gardens infested with *H. halys*, more bugs and greater damage were noted on tomato fruits near the traps (Sargent et al. 2014). Thus, judicious placement of traps some distance away from susceptible crops may be essential to prevent these unwanted effects.

15.2.10 *Murgantia histrionica* (Hahn) [Pentatomidae: Pentatominae: Strachiini]

The harlequin bug, *Murgantia histrionica*, is a pest of cole crops and related plants in the mustard family in the southern United States. Zahn et al. (2008, 2012) reported that sexually mature male *M. histrionica* produce an aggregation pheromone (dubbed murgantiol), which they identified as one of the stereoisomers of 10,11-epoxy-1-bisabolen-3-ol (**51**). ^{1}H and ^{13}C NMR spectra of murgantiol isolated from extracts of volatiles from males confirmed the proposed structure but did not determine the relative and absolute configurations of the four chiral centers in murgantiol. A relatively short and nonstereoselective synthesis of murgantiol from (*S*)-citronellal (Zahn et al. 2008) is shown in **Figure 15.8**.

Thus, (*S*)-citronellal was converted to cyclohexenone **48** via a tandem Michael addition/Robinson ring annulation (~1:1 mixture of diastereomers) following Chavan et al. (1997). Addition of methyllithium to enone **48** afforded a 38:62 mixture of less/more polar diastereomers **51**, which were separated by silica gel chromatography. Each pair of stereoisomers then was epoxidized with *m*-chloroperbenzoic acid to afford binary mixtures of less and more polar diastereomeric epoxybisabolenols **52**. The synthesis was repeated with (*R*)-citronellal to afford the four diastereomeric epoxybisabolenols **52** with the (7*R*)-configuration. The GC retention time of the male-specific compound isolated from *M. histrionica* matched that of one of the less polar stereoisomers of **52**, which were subsequently shown to have the *cis* relative configuration for the 1,4 substituents on the cyclohexene ring (Khrimian et al. 2014a). In Y-tube bioassays, female *M. histrionica* responded to compounds derived from both enantiomers of citronellal, further complicating the stereochemical assignment (Zahn et al. 2008). Khrimian et al. (2014b) repeated the collection of the male-produced volatiles of *M. histrionica* and discovered that they consisted of not one but two diastereomers of 10,11-epoxy-1-bisabolen-3-ol using two enantioselective columns, Chiraldex G-TA and Hydrodex β-6TBDM. Synthesis of all sixteen stereoisomers of 10,11-epoxy-1-bisabolen-3-ol allowed structural determination of the two components of the *M. histrionica* male-produced volatiles as the (3*S*,6*S*,7*R*,10*S*)- and (3*S*,6*S*,7*R*,10*R*)-stereoisomers **37** and **44** produced in a 1.4:1 ratio (see above under *H. halys*, and **Figure 15.7**) (Khrimian et al. 2014b).

In field trials, Weber et al. (2014b) demonstrated significant attraction of both sexes and nymphs to collard (*Brassica oleracea* L., acephala group) plants baited with either pheromone component compared to plants with no lure added. The combination of the two stereoisomers **37** and **44** was most attractive to

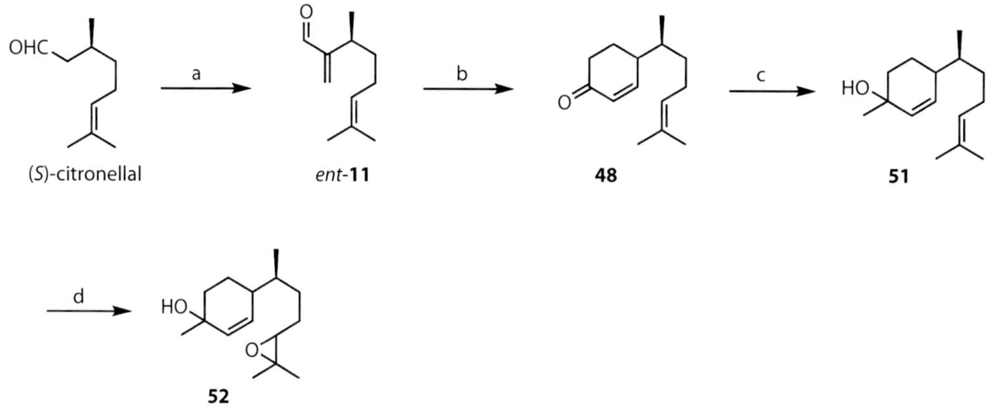

FIGURE 15.8 Synthesis of *Murgantia histrionica* pheromone (murgantiol) as a mixture of diastereomers from (*S*)-citronellal. a) HCHO/Piperidine; b) CH$_3$COCH$_2$COOCH$_3$/CH$_3$ONa; c) CH$_3$Li; d) *m*-CPBA/NaHCO$_3$.

adults and nymphs in the naturally occurring ratio of 1.4:1, and all blends generally were more attractive than either single component (Weber et al. 2014b). Mixed-isomer preparations of 10,11-epoxy-1-bisabolen-3-ol synthesized from (*R*)-citronellal (identical to lures for *H. halys*, see Leskey et al. 2015b) also were attractive to both adults and nymphs, and presence of the collard host plant increased attraction (Weber et al. 2014b). Cabrera Walsh et al. (2015), using marked adults, showed that whereas collard plants baited with mixed-isomer preparations of 10,11-epoxy-1-bisabolen-3-ol synthesized from (*R*)-citronellal attracted more bugs than unbaited plants, they did not retain them any longer than unbaited plants. Given the importance of retention of target insects in a trap crop (Holden et al. 2012), these findings have important implications for use of *M. histrionica* aggregation pheromone in trap-cropping schemes.

Murgantia histrionica is a specialist on plants in the families Brassicaceae and Capparaceae. Isothiocyanates are produced by breakdown of glucosinolate compounds within these plants and, also are sequestered by the bugs, rendering them distasteful and repellent to predators as reflected by the bugs in their aposematic coloration (Aliabadi et al. 2002). At least two common isothiocyanates—allyl and benzyl—are attractive to *M. histrionica* adults, and these plant volatiles may add to the attractiveness of the mixed-isomer murgantiol lures (Weber et al. unpublished).

15.2.11 *Nezara viridula* (L.) and *Nezara antennata* Scott [Pentatomidae: Pentatominae: Nezarini]

Nezara viridula is an important agricultural pest in warmer regions worldwide. It is thought to have originated from east Africa, with host plants in more than 30 plant families (Capinera 2001). The major pheromone components were identified as a mixture of *trans*-(*Z*)-bisabolene epoxide (**2**, **Figure 15.1**) and the corresponding *cis*-isomer **1** (**Figure 15.1**) (Aldrich et al. 1987, Baker et al. 1987). Cribb et al. (2006) described the unicellular glands on the abdominal sternites from which the pheromone is released. The pheromone attracts adults of both sexes and nymphs in the field (reviewed in Millar 2005). Females are attracted from a distance by the male-produced pheromone and then once on the same substrate, both sexes use substrate-borne vibrational signals and visual signals in shorter-range mate location and courtship (Čokl et al. 2007, Zgonik and Čokl 2014). Tillman et al. (2010), using yellow pyramid traps topped with clear chambers baited with a 3:1 *trans*- to *cis*-(*Z*)-bisabolene epoxide blend and containing a pyrethroid insecticide, showed that *N. viridula* can be trapped under field conditions, and that this attraction is dose-dependent. Shimizu and Tsutsumi (2011), using caged male bugs as lures with water pan traps, showed that *N. viridula* and *N. antennata* males attracted males and females of both species. *Nezara antennata* is native to Asia and is known to mate with *N. viridula* where they are sympatric. Dissections of *N. antennata* males and females attracted to *N. antennata* males showed that almost all were prereproductive and not well-fed, leading Shimizu and Tsutsumi (2012) to conclude that the adults might use the volatiles emitted by adult males as cues to find food.

15.2.12 *Oebalus poecilus* (Dallas) [Pentatomidae: Pentatominae: Carpocorini]

The small rice stink bug, *Oebalus poecilus*, is one of the most important rice pests in South America (Panizzi 1997) along with other *Oebalus* spp. and *Tibraca limbativentris* (see below). De Oliveira et al. (2013) showed that *O. poecilus* males produced an attractant pheromone, which was identified as one of the isomers of 1,10-bisaboladien-3-ol (= zingiberenol) by comparison with authentic standards synthesized from both enantiomeric citronellals using Hagiwara's method (Hagiwara et al. 2002). To determine the absolute configuration, the male-produced compound was dehydrated to zingiberene, the retention time of which on a chiral stationary phase β-DEX 325 column matched that of the known (-)-(6*R*,7*S*)-zingiberene (**Figure 15.9**) from ginger oil (Soffer and Burk 1985, Bhonsle et al. 1994,), proving the (6*R*,7*S*)-configuration. De Oliveira et al. (2013) also found that zingiberenol from *O. poecilus* coeluted with less polar diastereomers of (7*S*)-1,10-bisaboladien-3-ol synthesized from (*S*)-citronellal on a DB-5 column. Thus, de Oliveira et al. (2013) recorded the ^{13}C NMR spectrum of a two-component mixture of the less polar (7*S*)-1,10-bisaboladien-3-ols and found that the resonances of the quaternary carbons bearing hydroxyl and methyl groups were very similar to those in the ^{13}C NMR spectrum of (1*R*,4*R*)-menthenol (**Figure 15.9**), suggesting that the relative configurations of the two molecules were the same.

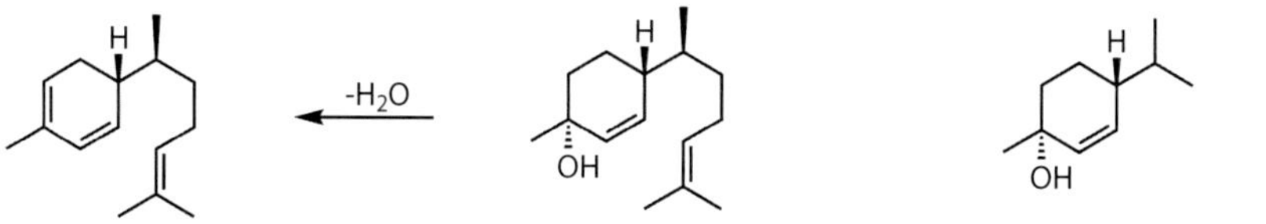

(-)-Zingiberene   *Oebalus poecilus* pheromone **50**   (1*R*,4*R*)-Menthenol

FIGURE 15.9   Structure of *Oebalus poecilus* pheromone **50** correlated with (-)-zingiberene and (1*R*,4*R*)-menthenol.

Given that the insect-produced compound had been shown to have the (6*R*,7*S*)-configuration as described above, the correlation with (1*R*,4*R*)-menthenol proved that the *O. poecilus* had to be (3*R*,6*R*,7*S*)-1,10-bisaboladien-3-ol (**50**) (**Figure 15.9**). Terhune et al. (1975) identified a zingiberenol in the essential oil of ginger (*Zingiber officinale* Roscoe) rhizomes without determining its absolute configuration. Khrimian et al. (2015) found not one but two stereoisomeric zingiberenols in ginger, bisaboladienol **50** and (3*S*,6*R*,7*S*)-1,10-bisaboladien-3-ol.

Using Y-tube olfactometer bioassays, de Oliveira et al. (2013) showed that the male-produced pheromone was attractive only to females. Also, females responded to odors from males only in the morning hours, and they responded as well to synthetic mixtures of isomers containing the identified pheromone **50** as to the odors from males. Results from field trials of zingiberenol **50**, either as a pure compound or as a mixture of isomers, have not yet been reported.

15.2.13 *Pallantia macunaima* Grazia [Pentatomidae: Pentatominae: Pentatomini]

Fávaro et al. (2013) identified the first ketone pentatomid pheromone, the male-produced sex pheromone of the important Brazilian soybean pest, *Pallantia macunaima*, as (6*R*,10*S*)-6,10,13-trimethyltetradecan-2-one ((6*R*,10*S*)-**53**). This compound elicited strong electroantennogram responses from antennae of females but not males. Male-produced volatiles, as well as the synthetic pheromone, were attractive to females, but not to males, in Y-tube olfactometer bioassays. The trivial name, pallantione, was assigned to this novel pheromone.

The chemical structure of pallantione was elucidated by a combination of GC-FTIR, GC-MS, and microchemical reactions (hydrogenation, LiAlH$_4$ reduction/silylation, formation of tosylhydrazone, then LiAlH$_4$ and LiAlD$_4$ reductions) and confirmed by synthesis. The mixture of all four stereoisomers of 6,10,13-trimethyltetradecan-2-one was synthesized from geranyl acetone through intermediate aldehyde **54** via Wittig olefination and hydrogenation (**Figure 15.10**) (Fávaro et al. 2013).

The individual stereoisomers of pallantione were prepared (**Figure 15.11**) from commercial (*R*)- and (*S*)-propylene oxides (Muraki et al. 2013). The key steps in these syntheses were S$_N$2 displacements of secondary tosylates with carbanions that proceeded with almost complete inversion at the chiral centers. Thus, to achieve chain elongation by two carbons and install a methyl group at the 3-position, tosylate **55** was treated with the enolate of dimethyl malonate to afford **56**, which was decarboxylated to give the ester **57** with a chiral methyl branch. Ester **57** then was reduced to the corresponding primary alcohol, and the latter was converted to iodide (*R*)-**58**. Analogously, (*R*)-propylene oxide was converted to iodide (*S*)-**58**. The other chiral building blocks were also made from (*R*)- and (*S*)-propylene oxides. Thus, reaction between (*R*)-propylene oxide and 1-butenylmagnesium bromide produced a secondary alcohol, which was then tosylated to give **59**. S$_N$2 reaction between **59** and the anion derived from methyl phenyl sulfone then gave (*S*)-**60** with a chiral methyl. Sulfone (*S*)-**60** was alkylated with iodide (*R*)-**58** to complete the carbon skeleton, followed by removal of the sulfone from **61** and conversion of the

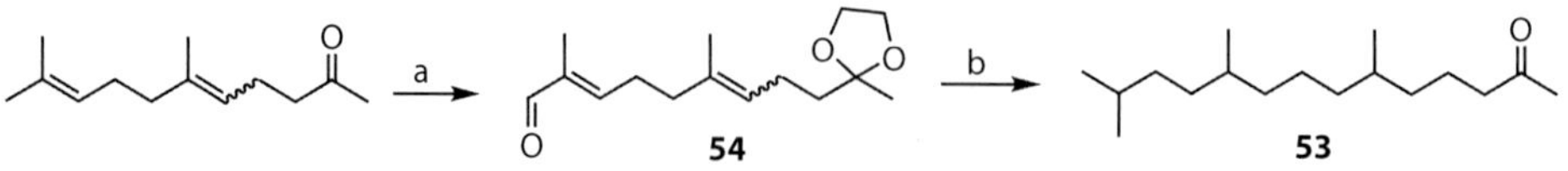

FIGURE 15.10   Non-stereoselective synthesis of *Pallantia macunaima* pheromone **53**. a) 1. HOCH$_2$CH$_2$OH/H$^+$, 2. SeO$_2$/t-BuOOH, 3. PCC; b) (CH$_3$)$_2$CHCH$_2$Ph$_3$Br/n-BuLi, 2. H$_2$/Pd-C; 3. HO$_2$CCO$_2$H/MeOH.

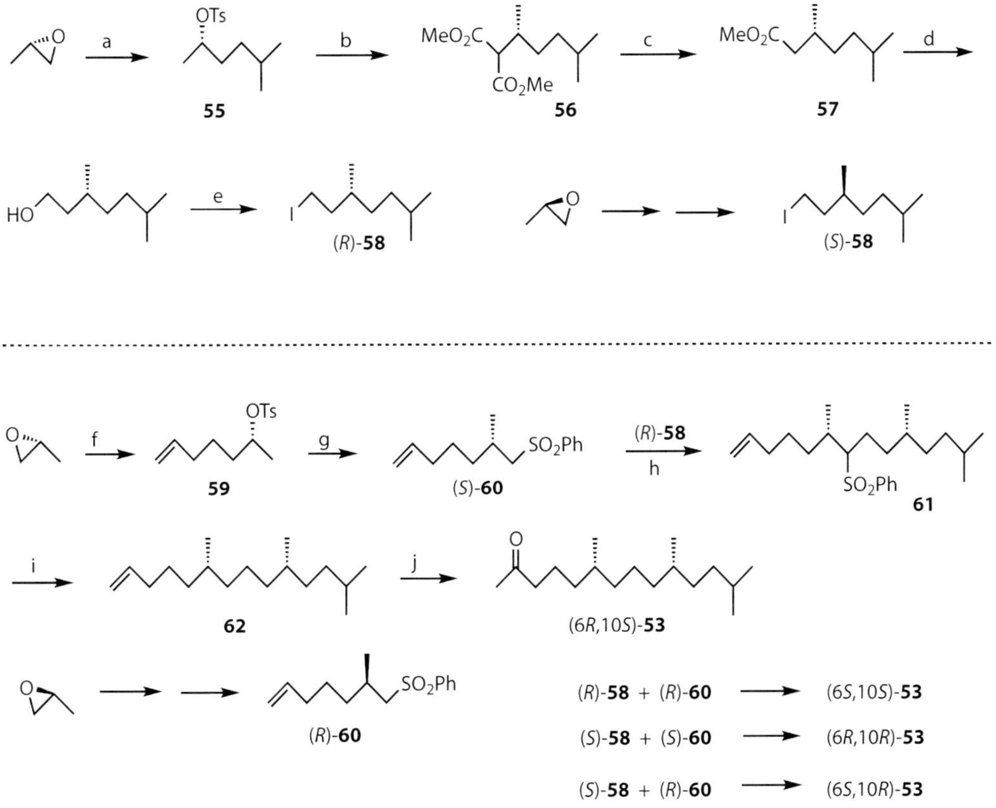

FIGURE 15.11 Syntheses of *Pallantia macunaima* sex pheromone (6R,10S)-**53** and its stereoisomers. a) 1. (CH$_3$)$_2$CHCH$_2$MgBr/Li$_2$CuCl$_4$, 2. TsCl/Et$_3$N; b) NaCH(CO$_2$Me)$_2$; c) LiCl/DMSO/H$_2$O, 150-160°C; d) LiAlH$_4$; e) I$_2$/PPh$_3$; f) 1. CH$_2$=CHCH$_2$CH$_2$MgBr/Li$_2$CuCl$_4$, 2. TsCl/Et$_3$N; g) PhSO$_2$Me/BuLi; h) BuLi/THF/DMI; i) Mg/MeOH; j) O$_2$/PdCl$_2$/CuCl.

terminal double bond in **62** to the desired methyl ketone, (6R,10S)-**53**. Analogously, sulfone (R)-**60** was prepared from (S)-propylene oxide (**Figure 15.11**), and the remaining three diastereomers [i.e., (6S,10S)-**53**, (6R,10R)-**53**, and (6S,10R)-**53**] were synthesized using the appropriate combinations of iodide **58** and sulfone **60** chiral building blocks (Muraki et al. 2013).

A partial resolution of the mixture of all four stereoisomers of 6,10,13-trimethyltetradecan-2-one **53** was achieved on a β-DEX chiral stationary phase GC column limiting the possible configurations of natural pallantione to the (6R,10S)- or (6S,10R)- isomers. Lastly, introducing an additional chiral center into the molecule by reducing the ketone and acetylating the resulting alcohol, then analyzing the acetates on the β-DEX column, allowed differentiation of the enantiomers. Based on these findings, the absolute configuration of the male-produced compound was determined to be (6R,10S)-6,10,13-trimethyltetradecan-2-one ((6R,10S)-**53**)(Fávaro et al. 2013).

15.2.14 *Piezodorus guildinii* (Westwood) and *Piezodorus hybneri* (Gmelin) [Pentatomidae: Pentatominae: Piezodorini]

15.2.14.1 Sex Pheromone of *Piezodorus guildinii*

Piezodorus guildinii, commonly known as the redbanded stink bug, is a serious pest of soybeans in South and Central America and, in the last decade, also the southern United States. Borges et al. (2007) isolated and identified its male-produced sex pheromone from volatiles collected from male bugs. Live male bugs or the synthetic pheromone attracted only females in Y-tube bioassays. Although males

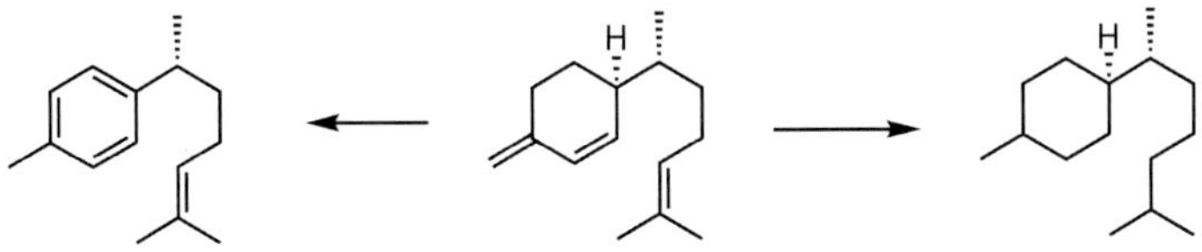

FIGURE 15.12 Sex pheromone of *Piezodorus guildinii*, (6*S*,7*R*)-(+)-β-sesquiphellandrene (**63**) correlated to curcumene **64** and bisabolane **65**.

produced the pheromone (mean 40 ng/day) from morning through evening, females only responded during the evening (1700-2000 h). The compound was identified as (6*S*,7*R*)-(+)-β-sesquiphellandrene (**63**), the general structure of which was confirmed by comparison with an authentic standard by GC-MS. The absolute configuration was determined by: (1) conversion to α-curcumene **64** and comparison of its GC retention time on a chiral Cyclodex-B column with those of standards prepared by dehydrogenation of (-)- and (+)-β-sesquiphellandrene; and (2) hydrogenation to a bisabolane **65** and analysis on a Cyclodex-B column along with authentic standards (McBrien et al. 2002) (**Figure 15.12**). It is noteworthy that β-sesquiphellandrene produced by *P. guildinii* has the opposite configuration to the (6*R*,7*S*)-β-sesquiphellandrene present in ginger oil and also is produced as a pheromone component by two other stink bug species, *Thyanta pallidovirens* and *Thyanta custator accerra* (McBrien et al. 2002).

15.2.14.2 Aggregation Pheromone of *Piezodorus hybneri:* Functions and Variation

Piezodorus hybneri is a major soybean pest in southern Japan. Leal et al. (1998) identified the pheromone blend and showed that both males and females were attracted to the 3-component blend of β-sesquiphellandrene, (*R*)-15-hexadecanolide, and methyl (*Z*)-8-hexadecenoate (ratio: 10:4:1). Males "also displayed a clear precopulatory behavior," even eliciting male-male sexual behavior. Leal and coworkers argued that this pheromone therefore functioned primarily as a sex pheromone rather than an aggregation pheromone. In field trials, the pheromone blend attracted predominantly females, and the individual components were not attractive (Endo et al. 2010). The pheromone attracted similar numbers of both sexes late in the fall, even though males would not be expected to produce it at this time (Endo et al. 2010). Endo et al. (2007, 2012) also studied the effects of age and diapause on pheromone emission by male *P. hybneri*. Only non-diapausing males produced pheromone; production started 3-6 days after adult emergence, was quantitatively and qualitatively different from individual to individual, and also changed within individuals over time. This led Endo et al. (2012) to question whether it was realistic to consider that the pheromone consisted of a set ratio of components. Whereas other studies have shown variation in pheromone component ratios in stink bugs (e.g., Miklas et al. 2000, for *Nezara viridula*), this is the only study to date that demonstrates changes in pheromone emission over the lifetime of individual bugs.

15.2.14.3 Attraction of *Piezodorus hybneri* to the Pheromone of *Riptortus clavatus* Thunberg (Heteroptera: Coreoidea: Alydidae)

It was discovered serendipitously in field trials that *Piezodorus hybneri* was attracted to one component, (*E*)-2-hexenyl (*E*)-2-hexenoate, of the 3-component pheromone of *Riptortus clavatus*, even though *P. hybneri* does not produce this compound (Endo et al. 2006, Huh et al. 2006). The *R. clavatus* component appeared to be most attractive to unfed adults of both sexes, suggesting a kairomonal role in plant host finding (Endo et al. 2010). It was also attractive to nymphs (Endo et al. 2010). The pheromone of *P. hybneri* and (*E*)-2-hexenyl (*E*)-2-hexenoate were not tested together.

15.2.15 *Plautia stali* Scott [Pentatomidae: Pentatominae: Antestiini]

The brown-winged green stink bug, *Plautia stali*, is a serious pest of tree fruits in Asia and also was discovered in Hawaii in 1967 (Mau and Mitchell 1978). The male-produced aggregation pheromone, attractive to conspecific males and females, was identified as methyl (2*E*,4*E*,6*Z*)-2,4,6-decatrienoate (**66**) (Sugie et al. 1996). Because of the pest status of this bug, the synthesis of ester **66** as an attractant for *P. stali* was patented in Japan (Yamashita et al. 1997) using controlled hydrogenation of methyl (2*E*,4*E*)-2,4-decadien-6-ynoate with Lindlar catalyst in a key step. However, this intermediate took several steps to make, and the reported yield of **65** from the reduction step was only 12%. Because **65** has been developed for commercial use in Japan, it seems likely that a better synthesis has been developed but has not been disclosed.

Khrimian (2005) reported syntheses of all eight geometric isomers of methyl 2,4,6-decatrienoate including the aggregation pheromone of *P. stali*, compound **66** (**Figure 15.13**) and a component of the sex pheromone of several *Thyanta* spp., compound **73** (**Figure 15.14**).

Syntheses of the four (4*E*)-isomers (**Figure 15.13**) were accomplished from the same precursor, (*E*)-4,4-dimethoxy-2-butenal (**67**), by fully exploiting Wittig-type olefinations, and taking advantage of an easy separation of (2*E*)- and (2*Z*)-unsaturated esters by silica gel chromatography. The starting aldehyde **67** can be prepared readily by a controlled deacetalization of fumaraldehyde bis(dimethylacetal), which is commercially available or can be prepared from furan (Gree et al. 1986). The *Plautia stali* pheromone **66** was prepared expeditiously in 55% overall yield and 96:4 *EEZ/EEE* ratio from **67** using consecutive *cis*-Wittig olefination to acetal **68** followed by acid-catalyzed hydrolysis and *trans*-Horner–Wadsworth–Emmons olefination of the (2*E*,4*Z*)-2,4-octadienal formed to give a 96:4 mixture of **66:71**. The method has been adopted by several semiochemical companies for commercial production of the *P. stali* pheromone (A.K., pers. comm.). *trans*-Wittig olefination of **67** yielded a diene acetal intermediate **70**, which was converted to the all-*E* stereoisomer **71** analogous to the synthesis of **66**. *cis*-Horner-Wadsworth-Emmons olefination with strongly electrophilic (CF$_3$CH$_2$O)$_2$P(O)CH$_2$COOCH$_3$ and a highly dissociated base, KN(TMS)$_2$/18-crown-6 (Still and Gennari 1983) was utilized to make stereoisomers **69** and **72** in good stereoisomeric purities from acetals **68** and **70**, respectively (**Figure 15.13**) (Khrimian 2005).

All four (4*Z*)-isomers of methyl 2,4,6-decatrienoate (**73**, **76**, **79**, and **80**) were synthesized by combining Horner–Wadsworth–Emmons olefinations and controlled stereoselective reductions of carbon-carbon triple bonds to introduce the required *Z*-double bonds (**Figure 15.14**) and, as previously mentioned, exploiting the easy separation of 2*E*- and 2*Z*-unsaturated esters on silica gel (Khrimian 2005). The thermally and chemically unstable (2*E*,4*Z*,6*Z*)-isomer **73**, the sex pheromone of several *Thyanta* spp. (Millar 1997, McBrien et al. 2002, Moraes et al. 2005a), was assembled by a one-pot oxidation of dienol **75** with

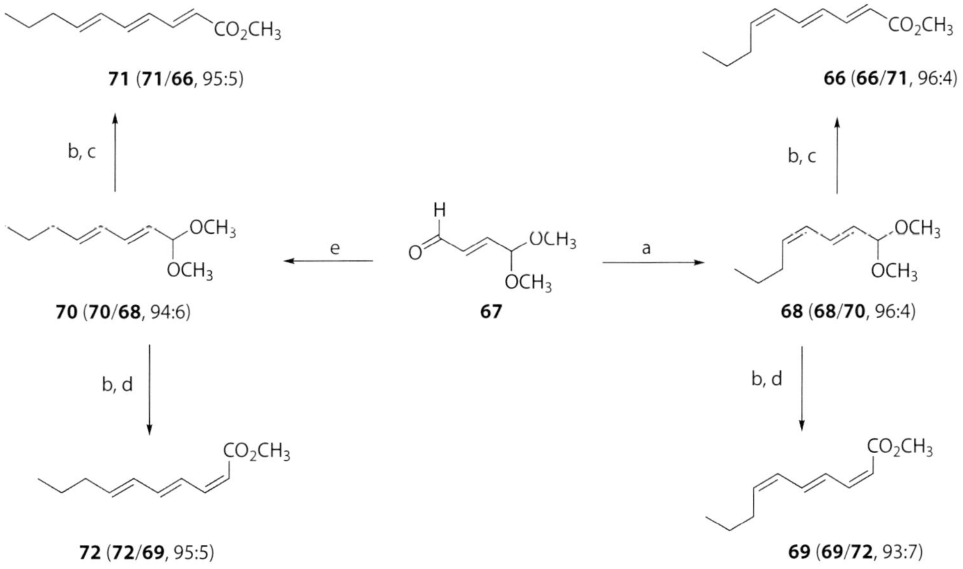

FIGURE 15.13 Syntheses of four methyl (4*E*)-2,4,6-decatrienoates including *Plautia stali* pheromone **66**. a) Ph$_3$PC$_4$H$_9$Br/ [(CH$_3$)$_3$Si]$_2$NNa; b) PTSA, acetone-H$_2$O; c) (CH$_3$O)$_2$P(O)CH$_2$CO$_2$CH$_3$, K$_2$CO$_3$-H$_2$O; d) (CF$_3$CH$_2$O)$_2$P(O)CH$_2$CO$_2$CH$_3$/ [(CH$_3$)$_3$Si]$_2$NK, 18-crown-6; e) Ph$_3$PC$_4$H$_9$Br/ 2 BuLi, THF.

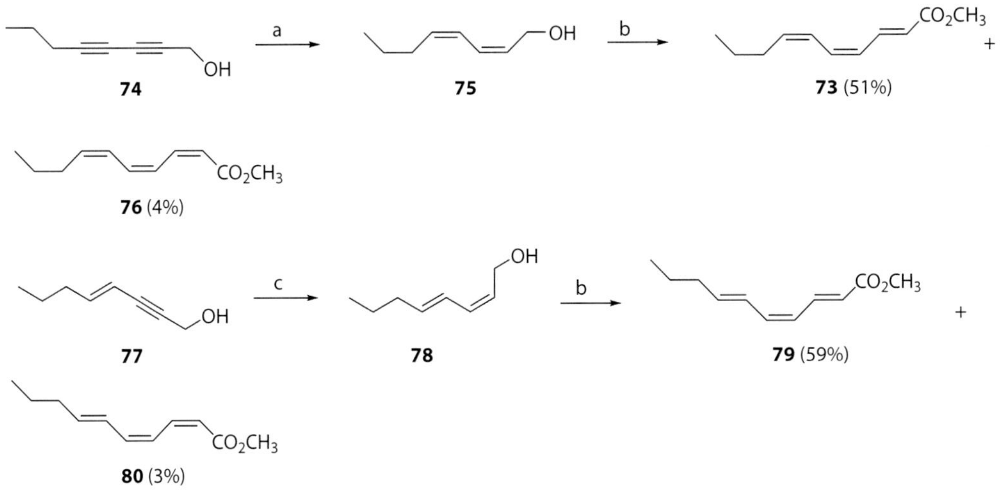

FIGURE 15.14 Syntheses of four methyl (4*Z*)-2,4,6-decatrienoates including ester **73**, a pheromone component of *Thyanta* spp. a) 1. DHP/PPTS, 2. dicyclohexylborane/AcOH/MeOH, 3. PPTS; b) 1. MnO_2, CH_2Cl_2, 2. Ph_3P=$CHCO_2Me$; (c) Zn(Cu/Ag), MeOH–H_2O.

in situ olefination of the intermediate aldehyde, in 51% overall yield. The original synthesis of ester **73** had used a one-pot double-carbocupration of acetylene with subsequent addition to methylpropiolate (Millar 1997).

With the *Plautia stali* pheromone commercially available from Shin-Etsu Chemical Co., Japanese researchers optimized pheromone-based trapping protocols to monitor this species in Asian pears and other tree fruit (Kakogawa et al. 2007). The traps are pyramidal, and first were yellow before being changed to clear in color, and are topped with a chamber with the pheromone lure and dichlorvos as a killing agent (Adachi et al. 2007, Katase et al. 2007). The lure also attracted *H. halys* (*vide supra*) and *Glaucias subpunctatus* (Walker) (Pentatomidae), both of which are significant fruit pests. The former now is known to emit a completely different male-produced pheromone as described above, whereas no pheromones have been identified yet for the latter species (Adachi et al. 2007, Katase et al. 2007).

The widespread planting of conifers, especially sugi [*Cryptomeria japonica* D. Don (Taxodiaceae)] and hinoki cypress [*Chamaecyparis obtusa* (Siebold & Zucc.) (Cuppressaceae)], has resulted in an abundant food source for all three bug species (i.e., *G. subpunctatus, H. halys, P. stali*), which feed on the cones and produce mass dispersals from this food source to fruit orchards up to 4 km distant (Kiritani 2007, Taki et al. 2014). To address these outbreaks, two pheromone-based suppression methods were tested: autodissemination of the entomopathogen *Beauveria bassiana* E-9102 (Tsutsumi et al. 2003), and use of trap plants poisoned with a systemic insecticide (Katase et al. 2007, Yamanaka et al. 2011). The autodissemination experiment increased mortality in the bugs from near zero in controls to 28-35% in the treated area to 70-75% in adults collected at the inoculation station (Tsutsumi et al. 2003). However, this method apparently has not been applied on a commercial scale. In a pilot test of systemic insecticides combined with pheromones, Katase et al. (2007) injected a potted sugi tree with the organophosphate insecticide acephate and baited it with *P. stali* pheromone, attracting and killing ~5000 bugs in 3 days. In a more ambitious trial, Yamanaka et al. (2011) used large potted (>1m tall) *Solanum torvum* Swartz (wild eggplants) baited with 34 mg of *P. stali* pheromone and poisoned with 1 g of imidacloprid per pot (~130-200 times the United States labeled field rate for eggplant). Poisoned eggplants were stationed at 50 m intervals around each of three persimmon orchards, with three control orchards. The border trap crop failed to reduce overall damage in the orchard. Damage was highest closest to the trap plants and, in related experiments, losses were elevated within a so-called spillover area of up to 150 m around the treated plants, as measured by fruit dropped within 7 days. The authors concluded that trap plants should be at least 100 m away from the crop that needed protection, and that attract-and-kill or push-pull strategies would have to adhere to this spatial limitation for any chance of success.

15.2.16 *Sehirus cinctus* (Palisot de Beauvois) [Cydnidae: Sehirinae]

The white-margined burrowing bug (*Sehirus cinctus*) exhibits brooding behavior in which the female provisions the nymphs with food in response to "solicitation pheromones." Kölliker et al. (2006) found that mother bugs already caring for offspring, and then exposed to volatiles from artificially-reared but poorly-fed nymphs, fed these nymphs significantly more compared to no exposure, or to exposure to volatiles from well-fed nymphs. At least eight compounds were identified in these volatiles, including α-pinene and camphene; however, these two chemicals did not elicit the feeding response evoked by the full blend. Interestingly, in the closely-related *Parastrachia japonensis* Scott (Heteroptera: Parastrachiidae), mothers signal their nymphal offspring with a distinct provisioning song when arriving with drupe fruits of the plant host (Nomakuchi et al. 2012).

15.2.17 *Tessaratoma papillosa* (Drury) [Tessaratomidae: Tessaratominae]

The litchi stink bug, *Tessaratoma papillosa*, is a serious pest of litchi, longan, and rambutan fruit in tropical Asia. Zhao et al. (2012) and Wang et al. (2012) described metathoracic glands and their contents from males and females, including differences in presumably allomonal chemical contents, along with electroantennogram and behavioral responses. Unfortunately, as in the case of several species of Scutelleridae (Millar 2005), behavioral evidence is not clear for what Wang et al. (2012) call "a complex pheromone system in *T. papillosa*." The authors state that males are attracted to secretions from females, but the pheromone has not yet been characterized properly.

15.2.18 *Thyanta custator* (F.), *Thyanta pallidovirens* (Stål), and *Thyanta perditor* (F.) [Pentatomidae: Pentatominae: Antestiini]

Moraes et al. (2005a) identified a male-produced sex pheromone of the neotropical red-shouldered stink bug, *Thyanta perditor*, one of a complex of pentatomids attacking soybeans in Brazil, as methyl (2E,4Z,6Z)-2,4,6-decatrienoate (**73**, **Figure 15.14**). The structure was confirmed by comparisons with retention times and spectra of an authentic standard (Millar 1997). It was attractive to females but not males in Y-tube bioassays. *Thyanta perditor* is the third *Thyanta* species, in addition to *T. pallidovirens* and *T. custator accerra* (McBrien et al. 2002), shown to use methyl (2E,4Z,6Z)-2,4,6-decatrienoate as a pheromone component. However, this compound appears to be the only pheromone component in the case of *T. perditor*, attracting conspecific females as a single component, whereas the pheromones of *T. pallidovirens* and *T. custator accerra* are more complex, and at least one of the three sesquiterpenes, (+)-α-curcumene, (-)-zingiberene, or (-)-β-sesquiphellandrene, produced by the males, is required in order for the latter two species to be attracted (McBrien et al. 2002). The pheromones of the latter two species were reviewed in Millar (2005), including syntheses of the various pheromone components.

All stereoisomers of methyl 2,4,6-decatrienoates have strong UV absorptions at ~ 300 nm (Millar 1997, Khrimian 2005, Moraes et al. 2005a) and, hence, are unstable in sunlight (Khrimian 2005, Khrimian et al. 2008). Thus, a solution of *Plautia stali* pheromone **66** (**Figure 15.13**) left unprotected under ambient conditions isomerized to a mixture of stereoisomers in which isomers **69** and **71** were the main byproducts (Khrimian et al. 2008). A similar trend was noticed when rubber septa impregnated with **66** were exposed to sunlight. The pheromone of *Thyanta pallidovirens*, ester **73**, both in a hexane solution and impregnated into rubber septa, isomerized in sunlight producing a mixture of geometric isomers, among which esters **66** and **71** were most abundant. Thus, at first glance the instability of these methyl 2,4,6-decatrienoate pheromones might seem problematic for their practical application in monitoring various target pests. However, Laumann et al. (2011) showed that traps baited with methyl (2E,4Z,6Z)-2,4,6-decatrienoate (**73**), protected from direct sunlight or not, attracted *T. perditor* females in a field trial in Brazil. Interestingly, even methyl (2E,4E,6Z)-2,4,6-decatrienoate (**66**) was somewhat attractive to *T. perditor* (Laumann et al. 2011), but whether or not this isomer is intrinsically attractive or whether it isomerized to the (2E,4Z,6Z)-isomer under field conditions, remains unanswered and, for practical purposes, does not matter. Methyl (2E,4Z,6Z)-2,4,6-decatrienoate also cross-attracted other species of stink bugs to some extent, including *Euschistus heros*, *Edessa meditabunda*, *Piezodorus*

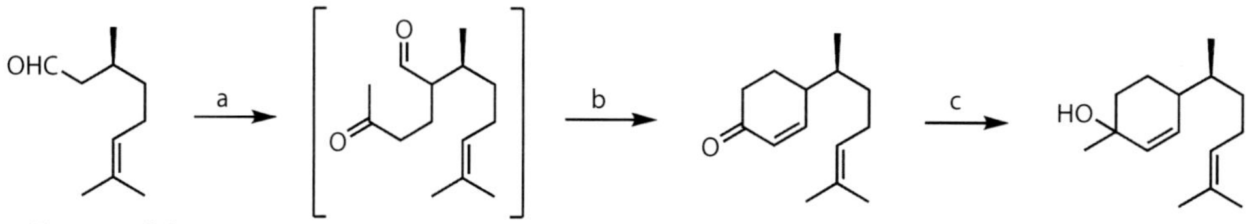

FIGURE 15.15 Non-stereoselective synthesis of zingiberenol II mix of isomers (**49**), a sex pheromone of *Tibraca limbativentris*, from (*S*)-citronellal. a) CH_2=CHC(O)CH_3/ Et_2NTMS; b) KOH/Bu_4NOH/THF-Et_2O; c) CH_3Li/Et_2O.

guildinii, and *Nezara viridula*, although statistical comparisons to unbaited traps were not presented (Laumann et al. 2011).

15.2.19 *Tibraca limbativentris* (Stål) [Pentatomidae: Pentatominae: Carpocorini]

Borges et al. (2006) isolated a male-produced sex attractant pheromone from the rice stalk stink bug, *Tibraca limbativentris*, which along with *Oebalus poecilus* (see above), is one of two major stink bug pests of rice in South America. Olfactometer bioassays showed that males attracted only females and primarily at night. GC-MS analyses of headspace volatiles from males showed two male-specific compounds on HP-1 and DB-WAX columns. The mass spectral fragmentation patterns along with matches to spectra in the NIST MS database suggested that these compounds were stereoisomers of 1,10-bisaboladien-3-ol (= zingiberenol), which was confirmed by non-stereoselective syntheses of mixtures of isomers (**Figure 15.15**) starting from (*R*)- and (*S*)-citronellal following Hagiwara et al. (2002) in an interesting variation of the synthesis shown in **Figure 15.8**.

Each synthesis produced a mixture of four stereoisomers, zingiberenol I mix from (*R*)-citronellal and zingiberenol II mix from (*S*)-citronellal. Trimethylsilyl derivatives of all four diastereomers of zingiberenol were partially separated on an HP-1 GC column, but the lack of authentic standards of the individual stereoisomers of known configuration prevented the assignment of the absolute configurations of the two male-specific compounds. In light of more recent findings (Khrimian et al. 2014a), the faster eluting (from an HP-1 GC column) and more abundant male-specific compound in *T. limbativentris* can be assigned the *cis* relative configuration, and the slower eluting minor compound *trans*. When the zingiberenol I and zingiberenol II mixtures were tested against each other in olfactometer bioassays, both were equally attractive, whereas when the mixtures were compared individually with a neutral control, only the zingiberenol II mixture derived from (*S*)-citronellal was attractive (Borges et al. 2006). It remains to be seen whether *T. limbativentris* and *Oebalus poecilus* share the same diastereomer of 1,10-bisaboladien-3-ol **50** with the (*7S*)-configuration (**Figure 15.9**) as their sex pheromone.

15.3 Allomones: Defensive Chemicals

15.3.1 Range of Chemistry

Defensive compounds of pentatomoid bugs are produced and stored in the dorsal abdominal and metathoracic glands. These organs have been studied in several pentatomoid species (Pavis 1987, Aldrich 1995a, Moraes et al. 2008b, Fávaro et al. 2011) and the secretions have been analyzed from many species. Whereas these chemicals have been implied to have defensive functions in numerous species, their putative functional roles as allomones has actually been tested in bioassays in only a few species.

The intraspecific variability in the chemistry of the secretions from nymphs and adults respectively is generally low, and the morphology of the glands is quite conserved across species (Aldrich et al. 1978, Pavis 1987, Aldrich 1988). In nymphs, the compounds are produced and stored in dorsal abdominal glands (DAGs) distributed in tergites 3 and 4, 4 and 5, and 5 and 6 (Aldrich et al. 1978, Aldrich 1988). In contrast, adults produce these compounds in the metathoracic glands (MTGs) and store them in a large orange reservoir (Aldrich 1988, Pavis 1987). The DAGs are not functional in most adults. There is no apparent sexual dimorphism in the MTGs or DAGs of phytophagous pentatomids.

The chemistry of the allomones produced by pentatomoids is in most cases rather simple. In nymphs, the principal components are short chain (*E*)-2-alkenals, 4-oxo-(*E*)-2-alkenals (C_6, C_8, C_{10}) and linear hydrocarbons (mainly C_{11} to C_{13}), and the production of these compounds changes with the nymphal instar. Usually, the first instars produce 4-oxo-(*E*)-2-decenal in relatively large proportions, whereas production in the second to fifth instars is reduced or even eliminated, and, instead, 4-oxo-(*E*)-2-hexenal and 4-oxo-(*E*)-2-octenal are produced in larger quantities (Borges and Aldrich 1992). The (*Z*)-isomers of 4-oxo-(*E*)-2-alkenals and (*E*)-2-alkenals also frequently are found in trace quantities in the secretions of nymphs and adults (Pareja et al. 2007). In the adults, the allomones are composed of the same compounds found in the later nymphal instars, with tridecane being the major compound. Adults also produce short-chain alcohols (C_6-C_8) and their esters, such as (*E*)-2-hexenyl, (*E*)-2-octenyl, and (*E*)-2-decenyl acetates and, depending on species, a variety of other minor compounds including monoterpenes, long-chain saturated and unsaturated aldehydes, diols, and pyrazines.

The stink bugs *Edessa rufomarginata* (De Geer) and *E. meditabunda* (subfamily Edessinae), differ from this general pattern. In these species, the major hydrocarbon is undecane instead of tridecane in both nymphs and adults, and the first instar does not produce 4-oxo-(*E*)-2-decenal (Borges and Aldrich 1992, Zarbin et al. 2012). Other pentatomids, such as *Tibraca limbativentris* and *Euschistus heros*, produce aldehydes with longer chains (e.g., tetradecanal) and *E. heros* also has the monoterpene linalool in its MTG secretions (Pareja et al. 2007). Linalool also has been found in DAG extracts of adult *Dichelops melacanthus* (Dallas) of both sexes (M.C.B.M., unpublished data).

Although stink bugs share a number of compounds in their allomonal blends, careful analytical studies with robust statistical analyses have shown that these blends can be quite complex, with species-specific differences in minor compounds and significant differences in ratios among compounds (Aldrich et al. 1996, Pareja et al. 2007, Fávaro et al. 2011). For example, for five neotropical pentatomid species (*Euschistus heros, Dichelops melacanthus, Chinavia ubica, Chinavia impicticornis,* and *Piezodorus guildinii*), the fifth instars and adults shared several allomonal compounds. However, a multivariate analysis was able to group the species according to their taxonomic relationships, with the two *Chinavia* species grouped together, and *D. melacanthus* and *E. heros* grouping together (Pareja et al. 2007). For these five species, the main compounds responsible for the separation of the adults were aldehydes and esters, with the *Chinavia* species producing higher amounts of (*E*)-2-decenal and the analogous ester (*E*)-2-decenyl acetate, whereas (*E*)-2-octenal and (*E*)-2-octenyl acetate were relatively more abundant in *E. heros* and *D. melacanthus* (Pareja et al. 2007).

Some pentatomids produce defensive compounds with more complex chemistry. For example, *Murgantia histrionica* adults produce (*E*)-2-octenal, (*2E,6E*)-2,6-octadienal and (*2E,6E*)-2,6-octadien-1,8-diol from the MTGs and 2-*sec*-butyl-3-methoxypyrazine and 2-isopropyl-3-methoxypyrazine from the prothoracic fluid (Aldrich et al. 1996). Diols and unsaturated aldehydes also were identified from the MTG extracts of *Eurydema ventralis* (Kolenati), *Eocanthecona furcellata*, *Euschistus heros*, *Chinavia impicticornis*, *Chinavia ubica*, *Dichelops melacanthus*, and *Pallantia macunaima* (Aldrich et al. 1996, Ho et al. 2003, Pareja et al. 2007, Fávaro et al. 2011) but in tiny quantities. The roles of these compounds as possible mediators of behavior remain to be determined.

Fávaro et al. (2012) conducted a study of the possible defensive compounds from *Agroecus griseus* of different life stages. The compounds found in the DAGs of nymphs were similar to those of other pentatomids and included (*E*)-2-hexenal, 4-oxo-(*E*)-2-hexenal, (*E*)-2-octenal, dodecane, (*E*)-2-decenal, 1-tridecene, (*Z*)-4-tridecene, tridecane, 4-oxo-(*E*)-2-decenal, and tetradecanal (Fávaro et al. 2012, 2013). A significant difference between exuvial extracts of different nymphal stadia was the presence of (*E*)-2-decenal and (*E*)-4-oxo-2-hexenal in the first instars and the absence of these chemicals in later instars, which parallels DAG contents of other pentatomids (Borges and Aldrich 1992). 4-Oxo-(*E*)-2-decenal also is known to mediate the aggregation behavior of first instars of several pentatomid species (Pavis et al. 1994, Fucarino et al. 2004). Analysis of the MTG content of *A. griseus* revealed aldehydes, hydrocarbons, esters, oxoalkenals, and two new compounds identified as (*S*)-2-methylbutyl acetate and 3-methyl-2-butenyl acetate (Fávaro et al. 2012). The absolute configuration of the former was established by GC analysis on an enantioselective column. The biological functions of these compounds have not been determined.

15.3.2 Effect of MTG and DAG Secretions against Natural Enemies

The deterrent effects of the chemicals produced by the DAGs and MTGs of stink bugs have been demonstrated in a few cases. For example, Noge et al. (2012) evaluated the effect of some possible defensive compounds of *Euschistus biformis*, including (*E*)-2-hexenal, (*E*)-2-octenal, (*E*)-2-octenyl acetate, 4-oxo-(*E*)-2-hexenal and (*E*)-2-hexenyl acetate, against the praying mantis *Tenodera aridifolia sinensis* (Saussure) (Mantodea: Mantidae). The compounds were evaluated individually and as blends in arena bioassays, and it was reported that (*E*)-2-hexenal, (*E*)-2-octenal, and (*E*)-2-octenyl acetate were repellent to the mantis (Noge et al. 2012). Eliyahu et al. (2012) evaluated the function of several possible defensive compounds from pentatomid nymphs against the jumping ant *Harpegnathos saltator* (Hymenoptera: Formicidae), hypothesizing that 4-oxo-(*E*)-2-decenal and (*E*)-2-decenal might be less effective than their more volatile, shorter chain homologs, 4-oxo-(*E*)-2-hexenal and (*E*)-2-hexenal, and that tridecane might have a synergistic effect. In arena bioassays, a blend of 4-oxo-(*E*)-2-hexenal, tridecane, and 4-oxo-(*E*)-2-decenal deterred ant attacks (Eliyahu et al. 2012). Another study with compounds from the MTG of *Coridius janus* (Hemiptera: Pentatomidae) reported that (*E*)-2-hexenal and tridecane deterred attacks by *Anoplolepis longipes* (Formicidae), a natural enemy of *C. janus* (Gunawardena and Herath 1991). Overall, these studies suggested that the shorter chain aldehydes such as (*E*)-2-hexenal and 4-oxo-(*E*)-2-hexenal indeed were more effective against predators than their longer chain homologs, and that tridecane had an important role in enhancing the efficacy of these aldehydes (Gunawardena and Herath 1991, Eliyahu et al. 2012).

The aldehydes produced by pentatomid bugs provide protection against at least some entomopathogenic fungi. Borges et al. (1993) showed inhibition of the entomopathogen *Metarhizium anisopliae* (Metschnikoff) Sorokin (Hypocreales: Clavicipitaceae) by two aldehyde allomones, (*E*)-2-decenal and (*E*)-2-hexenal, produced by *Nezara viridula*. They reported that solutions containing the aldehydes and tested individually had the same effect when they were combined, and (*E*)-2-decenal (25 µg) alone was able to completely inhibit fungal germination. Sosa-Gomez (1997) also showed that (*E*)-2-decenal was strongly fungistatic on *N. viridula* cuticle against *M. anisopliae* but not against *Beauveria* and *Paecilomyces* spp. Da Silva et al. (2015) found that extracts of MTGs and DAGs, respectively, of adult and nymphal *Tibraca limbativentris*, and synthetic versions of the major components, (*E*)-2-hexenal, (*E*)-2-octenal, and (*E*)-2-decenal, inhibited spore germination, growth, and sporulation of *M. anisopliae* strain CG168, which is expected to be registered for commercial use against this pentatomid in Brazil. The decreased susceptibility of later instars and adults to this fungal infection was attributed at least in part to the greater quantities of the three aldehydes present in these more resistant stages (da Silva et al. 2015). Ulrich et al. (2015) demonstrated the dose-dependent contact and fumigant activity of (*E*)-2-hexenal and (*E*)-2-octenal against *M. anisopliae* (sensu lato, ARSEF 1548) and a protective effect of (*E*)-2-octenal against mortality from *M. anisopliae* in the bed bug *Cimex lectularius* L. (Cimicidae). Preliminary results with *Halyomorpha halys* suggest that adult bugs in groups are less susceptible to mortality from *M. anisopliae*, but the mechanisms for this apparent effect are still being pursued (R. St. Leger, pers. comm.).

15.3.3 Other Behavioral Roles: Within Species

Volatile aldehydes and esters are primarily responsible for the strong odor that stink bugs emit when stressed, and in addition to deterring predation, these and other gland components also may have intraspecific roles in communication. Also, different effects can be elicited by different doses: in low quantities, compounds can act as aggregation pheromones, and, in larger quantities, as alarm pheromones (Ishiwatari 1976). Similarly, Lockwood and Story (1985) reported that tridecane acts as an aggregation pheromone at low doses and as an alarm pheromone at high doses in *Nezara viridula*. However, given more recent studies indicating that the aggregation pheromone of *N. viridula* consists of bisabolene epoxides (*vide supra*), this earlier report must be treated with caution. Furthermore, there also are contradictory reports on the biological roles of hydrocarbons normally present in stink bug secretions. For example, Fucarino et al. (2004) were not able to reproduce the results of Lockwood and Story (1987), likely because the doses used in the Lockwood and Story study were well above the biologically relevant range. Thus, it remains equivocal as to whether the reported responses to linear hydrocarbons represent a true, biologically relevant aggregation effect. Aldrich (1988) proposed that one of the functions of the aliphatic hydrocarbons in the gland secretions is to serve as solvents or carriers, rather than as inherently bioactive compounds, to modulate the evaporation of the other compounds, suggesting a reason for the relatively large quantities of hydrocarbons found in stink bug glands.

Most pentatomids aggregate upon hatching and then disperse during later stadia. Several authors have proposed that 4-oxo-(E)-2-decenal might be involved in the aggregation behavior of first instars (Borges and Aldrich 1992, Pavis et al. 1994, Fucarino et al. 2004). This compound has been identified from the DAGs of first instar nymphs for *Euschistus heros*, *E. conspersus*, *E. tristigmus*, *Thyanta perditor*, *T. pallidovirens*, *Nezara viridula*, and *Chinavia aseada* (=*Acrosternum aseadum*), and its potential role as an aggregation pheromone has been demonstrated for *Nezara viridula* (Pavis et al. 1994, Fucarino et al. 2004). Among insects, 4-oxo-(E)-2-alkenals have only been reported from the Heteroptera to date.

15.4 Kairomones for Natural Enemies

Parasitoids and predators can exploit the volatile chemicals produced by their hosts as kairomones to locate their hosts (**Tables 15.2** and **15.3**). The compounds exploited can include the host's defensive secretions (**Table 15.2**), or compounds produced by the host for intraspecific signaling such as sex, alarm, or other pheromones (**Table 15.3**). Alternatively, natural enemies may exploit compounds associated with products derived from the host such as feces/frass, exuvia, scales, honeydew, or host "footprints", where the host has no control over the production of the compounds (Vinson 1985, Vet and Dicke 1992, Godfray 1994, Borges et al. 2003, Afsheen et al. 2008, Fatouros et al. 2008, Laumann et al. 2009, Conti and Colazza 2012). Each of these types of kairomonal interactions is considered in more detail below.

Natural enemies often have short adult life spans, and hosts may have only a short time window during which they can be parasitized; therefore, parasitoids need to locate suitable hosts quickly. Platygastrid wasp parasitoids, for example, prefer to parasitize pentatomid eggs that are less than 72 hours old (Bin et al. 1993). Therefore, natural enemies need reliable information about the presence of their hosts, and semiochemicals from hosts or host-related cues can provide such reliable information.

TABLE 15.2

Allomones from Pentatomoidea That Are Exploited as Kairomones by Natural Enemies

Bug Species	Chemicals	Natural Enemy Species	Reference
Coridius janus (**Dinidoridae**)	(E)-2-Hexenal, C_8, C_{10} - C_{14} and C_{16} linear hydrocarbons	Hymenoptera: Formicidae: *Anoplolepis longipes*	Gunawardena and Herath 1991
Euschistus biformis	(E)-2-Hexenal, (E)-2-octenal, (E)-2-octenyl acetate, 4-oxo-(E)-2-hexenal, (E)-2-hexenyl acetate	Mantodea: Mantida *Tenedora aridofolia*	Noge et al. 2012
Euschistus heros	Footprints (the bioactive compounds were not identified)	Hymenoptera: Platygastridae: *Telenomus podisi*	Borges et al. 2003
Euschistus heros, Dichelops melacanthus, Nezara viridula	Tridecane, (E)-2-hexenal, (E)-2-decenal, undecane, 4-oxo-(E)-2-hexenal	Hymenoptera: Platygastridae: *Telenomus podisi* and *Trissolcus basalis*	Mattiacci et al. 1993, Laumann et al. 2009
Nezara viridula	(E)-2-Decenal and (E)-2-octenal	Hymenoptera: Platygastridae: *Telenomus podisi*	Borges and Aldrich 1994
Nezara viridula	Nonadecane	Hymenoptera: Platygastridae: *Trissolcus basalis*	Colazza et al. 2007
Pentatomidae	4-Oxo-(E)-2-hexenal, 4-oxo-(E)-2-decenal, (E)-2-hexenal, (E)-2-decenal, tridecane	Hymenoptera: Formicidae: *Harpegnathos saltator*	Eliyahu et al. 2012

TABLE 15.3

Volatile Pheromones of Pentatomoidea That Are Exploited as Kairomones by Natural Enemies

Bug Species	Chemicals	Natural Enemy Species	References
Euschistus conspersus	Methyl (2*E*,4*Z*)-2,4-decadienoate	Diptera: Tachinidae: *Gymnoclytia occidentalis*	Krupke and Brunner 2003
Euschistus heros	Methyl 2,6,10-trimethyltridecanoate	Hymenoptera: Platygastridae: *Telenomus podisi*	Borges et al. 1998, 1999; Silva et al. 2006
Euschistus spp.	Methyl (2*E*,4*Z*)-2,4-decadienoate	Diptera: Tachinidae: *Gymnosoma par*	Aldrich et al. 2006, 2007
Murgantia histrionica	*Extracts from adults, nymphs, and eggs*	Hymenoptera: Platygastridae: *Trissolcus brochymenae*	Conti et al. 2003
Nezara viridula	One or more of *(Z)*-α-bisabolene, (4*S*)-*cis* and (4*S*)-*trans*-bisabolene epoxides; *Live males and females as odor sources*	Diptera: Tachinidae: *Trichopoda pennipes* Hymenoptera: Encyrtidae: *Ooencyrtus telenomicida* Hymenoptera: Platygastridae: *Trissolcus basalis*	Aldrich et al. 1987, 2006, 2007; Colazza et al. 1999; Mitchell and Mau 1971; Harris and Todd 1980
Plautia stali	Methyl (2*E*,4*E*,6*Z*)-2,4,6-decatrienoate	Diptera: Tachinidae: *Gymnosoma rotundatum, G. par, Euclytia flava, Euthera tentatrix*; Hymenoptera: Sphecidae: *Astata occidentalis*	Moriya and Shiga 1984; Aldrich et al. 2006, 2007; Higaki and Adachi 2011; Cottrell et al. 2014
Podisus fretus	One or more of (*E*)-2-hexenal + α-terpineol + linalool + terpinen-4-ol + benzyl alcohol	Hymenoptera: Vespidae: *Vespula maculifrons*	Aldrich et al. 1986a
Podisus maculiventris	One or more of (*E*)-2-hexenal + α-terpineol + linalool + terpinen-4-ol + benzyl alcohol	Diptera: Tachinidae: *Euclytia flava, Hemyda aurata, Cylindromyia fumipennis*; Diptera: Ceratopogonidae: *Forcipomyia crinita*; Hymenoptera: Platygastridae: *Telenomus* spp. Hymenoptera: Vespidae: *Vespula maculifrons*	Aldrich et al. 1984, 1986b; Bruni et al. 2000; Aldrich and Zhang 2002; Aldrich et al. 2006, 2007
Podisus neglectus	One or more of (*E*)-2-hexenal + α-terpineol + linalool + terpinen-4-ol + benzyl alcohol + (*E*)-2-octenal	Diptera: Tachinidae: *Euclytia flava, Hemyda aurata* Diptera: Ceratopogonidae: *Forcipomyia crinita* Hymenoptera: Platygastridae: *Telenomus* spp.	Aldrich et al. 1984, 1986b; Aldrich and Zhang 2002
Thyanta pallidovirens	One or more of methyl (2*E*,4*Z*,6*Z*)-2,4,6-decatrienoate, methyl (2*E*,4*E*,6*Z*)-2,4,6-decatrienoate, zingiberene, α-curcumene and sesquiphellandrene	Diptera: Tachinidae: *Euclytia flava*; Hymenoptera: Sphecidae: *Astata occidentalis*	Aldrich et al. 2006, 2007; Millar et al. 2001
Tibraca limbativentris	Zingiberenol	Hymenoptera: Platygastridae: *Telenomus podisi*	Tognon et al. 2014

15.4.1 Pentatomid Compounds as Kairomones for Egg Parasitoids

Studies conducted with platygastrid egg parasitoids demonstrate the complexity and diversity of cues and behavioral interactions between an egg parasitoid and its stink bug hosts. For example, in Y-tube olfactometer bioassays, the egg parasitoid *Telenomus podisi* (F.) was attracted to a defensive compound, (*E*)-2-hexenal, from its preferred host, *Euschistus heros* (Vieira et al. 2014). However, (*E*)-2-hexenal is a common plant volatile and is produced in large amounts by soybean, a major host plant of *E. heros*

(Moraes et al. 2008a). Thus, it remains uncertain whether the parasitoid responds to this compound as a cue directly associated with its host or as a means of finding the habitat of its host. Analogously, *Trissolcus basalis* (Wollaston) was attracted to (*E*)-2-decenal, a compound produced by its preferred host, *Nezara viridula*, in laboratory bioassays (Mattiacci et al. 1993). In another example, *T. podisi* and *Tr. basalis* are generalist parasitoids, but field observations in soybean fields showed high rates of *Tr. basalis* parasitism of *N. viridula* eggs, whereas *T. podisi* preferred *E. heros* eggs (Corrêa-Ferreira and Moscardi 1995, Medeiros et al. 1998, Pacheco and Corrêa-Ferreira 2000). Laboratory experiments using a multiple-host choice bioassay confirmed these field observations (Sujii et al. 2002). The allomones produced by the two stink bug species may be involved in this preference. However, when individuals of *Telenomus* sp. from an American population were offered two different odor sources, eggs treated with (*E*)-2-decenal or (*E*)-2-octenal, of which the latter is produced by *E. heros*, the parasitoids preferred the former odor (Borges and Aldrich, 1994). A subsequent study with Brazilian populations of *Tr. basalis* and *T. podisi* showed that each species recognized the allomones produced by their preferred hosts. Thus, in Y-olfactometer bioassays, *Tr. basalis* responded preferentially to (*E*)-2-decenal and 4-oxo-(*E*)-2-hexenal, whereas *T. podisi* preferred (*E*)-2-hexenal and 4-oxo-(*E*)-2-hexenal, as predicted. The lack of response of the *Telenomus* sp. American population to (*E*)-2-octenal remains to be clarified. One hypothesis is that *Telenomus* sp. learned to recognize *E. heros* through (*E*)-2-hexenal, which is produced by this stink bug, rather than through (*E*)-2-octenal. Despite some uncertainties as to the exact compounds which are attractive to the various species and populations, these results overall support the hypothesis that (*E*)-2-alkenals are important cues used by platygastrid parasitoids and other natural enemies to locate their preferred hosts (Borges and Aldrich 1992, Aldrich 1994b, Pareja et al. 2007, Laumann et al. 2009). In fact, recent field experiments showed that soybean crops treated with (*E*)-2-hexenal attracted higher numbers of natural enemies, mainly *T. podisi,* compared with untreated areas (Vieira et al. 2014).

Aldrich (1985) reported that an egg parasitoid can use the sex pheromone produced by male stink bugs to locate females and, indirectly, the stink bug eggs. This strategy apparently is used by *Telenomus calvus* (Ashmead), which exploits pheromones produced by male *Podisus maculiventris* and *P. neglectus* (Westwood) as an indirect mechanism to locate *Podisus* eggs (Aldrich 1995a). This parasitoid requires eggs less than 12 hours old to successfully develop, and so it needs a reliable method of locating fresh eggs. The pheromone blends of male *Podisus maculiventris* and *P. nigrispinus* (Dallas) (= *P. connexivus*) both contain (*E*)-2-hexenal, benzyl alcohol, and α-terpineol, providing a reliable cue for the wasp.

Telenomus podisi is one the main egg parasitoids of pentatomids, using different host-related cues to locate host eggs. In Brazil, *T. podisi* is the most abundant natural enemy of *E. heros* in soybean crops, with >80% parasitism of host eggs (Pacheco and Corrêa-Ferreira 2000, Michereff et al. 2014). In field experiments using traps baited with *E. heros* pheromone (1 mg per trap of racemic methyl 2,6,10-trimethyltridecanoate), two egg parasitoids were attracted, *T. podisi* and *Trissolcus urichi* (Crawford) (Borges et al. 1998, 2011). Laboratory experiments confirmed that *T. podisi* was attracted to synthetic racemic methyl 2,6,10-trimethyltridecanoate (Silva et al. 2006). Recently, Tognon and co-workers (2014) reported that *T. podisi* also is attracted to the male-produced pheromone of the rice stink bug *Tibraca limbativentris*, which consists of the sesquiterpenoid zingiberenol, with a completely different structure than the *E. heros* pheromone. Thus, *T. podisi* appears to possess a high degree of plasticity in its recognition of a variety of chemical structures associated with hosts, responding to the simple molecules found in the defensive secretions, such as (*E*)-2-alkenals and 4-oxo-(*E*)-2-alkenals, as well as to more complex structures such as the long-chain methyl esters (C_{10} to C_{15}) and sesquiterpenoids that constitute the male-produced aggregation pheromones (Aldrich 1985, Borges et al. 1998, Laumann et al. 2009, Michereff et al. 2013).

In addition to volatile allomones from DAGs and MTGs, and pheromones, natural enemies may also exploit less volatile and nonvolatile components of the cuticular lipids of stink bugs, such as long-chain hydrocarbons (Blomquist and Bagnères 2010). For example, Colazza et al. (2007) reported that nonadecane is present in the cuticular profile of male but not female *Nezara viridula*. The authors suggested that this information may be used by egg parasitoids to distinguish residues left by male or female bugs. In addition, Borges et al. (2003) showed that the egg parasitoid *T. podisi* from a Brazilian population can recognize "footprints" from *Euschistus heros* females, whereas *T. podisi* from an American population cannot. More recently, Salerno et al. (2009) showed that *Trissolcus brochymenae* (Ashmead) prefers the traces left by mated females of *Murgantia histrionica*, compared with traces from virgin females, males, or parous host females. However, the chemicals that mediate this interaction have not been identified, and, in general, there is almost no information about the cuticular lipids of pentatomoids and their possible functions as intra- or interspecific semiochemicals.

15.4.2 Pentatomid Compounds as Kairomones for Diptera

The dipteran family Tachinidae has approximately 10,000 described species, and tachinid flies are the largest and most important groups of parasitoids within the Diptera. All tachinids are endoparasitoids of other arthropods, mainly insects (Aldrich et al. 2006, Nakamura et al. 2013). Chemically mediated interactions between tachinids and their hosts have been studied in a number of systems (Feener and Brown 1997), including the exploitation of pentatomid pheromones as kairomones by tachinid flies (Mitchell and Mau 1971; Harris and Todd 1980; Moriya and Shiga 1984; Aldrich et al. 1984, 1987, 2007; Aldrich 1995a,b; Mishiro and Ohira 2002; Adachi et al. 2007; Jang and Park 2010; Higaki and Adachi 2011). The topic also has been reviewed recently (Nakamura et al. 2013).

Analogous to the platygastrid egg parasitoids, tachinid flies use multiple strategies to locate their hosts, including exploitation of the pheromones of their prey. Similarly to *Telenomus podisi*, tachinid flies can be attracted to the volatile pheromones of a variety of stink bug species, even though the molecular structures of the pheromones are dissimilar. For example, in field experiments, *Euclytia flava* (Townsend) was caught in traps baited with several different pentatomid pheromones, including (*E*)-2-hexenal + α-terpineol (*Podisus maculiventris*), methyl (2*E*,4*Z*,6*Z*)-2,4,6-decatrienoate (*Thyanta* spp.), and methyl (2*E*,4*E*,6*Z*)-2,4,6-decatrienoate (*Plautia stali*) (Aldrich et al. 2007). The responses to different host pheromones may be an adaptation to the different seasonal activity periods of its various hosts because *E. flava* is seldom captured in traps baited with *P. maculiventris* pheromone in late summer or early fall when *P. maculiventris* would not be producing pheromone, and this fly almost never emerges from overwintering *P. maculiventris* in the spring. Instead, during the later part of the growing season, *E. flava* were collected primarily from *Thyanta* spp. hosts. Therefore, these tachinid flies may have adapted to the different seasonal activity periods of their hosts, so that they can exploit a succession of host species throughout the growing season (Aldrich et al. 2006).

Aldrich (1995b) reported that it is occasionally possible to find *Podisus* spp. nymphs parasitized by tachinid flies in the field. Field experiments using reconstructed nymphal gland secretions containing 4-oxo-(*E*)-2-hexanal + linalool + tridecane supported this observation, with two species of tachinid flies, *Euclytia flava* and *Hemyda aurata* (Robineau-Desvoid), being attracted to the lures (Aldrich 1988).

In general, there have been relatively few studies demonstrating attraction of tachinid flies to stink bug secretions, in part because the parasitoids are laborious to rear. However, field observations have shown that these flies generally parasitize adults of both sexes, rather than nymphs, suggesting that the pheromones or other secretions of adults likely mediate these interactions. For example, a field experiment using (*E*)-2-octenal, a common stink bug allomone, combined with the sex pheromones of *Podisus* spp. and *Euschistus* spp., showed that attraction of the tachinids was not influenced by the presence of (*E*)-2-octenal, suggesting that the pheromones were the more important part of the volatile cues used by the flies to locate hosts (Aldrich and Zhang 2002).

Chloropid and milichiid flies also have been reported to be attracted to defensive compounds produced by stink bugs. These small flies are scavengers, so they are probably attracted by the scent of injured or dead bugs to feed on hemolymph, for example. Eisner et al. (1991) showed that milichiid flies (almost exclusively females) from four genera (*Paramyia*, *Neophylomyza*, *Milichiella*, and *Desmometopa*) were attracted to the stink bugs *Piezodorus guildinii*, *Nezara viridula*, and *Euschistus* sp. when these stink bugs were being attacked by the spider *Nephila clavipes* (L.) (Araneae: Nephilidae), and also were attracted to their common defensive compounds, (*E*)-2-hexenal and hexanal. Kondo et al. (2010) demonstrated the rapid attraction (typically in about 2 seconds) of female *Milichiella lacteipennis* (Loew) to crushed adults of Pentatomidae and Coreidae in Colombia, but did not test chemical attractants separately. Aldrich and Barros (1995) showed the field attraction of female *Milichiella arcuata* (Loew) to (*E*)-2-hexenal, and the attraction of chloropids (15 species total, predominantly females) to (*E*)-2-hexenal, (*E*)-2-octenal, and (*E*)-2-decenal, common defensive compounds of pentatomids and coreids.

15.4.3 Pentatomid Compounds as Kairomones for Arthropod Predators

Millar et al. (2001) reported that *Astata occidentalis* (Cresson), a predatory sphecid wasp that provisions its nest with stink bugs, was attracted to the sex pheromone of *Thyanta pallidovirens*. In field experiments, only female wasps were caught in traps baited with methyl (2*E*,4*Z*,6*Z*)-2,4,6-decatrienoate, the main pheromone component of *T. pallidovirens*, and blends of this compound with the sesquiterpenoid minor components of the pheromone (zingiberene, sesquiphellandrene, and α-curcumene) (Millar et al. 2001). Recently, Cottrell et al. (2014) caught substantial numbers of *A. occidentalis* in traps baited with the *Plautia stali* pheromone, methyl (2*E*,4*E*,6*Z*)-2,4,6-decatrienoate, deployed to monitor *Halyomorpha halys* and other pentatomids in Georgia and Washington state, United States. The authors noted that this could have been due to isomerization of methyl (2*E*,4*E*,6*Z*)-2,4,6-decatrienoate to the *Thyanta* pheromone, methyl (2*E*,4*Z*,6*Z*)-2,4,6-decatrienoate (Khrimian et al. 2008), or it may be that the wasp is attracted to several of the isomers of this compound.

Aldrich et al. (1986a) found that yellowjacket wasps (*Vespula maculifrons* (Buysson) (Hymenoptera: Vespidae)) were attracted to synthetic pheromone mixtures for *Podisus* spp., as were honey bees (*Apis mellifera* L.), which may have been incidentally attracted because the *Podisus* pheromone contains components also found in floral volatiles. Aldrich and Barros (1995) demonstrated attraction of male crab spiders *Xysticus ferox* (Hentz) (Araneae: Thomisidae) to (*E*)-2-octenal and (*E*)-2-decenal, but suggested that this probably represented semiochemical overlap with a putative spider sex pheromone rather than any indication of kairomonal prey detection.

15.4.4 Predatory Stink Bugs Exploiting Semiochemicals of Their Prey

Several predatory stink bug species (Asopinae) have been studied in relation to semiochemically mediated interactions with their prey and the prey's host plants. For example, Yasuda (1997) showed that *Eocanthecona furcellata* was attracted to volatiles, including several saturated hydrocarbons, from caterpillars of its noctuid prey *Spodoptera litura* (F.). Conversely, in a recent study, larvae of the Colorado potato beetle, *Leptinotarsa decemlineata* Say, were shown to reduce their feeding when predatory *Podisus maculiventris* were present. The evidence suggested that this effect was mediated by olfaction and, to a lesser extent, visual cues (Hermann and Thaler 2014). The effect was stronger for male bugs, leading the authors to suggest that application of the male-produced *P. maculiventris* aggregation pheromone might be useful in indirectly reducing pest feeding damage, in addition to recruiting predatory bugs (Hermann and Thaler 2014).

15.4.5 Tritrophic Interactions Involving Pentatomoid Bugs

Tritrophic interactions with Pentatomoidea may involve several different scenarios, with stink bugs being either the top level (predator), or the middle level (prey). Plants produce a great diversity of volatile and nonvolatile compounds, some of which are induced by herbivores. Both constitutive and induced phytochemicals influence herbivores directly in terms of host preference and feeding choices, and indirectly through attraction of their natural enemies (Poelman et al. 2008). Such "herbivore-induced plant volatiles" (HIPVs) have been suggested to be exploited by predators and parasitoids to locate prey or hosts, respectively, because these induced volatiles are reliable indicators of host presence, and because these compounds may be emitted in much higher amounts than volatiles from the herbivore itself (Vet and Dicke 1992).

The predatory bug *Perillus bioculatus* (F.), which specializes on chrysomelid beetles, is attracted to plant volatiles induced by Colorado potato beetle, *Leptinotarsa decemlineata*, feeding on potato, and less strongly to volatiles induced by mechanical damage to the plant (van Loon et al. 2000; Weissbecker et al. 1999, 2000). Two generalist predatory pentatomids, *Podisus maculiventris* and *P. nigrispinus* (Dallas), also are attracted to herbivore-induced plant volatiles (Sant'Ana and Dickens 1998, Greenstone and Dickens 2005). In a recent field study of potential tritrophic interactions, Kelly et al. (2014) tested responses of *Podisus maculiventris* to lures releasing either methyl salicylate (an herbivore-induced plant volatile) or *P. maculiventris* aggregation pheromone. The pheromone enhanced retention of bugs, but only in dry weather, and both lures caused localized aggregations of the bugs around the lures, drawing them away from areas in which larval sphingid prey were concentrated, consistent with Kaplan's (2012) suggestion that HIPVs might produce undesirable results under some circumstances of field deployment. Dickens (1999) showed that *P. maculiventris* and *Perillus bioculatus* were not attracted to common green leaf volatiles such as (Z)-3-hexen-1-ol and (E)-2-hexenal, proposing that unlike several other HIPVs tested, these are not reliable cues of prey presence for these predators.

In contrast to predatory species, herbivorous pentatomids apparently are not attracted to HIPVs, and indeed, they may avoid plants damaged by feeding to limit competition with conspecifics or to avoid attack by natural enemies that may be attracted to HIPVs. For example, in Y-olfactometer bioassays, *Euschistus heros* was not attracted to HIPVs emitted from soybean plants damaged by conspecifics but was attracted to constitutive volatiles from soybean. Similarly, *Tibraca limbativentris* females (but not males) preferred volatiles emitted by undamaged rice plants compared to volatiles emitted by rice plants fed upon by conspecifics (Melo Machado et al. 2014).

Several studies have shown that egg parasitoids exploit HIPVs indirectly to locate eggs of their hosts. For example, the egg parasitoid *Telenomus podisi* exploited HIPVs indirectly to locate *Euschistus heros* eggs (Moraes et al. 2005b, 2008a; Michereff et al. 2011) or eggs of another host, *Tibraca limbativentris* (Melo Machado et al. 2014). In contrast, another egg parasitoid, *Trissolcus basalis*, showed no preference for constitutive versus induced volatiles produced by rice plants fed on by *T. limbativentris* (Melo Machado et al. 2014). Recently, Michereff et al. (2014) reported that *T. podisi* was attracted to a soybean cultivar that released higher amounts of constitutive and herbivore-induced volatiles compared to other cultivars, and, in this cultivar, *Euschistus heros* eggs were more heavily parasitized than in other cultivars. In addition, parasitism occurred during the R5-R7 stage of soybean cultivation when *E. heros* populations are high, so the effect of the parasitoids was maximized (Borges et al. 2011, Michereff et al. 2014). Colazza et al. (2004) reported that *Nezara viridula* feeding on beans also elicited HIPVs that attracted the egg parasitoid *Tr. basalis*.

Oviposition by herbivores also may induce production of plant volatiles that attract parasitoids, and/or nonvolatiles that arrest them. Conti et al. (2010) demonstrated that with *Murgantia histrionica* hosts, a complex of host- and plant-produced compounds aided host egg location by the egg parasitoid *Trissolcus brochymenae*. Plant volatiles were apparently adsorbed into the plant epicuticular wax layer (Frati et al. 2013). Oviposition-induced plant volatiles now have been demonstrated for multiple herbivore-parasitoid pairs (Hilker and Fatouros 2015), but, to date, this is the only known system involving Pentatomoidea.

15.5 Overview: Semiochemicals in Life History of Pentatomoids, and Practical Applications

15.5.1 Patterns of Production and Response to Pheromones

All of the pheromones described from pentatomoid adults to date are produced by male bugs (**Table 15.1**). This pattern of pheromone emission by males, shared with diverse Coleoptera such as Curculionidae, Chrysomelidae, and some Cerambycidae, may be correlated to males of these species being the risk-taking sex that disperses to new habitats and locates new resources, especially in environments where necessary resources are ephemeral in space and time (Aldrich et al. 1984, Millar 2005).

The response pattern to the male-produced pheromones is divided roughly equally between those that appear to attract only females, and those that attract both sexes (and nymphs, when tested). Aggregation pheromones are less obvious in function than sex pheromones, but ecologists have postulated several fitness advantages associated with aggregation to produce densities higher than would occur without a semiochemical signal (Wyatt 2014). First, mates may be easier to find at higher densities, which overlaps with a strictly sexual function but also includes the phenomenon of attraction of males to signals of other males, which also are attractive to females. This exploitation of the signals of others allows a cheating male a chance to procure a mate without incurring the costs of producing and emitting pheromone (e.g., biosynthetic cost, becoming more apparent to natural enemies) (Wertheim et al. 2005, Cardé 2014). A second possible driving force for formation and persistence of aggregations is the more efficient use of resources such as host plants, based on nutritional benefits of gregarious feeding (Karban and Agrawal 2002). For piercing-sucking insects, this translates to the preferential supply of nutrients by the host plant, increasing fecundity (Dixon and Wratten 1971, Lopez et al. 1989, Awmack and Leather 2002). Third, aggregations can result in increased protection from natural enemies by providing more apparent aposematic visual or allomonal chemical signals. Benefits also can arise from being in groups where natural enemies are not density-dependent in their attack, causing a dilution of risk of attack, especially to individuals surrounded by other conspecifics (Wertheim et al. 2005). Finally, there may be some fitness improvement of abiotic environmental conditions under higher density; for example, Lockwood and Story (1986) found acceleration of development with higher density under cool (but not warm) conditions for early instars of *Nezara viridula*, as well as better protection from most predators. This effect is not only weather-dependent, but instar-dependent; mortality increased with density for later instars (Kiritani 1964). These potential mechanisms in favor of aggregation are not mutually exclusive, and may be quite complex, raising important questions of function and teleology (Cardé 2014). Nevertheless, to be selected for, there must be a net benefit, and there are clearly potential negative effects of higher density, including competition for food and mates, increased disease transmission, increased apparency to natural enemies, and the metabolic costs of aggregating, including pheromone production and the energetic costs of flying or walking to join an aggregation.

With pheromone emission under control of the sender, excessive densities also can be discouraged, with an inverse density response observed, for instance, with *Murgantia histrionica* males in rates of emission of aggregation pheromone (Zahn et al. 2008). This is likely a mechanism to avoid aggregation of too many bugs on host plants or patches. For this same species, Cabrera Walsh et al. (2015) found that although a synthetic mixture of the stereoisomers of murgantiol containing the aggregation pheromone was highly attractive to bugs, it did not retain bugs on host plants, and numbers of *M. histrionica* per baited plant reached a plateau due to a net balance of immigration and emigration.

Many factors are potentially significant in selection for aggregation pheromone production, release, and response. There must be a net benefit within species both to the sender and receiver, which accrues frequently enough, and also at low enough densities, for aggregation pheromones to be selected for over evolutionary time, presumably from pre-existing odor production and detection mechanisms (Wertheim et al. 2005). However, given the flexibility in production, release, and response to pentatomoid pheromones, evolution of this capability does not oblige individuals to send or to respond to the signals at any given time, particularly as adult bugs are relatively long-lived and can replenish their energy stores by feeding. This flexibility is an asset to stink bugs, even as it makes studying their chemical ecology more difficult!

15.5.1.1 Cross-Species Attraction

One of the most curious and still unexplained phenomena in pentatomid semiochemistry is the cross-attraction of a number of species to the pheromones of other species. This has been reported for both adults and nymphs (**Table 15.4**). In some cases, this may represent overlap in chemistry of the pheromones of the respective species, but, in other cases, it clearly does not. One of the better-known examples is the attraction of *Halyomorpha halys* to the pheromone of *Plautia stali*, methyl (2*E*,4*E*,6*Z*)-2,4,6-decatrienoate. This compound also attracts the Asian species *Glaucias subpunctatus*, for which no pheromone is known, and the North American species *Chinavia hilaris*, which, like *H. halys*, does not produce this compound (Aldrich et al. 2009, Tillman et al. 2010). There are other examples attributable to chemical overlap, for instance the Japanese native *Nezara antennata* and the exotic *N. viridula*, which share their two pheromone components (Moraes et al. 2008b, Shimizu and Tsutsumi 2011). There are still other cases in which either imprecise stereochemistry or field isomerization of pheromone components (Khrimian et al. 2008) may account for cross-attraction, leaving open the possibility that there may or may not be cross-attraction between living bugs in the field. Depending on the diel pattern of natural emissions, UV-light-driven isomerization may occur during daylight yet be undetected by laboratory collections in which pheromone-emitting substrates are subjected to lower light intensity and aerations often are conducted in UV-opaque collection vessels. Diel cycles of pheromone emission and response may make cross-attraction more or less likely in the field (discussion in Moraes et al. 2005a).

TABLE 15.4

Cross-Attractancy of Pheromones in the Pentatomidae

Chapter Section	Species	Cross-Attracted to Pheromones of:	Pheromone of This Species Cross-Attracts:
15.2.2	*Bagrada hilaris*	? *M. histrionica,*? *H. halys*	
15.2.3	*Chinavia hilaris*	*P. stali*	
15.2.4	*Chlorochroa ligata*	*C. uhleri*	*C. uhleri*
15.2.4	*Chlorochroa uhleri*	*C. ligata*	*C. ligata*
15.2.5	*Edessa meditabunda*	*T. perditor, E. heros*	
15.2.7	*Euschistus heros*	*T. perditor*	*P. guildinii, E. meditabunda*
15.2.9	*Halyomorpha halys*	*P. stali, M. histrionica*	? *B. hilaris, M. histrionica, P. stali*
15.2.10	*Murgantia histrionica*	*H. halys*	? *B. hilaris, H. halys, P. stali*
15.2.11	*Nezara viridula*	*N. antennata, T. perditor*	*N. antennata*
15.2.11	*Nezara antennata*	*N. viridula*	*N. viridula*
15.2.14	*Piezodorus guildinii*	*E. heros, T. perditor*	
15.2.14	*Piezodorus hybneri*	*Riptortus clavatus* (Alydidae)	
15.2.15	*Plautia stali*	*H. halys, M. histrionica*	*H. halys, C. hilaris, Glaucias subpunctatus* (Walker) (Pentatomidae)
15.2.18	*Thyanta perditor*		*E. heros, E. meditabunda, P. guildinii, N. viridula*

It is important to bear in mind that the male-produced pheromones of pentatomids are used for long-distance attraction, whereas vibrational and visual signals mediate short-range mate location (Čokl and Millar 2009, Millar et al. 2010). Thus, there may be minimal cost to the heterospecific sender or the receiver in the olfactory attraction, because a potential sexual interaction is rapidly and efficiently determined as heterospecific once the sender and receiver are on the same substrate. Thus, effective short-range communication mechanisms may mitigate selection pressures to avoid cross-attraction to long-range pheromones (see Symonds and Elgar 2008).

Analogous to the potential intraspecific benefits of aggregation pheromones outlined above, heterospecific benefits have also been proposed to account for cross-species attraction. Three benefits to the heterospecific pheromone receiver have been proposed specifically for the Pentatomoidea: (1) location of food plants (Tada et al. 2001a, Endo et al. 2006), (2) aggregation at suitable overwintering locations (Khrimian et al. 2008), and (3) density-based protection from tachinid fly parasites, which are attracted to many pentatomid pheromones (Aldrich et al. 2007). Nymphal aggregations also may be heterospecific, probably involving improvement of the physical environment and protection from natural enemies, analogous to intraspecific benefits (Lockwood and Story 1986), and these are mediated chemically through common allomonal chemicals (Fucarino et al. 2004). Benefits also may accrue to the sender, rendering the interaction synomonal, under each of the three reasons for receiver attraction cited above.

15.5.1.2 Multiple Components: Ratios and Variability

Most stink bug pheromones probably consist of at least two components: of the 46 species listed in **Table 15.1**, 14 are undefined, whereas 32 have one or more components that have been shown to mediate behavioral responses. Of these 32 species, 14 clearly require multiple components, and only seven species appear to have pheromones consisting of single components. The remaining 11 may have single or multiple components, which in these cases can only be determined by further stereochemical analysis and/or synthesis, coupled with behavioral bioassays.

Component ratios may vary among geographic populations (e.g., *Nezara viridula*, Aldrich et al. 1987, 1993, but disputed by Ryan et al. 1995), with age (*Piezodorus hybneri*, Endo et al. 2012), and with nutrition (*Euschistus heros*; Moraes et al. 2008c). Among individual males of *P. hybneri*, Endo et al. (2012) found that ratios of the three components in pheromone emissions were highly variable among individuals and over time. In contrast, Miklas et al. (2000) found that although individual male *N. viridula* varied as to the ratio of their two pheromone components, this ratio was almost constant for any given male in spite of large variations in overall quantities produced. The only three studies to look closely at pheromone production by individuals (Ryan et al. 1995, Miklas et al. 2000, Endo et al. 2012) all demonstrated substantial variability in both quantities produced and blend ratios, and this pattern likely is widespread among pentatomids. Moraes et al. (2008c) found that well-fed male *E. heros* produced all three components of the attractant pheromone, but those provided only water or neither food nor water, stopped production of two of the three components while increasing production of allomonal compounds such as (*E*)-2-hexenal. For all of the above reasons, researchers must be cautious in designating any set ratio of multiple components as "the" pheromone ratio for a particular stink bug species.

The fact that the ratio of components and/or the presence of one or more unnatural isomers in synthetic pheromone blends apparently is not critical to pheromone-induced responses for many stink bugs is contrary to findings for the pheromone blends of most other insects. This anomaly suggests that there may be more subtle information encoded in the blends, in addition to the attractant signal, and so more careful study of the variability in stink bug pheromone blends may prove fruitful. The variation in semiochemical emissions, and the responses to them, must be considered when optimizing attractants for monitoring and management purposes. Practical optimization of pheromone synthesis may entail use of more cost-efficient but non-stereoselective syntheses that produce mixtures of products that differ from the natural ratio(s). For instance, as described above, the pheromone blend of *Halyomorpha halys* was found to be a 3.5:1 mixture of two stereoisomers, (3*S*,6*S*,7*R*,10*S*)-and (3*R*,6*S*,7*R*,10*S*)-10,11-epoxy-1-bisabolen-3-ols (Khrimian et al. 2014a). Making sufficient quantities of these pure isomers for practical applications would be prohibitively expensive. However, a cost-effective synthesis from racemic citronellal yielded these two isomers in 1:1.5-2.0 ratio (Leskey et al. 2015b) admixed with the other 14 possible isomers. Remarkably, based on the quantity of the major component, the SSRS isomer, this mixture of 16 isomers of 10,11-epoxy-1-bisabolen-3-ol was as attractive in field tests as the 3.5:1 mixture of the two pure isomers (Leskey et al. 2015b), illustrating that off-ratios of pheromone components and presence of unnatural isomers may be unimportant in some cases. In contrast, in tests with the two components of *Murgantia histrionica* pheromone, Weber et al. (2014b) found that bugs were most strongly attracted to lures containing the natural 1.4:1 blend of (3*S*,6*S*,7*R*,10*S*)- and (3*S*,6*S*,7*R*,10*R*)-10,11-epoxy-1-bisabolen-3-ols, with decreased attraction to off-ratio lures, which in turn were superior to either single component.

15.5.1.3 Dose and Release Rates of Stink Bug Semiochemicals

Studies quantifying pheromone release by male pentatomids are much less numerous than those documenting identities of components and their proportions. Studies have demonstrated release rates of active components in the range of a few nanograms to a few micrograms per day per bug, usually with distinct diurnal rhythms. Also, when density effects were studied, in at least some species, males were found to produce less pheromone per bug at higher densities. This may account for some of the variability in quantities reported within species.

For example, Zahn et al. (2008) found that males of *Murgantia histrionica* release a blend of 10,11-epoxy-1-bisabolen-3-ols at ≈60 ng/bug/hour during peak release during the early afternoon, for a daily total of approximately 1 µg. The release rate was attenuated significantly (≈6-fold) when 10 or more male bugs were held in a 400 ml aeration chamber compared to individual bugs or groups of five bugs. Borges et al. (2007) found that *Piezodorus guildinii* males release sesquiphellandrene **63** at a mean rate of ≈40 ng/bug/day, at a density of 20 bugs within a 1-liter glass vessel, but aerations were not performed at other densities. Endo et al. (2012) measured the 3-component pheromone emissions by individual male *Piezodorus hybneri* each day from 1 to 16 days of age. Emission started 3 to 6 days after the final molt, and typically peaked at ~1 µg/bug/day on day 11 (Endo et al. 2012), exceeding the original estimate of ~200ng/bug/day of Leal et al. (1998). Bugs that were solvent extracted had a mean of ~10 µg/bug of total pheromone constituents (Endo et al. 2007). Ho & Millar (2001a, b) found that male *Chlorochroa* spp. (10 to 20 virgin males in 1 liter vessels) emitted means of ~320, ~240, and ~77 ng/bug/day for *C. ligata, C. sayi, and C. uhleri* respectively, each with an evening peak emission, of their respective major pheromone components. Male *Thyanta perditor* bugs emitted about 200 ng/bug/day of their sex pheromone **73** (Moraes et al. 2005a). *Chinavia ubica* males emitted a mean of ~3.0 and

~0.2 µg/bug/day of bisabolene epoxides **1** and **2,** respectively (Blassioli-Moraes et al. 2012). In contrast, the quantities of defensive compounds produced by both sexes of Pentatomoidea typically are much higher than those of the male-produced pheromones (Aldrich 1988; see, for example, quantification by de Oliveira et al. [2013]).

Dose-response studies testing male-produced aggregation pheromones in field trials typically have shown increasing attraction and trap captures up to very high doses and, presumably, release rates. Leskey et al. (2015b) quantified captures of *Halyomorpha halys* in pheromone-baited traps as well as attraction to the vicinity of lures for four orders of magnitude of lure loading, from 0.02 to 200 mg of **37** and 0.008 to 80 mg of **38**, with a clear dose response, with or without the addition of **66**, the *Plautia stali* pheromone, the latter in a commercial lure reported to contain 66 mg. Weber et al. (2014b), using rubber septum lures loaded with 0.02 to 20 mg each of **37** and **44**, showed increased attraction of *Murgantia histrionica* adults with increasing doses of the blend of components, with or without host plants. However, the presence of plants strongly enhanced overall captures. Tests of the interaction between component ratios and dose, as well as quantification of release rates, is ongoing for *H. halys* (T. Leskey, pers. comm.) and *M. histrionica* (D. W., unpublished data). Plastic pellets formulated by Fuji Flavor Co. (Japan) with different doses of methyl-2,6,10-trimethyltridecanoate (0.004, 0.01 and 1 mg) were evaluated in laboratory bioassays with *E. heros* females, and all doses tested attracted significantly more females than the clean air controls (M. Borges, pers. comm.).

In field bioassays, most studies have tested lures loaded with 10 or more bug equivalents per lure, but there have been few attempts to quantify release rates and the ratios actually emitted. However, Millar et al. (2010) showed that differing release rates of *Chlorochroa sayi* pheromone components from rubber septa resulted in changes in emitted blend ratios over time. Borges et al. (2011), using a lure formulated by Fuji Flavor (Japan) containing 1 mg of methyl-2,6,10-trimethyltridecanoate, showed that from day 12 until day 50, the release rate of the pheromone was ≈10 ng/day. This amount is approximately 1,000 times less than that produced by male *E. heros* per day (Zhang et al. 2003, Moraes et al. 2008c), but, nevertheless, the pheromone traps attracted females until day 49 (Borges et al. 2011). A variety of slow-release technologies may be required to optimize release rates and field longevities of stink bug pheromones of different chemistries.

15.5.2 Current Role of Stink Bug Semiochemicals in Pest Management

Use of stink bug pheromones in pest management currently is limited to monitoring the phenology and abundance of populations of a few major pest species to determine the need for suppressive treatments. Examples include *Euschistus conspersus* in California tomatoes (Cullen and Zalom 2007), *E. heros* in Brazilian soybeans (Borges et al. 2011), and *Plautia stali* and other fruit-piercing bugs in tree fruit in Japan (Adachi et al. 2007, Yamanaka et al. 2011). Furthermore, Leskey et al. (2015a) employed hundreds of traps to monitor the abundance and phenology of *Halyomorpha halys* across ten states in the United States as part of a continuing effort to define infestations and damage to crops and to monitor for this invasive pest.

Use of stink bug pheromones in mass trapping, trap-cropping, or other suppression tactics has not been implemented although there is continuing interest in development of such methods. Several experiments have tested "trap plants" treated with aggregation pheromones to attract pest species, and, in some cases, the plants also were treated with systemic insecticides to kill bugs that were attracted (Katase et al. 2007 and Yamanaka et al. 2011 for *P. stali*; D. W., unpublished data, for *Murgantia histrionica*). The "trap tree" concept currently is being tested, using the combined attractants for *Halyomorpha halys* and other stink bug pests of tree fruit in North America (T. Leskey, pers. comm.). None of these methods has yet advanced to commercial implementation. Successful implementation of management of stink bugs based on semiochemicals will require careful evaluation of each species for the likelihood of success, taking into account the pheromone chemistry, the life history and host specificity of the target, ecological (abiotic and biotic) characteristics of the agroecosystem (Millar 2007), and the costs of synthesis and deployment of pheromone devices.

Despite the frequently noted effects of pentatomoid semiochemicals on natural enemies, particularly tachinid flies and hymenopteran egg parasitoids, no methods based on host pheromones or allomones have been developed to manage or augment these natural enemies to improve the biological control of stink bugs. However, for a related coreoid bug species, Alim and Lim (2011) used the aggregation phero-mone of the soybean pest *Riptortus pedestris* to trap the adult bugs. They also used the same pheromone to attract parasitoids, particularly *Ooencyrtus nezarae* Ishii (Hymenoptera: Encyrtidae), to host egg masses, that were attached to the trap within screening that allowed parasitoids to freely enter and exit. The design prevented the entrapment of the parasitoids that were attracted, which can be a problem with less selective trap designs. Treated plots showed higher egg parasitism, marginally lower bug numbers, and significantly lower bug damage to soybean pods (Alim and Lim 2011).

15.5.3 Future Directions in Research and Applications

The proper identification of semiochemicals and natural semiochemical blends requires the sequential application of:

(1) rigorous analytical chemistry to identify all of the components in the volatiles released by the target species;

(2) stereospecific syntheses to produce libraries of pure compounds for verification of identifica-tions and for bioassays testing blends of precisely known compositions; and

(3) laboratory and field bioassays to work out the minimum blend of components that is both nec-essary and sufficient to elicit the same behavioral responses as the natural pheromone.

Because of the cost and difficulty associated with production of multigram or larger quantities of single isomers of semiochemicals with relatively complex structures, it may not be possible to use pure isomers for practical applications or even for experimental trials. However, if less expensive mixtures of isomers are to be used, they must be rigorously tested under field conditions to ensure that the unnatu-ral isomers and other components present in the mixtures do not antagonize the desired behavioral responses. For at least some pentatomoids, ratios of pheromone components and even the presence of multiple unnatural isomers and other components have not proven critical to obtaining good responses in the field. However, in other cases, pheromones of different species may interfere with each other (Millar et al. 2010).

The conservation of the basic biosynthetic pathways that produce classes of compounds such as terpenoids and acetogenins in insects and other organisms may provide opportunities to use engi-neered organisms to produce pentatomid and other insect semiochemicals. Synthetic chemists already exploit chiral phytochemicals such as citronellal as building blocks to synthesize semiochemicals. Genomic and transcriptomic research on stink bugs (i5K Consortium 2013, Sparks et al. 2014) will help to elucidate the details of semiochemical biosynthesis and also may provide the opportunity to use genetically transformed organisms such as yeasts or plants as "factories" for pheromone produc-tion (e.g., Ding et al. 2014). Such transformed plants also could be deployed in the field as possible dead-end trap crops, as Møldrup et al. (2012) demonstrated with a *Nicotiana* sp. that was attractive to ovipositing female diamondback moths (*Plutella xylostella* (L.)(Lepidoptera: Plutellidae) but lethal to their offspring.

Regardless of the methods used to produce pentatomoid semiochemicals, to paraphrase Millar (2007, 2014), the devil is in the deployment. Use in large scale, real-world pest management will require easy-to-use and effective tactics – be it monitoring, mass trapping, attract-and-kill, or trap-cropping – to deploy pheromones or other semiochemicals for stink bug management. Furthermore, these methods must be economically competitive with alternative methods that are available for managing a particu-lar pest in order to have any chance of being adopted for commercial use. Effective, environmentally friendly pest management takes place in the context of the agroecosystem and its surroundings, with one being conscious of the diverse and mobile nature of Pentatomoidea and other pest and beneficial species. Special attention must be paid to the known and powerful attraction of natural enemies to stink bug semiochemicals that, depending on the exact method of deployment, could have effects rang-ing from effective augmentation to severe depression of natural enemies that help to control stink bug populations.

15.6 Acknowledgments

We thank Miguel Borges (Embrapa Recursos Genéticos e Biotecnologia, Brasília, Brazil) for providing comments on the allomones section. Miguel Borges, Tracy Leskey (USDA ARS, Kearneysville, WV), and Ray St. Leger (University of Maryland Department of Entomology) graciously provided personal communications regarding ongoing research.

15.7 References Cited

Adachi, I., K. Uchino, and F. Mochizuki. 2007. Development of a pyramidal trap for monitoring fruit-piercing stink bugs baited with *Plautia crossota stali* (Hemiptera: Pentatomidae) aggregation pheromone. Applied Entomology and Zoology 42: 425-431.

Afsheen, S., X. Wang, R. Li, C. S. Zhu, and Y. G. Lou. 2008. Differential attraction of parasitoids in relation to specificity of kairomones from herbivores and their by-products. Insect Science 15: 381–397.

Aldrich, J. R. 1985. Pheromone of a true bug (Hemiptera-Heteroptera): Attractant for the predator, *Podisus maculiventris*, and kairomonal effects, pp. 95-119. *In* T. E. Acree and D. M. Soderlund (Eds.), Semiochemistry: flavors and pheromones. de Gruyter Press, Berlin. 289 pp.

Aldrich, J. R. 1988. Chemical ecology of the Heteroptera. Annual Review of Entomology 33: 211–238.

Aldrich, J. R. 1995a. Chemical communication in the true bugs and parasitoid exploitation, pp. 318-363. *In* R. T. Cardé and W. J. Bell (Eds.), Chemical Ecology of Insects II, Chapman & Hall, New York. 433 pp.

Aldrich, J. R. 1995b. Testing the "new associations" biological control concept with a tachinid parasitoid (*Euclytia flava*). Journal of Chemical Ecology 211: 1031–1042.

Aldrich, J. R., and T. M. Barros. 1995. Chemical attraction of male crab spiders (Araneae, Thomisidae) and kleptoparasitic flies (Diptera, Milichiidae and Chloropidae). Journal of Arachnology 23: 212–214.

Aldrich, J. R., and W. R. Lusby. 1986. Exocrine chemistry of beneficial insects: Male-specific secretions from predatory stink bugs (Hemiptera: Pentatomidae). Comparative Biochemistry and Physiology Part B: Comparative Biochemistry 85: 639–642.

Aldrich, J. R., and P. Puapoomchareon. 1996. Management of predaceous hemipterans with semiochemicals: Practice and potential. *In* Proceedings of the XX International Congress of Entomology, Firenze, Italy, p. 625.

Aldrich, J. R., and A. Zhang. 2002. Kairomone strains of *Euclytia flava* (Townsend), a parasitoid of stink bugs. Journal of Chemical Ecology 28: 1565–1582.

Aldrich, J. R., M. S. Blum, H. A. Lloyd, and H. M. Fales. 1978. Pentatomid natural products. Chemistry and morphology of the III-IV dorsal abdominal glands of adults. Journal of Chemical Ecology 4: 161–172.

Aldrich, J. R., J. P. Kochansky, and C. B. Abrams. 1984. Attractant for a beneficial insect and its parasitoids: pheromone of the predatory spined soldier bug, *Podisus maculiventris* (Hemiptera: Pentatomidae). Environmental Entomology 13: 1031–1036.

Aldrich, J. R., W. R. Lusby, and J. P. Kochansky. 1986a. Identification of a new predaceous stink bug pheromone and its attractiveness to the eastern yellowjacket. Experientia 42: 583–585.

Aldrich, J. R., J. E. Oliver, W. R. Lusby, and J. P. Kochansky. 1986b. Identification of male-specific exocrine secretions from predatory stink bugs (Hemiptera, Pentatomidae). Archives of Insect Biochemistry and Physiology 3: 1–12.

Aldrich, J. R., J. E. Oliver, W. R. Lusby, J. P. Kochansky, and J. A. Lockwood. 1987. Pheromone strains of the cosmopolitan pest *Nezara viridula* (Heteroptera: Pentatomidae). Journal of Experimental Zoology 244: 171–175.

Aldrich, J. R., W. R. Lusby, B. E. Marron, K. C. Nicolaou, M. P. Hoffmann, and L. T. Wilson. 1989. Pheromone blends of green stink bugs and possible parasitoid selection. Naturwissenschaften 76: 173–175.

Aldrich, J. R., M. P. Hoffmann, J. P. Kochansky, W. R. Lusby, J. E. Eger, and J. A. Payne. 1991. Identification and attractiveness of a major pheromone component for Nearctic *Euschistus* spp. stink bugs (Heteroptera: Pentatomidae). Environmental Entomology 20: 477–483.

Aldrich, J. R., H. Numata, M. Borges, F. Bin, G. K. Waite, and W. R. Lusby. 1993. Artifacts and pheromone blends from *Nezara* spp. and other stink bugs (Heteroptera: Pentatomidae). Zeitschrift für Naturforschung (Ser: C) 48: 73–79.

Aldrich, J. R., J. W. Avery, C. J. Lee, J. C. Graf, D. J. Harrison, and F. Bin. 1996. Semiochemistry of cabbage bugs (Heteroptera: Pentatomidae: *Eurydema* and *Murgantia*). Journal of Entomological Science 31: 172–182.

Aldrich, J. R., A. Khrimian, A. Zhang, and P. W. Shearer. 2006. Bug pheromones (Hemiptera, Heteroptera) and tachinid fly host-finding. Denisia 19: 1015–1031.

Aldrich, J. R., A. Khrimian, and M. J. Camp. 2007. Methyl 2,4,6-decatrienoates attract stink bugs and tachinid parasitoids. Journal of Chemical Ecology 33: 801–815.

Aldrich, J. R., A. Khrimian, M. J. Camp, and X. Chen. 2009. Semiochemically based monitoring of the invasion of the brown marmorated stink bug and unexpected attraction of the native green stink bug (Heteroptera: Pentatomidae) in Maryland. Florida Entomologist 92: 483–491.

Aliabadi, A., J. A. Renwick, and D. W. Whitman. 2002. Sequestration of glucosinolates by harlequin bug *Murgantia histrionica*. Journal of Chemical Ecology 28: 1749–1762.

Alim, M. A., and U. T. Lim. 2011. Refrigerated eggs of *Riptortus pedestris* (Hemiptera: Alydidae) added to aggregation pheromone traps increase field parasitism in soybean. Journal of Economic Entomology 104: 1833–1839.

Alizadeh, B. H., S. Kuwahara, W. S. Leal, and H.-C. Men. 2002. Synthesis of the racemate of (Z)-*exo*-α-bergamotenal, a pheromone component of the white-spotted spined bug, *Eysarcoris parvus* Uhler. Bioscience, Biotechnology, and Biochemistry 66: 1415–1418.

Ando, K. 1998. (Z)-Selective Horner-Wadsworth-Emmons reaction of α-substituted ethyl (diarylphosphono)-acetates with aldehydes. Journal of Organic Chemistry 63: 8411–8416.

Awmack, C. S., and S. R. Leather. 2002. Host plant quality and fecundity in herbivorous insects. Annual Review of Entomology 47: 817–844.

Baker, R., M. Borges, N. G. Cooke, and R. H. Herbert. 1987. Identification and synthesis of (Z)-(1'S,3'R,4'S)-(-)-2-(3',4'-epoxy-4'-methylcyclohexyl)-6-methylhepta-2,5-diene, the sex pheromone of the southern green stink bug, *Nezara viridula* (L.). Journal of the Chemical Society, Chemical Communications 6: 414–416.

Bhonsle, J. B., V. H. Deshpande, and T. Ravindranathan. 1994. Synthesis of (+)-zingiberene. Indian Journal of Chemistry 33B: 313–316.

Bin, F., S. B. Vinson, M. R. Strand, S. Colazza, and W. A. Jones Jr. 1993. Source of an egg kairomone for *Trissolcus basalis*, a parasitoid of *Nezara viridula*. Physiological Entomology 18: 7-15.

Biovision Foundation. 2014. Bagrada Bug. In Infonet-Biovision. Accessed 1 November 2014, from http://www.infonet-biovision.org/default/ct/103/pests.

Blassioli-Moraes, M. C., R. A. Laumann, M. W. M. Oliveira, C. M. Woodcock, P. Mayon, A. Hooper, J. A. Pickett, M. A. Birkett, and M. Borges. 2012. Sex pheromone communication in two sympatric Neotropical stink bug species *Chinavia ubica* and *Chinavia impicticornis*. Journal of Chemical Ecology 38: 836–845.

Blomquist, G. J., and A.-G. Bagnères. 2010. Introduction: History and overview of insect hydrocarbons, pp. 11-18. *In* G. J. Blomquist. and A.-G. Bagnères (Eds.), Insect Hydrocarbons: Biology, Biochemistry, and Chemical Ecology. Cambridge University Press UK. 492 pp.

Borges, M., and J. R. Aldrich. 1992. Instar-specific defensive secretions of stink bugs (Heteroptera: Pentatomidae). Experientia 48: 893–896.

Borges, M., and J. R. Aldrich. 1994. Attractant pheromone for Nearctic stink bug, *Euschistus obscurus* (Heteroptera: Pentatomidae): insight into a Neotropical relative. Journal of Chemical Ecology 20: 1095–1102.

Borges, M., S. C. M. Leal, M. S. Tigano-Milani, and M. C. C. Valadares. 1993. Efeito do feromônio de alarme do percevejo verde, *Nezara viridula* (L.) (Hemiptera: Pentatomidae), sobre o fungo entomopatogênico *Metarhizium anisopliae* (Metsch.) Sorok. Anais da Sociedade Entomologica do Brasil 22: 505–512.

Borges, M., F. G. V. Schmidt, E. R. Sujii, M. A. Medeiros, K. Mori, P. H. G. Zarbin, and T. B. Ferreira. 1998. Field responses of stink bugs to the natural and synthetic pheromone of the neotropical brown stink bug, *Euschistus heros* (Heteroptera: Pentatomidae). Physiological Entomology 23: 202–207.

Borges, M., M. L. M. Costa, E. R. Sujii, M. Das G. Cavalcanti, G. F. Redígolo, I. S. Resck, and E. F. Vilela. 1999. Semiochemical and physical stimuli involved in host recognition by *Telenomus podisi* (Hymenoptera: Scelionidae) toward *Euschistus heros* (Heteroptera: Pentatomidae). Physiological Entomology 24: 227–233.

Borges, M., S. Colazza, P. Ramirez-Lucas, K. R. Chauhan, J. R. Aldrich, and M. C. B. Moraes. 2003. Kairomonal effect of walking traces from *Euschistus heros* (Heteroptera: Pentatomidae) on two strains of *Telenomus podisi* (Hymenoptera: Scelionidae). Physiological Entomology 28: 349–355.

Borges, M., M. Birkett, J. R. Aldrich, J. E. Oliver, M. Chiba, Y. Murata, R. A. Laumann, J. A. Barrigossi, J. A. Pickett, and M. C. B. Moraes. 2006. Sex attractant pheromone from the rice stalk stink bug, *Tibraca limbativentris* Stal. Journal of Chemical Ecology 32: 2749–2761.

Borges, M., J. G. Millar, R. A. Laumann, and M. C. B. Moraes. 2007. A male-produced sex pheromone from the neotropical redbanded stink bug, *Piezodorus guildinii* (W.). Journal of Chemical Ecology 33: 1235–1248.

Borges, M., M. C. B. Moraes, M. F. Peixoto, C. S. S. Pires, E. R. Sujii, and R. A. Laumann. 2011. Monitoring the neotropical brown stink bug *Euschistus heros* (F.) (Hemiptera: Pentatomidae) with pheromone-baited traps in soybean fields. Journal of Applied Entomology 135: 68–80.

Bruni, R., J. Sant'Ana, J. R. Aldrich, and F. Bin. 2000. Influence of host pheromone on egg parasitism by scelionid wasps: comparison of phoretic and nonphoretic parasitoids. Journal of Insect Behaviour 13: 165–173.

Cabrera Walsh, G., A. S. DiMeglio, A. Khrimian, and D. C. Weber. 2015. Marking and retention of harlequin bug, *Murgantia histrionica* (Hahn) (Hemiptera: Pentatomidae), on pheromone baited and unbaited plants. Journal of Pest Science, DOI 10.1007/s10340-015-0663-1.

Capinera, J. 2001. Handbook of Vegetable Pests. Academic Press, New York. 800 pp.

Cardé, R.T. 2014. Defining attraction and aggregation pheromones: teleological versus functional perspectives. Journal of Chemical Ecology 40: 519–520.

Chavan, S. P., V. D. Dhondge, S. S. Patil, Y. T. S. Rao, and C. A. Govande. 1997. Enantioselective total synthesis of (+)-laevigatin. Tetrahedron: Asymmetry 8: 2517–2518.

Čokl, A., and J. G. Millar. 2009. Manipulation of insect signaling for monitoring and control of pest insects, pp. 279-316. *In* I. Ishaaya and A. R. Horowitz [Eds.], Biorational control of arthropod pests: application and resistance management. Springer, New York. 408 pp.

Čokl, A., M. Zorović, and J. G. Millar. 2007. Vibrational communication along plants by the stink bugs *Nezara viridula* and *Murgantia histrionica*. Behavioural Processes 75: 40–54.

Colazza, S., G. Salerno, and E. Wajnberg. 1999. Volatile and contact chemicals released by *Nezara viridula* (Heteroptera: Pentatomidae) have a kairomonal effect on the egg parasitoid *Trissolcus basalis* (Hymenoptera: Scelionidae). Biological Control 16: 310–317.

Colazza, S., J. S. McElfresh, and J. G. Millar. 2004. Identification of volatile synomones, induced by *Nezara viridula* feeding and oviposition on bean spp., that attract the egg parasitoid *Trissolcus basalis*. Journal of Chemical Ecology 30: 945–964.

Colazza, S., G. Aquila, C. De Pasquale, E. Peri, and J. G. Millar. 2007. The egg parasitoid *Trissolcus basalis* uses *n*-nonadecane, a cuticular hydrocarbon from stink bug host *Nezara viridula*, to discriminate between female and male hosts. Journal of Chemical Ecology 33: 1405–1420.

Conti, E., and S. Colazza. 2012. Chemical ecology of egg parasitoids associated with true bugs. Psyche. article ID 651015.

Conti, E., G. Salerno, F. Bin, H. J. Williams, and S. B. Vinson. 2003. Chemical cues from *Murgantia histrionica* eliciting host location and recognition in the egg parasitoid *Trissolcus brochymenae*. Journal of Chemical Ecology 29: 115–130.

Conti, E., G. Salerno, B. Leombruni, F. Frati, and F. Bin. 2010. Short-range allelochemicals from a plant–herbivore association: a singular case of oviposition-induced synomone for an egg parasitoid. Journal of Experimental Biology 213: 3911–3919.

Cottrell T. E., P. J. Landolt, Q.-H. Zhang, and R. S. Zack. 2014. A chemical lure for stink bugs (Hemiptera: Pentatomidae) is used as kairomone by *Astata occidentalis* (Hymenoptera: Sphecidae). Florida Entomologist 97: 233–237.

Corrêa-Ferreira, B. S., and F. Moscardi. 1995. Seasonal occurrence and host spectrum of egg parasitoids associated with soybean stink bugs. Biological Control 5: 196–202.

Cribb, B. W., K. N. Siriwardana, and G. H. Walter. 2006. Unicellular pheromone glands of the pentatomid bug *Nezara viridula* (Heteroptera: Insecta): ultrastructure, classification, and proposed function. Journal of Morphology 267: 831–840.

Cullen, E. M., and F. G. Zalom. 2005. Relationship between *Euschistus conspersus* (Hem., Pentatomidae) pheromone trap catch and canopy samples in processing tomatoes. Journal of Applied Entomology 129: 505–514.

Cullen, E. M., and F. G. Zalom. 2006. *Euschistus conspersus* female morphology and attraction to methyl (2*E*,4*Z*)-decadienoate pheromone–baited traps in processing tomatoes. Entomologia Experimentalis et Applicata 119: 163–173.

Cullen, E. M., and F. G. Zalom. 2007. On-farm trial assessing efficacy of three insecticide classes for management of stink bug and fruit damage on processing tomatoes. Plant Health Progress (March): 10 pp.

De Pasquale, C., S. Guarino, E. Peri, and S. Colazza. 2007. Investigation of cuticular hydrocarbons from *Bagrada hilaris* genders by SPME/GC-MS. Analytical and Bioanalytical Chemistry 389: 1259–1265.

Dickens, J. C. 1999. Predator-prey interactions: olfactory adaptations of generalist and specialist predators. Agricultural and Forest Entomology 1: 47–54.

Ding, B.-J., P. Hofvander, H.-L. Wang, T. P. Durrett, S. Stymne, and C. Löfstedt. 2014. A plant factory for moth pheromone production. Nature Communications 5: 3353 (7 pp.).

Dixon, A. F. G., and S. D. Wratten. 1971. Laboratory studies on aggregation, size and fecundity in the black bean aphid, *Aphis fabae* Scop. Bulletin of Entomological Research 61: 97–111.

Eisner, T., M. Eisner, and M. Deyrup. 1991. Chemical attraction of kleptoparasitic flies to heteropteran insects caught by orb-weaving spiders. Proceedings of the National Academy of Sciences of the United States of America 88: 8194–8197.

Eliyahu, D., R. A. Ceballos, V. Saeidi, and J. X. Becerra. 2012. Synergy versus potency in the defensive secretions from nymphs of two Pentatomorphan families (Hemiptera: Coreidae and Pentatomidae). Journal of Chemical Ecology 38: 1358–1365.

Endo, N., T. Wada, Y. Nishiba, and R. Sasaki, R. 2006. Interspecific pheromone cross-attraction among soybean bugs (Heteroptera): does *Piezodorus hybneri* (Pentatomidae) utilize the pheromone of *Riptortus clavatus* (Alydidae) as a kairomone? Journal of Chemical Ecology, 32: 1605–1612.

Endo, N., T. Yasuda, K. Matsukur, T. Wada, S. E. Muto, and R. Sasaki. 2007. Possible function of *Piezodorus hybneri* (Heteroptera: Pentatomidae) male pheromone: effects of adult age and diapause on sexual maturity and pheromone production. Applied Entomology and Zoology 42: 637–641.

Endo, N., R. Sasaki, and S. Muto. 2010. Pheromonal cross-attraction in true bugs (Heteroptera): attraction of *Piezodorus hybneri* (Pentatomidae) to its pheromone versus the pheromone of *Riptortus pedestris* (Alydidae). Environmental Entomology 39: 1973–1979.

Endo, N., T. Yasuda, T. Wada, S. E. Muto, and R. Sasaki. 2012. Age-related and individual variation in male *Piezodorus hybneri* (Heteroptera: Pentatomidae) pheromones. Psyche 2012: ID 609572, 4 pp.

Fatouros, N. E., M. Dicke, R. Mumm, T. Meiners, and M. Hilker. 2008. Foraging behavior of egg parasitoids exploiting chemical information. Behavioural Ecology 19: 677–689.

Fávaro, C. F., M. A. C. de M. Rodrigues, J. R. Aldrich, and P. H. G. Zarbin. 2011. Identification of semiochemicals in adults and nymphs of the stink bug *Pallantia macunaima* Grazia (Hemiptera: Pentatomidae). Journal of the Brazilian Chemical Society 22: 58–94.

Fávaro, C. F., T. B. Santos, and P. H. G. Zarbin. 2012. Defensive compounds and male-produced sex pheromone of the stink bug, *Agroecus griseus*. Journal of Chemical Ecology 38: 1124–1132.

Fávaro, C. F., R. A. Soldi, T. Ando, J. R. Aldrich, and P. H. G. Zarbin. 2013. (6*R*,10*S*)-Pallantione: the first ketone identified as a sex pheromone in stink bugs. Organic Letters 15: 1822–1825.

Feener, D. H. Jr., and B. V. Brown. 1997. Diptera as parasitoids. Annual Review of Entomology 42: 73–97.

Frati, F., G. Salerno, and E. Conti. 2013. Cabbage waxes affect *Trissolcus brochymenae* response to short-range synomones. Insect Science 20: 753–762.

Fucarino, A., J. G. Millar, J. S. McElfresh, and S. Colazza. 2004. Chemical and physical signals mediating conspecific and heterospecific aggregation behavior of first instar stink bugs. Journal of Chemical Ecology 30: 1257–1269.

Funayama, K. 2008. Seasonal fluctuations and physiological status of *Halyomorpha halys* (Stål) (Heteroptera: Pentatomidae) adults captured in traps baited with synthetic aggregation pheromone of *Plautia crossota stali* Scott (Heteroptera: Pentatomidae). Applied Entomology and Zoology 52: 69–75.

Fürstner, A., and A. Schlecker. 2008. A gold-catalyzed entry into the sesquisabinene and sesquithujene families of terpenoids and formal total syntheses of cedrene and cedrol. Chemistry—A European Journal 14: 9181–9191.

Godfray, H. C. J. 1994. Parasitoids: Behavioural and Evolutionary Ecology. Princeton University Press, Princeton, New Jersey. 473 pp.

Gree, R., H. Tourbah, and R. Carrie. 1986. Fumaraldehyde monodimethyl acetal: an easily accessible and versatile intermediate. Tetrahedron Letters 27: 4983–4986.

Greenstone, M. H., and J. C. Dickens. 2005. The production and appropriation of chemical signals among plants, herbivores and predators, pp. 139-165. *In* P. Barbosa and I. Catellanos (Eds.), Ecology of predator-prey interactions. Oxford University Press, Oxford, UK. 416 pp.

Guarino, S., C. De Pasquale, E. Peri, G. Alonzo and S. Colazza. 2008. Role of volatile and contact pheromones in the mating behavior of *Bagrada hilaris* (Heteroptera: Pentatomidae). European Journal of
Entomology 105: 613–617.

Gunawardena, N. E., and H. M. W. K. B. Herath. 1991. Significance of medium chain *n*-alkanes as accompanying compounds in hemipteran defensive secretions: an investigation based on the defensive secretion
of *Coridius janus*. Journal of Chemical Ecology 17: 2449–2557.

Hagiwara, H., T. Okabe, H. Ono, V. P. Kamat, T. Hoshi, T. Suzuki, and M. Ando. 2002. Total synthesis of
bisabolane sesquiterpenoids, α-bisabol-1-one, curcumene, curcuphenol and elvirol: utility of catalytic
enamine reaction in cyclohexenone synthesis. Journal of the Chemical Society, Perkin Transactions I:
895–900.

Harris, V. E., and J. W. Todd. 1980. Male-mediated aggregation of male, female, and 5th-instar southern
green stink bugs and concomitant attraction of a tachinid parasite, *Trichopoda pennipes*. Entomologia
Experimentalis et Applicata 27: 117–126.

Haye, T., T. Gariepy, K. Hoelmer, J.-P. Rossi, J.-C. Streito, X. Tassus, and N. Desneux. 2015. Range expansion
of the invasive brown marmorated stinkbug, *Halyomorpha halys*: an increasing threat to field, fruit and
vegetable crops worldwide. Journal of Pest Science, DOI 10.1007/s10340-015-0670-2.

Hermann, S. L., and J. S. Thaler. 2014. Prey perception of predation risk: volatile chemical cues mediate non-
consumptive effects of a predator on a herbivorous insect. Oecologia 176: 669–676.

Higaki, M., and I. Adachi. 2011. Response of a parasitoid fly, *Gymnosoma rotundatum* (Linnaeus) (Diptera:
Tachinidae) to the aggregation pheromone of *Plautia stali* Scott (Hemiptera: Pentatomidae) and its
parasitism of hosts under field conditions. Biological Control 58: 215–221.

Hilker, M., and N. E. Fatouros. 2015. Plant responses to insect egg deposition. Annual Review of Entomology
60: 493–515.

Ho, H-Y., and J. G. Millar. 2001a. Identification and synthesis of a male-produced sex pheromone from the
stink bug *Chlorochroa sayi*. Journal of Chemical Ecology 27: 1177–1201.

Ho, H-Y., and J. G. Millar. 2001b. Identification and synthesis of male-produced sex pheromone components of
the stink bugs *Chlorochroa ligata* and *Chlorochroa uhleri*. Journal of Chemical Ecology 27: 2067–2095.

Ho, H.-Y., R. Kou, and H.-K. Tseng. 2003. Semiochemicals from the predatory stink bug *Eocanthecona furcellata* (Wolff): Components of metathoracic gland, dorsal abdominal gland, and sternal gland secretions. Journal of Chemical Ecology 29: 2101–2114.

Ho, H.-Y., Y. C. Hsu, Y. C. Chuang, and Y. S. Chow. 2005. Effect of rearing conditions on production of sternal
gland secretion, and identification of minor components in the sternal gland secretion of the predatory
stink bug *Eocanthecona furcellata*. Journal of Chemical Ecology 31: 29–37.

Hodgson, D. M., Y.-K. Chung, I. Nuzzo, G. Freixas, K. K. Kulikiewicz, E. Cleator, and J.-M. Paris. 2007.
Intramolecular cyclopropanation of unsaturated terminal epoxides and chlorohydrins. Journal of the
American Chemical Society 129: 4456–4462.

Hoebeke, E. R., and M. E. Carter. 2003. *Halyomorpha halys* (Stal) (Heteroptera: Pentatomidae): A polyphagous plant pest from Asia newly detected in North America. Proceedings of the Entomological Society
of Washington 105: 225–237.

Holden, M. H., S. P. Ellner, D.-H. Lee, J. P. Nyrop, and J. P. Sanderson. 2012. Designing an effective trap
cropping strategy: the effects of attraction, retention and plant spatial distribution. Journal of Applied
Ecology 49: 715–722.

Huh, H. S., K. H. Park, H. Y. Choo, and C. G. Park. 2006. Attraction of *Piezodorus hybneri* to the aggregation
pheromone components of *Riptortus clavatus*. Journal of Chemical Ecology 32: 681–691.

i5K Consortium. 2013. The i5K Initiative: advancing arthropod genomics for knowledge, human health, agriculture, and the environment. Journal of Heredity 104: 595–600.

Ishiwatari, T. 1976. Studies on the scent of stink bugs (Hemiptera: Pentatomidae) II. Aggregation pheromone
activity. Applied Entomology and Zoology 11: 38–44.

James, D. G., K. Mori, J. R. Aldrich, and J. E. Oliver. 1994. Flight-mediated attraction of *Biprorulus bibax*
Breddin (Hemiptera: Pentatomidae) to natural and synthetic aggregation pheromone. Journal of
Chemical Ecology 20: 71–80.

James, D. G., R. Heffer, and M. Amaike. 1996. Field attraction of *Biprorulus bibax* Breddin (Hemiptera:
Pentatomidae) to synthetic aggregation pheromone and (*E*)-2-hexenal, a pentatomid defense chemical.
Journal of Chemical Ecology 22: 1697–1708.

Jang, S. A., and G. Park. 2010. *Gymnosoma rotundatum* (Diptera: Tachinidae) attracted to the aggregation pheromone of *Plautia stali* (Hemiptera: Pentatomidae). Journal of Asia-Pacific Entomology 13: 73–75.

Kakogawa, K., H. Kurihisa, T. Morita, K. Matsumoto, D. Mise, and T. Mizoguchi. 2007. Monitoring methods for population levels of the brown-winged green bug, *Plautia crossota stali* Scott (Hemiptera: Pentatomidae) and damage analysis on Japanese pear [*Pyrus pyrifolia*]. Bulletin of the Hiroshima Prefectural Technology Research Institute Agricultural Technology Research Center (Japan) (annual).

Kaplan, I. 2012. Attracting carnivorous arthropods with plant volatiles: the future of biocontrol or playing with fire? Biological Control 60: 77–89.

Karban, R., and A. A. Agrawal. 2002. Herbivore offense. Annual Review of Ecology and Systematics 33: 641–664.

Katase, M., K. Shimizu, H. Nagasaki, and I. Adachi. 2007. Application of synthetic aggregation pheromone of *Plautia crossota stali* Scott to the monitoring and mass trapping of fruit-piercing stink bugs. Bulletin of the Chiba Prefectural Agriculture Research Center (Japan) 4: 135–144 (In Japanese with a summary in English).

Kelly, J. L., J. R. Hagler, and I. Kaplan. 2014. Semiochemical lures reduce emigration and enhance pest control services in open-field predator augmentation. Biological Control 71: 70–77.

Khrimian A. 2005. The geometric isomers of methyl 2,4,6-decatrienoate, including pheromones of at least two species of stink bugs. Tetrahedron 61: 3651–3657.

Khrimian, A., P. W. Shearer, A. Zhang, G. C. Hamilton, and J. R. Aldrich. 2008. Field trapping of the invasive brown marmorated stink bug, *Halyomorpha halys*, with geometric isomers of methyl 2,4,6-decatrienoate. Journal of Agricultural and Food Chemistry 56: 196–203.

Khrimian, A., A. A. Cossé, and D. J. Crook. 2011. Absolute configuration of 7-epi-sesquithujene. Journal of Natural Products 74: 1414–1420.

Khrimian, A., A. Zhang, D. C. Weber, H.-Y. Ho, J. R. Aldrich, K. E. Vermillion, M. A. Siegler, S. Shirali, F. Guzman, and T. C. Leskey. 2014a. Discovery of the aggregation pheromone of the brown marmorated stink bug (*Halyomorpha halys*) through the creation of stereoisomeric libraries of 1-bisabolen-3-ols. Journal of Natural Products 77: 1708–1717.

Khrimian, A., S. Shirali, K. E. Vermillion, M. A. Siegler, F. Guzman, K. Chauhan, J. R. Aldrich, and D. C. Weber. 2014b. Determination of the stereochemistry of the aggregation pheromone of harlequin bug, *Murgantia histrionica*. Journal of Chemical Ecology 40: 1260–1268.

Khrimian, A., S. Shirali, and F. Guzman. 2015. Absolute configurations of zingiberenols isolated from ginger (*Zingiber officinale*) rhizomes. Journal of Natural Products 78: 3071–3074.

Kiritani, K. 1964. The effect of colony size upon the survival of larvae of the southern green stink bug, *Nezara viridula*. Japanese Journal of Applied Entomology and Zoology 8: 45–53.

Kiritani, K. 2007. The impact of global warming and land-use change on the pest status of rice and fruit bugs (Heteroptera) in Japan. Global Change Biology 13: 1586–1595.

Kochansky, J., J. R. Aldrich, and W. R. Lusby. 1989. Synthesis and pheromonal activity of 6, 10,13-trimethyl-1-tetradecanol for predatory stink bug, *Stiretrus anchorago* (Heteroptera: Pentatomidae). Journal of Chemical Ecology 15: 1717–1728.

Kölliker, M., J. P. Chuckalovcak, K. F. Haynes, and E. D. Brodie. 2006. Maternal food provisioning in relation to condition-dependent offspring odours in burrower bugs (*Sehirus cinctus*). Proceedings of the Royal Society B: Biological Sciences 273: 1523–1528.

Kondo, T., I. Brake, K. Imbachi López, and C. A. Korytkowski. 2010. Report of *Milichiella lacteipennis* Loew (Diptera: Milichiidae), attracted to various crushed bugs (Hemiptera: Coreidae & Pentatomidae). Boletin del Museo de Entomología de la Universidad del Valle 11: 16–20.

Krupke, C. H., and J. F. Brunner. 2003. Parasitoids of the consperse stink bug (Hemiptera: Pentatomidae) in north central Washington and attractiveness of a host-produced pheromone component. Journal of Entomological Science 38: 84–92.

Krupke, C. H., J. F. Brunner, M. D. Doerr, and A. D. Kahn. 2001. Field attraction of the stink bug *Euschistus conspersus* (Hemiptera: Pentatomidae) to synthetic pheromone-baited host plants. Journal of Economic Entomology 94: 1500–1505.

Krupke, C. H., V. P. Jones, and J. F. Brunner. 2006. Diel periodicity of *Euschistus conspersus* (Heteroptera: Pentatomidae) aggregation, mating, and feeding. Annals of the Entomological Society of America 99: 169–174.

Kuwahara, S., S. Hamade, W. S. Leal, J. Ishikawa, and O. Kodama. 2000. Synthesis of a novel sesquiter-pene isolated from the pheromone gland of a stink bug, *Tynacantha marginata* Dallas. Tetrahedron 56: 8111–8117.

Laumann, R. A., M. F. Aquino, M. C. B. Moraes, M. Pareja, and M. Borges. 2009. Response of egg parasitoids *Trissolcus basalis* and *Telenomus podisi* to compounds from defensive secretions of stink bugs. Journal of Chemical Ecology 35: 8–19.

Laumann, R. A., M. C. B. Moraes, A. Khrimian, and M. Borges. 2011. Field capture of *Thyanta perditor* with pheromone baited traps. Pesquisa Agropecuária Brasileira 46: 113–119.

Leal, W. S., Y. Ueda, and M. Ono. 1996. Attractant pheromone for male rice bug *Leptocorisa chinensis*: semiochemicals produced by both male and female. Journal of Chemical Ecology 22: 1429–1437.

Leal, W. S., S. Kuwahara, X. Shi, H. Higuchi, C. E. Marino, M. Ono, and J. Meinwald. 1998. Male-released sex pheromone of the stink bug *Piezodorus hybneri*. Journal of Chemical Ecology 24: 1817–1829.

Lee, D.-H., B. D. Short, S. V. Joseph, J. C. Bergh, and T. C. Leskey. 2013. Review of the biology, ecology, and management of *Halyomorpha halys* (Hemiptera: Pentatomidae) in China, Japan, and the Republic of Korea. Environmental Entomology 42: 627–641.

Lee, K.-C., C.-H. Kang, D. W. Lee, S. M. Lee, C. G. Park, and H. Y. Choo. 2002. Seasonal occurrence trends of hemipteran bug pests monitored by mercury light and aggregation pheromone traps in sweet persim-mon orchards. Korean Journal of Applied Entomology 41: 233–238.

Leskey T. C., G. C. Hamilton, A. L. Nielsen, D. F. Polk, C. Rodriguez-Saona, J. C. Berg, D. A. Herbert, T. P. Kuhar, D. Pfeiffer, G. P. Dively, C. Hooks, M. J. Raupp, P. M. Shrewsbury, G. Krawczyk, P. W. Shearer, J. Whalen, C. Koplinka-Loehr, E. Myers, D. Inkley, K. A. Hoelmer, D.-H. Lee, and S. E. Wright. 2012a. Pest status of the brown marmorated stink bug, *Halyomorpha halys* (Stål), in the USA. Outlooks on Pest Management 23: 218–226.

Leskey, T. C., S. E. Wright., B. D. Short and A. Khrimian. 2012b. Development of behaviorally based monitor-ing tools for the brown marmorated stink bug, *Halyomorpha halys* (Stål) (Heteroptera: Pentatomidae) in commercial tree fruit orchards. Journal of Entomological Science 47: 76–85.

Leskey, T. C., J. A. Agnello, C. Bergh, G. P. Dively, G. C. Hamilton, P. Jentsch, A. Khrimian, G. Krawczyk, T. P. Kuhar, D.-H. Lee, W. R. Morrison III, D. F. Polk, C. Rodriguez-Saona, P. W. Shearer, B. D. Short, P. M. Shrewsbury, J. F. Walgenbach, C. Welty, J. Whalen, D. C. Weber, and N. Wiman. 2015a. Attraction of the invasive *Halyomorpha halys* (Hemiptera: Pentatomidae) to traps baited with semiochemical stim-uli across the United States. Environmental Entomology 44: 746–756.

Leskey, T. C., A. Khrimian, D. C. Weber, J. R. Aldrich, B. D. Short, D.-H. Lee, and W. R. Morrison. 2015b. Behavioral responses of the invasive *Halyomorpha halys* (Stål) (Hemiptera: Pentatomidae) to traps baited with stereoisomeric mixtures of 10,11-epoxy-1-bisabolen-3-ol. Journal of Chemical Ecology 41: 418–429.

Lockwood, J. A., and R. N. Story. 1985. Bifunctional pheromone in the first instar of the southern green stink bug, *Nezara viridula* (L.) (Hemiptera: Pentatomidae): its characterization and interaction with other stimuli. Annals of the Entomological Society of America 78: 474–479.

Lockwood, J. A., and R. N. Story. 1986. Adaptive functions of nymphal aggregation in the southern green stink bug, *Nezara viridula* (L.) (Hemiptera: Pentatomidae). Environmental Entomology 15: 739–749.

Lockwood, J. A., and R. N. Story. 1987. Defensive secretion of the southern green stink bug (Hemiptera: Pentatomidae) as an alarm pheromone. Annals of the Entomological Society of America 80: 686–691.

Lopez, E. R., R. G. Van Driesche, and J. S. Elkinton. 1989. Influence of group size on daily per capita birth rates of the cabbage aphid (Homoptera: Aphididae) on collards. Environmental Entomology 18: 1086–1089.

Mattiacci, L., S. B. Vinson, H. J. Williams, J. R. Aldrich, and F. Bin. 1993. A long range attractant kairomone for egg parasitoid *Trissolcus basalis*, isolated from defensive secretion of its host, *Nezara viridula*. Journal of Chemical Ecology 19: 1167–1181.

Mau, R. F. L., and W. C. Mitchell. 1978. Development and reproduction of the oriental stink bug, *Plautia stali* (Hemiptera: Pentatomidae). Annals of the Entomological Society of America 71: 756–757.

McBrien, H. L., J. G. Millar, L. Gottlieb, X. Chen, and R. E. Rice. 2001. Male-produced sex attractant phero-mone of the green stink bug, *Acrosternum hilare* (Say). Journal of Chemical Ecology 27: 1821–1839.

McBrien, H. L., J. G. Millar, R. E. Rice, J. S. McElfresh, E. Cullen, and F. G. Zalom. 2002. Sex attractant pheromone of the red-shouldered stink bug *Thyanta pallidovirens*: a pheromone blend with multiple redundant components. Journal of Chemical Ecology 28: 1797–1818.

McPherson, J. E., and R. M. McPherson. 2000. Stink Bugs of Economic Importance in America North of Mexico. CRC Press, Boca Raton, FL. 272 pp.

Medeiros, M. A., M. S. Loiácono, M. Borges, and F. V. G. Schmidt. 1998. Incidência natural de parasitóides em ovos de percevejos (Hemiptera: Pentatomidae) encontrados na soja no Distrito Federal. Pesquisa Agropecuária Brasileira 33: 1431–1435.

Melo Machado, R. C., J. Sant'Ana, M. C. Blassioli-Moraes, R. A. Laumann, and M. Borges. 2014. Herbivory-induced plant volatiles from *Oryza sativa* and their influence on chemotaxis behaviour of *Tibraca limbativentris* Stål (Hemiptera: Pentatomidae) and egg parasitoids. Bulletin of Entomological Research 104: 347–356.

Michereff, M. F. F., R. A. Laumann, M. Borges, M. Michereff Filho, I. R. Diniz, A. Faria Neto, and M. C. B. Moraes. 2011. Volatiles mediating plant-herbivory-natural enemy interaction in resistant and susceptible soybean cultivars. Journal of Chemical Ecology 37: 273–285.

Michereff, M. F. F., M. Borges, I. R. Diniz, R. A. Laumann, and M. C. Blassioli-Moraes. 2013. Influence of volatile compounds from herbivore-damaged soybean plants on searching behavior of the egg parasitoid *Telenomus podisi*. Entomologia Experimentalis et Applicata 147: 9–17.

Michereff, M. F. F., M. Michereff Filho, M. C. Blassioli-Moraes, R. A. Laumann, I. R. Diniz, and M. Borges. 2014. Effect of resistant and susceptible soybean cultivars on the attraction of egg parasitoids under field conditions. Journal of Applied Entomology 139: 207–216.

Miklas, N., M. Renou, I. Malosse, and C. Malosse. 2000. Repeatability of pheromone blend composition in individual males of the southern green stink bug, *Nezara viridula*. Journal of Chemical Ecology 26: 2473–2485.

Millar, J. G. 1997. Methyl (2*E*,4*Z*,6*Z*)-deca-2,4,6-trienoate, a thermally unstable, sex-specific compound from the stink bug *Thyanta pallidovirens*. Tetrahedron Letters 38: 7971–7972.

Millar, J. G. 2005. Pheromones of true bugs. Topics in Current Chemistry 240: 37–84.

Millar, J. G. 2007. Insect pheromones for integrated pest management: promise versus reality. Redia 15: 51–55.

Millar, J. G. 2014. The devil is in the details. Journal of Chemical Ecology 40: 517–518.

Millar, J. G., R. E. Rice, S. A. Steffan, K. M. Daane, E. Cullen, and F. G. Zalom. 2001. Attraction of female digger wasps, *Astata occidentalis* Cresson (Hymenoptera: Sphecidae) to the sex pheromone of the stink bug *Thyanta pallidovirens* (Hemiptera:Pentatomidae). Pan-Pacific Entomologist 77: 244–248.

Millar, J. G., H. M. McBrien, and J. S. McElfresh. 2010. Field trials of aggregation pheromones for the stink bugs *Chlorochroa uhleri* and *Chlorochroa sayi* (Hemiptera: Pentatomidae). Journal of Economic Entomology 103: 1603–1612.

Mishiro, K., and Y. Ohira. 2002. Attraction of a synthetic aggregation pheromone of the brown-winged green bug, *Plautia crossota stali* Scott to its parasitoids, *Gymnosoma rotundata* and *Trissolcus plautiae*. Kyushu Plant Protection Research 48: 76–80.

Mitchell, W. C., and R. F. L. Mau. 1971. Response of female southern green stink bug and its parasite, *Trichopoda pennipes*, to male stink bug pheromones. Journal of Economic Entomology 64: 856–859.

Møldrup, M. E., F. Geu-Flores, M. De Vos, C. E. Olsen, J. Sun, G. Jander, and B. A. Halkier. 2012. Engineering of benzylglucosinolate in tobacco provides proof-of-concept for dead-end trap crops genetically modified to attract *Plutella xylostella* (diamondback moth). Plant Biotechnology Journal 10: 435–442.

Moraes, M. C. B., J. G. Millar, R. A. Laumann, E. R. Sujii, C. S. S. Pires, and M. Borges. 2005a. Sex attractant pheromone from the neotropical red-shouldered stink bug, *Thyanta perditor* (F.). Journal of Chemical Ecology 31: 1415–1427.

Moraes, M .C. B., R. A. Laumann, C. S. S. Pires, E. R. Sujii, and M. Borges. 2005b. Induced volatiles in soybean and pigeon pea plants artificially infested with the neotropical brown stink bug, *Euschistus heros*, and their effect on the egg parasitoid, *Telenomus podisi*. Entomologia Experimentalis et Applicata 115: 227–237.

Moraes, M. C .B., M. Pareja, R. A. Laumann, C. B. Hoffmann-Campo, and M. Borges. 2008a. Response of the parasitoid *Telenomus podisi* to induced volatiles from soybean damaged by stink bug herbivory and oviposition. Journal of Plant Interaction 3: 1742–1756

Moraes, M. C. B., M. Pareja, R. A. Laumann, and M. Borges. 2008b. The chemical volatiles (semiochemicals) produced by neotropical stink bugs (Hemiptera: Pentatomidae). Neotropical Entomology 37: 489–505.

Moraes, M. C. B., M. Borges, M. Pareja, H. G. Vieira, F. T. de Souza Sereno, and R. A. Laumann. 2008c. Food and humidity affect sex pheromone ratios in the stink bug, *Euschistus heros*. Physiological Entomology 33: 43–50.

Mori, K. 2007. Synthetic studies aimed at the elucidation of the stereostructure of the aggregation pheromone, 2-methyl-6-(4'-methylenebicyclo-[3.1.0]hexyl)hept-2-en-1-ol, produced by the male stink bug *Erysarcoris [sic] lewisi*. Tetrahedron: Asymmetry 18: 838–846.

Mori, K., T. Tashiro, T. Yoshimura, M. Takita, J. Tabata, S. Hiradate, and H. Sugie. 2008. Determination of the absolute configuration of the male aggregation pheromone, 2-methyl-6-(4'-methylenebicyclo[3.1.0] hexyl)hept-2-en-1-ol, of the stink bug *Erysarcoris [sic] lewisi* (Distant) as 2Z,6R,1'S,5'S by its synthesis. Tetrahedron Letters 49: 354–357.

Moriya, S., and M. Shiga. 1984. Attraction of the male brown-winged green bug, *Plautia stali* Scott (Heteroptera: Pentatomidae) for males and females of the same species. Applied Entomology and Zoology 32: 317–322.

Muraki, Y., T. Taguri, M. Yamamoto, P. H. G. Zarbin, and T. Ando. 2013. Synthesis of all four stereoisomers of 6,10,13-trimethyltetradecan-2-one, a sex pheromone component produced by males of the stink bug *Pallantia macunaima*. European Journal of Organic Chemistry 11: 2209–2215.

Nakamura, M., A. Hirai, M. Sogi, and E. Nakamura. 1998. Enantioselective addition of allylzinc reagent to alkynyl ketones. Journal of the American Chemical Society 120: 5846–5847.

Nakamura, S., R. T. Ichiki, and Y. Kainoh. 2013. Chemical ecology of tachinid parasitoids, pp. 145-167. *In* E. Wajnberg and S. Colazza (Eds.), Chemical ecology of insect parasitoids. Wiley-Blackwell, UK. 328 pp.

Noge, K., K. L. Prudic, and J. X. Becerra. 2012. Defensive roles of (*E*)-2-alkenals and related compounds in Heteroptera. Journal of Chemical Ecology 38: 1050–1056.

Nomakuchi, S., T. Yanagi, N. Baba, A. Takahira, M. Hironaka, and L. Filippi. 2012. Provisioning call by mothers of a subsocial shield bug. Journal of Zoology 288: 50–56.

Nordlund, D. A., and W. J. Lewis. 1976. Terminology of chemical releasing stimuli in intraspecific and interspecific interactions. Journal of Chemical Ecology 2: 211–220.

Northeastern IPM Center. 2015. Where is BMSB? http://www.stopbmsb.org/where-is-bmsb/host-plants/ Accessed 31 January 2015.

de Oliveira, M. W. M., M. Borges, C. K. Z Andrade, R. A. Laumann, J. A. F. Barrigossi, and M. C. Blassioli-Moraes. 2013. Zingiberenol, (1S,4R,1'S)-4-(1',5'-dimethylhex-4'-enyl)-1-methylcyclohex-2-en-1-ol, identified as the sex pheromone produced by males of the rice stink bug *Oebalus poecilus* (Heteroptera: Pentatomidae). Journal of Agricultural and Food Chemistry 61: 7777–7785.

Pacheco, D. J. P., and B. S. Corrêa-Ferreira. 2000. Parasitismo de *Telenomus podisi* Ashmead (Hymenoptera: Scelionidae) em populações de percevejos pragas da soja. Anais da Sociedade Entomológica do Brasil 29: 295–302.

Panizzi, A. R. 1997. Wild hosts of pentatomids: ecological significance and role in their pest status on crops. Annual Review of Entomology 42: 99–122.

Pareja, M., M. Borges, R. A. Laumann, and M. C. B. Moraes. 2007. Inter- and intraspecific variation in defensive compounds produced by five neotropical stink bug species (Hemiptera: Pentatomidae). Journal of Insect Physiology 53: 639–648.

Pavis, C. 1987. Les secretions exocrines des hétéroptères (allomones et pheromones). Une mise au point bibliographique. Agronomie, EDP Sciences 7: 547–561.

Pavis, C., C. Malosse, P. H. Ducrot, and C. Déscoins. 1994. Dorsal abdominal glands in nymphs of southern green stink bug, *Nezara viridula* (L.) (Heteroptera: Pentatomidae): chemistry of secretions of five instars and role of (*E*)-4-oxo-2-decenal, compound specific to first instars. Journal of Chemical Ecology 20: 2213–2227.

Poelman, E. H., J. J. A. van Loon, and M. Dicke. 2008. Consequences of variation in plant defense for biodiversity at higher trophic levels. Trends in Plant Science 13: 534–541.

Reed, D. A., J. C. Palumbo, T. M. Perring, and C. May. 2013. *Bagrada hilaris* (Hemiptera: Pentatomidae), an invasive stink bug attacking cole crops in the southwestern United States. Journal of Integrated Pest Management 4(3), C1–C7.

Rider, D.A. 2015. Pentatomoidea home page. www.ndsu.nodak.edu/ndsu/rider/Pentatomoidea/ accessed 31 May 2015.

Rider, D. A., and L. H. Rolston. 1987. Review of the genus *Agroecus* Dallas, with the description of a new species (Hemiptera: Pentatomidae). Journal of the New York Entomological Society 94: 428–439.

Ryan, M. A., C. J. Moore, and G. H. Walter. 1995. Individual variation in pheromone composition in *Nezara viridula* (Heteroptera: Pentatomidae): how valid is the basis for designating "pheromone strains"? Comparative Biochemistry and Physiology Part B 111: 189–193.

Salerno, G., F. Frati, E. Conti, C. de Pasquale, E. Peri, and S. Colazza. 2009. A finely tuned strategy adopted by an egg parasitoid to exploit chemical traces from host adults. Journal of Experimental Biology 212: 1825–1831.

Sant'Ana, J. and J. C. Dickens. 1998. Comparative electrophysiological studies of olfaction in predaceous bugs, *Podisus maculiventris* and *P. nigrispinus*. Journal of Chemical Ecology 24: 965-984.

Sargent, C., H. M. Martinson, and M. J. Raupp. 2014. Traps and trap placement may affect location of brown marmorated stink bug (Hemiptera: Pentatomidae) and increase injury to tomato fruits in home gardens. Environmental Entomology 43: 432–438.

Shimizu, N., and T. Tsutsumi. 2011. Cross-attraction of *Nezara viridula* (Linnaeus) and *N. antennata* Scott (Heteroptera: Pentatomidae) to male-produced pheromones. Japanese Journal of Applied Entomology and Zoology 55: 43–48.

Shimizu, N., and T. Tsutsumi. 2012. Physiological status of *Nezara antennata* Scott attracted to conspecific adult males. Japanese Journal of Applied Entomology and Zoology 56: 16-18.

Siewert, J., R. Sandmann, and P. von Zezschwitz. 2007. Rhodium-catalyzed enantioselective 1,2-addition of aluminum organyl compounds to cyclic enones. Angewandte Chemie International Edition 46: 7122–7124.

Silva, C. C., M. C. B. Moraes, R. A. Laumann, and M. Borges. 2006. Sensory response of the egg parasitoid *Telenomus podisi* to stimuli from the bug *Euschistus heros*. Pesquisa Agropecuária Brasileira 41: 1093–1098.

da Silva, R. A., E. D. Quintela, G. M. Mascarin, N. Pedrini, L. M. Lião and P. H. Ferri. 2015. Unveiling chemical defense in the rice stalk stink bug against the entomopathogenic fungus *Metarhizium anisopliae*. Journal of Invertebrate Pathology 127: 93–100.

Soffer, M. D., and L. A. Burk. 1985. The total stereostructure of (-)-isozingiberene dihydrochloride. Tetrahedron Letters 26: 3543–3546.

Sosa-Gomez, D. R., D. G. Boucias, and J. L. Nation. 1997. Attachment of *Metarhizium anisopliae* to the southern green stink bug *Nezara viridula* cuticle and fungistatic effect of cuticular lipids and aldehydes. Journal of Invertebrate Pathology 69: 31–39.

Sparks, M. E., K. S. Shelby, D. Kuhar, and D. E. Gundersen-Rindal. 2014. Transcriptome of the invasive brown marmorated stink bug, *Halyomorpha halys* (Stål)(Heteroptera: Pentatomidae). PloS One 9(11), e111646.

Still, W. C., and S. Gennari. 1983. Direct synthesis of Z-unsaturated esters. A useful modification of the Horner-Emmons olefination. Tetrahedron Letters 24: 4405–4408.

Sugie, H., M. Yoshida, K. Kawasaki, H. Noguchi, S. Moriya, K. Takagi, H. Fukuda, A. Fujiie, M. Yamanaka, Y. Ohira, T. Tsutsumi, K. Tsuda, K. Fukumoto, M. Yamashita, and H. Suzuki. 1996. Identification of the aggregation pheromone of the brown-winged green bug, *Plautia stali* Scott (Heteroptera: Pentatomidae). Applied Entomology and Zoology 31: 427–431.

Sujii, E. R., M. L. M. Costa, C. S. S. Pires, S. Colazza, and M. Borges. 2002. Inter- and intra-guild interactions in egg parasitoid species of the soybean stink bug complex. Pesquisa Agropecuária Brasileira 37: 1541–1549.

Symonds, M. R., and M. A. Elgar. 2008. The evolution of pheromone diversity. Trends in Ecology and Evolution 23: 220–228.

Tada, N., M. Yoshida, and Y. Sato. 2001a. Monitoring of forecasting for stink bugs in apple 1. Characteristics of attraction to aggregation pheromone in Iwate Prefecture. Annual Report of Plant Protection of North Japan 52: 224–226.

Tada, N., M. Yoshida, and Y. Sato. 2001b. Monitoring of forecasting for stink bugs in apple 2. The possibility of forecasting with aggregation pheromone. Annual Report of Plant Protection of North Japan 52: 227–229.

Taki, H., K. Tabuchi, H. Iijima, K. Okabe, and M. Toyama. 2014. Spatial and temporal influences of conifer planted forests on the orchard pest *Plautia stali* (Hemiptera: Pentatomidae). Applied Entomology and Zoology 49: 241–247.

Takita, M. 2007. Attraction of adults and nymphs by male adults of *Eysarcoris lewisi* (Distant) (Heteroptera: Pentatomidae) reared with a long photoperiod. Japanese Journal of Applied Entomology and Zoology 51: 231–233 (in Japanese).

Takita, M., H. Sugie, J. Tabata, S. Ishii, and S. Hiradate. 2008. Isolation and estimation of the aggregation pheromone from *Eysarcoris lewisi* (Distant) (Heteroptera: Pentatomidae). Applied Entomology and Zoology 43: 11–17.

Tashiro, T., and K. Mori. 2008. Synthesis and absolute configuration of the male aggregation pheromone of the stink bug *Erysarcoris* [*sic*] *lewisi* (Distant), (2*Z*,6*R*,1′*S*,5′*S*)-2-methyl-6-(4′-methylenebicyclo[3.1.0] hexyl)hept-2-en-1-ol. Tetrahedron: Asymmetry 19: 1215–1223.

Terhune, S. J., J. W. Hogg, A. C. Bromstein, and B. M. Lawrence. 1974. Four new sesquiterpene analogs of common monoterpenes. Canadian Journal of Chemistry 53: 3285–3293.

Tillman, P. G., J. R. Aldrich, A. Khrimian, and T. E. Cottrell. 2010. Pheromone attraction and cross-attraction of *Nezara, Acrosternum,* and *Euschistus* spp. stink bugs (Heteroptera: Pentatomidae) in the field. Environmental Entomology 39: 610–617.

Tillman, P. G., and T. E. Cottrell. 2012. Case study: trap crop with pheromone traps for suppressing *Euschistus servus* (Heteroptera: Pentatomidae) in cotton. Psyche 2012: ID 401703, 10 pp.

Tognon, R., J. Sant'Ana, and S. M. Jahnke. 2014. Influence of original host on chemotaxic behaviour and parasitism in *Telenomus podisi* Ashmead (Hymenoptera: Platygastridae). Bulletin of Entomological Research. 104: 781–787.

Tsutsumi, T., M. Teshiba, M. Yamanaka, Y. Ohira, and T. Higuchi. 2003. An autodissemination system for the control of brown winged green bug, *Plautia crossota stali* Scott (Heteroptera: Pentatomidae) by an entomopathogenic fungus, *Beauveria bassiana* E-9102 combined with aggregation pheromone. Japanese Journal of Applied Entomology and Zoology 47: 159–163 (in Japanese).

Ulrich, K. R., M. F. Feldlaufer, M. Kramer, and R. J. St. Leger. 2015. Inhibition of the entomopathogenic fungus *Metarhizium anisopliae* sensu lato in vitro by the bed bug defensive secretions (*E*)-2-hexenal and (*E*)-2-octenal. BioControl, DOI 10.1007/s10526-015-9667-2.

van Loon, J. A., E. W. Vos, and M. Dicke. 2000. Orientation behaviour of the predatory hemipteran *Perillus bioculatus* to plant and prey odours. Entomologia Experimentalis et Applicata 96: 51–58.

Vet, L. E. M., and M. Dicke. 1992. Ecology of infochemical use by natural enemies in a tritrophic context. Annual Review of Entomology 37: 141–172.

Vieira, C. R., M. C. Blassioli-Moraes, M. Borges, C. S. S. Pires, E. R. Sujii, and R. A. Laumann. 2014. Field evaluation of (*E*)-2-hexenal efficacy for behavioral manipulation of egg parasitoids in soybean. BioControl 59: 525–537.

Vinson, S. B. 1985. The behavior of parasitoids. Comprehensive Insect Physiology, Biochemistry, and Pharmacology 9: 417–469.

Wang, Y. J., D. X. Zhao, J. L. Gao, Z. Q. Peng, X. N. Li, and Y. Wang. 2012. Electrophysiological and behavioral activity of compounds in metathoracic glands of adults of *Tessaratoma papillosa* (Hemiptera: Pentatomidae). Applied Mechanics and Materials 108: 301–307.

Weber, D. C., T. C. Leskey, G. Cabrera Walsh, and A. Khrimian. 2014a. Synergy of aggregation pheromone with methyl (*E,E,Z*)-2,4,6-decatrienoate in attraction of *Halyomorpha halys* (Hemiptera: Pentatomidae). Journal of Economic Entomology 107: 1061–1068.

Weber, D. C., G. Cabrera Walsh, A. S. DiMeglio, M. M. Athanas, T. C. Leskey, and A. Khrimian. 2014b. Attractiveness of harlequin bug, *Murgantia histrionica* (Hemiptera: Pentatomidae), aggregation pheromone: field response to isomers, ratios and dose. Journal of Chemical Ecology 40: 1251–1259.

Weissbecker, B., J. J. Van Loon, and M. Dicke. 1999. Electroantennogram responses of a predator, *Perillus bioculatus,* and its prey, *Leptinotarsa decemlineata,* to plant volatiles. Journal of Chemical Ecology 25: 2313–2325.

Weissbecker, B., J. J. Van Loon, M. A. Posthumus, H. J. Bouwmeester, and M. Dicke. 2000. Identification of volatile potato sesquiterpenoids and their olfactory detection by the two-spotted stinkbug *Perillus bioculatus.* Journal of Chemical Ecology 26: 1433–1445.

Wertheim, B., E.-J. A. van Baalen, M. Dicke, and L. E. M. Vet. 2005. Pheromone-mediated aggregation in nonsocial arthropods: an evolutionary ecological perspective. Annual Review of Entomology 50: 321–346.

Wyatt, T. 2014. Pheromones and Animal Behaviour, 2nd ed. Cambridge University Press, Cambridge, UK. 419 pp.

Yamanaka, T., M. Teshiba, M. Tuda, and T. Tsutsumi. 2011. Possible use of synthetic aggregation pheromones to control stinkbug *Plautia stali* in kaki persimmon orchards. Agricultural and Forest Entomology 13: 321–331.

Yamashita, M., T. Fukumoto, and H. Suzuki. 1997. Preparation of methyl (3E,4E,6Z)-2,4,6-decatrienoate as attractant pheromone for *Plautia stali*. Japanese Patent No. 09176089.

Yasuda, T. 1997. Chemical cues from *Spodoptera litura* larvae elicit prey-locating behavior by the predatory stink bug, *Eocanthecona furcellata*. Entomologia Experimentalis et Applicata 82: 349–354.

Zahn, D. K., J. A. Moreira, and J. G. Millar. 2008. Identification, synthesis, and bioassay of a male-specific aggregation pheromone from the harlequin bug, *Murgantia histrionica*. Journal of Chemical Ecology 34: 238–251.

Zahn, D. K., J. A. Moreira, and J. G. Millar. 2012. Erratum to: Identification, synthesis, and bioassay of a male-specific aggregation pheromone from the harlequin bug, *Murgantia histrionica*. Journal of Chemical Ecology 38: 126.

Zarbin, P. H. G., C. F. Fávaro, D. M. Vidal, and M. A. C. M. Rodrigues. 2012. Male-produced sex pheromone of the stink bug *Edessa meditabunda*. Journal of Chemical Ecology 38: 825–835.

Zgonik, V., and A. Čokl. 2014. The role of signals of different modalities in initiating vibratory communication in *Nezara viridula*. Central European Journal of Biology 9: 200–211.

Zhang, A. J., M. Borges, J. R. Aldrich, and M. Camp. 2003. Stimulatory male volatiles for the neotropical brown stink bug, *Euschistus heros* (F.) (Heteroptera: Pentatomidae). Neotropical Entomology 32: 713–717.

Zhao, D., J. Gao, Y. Wang, J. Jiang, and R. Li. 2012. Morphology and volatile compounds of metathoracic scent gland in *Tessaratoma papillosa* (Drury) (Hemiptera: Tessaratomidae). Neotropical Entomology 41: 278–282.

Section X

Management

16

General Insect Management

Jeremy K. Greene, James A. Baum, Eric P. Benson, C. Scott Bundy,
Walker A. Jones, George G. Kennedy, J. E. McPherson, Fred R. Musser,
Francis P. F. Reay-Jones, Michael D. Toews, and James F. Walgenbach

CONTENTS

16.1   Management Tactics and Control 730
16.2   History 731
     16.2.1   Control Prior to Synthetic Organic Insecticides 731
         16.2.1.1   Brief Overview 731
         16.2.1.2   Ancient History 731
         16.2.1.3   Post-Classical Era 732
         16.2.1.4   Modern History 732
         16.2.1.5   Influence of Farming Practices on Control 734
     16.2.2   Age of Synthetic Organic Insecticides (Chlorinated Hydrocarbons,
         Organophosphates, Carbamates) 735
16.3   Modern Era 737
     16.3.1   Current Management Practices 737
         16.3.1.1   Cultural Control 737
             16.3.1.1.1   Crop Rotation or Fallowing 738
             16.3.1.1.2   Sanitation or Clean Culture 738
             16.3.1.1.3   Tillage 738
             16.3.1.1.4   Plant Density and Row Spacing 739
             16.3.1.1.5   Trap Cropping and Farmscape Influences 739
             16.3.1.1.6   Planting Date 739
             16.3.1.1.7   Water Management 739
             16.3.1.1.8   Nutrient Management 740
             16.3.1.1.9   Intercropping 740
             16.3.1.1.10   Harvest Management 740
             16.3.1.1.11   Host Plant Resistance (Non-GMOs) 741
         16.3.1.2   Physical/Mechanical Control 741
             16.3.1.2.1   Electricity 741
             16.3.1.2.2   Sound Waves 742
             16.3.1.2.3   Radiation: Visible, Ultraviolet, Ionizing 742
             16.3.1.2.4   Temperature 743
             16.3.1.2.5   Physical Barriers 743
             16.3.1.2.6   Hand Removal (Handpicking) 743
         16.3.1.3   Reproductive Control 744
             16.3.1.3.1   Sterilization Strategies 744
                 16.3.1.3.1.1   Chemosterilization 744
                 16.3.1.3.1.2   Radiation 744
                 16.3.1.3.1.3   Sterile-Insect Technique 745
             16.3.1.3.2   Mating Disruption 745

        16.3.1.4 Regulatory (Legal) Control 745
                  16.3.1.4.1 Quarantine...745
                  16.3.1.4.2 Eradication..746
                  16.3.1.4.3 Certification/Inspections 746
        16.3.1.5 Chemical Control..746
                  16.3.1.5.1 Economic Injury Levels and Action Thresholds.......................746
                  16.3.1.5.2 Insecticide Modes of Action.......................................747
                  16.3.1.5.3 Application Strategies 747
        16.3.1.6 Biological Control..748
                  16.3.1.6.1 Predators...748
                  16.3.1.6.2 Parasitoids..748
                  16.3.1.6.3 Entomopathogens and Nematodes 749
                  16.3.1.6.4 Classical Biological Control......................................749
                  16.3.1.6.5 Augmentation Biological Control..............................749
        16.3.1.7 Integrated Pest Management 749
                  16.3.1.7.1 Agricultural IPM 750
                              16.3.1.7.1.1 Stink Bug Control Programs for Selected Crops... 750
                                            16.3.1.7.1.1.1 Cotton...750
                                            16.3.1.7.1.1.2 Fruit and Vegetable Crops.............750
                                                           16.3.1.7.1.1.2.1 Damage 750
                                                           16.3.1.7.1.1.2.2 Chemical Control...751
                                                           16.3.1.7.1.1.2.3 Pheromone-
                                                                            Based Programs 752
                                                           16.3.1.7.1.1.2.4 Habitat
                                                                            Manipulation 752
                                                           16.3.1.7.1.1.2.5 Conclusions
                                                                            and Future
                                                                            Considerations.......753
                  16.3.1.7.2 Urban IPM 753
16.4 Future Management Practices 754
      16.4.1 Innovative Precision Agriculture Techniques.............................754
      16.4.2 Genetically Modified Organisms (GMOs).................................755
              16.4.2.1 Plant-Incorporated Proteins...755
              16.4.2.2 RNA Interference 756
16.5 Acknowledgments..757
16.6 References Cited 757

16.1 Management Tactics and Control

Insects have been competitors of humans for food and space and have served as vectors of animal and plant diseases for thousands of years. Therefore, it is not surprising that many tactics have been developed to help manage populations of these pests. These management tactics can be divided into four broad categories: (1) attacking the insects directly with insecticides, resistant crop varieties, and biological control agents; (2) avoiding or minimizing the peak population densities by using early or late maturing varieties and adjusting planting and harvesting dates; (3) disrupting life cycles by destroying weeds and other wild hosts; and (4) by not encouraging the build up of damaging populations by planting trap crops or other preferred hosts near the crop to be protected or using these hosts as ground cover in or near plants that are agriculturally important (McPherson and McPherson 2000).

These and other tactics fall into seven major types of control and are listed below (1–6 modified from Johansen 1978, 7 modified from Kogan 1998). Each will be discussed in detail later in this chapter. Not all of these have been used in control of pentatomoids, but it puts those that have been used into an overall framework.

1. Cultural Control: Reduction of insect populations using agricultural practices (e.g., rotation, trap crop, clean culture, resistant plant varieties, adjusting planting/harvesting dates).

2. Mechanical and Physical Control: Reduction of insect populations using devices that affect them directly or radically alter their physical environment.

    a. Mechanical control: Examples include handpicking, trapping, screens, barriers, sticky bands, and shading devices.

    b. Physical control: Examples include electricity, sound waves, infrared rays, X-rays, light, and heat or cold.

3. Regulatory (Legal) Control: Lawful regulation to eradicate, prevent, or control infestations or reduce damage by insects (e.g., quarantines).

4. Reproductive Control: Reduction of insect populations using physical treatments or substances that cause sterility, affect sexual behavior, or otherwise disrupt normal reproduction (e.g., insect sterilization, sex attractants, genetic manipulation).

5. Chemical Control: Reduction of insect populations or prevention of their injury by the using materials to poison them, attract them to particular devices, or repel them from specified areas.

6. Biological Control: Reduction of insect populations through the release, conservation, and establishment of living organisms that have been chosen for that purpose. This would include classical biocontrol (i.e., release of exotic natural enemies with the purpose of becoming established and controlling the targeted pests), conservation biocontrol (i.e., employing strategies to preserve and enhance the effectiveness of naturally occurring natural enemies), and augmentative biocontrol (i.e., release of natural enemies to augment natural populations of these enemies).

7. Integrated Pest Management (IPM): "A decision support system for the selection and use of pest control tactics, singly or harmoniously coordinated into a management strategy, based on cost/benefit analyses that take into account the interests of and impacts on producers, society, and the environment" (Kogan 1998).

16.2 History

16.2.1 Control Prior to Synthetic Organic Insecticides

16.2.1.1 Brief Overview

Stearns et al. (2010) divide history into three periods: Ancient History (3600 B.C. [B.C.E.[1]] – 500 A.D. [C.E.[2]]), Post-classical Era (500–1500 A.D.), and Modern History (1500 A.D. to present). Although these periods are somewhat arbitrary, they are convenient for discussing the progressive development of various control tactics over time. Much of this information is taken from Flint and van den Bosch (1981) and Dent (2000).

16.2.1.2 Ancient History

Management tactics, several of which still are used today, had their origins early in recorded history (Flint and van den Bosch 1981, Dent 2000) and included chemical, cultural, botanical, and biological control. Chemical control involved the use of inorganic compounds such as sulfur, arsenic, mercury, and oils (Flint and van den Bosch 1981, Dent 2000). The first records of chemical insecticides apparently are from the Sumerians in 2500 B.C., who used sulfur compounds to control insects and mites. In 1500 B.C., cultural controls were described and primarily involved manipulation of planting dates.

---

[1] Before Current Era, equivalent to B.C.
[2] Common Era, equivalent to A.D.

In 1200 B.C., botanical insecticides (particular plant species not given) were used by the Chinese for seed treatments and as fungicides. The Chinese also used mercury and arsenical compounds for control of body lice (Flint and van den Bosch 1981, Dent 2000). Homer, whether or not the individual actually existed (McCoy 2015), is the source of multiple pest control strategies, including the use of sulfur for pest control in 1000 B.C. (Hajek 2004) and wood ash in 750 B.C. (Stent 2006) (note the wide gap between years). In 324 B.C., the Chinese documented biological control when they placed ants [*Oecophylla smaragdina* (Fabricius)][3] in citrus trees to manage caterpillars and large boring insects (Hardy et al. 1996, Pedigo and Rice 2015). In 300 B.C., the Chinese recognized the connection between climate and periodic biological phenomena, which led to timing of planting (phenology) to avoid pest attacks (Pedigo and Rice 2015). In the First Century (0–99 A.D.), the Roman naturalist Pliny the Elder recommended arsenic as an insecticide (Rathore 2010). In 300 A.D., the Chinese recorded using biological control in citrus orchards to control caterpillar and beetle pests by setting up colonies of ants (*O. smaragdina*)[4] with bamboo bridges so that the ants could move between trees (Dent 2000, Pedigo and Rice 2015). In 400 A.D., a Chinese alchemist, Ko Hung, recommended treating the roots of rice (*Oryza sativa* L.) with white arsenic during transplanting to protect against insect pests (Flint and van den Bosch 1981).

Two compilations were published in the Post-classical Era (see below) but contained information on pest control and agricultural practices from the Ancient History Period. First, a 20-volume publication of agricultural literature called "Geoponika" was compiled in the 10th Century (and included a 6th Century publication of the same name by Cassianus Bassus) (Dalby 2011). It documented pest management practices from many cultures, including the Greeks and Romans. Originally written in Greek, it was translated into English by Owen in 1805 and 1806 and retranslated by Dalby in 2011.

In the 12th Century A.D., Ibn Al-Awam, a Spanish Moor, authored a classic book in Arabic on agriculture, "Kitab al-Felahah" ("Book of Agriculture"), which contained much information on pest management and agriculture, including techniques recorded from earlier civilizations and those tested by Al-Awam, himself (Olson and Eddy 1943, Orlob 1973).

16.2.1.3 Post-Classical Era

The Post-classical Era was comprised of several hundred years, including the "Dark Ages." During that time, advances in pest management occurred at different rates between the Eastern and Western Worlds. In the East, the Chinese first used arsenic to control garden insects in 900 A.D. (Singh 2012) and soap to control pests by 1101 A.D. (Jaglan and Singh 2007, Pedigo and Rice 2015). They continued to develop pest control strategies for the next several hundred years (Flint and van den Bosch 1981). However, in the Western World, there was little advancement in pest control, particularly after the fall of the Roman Empire (Flint and van den Bosch 1981). During that time, insect management strategies often were based on superstition, excommunication, and odd legal battles (Dethier 1976, Flint and van den Bosch 1981). For example, in 1476 A.D., cutworms in Berne, Switzerland were taken to court, found guilty, excommunicated by the Archbishop, and banished (Dethier 1976, Flint and van den Bosch 1981, Dent 2000). In 1485, caterpillars were ordered by the High Vicar of Valence to appear before him. He provided them with a defense council and, then, condemned them to leave the area (Dent 2000). During that time, legal action against insects was so common that a treatise was written on the rules by which grasshoppers could be tried in court (Dethier 1976).

16.2.1.4 Modern History

During the next 150–300 years, there apparently was little progress made in insect control although much progress was made in insect taxonomy and biological discoveries (1650–1780) (Orlob 1973). However, there were notable exceptions. The botanicals nicotine, pyrethrum, and rotenone were introduced into

---

[3] Listed as *Acephali amaragina* but most likely in error, some kind of misspelling of *Oecophylla smaragdina* (Ted R. Schultz, personal communication).

[4] Undoubtedly, a correct identification (Ted R. Schultz, personal communication).

the Western World during the 17th—19th Centuries but not always as an insecticide (i.e., rotenone) (see below). The discovery of these botanical insecticides apparently was promoted by an agricultural revolution during 1750–1880 that began in Europe and spread to North America and was due (in part) to various innovations in agricultural practices. Crop production became more extensive and pest outbreaks more severe (Flint and van den Bosch 1981). Finally, the propagation of insect resistant cultivars first was documented in the United States in the late 1700s and early 1800s, including wheat (*Triticum aestivum* L.) resistant to the Hessian fly [*Mayetiola destructor* (Say)] in 1792 and apple (*Malus domestica* Borkhausen) resistant to the wooly apple aphid [*Eriosoma lanigerum* (Hausmann)] in 1831 (Smith 2005).

Nicotine apparently was the first of the three botanicals noted above to be used as an insecticide in the Western World and is reported to have been introduced in the 1600s (Isman 2006, Singh 2012, Pedigo and Rice 2015). It is extracted from tobacco (*Nicotiana tabacum* L.) and related species within the Solanaceae (Isman 2006, Pedigo and Rice 2015). It was used in the late 1600s in France as a wash applied to pear (*Pyrus communis* L.) trees to control lace bugs (McIndoo 1943) and in the 1760s as an extract from tobacco leaves that was sprayed on vegetation, as "tobacco water and tobacco powder" against plant lice (McIndoo 1943), and as crushed tobacco leaves for control of aphids (Singh 2012). From 1763 into the 1890s, tobacco smoke was used indoors (greenhouses) and outdoors (under tents and hoods placed over trees) to kill aphids (McIndoo 1943). Tobacco first was used in the United States (Albany, NY) as an insecticide in 1814 in the form of "tobacco water" (McIndoo 1943). Finally, nicotine was the active ingredient in Black Leaf 40®, which commonly was used to control aphids and a variety of other insects until the early 1990s (Roberts and Reigert 2013). However, although nicotine was an effective insecticide, it also was found to be highly toxic to humans (Pedigo and Rice 2015).

Pyrethrum, extracted from flowers of several species of *Chrysanthemum* (or *Tanacetum*) (Asteraceae) (Glynne-Jones 2001, Isman 2006, Abivardi 2008), also has had a long history. The earliest record of which we are aware was in 400 B.C. during the reign of the Persian King Xerxes in which children were deloused with powder from dry flowers of pyrethrum (*Tanacetum cinerariifolium* Trevir. Sch. Bip.) (Silva-Aguayo 2013). Subsequently, in the First Century (0–99 A.D.), Persians used pyrethrum to protect stored grain (Unsworth 2010), and the Chinese, during the Chou Dynasty, noted the insecticidal properties of the compound (Mocatta 2003). Following, pyrethrum was traded along the Silk Route into Europe and the New World and was used to delouse troops from the Napoleonic Wars (1804–1815 A.D.) up through WWII (Mocatta 2003). Interestingly, pyrethrum became widely used as an insecticidal powder, known as Persian Insect Powder, in Europe and the United States by the early- to late-1800s (Abivardi 2008). In the early 1900s, a sprayable liquid extraction of pyrethrum first became available and is the form most commonly used today (Abivardi 2008).

Rotenone (*Derris*) was introduced in the 1600s (Singh 2012), but not as an insecticide. It originally was used as a fish poison in South America by the aborigines since at least 1649 and has been used as an insecticide since 1848 (Pedigo and Rice 2015). It is extracted from the roots of *Derris* spp. and related plants (*Lonchocarpus* spp. and *Tephrosia* spp.) within the Fabaceae (Isman 2006, Pedigo and Rice 2015). Rotenone products from *Lonchocarpus* spp. were called Cubé and those from *Derris* spp., derris or derris dust (Peairs 1947).

From the 1800s through the early part of the 1900s, not only were botanical insecticides introduced, but the variety of cultural and chemical controls increased, particularly evident in North America. Cultural controls commonly utilized included trap cropping, location of host plants, clean culture, timing of planting date and plant maturation, and others. Chemicals used as inorganic insecticides included such compounds as petroleum, kerosene, creosote, turpentine, arsenic, sulfur, phosphorus, mercury (Ware 1994, Singh 2012), and many others (Riley 1884). The need for a greater variety of controls in North America in the 1800s was associated with the invasion and/or spread of several pests including the San Jose scale [*Quadraspidiotus perniciosus* (Comstock)] and the Colorado potato beetle [*Leptinotarsa decemlineata* (Say)] (Singh 2012). In 1867, Paris green (a mixture of arsenic and copper sulfate) was introduced to control the Colorado potato beetle (O'Brien 1967) and, within a decade, it was also used against the codling moth [*Cydia pomonella* (L.)] (Singh 2012). In 1892, lead arsenate was introduced as one of the most effective inorganic insecticides for control of pests such as gypsy moth [*Lymantria dispar* (L.)], apple maggot [*Rhagoletis pomonella* (Walsh)], and various soil insects (Flint and van den Bosch 1981, Singh 2012). It was used primarily in orchards until the 1940s (Ware 1994) and resulted in some of the first concerns about the safety of its residue on fruits. Today, there still are orchard soils contaminated with lead and arsenic (Lah 2011).

During this same period (1800s–early 1900s), biological control received little attention even though it had been utilized as far back as 324 B.C. by the Chinese (see above), and its underutilization continued into the mid-1900s. However, there was one early success of biological control that was an extremely effective tactic in North America. The invasive cottony cushion scale (*Icerya purchasi* Maskell) was causing severe losses to the citrus industry in California by the 1880s. In an example of classical biological control, the United States sent an entomologist, Albert Koebele, to the scale's native Australia to search for natural enemies. One of the two beneficials he brought to California, the vedalia beetle [*Rodolia cardinalis* (Mulsant)] was extremely successful in controlling the scale, and California's citrus industry was protected by the early 1890s (Flint and van den Bosch 1981).

16.2.1.5 Influence of Farming Practices on Control

The increased emphasis on methods of insect control in North America was greatly influenced by the changing demographics of the human population and changes in farming practices that were occurring at about the same time. During this period, there was an increase in the United States population, with many immigrants having an agricultural background (Reinhardt and Ganzel 2003). Most farmers supported themselves by growing crops for their own consumption rather than selling to urban inhabitants. In fact, in 1790, 96.7% of the American population lived on farms, the remaining 3,3% in cities. Farmers, not unexpectedly, raised a wide variety of crops and livestock (Reinhardt and Ganzel 2003).

Between the late 1700s and early 1930s, catastrophic insect infestations often were isolated and, thus, pest problems generally were not severe because the plantings of various crops were smaller and interspersed (Reinhardt and Ganzel 2003). There were exceptions, of course (e.g., boll weevil, *Anthonomus grandis grandis* Boheman; Colorado potato beetle), but, for most farmers, handpicking (physical control) and several cultural control methods were recommended including clean culture, burning of weeds and rubbish piles in winter, and trap crops. Insecticides were not particularly effective and could be dangerous. Some of the most dangerous were arsenical compounds (i.e., Paris Green, Scheele's Green, London Purple). In fact, in 1925, a family of four in London became ill from arsenic poisoning after consuming apples from the western United States that had not been washed properly by the grower after spraying (Reinhardt and Ganzel 2003). Petroleum products, including kerosene, were not particularly appealing because, according to several entomologists at the time (e.g., Smith 1897; Paddock 1915, 1918; Jones 1918, Watson 1918), a solution strong enough to kill the insects also would injure or kill the plants. Nicotine sulfate generally was ineffective (Drake 1920). However, Riley (1884) did have good results using pyrethrum in a liquid solution.

As noted above, most farmers raised several crops annually on their farms from the late 1700s to the early 1930s. But, there were exceptions. Some farmers specialized in single crops (i.e., monoculture). An excellent example was cotton, which was grown in large plantations beginning in the 1800s, typically ranging from 500 to 1,000 acres (Vejnar 2011). Insect damage to cotton, particularly from the boll weevil, was of primary importance because of the high cash value of the crop. Not surprisingly, the damage caused by sucking insects was not appreciated fully until Morrill (1910) reported that these insects were capable of causing severe damage to the bolls. Cultural methods were the primary recommendations for control including timing, clean culture, handpicking (mechanical control), and spraying with kerosene.

A second example of the effects of monoculture on insect populations was that of potato and its primary pest, the Colorado potato beetle. This beetle first was noted as a major pest of this crop during an outbreak that occurred in 1859 in fields about 100 miles west of Omaha, Nebraska (Jacques 1988), undoubtedly encouraged by the cultivation of "considerable acreages" of this crop.

During the 1930s, several changes occurred in farming practices that had a significant effect on agricultural production (Reinhardt and Ganzel 2003). In 1930, farmers made up only 21% of the labor force.

Therefore, it made sense to produce more food than needed by the family and ship the excess to urban markets. This, increasingly, became easier because farmers were moving from horse-drawn cultivators to increasingly sophisticated tractors and could plow more acreage in shorter periods of time. Also, it became clearer that it was more profitable to concentrate on fewer cash crops such as corn and wheat. In fact, by the 1950s and 1960s, entire regions specialized in producing one cash crop (Recall, this practice actually had begun in the 1800s with cotton and potato). However, moving to a monoculture agricultural system meant that insects feeding on these crops now had vast areas of unbroken fields of their preferred hosts, obviously an unnatural ecosystem. What, was needed, therefore, was better insecticides. But, all that was available at that time were arsenicals, sulfur, kerosene, mineral oil emulsion, naphthalene, paradichlorobenze, nicotine sulphate, and rotenone, all of which were largely ineffective. This began to change in the late 1930s with the advent of synthetic organic insecticides (Reinhardt and Ganzel 2003).

16.2.2 Age of Synthetic Organic Insecticides (Chlorinated Hydrocarbons, Organophosphates, Carbamates)

The age of commercial use of synthetic organic insecticides began in the 1940s with the advent of DDT (Dichlorodiphenyltrichloroethane). DDT was first synthesized in 1873 by Othmar Zeidler, an Austrian chemist, who did not investigate its properties and, therefore, did not recognize its potential as an insecticide (Ware 1994, Pedigo and Rice 2015). This discovery was made approximately 65 years later by Paul Hermann Müller, a Swiss chemist, working for J. R. Geigy AG. He had been assigned to develop an insecticide and was looking for an ideal contact insecticide that would have a quick and powerful toxic effect upon as many insect species as possible (broad spectrum) but would cause little or no harm to plants or warm-blooded animals. He also wanted to develop an insecticide that had a long residual action with high chemical stability and was cheap to produce. He had two primary motivations: (1) a severe food shortage in Switzerland, which could be lessened by a better way to control insect infestation of crops, and (2) the presence of a typhus epidemic, carried by head and body lice, in Russia, the most extensive and lethal epidemic in history (1918–1922) (Patterson 1993). He began his search in 1935.

Müller worked for 4 years on his project, testing numerous compounds and failing 349 times. But, in September 1939, he found what he had been looking for, DDT. He placed a fly in a cage with this insecticide and, shortly thereafter, the fly died. Subsequently, DDT was tested by the Swiss government and the United States Department of Agriculture against the Colorado potato beetle and found it to be highly effective. Subsequent work showed that DDT was effective against a wide range of pests including sand flies, mosquitoes, fleas, and lice known to transmit, respectively, various tropical diseases, malaria, plague, and typhus. And, because DDT's discovery coincided with World War II and the presence of disease among the troops, DDT was used to protect the Allied Forces from these blood-sucking insect vectors (Pedigo and Rice 2015).

After the war, DDT was offered commercially for general use by the public on farms, orchards, and in and around homes. In fact, DDT was used in sprays, paints, wallpaper, dusts for pests, and many other formulations. It seemed to be the perfect insecticide because, at the time, all information about it was positive. For the next 20–25 years, DDT and other chlorinated hydrocarbons were used and overused indiscriminately. As Smith and Kennedy (2002) stated so succinctly, "Such optimism had a profound effect on the crop protection sciences. In entomology and weed sciences especially, research shifted focus away from pest biology on to pesticide technology. At this point, the birthright of pest control scientists as biologists became endangered." As reliance on pesticides increased, the simple question of whether or not a pesticide application was even necessary or justified economically led Doutt and Smith (1971) to coin the term "pesticide syndrome." Applications were based on the calendar rather than on the pest populations. Pesticide manufacturers encouraged farmers to view premature applications as protection from pest problems before they appeared and, typically, this was referred to as "preventive control."

By the late 1950s and the 1960s, concerns were emerging about the negative effects of pesticides on the environment. Examples included resurgence of pest populations because of elimination of their natural enemies, outbreaks of secondary pest populations that had been held in check by populations of natural enemies that were eliminated by insecticidal applications that targeted primary pest species, and appearance of resistance, which required increased applications per year and increased amounts per application. These developing problems were ignored or discounted by entomologists and the general public, particularly those involved with selling pesticides. In fact, the rates at which these negative effects were appearing were undoubtedly influenced significantly by the practice of preventive control.

Another problem during this period was the increasing expectations of consumers for high quality produce (i.e., no blemishes). As DeBach (1964) noted, the consumer had adopted the slogan that "the only good bug was a dead bug" and, in doing so, thoroughly accepted "the advertising idea that shiny clean fruit, etc., are better fruit. No thrips scars, not a scale insect, must be present. Quality and taste are really forgotten to a large extent; appearance is of prime importance."

How does one obtain perfect produce? With pesticides. Use of DDT and other chlorinated hydrocarbons produced perfect apples, tomatoes, and many other crops. The nonchemical methods (e.g., cultural control, biological control) then available could not produce the same high quality and generally were ignored (Turnbull and Chant 1961).

One characteristic of the chlorinated hydrocarbons that had been considered a positive attribute in the beginning was that of a long residual action. However, it began to be viewed negatively as the other concerns about pesticides mentioned above became more evident. Thus, the organophosphates and carbamates were developed, which had a short residual action. These could be applied just before crops were harvested and sent to market. Unfortunately, the organophosphates (OPs) generally are more toxic to humans than the chlorinated hydrocarbons (malathion probably is the least toxic of the OPs). Carbamates, as a class, generally are less toxic than the organophosphates, though both have a similar mode of action (synaptic poisons) (Ware 1994).

Biological control plays a major role in pest management today. But, research on this tactic already was being conducted when DDT was introduced to farmers in 1945 and had been shown to be effective since the late 1800s in some cases (see above). However, up into the 1960s, there was a reluctance by farmers to rely on biological control because it was much easier to use insecticides, and biological control could not provide 100% control. This is because an effective parasitoid rarely eliminates 100% of the host population. The result is there may be damage. The farmer, then, is faced with one of two choices: (1) accept an increased risk that there will be insect damage and a reduction in marketable yield and/or the price received for the harvested product or (2) continue to use insecticides.

Biological control varies in its effectiveness depending on the type of injury caused by the pest species, direct or indirect. Direct injury is damage to the part of the plant (e.g., fruit) targeted for harvest. Indirect injury is damage to non-harvested parts of the plant (e.g., twigs, leaves, bark). Biological control is most valuable for control of pests causing indirect injury because 100% control is not needed (Turnbull and Chant 1961).

The turning point in blind reliance on pesticides and belief that there was no downside to their usage was the 1962 publication of Rachel Carson's book, "Silent Spring," in which she dramatically and forcefully brought to the attention of the public the dangers associated with these insecticides. Not only were they hurting the environment, they could be deadly to those who used them. She not only discussed the harmful effects of pesticides on vertebrates and invertebrates but was particularly critical of the long-term effects of the chlorinated hydrocarbons because of their long residual characteristic and subsequent biomagnification in food chains. To this can be added the off-site movement of pesticides to surface waterways and ground water and pesticides on harvested products.

Rachel Carson's book was viewed by many, including entomologists, with much hostility. But, today, it is considered a landmark publication, which resulted in major changes in current pest control and how pesticides are viewed by the public. Numerous changes in public policy, including changes in the regulation of pesticides and their usage, occurred in both the United States and Europe. Today, pesticides are viewed by some as a necessary evil and should not be used if there is an alternative. If they are used, they should be used as infrequently as possible, exceptions being emergency situations (sudden outbreaks, particularly diseases).

In addition to the public awareness resulting from Rachel Carson's book, there was an ongoing push in the scientific community to deal with the impacts of the over use of insecticides by promoting alternate management tactics, including biological control. Interest in biological control had had a long history as evidenced by the early historical reports mentioned at the beginning of this chapter. Even just before and during the "golden age" of insecticides from 1939 into the 1960s, biological control was being studied. The earliest studies during this period were included in broader discussions of population dynamics (Smith 1935; Solomon 1949, 1957). However, there was a gradual shift to papers emphasizing biological control (e.g., DeBach 1951, DeBach and Bartlett 1951), leading to a seminal paper by Stern et al. (1959) that promoted and provided a conceptual framework for the integration of biological and chemical control. Fundamental to this framework were the economic injury level and economic threshold concepts, which recognize that insect pest populations often occur at levels below which the value of yield loss that they cause is less than the cost of preventing that loss. Together, these concepts provided a basis for determining when insecticide applications are justified. Although these two controls often were thought to be alternatives, they could be complementary and even augmentative. Stern et al. (1959) noted that biological control operates to permanently increase the general environmental resistance to the increase of a pest population, whereas chemical control constitutes short-term restricted pressure. Finally, they noted that one must "recognize the "'oneness'" of any environment, natural or man-made." The abiotic and biotic components together make up the ecosystem. "If an attempt is made to reduce the population level of one kind of animal (for example, a pest insect) by chemical treatment, modification of cultural practices, or by other means, other parts of the ecosystem will be affected as well. For this reason, the production of a given food or fiber must be considered in its entirety. This includes simultaneous consideration of insects, diseases, plant nutrition, plant physiology, and plant resistance, as well as the economics of the crop." Further, DeBach (1964) stated, "The prime requisite for integrated pest control is basic ecological knowledge of the entire complex involved including the extent of biological control of each host insect that occurs in the absence of treatment. Basic studies may take several years to reveal the best method of utilizing the chemicals that are found to be absolutely necessary, with the biological control that is known to occur if no chemicals are used." Finally, Doutt and Smith (1971) stated that there was a great deal of misunderstanding of what was meant by "integrated control." They further stated, "There are several unique aspects to the research upon which an integrated control program is built. One is philosophical since the researchers must believe that chemical treatments ought not to be applied until they are clearly needed. Another aspect is organizational in that the program is best developed by a team of specialists representing diverse disciplines but all of who are channeling their research toward a common goal. The third and fundamental aspect is that the program is based on a sophisticated understanding of the ecology of the ecosystem involved."

Today, integrated control has evolved into Integrated Pest Management (IPM), which is based on a more holistic approach. IPM frequently is utilized in crop protection with the goal of eliminating or reducing the amount of pesticide used. As noted at the beginning of this chapter, IPM is defined as "A decision support system for the selection and use of pest control tactics, singly or harmoniously coordinated into a management strategy, based on cost/benefit analyses that take into account the interests of and impacts on producers, society, and the environment" (Kogan 1998). Such management practices help control pests but affect the environment as little as possible.

16.3 Modern Era

16.3.1 Current Management Practices

16.3.1.1 Cultural Control

Cultural control is the use of agricultural practices to reduce crop injury from pests. Because these are agricultural practices, these practices already are used for the purpose of growing the crop. However, to be considered a cultural control for a pest, the choice of an agricultural practice must be for preventing or reducing pest damage. Therefore, use of cultural controls requires an intimate understanding of both the pest and the crop. Cultural controls that are widely adopted generally are consistent with agricultural recommendations for other purposes. The cost of implementing cultural controls depends on how closely the control strategy aligns with recommended agricultural practices. The following sections cover the primary cultural controls used in agriculture.

16.3.1.1.1 Crop Rotation or Fallowing

Crop rotation is a designed sequence of crops grown on a piece of land. The length of a rotation can be as short as two years or as long as 10 or more years, especially when a multi-year crop like alfalfa (*Medicago sativa* L.) is included in the rotation. Rotations can have benefits for soil fertility, disease management, and weed control in addition to insect management. The principle behind crop rotation for insect management is removing a requisite for insect development. This is most effective for pests with a narrow host range, those with a long generation time, and those that overwinter locally. One of the most widely adopted uses of crop rotation for insect control is rotating between corn (*Zea mays* L.) and soybean (*Glycine max* Merrill) in the American Midwest. Western corn rootworm (*Diabrotica virgifera virgifera* LeConte) only has one generation per year, larvae can only develop on corn roots, and eggs are the overwintering stage (Levine and Oloumi-Sadeghi 1991). By rotating between corn and soybean, corn rootworm eggs are laid in corn fields in the fall, but when the eggs hatch the following spring, the field now has soybean; this effectively starves the corn rootworm larvae with no damage to soybean roots.

Fallowing is similar to crop rotation except that a period of no plants is included in the sequence. Fallowing is normally practiced in arid climates to try and accumulate enough ground water to grow a crop. From a pest management perspective, it is the same as crop rotation because the habitat does not have a suitable host for a period of time, causing the insects to leave or die.

16.3.1.1.2 Sanitation or Clean Culture

Sanitation or clean culture involves practices that remove infested crop material or potential host habitat from the area. The principle of sanitation for pest control is to create adverse conditions for pest survival. Many pests survive between crops by feeding on culls, volunteer plants, and crop residue. By removing these pest havens through sanitation, surviving populations are reduced, thereby reducing the pest population in the subsequent crop. Examples where sanitation is a critical pest management component include controlling bark beetles (Curculionidae: Scolytinae) by debarking infested logs and burning or chipping the infested bark (Wermelinger 2004) and shredding cotton (*Gossypium* spp.) stalks at the end of the growing season to control the boll weevil (*Anthonomus grandis grandis*) (Summy et al. 1986). In both cases, the cultural control prevents the insects from completing their generation, thereby reducing the size of the next generation.

Sanitation also includes removing alternative crop hosts found on the periphery of fields. Johnsongrass [*Sorghum halepense* (L.)] and other wild sorghum species are key hosts for sorghum midge [*Contarinia sorghicola* (Coquillett)]. Reduction of Johnsongrass in the vicinity of sorghum [*Sorghum bicolor* (L.)] fields removes an early-season host of sorghum midge, reducing the size of the population available to infest sorghum later in the year (Young and Teetes 1977).

16.3.1.1.3 Tillage

Tillage is a common agricultural practice with many purposes. Weed control, incorporation of plant nutrients, and seedbed preparation are common reasons for tilling the soil. In addition to these benefits, some pests can be controlled effectively with tillage. Tillage can change the microclimate of soil-living insects, or it can disturb their habitat. European corn borer [*Ostrinia nubilalis* (Hübner)] is one of several corn borers that overwinter in the base of corn stalks. Mowing and tillage breaks apart their overwintering home, increasing mortality from predation and desiccation (Umeozor et al. 1985). Other insects such as corn earworm [*Helicoverpa zea* (Boddie)] overwinter as pupae in the soil. Tillage can either bury them so deeply that the moths cannot reach the surface, or expose them to the surface so they are eaten by predators or dessicate and die (Roach 1981).

16.3.1.1.4 Plant Density and Row Spacing

Plant density and row spacing are techniques that can be used to manage the micro-environment of insects and pathogens. High plant densities and narrower rows can create a cooler, more shaded environment, which has been shown to reduce citrus rust mite (Muma 1970). Wide soybean rows provide a more open canopy, which frequently results in higher densities of large larvae of *Helicoverpa zea* (Alston et al. 1991). This may not be from any direct effect on *H. zea*, but may be a result of pathogens, such as the fungus *Nomurea rileyi* (Farlow) Samson, which can provide more effective biological control of *H. zea* in a closed canopy with higher humidity (Sprenkel et al. 1979).

16.3.1.1.5 Trap Cropping and Farmscape Influences

The basic concept of trap cropping is to provide a highly attractive crop in proximity to the main crop so that pests move to the trap crop and spare the main crop. Insects that are in the trap crop can then be managed as needed to avoid later movement into the main crop. This technique has been shown to be effective in multiple systems (Hokkanen 1991, Shelton and Badenes-Perez 2006), and, yet, it rarely has been implemented on a commercial scale. Reasons for the rare implementation generally are related to logistical challenges of growing two crops in a field (e.g., herbicide compatibility, fertilizer requirements, planting dates, harvest dates) and the cost of taking land out of production to grow the trap crop, which may not have any market value. Stink bugs often prefer soybeans over cotton (Bundy and McPherson 2000) and prefer soybeans in the earliest maturing varieties. Therefore, a trap crop of early-maturing soybean can help protect a field of later-maturing soybean (McPherson and Newsom 1984).

In contrast to trap cropping to reduce pest damage, the location of fields within a farmscape can increase or minimize pest damage of generalist insect pests. In the southern United States, wheat dries down at the same time corn is in its early vegetative stages. Chinch bug [*Blissus leucopterus leucopterus* (Say)] can be found in maturing wheat (*Triticum aestivum* L.) at high densities but is seldom regarded as a pest at this time, so it is not managed. However, as the wheat begins to dry due to maturation, it becomes an unsuitable host for the bug, which then leaves the wheat. Young corn growing adjacent to the wheat is a suitable host, and damaging densities of chinch bugs can be found on the outside rows of corn planted beside wheat. Similarly, western tarnished plant bug, *Lygus hesperus* Knight, stays in alfalfa adjacent to cotton (Sevacherian and Stern 1974, Godfrey and Leigh 1994) unless all the alfalfa is mowed (Sevacherian and Stern 1975). Strategic arrangement of crops in the farmscape should seek to minimize interfaces that facilitate movement of a pest from one field to the next. The farther an insect needs to move to find a suitable host, the lower the likelihood it will find the host, which can be an effective cultural control strategy in itself.

16.3.1.1.6 Planting Date

Planting dates often are chosen for agronomic reasons, and planting before or after recommended planting dates can have a substantial economic cost. However, even within the recommended planting window, the choice of planting date can impact the likelihood of pest damage. Tarnished plant bug [*Lygus lineolaris* (Palisot de Beauvois)] densities generally increase throughout the growing season, so earlier cotton plantings require fewer insecticide applications and have less damage than later plantings (Adams et al. 2013). Another aspect of planting date is how it compares to neighboring fields of the same crop. For example, one of the most effective ways to minimize damage from *Contarinia sorghicola* is to plant all the sorghum at the same time so that all plants flower at the same time (Young and Teetes 1977). These midges only attack sorghum during pollination, so when all fields are pollinating at the same time, no fields suffer much damage because the midges are spread over many sorghum plants. Therefore, there are no opportunities for midges infesting an early-flowering field to complete a generation and then attack a late-planted field.

16.3.1.1.7 Water Management

Water management consists of draining water away from, or adding water to, a crop. The agronomic impact of water management is great, so using water management for pest management purposes is a reasonable option only when it is consistent with other agronomic needs. Flooding has been used to control several insects in crops that are tolerant of flooding. Cranberry, *Vaccinium macrocarpon* Aiton, is tolerant of flooded conditions, but the low dissolved oxygen content in flooded soil of a cranberry bog can provide 80–90% control of blackheaded fireworm, *Rhopobota naevana* (Hübner), within 9 days (Cockfield and Mahr 1992). Use of insecticide to control cranberry fruitworm, *Acrobasis vaccinii* Riley, during the summer growing season was reduced by 70% after implementing a 4-week flood during the spring in Massachusetts (Averill et al. 1997). Continuous flooding for up to 6 weeks killed many wireworms, *Melanotus communis* (Gyllenhal), in Florida (Hall and Cherry 1993). Mortality increased with warmer temperatures. In other situations, sprinkler irrigation is sufficient to reduce pest populations. Diamondback moth [*Plutella xylostella* (L.)] has reduced mating, dispersal, oviposition, and increased larval mortality when fields are irrigated with a sprinkler (Talekar and Shelton 1993). On the other hand, mosquito populations, many which develop in stagnant water, can be reduced by draining land that would otherwise hold pools of water (Keiser et al. 2005).

16.3.1.1.8 Nutrient Management

Several insect species, including aphids (Aphididae) and *Lygus lineolaris*, are known to concentrate in the most succulent parts of a field. Succulence may be a function of water management, but, often, it is a result of nutrient management, particularly nitrogen management. Excessive nitrogen often increases vegetative growth and makes foliage more succulent. This can increase the growth rates of insects such as aphids (Duffield et al. 1997, Nevo and Coll 2001), resulting in larger populations. Some nitrogen is essential for plant growth, but, often, there is a level where there is enough nitrogen to meet plant needs without impacting pest densities.

16.3.1.1.9 Intercropping

Intercropping is growing two dissimilar crops in the same space at the same time. This was common before agriculture became mechanized, but equipment limitations today prevent widespread use of this cultural control in many situations. Advantages of intercropping include increased yields (Li et al. 2001), reduced economic risk (Rao and Singh 1990), and improved non-chemical weed control (Liebman and Dyck 1993). For insect pest control, intercropping can be effective by making it harder for the pest to identify the crop, or the non-host plant may impede movement even after the pest arrives in the field (Trenbath 1993). Two commonly used intercrops are mixtures of alfalfa and a grass. Potato leafhopper [*Empoasca fabae* (Harris)] is a major pest of alfalfa, but intercropping alfalfa with a forage grass results in reduced densities of this insect (Roda et al. 1997a,b). Another pest impacted by intercropping is the alfalfa weevil, *Hypera postica* (Gyllenhal). This weevil is a specialist of alfalfa, so fields that are seeded with a forage hay experience reduced levels of damage by this insect (Roda et al. 1996). Because both alfalfa and forage grasses can be harvested together, this intercrop is not hampered by mechanized harvest.

16.3.1.1.10 Harvest Management

Harvest management can play a role in several ways. In concert with managing the proximity of a pest source field with potential sink fields (farmscape influences), the timing of the harvest of the source field can play an important role in the scale and timing of the pest movement from one field to another. In some cases, harvesting a crop early can be an alternative to spraying for the pest and then waiting until it is safe to harvest. Also, harvest management can be used to reduce pest damage by harvesting the crop in strips, thereby keeping the pest in the current crop rather than forcing it out of the field to feed in a nearby crop that is more easily injured by the insect. All of these management practices can be shown, using *Lygus hesperus* in an alfalfa-cotton landscape as an example. This plant bug prefers to feed in alfalfa but causes greater economic damage in cotton, so management strategies are designed to keep it from moving into cotton. Harvesting alfalfa early is one strategy whereby the alfalfa is harvested on a 28-day schedule rather than the more conventional 35-day schedule (Godfrey and Leigh 1994). The bugs can infest and oviposit in alfalfa as soon as new growth has emerged. By harvesting on a 28-day schedule, nymphs from the earliest-laid eggs cannot complete development to the adult stage, so the life cycle is cut short, and there are few adults available to migrate to nearby cotton fields. An alternative to the shorter harvest schedule is to harvest the alfalfa in strips, always maintaining a portion of the alfalfa field in a state that is attractive to the bugs, thereby minimizing any migration into cotton (Sevacherian and Stern 1975, Summers 1976).

16.3.1.1.11 Host Plant Resistance (Non-GMOs)

Selecting crop varieties based on their tolerance or resistance to pests can be considered another form of cultural control. Many crop-breeding programs have been focused on developing commercially acceptable varieties that are resistant to a wide range of pests. Differences in pest susceptibility among crop varieties can be based on pest evasion, antixenosis, antibiosis, or crop tolerance.

Pest evasion is a reduction in damage by not having the susceptible stage of the plant present when the pest is abundant. For example, an early-maturing variety of soybean can appear to be resistant to late-season migratory pests [e.g., fall armyworm, *Spodoptera frugiperda* (J. E. Smith); and soybean looper, *Chrysodeixis includens* (Walker)] because it matures before these pests arrive. However, if an early-maturing variety intended for a pest evasion approach is planted in a situation where the pest is present when the plant is susceptible, this variety is just as susceptible as other varieties.

Antixenosis, or non-preference, is a form of host plant resistance that has the biggest benefit when pests have a choice of hosts (Kogan and Ortman 1978). It has a genetic basis with less desirable morphological (e.g., thorns, leaf hairs) or physiological (chemical deterrents) features for the pest. These features do not actually kill the pest, but cause the pest to preferentially feed on other hosts when available. However, in large, monocrop situations, this type of host plant resistance has minimal impact because the pest has no choice, feeds on the plant, and can develop normally.

Antibiosis is the goal of most host plant resistance breeding programs. In this case, the pest is adversely impacted by feeding on the plant. This may be due to the presence of one or more toxic metabolites or a sub-optimal amount of some essential nutrient. Although this type of host plant resistance often provides the highest level of pest control, it also is the type of control that pests are most likely to overcome by developing resistance to it. Because those that are most susceptible to the resistance mechanism are killed, those that survive and mate are those that are most resistant, creating a situation where continued selection for resistance to a toxin eventually creates a population that can overcome the host plant resistance.

The last form of host plant resistance is tolerance to the pest. In this case, the plant has no impact on the pest, but the plant is able to compensate for pest damage and, therefore, the economic impact of the pest on the plant is reduced. For example, a variety of corn that grows extra corn roots may not be as susceptible to corn rootworm (*Diabrotica* spp.) feeding as other varieties (Owens et al. 1974, Ivezic et al. 2006).

16.3.1.2 Physical/Mechanical Control

Physical or mechanical control of insects involves altering the environment in some physical manner to make it undesirable or uninhabitable to insect pests (Banks 1976). Several strategies are discussed here generally and specifically when related to stink bugs or closely related groups.

16.3.1.2.1 Electricity

Ever since the invention of electricity, attempts at using electrical power to eliminate pests have been endless. One of the most successful strategies for killing insects with electricity involves the combined use of attractive light and energized metal wires. Traps designed to lure and kill insects with electricity in this manner have been in use for decades. Although most of these trap types primarily lure and kill flying insects, they are relatively non-selective, killing pestiferous and beneficial species, and are not generally effective in killing a targeted species (Heinen et al. 2003). Furthermore, electrocuting insect traps might actually release into the environment bacteria and viruses carried by insects when their bodies disintegrate (Broce 1993, Urban and Broce 2000). Finally, electric stink bug traps can be helpful for monitoring general population trends, but there is no evidence that enough bugs can be attracted to actually reduce population density in an outdoor setting.

16.3.1.2.2 Sound Waves

There is a plethora of sonic devices available that claim to deter or control animals. Research on devices that use sound waves as animal deterrents usually end with summaries that report poor performance or inconclusive effectiveness of tested gadgets (Bomford and O'Brien 1990). However, some potential utility has been observed with radio frequency (RF) and microwave energy (MW) in controlling pests of stored nuts (Wang and Tang 2001). The problems with using those energy waves are high costs, undesirable damage caused by heating to stored products, and non-uniform control of pests. Nelson (1996) provided a review of RF and MW energy for control of insects in stored grain, but the costs of alternative control measures using radiation were high compared with chemical control (Halverson et al. 1996). Early tests of sonic devices marketed for control of household insects, such as cockroaches, indicated that efficacy was very poor (Gold et al. 1984), and more recent reviews have reached the same conclusions about sonic pest repellents (Aflitto and DeGomez 2014). In a study testing sound repellency of *Helicoverpa zea*, high-frequency sounds did not repel this insect in sweet corn (Shorey et al. 1972). Most methods using sound waves are not effective in repelling or controlling insect pests, and some ultrasonic devices actually attract insects, such as mosquitoes (Andrade and Cabrini 2010). Additionally, ultrasonic waves do not penetrate into substrates making control negligible in complex field situations.

16.3.1.2.3 Radiation: Visible, Ultraviolet, Ionizing

Insects are greatly affected by light in its various forms. Photoperiod, the relative length of light and dark periods in a single day, determines when many insects prepare physiologically for overwintering in autumn or leaving overwintering sites in the spring. The simple pattern of sunlight presence or absence during the year has a dramatic and important impact on insect biology. Increasing amounts of sunlight during long summer days with much more light than dark generally trigger reproduction. Conversely, decreasing day length will eventually reach a critical threshold when newly emerged adults do not reproduce but rather accumulate fat bodies to survive during diapause. In a review of the impact of climate change on heteropterans, Musolin (2007) reported on the continuing evolution of the true bugs in the context of fluctuating ecological parameters. As global temperatures increase, insect geographic ranges may also shift resulting in exposure to different photoperiods. These photoperiodic cues can cause behavioral changes that are favorable or unfavorable to the species. Diapause in adults of the southern green stink bug, *Nezara viridula* (L.), is controlled primarily by a long-day photoperiodic response, and short-day conditions trigger a color change associated with induction of diapause (Harris et al. 1984, Musolin and Numata 2003) (see **Chapter 11** for further discussion of diapause). Clearly, light is important to insects in general, and to stink bugs specifically, and knowledge of how critical light is to insects has resulted in research to use light to manage insects (Ben-Yakir et al. 2012).

Ultraviolet (UV) radiation and colors in the visible spectrum can effect insect behavior. UV has been used for decades to attract insects to light traps (Frost 1957, Harding et al. 1966, Hendricks et al. 1975). Other specific wavelengths of light, such as yellow light, in the form of sticky cards, direct illumination, and other forms, have been used to count, control, or alter the behavior of insects (Shimoda and Honda 2013). When insects are attracted and move to light, the behavior is termed positive phototaxis; when light repels insects, the behavior is termed negative phototaxis (Jander 1963, Coombe 1981, Menzel and Greggers 1985, Reisenman et al. 1998, Kim et al. 2013). Negative phototaxis has been demonstrated with light reflected off mulching films on the ground between soybean rows, thereby delaying colonization of aphids (Kimura 1982), and the same has been observed with aphids, thrips, and whiteflies in tomatoes (Csizinszky et al. 1995). Other reflective ground surfaces have been used to prevent invasions of thrips and whiteflies in other crops (Nagatuka 2000, Simmons et al. 2010).

In addition to visible and UV light, other forms of electromagnetic radiation have been used in attempts to physically control insects. Historically, the most recognized example of this control involved significant reductions in reproductive potential of the screwworm, *Cochliomyia hominivorax* (Coquerel) (also see **Section 16.3.1.3.1.3**, Sterile-Insect Technique below), by release of millions of male flies sterilized by irradiation (Knipling 1955, Baumhover et al. 1955). Because a *C. hominivorax* female only mates once during her lifetime, copulation with a sterile male eliminates her reproductive potential. Use of this technique allowed researchers to eradicate this important lifestock pest from the southeastern

United States in 1959 (Meadows 1985). Ionizing radiation was used to kill or render nymphs of *Nezara viridula* sterile (Dyby and Sailer 1999). It was hoped that sterile individuals would promote autocidal control for the species after release of semisterile populations (Knipling 1969), but the results were limited.

16.3.1.2.4 Temperature

Temperature is the most important environmental factor for insects. Because most insects are ectothermic or poikilothermic, their life processes are affected by temperature. The general effects of temperature on insect physiology, behavior, and development were reviewed by Wigglesworth (1972). As climate change becomes more apparent, any rise in average global temperature undoubtedly will affect insects, and heteropterans will be no exception (Musolin 2007, Musolin et al. 2010). The influence of temperature on pentatomids has been studied, with most research focused on the impacts of temperature on rates of development of the insect (Ali and Ewiess 1977, Naresh and Smith 1983, Simmons and Yeargan 1988, Kotaki 1998, Musolin and Numata 2003). However, mutualistic gut-associated symbionts of stink bugs also can be affected negatively by temperature (Prado et al. 2010). Although few studies have addressed manipulations of temperature to manage hemipterans as pests of field crops, there are numerous applications for the strategy regarding insect and mite pests of stored products (Fields 1992). In a review of over 50 published papers on the effects of temperature on pests of stored products, the optimal range for growth and reproduction for stored-product insects and mites was defined as 25–33°C, with an extended range of 13–35°C acceptable for most development and reproduction, and <13 or >35°C eventually being lethal (Fields 1992). Heat treatments, increasing the ambient temperature to 50–60°C for a 24–36 hour period, are a widely used alternative to methyl bromide gas for management of stored product insects in food processing facilities (Mahroof et al. 2005). In response to rapid heating, studies show that the red flour beetle, *Tribolium castaneum* Herbst, produces heat-inducible proteins that confer theromotolerance (Mahroof et al. 2005). Heat treatments also are a residue-free management method for the bed bug, *Cimex lectularius* L., in residential spaces (Kells and Goblirsch 2011). Forced hot air has been used for quarantine treatments, such as West Indian fruit fly, *Anastrepha obliqua* (Macquart), on mango (Mangan and Ingle 1992).

16.3.1.2.5 Physical Barriers

Numerous types of physical barriers have been used to successfully control insects, and many of those methods for agriculturally important insects have been reviewed (Boiteau and Vernon 2001, Vincent et al. 2003). Trenches (Boiteau and Osborn 1999), fences (Bomford et al. 2000), mulches (Brust 1994), particle films (Glenn et al. 1999, Puterka et al. 2000), traps (Cohen and Yuval 2000), flooding (Averill et al. 1997), screening (Dobson 2015), barriers (Tillman 2014, Tillman et al. 2015), and various other physical control techniques have been used to control or deter insects. Materials that block UV light have been used to protect greenhouse crops from insect vectors of plant diseases (Diaz and Fereres 2007) by interfering with the vision of insects and limiting their ability to locate host plants. Although considered a cultural control strategy, trap cropping has been tested as a barrier strategy to concentrate pestiferous insects in a small area adjacent to the crop of main interest where they can be destroyed or held, preserving the protected commodity (Todd and Schumann 1988). This was shown to be unreliable for stink bugs in early season soybean production systems (Smith et al. 2009) that dominate most of the mid-southern United States. In other studies, the limitations and advantages of using trap cropping strategies for stink bugs were described (Mizell et al. 2008). Successful use of soybeans as a trap crop for protecting cotton was reported recently by Tillman et al. (2015). Use of a synthetic (1.83 m high black, polypropylene sheeting) physical barrier between peanuts and cotton did show promise in protecting cotton from stink bugs (Tillman et al. 2015) as did plantings of grain sorghum (Tillman 2014).

16.3.1.2.6 Hand Removal (Handpicking)

Manual removal of arthropod pests from undesired locations is a practice that dates back to ancient times. Removing parasites from animals, including humans, by hand was the original control tactic for pest control. This practice extended to plants of interest, including those that were cultivated. During early subsistence agriculture, manual removal of pests was part of the production of cotton (Bottrell and Adkisson 1977). Cultural practices for control of rice borers and other insects in rice were and continue to be used in Asia (Kiritani 1979). When combinations of insecticides, removal of weed hosts, and hand-picking of Old World bollworm (*Helicoverpa armigera* Hübner) are used in chickpea in Asia, effective control is achieved and yields are maximized (Wakil et al. 2009). In the United States, handpicking of insects continues to be in the recent recommendations for control of pests in home gardens producing vegetables (Sagers 2005). The harlequin bug, *Murgantia histrionica* (Hahn), and the *Bagrada* bug (or painted bug), *Bagrada hilaris* (Burmeister), are pests in home gardens and commercial operations of vegetable production, and guidelines for management include hand removal of these stink bug species as a cultural control technique (Bealmear et al. 2013, Kemble et al. 2015).

16.3.1.3 Reproductive Control

Reproductive pest control is a specialized pest management strategy whereby individuals are prevented from finding each other to mate, or individuals are genetically altered and rendered sterile before being released into the environment to mate with wild individuals. Due to inability to mate or production of sterile eggs, the pest populations decrease over time.

16.3.1.3.1 Sterilization Strategies

Partial or complete genetic sterilization of insects can be accomplished using ionizing radiation, che-mosterilants, or RNA interference (RNAi). The use of ionizing radiation and chemosterilants has been studied for many years across taxa, whereas gene silencing via RNAi is still in its infancy as evidenced by the fact that no RNAi technologies currently are available commercially. RNAi is used to suppress expression of target genes by post-transcriptional regulation. This technology has been shown to produce spermless males of *Anopheles gambiae* Giles that mate successfully with females that then lay sterile eggs (Thailayil et al. 2011). A few other (of many) examples of the unique RNAi strategies include release of individuals that carry a dominant female-killing allele (Schliekelman and Gould 2000), interference with salivary secretions necessary for feeding (Araujo et al. 2006), interference with genes expressed in the midgut (Ghanim et al. 2007), and suppression of genes that govern an insect's circadian clock (Ikeno et al. 2013). Deployment of a transgenic corn variety that incorporates RNAi for pest management of *Diabrotica virgifera virgifera* LeConte appears to be the most promising candidate for a commercial release of this technology (Baum et al. 2007, Murugesan and Siegfried 2012).

16.3.1.3.1.1 Chemosterilization   Chemosterilants are compounds that cause a treated organism to become irreversibly sterile. In agriculture, chemosterilants are used to reduce pest populations and associ-ated damage to a particular commodity. Using *Oncopeltus fasciatus* (Dallas) as a model, Economopoulos and Gordan (1971) used tretamine to sterilize 90% of the males released into a confined population. This treatment caused an immediate reduction in viable eggs produced by females that mated with the sterilized males, but the effect was short lived and became negligible after only 20 days. More recently, insect growth regulators have been used widely as chemosterilants to disrupt the development of imma-ture life stages by interfering with the endocrine mechanism. Some insect growth regulators, includ-ing pyriproxifen and triflumuron, have been used to sterilize *Glossina morsitans morsitans* Westwood (Hargrove and Langley 1990, Langley 1995). Similarly, cyromazine, diflubenzuron, and pyriproxyfen are effective against *Musca domestica* L. (Kocisova et al. 2004). Casana-Giner et al. (1999) reported that treatment of adults with the chitin synthesis inhibitor lufenuron prevented egg hatch in *Ceratitis capitata* (Wiedemann) and females that mated with lufenuron-treated males deposited nonviable eggs. Further, field deployment of a bait gel laced with lufenuron reduced *C. capitata* populations continuously over a 4-year period (Navarro-Llopis et al. 2007).

16.3.1.3.1.2 Radiation   Ionizing radiation has been used in an entomological context for pest manage-ment programs that utilize the sterile insect technique, quarantine purposes (disinfestation of commodi-ties at ports of entry), and for research projects on physiological interactions between living organisms.

The mode of action for radiation is that cells with a high mitotic rate are more radiosensitive; therefore, mitotically active reproductive cells are the most susceptible to killing and sterilization when exposed to ionizing radiation (Bakri et al. 2005). Research showed that adult hemipterans generally began to become sterile, at least partially, after exposure to 30 to 60100 Grays (Gy) of ionizing radiation (Mau et al. 1967, LaChance et al. 1970, LaChance and Riemann 1973). For example, *Nezara viridula* adult females that were irradiated with low level (<10 Gy) radiation as fourth instars produced large numbers of nonviable eggs and exhibited significantly lower fecundity than non-irradiated females (Dyby and Sailer 1999). Additional stresses, including diet and inbreeding, resulted in further declines in fecundity and fertility in the test population. When exposed to higher levels of radiation (20 Gy), only half of the individuals survived to the adult stage.

A frequent concern is that foods or insects exposed to radiation for quarantine or pest management may become radioactive. However, the quantum energies of acceptable radioactive sources for these purposes are insufficient to induce radioactivity (Bakri et al. 2005).

16.3.1.3.1.3 Sterile-Insect Technique   The sterile insect technique is a pest management strategy whereby an overwhelming number of sterile individuals (often males) are released into the environment to mate with wild individuals resulting in non-fertile offspring. The sterile insect technique is widely acclaimed for its role in the success of the area-wide integrated pest management program that eradicated *Cochliomyia hominivorax* from the southern United States (Knipling 1955, 1960). Unfortunately, the specific factors that made the *C. hominivorax* program successful are not present with other pests, such as stink bugs. For example, *C. hominivorax* individuals are easy to rear in large numbers, they mate only once in their lifetime, and adult males do not inflict economic injury. By comparison, it costs a minimum of several dollars per insect to rear stink bugs [e.g., the brown stink bug, *Euschistus servus* (Say); the brown marmorated stink bug, *Halyomorpha halys* (Stål); and the southern green stink bug, *Nezara viridula*) to adults (Michael D. Toews, personal observation). Rearing enough stink bugs to release thousands per hectare in an area-wide program would be cost prohibitive, not to mention that the sterilized adults would feed and inflict serious economic injury. Further, stink bugs mate multiple times thereby increasing the likelihood of mating with a feral (non-sterile) individual at some point.

16.3.1.3.2 Mating Disruption

Mating disruption is a pest management technique whereby large amounts of synthetic sex pheromone are released into the environment to interfere with the ability of males and females to find each other and mate. Unmated females fail to produce viable eggs, and pest populations quickly decline. Shorey et al. (1967) used mating disruption to control *Trichoplusia ni* (Hübner). Subsequent investigations and commercial applications have included disruption of agricultural and forest insect pests, generally moths (Cardé and Minks 1995). This technique has shown tremendous utility for reducing *Plodia interpunctella* Hübner populations in confined environments such as storage warehouses and food- and feed-processing mills (Trematerra et al. 2011). Mahroof and Phillips (2014) documented immediate trap shutdown and a significant long-term reduction of a beetle, *Lasioderma serricorne* (F.), in food- and feed-processing facilities. Mating disruption has not been studied for hemipterans with the exception of mealybug and scale insects (Walton et al. 2006, Vacas et al. 2011). Pheromone-based mating disruption has been suggested for *Euschistus servus* (Borges et al. 2001), but not tested empirically.

16.3.1.4 Regulatory (Legal) Control

Regulatory or legal control of insects can be defined as the approach to control insects with laws that mandate procedures, such as inspection, certification, quarantine, and eradication, to prevent or mitigate problems with offending species.

16.3.1.4.1 Quarantine

The economic impact of invasive species is an estimated $120 billion annually in losses in the United States, which includes $14.4 billion for arthropod crop pests and $2.1 billion for arthropod forest pests

(Pimentel et al. 2005). Pesticide costs are estimated at $500 million annually in the United States to control nonindigenous insect species (Pimentel et al. 2005). Quarantine methods can be used to prevent, detect, contain, and eradicate invasive species that may affect humans, animals, plants, and the natural environment (Mumford 2002). Once an invasive species becomes established, eradication can be used when the long-term cost of damage, control, or both are expected to exceed the short-term cost of eradication (Myers et al. 1998). A number of stink bug species have become established outside of their native distributions. Among these, *Nezara viridula* is found in Europe, Asia, Australasia, Africa, and the Americas, but is thought to be native to Africa (Todd 1989; see **Chapter 7**). *Bagrada hilaris* is native to Europe, Africa, and Asia, was found in 2008 in California and since has spread throughout the southwestern United States (Bundy et al. 2012, Reed et al. 2013; see **Chapter 3**). *Halyomorpha halys*, native to Asia, was found in 1996 in the United States and also has invaded parts of Europe, Australia, and New Zealand (Rice et al. 2014; see **Chapter 4**).

16.3.1.4.2 Eradication

Because successful eradication often is achieved when the target pest has a limited distribution, either through host or habitat specificity or is geographically isolated (Myers et al. 1998), eradication of stink bugs typically is not possible due to their polyphagous nature and high reproductive ability. Successful eradication has been achieved with the boll weevil, *Anthonomus grandis grandis* (Coleoptera: Curculionidae), in the southeastern United States in cotton by preventing diapause, decreasing reproduction, and reducing in-season survival by using insecticides and cultural practices (Smith 1998). Another example of a successful eradication is the tsetse fly, *Glossina austeni* Newstead, from Unguja Island, Zanzibar, using the sterile insect technique (Vreyson et al. 2000).

16.3.1.4.3 Certification/Inspections

Preventing the spread of invasive species is more challenging than ever, with on-going changes worldwide in climate, trade, and travel (Schwalbe and Hallman 2002). As an example of the threat to trade caused by a species closely related to stink bugs, the kudzu bug, *Megacopta cribraria* (F.), is native to Asia and has spread throughout the southern United States. In December 2011, inspectors in Honduras found two dead kudzu bugs in a shipping container from Georgia. This led to a temporary ban on agricultural shipments from the southern United States until all containers were inspected (Ruberson et al. 2013; see **Chapter 5**). Efforts in pest survey and detection methods will continue to be crucial in preventing the establishment of plant pests (McCullough et al. 2006).

16.3.1.5 Chemical Control

Chemical control is the use of natural or synthetic chemicals (pesticides) to reduce pest damage. Within the framework of integrated pest management of insects, insecticides should largely be reserved for controlling existing populations that are approaching economically damaging levels. Preventative use of insecticides generally is discouraged because of the potential for negative non-target impacts. The point at which an insecticide application can be economically justified is called the economic threshold.

16.3.1.5.1 Economic Injury Levels and Action Thresholds

The Economic Injury Level (EIL) for a given pest in a specific situation is the point where the expected cost from pest damage is equal to the cost of applying an insecticide to avoid the pest damage. The Action Threshold (AT), also known as Economic Threshold, Action Level, or simply Threshold, is a point that occurs before the EIL is reached where action (typically the application of a chemical insecticide) should be taken to ensure that an increasing pest population will not cause economic damage. To estimate the current and potential economic cost of an insect population requires (1) an estimate of the size of the existing pest population, (2) knowledge of the amount of commodity injury caused by a population of the insect, (3) knowledge of the amount of commodity damage caused by the plant injury observed, and (4) an estimate of insect population growth rate. Without all these components, an EIL and empirically based AT cannot be estimated. This practically limits the use of EIL and AT to situations where the pest can be sampled readily and enough research has been conducted to estimate injury, commodity damage, and insect growth rate. Furthermore, it requires that the lowest detectable amount of insects or injury does not cause an economic loss. When an insect has a cryptic lifestyle (e.g., soil-dwelling, stalk borer), or a single insect causes economic losses (e.g., disease vectors of humans and animals), a practical EIL and AT cannot be calculated. Fortunately, sampling stink bugs generally is feasible, and diseases vectored are limited in scope, so EILs and ATs are possible for many stink bug-host relationships (Hall and Teetes 1982, Negrón and Riley 1987, Pantoja et al. 2000, Greene et al. 2001, Musser et al. 2011).

The concept of EIL and AT was introduced by Stern et al. (1959). Later a mathematical calculation for EIL was developed by Pedigo et al. (1986). There is no general mathematical formula available for the AT because it is impacted by scouting frequency, time between scouting and taking an action, time between taking an action and getting results, the link between the insect stage sampled and the damaging life stage, and the population growth rate of the insect.

16.3.1.5.2 Insecticide Modes of Action

Insecticides have been developed that kill insects in numerous ways. These various modes of action provide growers with a variety of tools that have varying levels of specificity, risks to humans, and environmental risks. Each insecticide is categorized based on its mode of action (Insecticide Resistance Action Committee 2016) so that users can choose a product that best fits their needs. A common recommendation is for users to rotate among several modes of action in order to prevent or delay the development of resistance to the insecticides (Immaraju et al. 1990, Prabhaker et al. 1998). These insecticide modes of action can be grouped into four broad categories, namely neuromuscular toxins, insect growth regulators, cellular respiration disruptors, and midgut disrupters.

Neuromuscular toxins attack the nervous system or muscles of insects. These tend to be fast-acting products that control a broad spectrum of insects. Unfortunately, many aspects of the insect nervous system are similar to the human nervous system, so these products tend to be more dangerous to humans. Early insecticides and the majority of insecticides used currently belong to this category. Widely used classes of insecticides that work on the nervous system include the organophosphates, carbamates, pyrethroids, neonicotinoids, and diamides. Each of these insecticide classes works on a specific nervous system target (e.g., organophosphates inhibit acetylcholinesterase; pyrethroids lock axonic sodium channels open), so rotation of products even within this broad category can reduce the development of insecticide resistance.

Insect growth regulators impact growth and development, especially of immature insects. These insecticides generally mimic or inhibit essential hormones involved in the molting process. They generally have minimal human toxicity problems and are not lethal to adults of even the target species. Because they kill by disrupting molting, the impact is only evident following a molt, so newly molted insects may continue to live and feed for several days after application.

Cellular respiration disruptors affect the cellular respiration process, which shuts down normal physiological functions of the insect. Most products in this small category interfere with mitochondrial functions.

Midgut disruptors describe all *Bacillus thuringiensis* (Bt) products, both transgenic and conventional types. These products require ingestion by the insect. The insecticide binds to the midgut, creating pores that allow the contents of the midgut to mix with the hemolymph, thereby killing the insect (Gill et al. 1992). Bt commercialized products currently target numerous lepidopteran, dipteran, and coleopteran pests, while having little to no impact on other insect orders or other non-target species. However, many Bt toxins have been identified without knowing which species they control, so more diversity in insect targets of Bt products is expected in the future.

16.3.1.5.3 Application Strategies

The decision about how to apply chemical controls should be made based on the target pest, the host, and the environment. The first chemical control decision needs to be made before or at planting. Some soil-dwelling insects (e.g., white grubs, wireworms) feed on the seed or roots of seedlings, but, because they live below the soil surface, they cannot be controlled with foliar insecticide applications. Options include

(1) a broadcast application followed by tillage to incorporate the insecticide, (2) an in-furrow application where the insecticide is placed near the seed and buried there by the planter, and (3) a seed treatment where the seed is coated with an insecticide before planting. All three methods are applied before it is known if there is a problem, so although this is not an ideal situation from an integrated pest management approach, it is the only chemical control option for certain situations.

The most common method of applying insecticides to above ground pests is by foliar applications. These sprayers range from backpack single nozzle sprayers to large self-propelled sprayers to airplanes, but, in all cases, the goal is to apply a uniform layer of insecticide to the target. Insecticide labels, which accompany all insecticides sold in the United States, provide information on the amount of insecticide that can be applied legally to control a specific pest. Where possible, these applications should be made based on monitoring information that indicates that the pest is at an action threshold.

Considerations when making foliar applications should include whether the application needs to be broadcast over the entire area or whether it can be a spot treatment or banded application. Many insects, including many stink bugs, concentrate on the edge of a habitat (Tillman et al. 2009, Reay-Jones 2010, Reay-Jones et al. 2010). As a result, in some situations, an application around the edge of the habitat is sufficient to manage the pest. For young crops grown in rows, banding the insecticide over the row while leaving the space between the rows unsprayed is a way to reduce insecticide and environmental costs without reducing insect control. Another consideration of foliar insecticides is the timing of the application. The efficacy of an insecticide can be impacted by the physical environment, namely temperature (Musser and Shelton 2005, Satpute et al. 2007), time between application and rainfall (Nord and Pepper 1991, Willis et al. 1992), and wind (Smith et al. 2000). The time of application should also consider the growth stage of the target insect (young immatures are normally the easiest stage to kill), the level of crop damage, the potential impact on pollinators and other beneficial insects, and other planned crop production activities (e.g., irrigation, tillage, harvesting).

Regardless of the method or time of application, the goal of chemical control applications is to reduce insect densities to levels that will not cause economic loss.

16.3.1.6 Biological Control

16.3.1.6.1 Predators

The most important natural enemies of pentatomoids, as with many other insect taxa, are predators, parasitoids, and pathogens. Most predators of stink bugs are generalists. There are many records of predaceous pentatomids preying on invasive pentatomoids, but there are no reports that they have had a significant impact on prey numbers. Vertebrates, especially birds (Exnerová et al. 2003), often have been mentioned as eating stink bugs including economically important species. Bats have also been shown to consume pest stink bugs, sometimes in large numbers (e.g., Galorio and Nuñeza 2014). Two genera of Coccinellidae are specific to Plataspidae (Giorgi et al. 2009). Certain species in the genus *Synona* in India are specific to pest plataspids in the genus *Coptosoma* as well as the invasive kudzu bug, *Megacopta cribraria*, and could become candidates for classical biological control of *M. cribraria* in the United States (Subramanyam 1925 [cited in Poorani et al. 2008, p. 582]; Afroze and Shuja 1998).

16.3.1.6.2 Parasitoids

Parasitoids are the most effective biotic agents attacking the Pentatomoidea. All pentatomid species likely are attacked by tachinid flies, and many are attacked by several species, most of which are members of the subfamily Phasiinae. Tachinid species, by themselves, generally are not effective agents. Exceptions are *Trichopoda* spp. attacking *Nezara viridula* in the New World, possibly because these are new associations (Hokkanen and Pimentel 1989). The eggs are the most susceptible stage to parasitoids, and the genera *Trissolcus* and *Telenomus*, within the hymenopterous family Platygastridae, are the most important worldwide. Species in the families Encyrtidae and Eupelmidae are the next most common groups of egg parasitoids (Mills 2010). Efficiency of egg mass discovery can be fairly high in crops such as soybeans (up to 65% or more), with most of the individual eggs in a parasitized mass being attacked (e.g., Cingolani et al. 2014).

16.3.1.6.3 Entomopathogens and Nematodes

Several kinds of entomopathogens are reported from stink bugs, and some, especially certain fungi, have been observed to cause significant epizootics in invasive species that can have a significant local impact, either directly within crops or in overwintering sites (e.g., Seiter et al. 2014). Certain fungi can be specific to just a very few species of Pentatomidae, such as *Halyomorpha halys*, and, because of this specificity, they have been suggested for use as biological control agents (Sasaki et al. 2008).

Nematodes have been reported from several species of pentatomoids and have recently been reported from the redbanded stink bug, *Piezodorus guildinii* (Westwood), and the kudzu bug, *Megacopta cribraria*, but they do not account for significant mortality. Entomopathogens have been reported from several species. Trypansomatids recently have been identified in many Heteroptera and most apparently are specific at both the family and genus levels, but the effects on their hosts are unclear (Kozminsky et al. 2015). Sosa-Gomez (2006) reported species from *Nezara viridula* and *P. guildinii* in Brazil. Most pentatomoids studied have been shown to harbor bacteria and other endosymbionts, but the degree of mutualism in these associations is variable (e.g., Prado and Almeida 2009). Viruses have been recorded from certain stink bugs including *N. viridula* in South Africa (Williamson and von Wechmar 1995).

16.3.1.6.4 Classical Biological Control

Several projects have targeted invasive pest pentatomoids. The most widespread have been those against *Nezara viridula* (see **Chapter 7** for more details). Parasitoids have been imported and released against this pest nearly everywhere it has been found. The target areas have included Australia (Wilson 1960, Coombs and Sands 2000), Brazil (Kobayashi and Cosenza 1987), South Africa (Van Den Berg and Greenland 1996), the continental United States (Jones et al. 1983, 1995) and Hawaii (Davis 1964, 1966), and several Pacific islands (Rao et al. 1971). The parasitoids imported most often have been the hymenopterous egg parasitoid *Trissolcus basalis* (Wollaston) (Platygastridae) and the dipteran parasitoids *Trichopoda* spp. (Tachnidae).

16.3.1.6.5 Augmentation Biological Control

There has been much research on basic technology required to economically mass produce and release natural enemies against certain invasive stink bugs, including development of efficient host rearing, selection of the best candidate natural enemies, cold storage of host eggs and the use of pheromones. However, few actual programs have been put in place, primarily due to the cost of implementation. An exception was the Brazilian government-sponsored development and implementation program that utilized the egg parasitoid *Trissolcus basalis* for both inoculative and inundative releases to manage pest stink bugs in commercial soybean, including *Nezara viridula* and *Piezodorus guildinii* (Correa-Ferreira 1993, Correa-Ferreira and Panizzi 1999). These programs disappeared with the advent of changing farming methods (Panizzi 2013).

Although natural epizootics of fungal pathogens have been observed in invasive pentatomoids (e.g., Seiter et al. 2014), and shown promise for use against invasive species (e.g., Sosa-Gomez and Moscardi 1998, Gouli et al. 2012), no applied use has been undertaken as of yet. Inoculative applications of fungal pathogens might be effective in species such as *Megacopta cribraria* that overwinter gregariously in known habitats. Sedighi et al. (2013) showed that overwintering *Eurygaster integriceps* Puton (Heteroptera: Scutelleridae) were much more susceptible to mortality by *Metarhizium anisopliae* var. *major* (Metchnikoff) Sorokin (Hypocreales: Clavicipitaceae) than were active populations collected in spring.

16.3.1.7 Integrated Pest Management

In 1996, IPM was defined by the United States Congress in the Food Quality Protection Act as "a sustainable approach to managing pests by combining biological, chemical, cultural, mechanical, and physical tools in a way that minimizes economic, health, and environmental risks" (Congress 1996). A multitude of definitions for IPM can be found in the scientific literature (Bajwa and Kogan 2002), but most recommend an approach that deals with a balance of control tactics to achieve a greater permanency in pest management programs (Luckmann and Metcalf 1994).

For information on the history of IPM, see **Section 16.2.2**.

16.3.1.7.1 Agricultural IPM

Most of the aforementioned control strategies (cultural, physical/mechanical, reproductive, regulatory, chemical, and biological) can be used in agricultural IPM programs to manage pestiferous hemipterans. Although chemical control of pentatomids in the agricultural setting is predominant, other strategies can be used successfully, along with insecticides, to manage stink bugs in crops.

16.3.1.7.1.1 Stink Bug Control Programs for Selected Crops

16.3.1.7.1.1.1 Cotton
Programs for controlling economically important hemipterans in cropping systems can and do recommend various IPM strategies for managing insect pests. For example, in cotton, important stink bugs are controlled with insecticides after economic thresholds are met or exceeded (chemical control), but problems with the pest group can be alleviated to some degree if the crop is planted early (cultural control) or not adjacent to a crop, such as peanuts, associated with increased levels of injury from stink bugs (cultural control) (Tillman et al. 2009, Reay-Jones et al. 2010). Collaborative research efforts have been devoted to the establishment of treatment thresholds (chemical control) for stink bugs in cotton (Greene et al. 2001, 2009; Herbert et al. 2009), primarily because of the low-insecticide environment afforded by the widespread adoption of transgenic technology using genes/proteins from *Bacillus thuringiensis* (Bt) *kurstaki* Berliner (see "Genetically Modified Organisms and Plant-Incorporated Proteins" [see **Sections 16.4.2 and 16.4.2.1**] below). Because major lepidopteran pests currently are controlled by the in-plant, toxic proteins produced by Bt cotton (genetic control-advanced HPR), far fewer insecticides are used in the modern crop. This greatly reduces the coincidental control of stink bugs observed before Bt technology when most insecticide applications were applied for major lepidopteran pests, such as *Heliothis virescens* (F.) and *Helicoverpa zea*. By delaying use of insecticides in cotton until absolutely necessary, populations of natural enemies (primarily predaceous and parasitic arthropods) are allowed to build and help regulate populations of pests (biological control-conservation) until chemical control is justified. Also, habitats that promote population development of natural enemies can be planted near cotton or other crops of interest (biological-conservation and cultural control) (Landis et al. 2000). Recent research has indicated that natural and artificial barriers (physical/mechanical control) might deter stink bug movement into cotton (Tillman 2014).

16.3.1.7.1.1.2 Fruit and Vegetable Crops
The importance of stink bugs as pests of fruit and vegetable crops has increased substantially during the past decade due to changes in pest management programs used in these crops and the establishment and spread of several invasive species. Regulatory actions associated with the Food Quality Protection Act resulted in the cancellation of many older broad-spectrum insecticides (i.e., organophosphates, carbamates, organochlorines), several of which provided excellent stink bug control. Grower adoption of reduced-risk pest management programs that rely on narrow-spectrum insecticides with non-insecticide strategies such as pheromone-based mating disruption (Agnello et al. 2008) has elevated the importance of "non-targeted" pests such as stink bugs (Varela et al. 2011).

Stink bugs are potential pests of many fruit and vegetable crops, but the most severely affected in North America include stone and pome fruits, fruiting vegetables, and brassicas. Although there is a diversity of stink bug species that are potential pests of these crops (**Table 16.1**), those species that are the most common, based on the frequency of reports in the literature, are *Euschistus servus*, *E. conspersus* Uhler, *Chinavia hilaris* (Say), *Murgantia histrionica*, and *Nezara viridula*. In addition, two recently established invasive species, the polyphagous *Halyomorpha halys* (Hoebeke and Carter 2003, Leskey et al. 2012a) and *Bagrada hilaris* that attacks brassicas in the southwestern United States (Palumbo and Natwick 2010), have become major concerns.

16.3.1.7.1.1.2.1 Damage
Damage caused by stink bugs to fruit and vegetable crops is primarily cosmetic in nature and results in a downgrading or culling of fruit. It often is expressed as a surface and/or subsurface discoloration around the area where the bug inserted its proboscis when feeding.

TABLE 16.1

Species of Stink Bugs Most Frequently Reported as Pests of Tree Fruits and Vegetables in North America

Species	Common Name	Importance[1]
Chinavia hilaris (Say)	Green stink bug	Common on tree fruits and vegetables throughout North America, most common in southeast.
Bagrada hilaris (Burmeister)	Bagrada or painted bug	Invasive pest brassicas in southwestern US and Mexico.
Euschistus servus (Say)	Brown stink bug	Common on tree fruits and vegetables primarily in eastern North America, although it occurs in western regions.
Euschistus conspersus Uhler	Conspersus stink bug	Tree fruits and vegetables in western North America. Most common stink bug pest in western region.
Euschistus tristigmus (Say)	Dusky stink bug	Tree fruits and occasionally vegetables primarily in eastern North America. Less common than *Euschistus servus*.
Euschistus variolarius (Palisot de Beauvois)	Onespotted stink bug	Tree fruits, most common in northern areas of middle and eastern North America.
Halyomorpha halys (Stål)	Brown marmorated stink bug	Invasive pest common on tree fruits and vegetables in eastern and northwestern US range is expanding.
Murgantia histrionica (Hahn)	Harlequin bug	Brassicas throughout the southern half of North America.
Nezara viridula (L.)	Southern green stink bug	Primarily vegetables in southeastern North America, but its distribution extends to California.
Chlorochroa sayi (Stål)	Say stink bug	Occasional pest of vegetables primarily in western North America.
Thyanta custator custator (F.) (eastern U.S.) *Thyanta pallidovirens* (Stål) (western U.S.)	Redshouldered stink bug	Primarily vegetable across southern tier of North America.

[1] Information on crops affected and distribution obtained from McPherson and McPherson (2000), Campinera (2001), Palumbo and Natwick 2010, and www.StopBMSB.org (accessed 17 December 2015).

Stink bug feeding during petal fall on fruit trees results in increased amounts of aborted fruit (Nielsen and Hamilton 2009). On peaches, early season damage can result in a severe deformation commonly referred to as catfacing, which also can be caused by plant bugs and plum curculio (Rings 1957). Mid- and late season damage is expressed as gummosis (exudation of gummy substance at the probing site), a discolored depression with a water-soaked appearance and/or a discolored area surrounding the stylet sheath extending into the flesh of the fruit. Damage to apple can be difficult to diagnose because it closely resembles the physiological disorders cork spot and bitter pit (Brown 2003, Brown and Short 2010). The only consistent, definitive way to confirm damage by stink bugs is the presence of a feeding puncture (≈0.17 mm in diameter) observed under high magnification (Leskey et al. 2009). Damage to vegetables also is expressed as a discoloration on mature fruit near the feeding site. Feeding by *Nezara viridula* on immature green tomatoes can induce premature ripening of fruit, leading to reduced yields (Lye et al. 1988a). Symptoms of *Murgantia histrionica* and *Bagrada hilaris* feeding on brassicas can vary by crop, plant age, and plant structure attacked (Ludwig and Kok 2001, Haung et al. 2014). Symptoms can range from discoloration at the feeding site on mature leaves, to distorted growth when feeding on apical meristem tissue, to death of seedling plants. Finally, stink bug feeding also has been associated with fruit rot of tomato caused by the yeast *Eremothecium coryli* (Miyao et al. 2000, Brust and Rane 2011).

16.3.1.7.1.1.2.2 Chemical Control Management of stink bugs can be challenging on high-value fruit and vegetable crops due to consumer demand for high cosmetic standards and the resultant low tolerance levels for damage (Lye et al. 1988b, Zalom et al. 1997). Consequently, chemical control has been the primary means of managing stink bugs in these crops. Since the loss of many broad spectrum organophosphates, carbamates and organochlorines, management now is achieved largely by pyrethroid and neonicotinoid insecticides (Cullen and Zalom 2007, Kamminga et al. 2009, Walgenbach and Schoof 2011, Wallingford et al. 2012, Bergh 2013, Leskey et al. 2013, Kuhar et al. 2014, Palumbo et al. 2015). This reliance on insecticides was recently illustrated by Leskey et al. (2012b), who documented a four-fold increase in pesticide use, primarily pyrethroids, over a 1-year period in mid-Atlantic apple and peach orchards in response to increasing populations of *Halyomorpha halys*. Unfortunately, these groups of chemicals can negatively impact biological control agents (Atanassov et al. 2003, Villanueva and Walgenbach 2005) leading to secondary pest outbreaks, particularly in orchard crops (Hull and Starner 1983, Hardman et al. 1991, Hill and Foster 1998, Joseph et al. 2014).

16.3.1.7.1.1.2.3 Pheromone-Based Programs   Recent advances in understanding the biology and behavior of stink bugs in fruit and vegetable systems has led to progress in the development of monitoring programs, more efficient pesticide-use strategies, and behaviorally based management approaches. Identification of the *Euschistus* spp. aggregation pheromone, (2*E*,4*Z*)-decadienoate (Aldrich et al. 1991), has been used as a lure in conjunction with pyramid traps to monitor populations of *E. servus* and *E. tristigmus* in apple and peach orchards (Leskey and Hogmire 2005, Hogmire and Leskey 2006) and *E. conspersus* in California processing tomatoes (Cullen and Zalom 2000). Although these pheromone traps have been noted to attract many bugs to the vicinity of traps, only about 50% of *E. servus* attracted to within 10 meters were captured (Leskey and Hogmire 2005). In addition, pheromone traps placed adjacent to tomato fields for *E. conspersus* were female-biased and captures were poorly correlated with populations in the field (Cullen and Zalom 2005). Nonetheless, this female bias and the fact that those attracted to traps in the early season were reproductively active with mature eggs (Cullen and Zalom 2006), allowed trap captures to be used to set biofix for a developmental degree-day model to focus sampling efforts for early instars (Cullen and Zalom 2000).

Rapid progress is being made on development of monitoring programs for *Halyomorpha halys* since the discovery of its aggregation pheromone (Khrimian et al. 2014). The pheromone is similar to that of *Murgantia histrionica* (Zahn et al. 2008), the two differing in that they consist of different isomers of murgantiol. The addition of (2*E*,4*E*,6*Z*)-decatrienoate (commonly referred to as MDT) synergizes *H. halys*' response to its pheromone (Weber et al. 2014) and provides a lure that is attractive season-long to adults and nymphs (Leskey et al. 2015). Ongoing research is using baited traps to establish treatment thresholds in both orchard and vegetable crops.

Expanded knowledge of stink bug pheromones has provided opportunities for investigating semiochemical-based management approaches. For example, Krupke et al. (2001) found that when deploying the *Euschistus* aggregation pheromone on mullein plants, a host of *E. conspersus*, a large number of *E. conspersus* was quickly attracted to and retained on baited plants adjacent to an apple orchard. Combining the aggregation pheromone with host plants could potentially be used as a trap crop. Using this concept with *Halyomorpha halys* in apples, Morrison et al. (2015) were able to attract large numbers of stink bugs to individuals trees baited with *H. halys* pheromone and MDT synergist, and then used weekly pesticide sprays in baited trees to minimize stink bug damage in adjacent trees. These results suggest that an attract-and-kill strategy may be a viable option for managing this pest with reduced pesticide use.

16.3.1.7.1.1.2.4 Habitat Manipulation   One of the first instances of habitat manipulation to manage stink bugs was in peaches in North Carolina (Killian and Meyer 1984). Early season herbicide applications targeting winter annual weeds on the orchard floor was demonstrated to reduce the incidence of catfacing damage caused by stink bugs and *Lygus* bugs. This groundcover management strategy was combined successfully with mating disruption for the oriental fruit moth [*Grapholita molesta* (Busck)] as a reduced-risk approach to manage the complex of insects in New Jersey peaches, resulting in reduced stink bug populations as well as fewer organophosphate and carbamate insecticides compared to conventionally managed orchards (Atanassov et al. 2002). This system was later compromised by increasing populations of *Halyomorpha halys* in New Jersey. However, taking advantage of the border-arrestment behavior of *H. halys*, Blaauw et al. (2014) demonstrated that limiting insecticide sprays to the periphery of orchards reduced insecticide usage by 25–61% with equivalent or better control compared to whole orchard sprays.

Trap cropping has been a relatively little used practice for managing stink bugs in fruit or vegetable crops. Mustard [*Brassica juncea* (L.) and *B. kaber* de Candolle] and rapeseed (*B. napus* L.) have been evaluated as trap crops to reduce damage in collards and broccoli by *Murgantia histrionica* (Ludwig and Kok 1998, Bender et al. 1999, Wallingford et al. 2013). Failure of the trap crop to hold large densities of bugs and prevent them from moving to the cash crop has been an issue (Ludwig and Kok 1998), even when the trap crop was treated with a systemic neonicotinoid (Wallingford et al. 2013). The failure of soil-applied neonicotinoids likely was due to its short residual activity (Wallingford et al. 2012). Longer residual insecticides or frequent spraying of trap crops may provide more promising results.

Expanding our knowledge of the area-wide ecology of stink bugs offers promise in devising more large-scale farmscape approaches to managing stink bugs. Wooded areas and weedy borders are known sources of many species of stink bugs that infest adjacent crops (Pease and Zalom 2010, Bakken et al. 2015). Strategically altering these landscapes can reduce stink bug populations and increase natural enemy populations in nearby crops (Pease and Zalom 2010, Morandin et al. 2014).

16.3.1.7.1.1.2.5 Conclusions and Future Considerations The importance of stink bugs as pests of fruit and vegetable crops has increased considerably during the past decade due to pesticide regulatory actions and the establishment of new invasive species. Due to low damage tolerance levels on these crops, current management programs rely heavily on chemical control, primarily pyrethroid and neonicotinoid chemistry. These groups of insecticides negatively impact many beneficial arthropods and have led to outbreaks of secondary pests – clearly not a sustainable strategy. In view of society's aversion to genetically modified fresh fruits and vegetables, future advances in stink bug pest management likely will rely on a combination of behavioral, habitat manipulation, and chemical tactics. Greater efforts also will be directed to enhancing the role of biological control in managing stink bugs in fruit and vegetable systems, particularly classical biological control of invasive pests (Talamas et al. 2015).

16.3.1.7.2 Urban IPM

In general, some pentatomoids are considered either occasional invaders or only nuisance overwintering pests in and around structures. Bugs overwintering in leaf-litter sites or crevices on trees or shrubs around structures often are unnoticed by homeowners or property managers. Bugs that harbor in or on structures generally are more of a concern. Until recently, the boxelder bug [*Boisea trivittata* (Say)], in the family Rhopalidae, was far more common in structures than pentatomoids. On occasion, *Brochymena* spp. have been reported to overwinter in dwellings (Ruckes 1946, Scudder 1979). Increasingly, as photoperiod and temperatures decrease in the fall, overwintering aggregations of invasive *Megacopta cribraria* and *Halyomorpha halys* in and on buildings have become more common (Nielsen et al. 2008, Eger et al. 2010, Suiter et al. 2010). These same bug species also cause problems in the spring when they move from overwintering sites in structures as temperatures rise and daylight hours increase. In some situations, large numbers of *M. cribraria* or *H. halys* entering structures can pose more of a problem beyond just being a nuisance. Bugs harboring in sensitive areas around hospitals, nursing homes, dental offices, or food preparation areas can cause contamination concerns. Bug defecation or hemolymph secretions can stain fabric and wall coverings. Even physical injury may occur; secretions from *M. cribraria* have been recorded to cause skin discoloration, a mild burning irritation (Ruberson et al. 2013), and eye irritation (Seiter et al. 2013).

Management practices for many structure-invading pentatomoids, especially species that aggregate in large numbers, have been conducted largely by building owners or contract pest management professionals (PMP). Until recently, numerous applications were made with broad exterior sprays of pyrethroid insecticides (Nielsen et al. 2008, Seiter et al. 2013). Today, nearly all applications with pyrethroid sprays on outdoor structures are limited to spot or crack-and-crevice treatments. This includes treatments around windows, doors, and eaves. On the labels for most products, broad spray applications are limited to building foundations, up to a maximum height of three feet. Outdoor applications to other structural hard-scapes including driveways, patios, porches, sidewalks, also are limited to spot or crack-and-crevice applications.

To control structurally invading pentatomoids, insecticide label restrictions have enhanced the need for Urban IPM approaches to incorporate more non-chemical strategies and targeted chemical control applications. For non-chemical control, before temperature decreases to below 10°C on average at night-time in the fall, homeowners and property manager can reduce overwintering sites such as leaf-litter and excessive mulch where many bugs congregate. Property owners can reduce or remove plants for species like _Megacopta cribraria_ that have a known plant preference, such as kudzu or even wisteria (Benson and Greene 2012). Many individual or aggregating species of pentatomoids will first alight on structures in light colored areas or on the warmest side (Smith and Whitman 2007). Sealing as many cracks and crevices as possible on these structures, especially on sides prone to pest invasions, helps prevent bugs from entering (Meek 2011). Exclusion methods include caulking and weather stripping; and screening vents, windows, and doors. For species that are attracted to light such as _Halyomorpha halys_, lights can be turned off during periods when aggregation tend to occur or changed to yellow bulbs or sodium vapor lights that are less attractive to most insect pests (Smith and Whitman 2007).

For the pentatomoids that do invade structures, mechanical removal is a better strategy for control than insecticide treatments. If large number of bugs are killed and not removed, especially in inaccessible void areas, their dead bodies often create odor problems or serve as a food source for secondary pests such as ants or dermestid beetles (Meek 2011). For low numbers of invading bugs, sweeping them up or vacuuming often is a good option. However, bugs can leave a lingering odor in vacuum hoses (Smith and Whitman 2007). For large numbers of bugs, a wet/dry vacuum with some soapy water in the canister will kill the bugs; in a common household vacuum, it will reduce the potential of foul odors (Benson and Greene 2012). If a regular vacuum is used, the bag should be discarded. However, if bugs are removed, crushing them should be avoided as their body secretions can cause staining (Benson and Greene 2012).

Most insecticide active ingredients and formulations labeled for household and structural use will kill stink bugs and related families, especially pyrethroid-based spray products (Nielsen et al. 2008, Seiter et al. 2013). If used, insecticide treatments are best applied to outdoor areas to kill invading bugs before they enter structures. For the species that aggregate in large numbers, one treatment strategy is to spray the bugs directly or spot treat the structural areas where they tend to congregate. This is especially effective in the fall when night-time temperatures start to cool to approximately 10° C or below (Benson and Greene 2012). In the spring, it may be unnecessary to chemically treat bugs as they move from structures back to plants. However, if they need to be treated, the best strategy is to treat the bugs directly on sunny, cool mornings before they become active (Benson and Greene 2012).

In general, most pentatomoids are not structure-invading pests. The few that are tend to be incidental and seasonal, with most invasions occurring when seasonal temperatures begin to cool in the fall season. This provides the opportunity to plan and establish seasonal strategies for control rather than a year-round treatment program. Employing a seasonal-based plan enables homeowners, property managers, and PMPs the ability to employ many non-chemical strategies and targeted chemical treatments as part of an urban IPM program for structure-infesting pentatomoids.

16.4 Future Management Practices

16.4.1 Innovative Precision Agriculture Techniques

Management of insects using precision agriculture techniques continues to be a definitive goal of integrated pest management programs. Because precision agriculture is a measured and minimal expenditure of resources by definition, especially when referring to the judicious use of insecticides, it is completely compatible with the principles of IPM. Although advances with precision agriculture tactics have been made (Zhang et al. 2002), the biggest challenges to broader adoption center on the difficulties in developing decision-support systems for applying precision management inputs (McBratney et al. 2005). Despite complications related to decision making using data gathered for targeting inputs, the tools to precisely place inputs are in place and await proper decision-support systems. Global information systems and global positioning systems have made possible extremely accurate mapping of fields (Stafford 2000, Lamb and Brown 2001) and guidance equipment on tractors and sprayers (Bell 2000, Reid et al. 2000, Batte and Ehsani 2006). These advances have allowed for the creation of precision-input maps using soil characteristics, such as soil electrical conductivity (Lund et al. 1999, Adamchuk et al. 2004), yields

(Arslan and Colvin 2002), reflected light, such as normalized difference vegetation index (Basso et al. 2001), and other layers of collectable data. Precision placement of inputs using zone-management maps created from these layers of data is considered state-of-the-art precision agriculture today (Zhang et al. 2010a). Recently, the use of unmanned aerial vehicles has increased the pace exponentially at which data can be gathered by the pest management practitioner (Zhang and Kovacs 2012). Site-specific management is more challenging for insect pests than for soil fertility and weed pests, partially because of the high cost of obtaining sufficient data to spatially characterize insect populations (Krell et al. 2003). However, advantages include slowing the development of resistance and preserving natural enemies by maintaining unsprayed refuges in fields (Karimzadeh et al. 2011). Precision agriculture techniques targeting hemipterans have involved detecting important species or damage indices in data layers involving soil characteristics or hyperspectral imagery in a limited number of crops, such as cotton (Willers et al. 2005, Reisig and Godfrey 2007, Prabhakar et al. 2011), and suggesting gradient zones of management for future applications. Recent research incorporating olfactometry into precision agriculture has involved development of electronic nose (E-nose) technology to detect pests or crop injury. Successful research with E-nose detection of stink bug injury to cotton (Henderson et al. 2010, Lampson et al. 2014a,b) is encouraging for future work focusing on localized detection of pests with remote sensing equipment.

16.4.2 Genetically Modified Organisms (GMOs)

The use of genetically modified organisms (GMOs) in commercial agriculture has not been without controversy regarding misconceptions about health-related concerns (Stewart et al. 2000, Greenwell and Rughooputh 2004, McHughen and Wager 2010, Ammann 2014) and negative environmental impacts (Dale et al. 2002). However, the technology has proven to be generally beneficial to the environment by reducing the volume of foliar-applied insecticides used in the production of major crops, such as corn, soybeans, and cotton (Phipps and Park 2002, Huang et al. 2003, Benbrook 2012). One risk is the escape of transgenic crops or crop alleles, which has occurred with herbicide tolerant traits in canola found in feral populations and in non-transgenic populations (Schafer et al. 2011). However, alleged negative effects of GMOs on health have not been proven in reputable peer-reviewed research. Therefore, the benefits of GMO technology are numerous, particularly the segments that confer in-plant protection from insect pests.

16.4.2.1 *Plant-Incorporated Proteins*

Genetically modified cotton was made commercially available in 1996 with technology called 'Bollgard' cotton by Monsanto Company (Perlak et al. 2001). Genes from the naturally occurring bacterium *Bacillus thuringiensis* (Bt) *kurstaki* that produce proteins specifically toxic to caterpillar pests, were inserted into the cotton genome, allowing every cell in the transformed cotton plants to produce the same proteins. Specificity of the toxic proteins to lepidopterans is achieved through unique binding sites on the midgut epithelial lining of targeted pests. Plant material consumed by caterpillars feeding on Bt cotton contains the plant-incorporated proteins that bind to these sites after activation by a narrow range of pH (high – very basic) that also specifically exists in the digestive systems of lepidopterans. Incidentally and subsequently, the same type of transformations was made with genes that confer resistance to broad-spectrum herbicides, such as glyphosate (Roundup) (Nida et al. 1996). The in-plant protection from caterpillar insect pests revolutionized insect management in cotton, and the majority of cotton produced in the United States contains Bt technology (Williams 2015).

The widespread adoption of Bt cotton for control of major caterpillar pests, such as *Heliothis virescens* and *Helicoverpa zea*, did elevate the pest status of hemipteran pests, such as the tarnished plant bug, *Lygus lineolaris*, and phytophagous pentatomids because of the reduced use of insecticides in the crop (Turnipseed et al. 1995, Greene et al. 1999). Because foliar-applied insecticides have not been used as extensively for control of caterpillar pests in cotton as they were before 1996, coincidental control of hemipterans, historically considered secondary pests, has been essentially eliminated, allowing true bugs to become primary pests of the crop. Stink bugs and plant bugs continue to be major pests in Bt cotton because of the low-insecticide environment, but recent advances with transgenic technology have yielded proteins with activity on other groups of insects, including plant bugs (Baum et al. 2012). These advanced Bt genes remain under testing agreements as on-going research validates and qualifies performance in the field and laboratory.

16.4.2.2 RNA Interference

RNA interference (RNAi) could provide an alternative strategy for the management of insect pests (Gordon and Waterhouse 2007, Price and Gatehouse 2008). This post-transcriptional gene silencing phenomenon employs a conserved pathway found in eukaryotic cells by which exogenously applied and endogenously expressed double-stranded (ds) RNAs direct the degradation of complementary endogenous messenger RNA (mRNA) transcripts within the cell, resulting in sequence-specific gene suppression (Fire et al. 1998, Hannon 2002, Tomari and Zamore 2005). This conserved RNAi machinery includes RNAse III-like proteins referred to as Dicer or Dicer-like proteins that process long dsRNAs to 21- to 24-bp silencing (si) RNA duplexes. These siRNA duplexes are loaded into a multi-protein complex called the RNA-induced silencing complex (RISC) where the passenger (sense) strand is removed and the guide (antisense) strand remains to target mRNA for silencing. The guide strand in the RISC enables base pairing of the complex to complementary mRNA transcripts that are then subject to enzymatic cleavage by a class of proteins referred to as Argonaute proteins, resulting in arrest of mRNA translation.

RNAi-mediated suppression of essential genes in insects via ingestion of dsRNA can lead to increased mortality (Baum et al. 2007, Whyard et al. 2009). In sensitive insects, this environmental RNAi response requires a $\geq$ 21 nucleotide match between the dsRNA sequence and the target gene mRNA transcript (Whyard et al. 2009, Bolognesi et al. 2012, Bachman et al. 2013). In addition, multiple barriers exist that block the environmental RNAi response in non-target organisms, including mammals and other vertebrates (Petrick et al. 2013). The aggregate of these factors results in a technology that has potential for great selectivity towards susceptible insect species, more so than any insecticidal agent conceived to date. Commercial development of next-generation rootworm-protected corn hybrids that combine an RNAi-based trait with multiple *Bacillus thuringiensis* insecticidal protein-based traits for corn rootworm pest management is in progress (Kupferschmidt 2013).

The introduction of dsRNA into nymphs and adults via injection is generally effective in triggering an RNAi response in hemipteran species (e.g., Mutti et al. 2006, 2008; Walker and Allen, 2011; Zha et al. 2011; Yao et al. 2013). Target gene-specific silencing following ingestion of dsRNA has been reported in the triatomine bug, *Rhodnius prolixus* (Stål) (Araujo et al. 2006); the pea aphid, *Acyrthosiphon pisum* (Harris) (Shakesby et al. 2009; Whyard et al. 2009; Mao and Zeng 2012, 2014; Sapountzis et al. 2014); the peach aphid, *Myzus persicae* (Sulzer) (Mao and Zeng 2014); the cotton aphid, *Aphis gossypii* (Glover) (Gong et al. 2014); the brown planthopper, *Nilaparvata lugens* (Stål) (Chen et al. 2010; Li et al. 2011); the potato–tomato psyllid, *Bactericera cockerelli* (Šulc) (Wuriyanghan et al. 2011); the corn planthopper, *Peregrinus maidis* (Ashmead) (Yao et al. 2013); and the grain aphid, *Sitobion avenae* (F.) (Zhang et al. 2013) among others (Christiaens and Smagghe 2014). Dietary concentrations of dsRNA required for silencing and/or lethal phenotypes in hemipteran species via feeding vary widely and tend to be at least three orders of magnitude higher than effective concentrations observed with sensitive coleopteran species (Baum and Roberts 2014). Furthermore, different groups working with the same species and gene targets have reported conflicting results, suggesting that the response of hemipteran species to ingested dsRNAs is neither robust nor consistent (Christiaens et al. 2014). Among hemipteran species, the [sap-sucking] sweetpotato whitefly, *Bemisia tabaci* (Gennadius), appears to be one of the more responsive to ingested dsRNAs. Both siRNAs and dsRNAs exhibit activity in whitefly bioassays employing an artificial diet (Upadhyay et al. 2011), and transgenic tobacco plants expressing a *Bemisia* v-ATPase A dsRNA exhibit protection from whitefly feeding damage in controlled environment tests (Thakur et al. 2014). Recent studies, however, reinforce the conclusion that this is the exception to the rule and that hemipteran species, in general, are either unresponsive to ingested dsRNAs targeting essential genes or require high dietary dsRNA concentrations in the 50–1,000 ppm range for a significant effect on mortality, growth inhibition, or fecundity (Christiaens et al. 2014, Gong et al. 2014, Sapountzis et al. 2014, Wan et al. 2014). Rather than target essential genes, opportunities may exist to restore sensitivity to insecticides via suppression of resistance genes as has been reported in *Sitobion avenae* (Xu et al. 2014) and *Aphis gossypii* (Gong et al. 2014).

RNAi studies with stink bug species have not been reported in the literature, but some insights can be gained from published work on the plant bug *Lygus lineolaris*. Suppression of the *Inhibitor of Apoptosis* (*IAP*) gene via dsRNA injection resulted in increased *Lygus* mortality whereas feeding of the *IAP* dsRNA at a dietary concentration of 1,000 ppm had no phenotypic effect (Walker and Allen 2011, Allen and Walker 2012). The digestive physiology of hemipteran species accounts for one apparent barrier to oral delivery of dsRNAs. As with stink bugs, *Lygus* bugs engage in extra-oral digestion, a process that includes the injection of salivary secretions comprising lytic enzymes into plant feeding sites, creating a slurry of plant material that then is ingested. Salivary secretions from hemipteran species are known to contain digestive enzymes such as lipases, phosphatases, pectinases, and proteases (Kaloshian and Walling 2005). The report that *Lygus* salivary extracts also exhibit potent dsRNAse activity suggests that dsRNAs are likely to suffer degradation prior to ingestion (Allen and Walker 2012). Similarly, salivary- and hemolymph-nucleases have been identified as likely barriers to RNAi in the pea aphid (Christiaens et al. 2014).

An alternative strategy for management of hemipteran species relies on topical delivery of dsRNAs through the insect cuticle, a route that bypasses the harsh environment of the insect gut. This approach has been used with some success to deliver dsRNAs and achieve target gene silencing in lepidopteran (Wang et al. 2011), dipteran (Pridgeon et al. 2008), and hemipteran species (El-Shesheny et al. 2013, Killiny et al. 2014). In the case of the Asian citrus psyllid, *Diaphorina citri* (Kuwayama), deposition of dsRNA on the ventral side of the thorax resulted in suppression of five CYP4 (cytochrome P450 monooxygenase) genes and an increase in sensitivity to the neonicotinoid insecticide imidacloprid (Killiny et al. 2014). In the aforementioned studies, the concentrations of dsRNA required for a topical effect ranged from 50–200 ppm, a prohibitively high rate for actual field application on row crops impacted by stink bugs.

The identification of formulations that can package dsRNA, confer environmental stability, and improve delivery to insect cells may result in the lower use rates required for topical applications. Delivery agents including cationic lipids and dendrimers, cyclodextrin polymers, mesoporous silica, and various forms of polethyleneimine are being evaluated for RNAi human therapeutics (for review, see Zhou et al. 2013). Preliminary studies with dipteran and lepidopteran species suggest that some of these delivery agents can enable dsRNA delivery to otherwise recalcitrant insect species (Whyard et al. 2009, Zhang et al. 2010b, He et al. 2013). The requirements for effective delivery presumably would differ depending on whether the route of delivery is via topical exposure or ingestion. In the latter case, formulations would need to protect dsRNAs from stink bug salivary- and gut- nucleases and enable uptake into gut cells. In addition, because of the nature of stink bug feeding, these formulations must presumably penetrate plant tissue in order to be accessible for ingestion. More research is needed to evaluate the feasibility of this approach and the impact of dsRNA stabilization on both bioavailability and efficacy.

16.5 Acknowledgments

We thank Ted R. Schultz (National Museum of Natural History, Washington, DC) for his help concerning the identification of the ant species used by the Chinese in biological control during the Ancient History Period. We also thank Marlin E. Rice (Ames, IA) for his input on this chapter.

16.6 References Cited

Abivardi, C. 2008. Pyrethrum and Persian insect powder, pp. 3084–3090. *In* J. L. Capinera (Ed.), Encyclopedia of entomology (2nd edit.). Springer, Heidelberg, Germany. 4346 pp.

Adamchuk, V. I., J. W. Hummel, M. T. Morgan, and S. K. Upadhyaya. 2004. On-the-go sensors for precision agriculture. Computers and Electronics in Agriculture 44: 71–91.

Adams, B., A. Catchot, J. Gore, D. Cook, F. Musser, and D. Dodds. 2013. Impact of planting date and varietal maturity on tarnished plant bug (Hemiptera: Miridae) in cotton. Journal of Economic Entomology 106: 2378–2383.

Aflitto, N., and T. DeGomez. 2014. Sonic pest repellents. University of Arizona Extension. Publication AZ1639, p. 4 (October 2014).

Afroze, S., and U. Shuja. 1998. Bioecology of *Synia melanaria* Mulsant (Coleoptera: Coccinellidae), predating on *Coptosoma ostensum* Distant. Journal of Entomological Research 22(4): 329–336.

Agnello, A. M., A. Atanassov, J. C. Bergh, D. M. Biddinger, L. J. Gut, J. K Harper, J. J. Haas, H. W. Hogmire, L. A. Hull, L. F. Kime, G. Krawczyk, P. S. McGhee, J. P. Nyrop, W. H. Reissig, P. W. Shearer, R. W. Straub, R. T. Villanueva, and J. F. Walgenbach. 2008. Reduced-risk pest management programs for eastern U.S. apple and peach orchards: A 4-year project. American Entomologist 55: 190–203.

Aldrich, J. R., M. P. Hoffmann, J. P. Kochansky, W. R. Lusby, J. E. Eger, and J. A. Payne. 1991. Identification and attractiveness of a major pheromone component for Nearctic *Euschistus* spp. Stink bugs (Heteroptera: Pentatomidae). Environmental Entomologist 20: 477–483.

Ali, M., and M. A. Ewiess. 1977. Photoperiodic and temperature effects on rate of development and diapause in the green stink bug, *Nezara viridula* L. (Heteroptera: Pentatomidae). Journal of Applied Entomology 84: 256–264.

Allen, M. L., and W. B. Walker III. 2012. Saliva of *Lygus lineolaris* digests double stranded ribonucleic acids. Journal of Insect Physiology 58(3): 391–396.

Alston, D. G., J. R. Bradley, Jr., D. P. Schmitt, and H. D. Coble. 1991. Response of *Helicoverpa zea* (Lepidoptera: Noctuidae) populations to canopy development in soybean as influenced by *Heterodera glycines* (Nematoda: Heteroderidae) and annual weed population densities. Journal of Economic Entomology 84: 267–276.

Ammann, K. 2014. Genomic Misconception: a fresh look at the biosafety of transgenic and conventional crops. A plea for a process agnostic regulation. New Biotechnology 31: 1–17.

Andrade, C. F. S., and I. Cabrini. 2010. Electronic mosquito repellers induce increased biting rates in *Aedes aegypti* mosquitoes (Diptera: Culicidae). Journal of Vector Ecology 35(1): 75–78.

Araujo, R. N., A. Santos, F. S. Pinto, N. F. Gontijo, M. J. Lehane, and M. H. Pereira. 2006. RNA interference of the salivary gland nitrophorin 2 in the triatomine bug *Rhodnius prolixus* Hemiptera, Reduviidae by dsRNA ingestion or injection. Insect Biochemistry and Molecular Biology 36(9): 683–693.

Arslan, S., and T. S. Colvin. 2002. Grain yield mapping: yield sensing, yield reconstruction, and errors. Precision Agriculture 3: 135–154.

Atanassov, A., P.W. Shearer, G. Hamilton, and D. Polk. 2002. Development and implementation of a reduced risk peach arthropod management program in New Jersey. Journal of Economic Entomology 95: 803–812.

Atanassov, A., P.W. Shearer, and G.C. Hamilton. 2003. Peach pest management programs impact beneficial fauna abundance and *Grapholit molesta* (Lepidoptera: Tortricidae) egg parasitism and predation. Environmental Entomology 32: 780–788.

Averill, A. L., M. M. Sylvia, C. C. Kusek, and C. J. DeMoranville. 1997. Flooding in cranberry to minimize insecticide and fungicide inputs. American Journal of Alternative Agriculture 12(2): 50–54.

Bachman, P., R. Bolognesi, W. J. Moar, G. M., Mueller, M. S. Paradise, P. Ramaseshadri, J. Tan, J. P. Uffman, J. Warren, B. E. Wiggins, and S. L. Levine. 2013. Characterization of the spectrum of insecticidal activity of a double-stranded RNA with targeted activity against western corn rootworm *Diabrotica virgifera virgifera* LeConte. Transgenic Research 22(6): 1207–1222.

Bajwa, W. I., and M. Kogan. 2002. Compendium of IPM definitions (CID) – what is IPM and how is it defined in the worldwide literature? IPPC Publications No. 998, Integrated Plant Protection Center (IPPC), Oregon State University, Corvallis, OR 97331, USA.

Bakken, A. J., S. C. Schoof, M. Bickerton, K. L. Kamminga, J. C. Jenrette, S. Malone, M. A. Abney, D. A. Herbert, D. Reisig, T. P. Kuhar, and J. F. Walgenbach. 2015. Occurrence of brown marmorated stink bug (Hemiptera: Pentatomidae) on wild hosts in non-managed woodlands and soybean fields in North Carolina and Virginia. Environmental Entomology 44: 1011–1021.

Bakri, A., N. Heather, J. Hendrichs, and I. Ferris. 2005. Fifty years of radiation biology in entomology: lessons learned from IDIDAS. Annals of the Entomological Society of America 98: 1–12.

Banks, H. J. 1976. Physical control of insects – recent developments. Journal of Australia Entomology Society 15: 89–100.

Basso, B., J. T. Ritchie, F. J. Pierce, R. P. Braga, and J. W. Jones. 2001. Spatial validation of crop models for precision agriculture. Agricultural Systems 68: 97–112.

Batte, M. T., and M. R. Ehsani. 2006. The economics of precision guidance with auto-boom control for farmer-owned agricultural sprayers. Computers and Electronics in Agriculture 53: 28–44.

Baum, J. A., and J. K. Roberts. 2014. Progress towards RNAi-mediated insect pest management. Advances in Insect Physiology 47: 249–295.

Baum, J. A., T. Bogaert, W. Clinton, G. R. Heck, P. Feldmann, O. Ilagan, S. Johnson, G. Plaetinck, T. Munyikwa, M. Pleau, T. Vaughn, and J. Roberts. 2007. Control of coleopteran insect pests through RNA interference. Nature Biotechnology 25(11): 1322–1326.

Baum, J. A., U. R. Sukuru, S. R. Penn, S. E. Meyer, S. Subbarao, X. Shi, S. Flasinski, G. R. Heck, R. S. Brown, and T. L. Clark. 2012. Cotton plants expressing a hemipteran-active *Bacillus thuringiensis* crystal protein impact the development and survival of *Lygus hesperus* nymphs. Journal of Economic Entomology 105: 616–624.

Baumhover, A. H., A. J. Graham, D. E. Hopkins, F. H. Dudley. W. D. New, and R. C. Bushland. 1955. Control of screw-worms through release of sterilized flies. Journal of Economic Entomology 48: 462–466.

Bealmear, S., P. Warren, and K. Young. 2013. Bagrada bug: a new pest for Arizona gardeners. University of Arizona Extension. (Winter 2013). 4 pp.

Bell, T. 2000. Automatic tractor guidance using carrier-phase differential GPS. Computers and Electronics in Agriculture 25: 53–66.

Benbrook, C. M. 2012. Impacts of genetically engineered crops on pesticide use in the U.S. – the first sixteen years. Environmental Sciences Europe 24: 1–13. http://www.enveurope.com/content/24/1/24

Bender, D. A., W. P. Morrison, and R. E. Frisbie. 1999. Intercropping cabbage and Indian mustard for potential control of lepidopterous and other insects. Horticulture Science 34: 275–279.

Benson, E. P., and J. K. Greene. 2012. Kudzu bugs around the home. Clemson University EIIS/HS-50. http://www.clemson.edu/cafls/departments/esps/factsheets/household_structural/kudzu_bugs_hs50.html (accessed 4 December 2015).

Ben-Yakir, D., Y. Antignus, Y. offir, and Y. Shahak. 2012. Optical manipulations: an advance approach for reducing sucking insect pests, 249–267. *In* I. Ishaaya, S. R. Palli, and A. R. Horowitz (Eds.), Advanced technologies for managing insect pests. Springer, New York, London. 326 pp.

Bergh, J. C. 2013. Single insecticides targeting brown marmorated stink bug in apple, 2011. Arthropod Management Tests 38: A2.

Blaauw, B. R., D. Polk, and A. L. Nielsen. 2014. IPM-CPR for peaches: Utilizing behaviorally-based methods to manage key peach pests; brown marmorated stink bug (*Halyomorpha halys*) and Oriental fruit moth (*Grapholita molesta*). Pest Management Science 71: 1513–1522.

Boiteau, G., and W. P. L. Osborn. 1999. Comparison of plastic-lined trenches and extruded plastic traps for controlling *Leptinotarsa decemlineata* (Coleoptera: Chrysomelidae). The Canadian Entomologist 131: 567–572.

Boiteau, G., and R. S. Vernon. 2001. Physical barriers for the control of insect pests, pp. 224–247. In C. Vincent, B. Panneton, and F. Fleurat-Lessard (Eds.), Physical control methods in plant protection. Springer-Verlag, Berlin. (page numbers not readily available)

Bolognesi, R., P. Ramaseshadri, J. Anderson, P. Bachman, W. Clinton, R. Flannagan, O. Ilagan, C. Lawrence, S. Levine, W. Moar, and G. Mueller. 2012. Characterizing the mechanism of action of double-stranded RNA activity against western corn rootworm *Diabrotica virgifera virgifera* LeConte. PLoS ONE 7 (10): e47534.

Bomford, M., and P. H. O'Brien. 1990. Sonic deterrents in animal damage control: a review of device tests and effectiveness. Wildlife Society Bulletin 18: 411–422.

Bomford, M. K., R. S. Vernon, and P. Pats. 2000. Importance of overhangs on the efficacy of exclusion fences for managing cabbage flies (Diptera: Anthomyiidae). Environmental Entomology 29: 795–799.

Borges, M., A. Zhang, M. J. Camp, and J. R. Aldrich. 2001. Adult diapause morph of the brown stink bug, *Euschistus servus* (Say) (Heteroptera: Pentatomidae). Neotropical Entomology 30: 179–182.

Bottrell, D. G., and P. L. Adkisson. 1977. Cotton insect pest management. Annual Review of Entomology 22: 451–481.

Broce, A. B. 1993. Electrocuting and electronic insect traps: trapping efficiency and production of airborne particles. International Journal of Environmental Health Research 3: 47–48.

Brown, M. W. 2003. Characterization of stink bug (Heteroptera: Pentatomidae) damage to mid and late-season apples. Journal of Agricultural and Urban Entomology 20: 193–202.

Brown, M. W., and B. D. Short. 2010. Factors affecting appearance of stink bug (Hemiptera: Pentatomidae) injury to apple. Environmental Entomology 39: 134–139.

Brust, G. E. 1994. Natural enemies in straw mulch reduce Colorado potato beetle populations and damage in potato. Biological Control 4: 163–169.

Brust, G. E., and K. K. Rane. 2011. First report of the yeast *Eremothecium coryli* associated with brown marmorated stink bug feeding injury on tomato and apple. Phytopathology 101(6): 22.

Bundy, C. S., and R. M. McPherson. 2000. Dynamics and seasonal abundance of stink bugs (Heteroptera: Pentatomidae) in a cotton-soybean ecosystem. Journal of Economic Entomology 93: 697–706.

Bundy, C. S, T. R. Grasswitz, and C. Sutherland. 2012. First report of the invasive stink bug *Bagrada hilaris* (Burmeister) (Heteroptera: Pentatomidae) from New Mexico, with notes on its biology. Southwestern Entomologist 37: 411–414.

Campinera, J. L. 2001. Handbook of vegetable pests. San Diego, CA, Academic Press. 729 pp.

Cardé, R., and A. K. Minks. 1995. Control of moth pest by mating disruption, success and constraints. Annual Review of Entomology 40: 559–585.

Carson, R. 1962. Silent spring. Houghton-Mifflin Co., Boston. 368 pp.

Casana-Giner, V., A. Gandia-Balaguer, C. Mengod-Puerta, J. Primo-Millo, and E. Primo-Yufera. 1999. Insect growth regulators and chemosterilants for *Ceratitis capitata* (Diptera: Tephrididae). Journal of Economic Entomology 92: 303–308.

Chen, J., D. Zhang, Q. Yao, J. Zhang, X. Dong, H. Tian, and W. Zhang. 2010. Feeding-based RNA interference of a trehalose phosphate synthase gene in the brown planthopper, *Nilaparvata lugens*. Insect Molecular Biology 19(6): 777–786.

Christiaens, O., and G. Smagghe. 2014. The challenge of RNAi-mediated control of hemipterans. Current Opinion in Insect Science 6: 15–21.

Christiaens, O., L. Sweveres, and G. Smagghe. 2014. DsRNA degradation in the pea Aphid (*Acyrthosiphon pisum*) associated with lack of response in RNAi feeding and injection assay. Peptides 53: 307–314.

Cingolani, M. F., N. M. Greco, G. G. Liljesthröm. 2014. Egg parasitism of *Piezodorus guildinii* and *Nezara viridula* (Hemiptera: Pentatomidae) in soybean, alfalfa and red clover. Revista de la Facultad de Ciencias Agrarias. Universidad Nacional de Cuyo 46(1): 15–27.

Cockfield, S. D., and D. L. Mahr. 1992. Flooding cranberry beds to control blackheaded fireworm (Lepidoptera: Tortricidae). Journal of Economic Entomology 85: 2383–2388.

Cohen, H., and B. Yuval. 2000. Perimeter trapping strategy to reduce Mediterranean fruit fly (Diptera: Tephritidae) damage on different host species in Israel. Journal of Economic Entomology 93: 721–715.

Congress, U. S. 1996. Food Quality Protection Act of 1996. Public Law No. 104–170, 110 Stat. 1489.

Coombe, P. E. 1981. Visual behavior of the greenhouse whitefly, *Trialeurodes vaporariorum*. Physiological Entomology 7: 243–251.

Coombs, M., and D. P. A. Sands. 2000. Establishment in Australia of *Trichopoda giacomellii* (Blanchard) (Diptera: Tachinidae), a biological control agent for *Nezara viridula* (L.) (Hemiptera: Pentatomidae). Australian Journal of Entomology 39: 219–222.

Correa-Ferreira, B. S. 1993. Utilizacao do parasitoide de ovos *Trissolcus basalis* (Wollaston) no controle de percevejos da soja. Empresa Brasileira de Pesquisa Agropecuária, Centro Nacional de Pesquisa de Soja, Circular Técnica 11. 40 pp. (In Portuguese)

Correa-Ferreira, B. S., and A. R. Panizzi. 1999. Percevejos da soja e seu manejo. Empresa Brasileira de Pesquisa Agropecuária, Centro Nacional de Pesquisa de Soja, Circular Técnica 24. 45 pp. (In Portuguese)

Csizinszky, A. A., D. J. Schuster, and J. B. Kring. 1995. Color mulches influence yield and insect pest populations in tomatoes. Journal of the American Society for Horticultural Science 120(5): 778–784.

Cullen, E. M., and F. G. Zalom. 2000. Phenology-based field monitoring for conperse stink bug (Hemiptera: Pentatomidae) in processing tomatoes. Environmental Entomology 29: 560–567.

Cullen, E. M., and F. G. Zalom. 2005. Relationship between *Euschistus conspersus* (Hemiptera: Pentatomidae) pheromone trap catch and canopy samples in processing tomatoes. Journal of Applied Entomology 129: 505–514.

Cullen, E. M., and F. G. Zalom. 2006. *Euschistus conspersus* female morphology and attraction to methyl (2E,4Z)-decadienoate pheromone-baited traps in processing tomatoes. Entomologia Experimentalis Applicata 119: 163–173.

Cullen, E. M, and G. G. Zalom. 2007. On-farm trial assessing efficacy of three insecticide classes for management of stink bug and fruit damage on processing tomatoes. Plant Health Progress doi:10.1094/PHP-2007-0323-01-RS.

Dalby, A. 2011. Geoponika: farm work. A modern translation of the Roman and Byzantine farming handbook. Prospect Books, London, England. 368 pp.

Dale, P. J., B. Clarke, and E. M. G. Fontes. 2002. Potential for the environmental impact of transgenic crops. Nature Biotechnology 20: 567–574.

Davis, C. J. 1964. The introduction, propagation, liberation, and establishment of parasites to control *Nezara viridula* variety *smaragdula* (Fabricius) in Hawaii (Heteroptera: Pentatomidae). Proceedings of the Hawaiian Entomological Society 18: 369–375.

Davis, C. J. 1966. Progress in the biological control of the southern green stink bug, *Nezara viridula* variety *smaragdula* (Fabricius) in Hawaii (Heteroptera: Penttomidae). Mushi 39: 9–16.

DeBach, P. 1951. The necessity for an ecological approach to pest control on citrus in California. Journal of Economic Entomology 44: 443–447.

DeBach, P. 1964. Successes, trends, and future possibilities, pp. 673–713. *In* P. DeBach and E. I. Schlinger (Eds.), Biological control of insect pests and weeds. Reinhold Publishing Corporation, New York. 844 pp.

DeBach, P., and B. Bartlett. 1951. Effects of insecticides on biological control of insect pests of citrus. Journal of Economic Entomology 44: 372–383.

Dent, D. 2000. Insect pest management (2nd edit.). CABI Publishing, New York, NY. 410 pp.

Dethier, V. G. 1976. Man's plague? Insects and agriculture. Darwin Press, Princeton, NJ. 237 pp.

Diaz, B. M., and A. Fereres. 2007. Ultraviolet-blocking materials as a physical barrier to control insect pests and plant pathogens in protected crops. Pest Technology 1(2): 85–95.

Dobson, R. 2015. Mechanical exclusion and biological control strategies for the invasive brown marmorated stink bug, *Halyomorpha halys* (Hemiptera: Pentatomidae). Theses and Dissertations–Entomology. Paper 17. http://uknowledge.uky.edu/entomology_etds/17

Doutt, R. L., and R. F. Smith. 1971. The pesticide syndrome–diagnosis and suggested prophylaxis, pp. 3–15. *In* C. B. Huffaker (Ed.), Biological control. Plenum Press, New York, NY. 511 pp.

Drake, C. J. 1920. The southern green stink-bug in Florida. Florida State Plant Board of Florida, The Quarterly Bulletin 4: 41–94.

Duffield, S. J., R. J. Bryson, J. E. B. Young, R. Sylvester-Bradley, and R. K. Scott. 1997. The influence of nitrogen fertiliser on the population development of the cereal aphids *Sitobion avenae* (F.) and *Metopolophium dirhodum* (Wlk.) on field grown winter wheat. Annals of Applied Biology 130(1): 13–26.

Dyby, S. D., and R. I. Sailer. 1999. Impact of low-level radiation on fertility and fecundity of *Nezara viridula* (Hemiptera: Pentatomidae). Journal of Economic Entomology 92: 945–953.

Economopoulos, A. P., and H. T. Gordan. 1971. Chemosterilization of *Oncopeltus fasciatus* - Competition between normal and tretamine-sterilized insects. Journal of Economic Entomology 64: 1360–1364.

Eger, J. E., Jr., L. M. Ames, D. R. Suiter, T. M. Jenkins, D. A. Rider, and S. E. Halbert. 2010. Occurrence of the Old World bug *Megacopta cribraria* (Fabricius) (Heteroptera: Plataspidae) in Georgia: a serious home invader and potential legume pest. Insecta Mundi 0121: 1–11.

El-Shesheny, I., S. Hajeri, I. El-Hawary, S. Gowda, and N. Killiny. 2013. Silencing abnormal wing disc gene of the asian citrus psyllid, *Diaphorina citri* disrupts adult wing development and increases nymph mortality. PLoS ONE 8(5): e65392. doi:10.1371/journal.pone.006539.

Exnerová, A., A. Stys, A. Kristín, O. Volf, and M. Pudil. 2003. Birds as predators of true bugs (Heteroptera) in different habitats. Biologia 58: 253–264.

Fields, P. G. 1992. The control of stored-product insects and mites with extreme temperatures. Journal of Stored Products Research 28(2): 89–118.

Fire, A., S. Xu, M. K. Montgomery, S. A. Kostas, S. E. Driver, and C. C. Mello. 1998. Potent and specific genetic interference by double-stranded RNA in *Caenorhabditis elegans*. Nature 391: 806–811.

Flint, M. L., and R. van den Bosch. 1981. Introduction to integrated pest management. Plenum Press, New York, NY. 240 pp.

Frost, S. W. 1957. The Pennsylvania insect light trap. Journal of Economic Entomology 50: 287–292.

Galorio, A. H. N., and O. M. Nuñeza. 2014. Diet of cave-dwelling bats in Bukidnon and Davao Oriental, Philippines. Animal Biology and Animal Husbandry, Bioflux 6(2): 148–157.

Ghanim, M., H. Czosnek, and S. Kontsedalov. 2007. Tissue-specific gene silencing by RNA interference in the whitefly *Bemisia tabaci* (Gennadius). Insect Biochemistry and Molecular Biology 37: 732–738.

Gill, S. S., E. A. Cowles, and P. V. Pietrantonio. 1992. The mode of action of *Bacillus thuringiensis* endotoxins. Annual Review of Entomology 37: 615–634.

Giorgi, J. A., N. J. Vandenberg, J. V. McHugh, J. A. Forrester, A. Ślipiński, K. B. Miller, L. R. Shapiro, and M. F. Whiting. 2009. The evolution of food preferences in Coccinellidae. Biological Control 51: 215–231.

Glenn, D. M., G. M. Puterka, T. VanderZwet, R. E. Byers, and C. Feldhake. 1999. Hydrophobic particle films: a new paradigm for suppression of arthropod pests and plant diseases. Journal of Economic Entomology 92: 759–771.

Glynne-Jones, A. 2001. Pyrethrum. Pesticide Outlook 12: 195–198.

Godfrey, L. D., and T. F. Leigh. 1994. Alfalfa harvest strategy effect on *Lygus* bug (Hemiptera: Miridae) and insect predator population density: Implications for use as trap crop in cotton. Environmental Entomology 23: 1106–1118.

Gold, R. E., T. N. Decker, and A. D. Vance. 1984. Acoustical characterization and efficacy evaluation of ultrasonic pest control devises marketed for control of German coackroaches (Orthoptera: Blattellidae). Journal of Economic Entomology 77: 1507–1512.

Gong, Y.-H., X.-R. Yu, Q.-L. Shang, X.-Y. Shi, and X.-W. Gao. 2014. Oral delivery mediated RNA interference of a carboxylesterase gene results in reduced resistance to organophosphorus insecticides in the cotton aphid, *Aphis gossypii* Glover. PLoS ONE 9(8): e102823.

Gordon, K. H. J., and P. M. Waterhouse. 2007. RNAi for insect-proof plants. Nature Biotechnology 25(11): 1231–1232.

Gouli, V., S. Gouli, M. Skinner, G. Hamilton, J. S. Kim, and B. L. Parker. 2012. Virulence of select entomopathogenic fungi to the brown marmorated stink bug, *Halyomorpha halys* (Stål) (Heteroptera: Pentatomidae). Pest Management Science 68 (2): 155–157. doi10.1002/ps.2310.

Greene, J. K., S. G. Turnipseed, M. J. Sullivan, and G. A. Herzog. 1999. Boll damage by southern green stink bug (Hemiptera: Pentatomidae) and tarnished plant bug (Hemiptera: Miridae) caged on transgenic *Bacillus thuringiensis* cotton. Journal of Economic Entomology 92: 941–944.

Greene, J. K., S. G. Turnipseed, M. J. Sullivan, and O. L. May. 2001. Treatment thresholds for stink bugs (Hemiptera: Pentatomidae) in cotton. Journal of Economic Entomology 94: 403–409.

Greene, J. [K.], J. Bacheler, P. Roberts, M. Toews, J. Ruberson, F. Reay-Jones, D. Robinson, D. Mott, D. Morrison, T. Pegram, T. Walker, and C. Davis. 2009. Continued evaluations of internal boll-injury treatment thresholds for stink bugs in the Southeast, pp. 1091–1100. *In* Proceedings Beltwide Cotton Production Conferences, National Cotton Council of America, Memphis, TN.

Greenwell, P., and S. Rughooputh. 2004. Genetically modified food: good news but bad press. The Biomedical Scientist 48: 845–846.

Hajek, A. E. 2004. Natural enemies: an introduction to biological control. Cornell University Press, Ithaca, NY. 396 pp.

Hall, D. G., and R. H. Cherry. 1993. Effect of temperature in flooding to control the wireworm *Melanotus communis* (Coleoptera: Elateridae). The Florida Entomologist 76(1): 155–160.

Hall, D. G., and G. L. Teetes. 1982. Damage to grain sorghum by southern green stink bug, conchuela, and leaffooted bug. Journal of Economic Entomology 75: 620–625.

Halverson, S. L., W. E. Burkholder, T. S. Bigelow, E. V. Nordheim, and M. E. Misenheimer. 1996. High-power microwave radiation as an alternative insect control method for stored products. Journal of Economic Entomology 89: 1638–1648.

Hannon, G. J. 2002. RNA interference. Nature 418: 244–251.

Harding, Jr., W. C., J. G. Hartsock, and G. G. Rohwer. 1966. Blacklight trap standards for general insect surveys. Entomology Society of America 31–32. doi.org/10.1093/besa/12.1.31.

Hardman, J. M., R. E. L. Rogers, J. P. Nyrop, and T. Frisch. 1991. Effect of pesticide application son abundance of European red mite (Acari: Tetranychidae) and *Typhlodromus pyri* (Acari: Phytoseiidae) in Nova Scotian apple orchards. Journal of Economic Entomology 84: 570–580.

Hardy, W. F., R. N. Beachy, H. Browning, J. D. Caulder, R. Charudattan, P. Faulkner, F. L. Gould, M. K. Hinkle, B. A. Jaffee, M. K. Knudson, W. Joe Lewis, J. E. Loper, D. L. Mahr, and N. K. Van Alfen. 1996. Ecologically based pest management: new solutions for a new century. National Academies Press. Washington, D.C. 160 pp.

Hargrove, J. W. and P. A. Langley. 1990. Sterilizing tsetse (Diptera, Glossinidae) in the field – a successful trial. Bulletin of Entomological Research 80: 397–403.

Harris, V. E., J. W. Todd, and B. G. Mullinix. 1984. Color change as an indicator of adult diapause in the southern green stink bug, *Nezara viridula*. Journal of Agricultural Entomology 1: 82–91.

Haung, T. I., D. A. Reed, T. M. Perring, and J. C. Palumbo. 2014. Feeding damage by *Bagrada hilaris* (Hemiptera: Pentatomidae) and impact on growth and chlorophyll content of Brassicaceous plant species. Arthropod-Plant Interactions 8: 89–100.

He, B., Y. Chu, M. Yin, K. Müllen, C. An, and J. Shen. 2013. Fluorescent nanoparticle delivered dsRNA toward genetic control of insect pests. Advanced Materials 25(33): 4580–4584.

Heinen, J. T., J. Reznik, S. Hill, J. Kostrzewski, and A. Maziak. 2003. Factors affecting capture rates of insect taxa by retail electrocutors and eliminators in northern lower Michigan. The Great Lakes Entomologist 36(1–2): 61–69.

Henderson, W. G., A. Khalilian, Y. J. Han, J. K. Greene, and D. C. Degenhardt. 2010. Detecting stink bugs/damage in cotton utilizing a portable electronic nose. Computers and Electronics in Agriculture 70: 157–162.

Hendricks, D. E., P. D. Lingren, and J. P. Hollingsworth. 1975. Numbers of bollworms, tobacco budworms, and cotton leafworms caught in traps equipped with fluorescent lamps of five colors. Journal of Economic Entomology 68: 645–649.

Herbert, A., E. Blinka, J. Bacheler, J. Van Duyn, J. Greene, M. Toews, P. Roberts, and R. H. Smith. 2009. Managing stink bugs in cotton: research in the southeast region. Virginia Cooperative Extension, Virginia Polytechnic Institute and State University. Publication 444-390. 17 pp. http://pubs.ext.vt.edu/444/444-390/444-390.pdf

Hill, T. A., and R. E. Foster. 1998. Influence of selective insecticides on population dynamics of European red mite (Acari: Tetranychidae), apple rust mite (Acari: Eriophyidae), and their predator *Amblyesius fallacis* (Acari: Phytoseiidae) in apple. Journal of Economic Entomology 91: 191–198.

Hoebeke, E. R., and M. E. Carter. 2003. *Halyomorpha halys* (Stal) (Heteroptera: Pentatomidae): a polyphagous plant pest from Asia newly detected in North America. Proceedings of the Entomological Society of Washington 105: 225–237.

Hogmire, H. W., and T. C. Leskey. 2006. An improved trap for monitoring stink bugs (Heteroptera: Pentatomidae) in apple and peach orchards. Journal of Entomological Science 41: 9–21.

Hokkanen, H. M. T. 1991. Trap cropping in pest management. Annual Review of Entomology 36: 119–138.

Hokkanen, H. M. T., and D. Pimentel. 1989. New associations in biological control: theory and practice. The Canadian Entomologist 121: 829–840. doi.org/10.4039/Ent121829-10.

Huang, J., R. Hu, C. Pray, F. Qiao, and S. Rozelle. 2003. Biotechnology as an alternative to chemical pesticides: a case study of Bt cotton in China. Agricultural Economics 29: 55–67.

Hull, L. A., and V. A. Starner. 1983. Impact of four synthetic pyrethroids on major natural enemies and pests of apple in Pennsylvania. Journal of Economic Entomology 76: 122–130.

Ikeno, T., K. Ishikawa, H. Numata, and S. G. Goto. 2013. Circadian clock gene *Clock* is involved in the photoperiodic response of the bean bug *Riptortus pedestris*. Physiological Entomology 38: 157–162.

Immaraju, J. A., J. G. Morse, and R. F. Hobza. 1990. Field evaluation of insecticide rotation and mixtures as strategies for citrus thrips (Thysanoptera: Thripidae) resistance management in California. Journal of Economic Entomology 83: 306–314.

Insecticide Resistance Action Committee. 2016. IRAC mode of action classification scheme, April 2016. http://www.irac-online.org/documents/moa-classification/?ext=pdf.

Isman, M. B. 2006. Botanical insecticides, deterrents, and repellents in modern agriculture and an increasingly regulated world. Annual Review of Entomology 51: 45–66.

Ivezic, M., J. Tollefson, E. Raspudic, I. Brkic, M. Brmez, and B. Hibbard. 2006. Evaluation of corn hybrids for tolerance to corn rootworm (*Diabrotica virgifera virgifera* LeConte) larval feeding. Cereal Research Communications 34(2–3): 1101–1107.

Jacques, R. L. 1988. The potato beetles. The genus *Leptinotarsa* in North America (Coleoptera: Chrysomelidae). Flora and Fauna Handbook no. 3. E. J. Brill, New York, NY. 144 pp.

Jaglan, R. S., and R. Singh. 2007. History of integrated pest management, pp. 1–18. *In* P. C. Jain and M. C. Ghargava (Eds.), Entomology: novel approaches. New India Publishing Agency, New Delhi. 541 pp.

Jander, R. 1963. Insect orientation. Annual Review of Entomology 8: 95–114.

Johansen, C. 1978. Principles of insect control, pp. 189–208. *In* R. E. Pfadt, Fundamentals of applied entomology (3rd edit.). Macmillan Publishing Company, New York, NY. 798 pp.

Jones, T. H. 1918. The southern green plant-bug. United States Department of Agriculture Bulletin 689: 1–27.

Jones, W. A., Jr., S. Y. Young, M. Shepard, and W. H. Whitcomb. 1983. Use of imported natural enemies against insect pests of soybean, pp. 63–76. *In* H. Pitre (Ed.), Natural enemies of insect pests of soybean. Southern Cooperative Series Bulletin 285. (page numbers not readily available)

Jones, W. A., L. E. Ehler, M. P. Hoffmann, N. A. Davidson, L. T. Wilson, and J. W. Beardsley. 1995. Southern green stink bug. pp 81–83. *In* J. R. Nechols (Ed.), Biological Control in the Western United States. Accomplishments and Benefits of Regional Research Project W-84, 1964–1989. University of California Publication 3361. 336 pp.

Joseph, S. V., J. W. Stallings, T. C. Leskey, G. Krawczyk, D. Polk, B. Butler, and J. C. Bergh. 2014. Spatial distribution of brown marmorated stink bug (Hemiptera: Pentatomidae) injury at harvest in Mid-Atlantic apple orchards. Journal of Economic Entomology 107: 1839–1848.

Kaloshian, I., and L. L. Walling. 2005. Hemipterans as plant pathogens. Annual Review of Phytopathology 43(1): 491–521.

Kamminga, K. L., S. Malone, H. Doughty, D. A. Herbert, and T. P. Kuhar. 2009. Toxicity, feeding preference, and repellency associated with selected organic insecticides against *Acrosternum hilare* and *Euschistus servus* (Hemiptera: Pentatomidae. Journal of Economic Entomology 202: 1915–1921.

Karimzadeh, R., M. J. Hejazi, H. Helali, S. Iranipour, and S. A. Mohammadi. 2011. Assessing the impact of site-specific spraying on control of *Eurygaster integriceps* (Hemiptera: Scutelleridae) damage and natural enemies. Precision Agriculture 12: 576–593.

Keiser, J., B. H. Singer, and J. Utzinger. 2005. Reducing the burden of malaria in different eco-epidemiological settings with environmental management: a systematic review. The Lancet Infectious Diseases 5: 695–708.

Kells, S. A., and M. J. Goblirsch. 2011. Temperture and time requirements for controlling bed bugs (*Cimex lectularius*) under commercial heat treatment conditions. Insects 2: 412–422.

Kemble, J. M., L. M. Quesada-Ocampo, M. L. L. Ivey, K. M. Jennings, and J. F. Walgenbach (Eds.). 2015. 2015 Vegetable crop handbook for southeastern United States. 277 pp.

Khrimian, A., A. Zhang, D. C. Weber, H. Y. Ho, J. R. Aldrich, K. E. Vermillion, M. A. Siegler, S. Shirali, F. Guzman, and T. C. Leskey. 2014. Discovery of the aggregation pheromone of the brown marmorated stink bug (*Halyomorpha halys*) through the creation of stereoisomeeric libraries of 1-bisabolen-3-ols. Journal of Natural Products 77: 1708–1717.

Killian, J. C., and J. R. Meyer. 1984. Effect of orchard weed management on catfacing damage to peaches in North Carolina. Journal of Economic Entomology 77: 1596–1600.

Killiny, N., S. Hajeri, S. Tiwari, S. Gowda, and L. L. Stelinski. 2014. Double-stranded RNA uptake through topical application, mediates silencing of five CYP4 genes and suppresses insecticide resistance in *Diaphorina citri*. PLoS ONE 9(10): e110536.

Kim, M. G., J. Y. Yang, and H. S. Lee. 2013. Phototactic behavior: repellent effects of cigarette beetlc, *Lasioderma serricorne* (Coleoptera: Anobiidae), to light-emitting diodes. Journal of the Korean Society for Applied Biological Chemistry 56: 331–333.

Kimura, Y. 1982. Control of aphid infestation by mulching with silver-colored polyethylene films. Plant Protection 36: 469–473.

Kiritani, K. 1979. Pest management in rice. Annual Review of Entomology 24: 279–312.

Knipling, E. F. 1955. Possibilities of insect control or eradication through the use of sexually sterile males. Journal of Economic Entomology 48: 459–462.

Knipling, E. F. 1960. The eradication of the screwworm fly. Scientific American 203: 4–48.

Knipling, E. F. 1969. Concept and value of eradication or continuous suppression of insect populations, pp. 19–32. *In* Sterile-male technique for eradication or control of harmful insects (Proceedings, Panel Vienna, 1968). STI/PUB/224. IAEA, Vienna. (page numbers not readily available)

Kobayashi, T., and G. W. Cosenza. 1987. Integrated control of soybean stink bugs in the Cerrados. Japan Agricultural Research Quarterly 20(4): 307–376.

Kocisova, A., M. Petrovsky, J. Toporcak, and P. Novak. 2004. The potential of some insect growth regulators in housefly (*Musca domestica*) control. Biologia 59: 661–668.

Kogan, M. 1998. Integrated pest management: historical perspectives and contemporary developments. Annual Review of Entomology 43: 243–270.

Kogan, M., and E. F. Ortman. 1978. Antixenosis – a new term proposed to define Painter's "nonpreference" modality of resistance. Bulletin of the Entomological Society of America 24(2): 175–176.

Kotaki, T. 1998. Effects of low temperature on diapause termination and body colour change in adults of a stink bug, *Plautia stali*. Physiological Entomology 23: 53–61.

Kozminsky, E., N. Kraeva, A. Ishemgulova, E. Dobáková, J. Lukeš, P. Kment, V. Yurchenko, J. Vot´ypka, and D. A. Maslov. 2015. Host-specificity of monoxenous trypanosomatids: statistical snalysis of the distribution and transmission patterns of the parasites from Neotropical Heteroptera. Protist 166: 551–568. doi:10.1016/j.protis.2015.08.004.

Krell, R. K., L. P. Pedigo, and B. A. Babcock. 2003. Comparison of estimated costs and benefits of site-specific versus uniform management for the bean leaf beetle in soybean. Precision Agriculture 4: 401–411.

Krupke, C. H., J. F. Brunner, M. D. Doerr, and A. D. Kahn. 2001. Field attraction of the stink bug *Euschistus conspersus* (Hemiptera: Pentatomidae) to synthetic pheromone-baited host plants. Journal of Economic Entomology 94: 1500–1505.

Kuhar, T. P., H. Doughty, C. Philips, J. Aigner, L. Nottingham, and J. Wilson. 2014. Evaluation of foliar insecticides for the control of foliar insects in bell peppers in Virginia, 2013. Arthropod Management Tests 39: E19

Kupferschmidt, K. 2013. A lethal dose of RNA. Science 341: 732–733.

LaChance, L. E., and J. G. Riemann. 1973. Dominant lethal mutations in insects with holokinetic chromosomes. 1. Irradiation of Oncopeltus (Hemiptera: Lygaeidae) sperm and oocytes. Annals of the Entomological Society of America 66: 813–819.

LaChance, L. E., M. Begrugillier, and A. P. Leverich. 1970. Cytogenetics of inherited partial sterility in three generations of the large milkweed bug as related to holokinetic chromosomes. Chromosome (Berl.) 29: 20–41.

Lah, K. 2011. Pesticides – history. Toxipedia. http://www.toxipedia.org/display/toxipedia/Pesticides+-+History (accessed 25 March 2015).

Lamb, D. W., and R. B. Brown. 2001. PA – Precision agriculture: remote-sensing and mapping of weeds in crops. Journal of Agricultural Engineering Research 78: 117–125.

Lampson, B. D., A. Khalilian, J. K. Greene, Y. J. Han, and D. C. Degenhardt. 2014a. Development of a portable electronic nose for detection of cotton damaged by *Nezara viridula* (Hemiptera: Pentatomidae). Journal of Insects 2014: 1–8. doi:10.1155/2014/297219.

Lampson, B. D., Y. J. Han, A. Khalilian, J. K. Greene, D. C. Degenhardt, and J. O. Hallstrom. 2014b. Development of a portable electronic nose for detection of pests and plant damage. Computers and Electronics in Agriculture 108: 87–94.

Landis, D. A., S. D. Wratten, and G. M. Gurr. 2000. Habitat management to conserve natural enemies of arthropod pests in agriculture. Annual Review of Entomology 45: 175–201.

Langley, P. A. 1995. Evaluation of the chitin synthesis inhibitor triflumuron for controlling the tsetse *Glossina morsitans morsitans* (Diptera: Glossinidae). Bulletin of Entomological Research 85: 495–500.

Leskey, T. C., and H. W. Hogmire. 2005. Monitoring stink bugs (Hemiptera: Pentatomidae) in mid-Atlantic apple and peach orchards. Journal of Economic Entomology 98: 143–153.

Leskey, T. C., B. D. Short, S. E. Wright, and M. W. Brown. 2009. Diagnosis and variation in appearance of brown stink bug (Hemiptera: Pentatomidae) injury on apple. Journal of Entomological Science 44: 314–322.

Leskey, T. C., G. C. Hamilton, A. L. Nielsen, D. F. Polk, C. Rodriguez-Saona, J. C. Bergh, D. A. Herbert, T. P. Kuhar, D. Pfeiffer, G. Dively, C. R. R. Hooks, M. J. Raupp, P. M. Shrewsbury, G. Krawczyk, P. W. Shearer, J. Whalen, C. Koplinka-Loehr, E. Myers, D. Inkley, K. A. Hoelmer, D. H. Lee, and S. E. Wright. 2012a. Pest status of the brown marmorated stink bug, *Halyomorpha halys* (Stål), in the USA. Outlooks on Pest Management 23: 218–226.

Leskey T. C., B. D. Short., B. B. Butler, and S. E. Wright. 2012b. Impact of the invasive brown marmorated stink bug, *Halyomorpha halys* (Stål) in mid-Atlantic tree fruit orchards in the United States: case studies of commercial management. Psyche, Volume 2012 Article ID 535062, doi:10.1155/2012/535062.

Leskey, T. C., B. D. Short, and D. H. Lee. 2013. Efficacy of insecticide residues on adult *Halyomorpha halys* (Stål) (Hemiptera: Pentatomidae) mortality and injury in apple and peach orchards. Pest Management Science 70: 1097–1104.

Leskey, T. C., A. Agnello, J. C. Bergh, G. P. Dively, G. C. Hamilton, P. Jentsch, A. Khrimian, G. Drawczyk, T. P. Kuhar, D. H. Lee, W. R. Morrison, D. F. Polk, C. Rodriguez-Saona, P. W. Shearer, B. D. Short, P. M. Shrewsbury, J. F. Walgenbach, D. C. Weber, C. Welty, J. Whalen, N. Wiman, and F. Zaman. 2015. Attraction of the invasive *Halyomorpha halys* (Hemiptera: Pentatomidae) to traps baited with semiochemical stimuli across the United States. Environmental Entomology 44: 746–756.

Levine, E., and H. Oloumi-Sadeghi. 1991. Management of diabroticite rootworms in corn. Annual Review of Entomology 36: 229–255.

Li, J., Q. Chen, Y. Lin, T. Jiang, G. Wu, and H. Hua. 2011. RNA interference in *Nilaparvata lugens* (Homoptera, Delphacidae) based on dsRNA ingestion. Pest Management Science 67(7): 852–859.

Li, L., J. Sun, F. Zhang, X. Li, S. Yang, and Z. Rengel. 2001. Wheat/maize or wheat/soybean strip intercropping: I. Yield advantage and interspecific interactions on nutrients. Field Crops Research 71(2): 123–137.

Liebman, M., and E. Dyck. 1993. Crop rotation and intercropping strategies for weed management. Ecological Applications 3(1): 92–122.

Luckmann, W. H., and R. L. Metcalf. 1994. The pest management concept, pp. 1–34. *In* R. L. Metcalf and W. H. Luckmann (Eds.), Introduction to pest management. John Wiley and Sons, Inc., New York, NY, 661 pp.

Ludwig, S. W., and L. T. Kok. 1998. Evaluation of trap crops to manage harlequin bugs, *Murgantia histrionica* (Hahn) (Hemiptera: Pentatomidae) on broccoli. Crop Protection 17: 123–128.

Ludwig S. W., and L. T. Kok. 2001. Harlequin bug, *Murgantia histrionica* (Hahn) (Heteroptera: Pentatomidae), development on three crucifers and feeding damage on broccoli. Crop Protection 20: 247–251.

Lund, E. D., C. D. Christy, and P. E. Drummond. 1999. Practical applications of soil electrical conductivity mapping, pp. 771–779. *In* J. V. Safford (Ed.), Proceedings of the Second European Conference on Precision Agriculture, Sheffield Academic Press, Sheffield, UK. (page numbers not readily available)

Lye, B. H., R. N. Story, and V. L. Wright. 1988a. Southern green stink bug (Hemiptera: Pentatomidae) damage to fresh market tomatoes. Journal of Economic Entomology 81: 189–194.

Lye, B. H., R. N. Story, and V. L. Wright. 1988b. Damage threshold of the southern green stink bug, *Nezara viridula* (Hemiptera: Pentatomidae), on fresh market tomatoes. Journal of Entomological Science 23: 366–373.

Mahroof, R., and T. W. Phillips. 2014. Mating disruption of *Lasioderma serricorne* (Coleoptera: Anobiidae) in stored product habitats using the synthetic pheromone serricornin. Journal of Applied Entomology 138: 378–386.

Mahroof, R., K. Y. Zhu, and Bh. Subramanyam. 2005. Changes in expression of heat shock proteins in *Tribolium castaneum* (Coleoptera: Tenebrionidae) in relation to developmental stage, exposure time, and temperature. Annals of the Entomological Society of America 98: 100–107.

Mangan, R. L., and S. J. Ingle. 1992. Forced hot-air quarantine treatment for mangoes infested with West Indian fruit fly (Diptera: Tephritiddae). Journal of Economic Entomology 85: 1859–1864.

Mao, J., and F. Zeng. 2012. Feeding-based RNA interference of a gap gene is lethal to the pea aphid, *Acyrthosiphon pisum*. PLoS ONE 7 (11): e48718.

Mao, J., and F. Zeng. 2014. Plant-mediated RNAi of a gap gene-enhanced tobacco tolerance against the *Myzus persicae*. Transgenic Research 23(1): 145–152.

Mau, R., W. C., Mitchell, and M. Anwar. 1967. Preliminary studies on the effects of gamma irradiation of eggs and adults of the southern green stink bug, *Nezara viridula* (L.). Proceedings of the Hawaiian Entomological Society 19: 415–417.

McBratney, A., B. Whelan, and T. Ancev. 2005. Future directions of precision agriculture. Precision Agriculture 6: 7–23.

McCoy, T. 2015. More reasons why the Greek poet Homer may never have existed. The Washington Post, 6 January 2015. http://www.washingtonpost.com/news/morning-mix/wp/2015/01/06/more-reasons-why-the-greek-poet-homer-may-never-have-existed/ (accessed 24 March 2015).

McCullough, D.G., T.T. Work, J.F. Cavey, A.M. Liebhold, and D. Marshall. 2006. Interceptions of nonindigenous plant pests at US ports of entry and border crossings over a 17-year period. Biological Invasions. 8: 611–630.

McHughen, A., and R. Wager. 2010. Popular misconceptions: agricultural biotechnology. New Biotechnology 27: 724–728.

McIndoo, N. E. 1943. Insecticidal uses of nicotine and tobacco: a condensed summary of the literature, 1690–1934. USDA Agricultural Research Administration Bureau of Entomology and Plant Quarantine. E-597. 16 pp.

McPherson, J. E., and R. M. McPherson. 2000. Stink bugs of economic importance in America north of Mexico. CRC Press, Boca Raton, FL. 253 pp.

McPherson, R. M., and L. D. Newsom. 1984. Trap crops for control of stink bugs in soybean. Journal of the Georgia Entomological Society 19(4): 470–480.

Meadows, M. E. 1985. Eradication program in the southeast United States. Symposium on eradication of the screwworm from the United States and Mexico. Miscellaneous Publication of Entomological Society of America 62: 8–11.

Meek, F. 2011. Occasional invaders and overwintering pests, pp. 1250–1252. *In* S. A. Hedges (Ed.), Mallis handbook of pest control (10th edit.). Mallis Handbook Company, Valley View, Ohio. 1,599 pp.

Menzel, R., and U. Greggers. 1985. Natural phototaxis and its relationship to colour vision in honeybees. Journal of Comparative Physiology 157: 311–321.

Mills, N. 2010. Egg parasitoids in biological control and integrated pest management, pp. 389–406. *In* F. L. Consoli, J. R. P. Parra and R. A. Zucchi (Eds.), Egg parasitoids in agrosystems with emphasis on *Trichogramma*, Progress in Biological Control. Springer, New York. 479 pp.

Miyao, G. M., H. J. Phaff, and R. M. Davis. 2000. Outbreak of *Eremothecium coryli* fruit rot of tomato in California. Plant Disease 84: 594.

Mizell, R. F., T. C. Riddle, and A. S. Blount. 2008. Trap cropping system to suppress stink bugs in the Southern Coastal Plain. *In* Proceedings Florida State Horticulture Society 121: 377–382.

Mocatta, G. 2003. Pyrethrum – from ancient discovery to advanced agriculture. New Agriculturist (online). http://www.new-ag.info/03-6/develop/dev04.html (accessed 15 February 2015).

Morandin, L. A., R. F. Long, and C. Kremen. 2014. Hedgerows enhance beneficial insects on adjacent tomato fields in an intensive agricultural landscape. Agriculture Ecosystems and Environment 108: 164–170.

Morrill, A. W. 1910. Plant-bugs injurious to cotton bolls. United States Department of Agriculture Bulletin 86: 1–110.

Morrison, W. R., D. H. Lee, B. D. Short, A. Khrimian, and T. C. Leskey. 2015. Establishing the behavioral basis for an attract-and-kill strategy to manage the invasive *Halyomorpha halys* in apple orchards. Journal of Pest Science (in press). doi 10.1007/s10340-015-0679-6.

Muma, M. H. 1970. Preliminary studies on environmental manipulation to control injurious insects and mites in Florida citrus groves. Proceedings Tall Timbers Fire Ecology Conferences, Ecological Animal Control by Habitat Management 2: 23–40.

Mumford, J. D. 2002. Economic issues related to quarantine in international trade. European Review of Agricultural Economics 29: 329–348.

Murugesan, R., and B. D. Siegfried. 2012. Validation of RNA interference in western corn rootworm (*Diabrotica virgifera virgifera* LeConte (Coleoptera: Chrysomelidae) adults. Pest Management Science 68: 587–591.

Musolin, D. L. 2007. Insects in a warmer world: ecological, physiological, and life-history responses of true bugs (Heteroptera) to climate change. Global Change Biology 13: 1565–1585.

Musolin, D. L., and H. Numata. 2003. Photoperiodic and temperature control of diapause induction and colour change in southern green stink bug *Nezara viridula*. Physiological Entomology 28: 65–74.

Musolin, D. L., D. Tougou, and K. Fujisaki. 2010. Too hot to handle? Phenological and life-history responses to simulated climate change of the southern green stink bug *Nezara viridula* (Heteroptera: Pentatomidae). Global Change Biology 16: 73–87.

Musser, F. R., and A. M. Shelton. 2005. The influence of post-exposure temperature on the toxicity of insecticides to *Ostrinia nubilalis* (Lepidoptera: Crambidae). Pest Management Science 61: 508–510.

Musser, F. R., A. L. Catchot, B. K. Gibson, and K. S. Knighten. 2011. Economic injury levels for southern green stink bugs (Hemiptera: Pentatomidae) in R7 growth stage soybeans. Crop Protection 30(1): 63–69.

Mutti, N. S., Y. Park, J. C. Reese, and G. R. Reeck. 2006. RNAi knockdown of a salivary tran-script leading to lethality in the pea aphid, *Acyrthosiphon pisum*. Journal of Insect Science 6: 38.

Mutti, N. S., J. Louis, L. K. Pappan, K. Pappan, K. Begum, M.-S. Chen, Y. Park, N. Dittmer, J. Marshall, J. C. Reese, and G. R. Reeck. 2008. A protein from the salivary glands of the pea aphid, *Acyrthosiphon pisum*, is essential in feeding on a host plant. Proceedings of the National Academy of Sciences of the United States of America 105 (29): 9965–9969.

Myers, J. H., A. Savoie, and E. van Randen. 1998. Eradication and pest management. Annual Review of Entomology 43: 471–491.

Nagatuka, H. 2000. Effects of reflective sheet for whiteflies and thrips. Plant Protection 54. 359–362.

Naresh, J. S., and C. M. Smith. 1983. Development and survival of rice stink bugs (Hemiptera: Pentatomidae) reared on different host plants at four temperatures. Environmental Entomology 12: 1496–1499.

Navarro-Llopis, V., J. Sanchis, J. Primo-Millo, and E. Primo-Yufera. 2007. Chemosterilants as control agents of *Ceratitis capitata* (Diptera: Tephritidae) in field trials. Bulletin of Entomological Research 97: 359–368.

Negrón, J. F., and T. J. Riley. 1987. Southern green stink bug, *Nezara viridula* (Heteroptera: Pentatomidae), feeding in corn. Journal of Economic Entomology 80: 666–669.

Nelson, S. O. 1996. Review and assessment of radio-frequency and microwave energy for stored-grain insect control. Transactions of the American Society of Agricultural Engineers 39(4): 1475–1484.

Nevo, E., and M. Coll. 2001. Effect of nitrogen fertilization on *Aphis gossypii* (Homoptera: Aphididae): variation in size, color, and reproduction. Journal of Economic Entomology 94: 27–32.

Nida, D. L., K. H. Kolacz, R. E. Buehler, W. R. Deaton, W. R. Schuler, T. A. Armstrong, M. L. Taylor, C. C. Ebert, G. J. Rogan, S. R. Padgette, and R. L. Fuchs. 1996. Glyphosate-tolerant cotton: genetic characterization and protein expression. Journal of Agricultural and Food Chemistry 44: 1960–1966.

Nielsen, A. L., and G. C. Hamilton. 2009. Seasonal occurrence and impact of *Halyomorpha halys* (Hemiptera: Pentatomidae) in tree fruit. Journal of Economic Entomology 102: 1133–1140.

Nielsen, A. L., P. W. Shearer, and G. C. Hamilton. 2008. Toxicity of insecticides to *Halyomorpha halys* (Hemiptera: Pentatomidae) using glass-vial bioassays. Journal of Economic Entomology 101: 1439–1442.

Nord, J. C., and W. D. Pepper. 1991. Rain-fastness of insecticide deposits on loblolly pine foliage and the efficacy of adjuvants in preventing wash-off. Journal of Entomological Science 26(2): 287–298.

O'Brien, A. W. 1967. Insecticides. Action and metabolism. Academic Press, New York, NY. 332 pp.

Olson, L., and H. L. Eddy. 1943. Ibn-Al-Awam: a soil scientist in Moorish Spain. Geographical Review 33: 100–109.

Orlob, G. B. 1973. Ancient and medieval plant pathology. Pflanzenschutz-Nachrichten 26: 63–294.

Owen, T. 1805–1806. Geoponika: agricultural pursuits. 1805: Volume I: i +298 pp; Volume II: i + 301 pp.

Owens, J. C., D. C. Peters, and A. R. Hallauer. 1974. Corn rootworm tolerance in maize. Environmental Entomology 3: 767–772.

Paddock, F. B. 1915. The harlequin cabbage-bug. Texas Agricultural Experiment Station Bulletin 179: 1–9.

Paddock, F. B. 1918. Studies of the harlequin bug. Texas Agricultural Experiment Station Bulletin 227: 1–65.

Palumbo, J. C., and E. T. Natwick. 2010. The bagrada bug (Hemiptera: Pentatomidae): a new invasive pest of cole crops in Arizona and California. Plant Health Progress http://www.plantmanagementnetwork.org /pub/php/brief/2010/bagrada/

Palumbo, J. C., N. Prabhaker, D. A. Reed, T. M. Perring, S. J. Castle, and T. I Huang. 2015. Susceptibility of *Bagrada hilaris* (Hemiptera: Pentatomidae) to insecticides in laboratory and greenhouse bioassays. Journal of Economic Entomology 108: 672–682.

Panizzi, A. R. 2013. History and contemporary perspectives of the integrated pest management of soybean in Brazil. Neotropical Entomology 42(2): 119–127.

Pantoja, A., C. A. García, and M. C. Duque. 2000. Population dynamics and effects of *Oebalus ornatus* (Hemiptera: Pentatomidae) on rice yield and quality in southwestern Colombia. Journal of Economic Entomology 93(2): 276–279.

Patterson, K. D. 1993. Typhus and its control in Russia, 1870–1940. Medical history 37: 361–381.

Peairs, E. D. 1947. Insect pests of farm, garden, and orchard (4th edit.). John Wiley and Sons, Inc. New York, NY. 549 pp.

Pease, C. G., and F. G. Zalom. 2010. Influence of non-crop plants on stink bug (Hemiptera: Pentatomidae) and natural enemy abundance in tomatoes. Journal of Applied Entomology 134: 626–636.

Pedigo, L. P., and M. E. Rice. 2015. Entomology and pest management (6th edit.) Waveland Press. Long Grove, IL. 784 pp.

Pedigo, L. P., S. H. Hutchins, and L. G. Higley. 1986. Economic injury levels in theory and practice. Annual Review of Entomology 31: 341–368.

Perlak, F. J., M. Oppenhuizen, K. Gustafson, R. Voth, S. Sivasupramaniam, D. Heering, B. Carey, R. A. Ihrig, and J. K. Roberts. 2001. Development and commercial use of Bollgard® cotton in the USA – early promises versus today's reality. The Plant Journal 27: 489–501.

Petrick, J. S., B. Brower-Toland, A. L. Jackson, and L. D. Kier. 2013. Safety assessment of food and feed from biotechnology-derived crops employing RNA-mediated gene regulation to achieve desired traits: a scientific review. Regulatory Toxicology and Pharmacology 66: 167–176.

Phipps, R. H., and J. R. Park. 2002. Environmental benefits of genetically modified crops: global and European perspectives on their ability to reduce pesticide use. Journal of Animal and Feed Sciences 11: 1–18.

Pimentel, D., R. Zuniga, and D. Morrison. 2005. Update on the environmental and economic costs associated with alien invasive species in the United States. Ecological Economics 52: 273–288.

Poorani, J., A. Ślipiński, and R. G. A. Booth. 2008. A revision of the genus *Synona* (Coleoptera: Coccinellidae: Coccinellini). Annales Zoologici 58(3): 579–594. doi:10.3161/000345408X364427.

Prabhaker, N., N. C. Toscano, and T. J. Henneberry. 1998. Evaluation of insecticide rotations and mixtures as resistance management strategies for *Bemisia argentifolii* (Homoptera: Aleyrodidae). Journal of Economic Entomology 91: 820–826.

Prabhakar, M., Y. G. Prasad, M. Thirupathi, G. Sreedevi, B. Dharajothi, and B. Venkateswarlu. 2011. Use of ground based hyperspectral remote sensing for detection of stress in cotton caused by leafhopper (Hemiptera: Cicadellidae). Computers and Electronics in Agriculture 79: 189–198.

Prado, S. S., and R. P. P. Almeida. 2009. Role of symbiotic gut bacteria in the development of *Acrosternum hilare* and *Murgantia histrionica*. Entomologica Experimentalis et Applicata 132: 21–29.

Prado, S. S., K. Y. Hung, M. P. Daugherty, and R. P. P. Almeida. 2010. Indirect effects of temperature on stink bug fitness, via maintenance of gut-associated symbionts. Applied and Environmental Microbiology 76(4): 1261–1266.

Price, D. R. G., and J. A. Gatehouse. 2008. RNAi-mediated crop protection against insects. Trends in Biotechnology 26(7): 393–400.

Pridgeon, J. W., L. Zhao, J. J. Becnel, D. A. Strickman, G. G. Clark, and K. J. Linthicum. 2008. Topically applied AaeIAP1 double-stranded RNA kills female adults of *Aedes aegypti*. Journal Medical Entomology 45: 414–420.

Puterka, G. J., D. M. Glenn, D. G. Sekutowski, T. R. Unruh, and S. K. Jones. 2000. Progress toward liquid formulations of particle films for insect and disease control in pear. Environmental Entomology 29: 329–339.

Rao, M. R., and M. Singh. 1990. Productivity and risk evaluation in constrasting intercropping systems. Field Crops Research 23: 279–293.

Rao, V. P., M. A. Ghani, T. Sankaran, and K. C. Mathur. 1971. A Review of the Biological Control of Insects and other Pests in South-East Asia and the Pacific Region. Technical Communication No. 4. Commonwealth Agricultural Bureaux. Bangalore Press, Bangalore, India. 149 pp.

Rathore, H. S. 2010. Introduction, pp. 1–6. *In* L. M. L. Nollet and H. S. Rathore (Eds.), Handbook of pesticides: methods of pesticide residues analysis. CRC Press. Boca Raton, FL. 580 pp.

Reay-Jones, F. P. F. 2010. Spatial and temporal patterns of stink bugs (Hemiptera: Pentatomidae) in wheat. Environmental Entomology 39: 944–955.

Reay-Jones, F. P. F., M. D. Toews, J. K. Greene, and R. B. Reeves. 2010. Spatial dynamics of stink bugs (Hemiptera: Pentatomidae) and associated boll injury in southeastern cotton fields. Environmental Entomology 39: 956–969.

Reed, D. A., J. C. Palumbo, T. M. Perring, and C. May. 2013. *Bagrada hilaris* (Burmeister), a new stink bug attacking cole crops in the southwestern United States. Journal of Integrated Pest Management 4(3) http://dx.doi.org/10.1603/IPM13007.

Reid, J. F., Q. Zhang, N. Noguchi, and M. Dickson. 2000. Agricultural automatic guidance research in North America. Computers and Electronics in Agriculture 25: 155–167.

Reinhardt, C., and Ganzel, W. 2003. Living history farm. Farming in the 1930s. Wessels Living History Farm http://www.livinghistoryfarm.org/farminginthe30s/farminginthe1930s.html (accessed 12 February 2015).

Reisenman, C. E., C. R. Lazzari, and M. Giurfa. 1998. Circadian control of photonegative sensitivity in the haematophagous bug *Triatoma infestans*. Journal of Comparative Physiology 183: 522–541.

Reisig, D., and L. Godfrey. 2007. Spectral response of cotton aphid- (Homoptera: Aphididae) and spider mite- (Acari: Tetranychidae) infested cotton: controlled studies. Environmental Entomology 36: 1466–1474.

Rice, K. B., C. J. Bergh, E. J. Bergmann, D. J. Biddinger, C. Dieckhoff, G. Dively, H. Fraser, T. Gariepy, G. Hamilton, T. Haye, A. Herbert, K. Hoelmer, C. R. Hooks, A. Jones, G. Krawczyk, T. Kuhar, H. Martinson, W. Mitchell, A. L. Nielsen, D. G. Pfeiffer, M. J. Raupp, C. Rodriguez-Saona, P. Shearer, P. Shrewsbury, P. D. Venugopal, J. Whalen, N. G. Wiman, T. C. Leskey, and J. F. Tooker. 2014. Biology, ecology, and management of brown marmorated stink bug (Hemiptera: Pentatomidae). Journal of Integrated Pest Management 5(3): 1–13.

Riley, C. V. 1884. Report of the entomologist, pp. 285–418 + 2 pp., 10 plates. (Authors' edition. 1885. Annual report of Department of Agriculture for 1884). *In* Annual report of commissioner for the year 1884. United States Governing Printing Office, Washington, D. C. 581 pp.

Rings, R. W. 1957. Types and seasonal incidence of stink bug injury to peaches. Journal of Economic Entomology 50: 599–604.

Roach, S. H. 1981. Emergence of overwintered *Heliothis* spp. moths from three different tillage systems. Environmental Entomology 10: 817–818.

Roberts, J. R., and J. R. Reigart. 2013. Recognition and management of pesticide poisonings (6th edit.). http://www2.epa.gov/sites/production/files/documents/rmpp_6thed_titletoc_0.pdf (accessed 10 February 2015).

Roda, A. L., D. A. Landis, M. L. Coggins, E. Spandl, and O. B. Hesterman. 1996. Forage grasses decrease alfalfa weevil (Coleoptera: Curculionidae) damage and larval numbers in alfalfa-grass intercrops. Journal of Economic Entomology 89: 743–750.

Roda, A. L., D. A. Landis, and M. L. Coggins. 1997a. Forage grasses elicit emigration of adult potato leafhopper (Homoptera: Cicadellidae) from alfalfa–grass mixtures. Environmental Entomology 26: 745–753.

Roda, A. L., D. A. Landis, and J. R. Miller. 1997b. Contact-induced emigration of potato leafhopper (Homoptera: Cicadellidae) from alfalfa–forage grass mixtures. Environmental Entomology 26: 754–762.

Ruberson, J., K. Takasu, G. D. Buntin, J. Eger, W. Gardner, J. Greene, T. Jenkins, W. Jones, D. Olson, P. Roberts, D. Suiter, and M. D. Toews. 2013. From Asian curiosity to eruptive American pest: *Megacopta cribraria* (Hemiptera: Plataspidae) and prospects for its biological control. Applied Entomology and Zoology 48: 3–13.

Ruckes, H. 1946. Notes and keys on the genus *Brochymena* (Pentatomidae, Heteroptera). Entomologica Americana 26: 143–238.

Sagers, L. A. 2005. Vegetable crop pests. Paper 1525. Utah State University. http://digitalcommons.usu.edu/extension_histall/1525 (accessed 27 July 2016).

Sapountzis, P., G. Duport, S. Balmand, K. Gaget, S. Jaubert-Possamai, G. Febvay, H. Charles, Y. Rahbé, S. Colella, and F. Calevro. 2014. New insight into the RNA interference response against cathepsin-L gene in the pea aphid, *Acyrthosiphon pisum*: molting or gut phenotypes specifically induced by injection or feeding treatments. Insect Biochemistry and Molecular Biology 51: 20–32.

Sasaki, F., T. Miyamoto, A. Yamamoto, Y. Tamai, and T. Yajima. 2008. Morphological and genetic characteristics of the entomopathogenic fungus *Ophiocordyceps nutans* and its host insects. Mycological Research 112: 1241–1244.

Satpute, N. S., S. D. Deshmukh, N. G. V. Rao, S. N. Tikar, M. P. Moharil, and S. A. Nimbalkar. 2007. Temperature-dependent variation in toxicity of insecticides against *Earias vitella* (Lepidoptera: Noctuidae). Journal of Economic Entomology 100: 357–360.

Schafer, M. G., A. A. Ross, J. P. Londo, C. A. Burdick, E. H. Lee, S. E. Travers, P. K. Van de Water, and C. L. Sagers. 2011. The establishment of genetically engineered canola populations in the U.S. PLoS ONE 6: 1–4. doi:10.1371/journal.pone.0025736.

Schliekelman, P., and F. Gould. 2000. Pest control by the release of insects carrying a female-killing allele on multiple loci. Journal of Economic Entomology 93: 1566–1579.

Schwalbe, C. P., and G. J. Hallman. 2002. An appraisal of forces shaping the future of regulatory entomology, pp. 425–433. *In* G. J. Hallman and C. P Schwalbe (Eds.), Invasive arthropods in agriculture: problems and solutions. Science Publishers, Enfield, NH. 447 pp.

Scudder, G. G. E. 1979. Hemiptera, pp. 329–348. *In* H. V. Danks (Ed.), Canada and its insect fauna. Memoirs of the Entomological Society of Canada 111. doi:10.4039/entm111108329-1.

Sedighi, N., H. Abbasipour, A. S. Gorjan, and J. Karimi. 2013. Pathogenicity of the entomopathogenic fungus *Metarhizium anisopliae* var. *major* on different stages of the sunn pest, *Eurygaster integriceps*. Journal of Insect Science 13: 150. http://www.insectscience.org/13.150.

Seiter, N. J., E. P. Benson, F. P. F. Reay-Jones, J. K. Greene, and P. A. Zungoli. 2013. Residual efficacy of insecticides applied to exterior building material surfaces for control of nuisance infestations of *Megacopta cribraria* (Hemiptera: Plataspidae). Journal of Economic Entomology 106: 2448–2456.

Seiter, N. J., A. Grabke, J. K. Greene, J. L. Kerrigan, and F. P. F. Reay-Jones. 2014. *Beauveria bassiana* is a pathogen of *Megacopta cribraria* (Hemiptera: Plataspidae) in South Carolina. Journal of Entomological Science 49: 326–330.

Sevacherian, V., and V. M. Stern. 1974. Host plant preferences of *Lygus* bugs in alfalfa-interplanted cotton fields. Environmental Entomology 3: 761–766.

Sevacherian, V., and V. M. Stern. 1975. Movements of *Lygus* bugs between alfalfa and cotton. Environmental Entomology 4: 163–165.

Shakesby, A. J., I. S. Wallace, H. V. Isaacs, J. Pritchard, D. M. Roberts, and A. E. Douglas. 2009. A water-specific aquaporin involved in aphid osmoregulation. Insect Biochemistry and Molecular Biology 39: 1–10.

Shelton, A. M., and F. R. Badenes-Perez. 2006. Concepts and applications of trap cropping in pest management. Annual Review of Entomology 51: 285–308.

Shimoda, M., and K. Honda. 2013. Insect reactions to light and its applications to pest management. Applied Entomology and Zoology 48: 413–421.

Shorey, H. H., L. K. Gaston, and C. K. Saario. 1967. Sex pheromones of noctuid moths. XIV. Feasibility of behavioral control by disrupting pheromone communication in cabbage loopers. Journal of Economic Entomology 60: 1541–1545.

Shorey, H. H., T. L. Payne, and L. L. Sower. 1972. Evaluation of high-frequency sound for control of oviposition by corn earworm moths in sweet corn. Journal of Economic Entomology 65: 911–612.

Silva-Aguayo, G. 2013. Botanical insecticides. Radcliffe's IPM World Textbook. http://ipmworld.umn.edu /chapters/SilviaAguayo.htm (accessed 25 March 2015).

Simmons, A. M., and K. V. Yeargan. 1988. Development and survivorship of the green stink bug, *Acrosternum hilare* (Hemiptera: Pentatomidae) on soybean. Environmental Entomology 17: 527–532.

Simmons, A. M., C. S. Kousik, and A. Levi. 2010. Combining reflective mulch and host plant resistance for sweet potato whitefly (Hemiptera: Aleyrodidae) management in watermelon. Crop Protection 29: 898–902.

Singh, D. K. 2012. Pesticide chemistry and toxicology (Book Series: Toxicology: agriculture and environment), Volume. 1. Bentham Science Publishers (e-books). 213 pp.

Smith, C. M. 2005. Plant resistance to arthropods: molecular and conventional approaches. Springer. Dordrecht, Netherlands. 423 pp.

Smith, D. B., L. E. Bode, and P. D. Gerard. 2000. Predicting ground boom spray drift. Transactions of the American Society of Agricultural Engineers 43(3): 547–553.

Smith, E. H., and G. G. Kennedy. 2002. History of pesticides, pp. 376–380. *In* D. Pimentel (Ed.), Encyclopedia of pest management. Marcel Dekker, Inc., New York, NY. 931 pp.

Smith, E. H., and R. C. Whitman. 2007. Occasional Invaders, pp. 7.5.3.1–7.7.2. *In* National Pest Management Association Field Guide to Structural Pests (2nd edit.). (page numbers not readily available)

Smith, H. S. 1935. The role of biotic factors in the determination of population densities. Journal of Economic Entomology 28: 873–898.

Smith, J. E. 1897. The harlequin cabbage bug and the melon plant louse. New Jersey Agricultural Experiment Stations Bulletin 121: 3–14.

Smith, J. F., R. G. Luttrell, and J. K. Greene. 2009. Early-season soybean as a trap crop for stink bugs (Heteroptera: Pentatomidae) in Arkansas' changing system of soybean production. Environmental Entomology 38: 450–458.

Smith, J. W. 1998. Boll weevil eradication: area-wide pest management. Annals of the Entomological Society of America 91: 239–247.

Solomon, M. E. 1949. The natural control of animal populations. Journal of Animal Ecology 18: 1–35.

Solomon, M. E. 1957. Dynamics of insect populations. Annual Review of Entomology 2: 121–142.

Sosa-Gomez, D. R. 2006. Trypanosomatid prevalence in *Nezara viridula* (L.), *Euschistus heros* (Fabricius) and *Piezodorus guildinii* (Westwood) (Heteroptera: Pentatomidae) populations in northern Parana, Brazil. Neotropical Entomology 34: 341–347.

Sosa-Gomez, D. R., and F. Moscardi. 1998. Laboratory and field studies on the infection of stink bugs, *Nezara viridula, Piezodorus guildinii,* and *Euschistus heros* (Hemiptera: Pentatomidae) with *Metarhizium anisopliae* and *Beauveria bassiana* in Brazil. Journal of Invetebrate Pathology 71: 115–120.

Sprenkel, R. K., W. M. Brooks, J. W. Van Duyn, and L. L. Deitz. 1979. The effects of 3 cultural variables on the incidence of *Nomuraea rileyi,* phytophagous Lepidoptera and their predators on soybeans. Environmental Entomology 8: 334–339.

Stafford, J. V. 2000. Implementing precision agriculture in the 21st century. Journal of Agricultural Engineering Research 76: 267–275.

Stearns, P. N., M. Adas, S. B. Schwartz, and M. J. Gilbert. 2010. World civilizations: the global experience (6th edit). Pearson, Saddle River, NJ. 1080 pp.

Stent, V. 2006. The history of pest control. http://ezinearticles.com/?The-History-of-Pest-Control&id=133689 (accessed 24 March 2015).

Stern, V. M., R. F. Smith, R. van den Bosch, and K. S. Hagen. 1959. The integration of chemical and biological control of the spotted alfalfa aphid. The integrated control concept. Hilgardia 29: 81–101.

Stewart, C. N., H. A. Richards, and M. D. Halfhill. 2000. Transgenic plants and biosafety: science, misconceptions and public perceptions. BioTechniques 29: 832–843.

Subramanyam, T. V. 1925. *Coptosoma ostensum* Dist. and its enemy *Synia melanaria* Muls. Journal of the Bombay Natural History Society 30: 924–925.

Suiter, D. R., J. E. Eger, Jr., W. A. Gardner, R. C. Kemerait, J. N. All, P. M. Roberts, J. K. Greene, L. M. Ames, G. D. Buntin, T. M. Jenkins, and G. K. Douce. 2010. Discovery and distribution of *Megacopta cribraria* (Hemiptera: Heteroptera: Plataspidae) in northeast Georgia. Journal of Integrated Pest Management 1: F1–F4.

Summers, C. G. 1976. Population fluctuations of selected arthropods in alfalfa: influence of two harvesting practices. Environmental Entomology 5: 103–110.

Summy, K. R., W. G. Hart, M. D. Heilman, and J. R. Cate. 1986. Late-season boll weevil control: combined impact of stalk shredding and lethal soil temperatures, pp. 233–234. *In* Proceedings Beltwide Cotton Production Conferences, National Cotton Council, Memphis, TN. (page numbers not readily available)

Talamas, E. J., M. V. Herlihy, C. Dieckhoff, K. A. Hoelmer, M. L. Buffington, M. C. Bon, and D. C. Weber. 2015. *Trissolcus japonicus* (Ashmead) (Hymenoptera: Scelionidae) emerges in North America. Journal of Hymenoptera Research 43: 119–128.

Talekar, N. S., and A. M. Shelton. 1993. Biology, ecology, and management of the diamondback moth. Annual Review of Entomology 38: 275–301.

Thailayil, J., A. Crisanti, F. Catteruccia, K. Magnusson, and H. C. J. Godfray. 2011. Spermless males elicit large-scale female responses to mating in the malaria mosquito *Anopheles gambiae*. Proceedings of the National Academy of Sciences of the United States of America. 108(33): 13677–13681.

Thakur, N., S. K. Upadhyay, P. C. Verma, K. Chandrashekar, R. Tuli, and P. K. Singh. 2014. Enhanced whitefly resistance in transgenic tobacco plants expressing double stranded RNA of v-ATPase A Gene. PLoS ONE 9(3): e87235.

Tillman, P. G. 2014. Physical barriers for suppression of movement of adult stink bugs into cotton. Journal of Pest Science 87: 419–427.

Tillman, P. G., T. D. Northfield, R. F. Mizell, and T. C. Riddle. 2009. Spatiotemporal patterns and dispersal of stink bugs (Heteroptera: Pentatomidae) in peanut-cotton agri-scapes. Environmental Entomology 38: 1038–1052.

Tillman, P. G., A. Khrimian, T. E. Cottrell, X. Lou, R. R. Mizell, and C. J. Johnson. 2015. Trap cropping systems and a physical barrier for suppression of stink bugs (Hemiptera: Pentatomidae) in cotton. Journal of Economic Entomology 108: 2324–2334.

Todd, J. W. 1989. Ecology and behavior of *Nezara viridula*. Annual Review of Entomology 34: 273–292.

Todd, J. W., and F. W. Schumann. 1988. Combination of insecticide applications with trap crops of early maturing soybean and southern peas for population management of *Nezara viridula* in soybean (Hemiptera: Pentatomidae). Journal of Entomological Science 23: 192–199.

Tomari, Y., and P. D. Zamore. 2005. Perspective, machines for RNAi. Genes and Development 19(5): 517–529.

Trematerra, P., C. Athanassiou, V. Stejskal, A. Sciarretta, N. Kavallieratos and N. Palyvos. 2011. Large-scale mating disruption of *Ephestia* spp. and *Plodia interpunctella* in Czech Republic, Greece and Italy. Journal of Applied Entomology 135: 749–762.

Trenbath, B. R. 1993. Intercropping for the management of pests and diseases. Field Crops Research 34(3): 381–405.

Turnbull, A. L., and D. A. Chant. 1961. The practice and theory of biological control of insects in Canada. Canadian Journal of Zoology 39: 697–753.

Turnipseed, S. G., M. J. Sullivan, J. E. Mann, and M. E. Roof. 1995. Secondary pests in transgenic Bt cotton in South Carolina, pp. 768–769. *In* Proceedings Beltwide Cotton Conferences, National Cotton Council, Memphis, TN. (page numbers not readily available)

Umeozor, O. C., J. W. Van Duyn, J. R. Bradley, and G. G. Kennedy. 1985. Comparison of the effect of minimum-tillage treatments on the overwintering emergence of European corn borer (Lepidoptera: Pyralidae) in cornfields. Journal of Economic Entomology 78: 937–939.

Unsworth, J. 2010. History of pesticide use. International union of pure and applied chemistry. http://agrochemicals.iupac.org/index.php?option=com_sobi2&sobi2Task=sobi2Details&catid=3&sobi2Id=31 (accessed 20 March 2015).

Upadhyay, S. K., K. Chandrashekar, N. Thakur, P. C. Verma, J. F. Borgio, P. K. Singh, and R. Tuli. 2011. RNA interference for the control of whiteflies (*Bemisia tabaci*) by oral route. Journal of Biosciences 36: 153–161.

Urban, J. E., and A. Broce. 2000. Killing of flies in electrocuting insect traps releases bacteria and viruses. Current Microbiology 41: 267–270.

Vacas, S., P. Vanaclocha, C. Alfaro, J. Primo, M. J. Verdú, A. Urbaneja, and V. Navarro-Llopis. 2011. Mating disruption for control of *Aonidiella aurantii* Maskell (Hemiptera: Diaspididae) may contribute to increased effectiveness of natural enemies. Pest Management Science 68: 142–148.

Van Den Berg, M. A., and J. Greenland. 1996. Further releases of *Trichopoda pennipes*, parasitoid of the green stinkbug, *Nezara viridula*, in South Africa. Tachinid Times 9: 2.

Varela, L. G., R. A. Van Steenwyk, and R. B. Elkins. 2011. Evolution of secondary pests in California pear orchards under mating disruption for codling moth. Acta Horticulturae 909: 531–541.

Vejnar, R. J. 2011. Plantation agriculture. Encyclopedia of Alabama. http://www.encyclopediaofalabama.org/article/h-1832 (accessed 24 March 2015).

Villanueva, R. T., and J. F. Walgenbach. 2005. Development, oviposition and mortality of *Neoseiulus fallacis* (Phytoseiidae) in response to reduced-risk insecticides. Journal of Economic Entomology 98: 2114–2120.

Vincent, C., G. Hallman, B. Panneton, and F. Fleurat-Lessard. 2003. Management of agricultural insects with physical control methods. Annual Review of Entomology 48: 261–281.

Vreysen, M. J. B., K. M. Saleh, M. Y. Ali, A. M. Abdulla, Z. R. Zhu, K. G. Juma, V. A. Dyck, A. R. Msangi, P. A. Mkonyi, and H. U. Feldman. 2000. *Glossina austeni* (Diptera: Glossinidae) eradicated on the Island of Unguja, Zanzibar, using the sterile insect technique. Journal of Economic Entomology 93: 123–135.

Wakil, W., M. Ashfaq, M. U. Ghazanfar, M. Afzal, and T. Riasat. 2009. Phytoparasitica 37: 415–420.

Walgenbach, J. F., and S. C. Schoof. 2011. Flea beetle and harlequin bug control on cabbage, 2010. Arthropod Management Tests 36: E21.

Walker, W. B., and M. L. Allen. 2011. RNA interference-mediated knockdown of IAP in *Lygus lineolaris* induces mortality in adult and pre-adult life stages. Entomologia Experimentalis et Applicata 138: 83–92.

Wallingford, A. K., T. P. Kuhar, and P. B. Schultz. 2012. Toxicity and field efficacy of four neonicotinoids on harlequin bug (Hemiptera: Pentatomidae). Florida Entomologist 95: 1123–1126.

Wallingford, A. K., T. P. Kuhar, D. G. Pfeiffer, D. B. Tholl, J. H. Freeman, H. B. Doughty, and P. B. Schultz. 2013. Host plant preference of harlequin bug (Hemiptera: Pentatomidae) and evaluation of a trap cropping strategy for its control in Collard. Journal of Economic Entomology 106: 283–288.

Walton, V. M., T. E. Larsen, R. Malakar-Kuenen, J. G. Millar, K. M. Daane, and W. J. Bentley. 2006. Pheromone-based mating disruption of *Planococcus ficus* (Hemiptera: Pseudococcidae) in California vineyards. Journal of Economic Entomology 99: 1280–1290.

Wan, P.-J., S. Jia, N. Li, J.-M. Fan, and G.-Q. Li. 2014. RNA interference depletion of the Halloween gene disembodied implies its potential application for management of planthopper *Sogatella furcifera* and *Laodelphax striatellus*. PLos ONE 9(1): e86675.

Wang, S., and J. Tang. 2001. Radio frequency and microwave alternative treatments for insect control in nuts: a review. Journal of Agricultural Engineering 10(3–4): 105–120.

Wang, Y., H. Zhang, H.-C. Li, and X.-X. Miao. 2011. Second-generation sequencing supply an effective way to screen RNAi targets in large scale for potential application in pest insect control. PLoS ONE 6 (4): e18644.

Ware, G. W. 1994. The pesticide book (4th edit.). Thomson Publications, Fresno, CA. 386 pp.

Watson, J. R. 1918. Insects of a citrus grove. Florida Agricultural Experiment Station Bulletin 148: 165–267.

Weber, D. C., G. C. Walsh, A. S. DiMeglio, M. M. Athanas, T. C. Leskey, and A. Khrimian. 2014. Attractiveness of harlequin bug, *Murgantia histrionica*, aggregation pheromone. Field response to isomers, ratios, and dose. Journal of Chemical Ecology 40: 1251–1259.

Wermelinger, B. 2004. Ecology and management of the spruce bark beetle *Ips typographus* – a review of recent research. Forest Ecology and Management 202(1–3): 67–82.

Whyard, S., A. D. Singh, and S. Wong. 2009. Ingested double-stranded RNAs can act as species-specific insecticides. Insect Biochemistry and Molecular Biology 39(11): 824–832.

Wigglesworth, V. B. 1972. The principles of insect physiology. Chapman and Hall, London. 836 pp.

Willers, J. L., J. N. Jenkins, W. L. Ladner, P. D. Gerard, D. L. Boykin, K. B. Hood, P. L. McKibben, S. A. Samson, and M. M. Bethel. 2005. Site-specific approaches to cotton insect control. Sampling and remote sensing analysis techniques. Precision Agriculture 6: 431–452.

Williams, M. R. 2015. Cotton insect loss estimates 2014, pp. 494–506. *In* Proceedings of the Beltwide Cotton Conferences. National Cotton Council of America, Memphis, TN. (page numbers not readily available)

Williamson, C., and M. B. von Wechmar. 1995. The effects of two viruses on the metamorphosis, fecundity, and longevity of the green stinkbug, *Nezara viridula*. Journal of Invertebrate Pathology 65(2): 174–178.

Willis, G. H., L. L. McDowell, S. Smith, and L. M. Southwick. 1992. Foliar washoff of oil-applied malathion and permethrin as a function of time after application. Journal of Agricultural and Food Chemistry 6: 1086–1089.

Wilson, F. 1960. A review of the biological control of insects and weeds in Australia and Australian New Guinea. Technical Communication No. 1, Commonwealth Institute of Biological Control, Commonwealth Agricultural Bureaux, Ottawa, Canada. 101 pp.

Wuriyanghan, H., C. Rosa, and B. W. Falk. 2011. Oral delivery of double-stranded RNAs and siRNAs induces RNAi effects in the potato/tomato psyllid, *Bactericerca cockerelli*. PLoS ONE 6 (11): e27736.

Xu, L., X. Duan, Y. Lv, X. Zhang, Z. Nie, C. Xie, Z. Ni, and R. Liang. 2014. Silencing of an aphid carboxylesterase gene by use of plant-mediated RNAi impairs *Sitobion avenae* tolerance of phoxim insecticides. Transgenic Research 23(2): 389–396.

Yao, J., D. Rotenberg, A. Afsharifar, K. Barandoc-Alviar, and A. E. Whitfield. 2013. Development of RNAi methods for *Peregrinus maidis*, the corn planthopper. PLoS ONE 8(8): e370243.

Young, W. R., and G. L. Teetes. 1977. Sorghum entomology. Annual Review of Entomology 22: 193–218.

Zahn, D. K., J. A. Moreira, and J. G. Millar. 2008. Identification, synthesis, and bioassay of a male-specific aggregation pheromone from the harlequin bug, *Murgantia histrionica*. Journal of Chemical Ecology 34: 238–251.

Zalom, F. G., J. M. Smilanick, and L. E. Ehler. 1997. Fruit damage by stink bugs (Hemiptera: Pentatomidae) in bush-type tomatoes. Journal of Economic Entomology 90: 1300–1306.

Zha, W., X. Peng, R. Chen, B. Du, L. Zhu, and G. He. 2011. Knockdown of midgut genes by dsRNA-transgenic plant-mediated RNA interference in the hemipteran insect *Nilaparvata lugens*. PLoS ONE 6 (5): e20504.

Zhang, C., and J. M. Kovacs. 2012. The application of small unmanned aerial systems for precision agriculture: a review. Precision Agriculture 13: 693–712. doi 10.1007/s11119-012-9274-5.

Zhang, M., Y. Zhou, H. Wang, H. D. Jones, Q. Gao, D. Wang, Y. Ma, and L. Xia. 2013. Identifying potential RNAi targets in grain aphid *Sitobion avenae* F. based on transcriptome profiling of its alimentary canal after feeding on wheat plants. BMC Genomics 14: 560.

Zhang, N., M. Wang, and N. Wang. 2002. Precision agriculture – a worldwide overview. Computers and Electronics in Agriculture 36: 113–132.

Zhang, X., L. Shi, X. Jia, G. Seielstad, and C. Helgason. 2010a. Zone mapping application for precision-farming: a decision support tool for variable rate application. Precision Agriculture 11: 103–114. doi 10.1007/s11119-009-9130-4.

Zhang, X., J. Zhang, and K. Y. Zhu. 2010b. Chitosan/double-stranded RNA nanoparticle-mediated RNA interference to silence chitin synthase genes through larval feeding in the African malaria mosquito *Anopheles gambiae*. Insect Molecular Biology 19(5): 683–693.

Zhou, J., K.-T. Shum, J. Burnett, J. Rossi. 2013. Nanoparticle-based delivery of RNAi therapeutics, progress and challenges. Pharmaceuticals 61: 85–107.

Insects and Spiders Index

A

Abascantus, 58
Abeona, 75, 109
Ablaptus, 58
Ablerus, 307, 308
Acanthocephalonotum
    martinsnetoi, 58
Acanthocoris
    sordidus, 650, 655
Acanthosoma, 37
    denticauda japonica (syn. of *Acanthosoma*
        denticaudum), 503
    denticaudum, 503, 647
    firmatum, 37, 647, 655
    forficula, 647
    giganteum (syn. of *Acanthosoma firmatum*), 647
    haemorrhoidale, 647
    haemorrhoidale angulatum, 503
    labiduroides, 647
Acanthosomatidae, 3, 6, 12, 14, 28, 32, 34, 36–38, 44–47,
    112, 126, 136, 138, 501–503, 538, 568, 646, 647,
    654, 655, 656, 658, 678
Acanthosomatinae, 30, 32, 36–38, 184, 503
Acanthosomina, 30
Accarana, 85
Acephali
    amaragina, 732
Acesines, 67, 74, 75, 89
Acledra, 78, 79, 103
Acoloba, 49, 78, 83, 105
    lanceolata, 83, 190
Acoloba group, 100, 105
Acrobasis
    vaccinii, 740
Acroclisoides, 266
    africanus, 478, 483
Acrophyma, 37
Acrosternum (see *Chinavia*), 84, 91, 623
    heegeri, 578
    hilare (diff. comb. of *Chinavia hilaris*), 428, 429, 505,
        578, 614, 660, 682
    millierei, 65, 191
Acrozangis, 91
Actuarius, 89
Acyrthosiphon
    pisum, 756
Adelaidena, 98, 103
Adelocus, 105
Adevoplitus, 93, 94
Adomerus
    biguttatus, 503
    rotundus, 647, 658

    triguttulus, 40, 538, 647, 655, 658
    variegatus, 647
Adoxoplatys, 59, 108
Adria, 69, 83–85, 102
Adrisa, 39
    magna, 647, 655
Adustonotus, 79
Aedia, 49, 69, 105
    leucomelas, 513
Aednus, 58, 86, 89, 90
Aegaleus, 88, 89
Aegius, 109
Aeladria, 83
Aelia, 49, 69, 105
    acuminata, 63, 64, 197, 504, 511, 526, 528, 582, 583
    americana, 69
    fieberi, 504, 509, 510, 520, 526, 583, 584, 590
    furcula, 528
    melanota, 528
    rostrata, 504, 528, 537, 584, 597
    sibirica, 504, 582, 583
Aeliaria, 30
Aeliavuori, 69
    linnacostatus, 182
Aeliini, 50, 68, 69, 78, 83, 84, 102, 105, 188, 197, 504, 539
Aelioides, 69
Aeliomorpha, 69, 83, 105
Aeliomorpha group, 83, 100, 105
Aeliopsis, 69
    bucculata, 69
Aenaria, 87, 98
Aeptini, 50, 68, 69, 78, 83, 84, 89, 90, 92, 100–102, 182, 197
Aeptus, 69, 70, 83, 89, 102, 182
    singularis, 69, 197
Aeschrocoraria, 30, 70
Aeschrocorini, 50, 68, 70, 83, 100, 105, 188, 197
Aeschrocoris, 70, 71, 83
    obscurus, 197
Aeschrus, 70, 71
    inaequalis, 71
Aeschrus group, 100, 104
Afrania, 71
African cotton stainer, 657
Afrius, 56, 597
Afromenotes, 43
Agaclitus, 58
Agaeini, 50, 62, 71, 197
Agaeus, 71, 72, 85
    mimus, 72, 197
    pavimentatus, 12
    tessellatus, 72
Agatharchus, 79

Agathocles, 80, 96, 97
Agonosceliaria, 30
Agonoscelidini, 50, 66, 72, 100, 188, 197
Agonoscelis, 72, 85
    nubilis, 197
    puberula, 72
    rutila, 72
    versicoloratus, 72, 188, 484
Agonoscelis group, 100
Agonosoma, 130
    trilineatum, 195
Agroecus, 79
    griseus, 680, 681, 701
Aladoidae, 645
Alcaeorrhynchus, 48
    grandis, 186
Alcimocoris, 86
Alcimus, 86
Alciphron, 88
Alcippus, 58
Aleixus, 95
Aleria, 90
Alfalfa stink bug, 426
Alfalfa weevil, 740
Alitocoris, 59
    schraderi, 187
Alkindus, 137
Allophanurus
    indicus (syn. of *Trissolcus hyalinipennis*), 228, 231
Allopodops, 122, 123
Alophora
    indica, 227
    pusilla, 227
Alveostethus, 58
Alydidae, 307, 372, 590, 593, 646, 649, 655, 657, 658, 678,
      696, 710
Alydus
    calcaratus, 649, 655
    conspersus, 649
    tomentosus, 649
Amasenus, 96, 97
Amaurochrous, 119, 122, 123
    cinctipes, 119
    dubius, 53, 119, 187
Amaurocorinae, 32, 40, 42, 184
Amauromelpia, 79
Amauropepla, 123
Amberiana, 43
Amberianini, 32, 43
Ambiorix, 67, 83, 88, 94
    aenescens, 82
Amblybelus, 91
Amblycara, 75, 94, 109
Ambohicorypha, 72
Ambrosia beetle, 645
Amelanchier, 251
    laevis (syn. of *Amelanchier grandiflora*), 251
Amirantea, 94
Amnestinae, 31, 32, 39–41, 184, 503
Amnestus, 40
    championi, 184

    ficus, 40
    pusillus, 503
    pusio, 35, 42
Amphidexius, 101, 117
    suspensus, 193
Amphimachus, 88
Amyntaria, 30, 72
Amyntor, 72
Amyntorini, 50, 66, 72, 86, 197
Amyotea, 55, 56, 598
    malabarica, 56
Amyotinae, 56
Amyotini, 55, 56, 504, 597, 598
Anaca, 73
Anacanthocoris
    striicornis, 650
Anasa
    tristis, 616
Anasida, 55, 56
Anastatus, 266, 267, 275, 276
    antestiae, 478, 483–485
    bifasciatus, 275, 277
    gastopachae, 267
    mirabilis, 275
    pearsalli, 275
    reduvii, 275
Anastrepha
    obliqua, 743
Anaxarchus, 104
    reyi, 188
Anaxilaus, 74, 79
Anaximenes, 88
Anchesmus, 88
Anchises, 86
Ancyrosoma, 117, 123
    leucogrammes, 187
Andrallus, 56
    spinidens, 504, 513, 597
Anguleux, 30, 48
Anhanga, 59
Anoano, 99
Anopheles
    gambiae, 744
Anoplolepis
    longipes, 702, 703
Antestia, 73, 74, 81, 88, 109, 466, 467
    clymeneis (diff. comb. of *Antestiopsis clymeneis*), 469
    clymeneis var. *flaviventris* (syn. of *Antestiopsis*
      clymeneis), 469
    confusa (syn. of *Antestiopsis clymeneis*), 469
    cruciata (diff. comb. of *Antestiopsis cruciata*), 469
    faceta (misidentification of *Antestiopsis intricata* and
      A. facetoides), 468, 469
    intricata (diff. comb. of *Antestiopsis intricata*), 468
    lineaticollis (syn. of *Antestiopsis thunbergii*), 620
    trispinosa, 197
    variegata (misidentification of *Antestiopsis*
      facetoides), 469
Antestia bug, 466–472, 477–484, 486–489
Antestia group, 100
Antestiaria, 30, 73

Antestiini, 50, 62, 66, 73–75, 77, 81, 84, 88, 91, 94, 98–100, 106, 109, 117, 188, 197, 504, 680, 696, 699

Antestiopsis, ix, x, 74, 466–470, 478, 479, 487, 620, 621
   anchora, 188, 470
   clymeneis, 469
   clymeneis frappai, 469
   clymeneis galtiei, 469
   cruciata, 197, 467, 469, 487
   crypta, 470
   faceta (syn. of *Antestiopsis thunbergii*), 467
   facetoides, 468–470, 478
   falsa, 470
   intricata, 468–470, 472, 474–476, 478–485, 487, 488
   lepelleyi, 470
   lineaticollis (syn. of *Antestiopsis thunbergii*), 467, 468
   lineaticollis intricata (syn. of *Antestiopsis intricata*), 468
   littoralis, 470
   orbitalis (syn. of *Antestiopsis thunbergii*), 467
   orbitalis orbitalis (syn. of *Antestiopsis thunbergii*), 467
   thunbergii, 467, 468, 470–473, 475–480, 484–488, 620, 648, 655
   thunbergii bechuana, 468–470, 478, 486, 488
   thunbergii ghesquierei, 468, 470–472, 476–479, 484–486, 488
   transvaalia, 470
   variegata (syn. of *Antestiopsis thunbergii*), 469
Antheminia, 77
Anthocoridae, 266, 274, 309, 372, 439, 520
Anthonomus
   grandis, 399
   grandis grandis, 734, 738, 746
Antilochus
   coquebertii, 30, 308, 652
   nigripes, 652
Antiteuchus, 58
   costaricensis, 648, 655
   mixtus, 53, 186
   tripterus, 625
Antlion, 645
Ants, 31, 58, 61, 82, 92, 127, 271, 337, 341, 645, 663, 702, 732, 754
Apateticus, 56, 598
   bracteatus (diff. comb. of *Apoecilus bracteatus*), 570
   crocatus (syn. of *Apoecilus bracteatus*), 570
   cynicus (diff. comb. of *Apoecilus cynicus*), 502, 504, 570
Aphelinidae, 307, 308
Aphelocheiridae, 41, 501
Aphid, 5, 225, 229, 271, 300, 343, 613, 614, 617, 644, 645, 655, 657, 663, 666, 733, 756, 757
Aphididae, 300, 740
Aphis
   gossypii, 756, 757
Aphylinae, 28, 32, 34, 44, 49–51, 54, 126, 136, 186, 196
Aphylum, 54
   syntheticum, 52, 186, 196
Apines, 73, 88, 89

Apis
   mellifera, 707
Aplerotus, 82
Apodiphus, 86
   amygdali, 190
   robustus, 111
Apoecilus, 56, 598
   bracteatus, 570, 596
   crocatus (syn. of *Apoecilus bracteatus*)
   cynicus, 13, 502, 504, 568, 570, 596–598
Apple maggot, 734
Aprostocetus, 478
Aradidae, 43, 127, 501, 678
Aradoidea, 29
Araducta, 106
Araneida, 266
Arawacoris, 81
Archaeoditomotarsus, 37
Arctiidae, 590, 592
Aridelus, 483
   coffeae, 478
   egregius, 571, 572
   rufus, 478
   taylori, 478
Arilus, 628
   cristatus, 337, 342
Arma, 55, 56, 598
   chinensis, 266, 267, 575
   contusa, 57
   custos, 372, 504, 511, 515, 529, 532, 536, 574, 575, 577, 596
Arminae, 124
Armini, 55, 56
Armiventres, 30
Arniscus, 90, 103
   humeralis, 191
Arocatus
   longiceps, 651, 655
Arocera, 79, 80, 94, 110
   apta, 198
Arrow-headed bugs, 111
Artheneidae, 651, 655
Arvelius, 80, 81, 107, 455
   albopunctatus, x, 11, 64, 81, 456, 459, 460, 478, 626
   porrectispinus, 648
Asaroticus, 123, 124
Ascra, 59, 60
Asian citrus psyllid, 757
Asian lady beetle, 313
Asilidae, 274
Asolcus
   mopsus, 478
   suranus, 478, 483
Asopida, 48
Asopidae, 48
Asopides, 30, 55, 56
Asopina, 30, 56
Asopinae, 27, 30, 32, 47, 48, 50, 51, 54–57, 61, 100, 103, 104, 122, 124, 125, 136, 186, 196, 499, 504, 567, 568, 570, 597, 598, 645, 680, 684, 707
Asopini, 55, 56

Asopus, 55, 56
  argus, 56
  gibbus, 55
  puncticollis, 57
Asopus group, 100
Aspavia, 84, 87, 105
  armigera, 190
  hastator, 484
Aspidestrophus, 123
Aspideurus, 88
Aspongopus, 42
Assassin bug, 228, 337, 478, 645
Astata
  occidentalis, 704, 707
Atelides, 42
Atelocera, 85, 88
  serrata, 199
Auchenorrhyncha, 4, 111, 256, 612–614, 616
Augocoris, 14, 129, 131
Aulacetraires, 76
Aulacetrini, 77
Aulacetrus, 76, 77
Aurungabada, 134
Australojalla, 56
Auxentius, 85, 86
Avicenna, 96
Axiagastaria, 30
Axiagastini, 50, 67, 74, 86, 89, 97, 197
Axiagastus, 67, 74, 75, 86, 97
  cambelli, 75, 197

B

Babylas, 85, 86
Bachesua, 101, 102
Bactericera
  cockerelli, 756
Bagrada, 98
  cruciferarum (syn. of *Bagrada hilaris*), 207, 506
  hilaris, x, 5–7, 9, 10, 99, 201, 205–231, 506, 508,
      566, 681, 682, 710, 744, 746, 750, 751
  pictus (syn. of *Bagrada hilaris*), 207
Bagrada bug, 10, 206, 208, 506, 566, 681, 744
Bakerorandolotus, 114
Banasa, 87, 93, 94, 107, 110
  patagiata, 200
Bannacoris, 128
  arboreus, 129, 194
Banya, 106
  leplaei, 106
Banya group, 100, 106
Baryconus, 478
Basicryptus, 114
  distinctus, 35, 52, 196
Bathycoelia, 64, 75, 107, 109, 188
  alkyone, 197
  distincta, 76
  horvathi, 65, 75
  natalicola, 76
  thalassina, 76
Bathycoelia group, 100

Bathycoeliaria, 30, 75
Bathycoeliini, 50, 62, 75, 100, 107, 109, 188, 197
Bean bug, 353, 657
Bean plataspid, 7, 294
Bebaeus
  punctipes, 184
Bed bug, 55, 645, 702, 743
Beet armyworm, 343
Beewolf, 645
Belonochilus
  numenius, 651, 655
Belopis, 72
  unicolor, 197
Bemisia
  tabaci, 756
Benia, 76, 105, 106
Bergrothina, 73
Berry bug, 505, 537
Berytidae, 651, 655, 678
Bicyrtes
  quadrifasciata, 337
Bifurcipentatoma, 94
Biprorulus
  bibax, 96, 506, 528, 625, 681
Birch bug, 503
Birketsmithia, 87
Birna, 104
  griggae, 182
Bishop's mitre shield bug, 504, 526
Blachia, 55, 56
Black bug, 4
Black rice bug, 504, 510
Blackheaded fireworm, 740
Blaudinae, 37
Blaudusinae, 32, 37, 38, 184, 194
Blaudusini, 32, 37, 184
Blissidae, 651, 654, 655, 656, 658, 678
Blissus
  insularis, 651, 658
  leucopterus leucopterus, 739
Blue shield bug, 504
Bochartia, 228
Boea, 79, 96
Boerias, 76, 105
  victorini, 188
Bogosia, 267, 479, 483, 486
  rubens, 479, 483, 484, 486
Boisea
  trivittata, 753
Bolaca, 72, 86
  unicolor, 73
Bolbocoris, 121–124
  inaequalis, 187
  variolosus, 121
Boll weevil, 399, 440, 734, 738, 746
Bonacialus, 69, 70, 78, 83, 84, 102
Borrichias, 114
Boxelder bug, 753
Brachycerocorini, 116
Brachycerocoris, 121–124
  camelus, 119

Brachycerocoris group, 50, 118, 121–124
Brachycoris, 88
Brachyhalyini, 107
Brachyhalys, 107
    pilosum, 107
Brachymna, 98
    tenuis, 201
Brachynema, 91
    germari, 91
Brachyplatidae, 126
Brachyplatidinae, 127
Brachyplatys, 127
    hemisphaericus, 185
    subaeneus, 126, 295, 649, 658, 662
    vahlii, 126, 295, 649
Brachyplatys group, 127, 128, 185
Brachystethus, 59, 60, 106, 107
    geniculatus, 52, 187
Braconidae, 227, 478, 572
Brasilania, 107, 110
Brassica bug, 506
Braunus, 54, 79, 530
Brepholoxa, 95
Brévirostres, 30
Breviscutum, 128
Brizica, 85, 86
Brochymena, 54, 85, 86, 753
    quadripustulata, 190
Bromocoris, 85
Brontocoris, 598
Bronze orange bug, 508
Brown marmorated stink bug, ix, 7, 10, 244, 246, 248, 313, 397, 506, 584, 587, 590, 591, 688, 745, 751
Brown planthopper, 369, 756
Brown stink bug, 397, 428, 505, 745, 751
Brown-winged green stink bug, 691, 696
Bulbostethus, 56, 597
Burrower bug, 3, 41, 531, 538, 539
Burrus, 119
Byrsodepsini, 32, 43, 185
Byrsodepsus, 185

C

Cabbage bug, 339, 506, 515
Cabbage moth, 584
Cabbage worm, 339
Cachaniellus, 107
Cacoschistus, 111
Cahara, 86
Calagasma, 78, 164
Calidea, 130, 131
Calliphara, 130
    caesar, 195
Callostethus, 58
Canalirostres, 30
Candeocoris, 59
Canoparia, 38, 135
Canopidae, 27, 32, 34, 38, 39, 46, 126, 130, 135, 136, 185
Canopides, 30

Canopus, 38, 39, 135
    impressus, 185
    obtectus, 38
Cantao, 130, 131
    ocellatus, 649, 655
Cantharidae, 274
Canthecona, 55, 56, 597
Cantheconidea, 597
Canthophorus
    niveimarginatus, 40, 538, 647
Caonabo, 70
Cappaea, 76, 106
    halys, 243
    taprobanensis, 197, 624
Cappaearia, 30, 76
Cappaeini, 50, 67, 76, 78, 84, 96, 105, 106, 109, 134, 188, 197, 506, 680, 688
Carabidae, 274, 565
Carbula, 84, 87, 105, 110, 112
    humerigera, 502, 505, 511, 522, 537, 578, 597
    putoni, 190
Carbula group, 100, 105
Carenoplistus, 85, 86, 88
Caribo, 79
Caridophthalmus, 82, 92, 182, 190
Carpenter ant, 645, 655, 663
Carpocoraria, 30, 77
Carpocorates, 76
Carpocorini, 28, 50, 59, 61, 66, 69, 70, 72, 76–79, 82–84, 87, 93, 100, 101, 104, 105, 107–110, 117, 182, 188, 189, 198, 505, 597, 680, 681, 684, 693, 700
Carpocoris, 76, 77, 79, 112
    purpureipennis, 65, 198
Carpocoris group, 100, 104
Carpocoroides, 111, 112
Catacantharia, 30, 79
Catacanthini, 47, 50, 62, 65, 79–81, 93, 94, 96, 107, 110, 189, 198
Catacanthus, 47, 48, 60, 74, 79, 81
    incarnatus, 64, 80, 198
Catadipson
    aper, 184
Catalampusa, 107
    oenops, 107
Cataulax, 58, 109
Caternaultiella
    rugosa, 127, 528, 529
Catulona
    lucida, 187
Caura, 76, 105
Cavelerius
    saccharivorus, 651
Caystrini, 50, 58, 67, 70, 80, 86, 89, 90, 96, 100, 104, 182, 188, 198
Caystrus, 80, 104
    quadrimaculatus, 198
Caystrus group, 100
Cazira, 56
    horvathi, 196
Cenaeus
    carnifex, 652

Cephalocteinae, 32, 40, 41, 184
Cephalocteini, 32
Cephaloplatus, 58, 104
    granulatus, 188
Cephaloplatus group, 100, 104, 112
Cerambycidae, 708
Ceratitis
    capitata, 744
Ceratopogonidae, 704
Cerotoma
    ruficornis, 620
Chalcididae, 227
Chalcocoris, 74, 79
Chalcopis, 114
Chilacis
    typhae, 651, 655
Chiloxanthus
    pilosus, 501
Chinavia, 91, 427, 623
    aseada, 703
    collis, 199
    hilaris, 353, 368, 397, 399, 428, 429, 436, 437, 439, 505,
        508, 511, 578, 614, 620, 622, 648, 658, 660, 661,
        680, 682, 683, 709, 750, 751
    impicticornis, 682, 701
    marginata, 680
    pensylvanica, 680
    ubica, 682, 701, 711
    viridans, 64
Chinch bug, 739
Chinche verde, 353
Chlorochroa, 91, 191
    belfragei, 182
    ligata, 4, 6, 623, 648, 655, 661, 683, 710, 711
    sayi, 648, 661, 683, 711, 712, 751
    uhleri, 623, 641, 683, 710, 711
Chlorocorini, 50, 66, 80, 107, 110, 189, 190, 198
Chlorocoris, 80, 81
    complanatus, 198
Chloropepla, 80, 81, 110
Chloropid, 706
Chraesus, 70, 71
Chrysocoris
    purpureus, 507
Chrysodarecus, 108, 110
Chrysodeixis
    includens, 317, 741
Chrysomelid beetle, 56
Chrysomelidae, 592, 595, 708
Chrysoperla
    rufilabris, 309
Chrysopidae, 274, 309, 505
Cimex, 55
    anchora (diff. comb. of *Antestiopsis anchora*), 467
    cribraria (diff. comb. of *Megacopta cribraria*), 295
    cruciata (diff. comb. of *Antestiopsis cruciata*), 467, 469
    facetus (syn. of *Antestiopsis thunbergii*), 467
    lectularius, 702, 743
    olivaceus (syn. of *Antestiopsis thunbergii*), 467
    smaragdulus (syn. of *Nezara viridula*), 354, 361
    spirans (syn. of *Nezara viridula*), 366

    thunbergii (diff. comb. of *Antestiopsis thunbergii*), 467
    variegatus (syn. of *Antestiopsis thunbergii*), 467
    viridulus (diff. comb. of *Nezara viridula*), 354
Cimicidae, 645, 702
Cimicina, 30
Cimicinae, 55
Cimicomorpha, 613, 645
Cladrastis, 252
    kentukea (syn. of *Cladrastis lutea*), 252
Clavicorinae (syn. of Amnestinae), 41
Clavicoris, 41
Cleoqueria, 81
Cletus
    punctiger, 398, 650
    rusticus (syn. of *Cletus punctiger*)
    trigonus, 57, 650
Clown stink bug, 507, 585, 586
Cnephosa, 78
Cocalus, 89
Coccinellidae, 274, 309, 313, 439, 748
Cochliomyia
    hominivorax, 742, 745
Coctoteris, 85
Codling moth, 733
Coenus, 77, 79, 84
Coffee berry borer, 12, 467, 487
Coleoptera, 9, 41, 89, 229, 313, 399, 487, 565, 592, 595,
    708, 746
Coleorrhyncha, 4
Coleotichus
    borealis, 195
Collops
    vittatus, 228, 229
Colorado potato beetle, 517, 592, 707, 733–735
Colpocarena, 58, 182, 196
    complanatus, 196
Commius, 79
Compastaria, 30, 96
Compastes, 96
Coniscutes, 30, 42, 48
Conquistador, 598
Consperse stink bug, 505, 684
Contarinia
    sorghicola, 738, 739
Cooperocoris, 101
Coponia, 83
Coptosoma, 35, 126, 127, 748
    biguttulum, 126
    cribraria (syn. of *Megacopta cribraria*)
    duodecimpunctatum (diff. comb. of *Paracopta*
        duodecimpunctatum), 126, 295
    eocenica, 127
    japonicum, 649
    lyncea, 127
    mucronatum, 502, 507, 528, 571, 572
    parvipictum, 649, 658
    punctatissimum (diff. comb. of *Megacopta*
        punctatissima), 295
    scutellatum, 127, 185, 502, 507, 528, 529, 537, 571, 572
    sphaerula, 649
    xanthogramma, 126, 295

Coptosoma group, 127, 185
Coptosomatidae, 126
Coptosomatinae, 127
Coquerelia, 79, 81
    pectoralis, 198
Coquereliaria, 81
Coquerelidea, 81
Coquereliini, 50, 67, 80, 81, 198
Coracanthella, 123
Coreidae, 138, 205, 307, 613, 616, 618, 623, 624, 626, 650, 655
Coreoidea, 29, 645, 649, 654, 656, 658, 678, 696
Coridius, 42, 44
    chinensis, 44
    janus, 702, 703
    nepalensis, 44
    sanguinolentus, 194
    viduatus, 44, 45
Corienta
    transbaicalica, 47
Corimelaena, 39, 135, 137
    lateralis, 16, 136, 508
    lateralis lateralis, 508
    obscura, 508
    pulicaria, 137
Corimelaenidae (syn. of Thyreocoridae), 135, 136, 508
Corimelaeninae, 33, 39, 135–137
Corioxenidae, 479
Corioxenos
    antestiae, 479, 483, 484, 486
Corisseura, 84, 107
Corn earworm, 738
Corn planthopper, 756
Corn rootworm, 738, 741, 756
Cornegenapsylla
    sinica, 617
Coryzorhaphis, 56
Cosmopepla, 79
    bimaculata (syn. of *Cosmopepla lintneriana*)
    lintneriana, 622
Cosmopolitan stink bug, 353
Cotton aphid, 756
Cotton fleahopper, 370
Cotton green bug, 353
Cottony cushion scale, 734
Crab spider, 707
Crabronidae, 274
Cranberry fruitworm, 740
Crathis
    ansata, 195
Cresphontes, 88
Cressona, 114
Cressonini, 50, 109, 113–115
Cretacoris, 41
Cricket, 275
Critheus, 96, 97
    indicus, 97
    lineatifrons, 97
Crollius, 122, 123
Crypsinus, 123, 124
Cryptogamocoris, 121
    cornutus, 121

Curatia, 79
Curculionidae, 399, 708, 738, 746
Cuspicona, 91, 96, 106
    antica, 91
    simplex, 192
Cyclogastrina, 30
Cyclopelta
    parva, 44
Cydia
    pomonella, 733
Cydnidae, 3, 6, 7, 12, 15, 16, 28, 31, 32, 38–41, 43, 47, 126, 130, 135, 136, 501–503, 538, 568, 647, 654, 655, 658, 678, 681, 699
Cydnides, 30, 39
Cydnina, 30
Cydninae, 30, 32, 40, 41, 184, 194, 503
Cydnini, 32, 194
Cydnopsis
    affinis, 41
    nigromembranacea, 41
    ventralis, 41
Cydnus
    aterrimus, 194, 503
Cylindromyia
    fumipennis, 704
Cymidae, 678
Cyphostethus, 37
Cyptocephala, 108, 110
Cyptocoris, 57, 121, 136, 186, 196
Cyrtocoridae, 38
Cyrtocorinae, 28, 32, 49–51, 57, 136, 186, 196
Cyrtocoris
    egeris, 57
    gibbus, 186, 196
    trigonus, 57, 650
Cyrtomenus
    mirabilis, 35, 42, 184
Cyrtopeltis
    nicotianae, 617

D

Dabessus, 88, 106
Daclera
    levana, 649
Dalsira group, 114, 115
Dalleria, 84
Dalpada
    brevis (syn. of *Halyomorpha halys*), 243, 506
    brevis (var. *remota*) (syn. of *Halyomorpha halys*), 243
    remota (syn. of *Halyomorpha halys*), 243
Dalsira, 114, 115
Dalsira group, 114, 115
Damarius, 597
Dandinus, 101, 117, 123
Dardjilingia, 87, 108
Dasynus
    coccocinctus, 650
Decellella, 88
Degonetini, 50, 68, 81, 198

Degonetus, 81
    serratus, 81, 198
    sikkimensis, 81
Delegorguella
    ventralis, 191
Delocephalaria, 114
Delocephalini, 113–115
Delocephalus, 113, 114
    miniatus, 115
Dendrocoris, 95
    humeralis, 192
Deroplax, 130
    circumducta, 195
Deroploa, 123
    parva, 187
Deroploa group, 50, 117, 118, 120, 122, 123
Deroploini, 116
Deroploopsis, 123
    trispinosus, 119
Derula, 123
Deryeuma, 112
Desmometopa, 706
Diabrotica, 741
    virgifera virgifera, 738, 744
Diamondback moth, 343, 713, 740
Diaphorina
    citri, 757
Diaphyta, 96
Dichelops, 79
    furcatus, 12, 188
    melacanthus, 12, 505, 535, 661, 701, 703
Dichelorhinus, 114
Dicranocephalus
    agilis, 650
    albipes, 650
    lateralis, 651
    medius, 651
Dictyotus, 102, 103
    caenosus, 191
Dictyotus group, 100, 102, 103, 109
Diemenia, 82
    rubromarginata, 198
Diemenia group, 100, 103
Diemeniini, 50, 61, 82, 92, 100, 103, 111, 182, 190, 198
Dimorphopterus
    pallipes, 651, 655
Dindymus
    lanius, 652
Dinidor, 42, 43, 44
    mactabilis, 36, 652
Dinidorida, 42
Dinidoridae, 28, 32, 34, 42–44, 125, 133, 134, 136, 666, 703
Dinidorina, 30
Dinidorinae, 30, 32, 34, 43, 44, 194
Dinidorini, 32, 194
Dinidorites
    margiformis, 43
Dinocoris, 58
    lineatus, 186
Diplorhinus, 114
    furcatus, 187

Diplostira, 67, 82, 83
    valida, 82, 198
Diplostiraria, 30
Diplostirini, 50, 67, 82, 88, 94, 198
Diploxyaria, 30, 83
Diploxyini, 50, 68–71, 78, 83, 84, 92, 100, 102, 105, 190, 198
Diploxys, 83
    cordofana, 83
    fallax, 198
    orientalis, 83
    pulchricornis, 83
    punctata, 190
Diploxys group, 71, 100, 105
Dipsocoromorpha, 645
Diptera, 137, 227, 229, 401, 402, 499, 568, 569, 572, 625, 704, 706, 747, 749, 757
Discimita, 58, 90
Discocephala, 57
Discocephalidae, 48, 57
Discocephalina, 30, 48
Discocephalinae, 28, 30, 32, 48–50, 53, 57–59, 104, 108, 109, 145, 182, 183, 186, 187, 196, 626
Discocephalini, 50, 58, 59, 104, 182, 183, 186, 196
Discocera, 55, 56
    cayennensis, 55
Discocerini, 55
Discocoris
    drakei, 508
Disderia, 93, 107
Dismegistus, 39, 47, 48
    fimbriatus, 194
Dissolcus
    tetartus (misidentification of *Paratelenomus saccharalis*), 307
Ditomotarsinae, 32, 37, 38, 184
Ditomotarsini, 32, 37, 184
Ditomotarsus
    punctiventris, 184, 426, 437
Doesburgedessa, 60
Dolycorini, 77
Dolycoris, 72, 77
    baccarum, 35, 188, 505, 511, 526, 530, 537, 580–582, 584, 587, 622
    penicillatus, 528
Dorpiaria, 30
Dorpius (syn. of *Myrochea*), 103
Dorycoris, 56, 597
Drinostia, 98
Dryptocephala, 57, 58, 104
    spinosa, 52
Duadicus, 37
Dunnius, 89, 106
Durmia, 84, 85, 105, 110
Dusky stink bug, 505, 531, 751
Dybowskyia, 121, 123, 124
    reticulata, 119, 121, 504, 515, 520, 527, 539, 577, 590
Dymantaria, 30
Dymantis, 69, 70, 89, 90, 102, 112
Dymantiscus, 112

Dyroderes, 98
    umbraculatus, 193
Dysdercus
    andreae, 653
    cingulatus, 653
    decussatus, 653
    fasciatus, 653, 657, 658, 666
    olivaceus, 457, 653
    poecilus, 653
    suturellus, 621, 653
    voelkeri, 653
Dysnoetus, 101

E

Ealda, 83, 597
Earwig, 275
Ebony bug, 4
Ectenus, 85, 86
Edessa, 59, 60
    bugabensis, 648, 655
    eburatula, 648
    irrorata, 648
    jugata, 648
    junix, 648
    loxdalii, 626
    meditabunda, 12, 661, 666, 680, 683–685, 699, 701,
        710
    nigropunctata, 60
    protera, 60
    rufomarginata, 52, 196, 701
Edessidae, 48
Edessides, 30, 59
Edessinae, 28, 32, 34, 47, 49–51, 54, 59, 60, 93, 96, 106,
        107, 109, 187, 196, 680, 683, 701
Edessini, 59
Egg parasitoid, 7, 227–231, 266, 267, 275, 278, 307, 308,
        339, 366, 402, 430, 438, 439, 478, 483, 484,
        486, 488, 704–706, 708, 712, 748, 749
Eipeliella, 106, 117
Eipeliella group, 100, 106
Elanela, 88, 107
    hevera, 191
Elasmostethus, 37
    atricornis, 503
    brevis, 647
    humeralis, 528, 647, 658
    interstinctus, 521, 538
    ligatus placidus, 35, 36, 184
    nubilus, 647, 658
Elasmucha, 37
    dorsalis, 647
    grisea, 538, 539
    grisea grisea, 503
    lateralis, 503
    putoni, 646, 647
    signoreti, 647
Elemana, 86
Elsiella, 93
Eludocoris, 81
Elvisuraria, 30

Elvisurinae, 33, 130, 131, 195
Empiesta, 105
Empoasca
    fabae, 740
Encarsia, 308
Encarsiella
    boswelli, 307, 308
Encosternum, 185
    delegorguei, 134, 508
Encyrtidae, 227, 231, 275, 307, 308, 401, 478, 704, 713, 748
Engytatus
    nicotianae, 617
Enicocephalomorpha, 645
Ennius, 89
Eobanus, 122
Eocanthecona, 56, 597
    furcellata, 680, 684, 701, 707
Eonymia, 114
Ephynes, 132
Epipedus, 78
Epitoxicoris, 85, 86
Erachtheus, 89
Eretmocerus, 308
Eribotes, 69
Eriosoma
    lanigerum, 733
Erthesina, 85
    fullo, 190, 528, 531
Euclytia
    flava, 275, 704, 706
Eucorysses
    grandis, 649, 655
Eudryadocoris
    goniodes, 188
Eufrogattia, 123
Eulophidae, 478
Eumenotes, 43, 44
    obscura, 45, 185
Eumenotidae, 43
Eumenotini, 32, 43, 44, 185
Eupelmidae, 275, 478, 748
Euphorinae, 572
Eupolemus, 37
Eurinome, 78
European corn borer, 738
Eurus, 86
Euryasparia, 30, 83
Euryaspisaria, 30
Eurydema, 61, 98, 99, 112
    dominulus, 648, 661
    gebleri rugosa, 648, 655, 658
    oleracea, 506
    ornata, 201
    rugosa, 506, 515–517, 526, 587, 613, 661
    rugosum (misspelling of *Eurydema rugosa*), 506, 515, 587
    ventralis, 701
Eurydemaria, 30, 98
Eurydemini, 71, 98, 100, 506
Eurygaster, 130, 131
    integriceps, 507, 520, 528, 573, 749
    maura, 45, 528

Eurygastraria, 30
Eurygastridae, 129
Eurygastrides, 30, 129
Eurygastrinae, 33, 130, 132, 195, 507
Eurygastrini, 33
Eurysaspidini, 83
Eurysaspini, 50, 64, 75, 83, 95, 100, 106, 107, 190, 198
Eurysaspis, 83, 84, 106, 198
Eurysaspis group, 100
Eurystethus, 58, 186
   microlobatus, 58
Eusarcocoriaria, 30
Eusarcorini, 84
Euschistus, 61, 77, 79, 94, 111, 403, 539, 620, 622, 648, 706
   biformis, 702, 703
   conspersus, 276, 505, 531, 624, 680, 684, 703, 704,
      712, 750–752
   heros, 12, 189, 402, 436, 437, 505, 528, 535, 539,
      648, 655, 661, 680, 681, 683–685, 699, 701,
      703–705, 708, 710–712
   ictericus, 505, 680
   obscurus, 680, 681, 684
   politus, 680
   servus, 17, 248, 397, 399, 428, 429, 436, 437, 505, 531,
      614, 625, 684, 685, 745, 750–752
   servus euschistoides, 622
   servus servus, 622
   strenuus, 79
   tristigmus, 622, 680, 703, 751, 752
   tristigmus tristigmus, 505, 531, 535
   variolarius, 622, 751
   zopilotensis, 79
Euschistus group, 93, 117
Eusthenaria, 30
Eusthenina, 30
Eusthenini, 133
Euthera
   tentatrix, 439, 704
Euthyrhynchus, 57
   floridanus, 309, 502, 504
Eutrichopodopsis
   nitens, 439
Everardia
   picta, 193
Evoplitus, 93, 107
   humeralis, 63, 200
Evoplitus group, 62, 96
Exithemus, 80, 96, 97, 106
Eysarcocorini, 84
Eysarcoraires, 84
Eysarcoraria, 30
Eysarcoriaria, 30
Eysarcoriens, 84
Eysarcorini, 50, 57, 66, 69, 76, 83, 84, 86, 87, 89, 100,
      104–107, 110, 190, 199, 505, 597, 680, 685
Eysarcoris, 84, 85, 117
   aeneus, 53, 64, 199, 505
   lewisi, 505, 511, 525, 530, 680, 685–688
   mammillata, 57

   parvus, 680
   ventralis, 505, 520, 590
Eysarcoris group, 100, 104

F

Faizuda, 86
Fall armyworm, 741
Fecelia, 81
Fiarius, 597
Fire ant, 337, 341
Fire bug, 526, 529
Flaminia, 64, 83, 84
   natalensis, 190
Forcipomyia
   crinita, 704
Forest bug, 506, 572
Forficulidae, 274
Forget-me-not shield bug, 503
Formicidae, 9, 255, 309, 439, 702, 703
Frisimelica, 114

G

Galedanta, 59, 79
Galgupha, 35, 131, 135
   difficilis, 136
   ovalis, 508
Gambiana, 123
Garsauriinae, 32, 40, 42
Gastropod, 274
Gellia, 114
Geocoridae, 274, 309, 439, 678
Geocoris, 439
   punctipes, 309
   uliginosus, 309
Geomorpha, 70, 71
Geopeltus, 39
Geotomini, 32, 184
Gerridae, 228
Gerromorpha, 531, 645
Glaucias, 91
   subpunctatus, 265, 698, 709, 710
Glaucioides, 107
   englemani, 192
Globular stink bug, 294
Globuleux, 30, 48
Glossina
   austeni, 746
   morsitans morsitans, 744
Glottaspis, 86
Glyphepomis, 79, 117
Goilalaka, 58, 86
Gomphocranum, 69
Gonatocerus 308
Gonopsimorpha, 114
Gonopsis, 114
Gorse shield bug, 437
Grain aphid, 756

Grammedessa, 59, 60
Grapholita
    molesta, 752
Graphosoma, 117, 120, 123, 596
    lineatum, 119, 120, 196, 503, 508, 511, 521, 525, 526,
        532, 576, 577, 587
        rubrolineatum, 120, 397, 503, 577, 590
Graphosoma group, 50, 118, 120–123, 196
Graphosomaria, 30
Graphosomatinae, 30, 102, 106, 117
Graphosomatini, 116, 121, 122, 503, 539
Graphosominen, 117
Grazia, 107
Green bug, 353
Green cabbage bug, 506
Green peach aphid, 343, 506
Green plant bug, 353, 506
Green plant-bug, 353
Green shield bug, 505, 529, 572, 573
Green soldier bug, 353
Green soldier-bug, 353
Green stink bug, 353, 426, 428, 505, 506, 585, 751
Green vegetable bug, 353
Green-bug, 353
Green-bug of India, 353
Gryllidae, 274, 565
Gryon, 275, 439
    fulviventre, 478, 483–485
    gonikopalense, 230, 231
    karnalensis, 227
Gulielmus, 69, 70, 78, 79, 83, 84, 102
Gwea, 106
Gymnoclytia
    occidentalis, 704
Gymnosoma
    par, 704
    rotundatum, 704
Gyndes
    pallicornis, 652
Gynenica, 87, 105
    funerea, 199
Gypsy moth, 733

H

Hadronotus
    antestiae (syn. of *Gryon fulviventre*), 478
Hadrophanurus, 227
Hairy shield bug, 505
Halyabbas, 72, 73
    unicolor, 72, 73
Halyaria, 30, 112
Halydes, 30, 60, 85
Halydidae, 48
Halyini, 50, 58, 62, 66, 67, 71, 80, 82, 85–88, 90, 91, 94, 100,
    101, 103, 104, 107, 110, 112, 190, 191, 199
Halynoides, 111, 112
Halyomorpha, 76, 105, 106, 134, 246
    brevis (syn. of *Halyomorpha halys*), 246

halys, x, 7, 9, 10, 76, 197, 243–278, 308, 313, 397, 400,
    506, 511, 521, 525, 528–531, 539, 587, 590–592,
    618, 619, 625, 648, 657–659, 661, 666, 688–693,
    698, 702, 707, 709–712, 745, 746, 749–754
    mista (syn. of *Halyomorpha halys*), 244, 618
    picus (misidentification of *Halyomorpha halys*), 246,
        618
    remota (syn. of *Halyomorpha halys*), 246
    timorensis (misidentification, syn. of *Halyomorpha*
        halys), 244
Halyomorpha group, 76, 100, 105
Halys, 86, 112
    dentatus (syn. of *Halys fabricii*)
    fabricii, 528, 531
    sulcatus, 199
Halys group, 100, 103, 104
Harlequin bug, 10, 208, 228, 334, 341, 342, 506, 566, 692,
    744, 751
Harlequin cabbage bug, 339
Harmonia
    axyridis, 313, 439
Harpactor
    segmentarius (diff. comb. of *Rhynocoris*
        segmentarius), 228
Harpegnathos
    saltator, 702, 703
Haullevillea, 123
Hawthorn shield bug, 503
Hediocoris
    fasciatus, 478
Helicoverpa
    armigera, 744
    zea, 399, 738, 739, 742, 750, 755
Heliothis
    virescens, 399, 750, 755
Hellica, 37
Hemiptera, 4, 9, 29, 37, 229, 256, 266, 267, 612–615, 618,
    622, 623, 625, 629, 631, 632, 662, 678, 702
Hemitrochostoma
    lambirense, 127
Hemyda
    aurata, 704, 706
Hermolaus, 85
Hessian fly, 733
Heteroptera, xiii, 4, 7, 27, 29, 41, 54, 79, 111, 127,
    129, 137, 138, 227, 294, 307, 358, 369, 372,
    398, 427, 478, 498–502, 509, 520–522, 527,
    530, 531, 538, 539, 573, 593, 612–617, 624,
    625–626, 629, 645, 659, 661, 662, 666, 667,
    678, 696, 699, 703, 742, 743, 749
Heteroscelis, 56
Hexacladia, 227
Hippodamia
    convergens, 309, 439
Hoffmanseggiella, 96
Holcogaster, 77, 78
    exilis, 189
    fibulata, 77
Holcogastraria, 77

Holcostethus, 77, 79
  limbolarius, 189, 622
Homalogonia, 96, 109
Homoeocerus
  dilatatus, 650
  marginiventris, 650
  unipunctatus, 650
Homopterulum, 111
Hondocoris, 59
Honey bee, 707
Hoplistodera, 86
  pulchra, 199
Hoplistoderaria, 30, 86
Hoplistoderini, 50, 67, 84, 86, 199
Hoploxys, 56, 597
Horridipamera
  inconspicua, 652
  nietneri, 652
Hoteinae, 33, 130, 132, 195
House fly, 518
Humria, 90
Hyalomya
  pusilla, 227
Hybocoris, 123, 124
Hygia
  lativentris, 650
  opaca, 650
Hyllaria, 30, 73
Hyllini, 73
Hyllus, 73
Hymenarcys, 79
Hymenomaga, 105
Hymenoptera, 7, 9, 90, 137, 228, 229, 255, 337, 339, 345, 401, 402,
      430, 438, 488, 529, 565, 572, 702–704, 707, 712, 713
Hypanthracos, 79
Hypatropis, 79
Hypera
  postica, 740
Hyphantria
  cunea, 590, 592
Hypocreales, 267
Hypogomphus, 110
Hypothenemus
  hampei, 487
Hypsithocus, 78
Hyrmine, 79
  sexpunctata, 192

I

Icerya
  purchasi, 734
Idiostoloidea, 29
Indrapura, 74
  klossi, 74
Inermiventres, 30
Iphiarusa, 83, 93
Ippatha, 101, 102, 117
  angustilineata, 193

Ippatha group, 100, 102
Irochrotus
  lanatus, 572
  sibiricus, 195
Ischnodemus
  sabuleti, 501, 651, 655
Ischnopelta
  luteicornis, 186
Isoptera, 9
Italian striped bug, 503, 521, 575, 576
Izharocoris, 110

J

Jalla, 55, 56
Jallina, 56
Jallini, 55, 56, 504
Jalloides, 56
Janeirona, 93, 94, 107
Jascopidae, 111
Jayma, 114
Jeffocoris, 120, 123
Jostenicoris, 71
Jugalpada, 86
Jumiles, 79
Jumping ant, 702
Jurtina, 75, 109

K

Kaffraria, 114
Kafubu, 113, 114
Kahlamba, 78
Kalkadoona, 82
Kamaliana, 78
Kapunda, 97, 98, 103
Kapunda group, 100, 102, 103
Katongoplax, 114
Kayesia, 120, 122, 123
  parva, 45, 120
  setigera, 120
Kayesiina, 116, 122
Kelea, 106
Kelea group, 100, 106
Keleacoris, 88, 106
  congolensis, 106
Keriahana, 89
  elongata, 89
  bisignata, 88
Kermana, 107
Kitsonia, 101
Kitsonia group, 100, 101
Kitsoniocoris, 101
Kleidocerys
  resedae, 646, 651, 655
Kudzu bug, 7, 10, 294, 507, 644, 662, 664, 746, 748,
      749
Kumbutha group, 100–102
Kundelungua, 122

L

Lablab bug, 294, 316
Laccophorellini, 32, 37, 184
Lace bug, 618, 733
Ladeaschistus, 79
Lagynotomus
    elongatus (diff. comb. of *Niphe elongata*), 398
Lakhonia, 108
Lamprocoris, 130
Lamtoplax, 114, 115
Lanopini, 32, 37, 194
Lanopus
    rugosus, 36
Laprius, 89, 103
Largidae, 47, 652
Larix, 253
    kaempferi (syn. of *leptolepis*), 253
Lasioderma
    serricorne, 745
Latahcoris
    spectatus, 133
Lathraedoeus, 88
Latiscutella
    santosi, 41
Latiscutellidae (syn. of Amnestinae), 41
Leaf-cutting ant, 645
Leaffooted bug, 341, 623
Leafhopper, 616–618, 663
Leaf-miner, 482
Lelia, 93, 94, 106, 108
    octopunctata, 200
Leovitius, 108
Lepidoptera, 9, 229, 397, 499, 565, 584, 585, 590, 592, 595,
        679, 713
Leprosoma, 120, 123, 124
    inconspicuum, 119
Leptinotarsa
    decemlineata, 517, 592, 707, 733
Leptocorisa
    acuta, 649
    chinensis, 399, 649, 655, 682
    oratoria, 649
Leptoglossus
    australis (syn. of *Leptoglossus gonagra*)
    clypealis, 623
    gonagra, 624, 650
    phyllopus, 337, 341, 342
Leptolobus, 56, 597
Leptopodomorpha, 645
Lerida, 76, 105
Leridella, 105
Lesser streaked shield bug, 507
Lestonia, 44
    grossi, 44, 46
    haustorifera, 44–46, 194
Lestoniidae, 28, 29, 32, 36, 38, 44, 45, 126, 130,
        136, 194
Lestonocorini, 50, 66, 84, 87, 105, 108, 199
Lestonocoris, 87

Libertirostres, 30
Libyaspis, 127
    coccinelloides, 194
Libyaspis group, 127, 128
Liicoris, 97, 109
    tibetanus, 109
Lincus, 58, 59, 616, 626–628
    apollo, 626
    bipunctatus, 626, 627
    croupius (syn. of *Lincus bipunctatus*), 626, 627
    dentiger, 626
    hebes, 425, 628
    lamelliger, 626
    lethifer, 627
    lobuliger, 626–628
    malevolus, 628
    spathuliger, 626
    spurcus, 627, 628
    styliger, 626
    tumidifrons, 627
    vandoesburgi, 626
Lindbergicoris, 37
    gramineus, 647
Linea, 104
Lineostethus, 58
Linospa
    candida, 184
Liophanurus, 227
Litchi stink bug, 699
Lobepomis, 95
Lobopeltista, 114
Lobopeltista group, 113, 114
Lodosia, 90
Longirostres, 30
Lokaia, 76, 105
Longiscutes, 30, 48, 129
Lopadusa, 59, 60, 107
    augur, 60
    quinquedentata, 187
Loxa, 81
    deducta, 189, 661, 666
    viridis, 64
Lubentius, 101
Luridocimex, 79, 110
Lychee bug, 267
Lychee stink bug, 134
Lycipta, 79
Lygaeidae, 41, 134, 135, 138, 501, 626, 646, 651, 654,
        656, 678
Lygaeoidea, 29, 111, 645, 651, 654, 655, 656, 658, 678
Lygus
    hesperus, 739, 740
    lineolaris, 309, 739, 740, 755, 757
Lymantria
    dispar, 733

M

Mabusana, 76, 105, 106
Machaerotidae, 616

Macrina, 114
Macrocarenoides, 101
    scutellatus, 201
Macrocarenus, 101, 102
Macrocarenus group, 100–102
Macropeltidae, 48
Macropygium
    reticulare, 52, 53, 196
Macrorhaphis, 56, 597
    acuta, 484
Macroscytus
    brunneus, 184
    japonensis, 647
Madates, 98, 99
    limbata, 182
Madecorypha, 72
Magwamba, 114
Mahea, 37
Mamestra
    brassicae, 584
Man-faced stink bug, 80
Manoriana, 109
    pakistanensis, 109
Mantidae, 274, 478, 702
Mantodea, 478, 702
Marghita, 93, 94
Massocephalus, 76
Mataeoschistus (syn. of *Thnetoschistus*), 111
Mathiolus, 109
Mayetiola
    destructor, 733
Mayrinia, 81
    curvidens, 189
Mcphersonarcys, 79
Mealybug, 482, 645, 655, 745
Mecaphesa
    asperata, 439
Mecidaria, 30, 87
Mecidea, 49, 82, 87
    lindbergi, 189
    major, 63, 87
    minor, 5, 87
Mecideini, 50, 61, 82, 87, 100, 111, 199
Mecistorhinus, 58
Mecocephala, 79
Mecosoma, 56, 597
Megacopta, 127
    cribraria, x, 7, 9, 10, 13, 126, 293–321, 507, 520, 528, 644,
        649, 655, 658, 662–665, 746, 748, 749, 753, 754
    punctatissima (syn. of *Megacopta cribraria*), 295,
        296, 310, 311, 649, 657, 658, 661–663, 665
Megaedoeum, 134
Megalotomus
    costalis, 47, 649
    quinquespinosus, 650
Megarhaphis, 56
Megarididae, 27, 32, 33, 38, 46, 126, 130, 136, 185
Megaridinae, 46
Megaris, 27, 46, 185
    puertoricensis, 46
    semiamicta, 46

Megarrhamphini, 50, 113, 114, 196
Megarrhamphus, 113, 114
    hastatus, 196
Mégyménides, 30, 42
Megymeninae, 32, 42–44, 125, 185
Megymenini, 32, 43, 185
Megymenum, 42
    brevicorne, 185
Melampodius, 113–115
Melanaethus
    crenatus, 6
Melanocryptus, 114
Melanophara, 123
Melanotus
    communis, 740
Melyridae, 274
Memmia, 88
    femoralis, 199
Memmiini, 50, 62, 85, 86, 88, 90, 91, 199
Menaccarus, 97
Menecles, 94
    insertus, 198
Menestheus, 69, 102
Menestheus group, 68, 70, 100, 102
Menida, 88, 89
    disjecta, 505, 520, 528, 529, 563, 578
    scotti (syn. of *Menida disjecta*), 505, 520, 568, 573
    transversa, 191
    violacea, 64, 199
Menida group, 100, 106
Menidaria, 30, 75, 88
Menidini, 50, 63, 73, 75, 78, 88, 93, 95, 96, 100, 106, 107,
        191, 199, 505
Mercatus, 114
Mesohalys, 112
    munzenbergiana, 112
Mesopentacoridae, 32, 47, 111
Mesopentacoris
    costalis, 47
    orientalis, 47
Metatropis
    rufescens, 651
Metochus
    abbreviatus, 652
Metocryptus, 114
Metonymia, 114
Mezessea, 85
Milichiella
    arcuata, 706
    lacteipennis, 706
Milichiid, 706
Mimikana, 58, 86
Mimula, 79
Minchamia, 114
Minysporops
    dominicanus, 46
Miopygium
    cyclopeltoides, 53, 187
Miridae, 309, 370, 509, 613, 617
Misumena
    tricuspidata, 266

Mites, 229
Mitripus, 79
Modicia, 107
Moffartsia, 123
Molipteryx
   fuliginosa, 650
Monochrocerus, 81
Monteithiella, 78
Montrouzierellus, 56, 597
Morbora
   australis, 195
Mormidea, 79, 182
   collaris, 648
   v-luteum, 189
   ypsilon, 648
Mormidella, 109
Morna, 96
Multicolored Asian lady beetle, 313
Mulungua, 105
Murgantia, 54, 99, 369
   histrionica, 5, 9, 10, 65, 99, 201, 207–209,
      228, 333–337, 339–345, 506, 508, 566,
      648, 655, 660, 661, 681, 689, 690, 692,
      693, 701, 704, 706, 708–712, 744,
      750–753
Musca
   domestica, 518, 744
Muscanda, 109
   testacea, 109
Musgraveia
   sulciventris, 134, 502, 508
Mustha, 48, 61, 85
Mycoolona, 78, 104
   atricornis, 104
Mycoolona group, 100, 104
Mymaridae, 308
Myota, 93, 109
   aerea, 192
Myrocharia, 30
Myrochea, 89, 103
   cribrosa, 199
Myrocheina, 89
Myrocheini, 50, 58, 68–70, 78, 80, 84, 86, 89, 90, 92,
      100–103, 191, 199
Myzus
   persicae, 343, 756

N

Nabidae, 309, 520
Nabis
   roseipennis, 309
Nagusta
   puncticollis, 478
Natalicolinae, 33, 34, 133, 134, 185, 508
Natalicolini, 132, 133
Nazeeriana, 114
Neagenor, 85
Nealeria, 90
   asopoides, 199
Nealeriaria, 90

Nealeriini, 34, 50, 60, 68, 86, 90, 199
Necanicarbula, 112
Nene, 99
Neoaphylum, 54
Neocazira, 122
Neococalus, 78, 89, 90
Neocoquerelidea, 81
Neoderoploa, 95
Neodymantis, 89, 90
Neogynenica, 87, 108
Neojurtina, 109
   typica, 192
Neoleprosoma, 120, 123, 124
Neolethaeus
   assamensis, 652
   dallasi, 652
Neomazium, 61, 78, 82, 87
Neophylomyza, 706
Neoschyzops, 114
Neostrachia, 89
   bisignata, 89
Neotibilis, 94
Neotropical brown stink bug, 505
Neotropical red-shouldered stink bug, 699
Neottiglossa, 69
   cavifrons, 69
   undata, 188
Nephila
   clavipes, 706
Nephilidae, 706
Nepomorpha, 41, 531, 645
Nesocoris, 96
Nesogenes, 132
Neuroptera, 565
Nezara, 74, 91, 107
   antennata, 354, 355, 357, 359, 506, 537, 584, 585, 622,
      681, 693, 709
   viridula, 5, 7, 9, 11, 61, 91, 191, 351, 353–404,
      428–431, 435–439, 484, 506, 511, 518–523,
      529–535, 539, 566, 567, 584, 585, 587–589,
      612–617, 619–622, 624–626, 644, 648, 655,
      658–661, 666, 696, 700, 702, 703, 742, 743,
      745, 756, 748–751
   viridula (form *duyuna*), 354
   viridula (form *typica*), 354
   viridula (var. *aurantiaca*), 354, 355, 368
   viridula (var. *smaragdula*), 354, 355, 362, 368, 373
   viridula (var. *torquata*), 354, 355
   yunnana, 354, 355
Nezara group, 100
Nezaria, 30, 91, 109
Nezarini, 50, 63, 73, 74, 84, 88, 91, 93–96, 100, 107–109,
      182, 191, 199, 680, 682, 683, 693
Nilaparvata
   lugens, 370, 756
Nimboplax, 113–115
Niphe, 98
   elongata, 398
Nitilia, 98
Nocheta, 93, 107
Noctuidae, 317, 585, 595

Nopalis
  sulcatus, 36
Northern green soldier-bug, 353
Notius, 77, 78
Notobitus
  meleagris, 650
Noualhieridia, 37
  ornatula, 81
Novatilla, 74
Nudipèdes, 30
Nysius
  expressus, 651, 655
  plebeius, 651
Nyssonid wasp, 337

O

Ocellatocoris, 59
Ochisme, 101
Ochisme group, 100, 101, 102, 117
Ochlerini, 49, 50, 58, 59, 85, 86, 94, 187, 196, 626
Ochlerus, 59, 626
Ochrophara, 72, 98
Ochrorrhaca
  truncaticornis, 183
Ochyrotylus, 79
Ocirrhoe, 96, 104
  unimaculata, 193
Odiaria, 30
Odius, 80
Odmalea, 95
  concolor, 64, 95, 192
Odontoscelaria, 30
Odontoscelidae, 129
Odontoscelides, 30, 129
Odontoscelinae, 33, 132, 195, 507
Odontoscelis, 130, 135
  dorsalis, 502, 507, 572
  fuliginosa, 502, 507, 572
  lineola, 502, 507
Odontotarsaria, 30
Odontotarsinae, 33, 130, 132, 195, 507
Odontotarsini, 33
Odontotarsus, 130
  archaicus, 130
Oebalus, x, 11, 79, 455–457, 460, 693
  grisescens (syn. of *Oebalus ypsilongriseus*), 455, 459
  insularis, x, 11, 456–459
  poecilus, 459, 506, 511, 532, 680, 689, 690, 693, 694, 700
  pugnax, 11, 401, 456–458, 624
  pugnax pugnax, 456, 506, 624
  pugnax torrida, 456
  ypsilongriseus, x, 11, 456–459, 506, 531, 535
Oechalia, 55, 99, 598
  schellenbergi, 186
Oecophylla
  smaragdina, 732

Oedocoris, 108
Ogmocoris, 79
Olbia, 59
Old World bollworm, 744
Omyta, 49, 60, 85
Oncocoris, 82
  favillaceus, 190
Oncomerinae, 33, 34, 133, 134, 185, 508
Oncomerini, 133
Oncomeris
  flavicornis, 45
Oncomeraria, 30
Oncomerina, 30
Oncopeltus
  fasciatus, 613, 614, 744
Oncotropis, 67, 74
  carinatus, 74
Oncozygia, 122, 123
Oncozygidea, 123
Onebanded stink bug, 437
Onespotted stink bug, 751
Onylia, 84
Ooencyrtus, 266, 275, 308, 478
  johnsoni, 339, 341, 342, 345
  nezarae, 267, 307, 713
  telenomicida, 275, 401, 704
Ophiocordyceps
  nutans, 267
Oplistochilus, 123
Oplomus, 56
  dichrous, 680
  ebulinus, 680
Oploscelates, 97
Oploscelis, 97
Opsitoma, 92
  brunneus, 200
Opsitomaria, 92
Opsitomini, 34, 50, 60, 68, 92, 200
Orbiscutes, 30, 48, 129
Oriental fruit moth, 752
Oriental green stink bug, 537, 584
Origanaus, 109
Orius, 266
  insidiosus, 275, 309, 439, 684
  minutus, 372
Oropentatoma, 128
Orsillus
  depressus, 651
Ortholomus
  punctipennis, 651
Orthoptera, 565
Orthoschizops, 85
Ostrinia
  nubilalis, 738
Otantestia, 62, 74, 98
Ovalocoris
  parvis, 41
Oxynotidae, 48, 57

Oxynotides, 30, 57
Oxynotina, 30, 48
Oxynotus, 57
Oxyopes
    salticus, 309, 439
    shweta, 308
Oxyopidae, 308, 309, 439

P

Pachybrachius
    luridus, 652
Pachycoridae, 129
Pachycorides, 30, 129–132
Pachycorinae, 33, 130–132, 195, 507
Pachycoris, 131
    klugii, 507
    stallii, 507
    torridus, 131
Pachygrontha
    antennata, 651, 655
Pachygronthidae, 651, 654, 655
Pachymeridiidae, 47
Painted bug, 7, 10, 206–208, 566, 681, 744, 751
Palaeophloea
    monstrosa, 125
Pallantia, 107
    macula, 192
    macunaima, 681, 694, 695, 701
Palomena, 73, 91, 539, 596
    angulosa, 505, 568, 574, 596, 696
    prasina, 65, 199, 505, 529, 568, 572–574, 596
Palomenini, 91
Panaetius, 37
Panaorus
    albomaculatus, 652, 655
    japonicus, 652
Pangaeus
    bilineatus, 40
Pantochlora, 59
Pantochloraria, 59
Pantochlorini, 59
Parabrochymena, 85, 86
Parachinavia, 91
Paracopta
    duodecimpunctata, 295
Paracritheus, 86
Paradasynus
    spinosus, 650
Paradictyotus, 102
Paraedessa, 60
Paraleria, 90
Paralerida, 76, 105
Paralincus, 59
Parallophorella
    indica, 227
Paramecocephala, 79
Paramecocoris, 70, 89

Paramecus, 70, 89
Paramyia, 706
Paranevisanus, 86
Paranotius, 101
Parantiteuchus, 58
Paraplesius
    unicolor, 650
Parastalius, 59
Parastrachia, 39, 47
    japonensis, 6, 48, 194, 503, 528, 529, 538, 647, 655, 656, 658, 659, 699
Parastrachiidae, 6, 28, 31, 32, 40, 43, 47, 136, 194, 501–503, 538, 647, 654, 655, 656, 658, 659, 699
Parastrachiinae, 47
Paratelenomus
    saccharalis, 307, 308, 320, 321
    tetartus, 307
Paratibilis, 93
Parealda, 597
Parent bug, 3, 521, 528, 534, 538
Parentheca, 79
Parodmalea, 95
Paromius
    exiguus, 652
Parvacrena, 88, 93
Patanius
    vittatus, 110, 201
Pauropentacoris
    macruratus, 47
Pausias, 95
    pulverosus, 192
Pea aphid, 756, 757
Peach aphid, 343, 756
Pear bug, 246
Pediobius, 478, 483–485
Pegala, 96, 139
Pelidnocoris, 58, 104
Pellaea
    stictica, 52, 192, 661, 666
Penedalsira, 114
Pentamyrmecini, 31, 50, 61, 82, 92, 200
Pentamyrmex, 82
    spinosus, 92, 183, 200
Pentatoma, 77, 92–94, 105, 109
    confusa (syn. of *Antestiopsis clymeneis*), 469
    cruciata (diff. comb. of *Antestiopsis cruciata*), 469
    fibulata (syn. of *Holcogaster fibulata*), 77
    halys (diff. comb. of *Halyomorpha halys*), 243
    lineaticollis (syn. of *Antestiopsis thunbergii*), 467
    marginata (diff. comb. of *Chinavia marginata*), 354
    orbitalis (syn. of *Antestiopsis thunbergii*), 467
    pantherina (syn. of *Antestiopsis cruciata*), 469
    rufipes, 53, 65, 200, 502, 506, 508, 568, 572, 596
Pentatoma group, 74, 91, 100
Pentatomaria, 30
Pentatomiana, 79
Pentatomida, 18

Pentatomidae, ix, xiii, 3–7, 11–13, 15, 17, 27–34, 36–39,
    42–44, 46–51, 54–60, 66, 82, 85, 90, 92,
    94–96, 102, 103, 106, 108–113, 115, 117, 122,
    124–126, 129–132, 134–136, 138–140, 208,
    210, 212, 226–229, 231, 244, 256, 266, 267,
    272, 274, 275, 307–309, 353, 354, 371, 372, 381,
    397, 400, 403, 426–429, 435, 440, 466, 484,
    499, 501–503, 508–511, 513–517, 520, 525–532,
    535–537, 539, 567–570, 572–575, 577–580, 583,
    587, 588, 590, 592–598, 613–618, 620–624,
    626, 628–631, 644–646, 648, 654–661, 666,
    678–685, 688, 691–696, 698–711, 713, 743,
    748–750, 755
Pentatomides, 30, 60
Pentatomina, 30, 48
Pentatominae, 28, 30, 32, 34, 48–51, 53, 54, 57–61, 77, 94,
    102, 108, 111, 112, 124, 134, 136, 139, 182, 183,
    188–193, 197–201, 499, 504, 539, 567, 568, 597,
    680–685, 688, 692–696, 699, 700
Pentatomini, 50, 62, 64, 75, 77, 80, 81, 83, 90–94, 96, 97,
    100, 105, 107–110, 183, 192, 200, 505, 681, 694
Pentatomites, 111
Pentatomoidea, ix, x, xiii, 3–12, 17, 27–29, 32–34, 36, 38–40,
    43, 46–48, 50, 54, 61, 67, 78, 81, 92, 93, 100,
    102, 106, 111, 112, 116–118, 122, 125–129, 132,
    135–140, 206, 207, 231, 276, 296, 299, 300, 402,
    455, 497, 499–509, 511–513, 518–520, 525–532,
    537–539, 565–569, 571, 572, 575, 578, 579, 584,
    585, 587, 595–598, 616, 645, 647, 654, 656, 658,
    666, 678–681, 700, 701, 703, 704, 706–710, 712,
    713, 730, 748, 749, 753, 754
Pentatomoides, 111, 112
    atratus, 111
Pentatomomorpha, 29, 613, 648, 652, 654, 656, 666, 678
Peregrinus
    maidis, 756
Peribalus, 77, 79
Periboea, 96
Perillus, 55, 56
    bioculatus, 504, 508, 511, 517, 526, 592, 594, 595, 707,
        708
    strigipes, 680
Peromatus, 51, 59
    notatus, 52, 187
Petalaspis, 96
Peucetia
    viridans, 309
Phaeocoris, 90
Phalaecus, 34, 60, 93, 94
Pharypia, 80, 94
    pulchella, 200
Phasia
    hemiptera, 572
    robertsonii, 308
Phasiinae, 227, 568, 569, 748
Phimodera
    flori, 507
    humeralis, 507
Phimoderini, 33
Phiocordycipitaceae, 267
Phléides, 30

Phloea, 125, 267
    corticata, 194
    subquadrata, 539
Phloeidae, 28, 29, 33, 124, 125, 134, 185, 194, 539, 666
Phloeina, 30
Phloeophana, 125
    longirostris, 35, 185
Photinia (syn. of *Aronia*), 254
Phoeacia, 58, 139, 183, 186
Phricodini, 50, 61, 94, 100, 200
Phricodus, 61, 85, 94
    bessaci, 200
Phricodus group, 100
Phyllocephala, 113–115
Phyllocephala group, 100, 104, 114, 115
Phyllocephalida, 48
Phyllocephalidae, 48, 112
Phyllocéphalides, 30, 112
Phyllocephalina, 30
Phyllocephalinae, 30, 32, 34, 48, 50, 51, 100, 104, 109,
    110, 112–115, 187, 196
Phyllocephalini, 50, 113–115, 187, 196
Phymatocoris, 123, 124
Physopelta
    cincticollis, 652
    gutta, 265
    parviceps, 652
    slanbuschii, 652
Picromerus, 55
    bidens, 52, 186, 502, 504, 530, 537, 568–570, 596–598
Pied shield bug, 503
Piesma
    cinereum, 617
    quadratum, 617
Piezodoraria, 30, 94
Piezodorini, 50, 65, 84, 88, 93–96, 100, 107, 192, 200,
    506, 681, 695
Piezodorus, 88, 94, 95, 426, 430, 437
    grossi (syn. of *Piezodorus oceanicus*), 426
    guildinii, 9, 11, 63, 95, 397, 402, 425–440, 506, 531,
        532, 539, 622, 661, 666, 681, 685, 695, 696,
        700, 701, 706, 710, 711, 749
    hybneri, 426, 437, 506, 579, 580, 622, 696, 710
    lituratus, 95, 200, 426, 437
    oceanicus, 426, 437
    punctiventris, 426, 437
    purus, 426, 437
    rubrofasciatus (syn. of *Piezodorus hybneri*), 426, 437
Piezodorus group, 100
Piezosternini, 133
Piezosternum, 109, 133, 134
    subulatum, 12
    thunbergi, 36, 185
Placocoris, 108, 109, 110
    albovenosus, 110
    viridis, 110
Placosternum, 93
Planois
    gayi, 35
Planopsys, 597
Plant bugs, 353, 509, 613, 751, 755, 756

Planthoppers, 369, 617, 618, 756
Platacantha, 83, 84
　lutea, 618
Plataspidae, 4, 6, 7, 10, 12, 13, 27, 28, 33, 38, 44–46, 125–127, 130, 135, 136, 140, 185, 194, 294, 295, 307, 320, 501, 502, 507, 520, 528, 529, 537, 571, 572, 644, 649, 654–658, 661–665, 678, 682, 748
Plataspididae, 126
Plataspidinae, 30, 127
Plataspina, 30
Plataspinae, 127
Platencha, 93, 183
Platinopini, 504, 597
Platycarenus, 57
Platycoraria, 82
Platycorini, 82
Platycoris, 82, 86, 110
Platygastridae, 7, 137, 227, 228, 401, 402, 430, 438, 478, 748, 749
Platygastrids, 488
Platynopiellus, 56, 597
Platynopus, 56, 597
Platytatini, 133
Platytatus, 134
Plautia, 73, 74, 91
　affinis, 35
　picturata (syn. of *Antestiopsis cruciata*), 469
　stali, 265–267, 272, 504, 509–511, 522, 529, 531, 532, 536, 590, 648, 655, 658, 661, 696, 697, 707, 709, 712
Plautiaria, 30, 73
Pléniventres, 30
Plinachtus
　basalis, 650
　bicoloripes, 650
Plodia
　interpunctella, 745
Plum curculio, 751
Plutella
　xylostella, 343, 713, 740
Plutellidae, 713
Podisus, 7, 598, 705–707
　connexivus (syn. of *Podisus nigrispinus*), 705
　fretus (syn. of *Podisus neglectus*), 680, 704, 705
　maculiventris, 186, 228, 229, 276, 309, 502, 504, 511, 512, 514, 517, 521, 523–525, 530, 532, 566, 588, 592–594, 597, 598, 705–708
　neglectus, 680, 704, 705
　nigrispinus, 705, 707
Pododides, 30, 39, 60, 97
Pododus, 97
Podoparia, 30
Podopidae, 48
Podopides, 30, 60, 115
Podopina, 116, 122
Podopinae, 32, 48, 50, 53, 100–102, 104, 106, 115, 117–123, 129, 130, 136, 187, 196, 499, 503, 539, 567, 597, 598
Podopini, 116, 117, 122
Podops, 117, 119, 122, 123
　curvidens, 119

　inuncta, 119, 196
　nodicollis (syn. of *Phimodera humeralis*), 507
Podops group, 50, 102, 117–119, 122, 123, 196
Poecilocoris, 131
　lewisi, 502, 507, 511, 537, 585–587
　purpurascens, 195
Poecilometis, 85, 86
　mistus (syn. of *Halyomorpha halys*), 243
Poecilotoma, 78, 103
　grandicornis, 193
Poecilotoma group, 100, 103
Poliocoris, 112
Polioschistus, 111
Polycarmes, 86
Polyphyma
　koenigi, 195
Polytes, 131
Poriptus, 70, 78, 79, 83, 102
Porphyroptera, 48, 73, 74
Potato leafhopper, 740
Potato-tomato psyllid, 756
Poteschistus, 111
Praying mantis, 702
Priapismus, 58
Priassus, 93
Pricecoridae (syn. of Amnestinae), 41
Pricecoris
　beckerae, 41
Primipentatoma, 128
Primipentatomidae, 33, 111, 128
Prionaca, 81
Prionochilus, 108
Prionocompastes, 34, 97
Prionogastrina, 30
Prionogastrini, 33, 133
Prionosoma
　podopioides, 189
Prionotocoris, 78, 117
Proapateticus (nomen nudum), 598
Probergrothius
　angolensis, 653
　nigricornis, 653
　sanguinolens, 653
　sexpunctatus, 653
Procleticini, 50, 68, 93, 95, 117, 192, 200
Procleticus, 95
Promecocoris, 131
Protestrica, 123
Proxys
　albopunctulatus, 189
Prytanicoris, 76
Psacastini, 33
Pseudadoxoplatys
　mendacis, 187
Pseudanasida, 56
Pseudapines, 78
Pseudatelus, 85, 88
　spinulosus, 191
Pseudatomoscelis
　seriatus, 370
Pseudevoplitus, 93, 94

Pseudocromata, 59
Pseudolerida, 85, 105
Pseudopalomena, 111, 112
Pseudophonoctonus
    formosus, 478
Psix
    sp. nr. *striaticeps*, 227
Psyllid, 617, 618, 629, 645, 756, 757
Pteromalidae, 401, 478
Pugione, 96
Pumpkin bug, 353
Putonia, 123, 124
Pycanum, 133
Pygoplatys, 134
    tenangau, 194
Pyrrhocoridae, 47, 138, 308, 520, 529, 654–657, 678
Pyrrhocoris
    apterus, 526, 529, 653, 655, 658
    sibiricus, 653
Pyrrhocoroidea, 29, 652, 654, 656

Q

Quadraspidiotus
    perniciosus, 733
Quadrocoris, 128

R

Ragweed leaf beetle, 595
Ragweed moth, 595
Ramivena, 94
Ramosiana, 94
Randolotus, 114
Red flour beetle, 743
Redbanded shield bug, 426, 437
Redbanded stink bug, 11, 426, 532, 695, 749
Redshouldered stink bug, 426, 751
Reduviidae, 111, 127, 274, 308, 309, 478, 501, 645
Reduvius, 308
Rhacognathus, 55
Rhagoletis
    pomonella, 734
Rhaphigaster, 93, 94
    nebulosa, 53, 200
Rhaphigastrides, 30, 60, 354
Rhodnius
    prolixus, 756
Rhombocoris, 78
Rhomboidea, 111, 112
Rhopalidae, 678, 753
Rhopalimorpha, 37
Rhopobota
    naevana, 740
Rhynchocoraria, 30
Rhynchocorina, 96
Rhynchocorini, 50, 62, 84, 91, 93, 94, 96, 100, 104,
    106–109, 183, 192, 193, 200, 506, 681
Rhynchocoris, 96
    humeralis, 63, 200
    poseidon, 624

serratus (preoccupied, replaced by *Rhynchocoris*
    poseidon), 624
Rhynchocoris group, 100
Rhyncholepta, 66, 81, 110
    grandicallosa, 190
Rhynocoris
    albopunctatus, 478
    segmentarius, 228
Rhyparochromidae, 626, 646, 652, 655
Rhyssocephala, 79, 80
Riaziana, 110
    serrata, 110
Rice stalk stink bug, 700
Rice stink bug, 456, 689, 693, 705
Rio, 107
    punctatus, 191
Riptortus
    clavatus (syn. of *Riptortus pedestris*), 307, 372, 590,
        593, 622
    linearis, 650
    pedestris, 307, 372, 590, 593, 594, 622, 646, 650, 655,
        657, 658, 666, 713
Risbecella, 70, 71
Rodolia
    cardinalis, 734
Roebournea, 114
Roferta
    marginalis, 191
Rolstoniellini, 50, 67, 78, 80, 93, 96, 107, 201
Rolstoniellus, 34, 60, 96, 97
    boutanicus, 201
Rubiconia, 76, 78, 84
Rubiconiaires, 76, 84
Rubiconiini, 77
Ruckesona, 128
    vitrella, 129
Runibia, 79, 80
    perspicua, 189

S

Sabaeus, 94
Saceseurus, 88
Sagriva, 42
Saileriola, 128
Saileriolidae, 31, 33, 128, 129, 137, 194
Saileriolinae, 129
Saldidae, 501
Salvianus, 114
Sambirania, 107
San Jose scale, 733
Sand flies, 735
Sand wasps, 274
Sandehana, 114
Sangarius
    paradoxus, 184
Sarcodexia
    sternodontis, 337
Sarcophaga
    kempi, 227
Sarcophagidae, 227

Sarcophagid fly, 337
Sarju, 86, 110
Sastragala, 37
    esakii, 647, 658
    scutellata, 647
Say stink bug, 751
Scantius
    aegypticus, 653
    obscurus, 653
Scaptocorini, 32, 184
Scaptocoris, 40
    carvalhoi, 531, 539
    castaneus, 42, 184
Scarab shield bug, 508
Scelionidae, 7, 137, 275, 307, 308
Scelioninae, 478
Schiodtella, 39
    japonica, 647, 655
Schismatops, 114
Schlectendalia
    chinensis, 300
Schyzops, 114
Sciadiocoris, 134
Sciocoraria, 30
Sciocorides, 30, 60, 97
Sciocorini, 39, 50, 68, 90, 97, 98, 100, 101, 193, 201
Sciocoris, 97, 98, 103
    crassus, 97
    cursitans, 201
    longifrons, 97
Sciomenida, 88
Scolytinae, 487, 738
Scotinophara, 117, 120, 122, 123
    coarctata, 117
    horvathi, 120
    lurida, 117, 120, 398, 504, 510, 514, 521, 525, 537
    scotti, 120
    vermiculata, 117
Scotinopharina, 116, 122
Scotinopharini, 504
Screwworm, 742
Scribonia, 191
Scutata, 48
Scutellaria, 30
Scutelleraria, 30
Scutelleridae, 3, 6, 12, 14, 16, 28, 33, 34, 44, 48, 102, 115, 118, 126, 129–132, 136, 294, 307, 501, 502, 507, 527, 528, 538, 568, 584, 597, 649, 654, 655, 678, 699, 749
Scutellérides, 30, 129
Scutellerina, 30
Scutellerinae, 30, 33, 130–132, 195, 507
Scutellerini, 33, 132, 195
Scutelleroidea, 48, 129
Scylax, 70, 71, 83
Scymnus, 439
Séhirides, 30, 39
Sehirinae, 32, 40, 41–43, 47, 184, 503, 699
Sehirus
    cinctus, 35, 40, 42, 184, 503, 680, 681, 699

    cinctus cinctus, 40, 503
    luctuosus, 503, 538
Senectius, 110
    metallicus, 110
Sennertus, 72, 73
Sephela, 98
Sephela group, 100
Sephelaria, 98
Sephelini, 50, 68, 72, 78, 84, 87, 98, 100, 183, 201
Sepidiocoris, 123
Sepinini, 33, 133
Sepontia, 27, 34, 48, 60, 88
Sepontiella, 34, 60
Serbana, 125
    borneensis, 124, 196
Serbaninae, 28, 29, 32, 49–51, 124, 196
Serdia, 94
    concolor, 192
Severinina, 122, 123
Sharpshooter, 5, 614, 616, 645, 655
Shieldbacked bug, 4, 572, 584, 585
Sibaria, 79
    englemani, 648, 655, 658
Sinea, 309
    diadema, 228, 229
Sinopla
    humeralis, 194
    perpunctatus, 37, 508, 534
Sitobion
    avenae, 756, 757
Sloe bug, 505, 530, 580–582
Slug, 274
Small green stink bug, 426, 506
Small rice stink bug, 693
Soft-winged flower beetle, 228
Solenopsis
    geminata, 337, 341
    invicta, 309, 439
Solenosthedium
    rubropunctatum, 131
Solomonius, 86
Solomonius group, 86
Solubea (syn. of *Oebalus*), 455
    grisescens (syn. of *Oebalus ypsilongriseus*), 459
Sorghum midge, 738
Southern green plant-bug, 353
Southern green stink bug, ix, 7, 11, 227, 229, 353, 371, 428, 484, 518, 519, 522, 533–535, 566, 587, 589, 644, 659, 742, 745
Southern green stink-bug, 353
Southern green stinkbug, 353
Southern stinkbug, 353
Soybean looper, 317, 741
Sphaerocoraria, 30
Sphaerocorini, 33, 132, 195
Sphecidae, 704
Sphingid, 708
Sphyrocoris, 131
Spider, 274, 275, 308, 385, 706, 707
Spinalanx, 79

Spined assassin bug, 228
Spined soldier bug, 228, 229, 504, 512, 514, 524, 566, 592
Spined stink bug, 504
Spinidae, 111
Spinipèdes, 30, 39
Spinus, 111
Spissirostres, 30, 55
Spittlebug, 616
Spodoptera
    exigua, 343
    frugiperda, 741
    litura, 513, 707
Stachyomia, 86
Stagonomini, 84
Stagonomus, 84, 85, 87
    amoenus, 190
Stapecolis, 59
Staphylinid beetle, 645
Staria, 78, 84
Steganocerus
    multipunctatus, 195
Steleocoris, 78, 83
Stenocephalidae, 626, 650
Stenotus
    rubrovittatus, 399
Stenozygum, 71, 99
Sternodontus, 123
Sternorrhyncha, 4, 256, 612, 613
Stictochilus, 93, 94
    tripunctatus, 192
Stilbotes, 55, 56
Stilbotini, 55, 56
Stink bug, ix, 3–7, 9–12, 208, 210, 211, 213, 227–229, 231,
        256, 265–268, 271, 274–277, 313, 317, 334, 336,
        339, 343, 364–366, 368, 382, 396–401, 403,
        426, 428–431, 436–440, 456, 458–460, 486,
        501, 531, 578, 597, 617, 619–625, 631, 645, 656,
        658–661, 666, 667, 678, 679, 681, 682, 684,
        688, 701–713, 739, 741–755, 757
Stinkwood stink bug, 246
Stirétrides, 30, 55
Stiretrinae, 499, 504, 567
Stiretrinae (syn. of Asopinae), 499, 504, 567
Stiretrini, 55
Stiretrus, 55, 56
    anchorago, 680
    erythrocephalus, 186
Stirotarsinae, 32, 49–51, 124, 183, 196
Stirotarsini, 124
Stirotarsus, 124, 125
    abnormis, 124, 183, 196
Stollia (syn. of *Eysarcoris*), 84
Stolliini, 84
Storthecoris, 123
Storthogaster, 114
Strachia, 47, 98, 99
    crucigera, 201
    geometrica (syn. of *Antestiopsis cruciata*), 469
    pentatomoides (syn. of *Antestiopsis thunbergii*), 467
    velata (syn. of *Antestiopsis cruciata*), 469
Strachia group, 100

Strachiaires, 98
Strachiaria, 30
Strachiini, 47, 50, 62, 71, 72, 98–100, 182, 201, 228, 506,
        681, 692
Strepsiptera, 479, 483, 484, 486
Strombosoma, 135
    impictum, 136
Strongygaster
    triangulifera, 308
Stysiana, 79, 110
Subafrius, 56, 597
Sudan millet bug, 484
Sulcientres, 30
Sunn pest, 69, 131, 520, 528, 573
Supputius, 598
Surenus, 80, 86
Suspectocoris, 111, 112
    grandis, 37, 112
Sweetpotato whitefly, 756
Synona, 309, 748

T

Tachengia, 86
Tachinid flies, 7, 227, 275, 308, 337, 439, 486, 568, 569, 706,
        712, 748
Tachinidae, 7, 227, 307, 402, 479, 483, 568, 569, 572, 604, 706
Tahitocoridae, 122
Tahitocorini, 122
Tahitocoris, 122
    cheesmani, 122
Tamolia, 134
Tancreisca, 90
Tantia, 114
Tarachidia
    candefacta, 595
Tarisa, 54, 115, 121–124
Tarisa group, 50, 100–102, 118, 121, 123, 124
Tarisaria, 30
Tarisini, 539
Tarnished plant bug, 739, 755
Taubatecoris, 112
Taurocerus, 93, 94
Taurodes, 86
Tectocorinae, 33, 130–132, 195
Tectocoris, 130, 131
    diophthalmus, 195
Telenomine, 478
Telenomus, 266, 275, 438, 748
    calvus, 705
    chloropus (syn. of *Telenomus turesis*), 402
    mitsukurii (diff. comb. of *Trissolcus mitsukurii*), 267
    nakagawai (possible syn. of *Telenomus turesis*), 402
    persimilis, 275
    podisi, 228, 275, 276, 339, 341, 342, 345, 438, 439,
            703–706, 708
    samueli, 227
    sechellensis, 478, 483–485
    turesis, 402
    utahensis (diff. comb. of *Trissolcus utahensis*), 275
Teleocoris, 112

Teleoschistus, 111
Tenerva, 76
Tenodera
    aridifolia, 702
    aridifolia sinensis, 702
Ténuirostres, 30
Tepa, 108
Terania, 95
Tessaratoma
    papillosa, 45, 134, 267, 617, 677, 681, 699
Tessaratomaria, 30
Tessaratomidae, 6, 12, 14, 27, 28, 33, 34, 43, 59, 93, 109,
        110, 113, 115, 132–134, 136, 307, 501, 502, 508,
        538, 666, 678, 681, 699
Tessaratomina, 30
Tessaratominae, 30, 33, 34, 132–134, 194, 681, 699
Tessaratomini, 33, 133, 134, 194, 681, 699
Tessaratomoides,133
    maximus, 133
Testrica, 123
    vacca, 127
Tetradium, 255
    danielli (syn. of *Eudia*, syn. of *hupensis*), 255
Tetrisia
    vacca, 127
Tetroda, 114
Tetrodias, 114
Tetrodini, 50, 113–115
Tettigoniidae, 274
Tetyra
    bipunctata, 507
Tetyraria, 30
Tetyridae, 129
Tétyrides, 30, 129
Texas, 69
Thalmini, 32, 34
Thangnang, 89
Thasus
    neocalifornicus, 650, 655
Thaumastella, 39, 134
    aradoides, 135, 185
    elizabethae, 135
    namaquensis, 135
Thaumastellidae, 31, 33, 38, 134, 135, 185
Thaumastellinae, 39
Thaumastocoridae, 501, 502, 508
Theloris, 78
Theseus, 86
Thestral, 78, 103
Thlimmoschistus, 111
Thnetoschistus, 111
Tholagmus, 120, 123, 124
Tholosanus, 103
Tholosanus group, 90, 100, 103
Thomisidae, 266, 309, 439, 707
Thongolifha, 134
Thoreyella, 95
    brasiliensis, 200
Thoria, 122, 123
Thyanta, 108, 110
    acuminata, 63

    calceata, 506, 531
    custator, 426, 622, 696, 699
    custator accerra, 426, 696
    custator custator, 751
    pallidovirens, 623, 648, 655, 661, 680, 696, 699, 703,
        704, 707, 751
    perditor, 201, 601, 661, 666, 680, 691, 699, 703, 710, 711
Thyreocoraria, 135
Thyreocoridae, 4, 6, 7, 12, 15, 16, 28, 31, 33, 39, 126, 129,
        130, 135–137, 501, 502, 508
Thyréocorides, 30, 135
Thyreocorinae, 33, 39, 135–137, 194, 508
Thyreocoris, 39, 135, 136, 137
    scarabaeoides, 131, 136, 194, 508
Tibetocoris, 97
Tibilis, 93, 94, 107, 108
Tibraca, 79
    limbativentris, 12, 64, 189, 693, 700–702, 704, 705, 708
Timuria, 90
Tinganina, 86
Tiphodytes, 228
Tipulparra, 86
Tiridates, 131
Tiroschistus, 111
Togo
    hemipterus, 646, 652
Tolono, 37
Tolumnia, 76
Tomato and bean bug, 353
Tomato stink bug, 11, 459
Tornosia, 54, 120, 122, 123
    insularis, 119
Trachyops, 101
    australis, 201
Tribolium
    castaneum, 743
Trichopepla, 28, 72, 77, 79
    semivittata, 189
Trichophora, 29, 138
Trichoplusia
    ni, 745
Trichopoda, 402, 748, 749
    pennipes, 275, 337, 402, 439
Tricompastes, 99
Trigonosomaria, 30
Trigonotylus
    caelestialium, 399
Trincavellius, 97, 98
    galapagoensis, 193
Triozid, 629
Tripanda, 67, 76, 105
Triplatygini, 50, 58, 68, 99, 100, 201
Triplatyx, 99
    bilobatus, 100, 201
Triplatyxaria, 99
Trissolcus, 266, 275, 276, 307, 337, 438, 484, 485, 748
    basalis, 7, 228, 276, 401, 402, 430, 439, 529, 703–705,
        708, 749
    brochymenae, 228, 275, 339, 341, 342, 345, 439, 704,
        705, 708
    cultratus, 267, 276

edessae, 275
euschisti, 228, 275, 339, 341, 342, 345
flavipes, 267
halyomorphae (syn. of *Trissolcus japonicus*), 266
hullensis, 228, 275
hyalinipennis, 227, 228, 230
itoi, 267, 276
japonicus, 266, 267, 276, 278, 308
latisulcus, 307
mitsukurii, 267, 276, 402
mopsus, 478, 483–485
murgantiae (syn. of *Trissolcus brochymenae*), 339, 341, 342
plautiae, 267
podisi (syn. of *Trissolcus euschisti*), 339, 341, 342
suranus, 478, 483
teretis, 439
thyantae, 275
urichi, 439, 705
utahensis, 275
Tritomegas
   bicolor, 503, 538
Trochiscocaria, 98, 99
Trochiscocoris, 62, 98, 99
   hemipterus, 502, 506
Troilus, 598
Tropicaria, 30
Tropicoraria, 30, 81
Tropicorini, 110
Tropicoris, 109
Tropicorypha, 76
Tropidotylus
   minister, 127
   servus, 127
Tropicorypharia, 30, 76
True bug, 4, 294, 371, 373, 486, 498, 501, 502, 527, 531, 535, 537–539, 567, 572, 582, 588, 590, 592, 596, 597, 613, 617, 629, 678, 679, 742, 755
Tsetse fly, 655, 746
Tshibalaka, 114
Tshingisella, 123, 124
Turrubulana, 101
Twospotted stink bug, 504, 592
Tylospilus, 55, 598
Tynacantha, 598
   marginata, 680
Tyoma, 70, 71
   cryptorhyncha, 71, 188
   verrucosa, 71
Tyoma group, 100, 104
Tyomana, 70, 71
Typhodytes, 227
Tyrannocoris, 598

U

Uddmania, 109, 113, 114
Udonga, 88, 93
   montana, 89

Umgababa, 87
Urochela, 137
   luteovaria, 138, 649
   quadrinotata, 138, 529, 649
Urolabidina, 30
Urostylidae, 137
Urostylididae, 31, 33, 47, 127–129, 137, 138, 194, 649, 654, 655, 657, 658
Urostylina, 30
Urostylinae, 30
Urostylis
   annulicornis, 194, 649
   striicornis, 138, 649
   westwoodii, 138, 649, 655, 657, 658
Urothyreus
   horvathianus, 195
Utheria, 102

V

Variegated coffee bug, 466
Vedalia beetle, 734
Ventocoris, 123, 124
   trigonum, 119
Vespidae, 9, 255, 704, 707
Vespula
   maculifrons, 704, 707
Veterna, 76, 105
   pugionata, 188
Veterna group, 76, 100, 105, 106
Vidada, 107
Vilpianus, 123, 124
Vitellus, 96
Vitruvius, 110
   insignis, 110, 193
Vulsirea, 79
   violacea, 189

W

Weda, 122, 123
   parvula, 119
   stylata, 119
Weevil, 5, 399, 440, 645, 655, 734, 738, 740, 746
West Indian fruit fly, 743
Western corn rootworm, 738
Western tarnished plant bug, 739
Wheat stink bug, 504
Whiteflies, 229, 308, 742
White-margined burrower bug, 503
White-spotted stink bug, 520
Wooly apple aphid, 733

X

Xiengia, 48, 109
   elongata, 109

Xylastodorinae, 508
Xynocoris, 59
Xysticus, 309
    ferox, 707

Y

Yellow-brown stink bug, 246
Yellowjacket wasp, 707
Yemma
    exilis, 651, 655

Z

Zaplutus, 86
Zelus, 309
    renardii, 309
Zhengius, 97
Zicrona, 56
    caerulea, 52, 504
Zorcadium, 95
Zouicoris, 108
Zygogramma
    suturalis, 595

Plants Index

A

Abelia
   × *grandiflora*, 251
Abelmoschus
   esculentus, 217, 251, 390, 437, 471
Abutilon
   grandifolium, 390
Acacia, 260
Acacia, 251
Açaí, 626
Acanthaceae, 300, 301, 383, 395
Acephala group, 214, 344, 384, 692
Acer, 258
   buergerianum, 251
   campestre, 251
   circinatum, 251
   freemanii, 251
   griseum, 251
   japonicum, 251
   macrophyllum, 251
   negundo, 251
   palmatum, 251
   pensylvanicum, 251
   platanoides, 251
   rubrum, 251, 255
   saccharinum, 251
   saccharum, 251
   tegmentosum, 251
Achyranthes, 472
   aspera, 484
Adenocalymma
   comosum, 432
Admiral pea, 271, 274
Aeschynomene
   americana, 301, 306
Aesculus
   carnea, 251
   glabra, 251
African basil, 390
Agapanthus, 390
Agapanthus, 390
Agathi, 300, 304, 316
Ageratum
   conyzoides, 472, 486
Ailanthus
   altissima, 251, 255, 273
Aizoaceae, 135
Ajonjolí, 391
Albizia
   julibrissin, 301
Alcea
   rosea, 390

Alder, 538
Alfalfa (= Lucerne), 11, 216, 229, 303, 306, 388, 426, 431, 435, 439, 571, 683, 738, 739, 741
Algodón, 390
Algodonero, 390
Allegheny (apple) serviceberry, 251
Alligatorweed, 301
Alnus
   glutinosa, 538
Alpinia
   purpurata, 626
Alternanthera
   philoxeroides, 301
Althaea
   rosea (syn. of *Alcea rosea*)
Altingiaceae, 307
Amaranth (= Love-lies-bleeding), 251
Amaranthaceae, 121, 213, 214, 301, 382, 383, 395, 432, 472, 484
Amaranthus
   caudatus, 251
   hybridus, 383
   hypochondriacus, 383, 396
   leucocarpus (syn. of *Amaranthus hypochondriacus*)
   palmeri, 383
   spinosus, 383
   viridis, 383
Ambrosia
   artemisiifolia, 595
   trifida, 383
Amelanchier
   grandiflora (syn. of *Amelanchier laevis*)
   laevis, 251
American basswood, 255
American elm, 255
American joint vetch, 301, 306
American mountain ash, 254
American pokeweed, 254
American sycamore, 254
American wisteria, 304
American witchhazel, 253
American yellowwood, 252, 302
Amphicarpaea
   bracteata, 301
Amur (Japanese downy) maple, 251
Anacardiaceae, 214, 300, 301, 383, 395, 432
Anchietea
   pyrifolia, 435
   salutaris (syn. of *Anchietea pyrifolia*)
Annonaceae, 214, 471
Angiopteris
   evecta, 122

Antirrhinum
    majus, 251
Apiaceae, 120, 121, 137, 213, 214, 383, 395, 432
Apple, 244, 250, 253, 258, 260, 264, 265, 268–270, 273,
        277, 625
Apricot, 254, 268
Aquifoliaceae, 432
Arabica coffee, 12, 467, 469, 470, 472, 477, 478, 480, 483,
        486–488
Araceae, 57
Arachis
    glabrata, 301, 306
    hypogaea, 40, 301, 307, 386
Arborvitae (overwintering), 264, 386
Arctium
    minus, 251
Armoracia
    rusticana, 251
Aronia, 254
Arroz, 391
Artichoke, 214
Arugula, 213, 215, 220, 221, 224, 225
Asian pear, 244, 254
Asian spiderflower, 385
Asiatic (Japanese cornel) dogwood, 252
Asimina
    triloba, 214, 251
Asparagus, 251, 265
Asparagus
    officinalis, 251
Asteraceae, 121, 213, 214, 301, 307, 395, 396, 432, 472,
        484, 486, 595, 733
Astragalus
    sinicus, 301
Astrocaryum, 628
Asystasia
    coromandeliana (syn. of *Asystasia gangetica*)
    gangetica, 383
Autumn olive, 252
Avena
    sativa, 217, 391
Avocado, 76
Azalea, 313
Azuki bean, 304

B

Bagpod, 389
Balsam pear, 386
Bamboo, 73, 97, 113, 431
Banana, 129, 305, 396, 471, 486
Baptisia
    australis, 251, 301
Barbarea, 214
Barley, 217
Basswood, 258, 578
Bean, 253, 258, 260, 265, 270, 271, 302, 303, 334, 388, 620
Bee-bee tree (= Korean euodia), 255
Beet (overwintering), 385
Beggarweed, 388
Bell pepper, 251, 258, 274, 306

Berenjena, 394
Berenjena silvestre, 394
Bermuda grass, 217
Beta
    cicla (syn. of *Beta vulgaris* ssp. *cicla*)
    vulgaris, 214, 396
    vulgaris ssp. *cicla*, 251, 385
Betula
    nigra, 251
    papyrifera, 251
    pendula, 251, 538
Bidens
    pilosa, 214, 384, 432, 472, 486
Bigleaf maple, 251
Bignoniaceae, 72, 432
Birch, 538
Bird's-foot trefoil, 431, 435
Bitter cress, 334
Black cherry, 254, 392
Black gram, 217, 304, 389
Black lentil, 217, 304
Black locust, 217, 254, 303
Black medic, 307
Black mulberry, 391
Black mustard, 384
Black nightshade, 394
Black walnut, 253
Black willow, 306
Blackberry, 254, 270
Black-eyed pea, 300, 304, 306, 317
Blackgum (= Tupelo), 253
Blackhaw (= Vibernum), 255
Blackhaw cranberry, 255, 258
Black-jack, 214
Blue wild indigo, 251
Blueberry, 255, 270, 614
Boerhavia
    coccinea, 391
Bok choy, 215
Bombay hemp, 387
Boraginaceae, 72, 384, 395
Bougainvillea
    glabra, 434
Boxelder, 251, 753
Brassica, 215, 342, 344, 384
Brassica
    campestris (syn. of *Brassica rapa* var. *rapa*)
    juncea, 214 , 251, 344, 753
    kaber, 753
    napobrassica, 215
    napus, 211, 215, 336, 344, 384, 432, 516, 753
    nigra, 384
    oleracea, 210, 211, 214, 251, 344, 384
    oleracea var. *acephala*, 214 , 344, 384
    oleracea var. *botrytis*, 214 , 384
    oleracea var. *capitata*, 213, 215, 344
    oleracea var. *gemmifera*, 215, 384
    rapa, 209, 344
    rapa chinensis, 215
    rapa narinosa, 215
    rapa nipposinica, 215

rapa oleifera, 215
rapa var. *pekinensis*, 215
rapa var. *rapa*, 215
ruvo, 225
Brassicaceae, 99, 120, 206, 211, 213, 214, 224, 225, 384, 385, 396, 432, 682, 693
Bridelia
micrantha, 471
Broad bean, 304
Broccoli, 10, 210, 211, 213, 214, 219–224, 334, 336, 338, 344
Bromeliaceae, 385, 395
Broom, 571
Brown mustard, 516
Brussels sprouts, 215, 334, 384
Buckthorn, 578
Buddleja, 251
Buffalo gourd, 386
Bur grass, 391
Burclover, 388
Bush clover, 302
Butea
monosperma, 301
Butter bean (= Butterbean), 388
Butterfly bush, 251, 258
Butterfly pea, 387

C

Cabbage, 10, 211, 215, 220, 221, 225, 226, 251, 334, 335, 337–340, 344, 384
Cacao, 76, 394, 625
Cactaceae, 432
Caesalpinia
decapetala, 472
Cajanus
cajan, 300, 301, 387, 431, 432
indicus (syn. of *Cajanus cajan*)
Calendula, 383
Callery (Bradford) pear, 254
Callitris, 46
Calotropis
procera, 213
Camellia
sinensis, 218, 251
Canada goldenrod, 384
Canadian horseweed, 383
Canavalia
cathartica, 387
ensiformis, 387
gladiata (syn. of *Canavalia ensiformis*)
microcarpa (syn. of *Canavalia cathartica*)
Candytuft, 215
Cannabaceae, 216, 385, 395, 472
Cannabis, 385
Cannabis, 385
sativa, 216, 251, 385
Canola, 211, 215, 219, 755
Cantaloupe (= Melón), 216, 386
Canthium, 472
Cape gooseberry, 218

Cape mustard, 216
Cape pepper-grass, 215
Caper, 216, 682
Capitata group, 215, 344
Capparaceae, 216, 385, 395, 682, 693
Capparidaceae (as Capparaceae), 385
Capparis
spinosa, 216, 220
Capsella
bursa-pastoris, 215
Capsicum, 435
annuum, 213, 218, 251, 306, 393
Caragana
arborescens, 251
Carduus
nutans, 383
spinosissimus, 383
Carica
papaya, 216
Caricaceae, 216
Carpinus
betulus, 251
Carrot, 214
Carya
illinoinensis, 251, 304, 390
ovata, 252
pecan (syn. of *Carya illinoinensis*)
Cashew, 80
Cassava, 626
Cassia, 387
Cassia
fasciculata (syn. of *Chamaecrista fasciculata* var. *fasciculata*)
occidentalis (syn. of *Senna occidentalis*)
Castilleja
indivisa, 393
Castor bean, 386
Castor oil bean, 386
Castor oil plant, 216, 471
Castor oil tree, 472
Casuarina
equisetifolia, 472
Casuarinaceae, 472
Catalpa, 252, 258
Catalpa
bignonioides, 252, 268
Catharanthus
roseus, 252, 266
Cauliflower, 10, 214, 220, 221, 334, 344, 384
Cayenne pepper, 251
Cecropia
palmata, 626
Cedar, 252, 260
Cedrus, 252
Celastrus
orbiculatus, 252
Celosia, 252
Celtis
australis, 394
koraiensis, 252
occidentalis, 252

Cenchrus, 391
Cephalanthus
    occidentalis, 252
Cercidiphyllum
    japonicum, 252
Cercis
    canadensis, 252, 301
    canadensis var. *canadensis*, 301
    canadensis var. *texensis*, 252
Cereal plant, 383
Cereal rye, 254
Chamaecrista
    aeschinomene (syn. of *Chamaecrista nictitans*)
    fasciculata, 301, 432
    fasciculata var. *fasciculata*, 387
    nictitans, 432
Chamaecyparis
    obtusa, 698
Chayote, 386
Cheese weed, 390
Chenopodiaceae, 121, 214, 385, 395
Chenopodium, 385
    acuminatum (syn. of *Chenopodium album* var.
            acuminatum)
    album, 213, 214, 385
    album var. *acuminatum*, 385
    ambrosioides, 481
    berlandieri, 252
Cherry, 254
Cherry laurel, 254
Cherry plum, 254
Cherry tomato, 81, 459
Chícharo de vaca, 389
Chickling pea, 431
Chickpea, 302, 744
Chile morrón, 393
Chinese arborvitae, 254, 258, 264
Chinese cabbage, 215, 338, 344
Chinese elm, 255
Chinese fringetree, 252
Chinese lespedeza, 302
Chinese long bean, 390
Chinese milk vetch, 251, 258, 264, 301
Chinese (Asian) pear, 254
Chinese pistache, 254
Chinese privet, 253, 305
Chinese scholar tree, 255, 258, 260, 264, 266
Chinese sumac, 300, 301
Chinese tea, 251
Chinese violet, 383
Chinese wisteria, 304
Chinese-quince, 254
Chionanthus
    retusus, 252
    virginicus, 252
Chokeberry, 254
Chrysanthemoides
    monilifera, 472
Chrysanthemum, 214, 733
Chrysanthemum, 214, 733
    cinerariifolium, 481

Chual, 385
Cicer
    arietinum, 302, 387
Cirsium
    texanum, 383
Citrullus
    lanatus, 216
    lanatus var. *lanatus*, 386
    vulgaris (syn. of *Citrullus lanatus* var. *lanatus*)
Citrus, 96, 218, 252, 258, 265, 305, 393, 624
Citrus, 218, 305
    × *aurantiifolia* (= *medica* × sp. hybrid), 393
    × *limonia* (= *limon* × *reticulata* hybrid), 393
    × *paradisi* (= *maxima* × *sinensis* hybrid), 393
    × *sinensis* (= *maxima* × *reticulata* hybrid), 300, 305,
            393, 472
    acida (syn. of *Citrus* × *aurantiifolia*)
    aurantifolium (syn. of *Citrus* × *aurantiifolia*)
    aurantium (syn. of *Citrus* × *sinensis*)
    junos, 252, 265
    limonum (syn. of *Citrus* × *limonia*)
    maxima var. *uvacarpa* (syn. of *Citrus* × *paradisi*)
    nobilis (syn. of *Citrus reticulata*)
    reticulata, 305, 393
    unshiu, 252, 265, 305
Cladrastis
    kentukea, 252, 302
    lutea (syn. of *Cladrastis kentukea*)
Clammy locust, 578
Clasping coneflower, 383
Cleome
    graveolens (syn. of *Polanisia dodecandra*)
    gynandra, 385
    pungens (syn. of *Cleome spinosa*)
    spinosa, 385
    viscosa, 385
Clerodendron
    villosum, 72
Clitoria, 387
Clover, 11, 304, 428, 430, 439
Clusiaceae, 471
Cluster bean, 302
Cock's comb, 252
Cocklebur, 301
Cockspur weed, 391, 396
Cocoa, 394
Coconut, 616, 618, 625–627
Coconut palm, 75
Cocos
    nucifera, 626, 627
Coffea, 392
    arabica, 218, 435, 468, 472, 620
    canephora, 469, 472, 620
    liberica, 469, 472, 639
Coffee, 5, 12, 74, 218, 392, 466–472, 475–488, 620, 621,
            625, 626
Coffee Arabica, 466
Coffee senna, 389
Coix
    lacryma-jobi, 391
Col, 384

Cole crops, 99, 206, 207, 209, 213, 219, 222, 223, 692
Collards, 10, 99, 214, 220, 221, 251, 334, 335, 337, 338, 340, 344, 384
Comfrey, 255
Common buckthorn, 254
Common buttonbush, 252
Common dogwood, 252
Common hackberry, 252
Common hop, 253
Common millet, 253, 265
Common ragweed, 595
Common sowthistle, 384
Compositae (syn. of Asteraceae), 484
Convolvulaceae, 44, 216, 301, 386, 395
Convolvulus
    arvensis, 216
Conyza
    canadensis var. *canadensis*, 383
Corchorus
    capsularis, 304
Cordia
    millenii, 72
Coriander, 383
Coriandrum
    sativum, 213, 383
Corn, 10, 218, 258, 259, 265, 268, 270, 305, 334, 337, 338, 392, 625, 626
Corn sow thistle, 214
Cornaceae, 138, 585
Cornus
    controversa, 585
    drummondii, 622
    florida, 252
    kousa, 252
    macrophylla, 252
    officinalis, 252
    racemosa, 252
    sanguinea, 252, 268
    sericea, 252
Cornus Stellar series, 252
Corylus
    colurna, 252
Cotton, 5, 9–11, 217, 270, 294, 297, 300, 304, 309, 317, 319, 320, 334, 340, 390, 399–403, 435, 437, 439, 440, 616, 619, 621, 631
Cowitch, 303
Cowpea, 5, 71, 217, 274, 389, 399, 620, 622
Crab apple, 253
Cranberry, 740
Crapemyrtle, 253
Crataegus
    laevigata, 252
    monogyna, 252
    pinnatifida, 252
    viridis, 252
Creole cotton, 390
Crimson clover, 304, 389
Crimson fountaingrass, 391
Crimsoneyed rosemallow, 253
Crook-neck squash, 386

Crossandra
    infundibuliformis, 300, 301
    undulaefolia (syn. of *Crossandra infundibuliformis*)
Crotalaria, 431, 433
    brevidens, 432
    incana, 387
    juncea, 387
    lanceolata, 387, 431, 433
    pallida, 433
    spectabilis, 302, 387
    trichotoma, 387
    usaramoensis (syn. of *Crotalaria trichotoma*)
Croton
    zambesicus, 71
Cruciferae (syn. of Brassicaceae)
Cryptomeria
    japonica, 698
Cucumber, 265, 386
Cucumis
    melo, 213, 386
    melo subsp. *melo* var. *cantalupo*, 216
    sativus, 213, 252, 386
Cucurbit, 616, 618, 624
Cucurbita, 386
    foetidissima, 213, 355, 386
    maxima, 386
    moschata, 386
    pepo, 213, 252
Cucurbitaceae, 44, 213, 216, 386, 395, 396, 432
Cultivated grape, 394
Cupressaceae, 46, 386, 395
Curcuma
    longa, 306
Cuscuta
    campestris, 617
Cyamopsis
    psoralioides (syn. of *Cyamopsis tetragonoloba*)
    tetragonoloba, 302
Cymbidium, 252, 266
Cynara
    scolymus, 214, 383
Cynodon
    dactylon, 217
Cyperaceae, 120, 131, 386, 395
Cyperus
    esculentus, 386
    rotundus, 217

D

Dactyloctenum
    aegyptium, 217
Dahlia, 214
Dahlia, 214
Daucus
    carota, 137, 214
Dawn redwood, 253
Dendrobium, 391
Derris, 733
Descurainia
    sophia, 215

Desmodium, 387
   adscendens, 387
   intortum, 433
   ovalifolium (syn. of *Desmodium adscendens*)
   tortuosum, 302, 387, 396, 433
   triflorum, 387
   uncinatum, 433
Deutzia
   crenata, 304
Dhaincha, 303
Dimocarpus
   longan, 617
Diospyros
   kaki, 252
Dixie ticktrefoil, 387
Dodder, 617
Dog (native) rose, 254
Dogwood, 252, 622
Dolichos
   biflorus (syn. of *Vigna unguiculata* ssp.
      cylindrica)
   lablab (syn. of *Lablab purpureus*)
   unguiculatus (syn. of *Vigna unguiculata*)
Dracopis
   amplexicaulis, 383
Dwarf essex rape, 384
Dysphania
   ambrosioides, 481

E

Eastern hemlock, 255
Eastern redbud, 252
Eastern redcedar, 253
Echinochloa, 391
Edible fig, 252
Edible rose, 268
Egg-plant, 394
Eggplant, 71, 254, 258, 265, 270, 274, 394,
     459, 698
Egyptian clover, 217, 389
Egyptian crowfoot, 217
Elaeagnus
   angustifolia, 252
   umbellata, 252
Elaeis
   guineensis, 627, 628
Eleusine
   coracana, 391, 471
Elm, 255, 258, 260, 264
Empress tree, 618, 619
English (smoothleaf) elm, 255
English holly, 253
English oak, 254
Ericaceae, 432
Eriobotrya
   japonica, 305
Eruca
   sativa, 213, 215
Eucalyptus, 111, 431

Eucalyptus, 46
   camauldulensis, 54
Eugenia, 46
   uniflora, 434
Euodia
   hupehensis (syn. of *Tetradium daniellii*)
Euphorbia, 627
   hirta, 216
Euphorbiaceae, 57, 71, 125, 131, 216, 386, 395,
     432, 471
Euphorbs, 625
European ash, 252
European hornbeam, 251
European white birch, 251
Ever-lasting pea, 388

F

Faba bean, 389
Fabaceae, 57, 71, 121, 125, 127, 216, 264, 294, 299, 300,
     301, 306, 307, 320, 386–390, 395, 430–432,
     471, 472
Fagaceae, 138, 307
False mallow, 390
Fava bean, 304, 389
Ficus, 40
   carica, 252, 305
   colubrinae, 40
Field bean, 299, 302, 330
Field bindweed, 216
Field corn, 255, 256
Field mustard, 210, 215
Field pumpkin (= Summer squash), 252
Field rose, 268
Field sow thistle, 214
Fig, 305
Filaos, 472
Filbert (= Hazelnut), 252, 437
Finger millet, 391, 471
Firecracker plant, 300, 301
Firethorn, 254
Flame of the forest, 301
Flixweed, 215
Florida beggarweed, 302
Flowering dogwood, 252
Foeniculum
   vulgare, 432
Forest jasmine, 472
Forsythia
   suspensa, 252
Fragaria
   × *ananassa*, 435
Fraxinus
   americana, 250, 252
   excelsior, 252, 268
   pennsylvanica, 252
Freeman maple, 251
French bean, 388
Frijol, 388

Fuji cherry, 254
Fuzzy rattlepod, 387

G

Gaertnera, 469
Galiniera
    coffeoides, 472
Galinsoga
    parviflora, 472, 486
Garbanzo, 387
Garden cucumber, 252
Garden pea, 303
Garden snapdragon, 251
Garden tomato, 254
Garden vetch, 304
Geigeria
    alata, 214
Giant goldenrod, 384
Giant ragweed, 383
Ginger, 694
Ginkgo (= Maidenhair tree), 252
Ginkgo
    biloba, 252
Girasol, 383
Glaucias
    subpunctatus, 265
Gleditsia
    triacanthos var. *inermis*, 252
Gliricidia
    maculata, 471
Globe artichoke, 383
Glossy abelia, 251
Glycine
    hispida (syn. of *Glycine max*)
    max, 213, 253, 294, 302, 388, 426, 433, 620,
        662, 738
    wightii (syn. of *Neonotonia wightii* var. *wightii*)
Gmelina
    arborea, 72
Gochnatia
    polymorpha, 432
Goji, 268
Goldenrain tree, 253
Gossypium, 738
    barbadense, 390
    herbaceum (syn. of *Gossypium hirsutum* var.
        hirsutum)
    hirsutum, 217, 300, 304, 434, 435
    hirsutum var. *hirsutum*, 390
Grain sorghum, 274, 399, 401, 685, 743
Gramineae (syn. of Poaceae)
Grape, 264, 270
Grass (or grasses), 69, 70, 80, 83, 85, 87, 90, 98, 105, 113,
        115, 740
Grapefruit, 393
Gray dogwood, 252
Green ash, 252
Green bean, 258, 303, 368, 388, 430, 460, 625, 685
Green gram, 217

Green hawthorn, 252
Green pepper, 274, 625
Groundnut, 260
Gud-gad, 214
Gynandropsis
    gynandra (syn. of *Cleome gynandra*)
    pentaphylla (syn. of *Cleome gynandra*)

H

Hackberry, 394
Hairy indigo, 302
Hairy lespedeza, 302
Halesia
    tetraptera, 253
Hamamelis
    japonica, 253
    virginiana, 253
Haw, 260
Hazelnut (= Filbert), 252, 258, 437
Hebanthe
    erianthos, 432
Hedge maple, 251
Helianthus
    annuus, 383, 432
    tuberosus, 383, 432
Heliotrope, 384
Heliotropium, 384
Hemp, 216, 251, 385
Henbit, 390
Heptacodium
    miconioides, 253
Herb sophia, 215
Hibiscus (= Rose of Sharon), 253
Hibiscus
    cannabinus, 390
    esculentus (syn. of *Abelmoschus esculentus*)
    moscheutos, 253
    syriacus, 253
Highbush blueberry, 255
Highbush cranberry, 255, 258
Higuerilla, 386
Hinoki cypress, 698
Hirschfeldia
    incana, 215
Hoary pea, 389
Hog peanut, 301
Hollyhock, 217, 390
Hollyleaved barberry (= Oregon grape), 253
Honeysuckle, 253, 258
Hops, 270
Horse gram, 390
Horsenettle, 306
Horseradish, 251
Hot chili pepper, 258
Humulus
    lupulus, 253
Hybanthus
    atropurpureus, 435
Hydrangeaceae, 304

I

Iberis, 215
Ilex
    aquifolium, 253
    paraguariensis, 432
Indian beech tree, 303
Indian mustard, 211, 214, 219, 220, 225, 344
Indian paintbrush, 393
Indigo, 216, 302, 388, 428, 431
Indigofera, 302, 388, 428, 431
    endecaphylla, 433
    hirsuta, 302, 433
    suffruticosa, 433
    truxillensis, 433
Invasive witchhazel, 253
Ipomoea
    batatas, 301, 386
    purpurea, 386
Isomeris
    arborea, 336

J

Japanese flowering cherry, 254
Japanese larch, 253
Japanese maple, 251
Japanese pagoda tree, 254
Japanese persimmon, 252
Japanese privet (= Wax-leaf privet), 253
Japanese pumpkin, 386
Japanese radish, 225
Japanese rose, 582
Japanese snowbell, 255
Japanese stewartia, 255
Japanese wisteria, 304
Japanese yew, 255, 260
Jasmine, 470
Jasminum
    abyssinicum, 472
Jerusalem artichoke, 383
Jicama, 303
Jimson weed, 338
Jitomate, 394
Job's tears, 391
Johnson grass, 217, 305
Jointed charlock, 216
Juglandaceae, 304, 307, 390, 395
Juglans, 307
    nigra, 253
Juniperus, 386, 396
    virginiana, 253
Jute, 304

K

Kahi (= Kans grass), 217
Kale, 10, 214, 219–221, 223, 224, 334, 337, 338,
        340, 344
Kans grass (= Kahi), 217

Kantola, 216
Katsura tree, 252
Kentucky (American) yellowwood, 252
Key lime, 393
Kidney bean, 11, 303, 316, 388
Knol khol, 215
Knotweed, 218, 392
Koelreuteria
    paniculata, 253
Kohlrabi, 215, 334
Korean euodia (= Bee-bee tree), 255
Korean hackberry, 252
Korean stewartia, 255
Korean sun pear, 254
Kousa dogwood, 252
Kudzu, 7, 10, 294, 295, 297–300, 303, 306, 309, 310,
        312–315, 318–321, 662, 664, 665
Kumquat, 393

L

Labiatae (syn. of Lamiaceae)
Lablab bean, 299, 300, 302, 316
Lablab
    purpureus, 299, 302, 433
    purpureus var. *lignosus* (syn. of *Lablab purpureus*)
Lactuca
    sativa, 213, 214, 307, 383
Lagerstroemia
    indica, 253
Lamb's quarters, 214, 385
Lamiaceae, 81, 85, 390, 395
Lamium
    amplexicaule, 390
Lanceleaf crotalaria, 387, 431, 435
Lantana, 394
Lantana, 394
Large-leaf dogwood, 252
Larix
    kaempferi, 253
    leptolepis (syn. of *Larix kaempferi*)
Lathyrus
    latifolius, 388
Lauraceae, 434
Leaf beet, 385
Leafy wildparsley, 253
Leavenworth's goldenrod, 384
Legumes, 10, 79, 95, 126, 264, 267, 294, 295, 300, 316, 317,
        426, 428, 431, 435, 437–439, 618, 620, 621, 623
Leguminosae (syn. of Fabaceae)
Lemon, 393
Lens
    culinaris, 302, 433
    esculenta (syn. of *Lens culinaris*)
Lentil, 11, 302, 388
Lepidium
    alyssoides, 210, 215
    capense, 215
    latifolium, 215
    virginicum, 385
Lespedeza, 302, 303, 307

Lespedeza, 303, 307
  cuneata, 302
  cyrtobotrya, 302, 306
  hirta, 302
Lesser burdock, 251
Lettuce, 214, 307, 383
Leucosidea
  sericea, 472
Ligustrum
  japonicum, 253
  lucidum, 431, 434
  sinense, 253, 305
Liliaceae, 390, 395
Lima bean, 216, 253, 258, 300, 303, 306, 316, 317, 388, 399, 622
Linaceae, 434
Linden arrowwood, 255
Linum
  usitatissimum, 434
Liquidambar, 250
  styraciflua, 253, 307
Liriodendron
  tulipifera, 253
Litchi, 699
Litchi
  sinensis, 617
Littleleaf linden, 255
Loblolly pine (overwintering), 391
Lobularia
  maritima, 213, 216, 225
Locust, 257
Lolium
  perenne, 305
Lonchocarpus, 733
London planetree, 254
London rocket, 210, 216, 385
Longan, 617
Lonicera
  tatarica, 253
Loquat, 305
Lotononis
  bainesii, 433
Lotus
  corniculatus, 213, 431, 433, 448
Love-lies-bleeding (= Amaranth), 251
Lucerne (= Alfalfa), 388
Lupinus
  albus, 431, 433
  angustifolius, 433
  luteus, 433
Lychee, 134, 617
Lycium
  chinense, 253, 266
Lycopersicon
  esculentum (syn. of *Solanum lycopersicum* var. *lycopersicum*)
  lycopersicum (syn. of *Solanum lycopersicum* var. *lycopersicum*)
Lycopersicum
  esculentum (syn. of *Solanum lycopersicum* var. *lycopersicum*)
  melongena (syn. of *Solanum melongena*)

nigrum (syn. of *Solanum nigrum*)
tuberosum (syn. of *Solanum tuberosum*)
Lythrum
  salicaria, 253

M

Macadamia
  integrifolia, 392
  ternifolia (syn. of *Macadamia integrifolia*)
Macadamia nut, 76, 392, 659
Macroptilium
  lathyroides, 388, 433
Madras hemp, 387
Magnolia
  grandiflora, 253
  stellata, 253
Mahonia
  aquifolium, 253
Maidenhair tree (= Ginkgo), 252
Maize, 218, 220, 227, 471, 618, 625
Malpighia
  glabra, 434
Malpighiaceae, 434
Malus, 253, 255
  baccata, 253
  domestica, 253, 733
  pumila (syn. of *Malus domestica*), 253
  sargentii, 253
  zumi, 253
Malva
  parviflora, 390
Malvaceae, 57, 131, 217, 300, 304, 390, 395, 434, 471
Malvastrum
  coromandelianum, 390
Mamani, 389
Manchurian snakebark maple, 251
Mandarin, 264
Mangifera
  indica, 214, 301, 384
Mango, 214, 301, 383, 743
Maple, 251, 258
Mapouria, 469
Marattiaceae, 122
Marigold, 383
Matrimony vine, 253, 265, 266
Matthiola, 216
Mauna loa, 387
Mauritius thorn, 472
Medicago
  hispida (syn. of *Medicago polymorpha*)
  lupulina, 307
  polymorpha, 388, 433
  sativa, 216, 303, 306, 426, 431, 433, 738
Melastomataceae, 434
Melilotus
  alba, 303, 306
Melón (= Cantaloupe), 44, 216, 386
Mesa pepperwort, 210
Metasequoia
  glyptostroboides, 253

Metrosideros
  collina (syn. of *Metrosideros polymorpha* var.
      polymorpha)
  polymorpha var. *polymorpha*, 391
Mexican clover, 393
Mexican hat, 384
Miconea
  cinerascens, 434
Milk thistle, 384
Milkweed, 625
Millet, 217, 253, 265, 271
Millettia, 303
  ferruginea, 481
Mimosa (= Sensitive plant), 253, 301
Mimosa, 253
Mimosaceae, 125
Mizuna, 215
Momordica
  charantia, 386, 624
  dioica, 216
Moraceae, 40, 125, 305
Morella
  cerifera, 305
Morning glory, 386
Morus
  alba, 217, 253, 305, 391
  nigra, 391
Moth orchid, 253
Mountain (Carolina) silverbell, 253
Mountain ash, 254
Mucuna
  pruriens, 303, 316, 388
Mulberry, 217, 253, 258, 264, 391
Multiflora rose, 254
Mullein, 684, 752
Mung bean, 304, 306, 316, 317, 622
Musa, 396, 471, 486
  acuminata, 305
Musaceae, 305, 396, 471
Muscadine, 306
Musineon
  divaricatum, 253
Mustard, 209, 211, 213–215, 219, 220, 222, 225, 334, 336,
      338–340, 344, 384, 396, 753
Mustard greens, 214, 224, 225
Myrciaria
  tenella, 434
Myricaceae, 305
Myrtaceae, 46, 54, 125, 131, 391, 395, 434, 539

N

Nabo, 384
Napa cabbage, 215
Nasturtium, 216, 218
Nasturtium
  indicum, 216
  integrifolium, 218
  officinale, 385
Nectandra, 434
Nectarine, 245, 254, 267, 625

Needle bush, 304
Neem, 224, 481, 501
Neonotonia
  wightii var. *wightii*, 388
Nicotiana, 393
  glauca, 213
  tabacum, 393, 733
Nictaginaceae, 434
Nodding plumeless thistle, 383
Northern red oak, 254
Norway maple, 251
Nutgrass, 217, 386
Nutsedge, 386
Nyctaginaceae, 391
Nyssa
  sylvatica, 253

O

Oak, 254, 257, 321
Oats, 217, 391
Ocimum
  gratissimum, 390
  suave (syn. of *Ocimum gratissimum*)
Oenothera
  speciosa, 391
'Ohi'a lehua, 391
Ohio buckeye, 251
Oil palm, 5, 616, 625–628
Okame flowering cherry, 254
Okra, 217, 251, 258, 270, 271, 390, 437, 460, 471
Olacaceae, 529
Oleaceae, 305, 431, 434, 472
Onagraceae, 391, 395
Oneseed hawthorn, 252
Orange, 300, 305, 393, 472, 624
Orchid, 252, 265, 266, 391
Orchidaceae, 266, 391, 395
Oregon grape (= Hollyleaved barberry), 253
Oriental arborvitae, 260
Oriental bittersweet, 252
Ornamental junipers (overwintering), 386
Ornamental red ginger, 626
Oryza
  sativa, 217, 305, 391, 434, 732

P

Pachyrhyzus
  erosus, 303
Pak choy, 215
Palm, 58, 129, 626–628
Palmer amaranth, 383
Panicgrass (= Zacate carricillo), 391
Panicum
  miliaceum, 217, 253, 265
  turgidum, 113
Papaya, 216
Paper birch, 251
Paperbark maple, 251
Paradise apple, 253

Parrot tree, 301
Parthenium
    hysterophorus, 213, 383
Parthenocissus
    quinquefolia, 268
Partridge pea, 301, 387
Passiflora
    incarnata, 391
Passifloraceae, 391, 395
Passion flower, 391
Paulownia, 253, 258, 260, 264, 266, 618, 619
Paulownia
    tomentosa, 250, 253, 264, 266, 618
Pavetta
    elliottii, 472
Pawpaw, 214, 251
Pea, 217, 254, 265, 271, 303
Pea-vines, 353
Peach, 254, 256, 258, 260, 264, 265, 268–270, 273, 277, 392, 752
Peanut, 11, 250, 301, 307, 386, 397
Pear, 258, 260, 264, 268, 270, 392, 625, 733
Pearl millet, 217, 220, 254, 265, 274, 391
Pecan, 251, 304, 390, 399, 400, 625
Pedaliaceae, 391, 395
Peireskia
    aculeata, 432
Peking (Chinese) tree lilac, 255
Pennisetum
    glaucum, 217, 391
    setaceum, 391
    spicatum (syn. of *Pennisetum glaucum*)
    typhoides, 217
    typhoideum (syn. of *Pennisetum glaucum*)
Pepino, 386
Pepper, 218, 231, 258, 260, 270, 271, 274, 393, 459
Peppergrass, 334, 385
Pepperweed, 215, 385
Perennial peanut, 301, 306
Perennial ryegrass, 305
Periwinkle, 252, 266, 619
Persimmon, 258, 260, 264–266, 268, 698
Pfaffia
    paniculata (syn. of *Hebanthe erianthos*)
Phakosporaceae, 297
Phalaenopsis, 253
Pharnaceum
    aurantium, 135
Phaseolus, 303
    coccineus, 399
    lathyroides (syn. of *Macroptilium lathyroides*)
    limensis (syn. of *Phaseolus lunatus*)
    lunatus, 216, 253, 300, 303, 388, 396, 433
    mungo (syn. of *Vigna mungo*)
    vulgaris, 217, 299, 303, 306, 316, 388, 430, 433, 620
Photinia (syn. of *Aronia*)
Phyllanthaceae, 471
Phyllostachys, 431, 434
Physalis
    peruviana, 218

Physopelta
    gutta, 265
Phytolacca
    americana, 254, 391
    americana var. *americana*, 391
    decandra (syn. of *Phytolacca americana* var. *americana*)
    dioica, 434
Phytolaccaceae, 391, 395, 434
Pigeon pea, 300, 301, 306, 309, 316, 317, 387, 431
Pigweed, 338, 383
Pill-pod spurge, 216
Pinaceae, 73, 138, 305, 307, 391, 395
Pine, 260, 299, 305, 307, 313, 315, 318
Pinkladies, 391
Pinnate false threadleaf, 214
Pinto bean, 303, 306, 317, 320
Pinus, 305, 307
    taeda, 391, 396
    yunnanensis, 73
Piper, 434
Piperaceae, 57, 434
Pistachio, 5, 91, 622–624
Pistacia
    chinensis, 254
    vera, 622
Pisum, 389
    arvense (syn. of *Pisum sativum*)
    sativum, 217, 254, 303, 320, 389, 433
Pitseed goosefoot, 252
Pittosporaceae, 431, 434
Pittosporum
    tobira, 248
    undulatum, 431, 434
Platanus
    × *acerifolia*, 254, 255
    occidentalis, 254
Platycladus
    orientalis, 254
Plinia
    cauliflora, 539
Pluchea
    lanceolata, 213
Plum, 264, 268, 751
Poaceae, 73, 113, 120, 121, 131, 217, 305, 307, 395, 396, 434, 471
Pokeweed, 391
Polanisia
    dodecandra, 385
    viscosa (syn. of *Cleome viscosa*)
Pole bean, 303
Polygonaceae, 218, 392, 395
Polygonum, 392
    plebeium, 218
Pomegranate, 254, 260
Pongamia
    glabra (syn. of *Pongamia pinnata*)
    pinnata, 303
Popolo, 394
Portulaca
    oleracea, 392
Portulacaceae, 392, 395

Potato, 218, 220, 231, 306, 334, 394, 459, 460, 618, 707, 735
Princess tree, 253, 258
Proteaceae, 392, 395
Prunus
    armeniaca, 254
    avium, 254
    cerasifera, 254
    incam, 254
    incisa, 254
    laurocerasus, 254
    persica, 254, 255, 392
    serotina, 254, 392
    serrulata, 254
    subhirtella, 254
Pseudocydonia
    sinensis, 254
Psidium
    guajava, 434
Psychotria
    nairobiensis, 472
Pueraria
    lobata (syn. of *Pueraria montana* var. *lobata*)
    montana, 294, 297, 303, 662
    montana var. *lobata*, 294, 297, 303, 389
    phaseoloides, 303, 320, 389
    thungergiana (syn. of *Pueraria montana* var. *lobata*)
Puncturevine, 218
Punica
    granatum var. *granatum*, 254
Purple loosestrife, 253
Pyracantha, 254
Pyrethrum, 339, 341, 476, 477, 481, 483, 732, 733
Pyrostegia
    venusta, 432
Pyrus, 250, 258
    calleryana, 254
    communis, 733
    fauriei, 254
    pyrifolia, 250, 254, 719

Q

Quelito bledo, 383
Quelito espinoso, 383
Quelito morado, 383
Quercus, 257
    alba, 254
    coccinea, 254
    robur, 254
    rubra, 254, 307

R

Radish, 10, 213, 216, 220, 224, 225, 334, 339, 340, 344, 385
Ragweed, 338, 595
Rambutan, 699
Ranunculaceae, 120
Rape, 219, 240, 340, 516
Rapeseed, 211, 215, 344, 753
Raphanus
    raphanistrum, 216, 225, 385
    sativus, 213, 216, 344, 385, 431, 432
Rapistrum
    rugosum, 355, 385
Raspberry, 254, 270
Ratibida
    columnifera, 384
Rattlebox, 387
Rattlepod, 302
Raya, 214
Red cabbage, 213
Red clover, 304, 306, 389, 435
Red gram, 301
Red horse-chestnut, 251
Red maple, 251
Red oak, 254, 307
Red pepper, 393
Redbud, 252, 301
Redosier dogwood, 252
Redwhisker clammyweed, 385
Rhamnaceae, 121, 218
Rhamnus
    cathartica, 254, 578
Rheum
    rhabarbarum, 392
    rhaponticum (syn. of *Rheum rhabarbarum*)
Rhubarb, 392
Rhus
    chinensis, 300, 301
    javonica (syn. of *Rhus chinensis*)
Rice, 11, 79, 117, 217, 258, 305, 358, 391, 456–459, 470, 618, 624, 708
Richardia
    brasiliensis, 393
    scabra, 393
Richardsonia
    pilosa (syn. of *Richardia scabra*)
    scabra (syn. of *Richardia scabra*)
Ricinus
    communis, 216, 386, 399, 432, 471
River birch, 251
Riverbank wild grape, 255
River Red Gum, 54
Robinia, 257
    pseudoacacia, 217, 254, 303
    viscosa, 578
Robusta coffee, 469, 470, 488
Rosa, 248
    canina, 254
    multiflora, 254
    rugosa, 254
Rosaceae, 120, 125, 138, 255, 264, 305, 392, 395, 435, 472
Rosales, 255
Rose of Sharon (= Hibiscus), 253

Rosids, 265
Rough cocklebur, 384
Rubiaceae, 120, 218, 392, 393, 395, 469, 471, 472, 477, 488, 620
Rubus
    fruticosus, 305
    phoenicolasius, 254
Rugosa rose, 254, 258
Rumex, 392
Runner bean, 399
Russian olive, 252, 258
Ruta
    chalepensis, 481
Rutabaga, 215, 220, 334, 338, 384
Rutaceae, 218, 265, 300, 305, 393, 395, 472
Rye, 392

S

Saccharum
    officinarum, 217, 305, 392
    spontaneum, 217
Saldinia, 469
Salicaceae, 306
Salix
    nigra, 306
Salsola
    vermiculata, 121
Salvia, 390
Salvia, 390
Sandalwood, 73, 81
Sandbur, 391
Sandía, 386
Santalaceae, 73, 81
Santalum
    album, 81
Santa Maria feverfew, 383
Saphindales, 255
Sapindaceae, 435
Sargent's crab apple, 253
Sarson, 215
Sassafras, 254
Sassafras
    albidum, 254
Satsuma mandarin, 305
Satsuma orange, 393
Scarlet oak, 254
Scarlet spiderling, 391
Schinus
    terebinthifolius, 432
Schizachyrium
    condensatum, 435
Schkuhria
    pinnata, 214
Schoepfia
    jasminodora, 48, 529
Schoepfiaceae, 48
Scrophulariaceae, 393, 415
Secale, 392
    cereale, 254

Sechium
    edule, 386, 432
Senecio
    vulgaris, 213
Senna
    obtusifolia, 303
    occidentalis, 389
Sensitive plant (= Mimosa), 253, 301
Serjania
    fuscifolia, 435
Sesame, 391
Sesamum
    indicum (syn. of *Sesamum orientale*)
    orientale, 391
Sesbania, 389
    aculeata (syn. of *Sesbania bispinosa*)
    bispinosa, 303, 434
    grandiflora, 300, 304
    vesicaria, 389
Setaria
    italica, 254, 265
Seven sons flower, 253
Shadbush, 251, 258
Shagbark hickory, 252
Shepherd's purse, 334
Shortpod mustard, 215
Showy crotalaria, 387
Siberian crab apple, 253
Siberian pea shrub, 251, 258
Sicklepod, 303
Silver linden, 255
Silver maple, 251
Silverleaf nightshade, 393, 459
Silybum
    marianum, 384
Simaroubacea, 255
Sisymbrium
    capense, 216
    irio, 210, 385
    sophia, 215
Slender amaranth, 383
Slender deutzia, 304
Slim amaranth, 383
Smooth (English) hawthorn, 252
Snap bean, 217, 299, 303, 388
Solanaceae, 60, 71, 81, 213, 218, 266, 306, 307, 393–396,
    435, 459, 472, 733
Solanum, 81
    anomalum, 472
    carolinense, 306
    elaeagnifolium, 413
    lycopersicum, 213, 218, 254, 306, 307, 394
    lycopersicum var. *lycopersicum*, 218, 254, 306, 307, 394
    madrense, 394
    mauritianum, 435
    melongena, 254, 394
    nigrum, 213, 394
    torvum, 698
    tuberosum, 218, 306, 394

Solidago
  altissima, 251, 255, 273, 384
  canadensis var. *scabra* (syn. of *Solidago altissima*)
  gigantea, 384
  leavenworthii, 384
Sonchus
  arvensis, 214
  oleraceus, 384
Sophora
  chrysophylla, 389
  japonica, 254
Sorbus
  airia, 254
  americana, 254
  aucuparia, 254, 268
Sorghum, 217, 254, 258, 265, 270, 271, 274, 305, 307, 392,
      399, 403, 458, 471, 738, 739
Sorghum, 471
  bicolor, 217, 254, 265, 305, 307, 738
  bicolor ssp. *bicolor*, 392
  bicolor var. *sudanense*, 217
  halepense, 217, 305, 738
  vulgare (syn. of *Sorghum bicolor* ssp. *bicolor*)
Sorgo, 392
Sorrel, 392
Southern magnolia, 253
Soya, 388
Soybean, 5, 10, 11, 79, 95, 126, 244, 250, 253, 256,
      258, 259, 265, 270–273, 294–300, 302,
      306–313, 316–321, 337, 338, 366, 388,
      396–403, 426–431, 435–440, 460, 579,
      613, 620–622, 625, 659, 662–666, 683–
      685, 694–696, 699, 704, 705, 708, 712,
      713, 738, 739, 741–743, 748, 749, 755
Spanish moss (overwintering), 371, 385
Spanish needle, 384
Spiderwisp, 385
Spinach, 214, 385
Spinacia
  oleracea, 213, 214, 385
Spiny amaranth, 383
Spiny gourd, 216
Spiny spiderflower, 385
Spiraea, 254
Spirea, 254
Spondias
  mombin, 432
Spring raab, 225
Squash, 44, 386
Star magnolia, 253
Sterculiaceae, 394, 395
Stewartia
  koreana, 255
  pseudocamellia, 255
Stock, 216
Strawberry, 11
String bean, 303, 388
Striped maple, 251
Stylo, 304
Stylosanthes
  guianensis, 304

Styphnolobium
  japonicum, 255
Styrax
  japonicas, 255
Sudan grass, 217, 226
Sugar cane, 305, 392
Sugar maple, 251
Sugarbeet, 214
Sugarcane, 217, 392
Sugi, 698
Summer squash (= Field pumpkin), 252
Sunflower, 10, 253, 270, 271, 274, 383, 460
Sunn-hemp, 387
Sweet alyssum, 213, 216, 225
Sweet banana pepper, 258
Sweet bell pepper, 258
Sweet cherry, 254
Sweet corn, 255, 258, 270, 403, 742
Sweet gum, 307
Sweet persimmon, 252, 258, 265
Sweet potato, 301, 386, 459
Sweetgum, 253
Swiss chard, 251
Sword bean, 387
Symphonia
  globulifera, 471
Symphytum, 255
Syringa
  pekinensis, 255

T

Tabebuia
  rosea, 72
Taiwan-kudzu, 310
Tanacetum
  cinerariifolium, 481, 733
Tangerine, 305, 393
Tansy mustard, 215
Taramira, 215
Tat-soi, 215
Tatarian honeysuckle, 253
Taxodiaceae, 698
Taxus
  cuspidata, 255
Tea, 5, 218, 265
Teak, 81
Tecoma
  stans, 72
Tectona
  grandis, 81
Tephrosia, 389, 733
Tetradium
  daniellii, 255
  hupehensis (syn. of *Tetradium daniellii*)
Texas redbud, 252
Texas thistle, 383
Theaceae, 138, 218
Theobroma
  cacao, 394
Thistle, 214, 383, 384, 396

Thornless common honeylocust, 252
Threeflower ticktrefoil, 387
Thuja, 386, 396
Thymus
    vulgaris, 481
Tick trefoil, 387
Til, 390
Tilia
    americana, 255
    cordata, 255
    heteropholia, 578
    tomentosa, 255
Tiliaceae, 138
Tillandsia
    usneoides, 385, 396
Tobacco, 393, 399, 733
Tomate, 394
Tomato, 11, 81, 218, 254, 258, 259, 265, 268–271, 306, 307, 313,
        337, 338, 394, 398, 401, 459, 624–627, 684, 736, 751
Tree of heaven, 251, 256, 273
Trema
    orientalis, 472
Tribulus
    terrestris, 218
Trident maple, 251
Trifolium, 304
    alexandrinum, 217, 389
    incarnatum, 304, 389
    pratense, 304, 306, 389, 431, 434
    repens, 304, 306, 389, 434, 435
Trigo, 392
Triticum
    aestivum, 218, 305, 392, 435, 733, 739
    vulgare (syn. of *Triticum aestivum*)
Tropaeolaceae, 218
Tropaeolum
    majus, 216
Tropical kudzu, 303, 320, 389
Tropical Mexican clover, 383
Tropical soda apple, 459
Tsuga
    canadensis, 255
Tuliptree, 253
Tupelo (= Blackgum), 253
Turmeric, 306
Turnip, 215, 220, 224, 334, 336, 338, 340, 344, 384
Turnipweed, 385

U

Ulmaceae, 394, 395
Ulmus, 255, 258
    americana, 255
    minor, 255
    parvifolia, 255
    procera (syn. of *Ulmus minor*), 255
Umbelliferae (syn. of Apiaceae), 432
Upland cotton, 390
Upright prairie coneflower, 384
Urd bean, 304, 316
Urticaceae, 125

V

Vaccinium
    corymbosum, 255, 432
    macrocarpon, 740
Vachellia
    farnesiana, 304
Vangueria
    apiculata, 472
Velvet bean, 303, 316, 388
Verbenaceae, 72, 121, 394, 395
Verdolaga, 392
Vernonia
    amygdalina, 472
Vetch, 217, 255, 265, 304
Viburnum (= Blackhaw), 255
Viburnum
    × *burkwoodii*, 255
    dilatatum, 255
    edule, 255
    opulus var. *americanum*, 250, 255
    prunifolium, 250, 255
Vicia
    angustifolia, 304, 306
    faba, 213, 275, 276, 304, 389
    sativa, 304, 431
Vigna
    angularis, 304
    mungo, 217, 304, 316, 389
    radiata, 217, 304, 306
    sesquipedalis (syn. of *Vigna unguiculata* ssp.
        sesquipedalis)
    sinensis (syn. of *Vigna unguiculata*)
    unguiculata, 217, 300, 304, 389, 390, 434, 620
    unguiculata ssp. *cylindrica*, 390
    unguiculata ssp. *sesquipedalis*, 390
Vine maple, 251
Viola
    tricolor, 137
Violaceae, 137, 435
Vitaceae, 306, 394, 395
Vitis, 394
    riparia, 255
    vinifera, 255, 394
    rotundiflora, 306

W

Walnut, 258, 307
Watercress, 385
Watermelon, 216, 386
Wax myrtle, 305
Wax-leaf privet (= Japanese privet), 253
Weeping forsythia, 252
West Indian rattlebox, 387
Wheat, 69, 218, 220, 225, 228, 231, 270, 271, 305, 392,
        435, 733, 735, 739
White (American) ash, 252, 258
White basswood, 578
White clover, 304, 306, 389, 435
White fringetree, 252

White lupine, 431
White mulberry, 253, 260, 305, 391
White oak, 254
White sweet clover, 303, 306, 307
Wight's neonotonia, 388
Wild blackberry, 305, 392
Wild bushbean, 388
Wild grape, 255, 394
Wild indigo, 251, 301
Wild jujube, 218
Wild legume, 387, 389
Wild mustard, 251
Wild plum, 392
Wild radish, 216, 225, 385, 431
Wild spider flower, 385
Willow, 254, 258, 260, 264
Wine grape, 255, 270
Wine raspberry (= Wineberry), 254
Wineberry (= Wine raspberry), 254
Winter cress, 214
Winter pea, 303, 320
Winter-flowering (Higan) cherry, 254
Winterbeam, 254
Wisteria, 304, 754
Wisteria
   brachybotrys, 304
   floribunda, 304
   frutescens, 304
   sinensis, 304

X

Xanthium, 384
   strumarium, 301
Xpelón, 389

Y

Yardlong bean, 390
Yellow thistle, 383
Yuzu, 252, 258, 264, 265

Z

Zacate carricillo (= Panicgrass), 391
Zacate pinto, 391
Zarzabacoa galana, 387
Zea
   mays, 218, 255, 305, 392, 471, 738
Zingiber
   officinale, 694
Zingiberaceae, 306
Ziziphus
   rotundifolia, 218
   zuzuba, 213
Zygophyllaceae, 218

Microorganisms and Plant Diseases Index

A

Achromobacter, 311
Actinobacteria, 653, 655, 666
Actinomyces (Bacteria), 656
Actinomycete, 645
Aeromonas
    hydrophila (Bacterium), 655
Agamermis (Mermithid nematode), 308
Agrobacterium
    tumefaciens, 655
Alternaria, 622
    alternata (Fungus), 624
Archaea, 644
Ashbya
    gossypii (Fungus) (syn. of *Eremothecium gossypii*)
Asian soybean rust, 297
Aspergillus
    flavus (Fungus), 625
Aureobasidium
    pullulans (Fungus), 623

B

Bacilli (Bacteria), 616
Bacillus
    subtilis, 655
    thuringiensis (Bt), 747, 756
    thuringiensis (Bt) *kurstaki*, 396, 397, 750, 755
Bacterial blight of beans, 620
Bacterial blight of cowpea, 620
Bacterial symbionts, 6, 479, 480
Bacteroides
    fragilis (Bacterium), 655
Baumannia, 645, 663
    cicadellinicola (Bacterium), 655
Beauveria, 698
    bassiana (Fungus), 275, 309, 321, 698
Beet leafcurl disease (Virus), 617
Beet savoy disease (uncertain placement; possibly virus), 617
Betaproteobacteria, 654–656
Blochmannia, 645, 663
    floridanus (Bacterium), 655
Boll rot (Bacteria and Fungi), 5, 400, 619, 621
Botryosphaeria
    dothidea (Fungi), 623
Buchnera, 644, 656–659, 663, 664
    aphidicola (Bacterium), 644, 655
Burkholderia (Bacteria), 649–652, 656, 657, 659
    cepacia, 655
    fungorum, 655
    phymatum, 655
    pseudomallei, 655

C

Campylobacter
    concisus (Bacterium), 655
Candidatus (Bacteria)
    Benitsuchiphilus tojoi, 647, 655, 659
    Ishikawaella, 659, 661–665
    Ishikawaella capsulata, 310, 311, 649, 655, 657, 658, 661, 662
    Liberibacter, 629
    Pantoea carbekii, 256, 648, 655, 657–659
    Rohrkolberia cinguli, 651, 655
    Rosenkranzia clausaccus, 647, 655, 658
    Schneideria nysicola, 651, 655
    Tachikawaea, 659
    Tachikawaea gelatinosa, 649, 655, 657, 658
Carsonella (Bacteria), 645
Caulobacter
    crescentus (Bacterium), 655
Clavicipitaceae (Fungi), 309, 702, 749
Clostridium, 653
    perfringens (Bacterium), 655
Coconut lethal yellowing (Mollicutes), 618
Coconut root wilt (Mollicutes), 618
Coffee necrosis (Trypanosomatidae), 626
Colletotrichum
    truncatum (Fungi), 622
Common smut (Fungi), 625
Coriobacterium, 653
    glomerans, 655
Cornyebacterium (Bacteria), 620
Crebrothecium
    ashbyi (Fungus) (syn. of *Eremothecium ashbyi*)
Cucurbit yellow vine disease (Bacterium), 616, 618
Curvularia
    lunata (Fungus), 624

D

Deinococcus
    radiodurans (Bacterium), 655
Dinoflagellates, 644
Dry rot (Fungi), 623, 624

E

Endosymbiont (Mostly bacteria), 138, 256, 300, 309–311, 367, 368, 370, 430, 629, 661, 663–666, 749

Enterobacter
    cloacae (Bacterium), 655
Enterobacteriaceae, 370, 480, 654, 656
Eremothecium (Fungi), 479, 621
    ashbyi, 621, 622
    coryli, 370, 400, 479, 613, 616, 621–625, 751
    gossypii, 479, 621
Erwinia (Bacteria), 659, 661
    amylovora, 655
    rhapontici, 655
    tracheiphila, 655
Escherichia
    coli (Bacterium), 655, 658

F

Firmicutes, 653, 655, 656
Flagellates, 228, 626, 627
Fruit rot (Bacteria and Fungi), 623, 624, 627
Fruit rot of tomato (Bacteria and Fungi), 624, 751
Fungal pathogen (Fungi), 7, 267, 400, 620–625, 749
Fungi, 5, 39, 58, 127, 226, 228, 402, 479, 616, 617,
        620–625, 644, 659
Fusarium (Fungi), 622
Fusobacterium
    nucleatum (Bacterium), 655

G

Gammaproteobacteria, 309, 479, 654–656, 661, 662
Gammaproteobacterial symbionts, 309, 479, 654, 661
Gordonibacter (Bacteria), 653, 655

H

Hamiltonella
    defensa (Bacterium), 666
Hartrot (Trypanosomatidae), 5, 626, 627
Heart rot (see Hartrot)
Helicobacter
    pylori (Bacterium), 655
Herpetomonas (Trypanosomatidae), 625
Hexamermis (Mermithid nematode), 439
Hypocreales (Fungi), 267, 702, 749

I

Isaria (Fungi), 228
Ishikawaella
    capsulata (Bacterium), 310, 311, 649, 655, 657–659,
        661–665

K

Klebsiella, 370, 653
    pneumoniae (Bacterium), 370, 616
Kleidoceria
    schneideri (Bacterium), 651, 655

L

Lactococcus
    lactis (Bacterium), 655
Leptomonas (Trypanosomatidae), 625
Longan witches' broom disease (Virus), 617

M

Marchitez (Trypanosomatidae), 5, 627
Marchitez sorpresiva (Trypanosomatidae), 626
Mermis (Mermithid nematode), 439
Mermithid nematode, 308
    Agamermis, 308
    Hexamermis, 439
    Mermis, 439
    Pentatomimermis, 439
Metarhizium (Fungi)
    anisopliae, 275, 702, 749
    anisopliae major, 749
Mollicutes, 5, 618
Moniliasis (Fungi), 625
Moniliophthora
    roreri (Fungus), 625
Mycobacterium
    tuberculosis (Bacterium), 655
Mycoplasma (Mollicutes), 618
Mycoplasma-like organisms (Mollicutes), 618, 619

N

Nardonella (Bacteria), 645, 655
Neisseria
    gonorrhoeae (Bacterium), 655
Nematodes, 7, 228, 308, 439, 749
Nematospora (Fungi), 621
    coryli (syn. of *Eremothecium coryli*)
    gossypii (syn. of *Eremothecium gossypii*)
    phaseoli (syn. of *Eremothecium coryli*)
Nomurea
    rileyi (Fungus), 739

O

Ophiocordyceps
    nutans (Fungus), 267

P

Paecilomyces (Fungus), 702
Pandoraea
    pulmonicola (Bacterium), 655
Panicle and shoot blight (Fungi), 5, 623
Pantoea (Bacteria), 661
    agglomerans, 369, 370, 400, 403, 616, 619
    ananatis, 370, 616, 655
    coffeiphila, 480, 621
    stewartii, 655
Pathogenic bacteria, 479, 619, 620
Pathogenic fungi, 7, 267, 400, 479, 620–625

Paulownia witches' broom disease (Mollicutes), 267, 618, 619
Pecky rice (Fungi), 399, 458, 624
Penicillium (Fungi), 622, 625
Pentatomimermis (Mermithid nematode), 439
Phakopsora, 297, 440
  pachyrhizi (Fungus), 297
Phakopsoraceae, 297
Phoma (Fungi), 621
Phomopsis, 622
  sojae (Fungus), 622
Photorhabdus
  luminescens (Bacterium), 655
Phytomonas, 625–628
  leptovasorum (Trypanosomatidae), 626
  staheli (Trypanosomatidae), 626–628
Phytoplasma (Mollicutes), 5, 266, 267, 613, 616–619
Potato taste defect (PTD), 12, 467, 480, 620, 621
Proteobacteria, 647–652, 654–656
Protists, 644
Pseudomonas, 620
  aeruginosa (Bacterium), 655

R

Rhabdovirus (Virus), 617
Rhizobium (Bacteria), 644, 657
Rickettsia, 479
  typhi (Bacterium), 655

S

Salmonella Typhi (Bacterium), 655
Seed rot (Fungi), 5
Serratia (Bacteria)
  marcescens, 655
  symbiotica, 666
Sinorhizobium
  meliloti (Bacterium), 655
Slow wilt (Trypanosomatidae), 627
Sodalis, 479, 666
  glossinidius (Bacterium), 655, 666
Soybean rust, 297, 321, 440
Spiroplasmas (Mollicutes), 616–618
Spiroplasma, 479, 616–618

Stem canker (Fungus), 5, 620
Steinernematidae, 439
Stigmatomycosis (Fungi), 5, 621, 623, 624
Streptomyces
  coelicolor (Bacterium), 655
Sudden wilt (Trypanosomatidae), 626, 627
Symbionts, 5, 6, 7, 256, 643–667, 743

T

Tremblaya, 645
  princeps (Bacterium), 655
Trypanosomatid, 5, 81, 616, 617, 625–628

U

Ustilago
  maydis (Fungus), 625

V

Vein necrosis (Virus), 5, 620
Vibrio, 644, 657
  cholerae (Bacterium), 655
Viruses, 5, 612, 614, 616, 617, 741, 749

W

Wigglesworthia, 645, 657, 658, 663
  glossinidia (Bacterium), 655
Witches' broom (Mollicutes), 5, 264, 266, 267, 618
Wolbachia (Bacteria), 310, 311, 655, 666

X

Xanthomonads (Bacteria), 620
Xanthomonas, 620
  axonopodis, 620
  phaseoli, 620

Y

Yeast (Fungi), 616, 620–625, 659, 713, 751
Yeast spot (Fungi), 5, 621–623
Yersinia
  pestis (Bactcrium), 655